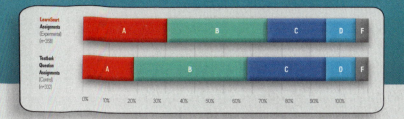

SMARTBOOK®

The first—and only—adaptive reading experience designed to transform the way students read.

> Engages students with a personalized reading experience
> Ensures students retain knowledge

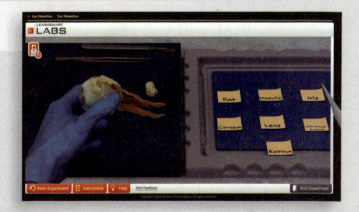

LEARNSMART®

The market-leading **adaptive study tool** proven to strengthen memory recall, increase class retention, and boost grades.

> Moves students beyond memorizing
> Allows instructors to align content with their goals
> Allows instructors to spend more time teaching higher-level concepts

LEARNSMART PREP®

An adaptive course preparation tool that quickly and efficiently helps students prepare for college-level work.

> Levels out student knowledge
> Keeps students on track

LEARNSMART ACHIEVE®

A learning system that continually adapts and provides learning tools to teach students the concepts they don't know.

> Adaptively provides learning resources
> A time management feature ensures students master course material to complete their assignments by the due date

LEARNSMART LABS®

LearnSmart Labs® is a super-adaptive simulated lab experience that brings meaningful scientific exploration to students. Through a series of adaptive questions, LearnSmart Labs identifies a student's knowledge gaps and provides resources to quickly and efficiently close those gaps. Once the student has mastered the necessary basic skills and concepts, they engage in a highly realistic simulated lab experience that allows for mistakes and the execution of the scientific method.

SEELEY'S
ESSENTIALS OF
ANATOMY & PHYSIOLOGY
NINTH EDITION

Cinnamon VanPutte
Southwestern Illinois College

Jennifer Regan
University of Southern Mississippi

Andrew Russo
University of Iowa

Mc
Graw
Hill
Education

SEELEY'S ESSENTIALS OF ANATOMY & PHYSIOLOGY, NINTH EDITION

Published by McGraw-Hill Education, 2 Penn Plaza, New York, NY 10121. Copyright © 2016 by McGraw-Hill Education. All rights reserved. Printed in the United States of America. Previous editions © 2013, 2010, and 2007. No part of this publication may be reproduced or distributed in any form or by any means, or stored in a database or retrieval system, without the prior written consent of McGraw-Hill Education, including, but not limited to, in any network or other electronic storage or transmission, or broadcast for distance learning.

Some ancillaries, including electronic and print components, may not be available to customers outside the United States.

This book is printed on acid-free paper.

1 2 3 4 5 6 7 8 9 0 DOW/DOW 1 0 9 8 7 6 5

ISBN 978-0-07-809732-4
MHID 0-07-809732-0

Senior Vice President, Products & Markets: *Kurt L. Strand*
Vice President, General Manager, Products & Markets: *Marty Lange*
Vice President, Content Design & Delivery: *Kimberly Meriwether David*
Managing Director: *Michael S. Hackett*
Director of Marketing: *James F. Connely*
Brand Manager: *Amy Reed*
Director, Product Development: *Rose Koos*
Product Developer: *Mandy C. Clark*
Marketing Manager: *Jessica Cannavo*
Director of Development: *Rose Koos*
Digital Product Developer: *Jake Theobold*
Program Manager: *Angela R. FitzPatrick*
Content Project Managers: *Jayne Klein/Sherry Kane*
Buyer: *Laura M. Fuller*
Design: *David W. Hash*
Content Licensing Specialists: *John Leland/Leonard Behnke*
Cover Images: *young woman sweating after working out ©Thomas_EyeDesign/Getty Images/RF;*
Two Staphylococcus epidermis Bacteria ©CDC/Janice Carr
Compositor: *ArtPlus Ltd.*
Typeface: *10/12 Minion*
Printer: *R. R. Donnelley*

All credits appearing on page or at the end of the book are considered to be an extension of the copyright page.

Library of Congress Cataloging-in-Publication Data

VanPutte, Cinnamon L.
    Seeley's essentials of anatomy & physiology / Cinnamon Van Putte, Jennifer Regan, Andrew Russo.–Ninth edition.
        pages cm
    Includes index.
    1. Human physiology. 2. Human anatomy. I. Regan, Jennifer L. II. Russo, Andrew F. III. Title.
        IV. Title: Essentials of anatomy and physiology.
    QP34.5.S418 2016
    612--dc23
                                                                        2014022213

The Internet addresses listed in the text were accurate at the time of publication. The inclusion of a website does not indicate an endorsement by the authors or McGraw-Hill Education, and McGraw-Hill Education does not guarantee the accuracy of the information presented at these sites.

www.mhhe.com

DEDICATION

This text is dedicated to our families. Without their uncompromising support and love, this effort would not have been possible. Our spouses and children have been more than patient while we've spent many nights at the computer surrounded by mountains of books. We also want to acknowledge and dedicate this edition to the previous authors as we continue the standard of excellence that they have set for so many years. For each of us, authoring this text is a culmination of our passion for teaching and represents an opportunity to pass knowledge on to students beyond our own classrooms; this has all been made possible by the support and mentorship we in turn have received from our teachers, colleagues, friends, and family.

About the Authors

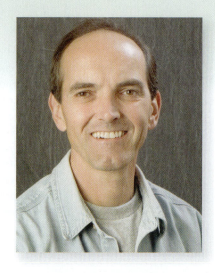

Cinnamon L. VanPutte
Professor of Biology
Southwestern Illinois College

Cinnamon has been teaching biology and human anatomy and physiology for almost two decades. At Southwestern Illinois College she is a full-time faculty member and the coordinator for the anatomy and physiology courses. Cinnamon is an active member of several professional societies, including the Human Anatomy & Physiology Society (HAPS). Her Ph.D. in zoology, with an emphasis in endocrinology, is from Texas A&M University. She worked in Dr. Duncan MacKenzie's lab, where she was indoctrinated in the major principles of physiology and the importance of critical thinking. The critical thinking component of *Seeley's Essentials of Human Anatomy & Physiology* epitomizes Cinnamon's passion for the field of human anatomy and physiology; she is committed to maintaining this tradition of excellence. Cinnamon and her husband, Robb, have two children: a daughter, Savannah, and a son, Ethan. Savannah is very creative and artistic; she loves to sing, write novels, and do art projects. Robb and Ethan have their black belts in karate and Ethan is one of the youngest black belts at his martial arts school. Cinnamon is also active in martial arts and is a competitive Brazilian Jiu-Jitsu practitioner. She has competed at both the Pan Jiu-Jitsu Championship and the World Jiu-Jitsu Championship.

Jennifer L. Regan
Instructor
University of Southern Mississippi

For over ten years, Jennifer has taught introductory biology, human anatomy and physiology, and genetics at the university and community college level. She has received the Instructor of the Year Award at both the departmental and college level while teaching at USM. In addition, she has been recognized for her dedication to teaching by student organizations such as the Alliance for Graduate Education in Mississippi and Increasing Minority Access to Graduate Education. Jennifer has dedicated much of her career to improving lecture and laboratory instruction at her institutions. Critical thinking and lifelong learning are two characteristics Jennifer hopes to instill in her students. She appreciates the Seeley approach to learning and is excited about contributing to further development of the textbook. She received her Ph.D. in biology at the University of Houston, under the direction of Edwin H. Bryant and Lisa M. Meffert. She is an active member of several professional organizations, including the Human Anatomy and Physiology Society. During her free time, Jennifer enjoys spending time with her husband, Hobbie, and two sons, Patrick and Nicholas.

Andrew F. Russo
Professor of Molecular Physiology and Biophysics
University of Iowa

Andrew has over 20 years of classroom experience with human physiology, neurobiology, molecular biology, and cell biology courses at the University of Iowa. He is a recipient of the Collegiate Teaching Award and is currently the course director for Medical Cell Biology and Director of the Biosciences Graduate Program. He is also a member of several professional societies, including the American Physiological Society and the Society for Neuroscience. Andrew received his Ph.D. in biochemistry from the University of California at Berkeley. His research interests are focused on the molecular neurobiology of migraine. His decision to join the author team for *Seeley's Essentials of Human Anatomy & Physiology* is the culmination of a passion for teaching that began in graduate school. He is excited about the opportunity to hook students' interest in learning by presenting cutting-edge clinical and scientific advances. Andy is married to Maureen, a physical therapist, and has three daughters Erilynn, Becky, and Colleen, now in college and graduate school. He enjoys all types of outdoor sports, especially bicycling, skiing, ultimate Frisbee and, before moving to Iowa, bodyboard surfing.

Brief Contents

Contents

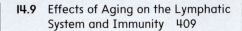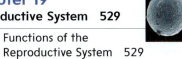

Help Your Students Prepare for Class

LearnSmartAdvantage.com

SmartBook is the first and only adaptive reading experience available for the higher education market. Powered by an intelligent diagnostic and adaptive engine, SmartBook facilitates the reading process by identifying what content a student knows and doesn't know through adaptive assessments. As the student reads, the reading material constantly adapts to ensure the student is focused on the content he or she needs the most to close any knowledge gaps.

LearnSmart Labs is a super adaptive simulated lab experience that brings meaningful scientific exploration to students. Through a series of adaptive questions, LearnSmart Labs identifies a student's knowledge gaps and provides resources to quickly and efficiently close those gaps. Once the student has mastered the necessary basic skills and concepts, they engage in a highly realistic simulated lab experience that allows for mistakes and the execution of the scientific method.

LearnSmart is the only adaptive learning program proven to effectively assess a student's knowledge of basic course content and help them master it. By considering both confidence level and responses to actual content questions, LearnSmart identifies what an individual student knows and doesn't know and builds an optimal learning path, so that they spend less time on concepts they already know and more time on those they don't. LearnSmart also predicts when a student will forget concepts and introduces remedial content to prevent this. The result is that LearnSmart's adaptive learning path helps students learn faster, study more efficiently, and retain more knowledge, allowing instructors to focus valuable class time on higher-level concepts.

The primary goal of LearnSmart Prep is to help students who are unprepared to take college level courses. Using super adaptive technology, the program identifies what a student doesn't know, and then provides "teachable moments" designed to mimic the office hour experience. When combined with a personalized learning plan, an unprepared or struggling student has all the tools they need to quickly and effectively learn the foundational knowledge and skills necessary to be successful in a college level course.

McGraw Hill Education | LEARNSMART®

— A Diagnostic, Adaptive Learning System to help you learn—smarter

LearnSmart is the only adaptive learning program proven to effectively assess a student's knowledge of basic course content and help them master it. By considering both confidence level and responses to actual content questions, LearnSmart identifies what an individual student knows and doesn't know and builds an optimal learning path, so that they spend less time on concepts they already know and more time on those they don't. LearnSmart also predicts when a student will forget concepts and introduces remedial content to prevent this. The result is that LearnSmart's adaptive learning path helps students learn faster, study more efficiently, and retain more knowledge, allowing instructors to focus valuable class time on higher-level concepts.

◀ Study with LearnSmart by working through modules and using LearnSmart's reporting to better understand your strengths and weaknesses.

Gauge your student's progress using reports in LearnSmart and Connect. Students can run these same reports in LearnSmart to track their own progress.

◀ Reports in Connect and LearnSmart help you monitor student assignments and performance, allowing for "just-in-time" teaching to clarify concepts that are more difficult for your students to understand.

▲ The Tree of Knowledge tracks your progress, reporting on short-term successes and long-term retention.

▲ Download the LearnSmart app from iTunes or Google Play and work on LearnSmart from anywhere!

McGraw-Hill Campus is an LMS integration service that offers instructors and students universal sign-on, automatic registration, and gradebook synchronization with our learning platforms and content. Gain seamless access to our full library of digital assets—1,500 e-texts and instructor resources that let you build richer courses from within your chosen LMS!

McGraw-Hill Connect®
Anatomy & Physiology

McGraw-Hill Connect® Anatomy & Physiology integrated learning platform provides auto-graded assessments, a customizable, assignable eBook, an adaptive diagnostic tool, and powerful reporting against learning outcomes and level of difficulty—all in an easy-to-use interface. Connect Anatomy & Physiology is specific to your book and can be completely customized to your course and specific learning outcomes, so you help your students connect to just the material they need to know.

Save time with auto-graded assessments and tutorials

Fully editable, customizable, auto-graded interactive assignments using high quality art from the textbook, animations, and videos from a variety of sources take you way beyond multiple choice. Assignable content is available for every Learning Outcome in the book.

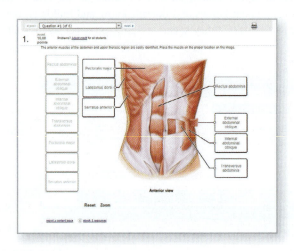

New! Clinical Question Types for Each Chapter!

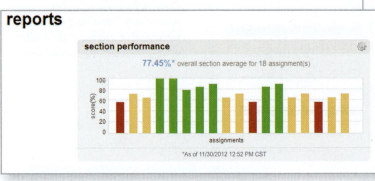

Gather assessment information

Generate powerful data related to student performance against learning outcomes, specific topics, level of difficulty, and more.

All **Connect** content is pre-tagged to Learning Outcomes for each chapter as well as topic, section, Bloom's Level, and Human Anatomy and Physiology Society (HAPS) Learning Outcomes to assist you in both filtering out unneeded questions for ease of creating assignments and in reporting on your students' performance against these points. This will enhance your ability to effectively assess student learning in your courses by allowing you to align your learning activities to peer-reviewed standards from an international organization.

Presentation Tools Allow Instructors to Customize Lecture

Everything you need, in one location

Enhanced Lecture Presentations contain lecture outlines, FlexArt, adjustable leader lines and labels, art, photos, tables, and animations embedded where appropriate. Fully customizable, but complete and ready to use, these presentations will enable you to spend less time preparing for lecture!

Animations—over 100 animations bringing key concepts to life, available for instructors and students.

Animation PPTs—animations are truly embedded in PowerPoint for ultimate ease of use! Just copy and paste into your custom slideshow and you're done!

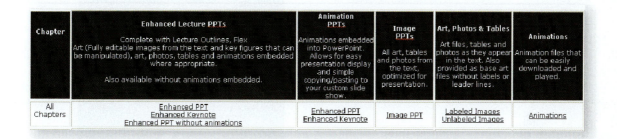

Chapter	Enhanced Lecture PPTs	Animation PPTs	Image PPTs	Art, Photos & Tables	Animations
	Complete with Lecture Outlines, Flex Art (Fully editable images from the text and key figures that can be manipulated), art, photos, tables and animations embedded where appropriate. Also available without animations embedded.	Animations embedded into PowerPoint. Allows for easy presentation display and simple copying/pasting to your custom slide show.	All art, tables and photos from the text, optimized for presentation.	Art files, tables and photos as they appear in the text. Also provided as base art files without labels or leader lines.	Animation files that can be easily downloaded and played.
All Chapters	Enhanced PPT Enhanced Keynote Enhanced PPT without animations	Enhanced PPT Enhanced Keynote	Image PPT	Labeled Images Unlabeled Images	Animations

Take your course online—*easily*—with one-click Digital Lecture Capture

McGraw-Hill Tegrity Campus™ records and distributes your lecture with just a click of a button. Students can view them anytime/anywhere via computer, iPod, or mobile device. Tegrity Campus indexes as it records your slideshow presentations and anything shown on your computer so **students can see keywords to find exactly what they want to study.**

Anatomy & Physiology | REVEALED® 3.0

An Interactive Cadaver Dissection Experience

This unique multimedia tool is designed to help you master human anatomy and physiology with:

- Content customized to your course
- Stunning cadaver specimens
- Vivid animations
- Lab practical quizzing

my Course Content
- Maximize efficiency by studying exactly what's required.
- Your instructor selects the content that's relevant to your course.

Dissection
- Peel layers of the body to reveal structures beneath the surface.

Animation
- Over 150 animations make anatomy and physiology easier to visualize and understand.

Histology
- Study interactive slides that simulate what you see in lab.

Imaging
- Correlate dissected anatomy with X-ray, MRI, and CT scans.

Quiz
- Gauge proficiency with customized quizzes and lab practicals that cover only what you need for your course.

WWW.APREVEALED.COM

Anatomy & Physiology Revealed® is now available in Cat and Fetal Pig versions!

Physiology Interactive Lab Simulations (Ph.I.L.S.) 4.0

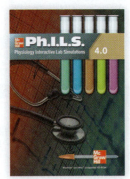

Ph.I.L.S. 4.0 is the perfect way to reinforce key physiology concepts with powerful lab experiments. Created by Dr. Phil Stephens at Villanova University, this program offers *42 laboratory simulations* that may be used to supplement or substitute for wet labs. All 42 labs are self-contained experiments—no lengthy instruction manual required. Users can adjust variables, view outcomes, make predictions, draw conclusions, and print lab reports. This easy-to-use software offers the flexibility to change the parameters of the lab experiment. There are no limits!

What Sets Seeley's Essentials Apart?

Seeley's Essentials of Anatomy & Physiology is designed to help students develop a solid, basic understanding of essential concepts in anatomy and physiology without an encyclopedic presentation of detail. Our goal as authors is to offer a textbook that provides enough information to allow students to understand basic concepts, and from that knowledge, make reasonable predictions and analyses. We have taken great care to select critically important information and present it in a way that maximizes understanding.

Emphasis on Critical Thinking—
Building a Knowledge Base for Solving Problems

An emphasis on critical thinking is integrated throughout this textbook. This approach can be found in questions starting each chapter and embedded within the narrative; in clinical material that is designed to bridge concepts explained in the text with real-life applications and scenarios; in end-of-chapter questions that go beyond rote memorization; and in a visual program that presents material in understandable, relevant images.

▶ Problem-solving perspective from the book's inception

▶ Pedagogy builds student comprehension from knowledge to application (Predict questions, Critical Thinking questions, and Learn to Predict Answer)

Predict 5

What combination of movements at the shoulder and elbow joints allows a person to perform a crawl stroke in swimming?

Predict Questions challenge students to use their understanding of new concepts to solve a problem. Answers to the questions are provided at the end of the book, allowing students to evaluate their responses and to understand the logic used to arrive at the correct answer.

⌁ CRITICAL THINKING

1. A friend tells you that an ECG revealed that her son has a slight heart murmur. Should you be convinced that he has a heart murmur? Explain.

2. Predict the effect on Starling's law of the heart if the parasympathetic (vagus) nerves to the heart are cut.

3. Predict the effect on heart rate if the sensory nerve fibers from the baroreceptors are cut.

4. An experiment is performed on a dog in which the arterial blood pressure in the aorta is monitored before and after the common carotid arteries are clamped. Explain the change in arterial blood pressure that would occur. (*Hint:* Baroreceptors are located in the internal carotid arteries, which are superior to the site of clamping of the common carotid arteries.)

5. Predict the consequences on the heart if a person took a large dose of a drug that blocks calcium channels.

6. What happens to cardiac output following the ingestion of a large amount of fluid?

7. At rest, the cardiac output of athletes and nonathletes can be equal, but the heart rate of athletes is lower than that of nonathletes. At maximum exertion, the maximum heart rate of athletes and nonathletes can be equal, but the cardiac output of athletes is greater than that of nonathletes. Explain these differences.

8. Explain why it is useful that the walls of the ventricles are thicker than those of the atria.

9. Predict the effect of an incompetent aortic semilunar valve on ventricular and aortic pressure during ventricular systole and diastole.

Answers in Appendix D

Critical Thinking These innovative exercises encourage students to apply chapter concepts to solve a problem. Answering these questions helps students build a working knowledge of anatomy and physiology while developing reasoning skills. Answers are provided in Appendix D.

A CASE IN POINT

Injections

Howey Stickum, a student nurse, learns three ways to give injections. An **intradermal injection** is administered by drawing the skin taut and inserting a small needle at a shallow angle into the dermis; an example is the tuberculin skin test. A **subcutaneous injection** is achieved by pinching the skin to form a "tent" and inserting a short needle into the adipose tissue of the subcutaneous tissue; an example is an insulin injection. An **intramuscular injection** is accomplished by inserting a long needle at a 90-degree angle to the skin into a muscle deep to the subcutaneous tissue. Intramuscular injections are used for most vaccines and certain antibiotics.

A Case in Point

These case studies explore relevant issues of clinical interest and explain how material just presented in the text can be used to understand important anatomical and physiological concepts, particularly in a clinical setting.

Clinical Emphasis—*Case Studies Bring Relevance to the Reader*

When problems in structure and/or function of the human body occur, this is often the best time to comprehend how the two are related. A Case in Point readings, untitled Clinical Asides, and Clinical Impact boxes all work to provide a thorough clinical education that fully supports the surrounding textual material. Systems Pathology and Systems Interactions vignettes provide a modern and system's interaction approach to clinical study of the materials presented.

▶ Clinical Impact boxes (placed at key points in the text)

▶ Chapter opening clinical scenarios/vignettes have been given a new look and many are revised

▶ Learn to Predict and chapter Predict questions with unique Learn to Predict Answers

▶ Clinical Asides

▶ Clinical Impact Essays

▶ Clinical Pathologies and Systems Interactions

CLINICAL IMPACT   Bone Fractures

Bone fractures can be classified as **open** (or *compound*), if the bone protrudes through the skin, and **closed** (or *simple*), if the skin is not perforated. Figure 6A illustrates some of the different types of fractures. If the fracture totally separates the two bone fragments, it is called **complete**; if it doesn't, it is called **incomplete**. An incomplete fracture that occurs on the convex side of the curve of a bone is called a **greenstick fracture**. A **comminuted** (kom'i-nū-ted; broken into small pieces) fracture is one in which the bone breaks into more than two fragments. An **impacted** fracture occurs when one of the fragments of one part of the bone is driven into the spongy bone of another fragment.

Fractures can also be classified according to the direction of the fracture line as **linear** (parallel to the long axis);

transverse (at right angles to the long axis); or **oblique** or **spiral** (at an angle other than a right angle to the long axis).

Figure 6A
Types of bone fractures.

Clinical Impact These in-depth essays explore relevant topics of clinical interest. Subjects covered include pathologies, current research, sports medicine, exercise physiology, pharmacology, and various clinical applications.

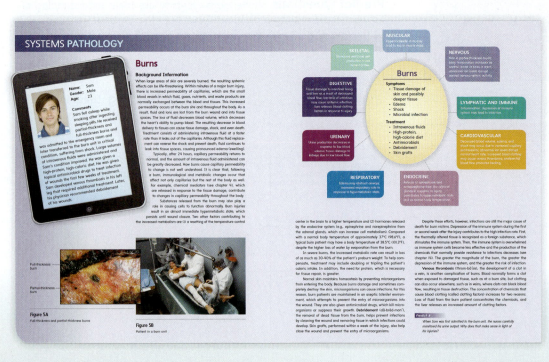

SYSTEMS PATHOLOGY

Burns

Systems Pathology Vignettes These spreads explore a specific condition or disorder related to a particular body system. Presented in a simplified case study format, each Systems Pathology vignette begins with a patient history followed by background information about the featured topic.

Exceptional Art—*Instructive Artwork Promotes Interest and Clarifies Ideas*

A picture is worth a thousand words—especially when you're learning anatomy and physiology. Brilliantly rendered and carefully reviewed for accuracy and consistency, the precisely labeled illustrations and photos provide concrete, visual reinforcement of important topics discussed throughout the text.

Realistic Anatomical Art The anatomical figures in *Seeley's Essentials of Anatomy & Physiology* have been carefully drawn to convey realistic, three-dimensional detail. Richly textured bones and artfully shaded muscles, organs, and vessels lend a sense of realism to the figures that helps students envision the appearance of actual structures within the body.

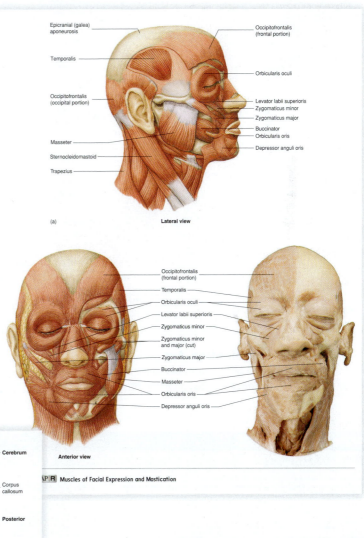

(a) **Lateral view**

Epicranial (galea) aponeurosis
Occipitofrontalis (frontal portion)
Temporalis
Orbicularis oculi
Occipitofrontalis (occipital portion)
Levator labii superioris
Zygomaticus minor
Zygomaticus major
Buccinator
Orbicularis oris
Masseter
Depressor anguli oris
Sternocleidomastoid
Trapezius

Anterior view

Occipitofrontalis (frontal portion)
Temporalis
Orbicularis oculi
Levator labii superioris
Zygomaticus minor
Zygomaticus minor and major (cut)
Zygomaticus major
Buccinator
Masseter
Orbicularis oris
Depressor anguli oris

AP|R Muscles of Facial Expression and Mastication

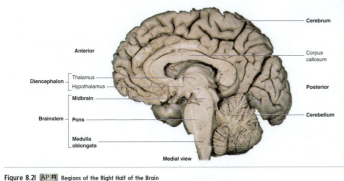

Cerebrum
Anterior
Corpus callosum
Diencephalon — Thalamus
Hypothalamus
Posterior
Midbrain
Brainstem — Pons
Cerebellum
Medulla oblongata
Medial view

Figure 8.21 **AP|R** Regions of the Right Half of the Brain

Atlas-quality cadaver images Clearly labeled photos of dissected human cadavers provide detailed views of anatomical structures, capturing the intangible characteristics of actual human anatomy that can be appreciated only when viewed in human specimens.

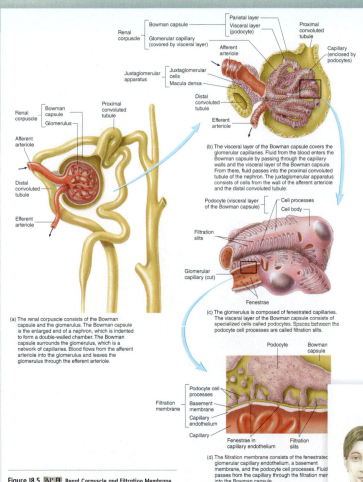

Bowman capsule
Parietal layer
Visceral layer (podocyte)
Proximal convoluted tubule
Renal corpuscle
Glomerular capillary (covered by visceral layer)
Capillary (enclosed by podocytes)
Afferent arteriole
Juxtaglomerular apparatus
Juxtaglomerular cells
Macula densa
Distal convoluted tubule
Efferent arteriole

Renal corpuscle
Bowman capsule
Glomerulus
Proximal convoluted tubule
Afferent arteriole
Distal convoluted tubule
Efferent arteriole

(a) The renal corpuscle consists of the Bowman capsule and the glomerulus. The Bowman capsule is the enlarged end of a nephron, which is indented to form a double-walled chamber. The Bowman capsule surrounds the glomerulus, which is a network of capillaries. Blood flows from the afferent arteriole into the glomerulus and leaves the glomerulus through the efferent arteriole.

(b) The visceral layer of the Bowman capsule covers the glomerular capillaries. Fluid from the blood enters the Bowman capsule by passing through the capillary walls and the visceral layer of the Bowman capsule. From there, fluid passes into the proximal convoluted tubule of the nephron. The juxtaglomerular apparatus consists of cells from the wall of the afferent arteriole and the distal convoluted tubule.

Podocyte (visceral layer of the Bowman capsule)
Cell processes
Cell body
Filtration slits
Glomerular capillary (cut)
Fenestrae

(c) The glomerulus is composed of fenestrated capillaries. The visceral layer of the Bowman capsule consists of specialized cells called podocytes. Spaces between the podocyte cell processes are called filtration slits.

Podocyte
Bowman capsule
Filtration membrane
Podocyte cell processes
Basement membrane
Capillary endothelium
Capillary
Fenestrae in capillary endothelium
Filtration slits

(d) The filtration membrane consists of the fenestrated glomerular capillary endothelium, a basement membrane, and the podocyte cell processes. Fluid passes from the capillary through the filtration membrane into the Bowman capsule.

Figure 18.5 **AP|R** Renal Corpuscle and Filtration Membrane

Multi-level Perspective Illustrations depicting complex structures or processes combine macroscopic and microscopic views to help students see the relationships between increasingly detailed drawings.

Combination Art Drawings are often paired with photographs to enhance the visualization of structures.

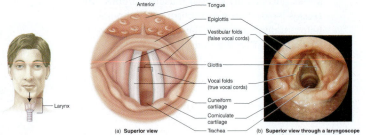

Anterior
Tongue
Epiglottis
Vestibular folds (false vocal cords)
Glottis
Vocal folds (true vocal cords)
Cuneiform cartilage
Corniculate cartilage
Larynx
Trachea

(a) Superior view
(b) Superior view through a laryngoscope

Figure 15.4 **AP|R** Vestibular and Vocal Folds
(Far left) The arrow shows the direction of viewing the vestibular and vocal folds. (a) The relationship of the vestibular folds to the vocal folds and the laryngeal cartilages. (b) Superior view of the vestibular and vocal folds as seen through a laryngoscope.

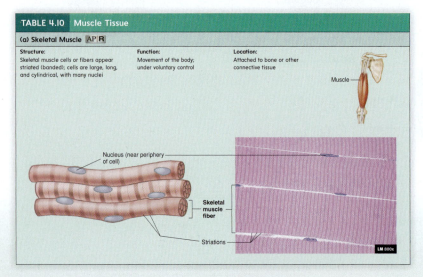

TABLE 4.10 Muscle Tissue

(a) Skeletal Muscle **AP|R**

Structure:	Function:	Location:
Skeletal muscle cells or fibers appear striated (banded); cells are large, long, and cylindrical, with many nuclei	Movement of the body; under voluntary control	Attached to bone or other connective tissue

Muscle

Nucleus (near periphery of cell)
Skeletal muscle fiber
Striations
LM 800x

Histology Micrographs Light micrographs, as well as scanning and transmission electron micrographs, are used in conjunction with illustrations to present a true picture of anatomy and physiology from the cellular level.

Specialized Figures Clarify Tough Concepts

Studying anatomy and physiology does not have to be an intimidating task mired in memorization. *Seeley's Essentials of Anatomy & Physiology* uses two special types of illustrations to help students not only learn the steps involved in specific processes, but also apply the knowledge as they predict outcomes in similar situations. Process Figures organize the key occurrences of physiological processes in an easy-to-follow format. Homeostasis Figures summarize the mechanisms of homeostasis by diagramming how a given system regulates a parameter within a narrow range of values.

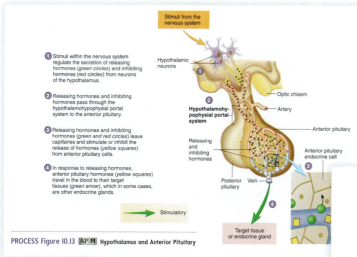

PROCESS Figure 10.13 **AP|R** Hypothalamus and Anterior Pituitary

Step-by-Step Process Figures

Process Figures break down physiological processes into a series of smaller steps, allowing readers to build their understanding by learning each important phase. Numbers are placed carefully in the art, permitting students to zero right in to where the action described in each step takes place.

Correlated with APR! Homeostasis Figures with in-art explanations and organ icons

▶ These specialized flowcharts illustrating the mechanisms that body systems employ to maintain homeostasis have been refined and improved in the ninth edition.

▶ More succinct explanations

▶ Small icon illustrations included in boxes depict the organ or structure being discussed.

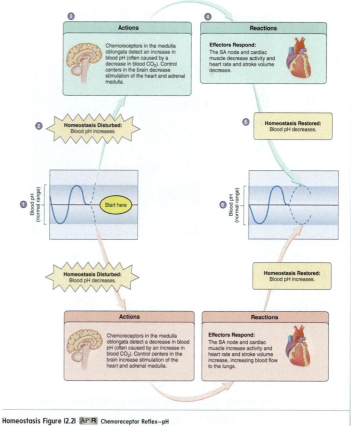

Homeostasis Figure 12.21 **AP|R** Chemoreceptor Reflex—pH

The chemoreceptor reflex maintains homeostasis in response to changes in blood concentrations of CO_2 and H^+ (or pH). (1) Blood pH is within its normal range. (2) Blood pH increases outside the normal range. (3) Chemoreceptors in the medulla oblongata detect increased blood pH. Control centers in the brain decrease sympathetic stimulation of the heart and adrenal medulla. (4) Heart rate and stroke volume decrease, reducing blood flow to lungs. (5) These changes cause blood pH to decrease (as a result of increase in blood CO_2). (6) Blood pH returns to its normal range, and homeostasis is restored.

Outstanding Instructor and Student Resources—
Focusing teaching and engaging students

▶ In-text Learning Outcomes are linked to section headers and Assessment Questions

▶ *McGraw-Hill Anatomy & Physiology REVEALED® (APR)* links to figures for eBook

▶ Learning Outcomes correlation guide between Predict, Learn to Predict, Review and Comprehension, and Critical Thinking Questions

▶ Correlation guide between APR and the textbook

▶ Enhanced Lecture PowerPoints with APR cadaver images

▶ Lecture PowerPoints with embedded animations

▶ McGraw-Hill Connect® Course Management system

▶ Access to media-rich eBooks

▶ McGraw-Hill LearnSmart® tailors study time and identifies at-risk students

▶ **NEW!** Clinical questions added to the Connect Question Bank based on the Clinical Features within each chapter

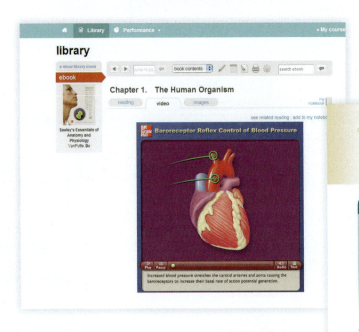

◀ The interactive eBook takes the reading experience to a new level with links to animations and interactive exercises that supplement the text.

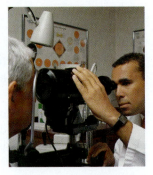

Functionality such as highlighting and post-it notes ▶ allow customizing for a personalized study guide.

Learn to Predict and Learn to Predict Answer—
Helping students learn how to think critically

▶ Part of the overall critical thinking Predict questions that appear throughout each chapter, a special Learn to Predict question now opens every chapter. This specifically written scenario links with the chapter opener photo and helps introduce the subject matter covered within the chapter.

LEARN TO PREDICT

Renzo, the dancer in the photo, is perfectly balanced, yet a slight movement in any direction would cause him to adjust his position. The human body adjusts its balance among all its parts through a process called homeostasis.

Let's imagine that Renzo is suffering from a blood sugar disorder. Earlier, just before this photo was taken, he'd eaten an energy bar. As an energy bar is digested, blood sugar rises. Normally, tiny collections of cells embedded in the pancreas respond to the rise in blood sugar by secreting the chemical insulin. Insulin increases the movement of sugar from the blood into the cells. However, Renzo did not feel satisfied from his energy bar. He felt dizzy and was still hungry, all symptoms he worried could be due to a family history of diabetes. Fortunately, the on-site trainer tested his blood sugar and noted that it was much higher than normal. After a visit to his regular physician, Renzo was outfitted with an insulin pump and his blood sugar levels are more consistent.

After reading about homeostasis in this chapter, create an explanation for Renzo's blood sugar levels before and after his visit to the doctor.

▶ A new Learn to Predict Answer box at the end of each chapter teaches students step-by-step how to answer the chapter-opening critical thinking question. This is foundational to real learning and is a crucial part of helping students put facts together to reach that "Aha" moment of true comprehension.

ANSWER TO LEARN TO PREDICT

The first Predict feature in every chapter of this text is designed to help you develop the skills to successfully answer critical thinking questions. The first step in the process is always to analyze the question itself. In this case, the question asks you to evaluate the mechanisms governing Renzo's blood sugar levels, and it provides the clue that there's a homeostatic mechanism involved. In addition, the question describes a series of events that helps create an explanation: Renzo doesn't feel satisfied after eating, has elevated blood sugar, and then is prescribed an insulin pump.

In chapter 1, we learn that homeostasis is the maintenance of a relatively constant internal environment. Renzo experienced hunger despite eating, and his blood sugar levels were higher than normal. In this situation, we see a disruption in homeostasis because his blood sugar stayed too high after eating. Normally, an increased blood sugar after a meal would return to the normal range by the activity of insulin secreted by the pancreas. When blood sugar returns to normal, insulin secretion stops. In Renzo's case, his pancreas has stopped making insulin. Thus, the doctor prescribed an insulin pump to take over for his pancreas. Now when Renzo eats, the insulin pump puts insulin into his blood and his blood sugar levels are maintained near the set point.

Answers to the rest of this chapter's Predict questions are in Appendix E.

PEDAGOGICAL FEATURES ENSURE SUCCESS

A major change you will notice in the ninth edition is the incorporation of Learning Outcomes that are closely linked with in-chapter Predict and Learn to Predict questions as well as the Summary, Critical Thinking, and Review and Comprehension questions. These carefully designed learning aids assist students in reviewing chapter content, evaluating their grasp of key concepts, and utilizing what they've learned.

3.4 MOVEMENT THROUGH THE CELL MEMBRANE

Learning Outcomes *After reading this section, you should be able to*

A. Define diffusion and concentration gradient.
B. Explain the role of osmosis and that of osmotic pressure in controlling the movement of water across the cell membrane. Compare hypotonic, isotonic, and hypertonic solutions.
C. Define carrier-mediated transport, and compare the processes of facilitated diffusion, active transport, and secondary active transport.
D. Describe endocytosis and exocytosis.

Studying Anatomy and Physiology does not have to be intimidating

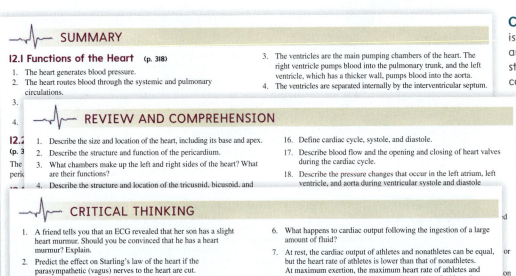

SUMMARY

12.1 Functions of the Heart   (p. 318)

1. The heart generates blood pressure.
2. The heart routes blood through the systemic and pulmonary circulations.
3.
4.

3. The ventricles are the main pumping chambers of the heart. The right ventricle pumps blood into the pulmonary trunk, and the left ventricle, which has a thicker wall, pumps blood into the aorta.
4. The ventricles are separated internally by the interventricular septum.

12.2
(p. 3

The
peri

REVIEW AND COMPREHENSION

1. Describe the size and location of the heart, including its base and apex.
2. Describe the structure and function of the pericardium.
3. What chambers make up the left and right sides of the heart? What are their functions?
4. Describe the structure and location of the tricuspid, bicuspid, and

16. Define cardiac cycle, systole, and diastole.
17. Describe blood flow and the opening and closing of heart valves during the cardiac cycle.
18. Describe the pressure changes that occur in the left atrium, left ventricle, and aorta during ventricular systole and diastole

CRITICAL THINKING

1. A friend tells you that an ECG revealed that her son has a slight heart murmur. Should you be convinced that he has a heart murmur? Explain.
2. Predict the effect on Starling's law of the heart if the parasympathetic (vagus) nerves to the heart are cut.
3. Predict the effect on heart rate if the sensory nerve fibers from the baroreceptors are cut.
4. An experiment is performed on a dog in which the arterial blood pressure in the aorta is monitored before and after the common carotid arteries are clamped. Explain the change in arterial blood pressure that would occur. (*Hint:* Baroreceptors are located in the internal carotid arteries, which are superior to the site of clamping of the common carotid arteries.)
5. Predict the consequences on the heart if a person took a large dose of a drug that blocks calcium channels.

6. What happens to cardiac output following the ingestion of a large amount of fluid?
7. At rest, the cardiac output of athletes and nonathletes can be equal, but the heart rate of athletes is lower than that of nonathletes. At maximum exertion, the maximum heart rate of athletes and nonathletes can be equal, but the cardiac output of athletes is greater than that of nonathletes. Explain these differences.
8. Explain why it is useful that the walls of the ventricles are thicker than those of the atria.
9. Predict the effect of an incompetent aortic semilunar valve on ventricular and aortic pressure during ventricular systole and diastole.

Answers in Appendix D

connect | ANATOMY & PHYSIOLOGY

Anatomy & Physiology **REVEALED** aprevealed.com

Chapter Summary   The summary is now conveniently linked by section and page number while it briefly states the important facts and concepts covered in each chapter.

Review and Comprehension

These multiple-choice practice questions cover the main points presented in the chapter. Completing this self-test helps students gauge their mastery of the material.

Critical Thinking Questions

These innovative exercises encourage students to apply chapter concepts to solve problems. Answering these questions helps build their working knowledge of anatomy and physiology while developing reasoning and critical thinking skills.

Answers to Predict Questions

These innovative critical thinking questions encourage students to become active learners as they read. Predict Questions challenge the understanding of new concepts needed to solve a problem. The questions are answered in Appendix E, allowing students to evaluate their responses and understand the logic used to arrive at the correct answer.

Ninth Edition Changes

WHAT'S NEW AND IMPROVED?

The ninth edition of *Seeley's Essentials of Anatomy & Physiology* is the result of extensive analysis of the text and evaluation of input from anatomy and physiology instructors who have thoroughly reviewed chapters. The result is a retention of the beloved features which foster student understanding, with an emphasis on a sharper focus within many sections, affording an even more logical flow within the text. Updating of content, along with revision of Homeostasis Figures and the addition of a new feature entitled Microbes In Your Body, make this an exciting edition.

Learning Outcomes and Assessment—
Helping instructors track student progress

UPDATED! Learning Outcomes are carefully written to outline expectations for each section

NEW! Microbes In Your Body feature discussing the many important and sometimes, little known roles of microbes and the physiology of homeostasis

UPDATED! Online student questions and test bank questions are correlated with Learning Outcomes to further scaffold and measure student progress and understanding

NEW! Online clinical study questions are based from clinical features within the text including Microbes In Your Body and System Pathologies, and are correlated with Learning Outcomes and HAPS Learning Objectives to further develop and measure higher level thinking and application of learned content

6.8 JOINTS

Learning Outcomes *After reading this section, you should be able to*

A. Describe the two systems for classifying joints.
B. Explain the structure of a fibrous joint, list the three types, and give examples of each type.
C. Give examples of cartilaginous joints.
D. Illustrate the structure of a synovial joint and explain the roles of the components of a synovial joint.
E. Classify synovial joints based on the shape of the bones in the joint and give an example of each type.
F. Demonstrate the difference between the following pairs of movements: flexion and extension; plantar flexion and dorsiflexion; abduction and adduction; supination and pronation; elevation and depression; protraction and retraction; opposition and reposition; inversion and eversion.

MICROBES IN YOUR BODY — Do our bacteria make us fat?

Obesity has increased at an alarming rate over the last three decades. It is estimated that over 150 billion adults worldwide are overweight or obese. In the United States, 1/3 of adults are obese. As obesity rates have increased, so have the rates of obesity-related health conditions such as insulin resistance, diabetes, and cardiovascular disease. Why this dramatic increase? There are two main reasons for obesity: diet/lifestyle and gut bacteria; and it seems these two may be related.

The most familiar cause of obesity is diet and lifestyle. The "typical" Western diet consists of frequent large meals high in refined grains, red meat, saturated fats, and sugary drinks. This is in sharp contrast to healthier diets rich in whole grains, vegetables, fruits, and nuts that help with weight control and prevention of chronic disease. From an evolutionary perspective, our bodies are adapted to conserve energy because food sources were scarce for ancient humans. Many of us now have easy access to energy-rich foods. Combined with a reduction in physical activity and less sleep for many Americans, the Western diet and lifestyle can lead to obesity and poor health.

Bacteroidetes than Firmicutes, while the opposite is true for obese people.

We now know that gut microbiota affect nutrient processing and absorption, hormonal regulation of nutrient use by body cells, and even our hunger level. In addition, our diet can influence the type of bacteria in our GI system. Studies of humans on carbohydrate-restricted or fat-restricted diets demonstrated that after weight loss, the number of Bacteroidetes ("lean person" bacteria) increased, while the number of Firmicutes ("obese person" bacteria) decreased. This makes sense in light of the fact that Firmicutes bacteria break down ingested food more completely than Bacteroidetes, which makes the food's energy easier to absorb by the human gut. Obese individuals store the absorbed energy in adipose tissue, which contributes to weight gain.

Furthermore, experiments with germ-free mice—mice lacking normal gut microbiota—have demonstrated just how important normal gut bacteria are for homeostasis. In the absence of normal gut microbiota, malfunctions in germ-free mice are widespread and significant. For example, when germ-free mice received gut microbiota transplants

pathogens. Finally, germ-free mice display an enhanced stress response, which is substantially reduced upon implantation of gut microbiota. Overall, these experiments demonstrate that there is a much greater correlation between bacteria, gut health, obesity, and anxiety than ever before realized.

Changes in gut microbiota also alter the hormonal regulation of nutrient use. Inflammation-promoting effects of an imbalanced gut microbiota is thought to induce obesity via promoting insulin resistance, a known autoimmune [...]. This obser[...] vation is support[...] betes symptoms[...] gery when pati[...] in gut microbio[...] well documented[...] metabolism is cri[...] anti-hunger horm[...] neurotransmitters[...] in normal gut mi[...] may very well disr[...] nals and gut perm[...] eating and inflam[...]

These obser[...] can we manipulat[...] people to cause [...]

◄ This feature helps students to understand the important role microbes play in helping various systems of the body to maintain homeostasis.

MICROBES IN YOUR BODY — Using Bacteria to Fight Bacteria

Acne (acne vulgaris) is the most common skin condition in the United States. Though 80% of all American adolescents develop acne, adults can also be affected by it. When considering all age groups, approximately 40 to 50 million Americans suffer from acne. Unfortunately, there is not a tried and true cure for acne; however, new research examining the skin microbiome may have found a natural and effective treatment to get healthy, clear skin. Unique species of bacteria, *Propionibacterium acnes* (*P. acnes*), are found in sebum-rich areas of the skin, such as the forehead, side of the nose, and back. Although it has been

difficult to study these bacteria, the inception of the Human Microbiome Project (see "Getting to Know Your Bacteria" in chapter 1) allowed scientists to determine specific genetic traits of skin microbiome bacteria. Using this technique, scientists have identified three unique strains of *P. acnes*. Of the three strains, one strain is more dominant in people with acne-free skin. Research has shown that this strain of *P. acnes* does not adversely affect the host. However, the other two strains of *P. acnes* are pathogenic to humans. So, how does this information help scientists learn how to prevent acne? It seems that the "good" *P. acnes* prevents

invasion of the skin by certain bacteria through a natural metabolic process. When *P. acnes* breaks down lipids, the skin pH is lowered to a level not tolerated by the invading bacteria. Scientists have proposed that the strain of *P. acnes* in healthy skin ("good" *P. acnes*) kills off the pathogenic strains of *P. acnes* ("bad" *P. acnes*) in a similar fashion. Since acne-affected people do not host the "good" strain, the "bad" strain can take over and cause the annoying skin eruptions of acne. Thus, perhaps in the future to prevent acne, affected people can apply the "good" *P. acnes* in a cream to prevent the "bad" *P. acnes* from taking over.

Chapter-by-Chapter Changes

Chapter 1

- Added figure legend to chapter opener photo to link photo more closely to the Learn to Predict for a complete story
- Throughout the entire textbook, dividing lines were added between the figures and the legends to help students clearly visualize the art concepts
- Systems figures were enhanced to increase clarity
- Homeostasis discussion was rewritten per reviewer feedback to: simplify, clarify, and make more accurate
- New predict #2 question and answer written to reflect changes in homeostasis discussion
- Figure 1.5 revised to match new homeostasis discussion
- Throughout the chapter and the entire textbook, adipose tissue replaces fat to be more accurate when referring to the material (adipose) where the chemical (fat) is stored
- Throughout the entire textbook, all homeostasis figures were revised for consistency and accuracy
- Figure 1.7 was updated to enhance students' comprehension of positive feedback, which is frequently misunderstood
- Figure 1.14 was updated by adding in organ art to help students relate the terms to actual organs
- New feature added: Microbes in Your Body, "Getting to Know your Bacteria." This helps the text to stay current in the field of biology where there is a greater focus on the microbiome and its importance in human health and homeostasis

Chapter 2

- Added a legend to the chapter opening photo to link better to the Learn to Predict
- Updated the discussion in section 2.1, "Ionic Bonding" for clarity
- Figure 2.2 was updated for better visibility and clarity
- Figure 2.3 was also updated for better visibility and clarity
- For consistency throughout the entire text, some symbols have replaced the words where appropriate (CO_2, O_2, and H_2O)
- Added table 2.3 to distinguish amongst chemical bond types
- Figure 2.13b was updated to represent unsaturated fatty acids in a more realistic way. Students need to see the molecule actually bent and not linear
- Figure 2.15 was updated to match other figures throughout the textbook
- Figure 2.17 was updated to match other graphs throughout the textbook
- A figure legend was added to figure 2.20 to explain why the bond between adjacent phosphates is represented differently than all the previous bonds shown in the chapter. Students without a chemistry background may be unfamiliar with this symbol

Chapter 3

- Increased size of figure 3.1 for better visual of organelles
- Image coloration changed for cytoplasm clarity
- Process figure 3.26 revised for clarity
- Osmosis discussion revised for clarity

Chapter 4

- Table 4.1 was updated to match the art in this chapter
- Table 4.2 was updated for consistency throughout the chapter
- In table 4.2a, the histology image was replaced with a clearer one of simple squamous epithelium
- In tables 4.4a, 4.6a, 4.7c, 4.9, and 4.10c a clearer histology image was used for clarity
- The terminology was changed from "respiratory passages" to "respiratory airways" for clarity
- The language in section 4.6, "Tissue Membranes" was clarified to indicate that the section describes tissue membranes and not cell membranes. "Fat" was changed to "adipose tissue" where appropriate

Chapter 5

- New Microbes in Your Body: Using Bacteria to Fight Bacteria

Chapter 6

- Throughout the chapter, the bone shading was lightened for realism
- A photo caption was added to the cover opener photo to link it to the Learn to Predict
- Figure 6.8 was updated to add an x-ray of a broken bone before and after callus formation
- In 6A, "greenstick fracture" was more clearly defined
- Throughout the chapter, the skull art's coloration was substantially brightened to help students more easily differentiate between the individual skull bones (figures 6.11 and 6.12a)
- In figure 6.12a, the nasal conchae drawings were clarified because in the former edition, the bones were not distinguishable from the background
- Figures 6.14, 6.15, 6.19a, 6.20, 6.25, 6.26, 6.31, 6.33 were revised to add photos of actual skulls, which share leader lines with the line art. This helps students conceptualize the anatomy more clearly
- Figures 6.24 and 6.28 were revised for accuracy of leader line placement
- The definition of flexion and extension was updated and corrected per reviewer feedback

Chapter 7

- Throughout the chapter, the actin and myosin myofilament line art was arranged so the myosin appears thicker than the actin
- A new Learn to Predict question was written that is more closely aligned with the chapter opening photo and muscle function
- A legend was added to the chapter opener photo to tie it in with the Learn to Predict

- The text for "Skeletal Muscle Structure" in section 7.2 was rewritten to flow logically from a macro view to a micro view
- Figure 7.2 was heavily revised so the art is oriented linearly and flows directly to the next, more magnified level of muscle structure
- Figure 7.3 was also heavily revised: Part a was added to show the logical flow from the macro to the micro; part b was cropped so the myofibrils are oriented linearly on the page and correlate more directly to part a; part c was added to provide a visual orientation of myofilament arrangement relative to each other
- Predict question #2 is new and covers muscle fiber electrical activity—a predict question topic that was missing in the previous edition
- Figure 7.11 was revised to better correlate a given response to its corresponding stimulus frequency
- Table 7.1 was added for clear distinction amongst fiber types
- The section on Energy Requirements for Muscle Contraction was updated to reflect the most up-to-date information about lactate fate. The definition of aerobic and anaerobic respiration in skeletal muscle was clarified
- The section on muscle fatigue was updated
- Figure 7.12 was heavily revised to visually differentiate between energy usage at rest vs. exercise
- A section on fiber type effect on activity level was added
- Table 7.3 on muscle nomenclature was added
- Figure 7.16 was updated to add a cadaver photo with shared leader lines with the line art. This helps students visualize the anatomy more clearly
- Table 7.13 was revised for consistent pronunciation of "teres"
- The Diseases and Disorders table was revised to accurately discuss ATP production and not lactic acid production

Chapter 8

- New figure 8.2 better represents the organization of the nervous system
- Revision of figure 8.11 more accurately represents saltatory conduction
- Dermatomal map added to figure 8.20

Chapter 9

- New figure 9.1 to present types of senses
- New section 9.2 describes types of receptors
- New figure 9.6 presents the pathways for the sense of taste
- Figure 9.16a revised for clarity
- New figure 9.20 presents the auditory pathway

Chapter 10

- Added a new Microbes in Your Body—"Do Our Bacteria Make Us Fat?" boxed reading

- Table 10.1 was updated to clarify the definition of autocrine chemical messengers
- The definition of paracrine chemical messengers in section 10.1 was updated
- Section 10.3 was updated to clarify hormones' sources as groups of cells as well as glands
- Section 10.4 was updated for clarity and accuracy
- Updated section 10.5 "Inhibition of Hormone Release by Hormonal Stimuli" for clarity
- Section 10.6 "Classes of Receptors" was revised to reflect newer research on membrane-bound receptor action by lipid-soluble hormones
- Figures 10.7a and 10.8 were updated for consistency with others for style
- Figure 10.8 was updated to reflect the information about membrane-bound receptor actions
- Figures 10.9 and 10.10 were updated to match the style of others throughout the textbook
- Section 10.6—"Membrane-bound Receptors and Signal Amplification" was revised for clarity and to incorporate lipid-soluble hormones
- Section 10.7—"Hormones of the Posterior Pituitary" was revised for clarity
- Figure 10.17 was revised for consistency throughout the textbook
- The term "intracellular receptor" was changed to "nuclear receptor" throughout the chapter
- Figure 10.19 was updated for clarity
- Figures 10.20 and 10.21 were updated for consistency with other figures
- Section 10.7—"Pancreas, Insulin, and Diabetes" was revised for accuracy. It now includes a definition of somatostatin
- Figure 10.22 was revised to include somatostatin
- Table 10.3 was updated to use adipose not fat where appropriate
- Figure 10.23 was revised for consistency

Chapter 11

- Figure 11.1 updated to show blood as % body weight
- Figure 11.2 revised to introduce myeloid and lymphoid stem cells
- Revision of 11.6 clarifies the relationship between transfusion reactions and kidney failure
- Figure 11.13 revised to better represent the relationship between the maternal blood and fetal blood

Chapter 12

- Revisions to figure 12.13 allow for better visualization of cardiac muscle cell structure
- Figure 12.14 revised to contrast skeletal muscle and cardiac muscle refractory period and resultant tension production
- Discussion of cardiac cycle revised to correlate with the descriptions of blood flow and the ECG. Figure 12.17 also updated according to these revisions

Chapter 13

- Figure 13.24 revised to better represent influences of blood pressure and osmosis on capillary exchange
- Clinical Impact "Circulatory Shock" updated to distinguish between septic shock and blood poisoning

Chapter 14

- Figure 14.11 revised to include plasma cells producing antibodies
- New Microbes in Your Body feature: Do Our Gut Bacteria Drive Immune Development and Function?

Chapter 15

- The Learn to Predict was updated to include questions about a ventilator to link it to the photo
- The chapter opener photo was updated
- Section 15.1 was revised to incorporate the term pathogen
- Figure 15.2 was updated to better represent the pharyngeal tonsils
- Figure 15.7 was updated for accuracy
- Section 15.3—"Pressure Changes and Airflow" was updated for clarity; Pleural Pressure to direct students attention to the boxed reading
- The term "aerobic respiration" was converted to "cellular respiration" for consistency throughout the textbook
- Figure 15.14 was updated for clarity and accuracy of legend text
- Section 15.6—"Generation of Rhythmic Breathing" was revised for accuracy regarding somatic nervous system regulation of breathing
- Section 15.6—"Chemical Control of Breathing" was revised to distinguish between CO_2 levels during exercise and hyperventilation
- Figure 15.7 was updated for consistency
- Section 15.6—"Effect of Exercise on Breathing" was updated for accuracy regarding lactic acid production
- Figure 15B legend was revised for better distinction between the two images and magnifications were added to the photos
- The Diseases and Disorders table was updated to correctly place text with the "Thrombosis of the pulmonary arteries"

Chapter 16

- A new feature "Microbes in Your Body—Fecal Implants" was added
- Section 16.3—"Anatomy of the Oral Cavity" was updated to include the lingual tonsils
- Figure 16.5 was updated for consistency throughout the text
- Section 16.3—"Salivary Glands" was updated for accuracy; "Saliva" for clarity
- Section 16.3—"Esophagus" was updated for accuracy with skeletal: smooth muscle proportions

- Section 16.6—"Liver" was updated for accuracy with bile production and gallstone formation
- Section 16.6—"Functions of the Pancreas" were updated for clarity
- Figures 16.22, 16.23, 16.24, 16.25, and 16.27 was updated for consistency throughout the text

Chapter 17

- Recommended fiber intake added to the discussion of carbohydrates
- FDA proposed changes to food labels added to figure 17.2

Chapter 18

- The Learn to Predict was revised to link more closely to the chapter opener photo
- Throughout the chapter, the term "Bowman's" was changed to "The Bowman" capsule
- Figure 18.3 was updated for clarity and labels for Renal Column were added to parts a and b
- Section 18.3 was revised to indicate "Production," which is a more active regulation term and an analogy for kidney function was added to help students conceptualize the mechanisms more clearly
- Figure 18.5 was edited to reflect term changes
- Table 18.1 was updated to give normal values for pH and specific gravity
- Section 18.3 was updated to give better filtration definition
- Section 18.3—"Filtration" was revised for clarity and accuracy
- Figures 18.11, 18.13, 18.17, 18.19 and 18.22 were edited for clarity and consistency throughout the textbook

Chapter 19

- The Learn to Predict questions were updated to compare meiosis in males and females
- A caption was added to the chapter opener photo to link more closely to the Learn to Predict
- The box on "Descent of the Testes" was updated to include a discussion of treatments
- The language was changed from "Sex Hormones" to "Reproductive Hormones" to reflect the more current style
- Figures 19.7 and 19.14 were updated for consistency
- Figure 19A was updated to have more modern photos

Chapter 20

- Revision of Respiratory and Circulatory Changes in the newborn to better explain the changes of oxygenated blood and deoxygenated blood flow through vessels before and after birth
- Discussion of segregation errors revised for clarity

List of Clinical Impact Essays

Acknowledgments

In today's world, no textbook is brought to fruition through the work of the authors alone. Without the support of friends, family, and colleagues, it would not have been possible for us to complete our work on this text. The final product is truly a team effort. We want to express sincere gratitude to the staff of McGraw-Hill for their help and encouragement. We sincerely appreciate Brand Manager Amy Reed and Product Developer Mandy Clark, for their hours of work, suggestions, patience, and undying encouragement. We also thank Content Project Manager Jayne Klein, Content Licensing Specialist John Leland, Buyer Laura Fuller, and Designer David Hash for their hours spent turning a manuscript into a book; Media Project Manager Sherry Kane for her assistance in building the various products that support our text; and Marketing Manager Jessica Cannavo for her enthusiasm in promoting this book. The McGraw-Hill Education employees with whom we have worked are extremely professional, but more than that, they are completely dedicated to their role as part of the content team.

Thanks are gratefully offered to Copy Editor Beth Bulger for carefully polishing our words.

We also thank the many illustrators who worked on the development and execution of the illustration program and who provided photographs and photomicrographs for the ninth edition of *Essentials of Anatomy & Physiology*. The art program for this text represents a monumental effort, and we appreciate their contribution to the overall appearance and pedagogical value of the illustrations.

Finally, we sincerely thank the reviewers and the teachers who have provided us with excellent constructive criticism. The remuneration they received represents only a token payment for their efforts. To conscientiously review a textbook requires a true commitment and dedication to excellence in teaching. Their helpful criticisms and suggestions for improvement were significant contributions that we greatly appreciate. We acknowledge them by name in the next section.

Cinnamon VanPutte
Jennifer Regan
Andy Russo

Reviewers

Malene Arnaud-Davis
Delgado Community College

Nahel Awadallah
Johnston Community College

Yavuz Cakir
Benedict College

Alan Crandall
Idaho State University

Lyndasu Crowe
Darton State College

Philias Denette
Delgado Community College

Lynn Gargan
Tarrant County College—NE Campus

Ronald Harris
Marymount College

Matt Lee
Diablo Valley College

Elizabeth Mays
Illinois Central College

Jose' E. Rivera
Indiana University of Pennsylvania

Edith Sarino
Arizona Western College

Alice Shapiro
Lane Community College

Scott Wersinger
University at Buffalo, SUNY

Paul Zillgitt
University of Wisconsin-Waukesha

The Human Organism

LEARN TO PREDICT

Renzo, the dancer in the photo, is perfectly balanced, yet a slight movement in any direction would cause him to adjust his position. The human body adjusts its balance among all its parts through a process called homeostasis.

Let's imagine that Renzo is suffering from a blood sugar disorder. Earlier, just before this photo was taken, he'd eaten an energy bar. As an energy bar is digested, blood sugar rises. Normally, tiny collections of cells embedded in the pancreas respond to the rise in blood sugar by secreting the chemical insulin. Insulin increases the movement of sugar from the blood into the cells. However, Renzo did not feel satisfied from his energy bar. He felt dizzy and was still hungry, all symptoms he worried could be due to a family history of diabetes. Fortunately, the on-site trainer tested his blood sugar and noted that it was much higher than normal. After a visit to his regular physician, Renzo was outfitted with an insulin pump and his blood sugar levels are more consistent.

After reading about homeostasis in this chapter, create an explanation for Renzo's blood sugar levels before and after his visit to the doctor.

Module 1 Body Orientation

1.1 ANATOMY

Learning Outcomes *After reading this section, you should be able to*

A. Define anatomy and describe the levels at which anatomy can be studied.
B. Explain the importance of the relationship between structure and function.

Human anatomy and physiology is the study of the structure and function of the human body. The human body has many intricate parts with coordinated functions maintained by a complex system of checks and balances. The coordinated function of all the parts of the human body allows us to detect changes or stimuli, respond to stimuli, and perform many other actions.

Knowing human anatomy and physiology also provides the basis for understanding disease. The study of human anatomy and physiology is important for students who plan a career in the health sciences because health professionals need a sound knowledge of structure and function in order to perform their duties. In addition, understanding anatomy and physiology prepares all of us to evaluate recommended treatments, critically review advertisements and reports in the popular literature, and rationally discuss the human body with health professionals and nonprofessionals.

Anatomy (ă-nat′ō-mē) is the scientific discipline that investigates the structure of the body. The word *anatomy* means to dissect, or cut apart and separate, the parts of the body for study.

Anatomy covers a wide range of studies, including the structure of body parts, their microscopic organization, and the processes by which they develop. In addition, anatomy examines the relationship between the structure of a body part and its function. Just as the structure of a hammer makes it well suited for pounding nails, the structure of body parts allows them to perform specific functions effectively. For example, bones can provide strength and support because bone cells secrete a hard, mineralized substance. Understanding the relationship between structure and function makes it easier to understand and appreciate anatomy.

Two basic approaches to the study of anatomy are systemic anatomy and regional anatomy. **Systemic anatomy** is the study of the body by systems, such as the cardiovascular, nervous, skeletal, and muscular systems. It is the approach taken in this and most introductory textbooks. **Regional anatomy** is the study of the organization of the body by areas. Within each region, such as the head, abdomen, or arm, all systems are studied simultaneously. This is the approach taken in most medical and dental schools.

Anatomists have two general ways to examine the internal structures of a living person: surface anatomy and anatomical imaging. **Surface anatomy** is the study of external features, such as bony projections, which serve as landmarks for locating deeper structures (for examples, see chapters 6 and 7). **Anatomical imaging** involves the use of x-rays, ultrasound, magnetic resonance imaging (MRI), and other technologies to create pictures of internal structures. Both surface anatomy and anatomical imaging provide important information for diagnosing disease.

1.2 PHYSIOLOGY

Learning Outcomes *After reading this section, you should be able to*

- **A.** Define physiology.
- **B.** State two major goals of physiology.

Physiology (fiz-ē-ol′ō-jē; the study of nature) is the scientific discipline that deals with the processes or functions of living things. It is important in physiology to recognize structures as dynamic rather than fixed and unchanging. The major goals of physiology are (1) to understand and predict the body's responses to stimuli and (2) to understand how the body maintains conditions within a narrow range of values in the presence of continually changing internal and external environments. **Human physiology** is the study of a specific organism, the human, whereas **cellular physiology** and **systemic physiology** are subdivisions that emphasize specific organizational levels.

1.3 STRUCTURAL AND FUNCTIONAL ORGANIZATION OF THE HUMAN BODY

Learning Outcomes *After reading this section, you should be able to*

- **A.** Describe the six levels of organization of the body, and describe the major characteristics of each level.
- **B.** List the eleven organ systems, identify their components, and describe the major functions of each system.

The body can be studied at six structural levels: chemical, cell, tissue, organ, organ system, and organism (figure 1.1).

Chemical Level

The structural and functional characteristics of all organisms are determined by their chemical makeup. The **chemical** level of organization involves how atoms, such as hydrogen and carbon, interact and combine into molecules. The function of a molecule is intimately related to its structure. For example, collagen molecules are strong, ropelike fibers that give skin structural strength and flexibility. With old age, the structure of collagen changes, and the skin becomes fragile and more easily torn. A brief overview of chemistry is presented in chapter 2.

Cell Level

Cells are the basic structural and functional units of organisms, such as plants and animals. Molecules can combine to form **organelles** (or′gă-nelz; little organs), which are the small structures that make up some cells. For example, the nucleus contains the cell's hereditary information, and mitochondria manufacture adenosine triphosphate (ATP), a molecule cells use for a source of energy. Although cell types differ in their structure and function, they have many characteristics in common. Knowledge of these characteristics and their variations is essential to a basic understanding of anatomy and physiology. The cell is discussed in chapter 3.

Tissue Level

A **tissue** (tish′ū) is a group of similar cells and the materials surrounding them. The characteristics of the cells and surrounding materials determine the functions of the tissue. The many tissues that make up the body are classified into four primary types: epithelial, connective, muscle, and nervous. Tissues are discussed in chapter 4.

Organ Level

An **organ** (ōr′găn; a tool) is composed of two or more tissue types that together perform one or more common functions. The urinary bladder, skin, stomach, and heart are examples of organs (figure 1.2).

Organ System Level

An **organ system** is a group of organs classified as a unit because of a common function or set of functions. For example, the urinary system consists of the kidneys, ureter, urinary bladder, and urethra. The kidneys produce urine, which is transported by the ureters to the urinary bladder, where it is stored until eliminated from the body by passing through the urethra. In this text, we consider eleven major organ systems: integumentary, skeletal, muscular, lymphatic, respiratory, digestive, nervous, endocrine, cardiovascular, urinary, and reproductive (figure 1.3).

The coordinated activity of the organ systems is necessary for normal function. For example, the digestive system takes in and processes food, which is carried by the blood of the cardiovascular system to the cells of the other systems. These cells use the food and produce waste products that are carried by the blood to the kidneys of the urinary system, which removes waste products from the blood. Because the organ systems are so interrelated, dysfunction in one organ system can have profound effects on other systems. For example, a heart attack can result in inadequate circulation of blood. Consequently, the organs of other systems, such as the brain and kidneys, can malfunction. Throughout this text, Systems Pathology essays consider the interactions of the organ systems.

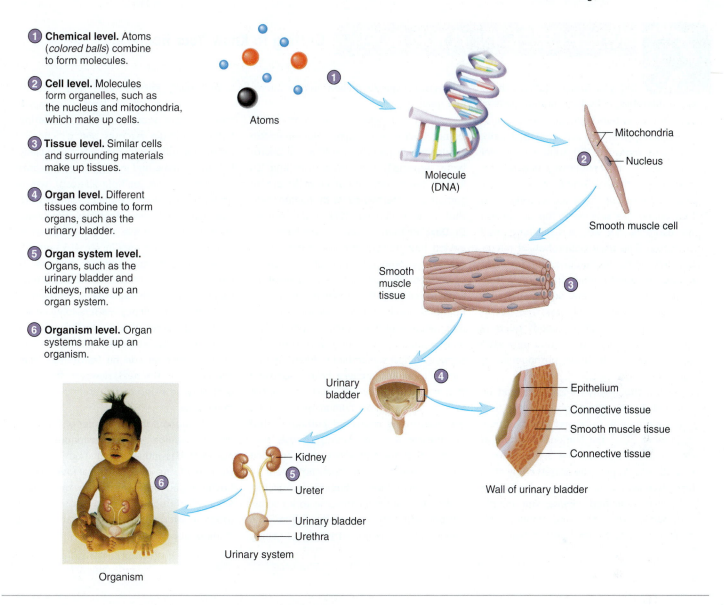

① **Chemical level.** Atoms (*colored balls*) combine to form molecules.

② **Cell level.** Molecules form organelles, such as the nucleus and mitochondria, which make up cells.

③ **Tissue level.** Similar cells and surrounding materials make up tissues.

④ **Organ level.** Different tissues combine to form organs, such as the urinary bladder.

⑤ **Organ system level.** Organs, such as the urinary bladder and kidneys, make up an organ system.

⑥ **Organism level.** Organ systems make up an organism.

Atoms

Molecule (DNA)

Mitochondria

Nucleus

Smooth muscle cell

Smooth muscle tissue

Urinary bladder

Epithelium

Connective tissue

Smooth muscle tissue

Connective tissue

Wall of urinary bladder

Kidney

Ureter

Urinary bladder

Urethra

Urinary system

Organism

PROCESS Figure 1.1 Levels of Organization for the Human Body

Organism Level

An **organism** is any living thing considered as a whole, whether composed of one cell, such as a bacterium, or of trillions of cells, such as a human. The human organism is a complex of organ systems that are mutually dependent on one another (figure 1.3).

1.4 CHARACTERISTICS OF LIFE

Learning Outcome *After reading this section, you should be able to*

A. List and define six characteristics of life.

Humans are organisms sharing characteristics with other organisms. The most important common feature of all organisms is life. This text recognizes six essential characteristics of life:

1. **Organization** refers to the specific interrelationships among the parts of an organism and how those parts interact to perform specific functions. Living things are highly organized. All organisms are composed of one or more cells. Some cells, in turn, are composed of highly specialized organelles, which depend on the precise functions of large molecules. Disruption of this organized state can result in loss of function and death.

2. **Metabolism** (mě-tab′ō-lizm) is the ability to use energy to perform vital functions, such as growth, movement, and reproduction. Plants capture energy from sunlight, and humans obtain energy from food.

3. **Responsiveness** is the ability of an organism to sense changes in the environment and make the adjustments that help maintain its life. Responses include movement toward food or water and away from danger or poor environmental conditions. Organisms can also make adjustments that maintain their internal environment. For example, if body temperature increases in a hot environment, sweat glands produce sweat, which can lower body temperature down to the normal level.

MICROBES IN YOUR BODY   Getting to Know Your Bacteria

Did you know that you have more microbial cells than human cells in your body? Astoundingly, for every cell in your body, there are ten microbial cells. That's as many as 100 trillion microbial cells, which can collectively account for anywhere between 2 and 6 pounds of your body weight! A microbe is any living thing that cannot be seen with the naked eye (for example, bacteria, viruses, fungi, and protozoa). The total population of microbial cells on the human body is referred to as the microbiota, while the combination of these microbial cells and their genes is known as the microbiome. The microbiota includes so-called "good" bacteria that do not cause disease and may even help us. It also includes pathogenic, or "bad" bacteria.

With that many microbes in and on our bodies, you might wonder how they affect our health. To answer that question, in October 2007 the National Institute of Health (NIH) initiated the 5-year Human Microbiome Project, the largest study of its kind. Five significant regions of the human body were examined: airway, skin, mouth, gastrointestinal tract, and vagina. This project identified over 5000 species and sequenced over 20 million unique microbial genes.

What did scientists learn from the Human Microbiome Project? Human health is dependent upon the health of our microbiota, especially the "good" bacteria. In fact, it seems that our microbiota are so completely intertwined with human cells that in a 2013 *New York Times* article, Dr. David Relman of Stanford University suggested that humans are like corals. Corals are marine organisms that are collections of different life forms all existing together. More specifically, the human microbiome is intimately involved in the development and maintenance of the immune system. And more evidence is mounting for a correlation between a host's microbiota, digestion, and metabolism. Researchers have suggested that microbial genes are more responsible for our survival than human genes. There are even a few consistent pathogens that are present without causing disease, suggesting that their presence may be good for us. However, there does not seem to be a universal healthy human microbiome. Rather, the human microbiome varies across lifespan, ethnicity, nationality, culture, and geographical location. Instead of being a detriment, this variation may actually be very useful for at least one major reason. There seems to be a correlation between certain diseases and a "characteristic microbiome community," especially for autoimmune and inflammatory diseases (Crohn's, asthma, multiple sclerosis), which have become more prevalent. Scientists are beginning to believe that any significant change in the profile of the microbiome of the human gut may increase a person's susceptibility to autoimmune diseases. It has been proposed that these changes may be associated with exposure to antibiotics, particularly in infancy. Fortunately, newer studies of microbial transplantations have shown that the protective and other functions of bacteria can be transferred from one person to the next. However, this work is all very new and much research remains to be done.

Throughout the remainder of this text, we will highlight specific instances where our microbes influence our body systems. In light of the importance of our body's bacteria and other microbes, the prevalence of antibacterial soap and hand gel usage in everyday life may be something to think about.

---

4. **Growth** refers to an increase in size of all or part of the organism. It can result from an increase in cell number, cell size, or the amount of substance surrounding cells. For example, bones become larger as the number of bone cells increases and they become surrounded by bone matrix.

5. **Development** includes the changes an organism undergoes through time; it begins with fertilization and ends at death. The greatest developmental changes occur before birth, but many changes continue after birth, and some continue throughout life. Development usually involves growth, but it also involves differentiation. **Differentiation** is change in cell structure and function from generalized to specialized. For example, following fertilization, generalized cells specialize to become specific cell types, such as skin, bone, muscle, or nerve cells. These differentiated cells form tissues and organs.

6. **Reproduction** is the formation of new cells or new organisms. Without reproduction of cells, growth and tissue repair are impossible. Without reproduction of the organism, the species becomes extinct.

1.5 HOMEOSTASIS

Learning Outcomes *After reading this section, you should be able to*

A. Define homeostasis, and explain why it is important for proper body function.

B. Describe a negative-feedback mechanism and give an example.

C. Describe a positive-feedback mechanism and give an example.

Homeostasis (hō′mē-ō-stā′sis; *homeo-*, the same) is the existence and maintenance of a relatively constant environment within the body despite fluctuations in either the external environment or the internal environment. Most body cells are surrounded by a small amount of fluid, and normal cell functions depend on the maintenance of the cells' fluid environment within a narrow range of conditions, including temperature, volume, and chemical content. These conditions are called **variables** because their values can change. For example, body temperature is a variable that can increase in a hot environment or decrease in a cold environment.

CLINICAL IMPACT   Cadavers and the Law

The study of human bodies is the foundation of medical education, and for much of history, anatomists have used the bodies of people who have died, called cadavers, for these studies. However, public sentiment has often made it difficult for anatomists to obtain human bodies for dissection. In the early 1800s, the benefits of human dissection for training physicians had become very apparent, and the need for cadavers increased beyond the ability to acquire them legally. Thus arose the resurrectionists, or body snatchers. For a fee and no questions asked, they removed bodies from graves and provided them to

medical schools. Because the bodies were not easy to obtain and were not always in the best condition, two enterprising men named William Burke and William Hare went one step further. Over a period of time, they murdered seventeen people in Scotland and sold their bodies to a medical school. When discovered, Hare testified against Burke and went free. Burke was convicted, hanged, and publicly dissected. Discovery of Burke's activities so outraged the public that sensible laws regulating the acquisition of cadavers were soon passed, and this dark chapter in the history of anatomy was closed.

Today, in the United States, it is quite simple to donate your body for scientific study. The Uniform Anatomical Gift Act allows individuals to donate their organs or entire cadaver by putting a notation on their driver's license. You need only to contact a medical school or private agency to file the forms that give them the rights to your cadaver. Once the donor dies, the family of the deceased usually pays only the transportation costs for the remains. After dissection, the body is cremated, and the cremains can be returned to the family.

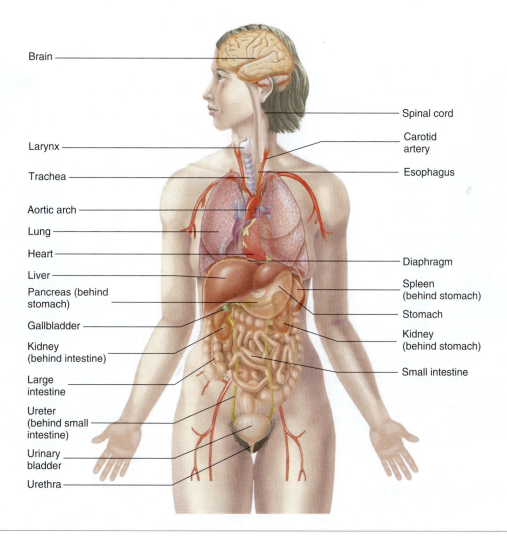

Figure I.2 Major Organs of the Body

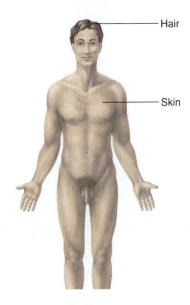

Integumentary System

Provides protection, regulates temperature, prevents water loss, and helps produce vitamin D. Consists of skin, hair, nails, and sweat glands.

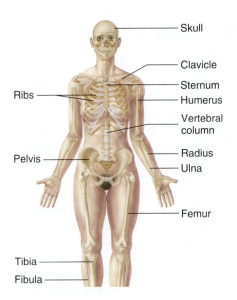

Skeletal System

Provides protection and support, allows body movements, produces blood cells, and stores minerals and adipose tissue. Consists of bones, associated cartilages, ligaments, and joints.

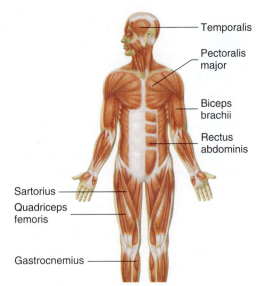

Muscular System

Produces body movements, maintains posture, and produces body heat. Consists of muscles attached to the skeleton by tendons.

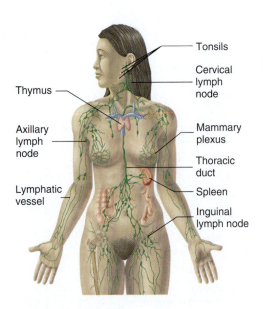

Lymphatic System

Removes foreign substances from the blood and lymph, combats disease, maintains tissue fluid balance, and absorbs dietary fats from the digestive tract. Consists of the lymphatic vessels, lymph nodes, and other lymphatic organs.

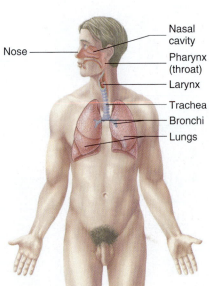

Respiratory System

Exchanges oxygen and carbon dioxide between the blood and air and regulates blood pH. Consists of the lungs and respiratory passages.

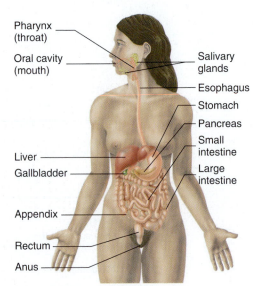

Digestive System

Performs the mechanical and chemical processes of digestion, absorption of nutrients, and elimination of wastes. Consists of the mouth, esophagus, stomach, intestines, and accessory organs.

Figure 1.3 AP|R Organ Systems of the Body

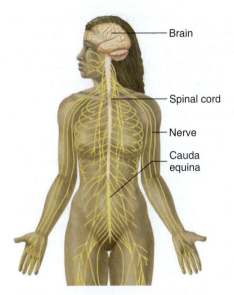

Nervous System

A major regulatory system that detects sensations and controls movements, physiological processes, and intellectual functions. Consists of the brain, spinal cord, nerves, and sensory receptors.

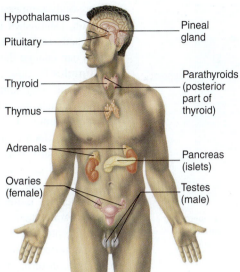

Endocrine System

A major regulatory system that influences metabolism, growth, reproduction, and many other functions. Consists of glands, such as the pituitary, that secrete hormones.

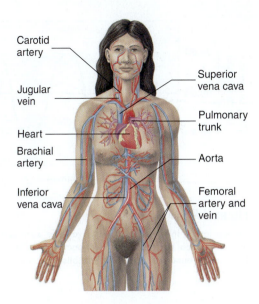

Cardiovascular System

Transports nutrients, waste products, gases, and hormones throughout the body; plays a role in the immune response and the regulation of body temperature. Consists of the heart, blood vessels, and blood.

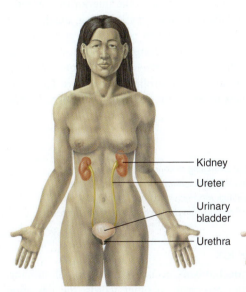

Urinary System

Removes waste products from the blood and regulates blood pH, ion balance, and water balance. Consists of the kidneys, urinary bladder, and ducts that carry urine.

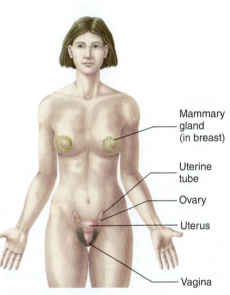

Female Reproductive System

Produces oocytes and is the site of fertilization and fetal development; produces milk for the newborn; produces hormones that influence sexual function and behaviors. Consists of the ovaries, uterine tubes, uterus, vagina, mammary glands, and associated structures.

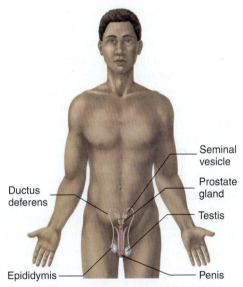

Male Reproductive System

Produces and transfers sperm cells to the female and produces hormones that influence sexual functions and behaviors. Consists of the testes, accessory structures, ducts, and penis.

Figure I.3 AP|R Organ Systems of the Body *(continued)*

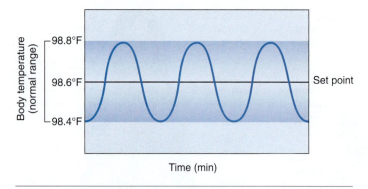

Figure 1.4 Homeostasis

Homeostasis is the maintenance of a variable, such as body temperature, around an ideal normal value, or set point. The value of the variable fluctuates around the set point to establish a normal range of values.

Homeostatic mechanisms, such as sweating or shivering, normally maintain body temperature near an ideal normal value, or **set point** (figure 1.4). Most homeostatic mechanisms are governed by the nervous system or the endocrine system. Note that homeostatic mechanisms are not able to maintain body temperature *precisely* at the set point. Instead, body temperature increases and decreases slightly around the set point, producing a **normal range** of values. As long as body temperatures remain within this normal range, homeostasis is maintained.

The organ systems help control the internal environment so that it remains relatively constant. For example, the digestive, respiratory, cardiovascular, and urinary systems function together so that each cell in the body receives adequate oxygen and nutrients and so that waste products do not accumulate to a toxic level. If the fluid surrounding cells deviates from homeostasis, the cells do not function normally and may even die. Disease disrupts homeostasis and sometimes results in death. Modern medicine attempts to understand disturbances in homeostasis and works to reestablish a normal range of values.

Negative Feedback

Most systems of the body are regulated by **negative-feedback mechanisms,** which maintain homeostasis. *Negative* means that any deviation from the set point is made smaller or is resisted. Negative feedback does not prevent variation but maintains variation within a normal range.

The maintenance of normal body temperature is an example of a negative-feedback mechanism. Normal body temperature is important because it allows molecules and enzymes to keep their normal shape so they can function optimally. An optimal body temperature prevents molecules from being permanently destroyed. Picture the change in appearance of egg whites as they are cooked; a similar phenomenon can happen to molecules in our body if the temperature becomes too high. Thus, normal body temperature is required to ensure that tissue homeostasis is maintained.

Many negative-feedback mechanisms, such as the one that maintains normal body temperature, have three components: (1) A **receptor** (rē-sep′tŏr, rē-sep′tōr) monitors the value of a variable,

such as body temperature; (2) a **control center,** such as part of the brain, establishes the set point around which the variable is maintained; and (3) an **effector** (ē-fek′tŏr), such as the sweat glands, can change the value of the variable. A changed variable is a **stimulus** because it initiates a homeostatic mechanism.

Normal body temperature depends on the coordination of multiple structures, which are regulated by the control center, or hypothalamus, in the brain. If body temperature rises, sweat glands (the effectors) produce sweat and the body cools. If body temperature falls, sweat glands do not produce sweat (figure 1.5). The stepwise process that regulates body temperature involves the interaction of receptors, the control center, and effectors. Often, there is more than one effector and the control center must integrate them. In the case of elevated body temperature, thermoreceptors in the skin and hypothalamus detect the increase in temperature and send the information to the hypothalamus control center. In turn, the hypothalamus stimulates blood vessels in the skin to relax and sweat glands to produce sweat, which sends more blood to the body's surface for radiation of heat away from the body. The sweat glands and skin blood vessels are the effectors in this scenario. Once body temperature returns to normal, the control center signals the sweat glands to reduce sweat production and the blood vessels constrict to their normal diameter. On the other hand, if body temperature drops, the control center does not stimulate the sweat glands. Instead, the skin blood vessels constrict more than normal and blood is directed to deeper regions of the body, conserving heat in the interior of the body. In addition, the hypothalamus stimulates shivering, quick cycles of skeletal muscle contractions, which generates a great amount of heat. Again, once the body temperature returns to normal, the effectors stop. In both cases, the effectors do not produce their responses indefinitely and are controlled by negative feedback. Negative feedback acts to return the variable to its normal range (figure 1.6).

> **Predict 2**
>
> *What effect would swimming in cool water have on body temperature regulation mechanisms? What would happen if a negative-feedback mechanism did not return the value of a variable, such as body temperature, to its normal range?*

Positive Feedback

Positive-feedback mechanisms occur when the initial stimulus further stimulates the response. In other words, the deviation from the set point becomes even greater. At times, this type of response is required to re-achieve homeostasis. For example, during blood loss, a chemical responsible for clot formation stimulates production of itself. In this way, a disruption in homeostasis is resolved through a positive-feedback mechanism. What prevents the entire vascular system from clotting? The clot formation process is self-limiting. Eventually, the components needed to form a clot will be depleted in the damaged area and more clot material cannot be formed (figure 1.7).

Birth is another example of a normally occurring positive-feedback mechanism. Near the end of pregnancy, the uterus is stretched by the baby's large size. This stretching, especially around the opening of the uterus, stimulates contractions of the

Humors and Homeostasis

The idea that the body maintains a balance (homeostasis) can be traced back to ancient Greece. Early physicians believed that the body supported four juices, or humors: the red juice of blood, the yellow juice of bile, the white juice secreted from the nose and lungs, and a black juice in the pancreas. They also thought that health resulted from a proper balance of these juices and that an excess of any one of them caused disease. Normally, they believed, the body would attempt to heal itself by expelling the excess juice, as when mucus runs from the nose of a person with a cold. This belief led to the practice of bloodletting to restore the body's normal balance of juices. Typically, physicians used sharp instruments to puncture the larger, external

vessels, but sometimes they applied leeches, blood-eating organisms, to the skin.

Tragically, in the eighteenth and nineteenth centuries, bloodletting went to extremes. During this period, a physician might recommend bloodletting, but barbers conducted the actual procedure. In fact, the traditional red-and-white-striped barber pole originated as a symbol for bloodletting. The brass basin on top of the pole represented the bowl for leeches, and the bowl on the bottom represented the basin for collecting blood. The stripes represented the bandages used as tourniquets, and the pole itself stood for the wooden staff patients gripped during the procedure. The fact that bloodletting did not improve the patient's condition was taken as evidence that not enough blood

had been removed to restore a healthy balance of the body's juices. Thus, the obvious solution was to let still more blood, undoubtedly causing many deaths. Eventually, the failure of this approach became obvious, and the practice was abandoned.

The modern term for bloodletting is **phlebotomy** (fle-bot′ō-mē), but it is practiced in a controlled setting and removes only small volumes of blood, usually for laboratory testing. There are some diseases in which bloodletting is still useful—for example, **polycythemia** (pol′ē-sī-thē′mē-ă), an overabundance of red blood cells. However, bloodletting in these patients does not continue until the patient faints or dies. Fortunately, we now understand more about how the body maintains homeostasis.

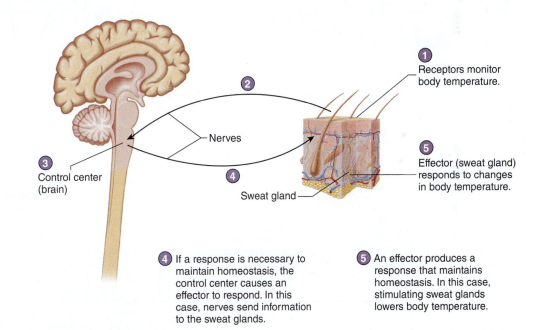

1 Receptors monitor the value of a variable. In this case, receptors in the skin monitor body temperature.

2 Information about the value of the variable is sent to a control center. In this case, nerves send information to the part of the brain responsible for regulating body temperature.

3 The control center compares the value of the variable against the set point.

Nerves

Control center (brain)

1 Receptors monitor body temperature.

5 Effector (sweat gland) responds to changes in body temperature.

Sweat gland

4 If a response is necessary to maintain homeostasis, the control center causes an effector to respond. In this case, nerves send information to the sweat glands.

5 An effector produces a response that maintains homeostasis. In this case, stimulating sweat glands lowers body temperature.

Figure I.5 Negative-Feedback Mechanism: Body Temperature

uterine muscles. The uterine contractions push the baby against the opening of the uterus, stretching it further. This stimulates additional contractions, which result in additional stretching. This positive-feedback sequence ends when the baby is delivered from the uterus and the stretching stimulus is eliminated.

On the other hand, occasionally a positive-feedback mechanism can be detrimental. One example of a detrimental positive-feedback mechanism is inadequate delivery of blood to cardiac

(heart) muscle. Contraction of cardiac muscle generates blood pressure and moves blood through the blood vessels to the tissues. A system of blood vessels on the outside of the heart provides cardiac muscle with a blood supply sufficient to allow normal contractions to occur. In effect, the heart pumps blood to itself. Just as with other tissues, blood pressure must be maintained to ensure adequate delivery of blood to the cardiac muscle. Following extreme blood loss, blood pressure decreases to the point that the delivery of blood

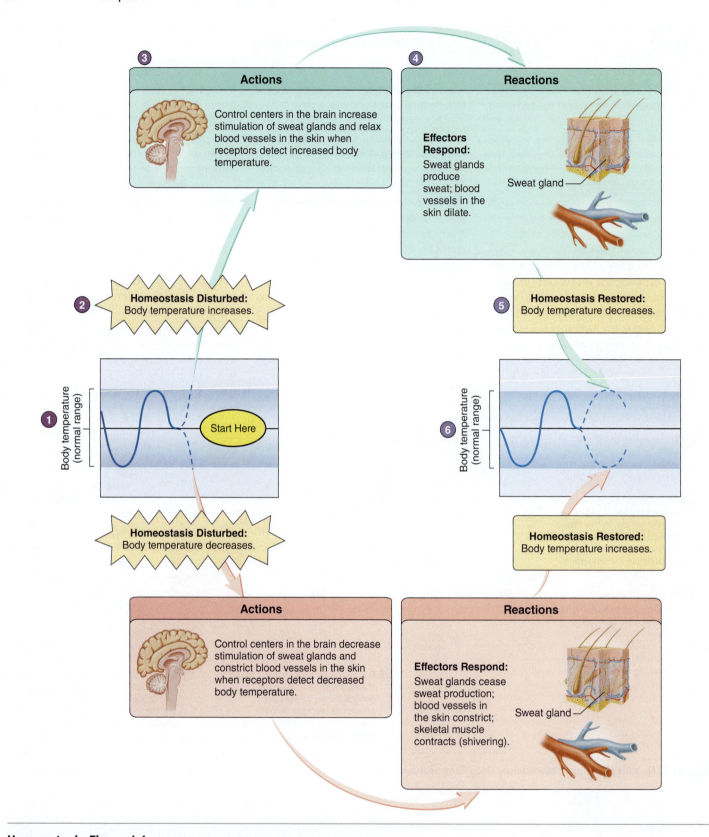

Homeostasis Figure 1.6 Negative-Feedback Control of Body Temperature

Throughout this book, all homeostasis figures have the same format as shown here. The changes caused by the increase of a variable outside the normal range are shown in the *green boxes,* and the changes caused by a decrease are shown in the *red boxes.* To help you learn how to interpret homeostasis figures, some of the steps in this figure are numbered. (1) Body temperature is within its normal range. (2) Body temperature increases outside the normal range, which causes homeostasis to be disturbed. (3) The body temperature control center in the brain responds to the change in body temperature. (4) The control center causes sweat glands to produce sweat and blood vessels in the skin to dilate. (5) These changes cause body temperature to decrease. (6) Body temperature returns to its normal range, and homeostasis is restored. Observe the responses to a decrease in body temperature outside its normal range by following the *red arrows.*

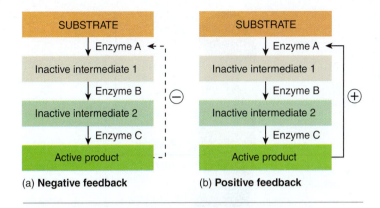

Figure I.7 Comparison of Negative-feedback and Positive-feedback Mechanisms

(*a*) In negative feedback, the response *stops* the effector. (*b*) In positive feedback, the response keeps the reaction going. For example, during blood clotting, the "active product" represents thrombin, which triggers, "enzyme A," the first step in the cascade that leads to the production of thrombin.

to cardiac muscle is inadequate. As a result, cardiac muscle homeostasis is disrupted, and cardiac muscle does not function normally. The heart pumps less blood, which causes the blood pressure to drop even lower. The additional decrease in blood pressure further reduces blood delivery to cardiac muscle, and the heart pumps even less blood, which again decreases the blood pressure. The process continues until the blood pressure is too low to sustain the cardiac muscle, the heart stops beating, and death results.

Following a moderate amount of blood loss (e.g., after donating a pint of blood), negative-feedback mechanisms result in an increase in heart rate that restores blood pressure. However, if blood loss is severe, negative-feedback mechanisms may not be able to maintain homeostasis, and the positive-feedback effect of an ever-decreasing blood pressure can develop.

A basic principle to remember is that many disease states result from the failure of negative-feedback mechanisms to maintain homeostasis. The purpose of medical therapy is to overcome illness by aiding negative-feedback mechanisms. For example, a transfusion can reverse a constantly decreasing blood pressure and restore homeostasis.

Predict 3

> Is the sensation of thirst associated with a negative- or a positive-feedback mechanism? Explain. (Hint: What is being regulated when you become thirsty?)

I.6 TERMINOLOGY AND THE BODY PLAN

Learning Outcomes *After reading this section, you should be able to*

A. Describe a person in anatomical position.
B. Define the directional terms for the human body, and use them to locate specific body structures.
C. Know the terms for the parts and regions of the body.
D. Name and describe the three major planes of the body and the body organs.
E. Describe the major trunk cavities and their divisions.
F. Describe the serous membranes, their locations, and their functions.

When you begin to study anatomy and physiology, the number of new words may seem overwhelming. Learning is easier and more interesting if you pay attention to the origin, or **etymology** (et′ĕ-mol′o-jē), of new words. Most of the terms are derived from Latin or Greek. For example, *anterior* in Latin means "to go before." Therefore, the anterior surface of the body is the one that goes before when we are walking.

Words are often modified by adding a prefix or suffix. For example, the suffix *-itis* means an inflammation, so *appendicitis* is an inflammation of the appendix. As new terms are introduced in this text, their meanings are often explained. The glossary and the list of word roots, prefixes, and suffixes on the inside back cover of the textbook also provide additional information about the new terms.

Body Positions

The **anatomical position** refers to a person standing erect with the face directed forward, the upper limbs hanging to the sides, and the palms of the hands facing forward (figure 1.8). A person is **supine** when lying face upward and **prone** when lying face downward.

The position of the body can affect the description of body parts relative to each other. In the anatomical position, the elbow is above the hand, but in the supine or prone position, the elbow and hand are at the same level. To avoid confusion, relational descriptions are always based on the anatomical position, no matter the actual position of the body.

Directional Terms

Directional terms describe parts of the body relative to each other (figure 1.8 and table 1.1). It is important to become familiar with these directional terms as soon as possible because you will see them repeatedly throughout the text. *Right* and *left* are used as directional terms in anatomical terminology. *Up* is replaced by **superior,** *down* by **inferior,** *front* by **anterior,** and *back* by **posterior.**

As previously mentioned, the word *anterior* means that which goes before; the word **ventral** means belly. Therefore, the anterior surface of the human body is also called the ventral surface, or belly, because the belly "goes first" when we are walking. The word *posterior* means that which follows, and **dorsal** means back. Thus, the posterior surface of the body is the dorsal surface, or back, which follows as we are walking.

Proximal means nearest, whereas **distal** means distant. These terms are used to refer to linear structures, such as the limbs, in which one end is near another structure and the other end is farther away. Each limb is attached at its proximal end to the body, and the distal end, such as the hand, is farther away.

Medial means toward the midline, and **lateral** means away from the midline. The nose is located in a medial position on the face, and the ears are lateral to the nose. The term **superficial** refers to a structure close to the surface of the body, and **deep** is toward the interior of the body. For example, the skin is superficial to muscle and bone.

Predict 4

> Provide the correct directional term for the following statement:
> When a boy is standing on his head, his nose is _____ to his mouth.

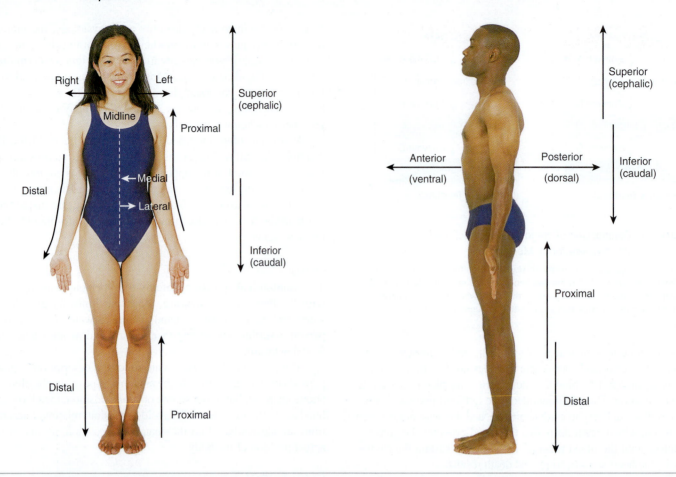

Figure I.8 AP|R **Directional Terms**

All directional terms are in relation to the body in the anatomical position: a person standing erect with the face directed forward, the arms hanging to the sides, and the palms of the hands facing forward.

TABLE I.I	Directional Terms for the Human Body		
Term	**Etymology**	**Definition***	**Example**
Right		Toward the body's right side	The right ear
Left		Toward the body's left side	The left ear
Inferior	Lower	Below	The nose is inferior to the forehead.
Superior	Higher	Above	The mouth is superior to the chin.
Anterior	To go before	Toward the front of the body	The teeth are anterior to the throat.
Posterior	*Posterus*, following	Toward the back of the body	The brain is posterior to the eyes.
Dorsal	*Dorsum*, back	Toward the back (synonymous with *posterior*)	The spine is dorsal to the breastbone.
Ventral	*Venter*, belly	Toward the belly (synonymous with *anterior*)	The navel is ventral to the spine.
Proximal	*Proximus*, nearest	Closer to a point of attachment	The elbow is proximal to the wrist.
Distal	*di + sto*, to be distant	Farther from a point of attachment	The knee is distal to the hip.
Lateral	*Latus*, side	Away from the midline of the body	The nipple is lateral to the breastbone.
Medial	*Medialis*, middle	Toward the middle or midline of the body	The bridge of the nose is medial to the eye.
Superficial	*Superficialis*, surface	Toward or on the surface	The skin is superficial to muscle.
Deep	*Deop*, deep	Away from the surface, internal	The lungs are deep to the ribs.

*All directional terms refer to a human in the anatomical position.

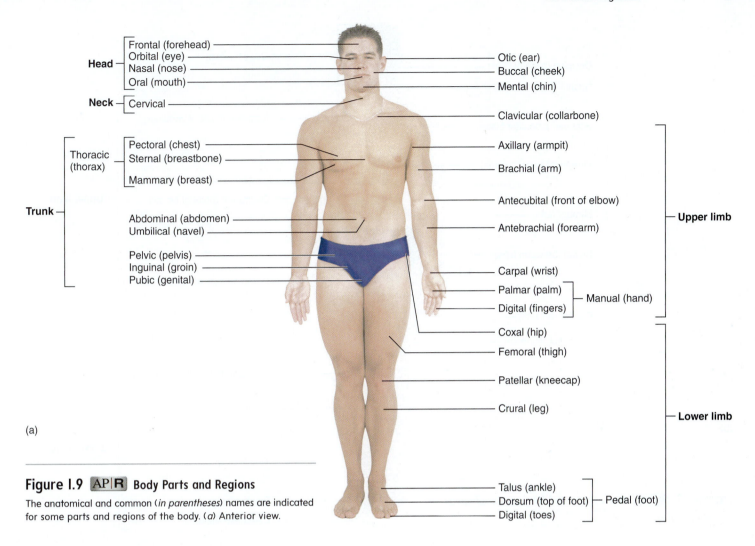

Head
- Frontal (forehead)
- Orbital (eye)
- Nasal (nose)
- Oral (mouth)

Neck
- Cervical

Trunk

Thoracic (thorax)
- Pectoral (chest)
- Sternal (breastbone)
- Mammary (breast)

- Abdominal (abdomen)
- Umbilical (navel)

- Pelvic (pelvis)
- Inguinal (groin)
- Pubic (genital)

- Otic (ear)
- Buccal (cheek)
- Mental (chin)
- Clavicular (collarbone)
- Axillary (armpit)
- Brachial (arm)
- Antecubital (front of elbow)
- Antebrachial (forearm)
- Carpal (wrist)
- Palmar (palm)
- Digital (fingers)
- Manual (hand)
- Coxal (hip)
- Femoral (thigh)
- Patellar (kneecap)
- Crural (leg)

Upper limb

Lower limb

- Talus (ankle)
- Dorsum (top of foot)
- Digital (toes)
- Pedal (foot)

(a)

Figure 1.9 AP|R **Body Parts and Regions**

The anatomical and common (*in parentheses*) names are indicated for some parts and regions of the body. (*a*) Anterior view.

Body Parts and Regions

Health professionals use a number of terms when referring to different regions or parts of the body. Figure 1.9 shows the anatomical terms, with the common terms in parentheses. The central region of the body consists of the **head, neck,** and **trunk.** The trunk can be divided into the **thorax** (chest), **abdomen** (region between the thorax and pelvis), and **pelvis** (the inferior end of the trunk associated with the hips). The upper limb is divided into the arm, forearm, wrist, and hand. The **arm** extends from the shoulder to the elbow, and the **forearm** extends from the elbow to the wrist. The lower limb is divided into the thigh, leg, ankle, and foot. The **thigh** extends from the hip to the knee, and the **leg** extends from the knee to the ankle. Note that, contrary to popular usage, the terms *arm* and *leg* refer to only a part of the respective limb.

The abdomen is often subdivided superficially into four sections, or **quadrants,** by two imaginary lines—one horizontal and one vertical—that intersect at the navel (figure 1.10*a*). The quadrants formed are the right-upper, left-upper, right-lower, and left-lower quadrants. In addition to these quadrants, the abdomen is sometimes subdivided into **regions** by four imaginary lines—two horizontal and two vertical. These four lines create an imaginary tic-tac-toe figure on the abdomen, resulting in nine regions: epigastric (ep-i-gas′trik), right and left hypochondriac

(hī-pō-kon′drē-ak), umbilical (ŭm-bil′i-kăl), right and left lumbar (lŭm′bar), hypogastric (hī-pō-gas′trik), and right and left iliac (il′ē-ak) (figure 1.10*b*). Clinicians use the quadrants or regions as reference points for locating the underlying organs. For example, the appendix is in the right-lower quadrant, and the pain of an acute appendicitis is usually felt there.

Predict 5

Using figures 1.2 and 1.10a, determine in which quadrant each of the following organs is located: spleen, gallbladder, kidneys, most of the stomach, and most of the liver.

A CASE IN POINT

Epigastric Pain

Wilby Hurtt has pain in the epigastric region (figure 1.10*b*), which is most noticeable following meals and at night when he is lying in bed. He probably has gastroesophageal reflux disease (GERD), in which stomach acid improperly moves into the esophagus, damaging and irritating its lining. Epigastric pain, however, can have many causes and should be evaluated by a physician. For example, gallstones, stomach or small intestine ulcers, inflammation of the pancreas, and heart disease can also cause epigastric pain.

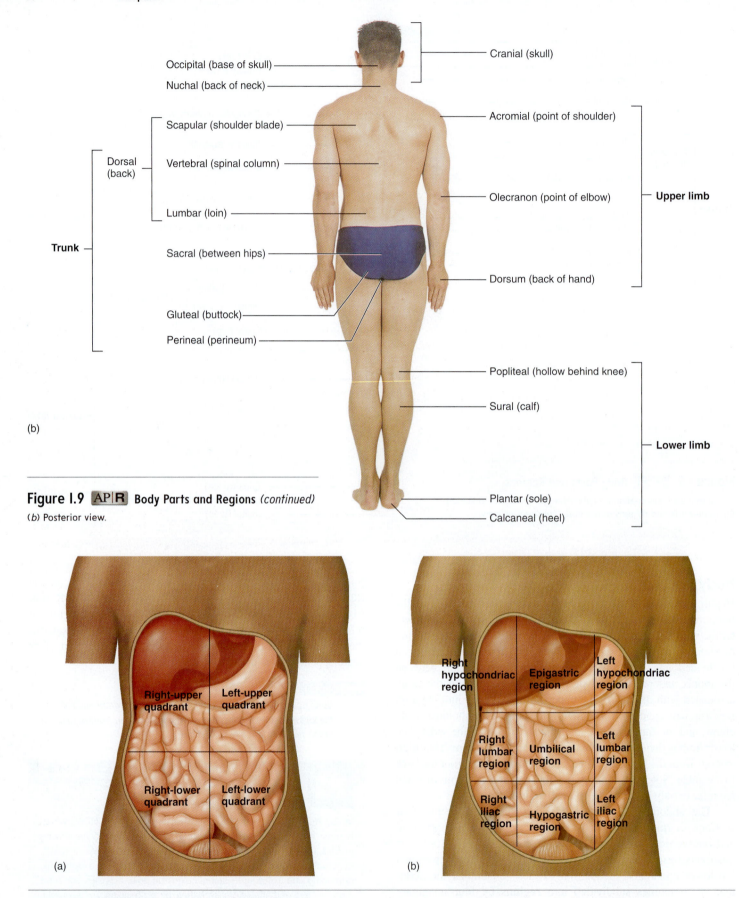

Occipital (base of skull)

Nuchal (back of neck)

Cranial (skull)

Scapular (shoulder blade)

Vertebral (spinal column)

Lumbar (loin)

Dorsal (back)

Acromial (point of shoulder)

Olecranon (point of elbow)

Dorsum (back of hand)

Upper limb

Trunk

Sacral (between hips)

Gluteal (buttock)

Perineal (perineum)

Popliteal (hollow behind knee)

Sural (calf)

Lower limb

Plantar (sole)

Calcaneal (heel)

(b)

Figure 1.9 AP|R **Body Parts and Regions** *(continued)*
(*b*) Posterior view.

Right-upper quadrant

Left-upper quadrant

Right-lower quadrant

Left-lower quadrant

(a)

Right hypochondriac region

Epigastric region

Left hypochondriac region

Right lumbar region

Umbilical region

Left lumbar region

Right iliac region

Hypogastric region

Left iliac region

(b)

Figure 1.10 AP|R Subdivisions of the Abdomen
Lines are superimposed over internal organs to demonstrate the relationship of the organs to the subdivisions.
(*a*) Abdominal quadrants consist of four subdivisions. (*b*) Abdominal regions consist of nine subdivisions.

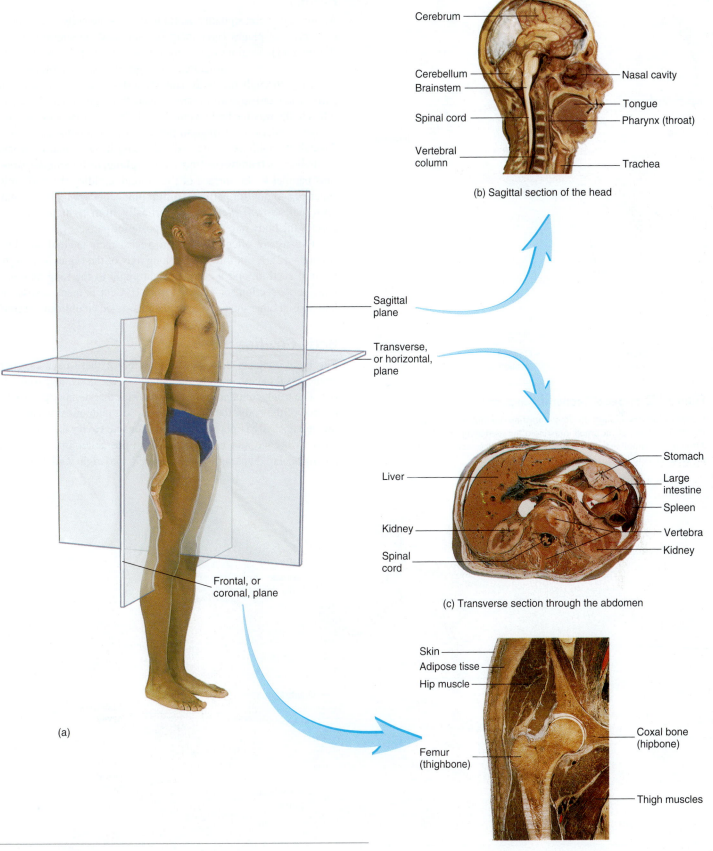

(b) Sagittal section of the head

Cerebrum

Cerebellum
Brainstem

Spinal cord

Vertebral
column

Nasal cavity

Tongue
Pharynx (throat)

Trachea

(c) Transverse section through the abdomen

Liver

Kidney

Spinal
cord

Stomach

Large
intestine

Spleen

Vertebra

Kidney

Sagittal
plane

Transverse,
or horizontal,
plane

Frontal, or
coronal, plane

(a)

(d) Frontal section through the right hip

Skin
Adipose tisse
Hip muscle

Femur
(thighbone)

Coxal bone
(hipbone)

Thigh muscles

Figure I.11 AP|R **Planes of Section of the Body**

(*a*) Planes of section through the body are indicated by "glass" sheets. Also shown
are actual sections through (*b*) the head (*viewed from the right*), (*c*) the abdomen
(*inferior view; liver is on the right*), and (*d*) the hip (*anterior view*).

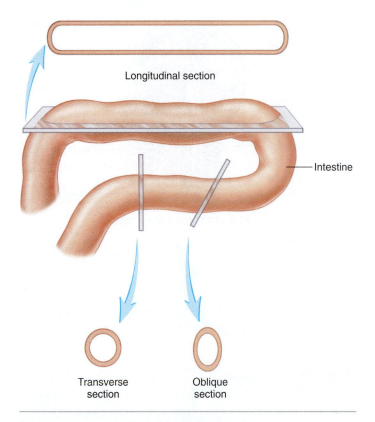

Longitudinal section

Intestine

Transverse section

Oblique section

Figure 1.12 Planes of Section Through an Organ

Planes of section through the small intestine are indicated by "glass" sheets. The views of the small intestine after sectioning are also shown. Although the small intestine is basically a tube, the sections appear quite different in shape.

Planes

At times, it is conceptually useful to discuss the body in reference to a series of planes (imaginary flat surfaces) passing through it (figure 1.11). Sectioning the body is a way to "look inside" and observe the body's structures. A **sagittal** (saj′i-tăl) **plane** runs vertically through the body and separates it into right and left parts. The word *sagittal* literally means the flight of an arrow and refers to the way the body would be split by an arrow passing anteriorly to posteriorly. A **median plane** is a sagittal plane that passes through the midline of the body, dividing it into equal right and left halves. A **transverse** (trans-vers′) **plane,** or *horizontal plane,* runs parallel to the surface of the ground, dividing the body into superior and inferior parts. A **frontal plane,** or *coronal* (kōr′ŏ-năl, kō-rō′nal; crown) *plane,* runs vertically from right to left and divides the body into anterior and posterior parts.

Organs are often sectioned to reveal their internal structure (figure 1.12). A cut through the long axis of the organ is a **longitudinal section,** and a cut at a right angle to the long axis is a **transverse section,** or *cross section.* If a cut is made across the long axis at other than a right angle, it is called an **oblique section.**

Body Cavities

The body contains many cavities. Some of these cavities, such as the nasal cavity, open to the outside of the body, and some do not. The trunk contains three large cavities that do not open to the outside of the body: the thoracic cavity, the abdominal cavity, and the pelvic cavity (figure 1.13). The **thoracic cavity** is surrounded by the rib cage and is separated from the abdominal cavity by the muscular diaphragm. It is divided into right and left parts

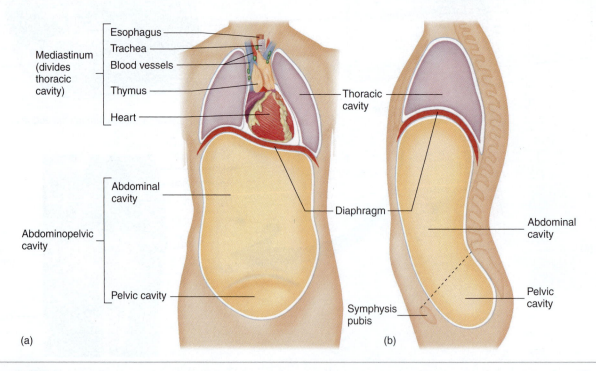

Mediastinum (divides thoracic cavity)

Esophagus

Trachea

Blood vessels

Thymus

Heart

Thoracic cavity

Abdominal cavity

Abdominopelvic cavity

Diaphragm

Abdominal cavity

Pelvic cavity

Pelvic cavity

Symphysis pubis

(a)

(b)

Figure 1.13 AP|R Trunk Cavities

(a) Anterior view showing the major trunk cavities. The diaphragm separates the thoracic cavity from the abdominal cavity. The mediastinum, which includes the heart, is a partition of organs dividing the thoracic cavity. (b) Sagittal section of the trunk cavities viewed from the left. The *dashed line* shows the division between the abdominal and pelvic cavities. The mediastinum has been removed to show the thoracic cavity.

by a median structure called the **mediastinum** (me′dē-as-tī′nŭm; wall). The mediastinum is a partition containing the heart, the thymus, the trachea, the esophagus, and other structures. The two lungs are located on each side of the mediastinum.

The **abdominal cavity** is bounded primarily by the abdominal muscles and contains the stomach, the intestines, the liver, the spleen, the pancreas, and the kidneys. The **pelvic** (pel′vik) **cavity** is a small space enclosed by the bones of the pelvis and contains the urinary bladder, part of the large intestine, and the internal reproductive organs. The abdominal and pelvic cavities are not physically separated and sometimes are called the **abdominopelvic** (ab-dom′i-nō-pel′vik) **cavity.**

Serous Membranes

Serous (sēr′ŭs) **membranes** line the trunk cavities and cover the organs of these cavities. To understand the relationship between serous membranes and an organ, imagine pushing your fist into an inflated balloon. The inner balloon wall in contact with your fist (organ) represents the **visceral** (vis′er-ăl; organ) **serous membrane,** and the outer part of the balloon wall represents the **parietal** (pă-rī′ĕ-tăl; wall) **serous membrane** (figure 1.14). The cavity, or space, between the visceral and parietal serous membranes is normally filled with a thin, lubricating film of serous fluid produced by the membranes. As an organ rubs against another organ or against the body wall, the serous fluid and smooth serous membranes reduce friction.

The thoracic cavity contains three serous membrane-lined cavities: a pericardial cavity and two pleural cavities. The **pericardial** (per-i-kar′dē-ăl; around the heart) **cavity** surrounds the heart (figure 1.15a). The visceral pericardium covers the heart, which is contained within a connective tissue sac lined with the parietal pericardium. The pericardial cavity, which contains pericardial fluid, is located between the visceral pericardium and the parietal pericardium.

A **pleural** (ploor′ăl; associated with the ribs) **cavity** surrounds each lung, which is covered by visceral pleura (figure 1.15b). Parietal pleura lines the inner surface of the thoracic wall, the lateral surfaces of the mediastinum, and the superior surface of the diaphragm. The pleural cavity is located between the visceral pleura and the parietal pleura and contains pleural fluid.

The abdominopelvic cavity contains a serous membrane-lined cavity called the **peritoneal** (per′i-tō-nē′ăl; to stretch over) **cavity** (figure 1.15c). Visceral peritoneum covers many of the organs of the abdominopelvic cavity. Parietal peritoneum lines the wall of the abdominopelvic cavity and the inferior surface of the diaphragm. The peritoneal cavity is located between the visceral peritoneum and the parietal peritoneum and contains peritoneal fluid.

The serous membranes can become inflamed—usually as a result of an infection. **Pericarditis** (per′i-kar-dī′tis) is inflammation of the pericardium, **pleurisy** (ploor′i-sē) is inflammation of the pleura, and **peritonitis** (per′i-tō-nī′tis) is inflammation of the peritoneum.

A CASE IN POINT

Peritonitis

May Day is rushed to the hospital emergency room. Earlier today, she experienced diffuse abdominal pain, but no fever. Then the pain became more intense and shifted to her right-lower quadrant. She also developed a fever. The examining physician concludes that May Day has **appendicitis**, an inflammation of the appendix that is usually caused by an infection. The appendix is a small, wormlike sac attached to the large intestine. The outer surface of the appendix is visceral peritoneum. An infection of the appendix can rupture its wall, releasing bacteria into the peritoneal cavity and causing peritonitis. Appendicitis is the most common cause of emergency abdominal surgery in children, and it often leads to peritonitis if not treated. May has her appendix removed, is treated with antibiotics, and makes a full recovery.

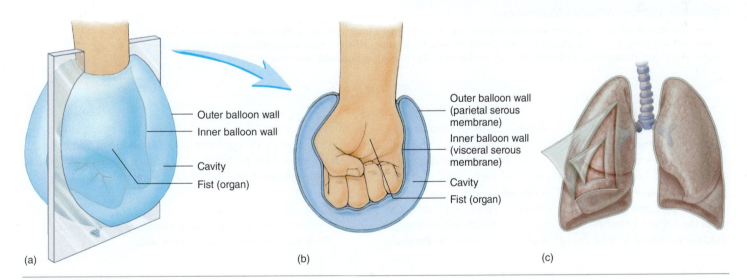

(a)   (b)   (c)

Outer balloon wall
Inner balloon wall
Cavity
Fist (organ)

Outer balloon wall (parietal serous membrane)
Inner balloon wall (visceral serous membrane)
Cavity
Fist (organ)

Figure 1.14 Serous Membranes

(a) Fist pushing into a balloon. A "glass" sheet indicates the location of a cross section through the balloon. (b) Interior view produced by the section in (a). The fist represents an organ, and the walls of the balloon represent the serous membranes. The inner wall of the balloon represents a visceral serous membrane in contact with the fist (organ). The outer wall of the balloon represents a parietal serous membrane. (c) View of the serous membranes surrounding the lungs. The membrane in contact with the lungs is the visceral pleura; the membrane lining the lung cavity is the parietal pleura.

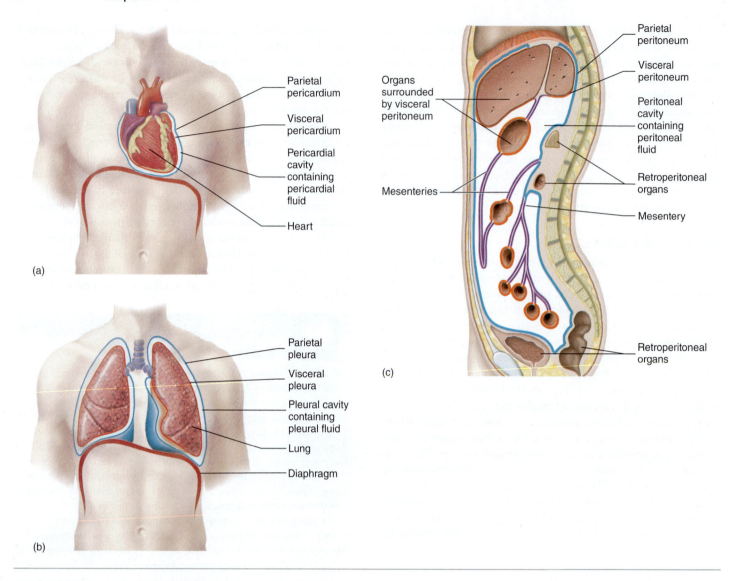

Figure 1.15 Location of Serous Membranes

(*a*) Frontal section showing the parietal pericardium (*blue*), visceral pericardium (*red*), and pericardial cavity. (*b*) Frontal section showing the parietal pleura (*blue*), visceral pleura (*red*), and pleural cavities. (*c*) Sagittal section through the abdominopelvic cavity showing the parietal peritoneum (*blue*), visceral peritoneum (*red*), peritoneal cavity, mesenteries (*purple*), and retroperitoneal organs.

Mesenteries (mes′en-ter-ēz), which consist of two layers of peritoneum fused together (figure 1.15*c*), connect the visceral peritoneum of some abdominopelvic organs to the parietal peritoneum on the body wall or to the visceral peritoneum of other abdominopelvic organs. The mesenteries anchor the organs to the body wall and provide a pathway for nerves and blood vessels to reach the organs. Other abdominopelvic organs are more closely attached to the body wall and do not have mesenteries. Parietal peritoneum covers these other organs, which are said to be **retroperitoneal** (re′trō-per′i-tō-nē′ăl; *retro*, behind). The retroperitoneal organs include the kidneys, the adrenal glands, the pancreas, parts of the intestines, and the urinary bladder (figure 1.15*c*).

Predict 6

Explain how an organ can be located within the abdominopelvic cavity but not be within the peritoneal cavity.

SUMMARY

Knowledge of anatomy and physiology can be used to predict the body's responses to stimuli when healthy or diseased.

1.1 Anatomy   (p. 1)

1. Anatomy is the study of the structures of the body.
2. Systemic anatomy is the study of the body by organ systems. Regional anatomy is the study of the body by areas.
3. Surface anatomy uses superficial structures to locate deeper structures, and anatomical imaging is a noninvasive method for examining deep structures.

1.2 Physiology   (p. 2)

Physiology is the study of the processes and functions of the body.

1.3 Structural and Functional Organization of the Human Body   (p. 2)

1. The human body can be organized into six levels: chemical, cell, tissue, organ, organ system, and organism.
2. The eleven organ systems are the integumentary, skeletal, muscular, lymphatic, respiratory, digestive, nervous, endocrine, cardiovascular, urinary, and reproductive systems (see figure 1.3).

1.4 Characteristics of Life   (p. 3)

The characteristics of life are organization, metabolism, responsiveness, growth, development, and reproduction.

1.5 Homeostasis   (p. 4)

1. Homeostasis is the condition in which body functions, body fluids, and other factors of the internal environment are maintained within a range of values suitable to support life.
2. Negative-feedback mechanisms maintain homeostasis.
3. Positive-feedback mechanisms make deviations from normal even greater. Although a few positive-feedback mechanisms normally exist in the body, most positive-feedback mechanisms are harmful.

1.6 Terminology and the Body Plan   (p. 11)

Body Positions

1. A human standing erect with the face directed forward, the arms hanging to the sides, and the palms facing forward is in the anatomical position.
2. A face-upward position is supine and a face-downward one is prone.

Directional Terms

Directional terms always refer to the anatomical position, regardless of the body's actual position (see table 1.1).

Body Parts and Regions

1. The body can be divided into the head, neck, trunk, upper limbs, and lower limbs.
2. The abdomen can be divided superficially into four quadrants or nine regions, which are useful for locating internal organs or describing the location of a pain.

Planes

1. A sagittal plane divides the body into left and right parts, a transverse plane divides the body into superior and inferior parts, and a frontal plane divides the body into anterior and posterior parts.
2. A longitudinal section divides an organ along its long axis, a transverse section cuts an organ at a right angle to the long axis, and an oblique section cuts across the long axis at an angle other than a right angle.

Body Cavities

1. The thoracic cavity is bounded by the ribs and the diaphragm. The mediastinum divides the thoracic cavity into two parts.
2. The abdominal cavity is bounded by the diaphragm and the abdominal muscles.
3. The pelvic cavity is surrounded by the pelvic bones.

Serous Membranes

1. The trunk cavities are lined by serous membranes. The parietal part of a serous membrane lines the wall of the cavity, and the visceral part covers the internal organs.
2. The serous membranes secrete fluid that fills the space between the parietal and visceral membranes. The serous membranes protect organs from friction.
3. The pericardial cavity surrounds the heart, the pleural cavities surround the lungs, and the peritoneal cavity surrounds certain abdominal and pelvic organs.
4. Mesenteries are parts of the peritoneum that hold the abdominal organs in place and provide a passageway for blood vessels and nerves to organs.
5. Retroperitoneal organs are found "behind" the parietal peritoneum. The kidneys, the adrenal glands, the pancreas, parts of the intestines, and the urinary bladder are examples of retroperitoneal organs.

REVIEW AND COMPREHENSION

1. Define anatomy, surface anatomy, anatomical imaging, and physiology.

2. List six structural levels at which the body can be studied.

3. Define tissue. What are the four primary tissue types?

4. Define organ and organ system. What are the eleven organ systems of the body and their functions?

5. Name six characteristics of life.

6. What does the term *homeostasis* mean? If a deviation from homeostasis occurs, what kind of mechanism restores homeostasis?

7. Describe a negative-feedback mechanism in terms of receptor, control center, and effector. Give an example of a negative-feedback mechanism in the body.

8. Define positive feedback. Why are positive-feedback mechanisms generally harmful? Give one example each of a harmful and a beneficial positive-feedback mechanism in the body.

9. Why is knowledge of the etymology of anatomical and physiological terms useful?

10. Describe the anatomical position. Why is it important to remember the anatomical position when using directional terms?

11. Define and give an example of the following directional terms: *inferior, superior, anterior, posterior, dorsal, ventral, proximal, distal, lateral, medial, superficial,* and *deep.*

12. List the subdivisions of the trunk, the upper limbs, and the lower limbs.

13. Describe the four-quadrant and nine-region methods of subdividing the abdomen. What is the purpose of these methods?

14. Define the sagittal, median, transverse, and frontal planes of the body.

15. Define the longitudinal, transverse, and oblique sections of an organ.

16. Define the following cavities: thoracic, abdominal, pelvic, and abdominopelvic. What is the mediastinum?

17. What is the difference between the visceral and parietal layers of a serous membrane? What function do serous membranes perform?

18. Name the serous membranes associated with the heart, lungs, and abdominopelvic organs.

19. Define mesentery. What does the term *retroperitoneal* mean? Give an example of a retroperitoneal organ.

CRITICAL THINKING

1. A male has lost blood as a result of a gunshot wound. Even though the bleeding has been stopped, his blood pressure is low and dropping, and his heart rate is elevated. Following a blood transfusion, his blood pressure increases and his heart rate decreases. Which of the following statement(s) is (are) consistent with these observations?
   a. Negative-feedback mechanisms can be inadequate without medical intervention.
   b. The transfusion interrupted a positive-feedback mechanism.
   c. The increased heart rate after the gunshot wound and before the transfusion is a result of a positive-feedback mechanism.
   d. Both a and b are correct.
   e. Answers a, b, and c are correct.

2. During physical exercise, the respiration rate increases. Two students are discussing the mechanisms involved. Student A claims they are positive-feedback mechanisms, and student B claims they are negative-feedback mechanisms. Do you agree with student A or student B, and why?

3. Complete the following statements using the correct directional terms for the human body.
   a. The navel is _____ to the nose.
   b. The heart is _____ to the breastbone (sternum).
   c. The forearm is _____ to the arm.
   d. The ear is _____ to the brain.

4. Describe, using as many directional terms as you can, the relationship between your kneecap and your heel.

5. According to "Dear Abby," a wedding band should be worn closest to the heart, and an engagement ring should be worn as a "guard" on the outside. Should a woman's wedding band be worn proximal or distal to her engagement ring?

6. In which quadrant and region is the pancreas located? In which quadrant and region is the urinary bladder located?

7. During pregnancy, which would increase more in size, the mother's abdominal cavity or her pelvic cavity? Explain.

8. A bullet enters the left side of a male, passes through the left lung, and lodges in the heart. Name in order the serous membranes and the cavities through which the bullet passes.

9. Can a kidney be removed without cutting through parietal peritoneum? Explain.

Answers in Appendix D

aprevealed.com

2 The Chemical Basis of Life

LEARN TO PREDICT

Rafael is playing the soccer match of his life. However, hot weather conditions are making this match one of his hardest yet. As the game began, Rafael noticed himself sweating, which was helping to keep him cool. However, he had been ill the night before and had not been able to hydrate as was normal for him. Towards the end of the second half, he realized he was no longer sweating and he began to feel overheated and dizzy.

After reading this chapter and learning how the properties of molecules contribute to their physiological functions, predict how Rafael's inability to sweat affects his internal temperature.

▲ The sweat on Rafael's skin will evaporate, cooling his body.

Module 2 Cells and Chemistry

2.1 BASIC CHEMISTRY

Learning Outcomes *After reading this section, you should be able to*

A. Define chemistry and state its relevance to anatomy and physiology.
B. Define matter, mass, and weight.
C. Distinguish between an element and an atom.
D. Define atomic number and mass number.
E. Name the subatomic particles of an atom, and indicate their location.
F. Compare and contrast ionic and covalent bonds.
G. Explain what creates a hydrogen bond and relate its importance.
H. Differentiate between a molecule and a compound.
I. Describe the process of dissociation.

Chemicals make up the body's structures, and the interactions of chemicals with one another are responsible for the body's functions. The processes of nerve impulse generation, digestion, muscle contraction, and metabolism can all be described in chemical terms. Many abnormal conditions and their treatments can also be explained in chemical terms, even though their symptoms appear as malfunctions in organ systems. For example, Parkinson disease, which causes uncontrolled shaking movements, results from a shortage of a chemical called dopamine in certain nerve cells in the brain. It can be treated by giving patients another chemical, which brain cells convert to dopamine.

A basic knowledge of chemistry is essential for understanding anatomy and physiology. **Chemistry** is the scientific discipline concerned with the atomic composition and structure of substances and the reactions they undergo. This chapter outlines some basic chemical principles and emphasizes their relationship to humans.

Matter, Mass, and Weight

All living and nonliving things are composed of **matter,** which is anything that occupies space and has mass. **Mass** is the amount of matter in an object, and **weight** is the gravitational force acting on an object of a given mass. For example, the weight of an apple results from the force of gravity "pulling" on the apple's mass.

Predict 2

The difference between mass and weight can be illustrated by considering an astronaut. How would an astronaut's mass and weight in outer space compare with that astronaut's mass and weight on the earth's surface?

The international unit for mass is the **kilogram (kg),** which is the mass of a platinum-iridium cylinder kept at the International Bureau of Weights and Measurements in France. The mass of all other objects is compared with this cylinder. For example, a 2.2-lb lead weight and 1 liter (L) (1.06 qt) of water each have a mass of approximately 1 kg. An object with 1/1000 the mass of the standard kilogram cylinder is said to have a mass of 1 **gram (g).**

Elements and Atoms

An **element** is the simplest type of matter having unique chemical properties. A list of the elements commonly found in the human body appears in table 2.1. About 96% of the body's weight results from the elements oxygen, carbon, hydrogen, and nitrogen. However, many other elements also play important roles in the human body. For example, calcium helps form bones, and sodium ions are essential for neuronal activity. Some of these elements are present in only trace amounts but are still essential for life.

An **atom** (at′ŏm; indivisible) is the smallest particle of an element that has the chemical characteristics of that element. An element is composed of atoms of only one kind. For example, the element carbon is composed of only carbon atoms, and the element oxygen is composed of only oxygen atoms.

An element, or an atom of that element, is often represented by a symbol. Usually the symbol is the first letter or letters of the element's name—for example, C for carbon, H for hydrogen, and Ca for calcium. Occasionally, the symbol is taken from the Latin, Greek, or Arabic name for the element; for example, the symbol for sodium is Na, from the Latin word *natrium.*

Atomic Structure

The characteristics of matter result from the structure, organization, and behavior of atoms. Atoms are composed of subatomic particles, some of which have an electrical charge. The three major types of subatomic particles are neutrons, protons, and electrons. **Neutrons** (noo′tronz) have no electrical charge, **protons** (prō′tonz) have positive charges, and **electrons** (e-lek′tronz) have negative charges. The positive charge of a proton is equal in magnitude to the negative charge of an electron. The number of protons and number of electrons in each atom are equal, and the individual charges cancel each other. Therefore, each atom is electrically neutral.

Protons and neutrons form the **nucleus** at the center of the atom, and electrons move around the nucleus (figure 2.1). The nucleus accounts for 99.97% of an atom's mass, but only 1-ten-trillionth of its volume. Most of the volume of an atom is occupied

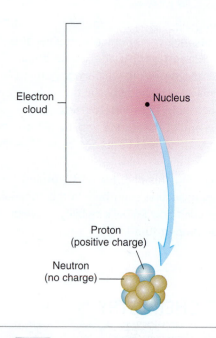

Figure 2.1 AP|R **Model of an Atom**

The tiny, dense nucleus consists of positively charged protons and uncharged neutrons. Most of the volume of an atom is occupied by rapidly moving, negatively charged electrons, which can be represented as an electron cloud.

			Atomic	Mass	Percent in Human	Percent in Human
TABLE 2.1	**Common Elements in the Human Body**					
Element	**Symbol**		**Atomic Number**	**Mass Number**	**Percent in Human Body by Weight**	**Percent in Human Body by Number of Atoms**
Hydrogen	H		1	1	9.5	63.0
Carbon	C		6	12	18.5	9.5
Nitrogen	N		7	14	3.3	1.4
Oxygen	O		8	16	65.0	25.5
Sodium	Na		11	23	0.2	0.3
Phosphorus	P		15	31	1.0	0.22
Sulfur	S		16	32	0.3	0.05
Chlorine	Cl		17	35	0.2	0.03
Potassium	K		19	39	0.4	0.06
Calcium	Ca		20	40	1.5	0.31
Iron	Fe		26	56	Trace	Trace
Iodine	I		53	127	Trace	Trace

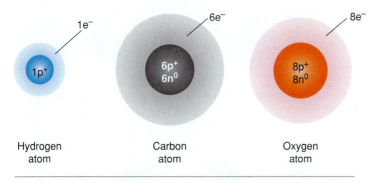

Figure 2.2 **Hydrogen, Carbon, and Oxygen Atoms**

Within the nucleus of an atom are specific numbers of positively charged protons (p⁺) and uncharged neutrons (n⁰). The negatively charged electrons (e⁻) are around the nucleus. The atoms depicted here are electrically neutral because the number of protons and number of electrons within each atom are equal.

by the electrons. Although it is impossible to know precisely where any given electron is located at any particular moment, the region where electrons are most likely to be found can be represented by an **electron cloud** (figure 2.1).

Each element is uniquely defined by the number of protons in the atoms of that element. For example, only hydrogen atoms have one proton, only carbon atoms have six protons, and only oxygen atoms have eight protons (figure 2.2; see table 2.1). The number of protons in each atom is called the **atomic number,** and because the number of electrons and number of protons are equal, the atomic number is also the number of electrons.

Protons and neutrons have about the same mass, and they are responsible for most of the mass of atoms. Electrons, on the other

hand, have very little mass. The **mass number** of an element is the number of protons plus the number of neutrons in each atom. For example, the mass number for carbon is 12 because it has 6 protons and 6 neutrons.

Predict 3

The atomic number of fluorine is 9, and the mass number is 19. What is the number of protons, neutrons, and electrons in an atom of fluorine?

Electrons and Chemical Bonding

The chemical behavior of an atom is determined largely by its outermost electrons. **Chemical bonding** occurs when the outermost electrons are transferred or shared between atoms. Two major types of chemical bonding are ionic bonding and covalent bonding.

Ionic Bonding

An atom is electrically neutral because it has an equal number of protons and electrons. If an atom loses or gains electrons, the numbers of protons and electrons are no longer equal, and a charged particle called an **ion** (ī′on) is formed. After an atom loses an electron, it has one more proton than it has electrons and is positively charged. For example, a sodium atom (Na) can lose an electron to become a positively charged sodium ion (Na⁺) (figure 2.3a). After an atom gains an electron, it has one more electron than it has protons and is negatively charged. For example, a chlorine atom (Cl) can accept an electron to become a negatively charged chloride ion (Cl⁻). Because oppositely charged ions are attracted to each other, positively charged ions tend to remain close to negatively charged ions. Thus, an **ionic** (ī-on′ik) **bond** occurs when electrons are transferred between atoms, creating oppositely charged ions. For example, Na⁺ and Cl⁻ are held together by ionic bonding to form an array of ions called sodium chloride (NaCl), or table salt (figure 2.3b,c).

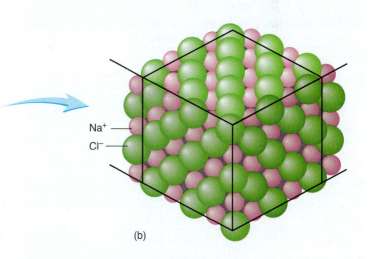

Figure 2.3 **AP|R** **Ionic Bonding**

(a) A sodium atom loses an electron to become a smaller, positively charged ion, and a chlorine atom gains an electron to become a larger, negatively charged ion. The attraction between the oppositely charged ions results in ionic bonding and the formation of sodium chloride. (b) The Na⁺ and Cl⁻ are organized to form a cube-shaped array. (c) A photomicrograph of salt crystals reflects the cubic arrangement of the ions.

Ions are denoted by using the symbol of the atom from which the ion was formed and adding a plus (+) or minus (−) superscript to indicate the ion's charge. For example, a sodium ion is Na^+, and a chloride ion is Cl^-. If more than one electron has been lost or gained, a number is used with the plus or minus sign. Thus, Ca^{2+} is a calcium ion formed by the loss of two electrons. Table 2.2 lists some ions commonly found in the body.

Predict 4

> *If an iron (Fe) atom loses three electrons, what is the charge of the resulting ion? Write the symbol for this ion.*

Covalent Bonding

A **covalent bond** forms when atoms share one or more pairs of electrons. The resulting combination of atoms is called a **molecule** (mol′ĕ-kūl). An example is the covalent bond between two hydrogen atoms to form a hydrogen molecule (figure 2.4). Each hydrogen atom has one electron. As the atoms get closer together, the positively charged nucleus of each atom begins to attract the electron of the other atom. At an optimal distance, the two nuclei mutually attract the two electrons, and each electron is shared by both nuclei. The two hydrogen atoms are now held together by a covalent bond.

The sharing of one pair of electrons by two atoms results in a **single covalent bond.** A single line between the symbols of the atoms involved (for example, H—H) represents a single covalent bond. A **double covalent bond** results when two atoms share two pairs of electrons. When a carbon atom combines with two oxygen atoms to form carbon dioxide, two double covalent bonds are formed. Double covalent bonds are indicated by a double line between the atoms (O=C=O).

The two hydrogen atoms do not interact because they are too far apart.

The positively charged nucleus of each hydrogen atom begins to attract the electron of the other.

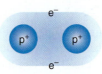

A covalent bond forms when the electrons are shared between the nuclei because the electrons are equally attracted to each nucleus.

Figure 2.4 Covalent Bonding

TABLE 2.2	Important Ions in the Human Body	
Ion	**Symbol**	**Significance***
Calcium	Ca^{2+}	Part of bones and teeth; functions in blood clotting, muscle contraction, release of neurotransmitters
Sodium	Na^+	Membrane potentials, water balance
Potassium	K^+	Membrane potentials
Hydrogen	H^+	Acid-base balance
Hydroxide	OH^-	Acid-base balance
Chloride	Cl^-	Water balance
Bicarbonate	HCO_3^-	Acid-base balance
Ammonium	NH_4^+	Acid-base balance
Phosphate	PO_4^{3-}	Part of bones and teeth; functions in energy exchange, acid-base balance
Iron	Fe^{2+}	Red blood cell function
Magnesium	Mg^2	Necessary for enzymes
Iodide	I^-	Present in thyroid hormones

*The ions are part of the structures or play important roles in the processes listed.

Electrons can be shared unequally in covalent bonds. When there is an unequal, asymmetrical sharing of electrons, the bond is called a **polar covalent bond** because the unequal sharing of electrons results in one end (pole) of the molecule having a partial electrical charge opposite to that of the other end. For example, two hydrogen atoms can share their electrons with an oxygen atom to form a water molecule (H_2O), as shown in figure 2.5. However, the hydrogen atoms do not share the electrons equally with the oxygen atom, and the electrons tend to spend more time around the oxygen atoms than around the hydrogen atoms. Molecules with this asymmetrical electrical charge are called **polar molecules.**

When there is an equal sharing of electrons between atoms, the bond is called a **nonpolar covalent bond.** Molecules with a symmetrical electrical charge are called **nonpolar molecules.**

Hydrogen Bonds

A polar molecule has a positive "end" and a negative "end." The positive end of one polar molecule can be weakly attracted to the negative end of another polar molecule. Although this attraction is called a **hydrogen bond,** it is not a chemical bond because electrons are not transferred or shared between the atoms of the different polar molecules. The attraction between molecules resulting from hydrogen bonds is much weaker than in ionic or covalent bonds. For example, the positively charged hydrogen of one water molecule is weakly attracted to a negatively charged oxygen of another water molecule (figure 2.6). Thus, the water molecules are held together by hydrogen bonds.

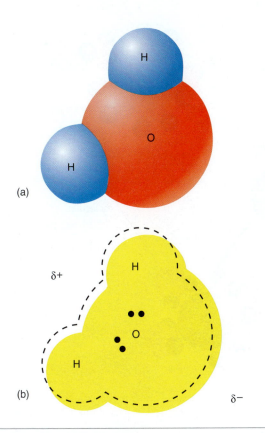

(a)

(b)

$\delta+$

$\delta-$

Figure 2.5 Polar Covalent Bonds

(*a*) A water molecule forms when two hydrogen atoms form covalent bonds with an oxygen atom. (*b*) The hydrogen atoms and oxygen atoms are sharing electron pairs (*indicated by the black dots*), but the sharing is unequal. The *dashed outline* shows the expected location of the electron cloud if the electrons are shared equally. But as the yellow area indicates, the actual electron cloud (*yellow*) is shifted toward the oxygen. Consequently, the oxygen side of the molecule has a slight negative charge (*indicated by* $\delta-$), and the hydrogen side of the molecule has a slight positive charge (*indicated by* $\delta+$).

Hydrogen bonds also play an important role in determining the shape of complex molecules. The bonds can occur between different polar parts of a single large molecule to hold the molecule in its normal three-dimensional shape (see "Proteins" and "Nucleic Acids: DNA and RNA" later in this chapter). Table 2.3 summarizes the important characteristics of chemical bonds (ionic and covalent) and forces between separate molecules (hydrogen bonds).

Molecules and Compounds

A molecule is formed when two or more atoms chemically combine to form a structure that behaves as an independent unit. Sometimes the atoms that combine are of the same type, as when two hydrogen atoms combine to form a hydrogen molecule. But more typically, a molecule consists of two or more different types of atoms, such as two hydrogen atoms and an oxygen atom combining to form water. Thus, a glass of water consists of a collection of individual water molecules positioned next to one another.

A **compound** (kom′pownd; to place together) is a substance resulting from the chemical combination of two or more *different* types of atoms. Water is an example of a substance that is a

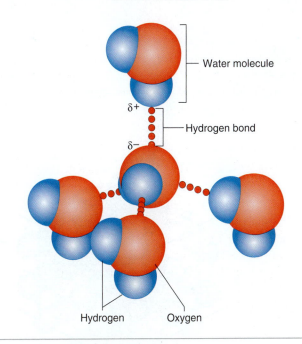

Water molecule

$\delta+$

Hydrogen bond

$\delta-$

Hydrogen   Oxygen

Figure 2.6 Hydrogen Bonds

The positive hydrogen part of one water molecule ($\delta+$) forms a hydrogen bond (*red dotted line*) with the negative oxygen part of another water molecule ($\delta-$). As a result, hydrogen bonds hold separate water molecules together.

TABLE 2.3	Comparison of Bonds
Bond	**Example**
Ionic Bond A complete transfer of electrons between two atoms results in separate positively charged and negatively charged ions.	Na^+Cl^- Sodium chloride
Polar Covalent Bond An unequal sharing of electrons between two atoms results in a slightly positive charge ($\delta+$) on one side of the molecule and a slightly negative charge (δ) on the other side of the molecule.	$\delta+$ O $\delta-$ Water
Nonpolar Covalent Bond An equal sharing of electrons between two atoms results in an even charge distribution among the atoms of the molecule.	H–C–H Methane
Hydrogen Bond The attraction of oppositely charged ends of one polar molecule to another polar molecule holds molecules or parts of molecules together.	O···H–O Water molecules

compound and a molecule. Not all molecules are compounds. For example, a hydrogen molecule is not a compound because it does not consist of different types of atoms.

Some compounds are molecules and some are not. (Remember that, to be a molecule, a structure must be an independent unit.) Covalent compounds, in which different types of atoms are held

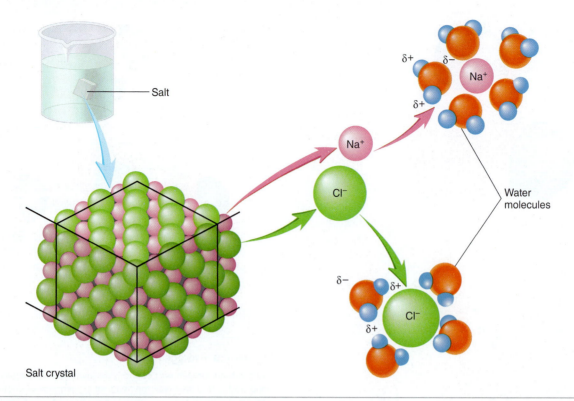

Figure 2.7 AP|R **Dissociation**

Sodium chloride (table salt) dissociates in water. The positively charged Na^+ are attracted to the negatively charged oxygen (*red*) end of the water molecule, and the negatively charged Cl^- are attracted to the positively charged hydrogen (*blue*) end of the water molecule.

together by covalent bonds, are molecules because the sharing of electrons results in distinct units. On the other hand, ionic compounds, in which ions are held together by the force of attraction between opposite charges, are not molecules because they do not consist of distinct units. Sodium chloride is an example of a substance that is a compound but not a molecule. A piece of sodium chloride does not consist of individual sodium chloride molecules positioned next to one another. Instead, it is an organized array of individual Na^+ and individual Cl^- in which each charged ion is surrounded by several ions of the opposite charge (see figure 2.3b).

Molecules and compounds can be represented by the symbols of the atoms forming the molecule or compound plus subscripts denoting the quantity of each type of atom present. For example, glucose (a sugar) can be represented as $C_6H_{12}O_6$, indicating that glucose is composed of 6 carbon, 12 hydrogen, and 6 oxygen atoms. Letter symbols represent most atoms and molecules. Throughout this chapter and the text, oxygen (O_2), carbon dioxide (CO_2), and other commonly discussed ions and molecules will be identified by their symbols when appropriate.

Dissociation

When ionic compounds dissolve in water, their ions **dissociate** (di-sō′sē-āt′), or separate, from each other because the positively charged ions are attracted to the negative ends of the water molecules, and the negatively charged ions are attracted to the positive ends of the water molecules. For example, when sodium chloride dissociates in water, the Na^+ and Cl^- separate, and water molecules surround and isolate the ions, keeping them in solution (figure 2.7). These

dissociated ions are sometimes called **electrolytes** (ē-lek′trō-lītz) because they have the capacity to conduct an electrical current, the flow of charged particles. For example, an electrocardiogram (ECG) is a recording of electrical currents produced by the heart. These currents can be detected by electrodes on the surface of the body because the ions in the body fluids conduct electrical currents.

When molecules dissolve in water, the molecules usually remain intact even though they are surrounded by water molecules. Thus, in a glucose solution, glucose molecules are surrounded by water molecules.

2.2 CHEMICAL REACTIONS

Learning Outcomes *After reading this section, you should be able to*

A. Summarize the characteristics of synthesis, decomposition, and exchange reactions.
B. Explain how reversible reactions produce chemical equilibrium.
C. Distinguish between chemical reactions that release energy and those that take in energy.
D. Describe the factors that can affect the rate of chemical reactions.

In a **chemical reaction,** atoms, ions, molecules, or compounds interact either to form or to break chemical bonds. The substances that enter into a chemical reaction are called the **reactants,** and the substances that result from the chemical reaction are called the **products.**

Classification of Chemical Reactions

For our purposes, chemical reactions can be classified as synthesis, decomposition, or exchange reactions.

Synthesis Reactions

When two or more reactants combine to form a larger, more complex product, the process is called a **synthesis reaction,** represented symbolically as

$$A + B \rightarrow AB$$

Examples of synthesis reactions include the synthesis of the complex molecules of the human body from the basic "building blocks" obtained in food and the synthesis of **adenosine triphosphate (ATP)** (ă-den′ō-sēn trī-fos′fāt) molecules. In ATP, *A* stands for adenosine, *T* stands for *tri-* (or three), and *P* stands for a phosphate group (PO_4^{3-}). Thus, ATP consists of adenosine and three phosphate groups. ATP is synthesized when adenosine diphosphate (ADP), which has two (*di-*) phosphate groups, combines with a phosphate group to form the larger ATP molecule. The phosphate group that reacts with ADP is often denoted as P_i, where the *i* indicates that the phosphate group is associated with an inorganic substance (see "Inorganic Molecules " later in this chapter).

$$\underset{\text{(ADP)}}{\text{A-P-P}} \quad + \quad \underset{\substack{\text{(Phosphate} \\ \text{group)}}}{P_i} \quad \rightarrow \quad \underset{\text{(ATP)}}{\text{A-P-P-P}}$$

All of the synthesis reactions that occur in the body are collectively referred to as **anabolism** (ă-nab′-ō-lizm). Growth, maintenance, and repair of the body could not take place without anabolic reactions.

Decomposition Reactions

In a **decomposition reaction,** reactants are broken down into smaller, less complex products. A decomposition reaction is the reverse of a synthesis reaction and can be represented in this way:

$$AB \rightarrow A + B$$

Examples of decomposition reactions include the breakdown of food molecules into basic building blocks and the breakdown of ATP to ADP and a phosphate group.

$$\underset{\text{(ATP)}}{\text{A-P-P-P}} \quad \rightarrow \quad \underset{\text{(ADP)}}{\text{A-P-P}} \quad + \quad \underset{\substack{\text{(Phosphate} \\ \text{group)}}}{P_i}$$

The decomposition reactions that occur in the body are collectively called **catabolism** (kă-tab′-ō-lizm). They include the digestion of food molecules in the intestine and within cells, the breakdown of fat stores, and the breakdown of foreign matter and microorganisms in certain blood cells that protect the body. All of the anabolic and catabolic reactions in the body are collectively defined as **metabolism.**

Exchange Reactions

An **exchange reaction** is a combination of a decomposition reaction and a synthesis reaction. In decomposition, the reactants are broken down. In synthesis, the products of the decomposition reaction are combined. The symbolic representation of an exchange reaction is

$$AB + CD \rightarrow AC + BD$$

An example of an exchange reaction is the reaction of hydrochloric acid (HCl) with sodium hydroxide (NaOH) to form table salt (NaCl) and water (H_2O):

$$HCl + NaOH \rightarrow NaCl + H_2O$$

Reversible Reactions

A **reversible reaction** is a chemical reaction that can proceed from reactants to products and from products to reactants. When the rate of product formation is equal to the rate of reactant formation, the reaction is said to be at **equilibrium.** At equilibrium, the amount of the reactants relative to the amount of products remains constant.

The following analogy may help clarify the concept of reversible reactions and equilibrium. Imagine a trough containing water. The trough is divided into two compartments by a partition, but the partition contains holes that allow the water to move freely between the compartments. Because the water can move in either direction, this is like a reversible reaction. The amount of water in the left compartment represents the amount of reactant, and the amount of water in the right compartment represents the amount of product. At equilibrium, the amount of reactant relative to the amount of product in each compartment is always the same because the partition allows water to pass between the two compartments until the level of water is the same in both compartments. If the amount of reactant is increased by adding water to the left compartment, water flows from the left compartment through the partition to the right compartment until the level of water is the same in both. Thus, the amounts of reactant and product are once again equal. Unlike this analogy, however, the amount of reactant relative to the amount of product in most reversible reactions is not one to one. Depending on the specific reversible reaction, there can be one part reactant to two parts product, two parts reactant to one part product, or many other possibilities.

An important reversible reaction in the human body occurs when carbon dioxide (CO_2) and water (H_2O) form hydrogen ions (H^+) and bicarbonate ions (HCO_3^-). The reversibility of the reaction is indicated by two arrows pointing in opposite directions:

$$CO_2 + H_2O \rightleftarrows H^+ + HCO_3^-$$

If CO_2 is added to H_2O, the amount of CO_2 relative to the amount of H^+ increases. However, the reaction of CO_2 with H_2O produces more H^+, and the amount of CO_2 relative to the amount of H^+ returns to equilibrium. Conversely, adding H^+ results in the formation of more CO_2, and the equilibrium is restored.

Maintaining a constant level of H^+ in body fluids is necessary for the nervous system to function properly. This level can be maintained, in part, by controlling blood CO_2 levels. For example, slowing the respiration rate causes blood CO_2 levels to increase, which causes an increase in H^+ concentration in the blood.

Predict 5

If the respiration rate increases, CO_2 is removed from the blood. What effect does this have on blood H^+ levels?

Energy and Chemical Reactions

Energy is defined as the capacity to do **work**—that is, to move matter. Energy can be subdivided into potential energy and kinetic energy. **Potential energy** is stored energy that could do work but

CLINICAL IMPACT Clinical Uses of Atomic Particles

Protons, neutrons, and electrons are responsible for the chemical properties of atoms. They also have other properties that can be useful in a clinical setting. For example, some of these properties have enabled the development of methods for examining the inside of the body.

Isotopes (ī′sō-tōpz; *isos,* equal + *topos,* part) are two or more forms of the same element that have the same number of protons and electrons but a different number of neutrons. Thus, isotopes have the same atomic number (i.e., number of protons) but different mass numbers (i.e., sum of the protons and neutrons). For example, hydrogen and its isotope deuterium each have an atomic number of 1 because they both have 1 proton. However, hydrogen has no neutrons, whereas deuterium has 1 neutron. Therefore, the mass number of hydrogen is 1, and that of deuterium is 2. Water made with deuterium is called heavy water because of the weight of the "extra" neutron. Because isotopes of the same atom have the same number of electrons, they are very similar in their chemical behavior. The nuclei of some isotopes are stable and do not change. Radioactive isotopes, however, have unstable nuclei that lose neutrons or protons. Several different kinds of radiation can be produced when neutrons and protons, or the products formed by their breakdown, are released from the nucleus of the isotope.

The radiation given off by some radioactive isotopes can penetrate and destroy tissues. Rapidly dividing cells are more sensitive to radiation than are slowly dividing cells. Radiation is used to treat cancerous (malignant) tumors because cancer cells divide rapidly. If the treatment is effective, few healthy cells are destroyed, but the cancerous cells are killed.

Radioactive isotopes are also used in medical diagnosis. The radiation can be detected, and the movement of the radioactive isotopes throughout the body can be traced. For example, the thyroid gland normally takes up iodine and uses it in the formation of thyroid hormones. Radioactive iodine can be used to determine if iodine uptake is normal in the thyroid gland.

Radiation can be produced in ways other than by changing the nucleus of atoms. X-rays are a type of radiation formed when electrons lose energy by moving from a higher energy state to a lower one. Health professionals use x-rays to examine bones to determine if they are broken and x-rays of teeth to see if they have caries (cavities). Mammograms, which are low-energy radiographs (x-ray films) of the breast, can reveal tumors because the tumors are slightly denser than normal tissue.

Computers can be used to analyze a series of radiographs, each made at a slightly different body location. The computer assembles these radiographic "slices" through the body to form a three-dimensional image. A **computed tomography** (tō-mog′ră-fē) **(CT) scan** is an example of this technique (figure 2A). CT scans are used to detect tumors and other abnormalities in the body.

Magnetic resonance imaging (MRI) is another method for looking into the body (figure 2B). The patient is placed into a very powerful magnetic field, which aligns the hydrogen nuclei. Radio waves given off by the hydrogen nuclei are monitored, and a computer uses these data to make an image of the body. Because MRI detects hydrogen, it is very effective for visualizing soft tissues that contain a lot of water. MRI technology can reveal tumors and other abnormalities.

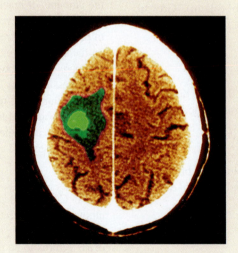

Figure 2A

CT scan of the brain with iodine injection, showing one large brain tumor (green area) that has metastasized (spread) to the brain from cancer in the large intestine.

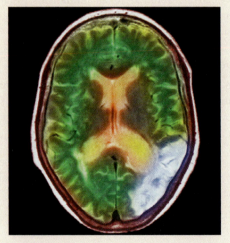

Figure 2B

Colorized MRI brain scan showing a stroke. The whitish area in the lower right part of the MRI is blood that has leaked into the surrounding tissue.

is not doing so. For example, a coiled spring has potential energy. It could push against an object and move the object, but as long as the spring does not uncoil, no work is accomplished. **Kinetic** (ki-net′ik; of motion) **energy** is energy caused by the movement of an object and is the form of energy that actually does work. An uncoiling spring pushing an object and causing it to move is an example. When potential energy is released, it becomes kinetic energy, thus doing work.

Potential and kinetic energy exist in many different forms: chemical energy, mechanical energy, heat energy, electrical energy, and electromagnetic (radiant) energy. Here we examine how chemical energy and mechanical energy play important roles in the body.

The **chemical energy** of a substance is a form of potential energy stored in chemical bonds. Consider two balls attached by a relaxed spring. In order to push the balls together and compress the spring, energy must be put into this system. As the spring is compressed, potential energy increases. When the compressed spring expands, potential energy decreases. In the same way, similarly charged particles, such as two negatively charged electrons or two positively charged nuclei, repel each other. As similarly charged particles move closer together, their potential energy increases, much like compressing a spring, and as they move farther apart, their potential energy decreases. Chemical bonding is a form of potential energy because of the charges and positions of the subatomic particles bound together.

Chemical reactions are important because of the products they form and the energy changes that result as the relative positions of subatomic particles change. If the products of a chemical reaction contain less potential energy than the reactants, energy is released. For example, food molecules contain more potential energy than waste products. The difference in potential energy between food and waste products is used by the human body to drive activities such as growth, repair, movement, and heat production.

An example of a reaction that releases energy is the breakdown of ATP to ADP and a phosphate group (figure 2.8*a*). The phosphate group is attached to the ADP molecule by a covalent bond, which has potential energy. After the breakdown of ATP, some of that energy is released as heat, and some is available to cells for activities such as synthesizing new molecules or contracting muscles:

$$ATP \rightarrow ADP + P_i + Energy \text{ (used by cells)}$$

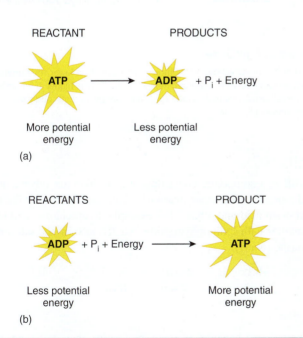

REACTANT　　　　　　PRODUCTS

ATP ⟶ **ADP** + P_i + Energy

More potential　　　Less potential
energy　　　　　　　energy

(a)

REACTANTS　　　　　　PRODUCT

ADP + P_i + Energy ⟶ **ATP**

Less potential　　　　More potential
energy　　　　　　　　energy

(b)

Figure 2.8 AP|R Energy and Chemical Reactions

In the two reactions shown here, the larger "sunburst" represents greater potential energy and the smaller "sunburst" represents less potential energy. (*a*) Energy is released as a result of the breakdown of ATP. (*b*) The input of energy is required for the synthesis of ATP.

According to the law of conservation of energy, the total energy of the universe is constant. Therefore, energy is neither created nor destroyed. However, one type of energy can be changed into another. Potential energy is converted into kinetic energy. Since the conversion between energy states is not 100% efficient, heat energy is released. For example, as a spring is released, its potential energy is converted to mechanical energy and heat energy. **Mechanical energy** is energy resulting from the position or movement of objects. Many of the activities of the human body, such as moving a limb, breathing, or circulating blood, involve mechanical energy.

Predict 6

Why does body temperature increase during exercise?

If the products of a chemical reaction contain more energy than the reactants (figure 2.8*b*), energy must be added from another source. The energy released during the breakdown of food molecules is the source of energy for this kind of reaction in the body. The energy from food molecules is used to synthesize molecules such as ATP, fats, and proteins:

$$ADP + P_i + Energy \text{ (from food molecules)} \rightarrow ATP$$

Rate of Chemical Reactions

The rate at which a chemical reaction proceeds is influenced by several factors, including how easily the substances react with one another, their concentrations, the temperature, and the presence of a catalyst.

Reactants

Reactants differ from one another in their ability to undergo chemical reactions. For example, iron corrodes much more rapidly than does stainless steel. For this reason, during the refurbishment of the Statue of Liberty in the 1980s, the iron bars forming the statue's skeleton were replaced with stainless steel bars.

Concentration

Within limits, the greater the concentration of the reactants, the greater the rate at which a chemical reaction will occur because, as the concentration increases, the reacting molecules are more likely to come in contact with one another. For example, the normal concentration of oxygen inside cells enables it to come in contact with other molecules, producing the chemical reactions necessary for life. If the oxygen concentration decreases, the rate of chemical reactions decreases. A decrease in oxygen in cells can impair cell function and even result in cell death.

Temperature

Because molecular motion changes as environmental temperature changes, the rate of chemical reactions is partially dependent on temperature. For example, reactions occur throughout the body at a faster rate when a person has a fever of only a few degrees. The result is increased activity in most organ systems, such as increased heart and respiratory rates. By contrast, the rate of reactions decreases when body temperature drops. The clumsy movement of very cold fingers results largely from the reduced rate of chemical reactions in cold muscle tissue.

Catalysts

At normal body temperatures, most chemical reactions would take place too slowly to sustain life if not for substances called catalysts. A **catalyst** (kat′ă-list) increases the rate of a chemical reaction, without itself being permanently changed or depleted. An enzyme is a protein molecule that acts as a catalyst. Many of the chemical reactions that occur in the body require enzymes. We consider them in greater detail later in this chapter, in the section titled "Enzymes."

2.3 ACIDS AND BASES

Learning Outcomes *After reading this section, you should be able to*

A. Describe the pH scale and its relationship to acidic and basic solutions.

B. Explain the importance of buffers in organisms.

The body has many molecules and compounds, called acids and bases, that can alter body functions. An **acid** is a proton donor. Because a hydrogen atom without its electron is a proton, any substance that releases hydrogen ions (H^+) in water is an acid. For example, hydrochloric acid (HCl) in the stomach forms H^+ and chloride ions (Cl^-):

$$HCl \rightarrow H^+ + Cl^-$$

A **base** is a proton acceptor. For example, sodium hydroxide (NaOH) forms sodium ions (Na^+) and hydroxide ions (OH^-). It is a base because the OH^- is a proton acceptor that binds with a H^+ to form water.

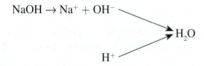

The pH Scale

The **pH scale** indicates the H^+ concentration of a solution (figure 2.9). The scale ranges from 0 to 14. A **neutral solution** has an equal number of H^+ and OH^- and thus a pH of 7.0. An **acidic solution** has a greater concentration of H^+ than of OH^- and thus a pH *less* than 7.0. A **basic,** or *alkaline* (al′kă-līn), **solution** has fewer H^+ than OH^- and thus a pH *greater* than 7.0. Notice that the pH number and the actual H^+ concentration are inversely related, meaning that the lower the pH number, the higher the H^+ concentration.

As the pH value becomes smaller, the solution becomes more acidic; as the pH value becomes larger, the solution becomes more basic. A change of one unit on the pH scale represents a 10-fold change in the H^+ concentration. For example, a solution with a pH of 6.0 has 10 times more H^+ than a solution with a pH of 7.0. Thus, small changes in pH represent large changes in H^+ concentration.

The normal pH range for human blood is 7.35 to 7.45. If blood pH drops below 7.35, a condition called **acidosis** (as-i-dō′sis) results. The nervous system is depressed, and the individual becomes disoriented and possibly comatose. If blood pH rises above 7.45, **alkalosis** (al-kă-lō′sis) results. The nervous system becomes overexcitable, and the individual can be extremely nervous or have convulsions. Both acidosis and alkalosis can result in death.

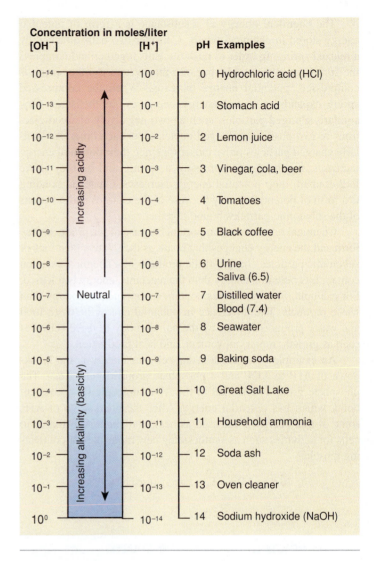

Figure 2.9 pH Scale

A pH of 7 is neutral. Values less than 7 are acidic (the lower the number, the more acidic). Values greater than 7 are basic (the higher the number, the more basic). This scale shows some representative fluids and their approximate pH values.

Salts

A **salt** is a compound consisting of a positive ion other than H^+ and a negative ion other than OH^-. Salts are formed by the reaction of an acid and a base. For example, hydrochloric acid (HCl) combines with sodium hydroxide (NaOH) to form the salt sodium chloride (NaCl):

| HCl | + | NaOH | → | NaCl | + | H_2O |
| (Acid) | | (Base) | | (Salt) | | (Water) |

Buffers

The chemical behavior of many molecules changes as the pH of the solution in which they are dissolved changes. The survival of an organism depends on its ability to maintain homeostasis by keeping body fluid pH within a narrow range. One way normal body fluid pH is maintained is through the use of buffers. A **buffer** (bŭf′er) is a chemical that resists changes in pH when either an

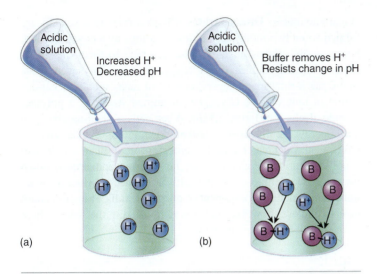

Figure 2.10 Buffers

(*a*) The addition of an acid to a nonbuffered solution results in an increase of H⁺ and a decrease in pH. (*b*) The addition of an acid to a buffered solution results in a much smaller change in pH. The added H⁺ bind to the buffer (symbolized by the letter *B*).

acid or a base is added to a solution containing the buffer. When an acid is added to a buffered solution, the buffer binds to the H⁺, preventing these ions from causing a decrease in the pH of the solution (figure 2.10).

Predict 7

> If a base is added to a solution, will the pH of the solution increase or decrease? If the solution is buffered, what response from the buffer prevents the change in pH?

2.4 INORGANIC MOLECULES

Learning Outcomes *After reading this section, you should be able to*

A. Distinguish between inorganic and organic molecules.
B. Describe how the properties of O_2, CO_2, and water contribute to their physiological functions.

Early scientists believed that inorganic substances came from non-living sources and that organic substances were extracted from living organisms. As the science of chemistry developed, however, it became apparent that the body also contains inorganic substances and that organic substances can be manufactured in the laboratory. As currently defined, **inorganic chemistry** deals with those substances that do not contain carbon, whereas **organic chemistry** is the study of carbon-containing substances. These definitions have a few exceptions. For example, CO_2 and carbon monoxide (CO) are classified as inorganic molecules, even though they contain carbon.

Inorganic substances play many vital roles in human anatomy and physiology. Examples include the O_2 and other gases we breathe, the calcium phosphate that makes up our bones, and the metals that are required for protein functions, such as iron in hemoglobin and zinc in alcohol dehydrogenase. In the next sections, we discuss the important roles of O_2, CO_2, and water—all inorganic molecules—in the body.

Oxygen and Carbon Dioxide

Oxygen (O_2) is a small, nonpolar, inorganic molecule consisting of two oxygen atoms bound together by a double covalent bond. About 21% of the gas in the atmosphere is O_2, and it is essential for most living organisms. Humans require O_2 in the final step of a series of chemical reactions in which energy is extracted from food molecules (see chapter 17).

Carbon dioxide (CO_2) consists of one carbon atom bound to two oxygen atoms. Each oxygen atom is bound to the carbon atom by a double covalent bond. Carbon dioxide is produced when food molecules, such as glucose, are metabolized within the cells of the body. Once CO_2 is produced, it is eliminated from the cell as a metabolic by-product, transferred to the lungs by the blood, and exhaled during respiration. If CO_2 is allowed to accumulate within cells, it becomes toxic.

Water

Water (H_2O) is an inorganic molecule that consists of one atom of oxygen joined by polar covalent bonds to two atoms of hydrogen. Water has many important roles in humans and all living organisms:

1. *Stabilizing body temperature.* Because heat energy causes not only movement of water molecules, but also disruption of hydrogen bonds, water can absorb large amounts of heat and remain at a stable temperature. Blood, which is mostly water, is warmed deep in the body and then flows to the surface, where the heat is released. In addition, water evaporation in the form of sweat results in significant heat loss from the body.
2. *Providing protection.* Water is an effective lubricant. For example, tears protect the surface of the eye from the rubbing of the eyelids. Water also forms a fluid cushion around organs, which helps protect them from damage. The fluid that surrounds the brain is an example.
3. *Facilitating chemical reactions.* Most of the chemical reactions necessary for life do not take place unless the reacting molecules are dissolved in water. For example, NaCl must dissociate in water into Na⁺ and Cl⁻ before those ions can react with other ions. Water also directly participates in many chemical reactions. For example, during the digestion of food, large molecules and water react to form smaller molecules.
4. *Transporting substances.* Many substances dissolve in water and can be moved from place to place as the water moves. For example, blood transports nutrients, gases, and waste products within the body.

2.5 ORGANIC MOLECULES

Learning Outcomes *After reading this section, you should be able to*

A. Describe the structural organization and major functions of carbohydrates, lipids, proteins, and nucleic acids.
B. Explain how enzymes work.

Carbon's ability to form covalent bonds with other atoms makes possible the formation of the large, diverse, complicated molecules necessary for life. Carbon atoms bound together by covalent bonds constitute the "framework" of many large molecules. Two mechanisms that allow the formation of a wide variety of molecules are variation in the length of the carbon chains and the combination of the atoms bound to the carbon framework. For example, proteins have thousands of carbon atoms bound by covalent bonds to one another and to other atoms, such as nitrogen, sulfur, hydrogen, and oxygen.

The four major groups of organic molecules essential to living organisms are carbohydrates, lipids, proteins, and nucleic acids. Each of these groups has specific structural and functional characteristics (table 2.4).

Carbohydrates

Carbohydrates are composed of carbon, hydrogen, and oxygen atoms. In most carbohydrates, for each carbon atom there are two hydrogen atoms and one oxygen atom. Note that this two-to-one ratio is the same as in water (H_2O). The molecules are called carbohydrates because each carbon (carbo-) is combined with the same atoms that form water (hydrated). For example, the chemical formula for glucose is $C_6H_{12}O_6$.

The smallest carbohydrates are **monosaccharides** (mon-ō-sak′ă-rīdz; one sugar), or simple sugars. Glucose (blood sugar) and fructose (fruit sugar) are important monosaccharide energy sources for many of the body's cells. Larger carbohydrates are formed by chemically binding monosaccharides together. For this reason, monosaccharides are considered the building blocks

of carbohydrates. **Disaccharides** (dī-sak′ă-rīdz; two sugars) are formed when two monosaccharides are joined by a covalent bond. For example, glucose and fructose combine to form the disaccharide sucrose (table sugar) (figure 2.11a). **Polysaccharides** (pol-ē-sak′ă-rīdz; many sugars) consist of many monosaccharides bound in long chains. Glycogen, or animal starch, is a polysaccharide of glucose (figure 2.11b). When cells containing glycogen need energy, the glycogen is broken down into individual glucose molecules, which can be used as energy sources. Plant starch, also a polysaccharide of glucose, can be ingested and broken down into glucose. Cellulose, another polysaccharide of glucose, is an important structural component of plant cell walls. Humans cannot digest cellulose, however, and it is eliminated in the feces, where the cellulose fibers provide bulk.

Lipids

Lipids are substances that dissolve in nonpolar solvents, such as alcohol or acetone, but not in polar solvents, such as water. Lipids are composed mainly of carbon, hydrogen, and oxygen, but other elements, such as phosphorus and nitrogen, are minor components of some lipids. Lipids contain a lower proportion of oxygen to carbon than do carbohydrates.

Fats, phospholipids, eicosanoids, and steroids are examples of lipids. **Fats** are important energy-storage molecules; they also pad and insulate the body. The building blocks of fats are **glycerol** (glis′er-ol) and **fatty acids** (figure 2.12). Glycerol is a 3-carbon

TABLE 2.4	Important Organic Molecules and Their Functions in the Body			
Molecule	**Elements**	**Building Blocks**	**Function**	**Examples**
Carbohydrate	C, H, O	Monosaccharides	Energy	Monosaccharides can be used as energy sources. Glycogen (a polysaccharide) is an energy-storage molecule.
Lipid	C, H, O (P, N in some)	Glycerol and fatty acids (for fats)	Energy	Fats can be stored and broken down later for energy; per unit of weight, fats yield twice as much energy as carbohydrates.
			Structure	Phospholipids and cholesterol are important components of cell membranes.
			Regulation	Steroid hormones regulate many physiological processes (e.g., estrogen and testosterone are responsible for many of the differences between males and females).
Protein	C, H, O, N (S in most)	Amino acids	Regulation	Enzymes control the rate of chemical reactions. Hormones regulate many physiological processes (e.g., insulin affects glucose transport into cells).
			Structure	Collagen fibers form a structural framework in many parts of the body.
			Energy	Proteins can be broken down for energy; per unit of weight, they yield the same energy as carbohydrates.
			Contraction	Actin and myosin in muscle are responsible for muscle contraction.
			Transport	Hemoglobin transports O_2 in the blood.
			Protection	Antibodies and complement protect against microorganisms and other foreign substances.
Nucleic acid	C, H, O, N, P	Nucleotides	Regulation	DNA directs the activities of the cell.
			Heredity	Genes are pieces of DNA that can be passed from one generation to the next.
			Gene expression	RNA is involved in gene expression.

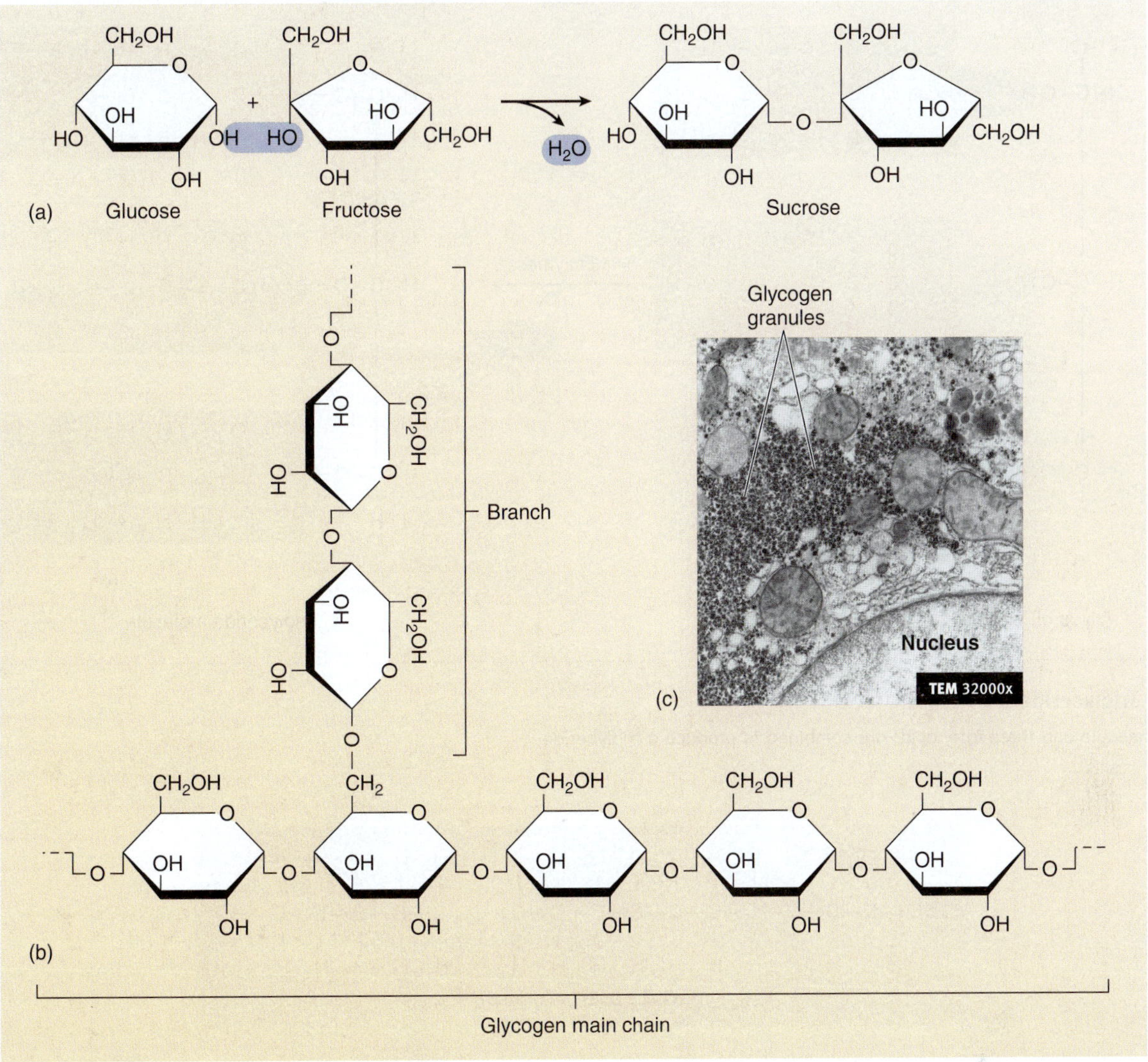

Figure 2.11 Carbohydrates

(*a*) Glucose and fructose are monosaccharides that combine to form the disaccharide sucrose. (*b*) Glycogen is a polysaccharide formed by combining many glucose molecules. (*c*) Transmission electron micrograph of stored glycogen in a human cell. Glycogen clusters together into particles called granules.

molecule with a **hydroxyl** (hī-drok′sil) **group** (—OH) attached to each carbon atom, and fatty acids consist of a carbon chain with a **carboxyl** (kar-bok′sil) **group** attached at one end. A carboxyl group consists of both an oxygen atom and a hydroxyl group attached to a carbon atom (—COOH):

$$
\begin{array}{ccc}
\overset{\displaystyle O}{\underset{\displaystyle \parallel}{}} & & \overset{\displaystyle O}{\underset{\displaystyle \parallel}{}} \\
-\text{C}-\text{OH} & \text{or} & \text{HO}-\text{C}-
\end{array}
$$

The carboxyl group is responsible for the acidic nature of the molecule because it releases H⁺ into solution. **Triglycerides** (trī-glis′er-īdz) are the most common type of fat molecules. Triglycerides have three fatty acids bound to a glycerol molecule.

Fatty acids differ from one another according to the length and degree of saturation of their carbon chains. Most naturally occurring fatty acids contain 14 to 18 carbon atoms. A fatty acid is **saturated** if it contains only single covalent bonds between the carbon atoms (figure 2.13*a*). Sources of saturated fats

include beef, pork, whole milk, cheese, butter, eggs, coconut oil, and palm oil. The carbon chain is **unsaturated** if it has one or more double covalent bonds (figure 2.13*b*). Because the double covalent bonds can occur anywhere along the carbon chain, many types of unsaturated fatty acids with an equal degree of unsaturation are possible. **Monounsaturated fats,** such as olive and peanut oils, have one double covalent bond between carbon atoms. **Polyunsaturated fats,** such as safflower, sunflower, corn, and fish oils, have two or more double covalent bonds between carbon atoms. Unsaturated fats are the best type of fats in the diet because, unlike saturated fats, they do not contribute to the development of cardiovascular disease.

Trans **fats** are unsaturated fats that have been chemically altered by the addition of H atoms. The process makes the fats more saturated and hence more solid and stable (longer shelf-life). However, the change in structure of these chemicals makes the consumption of *trans* fats an even greater factor than saturated fats in the risk for cardiovascular disease.

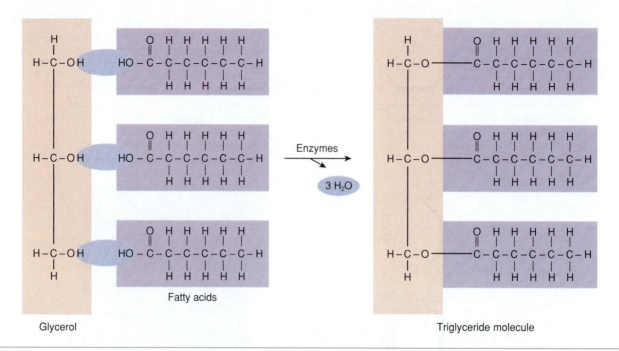

Figure 2.12 Triglyceride

One glycerol molecule and three fatty acids are combined to produce a triglyceride.

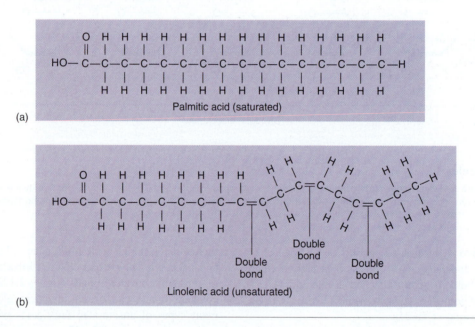

Figure 2.13 Fatty Acids

(a) Palmitic acid is a saturated fatty acid; it contains no double bonds between the carbons. (b) Linolenic acid is an unsaturated fatty acid; note the three double bonds between the carbons, which cause the molecule to have a bent shape.

Phospholipids are similar to triglycerides, except that one of the fatty acids bound to the glycerol is replaced by a molecule containing phosphorus (figure 2.14). A phospholipid is polar at the end of the molecule to which the phosphate is bound and nonpolar at the other end. The polar end of the molecule is attracted to water and is said to be **hydrophilic** (water-loving). The nonpolar end is repelled by water and is said to be **hydrophobic** (water-fearing).

Phospholipids are important structural components of cell membranes (see chapter 3).

The **eicosanoids** (ī′kō-să-noydz) are a group of important chemicals derived from fatty acids. Eicosanoids are made in most cells and are important regulatory molecules. Among their numerous effects is their role in the response of tissues to injuries. One example of eicosanoids is **prostaglandins** (pros′tă-glan′dinz), which have been implicated in regulating the secretion of some hormones, blood clotting, some reproductive functions, and many other processes. Many of the therapeutic effects of aspirin and other anti-inflammatory drugs result from their ability to inhibit prostaglandin synthesis.

Steroids are composed of carbon atoms bound together into four ringlike structures. Cholesterol is an important steroid because

other steroid molecules are synthesized from it. For example, bile salts, which increase fat absorption in the intestines, are derived from cholesterol, as are the reproductive hormones estrogen, progesterone, and testosterone (figure 2.15). In addition, cholesterol is an important component of cell membranes. Although high levels of cholesterol in the blood increase the risk of cardiovascular disease, a certain amount of cholesterol is vital for normal function.

Nitrogen

Phosphorus

Oxygen

Carbon

Hydrogen

Polar (hydrophilic) region (phosphate-containing region)

Nonpolar (hydrophobic) region (fatty acids)

(a)   (b)

Figure 2.14 Phospholipids

(*a*) Molecular model of a phospholipid. Note that one of the fatty acid tails is bent, indicating that it is unsaturated. (*b*) A simplified depiction of a phospholipid.

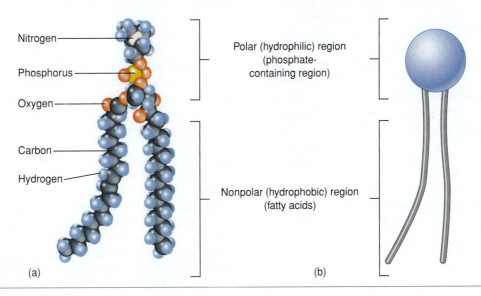

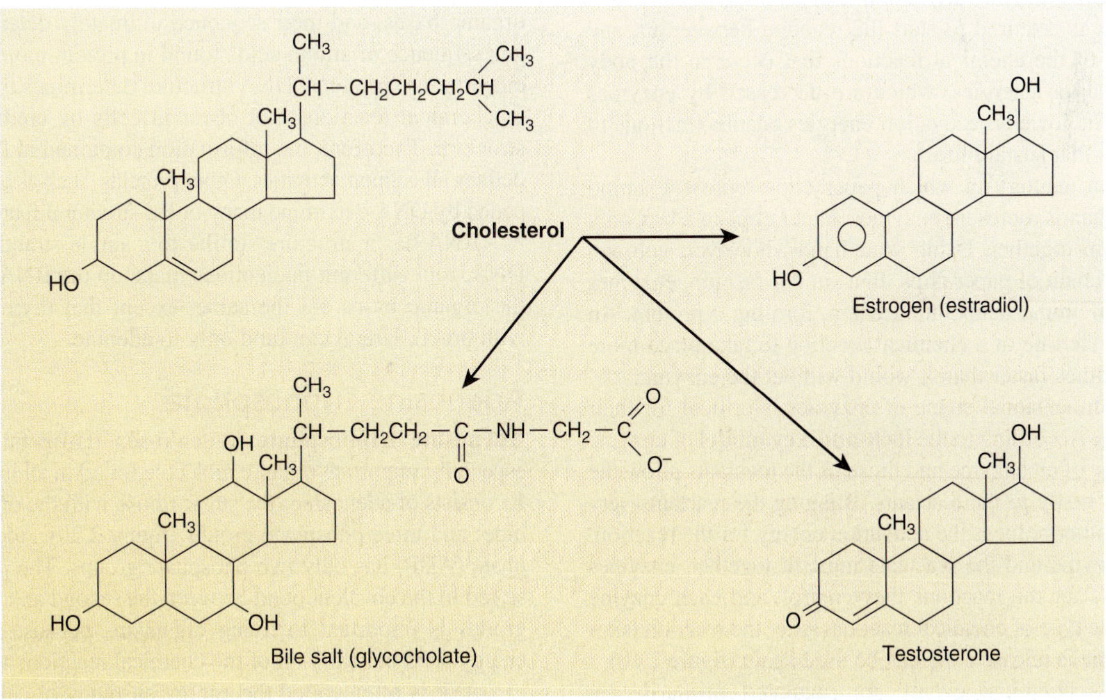

Cholesterol

Estrogen (estradiol)

Bile salt (glycocholate)

Testosterone

Figure 2.15 Steroids

Steroids are four-ringed molecules that differ from one another according to the groups attached to the rings. Cholesterol, the most common steroid, can be modified to produce other steroids.

Proteins

All **proteins** contain carbon, hydrogen, oxygen, and nitrogen, and most have some sulfur. The building blocks of proteins are **amino** (ă-mē′nō) **acids,** which are organic acids containing an **amine** (ă-mēn′) **group** (–NH$_2$) and a carboxyl group (figure 2.16*a*). There are 20 basic types of amino acids. Humans can synthesize 12 of them from simple organic molecules, but the remaining 8 so-called essential amino acids must be obtained in the diet.

A protein consists of many amino acids joined together to form a chain (figure 2.16*b,c*). Although there are only 20 amino acids, they can combine to form numerous types of proteins with unique structures and functions. Different proteins have different kinds and numbers of amino acids. Hydrogen bonds between amino acids in the chain cause the chain to fold or coil into a specific three-dimensional shape (figure 2.16*d,e*). The ability of proteins to perform their functions depends on their shape. If the hydrogen bonds that maintain the shape of the protein are broken, the protein becomes nonfunctional. This change in shape is called **denaturation,** and it can be caused by abnormally high temperatures or changes in pH.

Proteins perform many important functions. For example, enzymes are proteins that regulate the rate of chemical reactions, structural proteins provide the framework for many of the body's tissues, and muscles contain proteins that are responsible for muscle contraction.

Enzymes

An **enzyme** (en′zīm) is a protein catalyst that increases the rate at which a chemical reaction proceeds without the enzyme being permanently changed. Enzymes increase the rate of chemical reactions by lowering the **activation energy,** which is the energy necessary to start a chemical reaction. For example, heat in the form of a spark is required to start the reaction between O$_2$ and gasoline. Most of the chemical reactions that occur in the body have high activation energies, which are decreased by enzymes (figure 2.17). The lowered activation energies enable reactions to proceed at rates that sustain life.

Consider an analogy in which paper clips represent amino acids and your hands represent enzymes. Paper clips in a box only occasionally join together. Using your hands, however, you can rapidly make a chain of paper clips. In a similar fashion, enzymes can quickly join amino acids into a chain, forming a protein. An enzyme allows the rate of a chemical reaction to take place more than a million times faster than it would without the enzyme.

The three-dimensional shape of enzymes is critical for their normal function. According to the **lock-and-key model** of enzyme action, the shape of an enzyme and those of the reactants allow the enzyme to bind easily to the reactants. Bringing the reactants very close to one another reduces the activation energy for the reaction. Because the enzyme and the reactants must fit together, enzymes are very specific for the reactions they control, and each enzyme controls only one type of chemical reaction. After the reaction takes place, the enzyme is released and can be used again (figure 2.18).

The body's chemical events are regulated primarily by mechanisms that control either the concentration or the activity of enzymes. Either (1) the rate at which enzymes are produced in cells or (2) whether the enzymes are in an active or inactive form determines the rate of each chemical reaction.

Nucleic Acids: DNA and RNA

Deoxyribonucleic (dē-oks′ē-rī′bō-noo-klē′ik) **acid (DNA)** is the genetic material of cells, and copies of DNA are transferred from one generation of cells to the next. DNA contains the information that determines the structure of proteins. **Ribonucleic** (rī′bō-noo-klē′ik) **acid (RNA)** is structurally related to DNA, and three types of RNA also play important roles in gene expression or protein synthesis. In chapter 3, we explore the means by which DNA and RNA direct the functions of the cell.

The **nucleic** (noo-klē′ik, noo-klā′ik) **acids** are large molecules composed of carbon, hydrogen, oxygen, nitrogen, and phosphorus. Both DNA and RNA consist of basic building blocks called **nucleotides** (noo′klē-ō-tīdz). Each nucleotide is composed of a sugar (monosaccharide) to which a nitrogenous organic base and a phosphate group are attached (figure 2.19). The sugar is deoxyribose for DNA, ribose for RNA. The organic bases are thymine (thī′mēn, thī′min), cytosine (sī′tō-sēn), and uracil (ūr′ă-sil), which are single-ringed molecules, and adenine (ad′ĕ-nēn) and guanine (gwahn′ēn), which are double-ringed molecules.

DNA has two strands of nucleotides joined together to form a twisted, ladderlike structure called a double helix. The sides of the ladder are formed by covalent bonds between the sugar molecules and phosphate groups of adjacent nucleotides. The rungs of the ladder are formed by the bases of the nucleotides of one side connected to the bases of the other side by hydrogen bonds. Each nucleotide of DNA contains one of the organic bases: adenine, thymine, cytosine, or guanine. Adenine binds only to thymine because the structure of these organic bases allows two hydrogen bonds to form between them. Cytosine binds only to guanine because the structure of these organic bases allows three hydrogen bonds to form between them.

The sequence of organic bases in DNA molecules stores genetic information. Each DNA molecule consists of millions of organic bases, and their sequence ultimately determines the type and sequence of amino acids found in protein molecules. Because enzymes are proteins, DNA structure determines the rate and type of chemical reactions that occur in cells by controlling enzyme structure. Therefore, the information contained in DNA ultimately defines all cellular activities. Other proteins, such as collagen, that are coded by DNA determine many of the structural features of humans.

RNA has a structure similar to a single strand of DNA. Like DNA, four different nucleotides make up the RNA molecule, and the organic bases are the same, except that thymine is replaced with uracil. Uracil can bind only to adenine.

Adenosine Triphosphate

Adenosine triphosphate (ă-den′ō-sēn trī-fos′făt) **(ATP)** is an especially important organic molecule found in all living organisms. It consists of adenosine (the sugar ribose with the organic base adenine) and three phosphate groups (figure 2.20). Adenosine diphosphate (ADP) has only two phosphate groups. The potential energy stored in the covalent bond between the second and third phosphate groups is important to living organisms because it provides the energy used in nearly all of the chemical reactions within cells.

ATP is often called the energy currency of cells because it is capable of both storing and providing energy. The concentration of ATP is maintained within a narrow range of values, and essentially all energy-requiring chemical reactions stop when the quantity of ATP becomes inadequate.

(a) Two examples of amino acids. Each amino acid has an amine group (—NH₂) and a carboxyl group (—COOH).

(b) The individual amino acids are joined.

(c) A protein consists of a chain of different amino acids (represented by different-colored spheres).

(d) A three-dimensional representation of the amino acid chain shows the hydrogen bonds (*dotted red lines*) between different amino acids. The hydrogen bonds cause the amino acid chain to become folded or coiled.

(e) An entire protein has a complex three-dimensional shape.

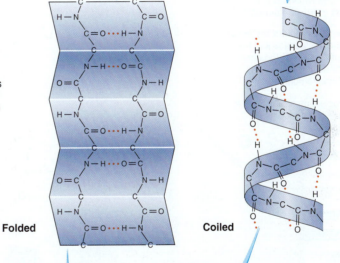

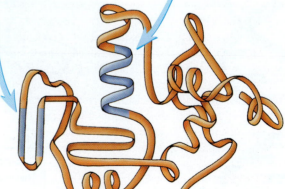

Figure 2.16 Protein Structure

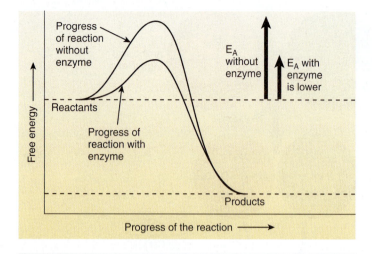

Figure 2.17 Activation Energy and Enzymes

Activation energy is required to initiate chemical reactions. Without an enzyme, a chemical reaction can proceed, but it needs more energy input. Enzymes lower the activation energy, making it easier for the reaction to proceed.

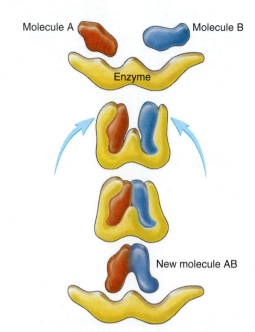

Figure 2.18 Enzyme Action

The enzyme brings two reacting molecules together. This is possible because the reacting molecules "fit" the shape of the enzyme (lock-and-key model). After the reaction, the unaltered enzyme can be used again.

① The building blocks of nucleic acids are nucleotides, which consist of a phosphate group, a sugar, and a nitrogen base.

② The phosphate groups connect the sugars to form two strands of nucleotides (*purple columns*).

③ Hydrogen bonds (*dotted red lines*) between the nucleotides join the two nucleotide strands together. Adenine binds to thymine, and cytosine binds to guanine.

④ The two nucleotide strands coil to form a double-stranded helix.

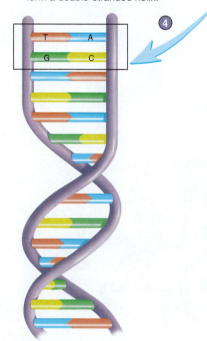

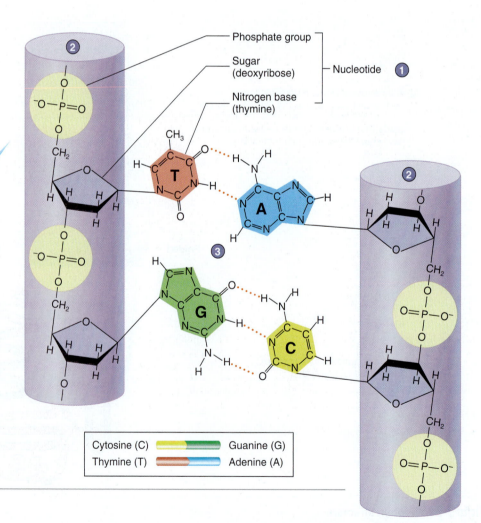

Figure 2.19 **AP|R** Structure of DNA

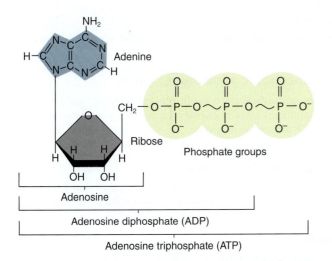

Figure 2.20 AP|R **Structure of ATP**

ATP releases energy when the bond between phosphate groups is broken. Notice that the bond between the oxygen atoms and phosphorus atoms of adjacent phosphate groups is drawn as a squiggly line. This symbolizes a less stable bond that is partly due to the negative charge on each phosphate and that, when broken, releases a great amount of energy. Consequently, this squiggly line is often referred to as a "high-energy" bond.

A CASE IN POINT

Cyanide Poisoning

Although Justin Hale was rescued from his burning home, he suffered from cyanide poisoning. Inhalation of smoke released by the burning of rubber and plastic during household fires is the most common cause of cyanide poisoning. Cyanide compounds can be lethal to humans because they interfere with the production of ATP in mitochondria (see chapter 3). Without adequate ATP, cells malfunction and can die. The heart and brain are especially susceptible to cyanide poisoning.

Cyanide poisoning by inhalation or absorption through the skin can also occur in certain manufacturing industries, and cyanide gas was used during the Holocaust to kill people. Deliberate suicide by ingesting cyanide is rare but has been made famous by the "suicide capsules" in spy movies. In 1982, seven people in the Chicago area died after taking Tylenol that someone had laced with cyanide. Subsequent copycat tamperings occurred and led to the widespread use of tamper-proof capsules and packaging.

ANSWER TO LEARN TO PREDICT

In this question, you learn that Rafael had not been able to properly hydrate the night before the soccer match and the weather is hot. The question also tells you that Rafael was cool when he was sweating and became overheated when he stopped sweating. Sweat consists mainly of water, and as you learned in section 2.1, water molecules are polar, allowing them to hydrogen bond together. Later, in section 2.4, you learned that hydrogen bonds stabilize water, allowing it to absorb large amounts of heat. Thus, sweating effectively creates a cooler layer of air next to your skin because the water molecules in sweat can absorb heat. Because Rafael was dehydrated, he could not continue sweating, became overheated, and suffered a mild heat stroke.

Answers to the rest of this chapter's Predict questions are in Appendix E.

 ## SUMMARY

Chemistry is the study of the composition and structure of substances and the reactions they undergo.

2.1 Basic Chemistry (p. 21)

Matter, Mass, and Weight

1. Matter is anything that occupies space and has mass.
2. Mass is the amount of matter in an object, and weight results from the gravitational attraction between the earth and an object.

Elements and Atoms

1. An element is the simplest type of matter having unique chemical and physical properties.
2. An atom is the smallest particle of an element that has the chemical characteristics of that element. An element is composed of only one kind of atom.

Atomic Structure

1. Atoms consist of neutrons, positively charged protons, and negatively charged electrons.
2. An atom is electrically neutral because the number of protons equals the number of electrons.
3. Protons and neutrons are in the nucleus, and electrons can be represented by an electron cloud around the nucleus.
4. The atomic number is the unique number of protons in each atom of an element. The mass number is the number of protons and neutrons.

Electrons and Chemical Bonding

1. An ionic bond results when an electron is transferred from one atom to another.
2. A covalent bond results when a pair of electrons is shared between atoms. A polar covalent bond is an unequal sharing of electron pairs.

Hydrogen Bonds

A hydrogen bond is the weak attraction between the oppositely charged regions of polar molecules. Hydrogen bonds are important in determining the three-dimensional structure of large molecules.

Molecules and Compounds

1. A molecule is two or more atoms chemically combined to form a structure that behaves as an independent unit.
2. A compound is two or more different types of atoms chemically combined. A compound can be a molecule (covalent compound) or an organized array of ions (ionic compound).

Dissociation

Dissociation is the separation of ions in an ionic compound by polar water molecules.

2.2 Chemical Reactions   (p. 26)

Classification of Chemical Reactions

1. A synthesis reaction is the combination of reactants to form a new, larger product.
2. A decomposition reaction is the breakdown of larger reactants into smaller products.
3. An exchange reaction is a combination of a decomposition reaction, in which reactants are broken down, and a synthesis reaction, in which the products of the decomposition reaction recombine.

Reversible Reactions

1. In a reversible reaction, the reactants can form products, or the products can form reactants.
2. The amount of reactant relative to the amount of product is constant at equilibrium.

Energy and Chemical Reactions

1. Energy is the capacity to do work. Potential energy is stored energy that could do work, and kinetic energy does work by causing the movement of an object.
2. Energy exists in chemical bonds as potential energy.
3. Energy is released in chemical reactions when the products contain less potential energy than the reactants. The energy can be lost as heat, used to synthesize molecules, or used to do work.
4. Energy must be added in reactions when the products contain more potential energy than the reactants.
5. Energy can be neither created nor destroyed, but one type of energy can be changed into another.

Rate of Chemical Reactions

1. The rate of a chemical reaction increases when the concentration of the reactants increases, the temperature increases, or a catalyst is present.
2. A catalyst (enzyme) increases the rate of a chemical reaction without being altered permanently.

2.3 Acids and Bases   (p. 30)

Acids are proton (H^+) donors, and bases are proton acceptors.

The pH Scale

1. A neutral solution has an equal number of H^+ and OH^- and a pH of 7.0.
2. An acidic solution has more H^+ than OH^- and a pH less than 7.0.
3. A basic solution has fewer H^+ than OH^- and a pH greater than 7.0.

Salts

A salt forms when an acid reacts with a base.

Buffers

Buffers are chemicals that resist changes in pH when acids or bases are added.

2.4 Inorganic Molecules   (p. 31)

1. Inorganic chemistry is mostly concerned with non-carbon-containing substances but does include such carbon-containing substances as CO_2 and carbon monoxide.
2. Some inorganic chemicals play important roles in the body.

Oxygen and Carbon Dioxide

1. Oxygen is involved with the extraction of energy from food molecules.
2. Carbon dioxide is a by-product of the breakdown of food molecules.

Water

1. Water stabilizes body temperature.
2. Water provides protection by acting as a lubricant or a cushion.
3. Water is necessary for many chemical reactions.
4. Water transports many substances.

2.5 Organic Molecules   (p. 31)

Organic molecules contain carbon atoms bound together by covalent bonds.

Carbohydrates

1. Carbohydrates provide the body with energy.
2. Monosaccharides are the building blocks that form more complex carbohydrates, such as disaccharides and polysaccharides.

Lipids

1. Lipids are substances that dissolve in nonpolar solvents, such as alcohol or acetone, but not in polar solvents, such as water. Fats, phospholipids, and steroids are examples of lipids.
2. Lipids provide energy (fats), serve as structural components (phospholipids), and regulate physiological processes (steroids).
3. The building blocks of triglycerides (fats) are glycerol and fatty acids.
4. Fatty acids can be saturated (have only single covalent bonds between carbon atoms) or unsaturated (have one or more double covalent bonds between carbon atoms).

Proteins

1. Proteins regulate chemical reactions (enzymes), serve as structural components, and cause muscle contraction.
2. The building blocks of proteins are amino acids.
3. Denaturation of proteins disrupts hydrogen bonds, which changes the shape of proteins and makes them nonfunctional.
4. Enzymes are specific, bind to reactants according to the lock-and-key model, and function by lowering activation energy.

Nucleic Acids: DNA and RNA

1. The basic unit of nucleic acids is the nucleotide, which is a monosaccharide with an attached phosphate and organic base.
2. DNA nucleotides contain the monosaccharide deoxyribose and the organic bases adenine, thymine, guanine, and cytosine. DNA occurs as a double strand of joined nucleotides and is the genetic material of cells.
3. RNA nucleotides are composed of the monosaccharide ribose. The organic bases are the same as for DNA, except that thymine is replaced with uracil.

Adenosine Triphosphate

ATP stores energy, which can be used in cell processes.

REVIEW AND COMPREHENSION

1. Define chemistry. Why is an understanding of chemistry important to the study of human anatomy and physiology?

2. Define matter. What is the difference between mass and weight?

3. Define element and atom. How many different kinds of atoms are present in a specific element?

4. List the components of an atom, and explain how they are organized to form an atom. Compare the charges of the subatomic particles.

5. Define the atomic number and the mass number of an element.

6. Distinguish among ionic, covalent, polar covalent, and hydrogen bonds. Define ion.

7. What is the difference between a molecule and a compound?

8. What happens to ionic and covalent compounds when they dissolve in water?

9. Define chemical reaction. Describe synthesis, decomposition, and exchange reactions, giving an example of each.

10. What is meant by the equilibrium condition in a reversible reaction?

11. Define potential energy and kinetic energy.

12. Give an example of a chemical reaction that releases energy and an example of a chemical reaction that requires the input of energy.

13. Name three ways that the rate of chemical reactions can be increased.

14. What is an acid and what is a base? Describe the pH scale.

15. Define a salt. What is a buffer, and why are buffers important?

16. Distinguish between inorganic and organic chemistry.

17. Why is O_2 necessary for human life? Where does the CO_2 we breathe out come from?

18. List four functions that water performs in the human body.

19. Name the four major types of organic molecules. Give a function for each.

20. Describe the action of enzymes in terms of activation energy and the lock-and-key model.

CRITICAL THINKING

1. If an atom of iodine (I) gains an electron, what is the charge of the resulting ion? Write the symbol for this ion.

2. For each of the following chemical equations, determine if a synthesis reaction, a decomposition reaction, or dissociation has taken place:
   a. $HCl \rightarrow H^+ + Cl^-$
   b. Glucose + Fructose → Sucrose (table sugar)
   c. $2 H_2O \rightarrow 2 H_2 + O_2$

3. In terms of the energy in chemical bonds, explain why eating food is necessary for increasing muscle mass.

4. Given that the H^+ concentration in a solution is based on the following reversible reaction,

$$CO_2 + H_2O \rightleftharpoons H^+ + HCO_3^-$$

what happens to the pH of the solution when $NaHCO_3$ (sodium bicarbonate) is added to the solution? (*Hint:* The sodium bicarbonate dissociates to form Na^+ and HCO_3^-.)

5. A mixture of chemicals is warmed slightly. As a consequence, although little heat is added, the solution becomes very hot. Explain what happens to make the solution hot.

6. Two solutions, when mixed together at room temperature, produce a chemical reaction. However, when the solutions are boiled and allowed to cool to room temperature before mixing, no chemical reaction takes place. Explain.

Answers in Appendix D

3 Cell Structures and Their Functions

LEARN to PREDICT

Carlos, the boy in the picture, has diabetes insipidus, an incurable disease that causes his kidneys to produce an unusually large volume of dilute urine. He always carries a water bottle, and he never likes to be too far from a restroom. To keep his body fluids in a state of homeostasis (see chapter I), Carlos has to drink enough water to replace what he loses as urine. Diabetes insipidus is a genetic disease in which the body does not respond normally to an important chemical signal called ADH, which regulates water loss from the kidneys.

After reading about cell structure and genetic expression in chapter 3, explain how Carlos's condition developed at the cellular level.

3.1 CELL STRUCTURE

Learning Outcome *After reading this section, you should be able to*

A. Explain how the structures of a cell contribute to its function.

The study of cells is an important link between the study of chemistry in chapter 2 and the study of tissues in chapter 4. Knowledge of chemistry makes it possible to understand cells because cells are composed of chemicals, and those chemicals are responsible for many of the cells' characteristics.

Cells, in turn, determine the form and functions of the human body. From cellular function, we can progress to the study of tissues. In addition, many diseases and other human disorders have a cellular basis. The human body is composed of trillions of cells and acts as a host to countless other organisms (see "Microbes In Your Body: Getting to Know Your Bacteria" in chapter 1). Because the body is made of many, many cells, we must first understand the anatomy and physiology of the cell before we can understand the anatomy and physiology of the human body.

The cell is the basic living unit of all organisms. The simplest organisms consist of single cells, whereas humans are composed of multiple cells. An average-sized cell is one-fifth the size of the smallest dot you can make on a sheet of paper with a sharp pencil! But despite their extremely small size, cells are complex living structures. Cells have many characteristics in common; however,

Module 2 Cells and Chemistry

most cells are also specialized to perform specific functions. The human body is made up of many populations of specialized cells. The coordinated functions of these populations are critical to the survival of humans and all other complex organisms. This chapter considers the structures of cells and how cells perform the activities necessary for maintaining homeostasis.

Each cell is a highly organized unit. Within cells, specialized structures called **organelles** (or′gă-nelz; little organs) perform specific functions (figure 3.1 and table 3.1). The nucleus is an organelle containing the cell's genetic material. The living material surrounding the nucleus is called **cytoplasm** (sī′tō-plazm), and it contains many types of organelles. The cytoplasm is enclosed by the **cell membrane,** or *plasma membrane.*

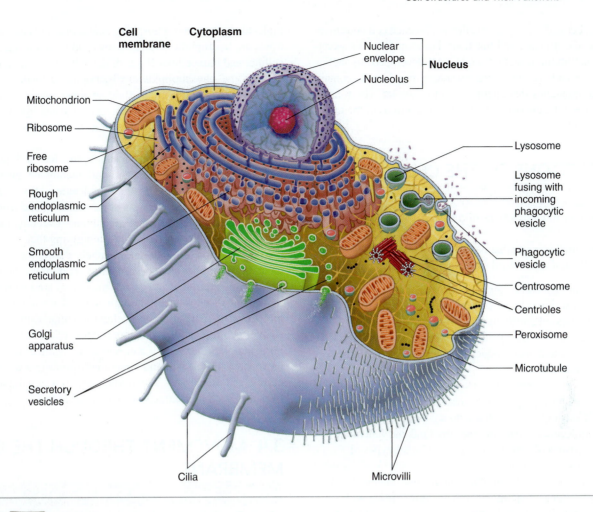

Figure 3.1 AP|R **Generalized Cell**

This generalized cell shows the major organelles contained in cells. However, no single cell contains all of these organelle types. Furthermore, one kind of cell may contain many organelles of one type, whereas another kind of cell may contain very few.

TABLE 3.1	Organelles and Their Locations and Functions	
Organelles	**Location**	**Function(s)**
Nucleus	Often near center of the cell	Contains genetic material of cell (DNA) and nucleoli; site of RNA synthesis and ribosomal subunit assembly
Ribosomes	In cytoplasm	Site of protein synthesis
Rough endoplasmic reticulum	In cytoplasm	Has many ribosomes attached; site of protein synthesis (rough ER)
Smooth endoplasmic reticulum	In cytoplasm	Site of lipid synthesis; participates in detoxification (smooth ER)
Golgi apparatus	In cytoplasm	Modifies protein structure and packages proteins in secretory vesicles
Secretory vesicle	In cytoplasm	Contains materials produced in the cell; formed by the Golgi apparatus; secreted by exocytosis
Lysosome	In cytoplasm	Contains enzymes that digest material taken into the cell
Mitochondrion	In cytoplasm	Site of aerobic respiration and the major site of ATP synthesis
Microtubule	In cytoplasm	Supports cytoplasm; assists in cell division and forms components of cilia and flagella
Centrioles	In cytoplasm	Facilitate the movement of chromosomes during cell division
Cilia	On cell surface with many on each cell	Move substances over surfaces of certain cells
Flagella	On sperm cell surface with one per cell	Propel sperm cells
Microvilli	Extensions of cell surface with many on each cell	Increase surface area of certain cells

The number and type of organelles within each cell determine the cell's specific structure and functions. For example, cells secreting large amounts of protein contain well-developed organelles that synthesize and secrete protein, whereas muscle cells contain proteins and organelles that enable them to contract. The following sections describe the structure and main functions of the major organelles in cells.

3.2 FUNCTIONS OF THE CELL

Learning Outcome *After reading this section, you should be able to*

A. List the four main functions of a cell.

Cells are the smallest units that have all the characteristics of life. Our body cells perform several important functions:

1. *Cell metabolism and energy use.* The chemical reactions that occur within cells are collectively called cell metabolism. Energy released during metabolism is used for cell activities, such as the synthesis of new molecules, muscle contraction, and heat production, which helps maintain body temperature.
2. *Synthesis of molecules.* Cells synthesize various types of molecules, including proteins, nucleic acids, and lipids. The different cells of the body do not all produce the same molecules. Therefore, the structural and functional characteristics of cells are determined by the types of molecules they produce.
3. *Communication.* Cells produce and receive chemical and electrical signals that allow them to communicate with one another. For example, nerve cells communicate with one another and with muscle cells, causing muscle cells to contract.
4. *Reproduction and inheritance.* Each cell contains a copy of the genetic information of the individual. Specialized cells (sperm cells and oocytes) transmit that genetic information to the next generation.

3.3 CELL MEMBRANE

Learning Outcome *After reading this section, you should be able to*

A. Describe the structure of the cell membrane.

The **cell membrane,** or *plasma* (plaz′mă) *membrane,* is the outermost component of a cell. The cell membrane encloses the cytoplasm and forms the boundary between material inside the cell and material outside it. Substances outside the cell are called **extracellular** substances, and those inside the cell are called **intracellular** substances. Besides enclosing the cell, the cell membrane supports the cell contents, acts as a selective barrier that determines what moves into and out of the cell, and plays a role in communication between cells. The major molecules that make up the cell membrane are phospholipids and proteins. In addition, the membrane contains other molecules, such as cholesterol and carbohydrates.

Studies of the arrangement of molecules in the cell membrane have given rise to a model of its structure called the **fluid-mosaic model** (figure 3.2). The phospholipids form a double layer of molecules. The polar, phosphate-containing ends of the phospholipids are hydrophilic (water-loving) and therefore face the extracellular and intracellular fluids of the cell. The nonpolar, fatty acid ends of the phospholipids are hydrophobic (water-fearing) and therefore face away from the fluid on either side of the membrane, toward the center of the double layer of phospholipids. The double layer of phospholipids forms a lipid barrier between the inside and outside of the cell.

The double layer of phospholipids has a liquid quality. Cholesterol within the phospholipid membrane gives it added strength and flexibility. Protein molecules "float" among the phospholipid molecules and, in some cases, extend from the inner to the outer surface of the cell membrane. Carbohydrates may be bound to some protein molecules, modifying their functions. The proteins function as membrane channels, carrier molecules, receptor molecules, enzymes, or structural supports in the membrane. **Membrane channels** and carrier molecules are involved with the movement of substances through the cell membrane. **Receptor molecules** are part of an intercellular communication system that enables cell recognition and coordination of the activities of cells. For example, a nerve cell can release a chemical messenger that moves to a muscle cell and temporarily binds to a receptor on the muscle cell membrane. The binding acts as a signal that triggers a response, such as contraction of the muscle cell.

3.4 MOVEMENT THROUGH THE CELL MEMBRANE

Learning Outcomes *After reading this section, you should be able to*

A. Define diffusion and concentration gradient.
B. Explain the role of osmosis and that of osmotic pressure in controlling the movement of water across the cell membrane. Compare hypotonic, isotonic, and hypertonic solutions.
C. Define carrier-mediated transport, and compare the processes of facilitated diffusion, active transport, and secondary active transport.
D. Describe endocytosis and exocytosis.

Cell membranes are **selectively permeable,** meaning that they allow some substances, but not others, to pass into or out of the cells. Intracellular material has a different composition than extracellular material, and the cell's survival depends on maintaining the difference. Substances such as enzymes, glycogen, and potassium ions (K^+) are found at higher concentrations intracellularly, whereas Na^+, Ca^{2+}, and Cl^- are found in greater concentrations extracellularly. In addition, nutrients must enter cells continually, and waste products must exit. Because of the permeability characteristics of cell membranes and their ability to transport certain molecules, cells are able to maintain proper intracellular concentrations of molecules. Rupture of the membrane, alteration of its permeability characteristics, or inhibition of transport processes can disrupt the normal intracellular concentration of molecules and lead to cell death.

Movement through the cell membrane may be passive or active. Passive membrane transport does not require the cell to expend energy. Active membrane transport does require the cell to expend energy, usually in the form of ATP. Passive membrane

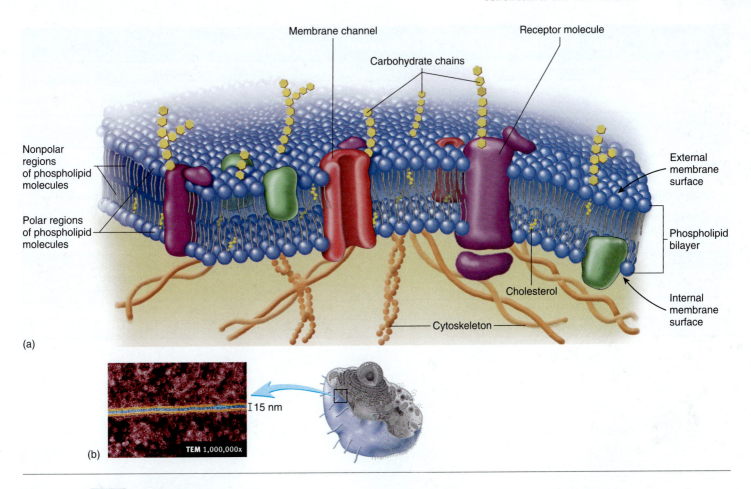

Membrane channel

Receptor molecule

Carbohydrate chains

Nonpolar
regions
of phospholipid
molecules

Polar regions
of phospholipid
molecules

External
membrane
surface

Phospholipid
bilayer

Cholesterol

Internal
membrane
surface

Cytoskeleton

(a)

I 15 nm

TEM 1,000,000x

(b)

Figure 3.2 AP|R **Cell Membrane**

(*a*) Fluid-mosaic model of the cell membrane. The membrane is composed of a bilayer of phospholipids and cholesterol with proteins "floating" in the membrane. The nonpolar hydrophobic region of each phospholipid molecule is directed toward the center of the membrane, and the polar hydrophilic region is directed toward the fluid environment either outside or inside the cell. (*b*) Colorized transmission electron micrograph showing the cell membrane of a single cell. Proteins at either surface of the lipid bilayer stain more readily than the lipid bilayer does and give the membrane the appearance of having three parts: The two yellow, outer parts are proteins and the phospholipid heads, and the blue, central part is the phospholipid tails and cholesterol.

transport mechanisms include diffusion, osmosis, and facilitated diffusion. Active membrane transport mechanisms include active transport, secondary active transport, endocytosis, and exocytosis. Table 3.2 lists the specific types of movement across cell membranes and we discuss each method in detail in the following sections.

Diffusion

A **solution** is generally composed of one or more substances, called **solutes,** dissolved in the predominant liquid or gas, which is called the **solvent.** Solutes, such as ions or molecules, tend to move from an area of higher concentration of a solute to an area of lower concentration of that same solute in solution. This process is called **diffusion** (figure 3.3, steps 1 and 2). An example of diffusion is the distribution of smoke throughout a room in which there are no air currents. Another example of diffusion is the gradual spread of salt throughout a beaker of still water.

Diffusion results from the natural, constant random motion of all solutes in a solution. More solute particles occur in an area of higher concentration than in an area of lower concentration. Because particles move randomly, the chances are greater that

solute particles will move from the higher toward the lower concentration than from the lower toward the higher concentration. At equilibrium, the net movement of solutes stops, although the random motion continues, and the movement of solutes in any one direction is balanced by an equal movement of solutes in the opposite direction (figure 3.3, step 3).

A **concentration gradient** is the difference in the concentration of a solute in a solvent between two points divided by the distance between the two points. The concentration gradient is said to be steeper when the concentration difference is large and/or the distance is small. When we say that a substance moves down (or with) the concentration gradient, we mean that solutes are diffusing from a higher toward a lower concentration of solutes. When we say that a solute moves up (or against) its concentration gradient, this means that the substance moves from an area of lower solute concentration to an area of higher solute concentration. This second type of movement does not occur by diffusion and requires energy in order to occur.

In the body, diffusion is an important means of transporting substances through the extracellular and intracellular fluids. In

TABLE 3.2	Types and Characteristics of Movement Across Membranes			
Types	**Transport**	**Requires ATP**	**Examples**	
Diffusion	With the concentration gradient through the lipid portion of the cell membrane or through membrane channels	No	O_2, CO_2, Cl^-, urea	
Osmosis	With the concentration gradient (for water) through the lipid portion of the cell membrane or through membrane channels	No	Water	
Carrier-mediated transport mechanisms				
Facilitated diffusion	With the concentration gradient by carrier molecules	No	Glucose in most cells	
Active transport	Against the concentration gradient* by carrier molecules	Yes	Na^+, K^+, Ca^{2+}, H^+, and amino acids	
Secondary active transport	Against the concentration gradient by carrier molecules; the energy for secondary active transport of one substance comes from the concentration gradient of another	Yes	Glucose, amino acids	
Endocytosis	Movement into cells by vesicles	Yes	Ingestion of particles by phagocytosis or receptor-mediated endocytosis and liquids by pinocytosis	
Exocytosis	Movement out of cells by vesicles	Yes	Secretion of proteins	

*Active transport normally moves substances *against* their concentration gradient, but it can also move substances *with* their concentration gradient.

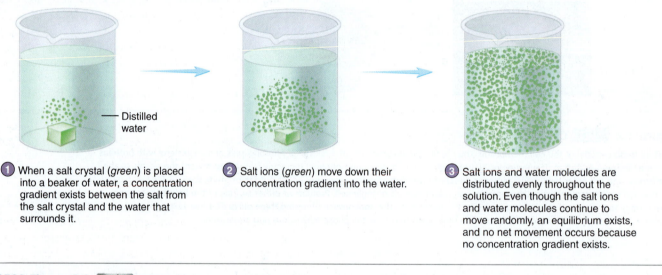

— Distilled water

1. When a salt crystal (*green*) is placed into a beaker of water, a concentration gradient exists between the salt from the salt crystal and the water that surrounds it.

2. Salt ions (*green*) move down their concentration gradient into the water.

3. Salt ions and water molecules are distributed evenly throughout the solution. Even though the salt ions and water molecules continue to move randomly, an equilibrium exists, and no net movement occurs because no concentration gradient exists.

PROCESS Figure 3.3 AP|R **Diffusion**

addition, substances, such as nutrients and some waste products, can diffuse into and out of the cell. The normal intracellular concentrations of many substances depend on diffusion. For example, if the extracellular concentration of O_2 is reduced, not enough O_2 diffuses into the cell, and the cell cannot function normally.

Predict 2

Urea is a toxic waste produced inside liver cells. It diffuses from those cells into the blood and is eliminated from the body by the kidneys. What would happen to the intracellular and extracellular concentrations of urea if the kidneys stopped functioning?

The phospholipid bilayer acts as a barrier to most water-soluble substances. However, certain small, water-soluble substances can diffuse between the phospholipid molecules of cell membranes. Other water-soluble substances can diffuse across the cell membrane only by passing through cell membrane **channels** (figure 3.4).

Molecules that are lipid-soluble, such as O_2, CO_2, and steroids, pass readily through the phospholipid bilayer.

Cell membrane channels consist of large protein molecules that extend from one surface of cell membranes to the other (figure 3.5). There are several channel types, each of which allows only certain substances to pass through. The size, shape, and charge of a molecule all determine whether it can pass through each kind of channel. For example, Na^+ passes through Na^+ channels, and K^+ and Cl^- pass through K^+ and Cl^- channels, respectively. Rapid movement of water across the cell membrane also occurs through membrane channels.

In addition, cell membrane channels differ in the degree to which ions pass through them. Some channels constantly allow ions to pass through. These channels are called **leak channels.** Other channels limit the movement of ions across the membrane by opening and closing. These channels are called **gated channels** (figure 3.5).

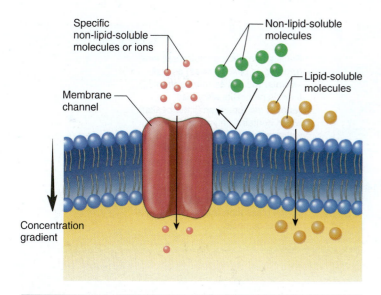

Specific
non-lipid-soluble
molecules or ions

Non-lipid-soluble
molecules

Membrane
channel

Lipid-soluble
molecules

Concentration
gradient

Figure 3.4 Diffusion Through the Cell Membrane

Non-lipid-soluble molecules (*red*) diffuse through membrane channels.
Lipid-soluble molecules (*orange*) diffuse directly through the cell membrane.

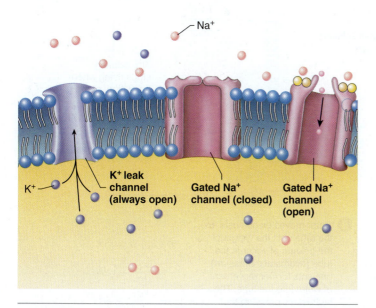

Na⁺

K⁺

**K⁺ leak
channel
(always open)**

**Gated Na⁺
channel (closed)**

**Gated Na⁺
channel
(open)**

Figure 3.5 Leak and Gated Membrane Channels

In this example, the K⁺ leak channel (*purple*) is always open, allowing K⁺ to
diffuse across the cell membrane. The gated Na⁺ channel (*pink*) regulates
the movement of Na⁺ across the membrane by opening and closing.

Osmosis

Osmosis (os-mō′sis) is the diffusion of water (a solvent) across
a selectively permeable membrane, such as the cell membrane,
from a region of higher water concentration to one of lower water
concentration (figure 3.6; see table 3.2). Even though water is a
polar molecule, it is small enough that it can move across the cell
membrane by passing either between the phospholipid molecules
or through water channels. Osmosis is important to cells because
large volume changes caused by water movement can disrupt normal cell functions. Osmosis occurs when the cell membrane is less

permeable, selectively permeable, or not permeable to solutes *and*
a concentration gradient for water exists across the cell membrane.
Water diffuses from a solution with a higher water concentration
across the cell membrane into a solution with a lower water concentration. The ability to predict the direction of water movement
across the cell membrane depends on knowing which solution on
either side of the membrane has the higher water concentration.

The concentration of a solution, however, is expressed not in
terms of water, but in terms of solute concentration. For example,
if sugar solution A is more concentrated than sugar solution B,
then solution A has more sugar (solute) than solution B. As the
concentration of a solution increases, the amount of water (solvent) proportionately decreases. Water diffuses from the less
concentrated solution B (less sugar, more water), into the more
concentrated solution A (more sugar, less water). In other words,
water diffuses toward areas of high solute concentration and
dilutes those solutes.

Osmotic pressure is the force required to prevent the movement of water across a selectively permeable membrane. Thus,
osmotic pressure is a measure of the tendency of water to move
by osmosis across a selectively permeable membrane. It can be
measured by placing a solution into a tube that is closed at one end
by a selectively permeable membrane and immersing the tube in
distilled water (see figure 3.6, step 1). Water molecules move by
osmosis through the membrane into the tube, forcing the solution
to move up the tube (see figure 3.6, step 2). As the solution rises,
its weight produces **hydrostatic pressure** (see figure 3.6, step 3),
which moves water out of the tube back into the distilled water
surrounding the tube. Net movement of water into the tube stops
when the hydrostatic pressure in the tube causes water to move out
of the tube at the same rate at which it diffuses into the tube by
osmosis. The osmotic pressure of the solution in the tube is equal
to the hydrostatic pressure that prevents net movement of water
into the tube.

The greater the concentration of a solution, the greater
its osmotic pressure, and the greater the tendency for water to
move into the solution. This occurs because water moves from
less concentrated solutions (less solute, more water) into more
concentrated solutions (more solute, less water). The greater the
concentration of a solution, the greater the tendency for water to
move into the solution, and the greater the osmotic pressure must
be to prevent that movement.

When placed into a solution, a cell may swell, remain
unchanged, or shrink, depending on the concentration gradient
between the solution and the cell's cytoplasm. A **hypotonic** (hī′pō-
ton′ik; *hypo*, under) solution usually has a lower concentration of
solutes and a higher concentration of water relative to the cytoplasm
of the cell. Thus, the solution has less tone, or osmotic pressure,
than the cell. Water moves by osmosis into the cell, causing it to
swell. If the cell swells enough, it can rupture, a process called
lysis (lī′sis) (figure 3.7*a*). When a cell is immersed in an **isotonic**
(ī′sō-ton′ik; *iso*, equal) solution, the concentrations of various solutes and water are the same on both sides of the cell membrane.
The cell therefore neither shrinks nor swells (figure 3.7*b*). When
a cell is immersed in a **hypertonic** (hī′per-ton′ik; *hyper*, above)
solution, the solution usually has a higher concentration of solutes
and a lower concentration of water relative to the cytoplasm of the
cell. Water moves by osmosis from the cell into the hypertonic

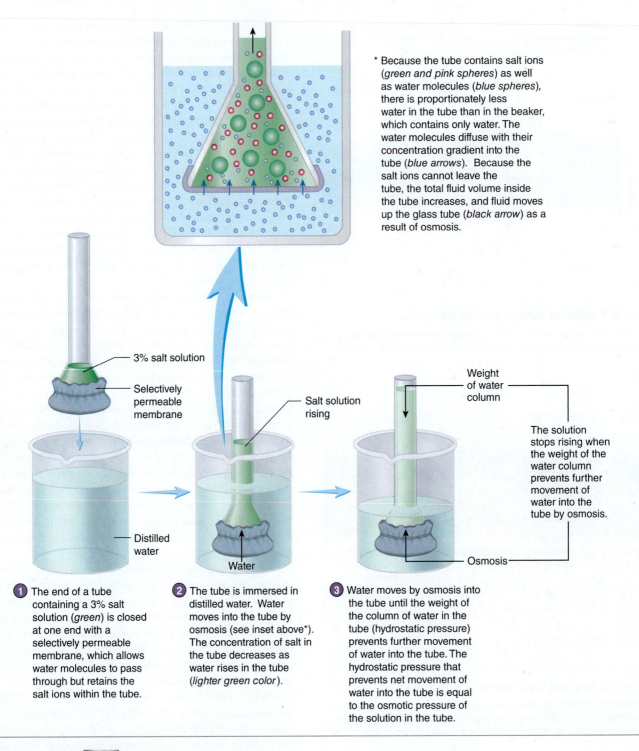

* Because the tube contains salt ions (*green and pink spheres*) as well as water molecules (*blue spheres*), there is proportionately less water in the tube than in the beaker, which contains only water. The water molecules diffuse with their concentration gradient into the tube (*blue arrows*). Because the salt ions cannot leave the tube, the total fluid volume inside the tube increases, and fluid moves up the glass tube (*black arrow*) as a result of osmosis.

3% salt solution

Selectively permeable membrane

Salt solution rising

Weight of water column

The solution stops rising when the weight of the water column prevents further movement of water into the tube by osmosis.

Distilled water

Water

Osmosis

1. The end of a tube containing a 3% salt solution (*green*) is closed at one end with a selectively permeable membrane, which allows water molecules to pass through but retains the salt ions within the tube.

2. The tube is immersed in distilled water. Water moves into the tube by osmosis (see inset above*). The concentration of salt in the tube decreases as water rises in the tube (*lighter green color*).

3. Water moves by osmosis into the tube until the weight of the column of water in the tube (hydrostatic pressure) prevents further movement of water into the tube. The hydrostatic pressure that prevents net movement of water into the tube is equal to the osmotic pressure of the solution in the tube.

PROCESS Figure 3.6 AP|R **Osmosis**

solution, resulting in cell shrinkage, or **crenation** (krē-nā′shŭn) (figure 3.7c). In general, solutions injected into blood vessels or into tissues must be isotonic to the body's cells because swelling or shrinking disrupts normal cell function and can lead to cell death.

Carrier-Mediated Transport Mechanisms

Many nutrient molecules, such as amino acids and glucose, cannot enter the cell by diffusion. Likewise, many substances produced in cells, such as proteins, cannot leave the cell by diffusion.

Carrier molecules, which are proteins within the cell membrane, are involved in **carrier-mediated transport mechanisms,** which move large, water-soluble molecules or electrically charged ions across the cell membrane. A molecule to be transported binds to a specific carrier molecule on one side of the membrane. The binding of the molecule to the carrier molecule in the cell membrane causes the three-dimensional shape of the carrier molecule to change, and the transported molecule is moved to the opposite side of the cell membrane. The transported molecule is then released by

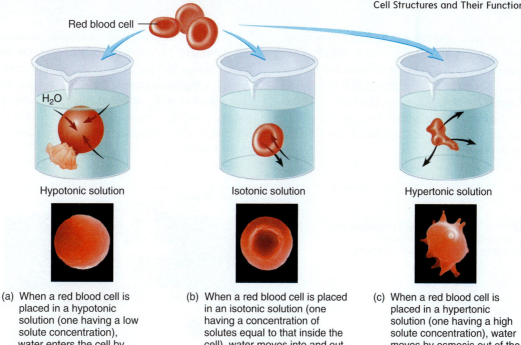

Red blood cell

H₂O

Hypotonic solution

Isotonic solution

Hypertonic solution

(a) When a red blood cell is placed in a hypotonic solution (one having a low solute concentration), water enters the cell by osmosis (*black arrows*), causing the cell to swell or even burst (lyse; *puff of red in lower part of cell*).

(b) When a red blood cell is placed in an isotonic solution (one having a concentration of solutes equal to that inside the cell), water moves into and out of the cell at the same rate (*black arrows*). No net water movement occurs, and the cell shape remains normal.

(c) When a red blood cell is placed in a hypertonic solution (one having a high solute concentration), water moves by osmosis out of the cell and into the solution (*black arrows*), resulting in shrinkage (crenation).

Figure 3.7 Effects of Hypotonic, Isotonic, and Hypertonic Solutions on Red Blood Cells

the carrier molecule, which resumes its original shape and is available to transport another molecule. Carrier-mediated transport mechanisms exhibit **specificity;** that is, only specific molecules are transported by the carriers. There are three kinds of carrier-mediated transport: facilitated diffusion, active transport, and secondary active transport.

Facilitated Diffusion

Facilitated diffusion is a carrier-mediated transport process that moves substances across the cell membrane from an area of higher concentration to an area of lower concentration of that substance (figure 3.8; see table 3.2). Because movement is with the concentration gradient, metabolic energy in the form of ATP is not required.

> **Predict 3**
>
> *The transport of glucose into most cells occurs by facilitated diffusion. Because diffusion occurs from a higher to a lower concentration, glucose cannot accumulate within these cells at a higher concentration than exists outside the cell. Once glucose enters a cell, it is rapidly converted to other molecules, such as glucose phosphate or glycogen. What effect does this conversion have on the cell's ability to transport glucose?*

Active Transport

Active transport is a carrier-mediated process that moves substances across the cell membrane from regions of lower concentration to those of higher concentration against a concentration gradient (see table 3.2). Consequently, active transport processes accumulate substances on one side of the cell membrane at concentrations many times greater than those on the other side. These dramatic

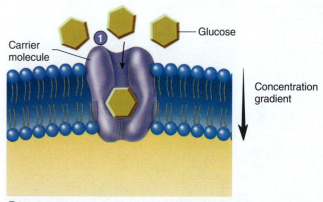

Carrier molecule

Glucose

Concentration gradient

① The carrier molecule binds with a molecule, such as glucose, on the outside of the cell membrane.

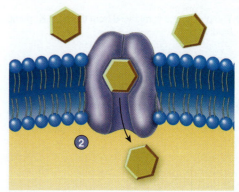

② The carrier molecule changes shape and releases the molecule on the inside of the cell membrane.

PROCESS Figure 3.8 AP|R **Facilitated Diffusion**

concentration differences are important for normal cell activity. Active transport requires energy in the form of ATP; if ATP is not available, active transport stops. One example of active transport is the movement of various amino acids from the small intestine into the blood. The malfunction of active transport can lead to serious health conditions. **Cystic fibrosis** is a genetic disorder that affects the active transport of Cl^- into cells, as described in the Clinical Impact essay later in this chapter.

In some cases, the active transport mechanism can exchange one substance for another. For example, the **sodium-potassium pump** moves Na^+ out of cells and K^+ into cells (figure 3.9). The result is a higher concentration of Na^+ outside the cell and a higher concentration of K^+ inside the cell. The concentration gradients for Na^+ and K^+, established by the sodium-potassium pump, are essential in maintaining the resting membrane potential (see chapter 8).

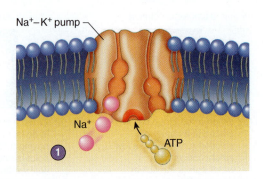

① Three sodium ions (Na^+) and adenosine triphosphate (ATP) bind to the sodium–potassium (Na^+–K^+) pump.

② The ATP breaks down to adenosine diphosphate (ADP) and a phosphate (P) and releases energy. That energy is used to power the shape change in the Na^+–K^+ pump.

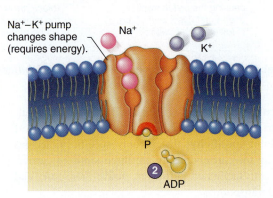

③ The Na^+–K^+ pump changes shape, and the Na^+ are transported across the membrane and into the extracellular fluid.

④ Two potassium ions (K^+) bind to the Na^+–K^+ pump.

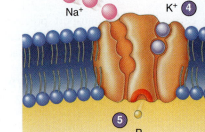

⑤ The phosphate is released from the Na^+–K^+ pump binding site.

⑥ The Na^+–K^+ pump changes shape, transporting K^+ across the membrane and into the cytoplasm. The Na^+–K^+ pump can again bind to Na^+ and ATP.

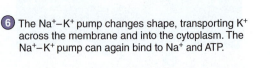
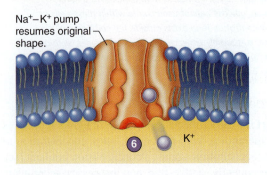

PROCESS Figure 3.9 AP|R **Active Transport: Sodium-Potassium Pump**

CLINICAL IMPACT Cystic Fibrosis

Cystic fibrosis is a genetic disorder that occurs at a rate of approximately 1 per 2000 births and currently affects 33,000 people in the United States. It is the most common lethal genetic disorder among caucasians. The diagnosis is based on the following signs: recurrent respiratory disease, increased Na⁺ in the sweat, and high levels of unabsorbed fats in the stool. Approximately 98% of all cases of cystic fibrosis are diagnosed before the patient is 18 years old.

At the molecular level, cystic fibrosis results from an abnormality in Cl^- channels. There are three types of cystic fibrosis. (1) In about 70% of cases, a defective channel protein fails to reach the cell membrane from its site of production inside the cell. (2) In a less common type, the channel protein is incorporated into the cell membrane but fails to bind ATP. (3) In the final category, the channel protein is incorporated into the cell membrane and ATP is bound to the channel protein, but the channel does not open. The result of any of these defects is that Cl^- does not exit cells at a normal rate.

Normally, as Cl^- moves out of cells that line body tubes, such as ducts or respiratory passages, water follows by osmosis. In cystic fibrosis, Cl^- does not exit these cells at normal rates, and therefore less water moves into the tubes. With less water present, the mucus produced by the cells lining those tubes is thick, and the cells' cilia cannot readily move the mucus over the cell surface. As a result, the tubes become clogged with mucus and lose much of their normal function.

The most critical effects of cystic fibrosis, accounting for 90% of deaths, are on the respiratory system. Cystic fibrosis also affects the secretory cells lining the ducts of the pancreas, sweat glands, and salivary glands.

In normal lungs, ciliated cells move a thin, fluid layer of mucus. In people with cystic fibrosis, the viscous mucus resists movement by cilia and accumulates in the lung passages. The accumulated mucus obstructs the passageways and increases the likelihood of infections. Infections are frequent and may result in chronic pneumonia. Chronic coughing occurs as the affected person attempts to remove the mucus.

At one time, cystic fibrosis was fatal during early childhood, but many patients are now surviving into young adulthood because of modern medical treatment. Currently, approximately 80% of people with cystic fibrosis live past age 20. Pulmonary therapy consists of supporting and enhancing existing respiratory functions, and infections are treated with antibiotics.

The buildup of thick mucus in the pancreatic and hepatic ducts blocks them so that pancreatic digestive enzymes and bile salts cannot reach the small intestine. As a result, fats and fat-soluble vitamins, which require bile salts for absorption and which cannot be adequately digested without pancreatic enzymes, are not taken up by intestinal cells in normal amounts. The patient suffers from deficiencies of vitamins A, D, E, and K, which result in conditions such as night blindness, skin disorders, rickets, and excessive bleeding. Therapy includes administering the missing vitamins to the patient and reducing dietary fat intake.

Future treatments for cystic fibrosis could include the development of drugs that correct or assist Cl^- transport. Alternatively, the disease may someday be cured through gene therapy—that is, inserting a functional copy of the defective gene into patients' cells.

Predict 4

Predict the effect of cystic fibrosis on the concentrations of Cl^- inside and outside the cell.

Secondary Active Transport

Secondary active transport involves the active transport of one substance, such as Na⁺, across the cell membrane, establishing a concentration gradient. The diffusion of that transported substance down its concentration gradient provides the energy to transport a second substance, such as glucose, across the cell membrane (figure 3.10). In **cotransport,** the diffusing substance moves in the same direction as the transported substance; in **countertransport,** the diffusing substance moves in a direction opposite to that of the transported substance.

A CASE IN POINT

Addison Disease

Lowe Blood experienced unexplained weight loss, fatigue, and low blood pressure. After running several tests, his physician concluded that he had **Addison disease,** which involves decreased aldosterone production by the adrenal cortex. When aldosterone production falls too low, certain kidney cells are unable to transport Na⁺. This results in Na⁺ and water loss in the urine, causing blood volume and blood pressure to drop.

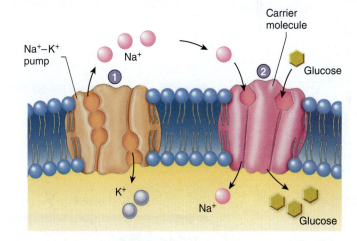

① A Na⁺–K⁺ pump maintains a concentration of Na⁺ that is higher outside the cell than inside.

② Na⁺ move back into the cell by a carrier molecule that also moves glucose. The concentration gradient for Na⁺ provides the energy required to move glucose, by cotransport, against its concentration gradient.

PROCESS Figure 3.10 Secondary Active Transport

Endocytosis and Exocytosis

Large water-soluble molecules, small pieces of matter, and even whole cells can be transported across cell membranes in membrane-bound sacs called **vesicles.** Because of the fluid nature of membranes, vesicles and cell membranes can fuse, allowing the contents of the vesicles to cross the cell membrane.

Endocytosis (en′dō-sī-tō′sis) is the uptake of material through the cell membrane by the formation of a vesicle (see table 3.2). The cell membrane invaginates (folds inward) to form a vesicle containing the material to be taken into the cell. The vesicle then moves into the cytoplasm.

Endocytosis usually exhibits specificity. The cell membrane contains specific receptor molecules that bind to specific substances. When a specific substance binds to the receptor molecule, endocytosis is triggered, and the substance is transported into the cell. This process is called **receptor-mediated endocytosis** (figure 3.11). Cholesterol and growth factors are examples of molecules that can be taken into a cell by receptor-mediated endocytosis. Bacterial phagocytosis is also receptor-mediated.

The term **phagocytosis** (fag′ō-sī-tō′sis; cell-eating) is often used for endocytosis when solid particles are ingested. A part of the cell membrane extends around a particle and fuses so that the particle is surrounded by the membrane. That part of the membrane then "pinches off" to form a vesicle containing the particle. The vesicle is now within the cytoplasm of the cell, and the cell membrane is left intact. White blood cells and some other cell types phagocytize bacteria, cell debris, and foreign particles. Phagocytosis is an important means by which white blood cells take up and destroy harmful substances that have entered the body. **Pinocytosis** (pin′ō-sī-tō′sis; cell-drinking) is distinguished from phagocytosis in that much smaller vesicles are formed, and they contain liquid rather than particles.

In some cells, membrane-bound sacs called **secretory vesicles** accumulate materials for *release* from the cell. The secretory vesicles move to the cell membrane, where the membrane of the vesicle fuses with the cell membrane, and the material in the vesicle is eliminated from the cell. This process is called **exocytosis** (ek′sō-sī-tō′sis; *exo*, outside) (figure 3.12; see table 3.2). Examples of exocytosis are the secretion of digestive enzymes by the pancreas and the secretion of mucus by the salivary glands.

To sum up, endocytosis results in the uptake of materials by cells, and exocytosis allows the release of materials from cells. Vesicle formation for both endocytosis and exocytosis requires energy in the form of ATP.

3.5 ORGANELLES

Learning Outcomes *After reading this section, you should be able to*

A. Describe the structure and function of the nucleus and the nucleoli.

B. Compare the structure and function of rough and smooth endoplasmic reticulum.

C. Describe the roles of the Golgi apparatuses and secretory vesicles in secretion.

D. Explain the role of lysosomes and peroxisomes in digesting material taken into cells by phagocytosis.

E. Describe the structure and function of mitochondria.

F. Describe the structure and function of the cytoskeleton.

G. Describe the structure and function of centrioles.

H. Compare the structure and function of cilia, flagella, and microvilli.

Nucleus

The **nucleus** (noo′klē-ŭs; a little nut or the stone of a fruit) is a large organelle usually located near the center of the cell (see figure 3.1). All cells of the body have a nucleus at some point in their life cycle, although some cells, such as red blood cells, lose their nuclei as they mature. Other cells, such as skeletal muscle cells, contain more than one nucleus.

The nucleus is bounded by a **nuclear envelope,** which consists of outer and inner membranes with a narrow space between them (figure 3.13). At many points on the surface of the nucleus, the inner and outer membranes come together to form **nuclear pores,** through which materials can pass into or out of the nucleus.

The nuclei of human cells contain 23 pairs of **chromosomes** (krō′mō′-somz), which consist of DNA and proteins. During most of a cell's life, the chromosomes are loosely coiled and collectively called **chromatin** (figure 3.14; see figure 3.13*b*). When a cell prepares to divide, the chromosomes become tightly coiled and are visible when viewed with a microscope (see "Cell Life Cycle" later in this chapter). The genes that influence the structural and functional features of every individual are portions of

Molecules to be transported

Receptor molecules

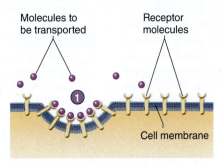

1 Receptor molecules on the cell surface bind to molecules to be taken into the cell.

Cell membrane

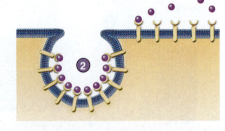

2 The receptors and the bound molecules are taken into the cell as a vesicle is formed.

Vesicle

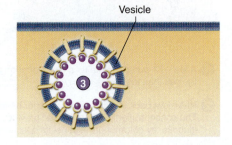

3 The vesicle membrane fuses and the vesicle separates from the cell membrane.

PROCESS Figure 3.11 Receptor-Mediated Endocytosis

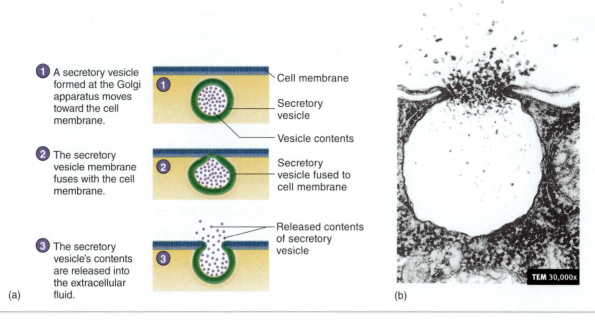

① A secretory vesicle formed at the Golgi apparatus moves toward the cell membrane.

Cell membrane

Secretory vesicle

Vesicle contents

② The secretory vesicle membrane fuses with the cell membrane.

Secretory vesicle fused to cell membrane

③ The secretory vesicle's contents are released into the extracellular fluid.

Released contents of secretory vesicle

(a)

(b)

TEM 30,000x

PROCESS Figure 3.12 Exocytosis

(*a*) Diagram of exocytosis. (*b*) Transmission electron micrograph of exocytosis.

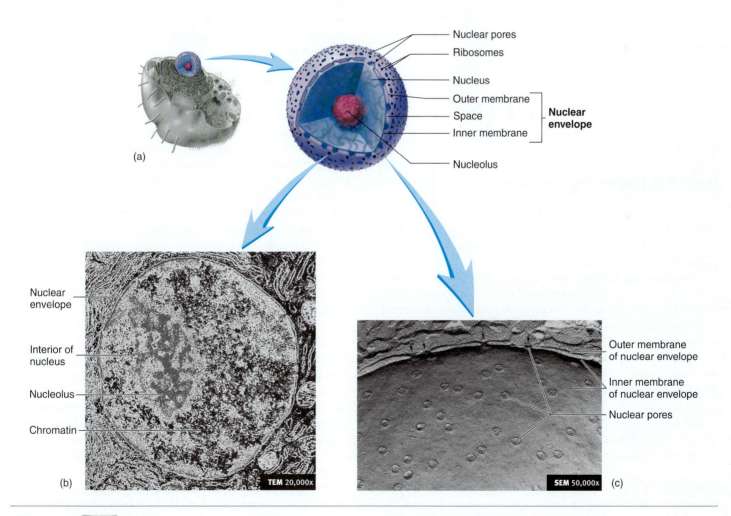

Nuclear pores
Ribosomes
Nucleus
Outer membrane
Space
Inner membrane
Nucleolus

Nuclear envelope

(a)

Nuclear envelope

Interior of nucleus

Nucleolus

Chromatin

(b)   TEM 20,000x

Outer membrane of nuclear envelope

Inner membrane of nuclear envelope

Nuclear pores

(c)   SEM 50,000x

Figure 3.13 AP|R Nucleus

(*a*) The nuclear envelope consists of inner and outer membranes, which become fused at the nuclear pores. The nucleolus is a condensed region of the nucleus not bounded by a membrane and consisting mostly of RNA and protein. (*b*) Transmission electron micrograph of the nucleus. (*c*) Scanning electron micrograph showing the membranes of the nuclear envelope and the nuclear pores.

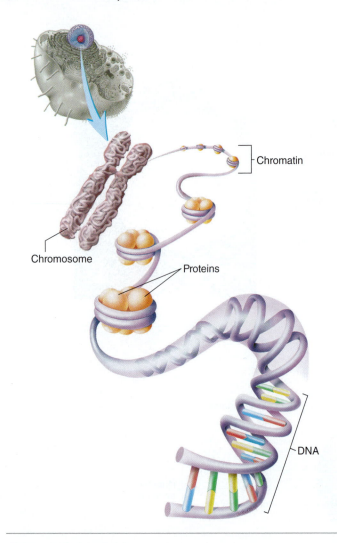

Chromatin

Chromosome

Proteins

DNA

Figure 3.14 Structure of a Chromosome

DNA molecules. These sections of DNA molecules determine the structure of proteins. By determining the structure of proteins, genes direct cell structure and function.

Nucleoli (noo-klē′ō-li; sing. nucleolus, little nucleus) are diffuse bodies with no surrounding membrane that are found within the nucleus (see figure 3.13). There are usually one to several nucleoli within the nucleus. The subunits of ribosomes, a type of cytoplasmic organelle, are formed within a nucleolus. Proteins produced in the cytoplasm move through the nuclear pores into the nucleus and to the nucleolus. These proteins are joined to **ribosomal ribonucleic** (rī′bō-noo-klē′ik) **acid (rRNA),** produced within the nucleolus, to form large and small ribosomal subunits (figure 3.15). The ribosomal subunits then move from the nucleus through the nuclear pores into the cytoplasm, where one large and one small subunit join to form a ribosome during protein synthesis.

Ribosomes

Ribosomes (rī′bō-sōmz) are the organelles where proteins are produced (see "Gene Expression" later in this chapter). Ribosomes may be attached to other organelles, such as the endoplasmic reticulum. Ribosomes that are not attached to any other organelle are called free ribosomes.

Rough and Smooth Endoplasmic Reticulum

The **endoplasmic reticulum** (en′dō-plas′mik re-tik′ū-lŭm; *rete,* a network) **(ER)** is a series of membranes forming sacs and tubules that extends from the outer nuclear membrane into the cytoplasm (figure 3.16). **Rough ER** is ER with ribosomes attached to it. A large amount of rough ER in a cell indicates that it is synthesizing large amounts of protein for export from the cell. On the other hand, ER without ribosomes is called **smooth ER.** Smooth ER is a site for lipid synthesis and participates in detoxification of chemicals within cells. In skeletal muscle cells, the smooth ER stores calcium ions.

Golgi Apparatus

The **Golgi** (gol′jē) **apparatus** consists of closely packed stacks of curved, membrane-bound sacs (figure 3.17). It collects, modifies, packages, and distributes proteins and lipids manufactured by the ER. For example, proteins produced at the ribosomes enter the Golgi apparatus from the ER. In some cases, the Golgi apparatus chemically modifies the proteins by attaching carbohydrate or lipid molecules to them. The proteins then are packaged into membrane sacs, called secretory vesicles, that pinch off from the margins of the Golgi apparatus (see the next section). The Golgi apparatus is present in larger numbers and is most highly developed in cells that secrete protein, such as those of the salivary glands or the pancreas.

Secretory Vesicles

A **vesicle** (ves′i-kl; a bladder) is a small, membrane-bound sac that transports or stores materials within cells. Secretory vesicles pinch off from the Golgi apparatus and move to the cell membrane (figure 3.17). The membrane of a secretory vesicle then fuses with the cell membrane, and the contents of the vesicle are released to the exterior of the cell. In many cells, secretory vesicles accumulate in the cytoplasm and are released to the exterior when the cell receives a signal. For example, nerve cells release substances called neurotransmitters from secretory vesicles to communicate with other cells. Also, secretory vesicles containing hormones remain in the cytoplasm of endocrine cells until signals stimulate their release. For example, insulin remains in the cytoplasm of pancreatic cells until rising blood glucose levels stimulate its secretion.

Lysosomes and Peroxisomes

Lysosomes (lī′sō-sōmz) are membrane-bound vesicles formed from the Golgi apparatus. They contain a variety of enzymes that function as intracellular digestive systems. Vesicles formed by endocytosis may fuse with lysosomes (figure 3.18). The enzymes within the lysosomes break down the materials in the endocytotic vesicle. For example, white blood cells phagocytize bacteria. Then enzymes within lysosomes destroy the phagocytized bacteria.

Peroxisomes (per-ok′si-sōmz) are small, membrane-bound vesicles containing enzymes that break down fatty acids, amino acids, and hydrogen peroxide (H_2O_2). Hydrogen peroxide is a by-product of fatty acid and amino acid breakdown and can be toxic to a cell. The enzymes in peroxisomes break down hydrogen peroxide to water and O_2. Cells active in detoxification, such as liver and kidney cells, have many peroxisomes.

1. Ribosomal proteins, produced in the cytoplasm, are transported through nuclear pores into the nucleolus.

2. rRNA, most of which is produced in the nucleolus, is assembled with ribosomal proteins to form small and large ribosomal subunits.

3. The small and large ribosomal subunits leave the nucleolus and the nucleus through nuclear pores.

4. The small and large subunits, now in the cytoplasm, combine with each other and with mRNA during protein synthesis.

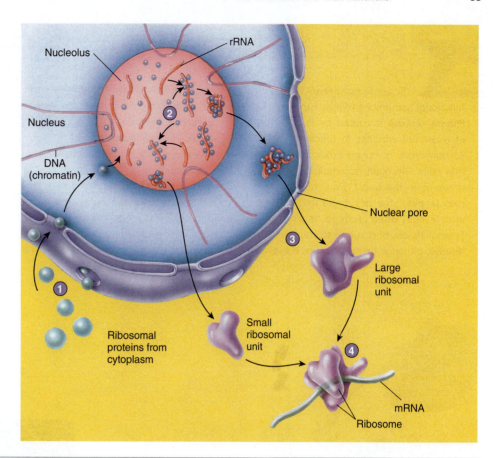

PROCESS Figure 3.15 **Production of Ribosomes**

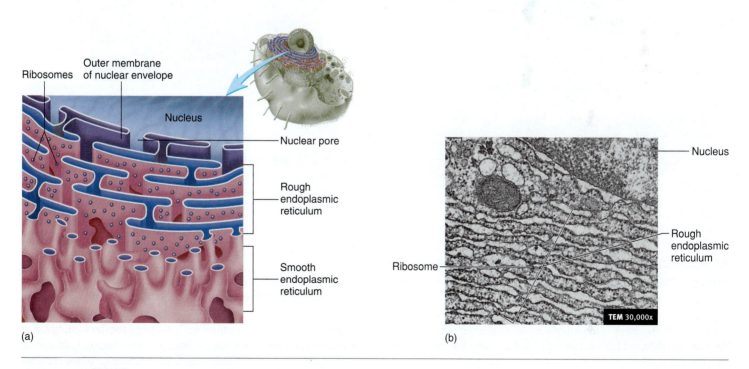

(a)

(b)

Figure 3.16 AP|R **Endoplasmic Reticulum**

(a) The endoplasmic reticulum is continuous with the nuclear envelope and can exist as either rough endoplasmic reticulum (with ribosomes) or smooth endoplasmic reticulum (without ribosomes). (b) Transmission electron micrograph of the rough endoplasmic reticulum.

CLINICAL IMPACT   Carbohydrate and Lipid Disorders

Some diseases result from nonfunctional lysosomal enzymes. For example, **Pompe disease** is caused by the inability of lysosomal enzymes to break down the carbohydrate glycogen produced in certain cells. Glycogen accumulates in large amounts in the heart, liver, and skeletal muscle cells; the accumulation in heart muscle cells often leads to heart failure. Specialists are best suited to treat Pompe disease, although there is no cure. However, clinical trials have shown that replacement of the nonfunctional enzymes can reduce the symptoms of Pompe disease.

Lipid-storage disorders are a group of hereditary diseases characterized by the accumulation of large amounts of lipids in cells that lack the enzymes necessary to break down the lipids. The accumulation of lipids leads to cell and tissue damage in the brain, spleen, liver, and bone marrow. **Tay-Sachs disease,** one such lipid-storage disorder, affects the nervous system. Tay-Sachs disease results from a lack of lysosomal enzymes that normally break down specific lipids in nerve cells of the brain. Buildup of these lipids in the brain leads to neurological problems, such as blindness, deafness, and paralysis, within the first several months of life. There is no successful treatment for Tay-Sachs disease; children with this disease usually die before age 4.

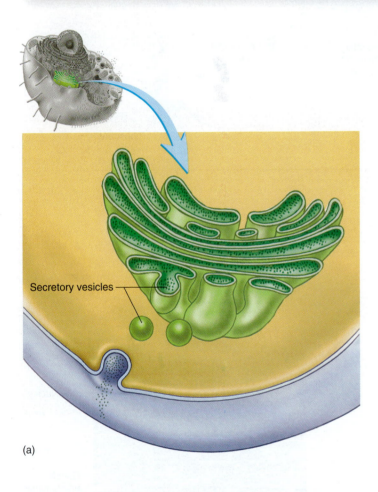

Secretory vesicles

(a)

Secretory vesicle

Golgi apparatus

TEM 96,000x

(b)

Figure 3.17 AP|R **Golgi Apparatus**

(*a*) The Golgi apparatus is composed of flattened, membranous sacs and resembles a stack of dinner plates or pancakes. (*b*) Transmission electron micrograph of the Golgi apparatus.

Mitochondria

Mitochondria (mī′tō-kon′drē-ă; sing. mitochondrion) are small organelles with inner and outer membranes separated by a space (figure 3.19; see figure 3.1). The outer membranes have a smooth contour, but the inner membranes have numerous folds, called **cristae** (kris′tē), which project like shelves into the interior of the mitochondria.

Mitochondria are the major sites of adenosine triphosphate (ATP) production within cells. Mitochondria carry out aerobic respiration (discussed in greater detail in chapter 17), a series of chemical reactions that require O_2 to break down food molecules to produce ATP. ATP is the main energy source for most chemical reactions within the cell, and cells with a large energy requirement have more mitochondria than cells that require less energy. For example, cells that carry out extensive active transport, described earlier in this chapter, contain many mitochondria. When muscles enlarge as a result of exercise, the mitochondria increase in number within the muscle cells and provide the additional ATP required for muscle contraction.

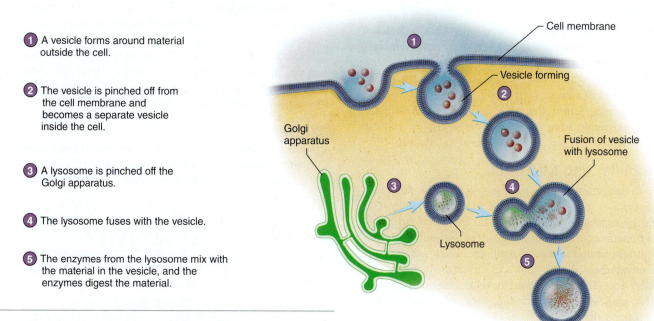

1. A vesicle forms around material outside the cell.

2. The vesicle is pinched off from the cell membrane and becomes a separate vesicle inside the cell.

3. A lysosome is pinched off the Golgi apparatus.

4. The lysosome fuses with the vesicle.

5. The enzymes from the lysosome mix with the material in the vesicle, and the enzymes digest the material.

PROCESS Figure 3.18 Action of Lysosomes

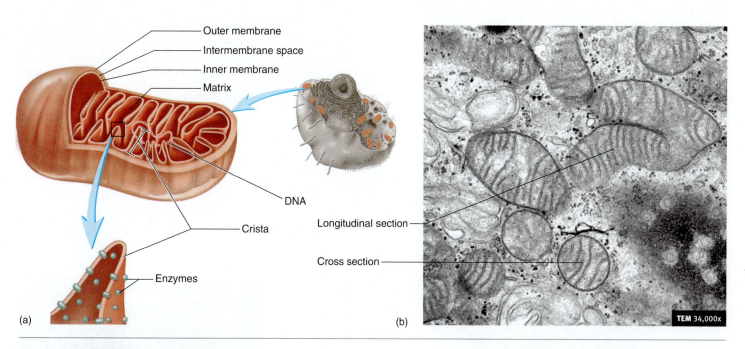

(a)   (b)

Figure 3.19 AP|R Mitochondrion

(a) Typical mitochondrion structure. (b) Transmission electron micrograph of mitochondria in longitudinal and cross sections.

Cytoskeleton

The **cytoskeleton** (sī-tō-skel′ĕ-ton) consists of proteins that support the cell, hold organelles in place, and enable the cell to change shape. These protein structures are microtubules, microfilaments, and intermediate filaments (figure 3.20).

Microtubules are hollow structures formed from protein subunits. They perform a variety of roles, such as helping support the cytoplasm of cells, assisting in cell division, and forming essential components of certain organelles, such as cilia and flagella.

Microfilaments are small fibrils formed from protein subunits that structurally support the cytoplasm. Some microfilaments are involved with cell movement. For example, microfilaments in muscle cells enable the cells to shorten, or contract.

Intermediate filaments are fibrils formed from protein subunits that are smaller in diameter than microtubules but larger in diameter than microfilaments. They provide mechanical support to the cell.

Centrioles

The **centrosome** (sen′trō-sōm) is a specialized zone of cytoplasm close to the nucleus, where microtubule formation occurs. It contains two **centrioles** (sen′trē-ōlz), which are normally oriented perpendicular to each other. Each centriole is a small, cylindrical organelle composed of nine triplets; each triplet consists of three parallel microtubules joined together (figure 3.21). Additional microtubules, extending from the area of the centrioles, play an important role in cell division, as we learn later in "Mitosis."

Cilia, Flagella, and Microvilli

Cilia (sĭl′ē-ă; sing. cilium, an eyelash) project from the surface of cells (see figure 3.1). They vary in number from none to thousands per cell and are capable of moving. Cilia are cylindrical structures that extend from the cell. Cilia are composed of microtubules, organized in a pattern similar to that of centrioles, which are enclosed by the cell membrane. Cilia are numerous on surface cells that line the respiratory tract. Their coordinated movement transports mucus, in which dust particles are embedded, upward and away from the lungs. This action helps keep the lungs clear of debris.

Flagella (flă-jel′ă; sing. flagellum, a whip) have a structure similar to that of cilia but are much longer, and they usually occur only one per cell. Sperm cells each have one flagellum, which propels the sperm cell.

Microvilli (mī′krō-vil′ī; *mikros,* small + *villus,* shaggy hair) are specialized extensions of the cell membrane that are supported by microfilaments (see figure 3.1), but they do not actively move as cilia and flagella do. Microvilli are numerous on cells that have them and they increase the surface area of those cells. They are abundant on the surface of cells that line the intestine, kidney, and other areas in which absorption is an important function.

Predict 5

List the organelles that are plentiful in cells that (a) synthesize and secrete proteins, (b) actively transport substances into cells, and (c) ingest foreign substances by endocytosis. Explain the function of each organelle you list.

3.6 WHOLE-CELL ACTIVITY

Learning Outcomes   *After reading this section, you should be able to*

A. Describe the process of gene expression.
B. Explain what is accomplished during mitosis.
C. Define differentiation, and explain how different cell types develop.

A cell's characteristics are ultimately determined by the types of proteins it produces. The proteins produced are in turn determined by the genetic information in the nucleus. In order to understand how a cell functions, we must consider the relationship between genes and proteins. For example, the transport of many food

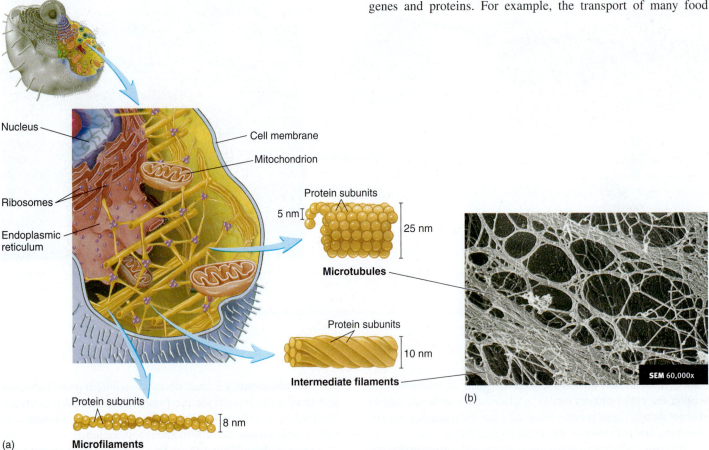

Nucleus

Cell membrane

Mitochondrion

Ribosomes

Endoplasmic reticulum

Protein subunits

5 nm

25 nm

Microtubules

Protein subunits

10 nm

Intermediate filaments

SEM 60,000x

(b)

Protein subunits

8 nm

(a)   **Microfilaments**

Figure 3.20 AP|R Cytoskeleton

(*a*) Diagram of the cytoskeleton. (*b*) Scanning electron micrograph of the cytoskeleton.

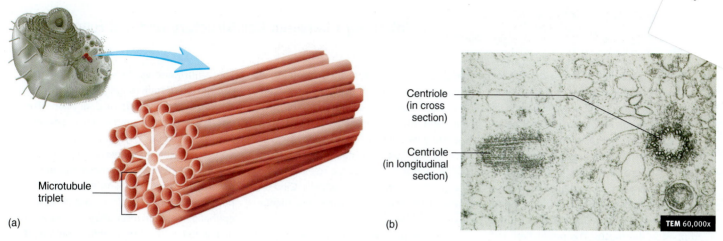

Centriole
(in cross
section)

Centriole
(in longitudinal
section)

TEM 60,000x

Microtubule
triplet

(a)

(b)

Figure 3.2l Centriole

(*a*) Structure of a centriole, which is composed of nine triplets of microtubules. Each triplet contains one complete microtubule fused to two incomplete microtubules. (*b*) Transmission electron micrograph of a pair of centrioles, which are normally located together near the nucleus. One is shown in cross section and one in longitudinal section.

molecules into the cell requires cell membrane proteins. Amino acids are assembled to synthesize proteins, including the transport proteins of the cell membrane. Information contained in DNA within the nucleus determines which amino acids are combined at ribosomes to form proteins. As you learned in the beginning of the chapter, the human body is composed of trillions of cells, with many different characteristics. Each human begins life as a single cell. Through cell division and cell differentiation, the cells that make up the human body are formed. The following sections illustrate the whole-cell activities that determine the characteristics of a functioning cell and the growth and maintenance of the human body.

Gene Expression

DNA contains the information that directs protein synthesis. This process is called **gene expression.** The proteins produced in a cell include those that serve as structural components inside the cell, proteins secreted to the outside of the cell, and enzymes that regulate chemical reactions in the cell. DNA influences the structural and functional characteristics of the entire organism because it directs protein synthesis. Whether an individual has blue eyes, brown hair, or other inherited traits is determined ultimately by DNA.

A DNA molecule consists of nucleotides joined together to form two nucleotide strands (see figures 2.19 and 3.23). The two strands are connected and resemble a ladder that is twisted around its long axis. The nucleotides function as chemical "letters" that form chemical "words." A **gene** is a sequence of nucleotides (making a word) that provides a chemical set of instructions for making a specific protein. Each DNA molecule contains many different genes.

Recall from chapter 2 that proteins consist of amino acids. The unique structural and functional characteristics of different proteins are determined by the kinds, numbers, and arrangement of their amino acids. The nucleotide sequence of a gene determines the amino acid sequence of a specific protein.

Gene expression involves two steps—transcription and translation (figure 3.22). This process can be illustrated with an analogy. Suppose a chef wants a cake recipe that is found only in a cookbook in the library. Because the book cannot be checked out, the chef makes a copy, or **transcription,** of the recipe. Later in the kitchen, the information contained in the copied recipe is used to make the cake. The changing of something from one form to another (from recipe to cake) is called **translation.**

In terms of this analogy, DNA (the cookbook) contains many genes (recipes) for making different proteins (food items). DNA, however, is too large a molecule to pass through the nuclear pores to the ribosomes (kitchen) where the proteins are made. Just as a book stays in the library, DNA remains in the nucleus. Therefore, through transcription, the cell makes a copy of the gene necessary to make a particular protein. The copy, which is called **messenger RNA (mRNA),** travels from the nucleus to the ribosomes in the cytoplasm, where the information in the copy is used to construct a protein by means of translation. Of course, the actual ingredients are needed to turn a recipe into a cake. The ingredients necessary to synthesize a protein are amino acids. Specialized molecules, called **transfer RNAs (tRNAs),** carry the amino acids to the ribosome (figure 3.22).

In summary, gene expression involves transcription (making a copy of a gene) and translation (converting that copied information into a protein). Next, we consider the details of transcription and translation.

Transcription

The first step in gene expression, transcription, takes place in the nucleus of the cell (figure 3.22, steps 1 and 2). DNA determines the structure of mRNA through transcription. The double strands of a DNA segment separate, and DNA nucleotides pair with RNA nucleotides (figure 3.23, steps 1 and 2). Each nucleotide of DNA contains one of the following organic bases: thymine, adenine, cytosine, or guanine; each nucleotide of mRNA contains uracil, adenine, cytosine, or guanine. The number and sequence

CLINICAL IMPACT   Relationships Between Cell Structure and Cell Function

Each cell is well adapted for the functions it performs, and the abundance of organelles in each cell reflects those functions. For example, epithelial cells lining the larger-diameter respiratory passages secrete mucus and transport it toward the throat, where it is either swallowed or expelled from the body by coughing. Particles of dust and other debris suspended in the air become trapped in the mucus. The production and transport of mucus from the respiratory passages keeps these passages clean. Cells of the respiratory system have abundant rough ER, Golgi apparatuses, secretory vesicles, and cilia. The ribosomes on the

rough ER are the sites where proteins, a major component of mucus, are produced. The Golgi apparatuses package the proteins and other components of mucus into secretory vesicles, which move to the surface of the epithelial cells. The contents of the secretory vesicles are released onto the surface of the epithelial cells. Cilia on the cell surfaces then propel the mucus toward the throat.

In people who smoke, prolonged exposure of the respiratory epithelium to the irritation of tobacco smoke causes the respiratory epithelial cells to change in structure and function. The cells flatten and form several layers of epithelial cells.

These flattened epithelial cells no longer contain abundant rough ER, Golgi apparatuses, secretory vesicles, or cilia. The altered respiratory epithelium is adapted to protect the underlying cells from irritation, but it can no longer secrete mucus, nor does it have cilia to move the mucus toward the throat to clean the respiratory passages. Extensive replacement of normal epithelial cells in respiratory passages is associated with chronic inflammation of the respiratory passages (bronchitis), which is common in people who smoke heavily. Coughing is a common manifestation of inflammation and mucus accumulation in the respiratory passages.

1. DNA contains the information necessary to produce proteins.

2. Transcription of one DNA strand results in mRNA, which is a complementary copy of the information in the DNA strand needed to make a protein.

3. The mRNA leaves the nucleus and goes to a ribosome.

4. Amino acids, the building blocks of proteins, are carried to the ribosome by tRNAs.

5. In the process of translation, the information contained in mRNA is used to determine the number, kinds, and arrangement of amino acids in the polypeptide chain.

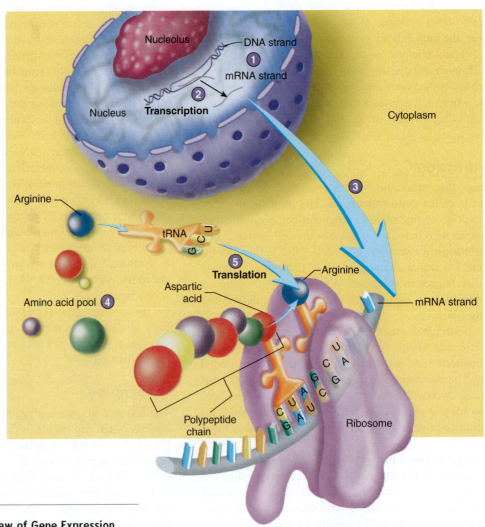

PROCESS Figure 3.22 Overview of Gene Expression

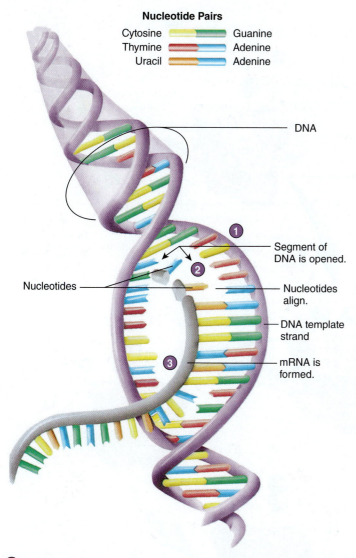

Nucleotide Pairs

Cytosine ▭▭ Guanine
Thymine ▭▭ Adenine
Uracil ▭▭ Adenine

DNA

① Segment of DNA is opened.

②

Nucleotides

Nucleotides align.

DNA template strand

③ mRNA is formed.

① The strands of the DNA molecule separate from each other. One DNA strand serves as a template for mRNA synthesis.

② Nucleotides that will form mRNA pair with DNA nucleotides according to the base-pair combinations shown in the key at the top of the figure. Thus, the sequence of nucleotides in the template DNA strand (*purple*) determines the sequence of nucleotides in mRNA (*gray*).

An enzyme (not shown) joins the nucleotides of mRNA together.

③ As nucleotides are added, an mRNA molecule is formed.

PROCESS Figure 3.23 **AP|R** **Formation of mRNA by Transcription of DNA**

of nucleotides in the DNA serve as a template to determine the number and sequence of nucleotides in the mRNA. DNA nucleotides pair only with specific RNA nucleotides: DNA's thymine with RNA's adenine, DNA's adenine with RNA's uracil, DNA's cytosine with RNA's guanine, and DNA's guanine with RNA's cytosine.

After the DNA nucleotides pair up with the RNA nucleotides, an enzyme catalyzes reactions that form chemical bonds between the RNA nucleotides to form a long mRNA segment (figure 3.23, step 3). Once the mRNA segment has been transcribed, portions of the mRNA molecule may be removed.

The information in mRNA is carried in groups of three nucleotides called **codons,** which specify a particular amino acid. For example, the nucleotide sequence uracil, cytosine, and adenine (UCA) specifies the amino acid serine. There are 64 possible mRNA codons, but only 20 amino acids. As a result, more than 1 codon can specify the same amino acid. For example, CGA, CGG, CGU, and CGC code for the amino acid arginine; UUU and UUC code for phenylalanine. Some codons do not specify a particular amino acid but perform other functions. For example, UAA does not code for an amino acid. It is called a stop codon because it acts as a signal to end the translation process.

Translation

Translation, the synthesis of proteins based on the information in mRNA, occurs at ribosomes (see figure 3.22, steps 3–5). The mRNA molecules produced by transcription pass through the nuclear pores to the ribosomes. Ribosomes consist of small and large subunits, which combine with mRNA during translation. The process of translation requires two types of RNA in addition to the mRNA: tRNA and **ribosomal RNA (rRNA).** There is one type of tRNA for each mRNA codon. The **anticodon,** a series of three nucleotides of tRNA, pairs with the codon of the mRNA. An amino acid is bound to another part of the tRNA. This ensures that the correct amino acid is matched with the codon of the mRNA. For example, the tRNA that pairs with the UUU codon of mRNA has the anticodon AAA and has the amino acid phenyl-alanine bound to it.

During translation, a ribosome binds to an mRNA. The ribosome aligns the mRNA with tRNA molecules so that the anti-codons of tRNA can pair with the appropriate codons on the mRNA (figure 3.24, steps 1 and 2). An enzyme associated with the ribosome causes the formation of a **peptide bond** between the amino acids bound to the tRNAs (figure 3.24, step 3). The ribosome moves down the mRNA one codon at a time, releasing one of the tRNA and allowing the next tRNA to move into position. As the process continues, a **polypeptide chain** is formed. Translation ends when the ribosome reaches the stop codon on the mRNA (figure 3.24, step 4). The polypeptide chain is released and becomes folded to form the three-dimensional structure of the protein molecule (see figure 2.16). A protein can consist of a single polypeptide chain or two or more polypeptide chains that are joined after each chain is produced on a separate ribosome.

Predict 6

Explain how changing one nucleotide within a DNA molecule of a cell can change the structure of a protein produced by that cell.

Cell Life Cycle

During growth and development, cell division allows for a dramatic increase in cell number after fertilization of an oocyte. The same process allows for the replacement and repair of damaged tissue. The cell life cycle includes two major phases: a nondividing phase, called **interphase,** and cell division. A cell spends most of its life cycle in interphase performing its normal functions. During interphase, the DNA (located in chromosomes in the cell's nucleus) is replicated. The two strands of DNA separate from each other, and each strand serves as a template for the production of

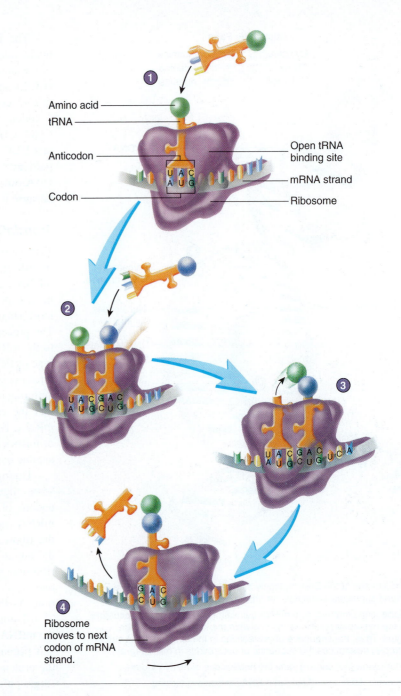

① To start protein synthesis, a ribosome binds to mRNA. The ribosome also has two binding sites for tRNA, one of which is occupied by a tRNA with its amino acid. Note that the codon of mRNA and the anticodon of tRNA are aligned and joined. The other tRNA binding site is open.

② By occupying the open tRNA binding site, the next tRNA is properly aligned with mRNA and with the other tRNA.

③ An enzyme within the ribosome catalyzes a synthesis reaction to form a peptide bond between the amino acids. Note that the amino acids are now associated with only one of the tRNAs.

④ The ribosome shifts position by three nucleotides. The tRNA without the amino acid is released from the ribosome, and the tRNA with the amino acids takes its position. A tRNA binding site is left open by the shift. Additional amino acids can be added by repeating steps 2 through 4. Eventually, a stop codon in the mRNA ends the addition of amino acids to the protein (polypeptide), which is released from the ribosome.

Amino acid
tRNA
Anticodon
Codon
Open tRNA binding site
mRNA strand
Ribosome

Ribosome moves to next codon of mRNA strand.

PROCESS Figure 3.24 AP|R **Translation of mRNA to Produce a Protein**

a new strand of DNA (figure 3.25). Nucleotides in the DNA of each template strand pair with nucleotides that are subsequently joined by enzymes to form a new strand of DNA. The sequence of nucleotides in the DNA template determines the sequence of nucleotides in the new strand of DNA because adenine pairs with thymine, and cytosine pairs with guanine. The two new strands of DNA combine with the two template strands to form two double strands of DNA.

At the end of interphase, a cell has two complete sets of genetic material. The DNA is dispersed throughout the nucleus as thin threads called chromatin (figure 3.26, step 1; see also figures 3.13*b* and 3.14).

Cell division is the formation of daughter cells from a single parent cell. The new cells necessary for growth and tissue repair are formed through mitosis (discussed next), and the sex cells necessary for reproduction are formed through meiosis (discussed in chapter 19).

Each cell of the human body, except for sex cells, contains 46 chromosomes. Sex cells have half the number of chromosomes as other cells (see "Meiosis" in chapter 19). The 46 chromosomes are the **diploid** (dip′loyd) number of chromosomes and are organized to form 23 pairs of chromosomes. Of the 23 pairs, 1 pair is the sex chromosomes, which consist of 2 **X chromosomes** if the person is a female or an X chromosome and a **Y chromosome** if the person is a

For convenience, mitosis is divided into four stages: prophase, metaphase, anaphase, and telophase (figure 3.26, steps 2–5). Although each stage represents certain major events, the process of mitosis is continuous. Learning each of the stages is helpful, but the most important concept to understand is how each of the two cells produced by mitosis obtains the same number and type of chromosomes as the parent cell.

- *Prophase.* During **prophase** (figure 3.26, step 2), the chromatin condenses to form visible chromosomes. After interphase, each chromosome is made up of two genetically identical strands of chromatin, called **chromatids** (krō′mă-tidz), which are linked at one point by a specialized region called the **centromere** (sen′trō-mēr; *kentron,* center + *meros,* part). Replication of the genetic material during interphase results in the two identical chromatids of each chromosome. Also during prophase, microtubules called spindle fibers extend from the centrioles to the centromeres (see figures 3.1 and 3.21). The centrioles divide and migrate to each pole of the cell. In late prophase, the nucleolus and nuclear envelope disappear.
- *Metaphase.* In **metaphase** (figure 3.26, step 3), the chromosomes align near the center of the cell.
- *Anaphase.* At the beginning of **anaphase** (figure 3.26, step 4), the chromatids separate. When this happens, each chromatid is then called a chromosome. At this point, two identical sets of 46 chromosomes are present in the cell. Each of the two sets of 46 chromosomes is moved by the spindle fibers toward the centriole at one of the poles of the cell. At the end of anaphase, each set of chromosomes has reached an opposite pole of the cell, and the cytoplasm begins to divide.
- *Telophase.* During **telophase** (tel′ō-fāz) (figure 3.26, step 5), the chromosomes in each of the daughter cells become organized to form two separate nuclei. The chromosomes begin to unravel and resemble the genetic material during interphase.

Following telophase, cytoplasm division is completed, and two separate daughter cells are produced (figure 3.26, step 6).

Differentiation

A sperm cell and an oocyte unite to form a single cell, and a new individual begins. The single cell formed during fertilization divides by mitosis to form two cells, which divide to form four cells, and so on (see chapter 20). The trillions of cells that ultimately make up the body of an adult, as a result, stem from that single cell. Therefore, all the cells in an individual's body contain the same amount and type of DNA. But even though the genetic information contained in cells is identical, not all cells look and function alike. Bone cells, for example, do not look like or function the same as muscle cells, nerve cells, or red blood cells (figure 3.27).

The process by which cells develop with specialized structures and functions is called **differentiation.** During differentiation of a cell, some portions of DNA are active, but others are inactive. The active and inactive sections of DNA differ with each cell type. For example, the portion of DNA responsible for the structure and function of a bone cell is different from that responsible for

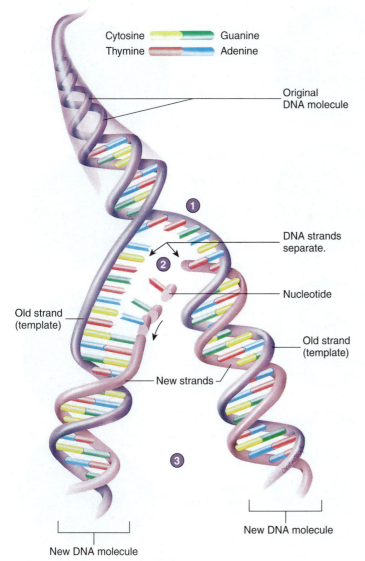

Cytosine ▭ Guanine
Thymine ▭ Adenine

Original DNA molecule

DNA strands separate.

Nucleotide

Old strand (template)

Old strand (template)

New strands

New DNA molecule

New DNA molecule

① The strands of the DNA molecule separate from each other.

② Each old strand (*dark purple*) functions as a template on which a new strand (*light purple*) is formed. The base-pairing relationship between nucleotides determines the sequence of nucleotides in the newly formed strands.

③ Two identical DNA molecules are produced.

PROCESS Figure 3.25 **AP|R** **Replication of DNA**

Replication during interphase produces two identical molecules of DNA.

male. The remaining 22 pairs of chromosomes are called **autosomes** (aw′tō-sōmz). The sex chromosomes determine the individual's sex, and the autosomes determine most other characteristics.

Mitosis

Most cells of the body, except those that give rise to sex cells, divide by **mitosis** (mī-tō′sis). During mitosis, a parent cell divides to form two daughter cells with the same amount and type of DNA as the parent cell. Because DNA determines the structure and function of cells, the daughter cells, which have the same DNA as the parent cell, can have the same structure and perform the same functions as the parent cell.

1 **Interphase** is the time between cell divisions. DNA is found as thin threads of chromatin in the nucleus. DNA replication occurs during interphase.

2 In **prophase**, the chromatin condenses into chromosomes. Each chromosome consists of two chromatids joined at the centromere. The centrioles move to the opposite ends of the cell, and the nucleolus and the nuclear envelope disappear.

3 In **metaphase**, the chromosomes align in the center of the cell in association with the spindle fibers.

4 In **anaphase**, the chromatids separate to form two sets of identical chromosomes. The chromosomes, assisted by the spindle fibers, move toward the centrioles at each end of the cell. The cytoplasm begins to divide.

5 In **telophase**, the chromosomes disperse, the nuclear envelopes and the nucleoli form, and the cytoplasm continues to divide to form two cells.

6 Mitosis is complete, and a new interphase begins. The chromosomes have unraveled to become chromatin. Cell division has produced two daughter cells, each with DNA that is identical to the DNA of the parent cell.

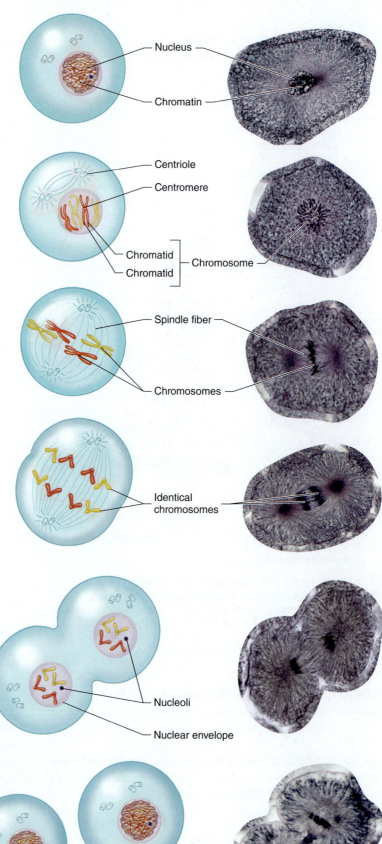

Nucleus

Chromatin

Centriole

Centromere

Chromatid

Chromatid

Chromosome

Spindle fiber

Chromosomes

Identical chromosomes

Nucleoli

Nuclear envelope

PROCESS Figure 3.26 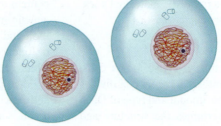 **Cell Life Cycle**

CLINICAL IMPACT   Cancer

A **tumor** (too′mōr) is any swelling that occurs within the body, usually involving cell proliferation. Tumors can be either **malignant** (mă-lig′nănt; with malice or intent to cause harm) or **benign** (bē-nīn′; kind).

Benign tumors are usually less dangerous than malignant ones. They are not inclined to spread, but they may increase in size. As a benign tumor enlarges, it can compress surrounding tissues and impair their functions. In some cases (e.g., brain tumors), the results are fatal.

Malignant tumors can spread by local growth and expansion or by **metastasis** (mē-tas′tă-sis; moving to another place), which results when tumor cells separate from the main tumor and are carried by the lymphatic system or blood vessels to a new site, where a second tumor forms.

Cancer (kan′ser) refers to a malignant, spreading tumor and the illness that results from it. Cancers lack the normal growth control that is exhibited by most other adult tissues. Cancer results when a cell or group of cells breaks away from the normal control of growth and differentiation. This breaking loose involves the genetic machinery and can be induced by viruses, environmental toxins, and other causes. The illness associated with cancer usually occurs as the tumor invades and destroys the healthy surrounding tissues, eliminating their functions.

Promising anticancer therapies are being developed in which the cells responsible for immune responses can be stimulated to recognize tumor cells and destroy them. A major advantage in such anticancer treatments is that the cells of the immune system can specifically attack the tumor cells and not other, healthy tissues. Other therapies currently under investigation include techniques to starve a tumor to death by cutting off its blood supply. Drugs that can inhibit blood vessel development are currently in use (bevacizumab, Avastin) or under investigation (thalidomide).

Predict 7

Cancer cells divide continuously. The normal mechanisms that regulate whether cell division occurs or ceases do not function properly in cancer cells. Cancer cells, such as breast cancer cells, do not look like normal, mature cells. Explain.

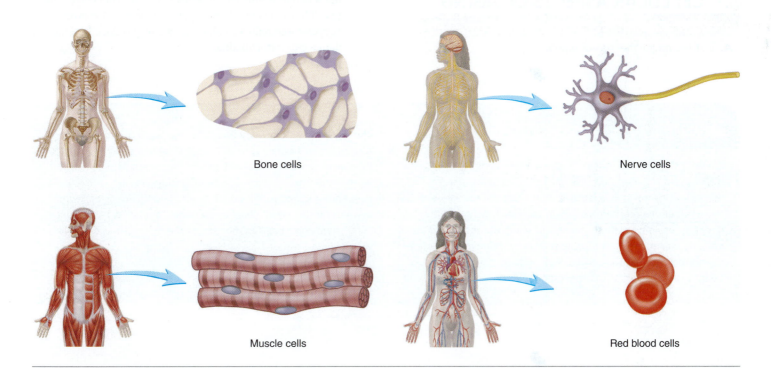

Bone cells

Nerve cells

Muscle cells

Red blood cells

Figure 3.27 Diversity of Cell Types

the structure and function of a muscle cell. Differentiation, then, results from the selective activation and inactivation of segments of DNA. The mechanisms that determine which portions of DNA are active in any one cell type are not fully understood, but the resulting differentiation produces the many cell types that function together to make a person. Eventually, as cells differentiate and mature, the rate at which they divide slows or even stops.

Apoptosis

Apoptosis (ăp′op-tō′sis), or *programmed cell death,* is a normal process by which cell numbers within various tissues are adjusted and controlled. In the developing fetus, apoptosis removes extra tissue, such as cells between the developing fingers and toes. In some adult tissues, apoptosis eliminates excess cells to maintain a constant number of cells within the tissue. Damaged or potentially dangerous cells, virus-infected cells, and potential cancer cells are also eliminated by apoptosis.

Apoptosis is regulated by specific genes. The proteins coded for by those genes initiate events within the cell that ultimately lead to the cell's death. As apoptosis begins, the chromatin within the nucleus condenses and fragments. This is followed by fragmentation of the nucleus and finally by death and fragmentation of the cell. Specialized cells called macrophages phagocytize the cell fragments.

3.7 CELLULAR ASPECTS OF AGING

Learning Outcome *After reading this section, you should be able to*

A. List the major theories of aging.

We are all familiar with the outward signs of aging, such as wrinkled skin, gray hair, and reduced vision. A number of cellular structures or events appear to be involved in causing these effects. The major hypotheses that attempt to explain how aging occurs concentrate on molecules within the cell, such as lipids, proteins, and nucleic acids. It is estimated that at least 35% of the factors affecting aging are genetic.

1. *Cellular clock.* One hypothesis of aging suggests the existence of a cellular clock that, after a certain passage of time or a certain number of cell divisions, results in the death of a given cell line.
2. *Death genes.* Another hypothesis suggests that there are "death genes," which turn on late in life, or sometimes prematurely, causing cells to deteriorate and die.
3. *DNA damage.* Other hypotheses suggest that, through time, DNA is damaged, resulting in cell degeneration and death.
4. *Free radicals.* DNA is also susceptible to direct damage, resulting in mutations that may result in cellular dysfunction and, ultimately, cell death. One of the major sources of DNA damage is apparently **free radicals,** which are atoms or molecules with an unpaired electron.
5. *Mitochondrial damage.* Mitochondrial DNA may be more sensitive to free-radical damage than is nuclear DNA. Mitochondrial DNA damage may result in loss of proteins critical to mitochondrial function. Because the mitochondria are the primary source of ATP, loss of mitochondrial function could lead to the loss of energy critical to cell function and, ultimately, to cell death. One proposal suggests that reduced caloric intake may reduce free-radical damage to mitochondria.

ANSWER TO LEARN TO PREDICT

Consider the important points made in the question. First, Carlos suffers from a genetic disease (diabetes insipidus) and second, this disease results in excessive water loss at the kidneys. In this chapter, we learned that DNA contains the information for directing protein synthesis. Since this is a genetic disease, one possible conclusion is that some type of protein is not functioning correctly. We also learned that proteins have several functions, including acting as receptor molecules and membrane channels. Malfunction of either of these types of proteins could disrupt water homeostasis. ADH is a chemical signal that regulates water loss from the kidneys. If the receptor molecules for ADH are altered and not functioning properly, the kidney cells could not respond to the signal. Another possibility is that water membrane channels, which allow water to move across the cell membrane, are not functioning properly and are disrupting water homeostasis.

Answers to the rest of this chapter's Predict questions are in Appendix E.

⎯⋀⎯ SUMMARY

3.1 Cell Structure (p. 42)

1. Cells are highly organized units containing organelles, which perform specific functions.
2. The nucleus contains genetic material, and cytoplasm is the living material between the nucleus and the cell membrane.

3.2 Functions of the Cell (p. 44)

1. Cells are the basic unit of life.
2. Cells protect and support the body, as well as provide for cell metabolism, communication, and inheritance.

3.3 Cell Membrane (p. 44)

1. The cell membrane forms the outer boundary of the cell. It determines what enters and leaves the cell.
2. The cell membrane is composed of a double layer of phospholipid molecules in which proteins float. The proteins function as membrane channels, carrier molecules, receptor molecules, enzymes, and structural components of the membrane.

3.4 Movement Through the Cell Membrane (p. 44)

Diffusion

1. Diffusion is the movement of a solute from an area of higher concentration to an area of lower concentration within a solvent. At equilibrium, the distribution of molecules is uniform.
2. A concentration gradient is the concentration of a solute at one point in a solvent minus the concentration of that solute at another point in the solvent divided by the distance between the points.
3. Lipid-soluble molecules pass through the cell membrane readily by dissolving in the lipid portion of the membrane. Small molecules and ions can pass through membrane channels.

Osmosis

1. Osmosis is the diffusion of water across a selectively permeable membrane.
2. Osmotic pressure is the force required to prevent movement of water across a selectively permeable membrane.
3. In a hypotonic solution, cells swell (and can undergo lysis); in an isotonic solution, cells neither swell nor shrink; in a hypertonic solution, cells shrink and undergo crenation.

Carrier-Mediated Transport Mechanisms

1. Carrier-mediated transport is the movement of a substance across a membrane by means of a carrier molecule. The substances transported tend to be large, water-soluble molecules or ions.
2. Facilitated diffusion moves substances from a higher to a lower concentration and does not require energy in the form of ATP.
3. Active transport can move substances from a lower to a higher concentration and requires ATP.
4. Secondary active transport uses the energy of one substance moving down its concentration gradient to move another substance across the cell membrane. In cotransport, both substances move in the same direction; in countertransport, they move in opposite directions.

Endocytosis and Exocytosis

1. Endocytosis is the movement of materials into cells by the formation of a vesicle. Receptor-mediated endocytosis involves cell receptors attaching to molecules, which are then transported into the cell. Phagocytosis is the movement of solid material into cells. Pinocytosis is similar to phagocytosis, except that the material ingested is much smaller and is in solution.
2. Exocytosis is the secretion of materials from cells by vesicle formation.

3.5 Organelles (p. 52)

Nucleus

1. The nuclear envelope consists of two separate membranes that form nuclear pores at many points on the surface of the nucleus.
2. DNA and associated proteins are found inside the nucleus as chromatin. DNA is the hereditary material of the cell and controls cell activities.
3. Nucleoli consist of RNA and proteins and are the sites of ribosomal subunit assembly.

Ribosomes

1. Ribosomes are the sites of protein synthesis.
2. A ribosome is composed of one large and one small subunit.

Rough and Smooth Endoplasmic Reticulum

1. Rough ER is ER with ribosomes attached. It is a major site of protein synthesis.
2. Smooth ER does not have ribosomes attached and is a major site of lipid synthesis.

Golgi Apparatus

The Golgi apparatus is a series of closely packed membrane sacs that collect, modify, package, and distribute proteins and lipids produced by the ER.

Secretory Vesicles

Secretory vesicles are membrane-bound sacs that carry substances from the Golgi apparatus to the cell membrane, where the vesicle contents are released.

Lysosomes and Peroxisomes

Lysosomes and peroxisomes are membrane-bound sacs containing enzymes. Within the cell, lysosomes break down phagocytized material. Peroxisomes break down fatty acids, amino acids, and hydrogen peroxide.

Mitochondria

Mitochondria are the major sites for the production of ATP, which cells use as an energy source. Mitochondria carry out aerobic respiration (requires O_2).

Cytoskeleton

1. The cytoskeleton supports the cytoplasm and organelles and is involved with cell movements.
2. The cytoskeleton is composed of microtubules, microfilaments, and intermediate filaments.

Centrioles

Centrioles, located in the centrosome, are made of microtubules. They facilitate chromosome movement during cell division.

Cilia, Flagella, and Microvilli

1. Cilia move substances over the surface of cells.
2. Flagella are much longer than cilia and propel sperm cells.
3. Microvilli increase the surface area of cells and thus aid in absorption.

3.6 Whole-Cell Activity   (p. 58)

A cell's characteristics are ultimately determined by the types of proteins it produces, which are determined by the genetic information in the nucleus. Understanding how genetic information is used in the cell and distributed to daughter cells is important for understanding basic cellular activity.

Gene Expression

1. Cell activity is regulated by enzymes (proteins), and DNA controls enzyme production.
2. During transcription, the sequence of nucleotides in DNA (a gene) determines the sequence of nucleotides in mRNA; the mRNA moves through the nuclear pores to ribosomes.
3. During translation, the sequence of codons in mRNA is used at ribosomes to produce proteins. Anticodons of tRNA bind to the codons of mRNA, and the amino acids carried by tRNA are joined to form a protein.

Cell Life Cycle

1. The life cycle of a cell consists of a series of events that produce new cells for growth and for tissue repair.
2. The two phases of the cell life cycle are interphase and cell division.
3. DNA replicates during interphase, the nondividing phase of the cell cycle.

4. Cell division occurs through mitosis, which is divided into four stages:
   • Prophase—each chromosome consists of two chromatids joined at the centromere.
   • Metaphase—chromosomes align at the center of the cell.
   • Anaphase—chromatids separate at the centromere and migrate to opposite poles.
   • Telophase—the two new nuclei assume their normal structure, and cell division is completed, producing two new daughter cells.

Differentiation

Differentiation, the process by which cells develop specialized structures and functions, results from the selective activation and inactivation of DNA sections.

Apoptosis

Apoptosis is the programmed death of cells. Apoptosis regulates the number of cells within various tissues of the body.

3.7 Cellular Aspects of Aging   (p. 66)

Aging may be due to the presence of "cellular clocks," the function of "death genes," DNA damage, free radicals, or mitochondrial damage.

REVIEW AND COMPREHENSION

1. Define cytoplasm and organelle.
2. List the functions of a cell.
3. Describe the structure of the cell membrane. What functions does it perform?
4. Define solution, solute, solvent, diffusion, and concentration gradient.
5. How do lipid-soluble molecules, small molecules that are not lipid-soluble, and large molecules that are not lipid-soluble cross the cell membrane?
6. Define osmosis and osmotic pressure.
7. What happens to cells that are placed in isotonic solutions? In hypertonic or hypotonic solutions? What are crenation and lysis?
8. What is carrier-mediated transport? How are facilitated diffusion and active transport similar, and how are they different?
9. How does secondary active transport work? Define cotransport and countertransport.
10. Describe receptor-mediated endocytosis, phagocytosis, pinocytosis, and exocytosis. What do they accomplish?
11. Describe the structure of the nucleus and the nuclear envelope. Name the structures in the nucleus, and give their functions.
12. Where are ribosomes assembled, and what kinds of molecules are found in them?

13. What is endoplasmic reticulum? Compare the functions of rough and smooth endoplasmic reticulum.
14. Describe the Golgi apparatus, and state its function.
15. Where are secretory vesicles produced? What are their contents, and how are they released?
16. What are the functions of lysosomes and peroxisomes?
17. Describe the structure and function of mitochondria.
18. Name the components of the cytoskeleton, and give their functions.
19. Describe the structure and function of centrioles.
20. Describe the structure and function of cilia, flagella, and microvilli.
21. Describe how proteins are synthesized and how the structure of DNA determines the structure of proteins.
22. Define autosome, sex chromosome, and diploid number.
23. How do the sex chromosomes of males and females differ?
24. Describe what happens during interphase and each stage of mitosis. What kinds of tissues undergo mitosis?
25. Define cell differentiation. In general terms, how does differentiation occur?
26. List the principal hypotheses of aging.

CRITICAL THINKING

1. The body of a man was found floating in the water of Grand Pacific Bay, which has a salt concentration slightly greater than that of body fluids. When examined during an autopsy, the cells in his lung tissue were clearly swollen. Choose the most logical conclusion.
   a. He probably drowned in the bay.
   b. He may have been murdered elsewhere.
   c. He did not drown.

2. Patients with kidney failure can be kept alive by dialysis, which removes toxic waste products from the blood. In a dialysis machine, blood flows past one side of a selectively permeable dialysis membrane, and dialysis fluid flows on the other side of the membrane. Small substances, such as ions, glucose, and urea, can pass through the dialysis membrane, but larger substances, such as proteins, cannot. If you wanted to use a dialysis machine to remove only the toxic waste product urea from blood, what could you use for the dialysis fluid?
   a. a solution that is isotonic and contains only protein
   b. a solution that is isotonic and contains the same concentration of substances as blood, except for having no urea in it
   c. distilled water
   d. blood

3. Secretory vesicles fuse with the cell membrane to release their contents to the outside of the cell. In this process, the membrane of the secretory vesicle becomes part of the cell membrane. Because small pieces of membrane are continually added to the cell membrane, we would expect the cell membrane to become larger and larger as secretion continues. However, the cell membrane stays the same size. Explain how this happens.

4. Suppose that a cell has the following characteristics: many mitochondria, well-developed rough ER, well-developed Golgi apparatuses, and numerous vesicles. Predict the major function of the cell. Explain how each characteristic supports your prediction.

5. The proteins (hemoglobin) in red blood cells normally organize relative to one another, forming "stacks" of proteins that are, in part, responsible for the normal shape of red blood cells. In sickle-cell anemia, proteins inside red blood cells do not stack normally. Consequently, the red blood cells become sickle-shaped and plug up small blood vessels. Sickle-cell anemia is hereditary and results from changing one nucleotide for a different nucleotide within the gene that is responsible for producing the protein. Explain how this change results in an abnormally functioning protein.

Answers in Appendix D

4 Tissues

LEARN to PREDICT

It is Matt's birthday, but he will not be eating any cake. Matt has gluten enteropathy, also called celiac disease. Celiac disease results from an inappropriate immune response to gluten, a group of proteins found in wheat and various other grains. After eating food containing gluten, such as most breads and cereals, Matt has bouts of diarrhea because his intestinal lining is unable to properly absorb water and nutrients. The poor absorption is due to a reduced number of villi, or folds of the intestinal lining, and reduced transport capacity of the remaining cells within the villi. In chapter 3 we learned that water and nutrients enter and exit the body's cells by osmosis and other transport processes. Chapter 4 describes how tissues are specialized to allow this flow of water and nutrients.

After reading this chapter, identify the type of tissue affected by Matt's disease and which parts of the cells in this tissue are damaged, thus reducing their ability to absorb water and nutrients. Then explain why Matt has diarrhea after eating food containing gluten.

▲ People with celiac disease must avoid foods containing wheat products.

4.1 TISSUES AND HISTOLOGY

Module 3 Tissue

Learning Outcome *After reading this section, you should be able to*

A. Describe the general makeup of a tissue.

A **tissue** (tish′ū) is a group of cells with similar structure and function that have similar extracellular substances located between them. The microscopic study of tissue structure is called **histology** (his-tol′ō-je; *histo-,* tissue + *-ology,* study). Knowledge of tissue structure and function is important in understanding how individual cells are organized to form tissues and how tissues are organized to form organs, organ systems, and the complete organism. The structure of each tissue type is related to its function, and the structure of the tissues in an organ is related to the organ's function.

Changes in tissues can result in development, growth, aging, trauma, or disease. For example, skeletal muscles enlarge because skeletal muscle cells increase in size in response to exercise. Reduced elasticity of blood vessel walls in aging people results from gradual changes in connective tissue. Many tissue abnormalities, including cancer, result from changes in tissues that can be identified by microscopic examination.

The four basic tissue types are epithelial, connective, muscle, and nervous. This chapter emphasizes epithelial and connective tissues. Later chapters consider muscle and nervous tissues in more detail.

4.2 EPITHELIAL TISSUE

Learning Outcomes *After reading this section, you should be able to*

A. List and explain the general characteristics of epithelial tissue.
B. Classify epithelial tissues based on the number of cell layers and the shape of the cells.

C. Name and describe the various types of epithelial tissue, including their chief functions and locations.

D. Relate the structural specializations of epithelial tissue with the functions they perform.

E. Differentiate between exocrine and endocrine glands, and unicellular and multicellular glands.

F. Categorize glands based on their structure and function.

Epithelium (ep-i-thē′lē-ŭm; pl. epithelia, ep-i-thē′lē-ă; *epi,* on + *thele,* covering or lining), or *epithelial tissue,* covers external and internal surfaces throughout the body. Surfaces of the body include the outer layer of the skin and the lining of cavities, such as the digestive tract, airways, and blood vessels. It also forms most glands. Epithelium consists almost entirely of cells with very little extracellular material between them. Although there are some exceptions, most epithelia have a **free surface,** which is not in contact with other cells, and a **basal surface** adjacent to a basement membrane, which attaches the epithelial cells to underlying tissues (figure 4.1). Epithelium may consist of a single layer of epithelial cells or multiple layers of epithelial cells between the free surface and the basement membrane.

The **basement membrane** is secreted partly by epithelial cells and partly by the cells of the underlying tissues. It consists of a meshwork of protein molecules with other molecules bound to them. Substances that cross the epithelium must also cross the basement membrane. It can function as a filter and as a barrier to the movement of cells. For example, if some epithelial cells are converted to cancer cells, the basement membrane can, for some time, help prevent the spread of the cancer into the underlying tissues.

Blood vessels do not extend from the underlying tissues into epithelium, so gases and nutrients that reach the epithelium must diffuse across the basement membrane from the underlying tissues, where blood vessels are abundant. Waste products produced by the epithelial cells diffuse across the basement membrane in the opposite direction, to reach the blood vessels.

Functions of Epithelia

The major functions of epithelia are

1. *Protecting underlying structures.* Examples include the outer layer of the skin and the epithelium of the oral cavity, which protect the underlying structures from abrasion.

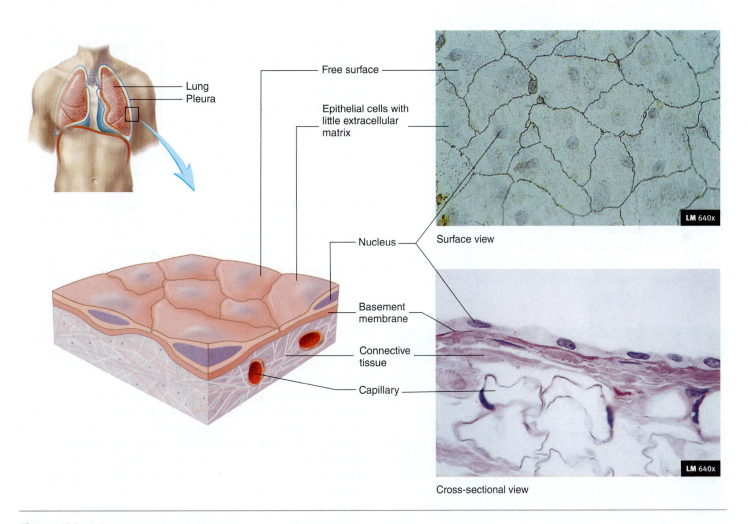

Figure 4.1 Characteristics of Epithelium

Surface and cross-sectional views of epithelium illustrate the following characteristics: little extracellular material between cells, a free surface, and a basement membrane attaching epithelial cells to underlying tissues. The capillaries in connective tissue do not penetrate the basement membrane. Nutrients, oxygen, and waste products diffuse across the basement membrane between the capillaries and the epithelial cells.

2. *Acting as a barrier.* Epithelium prevents many substances from moving through it. For example, the epithelium of the skin acts as a barrier to water and reduces water loss from the body. The epithelium of the skin also prevents many toxic molecules and microorganisms from entering the body.

3. *Permitting the passage of substances.* Epithelium also allows many substances to move through it. For example, oxygen and carbon dioxide are exchanged between the air and blood by diffusion through the epithelium in the lungs.

4. *Secreting substances.* Sweat glands, mucous glands, and the enzyme-secreting portion of the pancreas are all composed of epithelial cells.

5. *Absorbing substances.* The cell membranes of certain epithelial tissues contain carrier proteins (see chapter 3) that regulate the absorption of materials. For example, the epithelial cells of the intestines absorb digested food molecules, vitamins, and ions.

Classification of Epithelia

Epithelia are classified according to the number of cell layers and the shape of the cells (table 4.1). **Simple epithelium** consists of a single layer of cells. **Stratified epithelium** consists of more than one layer of epithelial cells, with some cells sitting on top of others. Categories of epithelium based on cell shape are **squamous** (skwā′mŭs; flat), **cuboidal** (cubelike), and **columnar** (tall and thin). In most cases, each epithelium is given two names, such as simple squamous, simple columnar, or stratified squamous epithelium. When epithelium is stratified, it is named according to the shape of the cells at the free surface.

Simple squamous epithelium is a single layer of thin, flat cells (table 4.2*a*). Some substances easily pass through this thin layer of cells, but other substances do not. For example, the airways end as small sacs called **alveoli** (al-vē′o-lī; sing. alveolus, hollow sac). The alveoli consist of simple squamous epithelium that allows oxygen from the air to diffuse into the body and carbon dioxide to diffuse out of the body into the air. Simple squamous epithelium in the filtration membranes of the kidneys forms thin barriers through which small molecules, but not large ones, can

pass. Small molecules, including water from blood, are filtered through these barriers as a major step in urine formation. Large molecules, such as proteins, and blood cells remain in the blood vessels of the kidneys.

Simple squamous epithelium also prevents abrasion between organs in the pericardial, pleural, and peritoneal cavities (see chapter 1). The outer surfaces of organs are covered with simple squamous epithelium that secretes a slippery fluid. The fluid lubricates the surfaces between the organs, preventing damage from friction when the organs rub against one another or the body wall.

Simple cuboidal epithelium is a single layer of cubelike cells (table 4.2*b*) that carry out active transport, facilitated diffusion, or secretion. Epithelial cells that secrete molecules such as proteins contain organelles that synthesize them. These cells have a greater volume than simple squamous epithelial cells and contain more cell organelles. The kidney tubules have large portions of their walls composed of simple cuboidal epithelium. These cells secrete waste products into the tubules and reabsorb useful materials from the tubules as urine is formed. Some cuboidal epithelial cells have cilia that move mucus over the free surface or microvilli that increase the surface area for secretion and absorption.

Simple columnar epithelium is a single layer of tall, thin cells (table 4.2*c*). These large cells contain organelles that enable them to perform complex functions. For example, the simple columnar epithelium of the small intestine produces and secretes mucus and digestive enzymes. **Mucus** (mū′kŭs) is a clear, viscous (thick) fluid. The mucus protects the lining of the intestine, and the digestive enzymes complete the process of digesting food. The columnar cells then absorb the digested foods by active transport, facilitated diffusion, or simple diffusion.

Pseudostratified columnar epithelium is a special type of simple epithelium (table 4.2*d*). The prefix *pseudo-* means false, so this type of epithelium appears stratified but is not. It consists of one layer of cells, with all the cells attached to the basement membrane. But it looks like two or more layers of cells because some of the cells are tall and reach the free surface, whereas others are short and do not reach the free surface. Pseudostratified columnar epithelium lines some glands and ducts, the auditory tubes, and some of the airways, such as the nasal cavity, nasal sinuses, pharynx, trachea, and bronchi. Pseudostratified columnar epithelium secretes mucus, which covers its free surface. Cilia located on the free surface move the mucus and the debris that accumulates in it. For example, cilia of the airways move mucus toward the throat, where it is swallowed.

Stratified squamous epithelium forms a thick epithelium because it consists of several layers of cells (table 4.3*a*). The deepest cells are cuboidal or columnar and are capable of dividing and producing new cells. As these newly formed cells are pushed to the surface, they become flat and thin. As the cells flatten, the cytoplasm of the epithelial cells is replaced by a protein called keratin, and the cells die. One type of stratified squamous epithelium forms the outer layer of the skin and is called keratinized squamous epithelium (see chapter 5). The dead cells provide protection against abrasion, and form a barrier that prevents microorganisms and toxic chemicals from entering the body, and reduces the loss of water from the body. If cells at the surface are damaged or rubbed away, they are replaced by cells formed in the deeper layers.

TABLE 4.1	Classification of Epithelium
Number of Layers or Category	**Shape of Cells**
Simple (single layer of cells)	Squamous Cuboidal Columnar

Stratified (more than one layer of cells)	Squamous
	   Nonkeratinized (moist)
	   Keratinized
	Cuboidal (very rare)
	Columnar (very rare)

Pseudostratified (modification of simple epithelium)	Columnar
Transitional (modification of stratified epithelium)	Roughly cuboidal to columnar when not stretched and squamouslike when stretched

TABLE 4.2 Simple Epithelium

(a) Simple Squamous Epithelium AP|R

Structure:
Single layer of flat, often hexagonal cells; the nuclei appear as bumps when viewed in cross section because the cells are so flat

Function:
Diffusion, filtration, some secretion, and some protection against friction

Location:
Lining of blood vessels and the heart, lymphatic vessels, alveoli of the lungs, portions of the kidney tubules, lining of serous membranes of body cavities (pleural, pericardial, peritoneal)

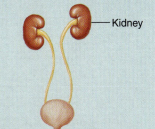

Kidney

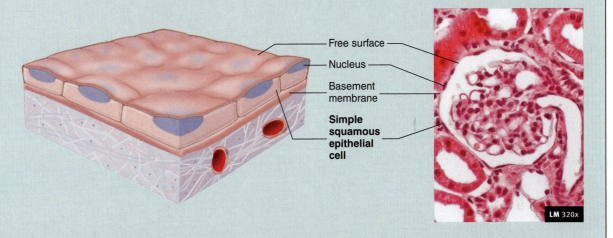

- Free surface
- Nucleus
- Basement membrane
- **Simple squamous epithelial cell**

LM 320x

(b) Simple Cuboidal Epithelium AP|R

Structure:
Single layer of cube-shaped cells; some cells have microvilli (kidney tubules) or cilia (terminal bronchioles of the lungs)

Function:
Secretion and absorption by cells of the kidney tubules; secretion by cells of glands and choroid plexuses; movement of particles embedded in mucus out of the terminal bronchioles by ciliated cells

Location:
Kidney tubules, glands and their ducts, choroid plexuses of the brain, lining of terminal bronchioles of the lungs, and surfaces of the ovaries

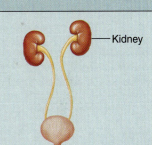

Kidney

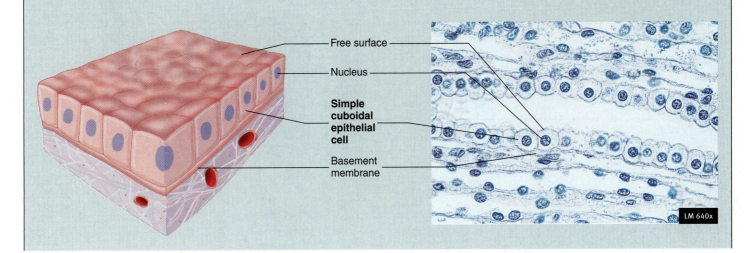

- Free surface
- Nucleus
- **Simple cuboidal epithelial cell**
- Basement membrane

LM 640x

TABLE 4.2   *continued*

(c) Simple Columnar Epithelium AP|R

Structure:
Single layer of tall, narrow cells; some cells have cilia (bronchioles of lungs, auditory tubes, uterine tubes, and uterus) or microvilli (intestines)

Function:
Movement of particles out of the bronchioles of the lungs by ciliated cells; partially responsible for the movement of oocytes through the uterine tubes by ciliated cells; secretion by cells of the glands, the stomach, and the intestine; absorption by cells of the intestine

Location:
Glands and some ducts, bronchioles of lungs, auditory tubes, uterus, uterine tubes, stomach, intestines, gallbladder, bile ducts, and ventricles of the brain

Lining of stomach and intestines

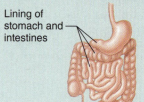

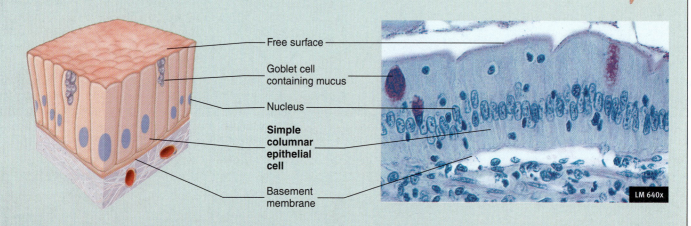

- Free surface
- Goblet cell containing mucus
- Nucleus
- **Simple columnar epithelial cell**
- Basement membrane

LM 640x

(d) Pseudostratified Columnar Epithelium AP|R

Structure:
Single layer of cells; some cells are tall and thin and reach the free surface, and others do not; the nuclei of these cells are at different levels and appear stratified; the cells are almost always ciliated and are associated with goblet cells that secrete mucus onto the free surface

Function:
Synthesize and secrete mucus onto the free surface and move mucus (or fluid) that contains foreign particles over the surface of the free surface and from passages

Location:
Lining of nasal cavity, nasal sinuses, auditory tubes, pharynx, trachea, and bronchi of lungs

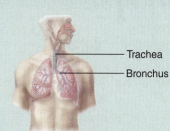

- Trachea
- Bronchus

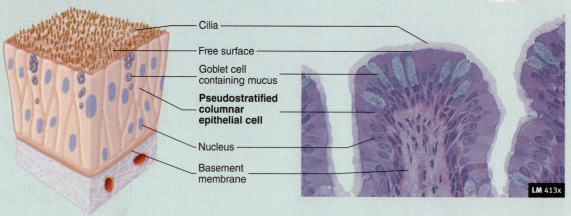

- Cilia
- Free surface
- Goblet cell containing mucus
- **Pseudostratified columnar epithelial cell**
- Nucleus
- Basement membrane

LM 413x

In contrast, stratified squamous epithelium of the mouth is composed of living cells with a moist surface. This nonkeratinized (moist) stratified squamous epithelium also provides protection against abrasion and acts as a mechanical barrier, preventing microorganisms from entering the body. Water, however, can move across it more readily than across the skin.

Stratified cuboidal epithelium consists of more than one layer of cuboidal epithelial cells. This epithelial type is relatively rare and is found in sweat gland ducts, ovarian follicular cells, and the salivary glands. It functions in absorption, secretion, and protection.

Stratified columnar epithelium consists of more than one layer of epithelial cells, but only the surface cells are columnar. The deeper layers are irregular or cuboidal in shape. Like stratified cuboidal epithelium, stratified columnar epithelium is relatively rare. It is found in the mammary gland ducts, the larynx, and a portion of the male urethra. This epithelium carries out secretion, protection, and some absorption.

Transitional epithelium is a special type of stratified epithelium that can be greatly stretched (table 4.3b). In the unstretched state, transitional epithelium consists of five or more layers of cuboidal or columnar cells that often are dome-shaped at the free surface. As transitional epithelium is stretched, the cells change to a low cuboidal or squamous shape, and the number of cell layers decreases. Transitional epithelium lines cavities that can expand greatly, such as the urinary bladder. It also protects underlying structures from the caustic effects of urine.

Structural and Functional Relationships

Cell Layers and Cell Shapes

The number of cell layers and the shape of the cells in a specific type of epithelium reflect the function the epithelium performs. Two important functions are controlling the passage of materials through the epithelium and protecting the underlying tissues. Simple epithelium, with its single layer of cells, is found in organs that primarily function to move materials. Examples include the diffusion of gases across the wall of the alveoli of the lungs, filtration of fluid across the filtration membranes in the kidneys, secretion from glands, and nutrient absorption by the intestines. The movement of materials through a stratified epithelium is hindered by its many layers. Stratified epithelium is well adapted for its protective function. As the outer cell layers are damaged, they are replaced by cells from deeper layers. Stratified squamous epithelium is found in areas of the body where abrasion can occur, such as in the skin, anal canal, and vagina.

Differences in function are also reflected in cell shape. Cells are normally flat and thin when the function is diffusion, such as in the alveoli of the lungs, or filtration, such as in kidney glomeruli. Cells with the major function of secretion or absorption are usually cuboidal or columnar. They are larger because they contain more organelles, which are responsible for the function of the cell. The stomach, for example, is lined with simple columnar epithelium. These cells contain many **secretory vesicles** (ves′i-klz) filled with mucus. The large amounts of mucus produced by the simple columnar epithelium protect the stomach lining against the digestive enzymes and acid produced in the stomach. An ulcer, or irritation in the stomach's epithelium and underlying tissue, can develop if

this protective mechanism fails. Simple cuboidal epithelial cells that secrete or absorb molecules, as occurs in the kidney tubules, contain many mitochondria, which produce the ATP required for active transport.

Predict 2

Explain the consequences of having (a) nonkeratinized stratified epithelium rather than simple columnar epithelium lining the digestive tract and (b) nonkeratinized stratified squamous epithelium rather than keratinized stratified squamous epithelium at the surface of the skin.

The shape and number of layers of epithelial cells can change if they are subjected to long-term irritation or other abnormal conditions. People who smoke cigarettes eventually experience changes in the epithelium of the larger airways. The delicate pseudostratified columnar epithelium, which performs a cleaning function by moving mucus and debris from the passageways, is replaced by stratified squamous epithelium, which is more resistant to irritation but does not perform a cleaning function. Also, lung cancer most often results from changes in epithelial cells in the lung passageways of smokers. The changes in cell structure are used to identify the cancer.

A CASE IN POINT

Detecting Cancer

Wanna Wonder was wary about having a Pap test. She insisted on knowing why she had to endure such a procedure. Her physician explained that the test could detect cervical cancer. Using a swab, a physician collects cells from the cervix of the uterus. The cells are stained and examined with a microscope by a skilled technician. Compared with normal cells, cancer cells have variable sizes and shapes, with large, unusually shaped nuclei. If cervical cancer is detected early, it can be treated before it spreads to other areas of the body.

Free Surfaces

Most epithelia have a free surface that is not in contact with other cells and faces away from underlying tissues. The characteristics of the free surface reflect its functions. The free surface can be smooth or lined with microvilli or cilia. A smooth free surface reduces friction as material moves across it. For example, the lining of blood vessels is simple squamous epithelium with a smooth surface, which reduces friction as blood flows through the vessels. **Microvilli** are cylindrical extensions of the cell membrane that increase the free surface area (see chapter 3). Normally, many microvilli cover the free surface of each cell involved in absorption or secretion, such as the cells lining the small intestine. **Cilia** (see chapter 3) propel materials along the free surface of cells. The nasal cavity and trachea are lined with pseudostratified columnar ciliated epithelium. Intermixed with the ciliated cells are specialized mucus-producing cells called **goblet cells** (see table 4.2d). Dust and other materials are trapped in the mucus that covers the epithelium, and movement of the cilia propels the mucus with its entrapped particles to the back of the throat, where it is swallowed or coughed up. The constant movement of mucus helps keep the airways clean.

TABLE 4.3   Stratified Epithelium

(a) Stratified Squamous Epithelium AP|R

Structure:
Several layers of cells that are cuboidal in the basal layer and progressively flattened toward the surface; the epithelium can be nonkeratinized (moist) or keratinized; in nonkeratinized stratified squamous epithelium, the surface cells retain a nucleus and cytoplasm; in keratinized stratified epithelium, the cytoplasm of cells at the surface is replaced by a protein called keratin, and the cells are dead

Function:
Protects against abrasion, forms a barrier against infection, and reduces loss of water from the body

Location:
Keratinized—outer layer of the skin; nonkeratinized—mouth, throat, larynx, esophagus, anus, vagina, inferior urethra, and corneas

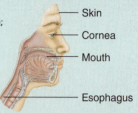

- Skin
- Cornea
- Mouth
- Esophagus

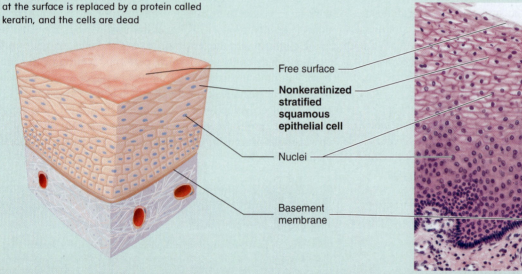

- Free surface
- **Nonkeratinized stratified squamous epithelial cell**
- Nuclei
- Basement membrane

LM 72x

(b) Transitional Epithelium AP|R

Structure:
Stratified cells that appear cuboidal when the organ or tube is not stretched and squamous when the organ or tube is stretched by fluid

Function:
Accommodates fluctuations in the volume of fluid in an organ or a tube; protects against the caustic effects of urine

Location:
Lining of urinary bladder, ureters, and superior urethra

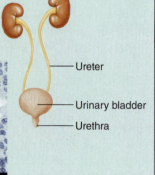

- Ureter
- Urinary bladder
- Urethra

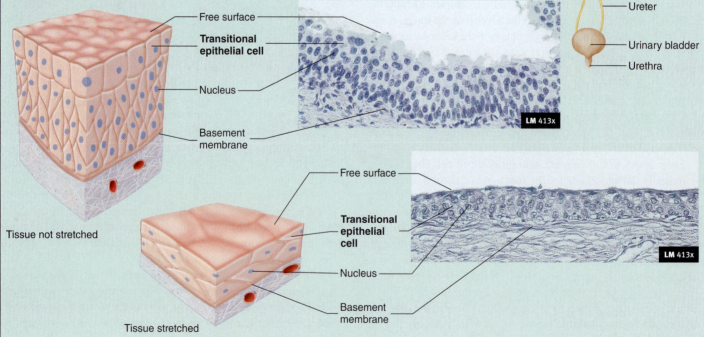

- Free surface
- **Transitional epithelial cell**
- Nucleus
- Basement membrane

LM 413x

Tissue not stretched

- Free surface
- **Transitional epithelial cell**
- Nucleus
- Basement membrane

LM 413x

Tissue stretched

Cell Connections

Epithelial cells are connected to one another in several ways (figure 4.2). **Tight junctions** bind adjacent cells together and form permeability barriers. Tight junctions prevent the passage of materials between epithelial cells because they completely surround each cell, similar to the way a belt surrounds the waist. Materials that pass through the epithelial layer must pass through the cells, so those cells regulate what materials can cross. Tight junctions are found in the lining of the intestines and in most other simple epithelia. **Desmosomes** (dez'mō-sōmz; *desmos*, a band + *soma*, body) are mechanical links that bind cells together. Modified desmosomes, called **hemidesmosomes** (hem-ē-dez'mō-sōmz; *hemi*, one-half), also anchor cells to the basement membrane. Many desmosomes are found in epithelia subjected to stress, such as the stratified squamous epithelium of the skin. **Gap junctions** are small channels that allow small molecules and ions to pass from one epithelial cell to an adjacent one. Most epithelial cells are connected to one another by gap junctions, and researchers believe that molecules or ions moving through the gap junctions act as communication signals to coordinate the activities of the cells.

Glands

A **gland** is a structure that secretes substances onto a surface, into a cavity, or into the blood. Most glands are composed primarily of epithelium and are multicellular. But sometimes single goblet cells

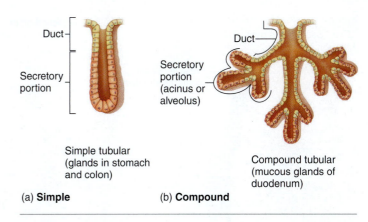

Simple tubular (glands in stomach and colon)

(a) **Simple**

Compound tubular (mucous glands of duodenum)

(b) **Compound**

Figure 4.3 Structure of Exocrine Glands

Based on the shapes of their ducts, exocrine glands may be called simple or compound. Based on their secretory units, they may be classed as tubular, acinar, or alveolar.

are classified as unicellular glands because they secrete mucus onto epithelial surfaces. Glands with ducts are called **exocrine** (ek'sō-krin; *exo*, outside + *krino*, to separate) glands (figure 4.3). The exocrine glands can be **simple,** with ducts that have no branches, or **compound,** with ducts that have many branches. The end of a duct can be **tubular.** Some tubular glands are straight, and others are coiled. Some glands have ends that are expanded into a saclike structure called an **acinus** (as'i-nŭs; grapelike) or an **alveolus** (al-vē'ō-lŭs; small cavity). Some compound glands have acini and tubules (tubuloacinar or tubuloalveolar glands) that secrete substances. Secretions from exocrine glands pass through the ducts onto a surface or into an organ. For example, sweat from sweat glands and oil from sebaceous glands flow onto the skin surface.

Exocrine glands can also be classified according to how products leave the cell. The most common type of secretion is **merocrine** (mer'ōkrin) **secretion.** In merocrine secretion, products are released, but no actual cellular material is lost (figure 4.4a). Secretions are either actively transported or packaged in vesicles and then released by the process of exocytosis at the free surface of the cell. Sweat and digestive enzymes produced by the pancreas are released by merocrine secretion. In **apocrine** (ap'ō-krin) **secretion,** the secretory products are released as fragments of the gland cell (figure 4.4b). Milk secretion by the mammary glands utilizes some apocrine secretion. **Holocrine** (hol'ō-krin) **secretion** involves the shedding of entire cells (figure 4.4c). Sebaceous (oil) glands of the skin utilize holocrine secretion.

Endocrine (en'dō-krin; *endo*, within) glands have no ducts and empty their secretions into the blood. These secretions, called **hormones** (hōr'mōnz), are carried by the blood to other parts of the body. Endocrine glands include the thyroid gland and the insulin-secreting portions of the pancreas. We discuss endocrine glands more fully in chapter 10.

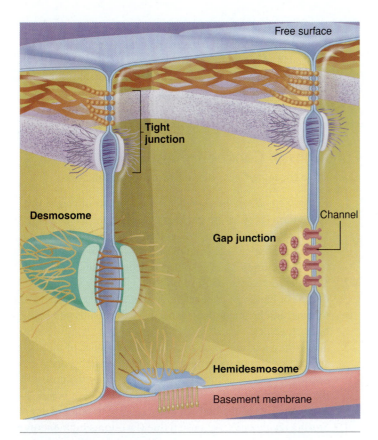

Free surface

Tight junction

Desmosome

Gap junction

Channel

Hemidesmosome

Basement membrane

Figure 4.2 Cell Connections

Desmosomes and tight junctions anchor cells to one another, and hemidesmosomes anchor cells to the basement membrane. Gap junctions allow adjacent cells to communicate with each other. Few cells have all of these different connections.

4.3 CONNECTIVE TISSUE

Learning Outcome   *After reading this section, you should be able to*

A. Describe the classification of connective tissue, and give examples of each major type.

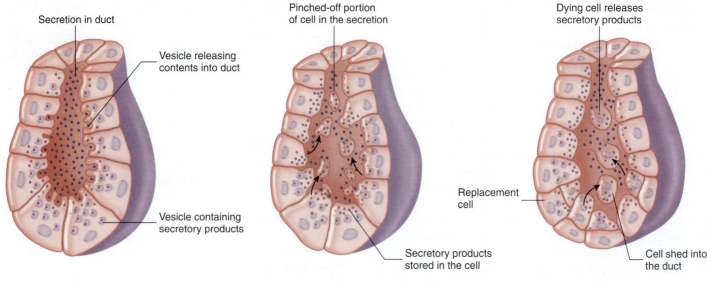

(a) Merocrine gland Cells of the gland produce secretions by active transport or produce vesicles that contain secretory products, and the vesicles empty their contents into the duct through exocytosis.	**(b) Apocrine gland** Secretory products are stored in the cell near the lumen of the duct. A portion of the cell near the lumen containing secretory products is pinched off the cell and joins secretions produced by a merocine process.	**(c) Holocrine gland** Secretory products are stored in the cells of the gland. Entire cells are shed by the gland and become part of the secretion. The lost cells are replaced by other cells deeper in the gland.

Figure 4.4 Exocrine Glands and Secretion Types
Exocrine glands are classified by type of secretion.

Connective tissue is found throughout the body. It is usually characterized by large amounts of extracellular material that separates cells from one another. The extracellular material, or **extracellular matrix** (mā′triks), has three major components: (1) protein fibers, (2) ground substance consisting of nonfibrous protein and other molecules, and (3) fluid.

Three types of protein fibers help form most connective tissues. **Collagen** (kol′lă-jen; glue-producing) **fibers,** which resemble microscopic ropes, are flexible but resist stretching. **Reticular** (rē-tik′ū-lăr) **fibers** are very fine, short collagen fibers that branch to form a supporting network. **Elastic fibers** have a structure similar to that of coiled metal bed springs; after being stretched, they can recoil to their original shape.

Ground substance is the shapeless background against which cells and collagen fibers can be seen when using a light microscope. Although ground substance appears shapeless, the molecules within it are highly structured. **Proteoglycans** (prō′tē-ō-glī′kanz; *proteo,* protein + *glycan,* polysaccharide) resemble the limbs of pine trees, with proteins forming the branches and polysaccharides forming the pine needles. This structure enables proteoglycans to trap large quantities of water between the polysaccharides.

Connective tissue cells are named according to their functions. Cells whose names contain the suffix -*blast* (germ) produce the matrix; cells ending in -*cyte* (cell) maintain it; and cells ending in -*clast* (break) break it down for remodeling. For example, **fibroblasts** (fī′bro-blasts; *fibra,* fiber) are cells that form fibers and ground substance in the extracellular matrix of fibrous connective tissue, and **fibroyctes** are cells that maintain it. **Osteoblasts** (os′tē-ō-blasts; *osteo,* bone) form bone, **osteocytes** (os′tē-ō-sītz) maintain bone, and **osteoclasts** (os′tē-ō-klasts) break down bone.

Also found in connective tissue are cells associated with the immune system. **Macrophages** (mak′rō-fāj-ez; *makros,* large + *phago,* to eat) are large white blood cells that are capable of moving about and ingesting foreign substances, including microorganisms, in the connective tissue. **Mast cells** are nonmotile cells that release chemicals, such as histamine, that promote inflammation.

Functions of Connective Tissue

Connective tissue performs the following major functions:

1. *Enclosing and separating other tissues.* Sheets of connective tissue form capsules around organs, such as the liver and the kidneys. Connective tissue also forms layers that separate tissues and organs. For example, connective tissues separate muscles, arteries, veins, and nerves from one another.
2. *Connecting tissues to one another.* Tendons are strong cables, or bands, of connective tissue that attach muscles to bone, and ligaments are connective tissue bands that hold bones together.
3. *Supporting and moving parts of the body.* Bones of the skeletal system provide rigid support for the body, and semirigid cartilage supports structures, such as the nose, the ears, and the surfaces of joints. Joints between bones allow one part of the body to move relative to other parts.
4. *Storing compounds.* Adipose tissue (fat) stores high-energy molecules, and bones store minerals, such as calcium and phosphate.
5. *Cushioning and insulating.* Adipose tissue cushions and protects the tissues it surrounds and provides an insulating layer beneath the skin that helps conserve heat.

6. *Transporting.* Blood transports gases, nutrients, enzymes, hormones, and cells of the immune system throughout the body.

7. *Protecting.* Cells of the immune system and blood provide protection against toxins and tissue injury, as well as against microorganisms. Bones protect underlying structures from injury.

Classification of Connective Tissue

Connective tissue types blend into one another, and the transition points cannot be precisely identified. As a result, connective tissue is somewhat arbitrarily classified by the type and proportions of cells and extracellular matrix.

Two major categories of connective tissue are embryonic and adult connective tissue. By eight weeks of development, most of the embryonic connective tissue has become specialized to form the types of connective tissue seen in adults.

Table 4.4 presents the classification of adult connective tissue used in this text. Adult connective tissue consists of three types: connective tissue proper (loose and dense), supporting connective tissue (cartilage and bone), and fluid connective tissue (blood).

Connective Tissue Proper

Loose Connective Tissue

Loose connective tissue (table 4.5) consists of relatively few protein fibers that form a lacy network, with numerous spaces filled with ground substance and fluid. Three subdivisions of loose connective tissue are areolar, adipose, and reticular. **Areolar** (a-re′ō-lar) has extracellular matrix consisting mostly of collagen fibers and a few elastic fibers. The most common cells in loose connective tissue are the fibroblasts, which are responsible for producing the matrix. Loose connective tissue is widely distributed throughout the body and is the "loose packing" material of most organs and other tissues; it attaches the skin to underlying tissues and provides nourishment for the structures with which it is associated (table 4.5a). The basement membranes of epithelia often rest on loose connective tissue.

Adipose (ad′i-pōs; fat) tissue consists of adipocytes, or fat cells, which contain large amounts of lipid for energy storage. Unlike other connective tissue types, adipose tissue is composed of large cells and a small amount of extracellular matrix, which consists of loosely arranged collagen and reticular fibers with some scattered elastic fibers. The individual cells are large and closely packed together (table 4.5b). Adipose tissue also pads and protects parts of the body and acts as a thermal insulator.

Reticular tissue forms the framework of lymphatic tissue, such as in the spleen and lymph nodes, as well as in bone marrow and the liver (table 4.5c).

Dense Connective Tissue

Dense connective tissue has a relatively large number of protein fibers that form thick bundles and fill nearly all of the extracellular space. These protein fibers are produced by fibroblasts. There are two major subcategories of dense connective tissue: collagenous and elastic.

Dense collagenous connective tissue has an extracellular matrix consisting mostly of collagen fibers (table 4.6a). Structures made up of dense collagenous connective tissue include tendons, which attach muscle to bone; many ligaments, which attach bones to other bones; and much of the dermis, which is the connective tissue of the skin. Dense collagenous connective tissue also forms many capsules that surround organs, such as the liver and kidneys. In tendons and ligaments, the collagen fibers are oriented in the same direction, and so the tissue is called **dense regular,** but in the dermis and in organ capsules, the fibers are oriented in many different directions, and so the tissue is called **dense irregular.**

Predict 3

In tendons, collagen fibers are oriented parallel to the length of the tendon. In the skin, collagen fibers are oriented in many directions. What are the functional advantages of the fiber arrangements in tendons and in the skin?

TABLE 4.4	Classification of Connective Tissues
Connective tissue proper	
Loose (fewer fibers, more ground substance)	
Areolar	
Adipose	
Reticular	
Dense (more fibers, less ground substance)	
Dense, regular collagenous	
Dense, regular elastic	
Dense, irregular collagenous (not shown)	
Dense, irregular elastic (not shown)	
Supporting connective tissue	
Cartilage (semisolid matrix)	
Hyaline	
Fibrocartilage	
Elastic	
Bone (solid matrix)	
Spongy	
Compact	
Fluid connective tissue	
Blood	
Hemopoietic Tissue	
Red marrow	
Yellow marrow	

A CASE IN POINT

Marfan Syndrome

Izzy Taller is a 6-foot, 10-inch-tall high school student. The basketball coach convinces Izzy to try out for the basketball team, which requires a physical exam. The physician determines that Izzy's limbs, fingers, and toes are disproportionately long in relation to the rest of his body; he has an abnormal heart sound, indicating a problem with a heart valve; and he has poor vision because the lenses of his eyes are positioned abnormally. The physician diagnoses Izzy's condition as Marfan syndrome, an autosomal-dominant (see chapter 20) genetic disorder that results in the production of an abnormal protein called fibrillin-l. This protein is necessary for the normal formation and maintenance of elastic fibers and the regulation of a growth factor called TGFß. Because elastic fibers are found in connective tissue throughout the body, several body systems can be affected. The major cause of death in Marfan syndrome is rupture of the aorta due to weakening of the connective tissue in its wall. Izzy Taller should not join the basketball team.

TABLE 4.5   Connective Tissue Proper: Loose Connective Tissue

(a) Areolar Connective Tissue AP|R

Structure:
A fine network of fibers (mostly collagen fibers with a few elastic fibers) with spaces between the fibers; fibroblasts, macrophages, and lymphocytes are located in the spaces

Function:
Loose packing, support, and nourishment for the structures with which it is associated

Location:
Widely distributed throughout the body; substance on which epithelial basement membranes rest; packing between glands, muscles, and nerves; attaches the skin to underlying tissues

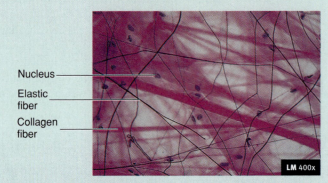

Nucleus
Elastic fiber
Collagen fiber

LM 400x

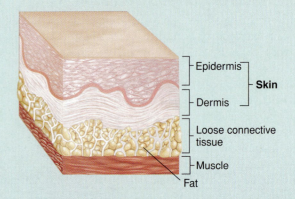

Epidermis ⎤
Dermis   ⎦ **Skin**
Loose connective tissue
Muscle
Fat

(b) Adipose Tissue AP|R

Structure:
Little extracellular matrix surrounding cells; the adipocytes, or fat cells, are so full of lipid that the cytoplasm is pushed to the periphery of the cell

Function:
Packing material, thermal insulator, energy storage, and protection of organs against injury from being bumped or jarred

Location:
Predominantly in subcutaneous areas, mesenteries, renal pelves, around kidneys, attached to the surface of the colon, mammary glands, and in loose connective tissue that penetrates into spaces and crevices

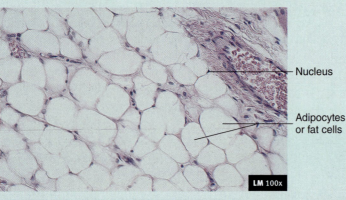

Nucleus

Adipocytes or fat cells

LM 100x

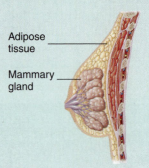

Adipose tissue

Mammary gland

(c) Reticular Tissue AP|R

Structure:
Fine network of reticular fibers irregularly arranged

Function:
Provides a superstructure for lymphatic and hemopoietic tissues

Location:
Within the lymph nodes, spleen, bone marrow

Leukocytes   Reticular fibers

LM 280x

Lymph node

Spleen

TABLE 4.6   Connective Tissue Proper: Dense Connective Tissue

(a) Dense Regular Collagenous Connective Tissue AP|R

Structure:
Matrix composed of collagen fibers running in somewhat the same direction in tendons and ligaments; collagen fibers run in several directions in the dermis of the skin and in organ capsules

Function:
Withstand great pulling forces exerted in the direction of fiber orientation due to great tensile strength and stretch resistance

Location:
Tendons (attach muscle to bone) and ligaments (attach bones to each other); also found in the dermis of the skin, organ capsules, and the outer layer of many blood vessels

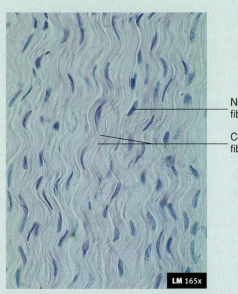

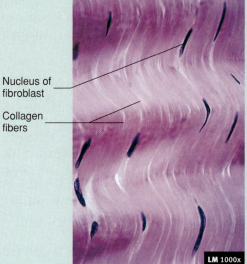

Nucleus of fibroblast

Collagen fibers

LM 165x

LM 1000x

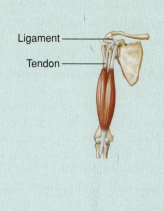

Ligament

Tendon

(b) Dense Regular Elastic Connective Tissue AP|R

Structure:
Matrix composed of collagen fibers and elastin fibers running in somewhat the same direction in elastic ligaments; elastic fibers run in connective tissue of blood vessel walls

Function:
Capable of stretching and recoiling like a rubber band with strength in the direction of fiber orientation

Location:
Elastic ligaments between the vertebrae and along the dorsal aspect of the neck (nucha) and in the vocal cords; also found in elastic connective tissue of blood vessel walls

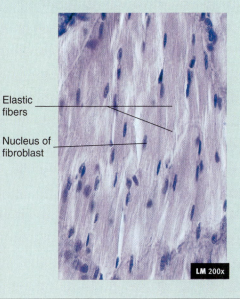

Elastic fibers

Nucleus of fibroblast

LM 100x

LM 200x

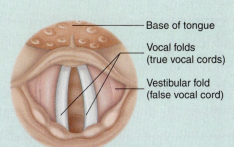

Base of tongue

Vocal folds (true vocal cords)

Vestibular fold (false vocal cord)

Dense elastic connective tissue has abundant elastic fibers among its collagen fibers. The elastic fibers allow the tissue to stretch and recoil. Examples include the dense elastic connective tissue in the vocal cords (table 4.6b), in elastic ligaments, and in the walls of large arteries. The elastic fibers are oriented in the same direction in elastic ligaments and in the vocal cords (dense regular tissue), but they are oriented in many different directions in the walls of arteries (dense irregular tissue; not pictured).

Predict 4

Scars consist of dense connective tissue made of collagen fibers. Vitamin C is required for collagen synthesis. Predict the effect of scurvy, a nutritional disease caused by vitamin C deficiency, on wound healing.

Supporting Connective Tissue

Cartilage

Cartilage (kar′ti-lij) is composed of **chondrocytes** (kon′drō-sītz), or cartilage cells, located in spaces called **lacunae** (lă-koo′nē; small spaces) within an extensive matrix (table 4.7). Collagen in the matrix gives cartilage flexibility and strength. Cartilage is resilient because the proteoglycans of the matrix trap water, which makes the cartilage relatively rigid and enables it to spring back after being compressed. Cartilage provides support, but if bent or slightly compressed, it resumes its original shape. Cartilage heals slowly after an injury because blood vessels do not penetrate it. Thus, the cells and nutrients necessary for tissue repair do not easily reach the damaged area.

There are three types of cartilage:

1. **Hyaline** (hī′ă-lin; clear or glassy) **cartilage** (see table 4.7a) is the most abundant type of cartilage and has many functions. It covers the ends of bones where they come together to form joints. In joints, hyaline cartilage forms smooth, resilient surfaces that can withstand repeated compression. Hyaline cartilage also forms the cartilage rings of the respiratory tract, the nasal cartilages, and the costal cartilages, which attach the ribs to the sternum (breastbone).

2. **Fibrocartilage** (table 4.7b) has more collagen than does hyaline cartilage, and bundles of collagen fibers can be seen in the matrix. In addition to withstanding compression, it is able to resist pulling or tearing forces. Fibrocartilage is found in the disks between the vertebrae (bones of the back) and in some joints, such as the knee and temporomandibular (jaw) joints.

3. **Elastic cartilage** (table 4.7c) contains elastic fibers in addition to collagen and proteoglycans. The elastic fibers appear as coiled fibers among bundles of collagen fibers. Elastic cartilage is able to recoil to its original shape when bent. The external ear, epiglottis, and auditory tube contain elastic cartilage.

Bone

Bone is a hard connective tissue that consists of living cells and a mineralized matrix (table 4.8). Osteocytes (*osteo,* bone), or bone cells, are located within lacunae. The strength and rigidity of the mineralized matrix enables bones to support and protect other tissues and organs. The two types of bone are **compact bone** and **spongy bone** (see chapter 6).

Fluid Connective Tissue

Blood

Blood is unique because the matrix is liquid, enabling blood cells to move through blood vessels (table 4.9). Some blood cells even leave the blood and wander into other tissues. The liquid matrix

TABLE 4.7	Supporting Connective Tissue: Cartilage

(a) Hyaline Cartilage AP|R

Structure:	Function:	Location:
Collagen fibers are small and evenly dispersed in the matrix, making the matrix appear transparent; the chondrocytes are found in spaces, or lacunae, within the firm but flexible matrix	Allows growth of long bones; provides rigidity with some flexibility in the trachea, bronchi, ribs, and nose; forms strong , smooth, yet somewhat flexible articulating surfaces; forms the embryonic skeleton	Growing long bones, cartilage rings of the respiratory system, costal cartilage of ribs, nasal cartilages, articulating surface of bones, and the embryonic skeleton

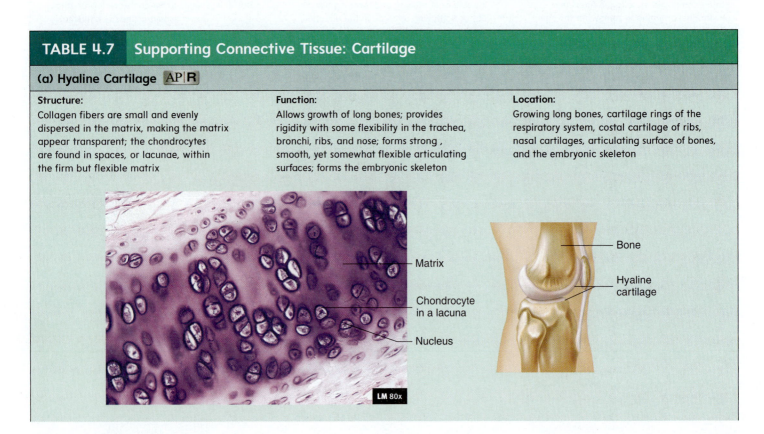

LM 80x

Matrix
Chondrocyte in a lacuna
Nucleus

Bone
Hyaline cartilage

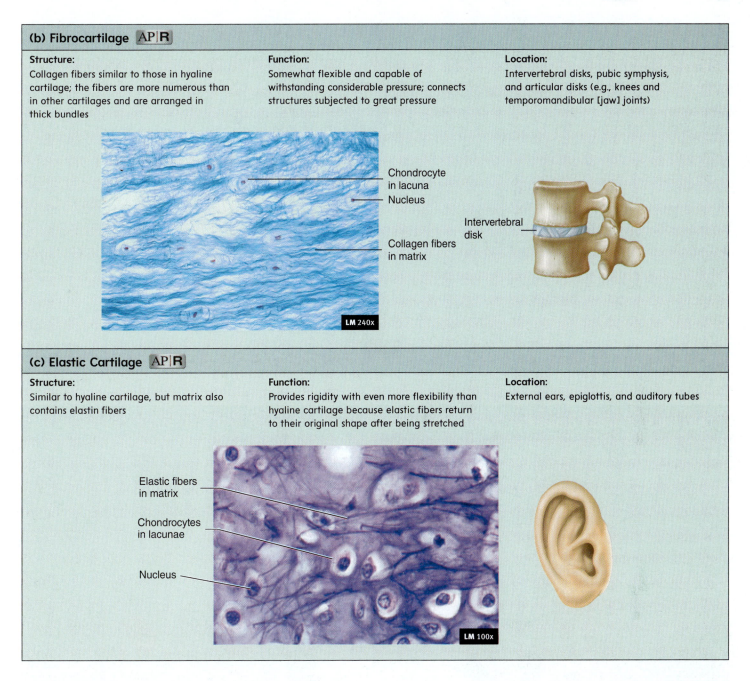

(b) Fibrocartilage AP|R

Structure:
Collagen fibers similar to those in hyaline cartilage; the fibers are more numerous than in other cartilages and are arranged in thick bundles

Function:
Somewhat flexible and capable of withstanding considerable pressure; connects structures subjected to great pressure

Location:
Intervertebral disks, pubic symphysis, and articular disks (e.g., knees and temporomandibular [jaw] joints)

Chondrocyte in lacuna
Nucleus
Collagen fibers in matrix
Intervertebral disk

LM 240x

(c) Elastic Cartilage AP|R

Structure:
Similar to hyaline cartilage, but matrix also contains elastin fibers

Function:
Provides rigidity with even more flexibility than hyaline cartilage because elastic fibers return to their original shape after being stretched

Location:
External ears, epiglottis, and auditory tubes

Elastic fibers in matrix
Chondrocytes in lacunae
Nucleus

LM 100x

enables blood to flow rapidly through the body, carrying nutrients, oxygen, waste products, and other materials. Blood is discussed more fully in chapter 11.

4.4 MUSCLE TISSUE

Learning Outcome *After reading this section, you should be able to*

A. Discuss the three types of muscle tissue by describing their general structures, their locations in the body, and their functions.

The main characteristic of **muscle tissue** is its ability to contract, or shorten, making movement possible. Muscle contraction results from contractile proteins located within the muscle cells (see chapter 7). The length of muscle cells is greater than the diameter. Muscle cells are sometimes called **muscle fibers** because they often resemble tiny threads.

The three types of muscle tissue are skeletal, cardiac, and smooth. **Skeletal muscle** is what we normally think of as "muscle" (table 4.10*a*). It is the meat of animals and constitutes about 40% of a person's body weight. As the name implies, skeletal muscle attaches to the skeleton and enables the body to move. Skeletal muscle is described as voluntary (under conscious control) because a person can purposefully cause skeletal muscle contraction to achieve specific body movements. However, the nervous system can cause skeletal muscles to contract without conscious involvement, as occurs during reflex movements and the maintenance of muscle tone. Skeletal muscle cells tend to be long and cylindrical, with several nuclei per cell. Some skeletal muscle cells extend the length of an entire muscle. Skeletal muscle

TABLE 4.8 Supporting Connective Tissue: Bone AP|R

Structure:
Hard, bony matrix predominates; many osteocytes (not seen in this bone preparation) are located within lacunae; the matrix is organized into layers called lamellae

Function:
Provides great strength and support and protects internal organs, such as the brain; bone also provides attachment sites for muscles and ligaments; the joints of bones allow movements

Location:
All bones of the body

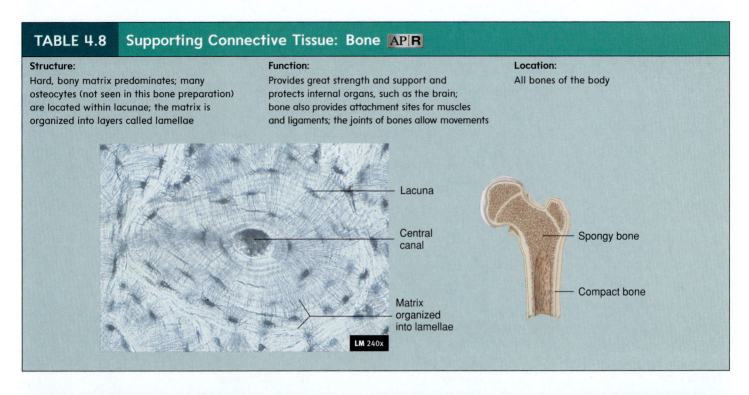

Lacuna

Central canal

Matrix organized into lamellae

LM 240x

Spongy bone

Compact bone

TABLE 4.9 Fluid Connective Tissue: Blood AP|R

Structure:
Blood cells and a fluid matrix

Function:
Transports oxygen, carbon dioxide, hormones, nutrients, waste products, and other substances; protects the body from infections and is involved in temperature regulation

Location:
Within the blood vessels; white blood cells frequently leave the blood vessels and enter the interstitial spaces

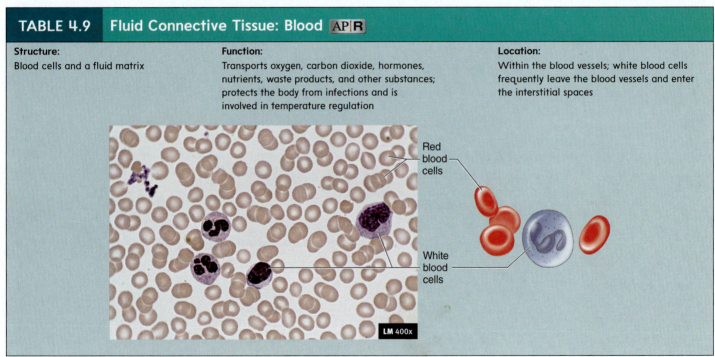

Red blood cells

White blood cells

LM 400x

cells are **striated** (strī′ăt-ed), or banded, because of the arrangement of contractile proteins within the cells (see chapter 7).

Cardiac muscle is the muscle of the heart; it is responsible for pumping blood (table 4.10b). Cardiac muscle is under involuntary (unconscious) control, although a person can learn to influence the heart rate by using techniques such as meditation and biofeedback. Cardiac muscle cells are cylindrical but much shorter than skeletal muscle cells. Cardiac muscle cells are striated and usually have one nucleus per cell. They are often branched and connected to one another by **intercalated** (in-ter′kă-lā-ted)

disks. The intercalated disks, which contain specialized gap junctions, are important in coordinating the contractions of the cardiac muscle cells (see chapter 12).

Smooth muscle forms the walls of hollow organs (except the heart); it is also found in the skin and the eyes (table 4.10c). Smooth muscle is responsible for a number of functions, such as moving food through the digestive tract and emptying the urinary bladder. Like cardiac muscle, smooth muscle is controlled involuntarily. Smooth muscle cells are tapered at each end, have a single nucleus, and are not striated.

TABLE 4.10   Muscle Tissue

(a) Skeletal Muscle AP|R

Structure:
Skeletal muscle cells or fibers appear striated (banded); cells are large, long, and cylindrical, with many nuclei

Function:
Movement of the body; under voluntary control

Location:
Attached to bone or other connective tissue

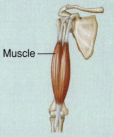

Muscle

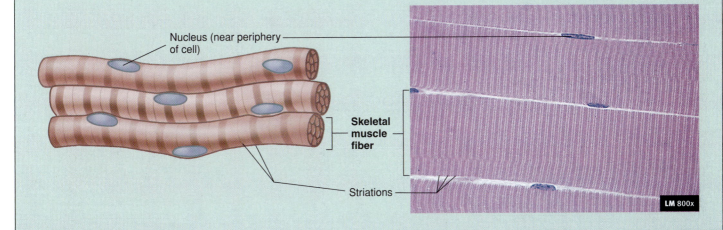

Nucleus (near periphery of cell)

Skeletal muscle fiber

Striations

LM 800x

(b) Cardiac Muscle AP|R

Structure:
Cardiac muscle cells are cylindrical and striated and have a single nucleus; they are branched and connected to one another by intercalated disks, which contain gap junctions

Function:
Pumps the blood; under involuntary (unconscious) control

Location:
In the heart

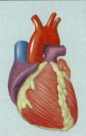

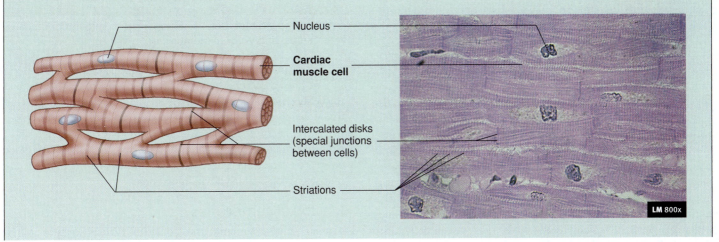

Nucleus

Cardiac muscle cell

Intercalated disks (special junctions between cells)

Striations

LM 800x

TABLE 4.10 *continued*

(c) Smooth Muscle AP|R

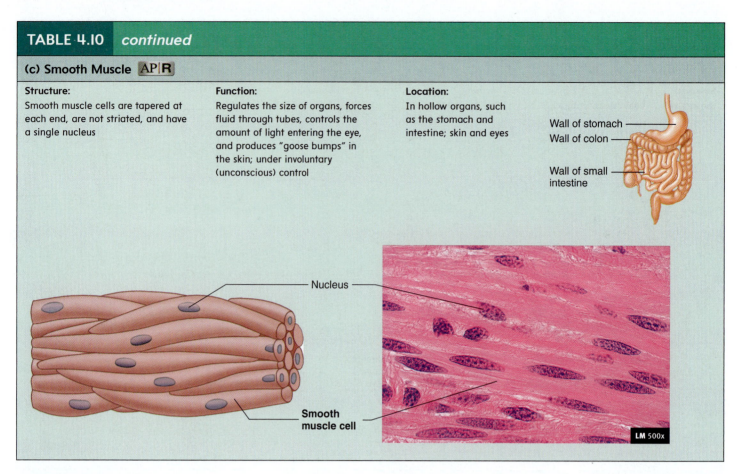

Structure:
Smooth muscle cells are tapered at each end, are not striated, and have a single nucleus

Function:
Regulates the size of organs, forces fluid through tubes, controls the amount of light entering the eye, and produces "goose bumps" in the skin; under involuntary (unconscious) control

Location:
In hollow organs, such as the stomach and intestine; skin and eyes

Wall of stomach
Wall of colon
Wall of small intestine

Nucleus

Smooth muscle cell

LM 500x

Predict 5

Make a table that summarizes the characteristics of the three major muscle types. The muscle types should form a column at the left side of the table, and the characteristics of muscle should form a row at the top of the table.

4.5 NERVOUS TISSUE

Learning Outcome *After reading this section, you should be able to*

A. Describe the functions of nervous tissue and the structure of a neuron.

Nervous tissue forms the brain, spinal cord, and nerves. It is responsible for coordinating and controlling many body activities. For example, the conscious control of skeletal muscles and the unconscious regulation of cardiac muscle are accomplished by nervous tissue. Awareness of ourselves and the external environment, emotions, reasoning skills, and memory are other functions performed by nervous tissue. Many of these functions depend on the ability of nervous tissue cells to communicate with one another and with the cells of other tissues by means of electrical signals called **action potentials.**

Nervous tissue consists of neurons and support cells. The **neuron** (noor′on), or *nerve cell,* is responsible for conducting action potentials. It is composed of three parts: a cell body, dendrites, and an axon (table 4.11). The **cell body** contains the nucleus and is the site of general cell functions. **Dendrites** (den′drītz; relating to a tree) and **axons** (ak′sonz) are nerve cell processes (extensions). Dendrites usually receive stimuli leading to electrical

changes that either increase or decrease action potentials in the neuron's axon. Action potentials usually originate at the base of an axon where it joins the cell body and travel to the end of the axon. **Neuroglia** (noo-rog′lē-ă; *glia,* glue) are the support cells of the nervous system; they nourish, protect, and insulate the neurons. We consider nervous tissue in greater detail in chapter 8.

4.6 TISSUE MEMBRANES

Learning Outcome *After reading this section, you should be able to*

A. List the structural and functional characteristics of mucous, serous, and synovial membranes.

A **membrane** is a thin sheet or layer of tissue that covers a structure or lines a cavity. Most membranes consist of epithelium and the connective tissue on which the epithelium rests. There are four tissue membranes in the body. The **skin,** or *cutaneous* (kū-tā′nē-ŭs, skin) *membrane,* is the external membrane. It is composed of stratified squamous epithelium and dense connective tissue (see chapter 5). The three major categories of internal membranes are mucous, serous, and synovial membranes.

Mucous Membranes

Mucous (mū′kŭs) **membranes** consist of various kinds of epithelium resting on a thick layer of loose connective tissue. They line cavities that open to the outside of the body, such as the digestive, respiratory, and reproductive tracts (figure 4.5a). Many, but not all, mucous membranes have mucous glands, which secrete

TABLE 4.11   Nervous Tissue AP|R

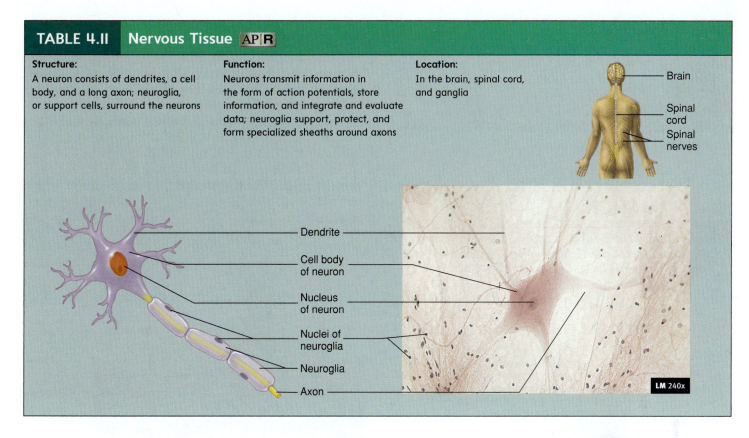

Structure:
A neuron consists of dendrites, a cell body, and a long axon; neuroglia, or support cells, surround the neurons

Function:
Neurons transmit information in the form of action potentials, store information, and integrate and evaluate data; neuroglia support, protect, and form specialized sheaths around axons

Location:
In the brain, spinal cord, and ganglia

Brain
Spinal cord
Spinal nerves

Dendrite
Cell body of neuron
Nucleus of neuron
Nuclei of neuroglia
Neuroglia
Axon

LM 240x

mucus. The functions of mucous membranes vary, depending on their location, but they include protection, absorption, and secretion. For example, the stratified squamous epithelium of the oral cavity (mouth) performs a protective function, whereas the simple columnar epithelium of the intestine absorbs nutrients and secretes digestive enzymes and mucus. Mucous membranes also line the nasal passages. When it becomes inflamed, we experience the "runny nose" characteristic of the common cold or an allergy.

Serous Membranes

Serous (sēr′ŭs; producing watery secretion) **membranes** consist of simple squamous epithelium resting on a delicate layer of loose connective tissue. Serous membranes line the trunk cavities and cover the organs within these cavities (figure 4.5*b*). Serous membranes secrete serous fluid, which covers the surface of the membranes. The smooth surface of the epithelial cells of the serous membranes combined with the lubricating qualities of the serous fluid prevent damage from abrasion when organs in the thoracic or abdominopelvic cavities rub against one another. Serous membranes are named according to their location: The **pleural** (ploor′ăl; a rib or the side) **membranes** are associated with the lungs; the **pericardial** (per-i-kar′dē-ăl; around the heart) **membranes** are associated with the heart; and the **peritoneal** (per′i-tō-nē′ăl; to stretch over) **membranes** are located in the abdominopelvic cavity (see figure 1.15). When the suffix -*itis* is added to the name of a structure, it means that the structure is inflamed (however, not all cases use the -*itis* suffix). Thus, **pericarditis** (per′i-kar-dī′tis) and **peritonitis** (per′i-tō-nī′tis) refer to inflammation of the pericardial membranes and peritoneal membranes, respectively. **Pleurisy** (ploor′i-sē) is inflammation of the pleural membranes.

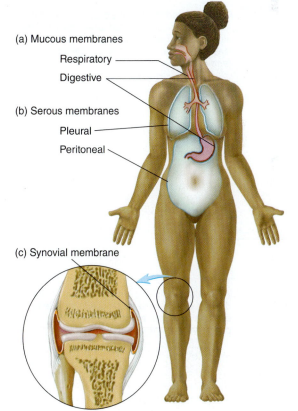

(a) Mucous membranes
Respiratory
Digestive

(b) Serous membranes
Pleural
Peritoneal

(c) Synovial membrane

Figure 4.5 Internal Membranes

(*a*) Mucous membranes line cavities that open to the outside and often contain mucous glands, which secrete mucus. (*b*) Serous membranes line cavities that do not open to the exterior, do not contain mucous glands, but do secrete serous fluid. (*c*) Synovial membranes line the cavities that surround synovial joints.

Synovial Membranes

Synovial (si-nō′vē-ăl) **membranes** are made up of only connective tissue. They line the inside of joint cavities (the space where bones come together within a movable joint) (figure 4.5c). Synovial membranes produce **synovial fluid,** which makes the joint very slippery, thereby reducing friction and allowing smooth movement within the joint. Synovial and other connective tissue membranes are discussed in chapter 6.

4.7 TISSUE DAMAGE AND INFLAMMATION

Learning Outcome *After reading this section, you should be able to*

A. Describe the process of inflammation in response to tissue damage, and explain how inflammation protects the body.

Inflammation (*flamma*, flame) occurs when tissues are damaged. For example, when viruses infect epithelial cells of the upper respiratory tract, inflammation and the symptoms of the common cold are produced. Inflammation can also result from the immediate and painful events that follow trauma, such as closing your finger in a car door or cutting yourself with a knife. Figure 4.6 illustrates the stages of the inflammatory response. Inflammation mobilizes the body's defenses and isolates and destroys microorganisms, foreign materials, and damaged cells so that tissue repair can proceed (see chapter 14). Inflammation produces five major symptoms: redness, heat, swelling, pain, and disturbance of function. Although unpleasant, the processes producing the symptoms are usually beneficial.

1. A splinter in the skin causes damage and introduces bacteria. Chemical mediators of inflammation are released or activated in injured tissues and adjacent blood vessels. Some blood vessels rupture, causing bleeding.

2. Chemical mediators cause capillaries to dilate and the skin to become red. Chemical mediators also increase capillary permeability, and fluid leaves the capillaries, producing swelling (*arrows*).

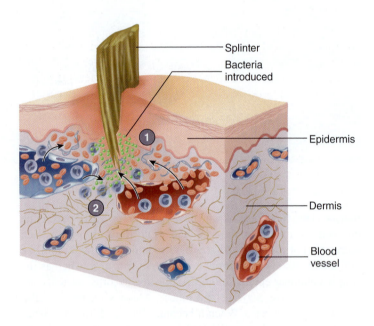

3. White blood cells (e.g., neutrophils) leave the dilated blood vessels and move to the site of bacterial infection, where they begin to phagocytize bacteria and other debris.

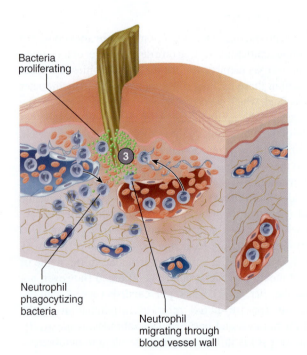

PROCESS Figure 4.6 Inflammation

Following an injury, substances called **chemical mediators** are released or activated in the injured tissues and adjacent blood vessels. The mediators include **histamine** (his′tă-mēn) and **prostaglandins** (pros-tă-glan′dinz). Some mediators cause dilation of blood vessels, which produces redness and heat, similar to what occurs when a person blushes. Dilation of blood vessels is beneficial because it increases the speed with which blood cells and other substances important for fighting infections and repairing the injury are brought to the injury site.

Chemical mediators of inflammation also increase the permeability of blood vessels, allowing materials and blood cells to move out of the vessels and into the tissue, where they can deal directly with the injury. **Edema** (e-dē′mă), or swelling, of the tissues results when water, proteins, and other substances from the blood move into the tissues. One of the proteins, fibrin, forms a fibrous network that "walls off" the site of injury from the rest of the body. This mechanism can help prevent the spread of infectious agents. One type of blood cell that enters the tissues is the **neutrophil** (noo′trō-fil), a phagocytic white blood cell that fights infections by ingesting bacteria. Neutrophils die after ingesting a small number of bacteria; the mixture of dead neutrophils, other cells, and fluid that can accumulate is called **pus.**

Pain associated with inflammation is produced in several ways. Nerve cell endings are stimulated by direct damage and by some chemical mediators to produce pain sensations. In addition, the increased pressure in the tissue caused by edema and pus accumulation can cause pain.

Pain, limitation of movement resulting from edema, and tissue destruction all contribute to the disturbance of function, which can be adaptive because it warns the person to protect the injured area from further damage.

Chronic Inflammation

Chronic, or prolonged, inflammation results when the agent responsible for an injury is not removed or something else interferes with the healing process. Infections of the lungs or kidneys usually result in a brief period of inflammation followed by repair. However, prolonged infections, or prolonged exposure to irritants, can result in chronic inflammation. Chronic inflammation caused by irritants, such as silica in the lungs, or abnormal immune responses can result in the replacement of normal tissue by fibrous connective tissue. Chronic inflammation of the stomach or small intestine may cause ulcers. The loss of normal tissue leads to the loss of normal organ function. Consequently, chronic inflammation of organs, such as the lungs, liver, or kidneys, can lead to death.

When the inflammatory response lasts longer or is more intense than is desirable, drugs are sometimes used to suppress the symptoms by inhibiting the synthesis, release, or actions of the chemical mediators of inflammation. For example, medications called antihistamines suppress the effects of histamine released in people with hay fever. Aspirin and related drugs, such as ibuprofen and naproxen, are effective anti-inflammatory agents that relieve pain by preventing the synthesis of prostaglandins and related substances.

Predict 6

In some injuries, tissues are so severely damaged that cells die and blood vessels are destroyed. For injuries such as these, where do the signs of inflammation, such as redness, heat, edema, and pain, occur?

4.8 TISSUE REPAIR

Learning Outcome *After reading this section, you should be able to*

A. Explain the major events involved in tissue repair.

Tissue repair is the substitution of viable cells for dead cells. Tissue repair can occur by regeneration or by fibrosis. In **regeneration,** the new cells are the same type as those that were destroyed, and normal function is usually restored. In **fibrosis,** or replacement, a new type of tissue develops that eventually causes scar production and the loss of some tissue function. The tissues involved and the severity of the wound determine the type of tissue repair that dominates.

Regeneration can completely repair some tissues, such as the skin and the mucous membrane of the intestine. In these cases, regeneration is accomplished primarily by stem cells. **Stem cells** are self-renewing, undifferentiated cells that continue to divide throughout life. With each division, there is a daughter stem cell and a second cell that can undergo differentiation. The differentiated cells are the same cell types as the dead cells. Regeneration can also involve division of differentiated cells in connective tissue and glands, such as the liver and pancreas. These cells do not normally divide, but retain the ability to divide after an injury.

Fibrosis is the predominant repair mechanism in some tissues. In the adult brain, heart, and skeletal muscle there are relatively few stem cells and the mature neurons, cardiac muscle, and skeletal muscle do not divide. While these cells cannot divide, they can recover from a limited amount of damage. For example, if the axon of a neuron is damaged, the neuron can grow a new axon, but it will die if the cell body is sufficiently damaged. If these cells are killed, they are often replaced by connective tissue.

Although neurons cannot form additional neurons, a small population of stem cells has been found in the adult brain. These stem cells can divide and form new neurons. It may be possible to develop treatments for some brain injuries that stimulate the stem cells. Researchers have identified a class of chemicals, called growth factors, that stimulate stem cells to divide and make injured neurons recover more rapidly. The new neurons may be incorporated with other functional neurons of the central nervous system.

In addition to the type of cells involved, the severity of an injury can influence whether repair is by regeneration or fibrosis. Generally, the more severe the injury, the greater the likelihood that repair involves fibrosis.

Repair of the skin is an illustration of tissue repair (figure 4.7). When the edges of a wound are close together, the wound fills with blood, and a clot forms (see chapter 11). The **clot** contains the threadlike protein fibrin, which binds the edges of the wound together and stops the bleeding. The surface of the clot dries to form a **scab,** which seals the wound and helps prevent infection.

An inflammatory response is activated to fight infectious agents in the wound and to help the repair process (see figure 4.6). Dilation of blood vessels brings blood cells and other substances to the injury area, and increased blood vessel permeability allows them to enter the tissue. The area is "walled off" by the fibrin, and neutrophils enter the tissue from the blood.

The epithelium at the edge of the wound undergoes regeneration and migrates under the scab while the inflammatory response

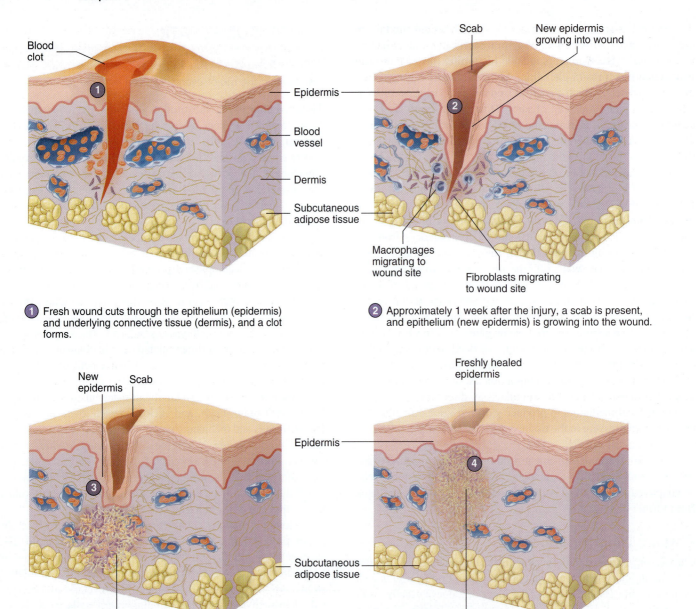

Blood clot

Epidermis

Blood vessel

Dermis

Subcutaneous adipose tissue

Scab   New epidermis growing into wound

Macrophages migrating to wound site

Fibroblasts migrating to wound site

① Fresh wound cuts through the epithelium (epidermis) and underlying connective tissue (dermis), and a clot forms.

② Approximately 1 week after the injury, a scab is present, and epithelium (new epidermis) is growing into the wound.

New epidermis   Scab

Granulation tissue (fibroblasts proliferating)

Freshly healed epidermis

Epidermis

Subcutaneous adipose tissue

Granulation tissue being replaced with new connective tissue

③ Approximately 2 weeks after the injury, the epithelium has grown completely into the wound, and fibroblasts have formed granulation tissue.

④ Approximately 1 month after the injury, the wound has completely closed, the scab has been sloughed, and the granulation tissue is being replaced by new connective tissue.

PROCESS Figure 4.7 Tissue Repair

proceeds. Eventually, the epithelial cells from the edges meet, and the epithelium is restored. After the epithelium is repaired, the scab is sloughed off (shed).

A second type of phagocytic cell, called a macrophage, removes the dead neutrophils, cellular debris, and the decomposing clot. Fibroblasts from the surrounding connective tissue migrate into the area, producing collagen and other extracellular matrix components. Capillaries grow from blood vessels at the edge of the wound and revascularize the area. The result is fibrosis, during which the clot is replaced with **granulation tissue,** a

delicate, granular-appearing connective tissue that consists of fibroblasts, collagen, and capillaries. Eventually, normal connective tissue replaces the granulation tissue. Sometimes a large amount of granulation tissue persists as a scar, which at first is bright red because numerous blood vessels are present. The scar turns from red to white as collagen accumulates and the blood vessels decrease in number.

When the wound edges are far apart, the clot may not completely close the gap, and it takes much longer for the epithelial cells to regenerate and cover the wound. With increased tissue damage, the degree of inflammation is greater, there is more cell debris for the phagocytes to remove, and the risk of infection increases. Much more granulation tissue forms, and **wound contracture,** a result of the contraction of fibroblasts in the granulation tissue, pulls the edges of the wound closer together. Although wound contracture reduces the size of the wound and speeds healing, it can lead to disfiguring and debilitating scars.

Predict 7

Explain why it is advisable to suture large wounds.

4.9 EFFECTS OF AGING ON TISSUES

Learning Outcome *After reading this section, you should be able to*

A. Describe the age-related changes that occur in cells and in extracellular matrix.

The consequences of some age-related changes are obvious, whereas others are subtle. For example, the appearance of skin changes as people age, and athletic performance declines, after approximately 30 to 35 years. With advanced age, the number of neurons and muscle cells decreases substantially. Reduced visual acuity, smell, taste, and touch occur, and the functional capacities of the respiratory and cardiovascular systems decline.

At the tissue level, age-related changes affect cells and the extracellular matrix produced by them. In general, cells divide more slowly in older than in younger people. The rate of red blood cell synthesis declines in the elderly. Injuries in the very young heal more rapidly and more completely than in older people, in part, because of the more rapid cell division. For example, a fracture in an infant's femur is likely to heal quickly and eventually leave no evidence in the bone. A similar fracture in an adult heals more slowly, and a scar, seen in radiographs of the bone, is likely to persist throughout life.

The consequences of changes in the extracellular matrix are important. Collagen fibers become more irregular in structure, even though they may increase in number. As a consequence, connective tissues with abundant collagen, such as tendons and ligaments, become less flexible and more fragile. Elastic fibers fragment, bind to Ca^{2+}, and become less elastic, causing elastic connective tissues to become less elastic. The reduced flexibility and elasticity of connective tissue is responsible for increased wrinkling of the skin, as well as an increased tendency for older people's bones to break.

The walls of arteries become less elastic because of changes in collagen and elastic fibers. Atherosclerosis results as plaques form in the walls of blood vessels, which contain collagen fibers, lipids, and calcium deposits. These changes result in reduced blood supply to tissues and increased susceptibility to blockage and rupture of arteries.

However, it is increasingly evident that many of the cell losses and functional declines can be slowed by physical and mental exercises.

ANSWER TO LEARN TO PREDICT

The question tells us that gluten enteropathy affects the intestinal lining, reducing its ability to absorb nutrients and water. It also reminds us that nutrient and water absorption occurs at the cellular level via several different transport processes.

Let us first identify the tissue type affected by gluten enteropathy. In this chapter we learned that epithelial tissue covers body surfaces, including the lining of the intestines. Further reading showed that the intestinal lining is composed of simple columnar epithelial tissue. Therefore, the tissue type affected by Matt's gluten enteropathy is simple columnar epithelium.

We are then asked to identify the specific cell parts affected by this disease. As stated in the question, the intestinal lining is organized into fingerlike projections called villi, which are covered by the simple columnar epithelium. We also learned that the epithelial cells of this tissue have microvilli. In chapter 3, we learned that microvilli are extensions of the plasma membrane that increase the surface area for absorption. Matt's gluten enteropathy reduced his ability to absorb nutrients and water, so we can conclude that the cell parts affected by the disease are the microvilli.

Finally, the question asks us to explain why Matt suffers from bouts of diarrhea after eating gluten. We know that gluten damages the intestinal lining by decreasing the number of villi and microvilli. This reduces the surface area for absorption. If the surface area decreases, fewer nutrients are absorbed. Chapter 3 showed us that water moves by osmosis to areas of higher solute concentration. The nutrient molecules are solutes in the intestines. Since the solutes are not being absorbed, the solute concentration remains high in the intestines, and water absorption decreases. As a result, the nutrients and water accumulate in the intestines, resulting in the watery feces of diarrhea.

Answers to the rest of this chapter's Predict questions are in Appendix E.

⎯⎻⎺⎻⎯ SUMMARY

4.1 Tissues and Histology (p. 70)

1. A tissue is a group of cells with similar structure and function, along with the extracellular substances located between the cells.
2. Histology is the study of tissues.

4.2 Epithelial Tissue (p. 70)

Epithelial tissue covers surfaces; it usually has a basement membrane, little extracellular material, and no blood vessels.

Functions of Epithelia

General functions of epithelia include protecting underlying structures, acting as a barrier, permitting the passage of substances, secreting substances, and absorbing substances.

Classification of Epithelia

1. Epithelia are classified according to the number of cell layers and the shape of the cells.
2. Simple epithelium has one layer of cells, whereas stratified epithelium has more than one.
3. Pseudostratified columnar epithelium is simple epithelium that appears to have two or more cell layers.
4. Transitional epithelium is stratified epithelium that can be greatly stretched.

Structural and Functional Relationships

1. Simple epithelium is involved with diffusion, secretion, or absorption. Stratified epithelium serves a protective role. Squamous cells function in diffusion or filtration. Cuboidal or columnar cells, which contain more organelles, secrete or absorb.
2. A smooth, free surface reduces friction. Microvilli increase surface area, and cilia move materials over the cell surface.
3. Tight junctions bind adjacent cells together and form a permeability barrier.
4. Desmosomes mechanically bind cells together, and hemidesmosomes mechanically bind cells to the basement membrane.
5. Gap junctions allow intercellular communication.

Glands

1. A gland is a single cell or a multicellular structure that secretes.
2. Exocrine glands have ducts, and endocrine glands do not.

4.3 Connective Tissue   (p. 77)

1. Connective tissue has an extracellular matrix consisting of protein fibers, ground substance, and fluid.
2. Collagen fibers are flexible but resist stretching; reticular fibers form a fiber network; and elastic fibers recoil.
3. Connective tissue cells that are blast cells form the matrix, cyte cells maintain it, and clast cells break it down.

Functions of Connective Tissue

Connective tissues enclose and separate other tissues, connect tissues to one another, help support and move body parts, store compounds, cushion and insulate the body, transport substances, and protect against toxins and injury.

Classification of Connective Tissue

1. Areolar connective tissue is the "packing material" of the body; it fills the spaces between organs and holds them in place.
2. Adipose tissue, or fat, stores energy. Adipose tissue also pads and protects parts of the body and acts as a thermal insulator.
3. Dense connective tissue has a matrix consisting of either densely packed collagen fibers (in tendons, ligaments, and the dermis of the skin) or densely packed elastic fibers (in elastic ligaments and the walls of arteries).
4. Reticular tissue forms a framework for lymphatic structures.

5. Cartilage provides support and is found in structures such as the disks between the vertebrae, the external ear, and the costal cartilages.
6. Bone has a mineralized matrix and forms most of the skeleton of the body.
7. Blood has a liquid matrix and is found in blood vessels.

4.4 Muscle Tissue   (p. 83)

1. Muscle tissue is specialized to shorten, or contract.
2. The three types of muscle tissue are skeletal, cardiac, and smooth muscle.

4.5 Nervous Tissue   (p. 86)

1. Nervous tissue is specialized to conduct action potentials (electrical signals).
2. Neurons conduct action potentials, and neuroglia support the neurons.

4.6 Tissue Membranes   (p. 86)

Mucous Membranes

Mucous membranes line cavities that open to the outside of the body (digestive, respiratory, and reproductive tracts). They contain glands and secrete mucus.

Serous Membranes

Serous membranes line trunk cavities that do not open to the outside of the body (pleural, pericardial, and peritoneal cavities). They do not contain mucous glands but do secrete serous fluid.

Synovial Membranes

Synovial membranes line joint cavities and secrete a lubricating fluid.

4.7 Tissue Damage and Inflammation   (p. 88)

1. Inflammation isolates and destroys harmful agents.
2. Inflammation produces redness, heat, swelling, pain, and disturbance of function.

Chronic Inflammation

Chronic inflammation results when the agent causing injury is not removed or something else interferes with the healing process.

4.8 Tissue Repair   (p. 89)

1. Tissue repair is the substitution of viable cells for dead cells by regeneration or fibrosis. In regeneration, stem cells, which can divide throughout life, and other dividing cells regenerate new cells of the same type as those that were destroyed. In fibrosis, the destroyed cells are replaced by different cell types, which causes scar formation.
2. Tissue repair involves clot formation, inflammation, the formation of granulation tissue, and the regeneration or fibrosis of tissues. In severe wounds, wound contracture can occur.

4.9 Effects of Aging on Tissues   (p. 91)

1. Cells divide more slowly as people age. Injuries heal more slowly.
2. Extracellular matrix containing collagen and elastic fibers becomes less flexible and less elastic. Consequently, skin wrinkles, elasticity in arteries is reduced, and bones break more easily.

REVIEW AND COMPREHENSION

1. Define tissue and histology.

2. In what areas of the body is epithelium located? What are four characteristics of epithelial tissue?

3. Explain how epithelial tissue is classified according to the number of cell layers and the shape of the cells. What are pseudostratified columnar and transitional epithelium?

4. What kinds of functions does a single layer of epithelium perform? A stratified layer? Give an example of each.

5. Contrast the functions performed by squamous cells with those of cuboidal or columnar cells. Give an example of each.

6. What is the function of an epithelial free surface that is smooth? of one that has microvilli? of one that has cilia?

7. Name the ways in which epithelial cells may be connected to one another, and give the function for each way.

8. Define gland. Distinguish between an exocrine and an endocrine gland.

9. Explain the differences among connective tissue cells that are termed blast, cyte, and clast cells.

10. What are the functions of connective tissues?

11. What are the major connective tissue types? How are they used to classify connective tissue?

12. Describe areolar connective tissue, and give an example.

13. How is adipose tissue different from other connective tissues? List the functions of adipose tissue.

14. Describe dense collagenous connective tissue and dense elastic connective tissue, and give two examples of each.

15. Describe the components of cartilage. Give an example of hyaline cartilage, fibrocartilage, and elastic cartilage.

16. Describe the components of bone.

17. Describe the cell types and matrix of blood, and list the functions of blood.

18. Functionally, what is unique about muscle? Which of the muscle types is under voluntary control? What tasks does each type perform?

19. Functionally, what is unique about nervous tissue? What do neurons and neuroglia accomplish? What is the difference between an axon and a dendrite?

20. Compare mucous and serous membranes according to the type of cavity they line and their secretions. Name the serous membranes associated with the lungs, heart, and abdominopelvic organs.

21. What is the function of the inflammatory response? Name the five symptoms of inflammation, and explain how each is produced.

22. Define tissue repair. What is the difference between regeneration and fibrosis?

23. Describe the process of tissue repair when the edges of a wound are close together versus when they are far apart.

24. Describe the effect of aging on cell division and the formation of extracellular matrix.

CRITICAL THINKING

1. What types of epithelium are likely to be found lining the trachea of a heavy smoker? Predict the changes that are likely to occur after he or she has stopped smoking for 1 or 2 years.

2. The blood-brain barrier is a specialized epithelium in capillaries that prevents many materials from passing from the blood into the brain. What kind of cell connections would you expect to find in the blood-brain barrier?

3. One of the functions of the pancreas is to secrete digestive enzymes that are carried by ducts to the small intestine. How many cell layers and what cell shape, cell surface, and type of cell-to-cell connections would you expect to be present in the epithelium that produces the digestive enzymes?

4. Explain the consequences
   a. if simple columnar epithelium replaced the nonkeratinized stratified squamous epithelium that lines the mouth
   b. if tendons were composed of dense elastic connective tissue instead of dense collagenous connective tissue
   c. if bones were made entirely of elastic cartilage

5. Some dense connective tissue has elastic fibers in addition to collagen fibers. This enables a structure to stretch and then recoil to its original shape. Examples are certain ligaments that hold together the vertebrae (bones of the back). When the back is bent (flexed), the ligaments are stretched. How does the elastic nature of these ligaments help the back function? How are the fibers in the ligaments organized?

6. The aorta is a large blood vessel attached to the heart. When the heart beats, blood is ejected into the aorta, which expands to accept the blood. The wall of the aorta is constructed with dense connective tissue that has elastic fibers. How do you think the fibers are arranged?

7. Antihistamines block the effect of a chemical mediator, histamine, which is released during the inflammatory response. When could taking an antihistamine be harmful, and when could it be beneficial?

8. Granulation tissue and scars consist of dense collagenous connective tissue. Vitamin C is required for collagen synthesis. Predict the effect of scurvy, which is a nutritional disease caused by vitamin C deficiency, on wound healing.

Answers in Appendix D

CHAPTER

5 Integumentary System

LEARN to PREDICT

Mia, Christine, and Landon are so excited about spending the day together. Their mothers are happy to see the children playing outside and getting some exercise. They spend the whole day running through the waves and playing in the sun. At the end of the day, Mia and Landon have tanned a bit, but poor Christine is sunburned, a condition that affects the integumentary system. By applying the information you learn from this chapter, explain what natural and artificial factors may have influenced the children's different reactions to the day of sun exposure.

5.1 FUNCTIONS OF THE INTEGUMENTARY SYSTEM

Learning Outcome *After reading this section, you should be able to*

A. Describe the general functions of the integumentary system.

The **integumentary** (in-teg-ū-men′tă-rē) **system** consists of the skin and accessory structures, such as hair, glands, and nails. Integument means covering, and the integumentary system is one of the more familiar systems of the body to everyone because it covers the outside of the body and is easily observed. We are also familiar with this system because we are concerned with the appearance of the integumentary system. Skin without blemishes is considered attractive, whereas acne is a source of embarrassment for many teenagers. The development of wrinkles and the graying or loss of hair are signs of aging that some people find unattractive. Because of these feelings, we invest much time, effort, and money on changing the appearance of the integumentary system. Many of us apply lotion to our skin, color our hair, and trim our nails. We also try to prevent sweating by using antiperspirants and to reduce or mask body odor by washing and by using deodorants and perfumes.

The appearance of the integumentary system can indicate physiological imbalances in the body. Some disorders, such as acne or warts, affect just the integumentary system. Other disorders affect different parts of the body but are reflected in the integumentary system, providing useful signs for diagnosis. For example, reduced blood flow through the skin during a heart

Module 4 Integumentary System

attack can cause a person to look pale, whereas increased blood flow as a result of fever can cause a flushed appearance. Also, some diseases cause skin rashes, such as those characteristic of measles, chicken pox, and allergic reactions. In addition, the integumentary system and the other systems often interact in complex ways in both healthy and diseased states (see Systems Pathology, "Burns," later in this chapter).

Although we are often concerned with how the integumentary system looks, it has many important functions that go beyond appearance. Major functions of the integumentary system include

1. *Protection.* The skin provides protection against abrasion and ultraviolet light. It also prevents microorganisms from entering the body and reduces water loss, thus preventing dehydration.

2. *Sensation.* The integumentary system has sensory receptors that can detect heat, cold, touch, pressure, and pain.

3. *Vitamin D production.* When exposed to ultraviolet light, the skin produces a molecule that can be transformed into vitamin D, an important regulator of calcium homeostasis.

4. *Temperature regulation.* The amount of blood flow beneath the skin's surface and the activity of sweat glands in the skin both help regulate body temperature.

5. *Excretion.* Small amounts of waste products are lost through the skin and in gland secretions.

5.2 SKIN

Learning Outcomes   *After reading this section, you should be able to*

A. Describe the structure and function of the epidermis.
B. Describe the epidermal strata, and relate them to the process of keratinization.
C. Describe the structure and discuss the function of the dermis.
D. Explain how melanin, blood, carotene, and collagen affect skin color.

The skin is made up of two major tissue layers: the epidermis and the dermis. The **epidermis** (ep-i-der′mis; upon the dermis) is the most superficial layer of skin. It is a layer of epithelial tissue that rests on the **dermis** (der′mis), a layer of dense connective tissue (figure 5.1). The thickness of the epidermis and dermis varies, depending on location, but on average the dermis is 10 to 20 times thicker than the epidermis. The epidermis prevents water loss and resists abrasion. The dermis is responsible for most of the skin's structural strength. The skin rests on the **subcutaneous** (sŭb-koo-tā′nē-ŭs; under the skin) **tissue,** which is a layer of connective tissue. The subcutaneous tissue is not part of the skin, but it does connect the skin to underlying muscle or bone. To give an analogy, if the subcutaneous tissue is the foundation on which a house rests, the dermis forms most of the house and the epidermis is its roof.

Epidermis

The epidermis is stratified squamous epithelium; in its deepest layers, new cells are produced by mitosis. As new cells form, they push older cells to the surface, where they slough, or flake off. The outermost cells protect the cells underneath, and the deeper, replicating cells replace cells lost from the surface. During their movement, the cells change shape and chemical composition. This process is called **keratinization** (ker′ă-tin-i-zā′shŭn) because the cells become filled with the protein **keratin** (ker′ă-tin), which makes them hard. As keratinization proceeds, epithelial cells eventually die and produce an outer layer of dead, hard cells that resists abrasion and forms a permeability barrier.

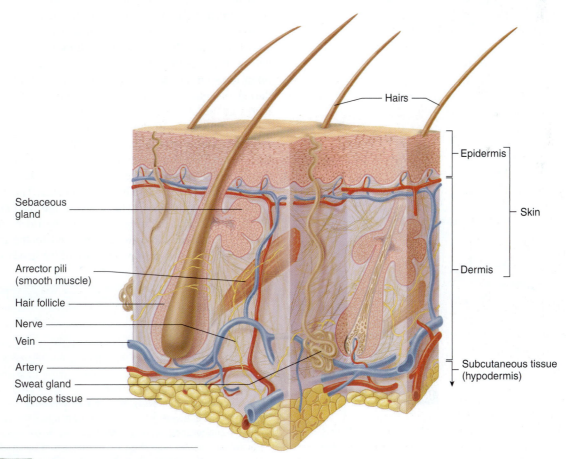

Figure 5.1 AP|R **Skin and Subcutaneous Tissue**

The skin, consisting of the epidermis and the dermis, is connected by the subcutaneous tissue to underlying structures. Note the accessory structures (hairs, glands, and arrector pili), some of which project into the subcutaneous tissue, and the large amount of adipose tissue in the subcutaneous tissue.

Although keratinization is a continuous process, distinct layers called **strata** (stra′tă; layer) can be seen in the epidermis (figure 5.2). The deepest stratum, the **stratum basale** (bā′să-lē; a base), consists of cuboidal or columnar cells that undergo mitotic divisions about every 19 days. One daughter cell becomes a new stratum basale cell and can divide again. The other daughter cell is pushed toward the surface, a journey that takes about 40–56 days. As cells move to the surface, changes in the cells produce intermediate strata.

The **stratum corneum** (kōr′nē-ŭm) is the most superficial stratum of the epidermis. It consists of dead squamous cells filled with keratin. Keratin gives the stratum corneum its structural strength. The stratum corneum cells are also coated and surrounded by lipids, which help prevent fluid loss through the skin.

Predict 2

What kinds of substances could easily pass through the skin by diffusion? What kinds would have difficulty?

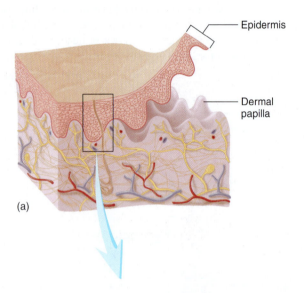

(a)

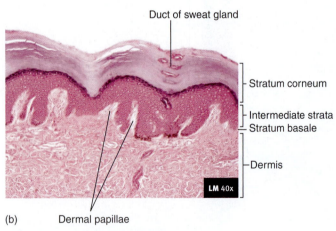

Duct of sweat gland

Stratum corneum

Intermediate strata
Stratum basale

Dermis

LM 40x

(b)   Dermal papillae

Figure 5.2 AP|R **Epidermis and Dermis**

(*a*) The epidermis rests on the dermis. Dermal papillae project toward the epidermis. (*b*) Photomicrograph of the epidermis resting on the dermis. Note the strata of the epidermis and the papillae of the dermis.

The stratum corneum is composed of 25 or more layers of dead squamous cells joined by desmosomes (see chapter 4). Eventually, the desmosomes break apart, and the cells are sloughed from the skin. Excessive sloughing of stratum corneum cells from the surface of the scalp is called dandruff. In skin subjected to friction, the number of layers in the stratum corneum greatly increases, producing a thickened area called a **callus** (kal′ŭs; hard skin). Over a bony prominence, the stratum corneum can thicken to form a cone-shaped structure called a **corn.**

Dermis

The dermis is composed of dense collagenous connective tissue containing fibroblasts, adipocytes, and macrophages. Nerves, hair follicles, smooth muscles, glands, and lymphatic vessels extend into the dermis (see figure 5.1).

Collagen and elastic fibers are responsible for the structural strength of the dermis. In fact, the dermis is the part of an animal hide from which leather is made. The collagen fibers of the dermis are oriented in many different directions and can resist stretch. However, more collagen fibers are oriented in some directions than in others. This produces **cleavage lines,** or *tension lines,* in the skin, and the skin is most resistant to stretch along these lines (figure 5.3). It is important for surgeons to be aware of cleavage lines. An incision made across the cleavage lines is likely to gap and produce considerable scar tissue, but an incision made parallel with the lines tends to gap less and produce less scar tissue (see chapter 4). If the skin is overstretched for any reason, the dermis can be damaged, leaving lines that are visible through the epidermis. These lines, called **stretch marks,** can develop when a person increases in size quite rapidly. For example, stretch marks often form on the skin of the abdomen and breasts of a woman during pregnancy or on the skin of athletes who have quickly increased muscle size by intense weight training.

The upper part of the dermis has projections called **dermal papillae** (pă-pil′e; nipple), which extend toward the epidermis (see figure 5.2). The dermal papillae contain many blood vessels that supply the overlying epidermis with nutrients, remove waste products, and help regulate body temperature. The dermal papillae in the palms of the hands, the soles of the feet, and the tips of the digits are arranged in parallel, curving ridges that shape the overlying epidermis into fingerprints and footprints. The ridges increase friction and improve the grip of the hands and feet.

A CASE IN POINT

Injections

Howey Stickum, a student nurse, learns three ways to give injections. An **intradermal injection** is administered by drawing the skin taut and inserting a small needle at a shallow angle into the dermis; an example is the tuberculin skin test. A **subcutaneous injection** is achieved by pinching the skin to form a "tent" and inserting a short needle into the adipose tissue of the subcutaneous tissue; an example is an insulin injection. An **intramuscular injection** is accomplished by inserting a long needle at a 90-degree angle to the skin into a muscle deep to the subcutaneous tissue. Intramuscular injections are used for most vaccines and certain antibiotics.

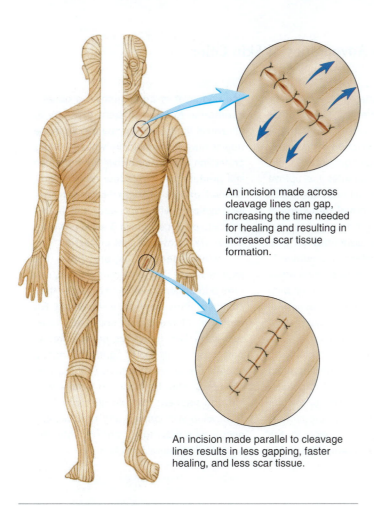

An incision made across cleavage lines can gap, increasing the time needed for healing and resulting in increased scar tissue formation.

An incision made parallel to cleavage lines results in less gapping, faster healing, and less scar tissue.

Figure 5.3 Cleavage Lines

The orientation of collagen fibers produces cleavage lines, or tension lines, in the skin.

Skin Color

Factors that determine skin color include pigments in the skin, blood circulating through the skin, and the thickness of the stratum corneum. **Melanin** (mel′ă-nin; black) is the group of pigments primarily responsible for skin, hair, and eye color. Most melanin molecules are brown to black pigments, but some are yellowish or reddish. Melanin provides protection against ultraviolet light from the sun.

Melanin is produced by **melanocytes** (mel′ă-nō-sītz; *melano*, black + *kytos*, cell), which are irregularly shaped cells with many long processes that extend between the epithelial cells of the deep part of the epidermis (figure 5.4). The Golgi apparatuses of the melanocytes package melanin into vesicles called **melanosomes** (mel′ă-nō-sōmz), which move into the cell processes of the mela- nocytes. Epithelial cells phagocytize the tips of the melanocyte cell processes, thereby acquiring melanosomes. Although all the epithelial cells of the epidermis can contain melanin, only the melanocytes produce it.

Large amounts of melanin form freckles or moles in some regions of the skin, as well as darkened areas in the genitalia, the nipples, and the circular areas around the nipples. Other areas, such as the lips, palms of the hands, and soles of the feet, contain less melanin.

Melanin production is determined by genetic factors, expo- sure to light, and hormones. Genetic factors are responsible for the amounts of melanin produced in different races. Since all races have about the same number of melanocytes, racial variations in skin color are determined by the amount, kind, and distribution of melanin. Although many genes are responsible for skin color, a single mutation can prevent the production of melanin. For exam- ple, **albinism** (al′bi-nizm) is a recessive genetic trait that causes a deficiency or an absence of melanin. Albinos have fair skin, white hair, and unpigmented irises in the eyes.

Exposure to ultraviolet light—for example, in sunlight— stimulates melanocytes to increase melanin production. The result is a suntan.

1. Melanosomes are produced by the Golgi apparatus of the melanocyte.

2. Melanosomes move into the melanocyte cell processes.

3. Epithelial cells phagocytize the tips of the melanocyte cell processes.

4. The melanosomes, which were produced inside the melanocytes, have been transferred to epithelial cells and are now inside them.

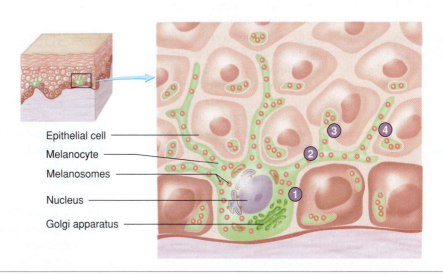

Epithelial cell
Melanocyte
Melanosomes
Nucleus
Golgi apparatus

PROCESS Figure 5.4 AP|R **Melanin Transfer to Epithelial Cells**

Melanocytes make melanin, which is packaged into melanosomes and transferred to many epithelial cells.

CLINICAL IMPACT Adaptive Advantages of Skin Color

The evolution of skin color in humans is intriguing because it helps explain certain modern health problems. During human evolution, the skeletal system of our ancestors changed, resulting in an upright posture and the ability to walk and run greater distances. As a result of increased physical activity, excess heat is produced, which can cause overheating. An increase in the number of sweat glands in the skin and a reduction in the amount of hair covering the skin help eliminate the excess heat.

With the reduction in hair over most of the body, prolonged exposure of the skin to sunlight can be harmful in two ways. First, it promotes the development of skin cancer (see "Skin Cancer" later in this chapter) by damaging DNA; second, it decreases the levels of the B vitamin folate in the blood by breaking it down through a photochemical reaction. Low folate levels are known to increase the risk

of abnormal development of the fetal nervous system (see chapter 20). On the other hand, exposure to ultraviolet light from the sun stimulates the production of vitamin D (see "Vitamin D Production" later in this chapter). Vitamin D promotes the uptake of calcium from the small intestine, which is important for the normal development of the skeletal system in the fetus and in children. Inadequate quantities of vitamin D can result in rickets, a condition in which the bones are soft, weak, and easily broken. Thus, increased skin pigmentation protects against skin cancer and abnormal development of the nervous system but impairs skeletal system development.

The optimal amount of melanin in the skin should be large enough to protect against the harmful effects of ultraviolet light but small enough to allow ultraviolet light to stimulate vitamin D production. Ultraviolet light intensity is high in the tropics but diminishes toward the poles.

The skin color of populations is a genetic adaptation to their different exposure to ultraviolet light. Dark-skinned people in the tropics have more melanin, which provides protection against ultraviolet light, but can still produce vitamin D year-round. Light-skinned people at higher latitudes have less melanin, which increases the body's ability to produce vitamin D while providing adequate ultraviolet light protection.

When people migrate from the regions where their ancestors evolved, cultural adaptations help them adjust to their changed ultraviolet light environment. For example, clothing and portable shade, such as tents or umbrellas, provide protection against ultraviolet light, and eating vitamin D–rich foods, such as fish, provides vitamin D. Still, light-skinned people who move to southern climates are more likely to develop skin cancer, and dark-skinned people who move to northern climates are more likely to develop rickets.

Certain hormones, such as estrogen and melanocyte-stimulating hormone, cause an increase in melanin production during pregnancy in the mother, darkening the nipples, the pigmented circular areas around the nipples, and the genitalia even more. The cheekbones and forehead can also darken, resulting in "the mask of pregnancy." Also, a dark line of pigmentation can appear on the midline of the abdomen.

Blood flowing through the skin imparts a reddish hue, and when blood flow increases, the red color intensifies. Examples include blushing and the redness resulting from the inflammatory response. A decrease in blood flow, as occurs in shock, can make the skin appear pale. A decrease in the blood O_2 content produces a bluish color of the skin, called **cyanosis** (sī-ă-nō′sis; dark blue). Birthmarks are congenital (present at birth) disorders of the blood vessels (capillaries) in the dermis.

Carotene (kar′ō-tēn) is a yellow pigment found in plants such as squash and carrots. Humans normally ingest carotene and use it as a source of vitamin A. Carotene is lipid-soluble; when consumed, it accumulates in the lipids of the stratum corneum and in the adipocytes of the dermis and subcutaneous tissue. If large amounts of carotene are consumed, the skin can become quite yellowish.

The location of pigments and other substances in the skin affects the color produced. If a dark pigment is located in the dermis or subcutaneous tissue, light reflected off the dark pigment can be scattered by collagen fibers of the dermis to produce a blue color. The deeper within the dermis or subcutaneous tissue any dark pigment is located, the bluer the pigment appears

because of the light-scattering effect of the overlying tissue. This effect causes the blue color of tattoos, bruises, and some superficial blood vessels.

Predict 3

Explain the differences in skin color between (a) the palms of the hands and the lips, (b) the palms of the hands of a person who does heavy manual labor and one who does not, and (c) the anterior and posterior surfaces of the forearm.

5.3 SUBCUTANEOUS TISSUE

Learning Outcome   After reading this section, you should be able to

A. Describe the structure and discuss the function of the subcutaneous tissue.

Just as a house rests on a foundation, the skin rests on the **subcutaneous tissue,** which attaches it to underlying bone and muscle and supplies it with blood vessels and nerves (see figure 5.1). The subcutaneous (sŭb-koo-tā′nē-ŭs; under the skin) tissue, which is not part of the skin, is sometimes called *hypodermis* (hī-pō-der′mis; under the dermis). The subcutaneous tissue is loose connective tissue, including adipose tissue that contains about half the body's stored lipids, although the amount and location vary with age, sex, and diet. Adipose tissue in the subcutaneous tissue functions as padding and insulation, and it is responsible for some of the differences in appearance between men and women as well as between individuals of the same sex.

The subcutaneous tissue can be used to estimate total body fat. The skin and subcutaneous tissue are pinched at selected locations, and the thickness of the fold is measured. The thicker the fold, the greater the amount of total body fat. The percentage of body fat varies in the population, but on average women have higher total body fat than do males. The acceptable percentage of body fat varies from 21% to 30% for females and from 13% to 25% for males. A body fat percentage above the acceptable range is an indicator of obesity.

5.4 ACCESSORY SKIN STRUCTURES

Learning Outcomes *After reading this section, you should be able to*

A. Describe the structure of a hair, and discuss the phases of hair growth.
B. Name the glands of the skin, and describe the secretions they produce.
C. Describe the parts of a nail, and explain how nails grow.

The accessory skin structures are hair, glands, and nails.

Hair

In humans, **hair** is found everywhere on the skin, except on the palms, the soles, the lips, the nipples, parts of the genitalia, and the distal segments of the fingers and toes.

Each hair arises from a **hair follicle,** an extension of the epidermis that originates deep in the dermis (figure 5.5*a*). The **shaft** of the hair protrudes above the surface of the skin, whereas the **root** and **hair bulb** are below the surface. A hair has a hard **cortex,** which surrounds a softer center, the **medulla** (me-dool´ă). The cortex is covered by the **cuticle** (kū´ti-kl; skin), a single layer of overlapping cells that holds the hair in the hair follicle. The hair follicle can play an important role in repair of the skin. If the surface epidermis is damaged, the epithelial cells within the hair follicle can divide and serve as a source of new epithelial cells.

Hair is produced in the hair bulb, which rests on the hair papilla (figure 5.5*b*). Blood vessels within the papilla supply the hair bulb with the nourishment needed to produce the hair. Hair is produced in cycles. During the growth stage, it is formed by epithelial cells within the hair bulb. These cells, like the cells of the stratum basale in the skin, divide and undergo keratinization. The hair grows longer as these cells are added to the base of the hair within the hair bulb. Thus, the hair root and shaft consist of columns of dead keratinized epithelial cells. During the resting stage, growth stops and the hair is held in the hair follicle. When the next growth stage begins, a new hair is formed and the old hair falls out. The duration of each stage depends on the individual hair. Eyelashes grow for about 30 days and rest for 105 days, whereas scalp hairs grow for 3 years and rest for 1–2 years. The loss of hair normally means that the hair is being replaced because the old hair falls out of the hair follicle when the new hair begins to grow. In some men, however, a permanent loss of hair results in "pattern baldness." Although many of the hair follicles are lost, some remain and produce a very short, transparent hair, which for practical purposes is invisible. These changes occur when male sex hormones act on the hair follicles of men who have the genetic predisposition for pattern baldness.

Hair color is determined by varying amounts and types of melanin. The production and distribution of melanin by melanocytes

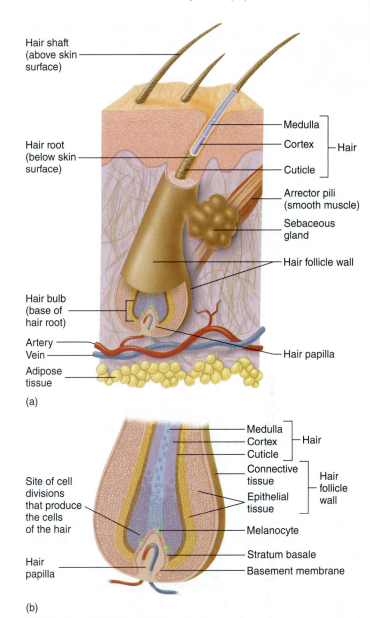

(a)

(b)

Figure 5.5 AP|R **Hair Follicle**

(*a*) Hair within a hair follicle. (*b*) Enlargement of the hair bulb and hair follicle wall.

occurs in the hair bulb by the same method as in the skin. With age, the amount of melanin in hair can decrease, causing the hair color to become faded, or the hair can contain no melanin and be white. Gray hair is usually a mixture of unfaded, faded, and white hairs.

Predict 4

Marie Antoinette's hair supposedly turned white overnight after she heard she would be sent to the guillotine. Explain why you believe or disbelieve this story.

Associated with each hair follicle are smooth muscle cells called the **arrector** (ă-rek´tōr; that which raises) **pili** (pī´li; hair) (see figure 5.5*a*). Contraction of the arrector pili causes the hair to become more perpendicular to the skin's surface, or to "stand on end," and it produces a raised area of skin called a "goose bump."

Glands

The major glands of the skin are the **sebaceous** (sē-bā′shŭs) **glands** and the **sweat glands** (figure 5.6). Sebaceous glands are simple, branched acinar glands (see chapter 4). Most are connected by a duct to the superficial part of a hair follicle. They produce **sebum,** an oily, white substance rich in lipids. The sebum is released by holocrine secretion (see chapter 4) and lubricates the hair and the surface of the skin, which prevents drying and protects against some bacteria.

There are two kinds of sweat glands: eccrine and apocrine. **Eccrine** (ek′rin) **sweat glands** are simple, coiled, tubular glands and release sweat by merocrine secretion. Eccrine glands are located in almost every part of the skin but most numerous in the palms and soles. They produce a secretion that is mostly water with a few salts. Eccrine sweat glands have ducts that open onto the surface of the skin through sweat pores. When the body temperature starts to rise above normal levels, the sweat glands produce sweat, which evaporates and cools the body. Sweat can also be released in the palms, soles, armpits, and other places because of emotional stress. Emotional sweating is used in lie detector (polygraph) tests because sweat gland activity usually increases when a person is nervous, such as when the person tells a lie. Such tests can detect even small amounts of sweat because the salt solution conducts electricity and lowers the electrical resistance of the skin.

Apocrine (ap′ō-krin) **sweat glands** are simple, coiled, tubular glands that produce a thick secretion rich in organic substances. These substances are released primarily by merocrine secretion, though some glands demonstrate holocrine secretion. They open into hair follicles, but only in the armpits and genitalia. Apocrine sweat glands become active at puberty because of the influence of sex hormones. The organic secretion, which is essentially odorless when released, is quickly broken down by bacteria into substances responsible for what is commonly known as body odor.

Nails

The **nail** is a thin plate, consisting of layers of dead stratum corneum cells that contain a very hard type of keratin. The visible part of the nail is the **nail body,** and the part of the nail covered by skin is the **nail root** (figure 5.7). The **cuticle,** or *eponychium* (ep-ō-nik′ē-ŭm), is stratum corneum that extends onto the nail body. The nail root extends distally from the **nail matrix.** The nail also attaches to the underlying **nail bed,** which is located distal

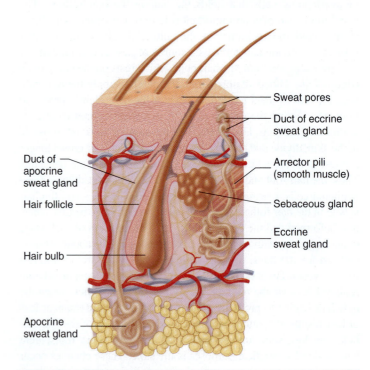

Sweat pores

Duct of eccrine sweat gland

Arrector pili (smooth muscle)

Sebaceous gland

Eccrine sweat gland

Duct of apocrine sweat gland

Hair follicle

Hair bulb

Apocrine sweat gland

Figure 5.6 Glands of the Skin

Sebaceous glands and apocrine sweat glands empty into a hair follicle. Eccrine sweat glands empty onto the surface of the skin.

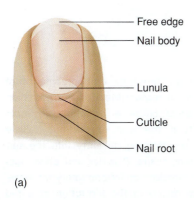

Free edge
Nail body
Lunula
Cuticle
Nail root

(a)

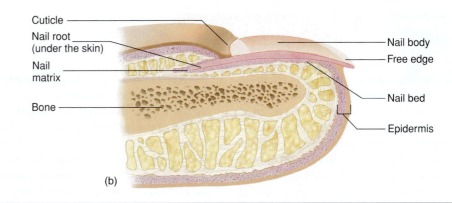

Cuticle
Nail root (under the skin)
Nail matrix
Bone

Nail body
Free edge
Nail bed
Epidermis

(b)

Figure 5.7 AP|R **Nail**

(*a*) Dorsal view of the exterior nail. (*b*) Lateral view of a sagittal section through the nail. Most of the epidermis is absent from the nail bed.

to the nail matrix. The nail matrix and bed are epithelial tissue with a stratum basale that gives rise to the cells that form the nail. The nail matrix is thicker than the nail bed and produces most of the nail. A small part of the nail matrix, the **lunula** (loo′noo-lă; moon), can be seen through the nail body as a whitish, crescent-shaped area at the base of the nail. Cell production within the nail matrix causes the nail to grow. Unlike hair, nails grow continuously and do not have a resting stage.

5.5 PHYSIOLOGY OF THE INTEGUMENTARY SYSTEM

Learning Outcome   *After reading this section, you should be able to*

A. Discuss the specific functions of skin, hair, glands, and nails.

Protection

The integumentary system performs many protective functions.

1. The intact skin plays an important role in reducing water loss because its lipids act as a barrier to the diffusion of water.
2. The skin prevents microorganisms and other foreign substances from entering the body. Secretions from skin glands also produce an environment unsuitable for some microorganisms.
3. The stratified squamous epithelium of the skin protects underlying structures against abrasion.
4. Melanin absorbs ultraviolet light and protects underlying structures from its damaging effects.
5. Hair provides protection in several ways: The hair on the head acts as a heat insulator, eyebrows keep sweat out of the eyes, eyelashes protect the eyes from foreign objects, and hair in the nose and ears prevents the entry of dust and other materials.
6. The nails protect the ends of the fingers and toes from damage and can be used in defense.

Sensation

Receptors in the epidermis and dermis can detect pain, heat, cold, and pressure (see chapter 9). Although hair does not have a nerve supply, sensory receptors around the hair follicle can detect the movement of a hair.

Vitamin D Production

When the skin is exposed to ultraviolet light, a precursor molecule of **vitamin D** is formed. The precursor is carried by the blood to the liver, where it is modified, and then to the kidneys, where the precursor is modified further to form active vitamin D. If exposed to enough ultraviolet light, humans can produce all the vitamin D they need. However, many people need to ingest vitamin D as well because clothing and indoor living reduce their exposure to ultraviolet light. Fatty fish (and fish oils) and vitamin D–fortified milk are the best sources of vitamin D. Eggs, butter, and liver contain small amounts of vitamin D but are not considered significant sources because too large a serving size is necessary to meet the daily vitamin D requirement. Adequate levels of vitamin D are necessary because active vitamin D stimulates the small intestine to absorb calcium and phosphate, the substances necessary for normal bone growth (see chapter 6) and normal muscle function (see chapter 7).

Temperature Regulation

Body temperature normally is maintained at about 37°C (98.6°F). Regulation of body temperature is important because the rate of chemical reactions within the body can be increased or decreased by changes in body temperature. Even slight changes in temperature can make enzymes operate less efficiently and disrupt the normal rates of chemical changes in the body.

Exercise, fever, and an increase in environmental temperature tend to raise body temperature. In order to maintain homeostasis, the body must rid itself of excess heat. Blood vessels in the dermis dilate and enable more blood to flow within the skin, thus transferring heat from deeper tissues to the skin (figure 5.8*a*), where the heat is lost by radiation (infrared energy), convection (air movement), or conduction (direct contact with an object). Sweat that spreads over the surface of the skin and evaporates also carries away heat and reduces body temperature.

If body temperature begins to drop below normal, heat can be conserved by the constriction of dermal blood vessels, which reduces blood flow to the skin (figure 5.8*b*). Thus, less heat is transferred from deeper structures to the skin, and heat loss is reduced. However, with smaller amounts of warm blood flowing through the skin, the skin temperature decreases. If the skin temperature drops below about 15°C (59°F), dermal blood vessels dilate.

CLINICAL IMPACT   Acne

Acne (ak′nē) is inflammation of the hair follicles and sebaceous glands. Four factors are believed to be responsible: hormones, sebum, abnormal production of cells, and the bacterium *Propionibacterium acnes*. During puberty, hormones, especially testosterone, stimulate the sebaceous glands, and sebum production increases. Because both the testes and the ovaries produce testosterone, the effect is seen in males and females.

The lesions of acne begin with the overproduction of epidermal cells in the hair follicle. These cells are shed from the wall of the hair follicle, and they stick to one another to form a mass of cells mixed with sebum that blocks the hair follicle. An accumulation of sebum behind the blockage produces a whitehead. A blackhead develops when the accumulating mass of cells and sebum pushes through the opening of the hair follicle.

Although there is general agreement that dirt is not responsible for the black color, the exact cause is disputed. A pimple results if the wall of the hair follicle ruptures, forming an entry into the surrounding tissue. *P. acnes* and other bacteria stimulate an inflammatory response, which results in the formation of a red pimple filled with pus. If tissue damage is extensive, scarring occurs.

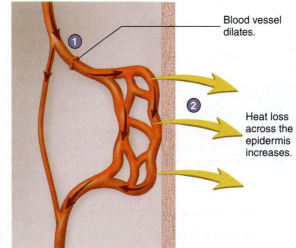

1 Blood vessel dilation results in increased blood flow toward the surface of the skin.

2 Increased blood flow beneath the epidermis results in increased heat loss (*gold arrows*).

Blood vessel dilates.

Heat loss across the epidermis increases.

(a)

3 Blood vessel constriction results in decreased blood flow toward the surface of the skin.

4 Decreased blood flow beneath the epidermis results in decreased heat loss.

Blood vessel constricts.

Heat loss across the epidermis decreases.

(b)

PROCESS Figure 5.8 AP|R Heat Exchange in the Skin

You may have noticed that on cold winter days, people's noses and ears turn red. Can you explain the advantage of this response?

Excretion

The integumentary system plays a minor role in **excretion,** the removal of waste products from the body. In addition to water and salts, sweat contains small amounts of waste products, such as urea, uric acid, and ammonia. Even though the body can lose large amounts of sweat, the sweat glands do not play a significant role in the excretion of waste products.

5.6 INTEGUMENTARY SYSTEM AS A DIAGNOSTIC AID

Learning Outcome   *After reading this section, you should be able to*

A. Explain how the integumentary system can be used as a diagnostic aid.

The integumentary system is useful in diagnosis because it is observed easily and often reflects events occurring in other parts of the body. For example, cyanosis, a bluish color to the skin caused by decreased blood O_2 content, is an indication of impaired circulatory or respiratory function. A yellowish skin color, called **jaundice** (jawn'dis), can occur when the liver is damaged by a disease, such as viral hepatitis. Normally, the liver secretes bile pigments, which are products of the breakdown of worn-out red blood cells, into the intestine. Bile pigments are yellow, and their buildup in the blood and tissues can indicate impaired liver function.

Rashes and lesions in the skin can be symptoms of problems elsewhere in the body. For example, scarlet fever results when bacteria infecting the throat release a toxin into the blood that causes a reddish rash on the skin. The development of a rash can also indicate an allergic reaction to foods or to drugs, such as penicillin.

The condition of the skin, hair, and nails is affected by nutritional status. In vitamin A deficiency, the skin produces excess keratin and assumes a characteristic sandpaper texture, whereas in iron-deficiency anemia the nails lose their normal contour and become flat or concave (spoon-shaped).

Hair concentrates many substances that can be detected by laboratory analysis, and comparison of a patient's hair with a "normal" hair can be useful in certain diagnoses. For example, lead poisoning results in high levels of lead in the hair. However, the use of hair analysis to determine the general health or nutritional status of an individual is unreliable.

5.7 BURNS

Learning Outcome   *After reading this section, you should be able to*

A. Classify burns on the basis of the amount of skin damage produced.

A **burn** is injury to a tissue caused by heat, cold, friction, chemicals, electricity, or radiation. Burns are classified according to their depth (figure 5.9). In **partial-thickness burns,** part of the stratum basale remains viable, and regeneration of the epidermis

occurs from within the burn area, as well as from the edges of the burn. Partial-thickness burns are divided into first- and second-degree burns.

First-degree burns involve only the epidermis and are red and painful. Slight **edema** (e-dē'mă), or swelling, may be present. They can be caused by sunburn or brief exposure to very hot or very cold objects, and they heal without scarring in about a week.

Second-degree burns damage both the epidermis and the dermis. If dermal damage is minimal, symptoms include redness, pain, edema, and blisters. Healing takes about 2 weeks, and no scarring results. However, if the burn goes deep into the dermis, the wound appears red, tan, or white; can take several months to heal; and might scar. In all second-degree burns, the epidermis regenerates from epithelial tissue in hair follicles and sweat glands, as well as from the edges of the wound.

In **full-thickness burns,** or *third-degree burns,* the epidermis and the dermis are completely destroyed, and recovery occurs from the edges of the burn wound. Third-degree burns often are surrounded by areas of first- and second-degree burns. Although the first- and second-degree burn areas are painful, the region of third-degree burn is usually painless because sensory receptors in the epidermis and dermis have been destroyed. Third-degree burns appear white, tan, brown, black, or deep cherry red.

Deep partial-thickness and full-thickness burns take a long time to heal, and they form scar tissue with disfiguring and debilitating wound contracture (see chapter 4). To prevent these complications and to speed healing, skin grafts are often performed. In a procedure called a split skin graft, the epidermis and part of

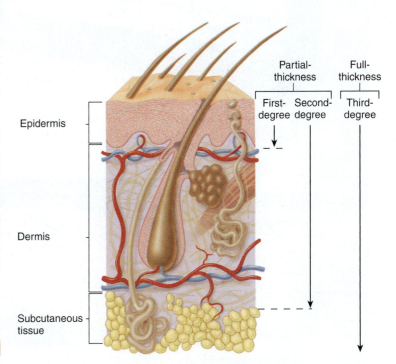

Figure 5.9 Burns

Burns are classified based on the extent of skin damage. Partial-thickness burns are subdivided into first-degree burns (damage to only the epidermis) and second-degree burns (damage to the epidermis and part of the dermis). Full-thickness, or third-degree, burns destroy the epidermis, the dermis, and sometimes deeper tissues.

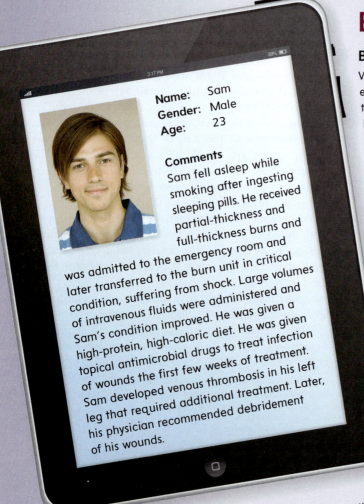

Name: Sam
Gender: Male
Age: 23

Comments

Sam fell asleep while smoking after ingesting sleeping pills. He received partial-thickness and full-thickness burns and was admitted to the emergency room and later transferred to the burn unit in critical condition, suffering from shock. Large volumes of intravenous fluids were administered and Sam's condition improved. He was given a high-protein, high-caloric diet. He was given topical antimicrobial drugs to treat infection of wounds the first few weeks of treatment. Sam developed venous thrombosis in his left leg that required additional treatment. Later, his physician recommended debridement of his wounds.

Burns

Background Information

When large areas of skin are severely burned, the resulting systemic effects can be life-threatening. Within minutes of a major burn injury, there is increased permeability of capillaries, which are the small blood vessels in which fluid, gases, nutrients, and waste products are normally exchanged between the blood and tissues. This increased permeability occurs at the burn site and throughout the body. As a result, fluid and ions are lost from the burn wound and into tissue spaces. The loss of fluid decreases blood volume, which decreases the heart's ability to pump blood. The resulting decrease in blood delivery to tissues can cause tissue damage, shock, and even death. Treatment consists of administering intravenous fluid at a faster rate than it leaks out of the capillaries. Although this fluid replacement can reverse the shock and prevent death, fluid continues to leak into tissue spaces, causing pronounced edema (swelling).

Typically, after 24 hours, capillary permeability returns to normal, and the amount of intravenous fluid administered can be greatly decreased. How burns cause capillary permeability to change is not well understood. It is clear that, following a burn, immunological and metabolic changes occur that affect not only capillaries but the rest of the body as well. For example, chemical mediators (see chapter 4), which are released in response to the tissue damage, contribute to changes in capillary permeability throughout the body.

Substances released from the burn may also play a role in causing cells to function abnormally. Burn injuries result in an almost immediate hypermetabolic state, which persists until wound closure. Two other factors contributing to the increased metabolism are (1) a resetting of the temperature control

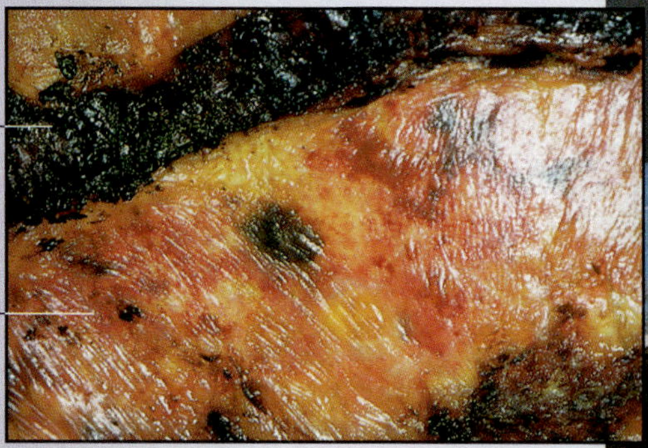

Full-thickness
burn

Partial-thickness
burn

Figure 5A
Full-thickness and partial-thickness burns

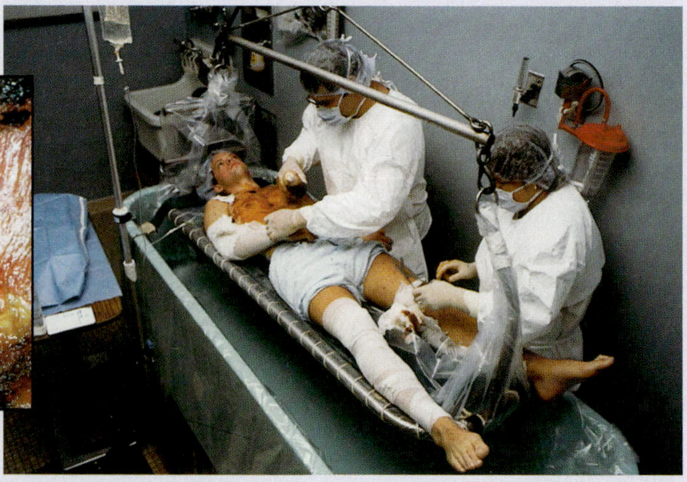

Figure 5B
Patient in a burn unit

Burns

MUSCULAR
Hypermetabolic state may lead to loss in muscle mass.

SKELETAL
Increased red blood cell production in red bone marrow.

NERVOUS
Pain in partial-thickness burns; body temperature increases as control center in brain is reset; abnormal ion levels disrupt normal nervous system activity.

DIGESTIVE
Tissue damage to intestinal lining and liver as a result of decreased blood flow; bacteria of intestines may cause systemic infection; liver releases blood-clotting factors in response to injury.

Symptoms
- Tissue damage of skin and possibly deeper tissue
- Edema
- Shock
- Microbial infection

Treatment
- Intravenous fluids
- High-protein, high-calorie diet
- Antimicrobials
- Debridement
- Skin grafts

LYMPHATIC AND IMMUNE
Inflammation; depression of immune system may lead to infection.

URINARY
Urine production decreases in response to low blood volume; tissue damage to kidneys due to low blood flow.

CARDIOVASCULAR
Decreased blood volume, edema, and shock may occur due to increased capillary permeability; abnormal ion levels disrupt normal heart rate; increased blood clotting may cause venous thrombosis; preferential blood flow promotes healing.

RESPIRATORY
Edema may obstruct airways; increased respiratory rate in response to hypermetabolic state.

ENDOCRINE
Release of epinephrine and norepinephrine from the adrenal glands in response to injury contributes to hypermetabolic state and increased body temperature.

center in the brain to a higher temperature and (2) hormones released by the endocrine system (e.g., epinephrine and norepinephrine from the adrenal glands, which can increase cell metabolism). Compared with a normal body temperature of approximately 37°C (98.6°F), a typical burn patient may have a body temperature of 38.5°C (101.3°F), despite the higher loss of water by evaporation from the burn.

In severe burns, the increased metabolic rate can result in loss of as much as 30–40% of the patient's preburn weight. To help compensate, treatment may include doubling or tripling the patient's caloric intake. In addition, the need for protein, which is necessary for tissue repair, is greater.

Normal skin maintains homeostasis by preventing microorganisms from entering the body. Because burns damage and sometimes completely destroy the skin, microorganisms can cause infections. For this reason, burn patients are maintained in an aseptic (sterile) environment, which attempts to prevent the entry of microorganisms into the wound. They are also given antimicrobial drugs, which kill microorganisms or suppress their growth. **Debridement** (dā-brēd-mon′), the removal of dead tissue from the burn, helps prevent infections by cleaning the wound and removing tissue in which infections could develop. Skin grafts, performed within a week of the injury, also help close the wound and prevent the entry of microorganisms.

Despite these efforts, however, infections are still the major cause of death for burn victims. Depression of the immune system during the first or second week after the injury contributes to the high infection rate. First, the thermally altered tissue is recognized as a foreign substance, which stimulates the immune system. Then, the immune system is overwhelmed as immune system cells become less effective and the production of the chemicals that normally provide resistance to infections decreases (see chapter 14). The greater the magnitude of the burn, the greater the depression of the immune system, and the greater the risk of infection.

Venous thrombosis (throm-bō′sis), the development of a clot in a vein, is another complication of burns. Blood normally forms a clot when exposed to damaged tissue, such as at a burn site, but clotting can also occur elsewhere, such as in veins, where clots can block blood flow, resulting in tissue destruction. The concentration of chemicals that cause blood clotting (called clotting factors) increases for two reasons: Loss of fluid from the burn patient concentrates the chemicals, and the liver releases an increased amount of clotting factors.

Predict 6

When Sam was first admitted to the burn unit, the nurses carefully monitored his urine output. Why does that make sense in light of his injuries?

the dermis are removed from another part of the body and placed over the burn. Interstitial fluid from the burn nourishes the graft until blood vessels can grow into the graft and supply it with nourishment. Meanwhile, the donor tissue produces new epidermis from epithelial tissue in the hair follicles and sweat glands in the same manner as in superficial second-degree burns.

When it is not possible or practical to move skin from one part of the body to a burn site, physicians sometimes use artificial skin or grafts from human cadavers. However, these techniques are often unsatisfactory because the body's immune system recognizes the graft as a foreign substance and rejects it. A solution to this problem is to grow some of the burn victim's own skin in the laboratory. A piece of healthy skin from the burn victim is removed and placed in a flask with nutrients and hormones that stimulate rapid growth. The new skin that results consists only of epidermis and does not contain glands or hair.

5.8 SKIN CANCER

Learning Outcome *After reading this section, you should be able to*

A. Name and define the types of skin cancer.

Skin cancer is the most common type of cancer. Although chemicals and radiation (x-rays) are known to induce cancer, the development of skin cancer most often is associated with exposure to ultraviolet (UV) light from the sun. Consequently, most skin cancers develop on the face, neck, or hands. The group of people most likely to have skin cancer are fair-skinned (they have less protection from the sun) or older than 50 (they have had long exposure to the sun).

There are three main types of skin cancer. **Basal cell carcinoma** (kar-si-nō′mă), the most frequent type, begins with cells in the stratum basale and extends into the dermis to produce an open ulcer (figure 5.10*a*). Surgical removal or radiation therapy cures this type of cancer. Fortunately, there is little danger that this type of cancer will spread, or metastasize, to other areas of the body. **Squamous cell carcinoma** develops from cells immediately superficial to the stratum basale. Normally, these cells undergo little or no cell division, but in squamous cell carcinoma, the cells continue to divide as they produce keratin. The typical result is a nodular, keratinized tumor confined to the epidermis (figure 5.10*b*). If untreated, the tumor can invade the dermis, metastasize, and

cause death. **Malignant melanoma** (mel′ă-nō′mă) is a rare form of skin cancer that arises from melanocytes, usually in a preexisting mole. A mole is an aggregation, or "nest," of melanocytes. The melanoma can appear as a large, flat, spreading lesion or as a deeply pigmented nodule (figure 5.10*c*). Metastasis is common, and unless diagnosed and treated early in development, this cancer is often fatal.

Limiting exposure to the sun and using sunscreens that block ultraviolet light can reduce the likelihood of developing skin cancer. Ultraviolet light is classified into two types based on their wavelengths: UVA has a longer wavelength than UVB. Exposure to UVA causes most tanning of the skin but is associated with the development of malignant melanoma. Exposure to UVB causes most burning of the skin and is associated with the development of basal cell and squamous cell carcinomas. It is advisable to use sunscreens that effectively block both UVA and UVB.

5.9 EFFECTS OF AGING ON THE INTEGUMENTARY SYSTEM

Learning Outcome *After reading this section, you should be able to*

A. List the changes the integumentary system undergoes with age.

As the body ages, the skin is more easily damaged because the epidermis thins and the amount of collagen in the dermis decreases. Skin infections are more likely, and repair of the skin occurs more slowly. A decrease in the number of elastic fibers in the dermis and loss of adipose tissue from the subcutaneous tissue cause the skin to sag and wrinkle. A decrease in the activity of sweat glands and in the blood supply to the dermis results in reduced ability to regulate body temperature. The skin becomes drier as sebaceous gland activity decreases. The number of melanocytes generally decreases, but in some areas the number of melanocytes increases to produce **age spots.** Note that age spots are different from **freckles,** which are caused by increased melanin production. Gray or white hair also results because of a decrease in or a lack of melanin production. Skin that is exposed to sunlight shows signs of aging more rapidly than nonexposed skin, so avoiding overexposure to sunlight and using sunscreen is advisable.

(a)

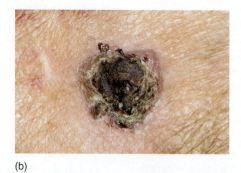
(b)

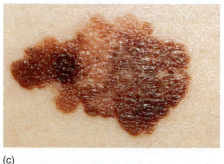
(c)

Figure 5.10 Cancer of the Skin
(*a*) Basal cell carcinoma. (*b*) Squamous cell carcinoma. (*c*) Malignant melanoma.

DISEASES AND DISORDERS: Skin

CONDITION	DESCRIPTION
Ringworm	Fungal infection that produces patchy scaling and inflammatory response in the skin
Eczema and dermatitis	Inflammatory conditions of the skin caused by allergy, infection, poor circulation, or exposure to chemical or environmental factors
Psoriasis	Chronic skin disease characterized by thicker than normal epidermal layer (stratum corneum) that sloughs to produce large, silvery scales; bleeding may occur if the scales are scraped away
Bacterial Infections	
Impetigo	Small blisters containing pus; easily rupture to form a thick, yellowish crust; usually affects children
Decubitus ulcers (bedsores or pressure sores)	Develop in people who are bedridden or confined to a wheelchair; compression of tissue and reduced circulation result in destruction of the subcutaneous tissue and skin, which later become infected by bacteria, forming ulcers
Viral Infections	
Rubeola (measles)	Skin lesions; caused by a virus contracted through the respiratory tract; may develop into pneumonia or infect the brain, causing damage
Rubella (German measles)	Skin lesions; usually mild viral disease contracted through the respiratory tract; may be dangerous if contracted during pregnancy because the virus can cross the placenta and damage the fetus
Chicken pox	Skin lesions; usually mild viral disease contracted through the respiratory tract
Shingles	Painful skin lesions that can recur when the dormant virus is activated by trauma, stress, or another illness; caused by the chicken pox virus after childhood infection
Cold sores (fever blisters)	Skin lesions; caused by herpes simplex I virus; transmitted by oral or respiratory routes; lesions recur
Genital herpes	Genital lesions; caused by herpes simplex II virus; transmitted by sexual contact

ANSWER TO LEARN TO PREDICT

In the section titled "Skin Color," we learned that melanocytes located in the epidermis produce melanin, a group of pigments that provides protection against ultraviolet light from the sun. We also know that the amount of melanin produced is determined by several natural factors, including our genes. We can assume that Mia and Landon naturally have darker skin than Christine because they each possess genes that allow them to produce more melanin when exposed to ultraviolet light. Recall also from the section titled "Skin Cancer" that we can artificially block ultraviolet light by applying sunscreens. Christine's mother may have forgotten to apply sunscreen on Christine during the day or used a sunscreen that didn't effectively block UVB, which commonly causes sunburn.

Answers to the rest of this chapter's Predict questions are in Appendix E.

SUMMARY

The integumentary system consists of the skin, hair, glands, and nails.

5.1 Functions of the Integumentary System
(p. 94)

The integumentary system consists of the skin, hair, glands, and nails. The integumentary system protects us from the external environment. Other functions include sensation, vitamin D production, temperature regulation, and excretion of small amounts of waste products.

5.2 Skin   (p. 95)

Epidermis

1. The epidermis is stratified squamous epithelium divided into strata.
   - New cells are produced in the stratum basale.
   - The stratum corneum consists of many layers of dead squamous cells containing keratin. The most superficial layers are sloughed.

2. Keratinization is the transformation of stratum basale cells into stratum corneum cells.
   - Structural strength results from keratin inside the cells and from desmosomes, which hold the cells together.
   - Lipids surrounding the stratum corneum cells help prevent fluid loss.

Dermis

1. The dermis is dense connective tissue.
2. Collagen and elastic fibers provide structural strength, and the blood vessels of the papillae supply the epidermis with nutrients.

Skin Color

1. Melanocytes produce melanin, which is responsible for different skin colors. Melanin production is determined genetically but can be modified by exposure to ultraviolet light and by hormones.

2. Carotene, a plant pigment ingested as a source of vitamin A, can cause the skin to appear yellowish.
3. Increased blood flow produces a red skin color, whereas decreased blood flow causes a pale skin color. Decreased blood O$_2$ results in the blue skin color of cyanosis.
4. Scattering of light by collagen produces a bluish color.

5.3 Subcutaneous Tissue (p. 98)

1. The subcutaneous tissue, which is not part of the skin, is loose connective tissue that attaches the skin to underlying tissues.
2. About half of the body's lipids are stored in the subcutaneous tissue.

5.4 Accessory Skin Structures (p. 99)

Hair

1. Hairs are columns of dead, keratinized epithelial cells. Each hair consists of a shaft (above the skin), a root (below the skin), and a hair bulb (site of hair cell formation).
2. Hairs have a growth phase and a resting phase.
3. Contraction of the arrector pili, which are smooth muscles, causes hair to "stand on end" and produces a "goose bump."

Glands

1. Sebaceous glands produce sebum, which oils the hair and the surface of the skin.
2. Eccrine sweat glands produce sweat, which cools the body.
3. Apocrine sweat glands produce an organic secretion that causes body odor when broken down by bacteria.

Nails

1. The nail consists of the nail body and the nail root.
2. The nail matrix produces the nail, which is composed of stratum corneum cells containing hard keratin.

5.5 Physiology of the Integumentary System (p. 101)

Protection

The skin reduces water loss, prevents the entry of microorganisms, and provides protection against abrasion and ultraviolet light; hair and nails also perform protective functions.

Sensation

The skin contains sensory receptors for pain, heat, cold, and pressure.

Vitamin D Production

1. Ultraviolet light stimulates the production of a precursor molecule in the skin that is modified by the liver and kidneys into vitamin D.
2. Vitamin D increases calcium uptake in the small intestine.

Temperature Regulation

1. Through dilation and constriction of blood vessels, the skin controls heat loss from the body.
2. Evaporation of sweat cools the body.

Excretion

Skin glands remove small amounts of waste products but are not important in excretion.

5.6 Integumentary System as a Diagnostic Aid (p. 103)

The integumentary system is easily observed and often reflects events occurring in other parts of the body (e.g., cyanosis, jaundice, rashes).

5.7 Burns (p. 103)

1. Partial-thickness burns damage only the epidermis (first-degree burn) or the epidermis and the dermis (second-degree burn).
2. Full-thickness burns (third-degree burns) destroy the epidermis, the dermis, and usually underlying tissues.

5.8 Skin Cancer (p. 106)

1. Basal cell carcinoma involves the cells of the stratum basale and is readily treatable.
2. Squamous cell carcinoma involves the cells immediately superficial to the stratum basale and can metastasize.
3. Malignant melanoma involves melanocytes, can metastasize, and is often fatal.

5.9 Effects of Aging on the Integumentary System (p. 106)

1. Blood flow to the skin is reduced, the skin becomes thinner, and elasticity is lost.
2. Sweat and sebaceous glands are less active, and the number of melanocytes decreases.

 REVIEW AND COMPREHENSION

1. Name the components of the integumentary system.
2. What kind of tissue is the epidermis? In which stratum of the epidermis are new cells formed? From which stratum are they sloughed?
3. Define keratinization. What structural changes does keratinization produce to make the skin resistant to abrasion and water loss?
4. What type of tissue is the dermis? What is responsible for its structural strength? How does the dermis supply the epidermis with blood?
5. Name the cells that produce melanin. What happens to the melanin after it is produced? What is the function of melanin?
6. Describe the factors that determine the amount of melanin produced in the skin.
7. How do melanin, blood, carotene, and collagen affect skin color?
8. What type of tissue is the subcutaneous tissue, and what are its functions?
9. What is a hair follicle? Define the root, shaft, and hair bulb of a hair. What kinds of cells are found in a hair?
10. Why is a hair follicle important in the repair of skin?
11. What part of a hair is the site of hair growth? What are the stages of hair growth?
12. What happens when the arrector pili of the skin contract?
13. What secretion do the sebaceous glands produce? What is the function of the secretion?
14. Which glands of the skin are responsible for cooling the body? Which glands are involved in producing body odor?
15. Name the parts of a nail. Where are the cells that make up the nail produced, and what kind of cells make up a nail? What is the lunula? Describe nail growth.
16. How do the components of the integumentary system provide protection?
17. List the types of sensations detected by receptors in the skin.

18. Describe the production of vitamin D by the body. What is the function of vitamin D?

19. How does the integumentary system help regulate body temperature?

20. Name the substances excreted by skin glands. Is the skin an important site of excretion?

21. Why is the skin a useful diagnostic aid? Give three examples of how the skin functions as a diagnostic aid.

22. Define the different categories of burns. How is repair accomplished after each type?

23. What is the most common cause of skin cancer? Describe three types of skin cancer and the risks of each type.

24. What changes occur in the skin as a result of aging?

 CRITICAL THINKING

1. A woman has stretch marks on her abdomen, yet she states that she has never been pregnant. Is this possible?

2. It has been several weeks since Tom has competed in a tennis match. After the match, he discovers that a blister has formed beneath an old callus on his foot and the callus has fallen off. When he examines the callus, it appears yellow. Can you explain why?

3. The lips are muscular folds forming the anterior boundary of the oral cavity. A mucous membrane covers the lips internally, and the skin of the face covers them externally. The red part of the lips (called the vermillion border) is covered by keratinized epithelium that is a transition between the epithelium of the mucous membrane and the facial skin. The vermillion border can become chapped (dry and cracked), whereas the mucous membrane and the facial skin do not. Propose as many reasons as you can to explain why the vermillion border is more prone to drying than the mucous membrane or the facial skin.

4. Pulling on hair can be quite painful, but cutting hair is not painful. Explain.

5. Given what you know about the cause of acne, propose some ways to prevent or treat it.

6. Consider the following statement: Dark-skinned children are more susceptible to rickets (insufficient calcium in the bones) than light-skinned children. Defend or refute this statement.

7. Harry, a light-skinned man, jogs on a cool day. What color would you expect his skin to be (a) after going outside and just before starting to run, (b) during the run, and (c) 5 minutes after the run?

Answers in Appendix D

6 Skeletal System: Bones and Joints

LEARN to PREDICT

Dr. Thomas Moore and Dr. Roberta Rutledge had worked together for almost two decades, and Roberta knew something was bothering Thomas. She noticed him wincing in pain whenever he bent down to retrieve something from a bottom shelf, and he was short-tempered rather than his usual happy self. Roberta knew Thomas would never admit to being injured if it meant he couldn't care for his patients, even if only for a few days. Finally, Roberta convinced him to let her x-ray his lower back. Right away, when Roberta showed Thomas the x-ray, he pointed to the cause of the pain he'd been suffering.

Using knowledge of the vertebral column, predict the source of Thomas's pain, based on his x-ray shown in this photo. Also, using your knowledge of vertebral anatomy, predict the region of the injury and explain why this region of the vertebral column is more prone to this type of injury than other regions.

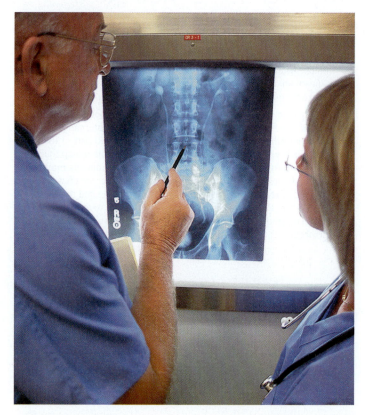

▲ Doctors examine an x-ray of the lumbar region of the spinal column to diagnose Thomas's ailment.

6.1 FUNCTIONS OF THE SKELETAL SYSTEM

Module 5 Skeletal System

Anatomy & Physiology REVEALED®
aprevealed.com

Learning Outcome *After reading this section, you should be able to*

A. Explain the functions of the skeletal system.

Sitting, standing, walking, picking up a pencil, and taking a breath all involve the skeletal system. Without the skeletal system, there would be no rigid framework to support the soft tissues of the body and no system of joints and levers to allow the body to move. The skeletal system consists of bones, such as those shown in figure 6.1, as well as their associated connective tissues, which include cartilage, tendons, and ligaments. The term *skeleton* is derived from a Greek word meaning dried. But the skeleton is far from being dry and nonliving. Rather, the skeletal system consists of dynamic, living tissues that are able to grow, detect pain stimuli, adapt to stress, and undergo repair after injury.

A joint, or an articulation, is a place where two bones come together. Many joints are movable, although some of them allow only limited movement; others allow no apparent movement. The structure of a given joint is directly correlated to its degree of movement.

Although the skeleton is usually thought of as the framework of the body, the skeletal system has many other functions in addition to support. The major functions of the skeletal system include

1. *Support.* Rigid, strong bone is well suited for bearing weight and is the major supporting tissue of the body. Cartilage provides firm yet flexible support within certain structures, such as the nose, external ear, thoracic cage, and trachea. Ligaments are strong bands of fibrous connective tissue that attach to bones and hold them together.
2. *Protection.* Bone is hard and protects the organs it surrounds. For example, the skull encloses and protects the brain, and the vertebrae surround the spinal cord. The rib cage protects the heart, lungs, and other organs of the thorax.

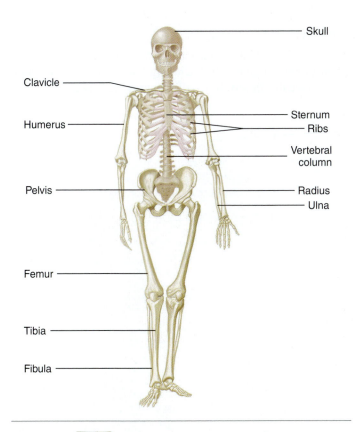

Figure 6.1 **AP|R** Major Bones of the Skeletal System

3. *Movement.* Skeletal muscles attach to bones by tendons, which are strong bands of connective tissue. Contraction of the skeletal muscles moves the bones, producing body movements. Joints, where two or more bones come together, allow movement between bones. Smooth cartilage covers the ends of bones within some joints, allowing the bones to move freely. Ligaments allow some movement between bones but prevent excessive movement.

4. *Storage.* Some minerals in the blood—principally, calcium and phosphorus—are stored in bone. Should blood levels of these minerals decrease, the minerals are released from bone into the blood. Adipose tissue is also stored within bone cavities. If needed, the lipids are released into the blood and used by other tissues as a source of energy.

5. *Blood cell production.* Many bones contain cavities filled with red bone marrow, which produces blood cells and platelets (see chapter 11).

6.2 EXTRACELLULAR MATRIX

Learning Outcome *After reading this section, you should be able to*

A. Describe the components of the extracellular matrix, and explain the function of each.

The bone, cartilage, tendons, and ligaments of the skeletal system are all connective tissues. Their characteristics are largely determined by the composition of their extracellular matrix. The matrix always contains collagen, ground substance, and other organic molecules, as well as water and minerals. But the types and quantities of these substances differ in each type of connective tissue.

Collagen (kol′lă-jen; *koila,* glue + *-gen,* producing) is a tough, ropelike protein. **Proteoglycans** (prō′tē-ō-glī′kanz; *proteo,* protein + *glycan,* polysaccharide) are large molecules consisting of polysaccharides attached to core proteins, similar to the way needles of a pine tree are attached to the tree's branches. The proteoglycans form large aggregates, much as pine branches combine to form a whole tree. Proteoglycans can attract and retain large amounts of water between their polysaccharide "needles."

The extracellular matrix of **tendons** and **ligaments** contains large amounts of collagen fibers, making these structures very tough, like ropes or cables. The extracellular matrix of **cartilage** (kar′ti-lij) contains collagen and proteoglycans. Collagen makes cartilage tough, whereas the water-filled proteoglycans make it smooth and resilient. As a result, cartilage is relatively rigid, but it springs back to its original shape after being bent or slightly compressed. It is an excellent shock absorber.

The extracellular matrix of bone contains collagen and minerals, including calcium and phosphate. The ropelike collagen fibers, like the reinforcing steel bars in concrete, lend flexible strength to the bone. The mineral component, like the concrete itself, gives the bone compression (weight-bearing) strength. Most of the mineral in bone is in the form of calcium phosphate crystals called **hydroxyapatite** (hī-drok′sē-ap-ă-tīt).

Predict 2

What would a bone be like if all of the minerals were removed? What would it be like if all of the collagen were removed?

 A CASE IN POINT

Brittle Bone Disease

May Trix is a 10-year-old girl who has a history of numerous broken bones. At first, physicians suspected she was a victim of child abuse, but eventually they determined that she has **brittle bone disease,** or *osteogenesis imperfecta,* which literally means imperfect bone formation. May is short for her age, and her limbs are short and bowed. Her vertebral column is also abnormally curved. Brittle bone disease is a rare disorder caused by any one of a number of faulty genes that results in either too little collagen formation or poor quality collagen. As a result, the bone matrix has decreased flexibility and is more easily broken than normal bone.

6.3 GENERAL FEATURES OF BONE

Learning Outcomes *After reading this section, you should be able to*

A. Explain the structural differences between compact bone and spongy bone.

B. Outline the processes of bone ossification, growth, remodeling, and repair.

There are four categories of bone, based on their shape: long, short, flat, and irregular. **Long bones** are longer than they are wide. Most of the bones of the upper and lower limbs are long bones. **Short bones** are approximately as wide as they are long; examples are the bones of the wrist and ankle. **Flat bones** have a relatively thin, flattened shape. Examples of flat bones are certain skull bones, the ribs, the scapulae (shoulder blades), and the sternum. **Irregular bones** include the vertebrae and facial bones, which have shapes that do not fit readily into the other three categories.

Structure of a Long Bone

A long bone serves as a useful model for illustrating the parts of a typical bone (figure 6.2). Each long bone consists of a central shaft, called the **diaphysis** (dī-af′i-sis; growing between), and two

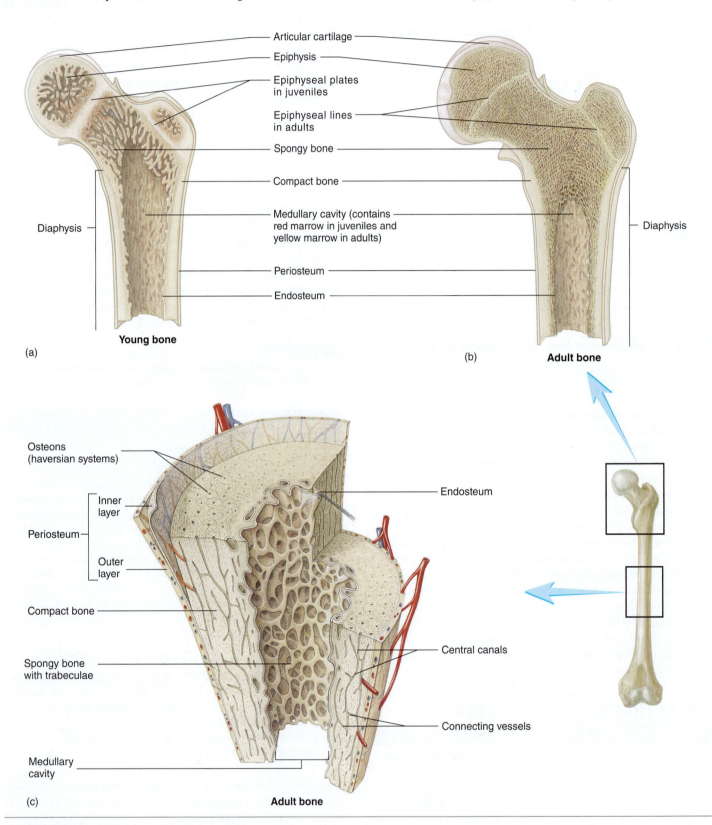

Figure 6.2 Structure of a Long Bone

(*a*) Young long bone (the femur) showing the epiphysis, epiphyseal plates, and diaphysis. (*b*) Adult long bone with epiphyseal lines.
(*c*) Internal features of a portion of the diaphysis in (*a*).

ends, each called an **epiphysis** (e-pif′i-sis; growing upon). A thin layer of **articular** (ar-tik′ū-lăr; joint) **cartilage** covers the ends of the epiphyses where the bone articulates (joins) with other bones. A long bone that is still growing has an **epiphyseal plate,** or *growth plate,* composed of cartilage, between each epiphysis and the diaphysis (figure 6.2*a*). The epiphyseal plate is where the bone grows in length. When bone growth stops, the cartilage of each epiphyseal plate is replaced by bone and becomes an **epiphyseal line** (figure 6.2*b*).

Bones contain cavities, such as the large **medullary cavity** in the diaphysis, as well as smaller cavities in the epiphyses of long bones and in the interior of other bones. These spaces are filled with soft tissue called **marrow. Yellow marrow** consists mostly of adipose tissue. **Red marrow** consists of blood-forming cells and is the only site of blood formation in adults (see chapter 11). Children's bones have proportionately more red marrow than do adult bones because, as a person ages, red marrow is mostly replaced by yellow marrow. In adults, red marrow is confined to the bones in the central axis of the body and in the most proximal epiphyses of the limbs.

Most of the outer surface of bone is covered by dense connective tissue called the **periosteum** (per-ē-os′tē-ŭm; *peri,* around + *osteon,* bone), which consists of two layers and contains blood vessels and nerves (figure 6.2*c*). The surface of the medullary cavity is lined with a thinner connective tissue membrane, the **endosteum** (en-dos′tē-ŭm; *endo,* inside).

Histology of Bone

The periosteum and endosteum contain **osteoblasts** (os′tē-ō-blasts; bone-forming cells), which function in the formation of bone, as well as in the repair and remodeling of bone. When osteoblasts become surrounded by matrix, they are referred to as **osteocytes** (os′tē-ō-sītz; bone cells). **Osteoclasts** (os′tē-ō-klastz; bone-eating cells) are also present and contribute to bone repair and remodeling by removing existing bone.

Bone is formed in thin sheets of extracellular matrix called **lamellae** (lă-mel′ē; plates), with osteocytes located between the lamellae within spaces called **lacunae** (lă-koo′nē; a hollows) (figure 6.3). Cell processes extend from the osteocytes across the extracellular matrix of the lamellae within tiny canals called **canaliculi** (kan-ă-lik′ū-lī; sing. canaliculus, little canal).

Bone tissue found throughout the skeleton is divided into two major types, based on the histological structure. **Compact bone** is mostly solid matrix and cells. **Spongy bone,** or *cancellous* (kan′sĕ-lŭs) *bone,* consists of a lacy network of bone with many small, marrow-filled spaces.

Compact Bone

Compact bone (figure 6.3) forms most of the diaphysis of a long bone and the thinner surfaces of all other bones. As you can see in figure 6.3, compact bone has a predictable pattern of repeating units. These units are called **osteons** (os′tē-onz). Each osteon consists of concentric rings of lamellae surrounding a **central canal**, or *Haversian* (ha-ver′shan) *canal.* As described earlier, osteocytes are located in lacunae between the lamellae of each osteon. Blood vessels that run parallel to the long axis of the bone are located in

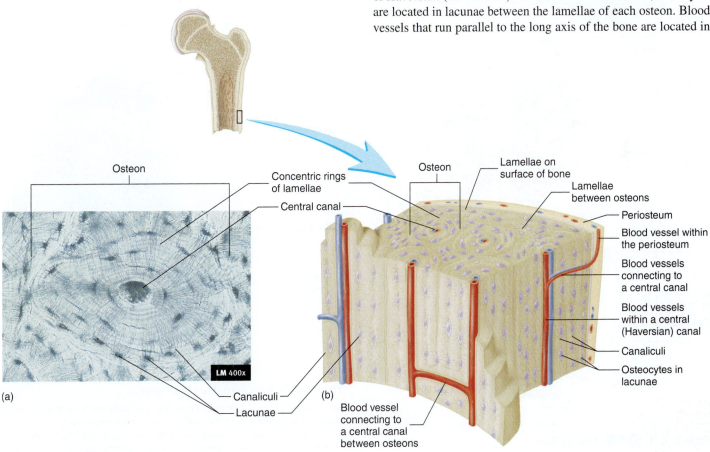

Osteon

Concentric rings of lamellae

Central canal

Canaliculi

Lacunae

LM 400x

(a)

Osteon

Lamellae on surface of bone

Lamellae between osteons

Periosteum

Blood vessel within the periosteum

Blood vessels connecting to a central canal

Blood vessels within a central (Haversian) canal

Canaliculi

Osteocytes in lacunae

(b)

Blood vessel connecting to a central canal between osteons

Figure 6.3 **AP|R** Structure of Bone Tissue

(*a*) Photomicrograph of compact bone. (*b*) Fine structure of compact bone.

the central canals. Osteocytes are connected to one another by cell processes in canaliculi. The canaliculi give the osteon the appearance of having tiny cracks within the lamellae.

Nutrients leave the blood vessels of the central canals and diffuse to the osteocytes through the canaliculi. Waste products diffuse in the opposite direction. The blood vessels in the central canals, in turn, are connected to blood vessels in the periosteum and endosteum.

Spongy Bone

Spongy bone, so called because of its appearance, is located mainly in the epiphyses of long bones. It forms the interior of all other bones. Spongy bone consists of delicate interconnecting rods or plates of bone called **trabeculae** (tră-bek′ū-lē; beams), which resemble the beams or scaffolding of a building (figure 6.4*a*). Like scaffolding, the trabeculae add strength to a bone without the added weight that would be present if the bone were solid mineralized matrix. The spaces between the trabeculae are filled with marrow. Each trabecula consists of several lamellae with osteocytes between them (figure 6.4*b*). Usually, no blood vessels penetrate the trabeculae, and the trabeculae have no central canals. Nutrients exit vessels in the marrow and pass by diffusion through canaliculi to the osteocytes of the trabeculae.

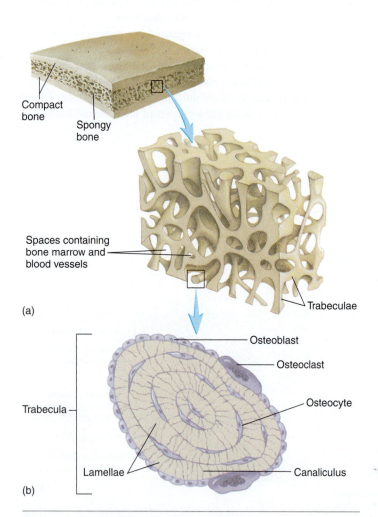

Figure 6.4 AP|R Spongy Bone

(*a*) Beams of bone, the trabeculae, surround spaces in the bone. In life, the spaces are filled with red or yellow bone marrow and with blood vessels. (*b*) Transverse section of a trabecula.

Bone Ossification

Ossification (os′i-fi-kā′shŭn; *os,* bone + *facio,* to make) is the formation of bone by osteoblasts. After an osteoblast becomes completely surrounded by bone matrix, it becomes a mature bone cell, or osteocyte. In the fetus, bones develop by two processes, each involving the formation of bone matrix on preexisting connective tissue (figure 6.5). Bone formation that occurs within connective tissue membranes is called intramembranous ossification, and bone formation that occurs inside cartilage is called endochondral ossification. Both types of bone formation result in compact and spongy bone.

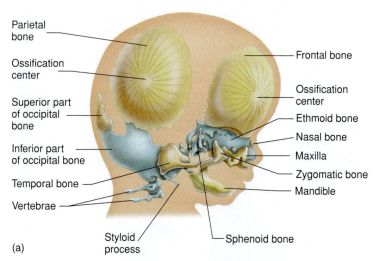

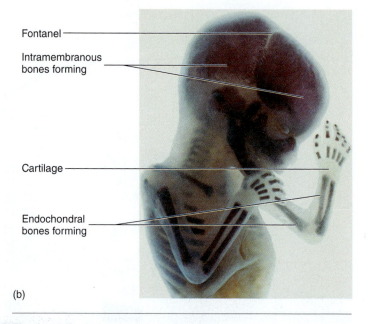

Figure 6.5 Bone Formation in a Fetus

(*a*) Intramembranous ossification occurs in a 12-week-old fetus at ossification centers in the flat bones of the skull (*yellow*). Endochondral ossification occurs in the bones forming the inferior part of the skull (*blue*). (*b*) Radiograph of an 18-week-old fetus, showing intramembranous and endochondral ossification. Intramembranous ossification occurs at ossification centers in the flat bones of the skull. Endochondral ossification has formed bones in the diaphyses of long bones. The epiphyses are still cartilage at this stage of development.

Intramembranous (in′tră-mem′bră-nŭs) **ossification** occurs when osteoblasts begin to produce bone in connective tissue membranes. This occurs primarily in the bones of the skull. Osteoblasts line up on the surface of connective tissue fibers and begin depositing bone matrix to form trabeculae. The process begins in areas called **ossification centers** (figure 6.5a), and the trabeculae radiate out from the centers. Usually, two or more ossification centers exist in each flat skull bone, and the skull bones result from fusion of these centers as they enlarge. The trabeculae are constantly remodeled after their initial formation, and they may enlarge or be replaced by compact bone.

The bones at the base of the skull and most of the remaining skeletal system develop through the process of **endochondral ossification** from cartilage models. The cartilage models have the general shape of the mature bone (figure 6.6, step 1). During endochondral ossification, cartilage cells, called **chondrocytes,** increase in number, enlarge, and die. Then the cartilage matrix becomes calcified (figure 6.6, step 2). As this process is occurring in the center of the cartilage model, blood vessels accumulate in the perichondrium. The presence of blood vessels in the outer surface of future bone causes some of the unspecified connective tissue cells on the surface to become osteoblasts. These osteoblasts then produce a collar of bone around part of the outer surface of the diaphysis, and the perichondrium becomes periosteum in that area. Blood vessels also grow into the center of the diaphyses, bringing in osteoblasts and stimulating ossification. The center part of the diaphysis, where bone first begins to appear, is called the **primary ossification center** (figure 6.6, step 3). Osteoblasts invade spaces in the center of the bone left by the dying cartilage cells. Some of the calcified cartilage matrix is removed by osteoclasts, and the osteoblasts line up on the remaining calcified matrix and begin to form bone trabeculae. As the bone develops, it

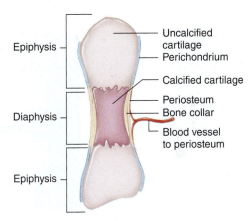

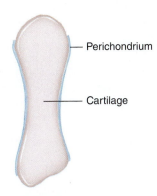

1 A cartilage model, with the general shape of the mature bone, is produced by chondrocytes. A perichondrium surrounds most of the cartilage model.

2 The chondrocytes enlarge, and cartilage is calcified. A bone collar is produced, and the perichondrium of the diaphysis becomes the periosteum.

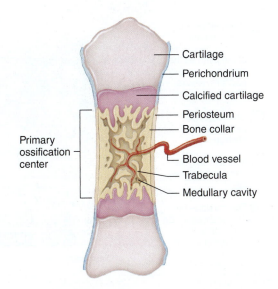

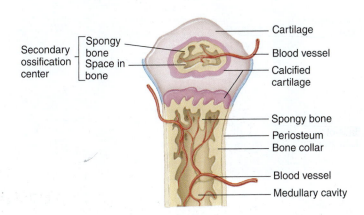

3 A primary ossification center forms as blood vessels and osteoblasts invade the calcified cartilage. The osteoblasts lay down bone matrix, forming trabeculae.

4 Secondary ossification centers form in the epiphyses of long bones.

PROCESS Figure 6.6 Endochondral Ossification of a Long Bone

is constantly changing. A medullary cavity forms in the center of the diaphysis as osteoclasts remove bone and calcified cartilage, which are replaced by bone marrow. Later, **secondary ossification centers** form in the epiphyses (figure 6.6, step 4).

Bone Growth

Bone growth occurs by the deposition of new bone lamellae onto existing bone or other connective tissue. As osteoblasts deposit new bone matrix on the surface of bones between the periosteum and the existing bone matrix, the bone increases in width, or diameter. This process is called appositional growth. Growth in the length of a bone, which is the major source of increased height in an individual, occurs in the epiphyseal plate. This type of bone growth occurs through endochondral ossification (figure 6.7). Chondrocytes increase in number on the epiphyseal side of the epiphyseal plate. They line up in columns parallel to the long axis of the bone, causing the bone to elongate. Then the chondrocytes

enlarge and die. The cartilage matrix becomes calcified. Much of the cartilage that forms around the enlarged cells is removed by osteoclasts, and the dying chondrocytes are replaced by osteoblasts. The osteoblasts start forming bone by depositing bone lamellae on the surface of the calcified cartilage. This process produces bone on the diaphyseal side of the epiphyseal plate.

Predict 3

Describe the appearance of an adult if cartilage growth did not occur in the long bones during childhood.

Bone Remodeling

Bone remodeling involves the removal of existing bone by osteoclasts and the deposition of new bone by osteoblasts. Bone remodeling occurs in all bone. Remodeling is responsible for changes in bone shape, the adjustment of bone to stress, bone repair, and calcium ion regulation in the body fluids. Remodeling is also involved in

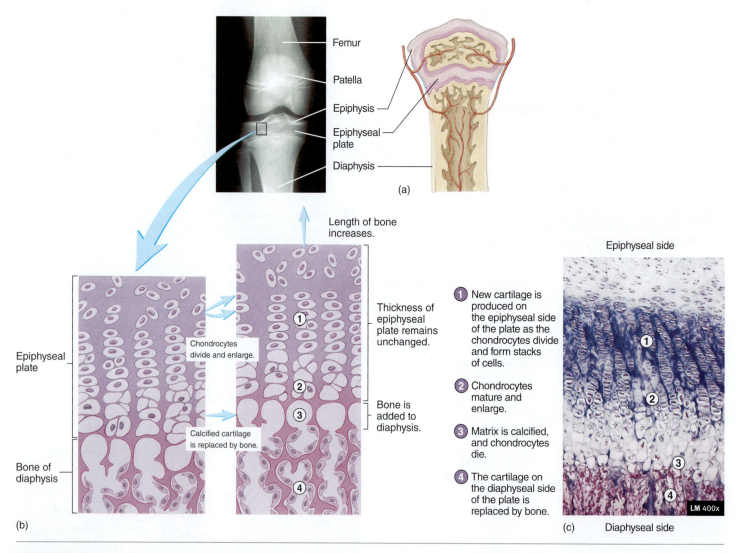

PROCESS Figure 6.7 Endochondral Bone Growth

(*a*) Location of the epiphyseal plate in a long bone. (*b*) As the chondrocytes of the epiphyseal plate divide and align in columns, the cartilage expands toward the epiphysis, and the bone elongates. At the same time, the older cartilage is calcified and then replaced by bone, which is remodeled, resulting in expansion of the medullary cavity of the diaphysis. The net result is an epiphyseal plate that remains uniform in thickness through time but is constantly moving toward the epiphysis, resulting in elongation of the bone. (*c*) Photomicrograph of an epiphyseal plate, demonstrating chondrocyte division and enlargement and the areas of calcification and ossification.

bone growth when newly formed spongy bone in the epiphyseal plate forms compact bone. A long bone increases in length and diameter as new bone is deposited on the outer surface and growth occurs at the epiphyseal plate. At the same time, bone is removed from the inner, medullary surface of the bone. As the bone diameter increases, the thickness of the compact bone relative to the medullary cavity tends to remain fairly constant. If the size of the medullary cavity did not also increase as bone size increased, the compact bone of the diaphysis would become thick and very heavy.

Because bone is the major storage site for calcium in the body, bone remodeling is important to maintain blood calcium levels within normal limits. Calcium is removed from bones when blood calcium levels decrease, and it is deposited when dietary calcium is adequate. This removal and deposition is under hormonal control (see "Bone and Calcium Homeostasis" in the next section).

If too much bone is deposited, the bones become thick or develop abnormal spurs or lumps that can interfere with normal function. Too little bone formation or too much bone removal, as occurs in osteoporosis, weakens the bones and makes them susceptible to fracture (see Systems Pathology, "Osteoporosis").

Bone Repair

Sometimes a bone is broken and needs to be repaired. When this occurs, blood vessels in the bone are also damaged. The vessels bleed, and a clot (hematoma) forms in the damaged area (figure 6.8, step 1). Two to three days after the injury, blood vessels and cells from surrounding tissues begin to invade the clot. Some of these cells produce a fibrous network of connective tissue between the broken bones, which holds the bone fragments together and fills the gap between them. Other cells produce islets of cartilage in the fibrous network. The network of fibers and islets of cartilage between the two bone fragments is called a *callus* (figure 6.8, step 2).

Osteoblasts enter the callus and begin forming spongy bone (figure 6.8, step 3). Spongy bone formation in the callus is usually complete 4–6 weeks after the injury. Immobilization of the bone is critical up to this time because movement can refracture the delicate

new matrix. Subsequently, the spongy bone is slowly remodeled to form compact and spongy bone, and the repair is complete (figure 6.8, step 4). Although immobilization at a fracture point is critical during the early stages of bone healing, complete immobilization is not good for the bone, the muscles, or the joints. Not long ago, it was common practice to immobilize a bone completely for as long as 10 weeks. But we now know that, if a bone is immobilized for as little as 2 weeks, the muscles associated with that bone may lose as much as half their strength. Furthermore, if a bone is completely immobilized, it is not subjected to the normal mechanical stresses that help it form. Bone matrix is reabsorbed, and the strength of the bone decreases. In experimental animals, complete immobilization of the back for 1 month resulted in up to a threefold decrease in vertebral compression strength. Modern therapy attempts to balance bone immobilization with enough exercise to keep muscle and bone from decreasing in size and strength and to maintain joint mobility. These goals are accomplished by limiting the amount of time a cast is left on the patient and by using "walking casts," which allow some stress on the bone and some movement. Total healing of the fracture may require several months. If a bone heals properly, the healed region can be even stronger than the adjacent bone.

6.4 BONE AND CALCIUM HOMEOSTASIS

Learning Outcomes   *After reading this section, you should be able to*

A. Explain the role of bone in calcium homeostasis.
B. Describe how parathyroid hormone and calcitonin influence bone health and calcium homeostasis.

Bone is the major storage site for calcium in the body, and movement of calcium into and out of bone helps determine blood calcium levels, which is critical for normal muscle and nervous system function. Calcium (Ca^{2+}) moves into bone as osteoblasts build new bone and out of bone as osteoclasts break down bone (figure 6.9). When osteoblast and osteoclast activity is balanced, the movements of calcium into and out of a bone are equal.

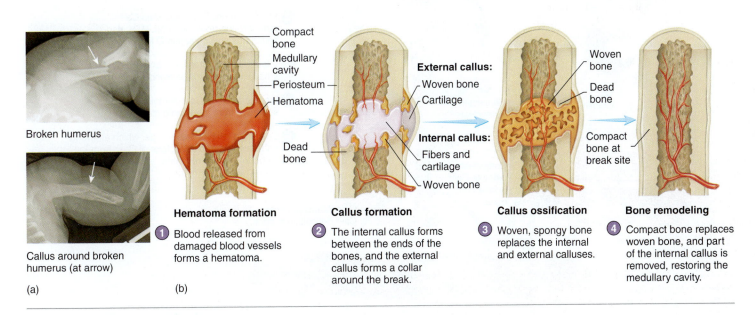

Broken humerus

Callus around broken humerus (at arrow)

(a)

Compact bone
Medullary cavity
Periosteum
Hematoma
Dead bone

External callus:
- Woven bone
- Cartilage

Internal callus:
- Fibers and cartilage
- Woven bone

Woven bone
Dead bone

Compact bone at break site

(b)

Hematoma formation
1. Blood released from damaged blood vessels forms a hematoma.

Callus formation
2. The internal callus forms between the ends of the bones, and the external callus forms a collar around the break.

Callus ossification
3. Woven, spongy bone replaces the internal and external calluses.

Bone remodeling
4. Compact bone replaces woven bone, and part of the internal callus is removed, restoring the medullary cavity.

PROCESS Figure 6.8 **Bone Repair**

CLINICAL IMPACT   Bone Fractures

Bone fractures can be classified as **open** (or *compound*), if the bone protrudes through the skin, and **closed** (or *simple*), if the skin is not perforated. Figure 6A illustrates some of the different types of fractures. If the fracture totally separates the two bone fragments, it is called **complete;** if it doesn't, it is called **incomplete.** An incomplete fracture that occurs on the convex side of the curve of a bone is called a **greenstick fracture.** A **comminuted** (kom'i-nū-ted; broken into small pieces) fracture is one in which the bone breaks into more than two fragments. An **impacted** fracture occurs when one of the fragments of one part of the bone is driven into the spongy bone of another fragment.

Fractures can also be classified according to the direction of the fracture line as **linear** (parallel to the long axis);

transverse (at right angles to the long axis); or **oblique** or **spiral** (at an angle other than a right angle to the long axis).

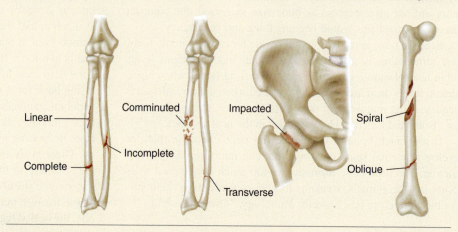

Figure 6A

Types of bone fractures.

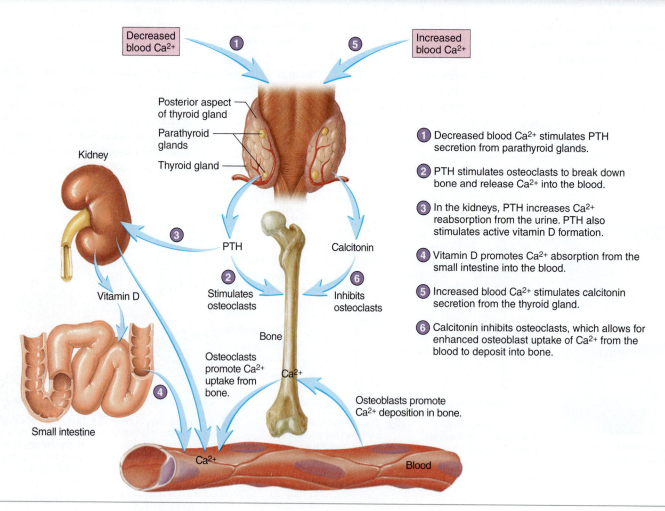

PROCESS Figure 6.9 Calcium Homeostasis

When blood calcium levels are too low, osteoclast activity increases, osteoclasts release calcium from bone into the blood, and blood calcium levels increase. Conversely, if blood calcium levels are too high, osteoclast activity decreases, osteoblasts remove calcium from the blood to produce new bone, and blood calcium levels decrease.

Calcium homeostasis is maintained by three hormones: **parathyroid hormone (PTH)** from the parathyroid glands, vitamin D from the skin or diet, and **calcitonin** (kal-si-tō′nin) from the thyroid gland. PTH and vitamin D are secreted when blood calcium levels are too low and calcitonin is secreted when blood calcium levels are too high.

PTH works through three simultaneous mechanisms to increase blood calcium levels.

1. PTH indirectly stimulates osteoclasts to break down bone, which releases stored calcium into the blood.
2. PTH stimulates the kidney to take up calcium from the urine and return it to the blood.
3. PTH stimulates the formation of active vitamin D, which, in turn, promotes increased calcium absorption from the small intestine.

PTH and vitamin D, therefore, cause blood calcium levels to increase, maintaining homeostatic levels. Decreasing blood calcium levels stimulate PTH secretion.

Calcitonin works to decrease blood calcium levels by inhibiting osteoclast activity. Even in the absence of osteoclast activity, osteoblast activity continues, removing calcium from the blood and depositing it into the bone. Thus, calcitonin maintains homeostatic blood calcium levels by decreasing calcium levels that are too high. In summary, PTH, vitamin D, and calcitonin work together to keep blood calcium levels within the homeostatic range and are described more fully in chapter 10.

6.5 GENERAL CONSIDERATIONS OF BONE ANATOMY

Learning Outcome   *After reading this section, you should be able to*

A. List and define the major features of a typical bone.

It is traditional to list 206 bones in the average adult skeleton (table 6.1), although the actual number varies from person to person and decreases with age as some bones fuse.

Anatomists use several common terms to describe the features of bones (table 6.2). For example, a hole in a bone is called a **foramen** (fō-rā′men; pl. foramina, fō-rā′min-ă; *foro,* to pierce). A foramen usually exists in a bone because some structure, such as a nerve or blood vessel, passes through the bone at that point. If

TABLE 6.1	Named Bones in the Adult Human Skeleton		
Bones			**Number**
Axial Skeleton			
Skull			
Braincase			
Paired	Parietal		2
	Temporal		2
Unpaired	Frontal		1
	Occipital		1
	Sphenoid		1
	Ethmoid		1
Face			
Paired	Maxilla		2
	Zygomatic		2
	Palatine		2
	Nasal		2
	Lacrimal		2
	Inferior nasal concha		2
Unpaired	Mandible		1
	Vomer		1
	TOTAL SKULL		22
Auditory Ossicles	Malleus		2
	Incus		2
	Stapes		2
	TOTAL		6
Hyoid			1
Vertebral Column			
Cervical vertebrae			7
Thoracic vertebrae			12
Lumbar vertebrae			5
Sacrum			1
Coccyx			1
	TOTAL VERTEBRAL COLUMN		26

Bones		**Number**
Thoracic Cage		
Ribs		24
Sternum (3 parts, sometimes considered 3 bones)		1
	TOTAL THORACIC CAGE	25
	TOTAL AXIAL SKELETON	80
Appendicular Skeleton		
Pectoral Girdle		
Scapula		2
Clavicle		2
Upper Limb		
Humerus		2
Ulna		2
Radius		2
Carpal bones		16
Metacarpal bones		10
Phalanges		28
	TOTAL GIRDLE AND UPPER LIMB	64
Pelvic Girdle		
Coxal bone		2
Lower Limb		
Femur		2
Tibia		2
Fibula		2
Patella		2
Tarsal bones		14
Metatarsal bones		10
Phalanges		28
	TOTAL GIRDLE AND LOWER LIMB	62
	TOTAL APPENDICULAR SKELETON	126
	TOTAL BONES	206

Skeletal

TABLE 6.2	Anatomical Terms for Features of Bones
Term	**Description**
Major Features	
Body, shaft	Main portion
Head	Enlarged (often rounded) end
Neck	Constricted area between head and body
Condyle	Smooth, rounded articular surface
Facet	Small, flattened articular surface
Crest	Prominent ridge
Process	Prominent projection
Tubercle, or tuberosity	Knob or enlargement
Trochanter	Large tuberosity found only on proximal femur
Epicondyle	Enlargement near or above a condyle
Openings or Depressions	
Foramen	Hole
Canal, meatus	Tunnel
Fissure	Cleft
Sinus	Cavity
Fossa	Depression

the hole is elongated into a tunnel-like passage through the bone, it is called a **canal** or a **meatus** (mē-ā′tus; a passage). A depression in a bone is called a **fossa** (fos′ă). A lump on a bone is called a **tubercle** (too′ber-kl; a knob) or a **tuberosity** (too′ber-os′i-tē), and a projection from a bone is called a **process.** Most tubercles and processes are sites of muscle attachment on the bone. Increased muscle pull, as occurs when a person lifts weights to build up muscle mass, can increase the size of some tubercles. The smooth, rounded end of a bone, where it forms a joint with another bone, is called a **condyle** (kon′dīl; knuckle).

The bones of the skeleton are divided into axial and appendicular portions (figure 6.10).

6.6 AXIAL SKELETON

Learning Outcomes *After reading this section, you should be able to*

A. Name the bones of the skull and describe their main features as seen from the lateral, frontal, internal, and inferior views.
B. List the bones that form the majority of the nasal septum.
C. Describe the locations and functions of the paranasal sinuses.
D. List the bones of the braincase and the face.
E. Describe the shape of the vertebral column, and list its divisions.
F. Discuss the common features of the vertebrae and contrast vertebrae from each region of the vertebral column.
G. List the bones and cartilage of the rib cage, including the three types of ribs.

The axial skeleton is composed of the skull, the vertebral column, and the thoracic cage.

Skull

The 22 bones of the skull are divided into those of the braincase and those of the face (see table 6.1). The **braincase,** which encloses the cranial cavity, consists of 8 bones that immediately surround and protect the brain; 14 **facial bones** form the structure of the face. Thirteen of the facial bones are rather solidly connected to form the bulk of the face. The mandible, however, forms a freely movable joint with the rest of the skull. There are also three auditory ossicles (os′i-klz) in each middle ear (six total).

Many students studying anatomy never see the individual bones of the skull. Even if they do, it makes more sense from a functional, or clinical, perspective to study most of the bones as they appear together in the intact skull because many of the anatomical features of the skull cannot be fully appreciated by examining the separate bones. For example, several ridges on the skull cross more than one bone, and several foramina are located between bones rather than within a single bone. For these reasons, it is more relevant to think of the skull, excluding the mandible, as a single unit. The major features of the intact skull are therefore described from four views.

Lateral View

The **parietal bones** (pă-rī′ĕ-tăl; wall) and **temporal** (tem′pō-răl) **bones** form a large portion of the side of the head (figure 6.11). (The word *temporal* refers to time, and the temporal bone is so named because the hairs of the temples turn white, indicating the passage of time.) These two bones join each other on the side of the head at the **squamous** (skwā′mŭs) **suture.** A suture is a joint uniting bones of the skull. Anteriorly, the parietal bone is joined to the **frontal** (forehead) **bone** by the **coronal** (kōr′ō-năl; *corona,* crown) **suture,** and posteriorly it is joined to the **occipital** (ok-sip′i-tăl; back of the head) bone by the **lambdoid** (lam′doyd) **suture.** A prominent feature of the temporal bone is a large opening, the **external auditory canal,** a canal that enables sound waves to reach the eardrum. The **mastoid** (mas′toyd) **process** of the temporal bone can be seen and felt as a prominent lump just posterior to the ear. Important neck muscles involved in rotation of the head attach to the mastoid process.

Part of the **sphenoid** (sfē′noyd) **bone** can be seen immediately anterior to the temporal bone. Although it appears to be two small, paired bones on the sides of the skull, the sphenoid bone is actually a single bone that extends completely across the skull. It resembles a butterfly, with its body in the center of the skull and its wings extending to the sides of the skull. Anterior to the sphenoid bone is the **zygomatic** (zī-gō-mat′ik) **bone,** or cheekbone, which can be easily felt. The **zygomatic arch,** which consists of joined processes of the temporal and zygomatic bones, forms a bridge across the side of the face and provides a major attachment site for a muscle moving the mandible.

The **maxilla** (mak-sil′ă; jawbone) forms the upper jaw, and **mandible** (man′di-bl; jaw) forms the lower jaw. The maxilla articulates by sutures to the temporal bone. The maxilla contains the superior set of teeth, and the mandible contains the inferior set of teeth.

Frontal View

The major structures seen from the frontal view are the frontal bone, the zygomatic bones, the maxillae, and the mandible (figure 6.12*a*). The teeth are very prominent in this view. Many bones of the face can be easily felt through the skin (figure 6.12*b*).

Skeletal

Axial Skeleton   **Appendicular Skeleton**   **Axial Skeleton**

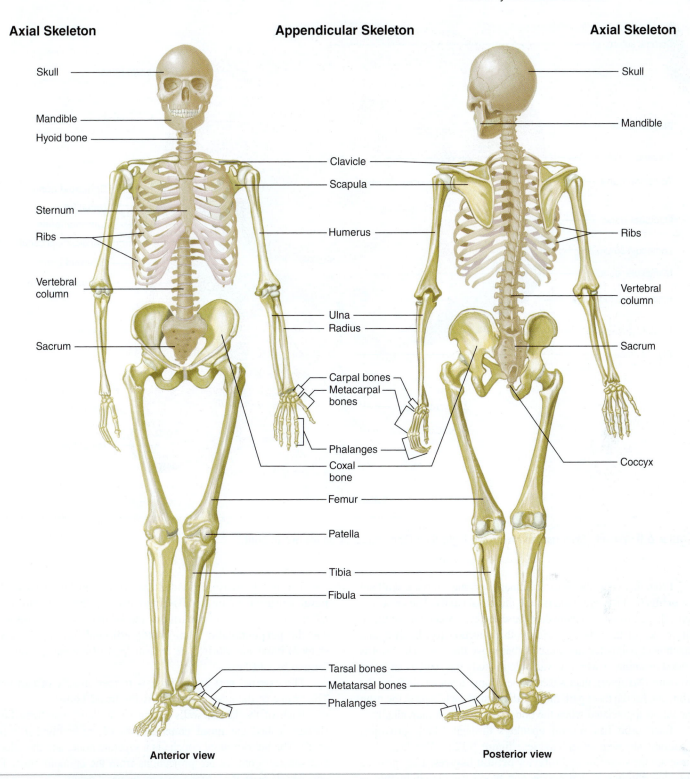

Skull

Mandible

Hyoid bone

Sternum

Ribs

Vertebral
column

Sacrum

Clavicle

Scapula

Humerus

Ulna

Radius

Carpal bones

Metacarpal
bones

Phalanges

Coxal
bone

Femur

Patella

Tibia

Fibula

Tarsal bones

Metatarsal bones

Phalanges

Skull

Mandible

Ribs

Vertebral
column

Sacrum

Coccyx

Anterior view   **Posterior view**

Figure 6.10 $\boxed{\text{AP|R}}$ **Complete Skeleton**

Bones of the axial skeleton are listed in the far left- and right-hand columns; bones of the appendicular skeleton are listed in the
center. (The skeleton is not shown in the anatomical position.)

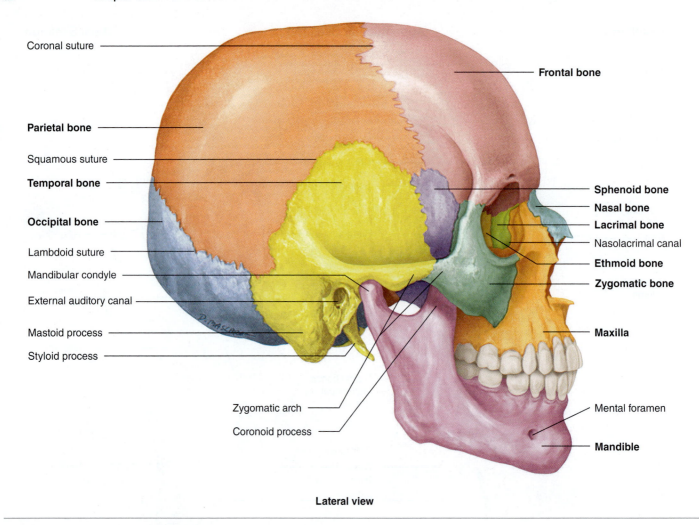

Coronal suture

Frontal bone

Parietal bone

Squamous suture

Temporal bone

Sphenoid bone

Nasal bone

Lacrimal bone

Nasolacrimal canal

Occipital bone

Ethmoid bone

Lambdoid suture

Zygomatic bone

Mandibular condyle

External auditory canal

Maxilla

Mastoid process

Styloid process

Zygomatic arch

Mental foramen

Coronoid process

Mandible

Lateral view

Figure 6.11 AP|R Skull as Seen from the Right Side (The names of bones are in bold.)

From this view, the most prominent openings into the skull are the **orbits** (ōr′bitz; eye sockets) and the **nasal cavity.** The orbits are cone-shaped fossae, so named because the eyes rotate within them. The bones of the orbits provide both protection for the eyes and attachment points for the muscles that move the eyes. The orbit is a good example of why it is valuable to study the skull as an intact structure. No fewer than seven bones come together to form the orbit, and for the most part, the contribution of each bone to the orbit cannot be appreciated when the bones are examined individually.

Each orbit has several openings through which structures communicate with other cavities (figure 6.12a). The largest of these are the **superior** and **inferior orbital fissures.** They provide openings through which nerves and blood vessels communicate with the orbit or pass to the face. The optic nerve, for the sense of vision, passes from the eye through the **optic foramen** and enters the cranial cavity. The **nasolacrimal** (nā-zō-lak′ri-măl; *nasus,* nose + *lacrima,* tear) **canal** (see figure 6.11) passes from the orbit into the nasal cavity. It contains a duct that carries tears from the eyes to the nasal cavity. A small **lacrimal** (lak′ri-măl) **bone** can be seen in the orbit just above the opening of this canal.

Predict 4

Why does your nose run when you cry?

The nasal cavity is divided into right and left halves by a **nasal septum** (sep′tŭm; wall) (figure 6.12a). The bony part of the nasal septum consists primarily of the **vomer** (vō′mer) inferiorly and the **perpendicular plate** of the **ethmoid** (eth′moyd; sieve-shaped) **bone** superiorly. The anterior part of the nasal septum is formed by cartilage.

The external part of the nose is formed mostly of cartilage. The bridge of the nose is formed by the **nasal bones.**

Each of the lateral walls of the nasal cavity has three bony shelves, called the **nasal conchae** (kon′kē; resembling a conch shell). The inferior nasal concha is a separate bone, and the middle and superior conchae are projections from the ethmoid bone. The conchae increase the surface area in the nasal cavity. The increased surface area of the overlying epithelium facilitates moistening and warming of the air inhaled through the nose (see chapter 15).

Several of the bones associated with the nasal cavity have large cavities within them, called the **paranasal** (par-ă-nā′săl; *para,* alongside) **sinuses** (figure 6.13), which open into the nasal cavity. The sinuses decrease the weight of the skull and act as resonating chambers during voice production. Compare a normal voice with the voice of a person who has a cold and whose sinuses are "stopped up." The sinuses are named for the bones where they are located and include the frontal, maxillary, ethmoidal, and sphenoidal sinuses.

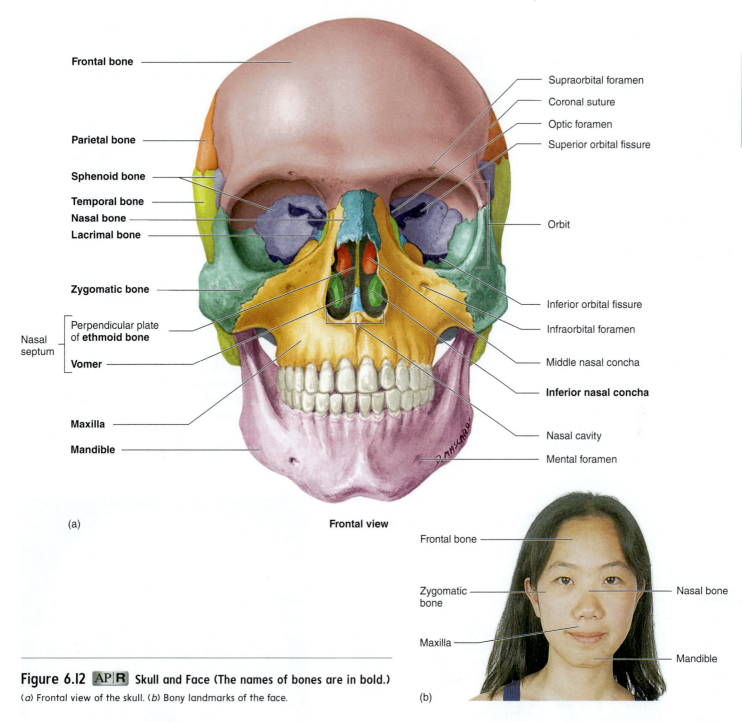

Frontal bone

Parietal bone

Sphenoid bone

Temporal bone

Nasal bone

Lacrimal bone

Zygomatic bone

Perpendicular plate
of **ethmoid bone**

Nasal
septum

Vomer

Maxilla

Mandible

Supraorbital foramen

Coronal suture

Optic foramen

Superior orbital fissure

Orbit

Inferior orbital fissure

Infraorbital foramen

Middle nasal concha

Inferior nasal concha

Nasal cavity

Mental foramen

(a)   **Frontal view**

Frontal bone

Zygomatic
bone

Maxilla

Nasal bone

Mandible

(b)

Figure 6.12 AP|R **Skull and Face (The names of bones are in bold.)**
(*a*) Frontal view of the skull. (*b*) Bony landmarks of the face.

The skull has additional sinuses, called the **mastoid air cells,** which are located inside the mastoid processes of the temporal bone. These air cells open into the middle ear instead of into the nasal cavity. An auditory tube connects the middle ear to the naso-pharynx (upper part of throat).

Interior of the Cranial Cavity

When the floor of the cranial cavity is viewed from above with the roof cut away (figure 6.14), it can be divided roughly into three cranial fossae (anterior, middle, and posterior), which are formed as the developing skull conforms to the shape of the brain. The

bones forming the floor of the cranial cavity, from anterior to posterior, are the frontal, ethmoid, sphenoid, temporal, and occipital bones. Several foramina can be seen in the floor of the middle fossa. These allow nerves and blood vessels to pass through the skull. For example, the foramen rotundum and foramen ovale transmit important nerves to the face. A major artery to the meninges (the membranes around the brain) passes through the foramen spinosum. The internal carotid artery passes through the carotid canal, and the internal jugular vein passes through the jugular foramen (see chapter 13). The large **foramen magnum,** through which the spinal cord joins the brain, is located in the posterior

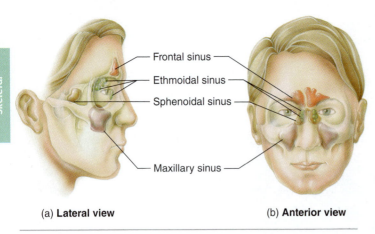

Frontal sinus

Ethmoidal sinus

Sphenoidal sinus

Maxillary sinus

(a) **Lateral view**   (b) **Anterior view**

Figure 6.13 Paranasal Sinuses

fossa. The central region of the sphenoid bone is modified into a structure resembling a saddle, the **sella turcica** (sel′ă tŭr′sĭ-kă; Turkish saddle), which contains the pituitary gland.

Base of Skull Viewed from Below

Many of the same foramina that are visible in the interior of the skull can also be seen in the base of the skull, when viewed from below, with the mandible removed (figure 6.15). Other specialized

structures, such as processes for muscle attachments, can also be seen. The foramen magnum is located in the occipital bone near the center of the skull base. **Occipital condyles** (ok-sip′i-tăl kon′dīlz), the smooth points of articulation between the skull and the vertebral column, are located beside the foramen magnum.

Two long, pointed **styloid** (stī′loyd; stylus or pen-shaped) **processes** project from the inferior surface of the temporal bone. The muscles involved in moving the tongue, the hyoid bone, and the pharynx (throat) originate from this process. The **mandibular fossa,** where the mandible articulates with the temporal bone, is anterior to the mastoid process.

The **hard palate** (pal′ăt) forms the floor of the nasal cavity and the roof of the mouth. The anterior two-thirds of the hard palate is formed by the maxillae, the posterior one-third by the **palatine** (pal′ă-tīn) **bones.** The connective tissue and muscles that make up the **soft palate** extend posteriorly from the hard, or bony, palate. The hard and soft palates separate the nasal cavity and nasopharynx from the mouth, enabling us to chew and breathe at the same time.

Hyoid Bone

The **hyoid bone** (figure 6.16) is an unpaired, U-shaped bone. It is not part of the skull (see table 6.1) and has no direct bony attachment to the skull. Muscles and ligaments attach it to the skull. The hyoid bone provides an attachment for some tongue muscles, and it is an attachment point for important neck muscles that elevate the larynx (voicebox) during speech or swallowing.

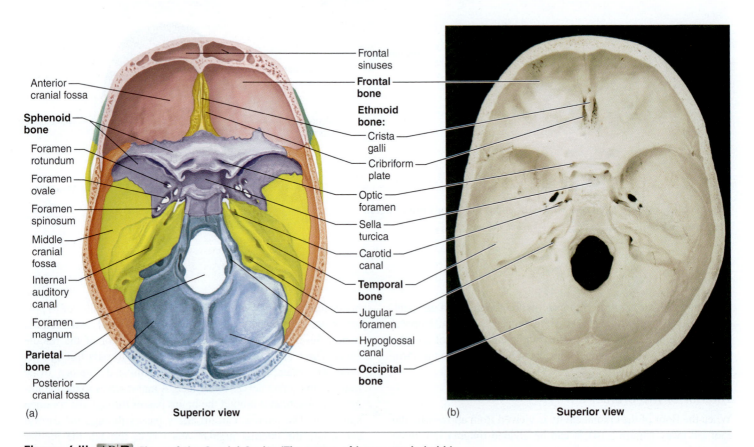

Anterior cranial fossa

Sphenoid bone

Foramen rotundum

Foramen ovale

Foramen spinosum

Middle cranial fossa

Internal auditory canal

Foramen magnum

Parietal bone

Posterior cranial fossa

(a)   **Superior view**

Frontal sinuses

Frontal bone

Ethmoid bone:
Crista galli
Cribriform plate

Optic foramen

Sella turcica

Carotid canal

Temporal bone

Jugular foramen

Hypoglossal canal

Occipital bone

(b)   **Superior view**

Figure 6.14 AP|R Floor of the Cranial Cavity (The names of bones are in bold.)

(a) Drawing of the floor of the cranial cavity. The roof of the skull has been removed, and the floor is viewed from above. (b) Photo of the same view.

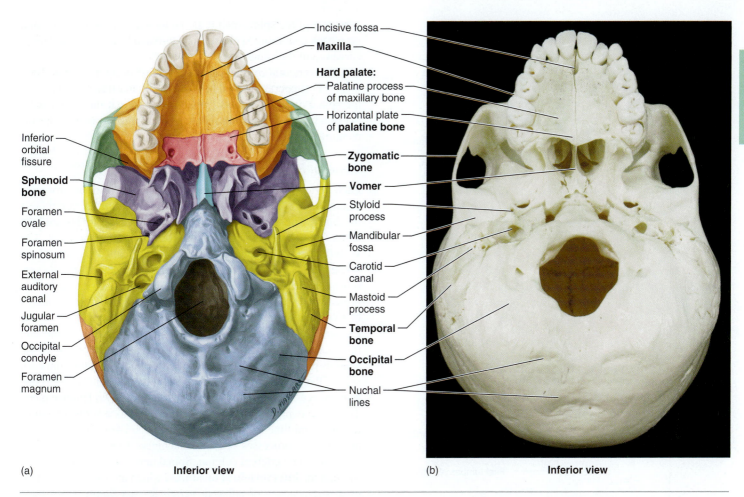

(a) Inferior view

(b) Inferior view

Labels for figure (a):
Incisive fossa
Maxilla
Hard palate:
Palatine process of maxillary bone
Horizontal plate of **palatine bone**
Zygomatic bone
Vomer
Styloid process
Mandibular fossa
Carotid canal
Mastoid process
Temporal bone
Occipital bone
Nuchal lines

Inferior orbital fissure
Sphenoid bone
Foramen ovale
Foramen spinosum
External auditory canal
Jugular foramen
Occipital condyle
Foramen magnum

Figure 6.15 AP|R **Base of the Skull as Viewed from Below (The names of bones are in bold.)**
(*a*) Drawing of skull's inferior with the mandible removed. (*b*) Photo of the same view.

Vertebral Column

The **vertebral column,** or backbone, is the central axis of the skeleton, extending from the base of the skull to slightly past the end of the pelvis. In adults, it usually consists of 26 individual bones, grouped into five regions (figure 6.17; see table 6.1): 7 **cervical** (ser′vǐ-kal; neck) **vertebrae** (ver′tě-brē; *verto,* to turn), 12 **thoracic** (thō-ras′ik) **vertebrae,** 5 **lumbar** (lŭm′bar) **vertebrae,** 1 **sacral** (sā′krǎl) **bone,** and 1 **coccyx** (kok′siks) **bone.** The adult sacral and coccyx bones fuse from 5 and 3–4 individual bones, respectively. For convenience, each of the five regions is identified by a letter, and the vertebrae within each region are numbered: C1–C7, T1–T12, L1–L5, S, and CO. You can remember the number of vertebrae in each region by remembering meal times: 7, 12, and 5.

The adult vertebral column has four major curvatures. The cervical region curves anteriorly, the thoracic region curves posteriorly, the lumbar region curves anteriorly, and the sacral and coccygeal regions together curve posteriorly.

Abnormal vertebral curvatures are not uncommon. **Kyphosis** (kī-fō′sis) is an abnormal posterior curvature of the spine, mostly in the upper thoracic region, resulting in a hunchback condition. **Lordosis** (lōr-dō′sis; curving forward) is an abnormal anterior

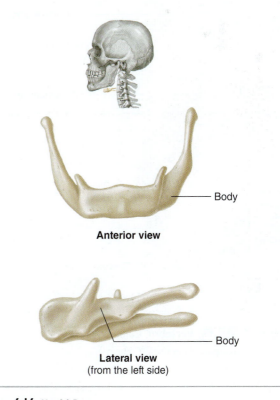

Anterior view
Body

Lateral view
(from the left side)
Body

Figure 6.16 Hyoid Bone

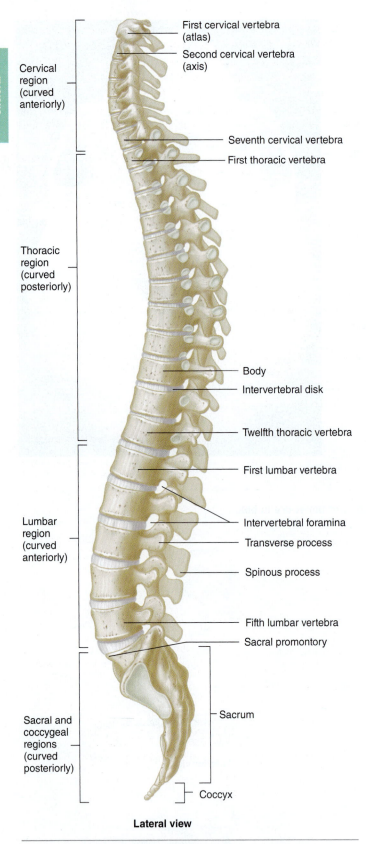

Cervical
region
(curved
anteriorly)

First cervical vertebra
(atlas)

Second cervical vertebra
(axis)

Seventh cervical vertebra

First thoracic vertebra

Thoracic
region
(curved
posteriorly)

Body

Intervertebral disk

Twelfth thoracic vertebra

First lumbar vertebra

Lumbar
region
(curved
anteriorly)

Intervertebral foramina

Transverse process

Spinous process

Fifth lumbar vertebra

Sacral promontory

Sacral and
coccygeal
regions
(curved
posteriorly)

Sacrum

Coccyx

Lateral view

Figure 6.17 AP|R **Vertebral Column**

Complete column viewed from the left side.

curvature of the spine, mainly in the lumbar region, resulting in a swayback condition. **Scoliosis** (skō-lē-ō′sis) is an abnormal lateral curvature of the spine.

The vertebral column performs the following five major functions: (1) supports the weight of the head and trunk; (2) protects the spinal cord; (3) allows spinal nerves to exit the spinal cord; (4) provides a site for muscle attachment; and (5) permits movement of the head and trunk.

General Plan of the Vertebrae

Each vertebra consists of a body, an arch, and various processes (figure 6.18). The weight-bearing portion of each vertebra is the **body.** The vertebral bodies are separated by **intervertebral disks** (see figure 6.17), which are formed by fibrocartilage. The **vertebral arch** surrounds a large opening called the **vertebral foramen.** The vertebral foramina of all the vertebrae form the **vertebral canal,** where the spinal cord is located. The vertebral canal protects the spinal cord from injury. Each vertebral arch consists of two **pedicles** (ped′ĭ-klz), which extend from the body to the transverse process of each vertebra, and two **laminae** (lam′ĭ-nē; thin plates), which extend from the transverse processes to the spinous process. A **transverse process** extends laterally from each side of the arch, between the pedicle and lamina, and a single **spinous process** projects dorsally from where the two laminae meet. The spinous processes can be seen and felt as a series of lumps down the midline of the back (see figure 6.24b). The transverse and spinous processes provide attachment sites for the muscles that move the vertebral column. Spinal nerves exit the spinal cord through the **intervertebral foramina,** which are formed by notches in the pedicles of adjacent vertebrae (see figure 6.17). Each vertebra has a superior and an inferior **articular process** where the vertebrae articulate with each other. Each articular process has a smooth "little face" called an **articular facet** (fas′et).

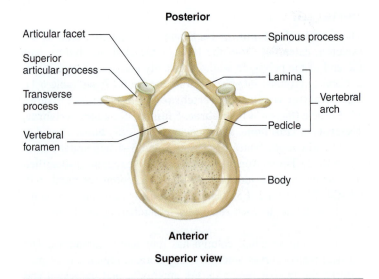

Posterior

Articular facet

Superior
articular process

Transverse
process

Vertebral
foramen

Spinous process

Lamina

Vertebral
arch

Pedicle

Body

Anterior

Superior view

Figure 6.18 **Vertebra**

Regional Differences in Vertebrae

The **cervical vertebrae** (figure 6.19*a–c*) have very small bodies, except for the atlas, which has no body. Because the cervical vertebrae are relatively delicate and have small bodies, dislocations and fractures are more common in this area than in other regions of the vertebral column. Each of the transverse processes has a transverse foramen through which the vertebral arteries pass toward the brain. Several of the cervical vertebrae also have partly split spinous processes. The first cervical vertebra (figure 6.19*a*) is called the **atlas** because it holds up the head, as Atlas in classical mythology held up the world. Movement between the atlas and the occipital bone is responsible for a "yes" motion of the head. It also allows a slight tilting of the head from side to side. The second

cervical vertebra (figure 6.19*b*) is called the **axis** because a considerable amount of rotation occurs at this vertebra, as in shaking the head "no." This rotation occurs around a process called the **dens** (denz), which protrudes superiorly from the axis.

The **thoracic vertebrae** (figure 6.19*d*) possess long, thin spinous processes that are directed inferiorly. The thoracic vertebrae also have extra articular facets on their lateral surfaces that articulate with the ribs.

The **lumbar vertebrae** (figure 6.19*e*) have large, thick bodies and heavy, rectangular transverse and spinous processes. Because the lumbar vertebrae have massive bodies and carry a large amount of weight, ruptured intervertebral disks are more common in this area than in other regions of the column. The superior

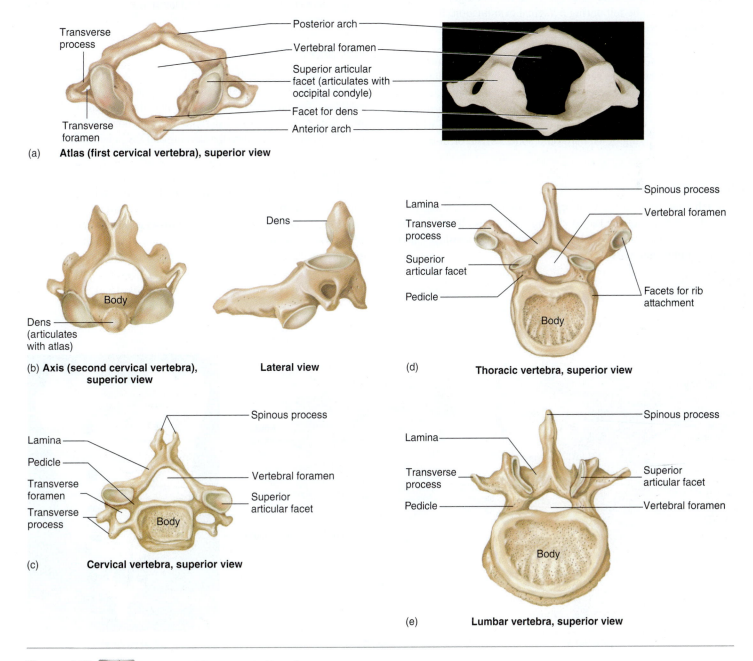

(a) **Atlas (first cervical vertebra), superior view**

(b) **Axis (second cervical vertebra), superior view**   **Lateral view**

(c) **Cervical vertebra, superior view**

(d) **Thoracic vertebra, superior view**

(e) **Lumbar vertebra, superior view**

Figure 6.19   AP|R   Regional Differences in Vertebrae

The posterior portion lies at the top of each illustration.

Skeletal

articular facets of the lumbar vertebrae face medially, whereas the inferior articular facets face laterally. This arrangement tends to "lock" adjacent lumbar vertebrae together, giving the lumbar part of the vertebral column more strength. The articular facets in other regions of the vertebral column have a more "open" position, allowing for more rotational movement but less stability than in the lumbar region.

The five sacral vertebrae are fused into a single bone called the **sacrum** (figure 6.20). The spinous processes of the first four sacral vertebrae form the **median sacral crest.** The spinous process of the fifth vertebra does not form, leaving a **sacral hiatus** (hī-ā-tŭs) at the inferior end of the sacrum, which is often the site of "caudal" anesthetic injections given just before childbirth. The anterior edge of the body of the first sacral vertebra bulges to form the **sacral promontory** (prom′on-tō-rē) (see figure 6.17), a landmark that can be felt during a vaginal examination. It is used as a reference point to determine if the pelvic openings are large enough to allow for normal vaginal delivery of a baby.

The coccyx, or tailbone, usually consists of four more-or-less fused vertebrae. The vertebrae of the coccyx do not have the typical structure of most other vertebrae. They consist of extremely reduced vertebral bodies, without the foramina or processes, usually fused into a single bone. The coccyx is easily broken when a person falls by sitting down hard on a solid surface or in women during childbirth.

Rib Cage

The **rib cage** protects the vital organs within the thorax and prevents the collapse of the thorax during respiration. It consists of the thoracic vertebrae, the ribs with their associated cartilages, and the sternum (figure 6.21).

A CASE IN POINT

Rib Fractures

Han D. Mann's ladder fell as he was working on his roof, and he landed chest-first on the ladder. Three ribs were fractured on his right side. It was difficult for Han to cough, laugh, or even breathe without severe pain in the right side of his chest. The middle ribs are those most commonly fractured, and the portion of each rib that forms the lateral wall of the thorax is the weakest and most commonly broken. The pain from rib fractures occurs because the broken ends move during respiration and other chest movements, stimulating pain receptors. Broken rib ends can damage internal organs, such as the lungs, spleen, liver, and diaphragm. Fractured ribs are not often dislocated, but dislocated ribs may have to be set for proper healing to occur. Binding the chest to limit movement can facilitate healing and lessen pain.

Ribs and Costal Cartilages

The 12 pairs of ribs can be divided into true ribs and false ribs. The **true ribs,** ribs 1–7, attach directly to the sternum by means of costal cartilages. The **false ribs,** ribs 8–12, do not attach directly to the sternum. Ribs 8–10 attach to the sternum by a common cartilage; ribs 11 and 12 do not attach at all to the sternum and are called **floating ribs.**

Sternum

The **sternum** (ster′nŭm), or breastbone (figure 6.21), is divided into three parts: the **manubrium** (mă-nū′brē-ŭm; handle), the **body,** and the **xiphoid** (zif′oyd, zī′foyd; sword) **process.** The sternum resembles a sword, with the manubrium forming the handle, the body forming the blade, and the xiphoid process forming the tip. At the

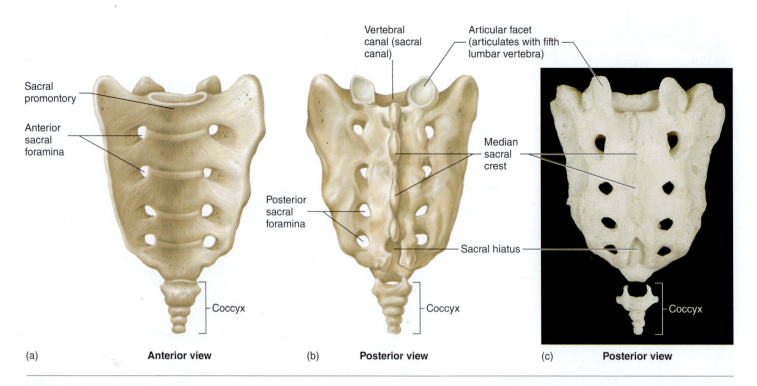

Sacral promontory

Anterior sacral foramina

Coccyx

(a)   **Anterior view**

Vertebral canal (sacral canal)

Articular facet (articulates with fifth lumbar vertebra)

Posterior sacral foramina

Coccyx

(b)   **Posterior view**

Median sacral crest

Sacral hiatus

Coccyx

(c)   **Posterior view**

Figure 6.20 AP|R Sacrum

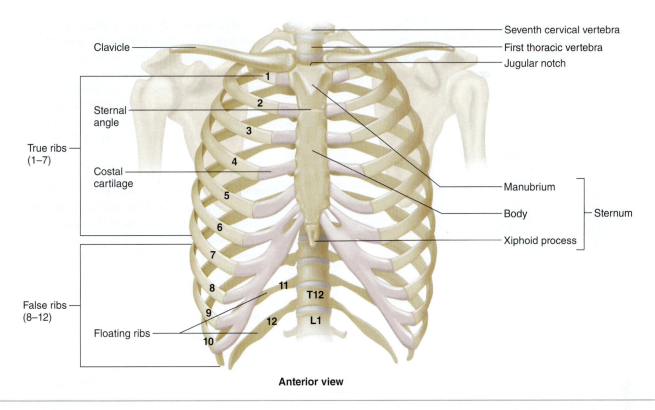

Clavicle

Sternal angle

True ribs (1–7)

Costal cartilage

False ribs (8–12)

Floating ribs

Seventh cervical vertebra

First thoracic vertebra

Jugular notch

Manubrium

Body

Sternum

Xiphoid process

T12

L1

Anterior view

Figure 6.21 AP|R **Rib Cage**

superior end of the sternum, a depression called the **jugular notch** is located between the ends of the clavicles where they articulate with the sternum. A slight elevation, called the **sternal angle,** can be felt at the junction of the manubrium and the body of the sternum. This junction is an important landmark because it identifies the location of the second rib. This identification allows the ribs to be counted; for example, it can help a health professional locate the apex of the heart, which is between the fifth and sixth ribs.

The xiphoid process is another important landmark of the sternum. During cardiopulmonary resuscitation (CPR), it is very important to place the hands over the body of the sternum rather than over the xiphoid process. Pressure applied to the xiphoid process can drive it into an underlying abdominal organ, such as the liver, causing internal bleeding.

6.7 APPENDICULAR SKELETON

Learning Outcomes *After reading this section, you should be able to*

A. Identify the bones that make up the pectoral girdle, and relate their structure and arrangement to the function of the girdle.

B. Name and describe the major bones of the upper limb.

C. Name and describe the bones of the pelvic girdle and explain why the pelvic girdle is more stable than the pectoral girdle.

D. Name the bones that make up the coxal bone. Distinguish between the male and female pelvis.

E. Identify and describe the bones of the lower limb.

The **appendicular** (ap'en-dik'ū-lăr; appendage) skeleton consists of the bones of the upper and lower limbs, as well as the girdles, which attach the limbs to the axial skeleton.

Pectoral Girdle

The **pectoral** (pek'tō-răl) **girdle,** or *shoulder girdle* (figure 6.22), consists of four bones, two scapulae and two clavicles, which attach the upper limb to the body. The **scapula** (skap'ū-lă), or *shoulder blade,* is a flat, triangular bone with three large fossae where muscles extending to the arm are attached (figures 6.23 and 6.24; see figure 6.10). A fourth fossa, the **glenoid** (glen'oyd) **cavity,** is where the head of the humerus connects to the scapula. A ridge, called the **spine,** runs across the posterior surface of the scapula. A projection, called the **acromion** (ă-krō'mē-on; *akron,* tip + *omos,* shoulder) **process,** extends from the scapular spine to form the point of the shoulder. The **clavicle** (klav'i-kl), or *collarbone,* articulates with the scapula at the acromion process. The proximal end of the clavicle is attached to the sternum, providing the only bony attachment of the scapula to the remainder of the skeleton. The clavicle is the first bone to begin ossification in the fetus. This relatively brittle bone may be fractured in the newborn during delivery. The bone remains slender in children and may be broken as a child attempts to take the impact of a fall on an outstretched hand. The clavicle is thicker in adults and is less vulnerable to fracture. Even though it is the first bone to begin ossification, it is the last to complete ossification. The **coracoid** (kōr'ă-koyd) **process** curves below the clavicle and provides for the attachment of arm and chest muscles.

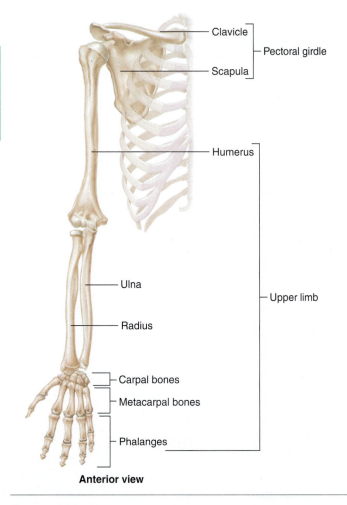

Figure 6.22 Bones of the Pectoral Girdle and Right Upper Limb

Upper Limb

The upper limb consists of the bones of the arm, forearm, wrist, and hand (see figure 6.22).

Arm

The **arm** is the region between the shoulder and the elbow; it contains the **humerus** (hū′mer-ŭs; shoulder) (figure 6.25). The proximal end of the humerus has a smooth, rounded **head,** which attaches the humerus to the scapula at the glenoid cavity. Around the edge of the humeral head is the anatomical neck. When the joint needs to be surgically replaced, this neck is not easily accessible. A more accessible site for surgical removal is at the surgical neck, located at the proximal end of the humeral shaft. Lateral to the head are two tubercles, a **greater tubercle** and a **lesser tubercle.** Muscles originating on the scapula attach to the greater and lesser tubercles and hold the humerus to the scapula. Approximately one-third of the way down the shaft of the humerus, on the lateral surface, is the **deltoid tuberosity,** where the deltoid muscle attaches. The size of the deltoid tuberosity can increase as the result of frequent and powerful pulls from the deltoid muscle. For example, in bodybuilders, the deltoid muscle and the deltoid tuberosity enlarge substantially. Anthropologists, examining ancient human remains, can use the presence of enlarged deltoid tuberosities as

evidence that a person was engaged in lifting heavy objects during life. If the humerus of a person exhibits an unusually large deltoid tuberosity for her age, it may indicate, in some societies, that she was a slave and was required to lift heavy loads. The distal end of the humerus is modified into specialized condyles that connect the humerus to the forearm bones. **Epicondyles** (ep′i-kon′dīlz; *epi,* upon) on the distal end of the humerus, just lateral to the condyles, provide attachment sites for forearm muscles.

Forearm

The **forearm** has two bones: the **ulna** (ŭl′nă) on the medial (little finger) side of the forearm and the **radius** on the lateral (thumb) side (figure 6.26). The proximal end of the ulna forms a **trochlear notch** that fits tightly over the end of the humerus, forming most of the elbow joint. Just proximal to the trochlear notch is an extension of the ulna, called the **olecranon** (ō-lek′ră-non; elbow) **process,** which can be felt as the point of the elbow (see figure 6.28). Just distal to the trochlear notch is a **coronoid** (kōr′ŏ-noyd) **process,** which helps complete the "grip" of the ulna on the distal end of the humerus. The distal end of the ulna forms a head, which articulates with the bones of the wrist, and a **styloid process** is located on its medial side. The ulnar head can be seen as a prominent lump on the posterior ulnar side of the wrist. The proximal end of the radius has a head by which the radius articulates with both the humerus and the ulna. The radius does not attach as firmly to the humerus as the ulna does. The radial head rotates against the humerus and ulna. Just distal to the radial head is a **radial tuberosity,** where one of the arm muscles, the biceps brachii, attaches. The distal end of the radius articulates with the wrist bones. A styloid process is located on the lateral side of the distal end of the radius. The radial and ulnar styloid processes provide attachment sites for ligaments of the wrist.

Wrist

The **wrist** is a relatively short region between the forearm and the hand; it is composed of eight **carpal** (kar′păl; wrist) **bones** (figure 6.27). These eight bones are the scaphoid (skaf′oyd), lunate (lū′nāt), triquetrum (trī-kwē′trŭm), pisiform (pis′i-fōrm), trapezium (tra-pē′zē-ŭm), trapezoid (trap′ě-zoyd), capitate (kap′i-tāt), and hamate (ha′māt). The carpal bones are arranged in two rows of four bones each and form a slight curvature that is concave anteriorly and convex posteriorly. A number of mnemonics have been developed to help students remember the carpal bones. The following one allows students to remember them in order from lateral to medial for the proximal row (top) and from medial to lateral (by the thumb) for the distal row: So Long Top Part, Here Comes The Thumb—that is, Scaphoid, Lunate, Triquetrum, Pisiform, Hamate, Capitate, Trapezoid, and Trapezium.

Hand

Five **metacarpal** (met′ă-kar′păl) **bones** are attached to the carpal bones and form the bony framework of the hand (figure 6.27). The metacarpal bones are aligned with the five **digits:** the thumb and fingers. They are numbered 1 to 5, from the thumb to the little finger. The ends, or heads, of the five metacarpal bones associated with the thumb and fingers form the knuckles (figure 6.28). Each finger consists of three small bones called **phalanges** (fă-lan′jēz;

CLINICAL IMPACT   Carpal Tunnel Syndrome

The bones and ligaments on the anterior side of the wrist form a **carpal tunnel,** which does not have much "give." Tendons and nerves pass from the forearm through the carpal tunnel to the hand. Fluid and connective tissue can accumulate in the carpal tunnel as a result of inflammation associated with overuse or trauma. The inflammation can also cause the tendons in the carpal tunnel to enlarge. The accumulated fluid and enlarged tendons can apply pressure to a major nerve passing through the tunnel. The pressure on this nerve causes **carpal tunnel syndrome,** characterized by tingling, burning, and numbness in the hand.

Treatments for carpal tunnel syndrome vary, depending on the severity of the syndrome. Mild cases can be treated nonsurgically with either anti-inflammatory medications or stretching exercises. However, if symptoms have lasted for more than 6 months, surgery is recommended. Surgical techniques involve cutting the carpal ligament to enlarge the carpal tunnel and ease pressure on the nerve.

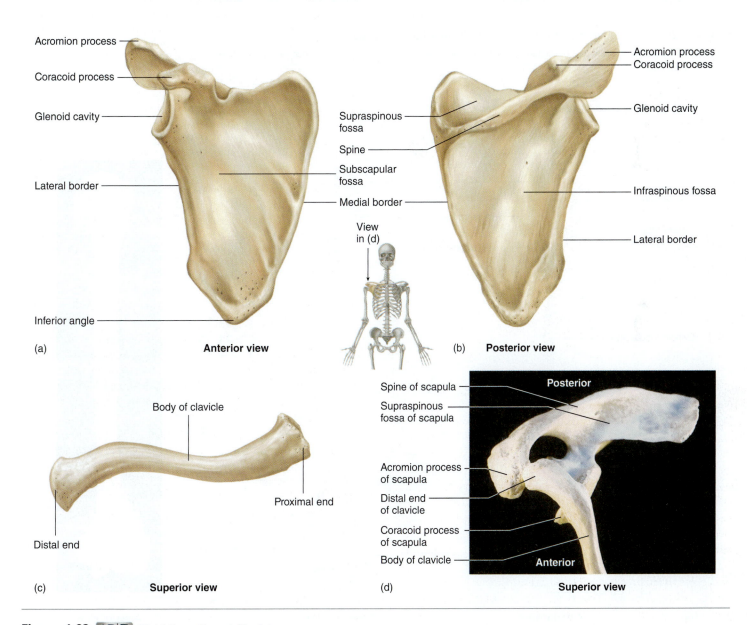

(a)   **Anterior view**

(b)   **Posterior view**

(c)   **Superior view**

(d)   **Superior view**

Figure 6.23 AP|R Right Scapula and Clavicle

(*a*) Right scapula, anterior view. (*b*) Right scapula, posterior view. (*c*) Right clavicle, superior view. (*d*) Photograph of the right scapula and clavicle from a superior view, showing the relationship between the distal end of the clavicle and the acromion process of the scapula.

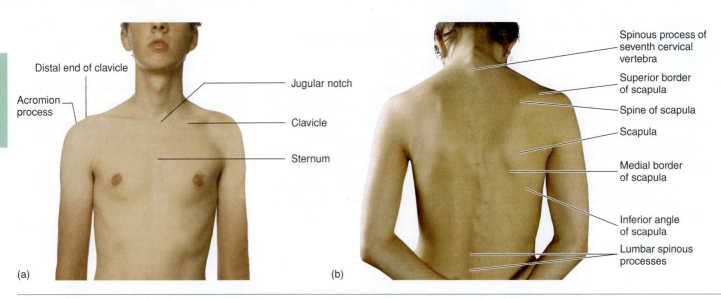

Distal end of clavicle

Acromion process

Jugular notch

Clavicle

Sternum

(a)

Spinous process of seventh cervical vertebra

Superior border of scapula

Spine of scapula

Scapula

Medial border of scapula

Inferior angle of scapula

Lumbar spinous processes

(b)

Figure 6.24 Surface Anatomy of the Pectoral Girdle and Thoracic Cage

(*a*) Bones of the pectoral girdle and the anterior thorax. (*b*) Bones of the scapula and the posterior vertebral column.

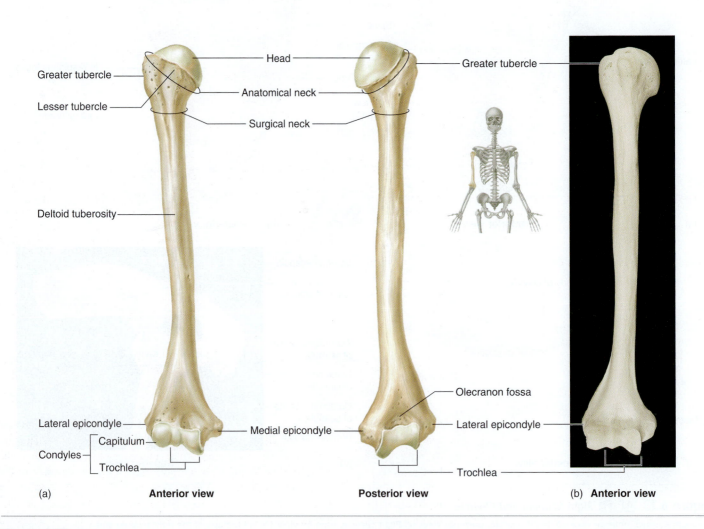

Greater tubercle

Lesser tubercle

Head

Anatomical neck

Surgical neck

Greater tubercle

Deltoid tuberosity

Lateral epicondyle

Condyles {
Capitulum

Trochlea

Medial epicondyle

Olecranon fossa

Lateral epicondyle

Trochlea

(a)   **Anterior view**

Posterior view

(b)   **Anterior view**

Figure 6.25 Right Humerus

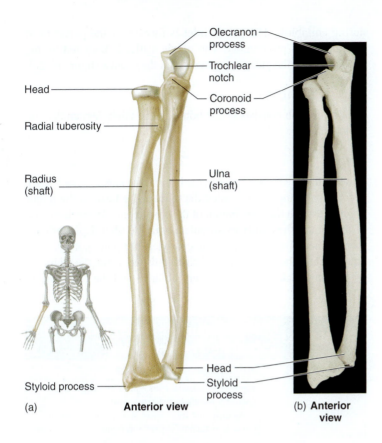

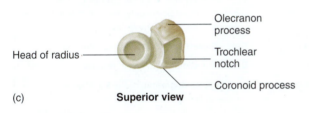

Figure 6.26 AP|R Right Ulna and Radius

(a) Anterior view of the right ulna and radius. (b) Photo of the same view of the right ulna and radius. (c) Proximal ends of the right ulna and radius.

sing. phalanx, fā′langks), named after the Greek phalanx, a wedge of soldiers holding their spears, tips outward, in front of them. The phalanges of each finger are called proximal, middle, and distal, according to their position in the digit. The thumb has two phalanges, proximal and distal. The digits are also numbered 1 to 5, starting from the thumb.

Pelvic Girdle

The **pelvic girdle** is the place where the lower limbs attach to the body (figure 6.29). The right and left **coxal** (kok′sul) **bones,** or hip bones, join each other anteriorly and the **sacrum** posteriorly to form a ring of bone called the **pelvic girdle.** The **pelvis** (pel′vis; basin) includes the pelvic girdle and the coccyx (figure 6.30). The sacrum and coccyx form part of the pelvis but are also part of the axial skeleton. Each coxal bone is formed by three bones fused to one another to form a single bone (figure 6.31). The **ilium** (il′ē-ŭm) is the most superior, the **ischium** (is′kē-ŭm) is inferior and posterior, and the **pubis** (pū′bis) is inferior and anterior. An **iliac crest** can be seen

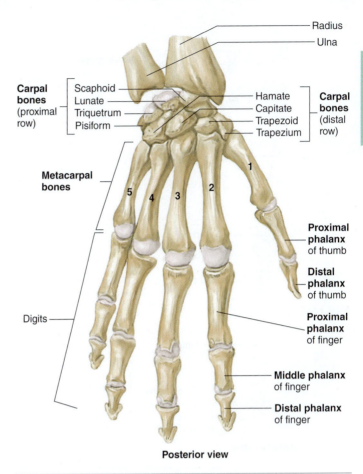

Posterior view

Figure 6.27 AP|R Bones of the Right Wrist and Hand

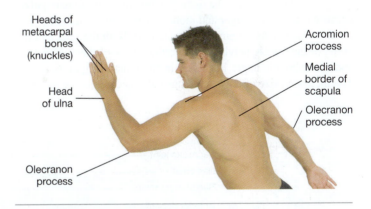

Figure 6.28 Surface Anatomy Showing Bones of the Pectoral Girdle and Upper Limb

along the superior margin of each ilium, and an **anterior superior iliac spine,** an important hip landmark, is located at the anterior end of the iliac crest. The coxal bones join each other anteriorly at the **pubic** (pū′bik) **symphysis** and join the sacrum posteriorly at the **sacroiliac** (sā-krō-il′ē-ak) **joints** (see figure 6.30). The **acetabulum** (as-ĕ-tab′ū-lŭm; vinegar cup) is the socket of the hip joint. The **obturator** (ob′too-rā-tŏr) **foramen** is the large hole in each coxal bone that is closed off by muscles and other structures.

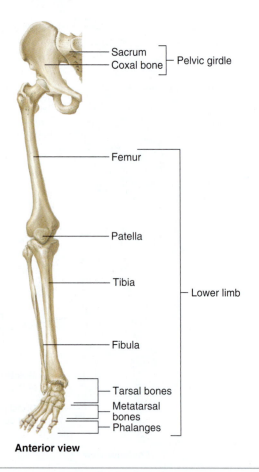

- Sacrum
- Coxal bone] — Pelvic girdle

- Femur

- Patella

- Tibia   — Lower limb

- Fibula

- Tarsal bones
- Metatarsal bones
- Phalanges

Anterior view

Figure 6.29 Bones of the Pelvic Girdle and Right Lower Limb

The male pelvis can be distinguished from the female pelvis because it is usually larger and more massive, but the female pelvis tends to be broader (figure 6.32; table 6.3). Both the inlet and the outlet of the female pelvis are larger than those of the male pelvis, and the subpubic angle is greater in the female (figure 6.32a,b). The increased size of these openings helps accommodate the fetus

during childbirth. The **pelvic inlet** is formed by the pelvic brim and the sacral promontory. The **pelvic outlet** is bounded by the ischial spines, the pubic symphysis, and the coccyx (figure 6.32c).

Lower Limb

The lower limb consists of the bones of the thigh, leg, ankle, and foot (see figure 6.29).

Thigh

The **thigh** is the region between the hip and the knee (figure 6.33a). It contains a single bone called the **femur.** The **head** of the femur articulates with the acetabulum of the coxal bone. At the distal end of the femur, the **condyles** articulate with the tibia. **Epicondyles,** located medial and lateral to the condyles, are points of ligament attachment. The femur can be distinguished from the humerus by its long neck, located between the head and the **trochanters**

TABLE 6.3	Differences Between Male and Female Pelvic Girdles	
Area	**Description of Difference**
General	Female pelvis somewhat lighter in weight and wider laterally but shorter superiorly to inferiorly and less funnel-shaped; less obvious muscle attachment points in female than in male
Sacrum	Broader in female, with the inferior portion directed more posteriorly; the sacral promontory projects less anteriorly in female
Pelvic inlet	Heart-shaped in male; oval in female
Pelvic outlet	Broader and more shallow in female
Subpubic angle	Less than 90 degrees in male; 90 degrees or more in female
Ilium	More shallow and flared laterally in female
Ischial spines	Farther apart in female
Ischial tuberosities	Turned laterally in female and medially in male (not shown in figure 6.32)

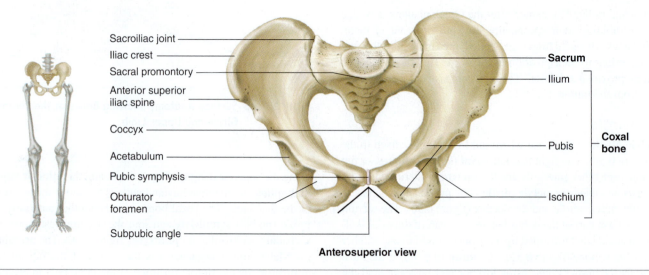

- Sacroiliac joint
- Iliac crest
- Sacral promontory
- Anterior superior iliac spine
- Coccyx
- Acetabulum
- Pubic symphysis
- Obturator foramen
- Subpubic angle

- Sacrum
- Ilium

- **Coxal bone**

- Pubis

- Ischium

Anterosuperior view

Figure 6.30 Pelvis

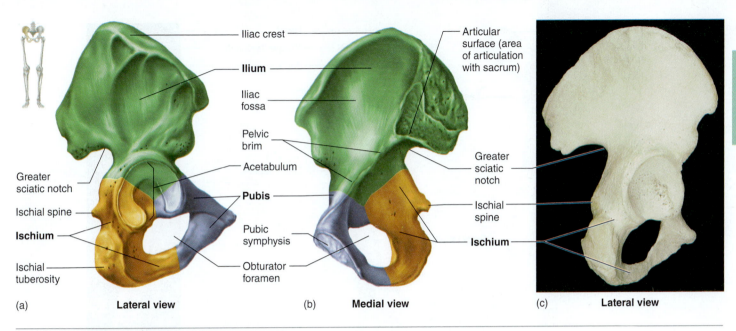

(a) **Lateral view**   (b) **Medial view**   (c) **Lateral view**

Figure 6.31 AP|R Right Coxal Bone

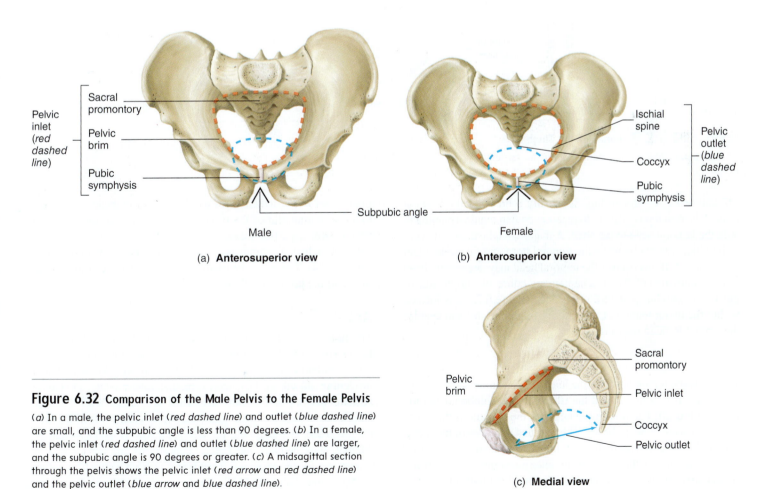

(a) **Anterosuperior view**   (b) **Anterosuperior view**

(c) **Medial view**

Figure 6.32 Comparison of the Male Pelvis to the Female Pelvis

(*a*) In a male, the pelvic inlet (*red dashed line*) and outlet (*blue dashed line*) are small, and the subpubic angle is less than 90 degrees. (*b*) In a female, the pelvic inlet (*red dashed line*) and outlet (*blue dashed line*) are larger, and the subpubic angle is 90 degrees or greater. (*c*) A midsagittal section through the pelvis shows the pelvic inlet (*red arrow* and *red dashed line*) and the pelvic outlet (*blue arrow* and *blue dashed line*).

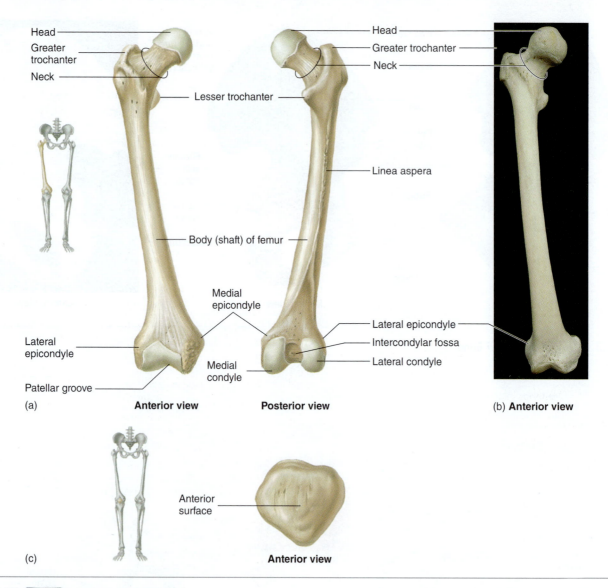

Figure 6.33 AP|R **Bones of the Thigh**

(a) Right femur. (b) Photo of anterior right femur. (c) Patella.

(trō′kan-terz). A "broken hip" is usually a break of the femoral neck. A broken hip is difficult to repair and often requires pinning to hold the femoral head to the shaft. A major complication can occur if the blood vessels between the femoral head and the acetabulum are damaged. If this occurs, the femoral head may degenerate from lack of nourishment. The trochanters are points of muscle attachment. The **patella** (pa-tel′ă), or kneecap (figure 6.33c), is located within the major tendon of the anterior thigh muscles and enables the tendon to bend over the knee.

Leg

The **leg** is the region between the knee and the ankle (figure 6.34). It contains two bones, called the **tibia** (tib′ē-ă; shinbone) and the **fibula** (fib′ū-lă). The tibia is the larger of the two and is the major weight-bearing bone of the leg. The rounded condyles of the femur rest on the flat condyles on the proximal end of the tibia. Just distal to the condyles of the tibia, on its anterior surface, is the **tibial tuberosity,** where the muscles of the anterior thigh attach. The

fibula does not articulate with the femur, but its head is attached to the proximal end of the tibia. The distal ends of the tibia and fibula form a partial socket that articulates with a bone of the ankle (the talus). A prominence can be seen on each side of the ankle (figure 6.34). These are the **medial malleolus** (mal-ē′ō-lŭs) of the tibia and the **lateral malleolus** of the fibula.

Ankle

The **ankle** consists of seven **tarsal** (tar′săl; the sole of the foot) **bones** (figure 6.35). The tarsal bones are the **talus** (tā′lŭs; ankle bone), **calcaneus** (kal-kā′nē-ŭs; heel), **cuboid** (kū′boyd), and **navicular** (nă-vik′yū-lăr), and the medial, intermediate, and lateral **cuneiforms** (kū′nē-i-fōrmz). The talus articulates with the tibia and fibula to form the ankle joint, and the calcaneus forms the heel (figure 6.36). A mnemonic for the distal row is **MILC**—that is, **M**edial, **I**ntermediate, and **L**ateral cuneiforms and the **C**uboid. A mnemonic for the proximal three bones is **N**o **T**hanks **C**ow—that is, **N**avicular, **T**alus, and **C**alcaneus.

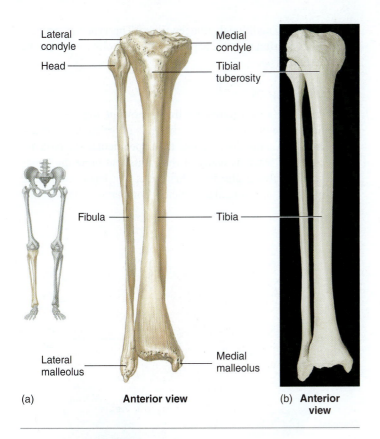

Figure 6.34 **AP|R** **Bones of the Leg**

The right tibia and fibula are shown.

Foot

The **metatarsal** (met′ă-tar′săl) **bones** and **phalanges** of the foot are arranged and numbered in a manner very similar to the metacarpal bones and phalanges of the hand (see figure 6.35). The metatarsal bones are somewhat longer than the metacarpal bones, whereas the phalanges of the foot are considerably shorter than those of the hand.

There are three primary **arches** in the foot, formed by the positions of the tarsal bones and metatarsal bones, and held in place by ligaments. Two longitudinal arches extend from the heel to the ball of the foot, and a transverse arch extends across the foot. The arches function similarly to the springs of a car, allowing the foot to give and spring back.

6.8 JOINTS

Learning Outcomes *After reading this section, you should be able to*

A. Describe the two systems for classifying joints.

B. Explain the structure of a fibrous joint, list the three types, and give examples of each type.

C. Give examples of cartilaginous joints.

D. Illustrate the structure of a synovial joint and explain the roles of the components of a synovial joint.

E. Classify synovial joints based on the shape of the bones in the joint and give an example of each type.

F. Demonstrate the difference between the following pairs of movements: flexion and extension; plantar flexion and dorsiflexion; abduction and adduction; supination and pronation; elevation and depression; protraction and retraction; opposition and reposition; inversion and eversion.

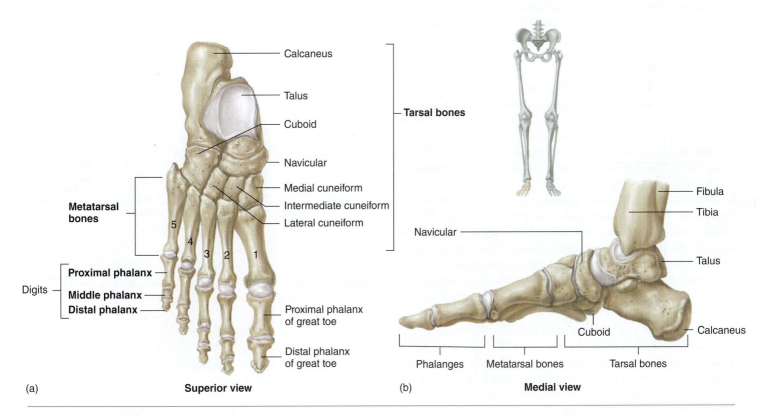

Figure 6.35 **AP|R** **Bones of the Right Foot**

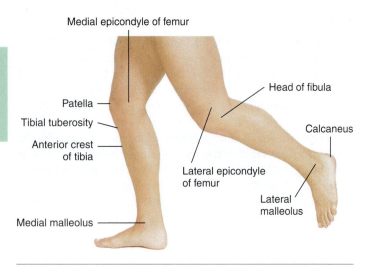

Figure 6.36 Surface Anatomy Showing Bones of the Lower Limb

A **joint,** or an *articulation,* is a place where two bones come together. A joint is usually considered movable, but that is not always the case. Many joints exhibit limited movement, and others are completely, or almost completely, immovable.

One method of classifying joints is a functional classification. Based on the degree of motion, a joint may be called a **synarthrosis** (sin′ar-thrō′sis; nonmovable joint), an **amphiarthrosis** (am′fi-ar-thrō′sis; slightly movable joint), or a **diarthrosis** (dī-ar-thrō′sis; freely movable joint). However, functional classification is somewhat restrictive and is not used in this text. Instead, we use a structural classification whereby joints are classified according to the type of connective tissue that binds the bones together and whether there is a fluid-filled joint capsule. The three major structural classes of joints are fibrous, cartilaginous, and synovial.

Fibrous Joints

Fibrous joints consist of two bones that are united by fibrous tissue and that exhibit little or no movement. Joints in this group are further subdivided on the basis of structure as sutures, syndesmoses, or gomphoses. **Sutures** (soo′choorz) are fibrous joints between the bones of the skull (see figure 6.11). In a newborn, some parts of the sutures are quite wide and are called **fontanels** (fon′tă-nelz′), or soft spots (figure 6.37). They allow flexibility in the skull during the birth process, as well as growth of the head after birth. **Syndesmoses** (sin′dez-mō′sēz) are fibrous joints in which the bones are separated by some distance and held together by ligaments. An example is the fibrous membrane connecting most of the distal parts of the radius and ulna. **Gomphoses** (gom-fō′sēz) consist of pegs fitted into sockets and held in place by ligaments. The joint between a tooth and its socket is a gomphosis.

Cartilaginous Joints

Cartilaginous joints unite two bones by means of cartilage. Only slight movement can occur at these joints. Examples are the cartilage in the epiphyseal plates of growing long bones and the cartilages between the ribs and the sternum. The cartilage of some cartilaginous joints, where much strain is placed on the joint, may

be reinforced by additional collagen fibers. This type of cartilage, called **fibrocartilage** (see chapter 4), forms joints such as the intervertebral disks.

Synovial Joints

Synovial (si-nō′vē-ăl) **joints** are freely movable joints that contain fluid in a cavity surrounding the ends of articulating bones. Most joints that unite the bones of the appendicular skeleton are synovial joints, whereas many of the joints that unite the bones of the axial skeleton are not. This pattern reflects the greater mobility of the appendicular skeleton compared to that of the axial skeleton.

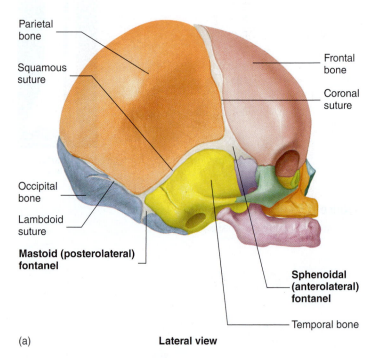

(a)   **Lateral view**

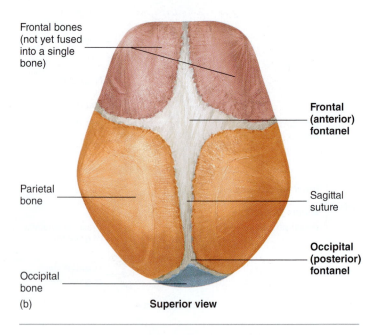

(b)   **Superior view**

Figure 6.37 Fetal Skull Showing Fontanels and Sutures

Several features of synovial joints are important to their function (figure 6.38). The articular surfaces of bones within synovial joints are covered with a thin layer of **articular cartilage,** which provides a smooth surface where the bones meet. The **joint cavity** is filled with fluid. The cavity is enclosed by a **joint capsule,** which helps hold the bones together and allows for movement. Portions of the fibrous part of the joint capsule may be thickened to form ligaments. In addition, ligaments and tendons outside the joint capsule contribute to the strength of the joint.

A **synovial membrane** lines the joint cavity everywhere except over the articular cartilage. The membrane produces **synovial fluid,** which is a complex mixture of polysaccharides, proteins, lipids, and cells. Synovial fluid forms a thin, lubricating film covering the surfaces of the joint. In certain synovial joints, the synovial membrane may extend as a pocket, or sac, called a **bursa** (ber′să; pocket). Bursae are located between structures that rub together, such as where a tendon crosses a bone; they reduce friction, which could damage the structures involved. Inflammation of a bursa, often resulting from abrasion, is called **bursitis.** A synovial membrane may extend as a **tendon sheath** along some tendons associated with joints (figure 6.38).

Types of Synovial Joints

Synovial joints are classified according to the shape of the adjoining articular surfaces (figure 6.39). **Plane joints,** or *gliding joints,* consist of two opposed flat surfaces that glide over each other. Examples of these joints are the articular facets between vertebrae.

Saddle joints consist of two saddle-shaped articulating surfaces oriented at right angles to each other. Movement in these joints can occur in two planes. The joint between the metacarpal bone and the carpal bone (trapezium) of the thumb is a saddle joint. **Hinge joints** permit movement in one plane only. They consist of a convex cylinder of one bone applied to a corresponding concavity of the other bone. Examples are the elbow and knee joints (figure 6.40a,b). The flat condylar surface of the knee joint is modified into a concave surface by shock-absorbing fibrocartilage pads called **menisci** (mĕ-nis′sī). **Pivot joints** restrict movement to rotation around a single axis. Each pivot joint consists of a cylindrical bony process that rotates within a ring composed partly of bone and partly of ligament. The rotation that occurs between the axis and atlas when shaking the head "no" is an example. The articulation between the proximal ends of the ulna and radius is also a pivot joint.

Ball-and-socket joints consist of a ball (head) at the end of one bone and a socket in an adjacent bone into which a portion of the ball fits. This type of joint allows a wide range of movement in almost any direction. Examples are the shoulder and hip joints (figure 6.40c,d). **Ellipsoid** (ē-lip′soyd) **joints,** or *condyloid* (kon′di-loyd) *joints,* are elongated ball-and-socket joints. The shape of the joint limits its range of movement nearly to that of a hinge motion, but in two planes. Examples of ellipsoid joints are the joint between the occipital condyles of the skull and the atlas of the vertebral column and the joints between the metacarpal bones and phalanges.

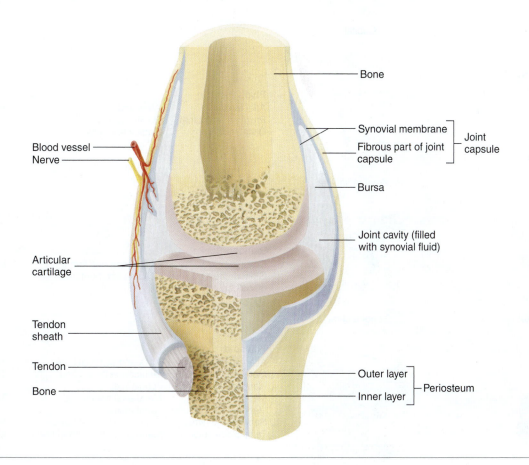

Figure 6.38 AP|R Structure of a Synovial Joint

Skeletal

Plane

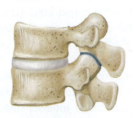

Intervertebral

Saddle

Carpometacarpal

Hinge

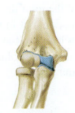

Cubital

Pivot

Proximal radioulnar

Ball-and-socket

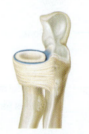

Glenohumeral

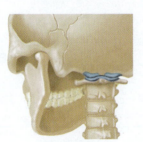

Ellipsoid

Atlantooccipital

Class and Example of Joint	Structures Joined	Movement
Plane		
Acromioclavicular	Acromion process of scapula and clavicle	Slight
Carpometacarpal	Carpals and metacarpals 2–5	Slight
Costovertebral	Ribs and vertebrae	Slight
Intercarpal	Between carpal bones	Slight
Intermetatarsal	Between metatarsal bones	Slight
Intertarsal	Between tarsal bones	Slight
Intervertebral	Between articular processes of adjacent vertebrae	Slight
Sacroiliac	Between sacrum and coxal bone (complex joint with several planes and synchondroses)	Slight
Tarsometatarsal	Tarsal bones and metatarsal bones	Slight
Saddle		
Carpometacarpal pollicis	Carpal and metacarpal of thumb	Two axes
Intercarpal	Between carpal bones	Slight
Sternoclavicular	Manubrium of sternum and clavicle	Slight
Hinge		
Cubital (elbow)	Humerus, ulna, and radius	One axis
Knee	Femur and tibia	One axis
Interphalangeal	Between phalanges	One axis
Talocrural (ankle)	Talus, tibia, and fibula	Multiple axes; one predominates
Pivot		
Atlantoaxial	Atlas and axis	Rotation
Proximal radioulnar	Radius and ulna	Rotation
Distal radioulnar	Radius and ulna	Rotation
Ball-and-Socket		
Coxal (hip)	Coxal bone and femur	Multiple axes
Glenohumeral (shoulder)	Scapula and humerus	Multiple axes
Ellipsoid		
Atlantooccipital	Atlas and occipital bone	Two axes
Metacarpophalangeal (knuckles)	Metacarpal bones and phalanges	Two axes
Metatarsophalangeal (ball of foot)	Metatarsal bones and phalanges	Two axes
Radiocarpal (wrist)	Radius and carpal bones	Multiple axes
Temporomandibular	Mandible and temporal bone	Multiple axes; one predominates

Figure 6.39 Types of Synovial Joints

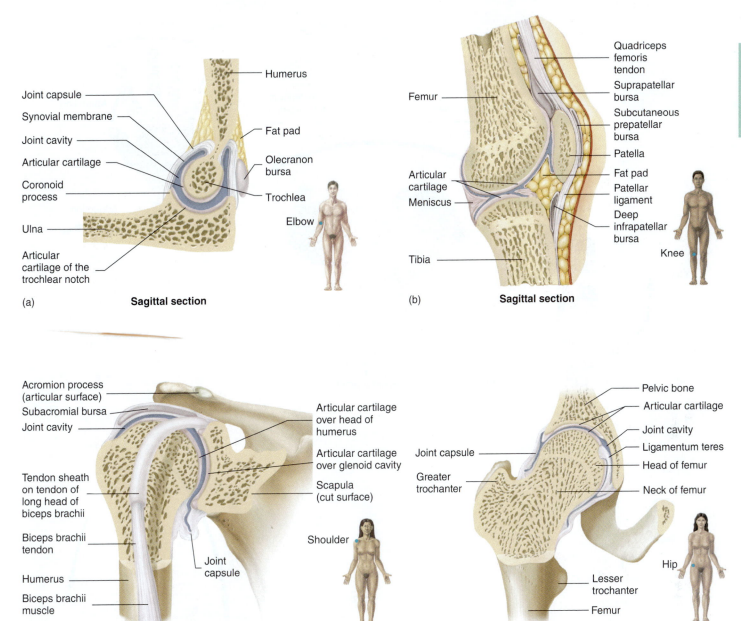

Figure 6.40 Examples of Synovial Joints

(a) Elbow. (b) Knee. (c) Shoulder. (d) Hip.

Types of Movement

The types of movement occurring at a given joint are related to the structure of that joint. Some joints are limited to only one type of movement, whereas others permit movement in several directions. All the movements are described relative to the anatomical position. Because most movements are accompanied by movements in the opposite direction, they are often illustrated in pairs (figure 6.41).

Flexion and **extension** are common opposing movements. The literal definitions are to bend (flex) and to straighten (extend). Flexion occurs when the bones of a particular joint are moved closer together, whereas extension occurs when the bones of a particular joint are moved farther apart, such that the bones are now arranged somewhat end-to-end (figure 6.41a,b). An example of flexion occurs when a person flexes the forearm to "make a muscle."

There are special cases of flexion when describing movement of the foot. Movement of the foot toward the plantar surface (sole of the foot), as when standing on the toes, is commonly called **plantar flexion.** Movement of the foot toward the shin, as when walking on the heels, is called **dorsiflexion.**

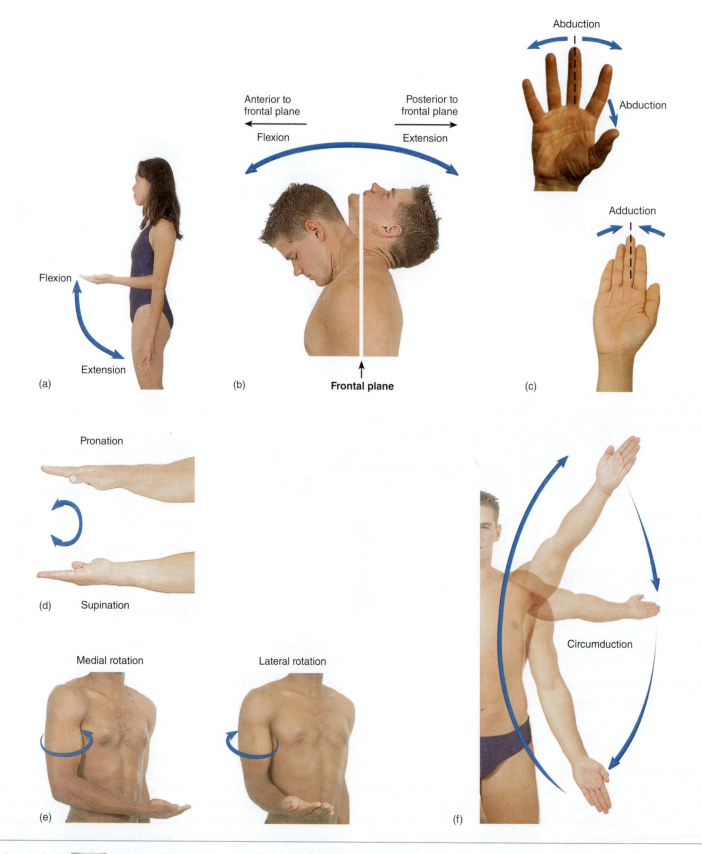

Figure 6.41 AP|R **Types of Movement**

(*a*) Flexion and extension of the elbow. (*b*) Flexion and extension of the neck. (*c*) Abduction and adduction of the fingers.
(*d*) Pronation and supination of the hand. (*e*) Medial and lateral rotation of the arm. (*f*) Circumduction of the arm.

A CASE IN POINT

Dislocated Shoulder

The shoulder joint is the most commonly dislocated joint in the body. Loosh Holder dislocated his shoulder joint while playing basketball. As a result of a "charging" foul, Loosh was knocked backward and fell. As he broke his fall with his extended right arm, the head of the right humerus was forced out of the glenoid cavity. While being helped up from the floor, Loosh felt severe pain in his shoulder, his right arm sagged, and he could not move his arm at the shoulder. Most dislocations result in stretching of the joint capsule and movement of the humeral head to the inferior, anterior side of the glenoid cavity. The dislocated humeral head is moved back to its normal position by carefully pulling it laterally over the inferior lip of the glenoid cavity and then superiorly into the glenoid cavity. Once the shoulder joint capsule has been stretched by a shoulder dislocation, the shoulder joint may be predisposed to future dislocations. Some individuals have hereditary "loose" joints and are more likely to experience a dislocated shoulder.

Abduction (ab-dŭk′shun; to take away) is movement away from the median or midsagittal plane; **adduction** (to bring together) is movement toward the median plane (figure 6.41*c*). Moving the legs away from the midline of the body, as in the outward movement of "jumping jacks," is abduction, and bringing the legs back together is adduction.

Pronation (prō-nā′shŭn) and **supination** (soo′pi-nā′shun) are best demonstrated with the elbow flexed at a 90-degree angle. When the elbow is flexed, pronation is rotation of the forearm so that the palm is down, and supination is rotation of the forearm so that the palm faces up (figure 6.41*d*).

Eversion (ē-ver′zhŭn) is turning the foot so that the plantar surface (bottom of the foot) faces laterally; **inversion** (in-ver′zhŭn) is turning the foot so that the plantar surface faces medially.

Rotation is the turning of a structure around its long axis, as in shaking the head "no." Rotation of the arm can best be demonstrated with the elbow flexed (figure 6.41*e*) so that rotation is not confused with supination and pronation of the forearm. With the elbow flexed, medial rotation of the arm brings the forearm against the anterior surface of the abdomen, and lateral rotation moves it away from the body.

Circumduction (ser-kŭm-dŭk′shŭn) occurs at freely movable joints, such as the shoulder. In circumduction, the arm moves so that it traces a cone where the shoulder joint is at the cone's apex (figure 6.41*f*).

In addition to the movements pictured in figure 6.41, several other movement types have been identified:

- **Protraction** (prō-trak′shŭn) is a movement in which a structure, such as the mandible, glides anteriorly.
- In **retraction** (rē-trak′shŭn), the structure glides posteriorly.
- **Elevation** is movement of a structure in a superior direction. Closing the mouth involves elevation of the mandible.
- **Depression** is movement of a structure in an inferior direction. Opening the mouth involves depression of the mandible.
- **Excursion** is movement of a structure to one side, as in moving the mandible from side to side.

- **Opposition** is a movement unique to the thumb and little finger. It occurs when the tips of the thumb and little finger are brought toward each other across the palm of the hand. The thumb can also oppose the other digits.
- **Reposition** returns the digits to the anatomical position.

Most movements that occur in the course of normal activities are combinations of movements. A complex movement can be described by naming the individual movements involved.

When the bones of a joint are forcefully pulled apart and the ligaments around the joint are pulled or torn, a *sprain* results. A *separation* exists when the bones remain apart after injury to a joint. A *dislocation* is when the end of one bone is pulled out of the socket in a ball-and-socket, ellipsoid, or pivot joint.

Hyperextension is usually defined as an abnormal, forced extension of a joint beyond its normal range of motion. For example, if a person falls and attempts to break the fall by putting out a hand, the force of the fall directed into the hand and wrist may cause hyperextension of the wrist, which may result in sprained joints or broken bones. Some health professionals, however, define hyperextension as the normal movement of a structure into the space posterior to the anatomical position.

Predict 5

What combination of movements at the shoulder and elbow joints allows a person to perform a crawl stroke in swimming?

6.9 EFFECTS OF AGING ON THE SKELETAL SYSTEM AND JOINTS

Learning Outcome   *After reading this section, you should be able to*

A. Describe the effects of aging on bone matrix and joints.

The most significant age-related changes in the skeletal system affect the joints as well as the quality and quantity of bone matrix. The bone matrix in an older bone is more brittle than in a younger bone because decreased collagen production results in relatively more mineral and less collagen fibers. With aging, the amount of matrix also decreases because the rate of matrix formation by osteoblasts becomes slower than the rate of matrix breakdown by osteoclasts.

Bone mass is at its highest around age 30, and men generally have denser bones than women because of the effects of testosterone and greater body weight. Race and ethnicity also affect bone mass. African-Americans and Latinos have higher bone masses than caucasians and Asians. After age 35, both men and women experience a loss of bone of 0.3–0.5% a year. This loss can increase 10-fold in women after menopause, when they can lose bone mass at a rate of 3–5% a year for approximately 5–7 years (see Systems Pathology, "Osteoporosis").

Significant loss of bone increases the likelihood of bone fractures. For example, loss of trabeculae greatly increases the risk of fractures of the vertebrae. In addition, loss of bone and the resulting fractures can cause deformity, loss of height, pain, and stiffness. Loss of bone from the jaws can also lead to tooth loss.

A number of changes occur within many joints as a person ages. Changes in synovial joints have the greatest effect and often present major problems for elderly people. With use, the cartilage covering articular surfaces can wear down. When a person

Name: Betty Williams
Gender: Female
Age: 65

Comments
Betty has smoked heavily for at least 50 years. She does not exercise, seldom goes outdoors, has a poor diet, and is slightly underweight.

While attending a family picnic, Betty tripped and fell. She reported severe right hip pain and was not able to put any weight on her right leg when she arrived in the ER. Radiographs revealed significant loss of bone density and a fractured femur neck (figures 6B, 6C).

Osteoporosis

Background Information

Osteoporosis, or porous bone, is a loss of bone matrix. This loss of bone mass makes bones so porous and weakened that they become deformed and prone to fracture (figure 6B). The occurrence of osteoporosis increases with age. In both men and women (although it is 2.5 times more common in women), bone mass starts to decrease at about age 40 and continually decreases thereafter. Women can eventually lose approximately one-half, and men one-quarter, of their spongy bone.

In women, decreased production of the female reproductive hormone estrogen can cause osteoporosis, mostly in spongy bone, especially in the vertebrae of the spine and the bones of the forearm. Collapse of the vertebrae can cause a decrease in height or, in more severe cases, kyphosis in the upper back (figure 6D). Estrogen levels decrease as a result of menopause; removal of the ovaries; amenorrhea (lack of menstrual cycle) due to extreme exercise or anorexia nervosa (self-starvation); or cigarette smoking.

In men, reduction in testosterone levels can cause loss of bone tissue. However, this is less of a problem in men than in women because men have denser bones than women, and testosterone levels generally don't decrease significantly until after age 65.

Inadequate dietary intake or absorption of calcium, sometimes due to certain medications, can also contribute to osteoporosis. Absorption of calcium from the small intestine decreases with age. Finally, too little exercise or disuse from injury can all cause osteoporosis. Significant amounts of bone are lost after only 8 weeks of immobilization.

Osteoporotic bone   Normal bone

Figure 6B
Photomicrograph of osteoporotic bone and normal bone.

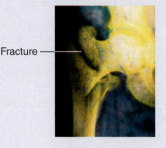

Fracture

Figure 6C
Radiograph

Kyphosis

Figure 6D

MUSCULAR
Muscle atrophy from reduced activity. Increased chance of falling and breaking a bone.

INTEGUMENTARY
Limited sun exposure lowers vitamin D production, which reduces Ca^{2+} absorption.

NERVOUS
Pain from injury may reduce activity level and therefore further injuries.

Osteoporosis

Symptoms
- Pain and stiffness especially in spine
- Easily broken bones
- Loss of height

Treatment
- Dietary calcium and vitamin D
- Exercise
- Calcitonin
- Alendronate

REPRODUCTIVE
Decreased estrogen following menopause contributes to osteoporosis.

ENDOCRINE
Calcitonin is used to treat osteoporosis.

DIGESTIVE
Inadequate calcium and vitamin D intake can cause insufficient Ca^{2+} absorption in small intestine.

CARDIOVASCULAR
If bone breakage has occurred, increased blood flow to that site removes debris. Blood carries nutrients necessary for repair.

RESPIRATORY
Excessive smoking lowers estrogen levels, which increases bone loss.

LYMPHATIC AND IMMUNE
Immune cells help prevent infection after surgery, such as hip replacement.

Early diagnosis of osteoporosis can lead to more preventative treatments. Instruments that measure the absorption of photons (particles of light) by bone are currently used; of these, dual-energy x-ray absorptiometry (DEXA) is considered the best. Supplementation of dietary calcium and vitamin D and exercise are the best preventative and rehabilitory measures to prevent bone loss or regain mild bone loss. Calcitonin (Miacalcin) inhibits osteoclast activity, and alendronate (Fosamax), which binds to hydroxyapatite, also inhibits osteoclasts. Although osteoporosis is linked to estrogen loss, estrogen therapy has been associated with many side effects, including breast cancer, and is no longer recommended as a treatment.

Predict 6

What advice should Betty give to her granddaughter so that the granddaughter will be less likely to develop osteoporosis when she is Betty's age?

Skeletal

DISEASES AND DISORDERS: Skeletal System

CONDITION	DESCRIPTION
Tumors	May be malignant or benign and cause a range of bone defects
SKELETAL DISORDERS	
Growth and Developmental Disorders	
Gigantism	Abnormally increased body size due to excessive growth at the epiphyseal plates
Dwarfism	Abnormally small body size due to improper growth at the epiphyseal plates
Osteogenesis imperfecta	Brittle bones that fracture easily due to insufficient or abnormal collagen
Rickets	Growth retardation due to nutritional deficiencies in minerals (Ca^{2+}) or vitamin D; results in bones that are soft, weak, and easily broken
Bacterial Infections	
Osteomyelitis	Bone inflammation often due to a bacterial infection that may lead to complete destruction of the bone
Tuberculosis	Typically, a lung bacterium that can also affect bone
Decalcification	
Osteomalacia	Softening of adult bones due to calcium depletion; often caused by vitamin D deficiency
Osteoporosis	Reduction in overall quantity of bone tissue; see Systems Pathology
JOINT DISORDERS	
Arthritis	Inflammation of a joint; causes include infectious agents, metabolic disorders, trauma, and immune disorders
Rheumatoid arthritis	General connective tissue autoimmune disease
Degenerative joint disease (osteoarthritis)	Gradual degeneration of a joint with advancing age; can be delayed with exercise
Gout	Increased production and accumulation of uric acid crystals in tissues, including joint capsules
Bursitis and bunions	
Bursitis	Inflammation of a bursa
Bunion	Most bunions are deformations of the first metatarsal (the great toe); bursitis may accompany this deformity; irritated by tight shoes
Joint replacement	Replacement of painful joints with artificial joints

is young, production of new, resilient matrix compensates for the wear. As a person ages, the rate of replacement declines, and the matrix becomes more rigid, thus adding to its rate of wear. The production rate of lubricating synovial fluid also declines with age, further contributing to the wear of the articular cartilage. Many people also experience arthritis, an inflammatory degeneration of the joints, with advancing age. In addition, the ligaments and tendons surrounding a joint shorten and become less flexible with age, resulting in decreased range of motion. Furthermore, many older people are less physically active, which causes the joints to become less flexible and decreases their range of motion.

The most effective preventative measure against the effects of aging on the skeletal system is the combination of increasing physical activity and taking dietary calcium and vitamin D supplements. Intensive exercise can even reverse loss of bone matrix.

Since the question tells us that Thomas suffers pain when bending over, it is most likely that he has a ruptured intervertebral disk in his lumbar region. When these disks rupture, a portion of the disk protrudes and pushes against the spinal cord or the spinal nerves, compromises their function, and produces pain. Herniation of the inferior lumbar disks is most common because, as you learned in section 6.6, the lumbar vertebrae support more weight than other regions of the vertebral column. Thomas' ruptured disk may be repaired with prolonged bed rest or it may require surgery. One such type of surgery involves removing the ruptured disk and replacing it with a piece of hipbone. Eventually, the adjacent vertebrae will become fused by bone across the gap.

Answers to the rest of this chapter's Predict questions are in Appendix E.

SUMMARY

The skeletal system consists of bone, cartilage, tendons, and ligaments.

6.1 Functions of the Skeletal System   (p. 110)

1. The skeletal system provides the major support for the body.
2. Bone protects internal organs.
3. Joints allow movement between bones.
4. Bones store and release minerals as needed by the body.
5. Bone marrow gives rise to blood cells and platelets.

6.2 Extracellular Matrix   (p. 111)

1. Bone, cartilage, tendons, and ligaments are connective tissues.
2. Varying amounts of collagen, proteoglycan, organic molecules, water, and minerals in the matrix determine the characteristics of connective tissue.

6.3 General Features of Bone   (p. 111)

There are four categories of bone: long, short, flat, and irregular.

Structure of a Long Bone

Long bones consist of a diaphysis (shaft), epiphyses (ends), and epiphyseal (growth) plates. The diaphysis contains a medullary cavity, which is filled with marrow, and the end of the epiphysis is covered by articular cartilage.

Histology of Bone

1. Osteoblasts are bone-forming cells.
2. Osteocytes are bone cells located between thin sheets of extracellular matrix called lamellae.
3. Compact bone tissue consists of osteons, which are composed of osteocytes organized into lamellae surrounding central canals.
4. Spongy bone tissue consists of trabeculae without central canals.

Bone Ossification

1. Intramembranous ossification occurs within connective tissue membranes.
2. Endochondral ossification occurs within cartilage.

Bone Growth

Bone elongation occurs at the epiphyseal plate as chondrocytes proliferate, enlarge, die, and are replaced by bone.

Bone Remodeling

Bone remodeling consists of removal of existing bone by osteoclasts and deposition of new bone by osteoblasts.

Bone Repair

During bone repair, cells move into the damaged area and form a callus, which is replaced by bone.

6.4 Bone and Calcium Homeostasis   (p. 117)

1. Osteoclasts remove calcium from bone, causing blood calcium levels to increase.
2. Osteoblasts deposit calcium into bone, causing blood calcium levels to decrease.
3. Parathyroid hormone increases bone breakdown, whereas calcitonin decreases bone breakdown.

6.5 General Considerations of Bone Anatomy   (p. 119)

There are 206 bones in the average adult skeleton.

6.6 Axial Skeleton   (p. 120)

The axial skeleton includes the skull, vertebral column, and thoracic cage.

Skull

1. The skull consists of 22 bones: 8 forming the braincase and 14 facial bones. The hyoid bone and 6 auditory ossicles are associated with the skull.
2. From a lateral view, the parietal, temporal, and sphenoid bones can be seen.
3. From a frontal view, the orbits and nasal cavity can be seen, as well as associated bones and structures, such as the frontal bone, zygomatic bone, maxilla, and mandible.
4. The interior of the cranial cavity contains three fossae with several foramina.
5. Seen from below, the base of the skull reveals numerous foramina and other structures, such as processes for muscle attachment.

Vertebral Column

1. The vertebral column contains 7 cervical, 12 thoracic, and 5 lumbar vertebrae, plus 1 sacral bone and 1 coccyx bone.
2. Each vertebra consists of a body, an arch, and processes.
3. Regional differences in vertebrae are as follows: Cervical vertebrae have transverse foramina; thoracic vertebrae have long spinous processes and attachment sites for the ribs; lumbar vertebrae have rectangular transverse and spinous processes, and the position of their facets limits rotation; the sacrum is a single, fused bone; the coccyx is 4 or fewer fused vertebrae.

Rib Cage

1. The rib cage consists of the thoracic vertebrae, the ribs, and the sternum.
2. There are 12 pairs of ribs: 7 true and 5 false (2 of the false ribs are also called floating ribs).
3. The sternum consists of the manubrium, the body, and the xiphoid process.

6.7 Appendicular Skeleton   (p. 129)

The appendicular skeleton consists of the bones of the upper and lower limbs and their girdles.

Pectoral Girdle

The pectoral girdle includes the scapulae and clavicles.

Upper Limb

The upper limb consists of the arm (humerus), forearm (ulna and radius), wrist (8 carpal bones), and hand (5 metacarpal bones, 3 phalanges in each finger, and 2 phalanges in the thumb).

Pelvic Girdle

The pelvic girdle is made up of the 2 coxal bones. Each coxal bone consists of an ilium, an ischium, and a pubis. The coxal bones, sacrum, and coccyx form the pelvis.

Lower Limb

The lower limb includes the thigh (femur), leg (tibia and fibula), ankle (7 tarsal bones), and foot (metatarsal bones and phalanges, similar to the bones in the hand).

6.8 Joints   (p. 137)

A joint is a place where bones come together.

Fibrous Joints

Fibrous joints consist of bones united by fibrous connective tissue. They allow little or no movement.

Cartilaginous Joints

Cartilaginous joints consist of bones united by cartilage, and they exhibit slight movement.

Synovial Joints

1. Synovial joints consist of articular cartilage over the uniting bones, a joint cavity lined by a synovial membrane and containing synovial fluid, and a joint capsule. They are highly movable joints.
2. Synovial joints can be classified as plane, saddle, hinge, pivot, ball-and-socket, or ellipsoid.

Types of Movement

The major types of movement are flexion/extension, abduction/adduction, pronation/supination, eversion/inversion, rotation, circumduction, protraction/retraction, elevation/depression, excursion, and opposition/reposition.

6.9 Effects of Aging on the Skeletal System and Joints   (p. 143)

1. Bone matrix becomes more brittle and decreases in total amount during aging.
2. Joints lose articular cartilage and become less flexible with age.
3. Prevention measures include exercise and calcium and vitamin D supplements.

REVIEW AND COMPREHENSION

1. What are the primary functions of the skeletal system?
2. Name the major types of fibers and molecules in the extracellular matrix of the skeletal system. How do they contribute to the functions of tendons, ligaments, cartilage, and bones?
3. Define diaphysis, epiphysis, epiphyseal plate, medullary cavity, articular cartilage, periosteum, and endosteum.
4. Describe the structure of compact bone. How do nutrients reach the osteocytes in compact bone?
5. Describe the structure of spongy bone. What are trabeculae? How do nutrients reach osteocytes in trabeculae?
6. Define and describe intramembranous and endochondral ossification.
7. How do bones grow in diameter? How do long bones grow in length?
8. What is accomplished by bone remodeling? How does bone repair occur?
9. Define the axial skeleton and the appendicular skeleton.
10. Name the bones of the braincase and the face.
11. Give the locations of the paranasal sinuses. What are their functions?
12. What is the function of the hard palate?
13. Through what foramen does the brain connect to the spinal cord?
14. How do the vertebrae protect the spinal cord? Where do spinal nerves exit the vertebral column?
15. Name and give the number of each type of vertebra. Describe the characteristics that distinguish the different types of vertebrae from one another.
16. What is the function of the thoracic cage? Name the parts of the sternum. Distinguish among true, false, and floating ribs.
17. Name the bones that make up the pectoral girdle, arm, forearm, wrist, and hand. How many phalanges are in each finger and in the thumb?
18. Define the pelvic girdle. What bones fuse to form each coxal bone? Where and with what bones do the coxal bones articulate?
19. Name the bones of the thigh, leg, ankle, and foot.
20. Define joint, or articulation. Name and describe the differences among the three major classes of joints.
21. Describe the structure of a synovial joint. How do the different parts of the joint permit joint movement?
22. On what basis are synovial joints classified? Describe the different types of synovial joints, and give examples of each. What movements does each type of joint allow?
23. Describe and give examples of flexion/extension, abduction/adduction, and supination/pronation.

CRITICAL THINKING

1. A 12-year-old boy fell while playing basketball. The physician explained that the head (epiphysis) of the femur was separated from the shaft (diaphysis). Although the bone was set properly, by the time the boy was 16, it was apparent that the injured lower limb was shorter than the normal one. Explain why this difference occurred.

2. Justin Time leaped from his hotel room to avoid burning to death in a fire. If he landed on his heels, what bone was likely to fracture? Unfortunately for Justin, a 240-pound fireman, Hefty Stomper, ran by and stepped heavily on the distal part of Justin's foot (not the toes). What bones now could be broken?

3. One day while shopping, Ms. Wantta Bargain picked up her 3-year-old son, Somm, by his right wrist and lifted him into a shopping cart. She heard a clicking sound, and Somm immediately began to cry and hold his elbow. Given that lifting the child caused a separation at the elbow, which is more likely: separation of the radius and humerus, or separation of the ulna and humerus?

4. Why are women knock-kneed more commonly than men?

5. A skeleton was discovered in a remote mountain area. The coroner determined that not only was the skeleton human, but it was female. Explain.

Answers in Appendix D

aprevealed.com

7 Muscular System

▲ Weight lifting can be used to develop muscle endurance as well as strength.

7.1 FUNCTIONS OF THE MUSCULAR SYSTEM

Learning Outcome *After reading this section, you should be able to*

A. List the functions of the muscular system.

A runner rounds the last corner of the track and sprints for the finish line. Her arms and legs are pumping as she tries to reach her maximum speed. Her heart is beating rapidly, and her breathing is rapid, deep, and regular. Blood is shunted away from her digestive organs, and a greater volume is delivered to her skeletal muscles to maximize their oxygen supply. These actions are accomplished by muscle tissue, the most abundant tissue of the body and one of the most adaptable.

You don't have to be running for the muscular system to be at work. Even when you aren't consciously moving, postural muscles keep you sitting or standing upright, respiratory muscles keep you breathing, the heart continuously pumps blood to all parts of your body, and blood vessels constrict or relax to direct blood to organs where it is needed.

In fact, movement within the body is accomplished in various ways: by cilia or flagella on the surface of certain cells, by the force of gravity, or by the contraction of muscles. But most of the body's movement results from muscle contraction. As described in chapter 4, there are three types of muscle tissue: skeletal,

Module 6 Muscular System

cardiac, and smooth (figure 7.1). This chapter focuses primarily on the structure and function of skeletal muscle; cardiac and smooth muscle are described only briefly. Following are the major functions of the muscular system:

1. *Movement of the body.* Contraction of skeletal muscles is responsible for the overall movements of the body, such as walking, running, and manipulating objects with the hands.
2. *Maintenance of posture.* Skeletal muscles constantly maintain tone, which keeps us sitting or standing erect.
3. *Respiration.* Muscles of the thorax carry out the movements necessary for respiration.
4. *Production of body heat.* When skeletal muscles contract, heat is given off as a by-product. This released heat is critical to the maintenance of body temperature.

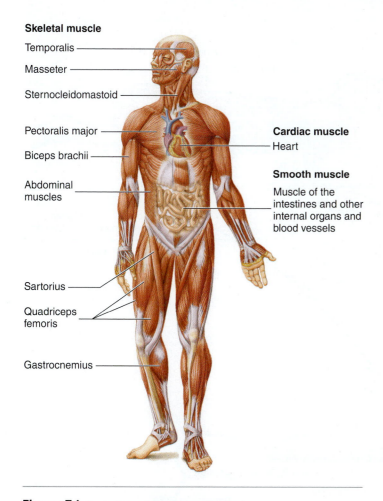

Skeletal muscle

Temporalis

Masseter

Sternocleidomastoid

Pectoralis major

Biceps brachii

Abdominal muscles

Sartorius

Quadriceps femoris

Gastrocnemius

Cardiac muscle
Heart

Smooth muscle
Muscle of the intestines and other internal organs and blood vessels

Figure 7.1 Overview of the Muscular System

E. Define muscle twitch, tetanus, recruitment, and summation.
F. Distinguish between fast-twitch and slow-twitch fibers, and explain the function for which each type is best adapted.
G. Distinguish between aerobic and anaerobic respiration.
H. Compare the mechanisms involved in the major types of fatigue.
I. Distinguish between isometric and isotonic contractions.
J. Define muscle tone.

Skeletal muscle, with its associated connective tissue, constitutes approximately 40% of body weight. Skeletal muscle is so named because most of the muscles are attached to the skeletal system. It is also called *striated muscle* because transverse bands, or striations, can be seen in the muscle under the microscope.

Skeletal muscle has four major functional characteristics: contractility, excitability, extensibility, and elasticity.

1. **Contractility** (kon-trak-til′i-tē) is the ability of skeletal muscle to shorten with force. When skeletal muscles contract, they cause the structures to which they are attached to move. Skeletal muscles shorten forcefully during contraction, but they lengthen passively. Either gravity or the contraction of an opposing muscle produces a force that pulls on the shortened muscle, causing it to lengthen.

2. **Excitability** (ek-sī′tă-bil′i-tē) is the capacity of skeletal muscle to respond to a stimulus. Normally, the stimulus is from nerves that we consciously control.

3. **Extensibility** (eks-ten′sī-bil′i-tē) means that skeletal muscles stretch. After a contraction, skeletal muscles can be stretched to their normal resting length and beyond to a limited degree.

4. **Elasticity** (ē-las-tis′i-tē) is the ability of skeletal muscles to recoil to their original resting length after they have been stretched.

Skeletal Muscle Structure

Connective Tissue Coverings of Muscle

Each skeletal muscle is surrounded by a connective tissue sheath called the **epimysium** (ep-i-mis′ē-ŭm), or *muscular fascia* (fash′ē-ă) (figure 7.2*a*). Each whole muscle is subdivided by a loose connective tissue called the **perimysium** (per′i-mis′ē-ŭm) into numerous visible bundles called muscle **fasciculi** (fă-sik′ū-lī). Each fascicle is then subdivided by a loose connective tissue called the **endomysium** (en′dō-mis′ē-ŭm) into separate muscle cells, called muscle fibers (figure 7.2*b*).

Muscle Fiber Structure

Examining the structure of a muscle fiber helps us understand the mechanism of muscle contraction. A muscle fiber is a single cylindrical fiber, with several nuclei located at its periphery. The largest human muscle fibers are up to 30 cm long and 0.15 mm in diameter. Such giant cells may contain several thousand nuclei. The cell membrane of the muscle fiber is called the **sarcolemma** (sar′kō-lem′ă; *sarco,* flesh) (figure 7.2*b*). The multiple nuclei of the muscle fiber are located just deep to the sarcolemma. Along the surface of the sarcolemma are many tubelike invaginations, called **transverse tubules,** or **T tubules,** which occur at regular intervals along the muscle fiber and extend inward into it. The

5. *Communication.* Skeletal muscles are involved in all aspects of communication, including speaking, writing, typing, gesturing, and facial expressions.

6. *Constriction of organs and vessels.* The contraction of smooth muscle within the walls of internal organs and vessels causes those structures to constrict. This constriction can help propel and mix food and water in the digestive tract, propel secretions from organs, and regulate blood flow through vessels.

7. *Contraction of the heart.* The contraction of cardiac muscle causes the heart to beat, propelling blood to all parts of the body.

7.2 CHARACTERISTICS OF SKELETAL MUSCLE

Learning Outcomes *After reading this section, you should be able to*

A. Describe the microscopic structure of a muscle, and produce diagrams that illustrate the arrangement of myofibrils, sarcomeres, and myofilaments.

B. Describe a resting membrane potential and an action potential.

C. Describe a neuromuscular junction.

D. Explain the events that occur in muscle contraction and relaxation.

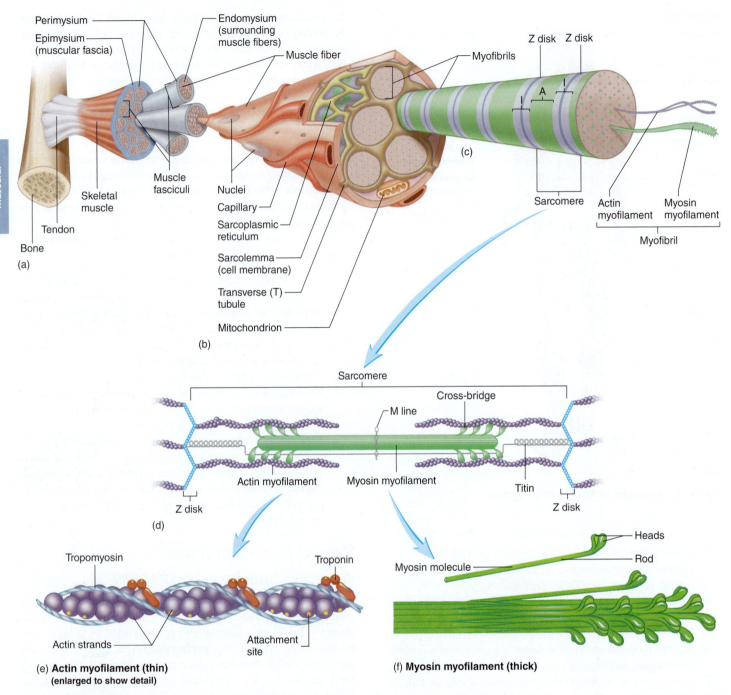

(a) Part of a muscle attached by a tendon to a bone. A muscle is composed of muscle fasciculi, each surrounded by perimysium. The fasciculi are composed of bundles of individual muscle fibers (muscle cells), each surrounded by endomysium. The entire muscle is surrounded by a connective tissue sheath called epimysium, or muscular fascia. (b) Enlargement of one muscle fiber containing several myofibrils. (c) A myofibril extended out the end of the muscle fiber, showing the banding patterns of the sarcomeres. (d) A single sarcomere of a myofibril is composed mainly of actin myofilaments and myosin myofilaments. The Z disks anchor the actin myofilaments, and the myosin myofilaments are held in place by the M line. (e) Part of an actin myofilament is enlarged. (f) Part of a myosin myofilament is enlarged.

Figure 7.2 AP|R Structure of a Muscle

T tubules are associated with a highly organized smooth endoplasmic reticulum called the **sarcoplasmic reticulum** (re-tik′ū-lŭm). T tubules connect the sarcolemma to the sarcoplasmic reticulum. The sarcoplasmic reticulum has a relatively high concentration of Ca^{2+}, which plays a major role in muscle contraction.

Inside each muscle fiber is the cytoplasm, called the **sarcoplasm** (sar′kō-plazm). It contains numerous **myofibrils** (mī-ō-fī′brilz; *myo,* muscle), threadlike structures that extend from one end of the muscle fiber to the other (figure 7.2c). Myofibrils consist of two major kinds of protein fibers: **actin** (ak′tin) **myofilaments** (mī-ō-fil′ă-ments) and **myosin** (mī′ō-sin) **myofilaments** (figure 7.2d). The actin and myosin myofilaments are arranged into highly ordered, repeating units called **sarcomeres** (sar′kō-mērz), which are joined end-to-end to form the myofibrils (figure 7.2c,d).

Actin and Myosin Myofilaments

Actin myofilaments, or thin filaments, are made up of three components: actin, troponin, and tropomyosin. The actin strands, which resemble two minute strands of pearls twisted together, have attachment sites for the myosin myofilaments (figure 7.2e). **Troponin** (trō′pō-nin) molecules are attached at specific intervals along the actin myofilaments. These molecules have binding sites for Ca^{2+}. **Tropomyosin** (trō-pō-mī′ō-sin) filaments are located along the groove between the twisted strands of actin myofilament subunits. The tropomyosin filaments block the myosin myofilament binding sites on the actin myofilaments in an unstimulated muscle. In other words, if no Ca^{2+} is present, the tropomyosin filaments cover the attachment sites on the actin myofilament. However, when Ca^{2+} is present, it binds to troponin, which causes the tropomyosin filaments to expose the attachment sites on the actin myofilaments.

Myosin myofilaments, or thick myofilaments, resemble bundles of minute golf clubs (figure 7.2f). The parts of the myosin molecule that resemble golf club heads are referred to as **myosin heads.** The myosin heads have three important properties: (1) The heads can bind to attachment sites on the actin myofilaments; (2) they can bend and straighten during contraction; and (3) they can break down ATP, releasing energy.

Sarcomeres

The sarcomere is the basic structural and functional unit of skeletal muscle because it is the smallest portion of skeletal muscle capable of contracting. The separate components of the sarcomere can slide past each other, causing the sarcomeres to shorten. When the sarcomeres shorten, the myofibrils shorten, which is the ultimate cause of contraction of the muscle fiber during a contraction. Each sarcomere extends from one Z disk to an adjacent Z disk. Each **Z disk** is a network of protein fibers forming an attachment site for actin myofilaments. The arrangement of the actin and myosin myofilaments in sarcomeres gives the myofibril a banded appearance (figure 7.3). A light **I band,** which consists only of actin myofilaments, spans each Z disk and ends at the myosin myofilaments. A darker, central region in each sarcomere, called an **A band,** extends the length of the myosin myofilaments. The actin and myosin myofilaments overlap for some distance at both ends of the A band. In the center of each sarcomere is a second light zone, called the **H zone,** which consists only of myosin myofilaments. The myosin myofilaments are anchored in the center of the sarcomere at a dark-staining band, called the **M line.** The alternating I bands and A bands of the sarcomeres are responsible for the striations in skeletal muscle fibers observed through the microscope (see table 4.10a). It is the close association of the sarcomeres, the T tubules, and the sarcoplasmic reticulum that enables a nerve stimulus to initiate contraction of the muscle fiber.

Excitability of Muscle Fibers

Muscle fibers, like other cells of the body, have electrical properties. This section describes the electrical properties of skeletal muscle fibers, and later sections illustrate their role in contraction.

Most cells in the body have an electrical charge difference across their cell membranes. The inside of the membrane is negatively charged while the outside of the cell membrane is positively charged. In other words, the cell membrane is **polarized** (figure 7.4, step 1). The charge difference, called the **resting membrane potential,** occurs because there is an uneven distribution of ions across the cell membrane. The resting membrane potential develops for three reasons: (1) The concentration of K^+ inside the cell membrane is higher than that outside the cell membrane; (2) the concentration of Na^+ outside the cell membrane is higher than that inside the cell membrane; and (3) the cell membrane is more permeable to K^+ than it is to Na^+. Recall from chapter 3 the different types of ion channels: nongated, or leak, channels, which are always open, and chemically gated channels, which are closed until a chemical, such as a neurotransmitter, binds to them and stimulates them to open (see figure 3.5). Because excitable cells have many K^+ leak channels, K^+ leaks out of the cell faster than Na^+ leaks into the cell. In other words, some K^+ channels are open, whereas other ion channels, such as those for Na^+, are closed. In addition, negatively charged molecules, such as proteins, are in essence "trapped" inside the cell because the cell membrane is impermeable to them. For these reasons, the inside of the cell membrane is more negatively charged than the outside of the cell membrane.

In addition to an outward concentration gradient for K^+, there exists an inward electrical gradient for K^+. The resting membrane potential results from the equilibrium of K^+ movement across the cell membrane. Because K^+ is positively charged, its movement from inside the cell to the outside causes the inside of the cell membrane to become even more negatively charged compared to the outside of the cell membrane. However, potassium diffuses down its concentration gradient only until the charge difference across the cell membrane is great enough to prevent any additional diffusion of K^+ out of the cell. The resting membrane potential is an equilibrium in which the tendency for K^+ to diffuse out of the cell is opposed by the negative charges inside the cell, which tend to attract the positively charged K^+ into the cell. At rest, the sodium-potassium pump transports K^+ from outside the cell to the inside and transports Na^+ from inside the cell to the outside. The active transport of Na^+ and K^+ by the sodium-potassium pump maintains the uneven distribution of Na^+ and K^+ across the cell membrane (see chapter 3).

A change in resting membrane potential is achieved by changes in membrane permeability to Na^+ or K^+ ions. A stimulation in a muscle fiber or nerve cell causes Na^+ channels to open

Muscular

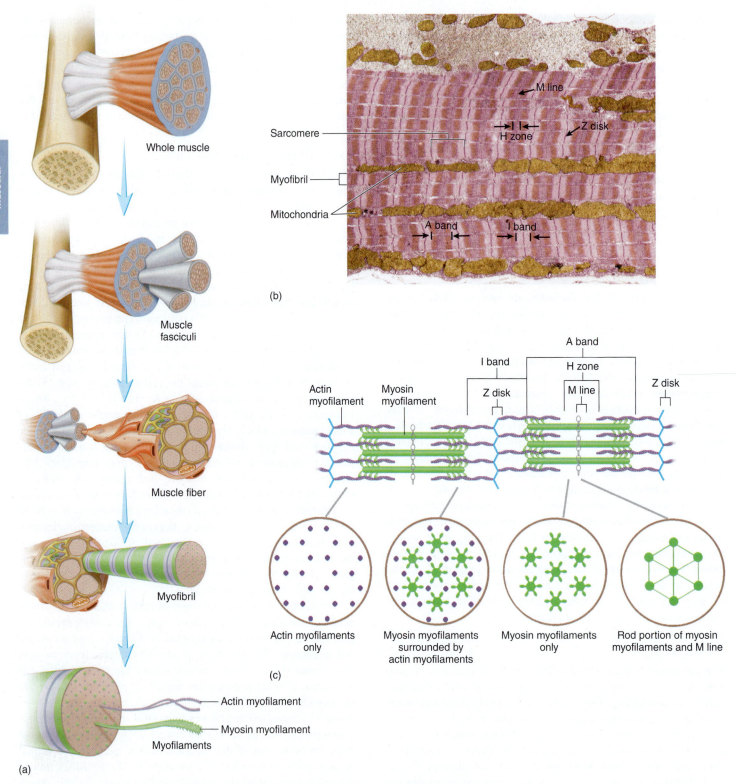

M line

Sarcomere

H zone

Z disk

Myofibril

Mitochondria

A band   I band

(b)

Whole muscle

Muscle
fasciculi

Muscle fiber

Myofibril

Actin myofilament

Myosin myofilament

Myofilaments

(a)

A band

I band   H zone

Z disk   M line   Z disk

Actin   Myosin
myofilament   myofilament

Actin myofilaments
only

Myosin myofilaments
surrounded by
actin myofilaments

Myosin myofilaments
only

Rod portion of myosin
myofilaments and M line

(c)

Figure 7.3 AP|R Skeletal Muscle

(a) Organization of skeletal muscle components. (b) Electron micrograph of skeletal muscle, showing several sarcomeres in
a muscle fiber. (c) Diagram of two adjacent sarcomeres, depicting the structures responsible for the banding pattern.

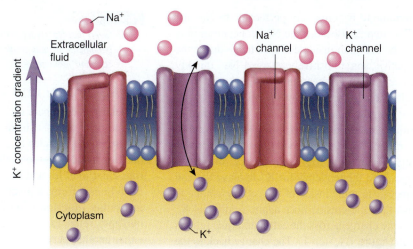

Charge Difference Across the Cell Membrane

1 **Resting membrane potential.** Na⁺ channels (*pink*) and some, but not all, K⁺ channels (*purple*) are closed. K⁺ diffuses down its concentration gradient through the open K⁺ channels, making the inside of the cell membrane negatively charged compared to the outside.

2 **Depolarization.** Na⁺ channels are open. Na⁺ diffuses down its concentration gradient through the open Na⁺ channels, making the inside of the cell membrane positively charged compared to the outside.

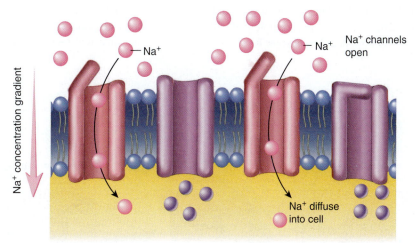

3 **Repolarization.** Na⁺ channels are closed, and Na⁺ movement into the cells stops. More K⁺ channels open. K⁺ movement out of the cell increases, making the inside of the cell membrane negatively charged compared to the outside, once again.

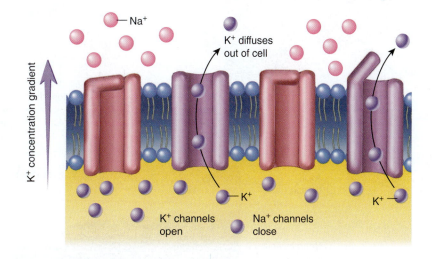

PROCESS Figure 7.4 AP|R **Ion Channels and the Action Potential**

Step 1 illustrates the status of Na⁺ and K⁺ channels in a resting cell. Steps 2 and 3 show how the channels open and close to produce an action potential. Next to each step, the charge difference across the plasma membrane is illustrated.

quickly and the membrane to become very permeable to Na⁺ for a brief time (figure 7.4, step 2). Because the Na⁺ concentration is much greater outside the cell than inside and the charge inside the cell membrane is negative, some positively charged Na⁺ quickly diffuses down its concentration gradient and toward the negative charges inside the cell, causing the inside of the cell membrane to become more positive than the outside of the cell. This change in the membrane potential is called **depolarization.** Near the end of depolarization, Na⁺ channels close, and additional K⁺ channels open (figure 7.4, step 3). Consequently, the tendency for Na⁺ to enter the cell decreases, and the tendency for K⁺ to leave the cell increases. These changes cause the inside of the cell membrane to become more negative than the outside once again. Additional K⁺ channels then close as the charge across the cell membrane returns to its resting condition (figure 7.4, step 1). The change back to the resting membrane potential is called **repolarization.** The rapid depolarization and repolarization of the cell membrane is called an **action potential.** In a muscle fiber, an action potential results in muscle contraction. The resting membrane potential and action potential are described in more detail in chapter 8.

Predict 2

Given that Na⁺ entry is the reason nerve cells or muscle fibers become activated, predict a mechanism for how some anesthetics "deaden" pain nerves.

Nerve Supply and Muscle Fiber Stimulation

Skeletal muscle fibers do not contract unless they are stimulated by motor neurons. **Motor neurons** are specialized nerve cells that stimulate muscles to contract. Motor neurons generate action potentials that travel to skeletal muscle fibers. Axons of these neurons enter muscles and send out branches to several muscle fibers. Each branch forms a junction with a muscle fiber, called a **neuromuscular junction** (figure 7.5). A more general term, **synapse** (sin′aps), refers to the cell-to-cell junction between a nerve cell and either another nerve cell or an effector cell, such as in a muscle or a gland. Neuromuscular junctions are located near the center of a muscle fiber. A single motor neuron and all the skeletal muscle fibers it innervates constitute a **motor unit.** A motor unit in a small, precisely controlled muscle, such as in the hand, may have only one or a few muscle fibers per unit, whereas the motor units of

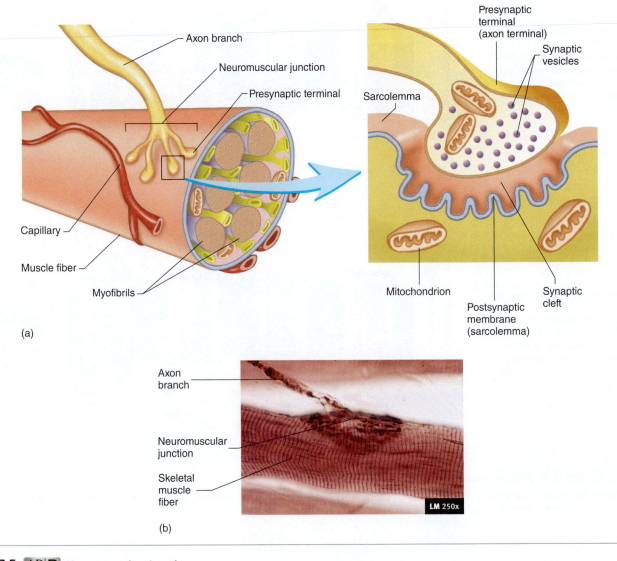

Figure 7.5 AP|R **Neuromuscular Junction**

(*a*) In a neuromuscular junction, several branches of an axon junction with a single muscle fiber. (*b*) Photomicrograph of neuromuscular junctions.

large thigh muscles may have as many as 1000 muscle fibers per motor unit. Therefore, the fewer fibers there are in the motor units of a muscle, the greater control you have over that muscle. Many motor units constitute a single muscle.

A neuromuscular junction is formed by a cluster of enlarged axon terminals resting in indentations of the muscle fiber's cell membrane. An enlarged axon terminal is the **presynaptic terminal;** the space between the presynaptic terminal and the muscle fiber membrane is the **synaptic cleft;** and the muscle fiber membrane is the **postsynaptic membrane.** Each presynaptic terminal contains many small vesicles, called **synaptic vesicles.** These vesicles contain **acetylcholine** (as-e-til-kō′lēn), or **ACh,** which functions as a **neurotransmitter,** a molecule released by a presynaptic nerve cell that stimulates or inhibits a postsynaptic cell.

When an action potential reaches the presynaptic terminal, it causes Ca^{2+} channels to open. Calcium ions enter the presynaptic terminal and cause several synaptic vesicles to release acetylcholine into the synaptic cleft by exocytosis (figure 7.6, steps 1 and 2). The acetylcholine diffuses across the synaptic cleft and binds to acetylcholine receptor sites on the Na^+ channels in the muscle fiber cell membrane. The combination of acetylcholine with its receptor opens Na^+ channels and therefore makes the cell membrane more permeable to Na^+. The resulting movement of Na^+ into the muscle fiber will initiate an action potential once threshold is reached. The action potential travels along the length of the muscle fiber and causes it to contract (figure 7.6, steps 3 and 4). The acetylcholine released into the synaptic cleft between the neuron and the muscle fiber is rapidly broken down by an enzyme, **acetylcholinesterase** (as′e-til-kō-lin-es′ter-ās). This enzymatic breakdown ensures that one action potential in the neuron yields only one action potential in the skeletal muscle fibers of that motor unit and only one contraction of each muscle fiber.

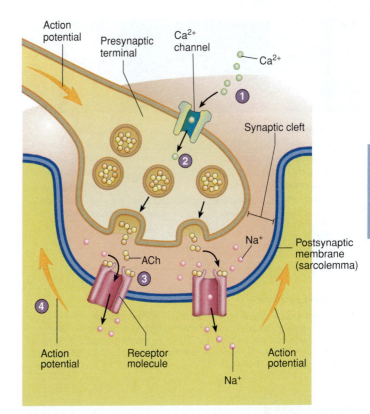

1. An action potential arrives at the presynaptic terminal, causing Ca^{2+} channels to open.

2. Calcium ions (Ca^{2+}) enter the presynaptic terminal and initiate the release of a neurotransmitter, acetylcholine (ACh), from synaptic vesicles into the presynaptic cleft.

3. Diffusion of ACh across the synaptic cleft and binding of ACh to ACh receptors on the postsynaptic muscle fiber membrane opens Na^+ channels.

4. Sodium ions (Na^+) diffuse down their concentration gradient, which results in depolarization of the muscle fiber membrane; once threshold has been reached, a postsynaptic action potential results.

PROCESS Figure 7.6 AP|R **Function of the Neuromuscular Junction**

ACh is released in response to an action potential at the neuromuscular junction.

A CASE IN POINT

Myasthenia Gravis

Les Strong was 45 years old when he first noticed muscle weakness in his face. While shaving, he could not tighten the left side of his face. Within a few days, he noticed the same problem on the right side of his face. He also experienced ptosis (drooping) of the left eyelid and slightly slurred speech. These changes worried him, and he scheduled an appointment with his physician. After a thorough examination, Les's physician referred him to a neurologist, who thought Les's muscular problems might be due to a condition called myasthenia gravis.

Myasthenia gravis is an autoimmune disorder in which antibodies are formed against acetylcholine receptors. As a result, acetylcholine receptors in the postsynaptic membranes of skeletal muscles are destroyed. With fewer receptors, acetylcholine is less likely to stimulate muscle contraction, resulting in muscle weakness and fatigue. In Les's case, a blood test revealed the presence of acetylcholine receptor antibodies, confirming the diagnosis of myasthenia gravis. Les was prescribed the anticholinesterase drug neostigmine (nē′ō-stig′min), which greatly reduced his symptoms and allowed him to continue a relatively normal life. Anticholinesterase drugs inhibit acetylcholinesterase activity, preventing the breakdown of acetylcholine. Consequently, acetylcholine levels in the synapse remain elevated, which increases the stimulation of the reduced number of functional acetylcholine receptors.

Muscle Contraction

Contraction of skeletal muscle tissue occurs as actin and myosin myofilaments slide past one another, causing the sarcomeres to shorten. Many sarcomeres are joined end-to-end to form myofibrils. Shortening of the sarcomeres causes myofibrils to shorten, thereby causing the entire muscle to shorten.

The sliding of actin myofilaments past myosin myofilaments during contraction is called the **sliding filament model** of muscle contraction. During contraction, neither the actin nor the myosin fibers shorten. The H zones and I bands shorten during contraction, but the A bands do not change in length (figure 7.7).

During muscle relaxation, sarcomeres lengthen. This lengthening requires an opposing force, such as that produced by other muscles or by gravity.

Figure 7.8 summarizes the sequence of events that occur as a stimulus causes a skeletal muscle to contract. Action potentials

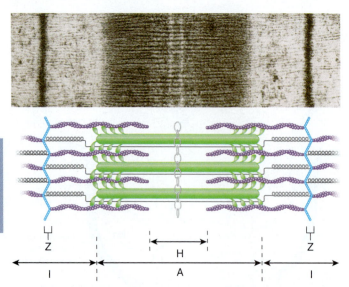

(a) Relaxed sarcomere

In a relaxed muscle, the actin and myosin myofilaments overlap slightly, and the H zone is visible. The sarcomere length is at its normal resting length. As a muscle contraction is initiated, actin myofilaments slide past the myosin myofilaments, the Z disks are brought closer together, and the sarcomere begins to shorten.

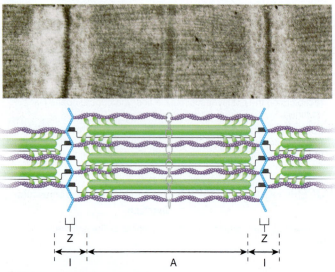

(b) Fully contracted sarcomere

In a contracted muscle, the A bands, which are equal to the length of the myosin myofilaments, do not narrow because the length of the myosin myofilaments does not change, nor does the length of the actin myofilaments. In addition, the ends of the actin myofilaments are pulled to and overlap in the center of the sarcomere, shortening it and the H zone disappears.

Figure 7.7 AP|R **Sarcomere Shortening**

produced in skeletal muscle fibers at the neuromuscular junction travel along the sarcolemma and the T tubule membranes (see figure 7.6). The action potentials cause the membranes of the sarcoplasmic reticulum adjacent to the T tubules to become more permeable to Ca^{2+}, and Ca^{2+} diffuses into the sarcoplasm. The Ca^{2+} binds to troponin molecules attached to the actin myofilaments.

This binding causes tropomyosin molecules to move into a groove along the actin molecule, exposing myosin attachment sites on the actin myofilament. The exposed attachment sites on the actin myofilament bind to the heads of the myosin myofilaments to form **cross-bridges** between the actin and myosin myofilaments.

Energy for muscle contraction is supplied to the muscles in the form of adenosine triphosphate (ATP), a high-energy molecule produced from the energy that is released during the metabolism of food (see chapters 3 and 17). The energy is released as ATP breaks down to adenosine diphosphate (ADP) and phosphate (P). During muscle contraction, the energy released from ATP is briefly stored in the myosin head. This energy is used to move the heads of the myosin myofilaments toward the center of the sarcomere, causing the actin myofilaments to slide past the myosin myofilaments. In the process, ADP and P are released from the myosin heads.

As a new ATP molecule attaches to the head of the myosin molecule, the cross-bridge is released, the ATP breaks down to ADP and P (which both remain bound to the myosin head), and the myosin head returns to its original position, where it can attach to the next site. As long as Ca^{2+} remains attached to troponin, and as long as ATP remains available, the cycle of cross-bridge formation, movement, and release repeats (figure 7.9). A new ATP must bind to myosin before the cross-bridge can be released. After a person dies, ATP is not available, and the cross-bridges that have formed are not released, causing the muscles to become rigid. This condition is called **rigor mortis** (rig′er mōr′tis; stiffness + death).

Part of the energy from ATP involved in muscle contraction is required for the formation and movement of the cross-bridges, and part is released as heat. The heat released during muscle contraction increases body temperature, which explains why a person becomes warmer during exercise. Shivering, a type of generalized muscle contraction, is one of the body's mechanisms for dealing with cold. The muscle movement involved in shivering produces heat, which raises the body temperature.

Muscle relaxation occurs as Ca^{2+} is actively transported back into the sarcoplasmic reticulum (a process that requires ATP). As a consequence, the attachment sites on the actin molecules are once again covered by tropomyosin so that cross-bridges cannot reform.

Muscle Twitch, Summation, Tetanus, and Recruitment

A **muscle twitch** is the contraction of a muscle fiber in response to a stimulus. Because most muscle fibers are grouped into motor units, a muscle twitch usually involves all the muscle fibers in a motor unit. A muscle twitch has three phases (figure 7.10). The **lag phase**, or *latent phase,* is the time between the application of a stimulus and the beginning of contraction. The **contraction phase** is the time during which the muscle contracts, and the **relaxation phase** is the time during which the muscle relaxes.

During the lag phase, action potentials are produced in one or more motor neurons. An action potential travels along the axon of a motor neuron to a neuromuscular junction (see figure 7.5). Once the stimulus reaches the neuromuscular junction, acetylcholine must be released from the presynaptic terminal, diffuse across the synaptic cleft, and bind to receptors that allow the entry of Na^+, which initiates an action potential on the postsynaptic

CLINICAL IMPACT   Acetylcholine Antagonists

Anything that affects the production, release, or degradation of acetylcholine or its ability to bind to its receptor on the muscle cell membrane can affect the transmission of action potentials across the neuromuscular junction. For example, some insecticides contain organophosphates that bind to and inhibit the function of acetylcholinesterase. As a result, acetylcholine is not degraded and accumulates in the synaptic cleft, where it acts as a constant stimulus to the muscle fiber. Insects exposed to such insecticides die partly because their respiratory muscles contract and cannot relax—a condition called **spastic paralysis** (spas'tik pa-ral'i-sis), which is followed by fatigue in the muscles.

Humans respond similarly to these insecticides. The skeletal muscles responsible for respiration cannot undergo their normal cycle of contraction and relaxation. Instead, they remain in a state of spastic paralysis until they become fatigued. Patients die of respiratory failure. Other organic poisons, such as **curare** (koo-ră'rē), the substance originally used by South American Indians in poison arrows, bind to the acetylcholine receptors on the muscle cell membrane and prevent acetylcholine from binding to them. Curare does not allow activation of the receptors; therefore, the muscle is incapable of contracting in response to nervous stimulation, a condition called **flaccid** (flak'sid, flas'id) **paralysis.** Curare is not a poison to which people are commonly exposed, but it has been used to investigate the role of acetylcholine in the neuromuscular synapse and is sometimes administered in small doses to relax muscles during certain kinds of surgery.

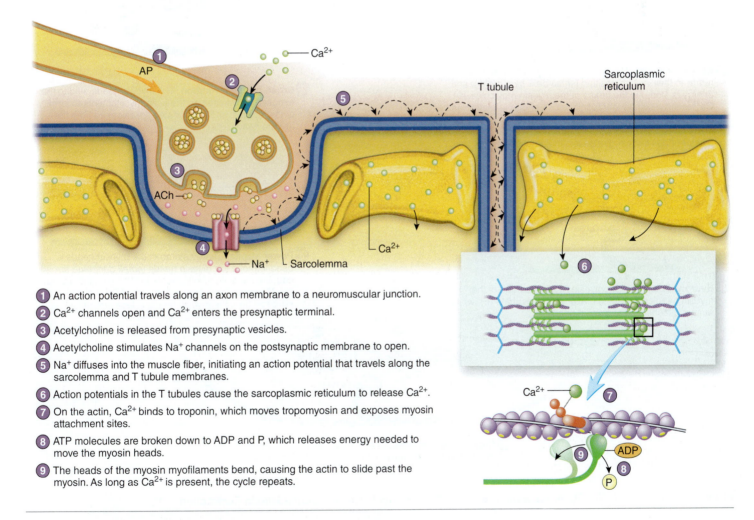

1. An action potential travels along an axon membrane to a neuromuscular junction.
2. Ca^{2+} channels open and Ca^{2+} enters the presynaptic terminal.
3. Acetylcholine is released from presynaptic vesicles.
4. Acetylcholine stimulates Na^+ channels on the postsynaptic membrane to open.
5. Na^+ diffuses into the muscle fiber, initiating an action potential that travels along the sarcolemma and T tubule membranes.
6. Action potentials in the T tubules cause the sarcoplasmic reticulum to release Ca^{2+}.
7. On the actin, Ca^{2+} binds to troponin, which moves tropomyosin and exposes myosin attachment sites.
8. ATP molecules are broken down to ADP and P, which releases energy needed to move the myosin heads.
9. The heads of the myosin myofilaments bend, causing the actin to slide past the myosin. As long as Ca^{2+} is present, the cycle repeats.

PROCESS Figure 7.8 AP|R Summary of Skeletal Muscle Contraction

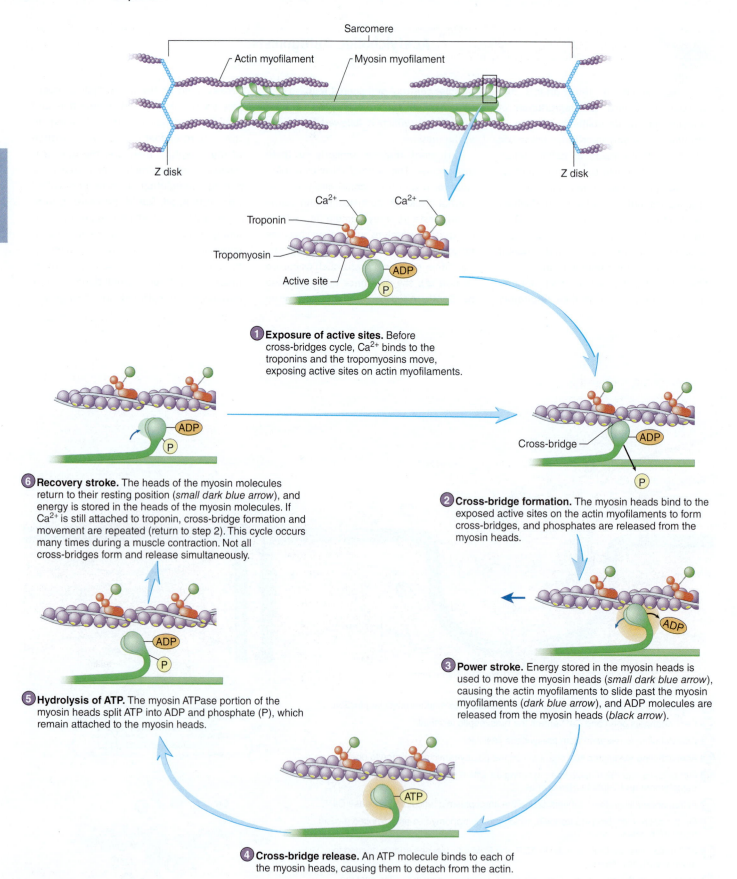

PROCESS Figure 7.9   AP|R   Breakdown of ATP and Cross-Bridge Movement During Muscle Contraction

membrane (see figure 7.6). Before the contraction phase can occur, the action potential must result in the release of Ca^{2+} from the sarcoplasmic reticulum and the formation of cross-bridges (see figure 7.9, steps 1–2).

The contraction phase results from cross-bridge movement and cycling (see figure 7.9, steps 3–6), which increases the tension produced by the muscle fibers (see figure 7.10).

During the relaxation phase, Ca^{2+} is actively transported back into the sarcoplasmic reticulum. As Ca^{2+} diffuses away from the troponin molecules, tropomyosin molecules once again block the attachment sites. Cross-bridge formation is prevented, and the tension produced by the muscle fibers decreases (see figure 7.10).

The strength of muscle contractions varies from weak to strong. For example, the force muscles generate to lift a feather is much less than the force required to lift a 25-pound weight. The force of contraction a muscle produces is increased in two ways: (1) Summation involves increasing the force of contraction of the muscle fibers within the muscle, and (2) recruitment involves increasing the number of muscle fibers contracting.

In **summation,** the force of contraction of individual muscle fibers is increased by rapidly stimulating them. When stimulus frequency, which is the number of times a motor neuron is stimulated per second, is low, there is time for complete relaxation of muscle fibers between muscle twitches (figure 7.11, stimulus frequency 1). As stimulus frequency increases (figure 7.11, stimulus frequencies 2–3), there is not enough time between contractions for muscle fibers to relax completely. Thus, one contraction *summates,* or is added onto, a previous contraction. As a result, the overall force of contraction increases. **Tetanus** (tet'a-nus; convulsive tension) is a sustained contraction that occurs when the frequency of stimulation is so rapid that no relaxation occurs (figure 7.11, stimulus frequency 4). It should be noted, however, that complete tetanus is rarely achieved under normal circumstances and is more commonly an experimentally induced muscular response. The increased force of contraction produced in summation and tetanus occurs because Ca^{2+} builds up in myofibrils, which promotes cross-bridge formation and cycling. The buildup of Ca^{2+} occurs because the rapid production of action potentials in muscle fibers causes Ca^{2+} to be released from the sarcoplasmic reticulum faster than it is actively transported back into the sarcoplasmic reticulum.

In **recruitment,** the number of muscle fibers contracting is increased by increasing the number of motor units stimulated, and the muscle contracts with more force. When only a few motor units are stimulated, a small force of contraction is produced because only a small number of muscle fibers are contracting. As the number of motor units stimulated increases, more muscle fibers are stimulated to contract, and the force of contraction increases. Maximum force of contraction is produced in a given muscle when all the motor units of that muscle are stimulated (recruited).

If all the motor units in a muscle could be stimulated simultaneously, the resulting motion would be quick and jerky. However, because the motor units are recruited gradually, some are stimulated and held in tetanus while additional motor units are recruited; thus, contractions are slow, smooth, and sustained. In the same way, smooth relaxation of muscle occurs because some motor units are held in tetanus while other motor units relax.

Fiber Types

Muscle fibers are sometimes classified as either slow-twitch or fast-twitch fibers (table 7.1). This classification is based on differences in the rod portion of the myosin myofilament (see figure 7.2). Slow-twitch fibers contain type I myosin as the predominant or even exclusive type. Fast-twitch fibers contain either type IIa or type IIb myosin myofilaments. Each of these three myosin types is the product of a different myosin gene.

The fast and slow names of these fibers indicate their contraction speeds. Fast-twitch muscle fibers contract quickly, whereas slow-twitch muscle fibers contract more slowly. Among the fast-twitch fibers, type IIb fibers are the fastest and type IIa fibers contract at an intermediate speed. The type IIb fibers can contract ten times faster than slow-twitch (type I) fibers. However, while slower, the slow-twitch fibers can sustain the contraction for longer than the fast-twitch fibers. Likewise, type IIa fibers can sustain contractions longer than type IIb, but not as long as type I.

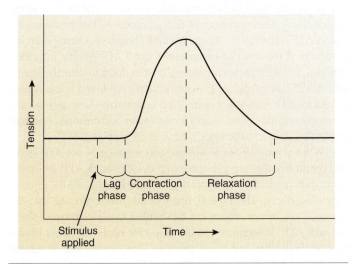

Figure 7.10 Phases of a Muscle Twitch

Hypothetical muscle twitch in a single muscle fiber. After the stimulus, there is a short lag phase, followed by a contraction phase and a relaxation phase.

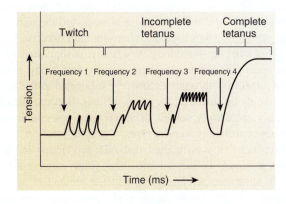

Figure 7.11 Multiple-Wave Summation

Stimuli 1–4 increase in frequency. For each stimulus, the arrow indicates the start of stimulation. Stimulus frequency 1 produces successive muscle twitches with complete relaxation between stimuli. Stimulus frequencies 2–3 do not allow complete relaxation between stimuli, resulting in incomplete tetanus. Stimulus frequency 4 allows no relaxation between stimuli, resulting in complete tetanus.

TABLE 7.1	Characteristics of Skeletal Muscle Fiber Types		
	Slow-Twitch (Type I)	Fast-Twitch Oxidative Glycolytic (Type IIa)	Fast-Twitch Glycolytic (Type IIb)
Fiber Diameter	Smallest	Intermediate	Largest
Myoglobin Content	High	High	Low
Mitochondria	Many	Many	Few
Metabolism	High aerobic capacity	High anaerobic capacity; intermediate aerobic capacity	Highest anaerobic capacity
Fatigue Resistance	High	Intermediate	Low
Myosin Head Activity	Slow	Fast	Fast
Glycogen Concentration	Low	High	High
Functions	Maintenance of posture; endurance activities	Endurance activities in endurance-trained muscles	Rapid, intense movement of short duration (sprinting)

Energy Requirements for Muscle Contraction

Muscle fibers are very energy-demanding cells whether at rest or during any form of exercise. This energy comes from either aerobic (with O_2) or anaerobic (without O_2) ATP production (see chapter 17).

Generally, ATP is derived from four processes in skeletal muscle:

1. Aerobic production of ATP during most exercise and normal conditions
2. Anaerobic production of ATP during intensive short-term work
3. Conversion of a molecule called **creatine** (krē′a-tēn) **phosphate** to ATP
4. Conversion of two ADP to one ATP and one AMP (adenosine monophosphate) during heavy exercise

Aerobic respiration, which occurs mostly in mitochondria, requires O_2 and breaks down glucose to produce ATP, CO_2, and H_2O. Aerobic respiration can also process lipids or amino acids to make ATP. **Anaerobic respiration,** which does not require O_2, breaks down glucose to produce ATP and lactate.

In general, slow-twitch fibers work aerobically, whereas fast-twitch fibers are more suited for working anaerobically. Low-intensity, long-duration exercise is supported through mainly aerobic pathways. High-intensity, short-duration exercise, such as sprinting or carrying something very heavy, is supported through partially anaerobic pathways. There are very few, if any, activities that are supported through exclusively anaerobic pathways and those can only be sustained for a few seconds. Because exercise is not usually exclusively aerobic or anaerobic, we see both muscle fiber types contributing to most types of muscle function.

Historically, it was thought that ATP production in skeletal muscle was clearly delineated into either purely aerobic activities or purely anaerobic activities, and that the product of anaerobic respiration was principally lactic acid. Lactic acid was considered to be a harmful waste product that must be removed from the body. However, it is now widely recognized that anaerobic respiration ultimately gives rise to lactic acid's alternate chemical form, lactate. Moreover, it is now known that lactate is a critical metabolic intermediate that is formed and utilized continuously even under fully aerobic conditions. Lactate is produced by skeletal muscle cells at all times, but particularly during exercise, and is subsequently broken down (70–75%) or used to make new glucose (30–35%). Thus, the aerobic and anaerobic mechanisms of ATP production are linked through lactate.

Aerobic respiration is much more efficient than anaerobic respiration, but takes several minutes. With aerobic respiration pathways, the breakdown of a single glucose molecule produces approximately 18 times more ATP than that through anaerobic respiration pathways. Additionally, aerobic respiration is more flexible than anaerobic respiration because of the ability to break down lipids and amino acids to form ATP, as noted earlier.

Anaerobic respiration produces far less ATP than aerobic respiration, but can produce ATP in a matter of a few seconds instead of a few minutes like aerobic respiration. However, ATP production rate by anaerobic respiration is too low to maintain activities for more than a few minutes.

Because muscle cells cannot store ATP, how do they generate enough ATP at a rate to keep pace with their high-energy demand? They store a different high-energy molecule called creatine phosphate. Creatine phosphate provides a means of storing energy that can be rapidly used to help maintain adequate ATP in contracting muscle fibers. During periods of rest, as excess ATP is produced, the excess ATP is used to synthesize creatine phosphate. During exercise, especially at the onset of exercise, the small ATP reserve is quickly depleted. Creatine phosphate is then broken down to directly synthesize ATP. Some of this ATP is immediately used, and some is used to restore ATP reserves. Figure 7.12 summarizes how aerobic and anaerobic respiration, lactic acid fermentation, and creatine phosphate production interact to produce a continuous supply of ATP.

When a muscle cell is working too strenuously for ATP stores and creatine phosphate to be able to provide enough ATP, anaerobic respiration predominates. Typically, the type II fibers are the primary anaerobic fibers. The type II fibers break down glucose into the intermediate, lactate, which can be shuttled to adjacent type I fibers to make ATP, or secreted into the blood for uptake by other tissues such as the liver to make new glucose. Thus, we see that in skeletal muscle, the type II fiber (anaerobic) pathways and the type I fiber (aerobic) pathways are not mutually exclusive. Rather, they work together, with lactate being the product of the type II fiber pathways that then serves as the starting point of the type I fiber pathways.

Rest

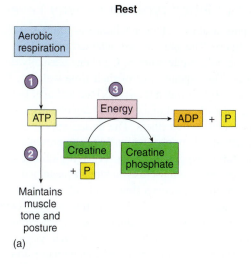

(a)

Exercise

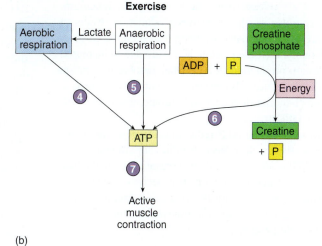

(b)

1 At rest, ATP is produced by aerobic respiration.

2 Small amounts of ATP are used in muscle contractions that maintain muscle tone and posture.

3 Excess ATP is used to produce creatine phosphate, an energy-storage molecule.

4 As exercise begins, ATP already in the cell is used first, but during moderate exercise, aerobic respiration provides most of the ATP necessary for muscle contraction.

5 During times of extreme exercise, anaerobic respiration provides small amounts of ATP that can sustain muscle contraction for brief periods.

6 Energy stored in creatine phosphate can also be used to produce ATP.

7 Throughout the time of exercise, ATP from all of these sources (4–6) provides energy for active muscle contraction.

PROCESS Figure 7.12 Fate of ATP in Resting and Exercising Muscle

Ultimately, if the use of ATP is greater than the production of ATP, the ATP:ADP ratio decreases, which interferes with the functioning of all of the major ATP-dependent enzymes in the muscle fibers. The ATP-dependent enzymes include the myosin head, the sarcoplasmic reticulum Ca^{2+} re-uptake pump, and the Na^+/K^+ pump for the resting membrane potential maintenance, all of which are required for proper muscle functioning. If the ATP:ADP ratio declines, an enzyme transfers one phosphate from one ADP to another ADP, generating one ATP and one AMP (adenosine monophosphate). The presence of AMP triggers a switch from anaerobic respiration to aerobic respiration of blood glucose and fatty acids. If this switch were not to occur, the muscles could not maintain their activity and could ultimately fail (see "Fatigue" in the next section). Figure 7.12 summarizes how aerobic and anaerobic respiration and creatine phosphate production interact to keep the muscles supplied with the ATP they need.

After intense exercise, the respiratory rate and volume remain elevated for a time, even though the muscles are no longer actively contracting. This increased respiratory activity provides the O_2 to pay back the *oxygen deficit*. The *recovery oxygen consumption* is the amount of O_2 needed in chemical reactions that occur to (1) convert lactate to glucose, (2) replenish the depleted ATP and creatine phosphate stores in muscle fibers, and (3) replenish O_2 stores in the lungs, blood, and muscles. After the lactate produced by anaerobic respiration is converted to glucose and creatine phosphate levels are restored, respiration rate returns to normal.

The magnitude of the oxygen deficit depends on the intensity of the exercise, the length of time it was sustained, and the physical condition of the individual. The metabolic capacity of an individual in poor physical condition is much lower than that of a well-trained athlete. With exercise and training, a person's ability to carry out both aerobic and anaerobic activities is enhanced.

Fatigue

Fatigue is a temporary state of reduced work capacity. Without fatigue, muscle fibers would be worked to the point of structural damage to them and their supportive tissues. Historically it was thought that buildup of lactic acid and the corresponding drop in pH (acidosis) was the major cause of fatigue. However, it is now established that there are multiple mechanisms underlying muscular fatigue.

These mechanisms include:

1. Acidosis and ATP depletion due to either an increased ATP consumption or a decreased ATP production
2. Oxidative stress, which is characterized by the buildup of excess reactive oxygen species (ROS; free radicals)
3. Local inflammatory reactions

Acidosis and ATP Depletion

Anaerobic respiration results in breakdown of glucose to lactate and protons, accounting for lowered pH. Lowered pH has several cellular effects, including decreased effectiveness of Ca^{2+} on actin and overall less Ca^{2+} release from the sarcoplasmic reticulum. Lactic acidosis can also result when liver dysfunction results in reduced clearance of lactate (such as using it to produce glucose, for example). Usually, increased lactate levels are due to increased anaerobic respiration production of ATP when aerobic respiration production of ATP is reduced. Increases in lactate are also seen in patients with mitochondrial disorders and chronic obstructive pulmonary disease (COPD).

However, to what extent ATP reductions are responsible for muscular fatigue is still not clear. Recent studies have demonstrated that cytoplasmic ATP levels stay relatively constant even in the face of decreasing muscle force production. But decreased ATP does cause fatigue. More specifically, it is the *localized* decreases in ATP levels or those associated with specific transport systems that are correlated with muscle fatigue.

Oxidative Stress

During intense exercise, increases in ROS production cause the breakdown of proteins, lipids, or nucleic acids. In addition, ROS trigger an immune system chemical called interleukin (IL)-6. IL-6 is a mediator of inflammation, which is the most likely cause of muscle soreness.

Inflammation

In addition to the stimulation of IL-6 by ROS, which causes inflammation, the immune system is directly activated by exercise. T lymphocytes, a type of white blood cell, migrate into heavily worked muscles. The presence of immune system intermediates increases the perception of pain, which most likely serves as a signal to protect those tissues from further damage.

An example of muscle fatigue occurs when a runner collapses on the track and must be helped off. The runner's muscle can no longer function regardless of how determined the runner is. Under conditions of extreme muscular fatigue, muscle may become incapable of either contracting or relaxing. This condition, called **physiological contracture,** occurs when there is too little ATP to bind to myosin myofilaments. Because binding of ATP to the myosin heads is necessary for cross-bridge release between the actin and myosin, the cross-bridges between the actin and myosin myofilaments cannot be broken, and the muscle cannot relax.

The most common type of fatigue, **psychological fatigue,** involves the central nervous system rather than the muscles themselves. The muscles are still capable of contracting, but the individual "perceives" that continued muscle contraction is impossible. A determined burst of activity in a tired runner in response to pressure from a competitor is an example of how psychological fatigue can be overcome.

Although fatigue reduces power output, the overall benefit is that it prevents complete exhaustion of ATP reserves, which could lead to severe damage of the muscle fibers.

Predict 3

After a 10-mile run with a sprint at the end, a runner continues to breathe heavily for a time. Indicate the type of respiration that is producing energy during the run, during the sprint, and after the run.

Effect of Fiber Type on Activity Level

The white meat of a chicken's breast is composed mainly of fast-twitch fibers. The muscles are adapted to contract rapidly for a short time but fatigue quickly. Chickens normally do not fly long distances. They spend most of their time walking. Ducks, on the other hand, fly for much longer periods and over greater distances. The red, or dark, meat of a chicken's leg or a duck's breast is composed of slow-twitch fibers. The darker appearance is due partly to a richer blood supply and partly to the presence of **myoglobin,** which stores oxygen temporarily. Myoglobin can continue to release oxygen in a muscle even when a sustained contraction has interrupted the continuous flow of blood.

Humans exhibit no clear separation of slow-twitch and fast-twitch muscle fibers in individual muscles. Most muscles have both types of fibers, although the number of each type varies in a given muscle. The large postural muscles contain more slow-twitch fibers, whereas muscles of the upper limb contain more fast-twitch fibers.

Average, healthy, active adults have roughly equal numbers of slow- and fast-twitch fibers in their muscles and over three times as many type IIa as type IIb fibers. In fact, athletes who are able to perform a variety of anaerobic and aerobic exercises tend to have a balanced mixture of fast-twitch and slow-twitch muscle fibers. However, a world-class sprinter may have over 80% type II fibers, with type IIa slightly predominating, whereas a world-class endurance athlete may have 95% type I fibers.

The ratio of muscle fiber types in a person's body apparently has a large hereditary component but can also be considerably influenced by training. Exercise increases the blood supply to muscles, the number of mitochondria per muscle fiber, and the number of myofibrils and myofilaments, thus causing muscle fibers to enlarge, or **hypertrophy** (hī-per′trō-fē). With weight training, type IIb myosin myofilaments can be replaced by type IIa myosin myofilaments as muscles enlarge. Muscle nuclei quit expressing type IIb genes and begin expressing type IIa genes, which are more resistant to fatigue. If the exercise stops, the type IIa genes turn off, and the type IIb genes turn back on. Vigorous exercise programs can cause a limited number of type I myofilaments to be replaced by type IIa myofilaments.

The number of cells in a skeletal muscle remains somewhat constant following birth. Enlargement of muscles after birth is primarily the result of an increase in the size of the existing muscle fibers. As people age, the number of muscle fibers actually decreases. However, there are undifferentiated cells just below the endomysium called **satellite cells.** When stimulated, satellite cells can differentiate and develop into a limited number of new, functional muscle fibers. These cells are stimulated by the destruction of existing muscle fibers, such as by injury or disease, or during intensive strength training.

Types of Muscle Contractions

Muscle contractions are classified as either isometric or isotonic. In **isometric** (equal distance) **contractions,** the length of the muscle does not change, but the amount of tension increases during the contraction process. Isometric contractions are responsible for the constant length of the body's postural muscles, such as the muscles of the back. On the other hand, in **isotonic** (equal tension) **contractions,** the amount of tension produced by the muscle is constant during contraction, but the length of the muscle decreases. Movements of the arms or fingers are predominantly isotonic contractions. Most muscle contractions are a combination of iso-metric and isotonic contractions in which the muscles shorten and the degree of tension increases.

Concentric (kon-sen′trik) **contractions** are isotonic contractions in which muscle tension increases as the muscle shortens. Many common movements are produced by concentric muscle contractions. **Eccentric** (ek-sen′trik) **contractions** are isotonic contractions in which tension is maintained in a muscle, but the opposing resistance causes the muscle to lengthen. Eccentric contractions are used when a person slowly lowers a heavy weight. Substantial force is produced in muscles during eccentric contractions, and muscles can be injured during repetitive eccentric contractions, as sometimes occurs in the hamstring muscles when a person runs downhill.

Muscle Tone

Muscle tone is the constant tension produced by body muscles over long periods of time. Muscle tone is responsible for keeping the back and legs straight, the head in an upright position, and the abdomen from bulging. Muscle tone depends on a small percentage of all the motor units in a muscle being stimulated at any point in time, causing their muscle fibers to contract tetanically and out of phase with one another.

7.3 SMOOTH MUSCLE AND CARDIAC MUSCLE

Learning Outcome   *After reading this section, you should be able to*

A. Distinguish among skeletal, smooth, and cardiac muscle.

Smooth muscle cells are small and spindle-shaped, usually with one nucleus per cell (table 7.2). They contain less actin and myosin than do skeletal muscle cells, and the myofilaments are not organized into sarcomeres. As a result, smooth muscle cells are not striated. Smooth muscle cells contract more slowly than skeletal muscle cells when stimulated by neurotransmitters from the nervous system and do not develop an oxygen deficit. The resting membrane potential of some smooth muscle cells fluctuates between slow depolarization and repolarization phases. As a result, smooth muscle cells can periodically and spontaneously generate action potentials that cause the smooth muscle cells to contract. The resulting periodic spontaneous contraction of smooth muscle is called **autorhythmicity.** Smooth muscle is under involuntary control, whereas skeletal muscle is under voluntary motor control. Some hormones, such as those that regulate the digestive system, can stimulate smooth muscle to contract.

Smooth muscle cells are organized to form layers. Most of those cells have gap junctions, specialized cell-to-cell contacts (see chapter 4), that allow action potentials to spread to all the smooth muscle cells in a tissue. Thus, all the smooth muscle cells tend to function as a unit and contract at the same time.

Cardiac muscle shares some characteristics with both smooth and skeletal muscle (table 7.2). Cardiac muscle cells are long, striated, and branching, with usually only one nucleus per cell. The actin and myosin myofilaments are organized into sarcomeres, but the distribution of myofilaments is not as uniform as in skeletal muscle. As a result, cardiac muscle cells are striated, but not as distinctly striated as skeletal muscle. When stimulated by neurotransmitters, the rate of cardiac muscle contraction is between

TABLE 7.2	Comparison of Muscle Types AP|R		
	Skeletal Muscle	**Cardiac Muscle**	**Smooth Muscle**
Location	Attached to bone	Heart	Walls or hollow organs, blood vessels, and glands
Appearance			
Cell Shape	Long, cylindrical	Branched	Spindle-shaped
Nucleus	Multiple, peripheral	Usually single, central	Single, central
Special Features	—	Intercalated disks	Cell-to-cell attachments
Striations	Yes	Yes	No
Autorhythmic	No	Yes	Yes
Control	Voluntary	Involuntary	Involuntary
Function	Move the whole body	Contract heart to propel blood through the body	Compress organs, ducts, tubes, and so on

those of smooth and skeletal muscle. Cardiac muscle contraction is autorhythmic. Cardiac muscle exhibits limited anaerobic respiration. Instead, it continues to contract at a level that can be sustained by aerobic respiration and consequently does not fatigue.

Cardiac muscle cells are connected to one another by **intercalated** (in-ter′ka-lā-ted) **disks.** Intercalated disks are specialized structures that include tight junctions and gap junctions and that facilitate action potential conduction between the cells. This cell-to-cell connection allows cardiac muscle cells to function as a unit. As a result, an action potential in one cardiac muscle cell can stimulate action potentials in adjacent cells, causing all to contract together. As with smooth muscle, cardiac muscle is under involuntary control and is influenced by hormones, such as epinephrine.

7.4 SKELETAL MUSCLE ANATOMY

General Principles

Most muscles extend from one bone to another and cross at least one joint. At each end, the muscle is connected to the bone by a **tendon.** Some broad, sheetlike tendons are called **aponeuroses** (ap′ō-noo-rō′sēz). A **retinaculum** (ret-i-nak′ū-lum; bracelet) is a band of connective tissue that holds down the tendons at each wrist and ankle. Muscle contraction causes most body movements by pulling one of the bones toward the other across the movable joint. Some muscles are not attached to bone at both ends. For example, some facial muscles attach to the skin, which moves as the muscles contract.

The two points of attachment of each muscle are its origin and insertion. The **origin,** also called the *head,* is the most stationary end of the muscle. The **insertion** is the end of the muscle attached to the bone undergoing the greatest movement. Origins are usually, but not always, proximal or medial to the insertion of a given muscle. The part of the muscle between the origin and the insertion is the **belly** (figure 7.13). Some muscles have multiple origins; for example, the biceps brachii has two, and the triceps brachii has three.

Muscles are typically grouped so that the action of one muscle or group of muscles is opposed by that of another muscle or group of muscles. For example, the biceps brachii flexes the elbow, and the triceps brachii extends the elbow. A muscle that accomplishes a certain movement, such as flexion, is called the **agonist** (ag′ō-nist). A muscle acting in opposition to an agonist is called an **antagonist** (an-tag′ō-nist). For example, when flexing the elbow, the biceps brachii is the agonist, whereas the triceps brachii, which relaxes and stretches to allow the elbow to bend, is the antagonist. When extending the elbow, the muscles' roles are reversed; the triceps brachii is the agonist, and the biceps brachii is the antagonist.

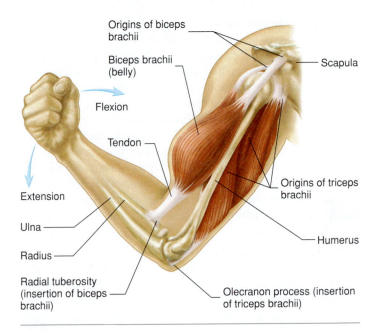

Figure 7.13 Muscle Attachment

Muscles are attached to bones by tendons. The biceps brachii has two origins that originate on the scapula. The triceps brachii has three origins that originate on the scapula and humerus. The biceps tendon inserts onto the radial tuberosity and onto nearby connective tissue. The triceps brachii inserts onto the olecranon of the ulna.

Muscles also tend to function in groups to accomplish specific movements. For example, the deltoid, biceps brachii, and pectoralis major all help flex the shoulder. Furthermore, many muscles are members of more than one group, depending on the type of movement being produced. For example, the anterior part of the deltoid muscle functions with the flexors of the shoulder, whereas the posterior part functions with the extensors of the shoulder. Members of a group of muscles working together to produce a movement are called **synergists** (sin′er-jistz). For example, the biceps brachii and the brachialis are synergists in elbow flexion. Among a group of synergists, if one muscle plays the major role in accomplishing the desired movement, it is called the **prime mover.** The brachialis is the prime mover in flexing the elbow. **Fixators** are muscles that hold one bone in place relative to the body while a usually more distal bone is moved. For example, the muscles of the scapula act as fixators to hold the scapula in place while other muscles contract to move the humerus.

Nomenclature

Most muscles have descriptive names (figure 7.14). Some muscles are named according to their location, such as the pectoralis (chest) muscles. Other muscles are named according to their origin and insertion, such as the brachioradialis (*brachio,* arm) muscle, which extends from the arm to the radius. Some muscles are named according to the number of origins, such as the biceps (*bi,* two + *ceps,* head) brachii, which has two origins, and some according to their function, such as the flexor digitorum, which flexes the digits (fingers). Other muscles are named according to their size (vastus, large), their shape (deltoid, triangular), or the orientation of their fasciculi (rectus, straight; table 7.3). Recognizing the descriptive nature of muscle names makes learning those names much easier.

Muscular

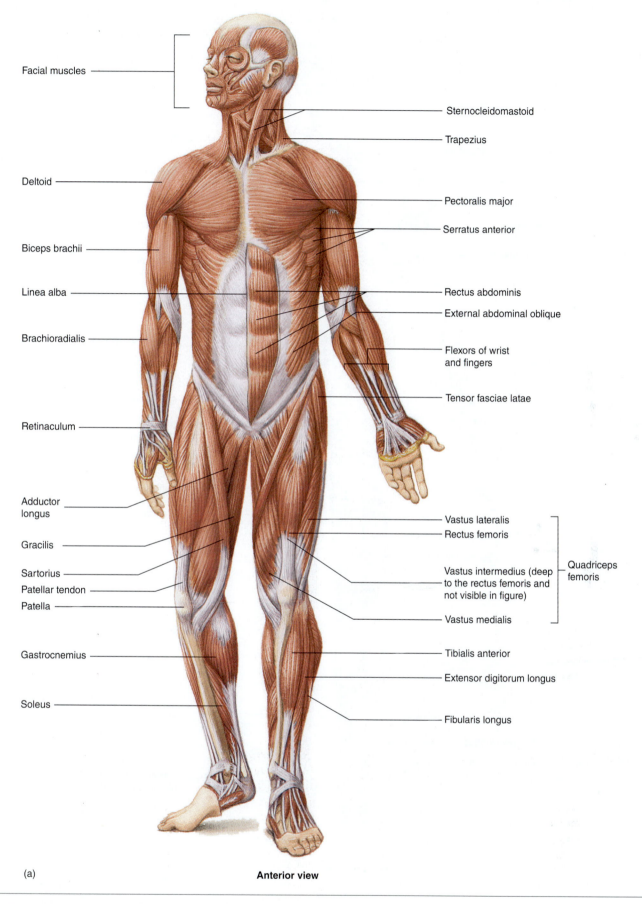

Facial muscles

Sternocleidomastoid

Trapezius

Deltoid

Pectoralis major

Serratus anterior

Biceps brachii

Linea alba

Rectus abdominis

External abdominal oblique

Brachioradialis

Flexors of wrist
and fingers

Tensor fasciae latae

Retinaculum

Adductor
longus

Vastus lateralis

Rectus femoris

Gracilis

Sartorius

Vastus intermedius (deep
to the rectus femoris and
not visible in figure)

Quadriceps
femoris

Patellar tendon

Patella

Vastus medialis

Gastrocnemius

Tibialis anterior

Extensor digitorum longus

Soleus

Fibularis longus

(a)

Anterior view

Figure 7.14 **Overview of the Superficial Body Musculature**

Red is muscle; *white* is connective tissue, such as tendons, aponeuroses, and retinacula.

Muscular

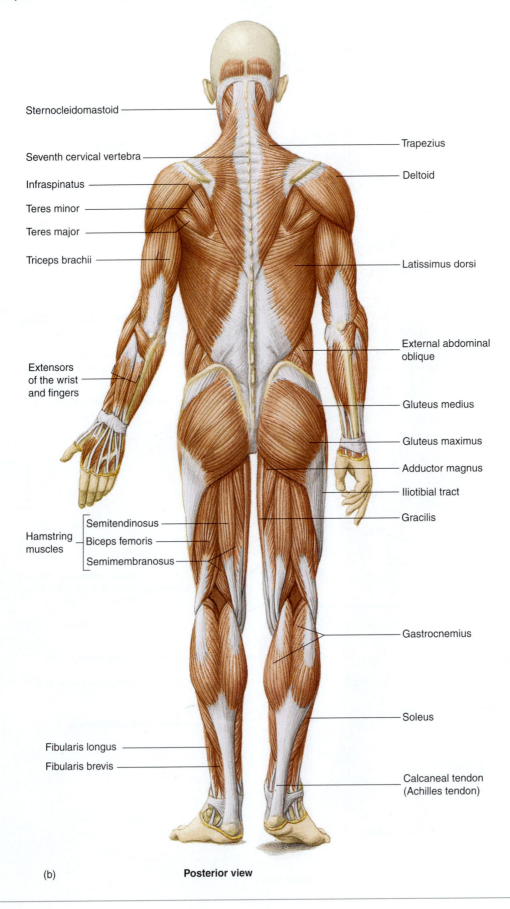

Sternocleidomastoid

Seventh cervical vertebra

Infraspinatus

Teres minor

Teres major

Triceps brachii

Extensors
of the wrist
and fingers

Hamstring
muscles

Semitendinosus

Biceps femoris

Semimembranosus

Fibularis longus

Fibularis brevis

Trapezius

Deltoid

Latissimus dorsi

External abdominal
oblique

Gluteus medius

Gluteus maximus

Adductor magnus

Iliotibial tract

Gracilis

Gastrocnemius

Soleus

Calcaneal tendon
(Achilles tendon)

(b)   **Posterior view**

Figure 7.14 **Overview of the Superficial Body Musculature** *(continued)*

TABLE 7.3   Fascicle Arrangement

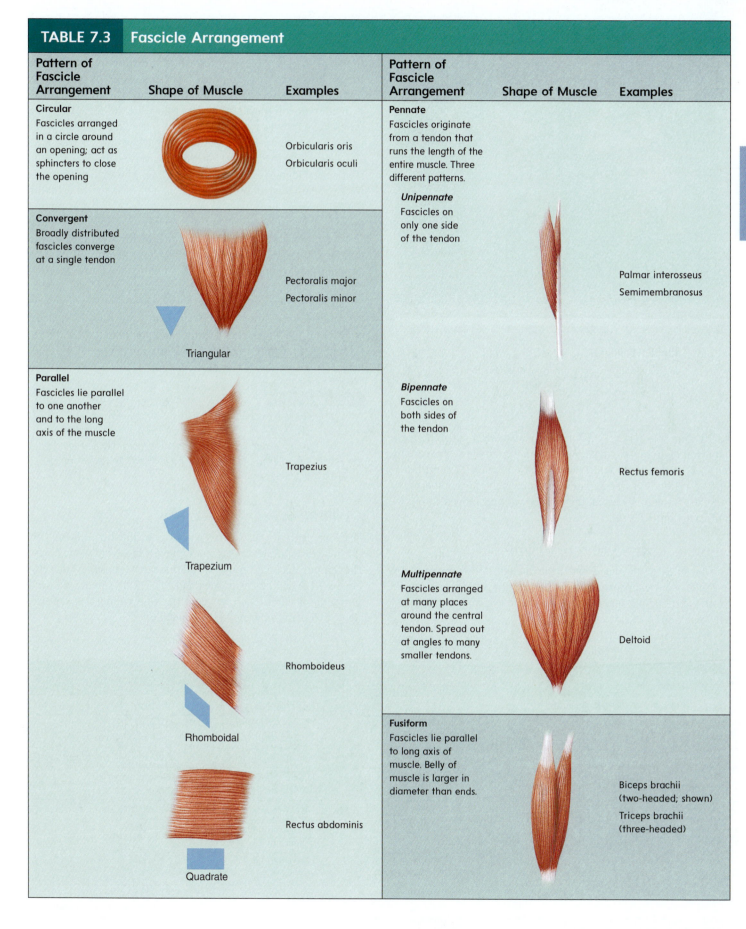

Pattern of Fascicle Arrangement	Shape of Muscle	Examples	Pattern of Fascicle Arrangement	Shape of Muscle	Examples
Circular Fascicles arranged in a circle around an opening; act as sphincters to close the opening		Orbicularis oris Orbicularis oculi	**Pennate** Fascicles originate from a tendon that runs the length of the entire muscle. Three different patterns.		
			Unipennate Fascicles on only one side of the tendon		Palmar interosseus Semimembranosus
Convergent Broadly distributed fascicles converge at a single tendon	Triangular	Pectoralis major Pectoralis minor			
Parallel Fascicles lie parallel to one another and to the long axis of the muscle	Trapezium	Trapezius	*Bipennate* Fascicles on both sides of the tendon		Rectus femoris
	Rhomboidal	Rhomboideus	*Multipennate* Fascicles arranged at many places around the central tendon. Spread out at angles to many smaller tendons.		Deltoid
	Quadrate	Rectus abdominis	**Fusiform** Fascicles lie parallel to long axis of muscle. Belly of muscle is larger in diameter than ends.		Biceps brachii (two-headed; shown) Triceps brachii (three-headed)

Muscular

Figure 7.15 Bodybuilders

Name as many muscles as you can in these photos. Compare these photos to the labeled muscles in figure 7.14.

Examining surface anatomy can also be a great help in understanding muscle anatomy. We have pointed out certain muscles of the upper and lower limbs that can be seen on the surface of the body, and figure 7.14 shows the most superficial muscles. Some muscles are especially well developed in bodybuilders (figure 7.15).

Muscles of the Head and Neck

The muscles of the head and neck include those involved in forming facial expressions, chewing, moving the tongue, swallowing, producing sounds, moving the eyes, and moving the head and neck.

Facial Expression

Several muscles act on the skin around the eyes and eyebrows (figure 7.16 and table 7.4). The **occipitofrontalis** (ok-sip′i-tō-frŭn-tā′lis) raises the eyebrows. The occipital and frontal portions of the muscle are connected by the epicranial aponeurosis. The **orbicularis oculi** (ōr-bik′ū-la′ris, circular + ok′ū-lī, eye) encircle the eyes, tightly close the eyelids, and cause "crow's feet" wrinkles in the skin at the lateral corners of the eyes.

Several other muscles function in moving the lips and the skin surrounding the mouth (figure 7.16). The **orbicularis oris** (ōr′is; mouth), which encircles the mouth, and the **buccinator** (buk′sĭ-nā′tōr; *bucca,* cheek) are sometimes called the kissing muscles because they pucker the mouth. The buccinator also flattens the cheeks as in whistling or blowing a trumpet and is therefore sometimes called the trumpeter's muscle. Smiling is accomplished primarily by the **zygomaticus** (zī′gō-mat′i-kŭs) muscles, which elevate the upper lip and corner of the mouth. Sneering is accomplished by the **levator labii superioris** (le-vā′ter lā′bē-ī soo-pēr′ē-ōr′is) because the muscle elevates one side of the upper lip. Frowning and pouting are largely performed by the **depressor anguli oris** (dē-pres′ŏr an′gū-lī ōr′ŭs), which depresses the corner of the mouth.

Predict 4

Harry Wolf, a notorious flirt, on seeing Sally Gorgeous, raises his eyebrows, winks, whistles, and smiles. Name the facial muscles he uses to carry out this communication. Sally, thoroughly displeased with this exhibition, frowns and sneers in disgust. What muscles does she use?

Mastication

The four pairs of muscles for chewing, or **mastication** (mas-ti-kā′shŭn), are some of the strongest muscles in the body (table 7.5). The **temporalis** (tem′pŏ-rā′lis) and **masseter** (ma-sē′ter) muscles (figure 7.16) can be easily seen and felt on the side of the head during mastication. The **pterygoid** (ter′ĭ-goyd) muscles, consisting of two pairs, are deep to the mandible.

TABLE 7.4	Muscles of Facial Expression (see figure 7.16)		
Muscle	**Origin**	**Insertion**	**Action**
Buccinator (buk′sĭ-nā′tōr)	Maxilla and mandible	Orbicularis oris at angle of mouth	Retracts angle of mouth; flattens cheek
Depressor anguli oris (dē-pres′ōr an′gū-lī ōr′ŭs)	Lower border of mandible	Skin of lip near angle of mouth	Depresses angle of mouth
Levator labii superioris (le-vā′ter lā′bē-ī soo-pēr′ē-ōr′is)	Maxilla	Skin and orbicularis oris of upper lip	Elevates upper lip
Occipitofrontalis (ok-sip′i-tō-frŭn′tā′lis)	Occipital bone	Skin of eyebrow and nose	Moves scalp; elevates eyebrows
Orbicularis oculi (ōr-bik′ū-lā′ris ok′ū-lī)	Maxilla and frontal bones	Circles orbit and inserts near origin	Closes eye
Orbicularis oris (ōr-bik′ū-lā′ris ōr′is)	Nasal septum, maxilla, and mandible	Fascia and other muscles of lips	Closes lip
Zygomaticus major (zī′gō-mat′i-kŭs)	Zygomatic bone	Angle of mouth	Elevates and abducts upper lip and corner of mouth
Zygomaticus minor (zī′gō-mat′i-kŭs)	Zygomatic bone	Orbicularis oris of upper lip	Elevates and abducts upper lip

Muscular

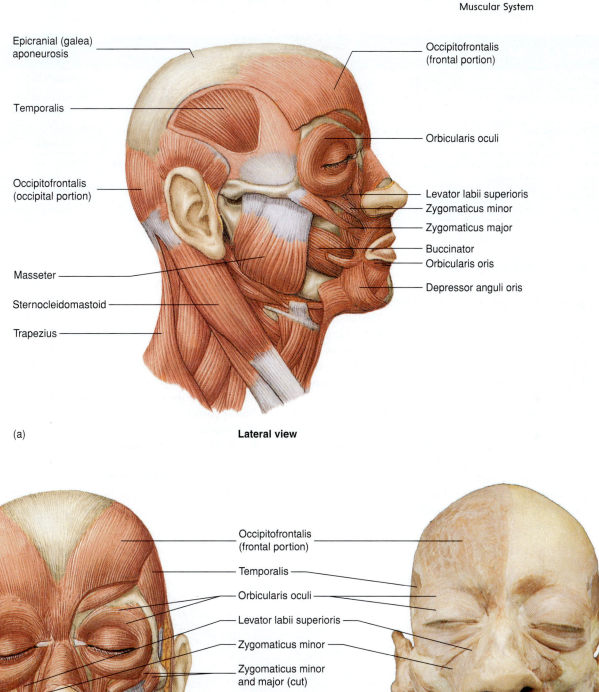

Epicranial (galea) aponeurosis

Occipitofrontalis (frontal portion)

Temporalis

Orbicularis oculi

Occipitofrontalis (occipital portion)

Levator labii superioris
Zygomaticus minor
Zygomaticus major
Buccinator
Orbicularis oris
Depressor anguli oris

Masseter

Sternocleidomastoid

Trapezius

(a)   **Lateral view**

Occipitofrontalis (frontal portion)

Temporalis

Orbicularis oculi

Levator labii superioris

Zygomaticus minor

Zygomaticus minor and major (cut)

Zygomaticus major

Buccinator

Masseter

Orbicularis oris

Depressor anguli oris

(b)   **Anterior view**

Figure 7.16 AP|R **Muscles of Facial Expression and Mastication**

TABLE 7.5   Muscles of Mastication (see figure 7.16)

Muscle	Origin	Insertion	Action
Temporalis (tem′pŏ-rā′lis)	Temporal fossa	Anterior portion of mandibular ramus and coronoid process	Elevates and retracts mandible; involved in excursion
Masseter (mă-sē′ter)	Zygomatic arch	Lateral side of mandibular ramus	Elevates and protracts mandible; involved in excursion
Lateral pterygoid (ter′i-goyd) (not shown in illustration)	Lateral pterygoid plate and greater wing of sphenoid	Condylar process of mandible and articular disk	Protracts and depresses mandible; involved in excursion
Medial pterygoid (not shown in illustration)	Lateral pterygoid plate of sphenoid and tuberosity of maxilla	Medial surface of mandible	Protracts and elevates mandible; involved in excursion

Tongue and Swallowing Muscles

The tongue is very important in mastication and speech. It moves food around in the mouth and, with the buccinator muscle, holds the food in place while the teeth grind the food. The tongue pushes food up to the palate and back toward the pharynx to initiate swallowing. The tongue consists of a mass of **intrinsic muscles,** which are located entirely within the tongue and change its shape. The **extrinsic muscles** are located outside the tongue but are attached to and move the tongue (figure 7.17 and table 7.6).

Swallowing involves a number of structures and their associated muscles, including the hyoid muscles, soft palate, pharynx (throat), and larynx (voicebox). The **hyoid** (hi′oyd) **muscles** are divided into a suprahyoid group (superior to the hyoid bone) and an infrahyoid group (inferior to the hyoid bone) (figure 7.17 and table 7.6). When the suprahyoid muscles hold the hyoid bone in place from above, the infrahyoid muscles can elevate the larynx.

To observe this effect, place your hand on your larynx (Adam's apple) and swallow.

The muscles of the soft palate close the posterior opening to the nasal cavity during swallowing, preventing food and liquid from entering the nasal cavity. When we swallow, muscles elevate the pharynx and larynx and then constrict the pharynx. Specifically, the **pharyngeal** (fă-rin′jē-ăl) **elevators** elevate the pharynx, and the **pharyngeal constrictors** constrict the pharynx from superior to inferior, forcing the food into the esophagus. Pharyngeal muscles also open the auditory tube, which connects the middle ear to the pharynx. Opening the auditory tube equalizes the pressure between the middle ear and the atmosphere. This is why it is sometimes helpful to chew gum or swallow when ascending or descending a mountain in a car or changing altitude in an airplane.

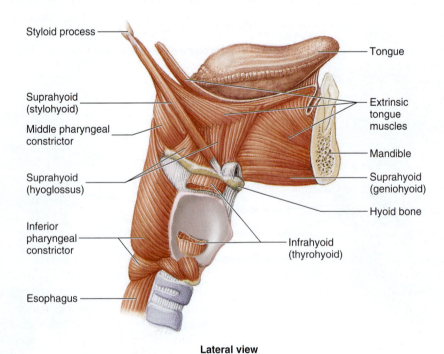

Lateral view

Figure 7.17 Tongue and Swallowing Muscles

Muscles of the tongue, hyoid, pharynx, and larynx as seen from the right.

TABLE 7.6   Tongue and Swallowing Muscles (see figure 7.17)

Muscle	Origin	Insertion	Action
Tongue muscles			
Intrinsic (not shown)	Inside tongue	Inside tongue	Changes shape of tongue
Extrinsic	Bones around oral cavity or soft palate	Onto tongue	Moves tongue
Hyoid muscles			
Suprahyoid (e.g., geniohyoid, stylohyoid, and hyoglossus)	Base of skull, mandible	Hyoid bone	Elevates or stabilizes hyoid
Infrahyoid (e.g., thyrohyoid)	Sternum, larynx	Hyoid bone	Depresses or stabilizes hyoid
Soft palate (not shown)	Skull or soft palate	Palate, tongue, or pharynx	Moves soft palate, tongue, or pharynx
Pharyngeal muscles			
Elevators (not shown)	Soft palate and auditory tube	Pharynx	Elevate pharynx
Constrictors	Larynx and hyoid	Pharynx	Constrict pharynx
Superior (not shown)			
Middle			
Inferior			

Muscular (side tab)

Neck Muscles

The deep neck muscles (figure 7.18 and table 7.7) include neck flexors, located along the anterior surfaces of the vertebral bodies, and neck extensors, located posteriorly. Rotation and lateral flexion of the head are accomplished by lateral and posterior neck muscles. The **sternocleidomastoid** (ster′nō-klī′dō-mas′toyd) muscle (see figure 7.16a), the prime mover of the lateral muscle group, is easily seen on the anterior and lateral sides of the neck. Contraction of only one sternocleidomastoid muscle rotates the head. Contraction of both sternocleidomastoids flexes the neck or extends the head, depending on what the other neck muscles are doing. **Torticollis** (tōr′ti-kol′is; a twisted neck), or wry neck, may result from injury to one of the sternocleidomastoid muscles. It is sometimes caused by damage to a baby's neck muscles during a difficult birth and usually can be corrected by exercising the muscle.

Predict 5

Shortening of the right sternocleidomastoid muscle rotates the head in which direction?

Trunk Muscles

Trunk muscles include those that move the vertebral column, the thorax and abdominal wall, and the pelvic floor.

Muscles Moving the Vertebral Column

In humans, the back muscles are very strong to maintain erect posture. The **erector spinae** (ē-rek′tōr spī′nē) group of muscles on each side of the back are primarily responsible for keeping the back straight and the body erect (table 7.8 and figure 7.18). **Deep back muscles,** located between the spinous and transverse processes of adjacent vertebrae, are responsible for several movements of the vertebral column, including extension, lateral flexion, and rotation. When the deep back muscles are stretched abnormally or torn, muscle strains and sprains of lumbar vertebral ligaments can occur, resulting in low back pain. Treatments include anti-inflammatory medication and RICE (*r*est, *i*ce, *c*ompression, and *e*levation). Low back exercises can also help the problem.

Thoracic Muscles

The muscles of the thorax (figure 7.19 and table 7.9) are involved almost entirely in the process of breathing. The **external intercostals** (in′ter-kos′tŭlz; between ribs) elevate the ribs during inspiration. The **internal intercostals** contract during forced expiration, depressing the ribs. However, the major movement produced in the thorax during quiet breathing is accomplished by the dome-shaped **diaphragm** (dī′ă-fram). When it contracts, the dome is flattened, causing the volume of the thoracic cavity to increase, resulting in inspiration.

Abdominal Wall Muscles

The muscles of the anterior abdominal wall (figure 7.20 and table 7.10) flex and rotate the vertebral column, compress the abdominal cavity, and hold in and protect the abdominal organs. In a relatively muscular person with little fat, a vertical indentation, extending from the sternum through the navel to the pubis, is visible. This tendinous area of the abdominal wall, called the **linea alba** (lin′ē-ă al′bă; white line), consists of white connective tissue rather than muscle. On each side of the linea alba is the **rectus abdominis** (rek′tŭs ab-dom′ĭ-nis; *rectus,* straight) muscle. **Tendinous intersections** cross the rectus abdominis at three or more locations, causing the abdominal wall of a lean, well-muscled person to appear segmented. Lateral to the rectus abdominis are three layers of muscle. From superficial to deep, these muscles are the **external abdominal oblique,** the **internal abdominal oblique,** and the **transversus abdominis** (transver′sŭs ab-dom′in-is) muscles. The fasciculi of these three muscle layers are oriented in different directions. When these muscles contract, they flex and rotate the vertebral column or compress the abdominal contents.

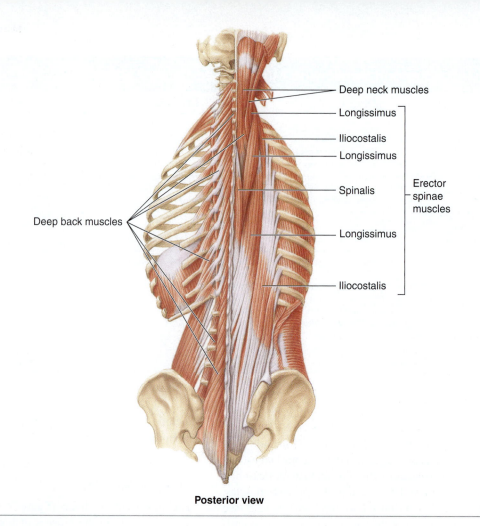

Posterior view

Figure 7.18 AP|R Deep Neck and Back Muscles

The upper limb, pectoral girdle, and associated muscles have been removed. On the right, the erector spinae muscles are shown.
On the left, these muscles are removed to reveal the deeper back muscles.

TABLE 7.7	Neck Muscles (see figures 7.14, 7.16, and 7.22)		
Muscle	**Origin**	**Insertion**	**Action**
Deep neck muscles			
Flexors (not shown)	Anterior side of vertebrae	Base of skull	Flex head and neck
Extensors	Posterior side of vertebrae	Base of skull	Extend head and neck
Sternocleidomastoid (ster′nō-klī′dō-mas′toyd)	Manubrium of sternum and medial part of clavicle	Mastoid process and nuchal line of skull	Individually rotate head; together flex neck
Trapezius (tra-pē′zē-ŭs)	Posterior surface of skull and upper vertebral column (C7–T12)	Clavicle, acromion process, and scapular spine	Extends and laterally flexes neck

TABLE 7.8	Muscles Acting on the Vertebral Column (see figure 7.18)		
Muscle	**Origin**	**Insertion**	**Action**
Superficial			
Erector spinae (ĕ-rek′tōr spī′nē) divides into three columns: Iliocostalis (il′ē-ō-kos-tā′lis)	Sacrum, ilium, vertebrae, and ribs	Ribs, vertebrae, and skull	Extends vertebral column
Longissimus (lon-gis′i-mŭs)			
Spinalis (spī-nā′lis)			
Deep back muscles	Vertebrae	Vertebrae	Extend vertebral column and help bend vertebral column laterally

Muscular

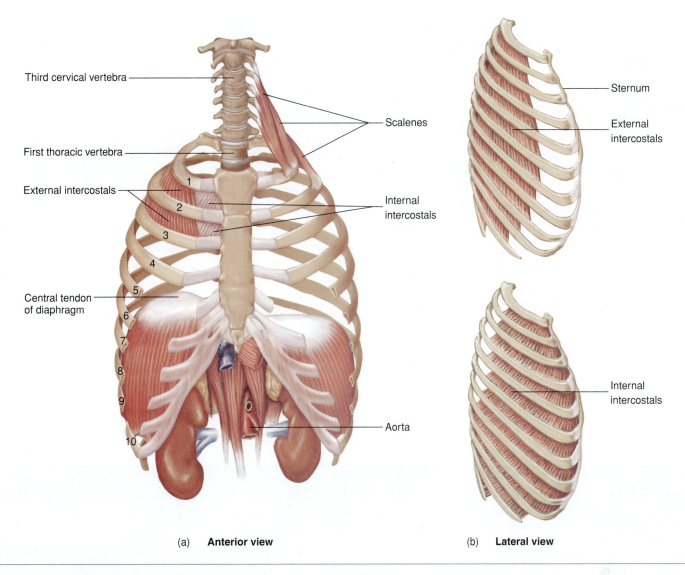

(a) **Anterior view**

(b) **Lateral view**

Figure 7.19 AP|R **Muscles of the Thorax**

(*a*) Anterior view shows a few selected intercostal muscles and the diaphragm. (*b*) Lateral view shows the external and internal intercostals.

TABLE 7.9	Muscles of the Thorax (see figures 7.19 and 7.22)		
Muscle	**Origin**	**Insertion**	**Action**
Scalenes (skā′lēnz)	Cervical vertebrae	First and second ribs	Inspiration; elevate ribs
External intercostals (in′ter-kos′tŭlz)	Inferior edge of each rib	Superior edge of next rib below origin	Inspiration; elevate ribs
Internal intercostals (in′ter-kos′tŭlz)	Superior edge of each rib	Inferior edge of next rib above origin	Forced expiration; depress ribs
Diaphragm (dī′ă-fram)	Inferior ribs, sternum, and lumbar vertebrae	Central tendon of diaphragm	Inspiration; depress floor of thorax

Pelvic Floor and Perineal Muscles

The pelvis is a ring of bone with an inferior opening that is closed by a muscular floor through which the anus and the openings of the urinary tract and reproductive tract penetrate. Most of the **pelvic floor,** also referred to as the *pelvic diaphragm,* is formed by the **levator ani** (le-vā′ter ā′nī) muscle. The area inferior to the pelvic floor is the **perineum** (per′i-nē′ŭm), which contains a number of muscles associated with the male or female reproductive structures (figure 7.21 and table 7.11). Several of these muscles help regulate urination and defecation.

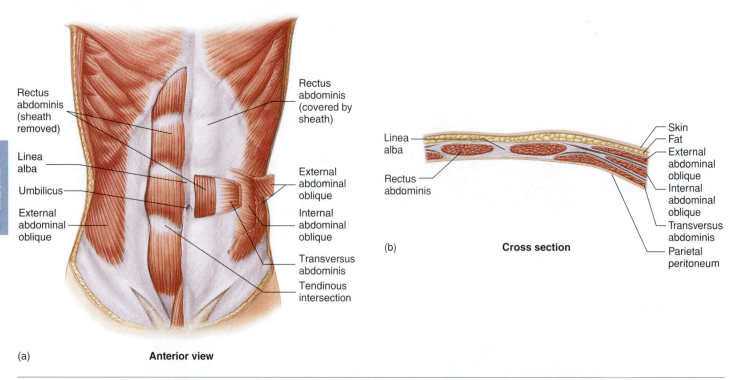

Rectus abdominis (sheath removed)

Linea alba

Umbilicus

External abdominal oblique

Rectus abdominis (covered by sheath)

External abdominal oblique

Internal abdominal oblique

Transversus abdominis

Tendinous intersection

(a)   **Anterior view**

Linea alba

Rectus abdominis

Skin
Fat
External abdominal oblique
Internal abdominal oblique
Transversus abdominis
Parietal peritoneum

(b)   **Cross section**

Figure 7.20 AP|R **Muscles of the Anterior Abdominal Wall**

(*a*) In this anterior view, windows reveal the various muscle layers. (*b*) A cross-sectional view of the muscle layers.

TABLE 7.10	Muscles of the Abdominal Wall (see figures 7.14, 7.20, 7.22)		
Muscle	**Origin**	**Insertion**	**Action**
Rectus abdominis (rek′tŭs ab-dom′i-nis)	Pubic crest and pubic symphysis	Xiphoid process and inferior ribs	Flexes vertebral column; compresses abdomen
External abdominal oblique	Ribs 5–12	Iliac crest, inguinal ligament, and fascia of rectus abdominis	Compresses abdomen; flexes and rotates vertebral column
Internal abdominal oblique	Iliac crest, inguinal ligament, and lumbar fascia	Ribs 10–12 and fascia of rectus abdominis	Compresses abdomen; flexes and rotates vertebral column
Transversus abdominis (trans-ver′sŭs ab-dom′in-is)	Costal cartilages 7–12, lumbar fascia, iliac crest, and inguinal ligament	Xiphoid process, fascia of rectus abdominis, and pubic tubercle	Compresses abdomen

Upper Limb Muscles

The muscles of the upper limb include those that attach the limb and pectoral girdle to the body and those in the arm, forearm, and hand.

Scapular Movements

The upper limb is primarily connected to the body by muscles. The muscles that attach the scapula to the thorax and move the scapula include the **trapezius** (tra-pē′zē-ŭs), the **levator scapulae** (le-vā′ter skap′ū-lē), the **rhomboids** (rom′boydz), the **serratus** (ser-ā′tŭs; serrated) **anterior,** and the **pectoralis** (pek′tō-ra′lis) **minor** (figure 7.22 and table 7.12). These muscles act as fixators to hold the scapula firmly in position when the muscles of the arm contract. The scapular muscles also move the scapula into different

positions, thereby increasing the range of movement of the upper limb. The trapezius forms the upper line from each shoulder to the neck. The origin of the serratus anterior from the first eight or nine ribs can be seen along the lateral thorax.

Arm Movements

The arm is attached to the thorax by the **pectoralis major** and **latissimus dorsi** (lă-tis′i-mŭs dōr′sī) muscles (figure 7.23*a* and table 7.13; see figure 7.22*c*). The pectoralis major adducts the arm and flexes the shoulder. It can also extend the shoulder from a flexed position. The latissimus dorsi medially rotates and adducts the arm and powerfully extends the shoulder. Because a swimmer uses these three motions during the power stroke of the crawl, the latissimus dorsi is often called the swimmer's muscle.

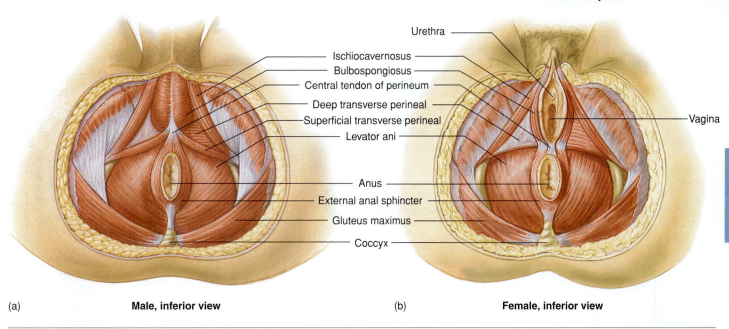

(a) **Male, inferior view**   (b) **Female, inferior view**

Figure 7.21 Muscles of the Pelvic Floor and Perineum

TABLE 7.11	Muscles of the Pelvic Floor and Perineum (see figure 7.21)		
Muscle	**Origin**	**Insertion**	**Action**
Pelvic Floor			
Levator ani (le′vā-ter ā′nī)	Posterior pubis and ischial spine	Sacrum and coccyx	Elevates anus; supports pelvic viscera
Perineum			
Bulbospongiosus (bul′bō-spŭn′jē-ō′sŭs)	Male—central tendon of perineum	Dorsal surface of penis and bulb of penis	Constricts urethra; erects penis
	Female—central tendon of perineum	Base of clitoris	Erects clitoris
Ischiocavernosus (ish′ē-ō-kav′er-nō′sŭs)	Ischial ramus	Corpus cavernosum	Compresses base of penis or clitoris
External anal sphincter	Coccyx	Central tendon of perineum	Keeps orifice of anal canal closed
Transverse perinei (pĕr′i-nē′ī)			
Deep	Ischial ramus	Midline connective tissue	Supports pelvic floor
Superficial	Ischial ramus	Central tendon of perineum	Fixes central tendon

Another group of four muscles, called the **rotator cuff muscles,** attaches the humerus to the scapula and forms a cuff or cap over the proximal humerus (table 7.13; see figure 7.22*b,c*). These muscles stabilize the joint by holding the head of the humerus in the glenoid cavity during arm movements, especially abduction. A rotator cuff injury involves damage to one or more of these muscles or their tendons. The **deltoid** (del′toyd) muscle attaches the humerus to the scapula and clavicle and is the major abductor of the upper limb. The pectoralis major forms the upper chest, and the deltoid forms the rounded mass of the shoulder (see figure 7.25). The deltoid is a common site for administering injections.

Forearm Movements

The arm can be divided into anterior and posterior compartments. The **triceps brachii** (trī′seps brā′kē-ī; three heads, arm), the primary extensor of the elbow, occupies the posterior compartment (figure 7.23*b* and table 7.14). The anterior compartment is occupied mostly by the **biceps** (bī′seps) **brachii** and the **brachialis** (brā′kē-ăl-is), the primary flexors of the elbow. The **brachioradialis** (brā′kē-ō-rā′dē-al′is), which is actually a posterior forearm muscle, helps flex the elbow.

Supination and Pronation

Supination of the forearm, or turning the flexed forearm so that the palm is up, is accomplished by the **supinator** (soo′pi-nā-ter) (figure 7.24 and table 7.15) and the biceps brachii, which tends to supinate the forearm while flexing the elbow. Pronation, turning the forearm so that the palm is down, is a function of two **pronator** (prō-nā′ter) muscles.

Wrist and Finger Movements

The twenty muscles of the forearm can also be divided into anterior and posterior groups. Only a few of these muscles, the most superficial, appear in table 7.15 and figure 7.24. Most of the

Muscular

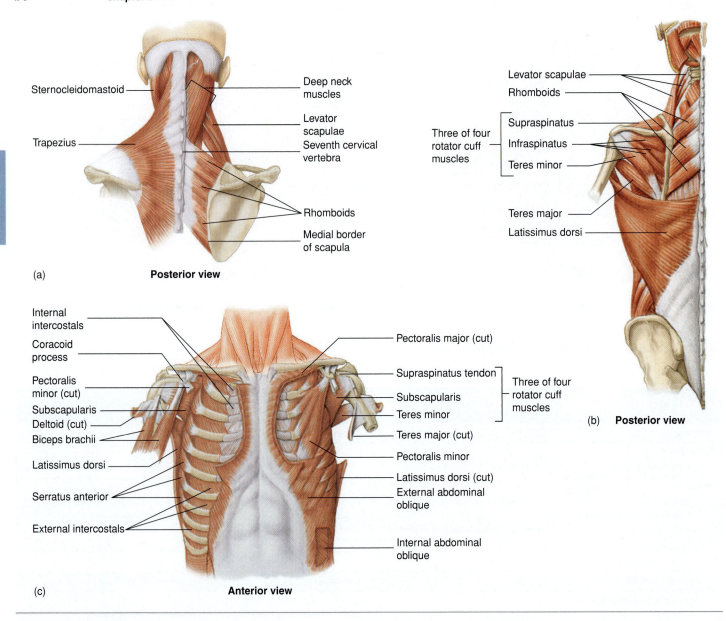

(a) **Posterior view**

Sternocleidomastoid
Trapezius
Deep neck muscles
Levator scapulae
Seventh cervical vertebra
Rhomboids
Medial border of scapula

(b) **Posterior view**

Levator scapulae
Rhomboids
Three of four rotator cuff muscles
Supraspinatus
Infraspinatus
Teres minor
Teres major
Latissimus dorsi

(c) **Anterior view**

Internal intercostals
Coracoid process
Pectoralis minor (cut)
Subscapularis
Deltoid (cut)
Biceps brachii
Latissimus dorsi
Serratus anterior
External intercostals
Pectoralis major (cut)
Supraspinatus tendon
Subscapularis
Teres minor
Three of four rotator cuff muscles
Teres major (cut)
Pectoralis minor
Latissimus dorsi (cut)
External abdominal oblique
Internal abdominal oblique

Figure 7.22 AP|R Muscles of the Shoulder

(*a*) Posterior view of the neck and upper shoulder. The left side shows the superficial muscles. On the right, the superficial muscles are removed to show the deep muscles. (*b*) Posterior view of the thoracic region, with the trapezius and deltoid muscles removed. (*c*) Anterior view of the thoracic region.

TABLE 7.12	Muscles Acting on the Scapula (see figures 7.14, 7.22, and 7.23)		
Muscle	**Origin**	**Insertion**	**Action**
Levator scapulae (le-vā′ter skap′ū-lē)	Transverse processes of C1–C4	Superior angle of scapula	Elevates, retracts, and rotates scapula; laterally flexes neck
Pectoralis minor (pek′tō-ra′lis)	Ribs 3–5	Coracoid process of scapula	Depresses scapula or elevates ribs
Rhomboids (rom′boydz) Major Minor	Spinous processes of T1–T4 Spinous processes of T1–T4	Medial border of scapula Medial border of scapula	Retracts, rotates, and fixes scapula Retracts, slightly elevates, rotates, and fixes scapula
Serratus anterior (ser-ā′tŭs)	Ribs 1–9	Medial border of scapula	Rotates and protracts scapula; elevates ribs
Trapezius (tra-pē′zē-ŭs)	Posterior surface of skull and spinous processes of C7–T12	Clavicle, acromion process, and scapular spine	Elevates, depresses, retracts, rotates, and fixes scapula; extends neck

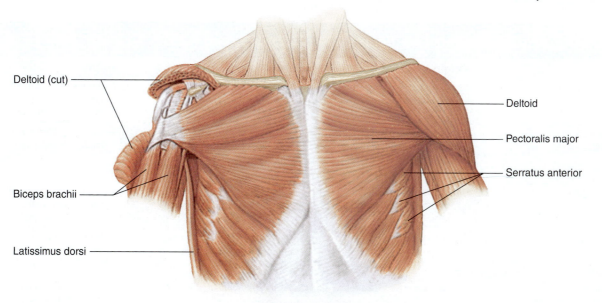

Deltoid (cut)

Biceps brachii

Latissimus dorsi

Deltoid

Pectoralis major

Serratus anterior

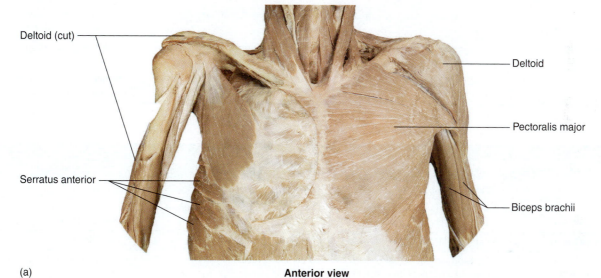

Deltoid (cut)

Serratus anterior

Deltoid

Pectoralis major

Biceps brachii

(a)   **Anterior view**

Muscular

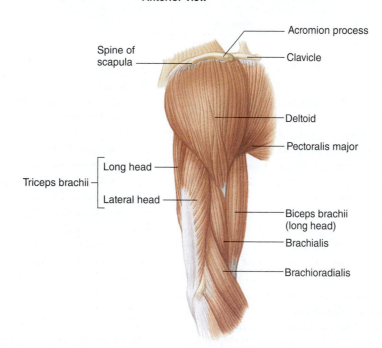

Spine of
scapula

Triceps brachii

Long head

Lateral head

Acromion process

Clavicle

Deltoid

Pectoralis major

Biceps brachii
(long head)

Brachialis

Brachioradialis

Figure 7.23 AP|R **Arm Muscles**   (b)   **Lateral view**

TABLE 7.13   Arm Movements (see figures 7.14, 7.22, 7.23, and 7.25)

Muscle	Origin	Insertion	Action
Deltoid (del′toyd)	Clavicle, acromion process, and scapular spine	Deltoid tuberosity	Flexes and extends shoulder; abducts and medially and laterally rotates arm
Latissimus dorsi (lă-tis′i-mŭs dōr′sī)	Spinous processes of T7–L5, sacrum and iliac crest, and inferior angle of scapula in some people	Medial crest of intertubercular groove	Extends shoulder; adducts and medially rotates arm
Pectoralis major (pek′tō-rā′lis)	Clavicle, sternum, superior six costal cartilages, and abdominal muscles	Lateral crest of intertubercular groove	Flexes shoulder; extends shoulder from flexed position; adducts and medially rotates arm
Teres major (ter′ēz)	Lateral border of scapula	Medial crest of intertubercular groove	Extends shoulder; adducts and medially rotates arm
Rotator Cuff			
Infraspinatus (in′fră-spī-nā′tŭs)	Infraspinous fossa of scapula	Greater tubercle of humerus	Stabilizes and extends shoulder and laterally rotates arm
Subscapularis (sŭb′skap-ū-lār′is)	Subscapular fossa of scapula	Lesser tubercle of humerus	Stabilizes and extends shoulder and medially rotates arm
Supraspinatus (sŭ′pră-spī-nā′tŭs)	Supraspinous fossa of scapula	Greater tubercle of humerus	Stabilizes shoulder and abducts arm
Teres minor (ter′ēz)	Lateral border of scapula	Greater tubercle of humerus	Stabilizes and extends shoulder; adducts and laterally rotates arm

TABLE 7.14   Arm Muscles (see figures 7.14, 7.22, 7.23, and 7.25)

Muscle	Origin	Insertion	Action
Biceps brachii (bī′seps brā′kē-ī)	Long head—supraglenoid tubercle Short head—coracoid process	Radial tuberosity and aponeurosis of biceps brachii	Flexes elbow; supinates forearm; flexes shoulder
Brachialis (brā′kē-al′is)	Anterior surface of humerus	Coronoid process of ulna	Flexes elbow
Triceps brachii (trī′seps brā′kē-ī)	Long head—lateral border of scapula Lateral head—lateral and posterior surface of humerus Medial head—posterior humerus	Olecranon process of ulna	Extends elbow; extends shoulder; adducts arm

anterior forearm muscles are responsible for flexion of the wrist and fingers, whereas most of the posterior forearm muscles cause extension. A strong band of fibrous connective tissue, the retinaculum (figure 7.24b), covers the flexor and extensor tendons and holds them in place around the wrist so that they do not "bowstring" during muscle contraction. Because the retinaculum does not stretch as a result of pressure, this characteristic is a contributing factor in carpal tunnel syndrome (see chapter 6).

The **flexor carpi** (kar′pī) muscles flex the wrist, and the **extensor carpi** muscles extend the wrist. The tendon of the flexor carpi radialis serves as a landmark for locating the radial pulse (figure 7.25a). The tendons of the wrist extensors are visible on the posterior surface of the forearm (figure 7.25b). Forceful, repeated contraction of the wrist extensor muscles, as occurs in a tennis backhand, may result in inflammation and pain where the extensor muscles attach to the lateral humeral epicondyle. This condition is sometimes referred to as "tennis elbow." Flexion of the fingers is the function of the **flexor digitorum** (dij′i-tōr′ŭm;

flexor of the digits, or fingers). Extension of the fingers is accomplished by the **extensor digitorum.** The tendons of this muscle are very visible on the dorsal surface of the hand (figure 7.25b). The thumb has its own set of flexors, extensors, adductors, and abductors. The little finger has some similar muscles.

Nineteen muscles, called **intrinsic hand muscles,** are located within the hand. **Interossei** (in′ter-os′ē-ī; between bones) muscles, located between the metacarpal bones, are responsible for abduction and adduction of the fingers. Other intrinsic hand muscles are involved in the many possible movements of the thumb and fingers. These muscles account for the fleshy masses at the base of the thumb and little finger and the fleshy region between the metacarpal bones of the thumb and index finger.

Lower Limb Muscles

The muscles of the lower limb include those located in the hip, the thigh, the leg, and the foot.

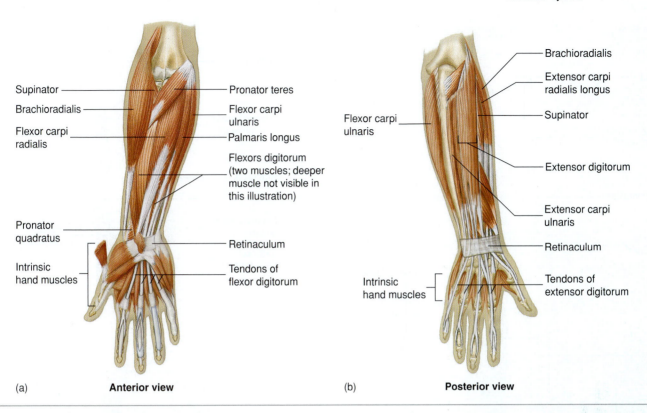

(a)   **Anterior view**

(b)   **Posterior view**

Figure 7.24 AP|R **Muscles of the Forearm**

(*a*) Anterior view. The flexor retinaculum has been removed. (*b*) Posterior view.

TABLE 7.15	Forearm Muscles (see figure 7.24)		
Muscle	**Origin**	**Insertion**	**Action**
Anterior Forearm			
Palmaris longus (pawl-mār′is lon′gus)	Medial epicondyle of humerus	Aponeurosis over palm	Tightens skin of palm
Flexor carpi radialis (kar′pī rā′dē-a-lĭs)	Medial epicondyle of humerus	Second and third metacarpal bones	Flexes and abducts wrist
Flexor carpi ulnaris (kar′pī ŭl-nār′is)	Medial epicondyle of humerus and ulna	Pisiform	Flexes and adducts wrist
Flexor digitorum profundus (dij′i-tōr′ŭm prō-fŭn′dŭs) (not shown)	Ulna	Distal phalanges of digits 2–5	Flexes fingers and wrist
Flexor digitorum superficialis (sū′per-fish′ē-a′lis)	Medial epicondyle of humerus, coronoid process, and radius	Middle phalanges of digits 2–5	Flexes fingers and wrist
Pronator			
Quadratus (prō′nā-tōr kwah-drā′tŭs)	Distal ulna	Distal radius	Pronates forearm
Teres (prō′nā-tōr te′rēz)	Medial epicondyle of humerus and coronoid process of ulna	Radius	Pronates forearm
Posterior Forearm			
Brachioradialis (brā′kē-ō-rā′dē-a′lis)	Lateral supracondylar ridge of humerus	Styloid process of radius	Flexes elbow
Extensor carpi radialis brevis (kar′pī rā′dē-a′lis brev′is) (not shown)	Lateral epicondyle of humerus	Base of third metacarpal bone	Extends and abducts wrist
Extensor carpi radialis longus (lon′gus)	Lateral supracondylar ridge of humerus	Base of second metacarpal bone	Extends and abducts wrist
Extensor carpi ulnaris (kar′pī ŭl-nār′is)	Lateral epicondyle of humerus and ulna	Base of fifth metacarpal bone	Extends and adducts wrist
Extensor digitorum (dij′i-tōr′ŭm)	Lateral epicondyle of humerus	Extensor tendon expansion over phalanges of digits 2–5	Extends fingers and wrist
Supinator (sū′pi-nā′tōr)	Lateral epicondyle of humerus and ulna	Radius	Supinates forearm (and hand)

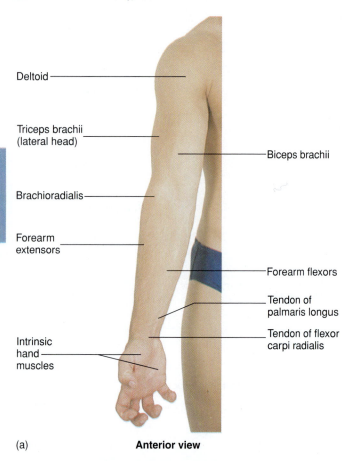

(a)   **Anterior view**

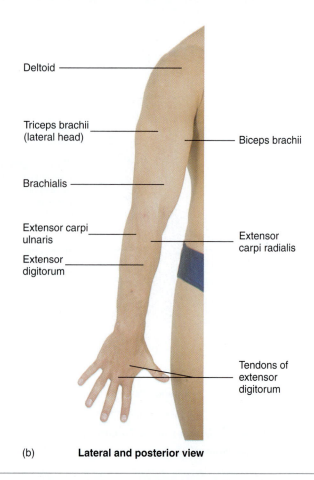

(b)   **Lateral and posterior view**

Figure 7.25 Surface Anatomy, Muscles of the Upper Limb

Thigh Movements

Several hip muscles originate on the coxal bone and insert onto the femur (figure 7.26 and table 7.16). The anterior muscle, the **iliopsoas** (il'ē-ō-sō'ŭs), flexes the hip (figure 7.26a). The posterior and lateral hip muscles consist of the **gluteal muscles** and the tensor fasciae latae. The **tensor fasciae latae** (ten'sŏr fa'shē-ă la'tē) is so named because it tenses a thick band of fascia on the lateral side of the thigh called the iliotibial tract. By so doing, it helps steady the femur on the tibia when a person is standing. The **gluteus** (glū'tē-us; buttock) **maximus,** which extends the hip and abducts and laterally rotates the thigh, contributes most of the mass that can be seen as the buttocks. The **gluteus medius,** which abducts and medially rotates the thigh, creates a smaller mass just superior and lateral to the maximus (see figure 7.27b). The gluteus maximus functions optimally to extend the hip when the thigh is flexed at a 45-degree angle. The gluteus medius is a common site for injections in the buttocks because the sciatic nerve lies deep to the gluteus maximus and could be damaged during an injection.

In addition to the hip muscles, some of the thigh muscles also attach to the coxal bone and can move the thigh. There are three groups of thigh muscles: the anterior thigh muscles, which flex the hip; the posterior thigh muscles, which extend the hip; and the medial thigh muscles, which adduct the thigh.

Predict 6

Consider a sprinter's stance and a bicyclist's racing posture, and explain why these athletes use these postures.

A CASE IN POINT

Groin Pull

Lowe Kikker was playing soccer on a very cold October day. He raced up to kick the ball, just as a player for the other team kicked the ball from the opposite direction. Initially, the ball did not move. Lowe felt a sudden, sharp pain in the medial side of his right thigh and fell to the ground, holding the injured area. His coach and team trainer ran onto the field. As the trainer pressed on the center part of his medial thigh, Lowe felt severe pain. He was able to stand and was helped off the field. On the sideline, the trainer applied ice to the adductor region of Lowe's right thigh. For the next few days, Lowe was told to rest, along with icing the region, compressing the muscles, and elevating the thigh (RICE). He was also given anti-inflammatory drugs. After a couple days of rest, Lowe began massage therapy. He missed the next game but was able to play in the finals. Lowe's injury was considered a grade 1 injury because there was pain and tenderness in the medial thigh muscles but no swelling. Grade 2 injury, in which the muscle is partially torn, is more painful and results in swelling of the area. Grade 3 injury, which involves a complete muscle tear, causes so much pain and swelling that the injured person cannot even walk. A groin pull involves one or more of the adductor muscles, most commonly the adductor longus. The damage usually occurs at the musculotendon junction, near its insertion.

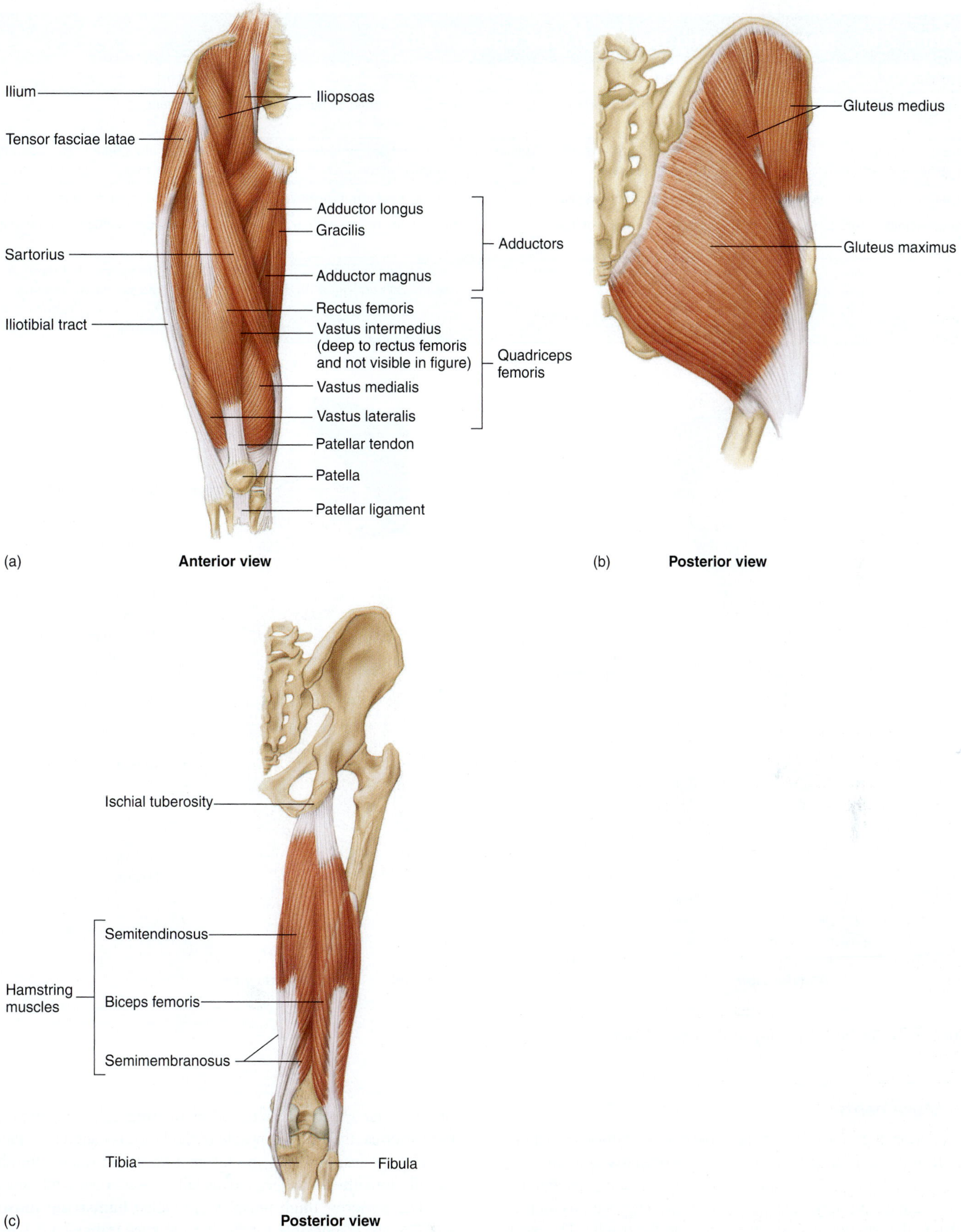

(a) **Anterior view**

Ilium

Iliopsoas

Tensor fasciae latae

Adductor longus

Gracilis

Adductors

Sartorius

Adductor magnus

Iliotibial tract

Rectus femoris

Vastus intermedius
(deep to rectus femoris
and not visible in figure)

Quadriceps
femoris

Vastus medialis

Vastus lateralis

Patellar tendon

Patella

Patellar ligament

(b) **Posterior view**

Gluteus medius

Gluteus maximus

(c) **Posterior view**

Ischial tuberosity

Semitendinosus

Hamstring
muscles

Biceps femoris

Semimembranosus

Tibia

Fibula

Figure 7.26 AP|R **Muscles of the Hip and Thigh**

(*a*) Anterior view. The vastus intermedius is labeled to allow for a complete listing of the quadriceps femoris muscles, but the muscle lies deep to the rectus femoris and cannot be seen in the figure. (*b*) Posterior view of the hip muscles. (*c*) Posterior view of the thigh muscles.

TABLE 7.16   Muscles Moving the Thigh (see figure 7.26)

Muscle	Origin	Insertion	Action
Iliopsoas (il′ē-ō-sō′ŭs)	Iliac fossa and vertebrae T12–L5	Lesser trochanter of femur and hip capsule	Flexes hip
Gluteus maximus (glū′tē-ŭs mak′si-mŭs)	Posterior surface of ilium, sacrum, and coccyx	Gluteal tuberosity of femur and iliotibial tract	Extends hip; abducts and laterally rotates thigh
Gluteus medius (glū′tē-ŭs mē′dē-ŭs)	Posterior surface of ilium	Greater trochanter of femur	Abducts and medially rotates thigh
Gluteus minimus (glū′tē-ŭs min′i-mŭs) (not shown)	Posterior surface of ilium	Greater trochanter of femur	Abducts and medially rotates thigh
Tensor fasciae latae (ten′sōr fa′shē-ē lā′tē)	Anterior superior iliac spine	Through lateral fascia of thigh to lateral condyle of tibia	Steadies femur on tibia through iliotibial tract when standing; flexes hip; medially rotates and abducts thigh

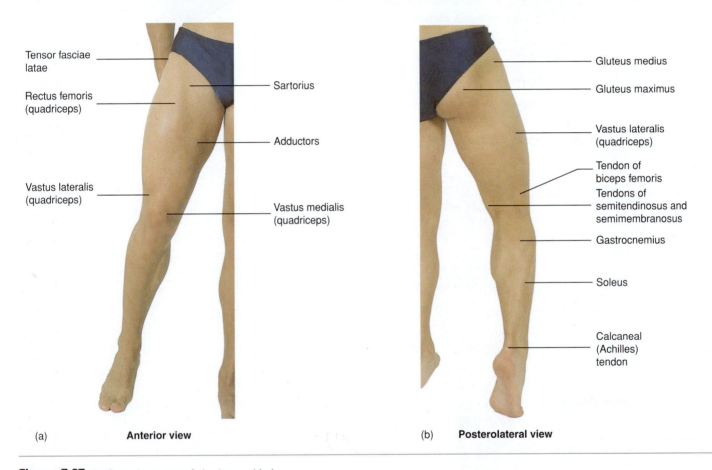

Tensor fasciae latae

Rectus femoris (quadriceps)

Vastus lateralis (quadriceps)

Sartorius

Adductors

Vastus medialis (quadriceps)

(a)   **Anterior view**

Gluteus medius

Gluteus maximus

Vastus lateralis (quadriceps)

Tendon of biceps femoris

Tendons of semitendinosus and semimembranosus

Gastrocnemius

Soleus

Calcaneal (Achilles) tendon

(b)   **Posterolateral view**

Figure 7.27 Surface Anatomy of the Lower Limb

Leg Movements

The anterior thigh muscles are the **quadriceps femoris** (kwah′dri-seps fe-mōr′is; four muscles) and the **sartorius** (sar-tōr′ē-ŭs) (table 7.17; figures 7.26a and 7.27a). The quadriceps femoris muscles are the primary extensors of the knee. They have a common insertion, the patellar tendon, on and around the patella. The patellar ligament is an extension of the patellar tendon onto the tibial tuberosity. The patellar ligament is tapped with a rubber hammer when testing the knee-jerk reflex in a physical examination (see figure 8.18). Health professionals often use one of the quadriceps muscles, the vastus lateralis, as an intermuscular injection site. The sartorius, the longest muscle in the body, is called the "tailor's muscle" because it flexes the hip and knee and rotates the thigh laterally for sitting cross-legged, as tailors used to sit while sewing.

The posterior thigh muscles are called **hamstring muscles,** and they are responsible for flexing the knee (table 7.17; see figures 7.14b and 7.26c). Their tendons are easily felt and seen on the medial and lateral posterior aspect of a slightly bent knee (figure 7.27b). The hamstrings are so named because these tendons in hogs or pigs could be used to suspend hams during curing.

TABLE 7.17	Leg Movements (see figures 7.14, 7.26, and 7.28)		
Muscle	**Origin**	**Insertion**	**Action**
Anterior Compartment			
Quadriceps femoris (kwah′dri-seps fem′ō-ris)			
Rectus femoris (rek′tŭs fem′ō-ris)	Ilium	Tibial tuberosity via patellar ligament	Extends knee; flexes hip
Vastus lateralis (vas′tus lat-er-ā′lis)	Greater trochanter and linea aspera of femur	Tibial tuberosity via patellar ligament	Extends knee
Vastus medialis (vas′tus mē′dē-ā′lis)	Linea aspera of femur	Tibial tuberosity via patellar ligament	Extends knee
Vastus intermedius (vas′tŭs in′ter-mē′dē-ŭs)	Body of femur	Tibial tuberosity via patellar ligament	Extends knee
Sartorius (sar-tōr′ē-ŭs)	Anterior superior iliac spine	Medial side of tibial tuberosity	Flexes hip and knee; laterally rotates thigh
Medial Compartment			
Adductor longus (a′dŭk-ter lon′gŭs)	Pubis	Linea aspera of femur	Adducts and laterally rotates thigh; flexes hip
Adductor magnus (a′dŭk-ter mag′nŭs)	Pubis and ischium	Femur	Adducts and laterally rotates thigh; extends knee
Gracilis (gras′i-lis)	Pubis near symphysis	Tibia	Adducts thigh; flexes knee
Posterior Compartment (Hamstring Muscles)			
Biceps femoris (bi′seps fem′ō-ris)	Long head—ischial tuberosity Short head—femur	Head of fibula	Flexes knee; laterally rotates leg; extends hip
Semimembranosus (se′mē-mem′bră-nō′sŭs)	Ischial tuberosity	Medial condyle of tibia and collateral ligament	Flexes knee; medially rotates leg; extends hip
Semitendinosus (se′mē-ten′di-nō′sŭs)	Ischial tuberosity	Tibia	Flexes knee; medially rotates leg; extends hip

Animals such as wolves often bring down their prey by biting through the hamstrings, thus preventing the prey animal from running. "To hamstring" someone is therefore to render him or her helpless. A "pulled hamstring" results from tearing one or more of these muscles or their tendons, usually where the tendons attach to the coxal bone.

The medial thigh muscles, the **adductor** (a′dŭk-ter) **muscles,** are primarily involved, as the name implies, in adducting the thigh (table 7.17).

Ankle and Toe Movements

The thirteen muscles in the leg, with tendons extending into the foot, can be divided into three groups: anterior, posterior, and lateral. As with the forearm, only the most superficial muscles are illustrated in figure 7.28 and listed in table 7.18. The anterior muscles (figure 7.28a) are extensor muscles involved in dorsiflexion (extension) of the foot and extension of the toes.

The superficial muscles of the posterior compartment of the leg (figure 7.28b), the **gastrocnemius** (gas′trok-nē′mē-us) and the **soleus** (sō′lē-ŭs), form the bulge of the calf (posterior leg; figure 7.28b). They join to form the common **calcaneal** (kal-kā′nē-ăl; heel) **tendon,** or *Achilles tendon.* These muscles are flexors and are involved in plantar flexion of the foot. The deep muscles of the posterior compartment plantar flex and invert the foot and flex the toes.

The lateral muscles of the leg, called the **fibularis** (fib-ū-lā′ris) **muscles** (figure 7.28c), are primarily everters (turning the lateral side of the foot outward) of the foot, but they also aid in plantar flexion.

The twenty muscles located within the foot, called the **intrinsic foot muscles,** flex, extend, abduct, and adduct the toes. They are arranged in a manner similar to the intrinsic muscles of the hand.

7.5 EFFECTS OF AGING ON SKELETAL MUSCLE

Learning Outcome *After reading this section, you should be able to*

A. Describe the effects of aging on skeletal muscle.

Aging skeletal muscle undergoes several changes that reduce muscle mass, increase the time a muscle takes to contract in response to nervous stimuli, reduce stamina, and increase recovery time. Loss of muscle fibers begins as early as 25 years of age, and by age 80 the muscle mass has been reduced by approximately 50%. Weight-lifting exercises help slow the loss of muscle mass but do not prevent the loss of muscle fibers. In addition, fast-twitch muscle fibers decrease in number more rapidly than slow-twitch fibers. Most of the loss of strength and speed is due to the loss of muscle fibers, particularly fast-twitch muscle fibers. The surface area of the neuromuscular junction decreases, and as a result, action potentials in neurons stimulate action potentials in muscle cells more slowly; thus, fewer action potentials are produced in muscle fibers. The number of motor neurons also decreases, and the remaining neurons innervate more muscle fibers. This decreases the number of motor units in

Muscular

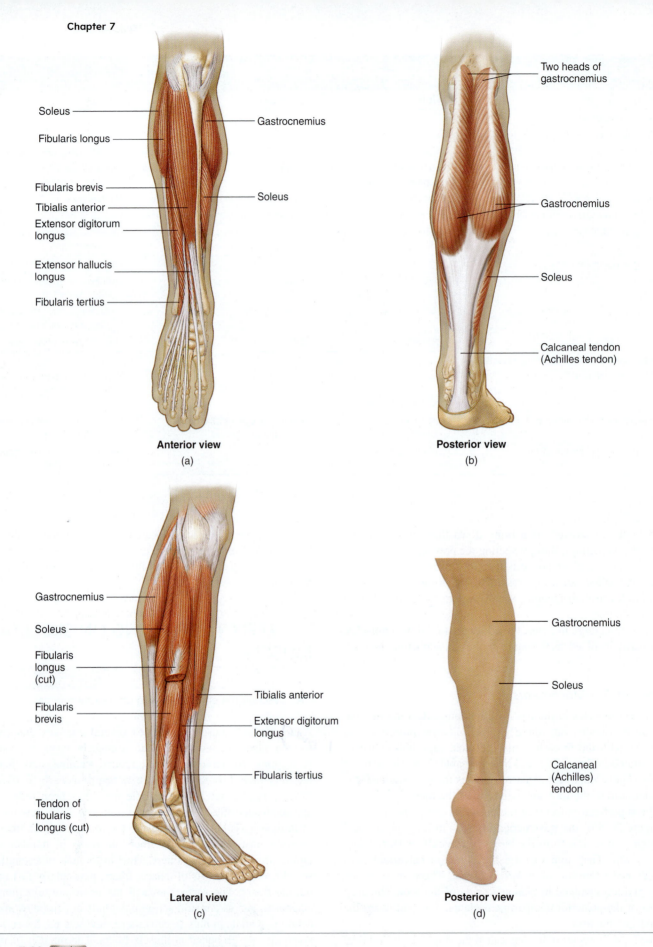

Anterior view
(a)

Soleus

Fibularis longus

Fibularis brevis

Tibialis anterior

Extensor digitorum longus

Extensor hallucis longus

Fibularis tertius

Gastrocnemius

Soleus

Posterior view
(b)

Two heads of gastrocnemius

Gastrocnemius

Soleus

Calcaneal tendon (Achilles tendon)

Gastrocnemius

Soleus

Fibularis longus (cut)

Fibularis brevis

Tendon of fibularis longus (cut)

Tibialis anterior

Extensor digitorum longus

Fibularis tertius

Lateral view
(c)

Gastrocnemius

Soleus

Calcaneal (Achilles) tendon

Posterior view
(d)

Figure 7.28 AP|R **Superficial Muscles of the Leg**

(a–c) Anterior, posterior, and lateral views. (d) Surface anatomy.

TABLE 7.18   Muscles of the Leg Acting on the Leg, Ankle, and Foot (see figures 7.14, 7.27, and 7.28)

Muscle	Origin	Insertion	Action
Anterior Compartment			
Extensor digitorum longus (dij′i-tōr′ŭm lon′gŭs)	Lateral condyle of tibia and fibula	Four tendons to phalanges of four lateral toes	Extends four lateral toes; dorsiflexes and everts foot
Extensor hallucis longus (hal′i-sis lon′gŭs)	Middle fibula and interosseous membrane	Distal phalanx of great toe	Extends great toe; dorsiflexes and inverts foot
Tibialis anterior (tib′ē-a′lis)	Tibia and interosseous membrane	Medial cuneiform and first metatarsal bone	Dorsiflexes and inverts foot
Fibularis tertius (peroneus tertius) (per′ō-nē′ŭs ter′shē-ŭs)	Fibula and interosseous membrane	Fifth metatarsal bone	Dorsiflexes and everts foot
Posterior Compartment			
Superficial			
Gastrocnemius (gas′trok-nē′mē-ŭs)	Medial and lateral condyles of femur	Through calcaneal (Achilles) tendon to calcaneus	Plantar flexes foot; flexes leg
Soleus (sō′lē-ŭs)	Fibula and tibia	Through calcaneal tendon to calcaneus	Plantar flexes foot
Deep			
Flexor digitorum longus (dij′i-tōr′ŭm lon′gŭs) (not shown)	Tibia	Four tendons to distal phalanges of four lateral toes	Flexes four lateral toes; plantar flexes and inverts foot
Flexor hallucis longus (hal′i-sis lon′gŭs) (not shown)	Fibula	Distal phalanx of great toe	Flexes great toe; plantar flexes and inverts foot
Tibialis posterior (tib′ē-a′lis) (not shown)	Tibia, interosseous membrane, and fibula	Navicular, cuneiforms, cuboid, and second through fourth metatarsal bones	Plantar flexes and inverts foot
Lateral Compartment			
Fibularis brevis (peroneus brevis) (fib-ū-lā′ris brev′is)	Fibula	Fifth metatarsal bone	Everts and plantar flexes foot
Fibularis longus (peroneus longus) (fib-ū-lā′ris lon′gŭs)	Fibula	Medial cuneiform and first metatarsal bone	Everts and plantar flexes foot

skeletal muscle, with a greater number of muscle fibers for each neuron, which may result in less precise muscle control. Aging is also associated with a decrease in the density of capillaries in skeletal muscles so that a longer recovery period is required after exercise.

Many of the age-related changes in skeletal muscle can be slowed dramatically if people remain physically active. As people age, they often assume a sedentary lifestyle. Studies show that elderly people who are sedentary can become stronger and more mobile in response to exercise.

DISEASES AND DISORDERS: Muscular System

CONDITION	DESCRIPTION
Cramps	Painful, spastic contractions of a muscle; usually due to a buildup of lactic acid
Fibromyalgia (fī-brō-mī-al′ja)	Non-life-threatening, chronic, widespread pain in muscles with no known cure; also known as chronic muscle pain syndrome
Hypertrophy	Enlargement of a muscle due to an increased number of myofibrils, as occurs with increased muscle use
Atrophy	Decrease in muscle size due to a decreased number of myofilaments; can occur due to disuse of a muscle, as in paralysis
Muscular dystrophy	Group of genetic disorders in which all types of muscle degenerate and atrophy
Duchenne muscular dystrophy	See Systems Pathology in this chapter
Myotonic muscular dystrophy	Muscles are weak and fail to relax following forceful contractions; affects the hands most severely; dominant trait in 1/20,000 births
Myasthenia gravis	See A Case in Point, "Myasthenia Gravis," earlier in this chapter
Tendinitis (ten-di-nī′tis)	Inflammation of a tendon or its attachment point due to overuse of the muscle

Muscular

Name: Greger
Gender: Male
Age: 3

Comments

A couple became concerned about their 3-year-old son, Greger, when they noticed that he was much weaker than other boys his age and his muscles appeared poorly developed. Over time, the differences between Greger and other boys his age became more obvious. Eventually, it was readily apparent that Greger had difficulty sitting, standing, climbing stairs, and even walking. He was clumsy and fell often. When Greger tried to stand, he would use his hands and arms to "climb up" his legs. Finally, the couple took Greger to his pediatrician, who, after several tests, informed them that Greger had Duchenne muscular dystrophy.

Duchenne Muscular Dystrophy

Background Information

Duchenne muscular dystrophy (DMD) is usually identified in children around 3 years of age when their parents notice slow motor development with progressive weakness and muscle wasting (atrophy). Typically, muscular weakness begins in the pelvic girdle, causing a waddling gait. Rising from the floor by "climbing up" the legs is characteristic and is caused by weakness of the lumbar and gluteal muscles (figure 7A). Within 3 to 5 years, the muscles of the shoulder girdle become involved. The replacement of muscle with connective tissue contributes to muscular atrophy and shortened, inflexible muscles called contractures. The contractures limit movements and can cause severe deformities of the skeleton. By 10 to 12 years of age, people with DMD are usually unable to walk, and few live beyond age 20. There is no effective treatment to prevent the progressive deterioration of muscles in DMD. Therapy primarily involves exercises to help strengthen muscles and prevent contractures.

Duchenne muscular dystrophy results from an abnormal gene on the X chromosome and is therefore a sex-linked (X-linked) condition. The gene is carried by females, but DMD affects males almost exclusively, at a frequency of 1 in 3000. The DMD gene is responsible for producing a protein called **dystrophin,** which plays a role in attaching myofibrils to other proteins in the cell membrane and regulating their activity. In a normal individual, dystrophin is thought to protect muscle cells against mechanical stress. In DMD patients, part of the gene is missing, and the protein it produces is nonfunctional, resulting in progressive muscular weakness and muscle contractures.

Figure 7A

DMD is characterized by progressive muscle weakness. Rising from the floor to a standing position becomes difficult, and the patient often has to use his or her arms to push against the thighs while rising.

SKELETAL

Shortened, inflexible muscles (contractures) cause severe deformities of the skeletal system. Kyphoscoliosis, severe curvature of the spinal column laterally and anteriorly, can be so advanced that normal respiratory movements are impaired. Surgery is sometimes required to prevent contractures from making it impossible for the individual to sit in a wheelchair.

URINARY

Reduced smooth muscle function and wheelchair dependency increase the frequency of urinary tract infections.

DIGESTIVE

DMD affects smooth muscle tissue, and the reduced ability of smooth muscle to contract can cause disorders of the digestive system. Digestive system disorders include an enlarged colon diameter, and twisting of the small intestine, resulting in intestinal obstruction, cramping, and reduced absorption of nutrients.

Duchenne Muscular Dystrophy

Symptoms
- Muscle weakness
- Muscle atrophy
- Contractures

Treatment
- Physical therapy to prevent contractures
- No effective treatment to prevent atrophy

NERVOUS

Some degree of mental retardation occurs in a large percentage of people with DMD, although the specific cause is unknown. Research suggests that dystrophin is important in the formation of synapses between nerve cells.

CARDIOVASCULAR

DMD affects cardiac muscle; consequently, heart failure occurs in a large number of people with advanced DMD. Cardiac involvement becomes serious in as many as 95% of cases and is one of the leading causes of death for individuals with DMD.

RESPIRATORY

Deformity of the thorax and increasing weakness of the respiratory muscles results in inadequate respiratory movements, which cause an increase in respiratory infections, such as pneumonia. Insufficient movement of air into and out of the lungs due to weak respiratory muscles is a major contributing factor in many deaths.

LYMPHATIC AND IMMUNE

No obvious direct effects occur to the lymphatic system, but damaged muscle fibers are phagocytized by macrophages.

Predict 7

A boy with advanced Duchenne muscular dystrophy developed pulmonary edema (accumulation of fluid in the lungs) and pneumonia caused by a bacterial infection. His physician diagnosed the condition in the following way: The pulmonary edema was the result of heart failure, and the increased fluid in the lungs provided a site where bacteria could invade and grow. The fact that the boy could not breathe deeply or cough effectively made the condition worse. How would the muscle tissues in a boy with advanced DMD with heart failure and ineffective respiratory movements differ from the muscle tissues in a boy with less advanced DMD?

ANSWER TO LEARN TO PREDICT

a. The first piece of information you are given is that Anthony increased the number of times he was contracting the same muscle group. Looking at figures 7.8 and 7.9, you see that for each cross-bridge formation event, a molecule of ATP is required. Therefore, being able to lift a particular weight once is different from being able to lift that weight many times in a row. Increasing the reps increases the amount of required ATP. You also learned in the "Fatigue" section that depletion of ATP along with decreased pH in muscle cells reduces their work capacity. Lifting the weight rapidly without resting would eventually lead to anaerobic ATP production. This is the reason that, in order to build muscle endurance, athletes reduce the weight but increase the number of reps.

b. Looking at figure 7.8, you see that when a muscle contraction is stimulated, Ca^{2+} is released from the sarcoplasmic reticulum. When Anthony contracted his muscles many times in a row, the level of Ca^{2+} inside his cells was higher than before he started exercising. In addition, the more times a muscle contracts, there is more Ca^{2+} released from the sarcoplasmic reticulum. So, the level of Ca^{2+} in his cells is higher after many reps than when he lifted each weight only once. This is why it is recommended that athletes warm up for at least 15–30 minutes before beginning any strenuous exercise—warming up helps the muscles function as efficiently and powerfully as possible (not to mention that warming up stretches muscles, tendons, and ligaments, which also helps to prevent injury!).

Answers to the rest of this chapter's Predict questions are in Appendix E.

SUMMARY

7.1 Functions of the Muscular System   (p. 150)

The muscular system produces body movement, maintains posture, causes respiration, produces body heat, performs movements involved in communication, constricts organs and vessels, and pumps blood.

7.2 Characteristics of Skeletal Muscle   (p. 151)

Skeletal muscle has contractility, excitability, extensibility, and elasticity.

Skeletal Muscle Structure

1. Muscle fibers are organized into fasciculi, and fasciculi are organized into muscles by associated connective tissue.
2. Each skeletal muscle fiber is a single cell containing numerous myofibrils.
3. Myofibrils are composed of actin and myosin myofilaments.
4. Sarcomeres are joined end-to-end to form myofibrils.

Excitability of Muscle Fibers

1. Cell membranes have a negative charge on the inside relative to a positive charge outside. This is called the resting membrane potential.
2. Action potentials are a brief reversal of the membrane charge. They are carried rapidly along the cell membrane.
3. Sodium ions (Na^+) move into cells during depolarization, and K^+ moves out of cells during repolarization.

Nerve Supply and Muscle Fiber Stimulation

1. Motor neurons carry action potentials to skeletal muscles, where the neuron and muscle fibers form neuromuscular junctions.
2. Neurons release acetylcholine, which binds to receptors on muscle cell membranes, stimulates an action potential in the muscle cell, and causes the muscle to contract.

Muscle Contraction

1. Action potentials are carried along T tubules to the sarcoplasmic reticulum, where they cause the release of calcium ions.
2. Calcium ions, released from the sarcoplasmic reticulum, bind to the actin myofilaments, exposing attachment sites.

3. Myosin forms cross-bridges with the exposed actin attachment sites.
4. The myosin molecules bend, causing the actin molecules to slide past; this is the sliding filament model. The H and I bands shorten; the A bands do not.
5. This process requires ATP breakdown.
6. A muscle twitch is the contraction of a muscle fiber in response to a stimulus; it consists of a lag phase, a contraction phase, and a relaxation phase.
7. Tetanus occurs when stimuli occur so rapidly that a muscle does not relax between twitches.

Energy Requirements for Muscle Contraction

1. Small contraction forces are generated when small numbers of motor units are recruited, and greater contraction forces are generated when large numbers of motor units are recruited.
2. Energy is produced by aerobic (with oxygen) and anaerobic (without oxygen) respiration.
3. After intense exercise, the rate of aerobic respiration remains elevated to repay the oxygen deficit.

Fatigue

1. Muscular fatigue occurs as ATP is depleted during muscle contraction. Physiological contracture occurs in extreme fatigue when a muscle can neither contract nor relax.

Effect of Fiber Type on Activity Level

1. Muscles contract either isometrically (tension increases, but muscle length stays the same) or isotonically (tension remains the same, but muscle length decreases).
2. Muscle tone consists of a small percentage of muscle fibers contracting tetanically and is responsible for posture.
3. Muscles contain a combination of slow-twitch and fast-twitch fibers.
4. Slow-twitch fibers are better suited for aerobic respiration, and fast-twitch fibers are adapted for anaerobic respiration.
5. Sprinters have more fast-twitch fibers, whereas distance runners have more slow-twitch fibers.

7.3 Smooth Muscle and Cardiac Muscle (p. 165)

1. Smooth muscle is not striated, has one nucleus per cell, contracts more slowly than skeletal muscle, can be autorhythmic, and is under involuntary control.
2. Cardiac muscle is striated, usually has one nucleus per cell, has intercalated disks, is autorhythmic, and is under involuntary control.

7.4 Skeletal Muscle Anatomy (p. 166)

General Principles

1. Most muscles have an origin on one bone, have an insertion onto another, and cross at least one joint.
2. A muscle causing a specific movement is an agonist. A muscle causing the opposite movement is an antagonist.
3. Muscles working together are synergists.
4. A prime mover is the muscle of a synergistic group that is primarily responsible for the movement.

Nomenclature

Muscles are named according to their location, origin and insertion, number of heads, or function.

Muscles of the Head and Neck

1. Muscles of facial expression are associated primarily with the mouth and eyes.
2. Four pairs of muscles are involved in mastication.
3. Tongue movements involve intrinsic and extrinsic muscles.
4. Swallowing involves the suprahyoid and infrahyoid muscles, plus muscles of the soft palate, pharynx, and larynx.
5. Neck muscles move the head.

Trunk Muscles

1. The erector spinae muscles hold the body erect.
2. Intercostal muscles and the diaphragm are involved in breathing.
3. Muscles of the abdominal wall flex and rotate the vertebral column, compress the abdominal cavity, and hold in and protect the abdominal organs.
4. Muscles form the floor of the pelvis.

Upper Limb Muscles

1. The upper limb is attached to the body primarily by muscles.
2. Arm movements are accomplished by pectoral, rotator cuff, and deltoid muscles.
3. The elbow is flexed and extended by anterior and posterior arm muscles, respectively.
4. Supination and pronation of the forearm are accomplished by supinators and pronators in the forearm.
5. Movements of the wrist and fingers are accomplished by most of the twenty forearm muscles and nineteen intrinsic muscles in the hand.

Lower Limb Muscles

1. Hip muscles flex and extend the hip and abduct the thigh.
2. Thigh muscles flex and extend the hip and adduct the thigh. They also flex and extend the knee.
3. Muscles of the leg and foot are similar to those of the forearm and hand.

7.5 Effects of Aging on Skeletal Muscle (p. 187)

Aging is associated with decreased muscle mass, slower reaction time, reduced stamina, and increased recovery time.

REVIEW AND COMPREHENSION

1. List the seven major functions of the muscular system.
2. Define contractility, excitability, extensibility, and elasticity.
3. List the connective tissue layers associated with muscles.
4. What are fasciculi?
5. What is a muscle fiber?
6. Explain the relevance of the structural relationship among sarcomeres, T tubules, and the sarcoplasmic reticulum.
7. What is a sarcomere?
8. Describe the composition of a myofibril. Describe the structure of actin and myosin myofilaments.
9. Explain the resting membrane potential and how it is produced.
10. Describe the production of an action potential.
11. What is a neuromuscular junction? What happens there?
12. Describe the sliding filament model of muscle contraction.
13. Explain how an action potential results in muscle contraction.
14. Define muscle twitch, tetanus, and recruitment.
15. Describe the two ways energy is produced in skeletal muscle.
16. Explain fatigue.
17. Compare isometric, isotonic, concentric, and eccentric contraction.
18. What is muscle tone?
19. Compare slow-twitch and fast-twitch muscle fibers.
20. How do smooth muscles and cardiac muscles differ from skeletal muscles?
21. Define origin, insertion, agonist, antagonist, synergist, prime mover, and fixator.
22. Describe the muscles of facial expression.
23. What is mastication? What muscles are involved?
24. What are the intrinsic and extrinsic tongue muscles?
25. What muscles are involved in swallowing?
26. What muscles are involved in respiration?
27. Describe the functions of the muscles of the anterior abdominal wall.
28. What is primarily responsible for attaching the upper limb to the body?
29. Describe, by muscle groups, movements of the arm, forearm, and hand.
30. Describe, by muscle groups, movements of the thigh, leg, and foot.

CRITICAL THINKING

1. Bob Canner improperly canned some home-grown vegetables and contracted botulism poisoning after eating them. Botulism results from a toxin, produced by bacteria, that prevents skeletal muscles from contracting. Symptoms include difficulty in swallowing and breathing. Eventually, Bob died of respiratory failure because his respiratory muscles relaxed and would not contract. Assuming that botulism toxin affects the neuromuscular junction, propose as many ways as you can to explain how botulism toxin produced the observed symptoms.

2. Harvey Leche milked cows by hand each morning before school. One morning, he overslept and had to hurry to get to school on time. While he was milking the cows as fast as he could, his hands became very tired, and then for a short time he could neither release his grip nor squeeze harder. Explain what happened.

3. A researcher was investigating the fast-twitch versus slow-twitch composition of muscle tissue in the gastrocnemius muscle (in the calf of the leg) of athletes. Describe the general differences this researcher would see when comparing the muscles from athletes who were outstanding in the following events: 100-m dash, weight lifting, and 10,000-m run.

4. Describe an exercise routine that would buildup each of the following groups of muscles: anterior arm, posterior arm, anterior forearm, anterior thigh, posterior leg, and abdomen.

5. Sherri Speedster started a 100-m dash but fell to the ground in pain. Examination of her right lower limb revealed the following symptoms: The knee was held in a slightly flexed position, but she could not flex it voluntarily; she could extend the knee with difficulty, but this caused her considerable pain; and there was considerable pain and bulging of the muscles in the posterior thigh. Explain the nature of her injury.

6. Javier has a number of prize apple trees in his backyard. To prevent them from becoming infested with insects, he sprayed them with an organophosphate insecticide. Being in a rush to spray the trees before leaving town on vacation, he failed to pay attention to the safety precautions on the packaging, and sprayed the trees without using any skin or respiratory protection. Soon he experienced severe stomach cramps, double vision, difficulty breathing, and spastic contractions of his skeletal muscles. Javier's wife took him to the emergency room, where he was diagnosed with organophosphate poisoning and given medication. Soon many of Javier's symptoms subsided.

   Organophosphate insecticides exert their effects by binding to the enzyme acetylcholinesterase within synaptic clefts, rendering it ineffective. Thus, the organophosphate poison and acetylcholine "compete" for the acetylcholinesterase and, as the organophosphate poison increases in concentration, the enzyme cannot effectively degrade acetylcholine. Organophosphate poisons will prolong acetylcholine actions at both skeletal and smooth muscle synapses. These include muscles needed for movement, breathing, and gastrointestinal contractions.

   Using the knowledge you gained about skeletal muscle contraction and figures 7.6 and 7.8, answer the following questions.
   a. Explain the spastic contractions that occurred in Javier's skeletal muscles.
   b. Refer to the Clinical Impact essay "Acetylcholine Antagonists," and propose a mechanism by which a drug could counteract the effects of organophosphate poisoning.

Answers in Appendix D

8 Nervous System

8.1 FUNCTIONS OF THE NERVOUS SYSTEM

Learning Outcome *After reading this section, you should be able to*

A. List the functions of the nervous system.

Module 7 Nervous System

The nervous system is involved in some way in nearly every body function. Some major functions of the nervous system are

1. *Receiving sensory input.* Sensory receptors monitor numerous external and internal stimuli. We are aware of sensations from some stimuli, such as vision, hearing, taste, smell, touch, pain, body position, and temperature. Other stimuli, such as blood pH, blood gases, and blood pressure, are processed at a subconscious level.

2. *Integrating information.* The brain and spinal cord are the major organs for processing sensory input and initiating responses. The input may produce an immediate response, be stored as memory, or be ignored.

3. *Controlling muscles and glands.* Skeletal muscles normally contract only when stimulated by the nervous system. Thus, by controlling skeletal muscle, the nervous system controls the major movements of the body. The nervous system also participates in controlling cardiac muscle, smooth muscle, and many glands.

4. *Maintaining homeostasis.* The nervous system plays an important role in maintaining homeostasis. This function depends on the nervous system's ability to detect, interpret, and respond to changes in internal and external conditions. In response, the nervous system can stimulate or inhibit the activities of other systems to help maintain a constant internal environment.

5. *Establishing and maintaining mental activity.* The brain is the center of mental activity, including consciousness, memory, and thinking.

8.2 DIVISIONS OF THE NERVOUS SYSTEM

Learning Outcome *After reading this section, you should be able to*

A. List the divisions of the nervous system, and describe the characteristics of each.

The nervous system can be divided into two major divisions: the central nervous system and the peripheral nervous system (figure 8.1). The **central nervous system (CNS)** consists of the brain and spinal cord. The **peripheral nervous system (PNS)** consists of all the nervous tissue outside the CNS (nerves and ganglia).

The PNS functions to link the CNS with the various parts of the body. The PNS carries information about the different tissues of the body to the CNS and carries commands from the CNS that alter body activities. The **sensory division**, or *afferent (toward) division*, of the PNS conducts action potentials from sensory receptors to the CNS (figure 8.2). The neurons that transmit action potentials from the periphery to the CNS are called **sensory neurons.** The **motor division**, or *efferent (away) division*, of the PNS conducts action potentials from the CNS to effector organs, such as muscles and glands. The neurons that transmit action potentials from the CNS toward the periphery are called **motor neurons.**

The motor division can be further subdivided based on the type of effector being innervated. The **somatic** (sō-mat′ik; bodily) **nervous system** transmits action potentials from the CNS to skeletal muscles, and the **autonomic** (aw-tō-nom′ik; self-governing) **nervous system (ANS)** transmits action potentials from the CNS to cardiac muscle, smooth muscle, and glands. The autonomic nervous system, in turn, is divided into sympathetic and parasympathetic divisions (see figure 8.2).

The **enteric nervous system (ENS)** is a unique subdivision of the peripheral nervous system. The ENS has both sensory and motor neurons contained wholly within the digestive tract. The ENS can function without input from the CNS or other parts of the PNS, although it is normally integrated with the CNS by sensory neurons and ANS motor neurons.

8.3 CELLS OF THE NERVOUS SYSTEM

Learning Outcomes *After reading this section, you should be able to*

A. Describe the structure of neurons and the function of their components. Describe the location, structure, and general function of glial cells.

B. Define and describe the structure of a nucleus, a ganglion, a nerve tract, and a nerve.

The two types of cells that make up the nervous system are neurons and glial cells.

Neurons

Neurons (noor′onz; nerve), or *nerve cells* (figure 8.3), receive stimuli, conduct action potentials, and transmit signals to other neurons or effector organs. There are three parts to a neuron: a cell body and two types of processes, called dendrites and axons.

Each neuron **cell body** contains a single nucleus. As with any other cell, the nucleus of the neuron is the source of information for

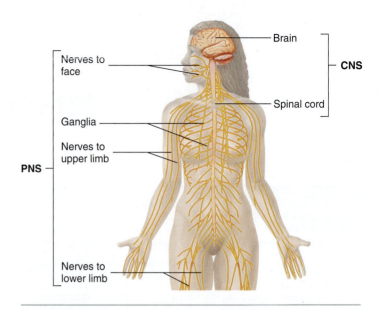

Figure 8.1 AP|R **Nervous System**
The central nervous system (CNS) consists of the brain and spinal cord. The peripheral nervous system (PNS) consists of nerves and ganglia.

gene expression. Extensive rough endoplasmic reticulum (rough ER), a Golgi apparatus, and mitochondria surround the nucleus. Large numbers of neurofilaments (intermediate filaments) and microtubules organize the cytoplasm into distinct areas.

Dendrites (den′drītz; trees) are short, often highly branching cytoplasmic extensions that are tapered from their bases at the neuron cell body to their tips. Most dendrites are extensions of the neuron cell body, but dendritelike structures also project from the peripheral ends of some sensory axons. Dendrites usually receive information from other neurons or from sensory receptors and transmit the information toward the neuron cell body.

Each neuron has an **axon**, a single long cell process extending from the neuron cell body. The area where the axon leaves the neuron cell body is called the **axon hillock.** Each axon has a uniform diameter and may vary in length from a few millimeters to more than a meter. Axons of sensory neurons conduct action potentials towards the CNS, and axons of motor neurons conduct action potentials away from the CNS. Axons also conduct action potentials from one part of the brain or spinal cord to another part. An axon may remain unbranched or may branch to form **collateral** (ko-lat′er-ăl) **axons.** Axons can be surrounded by a highly specialized insulating layer of cells called the myelin sheath (described in more detail in "Myelin Sheaths," later in this chapter).

Types of Neurons

Neurons can be classified on the basis of their structure or their function. For example, referring to a neuron as a "sensory neuron" indicates that it is carrying information to the CNS from the body. Alternatively, referring to a neuron as a "motor neuron" indicates that it is sending information to the body from the CNS. However, neurons are also distinguished from one another on the basis of their structure. In the structural classification, three categories of neurons exist, based on the arrangement of their processes (figure 8.4 and table 8.1). **Multipolar neurons** have many dendrites and a

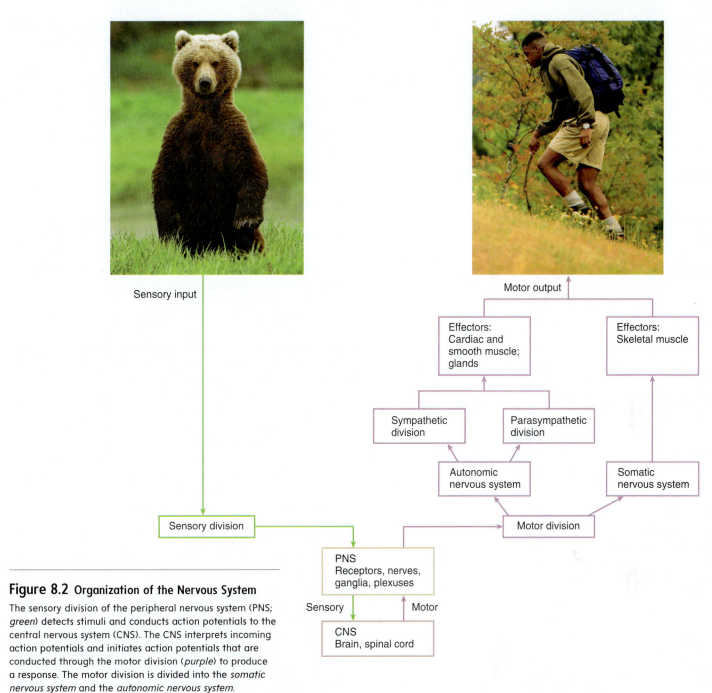

Figure 8.2 Organization of the Nervous System

The sensory division of the peripheral nervous system (PNS; *green*) detects stimuli and conducts action potentials to the central nervous system (CNS). The CNS interprets incoming action potentials and initiates action potentials that are conducted through the motor division (*purple*) to produce a response. The motor division is divided into the *somatic nervous system* and the *autonomic nervous system*.

single axon. Most of the neurons within the CNS and nearly all motor neurons are multipolar. **Bipolar neurons** have two processes: one dendrite and one axon. Bipolar neurons are located in some sensory organs, such as in the retina of the eye and in the nasal cavity. Most other sensory neurons are pseudo-unipolar. **Pseudo-unipolar neurons** have a single process extending from the cell body. This process divides into two processes a short distance from the cell body. One process extends to the periphery, and the other extends to the CNS. The two extensions function as a single axon with small, dendritelike sensory receptors at the periphery. The axon receives sensory information at the periphery and transmits that information in the form of action potentials to the CNS.

Glial Cells

Glial cells (glī′ăl, glē′ăl), or *neuroglia* (noo-rog′lē-ă; nerve glue), are the primarily supportive cells of the CNS and PNS, meaning these cells do not conduct action potentials. Neuroglia are far more numerous than neurons. Most neuroglia retain the ability to divide, whereas neurons do not. There are five types of glial cells. **Astrocytes** (as′trō-sītz) serve as the major supporting cells in the CNS. In this role, astrocytes can stimulate or inhibit the signaling activity of nearby neurons. In addition, astrocytes participate with the blood vessel endothelium to form a permeability barrier, called the **blood-brain barrier,** between the blood and the CNS. Astrocytes help limit damage to neural tissue; however, the repair

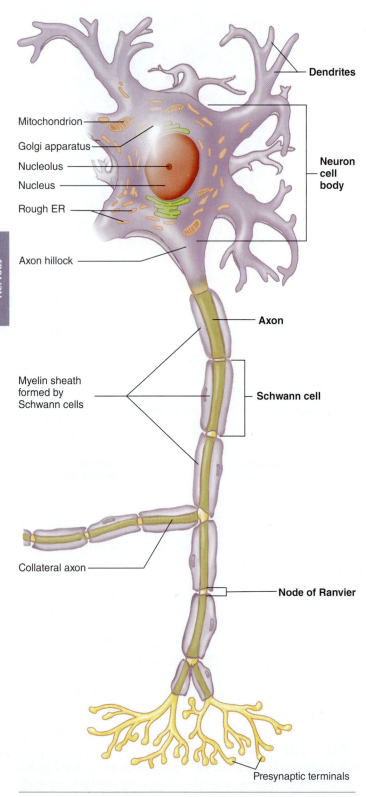

Mitochondrion

Golgi apparatus

Nucleolus

Nucleus

Rough ER

Axon hillock

Myelin sheath
formed by
Schwann cells

Collateral axon

Dendrites

Neuron
cell
body

Axon

Schwann cell

Node of Ranvier

Presynaptic terminals

Figure 8.3 AP|R **Typical Neuron**

Structural features of a typical motor neuron in the PNS include a cell body
and two types of cell processes: dendrites and an axon.

process can form a scar that blocks regeneration of damaged
axons. **Ependymal** (ep-en′di-măl) **cells** line the fluid-filled cavities
(ventricles and canals) within the CNS. Some ependymal cells
produce cerebrospinal fluid, and others, with cilia on the surface,
help move the cerebrospinal fluid through the CNS. **Microglia**
(mī-krog′lē-ă) act as immune cells of the CNS. They help protect
the brain by removing bacteria and cell debris. **Oligodendrocytes**
(ol′i-gō-den′drō-sītz) in the CNS and **Schwann cells** in the PNS
provide an insulating material that surrounds axons (figure 8.5
and table 8.1).

Myelin Sheaths

Myelin sheaths are specialized layers that wrap around the axons of
some neurons. These myelin sheaths are formed by the cell processes of oligodendrocytes in the CNS and Schwann cells in the PNS
(figure 8.5). Axons with these myelin sheaths are called **myelinated
axons** (figure 8.6*a*). Each oligodendrocyte process or Schwann cell
repeatedly wraps around a segment of an axon to form a series of
tightly wrapped cell membranes. Myelin is an excellent insulator
that prevents almost all ion movement across the cell membrane.
Gaps in the myelin sheath, called **nodes of Ranvier** (ron′vē-ā),
occur about every millimeter between the oligodendrocyte segments
or between individual Schwann cells. Ion movement can occur at
the nodes of Ranvier. Myelination of an axon increases the speed
and efficiency of action potential generation along the axon. The
details of action potentials are described fully in the next section.

Unmyelinated axons lack the myelin sheaths; however, these
axons rest in indentations of the oligodendrocytes in the CNS and
the Schwann cells in the PNS (figure 8.6*b*). A typical small nerve,
which consists of axons of multiple neurons, usually contains more
unmyelinated axons than myelinated axons.

Organization of Nervous Tissue

Both the CNS and the PNS contain areas of gray matter and areas
of white matter. **Gray matter** consists of groups of neuron cell
bodies and their dendrites, where there is very little myelin. In the
CNS, gray matter on the surface of the brain is called the **cortex,**
and clusters of gray matter located deeper within the brain are
called **nuclei.** In the PNS, a cluster of neuron cell bodies is called
a **ganglion** (gang′glē-on; pl. ganglia, a swelling or knot).

White matter consists of bundles of parallel axons with their
myelin sheaths, which are whitish in color. White matter of the
CNS forms **nerve tracts,** or *conduction pathways,* which propagate
action potentials from one area of the CNS to another. In the PNS,
bundles of axons and associated connective tissue form **nerves.**

8.4 ELECTRICAL SIGNALS AND NEURAL PATHWAYS

Learning Outcomes *After reading this section, you should be able to*

A. Describe a resting membrane potential, and explain how
an action potential is generated and propagated. Compare
the roles of leak and gated ion channels.

B. Describe the structure and function of a synapse.

C. List the parts of a reflex arc, and describe its function.

D. Describe a converging and a diverging circuit and the role
of summation in neural pathways.

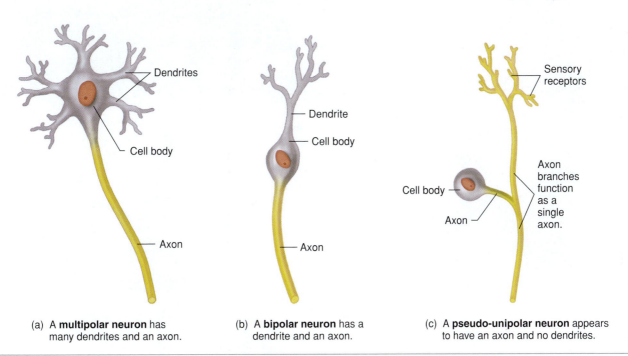

(a) A **multipolar neuron** has many dendrites and an axon.

(b) A **bipolar neuron** has a dendrite and an axon.

(c) A **pseudo-unipolar neuron** appears to have an axon and no dendrites.

Figure 8.4 Types of Neurons

TABLE 8.1	Cells of the Nervous System (see figures 8.4 and 8.5)	
Cell Type	**Description**	**Function**
Neuron		
Multipolar	Many dendrites and one axon	Most motor neurons and most CNS neurons
Bipolar	One dendrite and one axon	Found in special sense organs, such as eye and nose
Pseudo-unipolar	Appears to have a single axon	Most sensory neurons
Glial Cells		
Astrocytes	Highly branched	Provide structural support; regulate neuronal signaling; contribute to blood-brain barrier; help with neural tissue repair
Ependymal cells	Epithelial-like	Line ventricles of brain and central canal of the spinal cord, circulate cerebrospinal fluid (CSF); some form choroid plexuses, which produce CSF
Microglia	Small, mobile cells	Protect CNS from infection; become phagocytic in response to inflammation
Oligodendrocytes	Cells with processes that can surround several axons	Cell processes form myelin sheaths around axons or enclose unmyelinated axons in the CNS
Schwann cells	Single cells surrounding axons	Form myelin sheaths around axons or enclose unmyelinated axons in the PNS

Resting Membrane Potential

All cells exhibit electrical properties. The inside of most cell membranes is negatively charged compared to the outside of the cell membrane, which is positively charged (as discussed in chapter 7). This uneven distribution of charge means the cell membrane is **polarized.** In an unstimulated (or resting) cell, the uneven charge distribution is called the **resting membrane potential.** The outside of the cell membrane can be thought of as the positive pole of a battery and the inside as the negative pole. Thus, a small voltage difference, called a potential, can be measured across the resting cell membrane.

The resting membrane potential is generated by three main factors: (1) a higher concentration of K^+ immediately inside the cell membrane, (2) a higher concentration of Na^+ immediately outside the cell membrane, and (3) greater permeability of the cell membrane

to K^+ than to Na^+ (figure 8.7, step 1). Thus, the resting membrane potential results from differences in the concentration of ions across the membrane and the permeability characteristics of the membrane.

The difference in membrane permeability is due to the difference in the number of open ion channels. Recall from chapter 3 that ions cannot move freely across the cell membrane; instead, ions must flow through ion channels, which are proteins embedded in the cell membrane. Ions flow through channels due to the differences in their concentration across the membrane. There are two basic types of ion channels: leak channels and gated channels (see figure 3.5). **Leak channels** are always open. Thus, as the name suggests, ions can "leak" across the membrane down their concentration gradient. When a cell is at rest, the membrane potential is established by diffusion of ions through leak channels.

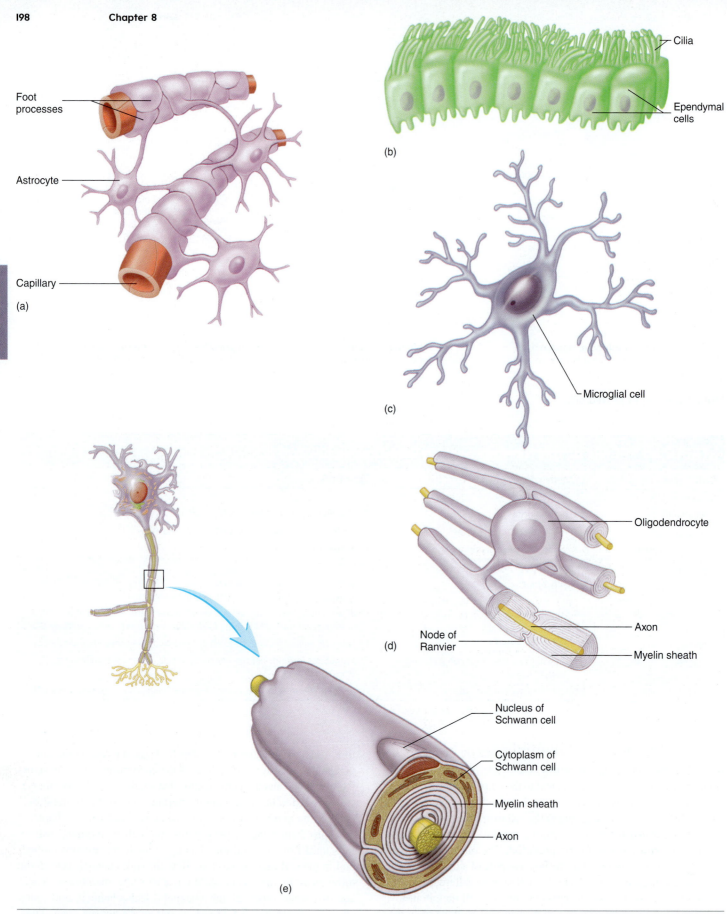

Nervous

Foot processes

Astrocyte

Capillary

(a)

Cilia

Ependymal cells

(b)

Microglial cell

(c)

Oligodendrocyte

Node of Ranvier

Axon

Myelin sheath

(d)

Nucleus of Schwann cell

Cytoplasm of Schwann cell

Myelin sheath

Axon

(e)

Figure 8.5 AP|R Types of Glial Cells

(*a*) Astrocytes, with foot processes surrounding a blood capillary. (*b*) Ependymal cells, with cilia extending from the surfaces. (*c*) Microglial cell. (*d*) Oligodendrocyte, forming a myelin sheath around parts of three axons within the CNS. (*e*) Schwann cell forming part of the myelin sheath of an axon in the PNS.

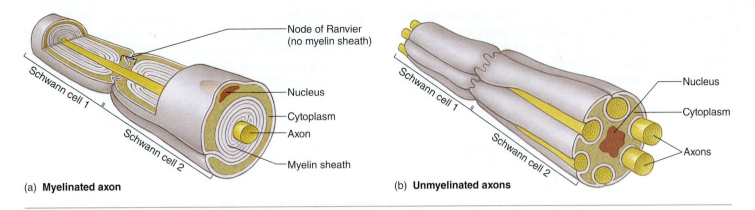

Figure 8.6 AP|R **Comparison of Myelinated and Unmyelinated Axons**
(*a*) Myelinated axon with two Schwann cells forming part of the myelin sheath around a single axon. Each Schwann cell surrounds part of one axon.
(*b*) Unmyelinated axons with two Schwann cells surrounding several axons in parallel formation. Each Schwann cell surrounds part of several axons.

(a) **Myelinated axon**

(b) **Unmyelinated axons**

Node of Ranvier
(no myelin sheath)

Nucleus

Cytoplasm

Axon

Myelin sheath

Nucleus

Cytoplasm

Axons

Schwann cell 1

Schwann cell 2

Nervous

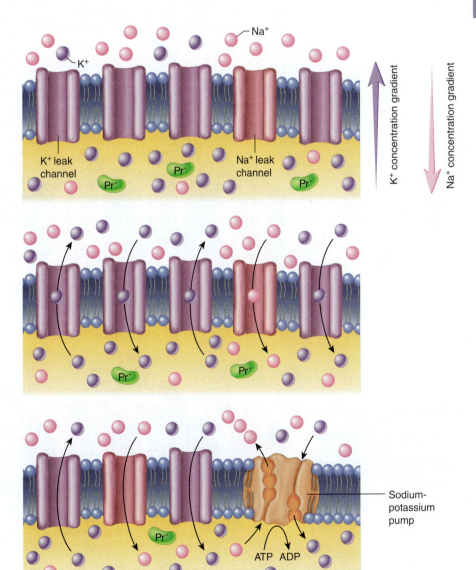

1 In a resting cell, there is a higher concentration of K⁺ (*purple circles*) inside the cell membrane and a higher concentration of Na⁺ (*pink circles*) outside the cell membrane. Because the membrane is not permeable to negatively charged proteins (*green*) they are isolated to inside of the cell membrane.

2 There are more K⁺ leak channels than Na⁺ leak channels. In the resting cell, only the leak channels are opened; the gated channels (not shown) are closed. Because of the ion concentration differences across the membrane, K⁺ diffuses out of the cell down its concentration gradient and Na⁺ diffuses into the cell down its concentration gradient. The tendency for K⁺ to diffuse out of the cell is opposed by the tendency of the positively charged K⁺ to be attracted back into the cell by the negative charge inside the cell.

3 The sodium-potassium pump helps maintain the differential levels of Na⁺ and K⁺ by pumping three Na⁺ out of the cell in exchange for two K⁺ into the cell. The pump is driven by ATP hydrolysis. The resting membrane potential is established when the movement of K⁺ out of the cell is equal to the movement of K⁺ into the cell.

PROCESS Figure 8.7 Generation of the Resting Membrane Potential

Because there are 50–100 times more K$^+$ leak channels than Na$^+$ leak channels, the resting membrane has much greater permeability to K$^+$ than to Na$^+$; therefore, the K$^+$ leak channels have the greatest contribution to the resting membrane potential. **Gated channels** are closed until opened by specific signals. **Chemically** **gated channels** are opened by neurotransmitters or other chemicals, whereas **voltage-gated channels** are opened by a change in membrane potential. When opened, the gated channels can change the membrane potential and are thus responsible for the action potential, described next.

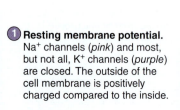

1 **Resting membrane potential.** Na$^+$ channels (*pink*) and most, but not all, K$^+$ channels (*purple*) are closed. The outside of the cell membrane is positively charged compared to the inside.

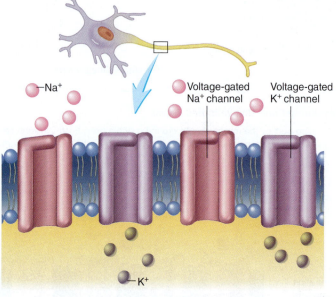

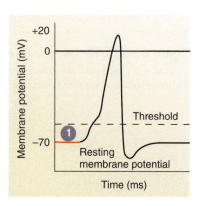

2 **Depolarization.** Na$^+$ channels open. K$^+$ channels begin to open. Depolarization results because the inward movement of Na$^+$ makes the inside of the membrane positive.

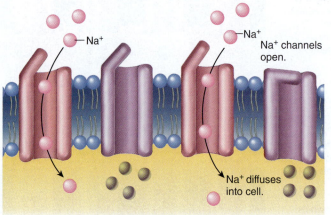

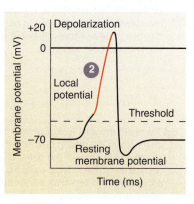

3 **Repolarization.** Na$^+$ channels close and additional K$^+$ channels open. Na$^+$ movement into the cell stops, and K$^+$ movement out of the cell increases, causing repolarization.

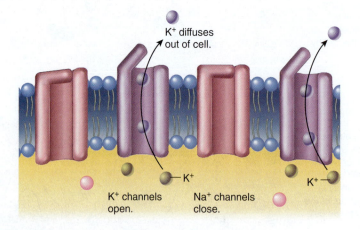

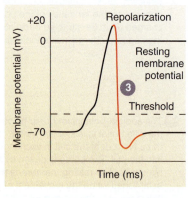

PROCESS Figure 8.8 AP|R **Voltage-Gated Ion Channels and the Action Potential**

Step 1 illustrates the status of voltage-gated Na$^+$ and K$^+$ channels in a resting cell. Steps 2 and 3 show how the channels open and close to produce an action potential. Next to each step, a graph shows in *red* the membrane potential resulting from the condition of the ion channels.

As mentioned earlier, there is a net negative charge inside and a net positive charge outside a resting cell. The primary source of negative charge inside the cell is a high concentration of negatively charged molecules, such as proteins, that cannot diffuse out of the cell because the cell membrane is impermeable to them (figure 8.7, step 2). Consequently, as positively charged K^+ leaks out of the cell via the leak channels, the charge inside the cell membrane becomes even more negative. The negatively charged molecules inside the cell tend to attract the positive K^+ back into the cell. The resting membrane potential is the point of equilibrium at which the tendency for K^+ to move down its concentration gradient out of the cell is balanced by the negative charge within the cell, which tends to attract the K^+ back into the cell (figure 8.7, step 2).

To compensate for the constant leakage of ions across the membrane, the **sodium-potassium pump** (Na^+–K^+ pump) is required to maintain the greater concentration of Na^+ outside the cell membrane and K^+ inside. The pump actively transports K^+ into the cell and Na^+ out of the cell (figure 8.7, step 3). The importance of this pump is indicated by the astounding amount of energy it consumes. It is estimated that the sodium-potassium pump consumes 25% of all the ATP in a typical cell and 70% of the ATP in a neuron.

Action Potentials

Muscle and nerve cells are **excitable cells,** meaning that the resting membrane potential changes in response to stimuli that activate gated ion channels. The opening and closing of gated channels can change the permeability characteristics of the cell membrane and hence change the membrane potential.

The channels responsible for the action potential are voltage-gated Na^+ and K^+ channels. When the cell membrane is at rest, the voltage-gated channels are closed (figure 8.8, step 1). When a stimulus is applied to a muscle cell or nerve cell, following neurotransmitter activation of chemically gated channels, Na^+ channels open very briefly, and Na^+ diffuses quickly into the cell (figure 8.8, step 2). This movement of Na^+, which is called a **local current,** causes the inside of the cell membrane to become positive, a change called **depolarization.** This depolarization results in a **local potential.** If depolarization is not strong enough, the Na^+ channels close again, and the local potential disappears without being conducted along the nerve cell membrane. If depolarization is large enough, Na^+ enters the cell so that the local potential reaches a **threshold** value. This threshold depolarization causes voltage-gated Na^+ channels to open. Threshold is most often reached at the axon hillock, near the cell body. The opening of these channels causes a massive, 600-fold increase in membrane permeability to Na^+. Voltage-gated K^+ channels also begin to open. As more Na^+ enters the cell, depolarization occurs until a brief reversal of charge takes place across the membrane—the inside of the cell membrane becomes positive relative to the outside of the cell membrane. The charge reversal causes Na^+ channels to close and more K^+ channels to open. Na^+ then stops entering the cell, and K^+ leaves the cell (figure 8.8, step 3). This repolarizes the cell membrane to its resting membrane potential. Depolarization and repolarization constitute an **action potential** (figure 8.9). At the end of repolarization, the charge on the cell membrane briefly becomes more negative than the resting membrane potential; this condition is called **hyperpolarization.** The elevated permeability to K^+ lasts only a very short time.

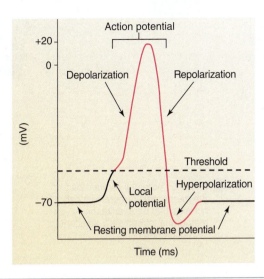

Figure 8.9 Action Potential

Once a local potential reaches threshold, an all-or-none action potential is started. During the depolarization phase, the voltage across the cell membrane changes from approximately −70 mV to approximately +20 mV. During the repolarization phase, the voltage across the cell membrane returns to −70 mV. There is a brief period of hyperpolarization at the end of repolarization before the membrane returns to its resting membrane potential. The entire process lasts 1 or 2 milliseconds (ms).

Predict 2

Predict the effect of a reduced extracellular concentration of Na⁺ on the generation of an action potential in an excitable cell.

In summary, the resting membrane potential is set by the activity of the leak channels. On stimulation, chemically gated channels are opened and initiate *local* potentials. If sufficiently strong, the local potentials activate voltage-gated channels to initiate an *action* potential.

Action potentials occur in an **all-or-none** fashion. That is, if threshold is reached, an action potential occurs; if the threshold is not reached, no action potential occurs. Action potentials in a cell are all of the same magnitude—in other words, the amount of charge reversal is always the same. Stronger stimuli produce a greater frequency of action potentials but do not increase the size of each action potential. Thus, neural signaling is based on the number of action potentials.

Action potentials are conducted slowly in unmyelinated axons and more rapidly in myelinated axons. In unmyelinated axons, an action potential in one part of a cell membrane stimulates local currents in adjacent parts of the cell membrane. The local currents in the adjacent membrane produce an action potential. By this means, the action potential is conducted along the entire axon cell membrane. This type of action potential conduction is called **continuous conduction** (figure 8.10).

In myelinated axons, an action potential at one node of Ranvier causes a local current to flow through the surrounding extracellular fluid and through the cytoplasm of the axon to the next node, stimulating an action potential at that node of Ranvier. By this means, action potentials "jump" from one node of Ranvier to the next along the length of the axon. This type of action potential conduction is called **saltatory** (sal′tă-tōr-ē; to leap) **conduction** (figure 8.11). Saltatory

Nervous

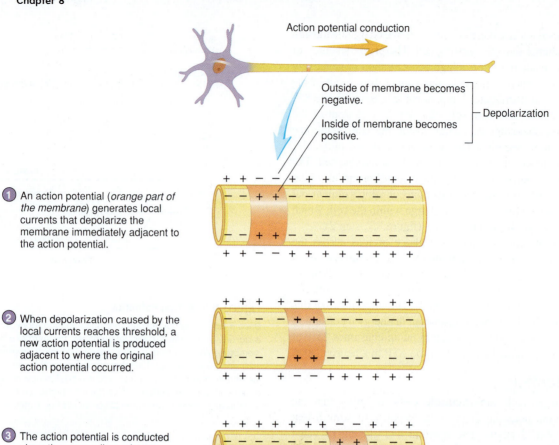

1. An action potential (*orange part of the membrane*) generates local currents that depolarize the membrane immediately adjacent to the action potential.

Action potential conduction

Outside of membrane becomes negative.

Inside of membrane becomes positive.

Depolarization

2. When depolarization caused by the local currents reaches threshold, a new action potential is produced adjacent to where the original action potential occurred.

3. The action potential is conducted along the axon cell membrane.

Site of next action potential

PROCESS Figure 8.10 AP|R **Continuous Conduction in an Unmyelinated Axon**

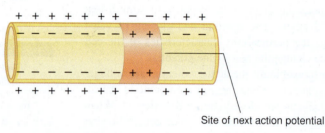

1. An action potential (*orange*) at a node of Ranvier generates local currents (*black arrows*). The local currents flow to the next node of Ranvier because the myelin sheath of the Schwann cell insulates the axon of the internode.

Node of Ranvier   Schwann cell   Internode

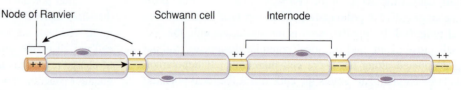

2. When the depolarization caused by the local currents reaches threshold at the next node of Ranvier, a new action potential is produced (*orange*).

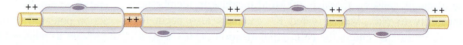

3. Action potential propagation is rapid in myelinated axons because the action potentials are produced at successive nodes of Ranvier (*1–5*) instead of at every part of the membrane along the axon.

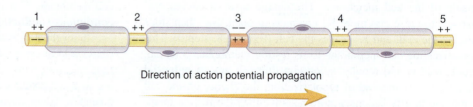

Direction of action potential propagation

PROCESS Figure 8.11 **Saltatory Conduction: Action Potential Conduction in a Myelinated Axon**

conduction greatly increases the conduction velocity because the nodes of Ranvier make it unnecessary for action potentials to travel along the entire cell membrane. Action potential conduction in a myelinated fiber is like a child skipping across the floor, whereas in an unmyelinated axon it is like a child walking heel to toe across the floor.

Medium-diameter, lightly myelinated axons, characteristic of autonomic neurons, conduct action potentials at the rate of about 3–15 meters per second (m/s), whereas large-diameter, heavily myelinated axons conduct action potentials at the rate of 15–120 m/s. These rapidly conducted action potentials, carried by sensory and motor neurons, allow for rapid responses to changes in the external environment. In addition, several hundred times fewer ions cross the cell membrane during conduction in myelinated cells than in unmyelinated cells. Much less energy is therefore required for the sodium-potassium pump to maintain the ion distribution.

The Synapse

A **synapse** (sin′aps) is a junction where the axon of one neuron interacts with another neuron or with cells of an effector organ, such as a muscle or gland (figure 8.12). The end of the axon forms a **presynaptic terminal.** The membrane of the dendrite or effector cell is the **postsynaptic membrane,** and the space separating the presynaptic and postsynaptic membranes is the **synaptic cleft.** Chemical substances called **neurotransmitters** (noor′ō-trans-mit′erz; *neuro-,* nerve + *transmitto,* to send across) are stored in **synaptic vesicles** in the presynaptic terminal. When an action potential reaches the presynaptic terminal, voltage-gated Ca^{2+} channels open, and Ca^{2+} moves into the cell. This influx of Ca^{2+} causes the release of neurotransmitters by exocytosis from the presynaptic terminal. The neurotransmitters diffuse across the synaptic cleft and bind to specific receptor molecules on the postsynaptic membrane. The binding of neurotransmitters to these membrane receptors causes chemically gated channels for Na^+, K^+, or Cl^- to open or close in the postsynaptic membrane, depending on the type of neurotransmitter in the presynaptic terminal and the type of receptors on the postsynaptic membrane. The response may be either stimulation or inhibition of an action potential in the postsynaptic cell. For example, if Na^+ channels open, the postsynaptic cell becomes depolarized, and an action potential will result if threshold is reached. If K^+ or Cl^- channels open, the inside of the postsynaptic cell tends to become more negative, or **hyperpolarized** (hī′per-pō′lăr-ī-zed), and an action potential is inhibited from occurring.

Of the many neurotransmitters or suspected neurotransmitter substances, the best known are **acetylcholine (ACh)** (as′e-til-kō′lēn) and **norepinephrine** (nōr′ep-i-nef′rin). Other neurotransmitters include serotonin (sēr-ō-tō′nin), dopamine (dō′pă-mēn), γ (gamma)-aminobutyric (gam′ă ă-mē′nō-bū-tēr′ik) acid (GABA), glycine, and endorphins (en′dōr-finz; endogenous morphine) (table 8.2). Neurotransmitter substances are rapidly broken down by enzymes within the synaptic cleft or are transported back into the presynaptic terminal. Consequently, they are removed from the synaptic cleft, so their effects on the postsynaptic membrane are very short-term. In synapses where acetylcholine is the neurotransmitter, such as in the neuromuscular junction (see chapter 7), an enzyme called **acetylcholinesterase** (as′e-til-kō-lin-es′-ter-ās) breaks down the acetylcholine. The breakdown products are then returned to the presynaptic terminal for reuse. Norepinephrine is

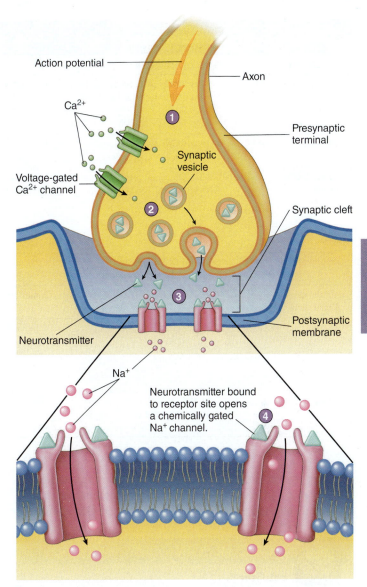

① Action potentials arriving at the presynaptic terminal cause voltage-gated Ca^{2+} channels to open.

② Ca^{2+} diffuses into the cell and causes synaptic vesicles to release neurotransmitter molecules.

③ Neurotransmitter molecules diffuse from the presynaptic terminal across the synaptic cleft.

④ Neurotransmitter molecules combine with their receptor sites and cause chemically gated Na^+ channels to open. Na^+ diffuses into the cell (*shown in illustration*) or out of the cell (*not shown*) and causes a change in membrane potential.

PROCESS Figure 8.12 AP|R **The Synapse**

A synapse consists of the end of a neuron (presynaptic terminal), a small space (synaptic cleft), and the postsynaptic membrane of another neuron or an effector cell, such as a muscle or gland cell.

either actively transported back into the presynaptic terminal or broken down by enzymes. The release and breakdown or removal of neurotransmitters occurs so rapidly that a postsynaptic cell can be stimulated many times a second. Drugs can modulate the action of neurotransmitters at the synapse. Cocaine and amphetamines increase the release and block the reuptake of norepinephrine,

TABLE 8.2   Neurotransmitters

Substance	Site of Release	Effect	Clinical Example
Acetylcholine (ACh)	CNS synapses, ANS synapses, and neuromuscular junctions	Excitatory or inhibitory	Alzheimer disease (a type of senile dementia) is associated with a decrease in acetylcholine-secreting neurons. Myasthenia gravis (weakness of skeletal muscles) results from a reduction in acetylcholine receptors.
Norepinephrine (NE)	Selected CNS synapses and some ANS synapses	Excitatory	Cocaine and amphetamines increase the release and block the reuptake of norepinephrine, resulting in overstimulation of postsynaptic neurons.
Serotonin	CNS synapses	Generally inhibitory	It is involved with mood, anxiety, and sleep induction. Levels of serotonin are elevated in schizophrenia (delusions, hallucinations, and withdrawal). Drugs that block serotonin transporters, such as Prozac, are used to treat depression and anxiety disorders.
Dopamine	Selected CNS synapses and some ANS synapses	Excitatory or inhibitory	Parkinson disease (depression of voluntary motor control) results from destruction of dopamine-secreting neurons.
Gamma-aminobutyric acid (GABA)	CNS synapses	Inhibitory	Drugs that increase GABA function have been used to treat epilepsy (excessive discharge of neurons).
Glycine	CNS synapses	Inhibitory	Glycine receptors are inhibited by the poison strychnine. Strychnine increases the excitability of certain neurons by blocking their inhibition. Strychnine poisoning results in powerful muscle contractions and convulsions. Tetanus of respiratory muscles can cause death.
Endorphins	Descending pain pathways	Inhibitory	The opiates morphine and heroin bind to endorphin receptors on presynaptic neurons and reduce pain by blocking the release of a neurotransmitter.

resulting in overstimulation of postsynaptic neurons and deleterious effects on the body. Drugs that block serotonin reuptake are particularly effective at treating depression and behavioral disorders.

A CASE IN POINT

Botulism

Ima Kannar removed a 5-year-old jar of green beans from the shelf in her food storage room. As she dusted off the lid, she noticed that it was a bit bowed out, but the beans inside looked and smelled okay when she opened the jar. She ate some of the beans for dinner, watched some television, and then went to bed. The next afternoon, as Ima was working in her garden, her throat felt dry. Then she started to experience double vision and difficulty breathing, so she dialed 911. By the time the paramedics arrived, 20 minutes later, Ima was not breathing. She was put on a respirator and rushed to the hospital, where physicians diagnosed a case of botulism. Ima recovered very slowly and was in the hospital for several months, much of that time on a respirator. Botulism is caused by the toxin of the bacterium *Clostridium botulinum,* which grows in anaerobic (without oxygen) environments, such as improperly processed canned food. The toxin blocks the release of acetylcholine in synapses. As a result, muscles cannot contract, and the person may die from respiratory failure.

Reflexes

A **reflex** is an involuntary reaction in response to a stimulus applied to the periphery and transmitted to the CNS. Reflexes allow a person to react to stimuli more quickly than is possible if conscious thought is involved. A **reflex arc** is the neuronal pathway by which a reflex occurs (figure 8.13). The reflex arc is the basic functional unit of the nervous system because it is the smallest, simplest pathway capable of receiving a stimulus and yielding a response. A reflex arc generally has five basic components: (1) a **sensory receptor;** (2) a **sensory neuron;** (3) in some reflexes, **interneurons,** which are neurons located between and communicating with two other neurons; (4) a **motor neuron;** and (5) an **effector organ** (muscles or glands). The simplest reflex arcs do not involve interneurons. Most reflexes occur in the spinal cord or brainstem rather than in the higher brain centers.

One example of a reflex occurs when a person's finger touches a hot stove. The heat stimulates pain receptors in the skin, and action potentials are produced. Sensory neurons conduct the action potentials to the spinal cord, where they synapse with interneurons. The interneurons, in turn, synapse with motor neurons in the spinal cord that conduct action potentials along their axons to flexor muscles in the upper limb. These muscles contract and pull the finger away from the stove. No conscious thought is required for this reflex, and withdrawal of the finger from the stimulus begins before the person is consciously aware of any pain.

Neuronal Pathways

Neurons are organized within the CNS to form pathways ranging from relatively simple to extremely complex. The two simplest pathways are converging and diverging pathways. In a **converging pathway,** two or more neurons synapse with (converge on) the same neuron (figure 8.14a). This allows information transmitted in more than one neuronal pathway to converge into a single pathway. In a **diverging pathway,** the axon from one neuron divides (diverges)

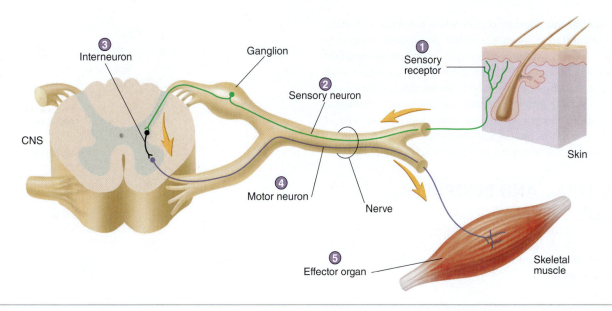

PROCESS Figure 8.13 AP|R **Reflex Arc**

The parts of a reflex arc are labeled in the order in which action potentials pass through them. The five components are the (I) sensory receptor, (2) sensory neuron, (3) interneuron, (4) motor neuron, and (5) effector organ. Simple reflex arcs have only four components because the reflex does not involve interneurons.

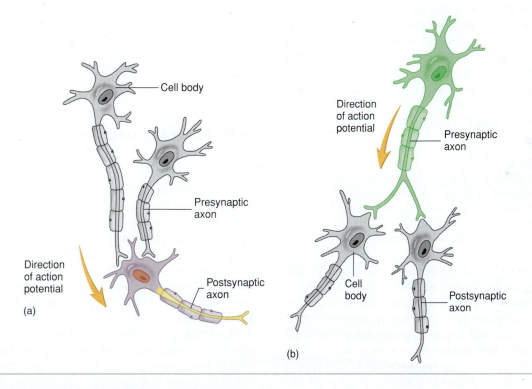

Figure 8.14 **Converging and Diverging Pathways in the Nervous System**

(a) General model of a converging pathway, showing two neurons converging onto one neuron. (b) General model of a diverging pathway, showing one neuron diverging onto two neurons.

and synapses with more than one other neuron (figure 8.14b). This allows information transmitted in one neuronal pathway to diverge into two or more pathways.

Within the CNS and in many PNS synapses, it takes more than a single action potential to have an effect. A single presynaptic action potential usually does not cause a sufficiently large postsynaptic local potential to reach threshold and produce an action potential in the target cell. Instead, many presynaptic action potentials are needed in a process called **summation.** Summation of signals in neuronal pathways allows integration of multiple subthreshold local potentials. Summation of the local potentials can bring the membrane potential to threshold and trigger an action potential. **Spatial summation** occurs when the local potentials originate from different locations on the postsynaptic neuron—for example, from converging pathways. **Temporal summation** occurs when local potentials overlap in time. This can occur from a single input that fires rapidly, which allows the resulting local potentials to overlap briefly. Spatial and temporal summation can lead to stimulation or inhibition, depending on the type of signal. Collectively, this integration of multiple inputs determines whether the postsynaptic neuron will fire an action potential.

8.5 CENTRAL AND PERIPHERAL NERVOUS SYSTEMS

Learning Outcome *After reading this section, you should be able to*

A. Compare and contrast the central and peripheral nervous systems.

As mentioned previously, the central nervous system (CNS) consists of the brain and spinal cord (see figure 8.1). The brain is housed within the braincase; the spinal cord is in the vertebral column.

The peripheral nervous system (PNS) consists of all the nerves and ganglia outside the brain and spinal cord. The PNS collects information from numerous sources both inside and on the surface of the body and relays it by way of sensory neurons to the CNS, where one of three results is possible: The information is ignored, triggers a reflex, or is evaluated more extensively. Motor neurons in the PNS relay information from the CNS to muscles and glands in various parts of the body, regulating activity in those structures. The nerves of the PNS can be divided into two groups: 12 pairs of cranial nerves and 31 pairs of spinal nerves.

8.6 SPINAL CORD

Learning Outcomes *After reading this section, you should be able to*

A. Describe the relationship between the spinal cord and the spinal nerves.

B. Describe a cross section of the spinal cord.

The **spinal cord** extends from the foramen magnum at the base of the skull to the second lumbar vertebra (figure 8.15). Spinal nerves communicate between the spinal cord and the body. The inferior end of the spinal cord and the spinal nerves exiting there resemble a horse's tail and are collectively called the **cauda equina** (kaw′dă, tail; ē-kwī′nă, horse).

A cross section reveals that the spinal cord consists of a superficial white matter portion and a deep gray matter portion (figure 8.16a). The white matter consists of myelinated axons, and the gray matter is mainly a collection of neuron cell bodies. The white matter in each half of the spinal cord is organized into three columns, called the **dorsal** (posterior), **ventral** (anterior), and **lateral columns.** Each column of the spinal cord contains ascending and descending tracts, or pathways. **Ascending tracts** consist of axons that conduct action potentials toward the brain, and **descending tracts** consist of axons that conduct action potentials away from the brain. Ascending tracts and descending tracts are discussed more fully in "Sensory Functions" and "Motor Functions" later in this chapter.

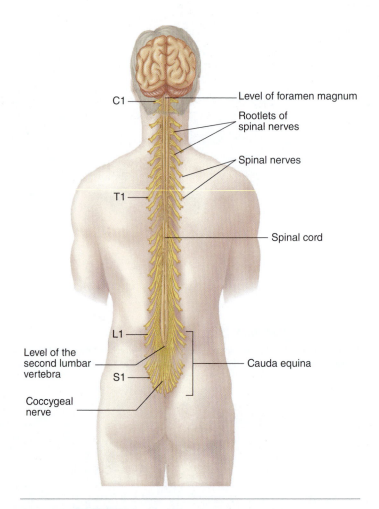

Figure 8.15 AP|R **Spinal Cord and Spinal Nerve Roots**

The gray matter of the spinal cord is shaped like the letter H, with **posterior horns** and **anterior horns.** Small **lateral horns** exist in levels of the cord associated with the autonomic nervous system. The **central canal** is a fluid-filled space in the center of the cord.

Spinal nerves arise from numerous rootlets along the dorsal and ventral surfaces of the spinal cord (figure 8.16a). The ventral rootlets combine to form a **ventral root** on the ventral (anterior) side of the spinal cord, and the dorsal rootlets combine to form a **dorsal root** on the dorsal (posterior) side of the cord at each segment. The ventral and dorsal roots unite just lateral to the spinal cord to form a spinal nerve. The dorsal root contains a ganglion, called the **dorsal root ganglion** (gang′glē-on; a swelling or knot).

The cell bodies of pseudo-unipolar sensory neurons are in the dorsal root ganglia (figure 8.16b). The axons of these neurons originate in the periphery of the body. They pass through spinal nerves and the dorsal roots to the posterior horn of the spinal cord gray matter. In the posterior horn, the axons either synapse with interneurons or pass into the white matter and ascend or descend in the spinal cord.

The cell bodies of motor neurons, which regulate the activities of muscles and glands, are located in the anterior and lateral horns of the spinal cord gray matter. Somatic motor neurons are in the anterior horn, and autonomic neurons are in the lateral

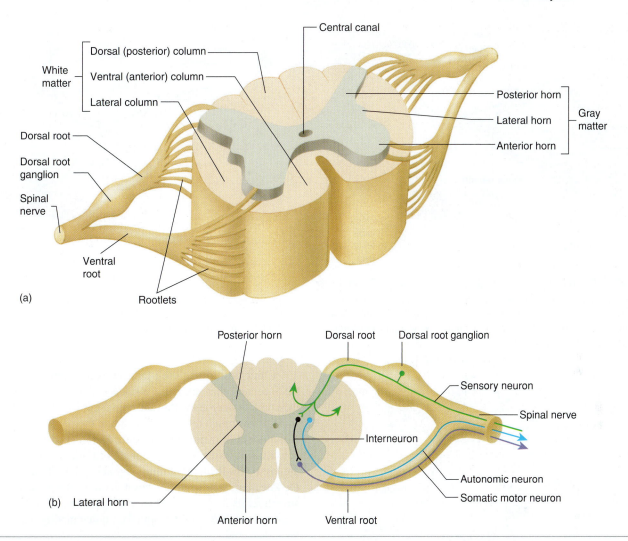

(a)

(b)

Figure 8.16 AP|R **Cross Section of the Spinal Cord**

(*a*) In each segment of the spinal cord, rootlets combine to form a dorsal root on the dorsal (posterior) side and a ventral root on the ventral (anterior) side. (*b*) Relationship of sensory and motor neurons to the spinal cord.

horn. Axons from the motor neurons form the ventral roots and pass into the spinal nerves. Thus, the dorsal root contains sensory axons, and the ventral root contains motor axons. Each spinal nerve therefore has both sensory and motor axons.

Predict 3

Describe the direction of action potential propagation in the spinal nerves, dorsal roots, and ventral roots.

An example of a converging pathway involves a motor neuron in the spinal cord that stimulates muscle contraction (figure 8.17*a*). Sensory fibers from pain receptors carry action potentials to the spinal cord and synapse with interneurons, which in turn synapse with a motor neuron. Neurons in the cerebral cortex, controlling conscious movement, also synapse with the same motor neuron by way of axons in descending tracts. Both the interneurons and the neurons in the cerebral cortex have axons that converge on the motor neuron, which can therefore be stimulated either through the reflex arc or by conscious thought.

An example of a diverging pathway involves sensory neurons within the spinal cord (figure 8.17*b*). The axon of a sensory neu-

ron carrying action potentials from pain receptors branches within the spinal cord. One branch produces a reflex response by synapsing with an interneuron. The interneuron, in turn, synapses with a motor neuron, which stimulates a muscle to withdraw the injured region of the body from the source of the pain. The other branch synapses with an ascending neuron that carries action potentials through a nerve tract to the brain, where the stimulation is interpreted as pain.

Spinal Cord Reflexes

Knee-Jerk Reflex

The simplest reflex is the **stretch reflex,** in which muscles contract in response to a stretching force applied to them. The **knee-jerk reflex,** or *patellar reflex* (figure 8.18), is a classic example of the stretch reflex. When the patellar ligament is tapped, the quadriceps femoris muscle tendon and the muscles themselves are stretched. Sensory receptors within these muscles are also stretched, and the stretch reflex is activated. Consequently, contraction of the muscles extends the leg, producing the characteristic knee-jerk response.

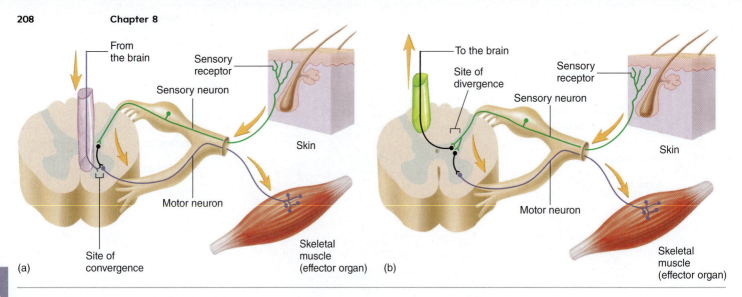

(a) Site of convergence

(b)

Figure 8.17 Converging and Diverging Pathways in the Spinal Cord
(*a*) In a converging circuit, a sensory neuron from the periphery (*green*), by way of an interneuron (*black*), and a descending neuron from the brain (*purple*) converge on a single motor neuron (*purple*). (*b*) In a diverging circuit, a sensory neuron from the periphery (*green*) diverges and sends information to a motor neuron by way of an interneuron (*black*) and sends information to the brain.

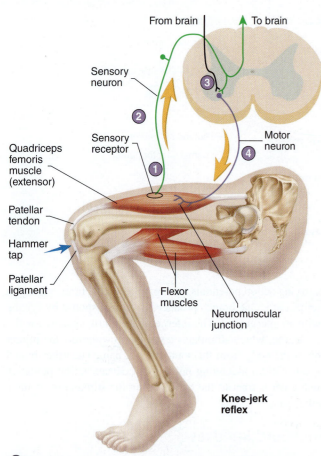

Knee-jerk reflex

1. Sensory receptors in the muscle detect stretch of the muscle.

2. Sensory neurons conduct action potentials to the spinal cord.

3. Sensory neurons synapse with motor neurons. Descending neurons (*black*) within the spinal cord also synapse with the neurons of the stretch reflex and modulate their activity.

4. Stimulation of the motor neurons causes the muscle to contract and resist being stretched.

PROCESS Figure 8.18 Knee-Jerk Reflex

Clinicians use the knee-jerk reflex to determine if the higher CNS centers that normally influence this reflex are functional. Descending neurons within the spinal cord synapse with the neurons of the stretch reflex and modulate their activity. This activity is important in maintaining posture and in coordinating muscular activity. Following a severe spinal cord injury, all spinal reflexes are lost below the level of injury. By about 2 weeks after the injury, the knee-jerk reflex returns, but it is often exaggerated. When the stretch reflex is absent or greatly exaggerated, it indicates that the neurons within the brain or spinal cord that modify this reflex have been damaged.

Withdrawal Reflex

The function of the **withdrawal reflex,** or *flexor reflex,* is to remove a limb or another body part from a painful stimulus. The sensory receptors are pain receptors (see chapter 9). Following painful stimuli, sensory neurons conduct action potentials through the dorsal root to the spinal cord, where the sensory neurons synapse with interneurons, which in turn synapse with motor neurons (figure 8.19). These neurons stimulate muscles, usually flexor muscles, that remove the limb from the source of the painful stimulus.

8.7 SPINAL NERVES

Learning Outcome   *After reading this section, you should be able to*

A. Define plexus, and describe the three primary plexuses, including their branches.

The **spinal nerves** arise along the spinal cord from the union of the dorsal roots and ventral roots (see figure 8.16). All the spinal nerves contain axons of both sensory and somatic motor neurons and thus are called **mixed nerves.** Some spinal nerves also contain parasympathetic or sympathetic axons. Most of the spinal nerves exit the vertebral column between adjacent vertebrae. Spinal nerves are categorized by the region of the vertebral column from which they emerge—cervical (C), thoracic (T), lumbar (L), sacral (S),

CLINICAL IMPACT   Spinal Cord Injury

About 10,000 new cases of **spinal cord injury** occur each year in the United States. Automobile and motorcycle accidents are the leading causes, followed by gunshot wounds, falls, and swimming accidents. Most spinal cord injuries are acute contusions (kon-too′shŭnz; bruising) of the cervical portion of the cord and do not completely sever the spinal cord.

Spinal cord injuries can interrupt ascending and/or descending tracts. Reflexes can still function below the level of injury, but sensations and/or motor functions and reflex modulation may be disrupted.

At the time of spinal cord injury, two types of tissue damage occur: (1) primary mechanical damage and (2) secondary tissue damage extending into a much larger region of the cord than the primary damage. The only treatment for primary damage is prevention, such as wearing seat belts when riding in automobiles and not diving in shal-

low water. Once an accident occurs, however, little can be done about primary damage.

Secondary spinal cord damage, which begins within minutes of the primary damage, is caused by ischemia (lack of blood supply), edema (fluid accumulation), ion imbalances, the release of "excitotoxins" (such as glutamate), and inflammatory cell invasion. Unlike primary damage, much secondary damage can be prevented or reversed if treatment is prompt. Giving the patient large doses of methylprednisolone, a synthetic anti-inflammatory steroid, within 8 hours of the injury can dramatically reduce the secondary damage to the cord. Other treatments include anatomical realignment and stabilization of the vertebral column and decompression of the spinal cord. Rehabilitation is based on retraining the patient to use whatever residual connections exist across the site of damage.

Although the spinal cord was formerly considered incapable of regeneration following severe damage, researchers have learned that most neurons of the adult spinal cord survive an injury and begin to regenerate, growing about 1 mm into the site of damage, before regressing to an inactive, atrophic state. The major block to adult spinal cord regeneration is the formation of a scar, consisting mainly of astrocytes, at the site of injury. Myelin in the scar apparently inhibits regeneration. However, research has shown that implantation of stem cells or other cell types, such as Schwann cells, can bridge the scar and stimulate regeneration. Certain growth factors can also stimulate regeneration. Current research continues to look for the right combination of chemicals and other factors to stimulate regeneration of the spinal cord following injury.

1. Pain receptors detect a painful stimulus.

2. Sensory neurons conduct action potentials to the spinal cord.

3. Sensory neurons synapse with interneurons that synapse with motor neurons.

4. Excitation of the motor neurons results in contraction of the flexor muscles and withdrawal of the limb from the painful stimulus.

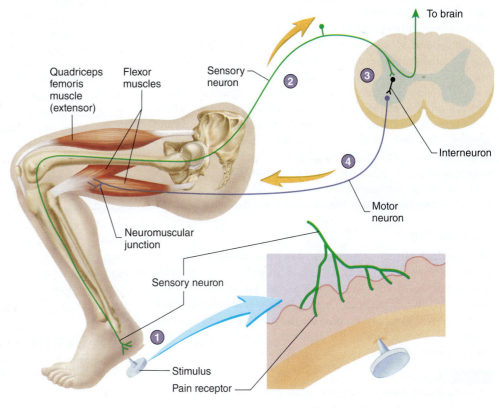

To brain

Quadriceps femoris muscle (extensor)

Flexor muscles

Sensory neuron

Interneuron

Motor neuron

Neuromuscular junction

Sensory neuron

Stimulus

Pain receptor

Withdrawal reflex

PROCESS Figure 8.19 Withdrawal Reflex

CLINICAL IMPACT   Radial Nerve Damage

The radial nerve lies very close to the medial side of the humerus in the proximal part of the arm and is susceptible to damage in that area. If a person uses crutches improperly so that the weight of the body is borne in the axilla and upper arm rather than by the hands, the top of the crutch can compress the radial nerve against the humerus. This compression can cause dysfunction of the radial nerve, resulting in paralysis of the posterior arm and forearm muscles and loss of sensation over the back of the forearm and hand. The condition, known as "crutch paralysis," is usually temporary as long as the patient begins to use the crutches correctly.

The radial nerve can be permanently damaged by a fracture of the humerus in the proximal part of the arm. A sharp edge of the broken bone may cut the nerve, resulting in permanent paralysis unless the nerve is surgically repaired. Because of potential damage to the radial nerve, a broken humerus should be treated very carefully.

and coccygeal (Co). The spinal nerves are also numbered (starting superiorly) according to their order within that region. The 31 pairs of spinal nerves are therefore C1 through C8, T1 through T12, L1 through L5, S1 through S5, and Co (figure 8.20a).

The nerves arising from each region of the spinal cord and vertebral column supply specific regions of the body. A **derma-tome** is the area of skin supplied with sensory innervation by a pair of spinal nerves. Each of the spinal nerves except C1 has a specific cutaneous sensory distribution. Figure 8.20b illustrates the dermatomal (der-mă-tō′măl) map for the sensory cutaneous distribution of the spinal nerves.

Most of the spinal nerves are organized into three major **plexuses** (plek′sŭs-ēz; braids) where neurons of several spinal nerves come together and intermingle. This reorganizes the neurons so that branches of nerves extending from each plexus contain neurons from different spinal segments. The three major plexuses are the cervical plexus, the brachial plexus, and the lumbosacral plexus (table 8.3). The major nerves of the neck and limbs are branches of these plexuses. Spinal nerves T2 through T11 do not join a plexus. Instead, these nerves extend around the thorax between the ribs, giving off branches to muscles and skin. Motor nerve fibers derived from plexuses innervate skeletal muscles, and sensory nerve fibers in those plexuses supply sensory innervation to the skin overlying those muscles (table 8.3). In addition to the major plexuses, the small **coccygeal plexus** supplies motor innervation to the muscles of the pelvic floor and sensory cutaneous innervation to the skin over the coccyx (figure 8.20b).

Cervical Plexus

The **cervical plexus** originates from spinal nerves C1 to C4. Branches from this plexus innervate several of the muscles attached to the hyoid bone, as well as the skin of the neck and posterior portion of the head. One of the most important branches of the cervical plexus is the **phrenic nerve,** which innervates the diaphragm. Contraction of the diaphragm is largely responsible for our ability to breathe (see chapter 15).

Predict 4

The phrenic nerve may be damaged where it descends along the neck or during open thorax or open heart surgery. Explain how damage to the right phrenic nerve affects the diaphragm. Describe the effect on the diaphragm of completely severing the spinal cord in the thoracic region versus in the upper cervical region.

Brachial Plexus

The **brachial plexus** originates from spinal nerves C5 to T1. Five major nerves emerge from the brachial plexus to supply the upper limb and shoulder. The **axillary nerve** innervates two shoulder muscles and the skin over part of the shoulder. The **radial nerve** innervates all the muscles in the posterior arm and forearm as well as the skin over the posterior surface of the arm, forearm, and hand. The **musculocutaneous** (mŭs′kū-lō-kū-tā′nē-ŭs; muscle + skin) **nerve** innervates the anterior muscles of the arm and the skin over the radial surface of the forearm. The **ulnar nerve** innervates two anterior forearm muscles and most of the intrinsic hand muscles. It also innervates the skin over the ulnar side of the hand. The ulnar nerve can be easily damaged where it passes posterior to the medial side of the elbow. The ulnar nerve at this location is called the "funny bone." The **median nerve** innervates most of the anterior forearm muscles and some of the intrinsic hand muscles. It also innervates the skin over the radial side of the hand.

Lumbosacral Plexus

The **lumbosacral** (lŭm′bō-sā′krăl) **plexus** originates from spinal nerves L1 to S4. Four major nerves exit the plexus to supply the lower limb. The **obturator** (ob′tū-rā-tŏr) **nerve** innervates the muscles of the medial thigh and the skin over the same region. The **femoral nerve** innervates the anterior thigh muscles and the skin over the anterior thigh and medial side of the leg. The **tibial nerve** innervates the posterior thigh muscles, the anterior and posterior leg muscles, and most of the intrinsic foot muscles. It also innervates the skin over the sole of the foot. The **common fibular** (fib′ū-lăr) **nerve** innervates the muscles of the lateral thigh and leg and some intrinsic foot muscles. It also innervates the skin over the anterior and lateral leg and the dorsal surface (top) of the foot. The tibial and common fibular nerves are bound together within a connective tissue sheath and together are called the **sciatic** (sī-at′ik) **nerve.**

8.8 BRAIN

Learning Outcomes   *After reading this section, you should be able to*

A. List the parts of the brain.
B. List the parts of the brainstem, and state their functions.
C. State where the cerebellum is located.
D. List the parts of the diencephalon, and state their functions.
E. List the lobes of the cerebrum, and state a function for each.

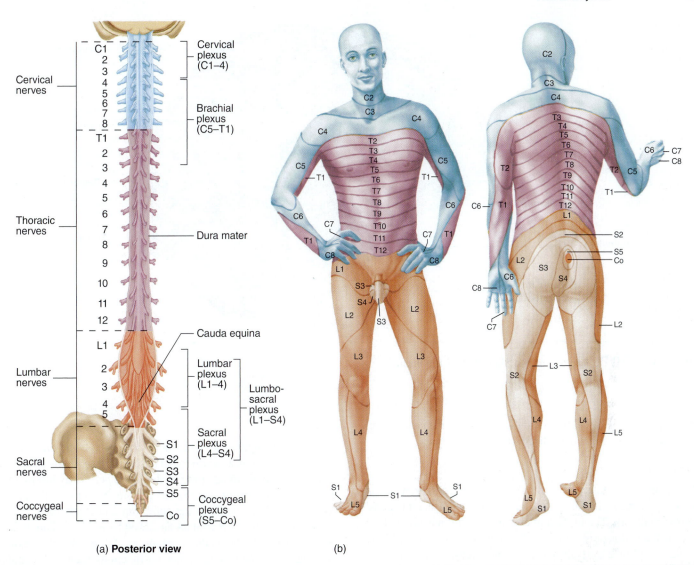

(a) **Posterior view**

(b)

Figure 8.20 AP|R Plexuses and Dermatomal Map

TABLE 8.3	Plexuses of the Spinal Nerves (see figure 8.20)			
Plexus	**Origin**	**Major Nerves**	**Muscles Innervated**	**Skin Innervated**
Cervical	C1–C4		Several neck muscles	Neck and posterior head
		Phrenic	Diaphragm	
Brachial	C5–T1	Axillary	Two shoulder muscles	Part of shoulder
		Radial	Posterior arm and forearm muscles (extensors)	Posterior arm, forearm, and hand
		Musculocutaneous	Anterior arm muscles (flexors)	Radial surface of forearm
		Ulnar	Two anterior forearm muscles (flexors), most intrinsic hand muscles	Ulnar side of hand
		Median	Most anterior forearm muscles (flexors), some intrinsic hand muscles	Radial side of hand
Lumbosacral	L1–S4	Obturator	Medial thigh muscles (adductors)	Medial thigh
		Femoral	Anterior thigh muscles (extensors)	Anterior thigh, medial leg, and foot
		Tibial	Posterior thigh muscles (flexors), anterior and posterior leg muscles, most foot muscles	Posterior leg and sole of foot
		Common fibular	Lateral thigh and leg, some foot muscles	Anterior and lateral leg, dorsal (top) part of foot
Coccygeal	S5 & Co		Pelvic floor muscles	Skin over coccyx

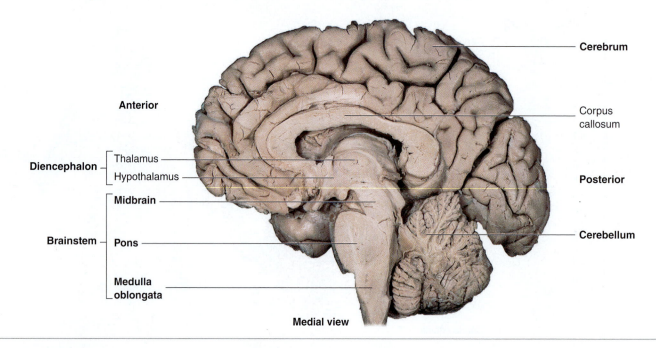

Cerebrum

Corpus
callosum

Anterior

Posterior

Diencephalon — Thalamus
Hypothalamus

Cerebellum

Midbrain

Brainstem — Pons

Medulla
oblongata

Medial view

Figure 8.21 AP|R **Regions of the Right Half of the Brain**

The major regions of the brain are the brainstem, the cerebellum, the diencephalon, and the cerebrum (figure 8.21).

Brainstem

The **brainstem** connects the spinal cord to the remainder of the brain (figure 8.22). It consists of the medulla oblongata, the pons, and the midbrain. The brainstem contains several nuclei involved in vital body functions, such as the control of heart rate, blood pressure, and breathing. Damage to small areas of the brainstem can cause death, whereas damage to relatively large areas of the cerebrum or cerebellum often do not. Nuclei for all but the first two cranial nerves are also located in the brainstem.

Medulla Oblongata

The **medulla oblongata** (ob′long-gă′tă) is the most inferior portion of the brainstem (figure 8.22) and is continuous with the spinal cord. It extends from the level of the foramen magnum to the pons. In addition to ascending and descending nerve tracts, the medulla oblongata contains discrete nuclei with specific functions, such as regulation of heart rate and blood vessel diameter, breathing, swallowing, vomiting, coughing, sneezing, balance, and coordination.

On the anterior surface, two prominent enlargements called **pyramids** extend the length of the medulla oblongata (figure 8.22a). The pyramids consist of descending nerve tracts, which transmit action potentials from the brain to motor neurons of the spinal cord and are involved in the conscious control of skeletal muscles.

Predict 5

*A large tumor or **hematoma** (hē-mă-tō′mă), a mass of blood that occurs as the result of bleeding into the tissues, can cause increased pressure within the skull. This pressure can force the medulla oblongata downward toward the foramen magnum. The displacement can compress the medulla oblongata and lead to death. Give two likely causes of death, and explain why they would occur.*

Pons

Immediately superior to the medulla oblongata is the **pons** (ponz; bridge). It contains ascending and descending nerve tracts, as well as several nuclei. Some of the nuclei in the pons relay information between the cerebrum and the cerebellum. Not only is the pons a functional bridge between the cerebrum and cerebellum, but on the anterior surface it resembles an arched footbridge (figure 8.22a). Several nuclei of the medulla oblongata, described earlier, extend into the lower part of the pons, so functions such as breathing, swallowing, and balance are controlled in the lower pons, as well as in the medulla oblongata. Other nuclei in the pons control functions such as chewing and salivation.

Midbrain

The **midbrain,** just superior to the pons, is the smallest region of the brainstem (figure 8.22b). The dorsal part of the midbrain consists of four mounds called the **colliculi** (ko-lik′ū-lī; sing. colliculus, hill). The two inferior colliculi are major relay centers for the auditory nerve pathways in the CNS. The two superior colliculi are involved in visual reflexes and receive touch and auditory input. Turning the head toward a tap on the shoulder, a sudden loud noise, or a bright flash of light is a reflex controlled in the superior colliculi. The midbrain contains nuclei involved in coordinating eye movements and controlling pupil diameter and lens shape. The midbrain also contains a black nuclear mass, called the **substantia nigra** (sŭb-stan′shē-ă nī′gră; black substance), which is part of the basal nuclei (see "Basal Nuclei" later in this chapter) and is involved in regulating general body movements. The rest of the midbrain consists largely of ascending tracts from the spinal cord to the cerebrum and descending tracts from the cerebrum to the spinal cord or cerebellum.

Nervous

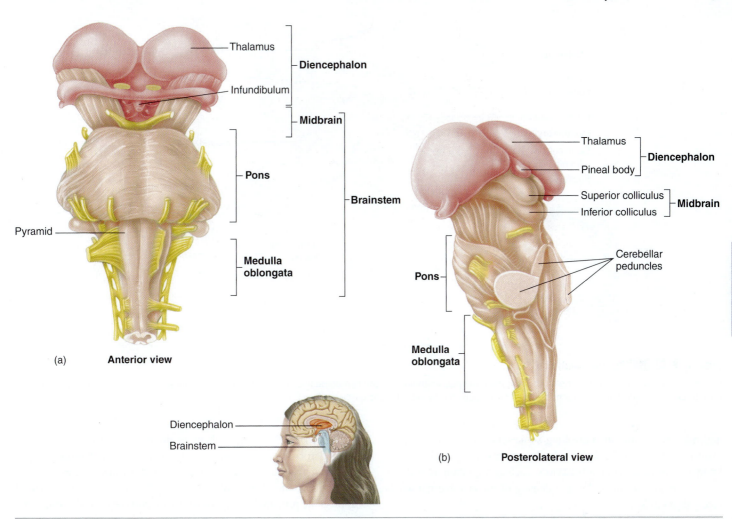

Figure 8.22 AP|R Brainstem and Diencephalon

Reticular Formation

Scattered throughout the brainstem is a group of nuclei collectively called the **reticular formation.** The reticular formation plays important regulatory functions. It is particularly involved in regulating cyclical motor functions, such as respiration, walking, and chewing. The reticular formation is a major component of the **reticular activating system,** which plays an important role in arousing and maintaining consciousness and in regulating the sleep-wake cycle. Stimuli such as a ringing alarm clock, sudden bright lights, smelling salts, or cold water splashed on the face can arouse consciousness. Conversely, removal of visual or auditory stimuli may lead to drowsiness or sleep. General anesthetics suppress the reticular activating system. Damage to cells of the reticular formation can cause coma.

Cerebellum

The **cerebellum** (ser-e-bel′ŭm; little brain), shown in figure 8.21, is attached to the brainstem by several large connections called **cerebellar peduncles** (pe-dŭng′klz; *pes,* foot). These connections provide routes of communication between the cerebellum and other parts of the CNS. The structure and function of the cerebellum are discussed in "Motor Functions" later in this chapter.

Diencephalon

The **diencephalon** (dī′en-sef′ă-lon) is the part of the brain between the brainstem and the cerebrum (figure 8.23). Its main components are the thalamus, the epithalamus, and the hypothalamus.

Thalamus

The **thalamus** (thal′a-mŭs) is by far the largest part of the diencephalon. It consists of a cluster of nuclei and is shaped somewhat like a yo-yo, with two large, lateral parts connected in the center by a small **interthalamic adhesion** (figure 8.23). Most sensory input that ascends through the spinal cord and brainstem projects to the thalamus, where ascending neurons synapse with thalamic neurons. Thalamic neurons, in turn, send their axons to the cerebral cortex. The thalamus also influences mood and registers an unlocalized, uncomfortable perception of pain.

Epithalamus

The **epithalamus** (ep′i-thal′ă-mŭs; *epi,* upon) is a small area superior and posterior to the thalamus (figure 8.23). It consists of a few small nuclei, which are involved in the emotional and visceral response to odors, and the pineal gland. The **pineal** (pin′ē-ăl; pinecone-shaped) **gland** is an endocrine gland that may influence the onset of puberty

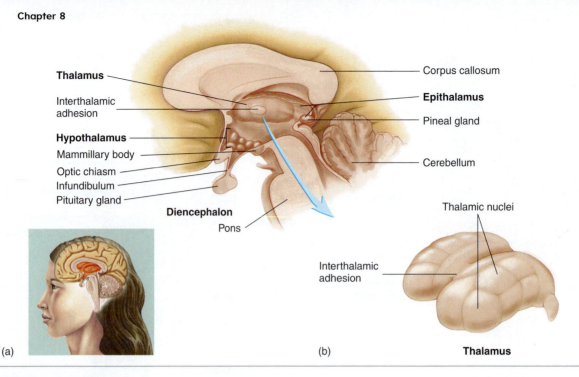

Figure 8.23 AP|R Diencephalon

(*a*) Median section of the diencephalon showing the thalamus, epithalamus, and hypothalamus. (*b*) Both halves of the thalamus as seen from an anterior, dorsolateral view with the separations between nuclei depicted by indentations on the surface.

and may play a role in controlling some long-term cycles that are influenced by the light-dark cycle. In animals, the pineal gland is known to influence annual behaviors, such as migration in birds, as well as changes in the color and density of fur in some mammals (see chapter 10).

Hypothalamus

The **hypothalamus** is the most inferior part of the diencephalon; it contains several small nuclei that are very important in maintaining homeostasis. The hypothalamus plays a central role in the control of body temperature, hunger, and thirst. Sensations such as sexual pleasure, rage, fear, and relaxation after a meal are related to hypothalamic functions. Emotional responses that seem inappropriate to the circumstances, such as "nervous perspiration" in response to stress or hunger as a result of depression, also involve the hypothalamus. A funnel-shaped stalk, the **infundibulum** (in-fŭn-dib′ū-lŭm; a funnel), extends from the floor of the hypothalamus to the pituitary gland. The hypothalamus plays a major role in controlling the secretion of hormones from the pituitary gland (see chapter 10). The **mammillary** (mam′i-lār-ē; *mamilla,* nipple) **bodies** form externally visible swellings on the posterior portion of the hypothalamus and are involved in emotional responses to odors and in memory.

Cerebrum

The **cerebrum** (ser′ĕ-brŭm, sĕ-rē′brŭm; brain) is the largest part of the brain (figure 8.24). It is divided into left and right hemispheres by a **longitudinal fissure.** The most conspicuous features on the surface of each hemisphere are numerous folds called **gyri** (jī′rī; sing. gyrus, *gyros,* circle), which greatly increase the surface area of the cortex, and intervening grooves called **sulci** (sŭl′sī; sing. sulcus, a furrow or ditch).

Each cerebral hemisphere is divided into lobes, which are named for the skull bones overlying them. The **frontal lobe** is important in the control of voluntary motor functions, motivation, aggression, mood, and olfactory (smell) reception. The **parietal lobe** is the principal center for receiving and consciously perceiving most sensory information, such as touch, pain, temperature, and balance. The frontal and parietal lobes are separated by the **central sulcus.** The **occipital lobe** functions in receiving and perceiving visual input and is not distinctly separate from the other lobes. The **temporal lobe** (figure 8.24*b*) is involved in olfactory (smell) and auditory (hearing) sensations and plays an important role in memory. Its anterior and inferior portions, called the "psychic cortex," are associated with functions such as abstract thought and judgment. Most of the temporal lobe is separated from the rest of the cerebrum by the **lateral fissure,** and deep within the fissure is the **insula** (in′soo-lă), often referred to as the **fifth lobe.**

8.9 SENSORY FUNCTIONS

Learning Outcomes *After reading this section, you should be able to*

- **A.** List the major ascending tracts, and state a function for each.
- **B.** Describe the sensory and association areas of the cerebral cortex, and discuss their interactions.

The CNS constantly receives a variety of stimuli originating both inside and outside the body. We are unaware of a large part of this sensory input, but it is vital to our survival and normal functions. Sensory input to the brainstem and diencephalon helps maintain homeostasis. Input to the cerebrum and cerebellum keeps us informed about our environment and allows the CNS to control motor functions. A small portion of the sensory input results in **perception,** the conscious awareness of stimuli (see chapter 9).

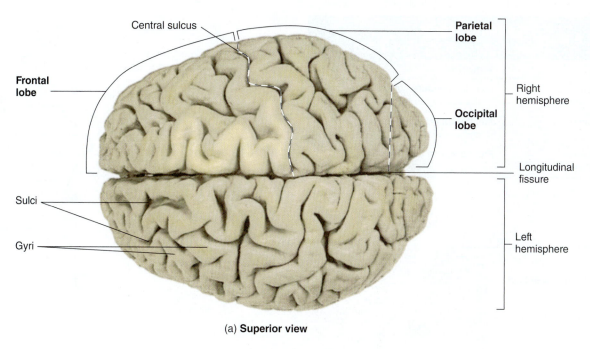

(a) **Superior view**

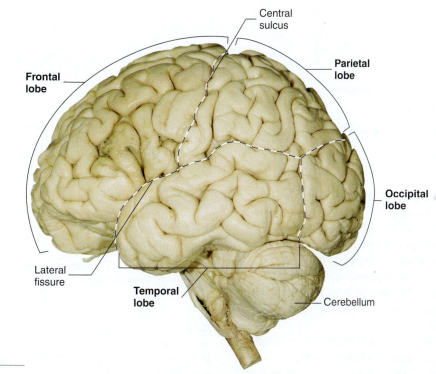

Figure 8.24 AP|R Brain

(b) **Lateral view**

Ascending Tracts

The spinal cord and brainstem contain a number of **ascending** (sensory) **tracts,** or pathways, that transmit information via action potentials from the periphery to various parts of the brain (table 8.4 and figure 8.25). Each tract is involved with a limited type of sensory input, such as pain, temperature, touch, position, or pressure, because each tract contains axons from specific sensory receptors specialized to detect a particular type of stimulus (see chapter 9).

Tracts are usually given composite names that indicate their origin and termination. The names of ascending tracts usually begin with the prefix *spino-,* indicating that they begin in the spinal cord. For example, the spinothalamic tract begins in the spinal cord and terminates in the thalamus.

Most ascending tracts consist of two or three neurons in sequence, from the periphery to the brain. Almost all neurons relaying information to the cerebrum terminate in the thalamus. Another neuron then relays the information from the thalamus to the

TABLE 8.4	Ascending Tracts (see figures 8.25 and 8.26)
Pathway	**Function**
Spinothalamic	Pain, temperature, light touch, pressure, tickle, and itch sensations
Dorsal column	Proprioception, touch, deep pressure, and vibration
Spinocerebellar (anterior and posterior)	Proprioception to cerebellum

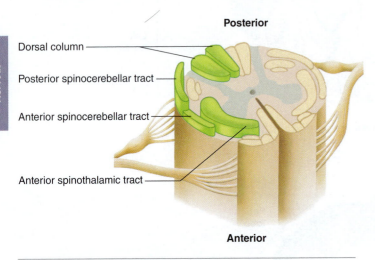

Figure 8.25 Ascending Tracts of the Spinal Cord

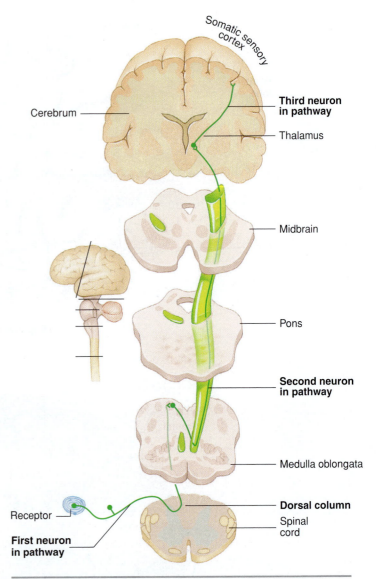

Figure 8.26 Dorsal Column

The dorsal column transmits action potentials dealing with touch, position, and pressure. Lines on the inset indicate levels of section.

cerebral cortex. The **spinothalamic tract,** which transmits action potentials dealing with pain and temperature to the thalamus and on to the cerebral cortex, is an example of an ascending tract. The **dorsal column,** which transmits action potentials dealing with touch, position, and pressure, is another example (figure 8.26).

Sensory tracts typically cross from one side of the body in the spinal cord or brainstem to the other side of the body. Thus, the left side of the brain receives sensory input from the right side of the body, and vice versa.

Ascending tracts also terminate in the brainstem or cerebellum. The anterior and posterior **spinocerebellar tracts,** for example, transmit information about body position to the cerebellum (see "Cerebellum" later in this chapter).

Sensory Areas of the Cerebral Cortex

Figure 8.27 depicts a lateral view of the left cerebral cortex with some of the sensory and motor areas indicated. The terms *area* and *cortex* are often used interchangeably for these regions of the cerebral cortex. Ascending tracts project to specific regions of the cerebral cortex, called **primary sensory areas,** where sensations are perceived. The **primary somatic sensory cortex,** or *general sensory area,* is located in the parietal lobe posterior to the central sulcus. Sensory fibers carrying general sensory input, such as pain, pressure, and temperature, synapse in the thalamus, and thalamic neurons relay the information to the primary somatic

sensory cortex. Sensory fibers from specific parts of the body project to specific regions of the primary somatic sensory cortex so that a topographic map of the body, with the head most inferior, exists in this part of the cerebral cortex (figure 8.27). Other primary sensory areas include the visual cortex in the occipital lobe, the primary auditory cortex in the temporal lobe, and the taste area in the insula.

Cortical areas immediately adjacent to the primary sensory areas, called **association areas,** are involved in the process of recognition. For example, sensory action potentials originating in the retina of the eye reach the visual cortex, where the image is perceived. Action potentials then pass from the visual cortex to the visual association area, where the present visual information is compared to past visual experience ("Have I seen this before?"). On the basis of this comparison, the visual association area "decides" whether the visual input is recognized and judges whether the input is significant. For example, if you pass a man walking down

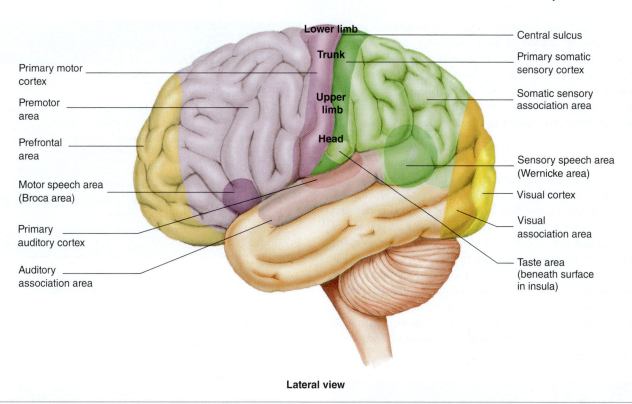

Lower limb

Trunk

Upper limb

Head

Central sulcus

Primary somatic sensory cortex

Somatic sensory association area

Sensory speech area (Wernicke area)

Visual cortex

Visual association area

Taste area (beneath surface in insula)

Primary motor cortex

Premotor area

Prefrontal area

Motor speech area (Broca area)

Primary auditory cortex

Auditory association area

Lateral view

Figure 8.27 AP|R **Sensory and Motor Areas of the Lateral Side of the Left Cerebral Cortex**

a street, you usually pay less attention to him if you've never seen him before than if you know him, unless a unique characteristic of the unknown person draws your attention. Other examples of association areas are the auditory association area, adjacent to the primary auditory cortex, and the somatic sensory association area, adjacent to the primary somatic sensory cortex.

8.10 MOTOR FUNCTIONS

Learning Outcomes *After reading this section, you should be able to*

A. Describe the motor area of the cerebral cortex, and discuss how it interacts with other parts of the frontal lobe.

B. Distinguish between upper and lower motor neurons and between direct and indirect tracts.

C. Discuss how the basal nuclei and cerebellum regulate motor functions.

The motor system of the brain and spinal cord is responsible for maintaining the body's posture and balance, as well as moving the trunk, head, limbs, tongue, and eyes and communicating through facial expressions and speech. Reflexes mediated through the spinal cord and brainstem are responsible for some body movements. These are called **involuntary movements** because they occur without conscious thought. **Voluntary movements,** on the other hand, are consciously activated to achieve a specific goal, such as walking or typing. Although consciously activated, the details of most voluntary movements occur automatically. For example, once you start walking, you don't have to think about the moment-to-moment control of every muscle because neural circuits in the

reticular formation and spinal cord automatically control your limbs. Once learned, complex tasks, such as typing, can be performed relatively automatically.

Voluntary movements result from the stimulation of upper and lower motor neurons. **Upper motor neurons** have cell bodies in the cerebral cortex. The axons of upper motor neurons form descending tracts that connect to lower motor neurons. **Lower motor neurons** have cell bodies in the anterior horn of the spinal cord gray matter or in cranial nerve nuclei. Their axons leave the central nervous system and extend through spinal or cranial nerves to skeletal muscles. Lower motor neurons form the motor units described in chapter 7.

Motor Areas of the Cerebral Cortex

The **primary motor cortex** is located in the posterior portion of the frontal lobe, directly anterior to the central sulcus (figure 8.27). Action potentials initiated in this region control voluntary movements of skeletal muscles. Upper motor neuron axons project from specific regions of this cortex to specific parts of the body so that a topographic map of the body exists in the primary motor cortex, with the head inferior, analogous to the topographic map of the primary somatic sensory cortex (figure 8.27). The **premotor area** of the frontal lobe is where motor functions are organized before they are actually initiated in the primary motor cortex. For example, if a person decides to take a step, the neurons of the premotor area are first stimulated, and the determination is made there as to which muscles must contract, in what order, and to what degree. Action potentials are then passed to the upper motor neurons of the primary motor cortex, which initiate each planned movement.

Nervous

The motivation and foresight to plan and initiate movements occur in the anterior portion of the frontal lobes, called the **prefrontal area.** This region of association cortex is well developed only in primates, especially humans. It is involved in motivation and regulation of emotional behavior and mood. The large size of this area in humans may account for our emotional complexity and our relatively well-developed capacity to think ahead and feel motivated.

Descending Tracts

The names of the descending tracts are based on their origin and termination (table 8.5 and figure 8.28). For example, the corticospinal tracts are so named because they begin in the cerebral cortex and terminate in the spinal cord. The corticospinal tracts are considered **direct** because they extend directly from upper motor neurons in the cerebral cortex to lower motor neurons in the spinal cord (a similar direct tract extends to lower motor neurons in the brainstem). Other tracts are named after the part of the brainstem from which they originate. Although they originate in the brainstem, these tracts are indirectly controlled by the cerebral cortex, basal nuclei, and cerebellum. These tracts are called **indirect** because no direct connection exists between the cortical and spinal neurons.

A CASE IN POINT

Recovery from Spinal Cord Injury

Harley Chopper was taking a curve on his motorcycle when he hit a patch of loose gravel, and his motorcycle slid onto its side. Harley fell off and was spun, head-first, into a retaining wall. Even though Harley was wearing a helmet, the compression of hitting the wall with great force injured his neck. Examination at the hospital revealed that Harley had suffered a fracture of the fifth and sixth cervical vertebrae. Physicians administered methylprednisolone to reduce inflammation around the spinal cord, and Harley's neck vertebrae were realigned and stabilized. Initially, Harley had no feeling, or movement, in his lower limbs and some feeling, but little movement, in his upper limbs. He was hospitalized for several weeks. Sensation and motor function gradually returned in all four limbs, but he required many months of physical therapy. Three years later, Harley had still not fully regained his ability to walk.

Immediately following Harley's injury, both ascending and descending tracts experienced a lack of function. This is not uncommon with a spinal cord injury, as the entire cord experiences the "shock" of the trauma. Without rapid treatment, that initial shock can become permanent loss. Sometimes, even with treatment, the damage is so severe that permanent loss of function results. In Harley's case, all ascending tracts regained normal function, but some functional deficit still remained in the lateral corticospinal tract to the lower limbs after 3 years.

The descending tracts control different types of movements (table 8.5). Tracts in the lateral columns (see figure 8.16; figure 8.28) are most important in controlling goal-directed limb movements, such as reaching and manipulating. The **lateral corticospinal tracts** are especially important in controlling the speed and precision of skilled movements of the hands. Tracts in the ventral columns, such as the reticulospinal tract, are most important for maintaining posture, balance, and limb position through their control of neck, trunk, and proximal limb muscles.

TABLE 8.5	Descending Tracts (see figures 8.28 and 8.29)
Pathway	**Function**
Direct	
Lateral corticospinal	Muscle tone and skilled movements, especially of hands
Anterior corticospinal	Muscle tone and movement of trunk muscles
Indirect	
Rubrospinal	Movement coordination
Reticulospinal	Posture adjustment, especially during movement
Vestibulospinal	Posture and balance
Tectospinal	Movement in response to visual reflexes

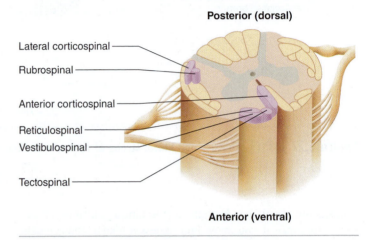

Figure 8.28 Descending Tracts of the Spinal Cord

The lateral corticospinal tract serves as an example of how descending pathways function. It begins in the cerebral cortex and descends into the brainstem (figure 8.29). At the inferior end of the pyramids of the medulla oblongata, the axons cross over to the opposite side of the body and continue into the spinal cord. Crossover of axons in the brainstem or spinal cord to the opposite side of the body is typical of descending pathways. Thus, the left side of the brain controls skeletal muscles on the right side of the body, and vice versa. The upper motor neuron synapses with interneurons that then synapse with lower motor neurons in the brainstem or spinal cord. The axon of the lower motor neuron extends to the skeletal muscle fiber.

Basal Nuclei

The **basal nuclei** are a group of functionally related nuclei (figure 8.30). Two primary nuclei are the **corpus striatum** (kōr´-pŭs strī-ā´tŭm), located deep within the cerebrum, and the **substantia nigra,** a group of darkly pigmented cells in the midbrain.

The basal nuclei are important in planning, organizing, and coordinating motor movements and posture. Complex neural circuits link the basal nuclei with each other, with the thalamus, and with the cerebral cortex. These connections form several feedback loops, some of which are stimulatory and others inhibitory. The stimulatory circuits facilitate muscle activity, especially at the

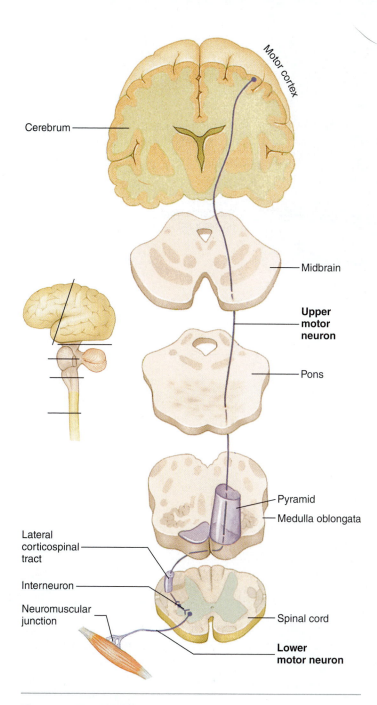

Figure 8.29 Example of a Direct Tract
The lateral corticospinal tract is responsible for movement below the head. Lines on the inset indicate levels of section.

person is not performing a task. Parkinson disease, Huntington disease, and cerebral palsy are basal nuclei disorders (see the Diseases and Disorders table at the end of this chapter).

Cerebellum

The **cerebellum** is attached by cerebellar peduncles to the brainstem (see figures 8.22 and 8.24b). The cerebellar cortex is composed of gray matter and has gyri and sulci, but the gyri are much smaller than those of the cerebrum. Internally, the cerebellum consists of gray nuclei and white nerve tracts. The cerebellum is involved in maintaining balance and muscle tone and in coordinating fine motor movement. If the cerebellum is damaged, muscle tone decreases, and fine motor movements become very clumsy.

A major function of the cerebellum is that of a **comparator** (figure 8.31). A comparator is a sensing device that compares the data from two sources—in this case, the motor cortex and peripheral structures. Action potentials from the cerebral motor cortex descend into the spinal cord to initiate voluntary movements. Collateral branches are also sent from the motor cortex to the cerebellum, giving information representing the intended movement. In addition, simultaneously, reaching the cerebellum are action potentials from **proprioceptive** (prō-prē-ō-sep′tiv) **neurons,** which innervate joints, tendons, and muscles and provide information about the position of body parts. The cerebellum compares information about the intended movement from the motor cortex to sensory information from the moving structures. If a difference is detected, the cerebellum sends action potentials to motor neurons in the motor cortex and the spinal cord to correct the discrepancy. The result is smooth and coordinated movements. For example, if you close your eyes, the cerebellar comparator function allows you to touch your nose smoothly and easily with your finger. If the cerebellum is not functioning, your finger tends to overshoot the target. One effect of alcohol is to inhibit the function of the cerebellum.

Another function of the cerebellum involves participating with the cerebrum in learning motor skills, such as playing the piano. Once the cerebrum and cerebellum "learn" these skills, the specialized movements can be accomplished smoothly and automatically.

8.11 OTHER BRAIN FUNCTIONS

Learning Outcomes *After reading this section, you should be able to*

A. Discuss the right and left cerebral hemispheres and speech.
B. Compare and contrast the features of working, short-term, and the two types of long-term memory.

Communication Between the Right and Left Hemispheres

The right cerebral hemisphere receives sensory input from and controls muscular activity in the left half of the body. The left cerebral hemisphere receives input from and controls muscles in the right half of the body. Sensory information received by one hemisphere is shared with the other through connections between the two hemispheres called **commissures** (kom′ĭ-shŭrz; joining together). The largest of these commissures is the **corpus callosum** (kōr′pus kă-lō′sŭm), a broad band of nerve tracts at the base of the longitudinal fissure (see figures 8.21 and 8.23).

beginning of a voluntary movement, such as rising from a sitting position or beginning to walk. The inhibitory circuits facilitate the actions of the stimulatory circuits by inhibiting muscle activity in antagonist muscles. In addition, inhibitory circuits inhibit random movements of the trunk and limbs. Inhibitory circuits also decrease muscle tone when the body, limbs, and head are at rest. Disorders of the basal nuclei result in difficulty rising from a sitting position and difficulty initiating walking. People with basal nuclei disorders exhibit increased muscle tone and exaggerated, uncontrolled movements when they are at rest. A specific feature of some basal nuclei disorders is "resting tremor," a slight shaking of the hands when a

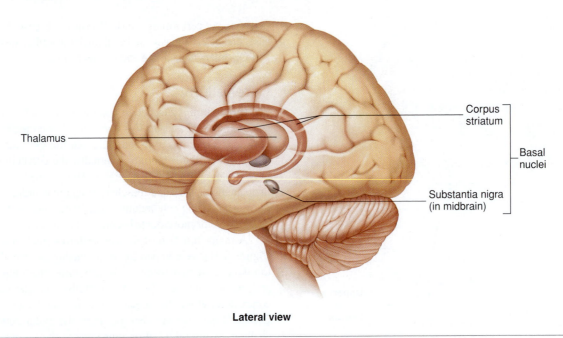

Lateral view

Figure 8.30 Basal Nuclei

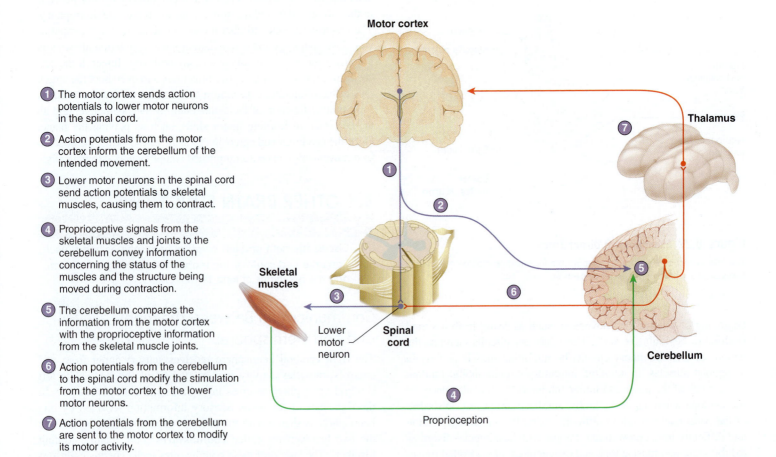

1. The motor cortex sends action potentials to lower motor neurons in the spinal cord.

2. Action potentials from the motor cortex inform the cerebellum of the intended movement.

3. Lower motor neurons in the spinal cord send action potentials to skeletal muscles, causing them to contract.

4. Proprioceptive signals from the skeletal muscles and joints to the cerebellum convey information concerning the status of the muscles and the structure being moved during contraction.

5. The cerebellum compares the information from the motor cortex with the proprioceptive information from the skeletal muscle joints.

6. Action potentials from the cerebellum to the spinal cord modify the stimulation from the motor cortex to the lower motor neurons.

7. Action potentials from the cerebellum are sent to the motor cortex to modify its motor activity.

PROCESS Figure 8.31 Cerebellar Comparator Function

Language and perhaps other functions, such as artistic activities, are not shared equally between the two hemispheres. Researchers believe the left hemisphere is the more analytical hemisphere, emphasizing such skills as mathematics and speech, whereas the right hemisphere is more involved in functions such as three-dimensional or spatial perception and musical ability.

Speech

In most people, the speech area is in the left cerebral cortex. Two major cortical areas are involved in speech: The **sensory speech area** (Wernicke area), located in the parietal lobe, functions in understanding and formulating coherent speech. The **motor speech area** (Broca area), located in the frontal lobe, controls the movement necessary for speech (see figure 8.27). Damage to these parts of the brain or to associated brain regions may result in **aphasia** (ă-fā′zē-ă; *a-,* without + *phasis,* speech), absent or defective speech or language comprehension. The most common cause is a stroke. It is estimated that 25–40% of stroke survivors exhibit aphasia.

Speech-related functions involve both sensory and motor pathways. For example, to repeat a word that you *hear* involves the following pathway: Action potentials from the ear reach the primary auditory cortex, where the word is perceived; the word is recognized in the auditory association area and comprehended in portions of the sensory speech area. Action potentials representing the word are then conducted through nerve tracts that connect the sensory and motor speech areas. In the motor speech area, the muscle activity needed to repeat the word is determined. Action potentials then go to the premotor area, where the movements are programmed, and finally to the primary motor cortex, where specific movements are triggered.

Speaking a *written* word involves a slightly different pathway: The information enters the visual cortex, then passes to the visual association area, where it is recognized. The information continues to the sensory speech area, where it is understood and formulated as it is to be spoken. From the sensory speech area, it follows the same route for repeating words that you hear: through nerve tracts to the motor speech area, to the premotor area, and then to the primary motor cortex.

Predict 6

> Describe a neural pathway that allows a blindfolded person to name an object placed in her hand.

Brain Waves and Consciousness

Different levels of consciousness can be revealed by different patterns of electrical activity in the brain. Electrodes placed on a person's scalp and attached to a recording device can record the brain's electrical activity, producing an **electroencephalogram** (ē-lek′trō-en-sef′ă-lō-gram) (**EEG**) (figure 8.32). These electrodes are not positioned so that they can detect individual action potentials, but they can detect the simultaneous action potentials in large numbers of neurons. Most of the time, EEG patterns are irregular, with no particular pattern, because the brain's electrical activity is normally not synchronous. At other times, however, EEG patterns can be detected as wavelike patterns known as **brain waves.** Their intensity and frequency differ, based on the state of brain activity. The different levels of consciousness in an awake and a sleeping person are marked by different types of brain waves. **Alpha waves** are observed in a normal person who is awake but in a quiet, resting state with the eyes closed. **Beta waves** have a higher frequency than alpha waves and occur during intense mental activity. During the beginning of sleep, a rapid transition takes place from a beta rhythm to an alpha rhythm. As sleep deepens, progressively more delta waves occur. **Delta waves** occur during deep sleep, in infants, and in patients with severe brain disorders. A fourth type of brain waves, **theta waves,** are usually observed in children, but they can also occur in adults who are experiencing frustration or who have certain brain disorders. Neurologists use these patterns to diagnose and determine the treatment for the disorders.

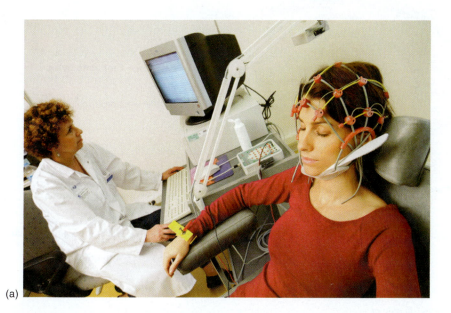

(a)

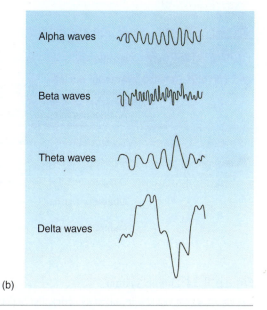

(b)

Alpha waves

Beta waves

Theta waves

Delta waves

Figure 8.32 Electroencephalogram
(*a*) Patient with electrodes attached to her head. (*b*) Four electroencephalographic tracings.

Memory

The storage of memory can be divided into three stages: working, short-term, and long-term. Long-term memories can also be subdivided based on the type of the memory: those dealing with facts (*declarative*) and those dealing with skills (*procedural*).

The brain briefly stores information required for the immediate performance of a task. This task-associated memory is called **working memory.** It lasts only a few seconds to minutes and occurs mostly in the frontal cortex. Working memory is limited primarily by the number of bits of information (about seven) that can be stored at any one time. When new information is presented, old information, previously stored in working memory, is eliminated. What happens to a telephone number you just looked up if you are distracted?

Short-term memory lasts longer than working memory and can be retained for a few minutes to a few days. Short-term memories are stored by a mechanism involving increased synaptic transmission. Short-term memory is susceptible to brain trauma, such as physical injury or decreased oxygen, and to certain drugs that affect neural function, such as general anesthetics.

Short-term memory is transferred to **long-term memory,** where it may be stored for only a few minutes or become permanent, by **consolidation,** a gradual process involving the formation of new and stronger synaptic connections. The length of time memory is stored may depend on how often it is retrieved and used. **Declarative memory,** or *explicit memory,* involves the retention of facts, such as names, dates, and places, as well as related emotional undertones. Emotion and mood apparently serve as gates in the brain and determine what is stored in long-term declarative memory. **Procedural memory,** or *reflexive memory,* involves the development of motor skills, such as riding a bicycle. Only a small amount of procedural memory is lost over time.

Long-term memory involves structural and functional changes in neurons that lead to long-term enhancement of synaptic transmission. A whole series of neurons, called **memory engrams,** or *memory traces,* are probably involved in the long-term retention of a given piece of information, a thought, or an idea. Repeating the information and associating it with existing memories help us transfer information from short-term to long-term memory.

Limbic System and Emotions

The olfactory cortex and certain deep cortical regions and nuclei of the cerebrum and the diencephalon are grouped together under the title **limbic** (lim′bik; a boundary) **system** (figure 8.33). The limbic system influences long-term declarative memory, emotions, visceral responses to emotions, motivation, and mood. A major source of sensory input to the limbic system is the olfactory nerves. The limbic system responds to olfactory stimulation by initiating responses necessary for survival, such as hunger and thirst. The limbic system is connected to, and functionally associated with, the hypothalamus. Lesions in the limbic system can result in voracious appetite, increased (often perverse) sexual activity, and docility (including loss of normal fear and anger responses).

8.12 MENINGES, VENTRICLES, AND CEREBROSPINAL FLUID

Learning Outcome *After reading this section, you should be able to*

A. Describe the three meningeal layers surrounding the central nervous system, the four ventricles of the brain, and the origin and circulation of the cerebrospinal fluid.

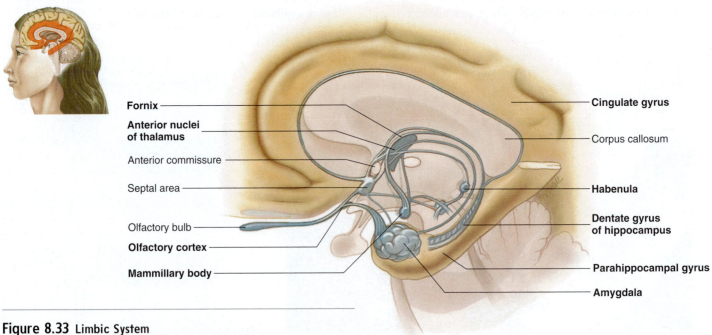

Fornix

Anterior nuclei of thalamus

Anterior commissure

Septal area

Olfactory bulb

Olfactory cortex

Mammillary body

Cingulate gyrus

Corpus callosum

Habenula

Dentate gyrus of hippocampus

Parahippocampal gyrus

Amygdala

Medial view

Figure 8.33 Limbic System

The limbic system includes the olfactory cortex, the cingulate gyrus (an area of the cerebral cortex on the medial side of each hemisphere), nuclei such as those of the hypothalamus and thalamus, the hippocampus (a mass of neuron cell bodies in the temporal lobe), and connecting nerve tracts, such as the fornix.

Meninges

Three connective tissue membranes, the **meninges** (mĕ-nin′jēz; *meninx,* membrane) (figure 8.34), surround and protect the brain and spinal cord. The most superficial and thickest of the meninges is the **dura mater** (doo′ră mā′ter; tough mother).

The dura mater around the brain consists of two layers, which function as a single layer but are physically separated into several regions to form dural folds and dural venous sinuses. Folds of dura mater extend into the longitudinal fissure between the two cerebral hemispheres and between the cerebrum and the cerebellum. The folds help hold the brain in place within the skull. The dural venous sinuses collect blood from the small veins of the brain and empty into the internal jugular veins, which exit the skull.

Within the skull, the dura mater adheres tightly to the cranial bones. In contrast, within the vertebral canal is an **epidural space** between the dura mater and the vertebrae (figure 8.34*c*). The epidural space is clinically important as the injection site for **epidural anesthesia** of the spinal nerves, which is often given to women during childbirth.

The second meningeal membrane is the very thin, wispy **arachnoid** (ă-rak′noyd; spiderlike, as in cobwebs) **mater.** The space between the dura mater and the arachnoid mater is the **subdural space,** which is normally only a potential space containing a very small amount of serous fluid.

A CASE IN POINT

Subdural Hematoma

May Fall is a 70-year-old woman who fell in her living room and struck her head on the coffee table. Some days later, she complained to her son, Earl E., that she was experiencing headaches and that her left hand seemed weak. Earl also noticed that his mother, who was normally very happy and positive, seemed irritable and short-tempered. Earl convinced his mother to see a physician. Based on May's age and history, her physician ordered a CT scan. The CT scan revealed an image consistent with a subdural hematoma over the right hemisphere, and May was scheduled for surgery. After making a large surgical flap, the surgeon opened the dura mater and removed the hematoma by suction. After the surgery, May required several months of physical therapy, but she recovered.

Damage to the veins crossing between the cerebral cortex and the dural venous sinuses can cause bleeding into the subdural space, resulting in a subdural hematoma, which can put pressure on the brain. The pressure can result in decreased brain function in the affected area. For example, the primary motor cortex is located in the posterior portion of the frontal lobe. Therefore, pressure on the portion of the right primary motor cortex involved in hand movements can cause decreased function in the left hand. The frontal lobe is also involved with mood; thus, pressure in that area can result in mood changes—for example, a normally happy person may become irritable. Subdural hematoma is more common in people over age 60 because their veins are less resilient and more easily damaged.

The spinal cord extends only to approximately the level of the second lumbar vertebra. Spinal nerves surrounded by meninges extend to the end of the vertebral column. Because there is no spinal cord in the inferior portion of the vertebral canal, a needle can be introduced into the subarachnoid space at that level without damaging the spinal cord. Health professionals use such a needle to inject anesthetic into the area as a **spinal block** or to take a sample of cerebrospinal fluid in a **spinal tap.** The cerebrospinal fluid can then be examined for infectious agents (meningitis) or for blood (hemorrhage).

The third meningeal membrane, the **pia mater** (pī′ă, pē′ă; affectionate mother), is very tightly bound to the surface of the brain and spinal cord. Between the arachnoid mater and the pia mater is the **subarachnoid space,** which is filled with cerebrospinal fluid and contains blood vessels.

Ventricles

The CNS contains fluid-filled cavities, called **ventricles,** which are quite small in some areas and large in others (figure 8.35). Each cerebral hemisphere contains a relatively large cavity called the **lateral ventricle. The third ventricle** is a smaller, midline cavity located in the center of the diencephalon between the two halves of the thalamus and connected by foramina (holes) to the lateral ventricles. The **fourth ventricle** is located at the base of the cerebellum and connected to the third ventricle by a narrow canal, called the **cerebral aqueduct.** The fourth ventricle is continuous with the **central canal** of the spinal cord. The fourth ventricle also opens into the subarachnoid space through foramina in its walls and roof.

Cerebrospinal Fluid

Cerebrospinal fluid (CSF) bathes the brain and spinal cord, providing a protective cushion around the CNS. It is produced by the **choroid** (kō′royd; lacy) **plexuses,** specialized structures made of ependymal cells, which are located in the ventricles (figure 8.36). CSF fills the brain ventricles, the central canal of the spinal cord, and the subarachnoid space. The CSF flows from the lateral ventricles into the third ventricle and then through the cerebral aqueduct into the fourth ventricle. A small amount of CSF enters the central canal of the spinal cord. The CSF exits the fourth ventricle through small openings in its walls and roof and enters the subarachnoid space. Masses of arachnoid tissue, called **arachnoid granulations,** penetrate the superior sagittal sinus, a dural venous sinus in the longitudinal fissure, and CSF passes from the subarachnoid space into the blood through these granulations.

Blockage of the openings in the fourth ventricle or the cerebral aqueduct can cause CSF to accumulate in the ventricles, a condition known as **hydrocephalus.** The accumulation of fluid creates increased pressure that dilates the ventricles and compresses the brain tissue, which usually results in irreversible brain damage. If the skull bones are not completely ossified when the hydrocephalus occurs, as in a fetus or newborn, the pressure can also cause severe enlargement of the head. Hydrocephalus is treated by placing a drainage tube (shunt) from the ventricles to the abdominal cavity to eliminate the high internal pressures.

8.13 CRANIAL NERVES

Learning Outcome   *After reading this section, you should be able to*

A. List the various types of cranial nerves, and briefly describe their functions.

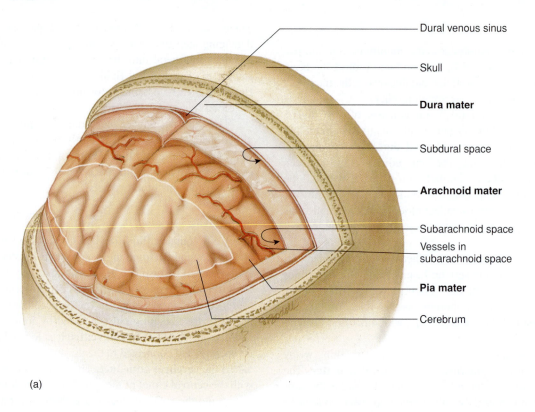

Dural venous sinus

Skull

Dura mater

Subdural space

Arachnoid mater

Subarachnoid space

Vessels in subarachnoid space

Pia mater

Cerebrum

(a)

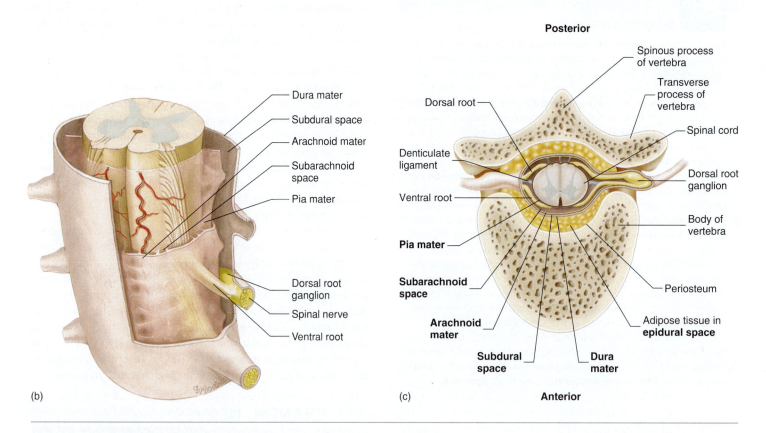

Dura mater

Subdural space

Arachnoid mater

Subarachnoid space

Pia mater

Dorsal root ganglion

Spinal nerve

Ventral root

(b)

Posterior

Spinous process of vertebra

Transverse process of vertebra

Dorsal root

Spinal cord

Denticulate ligament

Dorsal root ganglion

Ventral root

Body of vertebra

Pia mater

Subarachnoid space

Periosteum

Arachnoid mater

Adipose tissue in **epidural space**

Subdural space

Dura mater

(c)

Anterior

Figure 8.34 AP|R Meninges

(*a*) Anterior superior view of the head to show the meninges. (*b*) Meningeal membranes surrounding the spinal cord. (*c*) Cross section of a vertebra and the spinal cord.

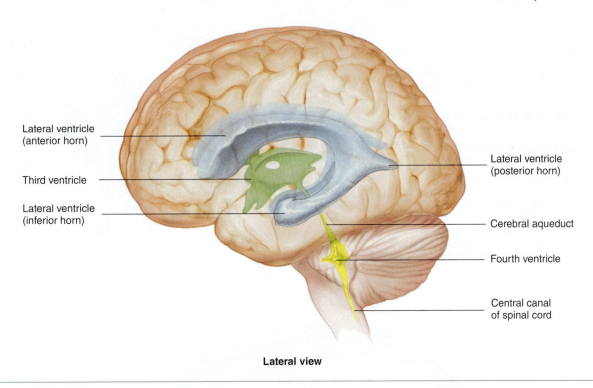

Lateral ventricle
(anterior horn)

Third ventricle

Lateral ventricle
(inferior horn)

Lateral ventricle
(posterior horn)

Cerebral aqueduct

Fourth ventricle

Central canal
of spinal cord

Lateral view

Figure 8.35 AP|R **Ventricles of the Brain Viewed from the Left**

The 12 pairs of **cranial nerves** are listed in table 8.6 and pictured in figure 8.37. They are designated by Roman numerals from I to XII. There are two general categories of cranial nerve function: sensory and motor. Sensory functions can be divided into the special senses, such as vision, and the more general senses, such as touch and pain in the face. Motor functions are subdivided into somatic motor and parasympathetic. Somatic motor cranial nerves innervate skeletal muscles in the head and neck. Parasympathetic cranial nerves innervate glands, smooth muscle throughout the body, and cardiac muscle of the heart.

Some cranial nerves are only sensory, and some are only somatic motor, whereas other cranial nerves are mixed nerves with sensory, somatic motor, and parasympathetic functions. Three cranial nerves—the olfactory (I), optic (II), and vestibulocochlear (VIII) nerves—are sensory only. Four other cranial nerves—the trochlear (IV), abducens (VI), accessory (XI), and hypoglossal (XII) nerves—are considered somatic motor only (although these motor nerves do provide proprioceptive information).

The trigeminal nerve (V) has sensory and somatic motor functions. It has the greatest general sensory distribution of all the cranial nerves and is the only cranial nerve supplying sensory information to the brain from the skin of the face. Sensory information from the skin over the rest of the body is carried to the CNS by spinal nerves. Injections of anesthetic by a dentist are designed to block sensory transmission through branches of the trigeminal nerve from the teeth. These dental branches of the trigeminal nerve are probably anesthetized more often than any other nerves in the body.

The oculomotor nerve (III) is somatic motor and parasympathetic. The facial (VII), glossopharyngeal (IX), and vagus (X) nerves have all three functions: sensory, somatic motor, and para-

sympathetic (table 8.6). The vagus nerve is probably the most important parasympathetic nerve in the body. It helps regulate the functions of the thoracic and abdominal organs, such as heart rate, respiration rate, and digestion.

As with the spinal nerves, the sensory and motor functions of many cranial nerves cross from one side of the head or body to the opposite side of the cerebral cortex. For example, sensory input from the right side of the face via the trigeminal nerve (V) crosses in the brainstem and projects to the left cerebral cortex. Motor output from the left cerebral cortex crosses in the brainstem to the right side of the face via the facial nerve (VII).

8.14 AUTONOMIC NERVOUS SYSTEM

Learning Outcomes   *After reading this section, you should be able to*

A. Contrast the structure of the autonomic nervous system and the somatic nervous system.

B. Name the two divisions of the autonomic nervous system, and describe the anatomical and neurotransmitter differences between them.

C. Compare and contrast the general functions of the parasympathetic and sympathetic nervous systems.

The autonomic nervous system (ANS) comprises motor neurons that carry action potentials from the CNS to the periphery. The autonomic neurons innervate smooth muscle, cardiac muscle, and glands. Autonomic functions are largely controlled unconsciously.

Axons from autonomic neurons do not extend all the way from the CNS to target tissues. This is in contrast to somatic motor neurons, which extend axons from the CNS to skeletal muscle. In the autonomic nervous system, two neurons in series extend from

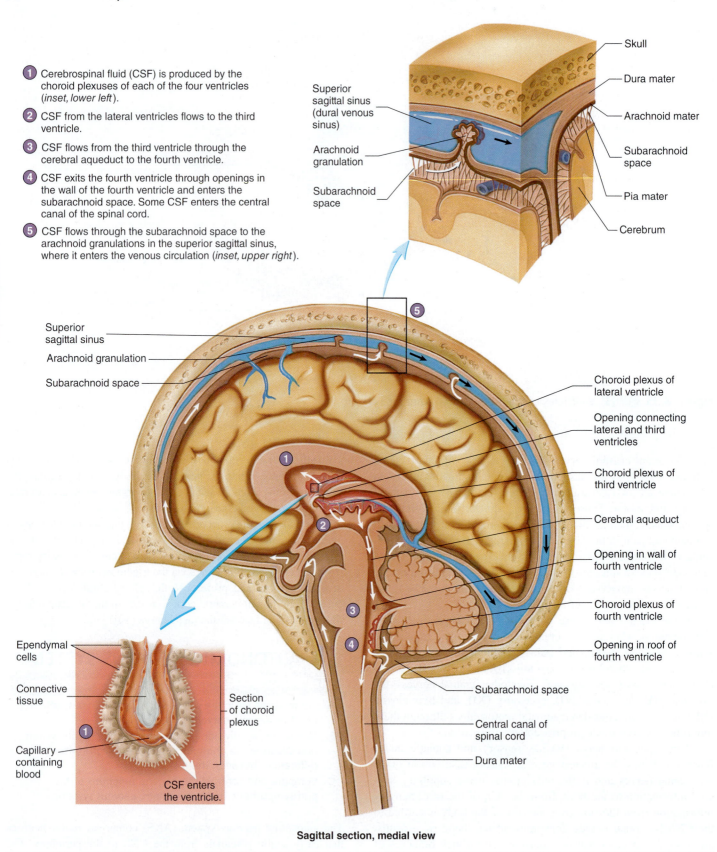

1. Cerebrospinal fluid (CSF) is produced by the choroid plexuses of each of the four ventricles (*inset, lower left*).

2. CSF from the lateral ventricles flows to the third ventricle.

3. CSF flows from the third ventricle through the cerebral aqueduct to the fourth ventricle.

4. CSF exits the fourth ventricle through openings in the wall of the fourth ventricle and enters the subarachnoid space. Some CSF enters the central canal of the spinal cord.

5. CSF flows through the subarachnoid space to the arachnoid granulations in the superior sagittal sinus, where it enters the venous circulation (*inset, upper right*).

Skull

Dura mater

Arachnoid mater

Subarachnoid space

Pia mater

Cerebrum

Superior sagittal sinus (dural venous sinus)

Arachnoid granulation

Subarachnoid space

Superior sagittal sinus

Arachnoid granulation

Subarachnoid space

Choroid plexus of lateral ventricle

Opening connecting lateral and third ventricles

Choroid plexus of third ventricle

Cerebral aqueduct

Opening in wall of fourth ventricle

Choroid plexus of fourth ventricle

Opening in roof of fourth ventricle

Subarachnoid space

Central canal of spinal cord

Dura mater

Ependymal cells

Connective tissue

Capillary containing blood

Section of choroid plexus

CSF enters the ventricle.

Sagittal section, medial view

PROCESS Figure 8.36 AP|R Flow of CSF

White arrows indicate the direction of CSF flow, and *black arrows* indicate the direction of blood flow in the sinuses.

Nervous

TABLE 8.6   Cranial Nerves and Their Functions (see figure 8.37)

Number	Name	General Function*	Specific Function
I	Olfactory	S	Smell
II	Optic	S	Vision
III	Oculomotor	M, P	Motor to four of six extrinsic eye muscles and upper eyelid; parasympathetic: constricts pupil, thickens lens
IV	Trochlear	M	Motor to one extrinsic eye muscle
V	Trigeminal	S, M	Sensory to face and teeth; motor to muscles of mastication (chewing)
VI	Abducens	M	Motor to one extrinsic eye muscle
VII	Facial	S, M, P	Sensory: taste; motor to muscles of facial expression; parasympathetic to salivary and tear glands
VIII	Vestibulocochlear	S	Hearing and balance
IX	Glossopharyngeal	S, M, P	Sensory: taste and touch to back of tongue; motor to pharyngeal muscles; parasympathetic to salivary glands
X	Vagus	S, M, P	Sensory to pharynx, larynx, and viscera; motor to palate, pharynx, and larynx; parasympathetic to viscera of thorax and abdomen
XI	Accessory	M	Motor to two neck and upper back muscles
XII	Hypoglossal	M	Motor to tongue muscles

*S, sensory; M, somatic motor; P, parasympathetic

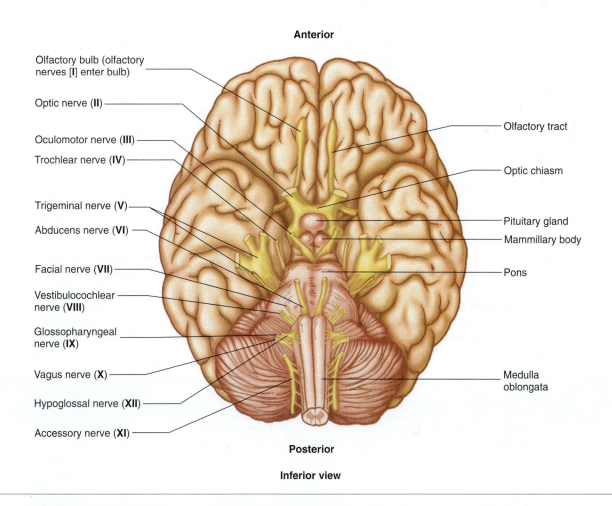

Anterior

Olfactory bulb (olfactory nerves [I] enter bulb)

Optic nerve (II)

Oculomotor nerve (III)

Trochlear nerve (IV)

Trigeminal nerve (V)

Abducens nerve (VI)

Facial nerve (VII)

Vestibulocochlear nerve (VIII)

Glossopharyngeal nerve (IX)

Vagus nerve (X)

Hypoglossal nerve (XII)

Accessory nerve (XI)

Olfactory tract

Optic chiasm

Pituitary gland

Mammillary body

Pons

Medulla oblongata

Posterior

Inferior view

Figure 8.37 AP|R Inferior Surface of the Brain Showing the Origin of the Cranial Nerves

the CNS to the effector organs. The first is called the **preganglionic neuron;** the second is the **postganglionic neuron** (figure 8.38). The neurons are so named because preganglionic neurons synapse with postganglionic neurons in **autonomic ganglia** outside the CNS. An exception is the preganglionic neuron that extends to the adrenal gland. There, the postganglionic neurons are actually the hormone-secreting cells of the adrenal medulla.

The autonomic nervous system is composed of the **sympathetic** (sim-pă-thet′ik; *sympatheo,* to feel with + *pathos,* suffering) **division** and the **parasympathetic** (par-ă-sim-pa-thet′ik; *para,* alongside of) **division** (table 8.7 and figure 8.39). Increased activity in sympathetic neurons generally prepares the individual for physical activity, whereas parasympathetic stimulation generally activates involuntary functions, such as digestion, that are normally associated with the body at rest.

Anatomy of the Sympathetic Division

Cell bodies of sympathetic preganglionic neurons are in the lateral horn of the spinal cord gray matter (see figure 8.16) between the first thoracic (T1) and the second lumbar (L2) segments. The axons of the preganglionic neurons exit through ventral roots and project to either sympathetic chain ganglia or collateral ganglia (figure 8.39). The **sympathetic chain ganglia** are connected to one another and are so named because they form a chain along both sides of the spinal cord. Some preganglionic fibers synapse with postganglionic fibers in the sympathetic chain ganglia. The postganglionic axons form sympathetic nerves that innervate structures of the thoracic cavity. The axons of those preganglionic fibers that do not synapse in the sympathetic chain ganglia form **splanchnic nerves** that extend to collateral ganglia. **Collateral ganglia** are located nearer target organs and consist of the celiac, superior mesenteric, and inferior mesenteric ganglia. Preganglionic neurons synapse with postganglionic neurons in the collateral ganglia. Postganglionic neurons in the collateral ganglia project to target tissues in the abdominal and pelvic regions.

Anatomy of the Parasympathetic Division

Preganglionic cell bodies of the parasympathetic division are located either within brainstem nuclei of the oculomotor nerve (III), facial nerve (VII), glossopharyngeal nerve (IX), or vagus nerve (X) or within the lateral part of the central gray matter of the spinal cord in the regions that give rise to spinal nerves S2 through S4.

Axons of the preganglionic neurons extend through spinal nerves or cranial nerves to **terminal ganglia** either located near effector organs in the head (figure 8.39) or embedded in the walls of effector organs in the thorax, abdomen, and pelvis. The axons

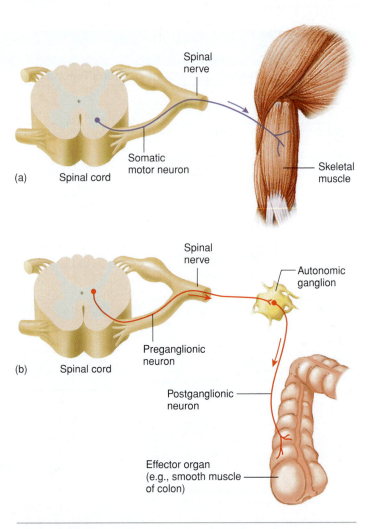

Figure 8.38 **Organization of Somatic and Autonomic Nervous System Neurons**

(*a*) The cell body of the somatic motor neuron is in the CNS, and its axon extends to the skeletal muscle. (*b*) The cell body of the preganglionic neuron is in the CNS, and its axon extends to the autonomic ganglion and synapses with the postganglionic neuron. The postganglionic neuron extends to and synapses with its effector organ.

of the postganglionic neurons extend a relatively short distance from the terminal ganglia to the target organ. Most of the thoracic and abdominal organs are supplied by preganglionic neurons of the **vagus nerve** extending from the brainstem. The vagus nerve branches to provide parasympathetic innervation to the heart, the lungs, the liver, and the stomach and other digestive organs.

TABLE 8.7	Sympathetic and Parasympathetic Divisions of the Autonomic Nervous System (see figure 8.39)		
Division	**Location of Preganglionic Cell Body**	**Location of Postganglionic Cell Body**	**General Function**
Sympathetic	T1–L2	Sympathetic chain ganglia or collateral ganglia	"Fight-or-flight"; prepares the body for physical activity
Parasympathetic	Cranial nerves III, VII, IX, X; S2–S4 spinal nerves	Terminal ganglia near or embedded in the walls of target organs	"Rest-and-digest"; stimulates involuntary activities of the body at rest

Nervous

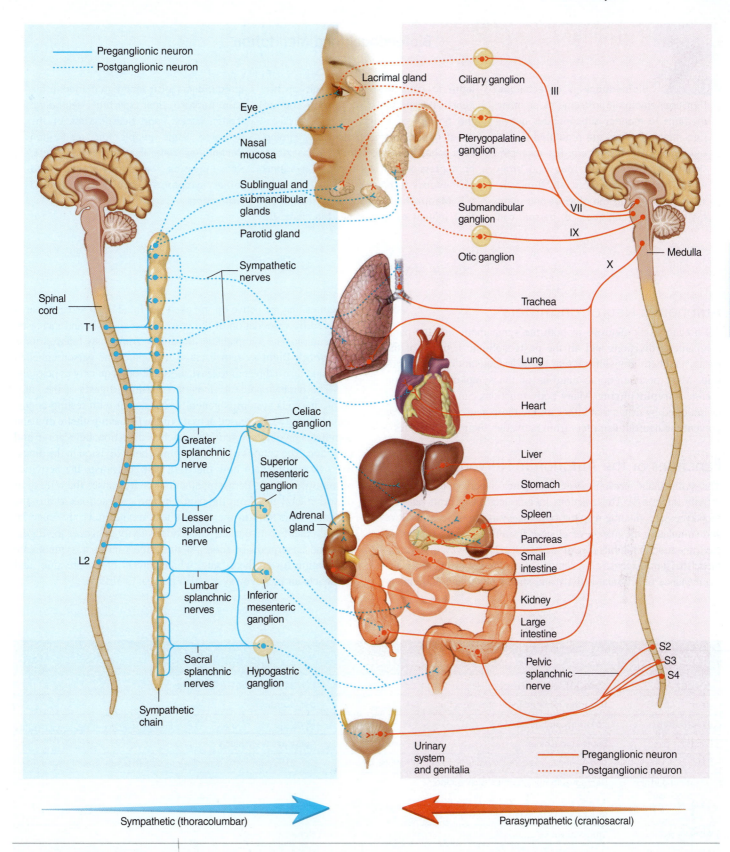

Preganglionic neuron
Postganglionic neuron

Lacrimal gland
Eye
Nasal mucosa
Sublingual and submandibular glands
Parotid gland
Sympathetic nerves
Spinal cord
T1
Greater splanchnic nerve
Superior mesenteric ganglion
Celiac ganglion
Lesser splanchnic nerve
Adrenal gland
L2
Lumbar splanchnic nerves
Inferior mesenteric ganglion
Sacral splanchnic nerves
Hypogastric ganglion
Sympathetic chain

Ciliary ganglion
Pterygopalatine ganglion
Submandibular ganglion
Otic ganglion
III
VII
IX
X
Medulla
Trachea
Lung
Heart
Liver
Stomach
Spleen
Pancreas
Small intestine
Kidney
Large intestine
Pelvic splanchnic nerve
S2
S3
S4
Urinary system and genitalia

Preganglionic neuron
Postganglionic neuron

Sympathetic (thoracolumbar)   Parasympathetic (craniosacral)

Figure 8.39 AP|R **Innervation of Organs by the ANS**
Preganglionic fibers are indicated by *solid lines,* and postganglionic fibers are indicated by *dashed lines.*

CLINICAL IMPACT   Biofeedback and Meditation

Biofeedback is a technique that uses electronic instruments or other methods to monitor subconscious activities, many of which are regulated by the autonomic nervous system. For example, skin temperature, heart rate, and brain waves may be monitored electronically. Then, by watching the monitor and using biofeedback techniques, a person can learn to consciously reduce his or her heart rate and blood pressure or change the pattern of brain waves. The severity of peptic (stomach) ulcers, migraine headaches, high blood pressure, anxiety, or depression can be reduced by using biofeedback techniques.

Meditation is another technique that influences autonomic functions. Its practitioners also claim that meditation can improve their spiritual well-being, consciousness, and holistic view of the universe. Some people find meditation techniques useful in reducing heart rate, blood pressure, the severity of ulcers, and other symptoms frequently associated with stress.

Autonomic Neurotransmitters

All preganglionic neurons of both the sympathetic and the parasympathetic divisions and all the postganglionic neurons of the parasympathetic division secrete the neurotransmitter **acetylcholine.** Most postganglionic neurons of the sympathetic division secrete **norepinephrine.** Many body functions can be stimulated or inhibited by drugs that either mimic these neurotransmitters or prevent the neurotransmitters from activating their target tissues.

Functions of the Autonomic Nervous System

The sympathetic nervous system prepares a person for physical activity (table 8.8). These actions include increasing heart rate and blood pressure, dilating respiratory passageways to increase airflow, and stimulating the release of glucose from the liver for energy. At the same time, it inhibits digestive activities. In this way, the sympathetic division decreases the activity of organs not essential for the maintenance of physical activity and shunts blood and nutrients to structures that are active during exercise. In addition, excess heat is removed by vasodilation of the vessels near the skin and increased perspiration. The sympathetic division is sometimes referred to as the **fight-or-flight system** because it prepares the person either to stand and face a threat or to leave the situation as quickly as possible.

The parasympathetic division (**rest-and-digest**) of the autonomic nervous system is generally consistent with resting conditions (table 8.8). Increased activity of the parasympathetic division stimulates involuntary activities, such as digestion, defecation, and urination. The actions of the parasympathetic division in the digestive system illustrate how the ANS can coordinate the activities of multiple organs. Parasympathetic activity causes the release of digestive enzymes from the pancreas and contractions to mix the enzymes with food in the small intestine and move the material through the digestive tract. This cooperativity enhances the digestion and absorption of food. At the same time, parasympathetic stimulation lowers heart rate, which lowers blood pressure, and constricts air passageways, which decreases airflow.

TABLE 8.8	Effects of the ANS on Various Tissues	
Target	**Sympathetic Effects**	**Parasympathetic Effects**
Heart	Increases rate and force of contraction	Decreases rate
Blood vessels	Constricts and dilates	None
Lungs	Dilates bronchioles	Constricts bronchioles
Eyes	Dilates pupil, relaxes ciliary muscle to adjust lens for far vision	Constricts pupil, contracts ciliary muscle to adjust lens for near vision
Intestinal and stomach walls	Decreases motility, contracts sphincters	Increases motility, relaxes sphincters
Liver	Breaks down glycogen, releases glucose	Synthesizes glycogen
Adipose tissue	Breaks down lipids	None
Adrenal gland	Secretes epinephrine, norepinephrine	None
Sweat glands	Secretes sweat	None
Salivary glands	Secretes thick saliva	Secretes watery saliva
Urinary bladder	Relaxes muscle, constricts sphincter	Contracts muscle, relaxes sphincter
Pancreas	Decreases secretion of digestive enzymes and insulin	Increases secretion of digestive enzymes and insulin

CLINICAL IMPACT   Autonomic Dysfunctions

Raynaud (rā-nō′) **disease** involves the spasmodic contraction of blood vessels in the periphery, especially in the digits, and results in pale, cold hands that are prone to ulcerations and gangrene as a result of poor circulation.

This condition may be caused by exaggerated sensitivity of the blood vessels to sympathetic stimulation. Occasionally, a surgeon may cut the preganglionic neurons to alleviate the condition. **Dysautonomia** (dis′aw-tō-nō′mē-ă) is an inherited condi-

tion characterized by reduced tear secretion, poor vasomotor control, dry mouth and throat, and other symptoms. It results from poorly controlled autonomic reflexes.

The sympathetic and parasympathetic divisions can each produce both stimulatory and inhibitory effects, depending on the target tissue. For example, the sympathetic division stimulates smooth muscle contraction in blood vessel walls and inhibits smooth muscle contraction in lung airway walls. Likewise, the parasympathetic division stimulates contraction of the urinary bladder and inhibits contraction of the heart.

Most organs that receive autonomic neurons are innervated by both the parasympathetic and the sympathetic division. In most cases, the influences of the two autonomic divisions are opposite. For example, sympathetic stimulation of the heart causes an increase in heart rate, whereas parasympathetic stimulation causes a decrease in heart rate. When both divisions innervate a single organ, the sympathetic division tends to play a major role during physical activity or stress, whereas the parasympathetic division has more influence during resting conditions. Despite the general opposing actions of the two divisions, in some situations, both divisions can act together to coordinate the activity of multiple targets. For example, in males, the parasympathetic division initiates erection of the penis, and the sympathetic division stimulates the release of secretions and helps initiate ejaculation. Not all organs receive dual innervation. Sweat glands and blood vessels are innervated by sympathetic neurons almost exclusively, whereas the smooth muscles associated with the lens of the eye are innervated primarily by parasympathetic neurons.

Predict 7

> List some of the responses stimulated by the autonomic nervous system in (a) a person who is extremely angry and (b) a person who has just finished eating and is now relaxing.

8.15 ENTERIC NERVOUS SYSTEM

Learning Outcome   *After reading this section, you should be able to*

A. Discuss how the enteric nervous system can act independently of the CNS.

The **enteric nervous system (ENS)** consists of plexuses within the wall of the digestive tract (see figure 16.2). The plexuses include (1) sensory neurons that connect the digestive tract to the CNS; (2) sympathetic and parasympathetic neurons that connect the CNS to the digestive tract; and (3) enteric neurons, located entirely within the enteric plexuses. A unique feature of enteric neurons is

that they are capable of monitoring and controlling the digestive tract independently of the CNS through local reflexes. For example, stretching of the digestive tract is detected by enteric sensory neurons, which stimulate enteric interneurons. The enteric interneurons stimulate enteric motor neurons, which stimulate glands to secrete. Although the ENS is capable of controlling the activities of the digestive tract completely independently of the CNS, normally the two systems work together. CNS control of parasympathetic branches of the vagus nerve and sympathetic nerves (primarily, the splanchnic nerves) can override the actions of enteric neurons. Hence, the ENS is an independent subdivision of the PNS that is integrated with the ANS.

8.16 EFFECTS OF AGING ON THE NERVOUS SYSTEM

Learning Outcome   *After reading this section, you should be able to*

A. Describe the changes that occur in the nervous system with advancing age.

As a person ages, sensory function gradually declines because of decreases in the number of sensory neurons, the function of the remaining neurons, and CNS processing. As a result of decreases in the number of skin receptors, elderly people are less conscious of anything touching or pressing on the skin and have more difficulty identifying objects by touch. These changes leave elderly people more prone to skin injuries.

A decreased sense of position of the limbs and in the joints can affect balance and coordination. Information about the position, tension, and length of tendons and muscles also decreases, further reducing the senses of movement, posture, and position, as well as the control and coordination of movement.

Other sensory neurons with reduced function include those that monitor blood pressure, thirst, objects in the throat, the amount of urine in the urinary bladder, and the amount of feces in the rectum. As a result, elderly people are more prone to high blood pressure, dehydration, swallowing and choking problems, urinary incontinence, and constipation or bowel incontinence.

A general decline in the number of motor neurons also occurs. As many as 50% of the lower motor neurons in the lumbar region of the spinal cord may be lost by age 60. Muscle fibers innervated by the lost motor neurons are also lost, resulting in a

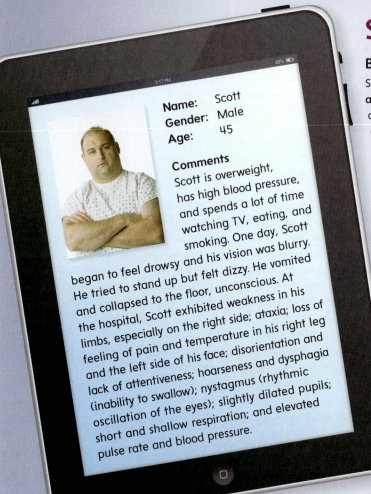

Name: Scott
Gender: Male
Age: 45

Comments
Scott is overweight, has high blood pressure, and spends a lot of time watching TV, eating, and smoking. One day, Scott began to feel drowsy and his vision was blurry. He tried to stand up but felt dizzy. He vomited and collapsed to the floor, unconscious. At the hospital, Scott exhibited weakness in his limbs, especially on the right side; ataxia; loss of feeling of pain and temperature in his right leg and the left side of his face; disorientation and lack of attentiveness; hoarseness and dysphagia (inability to swallow); nystagmus (rhythmic oscillation of the eyes); slightly dilated pupils; short and shallow respiration; and elevated pulse rate and blood pressure.

Stroke

Background Information

Scott had suffered a "stroke," also referred to as a **cerebrovascular accident (CVA).** The term **stroke** describes a heterogeneous group of conditions involving the death of brain tissue due to disruption of its vascular supply. There are two types of stroke: **Hemorrhagic stroke** results from bleeding of arteries supplying brain tissue, and **ischemic** (is-kē′mik) **stroke** results when arteries supplying brain tissue are blocked (figure 8A). The blockage in ischemic stroke can result from a thrombus, which is a clot that develops in place within an artery, or an embolism, which is a plug composed of a detached thrombus or other foreign body, such as a fat globule or gas bubble, that becomes lodged in an artery, blocking it. Scott was at high risk of developing a stroke. He was approaching middle age, was overweight, did not exercise enough, smoked, was under stress, and had a poor diet.

The combination of motor loss (as exhibited by weakness in the limbs) and sensory loss (evidenced by loss of pain and temperature sensation in the left lower limb and loss of all sensation in the right side of the face), along with the ataxia, dizziness, nystagmus, and hoarseness, suggests that the stroke affected the brainstem and cerebellum. Blockage of the vertebral artery, a major artery supplying the brain or its branches, can result in an area of dead tissue called a lateral medullary infarction. Damage to the descending motor tracts in that area, above the medullary crossover point, causes muscle weakness. Damage to ascending tracts can result in loss of pain and temperature sensation or other sensory modalities, depending on the affected tract.

Figure 8A

(*a*) MRI (magnetic resonance imaging) of a massive stroke in the left side (the viewer's right) of the brain. (*b*) Colorized NMR (nuclear magnetic resonance) showing disruption of blood flow to the left side of the brain (*yellow*). This disruption could cause a stroke.

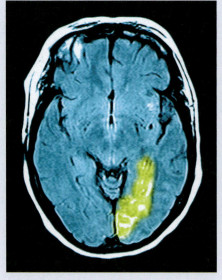

(a)

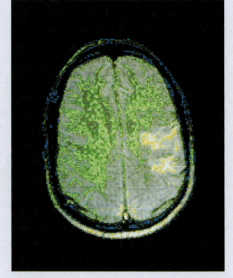

(b)

SKELETAL
Loss of bone mass as osteoclasts reabsorb bone due to lack of muscle activity.

INTEGUMENTARY
Decubitus (dē-kū′bi-tŭs) ulcers (bedsores) result from immobility due to loss of motor function; compression of blood vessels in areas of the skin leads to death of the skin in areas supplied by those vessels.

MUSCULAR
Atrophy of muscles due to decreased muscle function.

REPRODUCTIVE
There may be loss of libido (sex drive); innervation of the reproductive organs is often affected.

ENDOCRINE
Strokes in other parts of the brain could involve the functions of the hypothalamus, pineal gland, or pituitary gland.

Stroke

Symptoms
- Severe headache
- Loss of motor skills
- Weakness, numbness of face and limbs
- Difficulty with speech and swallowing
- Vision changes
- Confusion, inattentiveness, drowsiness, or unconsciousness
- Seizures

Treatment
- Varies depending on the type and severity of stroke
- Restore blood flow or stop blood loss
- Long-term treatment may include hypertension medication, physical and speech therapy

URINARY
Motor innervation of the bladder is often affected. Urinary tract infection may result from catheter implantation or from urine retention.

CARDIOVASCULAR
Risks: Phlebothrombosis (fleb′-ō-throm-bō′sis; blood clot in a vein) can occur from inactivity. Edema around the brain can increase the pressure inside the skull, activating reflexes that increase the blood pressure. If the cardioregulatory center in the brain is damaged, death may occur rapidly because of dramatically decreased blood pressure. Bleeding may result from the use of anticoagulants, and hypotension may result from the use of antihypertensives.

DIGESTIVE
Vomiting or dysphagia (dis-fā′jē-ă; difficulty swallowing) may occur if the brain centers controlling these functions are damaged. Hypovolemia (decreased blood volume) results from decreased fluid intake because of dysphagia; bowel control may be lost.

RESPIRATORY
Pneumonia may result from aspiration (fluid from the digestive tract entering the lungs when the patient vomits). Breathing may slow or stop completely if the respiratory center is damaged.

Damage to cranial nerve nuclei causes the loss of pain and temperature sensation in the face, dizziness, blurred vision, nystagmus, vomiting, and hoarseness. These signs and symptoms are not observed unless the lesion is in the brainstem, where these nuclei are located. Some damage to the cerebellum, also supplied by branches of the vertebral artery, can account for the ataxia.

General responses to neurological damage include drowsiness, disorientation, inattentiveness, loss of consciousness, and even seizures. Depression, due to either neurological damage or discouragement, is also common. Slight dilation of the pupils, short and shallow respiration, and increased pulse rate and blood pressure are all signs of Scott's anxiety about his current condition and his immedi-

ate future. Because he lost consciousness, Scott would not remember the last few minutes of what he was watching on television when he had his stroke. People in these circumstances are often worried about how they are going to deal with work tomorrow. They often have no idea that the motor and sensory losses may be permanent or that a long stretch of therapy and rehabilitation lies ahead.

Predict 8

Given that Scott exhibited weakness in his right limbs and loss of pain and temperature sensation in his right lower limb and the left side of his face, which side of the brainstem was most severely affected by the stroke? Explain your answer.

DISEASES AND DISORDERS: Nervous System

CONDITION	DESCRIPTION
	CENTRAL NERVOUS SYSTEM DISORDERS
Infections of the CNS	
Encephalitis	Inflammation of the brain caused by a virus and less often by bacteria or other agents; symptoms include fever, coma, and convulsions
Meningitis	Inflammation of meninges caused by viral or bacterial infection; symptoms include stiffness in the neck, headache, and fever; severe cases can cause paralysis, coma, or death
Rabies	Viral disease transmitted by infected animal; brain infection results in abnormal excitability, aggression, paralysis, and death
Tetanus	Caused by bacterial neurotoxin; affects lower motor neurons in spinal cord and brainstem, leading to muscle contraction; prevents muscle relaxation; body becomes rigid, including "lockjaw"; death results from spasms in respiratory muscles
Multiple sclerosis	Autoimmune condition; may be initiated by viral infection; inflammation in brain and spinal cord with demyelination and sclerotic (hard) sheaths results in poor conduction of action potentials; symptoms include exaggerated reflexes, tremor, and speech defects
CNS Movement Disorders	
Dyskinesias	Group of disorders, involving basal nuclei, that result in a resting tremor and brisk, jerky movements
Huntington disease	Dominant hereditary disorder; causes progressive degeneration of basal nuclei; characterized by involuntary movements
Cerebral palsy	General term for defects in motor functions or coordination due to brain damage caused by abnormal development or injury; symptoms include increased muscle tone, resting tremors, difficulty speaking and swallowing, and slow, writhing, aimless movements
Parkinson disease	Caused by a lesion in basal nuclei; characterized by muscular rigidity, resting tremor, general lack of movement, and a slow, shuffling gait
Other CNS Disorders	
Stroke	Caused by bleeding in the brain or a clot or spasm blocking cerebral blood vessels that results in a local area of cell death; symptoms include loss of speech, numbness, or paralysis
Alzheimer disease	Mental deterioration, or dementia; usually affects older people; involves loss of neurons in cerebral cortex; symptoms include general intellectual deficiency, memory loss, short attention span, moodiness, disorientation, and irritability
Tay-Sachs disease	Hereditary lipid-storage disorder of infants; primarily affects CNS; symptoms include paralysis, blindness, and death
Epilepsy	Seizures involving a sudden massive neuronal discharge, which may result in involuntary muscle contractions (convulsions)
Headaches	May be due to inflammation, dental irritations, eye disorders, tension in head and neck muscles, or unknown causes
	PERIPHERAL NERVOUS SYSTEM DISORDERS
General PNS Disorders	
Anesthesia	Loss of sensation; may be a pathological condition or may be induced temporarily to facilitate medical action
Neuritis	Inflammation of a nerve from a number of causes; in motor nerves, can result in loss of motor function; in sensory nerves, can result in anesthesia or neuralgia
Neuralgia	Nerve pain; involves severe spasms of throbbing or stabbing pain along the pathway of a nerve; can result from inflammation, nerve damage, or unknown cause
Infections of the PNS	
Leprosy	Bacterial disease that kills skin and PNS cells; characterized by disfiguring nodules and tissue necrosis
Herpes	Family of diseases characterized by skin lesions due to herpes viruses in sensory ganglia; different viruses cause oral lesions (cold sores), sexually transmitted disease with lesions on genitalia, or chickenpox in children (shingles in adults)
Poliomyelitis	Viral infection of the CNS; damages somatic motor neurons, leaving muscles without innervation, and leads to paralysis and atrophy
Other PNS Disorders	
Myotonic dystrophy	Dominant hereditary disease characterized by muscle weakness, dysfunction, and atrophy
Neurofibromatosis	Genetic disorder; neurofibromas (benign tumors along peripheral nerve tract) occur in early childhood and result in skin growths
Myasthenia gravis	Autoimmune disorder affecting acetylcholine receptors; makes the neuromuscular junction less functional; muscle weakness and increased fatigue lead to paralysis

general decline in muscle mass. Loss of motor units leads to more rapid fatigue as the remaining units must perform compensatory work. However, regular physical exercise can forestall and reduce these declines.

Reflexes slow as people age because both the generation and the conduction of action potentials and synaptic functions slow. The number of neurotransmitters and receptors declines. As reflexes slow, older people are less able to react automatically, quickly, and accurately to changes in internal and external conditions.

The size and weight of the brain decrease as a person ages. At least part of these changes results from the loss of neurons within the cerebrum. The remaining neurons can apparently compensate for much of this loss. Structural changes also occur in neurons. Neuron cell membranes become more rigid, the endoplasmic reticulum becomes more irregular in structure, neurofibrillar tangles develop in the cells, and amyloid plaques form in synapses. All these changes decrease the ability of neurons to function.

Short-term memory is decreased in most older people. This change varies greatly among individuals, but in general, such changes are slow until about age 60 and then become more rapid, especially after age 70. However, the total amount of memory loss is normally not great for most people. Older people have the most difficulty assimilating information that is unfamiliar and presented verbally and rapidly. Long-term memory appears to be unaffected or even improved in older people.

As with short-term memory, thinking, which includes problem solving, planning, and intelligence, generally declines slowly to age 60 but more rapidly thereafter. However, these changes are slight and quite variable. Many older people show no change, and about 10% show an increase in thinking ability. Many of these changes are affected by a person's background, education, health, motivation, and experience. It appears that continued mental activity can decrease the loss of mental skills with aging.

Older people tend to require more time to fall asleep and experience more periods of waking during the night, which are also of greater duration. Factors that can affect sleep include pain, indigestion, rhythmic leg movements, sleep apnea, decreased urinary bladder capacity, and poor peripheral circulation.

ANSWER TO LEARN TO PREDICT

James reacted to the pain of stepping on the toy cars not once, but twice. We know that each time he stepped on a toy, he switched the leg that supported his weight, first from the right to the left and then from the left to the right. Reading this chapter, we learned that the withdrawal reflex moves a limb away from a painful stimulus. James demonstrated this reflex when he stepped on the toys.

The withdrawal reflex, displayed by James, involves sensory neurons, interneurons, and motor neurons. The sensory neurons delivered the painful stimulus to the central nervous system; the interneurons relayed the information between the sensory and motor neurons; and the motor neurons regulated the skeletal muscle activity. When James stepped on the toy car with his right foot, he had transferred nearly all of his weight to his right leg before the withdrawal reflex was activated. The withdrawal reflex caused James to pick up his right leg and extend his left leg to support his weight. However, as James extended his left leg, he stepped on another toy car, activating the withdrawal reflex in his left leg. The repeated sequence of withdrawing one leg and stepping down on the other foot prevented James from falling down.

Answers to the rest of this chapter's Predict questions are in Appendix E.

SUMMARY

8.1 Functions of the Nervous System (p. 193)

The functions of the nervous system include receiving sensory input, integrating information, controlling muscles and glands, maintaining homeostasis, and serving as the center of mental activity.

8.2 Divisions of the Nervous System (p. 194)

1. The central nervous system (CNS) consists of the brain and spinal cord, whereas the peripheral nervous system (PNS) consists of all the nervous tissue outside the CNS.
2. The sensory division of the PNS transmits action potentials to the CNS; the motor division carries action potentials away from the CNS.
3. The motor division is divided into somatic and autonomic systems. The somatic nervous system innervates skeletal muscle and is mostly under voluntary control. The autonomic nervous system innervates cardiac muscle, smooth muscle, and glands and is mostly under involuntary control.

4. The autonomic nervous system is divided into sympathetic and parasympathetic divisions.
5. The enteric nervous system contains both sensory and motor neurons, which can function independently of the CNS.

8.3 Cells of the Nervous System (p. 194)

Neurons

1. Neurons receive stimuli and conduct action potentials. A neuron consists of a cell body, dendrites, and an axon.
2. Neurons are multipolar, bipolar, or pseudo-unipolar.

Glial Cells

Glial cells are the support cells of the nervous system. They include astrocytes, ependymal cells, microglia, oligodendrocytes, and Schwann cells.

Myelin Sheaths

Axons are either unmyelinated or myelinated.

Nervous

Organization of Nervous Tissue

Nervous tissue consists of white matter and gray matter. Gray matter forms the cortex and nuclei in the brain and ganglia in the PNS. White matter forms nerve tracts in the CNS and nerves in the PNS.

8.4 Electrical Signals and Neural Pathways   (p. 196)

Resting Membrane Potential

1. A resting membrane potential results from the charge difference across the membrane of cells.
2. The resting membrane potential is set by leak ion channels and the sodium-potassium pump.

Action Potentials

1. An action potential occurs when the charge across the cell membrane is briefly reversed.
2. Chemically gated and voltage-gated ion channels generate the action potential.

The Synapse

1. A synapse is a point of contact between two neurons or between a neuron and another cell, such as a muscle or gland cell.
2. An action potential arriving at the synapse causes the release of a neurotransmitter from the presynaptic terminal, which diffuses across the synaptic cleft and binds to the receptors of the postsynaptic membrane.

Reflexes

1. Reflex arcs are the functional units of the nervous system.
2. A complex reflex arc consists of a sensory receptor, a sensory neuron, interneurons, a motor neuron, and an effector organ; the simplest reflex arcs do not involve interneurons.

Neuronal Pathways

1. Neuronal pathways are either converging or diverging.
2. Spatial and temporal summation occur in neuronal pathways.

8.5 Central and Peripheral Nervous Systems (p. 206)

The CNS consists of the brain and spinal cord. The PNS consists of nerves and ganglia outside the CNS.

8.6 Spinal Cord   (p. 206)

1. The spinal cord extends from the foramen magnum to the second lumbar vertebra; below that is the cauda equina.
2. The spinal cord has a central gray part organized into horns and a peripheral white part forming nerve tracts.
3. Roots of spinal nerves extend out of the cord.

Spinal Cord Reflexes

1. The knee-jerk reflex occurs when the quadriceps femoris muscle is stretched.
2. The withdrawal reflex removes a body part from a painful stimulus.

8.7 Spinal Nerves   (p. 208)

1. The spinal nerves exit the vertebral column at the cervical, thoracic, lumbar, and sacral regions.
2. The nerves are grouped into plexuses.
3. The phrenic nerve, which supplies the diaphragm, is the most important branch of the cervical plexus.
4. The brachial plexus supplies nerves to the upper limb.
5. The lumbosacral plexus supplies nerves to the lower limb.

8.8 Brain   (p. 210)

Brainstem

1. The brainstem contains several nuclei.
2. The medulla oblongata contains nuclei that control activities such as heart rate, breathing, swallowing, and balance.
3. The pons contains relay nuclei between the cerebrum and cerebellum.
4. The midbrain is involved in hearing and in visual reflexes.
5. The reticular formation is scattered throughout the brainstem and is important in regulating cyclical motor functions. It is also involved in maintaining consciousness and in the sleep-wake cycle.

Cerebellum

The cerebellum is attached to the brainstem.

Diencephalon

The diencephalon consists of the thalamus (main sensory relay center), the epithalamus (the pineal gland may play a role in sexual maturation), and the hypothalamus (important in maintaining homeostasis).

Cerebrum

The cerebrum has two hemispheres, each divided into lobes: the frontal, parietal, occipital, and temporal lobes.

8.9 Sensory Functions   (p. 214)

1. The CNS constantly receives sensory input.
2. We are unaware of much of the input, but it is vital to our survival.
3. Some sensory input results in sensation.

Ascending Tracts

1. Ascending tracts transmit action potentials from the periphery to the brain.
2. Each tract carries a specific type of sensory information.

Sensory Areas of the Cerebral Cortex

1. Ascending tracts project to primary sensory areas of the cerebral cortex.
2. Association areas are involved in recognizing the sensory input.

8.10 Motor Functions   (p. 217)

1. Motor functions include involuntary and voluntary movements.
2. Upper motor neurons in the cerebral cortex connect to lower motor neurons in the spinal cord or cranial nerve nuclei.

Motor Areas of the Cerebral Cortex

1. Upper motor neurons are located in the primary motor cortex.
2. The premotor and prefrontal areas regulate movements.

Descending Tracts

Descending tracts project directly from upper motor neurons in the cerebral cortex to lower motor neurons in the spinal cord and brainstem; indirectly, they project from basal nuclei, the cerebellum, or the cerebral cortex through the brainstem to lower motor neurons in the spinal cord.

Basal Nuclei

1. Basal nuclei help plan, organize, and coordinate motor movements and posture.
2. People with basal nuclei disorders exhibit increased muscle tone and exaggerated, uncontrolled movements when at rest.

Cerebellum

1. The cerebellum is involved in balance, muscle tone, and muscle coordination.

2. Through its comparator function, the cerebellum compares the intended action to what is occurring and modifies the action to eliminate differences.

3. If the cerebellum is damaged, muscle tone decreases and fine motor movements become very clumsy.

8.11 Other Brain Functions   (p. 219)

Communication Between the Right and Left Hemispheres

1. Each hemisphere controls the opposite half of the body.
2. Commissures connect the two hemispheres.
3. The left hemisphere is thought to be the dominant analytical hemisphere, and the right hemisphere is thought to be dominant for spatial perception and musical ability.

Speech

Speech involves the sensory speech area, the motor speech area, and the interactions between them and other cortical areas.

Brain Waves and Consciousness

An EEG monitors brain waves, which are a summation of the electrical activity of the brain.

Memory

1. The types of memory are working (lasting a few seconds to minutes), short-term (lasting a few minutes), and long-term (permanent) memory.
2. Long-term memory includes declarative and procedural memories.

Limbic System and Emotions

1. The limbic system includes the olfactory cortex, deep cortical regions, and nuclei.
2. The limbic system is involved with memory, motivation, mood, and other visceral functions. Olfactory stimulation is a major influence.

8.12 Meninges, Ventricles, and Cerebrospinal Fluid   (p. 222)

Meninges

Three connective tissue meninges cover the CNS: the dura mater, arachnoid mater, and pia mater.

Ventricles

1. The brain and spinal cord contain fluid-filled cavities: the lateral ventricles in the cerebral hemispheres, a third ventricle in the diencephalon, a cerebral aqueduct in the midbrain, a fourth ventricle at the base of the cerebellum, and a central canal in the spinal cord.
2. The fourth ventricle has openings into the subarachnoid space.

Cerebrospinal Fluid

Cerebrospinal fluid is formed in the choroid plexuses in the ventricles. It exits through the fourth ventricle and reenters the blood through arachnoid granulations in the superior sagittal sinus.

8.13 Cranial Nerves   (p. 223)

1. There are 12 pairs of cranial nerves: 3 with only sensory function (S), 4 with only somatic motor function (M), 1 with somatic motor (M) and sensory function (S), 1 with somatic motor and parasympathetic (P) function, and 3 with all three functions. Four of the cranial nerves have parasympathetic function.
2. The cranial nerves are olfactory (I; S), optic (II; S), oculomotor (III; M, P), trochlear (IV; M), trigeminal (V; S, M), abducens (VI; M), facial (VII; S, M, P), vestibulocochlear (VIII; S), glossopharyngeal (IX; S, M, P), vagus (X; S, M, P), accessory (XI; M), and hypoglossal (XII; M).

8.14 Autonomic Nervous System   (p. 225)

1. The autonomic nervous system contains preganglionic and postganglionic neurons.
2. The autonomic nervous system has sympathetic and parasympathetic divisions.

Anatomy of the Sympathetic Division

1. Preganglionic cell bodies of the sympathetic division lie in the thoracic and upper lumbar regions of the spinal cord.
2. Postganglionic cell bodies are located in the sympathetic chain ganglia or in collateral ganglia.

Anatomy of the Parasympathetic Division

1. Preganglionic cell bodies of the parasympathetic division are associated with some of the cranial and sacral nerves.
2. Postganglionic cell bodies are located in terminal ganglia, either near or within target organs.

Autonomic Neurotransmitters

1. All autonomic preganglionic and parasympathetic postganglionic neurons secrete acetylcholine.
2. Most sympathetic postganglionic neurons secrete norepinephrine.

Functions of the Autonomic Nervous System

1. The sympathetic division prepares a person for action by increasing heart rate, blood pressure, respiration, and release of glucose for energy.
2. The parasympathetic division is involved in involuntary activities at rest, such as the digestion of food, defecation, and urination.

8.15 Enteric Nervous System   (p. 231)

1. The enteric nervous system forms plexuses in the digestive tract wall.
2. Enteric neurons are sensory, motor, or interneurons; they receive CNS input but can also function independently.

8.16 Effects of Aging on the Nervous System (p. 231)

1. In general, sensory and motor functions decline with age.
2. Mental functions, including memory, may decline with age, but this varies from person to person.

 ## REVIEW AND COMPREHENSION

1. List the functions of the nervous system.
2. Distinguish between the CNS and PNS.
3. Define the sensory and motor divisions of the PNS and the somatic and autonomic nervous systems.
4. What are the functions of neurons? Name the three parts of a neuron.
5. List the three types of neurons based on the organization of their processes.

6. Define glial cell. Name and describe the functions of the different glial cells.
7. What are the differences between unmyelinated and myelinated axons? Which conduct action potentials more rapidly? Why?
8. For nuclei, ganglia, nerve tracts, and nerves, name the cells or parts of cells found in each, state if they are white or gray matter, and name the part of the nervous system (CNS or PNS) in which they are found.

9. Explain the resting membrane potential and the roles of the K⁺ leak channels and the sodium-potassium pump.

10. List the series of events at the membrane that generate an action potential, including the contributions of chemically gated and voltage-gated ion channels.

11. Describe the sequence of events at a synapse, starting with an action potential in the presynaptic neuron and ending with the generation of an action potential in the postsynaptic neuron.

12. Describe the primary function of neurotransmitters at the synapse.

13. Define a reflex. Name the five components of a reflex arc, and explain its operation.

14. Compare and contrast the two major types of neuronal pathways. Give an example of each.

15. Describe the importance of spatial and temporal summation.

16. Describe the spinal cord gray matter. Where are sensory and motor neurons located in the gray matter?

17. Differentiate between the ventral root and the dorsal root in relation to a spinal nerve. Which contains sensory fibers, and which contains motor fibers?

18. Describe the knee-jerk and withdrawal reflexes.

19. List the spinal nerves by name and number.

20. Name the three main plexuses and the major nerves derived from each.

21. Name the four parts of the brainstem, and describe the general functions of each.

22. Name the three main components of the diencephalon, describing their functions.

23. Name the four lobes of the cerebrum, and describe the location and function of each.

24. List the ascending tracts, and state their functions.

25. Describe the locations in the cerebral cortex of the primary sensory areas and their association areas. How do the association areas interact with the primary areas?

26. Distinguish between upper and lower motor neurons.

27. Distinguish among the functions of the primary motor cortex, premotor area, and prefrontal area.

28. List the descending tracts, and state their functions.

29. Describe the functions of the basal nuclei.

30. Describe the comparator activities of the cerebellum.

31. What are the differences in function between the right and left cerebral hemispheres?

32. Describe the process required to speak a word that is seen compared to one that is heard.

33. Name the three types of memory, and describe the processes that result in long-term memory.

34. What is the function of the limbic system?

35. Name and describe the three meninges that surround the CNS.

36. Describe the production and circulation of the cerebrospinal fluid. Through what structures does the cerebrospinal fluid return to the blood?

37. What are the three principal functional categories of the cranial nerves? List a specific function for each cranial nerve.

38. Define preganglionic neuron and postganglionic neuron.

39. Compare the structure of the somatic nervous system and the autonomic nervous system in terms of the number of neurons between the CNS and the effector organs and the types of effector organs.

40. Contrast the functions of the sympathetic and parasympathetic divisions of the autonomic nervous system.

41. What kinds of neurons (sympathetic or parasympathetic, preganglionic or postganglionic) are found in the following?
    a. cranial nerve nuclei
    b. lateral horn of the thoracic spinal cord gray matter
    c. lateral portion of the sacral spinal cord gray matter
    d. chain ganglia
    e. ganglia in the wall of an organ

42. List the three parts of the enteric nervous system.

 CRITICAL THINKING

1. Given two series of neurons, explain why action potentials could be propagated along one series more rapidly than along the other series.

2. The left lung of a cancer patient was removed. To reduce the empty space left in the thorax after the lung was removed, the diaphragm on the left side was paralyzed so that the abdominal viscera would push the diaphragm upward into the space. What nerve should be cut to paralyze the left half of the diaphragm?

3. Name the nerve that, if damaged, produces the following symptoms:
   a. The elbow and wrist on one side are held in a flexed position and cannot be extended.
   b. The patient is unable to flex the right hip and extend the knee (as in kicking a ball).

4. A patient suffered brain damage in an automobile accident. Physicians suspected that the cerebellum was affected. On the basis of what you know about cerebellar function, how could you determine that the cerebellum was involved? What symptoms would you expect to see?

5. Louis Ville was accidentally struck in the head with a baseball bat. He fell to the ground, unconscious. Later, when he had regained consciousness, he was unable to remember any of the events that happened during the 10 minutes before the accident. Explain.

6. Name the cranial nerve that, if damaged, produces the following symptoms:
   a. The patient is unable to move the tongue.
   b. The patient is unable to see out of one eye.
   c. The patient is unable to feel one side of the face.
   d. The patient is unable to move the facial muscles on one side.
   e. The pupil of one eye is dilated and does not constrict.

7. Why doesn't injury to the spinal cord at the level of C6 significantly interfere with nervous system control of the digestive system?

Answers in Appendix D

CHAPTER

9 Senses

LEARN TO PREDICT

Freddy is an older man but he has never needed glasses. He has several family members that are near-sighted, meaning they have problems seeing things at a distance, and require corrective lenses. Freddy, on the other hand, has had 20/20 vision his whole life. Lately, though, he has noticed that he can't see quite so well when he is reading. He jokes with his friends that his "arms seem to be getting shorter."

After reading about the process of vision, explain what type of vision problem Freddy is experiencing and why his joke about his arms getting shorter relates to his visual problem.

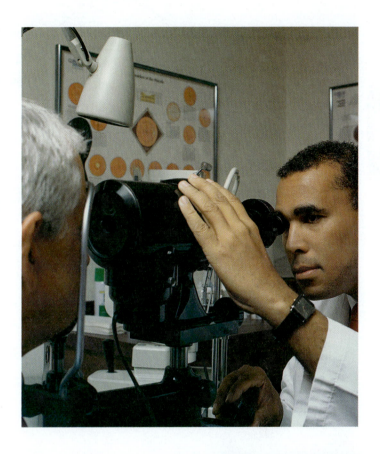

Module 7 Nervous System · Anatomy & Physiology **REVEALED**® aprevealed.com

9.1 SENSATION

Learning Outcomes *After reading this section, you should be able to*

A. Define sensation.

B. Distinguish between general senses and special senses.

Sense is the ability to perceive stimuli. The senses are the means by which the brain receives information about the environment and the body. **Sensation** is the process initiated by stimulating sensory receptors and **perception** is the conscious awareness of those stimuli. The brain constantly receives a wide variety of stimuli from both inside and outside the body, but stimulation of sensory receptors does not immediately result in perception. Sensory receptors respond to stimuli by generating action potentials that are propagated to the spinal cord and brain. Perception results when action potentials reach the cerebral cortex. Some other parts of the brain are also involved in perception. For example, the thalamus plays a role in the perception of pain.

Historically, five senses were recognized: smell, taste, sight, hearing, and touch. Today we recognize many more senses and divide them into two basic groups: general and special senses (figure 9.1). The **general senses** have receptors distributed over a large part of the body. They are divided into two groups: the somatic senses and the visceral senses. The **somatic senses** provide sensory information about the body and the environment. The **visceral senses** provide information about various internal organs, primarily involving pain and pressure.

Special senses are more specialized in structure and are localized to specific parts of the body. The special senses are smell, taste, sight, hearing, and balance.

9.2 SENSORY RECEPTORS

Learning Outcome *After reading this section, you should be able to*

A. List and describe five types of sensory receptors.

Sensory receptors are sensory nerve endings or specialized cells capable of responding to stimuli by developing action potentials. Several types of receptors are associated with both the general and the special senses, and each responds to a different type of stimulus:

Mechanoreceptors (mek′ă-nō-rē-sep′tŏrz) respond to mechanical stimuli, such as the bending or stretching of receptors.

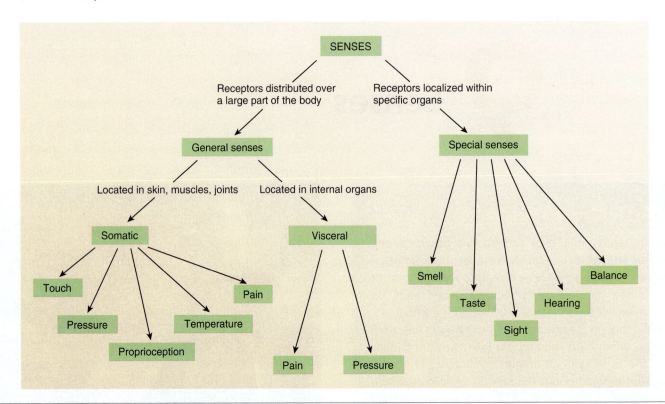

Figure 9.1 Classification of the Senses

Chemoreceptors (kem′ō-rē-sep′tŏrz) respond to chemicals, such as odor molecules.
Photoreceptors (fō′tō-rē-sep′tŏrz) respond to light.
Thermoreceptors (ther′mō-rē-sep′tŏrz) respond to temperature changes.
Nociceptors (nō′si-sep′tŏrs; *noceo,* to injure) respond to stimuli that result in the sensation of pain.

9.3 GENERAL SENSES

Learning Outcomes   *After reading this section, you should be able to*

A. List the general senses and the receptor type associated with each.
B. Define and describe pain and referred pain.

The general senses have sensory receptors that are widely distributed throughout the body. The general senses include the senses of touch, pressure, pain, temperature, vibration, itch, and proprioception (prō-prē-ō-sep′shun), which is the sense of movement and position of the body and limbs.

Many of the receptors for the general senses are associated with the skin (figure 9.2); others are associated with deeper structures, such as tendons, ligaments, and muscles. Structurally, the simplest and most common receptors are **free nerve endings,** which are relatively unspecialized neuronal branches similar to dendrites. Free nerve endings are distributed throughout almost all parts of the body. Some free nerve endings respond to painful stimuli, some to temperature, some to itch, and some to movement. Receptors for temperature are either **cold receptors** or **warm receptors.** Cold

receptors respond to decreasing temperatures but stop responding at temperatures below 12°C (54°F). Warm receptors respond to increasing temperatures but stop responding at temperatures above 47°C (117°F). It is sometimes difficult to distinguish very cold from very warm objects touching the skin because only pain receptors are stimulated at temperatures below 12°C or above 47°C.

Touch receptors are structurally more complex than free nerve endings, and many are enclosed by capsules. There are several types of touch receptors (figure 9.2). **Merkel disks** are small, superficial nerve endings involved in detecting light touch and superficial pressure. **Hair follicle receptors,** associated with hairs, are also involved in detecting light touch. Light touch receptors are very sensitive but not very discriminative, meaning that the point being touched cannot be precisely located. Receptors for fine, discriminative touch, called **Meissner corpuscles,** are located just deep to the epidermis. These receptors are very specific in localizing tactile sensations. Deeper tactile receptors, called **Ruffini corpuscles,** play an important role in detecting continuous pressure in the skin. The deepest receptors are associated with tendons and joints and are called **pacinian corpuscles.** These receptors relay information concerning deep pressure, vibration, and position (proprioception).

Pain

Pain is characterized by a group of unpleasant perceptual and emotional experiences. There are two types of pain sensation: (1) localized, sharp, pricking, or cutting pain resulting from rapidly conducted action potentials, and (2) diffuse, burning, or aching pain resulting from action potentials that are propagated more slowly.

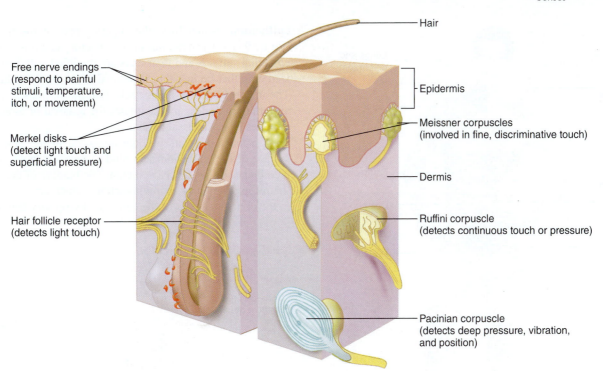

Figure 9.2 Sensory Receptors in the Skin

Superficial pain sensations in the skin are highly localized as a result of the simultaneous stimulation of pain receptors and tactile receptors. Deep or visceral pain sensations are not highly localized because of the absence of tactile receptors in the deeper structures. Visceral pain stimuli are normally perceived as diffuse pain.

Action potentials from pain receptors in local areas of the body can be suppressed by **local anesthesia,** a treatment where chemical anesthetics are injected near a sensory receptor or nerve, resulting in reduced pain sensation. Pain sensations can also be suppressed if loss of consciousness is produced. This is usually accomplished by **general anesthesia,** a treatment where chemical anesthetics that affect the reticular formation are administered.

Pain sensations can also be influenced by inherent control systems. Sensory axons from tactile receptors in the skin have collateral branches that synapse with neurons in the posterior horn of the spinal cord. Those neurons, in turn, synapse with and inhibit neurons that give rise to the spinothalamic tract, a sensory pathway that relays pain sensations to the brain (see table 8.4). For example, rubbing the skin in the area of an injury stimulates the tactile receptors, which send action potentials along the sensory axons to the spinal cord. According to the **gate control theory,** these action potentials "close the gate" and inhibit action potentials carried to the brain by the spinothalamic tract.

The gate control theory may explain the physiological basis for several techniques that have been used to reduce the intensity of pain. Action potentials carried by the spinothalamic tract can be inhibited by action potentials carried by descending neurons of the dorsal column system (see chapter 8). These neurons are stimulated by mental or physical activity, especially involving movement of the limbs. The descending neurons synapse with and inhibit neurons in the posterior horn that give rise to the spinothalamic tract. Vigorous mental or physical activity increases the rate of action potentials in neurons of the dorsal column and can reduce the sensation of pain. Exercise programs are important components in the clinical management of chronic pain. Acupuncture and acupressure procedures may also decrease the sensation of pain by stimulating descending dorsal column neurons, which inhibit action potentials in the spinothalamic tract neurons. The gate control theory also explains why the intensity of pain is decreased by diverting a person's attention.

Referred Pain

Referred pain is perceived to originate in a region of the body that is not the source of the pain stimulus. Most commonly, we sense referred pain when deeper structures, such as internal organs, are damaged or inflamed (figure 9.3). This occurs because sensory neurons from the superficial area to which the pain is referred and the neurons from the deeper, visceral area where the pain stimulation originates converge onto the same ascending neurons in the spinal cord. The brain cannot distinguish between the two sources of pain stimuli, and the painful sensation is referred to the most superficial structures innervated, such as the skin.

Referred pain is clinically useful in diagnosing the actual cause of the painful stimulus. For example, during a heart attack, pain receptors in the heart are stimulated when blood flow is blocked to some of the heart muscle. Heart attack victims, however, often may not feel the pain in the heart but instead perceive cutaneous pain radiating from the left shoulder down the arm (figure 9.3).

Predict 2

A man has constipation that causes distention and painful cramping in the colon (part of the large intestine). What kind of pain does he experience (localized or diffuse), and where does he perceive it? Explain.

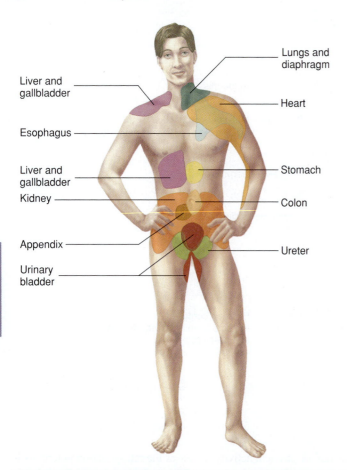

Figure 9.3 Areas of Referred Pain on the Body Surface
Pain from the indicated internal organs is referred to the surface areas shown.

9.4 SPECIAL SENSES

Learning Outcome *After reading this section, you should be able to*

A. List the special senses.

The senses of smell, taste, sight, hearing, and balance are associated with very specialized, localized sensory receptors. The sensations of smell and taste are closely related, both structurally and functionally, and both are initiated by the interaction of chemicals with sensory receptors. The sense of vision is initiated by the interaction of light with sensory receptors. Both hearing and balance function in response to the interaction of mechanical stimuli with sensory receptors. Hearing occurs in response to sound waves, and balance occurs in response to gravity or motion.

9.5 OLFACTION

Learning Outcome *After reading this section, you should be able to*

A. Describe olfactory neurons, and explain how airborne molecules can stimulate action potentials in the olfactory nerves.

The sense of smell, called **olfaction** (ol-fak´shŭn), occurs in response to airborne molecules, called **odorants,** that enter the nasal cavity. **Olfactory neurons** are bipolar neurons within the **olfactory**

epithelium, which lines the superior part of the nasal cavity (figure 9.4*a*). The dendrites of the olfactory neurons extend to the epithelial surface, and their ends are modified with long, specialized cilia that lie in a thin mucous film on the epithelial surface. The mucus keeps the nasal epithelium moist, traps and dissolves airborne molecules, and facilitates the removal of molecules and particles from the nasal epithelium.

Airborne odorants become dissolved in the mucus on the surface of the epithelium and bind to receptor molecules on the membranes of the specialized cilia. The binding of the odorant to the receptor initiates action potentials, which are then conducted to the olfactory cortex of the cerebrum by sensory neurons. There

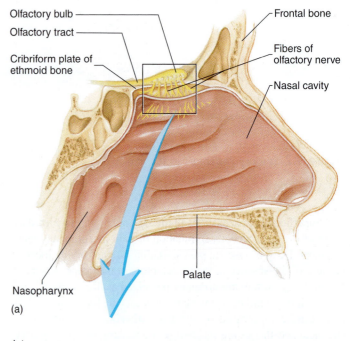

(a)

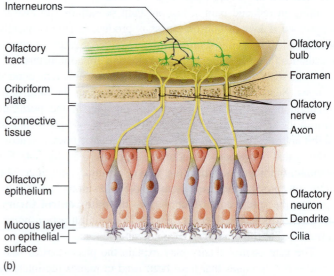

(b)

Figure 9.4 **AP|R** **Olfactory Epithelium and Olfactory Bulb**

(*a*) A sagittal section through the lateral wall of the nasal cavity shows the olfactory nerves, olfactory bulb, and olfactory tract. (*b*) The olfactory neurons lie within the olfactory epithelium. The axons of olfactory neurons pass through the cribriform plate to the olfactory bulb.

are at least 400 functional **olfactory receptors** in humans. Unlike most other receptors in the body, each olfactory receptor can bind multiple types of odorants; conversely, each type of odorant can bind to multiple olfactory receptors. These multiple combinations of odorants and receptors allow us to detect an estimated 10,000 different smells. Once an odorant has bound to its receptor, that receptor is desensitized and does not respond to another odor molecule for some time, which helps with adaptation (described next) to a particular odor. The threshold for the detection of odors is extremely low, so very few odorants bound to an olfactory neuron can initiate an action potential. The olfactory range and sensitivity is even greater in some animals than in humans, due to a larger number and more types of olfactory receptors. For example, dogs are often used to detect small traces of explosives and other chemicals that people cannot detect.

Neuronal Pathways for Olfaction

Axons from olfactory neurons form the olfactory nerves (cranial nerve I), which pass through foramina of the cribriform plate and enter the **olfactory bulb** (figure 9.4*b*). There they synapse with interneurons that relay action potentials to the brain through the **olfactory tracts.** Each olfactory tract terminates in an area of the brain called the **olfactory cortex,** located within the temporal and frontal lobes. Olfaction is the only major sensation that is relayed directly to the cerebral cortex without first passing through the thalamus. This is a reflection of the older, more primitive origin of the olfactory cortex. The olfactory cortex is involved with both the conscious perception of smell and the visceral and emotional reactions that are often linked to odors.

Within the olfactory bulb and olfactory cortex are feedback loops that tend to inhibit transmission of action potentials resulting from prolonged exposure to a given odorant. This feedback, plus the temporary decreased sensitivity at the level of the receptors, results in **adaptation** to a given odor. For example, if you enter a room that has an odor, such as a movie theater that smells like popcorn, you adapt to the odor and cannot smell it as well after the first few minutes. If you leave the room for some time and then reenter the room, the odor again seems more intense.

9.6 TASTE

Learning Outcome *After reading this section, you should be able to*

A. Outline the structure and function of a taste bud.

The sensory structures that detect taste stimuli are the **taste buds.** Taste buds are oval structures located on the surface of certain **papillae** (pă-pil′ē; nipples), which are enlargements on the surface of the tongue (figure 9.5). Taste buds are also distributed throughout other areas of the mouth and pharynx, such as on the palate, the root of the tongue, and the epiglottis. Each taste bud consists of two types of cells. Specialized epithelial cells form the exterior supporting capsule of each taste bud, and the interior consists of about 40 **taste cells.** Each taste cell contains hairlike processes, called **taste hairs,** that extend into a tiny opening in the surrounding stratified epithelium, called a **taste pore.** Dissolved molecules or ions bind to receptors on the taste hairs and initiate action potentials, which sensory neurons carry to the insula of the cerebral cortex.

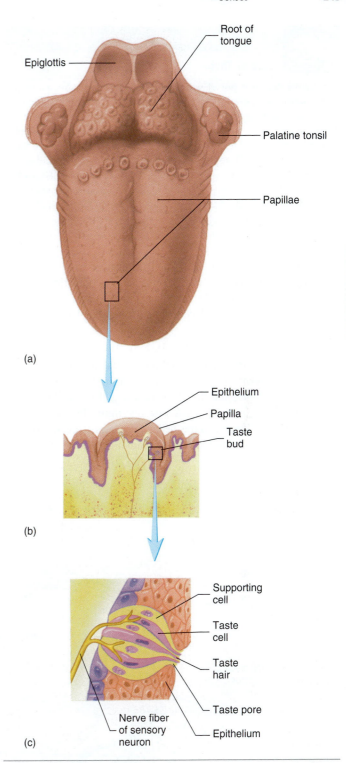

(a)

(b)

(c)

Figure 9.5 AP|R Tongue

(*a*) Dorsal surface of the tongue. (*b*) Section through a papilla showing the location of taste buds. (*c*) Enlarged view of a section through a taste bud.

Taste sensations are divided into five basic types: sour, salty, bitter, sweet, and umami (ū-ma′mē; savory). Although all taste buds are able to detect all five of the basic taste sensations, each taste bud is usually most sensitive to one class of taste stimuli. Presumably, our ability to perceive many different tastes is achieved through various combinations of these five types.

A CASE IN POINT

Loss of Taste

T. Burnes boiled a pot of water to make some tea. After pouring the water over the tea bag in her teacup, she absentmindedly took a large gulp of hot tea, severely burning her tongue. For the next several days, sensation in her tongue, including her sense of touch and sense of taste, was greatly reduced. Heat damage to the tongue epithelial tissue can cause injury or even death to epithelial cells, including taste cells in the taste buds. If the epithelial cells are only damaged, taste sensation returns within a few hours to a few days. If the cells die, it takes about 2 weeks for the epithelial cells to be replaced.

Many taste sensations are strongly influenced by olfactory sensations. This influence can be demonstrated by comparing the taste of some food before and after pinching your nose. It is easy to detect that the sense of taste is reduced while the nose is pinched.

Neuronal Pathways for Taste

Taste sensations are carried by three cranial nerves. The facial nerve (VII) transmits taste sensations from the anterior two-thirds of the tongue, and the glossopharyngeal nerve (IX) carries taste sensations from the posterior one-third. In addition, the vagus nerve (X) carries some taste sensations from the root of the tongue. Axons from these three cranial nerves synapse in the gustatory (taste) portion of brainstem nuclei. Axons of neurons in these brainstem nuclei synapse in the thalamus, and axons from neurons in the thalamus project to the taste area in the insula of the cerebrum (figure 9.6).

Predict 3

Why doesn't food taste as good when you have a cold?

9.7 VISION

Learning Outcomes   *After reading this section, you should be able to*

A. List the accessory structures of the eye, and explain their functions.
B. Name the tunics of the eye, list the parts of each tunic, and describe the functions of each part.
C. Explain the differences in function between rods and cones.
D. Describe the chambers of the eye and the fluids they contain.
E. Explain how images are focused on the retina.

The visual system includes the eyes, the accessory structures, and sensory neurons. The eyes are housed within bony cavities called orbits. Action potentials convey visual information from the eyes to the brain. We obtain much of our information about the world through the visual system. For example, education is largely based on visual input and depends on our ability to read words and numbers. Visual input includes information about light and dark, movement and color.

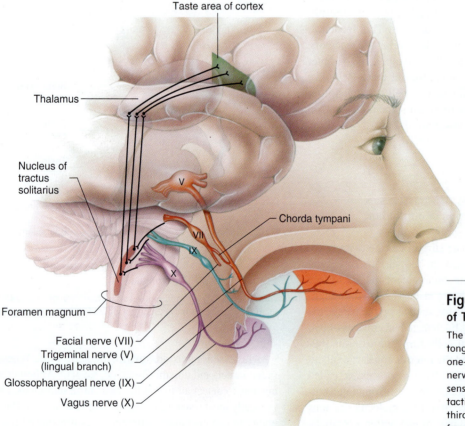

Figure 9.6 Pathways for the Sense of Taste

The facial nerve (anterior two-thirds of the tongue), glossopharyngeal nerve (posterior one-third of the tongue), and vagus nerve (root of the tongue) all carry taste sensations. The trigeminal nerve carries tactile sensations from the anterior two-thirds of the tongue. The chorda tympani from the facial nerve (carrying taste input) joins the trigeminal nerve.

Accessory Structures of the Eye

Accessory structures protect, lubricate, and move the eye. They include the eyebrows, eyelids, conjunctiva, lacrimal apparatus, and extrinsic eye muscles (figures 9.7 and 9.8).

Eyebrows

The **eyebrows** protect the eyes by preventing perspiration from running down the forehead and into the eyes, causing irritation. They also help shade the eyes from direct sunlight (figure 9.7*a*).

Eyelids

The **eyelids,** with their associated lashes, protect the eyes from foreign objects (figure 9.7*a,b*). If an object suddenly approaches the eye, the eyelids protect the eye by closing and then opening quite rapidly (blink reflex). Blinking, which normally occurs about 20 times per minute, also helps keep the eyes lubricated by spreading tears over the surface.

Conjunctiva

The **conjunctiva** (kon-jŭnk-tī′vă) is a thin, transparent mucous membrane covering the inner surface of the eyelids and the anterior surface of the eye (figure 9.7*b*). The secretions of the conjunctiva help lubricate the surface of the eye. Conjunctivitis is an inflammation of the conjunctiva (see the Diseases and Disorders table at the end of this chapter).

Lacrimal Apparatus

The **lacrimal** (lak′ri-măl; tear) **apparatus** consists of a lacrimal gland situated in the superior lateral corner of the orbit and a nasolacrimal duct and associated structures in the inferior medial corner of the orbit (figure 9.7*c*). The **lacrimal gland** produces tears, which pass over the anterior surface of the eye. Most of the fluid produced by the lacrimal glands evaporates from the surface of the eye, but excess tears are collected in the medial angle of the eyes by small ducts called **lacrimal canaliculi** (kan-ă-lik′ū-lī; little canals). These canaliculi open into a **lacrimal sac,** an enlargement of the **nasolacrimal** (nā-zō-lak′ri-măl) **duct,** which opens into the nasal cavity. Tears lubricate and cleanse the eye. They also contain an enzyme that helps combat eye infections.

Predict 4

Explain why it is often possible to smell (or "taste") medications, such as eyedrops, that have been placed into the eyes.

Extrinsic Eye Muscles

Movement of each eyeball is accomplished by six skeletal muscles called the **extrinsic eye muscles** (figure 9.8). Four of these muscles run more or less straight from their origins in the posterior portion of the orbit to their insertion sites on the eye, to attach to the four quadrants of the eyeball. They are the superior, inferior, medial, and lateral **rectus muscles.** Two muscles, the superior and inferior **oblique muscles,** are located at an angle to the long axis of the eyeball.

Anatomy of the Eye

The eyeball is a hollow, fluid-filled sphere. The wall of the eye is composed of three tissue layers, or **tunics** (figure 9.9). The outer, **fibrous tunic** consists of the sclera and cornea. The middle, **vascular tunic** consists of the choroid, ciliary body, and iris. The inner **nervous tunic** consists of the retina.

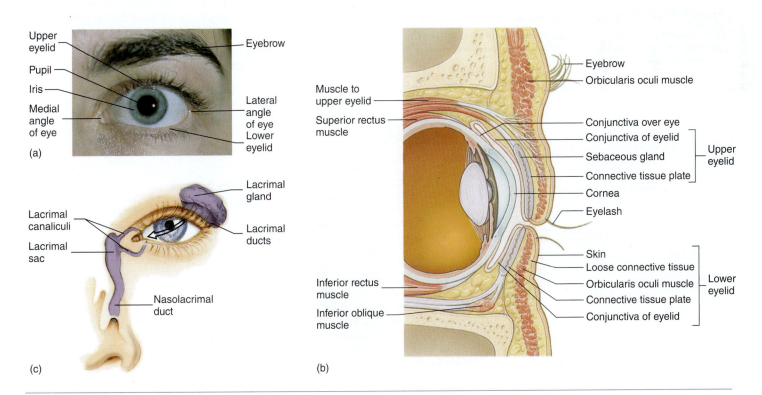

Figure 9.7 AP|R **The Eye and Its Accessory Structures**
(*a*) Photo of left eye. (*b*) Sagittal section through the eye. (*c*) Lacrimal apparatus.

CLINICAL IMPACT   Corneal Transplants

The cornea was one of the first organs to be successfully transplanted. Corneal transplants are often performed when corneal damage due to injury or infection has severely altered the surface of the cornea. Several characteristics make it relatively easy to transplant: It is easily accessible and relatively easily removed from the donor and grafted to the recipient; it does not have blood vessels and therefore does not require the growth of extensive circulation into the tissue after grafting; and it is less likely to stimulate the immune system and therefore less likely to be rejected than are other tissues.

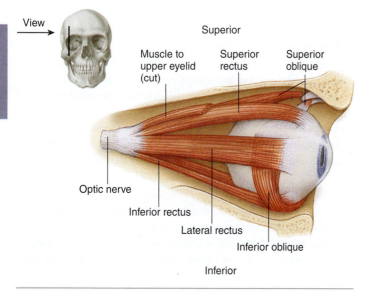

Figure 9.8 [AP|R] **Extrinsic Eye Muscles**

Extrinsic muscles of the right eye as seen from a lateral view with the lateral wall of the orbit removed. The medial rectus muscle cannot be seen from this view.

Fibrous Tunic

The **sclera** (sklēr′ă; hard) is the firm, white, outer connective tissue layer of the posterior five-sixths of the fibrous tunic. The sclera helps maintain the shape of the eye, protects the internal structures, and provides attachment sites for the extrinsic eye muscles. A small portion of the sclera can be seen as the "white of the eye."

The **cornea** (kōr′nē-ă) is the transparent anterior sixth of the eye, which permits light to enter. As part of the focusing system of the fibrous tunic, the cornea also bends, or refracts, the entering light.

Vascular Tunic

The middle tunic of the eye is called the **vascular tunic** because it contains most of the blood vessels of the eye. The posterior portion of the vascular tunic, associated with the sclera, is the **choroid** (kō′royd). This very thin structure consists of a vascular network and many melanin-containing pigment cells, causing it to appear black. The black color absorbs light, so that it is not reflected inside the eye. If light were reflected inside the eye, the reflection would interfere with vision. The interiors of cameras are black for the same reason.

Anteriorly, the vascular tunic consists of the ciliary body and the iris. The **ciliary** (sil′ē-ar-ē) **body** is continuous with the anterior margin of the choroid. The ciliary body contains smooth

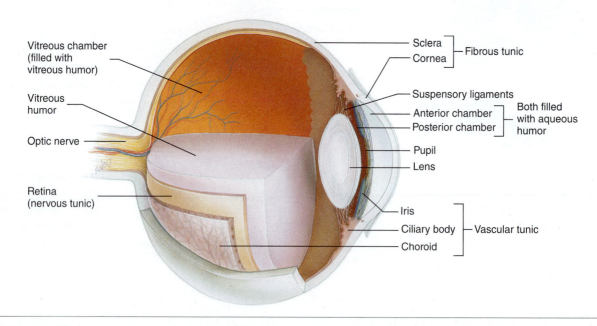

Figure 9.9 [AP|R] **Sagittal Section of the Eye, Demonstrating Its Tunics**

muscles called **ciliary muscles,** which attach to the perimeter of the lens by **suspensory ligaments** (figure 9.10). The **lens** is a flexible, biconvex, transparent disc (see figure 9.9).

The **iris** is the colored part of the eye. It is attached to the anterior margin of the ciliary body, anterior to the lens. The iris is a contractile structure consisting mainly of smooth muscle surrounding an opening called the **pupil.** Light passes through the pupil, and the iris regulates the diameter of the pupil, which controls the amount of light entering the eye. Parasympathetic stimulation from the oculomotor nerve (III) causes the circular smooth muscles of the iris to contract, constricting the pupil, whereas sympathetic stimulation causes radial smooth muscles of the iris to contract, dilating the pupil (figure 9.11). As light intensity increases, the pupil constricts; as light intensity decreases, the pupil dilates.

Nervous Tunic

The **nervous tunic** is the innermost tunic and consists of the **retina.** The retina covers the posterior five-sixths of the eye and is composed of two layers: an outer **pigmented retina** and an inner

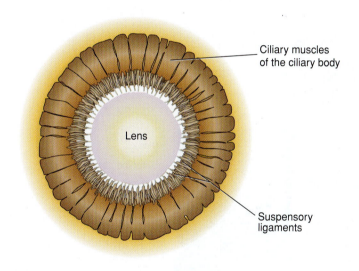

Figure 9.10 Lens and Ciliary Body

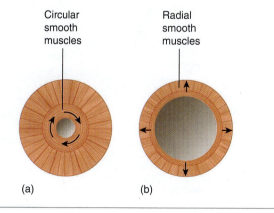

Figure 9.11 Iris

(*a*) Circular smooth muscles of the iris constrict the pupil. (*b*) Radial smooth muscles of the iris dilate the pupil.

sensory retina (figure 9.12*a*). The pigmented retina, with the choroid, keeps light from reflecting back into the eye. The sensory retina contains photoreceptor cells, called **rods** and **cones,** which respond to light. The sensory retina also contains numerous interneurons, some of which are named in figure 9.12. Over most of the retina, rods are 20 times more common than cones. Rods are very sensitive to light and can function in dim light, but they do not provide color vision. Cones require much more light, and they do provide color vision. There are three types of cones, each sensitive to a different color: blue, green, or red.

Predict 5

In dim light, colors seem to fade and objects appear in shades of gray. Explain this phenomenon.

The outer segments of rod and cone cells are modified by numerous foldings of the cell membrane to form discs (figure 9.12*c–e*). Rod cells contain a photosensitive pigment called **rhodopsin** (rō-dop′sin; purple pigment). Rhodopsin consists of a protein **opsin** (op′sin) loosely bound to a yellow pigment called **retinal** (ret′i-năl) (figure 9.12*f,* figure 9.13, step 1). When exposed to light, retinal changes shape, which then changes the activity of the entire rhodopsin molecule. This change in rhodopsin stimulates a response in the rod cell, resulting in vision (figure 9.13, steps 2 and 3*).* Retinal then completely detaches from opsin. Energy (ATP) is required to reattach retinal to opsin and return rhodopsin to the form it had before it was stimulated by light (figure 9.13, steps 4–6).

The manufacture of retinal in rod cells takes time and requires vitamin A. In bright light, much of the rhodopsin in rod cells is dissociated (opsin and retinal are separated). For example, suppose you go into a dark building on a bright day. It will take several seconds for your eyes to adjust to the dark as opsin and retinal reassociate to form rhodopsin in the rod cells, which can then react to the dim light. A person with a vitamin A deficiency may have a condition called **night blindness,** characterized by difficulty seeing in dim light. Night blindness can also result from **retinal detachment,** which is the separation of the sensory retina from the pigmented retina. Retinal detachment affects the periphery of the retina, where the rods are located, more than the center of the retina, where the cones are located. Because the rods are more sensitive than the cones to light, retinal detachment affects vision in low light to a greater extent than vision in bright light.

The photosensitive pigments in cone cells are slightly different from those in rod cells. The pigments in cone cells are sensitive to colors. Each color results from stimulation by a certain wavelength of light. Three major types of color-sensitive opsin exist; they are sensitive to blue, red, or green. The many colors that we can see result from the stimulation of combinations of these three types of cones.

The rod and cone cells synapse with bipolar cells of the sensory retina (see figure 9.12). These and the horizontal cells of the retina modify the output of the rod and cone cells. For example, this modification helps us perceive the borders between objects of contrasting brightness. The bipolar and horizontal cells synapse with ganglion cells, whose axons converge at the posterior of the eye to form the **optic nerve** (II; see figures 9.9 and 9.12*a*).

When the posterior region of the retina is examined with an ophthalmoscope (of-thal′mō-skōp), two major features can be observed: the macula and the optic disc (figure 9.14*a*). The **macula**

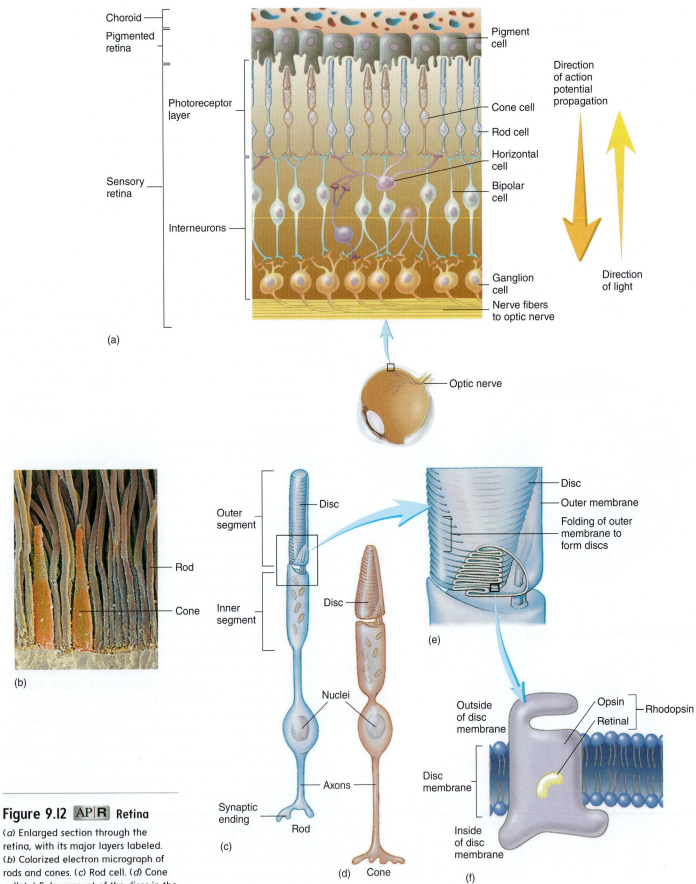

Choroid
Pigmented retina
Photoreceptor layer
Sensory retina
Interneurons

Pigment cell
Cone cell
Rod cell
Horizontal cell
Bipolar cell
Ganglion cell
Nerve fibers to optic nerve

Direction of action potential propagation
Direction of light

(a)

Optic nerve

Outer segment
Inner segment

Disc
Disc

Nuclei

Axons

Synaptic ending

Rod

(b)
Rod
Cone

(c)

(d)   Cone

Disc
Outer membrane
Folding of outer membrane to form discs

(e)

Outside of disc membrane
Disc membrane
Inside of disc membrane

Opsin
Retinal
Rhodopsin

(f)

Figure 9.12 AP|R Retina

(a) Enlarged section through the retina, with its major layers labeled. (b) Colorized electron micrograph of rods and cones. (c) Rod cell. (d) Cone cell. (e) Enlargement of the discs in the outer segment. (f) Enlargement of one of the discs, showing the relationship of rhodopsin to the membrane.

Nervous

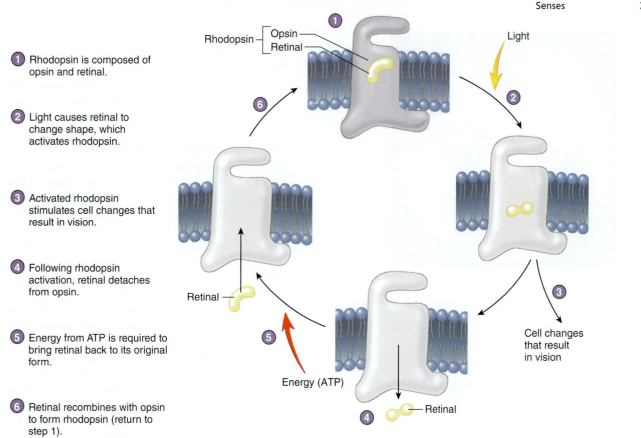

1. Rhodopsin is composed of opsin and retinal.

2. Light causes retinal to change shape, which activates rhodopsin.

3. Activated rhodopsin stimulates cell changes that result in vision.

4. Following rhodopsin activation, retinal detaches from opsin.

5. Energy from ATP is required to bring retinal back to its original form.

6. Retinal recombines with opsin to form rhodopsin (return to step 1).

Rhodopsin — Opsin Retinal

Light

Cell changes that result in vision

Retinal

Energy (ATP)

Retinal

PROCESS Figure 9.13 Effect of Light on Rhodopsin

(mak′ū-lā) is a small spot near the center of the posterior retina. In the center of the macula is a small pit, the **fovea** (fō′vē-ă; pit) **centralis.** The fovea centralis is the part of the retina where light is most focused when the eye is looking directly at an object. The fovea centralis contains only cone cells, and the cells are more tightly packed there than anywhere else in the retina. Hence, the fovea centralis is the region with the greatest ability to discriminate fine images, which explains why objects are best seen straight ahead.

The **optic disc** is a white spot just medial to the macula, through which a number of blood vessels enter the eye and spread over the surface of the retina. This is also the spot at which axons from the retina meet, pass through the two outer tunics, and exit the eye as the optic nerve. The optic disc contains no photoreceptor cells and does not respond to light; it is therefore called the **blind spot** of the eye. A small image projected onto the blind spot cannot be seen (figure 9.14*b*).

Chambers of the Eye

The interior of the eye is divided into the **anterior chamber,** the **posterior chamber,** and the **vitreous** (vit′rē-ŭs; glassy) **chamber** (see figure 9.9). The anterior and posterior chambers are located between the cornea and the lens. The iris separates the anterior and the posterior chambers, which are continuous with each other through the pupil. The much larger vitreous chamber is posterior to the lens.

The anterior and posterior chambers are filled with **aqueous humor** (watery fluid), which helps maintain pressure within the eye, refracts light, and provides nutrients to the inner surface of the eye. Aqueous humor is produced by the ciliary body as a blood filtrate and is returned to the circulation through a venous ring that surrounds the cornea. The presence of aqueous humor keeps the eye inflated, much like the air in a basketball. If flow of the aqueous humor from the eye through the venous ring is blocked, the pressure in the eye increases, resulting in a condition called **glaucoma** (see the Diseases and Disorders table at the end of this chapter). Glaucoma can eventually lead to blindness because the fluid compresses the retina, thereby restricting blood flow through it.

The vitreous chamber of the eye is filled with a transparent, jellylike substance called **vitreous humor.** The vitreous humor helps maintain pressure within the eye and holds the lens and the retina in place. It also refracts light. Unlike the aqueous humor, the vitreous humor does not circulate.

Functions of the Eye

The eye functions much like a camera. The iris allows light into the eye, which is focused by the cornea, lens, and humors onto the retina. The light striking the retina produces action potentials that are relayed to the brain.

Light Refraction

An important characteristic of light is that it can be refracted (bent). As light passes from air to some other, denser transparent substance, the light rays are refracted. If the surface of a lens is concave, the light rays are bent, so that they diverge as they pass

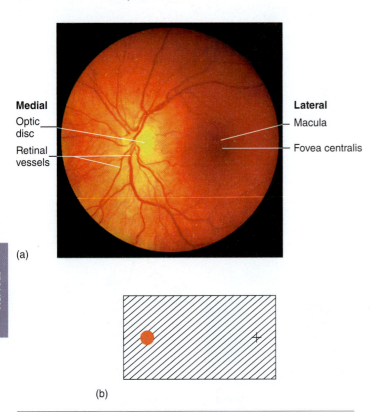

Medial

Optic disc

Retinal vessels

Lateral

Macula

Fovea centralis

(a)

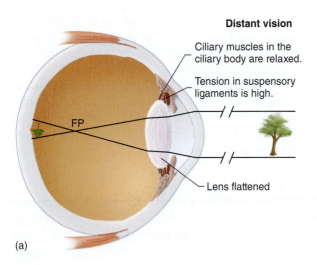

(b)

Figure 9.14 AP|R Ophthalmoscopic View of the Retina

(*a*) This view shows the posterior wall of the left eye as seen through the pupil. Notice the vessels entering the eye through the optic disc. The macula, with the fovea centralis in the center, is located lateral to the optic disc. (*b*) Demonstration of the blind spot. Close your right eye. Hold the drawing in front of your left eye and stare at the +. Move the drawing toward your eye. At a certain point, when the image of the spot is over the optic disc, the red spot seems to disappear.

through the lens; if the surface is convex, they converge. As the light rays converge, they finally reach a point at which they cross. The crossing point is called the **focal point (FP)** (figure 9.15), and causing light to converge is called **focusing.** The focal point in the eye occurs just anterior to the retina, and the tiny image that is focused on the retina is inverted compared to the actual object.

Focusing Images on the Retina

The cornea is a convex structure, and as light rays pass from the air through the cornea, they converge (figure 9.15). Additional convergence occurs as light passes through the aqueous humor, lens, and vitreous humor. The greatest contrast in media density is between the air and the cornea. The greatest amount of convergence therefore occurs at that point. However, the shape of the cornea and its distance from the retina are fixed, so the cornea cannot make any adjustment in focus. Fine adjustments in focus are accomplished by changing the shape of the lens.

When the ciliary muscles are relaxed, the suspensory ligaments of the ciliary body maintain elastic pressure on the perimeter of the lens, keeping it relatively flat and allowing for distant vision (figure 9.15*a*). When an object is brought closer than 20 feet (about 6½ m) from the eye, the ciliary muscles contract as a result of parasympathetic stimulation, pulling the ciliary body toward the lens. This reduces the tension on the suspensory ligaments of the lens and allows the lens to assume a more spherical form because of its own internal elastic nature (figure 9.15*b*). The spherical lens then has a more convex surface, causing greater refraction of light. This process is called **accommodation** (ă-kom′ō-dā′shŭn), and it enables the eye to focus on images closer than 20 feet from the eye.

Predict 6

As you are driving a car, what changes occur when you look down at the speedometer and then back up at the road?

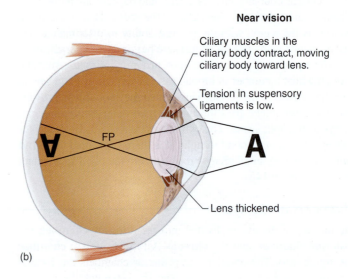

Distant vision

Ciliary muscles in the ciliary body are relaxed.

Tension in suspensory ligaments is high.

FP

Lens flattened

(a)

Near vision

Ciliary muscles in the ciliary body contract, moving ciliary body toward lens.

Tension in suspensory ligaments is low.

FP

Lens thickened

(b)

Figure 9.15 Focus by the Eye

The focal point (FP) is where light rays cross. (*a*) When viewing a distant image, the lens is flattened, and the image is focused on the retina. (*b*) In accommodation for near vision, the lens becomes more rounded, allowing the image to be focused on the retina.

CLINICAL IMPACT   Color Blindness

Color **blindness** is the absence of perception of one or more colors (figure 9A). Color perception may be decreased or completely lost. The loss may involve perception of all three colors or of one or two colors. Most forms of color blindness occur more frequently in males and are X-linked genetic traits (see chapter 20). In Western Europe, about 8% of all males have some form of color blindness, whereas only about 1% of the females are color blind.

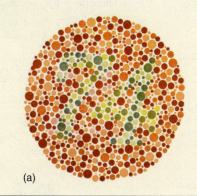
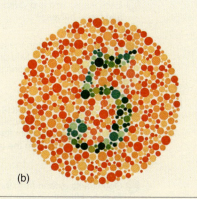

(a)   (b)

Figure 9A

These color blindness charts demonstrate the differences in color perception associated with some forms of color blindness. (*a*) A person with normal vision can see the number 74, whereas a person with red-green color blindness sees the number 21. (*b*) A person with normal vision can see the number 5. A person with red-green color blindness sees the number 2.

Reproduced from *Ishihara's Tests for Colour Deficiency* published by Kanehara Trading, Inc., Tokyo, Japan. Tests for color deficiency cannot be conducted with this material. For accurate testing, the original plates should be used.

When a person's vision is tested, a chart is placed 20 feet from the eye, and the person is asked to read a line of letters that has been standardized for normal vision. If the person can read the line, he or she has 20/20 vision, which means that the person can see at 20 feet what people with normal vision see at 20 feet. On the other hand, if the person can only read at 20 feet the line that people with normal vision see at 40 feet, the person's eyesight is 20/40, and corrective lenses are probably needed.

Neuronal Pathways for Vision

Figure 9.16 shows the neuronal pathways that transmit signals generated by light from the time light enters the eye until it reaches the area of the brain where vision is perceived. The **optic nerve** leaves the eye and exits the orbit through the optic foramen to enter the cranial cavity. Just inside the cranial cavity, the two optic nerves connect to each other at the **optic chiasm** (kī′azm; crossing). Axons from the nasal (medial) part of each retina cross through the optic chiasm and project to the opposite side of the brain. Axons from the temporal (lateral) part of each retina pass through the optic nerves and project to the brain on the same side of the body without crossing.

Beyond the optic chiasm, the route of the ganglionic axons is through the two **optic tracts** (figure 9.16). Most of the optic tract axons terminate in the thalamus. Some axons do not terminate in the thalamus but separate from the optic tracts to terminate in the superior colliculi, the center for visual reflexes. An example of a visual reflex is turning the head and eyes toward a stimulus, such as a sudden noise or flash of light. Neurons from the thalamus form the fibers of the **optic radiations,** which project to the **visual cortex** in the occipital lobe of the cerebrum

(figure 9.16). The visual cortex is the area of the cerebrum where vision is perceived.

The image seen by each eye is the **visual field** of that eye (figure 9.16*a*). Depth perception (three-dimensional, or binocular, vision) requires both eyes and occurs where the two visual fields overlap (figure 9.16*c*). Each eye sees a slightly different (monocular) view of the same object. The brain then processes the two images into a three-dimensional view of the object. If only one eye is functioning, the view of the object is flat, much like viewing a photograph.

A CASE IN POINT

Double Vision

I.C. Double awoke one morning and discovered that her vision was blurred. However, when she closed one eye, her vision was clear. She tried to ignore the problem and went about her work, but the condition persisted. Eventually, she consulted an optometrist, who found that her blurred vision resulted from double vision, or **diplopia.** The most common cause of diplopia is misalignment of the two eyes (binocular diplopia). This often results from weakness of the muscles moving the eyes. Most adults cannot easily ignore the double image and may need to wear a patch over one eye to eliminate one image. In children with diplopia, the brain may compensate for the two discordant images by ignoring one of the images; thus, the problem appears to go away. Double vision should not be ignored because it can be a symptom of a serious neurological problem, such as an expanding brain tumor compressing the nerves to the eye muscles, and decreasing their function. A physician should be consulted as soon as possible.

1 Each visual field is divided into temporal and nasal halves.

2 After passing through the lens, light from each half of a visual field projects to the opposite side of the retina stimulating receptors.

3 Axons from the retina pass through the optic nerve to the optic chiasma, where some cross. Axons from the nasal retina cross, and those from the temporal retina do not.

4 Optic tracts extend from the optic chiasm (with or without crossing) to the thalamus. Collateral branches of the axons in the optic tract synapse in the superior colliculi of the midbrain.

5 Optic radiations extend from the thalamus to the visual cortex of the occipital lobe.

6 The right part of each visual field (*dark green* and *light blue*) projects to the left side of the brain, and the left part of each visual field (*light green* and *dark blue*) projects to the right side of the brain.

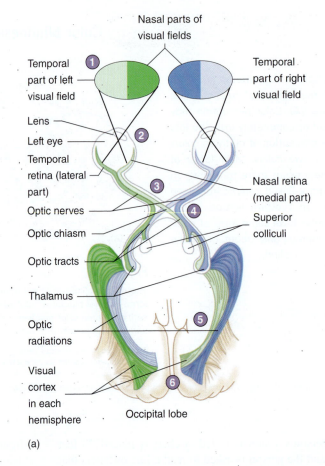

(a)

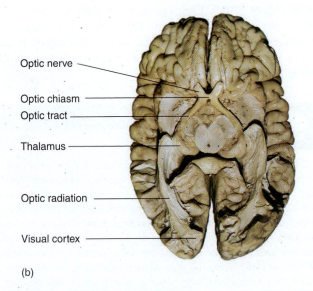

(b)

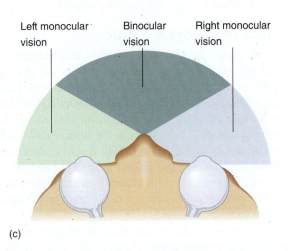

(c)

PROCESS Figure 9.16 AP|R Visual Pathways

(*a*) Pathways for both eyes (*superior view*). (*b*) Photograph of transverse section of the brain showing the visual nerves, tracts, and pathways (*inferior view*). (*c*) Overlap of the fields of vision (*superior view*).

9.8 HEARING AND BALANCE

Learning Outcomes *After reading this section, you should be able to*

A. Describe the structures of the outer and middle ears, and state the function of each.
B. Describe the anatomy of the cochlea, and explain how sounds are detected.
C. Explain how the structures of the vestibule and semicircular canals function in static and dynamic equilibrium.

The organs of hearing and balance are apportioned into three areas: the external, middle, and inner ears (figure 9.17). The external ear is the part extending from the outside of the head to the tympanic membrane, which is also called the eardrum. The middle ear is an air-filled chamber medial to the tympanic membrane. The inner ear is a set of fluid-filled chambers medial to the middle ear. The external and middle ears are involved in hearing only, whereas the inner ear functions in both hearing and balance.

Anatomy and Function of the Ear

External Ear

The **auricle** (aw′ri-kl; ear) is the fleshy part of the external ear on the outside of the head. The auricle opens into the **external auditory canal,** a passageway that leads to the eardrum. The auricle collects sound waves and directs them toward the external auditory canal, which transmits them to the tympanic membrane. The auditory canal is lined with hairs and **ceruminous** (sĕ-roo′mi-nŭs; *cera,* wax) **glands,** which produce **cerumen** (sĕ-roo′men), a modified sebum commonly called earwax. The hairs and cerumen help prevent foreign objects from reaching the delicate tympanic membrane.

The **tympanic** (tim-pan′ik; drumlike) **membrane,** or *eardrum,* is a thin membrane that separates the external ear from the middle ear. It consists of a thin layer of connective tissue sandwiched between two epithelial layers. Sound waves reaching the tympanic membrane cause it to vibrate.

Middle Ear

Medial to the tympanic membrane is the air-filled cavity of the middle ear. Two covered openings on the medial side of the middle ear, the **oval window** and the **round window,** connect the middle ear with the inner ear. The middle ear contains three **auditory ossicles** (os′i-klz; ear bones): the **malleus** (mal′ē-ŭs; hammer), the **incus** (ing′kŭs; anvil), and the **stapes** (stā′pēz; stirrup). These bones transmit vibrations from the tympanic membrane to the oval window. The malleus is attached to the medial surface of the tympanic membrane. The incus connects the malleus to the stapes. The base of the stapes is seated in the oval window, surrounded by a flexible ligament. As vibrations are transmitted from the malleus to the stapes, the force of the vibrations is amplified about 20-fold because

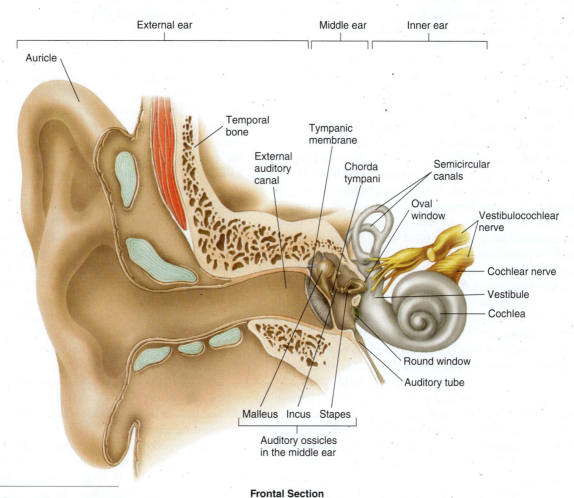

External ear   Middle ear   Inner ear

Auricle

Temporal bone

External auditory canal

Tympanic membrane

Chorda tympani

Semicircular canals

Oval window

Vestibulocochlear nerve

Cochlear nerve

Vestibule

Cochlea

Round window

Auditory tube

Malleus   Incus   Stapes

Auditory ossicles in the middle ear

Frontal Section

Figure 9.17 AP|R **Structure of the Ear**

the area of the tympanic membrane is about 20 times that of the oval window. Two small muscles in the middle ear, one attached to the malleus and the other to the stapes, help dampen vibrations caused by loud noises, thus protecting the delicate inner ear structures.

There are two unblocked openings into the middle ear. One opens into the mastoid air cells in the mastoid process of the temporal bone. The other, called the **auditory tube,** or *eustachian* (ū-stā′shŭn) *tube,* opens into the pharynx and enables air pressure to be equalized between the outside air and the middle ear cavity. Unequal pressure between the middle ear and the outside environment can distort the tympanic membrane, dampen its vibrations, and make hearing difficult. Distortion of the tympanic membrane also stimulates pain receptors associated with that structure. That distortion is why, as a person changes altitude, sounds seem muffled and the tympanic membrane may become painful. These symptoms can be relieved by opening the auditory tube to allow air to enter or exit the middle ear, such as by swallowing, yawning, chewing, or holding the nose and mouth shut while gently forcing air out of the lungs.

Inner Ear

The inner ear consists of interconnecting tunnels and chambers within the temporal bone, called the **bony labyrinth** (lab′i-rinth; maze) (figure 9.18*a*). Inside the bony labyrinth is a smaller set of membranous tunnels and chambers called the **membranous labyrinth** (figure 9.18*b*). The membranous labyrinth is filled with a clear fluid called **endolymph** (en′dō-limf), and the space between the membranous and bony labyrinths is filled with a fluid called **perilymph** (per′i-limf). The bony labyrinth can be divided into three regions: the cochlea, the vestibule, and the semicircular canals. The cochlea is involved in hearing. The vestibule and semicircular canals are involved primarily in balance.

The **cochlea** (kok′lē-ă; snail shell) is shaped like a snail shell (figure 9.18*a*) and contains a bony core shaped like a screw. The threads of this screw are called the **spiral lamina.** The cochlea is divided into three channels: the scala vestibuli, the scala tympani, and the cochlear duct (figure 9.18*b*). The **scala vestibuli** (skā′lă vestib′ū-lī; *scala,* stairway) extends from the oval window to the apex of the cochlea. The **scala tympani** (tim-pa′nē) extends in parallel with the scala vestibuli from the apex back to the round window. These two channels are perilymph-filled spaces between the walls of the bony and membranous labyrinths. The wall of the membranous labyrinth that lines the scala vestibuli is called the **vestibular** (ves-tib′ū-lār) **membrane;** the wall of the membranous labyrinth that lines the scala tympani is the **basilar membrane.** The **cochlear duct** is formed by the space between the vestibular membrane and the basilar membrane and is filled with endolymph.

Inside the cochlear duct is a specialized structure called the **spiral organ,** or *organ of Corti* (figure 9.18*c*). The spiral organ contains specialized sensory cells called **hair cells,** which have hairlike microvilli, often referred to as *stereocilia,* on their surfaces (figure 9.18*c,d,e*). The microvilli are stiffened by actin filaments. The hair tips are embedded within an acellular gelatinous shelf called the **tectorial** (tek-tōr′ē-ăl; a covering) **membrane,** which is attached to the spiral lamina (figure 9.18*b,c*).

Hair cells have no axons of their own, but each hair cell is associated with axon terminals of sensory neurons, the cell bodies of which are located within the **cochlear ganglion,** or *spiral ganglion.*

Axons of the sensory neurons join to form the cochlear nerve. This nerve joins the vestibular nerve to become the **vestibulocochlear nerve** (VIII), which carries action potentials to the brain.

Hearing

Vibrations create sound waves. Sound waves are collected by the auricle and conducted through the external auditory canal toward the tympanic membrane. Sound waves strike the tympanic membrane and cause it to vibrate. This vibration causes vibration of the three ossicles of the middle ear, and by this mechanical linkage, the force of vibration is amplified and transferred to the oval window (figure 9.19, steps 1–3).

Predict 7

When you hear a faint sound, why do you turn your head toward it?

Vibrations of the base of the stapes, seated in the oval window, produce waves in the perilymph of the cochlea. The two scalae can be thought of as a continuous, U-shaped tube, with the oval window at one end of the scala vestibuli and the round window at the other end of the scala tympani. The vibrations of the stapes in the oval window cause movement of the perilymph, which pushes against the membrane covering the round window (figure 9.19 step 4). This phenomenon is similar to pushing against a rubber diaphragm on one end of a fluid-filled glass tube. If the tube has a rubber diaphragm on each end, the fluid can move. If one end of the glass tube or of the cochlear tubes were solid, no fluid movement would occur.

The waves produced in the perilymph pass through the vestibular membrane and cause vibrations of the endolymph. Waves in the endolymph, within the cochlear duct, cause displacement of the basilar membrane. As the basilar membrane is displaced, the hair cells, seated on the basilar membrane, move with the movements of the membrane. The microvilli of the hair cells are embedded in the tectorial membrane, which is a rigid shelf that does not move. Because one end of the microvilli moves with the hair cells and their other ends are embedded in the nonmoving tectorial membrane, the microvilli bend. The bending of the microvilli stimulates the hair cells, which induces action potentials in the cochlear nerves (figure 9.19, steps 5–6).

The basilar membrane is not uniform throughout its length. The membrane is narrower and denser near the oval window and wider and less dense near the tip of the cochlea. The various regions of the membrane can be compared to the strings in a piano (i.e., some are short and thick, and others are longer and thinner). As a result of this organization, sounds with higher **pitches** cause maximum distortion of the basilar membrane nearer the oval window, whereas sounds with lower pitches cause maximum distortion nearer the apex of the cochlea. In each case, different hair cells are stimulated, and because of the differences in which hair cells are maximally stimulated, a person is able to detect variations in pitch. Sound **volume** is a function of sound wave amplitude, which causes the basilar membrane to distort more intensely and the hair cells to be stimulated more strongly.

Hearing impairment can have many causes. In general, there are two categories of hearing impairment: conduction deafness and sensorineural hearing loss (see the Diseases and Disorders table at the end of this chapter). **Conduction deafness** results from mechanical deficiencies—for example, destruction of the ligament that holds the malleus and incus together. **Sensorineural**

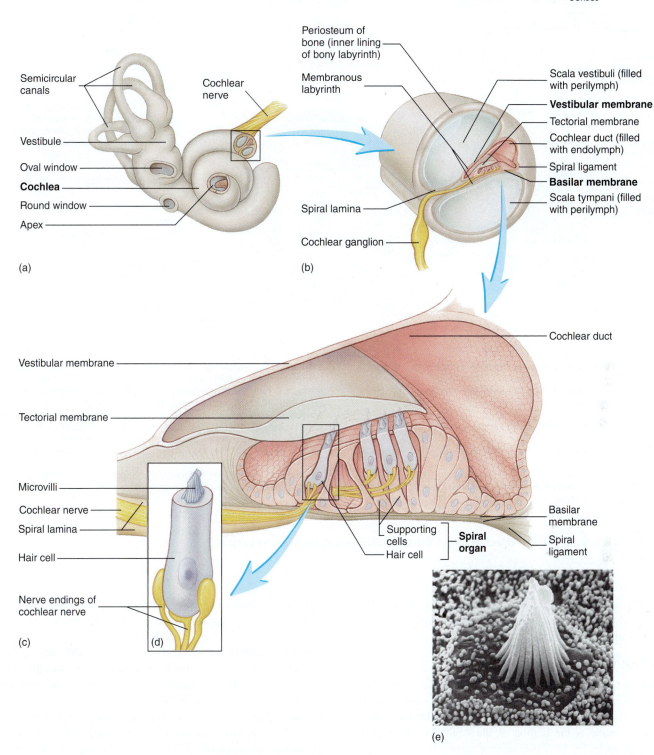

Periosteum of
bone (inner lining
of bony labyrinth)

Semicircular
canals

Cochlear
nerve

Vestibule

Oval window

Cochlea

Round window

Apex

(a)

Membranous
labyrinth

Spiral lamina

Cochlear ganglion

(b)

Scala vestibuli (filled
with perilymph)

Vestibular membrane

Tectorial membrane

Cochlear duct (filled
with endolymph)

Spiral ligament

Basilar membrane

Scala tympani (filled
with perilymph)

Cochlear duct

Vestibular membrane

Tectorial membrane

Microvilli

Cochlear nerve

Spiral lamina

Hair cell

Nerve endings of
cochlear nerve

(c)

(d)

Supporting
cells

Hair cell

**Spiral
organ**

Basilar
membrane

Spiral
ligament

(e)

Figure 9.18 AP|R Structure of the Inner Ear

(a) Bony labyrinth. The outer surface (gray) is the periosteum lining the inner surface of the bony labyrinth. (b) In this cross section of the cochlea, the outer layer is the periosteum lining the inner surface of the bony labyrinth. The membranous labyrinth is very small in the cochlea and consists of the vestibular and basilar membranes. The space between the membranous and bony labyrinths consists of two parallel tunnels: the scala vestibuli and the scala tympani. (c) An enlarged section of the cochlear duct (membranous labyrinth). (d) A greatly enlarged individual sensory hair cell. (e) Scanning electron micrograph of the microvilli of a hair cell.

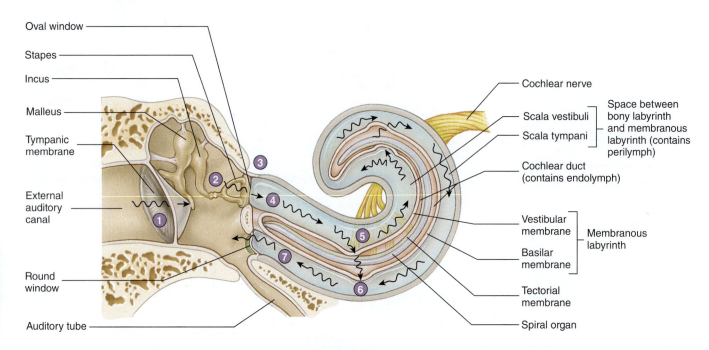

1 Sound waves strike the tympanic membrane and cause it to vibrate.

2 Vibration of the tympanic membrane causes the malleus, the incus, and the stapes to vibrate.

3 The base of the stapes vibrates in the oval window.

4 Vibration of the base of the stapes causes the perilymph in the scala vestibuli to vibrate.

5 Vibration of the perilymph causes the vestibular membrane to vibrate, which causes vibrations in the endolymph.

6 Vibration of the endolymph causes displacement of the basilar membrane. Short waves (high pitch) cause displacement of the basilar membrane near the oval window, and longer waves (low pitch) cause displacement of the basilar membrane some distance from the oval window. Movement of the basilar membrane is detected in the hair cells of the spiral organ, which are attached to the basilar membrane. Vibrations of the perilymph in the scala vestibuli and of the basilar membrane are transferred to the perilymph of the scala tympani.

7 Vibrations in the perilymph of the scala tympani are transferred to the round window, where they are dampened.

PROCESS Figure 9.19 AP|R **Effect of Sound Waves on Middle and Inner Ear Structures**

hearing loss is caused by deficiencies in the spiral organ or nerves; for example, loud sounds can damage the delicate microvilli of the hair cells, leading to destruction of the spiral organ.

Neuronal Pathways for Hearing

The senses of hearing and balance are both transmitted by the vestibulocochlear nerve (VIII). This nerve functions as two separate nerves, carrying information from two separate but closely related structures. The cochlear nerve is the portion of the vestibulocochlear nerve involved in hearing; the vestibular nerve is involved in balance. The cochlear nerve sends axons to the **cochlear nucleus** in the brainstem. Neurons in the cochlear nucleus project to other areas of the brainstem and to the **inferior colliculus** in the midbrain. Neurons from the inferior colliculus also project to the superior colliculus, where reflexes that turn the head and eyes in response to loud sounds are initiated. From the inferior colliculus, fibers project to the thalamus and from there to the auditory cortex of the cerebrum (figure 9.20).

Balance

The sense of balance, or equilibrium, has two components: static equilibrium and dynamic equilibrium. **Static equilibrium** is associated with the vestibule and is involved in evaluating the position

of the head relative to gravity. **Dynamic equilibrium** is associated with the semicircular canals and is involved in evaluating changes in the direction and rate of head movements.

The **vestibule** (ves'ti-bool) of the inner ear can be divided into two chambers: the **utricle** (ū'tri-kl) and the **saccule** (sak'ūl) (figure 9.21a). Each chamber contains specialized patches of epithelium called the **maculae** (mak'ū-lē), which are surrounded by endolymph. The maculae, like the spiral organ, contain hair cells. The tips of the microvilli of these cells are embedded in a gelatinous mass, often called the *otolithic membrane*, weighted by **otoliths** (ō'tō-liths; ear stones), particles composed of protein and calcium carbonate. The weighted gelatinous mass moves in response to gravity, bending the hair cell microvilli (figure 9.21c) and initiating action potentials in the associated neurons. The action potentials from these neurons are carried by axons of the vestibular portion of the vestibulocochlear nerve (VIII) to the brain, where they are interpreted as a change in the position of the head. For example, when a person bends over, the maculae are displaced by gravity, and the resultant action potentials provide information to the brain concerning the position of the head (figure 9.22).

Three **semicircular canals** are involved in dynamic equilibrium. The canals are placed at nearly right angles to one another,

Nervous

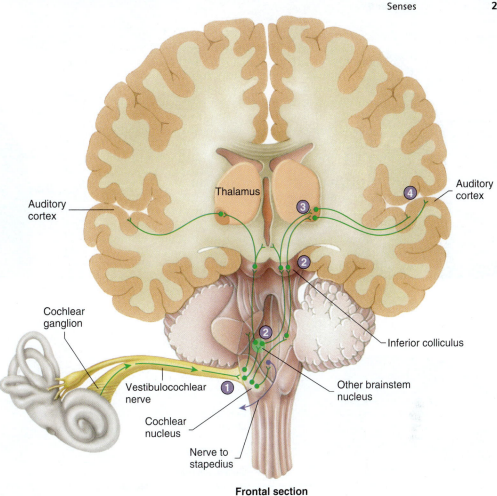

1. Sensory axons from the cochlear ganglion terminate in the cochlear nucleus in the brainstem.

2. Axons from the neurons in the cochlear nucleus project to other brainstem nuclei or to the inferior colliculus.

3. Axons from the inferior colliculus project to the thalamus.

4. Thalamic neurons project to the auditory cortex.

Thalamus

Auditory cortex

Auditory cortex

Cochlear ganglion

Inferior colliculus

Vestibulocochlear nerve

Other brainstem nucleus

Cochlear nucleus

Nerve to stapedius

Frontal section

PROCESS Figure 9.20 Central Nervous System Pathways for Hearing

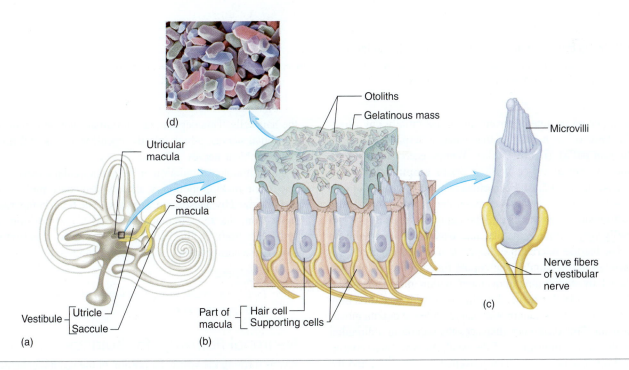

Otoliths

Gelatinous mass

Microvilli

(d)

Utricular macula

Saccular macula

Nerve fibers of vestibular nerve

(c)

Vestibule — Utricle / Saccule

(a)

Part of macula — Hair cell / Supporting cells

(b)

Figure 9.21 AP|R Location and Structure of the Macula

(a) Location of the utricular and saccular maculae within the vestibule. (b) Enlargement of the utricular macula, showing hair cells and otoliths. (c) Enlarged hair cell, showing the microvilli. (d) Colorized scanning electron micrograph of otoliths.

Nervous

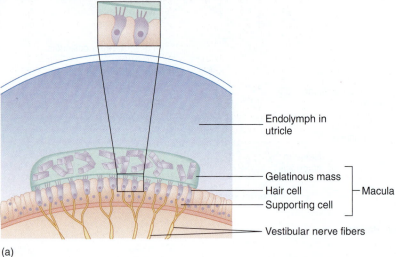

Endolymph in
utricle

Gelatinous mass
Hair cell   — Macula
Supporting cell

Vestibular nerve fibers

(a)

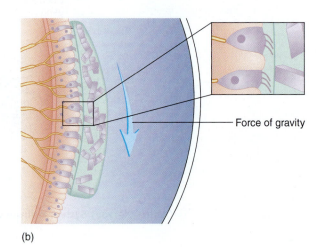

Force of gravity

(b)

Figure 9.22 Function of the Vestibule in Maintaining Balance

(*a*) In an upright position, the maculae don't move. (*b*) When the position of the head changes, as when a person bends over, the maculae respond by moving in the direction of gravity.

enabling a person to detect movements in essentially any direction. The base of each semicircular canal is expanded into an **ampulla** (am-pul′ă) (figure 9.23*a*). Within each ampulla, the epithelium is specialized to form a **crista ampullaris** (kris′tă am-pūl′ar′is) (figure 9.23*b*). Each crista consists of a ridge of epithelium with a curved, gelatinous mass, the **cupula** (koo′poo-lă; a tub), suspended over the crest. The cupula is structurally and functionally very similar to the maculae, except that it contains no otoliths. The hairlike microvilli of the crista hair cells (figure 9.23*c*) are embedded in the cupula. The cupula functions as a float that is displaced by endolymph movement within the semicircular canals (figure 9.24). As the head begins to move in one direction, the endolymph tends to remain stationary, while the cupula moves with the head. This difference displaces the cupula in a direction opposite that of the movement of the head. As movement continues, the fluid "catches up." When movement of the head and the cupula stops, the fluid continues to move, displacing the cupula in the direction of the movement. Movement of the cupula causes the hair cell microvilli to bend, which initiates depolarization in

the hair cells. This depolarization initiates action potentials in the vestibular nerves, which join the cochlear nerves to form the vestibulocochlear nerves.

Continuous stimulation of the semicircular canals—as occurs due to the rocking motion of a boat—can cause **motion sickness,** characterized by nausea and weakness. The brain compares sensory input from the semicircular canals, eyes, and position receptors (proprioceptors) in the back and lower limbs. Conflicting input from these sources can lead to motion sickness.

Predict 8

Explain why closing your eyes or looking at the horizon can help decrease motion sickness.

Neuronal Pathways for Balance

Axons forming the vestibular portion of the vestibulocochlear nerve (VIII) project to the vestibular nucleus in the brainstem. Axons run from this nucleus to numerous areas of the CNS, such as the cerebellum and cerebral cortex. Balance is a complex sensation

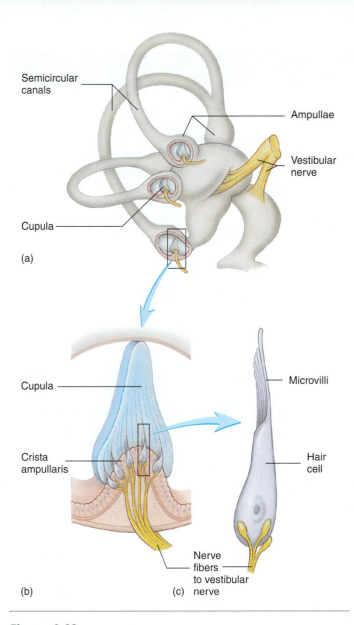

Semicircular canals

Ampullae

Vestibular nerve

Cupula

(a)

Cupula

Microvilli

Crista ampullaris

Hair cell

Nerve fibers to vestibular (c) nerve

(b)

Figure 9.23 Semicircular Canals

(*a*) Location of the ampullae of the semicircular canals. (*b*) Enlargement of the crista ampullaris, showing the cupula and hair cells. (*c*) Enlargement of a hair cell.

(a)

Figure 9.24 Function of the Crista Ampullaris

(*a*) As a person begins to tumble, the semicircular canals (*b*) move in the same direction as the body (*blue arrow*). The endolymph in the semicircular canals tends to stay in place as the body and the crista ampullaris begin to move. As a result, the cupula is displaced by the moving endolymph (*red arrow*) in a direction opposite the direction of movement.

involving sensory input to the vestibular nucleus not only from the inner ear but also from the limbs (proprioception) and visual system as well. In sobriety tests, people are asked to close their eyes while their balance is evaluated because alcohol affects the proprioceptive and vestibular components of balance to a greater extent than the visual component of balance.

A CASE IN POINT

Seasickness

Earl E. Fisher booked his first trip on a charter fishing boat. After boarding the boat, Earl was surprised that all those warnings about passengers becoming seasick did not seem to apply to him. At last, the boat arrived at the fishing site, the engine was cut, and the sea anchor was set. As the boat drifted, it began to roll and pitch. Earl noticed, for the first time, that the smell of the bait mixed unpleasantly with the diesel fumes. Earl felt a little light-headed and a bit drowsy. He then noticed that he was definitely nauseated. "I'm seasick," he realized. Seasickness is a form of motion sickness, which is caused by conflicting information reaching the brain from different sensory sources, such as the eyes and the semicircular canals of the inner ear. The brain reacts with a feeling of vertigo (a feeling of spinning) and nausea. In Earl's case, trying to fish seemed to worsen his condition. Eventually, his nausea intensified, and he leaned over the boat rail and vomited into the ocean. "Improves the fishing," the ship owner shouted to him cheerily. Earl noticed that he felt somewhat better, and he found that looking at the horizon rather than at the water helped his condition even more. He enjoyed the rest of the trip and even caught a couple of very nice fish.

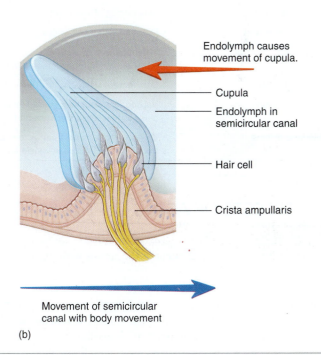

Endolymph causes movement of cupula.

Cupula

Endolymph in semicircular canal

Hair cell

Crista ampullaris

Movement of semicircular canal with body movement

(b)

Nervous

DISEASES AND DISORDERS: Vision, Hearing, and Balance

CONDITION	DESCRIPTION
	EYE DISORDERS
Infections	
Conjunctivitis (kon-jŭnk-ti-vi′tis)	Inflammation of conjunctiva, usually from bacterial infection; one form, pinkeye, occurs primarily in children
Trachoma (tră-kō′mă)	Type of conjunctivitis caused by *Chlamydia*; leading cause of infectious blindness in the world; transmitted by contact or flies
Stye (stī)	Infection of eyelash hair follicle
Defects of Focus, Alignment, or Color Vision	
Myopia (mī-ō′pē-ă)	Nearsightedness—ability to see close but not distant objects; caused when refractive power of cornea and lens is too great relative to length of eye
Hyperopia (hī-per-ō′pē-ă)	Farsightedness—ability to see distant but not close objects; caused when cornea is too flat or lens has too little refractive power relative to length of eye
Presbyopia (prez-bē-ō′pē-ă)	Decrease in near vision, due to reduced flexibility of lens and reduction in accommodation; a normal part of aging
Astigmatism (ă-stig′ mă-tizm)	Cornea or lens is not uniformly curved, so image is not sharply focused
Strabismus (stra-biz′mŭs)	One or both eyes are misdirected; can result from weak eye muscles
Diplopia (di-plō′pē-ă)	Double vision
Color blindness	Complete or partial absence of perception of one or more colors (see figure 9A); most forms are more frequent in males
Blindness	
Cataract (kat′ă-rakt)	Clouding of lens as a result of advancing age, infection, or trauma; most common cause of blindness in the world
Macular degeneration	Loss of sharp central vision, peripheral vision maintained; leading cause of legal blindness in older Americans; most causes not known
Glaucoma (glaw-kō′mă)	Excessive pressure buildup in aqueous humor; may destroy retina or optic nerve, resulting in blindness
Diabetic retinopathy	Involves optic nerve degeneration, cataracts, retinal detachment; often caused by blood vessel degeneration and hemorrhage
Retinal detachment	Separation of sensory retina from pigmented retina; relatively common problem; may result in vision loss
	EAR DISORDERS
Conduction deafness	Mechanical deficiency in transmission of sound waves from outer ear to spiral organ
Sensorineural hearing loss	Deficiencies of spiral organ or nerve pathways
Tinnitus (ti-nī′tus)	Phantom sound sensations, such as ringing in ears; common problem
Middle ear infection	Symptoms are low-grade fever, lethargy, irritability, and pulling at ear; in extreme cases, can damage or rupture tympanic membrane; common in young children
Inner ear infection	Can decrease detection of sound and maintenance of balance; may be caused by chronic middle ear infections
Motion sickness	Nausea and weakness caused when information to brain from semicircular canals conflicts with information from eyes and position sensors in back and lower limbs
Meniere disease	Vertigo, hearing loss, tinnitus, and a feeling of fullness in the affected ear; most common disease involving dizziness from inner ear; cause unknown but may involve a fluid abnormality in ears

9.9 EFFECTS OF AGING ON THE SENSES

Learning Outcome *After reading this section, you should be able to*

A. Describe changes that occur in the senses with aging.

As a person ages, both the general and the special sensory functions gradually decline. Among the general senses, free nerve endings and hair follicle receptors in the skin remain largely unchanged. However, the numbers of Meissner corpuscles and pacinian corpuscles decrease with age, and those that remain are often structurally distorted and less functional. As a result, elderly people are less conscious of something touching or pressing the skin, which increases the risk of skin injuries. The sense of two-point discrimination decreases, and the elderly have a more difficult time identifying objects by touch. A loss of pacinian corpuscles also decreases their awareness of limb and joint positions, which can affect balance and coordination. The functions of receptors for proprioception also decline with age, which decreases information on the position, tension, and length of tendons and muscles. This can further reduce coordination and control of movements.

Among the special senses, elderly people experience only a slight loss in the ability to detect odors. However, their ability to correctly identify specific odors decreases, especially in men over age 70.

In general, the sense of taste decreases as people age. The number of sensory receptors decreases, and the brain's ability to interpret taste sensations declines.

The lenses of the eyes lose flexibility as a person ages because the connective tissue of the lenses becomes more rigid. Consequently, the lenses' ability to change shape initially declines and eventually is lost. This condition, called **presbyopia,** is the most common age-related change in the eyes. In addition, the number of cones decreases, especially in the fovea centralis, resulting in a gradual decline in visual acuity and color perception.

The most common visual problem in older people requiring medical treatment, such as surgery, is the development of cataracts. Following cataracts in frequency are macular degeneration, glaucoma, and diabetic retinopathy, in that order (see the Diseases and Disorders table earlier in this chapter).

As people age, the number of hair cells in the cochlea decreases, resulting in age-related sensorineural hearing loss, called **presbya-**

cusis. This decline doesn't occur equally in both ears. Therefore, because direction is determined by comparing sounds coming into each ear, elderly people may experience a decreased ability to localize the origin of certain sounds. This may lead to a general sense of disorientation. In addition, CNS defects in the auditory pathways can lead to difficulty understanding sounds when echoes or background noises are present. This deficit makes it difficult for elderly people to understand rapid or broken speech.

With age, the number of hair cells in the saccule, utricle, and ampullae decreases. The number of otoliths also declines. As a result, elderly people experience a decreased sensitivity to gravity, acceleration, and rotation, which may lead to disequilibrium (instability) and vertigo (a feeling of spinning). Some elderly people feel that they can't maintain posture and are prone to fall.

ANSWER TO LEARN TO PREDICT

Since the question emphasized that Freddy is an older man, we can assume that his vision problems are age related. In the discussion of effects of aging on vision, we learned that presbyopia is the most common age-related change in the eye and results from a decrease in the flexibility of the lenses of the eyes. We also learned that we are able to focus images on the retina by changing the shape of the lenses. So, Freddy's lenses

are less flexible and that has reduced his ability to focus. But what objects is he having difficulty seeing? He can't focus on objects that are close to his eyes. Recall that the ciliary muscles contract to change the shape of the lenses when viewing close objects. Freddy's joke that his arms are getting shorter just means that he can't hold objects far enough from his face so the lens shape allows for proper focusing.

Answers to the rest of this chapter's Predict questions are in Appendix E.

SUMMARY

9.1 Sensation   (p. 239)

Perception of sensations results only from those stimuli that reach the cerebral cortex and are consciously perceived. Senses can be defined as general or special.

9.2 Sensory Receptors   (p. 239)

Sensory receptors are sensory nerve endings or specialized cells capable of responding to stimuli by developing action potentials.

9.3 General Senses   (p. 240)

Receptors for general senses, such as pain, temperature, touch, pressure, and proprioception, are scattered throughout the body.

Pain

1. Pain is an unpleasant sensation that may be either localized or diffuse.
2. Pain can be reduced or controlled by "gating mechanisms" involving the gate control theory.
3. Pain from deeper structures may be referred to more superficial structures, such as the skin.

9.4 Special Senses   (p. 242)

Smell and taste respond to chemical stimulation, vision to light stimulation, and hearing and balance to mechanical stimulation.

9.5 Olfaction   (p. 242)

1. Olfactory neurons have enlarged distal ends with long cilia. The cilia have receptors that respond to dissolved substances in the nasal mucus.

2. The wide range of detectable odors results from combinations of receptor responses.

Neuronal Pathways for Olfaction

Axons of the olfactory neurons form the olfactory nerves, which enter the olfactory bulb. Olfactory tracts carry action potentials from the olfactory bulbs to the olfactory cortex of the brain.

9.6 Taste   (p. 243)

1. Taste buds contain taste cells with hairs that extend into taste pores. Receptors on the hairs detect dissolved substances.
2. There are five basic types of taste: sour, salty, bitter, sweet, and umami.

Neuronal Pathways for Taste

The facial nerve carries taste from the anterior two-thirds of the tongue; the glossopharyngeal nerve carries taste from the posterior one-third of the tongue; and the vagus nerve carries taste from the root of the tongue.

9.7 Vision   (p. 244)

Accessory Structures of the Eye

1. The eyebrows prevent perspiration from entering the eyes.
2. The eyelids protect the eyes from foreign objects.
3. The conjunctiva covers the inner eyelids and the anterior surface of the eye.
4. Lacrimal glands produce tears that flow across the eye surface to lubricate and protect the eye. Excess tears pass through the nasolacrimal duct into the nasal cavity.

5. The extrinsic eye muscles move the eyeball.

Anatomy of the Eye

1. The fibrous tunic is the outer layer of the eye. It consists of the sclera and cornea.
2. The vascular tunic is the middle layer of the eye. It consists of the choroid, ciliary body, and iris.
3. The lens is held in place by the suspensory ligaments, which are attached to the smooth muscles of the ciliary body.
4. The nervous tunic (retina) is the inner layer of the eye and contains neurons sensitive to light.
5. Rods are responsible for vision in low illumination (night vision).
6. Cones are responsible for color vision.
7. Light causes retinal to change shape, which causes opsin to change shape, leading eventually to cellular changes that result in vision.
8. The fovea centralis in the center of the macula has the highest concentration of cones and is the area where images are detected most clearly.
9. The optic disc, or blind spot, is where the optic nerve exits the eye and blood vessels enter.
10. The anterior and posterior chambers of the eye are anterior to the lens and are filled with aqueous humor. The vitreous chamber is filled with vitreous humor. The humors keep the eye inflated, refract light, and provide nutrients to the inner surface of the eye.

Functions of the Eye

1. Light passing through a concave surface diverges. Light passing through a convex surface converges.
2. Converging light rays cross at the focal point and are said to be focused.
3. The cornea, aqueous humor, lens, and vitreous humor all refract light. The cornea is responsible for most of the convergence, whereas the lens can adjust the focus by changing shape (accommodation).

Neuronal Pathways for Vision

1. Axons pass through the optic nerves to the optic chiasm, where some cross. Axons from the nasal retina cross, and those from the temporal retina do not.
2. Optic tracts from the chiasm lead to the thalamus.
3. Optic radiations extend from the thalamus to the visual cortex in the occipital lobe.

9.8 Hearing and Balance (p. 253)

Anatomy and Function of the Ear

1. The external ear consists of the auricle and the external auditory canal.
2. The middle ear contains the three auditory ossicles.
3. The tympanic membrane (eardrum) is stretched across the external auditory canal.

4. The malleus, incus, and stapes connect the tympanic membrane to the oval window of the inner ear.
5. The auditory, or eustachian, tube connects the middle ear to the pharynx and equalizes pressure. The middle ear is also connected to the mastoid air cells.
6. The inner ear has three parts: the semicircular canals, the vestibule, and the cochlea.
7. The cochlea is a canal shaped like a snail's shell.
8. The cochlea is divided into three compartments by the vestibular and basilar membranes.
9. The spiral organ consists of hair cells that attach to the basilar and tectorial membranes.

Hearing

1. Sound waves are funneled through the auricle down the external auditory canal, causing the tympanic membrane to vibrate.
2. The tympanic membrane vibrations are passed along the ossicles to the oval window of the inner ear.
3. Movement of the stapes in the oval window causes the perilymph to move the vestibular membrane, which causes the endolymph to move the basilar membrane. Movement of the basilar membrane causes the hair cells in the spiral organ to move and generate action potentials, which travel along the vestibulocochlear nerve.

Neuronal Pathways for Hearing

From the vestibulocochlear nerve, action potentials travel to the cochlear nucleus and on to the cerebral cortex.

Balance

1. Static equilibrium evaluates the position of the head relative to gravity.
2. Maculae, located in the vestibule, consist of hair cells with the microvilli embedded in a gelatinous mass that contains otoliths. The gelatinous mass moves in response to gravity.
3. Dynamic equilibrium evaluates movements of the head.
4. The inner ear contains three semicircular canals, arranged perpendicular to each other. The ampulla of each semicircular canal contains a crista ampullaris, which has hair cells with microvilli embedded in a gelatinous mass, the cupula.

Neuronal Pathways for Balance

Axons in the vestibular portion of the vestibulocochlear nerve project to the vestibular nucleus and on to the cerebral cortex.

9.9 Effects of Aging on the Senses (p. 260)

Elderly people experience a general decline in some general senses and in taste, vision, hearing, and balance.

 # REVIEW AND COMPREHENSION

1. Describe how a stimulus becomes a sensation.
2. Contrast the features of the receptors associated with the general senses.
3. Explain how pain is reduced by analgesics and how it can be modified, according to the gate control theory.
4. Explain referred pain, and give an example.
5. Describe the process by which airborne molecules produce the sensation of smell.
6. How is the sense of taste related to the sense of smell?
7. What are the five primary tastes? How do they produce many different kinds of taste sensations?

8. Describe the following structures and state their functions: eyebrows, eyelids, conjunctiva, lacrimal apparatus, and extrinsic eye muscles.
9. Name the three tunics of the eye. Describe the structures composing each layer, and explain the functions of these structures.
10. Describe the three chambers of the eye, the substances that fill each, and the functions of these substances.
11. Describe the lens of the eye and how it is held in place.
12. Describe the arrangement of rods and cones in the fovea centralis and in the periphery of the eye.
13. What causes the pupil to constrict and dilate?
14. What is the blind spot of the eye, and what causes it?

15. What causes light to refract? What is a focal point?

16. Define accommodation. What does accommodation accomplish?

17. Name the three regions of the ear, name the structures in each region, and state the functions of each structure.

18. Describe the relationship among the tympanic membrane, the auditory ossicles, and the oval window of the inner ear.

19. Describe the structure of the cochlea.

20. Starting with the auricle, trace sound into the inner ear to the point at which action potentials are generated in the vestibulocochlear nerve.

21. Describe the maculae in the vestibule of the ear and their function in balance.

22. What is the function of the semicircular canals? Describe the crista ampullaris and its mode of operation.

CRITICAL THINKING

1. An elderly male with normal vision developed cataracts. A surgeon treated his condition by removing the lenses of his eyes. What kind of glasses do you recommend to compensate for the removal of his lenses?

2. On a camping trip, Starr Gazer was admiring the stars in the night sky. She noticed a little cluster of dim stars at the edge of her vision, but when she looked directly at that part of the sky, she could not see the cluster. On the other hand, when she looked toward the stars but not directly at them, she could see them. Explain what was happening.

3. SCUBA divers are subject to increased pressure as they descend toward the bottom of the ocean. Sometimes this pressure can lead to damage to the ear and loss of hearing. Describe the normal mechanisms that adjust for changes in pressure. Explain how increased pressure might cause reduced hearing, and suggest at least one other common condition that might interfere with this pressure adjustment.

4. If a vibrating tuning fork were placed against the mastoid process of your temporal bone, you would perceive the vibrations as sound, even if the external auditory canal were plugged. Explain how this happens.

5. The main way that people "catch" colds is through their hands. After touching an object contaminated with the cold virus, the person transfers the virus to the nasal cavity, where it causes an infection. Other than the obvious entry of the virus through the nose, how could the virus get into the nasal cavity?

Answers in Appendix D

10

Endocrine System

10.1 PRINCIPLES OF CHEMICAL COMMUNICATION

Module 8 Endocrine System

Learning Outcome *After reading this section, you should be able to*

A. Describe the four classes of chemical messengers.

The body has a remarkable capacity for maintaining homeostasis despite having to coordinate the activities of over 75 trillion cells. The principal means by which this coordination occurs is through chemical messengers, some produced by the nervous system and others produced by the endocrine system. **Chemical messengers** allow cells to communicate with each other to regulate body activities.

Most chemical messengers are produced by a specific collection of cells or by a gland. Recall from chapter 4 that a gland is an organ consisting of epithelial cells that specialize in **secretion,** which is the controlled release of chemicals from a cell. This text identifies four classes of chemical messengers based on the source of the chemical messenger and its mode of transport in the body (table 10.1). In this section, we describe chemical messengers in terms of how they function. But it is important to note that some chemical messengers fall into more than one functional category. For example, prostaglandins are listed in multiple categories because they have diverse functions and cannot be categorized in just one class. Therefore, the study of the endocrine system includes several of the following categories:

1. *Autocrine chemical messengers.* An autocrine chemical messenger stimulates the cell that originally secreted it, and sometimes nearby cells of the same type. Good examples of autocrine chemical messengers are those secreted by white blood cells during an infection. Several types of white blood cells can stimulate their own replication so that the total number of white blood cells increases rapidly (see chapter 14).

TABLE 10.1	Classes of Chemical Messengers	
Chemical Messenger	**Description**	**Example**
Autocrine	Secreted by cells in a local area; influences the activity of the same cell or cell type from which it was secreted	Eicosanoids (prostaglandins, thromboxanes, prostacyclins, leukotrienes)
Paracrine	Produced by a wide variety of tissues and secreted into extracellular fluid; has a localized effect on other tissues	Somatostatin, histamine, eicosanoids
Neurotransmitter	Produced by neurons; secreted into a synaptic cleft by presynaptic nerve terminals; travels short distances; influences postsynaptic cells	Acetylcholine, epinephrine
Endocrine	Secreted into the blood by specialized cells; travels some distance to target tissues; results in coordinated regulation of cell function	Thyroid hormones, growth hormone, insulin, epinephrine, estrogen, progesterone, testosterone, prostaglandins

2. *Paracrine chemical messengers.* Paracrine chemical messengers act locally on nearby cells. These chemical messengers are secreted by one cell type into the extracellular fluid and affect surrounding cells of a different type. An example of a paracrine chemical messenger is histamine, released by certain white blood cells during allergic reactions. Histamine stimulates vasodilation in nearby blood vessels.

3. *Neurotransmitters.* Neurotransmitters are chemical messengers secreted by neurons that activate an adjacent cell, whether it is another neuron, a muscle cell, or a glandular cell. Neurotransmitters are secreted into a synaptic cleft, rather than into the bloodstream (see chapter 8). Therefore, in the strictest sense neurotransmitters are paracrine messengers, but for our purposes it is most appropriate to consider them as a separate category.

4. *Endocrine chemical messengers.* Endocrine chemical messengers are secreted into the bloodstream by certain glands and cells, which together constitute the endocrine system. These chemical messengers affect cells that are distant from their source.

10.2 FUNCTIONS OF THE ENDOCRINE SYSTEM

Learning Outcome *After reading this section, you should be able to*

A. Describe the ten regulatory functions of the endocrine system.

The main regulatory functions of the endocrine system are the following:

1. *Metabolism.* The endocrine system regulates the rate of metabolism, the sum of the chemical changes that occur in tissues.

2. *Control of food intake and digestion.* The endocrine system regulates the level of satiety (fullness) and the breakdown of food into individual nutrients.

3. *Tissue development.* The endocrine system influences the development of tissues, such as those of the nervous system.

4. *Ion regulation.* The endocrine system regulates the solute concentration of the blood.

5. *Water balance.* The endocrine system regulates water balance by controlling solutes in the blood.

6. *Heart rate and blood pressure regulation.* The endocrine system helps regulate the heart rate and blood pressure and helps prepare the body for physical activity.

7. *Control of blood glucose and other nutrients.* The endocrine system regulates the levels of blood glucose and other nutrients in the blood.

8. *Control of reproductive functions.* The endocrine system controls the development and functions of the reproductive systems in males and females.

9. *Uterine contractions and milk release.* The endocrine system regulates uterine contractions during delivery and stimulates milk release from the breasts in lactating females.

10. *Immune system regulation.* The endocrine system helps control the production and functions of immune cells.

10.3 CHARACTERISTICS OF THE ENDOCRINE SYSTEM

Learning Outcomes *After reading this section, you should be able to*

A. Define hormone and target tissue.

B. Distinguish between endocrine and exocrine glands.

The **endocrine system** is composed of **endocrine glands** and specialized endocrine cells located throughout the body (figure 10.1). Endocrine glands and cells secrete minute amounts of chemical messengers called **hormones** (hor′mōnz) into the bloodstream, rather than into a duct. Hormones then travel through the general blood circulation to specific sites called **target tissues** or *effectors,* where they

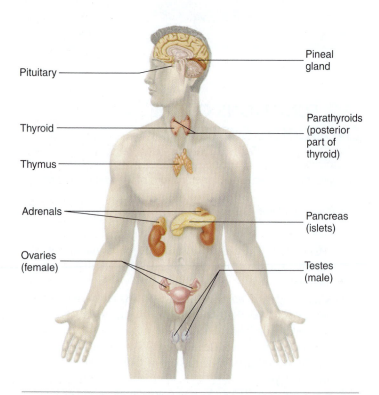

Pituitary

Thyroid

Thymus

Adrenals

Ovaries (female)

Pineal gland

Parathyroids (posterior part of thyroid)

Pancreas (islets)

Testes (male)

Figure 10.1 AP|R **Major Endocrine Glands and Their Locations**

produce a coordinated response of the target tissues. Thus, the term **endocrine** (en′dō-krin), derived from the Greek words *endo,* meaning within, and *krino,* to secrete, appropriately describes this system.

Endocrine glands are not to be confused with **exocrine glands.** Exocrine glands have ducts that carry their secretions to the outside of the body, or into a hollow organ, such as the stomach or intestines. Examples of exocrine secretions are saliva, sweat, breast milk, and digestive enzymes.

The study of the endocrine system, known as **endocrinology,** is the topic of this chapter. In this chapter, we present the general principles of hormones and we discuss specific hormones and their functions.

10.4 HORMONES

Learning Outcomes *After reading this section, you should be able to*

A. Describe the common characteristics of all hormones.

B. List and describe the two chemical categories of hormones.

C. Explain the influence of the chemical nature of a hormone on its transport in the blood, its removal from circulation, and its lifespan.

D. Describe the three main patterns of hormone secretion.

The word *hormone* is derived from the Greek word *hormon,* which means to set into motion. Hormones regulate almost every physiological process in our body.

Chemical Nature of Hormones

Hormones fit into one of two chemical categories: lipid-soluble hormones and water-soluble hormones, a distinction based on their chemical composition, which influences their chemical behavior. Recall from chapter 3 that the plasma membrane is a selectively permeable phospholipid bilayer that excludes most water-soluble molecules but allows lipid-soluble molecules to pass through. Therefore, the entire basis of a hormone's metabolism—its transport in the blood, its interaction with its target, and its removal from the body—is dependent on the hormone's chemical nature.

Within the two chemical categories, hormones can be subdivided into groups based on their chemical structures. Steroid hormones are those derived from cholesterol, thyroid hormones are derived from the amino acid tyrosine, and other hormones are categorized as amino acid derivatives, peptides, or proteins.

Lipid-Soluble Hormones

Lipid-soluble hormones are nonpolar, and include steroid hormones, thyroid hormones, and fatty acid derivative hormones, such as certain eicosanoids.

Transport of Lipid-Soluble Hormones

Because of their small size and low solubility in aqueous fluids, lipid-soluble hormones travel in the bloodstream attached to binding proteins, proteins that transport the hormones. As a result, the rate at which lipid-soluble hormones are degraded or eliminated from the circulation is greatly reduced and their lifespans range from a few days to as long as several weeks.

Without the binding proteins, the lipid-soluble hormones would quickly diffuse out of capillaries and be degraded by enzymes of the liver and lungs or be removed from the body by the

CLINICAL IMPACT   Lipid- and Water-Soluble Hormones in Medicine

Specific hormones are administered to treat certain illnesses. Hormones that are soluble in lipids, such as steroids, may be taken orally because they can diffuse across the wall of the stomach and intestine into the circulatory system. Examples include the synthetic estrogen and progesterone-like hormones in birth control pills and steroids that reduce the severity of inflammation, such as prednisone (pred'ni-sōn). In contrast to lipid-soluble hormones, protein hormones cannot diffuse across the wall of the intestine. Furthermore, protein hormones are broken down to individual amino acids before they are transported across the wall of the digestive system. The normal structure of a protein hormone is therefore destroyed, and its physiological activity is lost. Consequently, protein hormones must be injected rather than taken orally. The most commonly administered protein hormone is insulin, which is prescribed to treat diabetes mellitus.

---

kidneys. Circulating hydrolytic enzymes can also metabolize free lipid-soluble hormones. The breakdown products are then excreted in the urine or the bile.

Water-Soluble Hormones

Water-soluble hormones are polar molecules; they include protein hormones, peptide hormones, and most amino acid derivative hormones.

Transport of Water-Soluble Hormones

Because water-soluble hormones can dissolve in blood, many circulate as free hormones, meaning that most of them dissolve directly into the blood and are delivered to their target tissue without attaching to a binding protein. Because many water-soluble hormones are quite large, they do not readily diffuse through the walls of all capillaries; therefore, they tend to diffuse from the blood into tissue spaces more slowly. The capillaries of organs that are regulated by protein hormones are usually very porous, or *fenestrated* (see chapter 13). On the other hand, other water-soluble hormones are quite small and require attachment to a larger protein to avoid being filtered out of the blood.

All hormones are destroyed either in the circulation or at their target cells. The destruction and elimination of hormones limit the length of time they are active. When hormones are secreted that remain functional for only short periods, the body processes regulated by them tend to change quickly.

Water-soluble hormones have relatively short half-lives because they are rapidly degraded by enzymes, called proteases, within the bloodstream. The kidneys then remove the hormone breakdown products from the blood. Target cells also destroy water-soluble hormones when the hormones are internalized via endocytosis. Once the hormones are inside the target cell, lysosomal enzymes degrade them. Often, the target cell recycles the amino acids of peptide and protein hormones and uses them to synthesize new proteins. Hormones with short half-lives normally have concentrations that change rapidly within the blood and tend to regulate activities that have a rapid onset and short duration.

However, some water-soluble hormones are more stable in the circulation than others. In many instances, protein and peptide hormones have a carbohydrate attached to them, or their terminal ends are modified. These modifications protect them from protease activity to a greater extent than water-soluble hormones lacking such modifications. In addition, some water-soluble hormones also attach to binding proteins and therefore circulate in the plasma longer than free water-soluble hormones do.

10.5 CONTROL OF HORMONE SECRETION

Learning Outcomes   *After reading this section, you should be able to*

A. List and describe the three stimulatory influences on hormone secretion and give examples of each.

B. List and describe the three inhibitory influences on hormone secretion and give examples of each.

C. Describe the major mechanisms that maintain blood hormone levels.

Three types of stimuli regulate hormone release: humoral, neural, and hormonal. No matter what stimulus releases the hormone, however, the blood level of most hormones fluctuates within a homeostatic range through negative-feedback mechanisms (see chapter 1). In a few instances, positive-feedback systems also regulate blood hormone levels.

Stimulation of Hormone Release

Control by Humoral Stimuli

Blood-borne chemicals can directly stimulate the release of some hormones. These chemicals are referred to as **humoral stimuli** because they circulate in the blood, and the word *humoral* refers to body fluids, including blood. These hormones are sensitive to the blood levels of a particular substance, such as glucose, calcium, or sodium. Figure 10.2 illustrates that when the blood level of a particular chemical changes (calcium), the hormone (PTH) is released in response to the chemical's concentration. As another example, if a runner has just finished a long race during hot weather, he may not produce urine for up to 12 hours after the race because his elevated concentration of blood solutes stimulates the release of a water-conservation hormone called antidiuretic hormone (ADH). Similarly, elevated blood glucose levels directly stimulate insulin secretion by the pancreas, and elevated blood potassium levels directly stimulate aldosterone release by the adrenal cortex.

Control by Neural Stimuli

The second type of hormone regulation involves **neural stimuli** of endocrine glands. Following action potentials, neurons release a neurotransmitter into the synapse with the cells that produce the

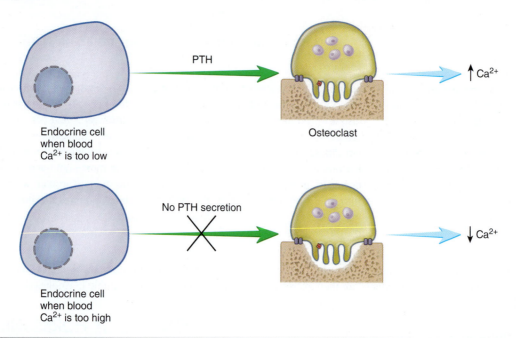

Figure 10.2 [AP|R] **Humoral Regulation of Hormone Secretion**

hormone. In some cases, the neurotransmitter stimulates the cells to increase hormone secretion. Figure 10.3 illustrates the neural control of hormone secretion from cells of an endocrine gland. For example, in response to stimuli, such as stress or exercise, the sympathetic division of the autonomic nervous system (see chapter 8) stimulates the adrenal gland to secrete epinephrine and norepinephrine, which help the body respond to the stimulus. Responses include an elevated heart rate and increased blood flow through the exercising muscles. When the stimulus is no longer present, the neural stimulation declines and the secretion of epinephrine and norepinephrine decreases.

Some neurons secrete chemical messengers directly into the blood when they are stimulated, making these chemical messengers hormones, which are called neuropeptides. Specialized neuropeptides stimulate hormone secretion from other endocrine cells and are called **releasing hormones,** a term usually reserved for hormones from the hypothalamus.

Control by Hormonal Stimuli

The third type of regulation uses **hormonal stimuli.** It occurs when a hormone is secreted that, in turn, stimulates the secretion of other hormones (figure 10.4). The most common examples are hormones from the anterior pituitary gland, called **tropic hormones.** Many tropic hormones are part of a complex process in which a releasing hormone from the hypothalamus stimulates the release of a tropic hormone from the pituitary gland. The pituitary tropic hormone then travels to a third endocrine gland and stimulates the release of a third hormone. For example, hormones from the hypothalamus and anterior pituitary regulate the secretion of thyroid hormones from the thyroid gland.

Inhibition of Hormone Release

Stimulating hormone secretion is important, but inhibiting hormone release is also important. This process involves the same three types of stimuli: humoral, neural, and hormonal.

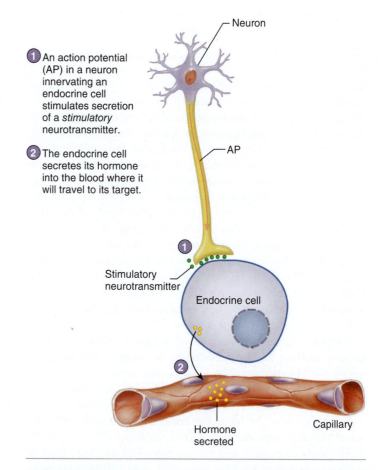

1. An action potential (AP) in a neuron innervating an endocrine cell stimulates secretion of a *stimulatory* neurotransmitter.

2. The endocrine cell secretes its hormone into the blood where it will travel to its target.

PROCESS Figure 10.3 **Control of Hormone Secretion by Direct Neural Innervation**

Inhibition of Hormone Release by Humoral Stimuli

Often when a hormone's release is sensitive to the presence of a humoral stimulus, there exists a companion hormone whose release is inhibited by the same humoral stimulus. Usually, the companion

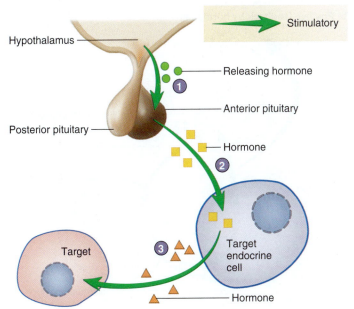

1. Neurons in the hypothalamus release stimulatory hormones, called releasing hormones. Releasing hormones travel in the blood to the anterior pituitary gland.

2. Releasing hormones stimulate the release of hormones from the anterior pituitary, which travel in the blood to their target endocrine cell.

3. The target endocrine cell secretes its hormone into the blood, where it travels to its target and produces a response.

PROCESS Figure 10.4 Hormonal Regulation of Hormone Secretion

hormone's effects oppose those of the secreted hormone and counteract the secreted hormone's action. For example, to raise blood pressure, the adrenal cortex secretes the hormone aldosterone in response to low blood pressure. However, if blood pressure goes up, the atria of the heart secrete the hormone atrial natriuretic peptide (ANP), which lowers blood pressure. Therefore, aldosterone and ANP work together to maintain homeostasis of blood pressure.

Inhibition of Hormone Release by Neural Stimuli

Neurons inhibit targets just as often as they stimulate targets. If the neurotransmitter is inhibitory, the target endocrine gland does not secrete its hormone.

Inhibition of Hormone Release by Hormonal Stimuli

Some hormones prevent the secretion of other hormones, which is a common mode of hormone regulation. For example, hormones from the hypothalamus that prevent the secretion of tropic hormones from the pituitary gland are called **inhibiting hormones.** Thyroid hormones can control their own blood levels by inhibiting their pituitary tropic hormone. Without the original stimulus, less thyroid hormone is released.

Regulation of Hormone Levels in the Blood

Two major mechanisms maintain hormone levels in the blood within a homeostatic range: negative feedback and positive feedback (see chapter 1).

1. *Negative feedback.* Most hormones are regulated by a negative-feedback mechanism, whereby the hormone's secretion is inhibited by the hormone itself once blood levels have reached a certain point and there is adequate hormone to activate the target cell. The hormone may inhibit the action of other, stimulatory hormones to prevent the secretion of the hormone in question. Thus, it is a self-limiting system (figure 10.5*a*). For example, thyroid hormones inhibit the secretion of their releasing hormone from the hypothalamus and their tropic hormone from the anterior pituitary.

2. *Positive feedback.* Some hormones, when stimulated by a tropic hormone, promote the synthesis and secretion of the tropic hormone in addition to stimulating their target cell. In turn, this stimulates further secretion of the original hormone. Thus, it is a self-propagating system (figure 10.5*b*). For example, prolonged estrogen stimulation promotes a release of the anterior pituitary hormone responsible for stimulating ovulation.

10.6 HORMONE RECEPTORS AND MECHANISMS OF ACTION

Learning Outcomes *After reading this section, you should be able to*

A. Describe the general properties of a receptor.
B. Explain the mechanisms of action for the two types of receptor classes.
C. Define amplification, and explain how, despite small hormone concentrations, water-soluble hormones can cause rapid responses.

Hormones exert their actions by binding to proteins called **receptors.** A hormone can stimulate only the cells that have the receptor for that hormone. The portion of each receptor molecule where a hormone binds is called a **receptor site,** and the shape and chemical characteristics of each receptor site allow only a specific type of hormone to bind to it. The tendency for each type of hormone to bind to one type of receptor, and not to others, is called **specificity** (figure 10.6). For example, insulin binds to insulin receptors, but not to receptors for thyroid hormones. However, some hormones, such as epinephrine, can bind to a "family" of receptors that are structurally similar. Because hormone receptors have a high affinity for the hormones that bind to them, only a small concentration of a given hormone is needed to activate a significant number of its receptors.

Classes of Receptors

Lipid-soluble and water-soluble hormones bind to their own classes of receptors. Figure 10.7 provides an overview of receptor type and mechanism of action.

1. *Lipid-soluble hormones bind to nuclear receptors.* Lipid-soluble hormones tend to be relatively small. They diffuse through the plasma membrane and bind to **nuclear receptors,** which are most often found in the cell nucleus (figure 10.7*a*). Nuclear receptors can also be located in the cytoplasm, but then move to the nucleus when activated. When hormones bind to nuclear receptors, the hormone-receptor complex interacts with DNA in the nucleus or with cellular enzymes to regulate the transcription of particular

(a) Negative feedback by hormones

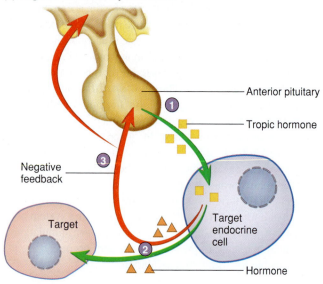

Anterior pituitary

Tropic hormone

Negative feedback

Target

Target endocrine cell

Hormone

1. The anterior pituitary gland secretes a tropic hormone, which travels in the blood to the target endocrine cell.

2. The hormone from the target endocrine cell travels to its target.

3. The hormone from the target endocrine cell also has a negative-feedback effect on the anterior pituitary and hypothalamus and decreases secretion of the tropic hormone.

(b) Positive feedback by hormones

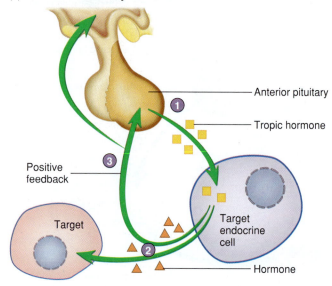

Anterior pituitary

Tropic hormone

Positive feedback

Target

Target endocrine cell

Hormone

1. The anterior pituitary gland secretes a tropic hormone, which travels in the blood to the target endocrine cell.

2. The hormone from the target endocrine cell travels to its target.

3. The hormone from the target endocrine cell also has a positive-feedback effect on the anterior pituitary and increases secretion of the tropic hormone.

PROCESS Figure 10.5 Negative and Positive Feedback

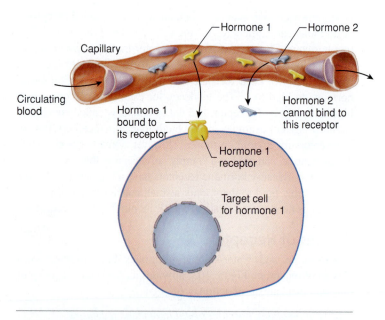

Capillary

Circulating blood

Hormone 1

Hormone 2

Hormone 1 bound to its receptor

Hormone 2 cannot bind to this receptor

Hormone 1 receptor

Target cell for hormone 1

Figure 10.6 AP|R **Target Tissue Specificity and Response**

Hormones bind to receptor proteins. The shape and chemical nature of each receptor site allow only certain hormones to bind. This relationship is called specificity. Additionally, in order for a target cell to respond to its hormone, the hormone must bind to its receptor.

genes in the target tissue, a process that takes several minutes to several hours. Thyroid hormones and steroid hormones (testosterone, estrogen, progesterone, aldosterone, and cortisol) generally bind to nuclear receptors.

In addition to modulating gene transcription, it is now recognized that lipid-soluble hormones have rapid effects (less than 1 minute) on target cells. These effects are most likely mediated through membrane-bound receptors (see next section).

2. *Water-soluble hormones bind to membrane-bound receptors.* Water-soluble hormones are polar molecules and cannot pass through the plasma membrane. Instead, they interact with **membrane-bound receptors,** which are proteins that extend across the plasma membrane, with their hormone-binding sites exposed on the plasma membrane's outer surface (figure 10.7b). When a hormone binds to a receptor on the outside of the plasma membrane, the hormone-receptor complex initiates a response inside the cell. Hormones that bind to membrane-bound receptors include proteins, peptides, some amino acid derivatives, such as epinephrine and norepinephrine, and some lipid-soluble hormones.

Action of Nuclear Receptors

After lipid-soluble hormones diffuse across the plasma membrane and bind to their receptors, the hormone-receptor complex binds to DNA to produce a response (figure 10.8). The receptors that bind to DNA have fingerlike projections that recognize and bind

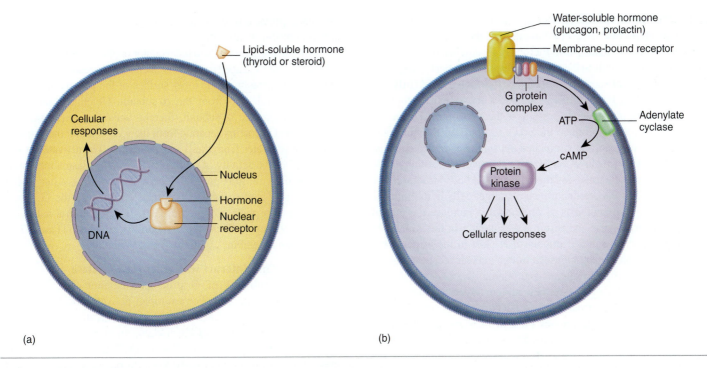

Figure 10.7 General Comparison of Nuclear and Membrane-Bound Receptors

(a) *Lipid-soluble* hormones diffuse through the plasma membrane of its target cell and bind to a cytoplasmic receptor or a nuclear receptor. In the nucleus, the combination of the hormone and the receptor initiates protein synthesis, described later in this chapter. (b) *Water-soluble* hormones bind to the external portion of membrane-bound receptors, which are integral membrane proteins on its target cell.

1. Lipid-soluble hormones diffuse through the plasma membrane.

2. Lipid-soluble hormones either bind to cytoplasmic receptors and travel to the nucleus or bind to nuclear receptors.

3. The hormone-receptor complex binds to a hormone-response element on the DNA, acting as a transcription factor.

4. The binding of the hormone-receptor complex to DNA stimulates the synthesis of messenger RNA (mRNA), which codes for specific proteins.

5. The mRNA leaves the nucleus, passes into the cytoplasm of the cell, and binds to ribosomes, where it directs the synthesis of specific proteins.

6. The newly synthesized proteins produce the cell's response to the lipid-soluble hormones—for example, the secretion of a new protein.

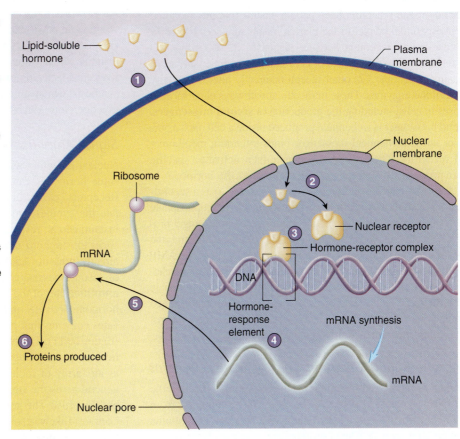

PROCESS Figure 10.8 [AP|R] **Nuclear Receptor Model**

to specific nucleotide sequences in the DNA called **hormone-response elements.** The combination of the hormone and its receptor forms a **transcription factor** because, when the hormone-receptor complex binds to the hormone-response element, it regulates the transcription of specific **messenger ribonucleic acid (mRNA)** molecules. Newly formed mRNA molecules move to the cytoplasm to be translated into specific proteins at the ribosomes. The newly synthesized proteins produce the cell's response to the hormone. For example, testosterone stimulates the synthesis of proteins that are responsible for male secondary sex characteristics, such as the formation of muscle mass and the typical male body structure. The steroid hormone aldosterone affects its target cells in the kidneys by stimulating the synthesis of proteins that increase the rate of Na^+ and K^+ transport. The result is a reduction in the amount of Na^+ and an increase in the amount of K^+ lost in the urine. Other hormones that produce responses through nuclear receptor mechanisms include thyroid hormones and vitamin D.

Target cells that synthesize new protein molecules in response to hormonal stimuli normally have a latent period of several hours between the time the hormones bind to their receptors and the time responses are observed. During this latent period, mRNA and new proteins are synthesized. Hormone-receptor complexes are eventually degraded within the cell, limiting the length of time hormones influence the cells' activities, and the cells slowly return to their previous functional states.

Membrane-Bound Receptors and Signal Amplification

Membrane-bound receptors have peptide chains that are anchored in the phospholipid bilayer of the plasma membrane (see chapter 3). Membrane-bound receptors activate responses in two ways: (1) Some receptors alter the activity of G proteins at the inner surface of the plasma membrane; (2) other receptors directly alter the activity of intracellular enzymes. These intracellular pathways elicit specific responses in cells, including the production of **second messengers.** A second messenger is a chemical produced inside a cell once a hormone or another chemical messenger binds to certain membrane-bound receptors. The second messenger then activates specific cellular processes inside the cell in response to the hormone. In some cases, this coordinated set of events is referred to as a **second-messenger system.** For example, cyclic adenosine monophosphate (cAMP) (the second messenger) is a common second messenger produced when a ligand binds to its receptor. Rather than the ligand (the first messenger) entering the cell to activate a cellular process, cAMP (the second messenger) stimulates the cellular process. This mechanism is usually employed by water-soluble hormones that are unable to cross the target cell's membrane. It has also been demonstrated that some lipid-soluble hormones activate second messenger systems, which is consistent with actions via membrane-bound receptors.

Membrane-Bound Receptors That Activate G Proteins

Many membrane-bound receptors produce responses through the action of G proteins. G proteins consist of three subunits; from largest to smallest, they are called alpha (α), beta (β), and gamma (γ) (figure 10.9, step 1). The G proteins are so named because one of the subunits binds to guanine nucleotides. In the inactive state, a guanine diphosphate (GDP) molecule is bound to the α subunit

of each G protein. In the active state, guanine triphosphate (GTP) is bound to the α subunit.

After a hormone binds to the receptor on the outside of a cell, the receptor changes shape (figure 10.9, step 2). As a result, the receptor binds to a G protein on the inner surface of the plasma membrane, and GDP is released from the α subunit. Guanine triphosphate (GTP) binds to the α subunit, thereby activating it (figure 10.9, step 3). The G proteins separate from the receptor, and the activated α subunit separates from the β and γ subunits. The activated α subunit can alter the activity of molecules within the plasma membrane or inside the cell, thus producing cellular responses. After a short time, the activated α subunit is turned off because the G protein removes a phosphate group from GTP, which converts it to GDP (figure 10.9, step 4). Thus, the α subunit is called a GTPase. The α subunit then recombines with the β and γ subunits.

G Proteins That Interact with Adenylate Cyclase

Activated α subunits of G proteins can alter the activity of enzymes inside the cell. For example, activated α subunits can influence the rate of cAMP formation by activating or inhibiting **adenylate cyclase** (a-den′i-lāt sī′klās), an enzyme that converts ATP to cAMP (figure 10.10). Cyclic AMP functions as a second messenger. For example, cAMP binds to protein kinases and activates them. **Protein kinases** are enzymes that, in turn, regulate the activity of other enzymes. Depending on the other enzyme, protein kinases can increase or decrease its activity. The amount of time cAMP is present to produce a response in a cell is limited. An enzyme in the cytoplasm, called **phosphodiesterase** (fos′fō-dī-es′ter-ās), breaks down cAMP to AMP. Once cAMP levels drop, the enzymes in the cell are no longer stimulated.

Cyclic AMP can elicit many different responses in the body because each cell type possesses a unique set of enzymes. For example, the hormone glucagon binds to receptors on the surface of liver cells, activating G proteins and causing an increase in cAMP synthesis, which stimulates the activity of enzymes that break down glycogen into glucose for release from liver cells (figure 10.10).

Signal Amplification

Nuclear receptors work by activating protein synthesis, which for some hormones can take several hours (see "Action of Nuclear Receptors" earlier in this chapter). However, hormones that stimulate the synthesis of second messengers can produce an almost instantaneous response because the second messenger influences existing enzymes. In other words, the response proteins are already present. Additionally, each receptor produces thousands of second messengers, leading to a cascade effect and ultimately **amplification** of the hormonal signal. With amplification, a single hormone activates many second messengers, each of which activates enzymes that produce an enormous amount of final product (figure 10.11). The efficiency of this second-messenger amplification is virtually unparalleled in the body and can be thought of as an "army of molecules" launching an offensive. In a war, the general gives the signal to attack, and thousands of soldiers carry out the order. The general alone could not eliminate thousands of enemies. Likewise, one hormone could not single-handedly produce millions of final products within a few seconds. However, with amplification, one hormone has an army of molecules working simultaneously to produce the final products.

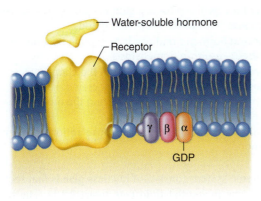

Water-soluble hormone
Receptor
γ β α
GDP

① Before the hormone binds to its receptor, the G protein consists of three subunits, with GDP attached to the α subunit, and freely floats in the plasma membrane.

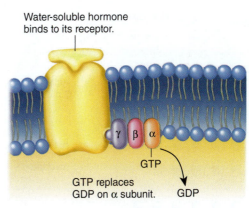

Water-soluble hormone binds to its receptor.
γ β α
GTP
GDP
GTP replaces GDP on α subunit.

② After the hormone binds to its membrane-bound receptor, the receptor changes shape, and the G protein binds to it. GTP replaces GDP on the α subunit of the G protein.

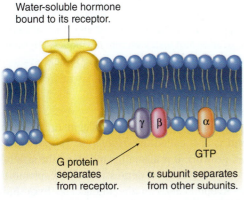

Water-soluble hormone bound to its receptor.
γ β α
GTP
G protein separates from receptor.
α subunit separates from other subunits.

③ The G protein separates from the receptor. The GTP-linked α subunit activates cellular responses, which vary among target cells.

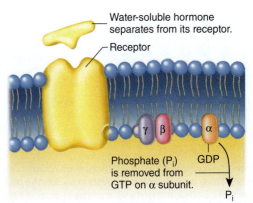

Water-soluble hormone separates from its receptor.
Receptor
γ β α
Phosphate (Pi) is removed from GTP on α subunit.
GDP
Pi

④ When the hormone separates from the receptor, additional G proteins are no longer activated. Inactivation of the α subunit occurs when phosphate (Pi) is removed from the GTP, leaving GDP bound to the α subunit.

PROCESS Figure 10.9 AP|R **Membrane-Bound Receptors Activating G Proteins**

① After a water-soluble hormone binds to its receptor, the G protein is activated.

② The activated α subunit, with GTP bound to it, binds to and activates an adenylate cyclase enzyme so that it converts ATP to cAMP.

③ The cAMP can activate protein kinase enzymes, which phosphorylate specific enzymes activating them. The chemical reactions catalyzed by the activated enzymes produce the cell's response.

④ Phosphodiesterase enzymes inactivate cAMP by converting cAMP to AMP.

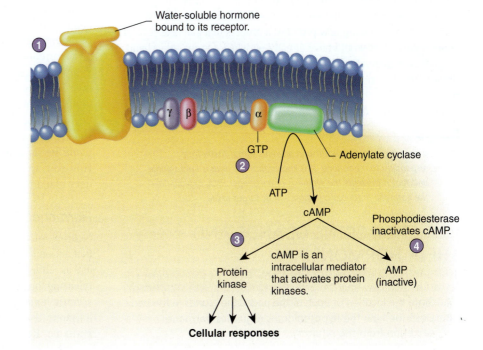

Water-soluble hormone bound to its receptor.
γ β α
GTP
Adenylate cyclase
ATP
cAMP
Phosphodiesterase inactivates cAMP.
Protein kinase
cAMP is an intracellular mediator that activates protein kinases.
AMP (inactive)
Cellular responses

PROCESS Figure 10.10 AP|R
Membrane-Bound Receptors Activating G Proteins to Increase the Synthesis of cAMP

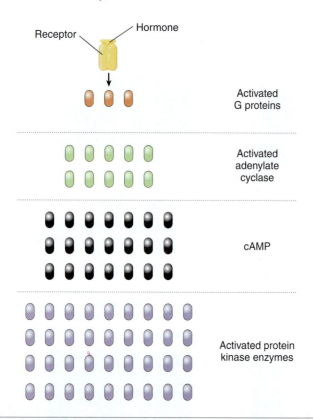

Receptor — Hormone

Activated
G proteins

Activated
adenylate
cyclase

cAMP

Activated protein
kinase enzymes

Figure 10.11 Cascade Effect

The combination of a hormone with a membrane-bound receptor activates several G proteins. The G proteins, in turn, activate many inactive adenylate cyclase enzymes, which cause the synthesis of a large number of cAMP molecules. The large number of cAMP molecules, in turn, activate many inactive protein kinase enzymes, which produce a rapid and amplified response.

Both nuclear receptor and membrane-bound receptor hormone systems are effective, but each is more suited to one type of response than another. For example, the reason epinephrine is effective in a fight-or-flight situation is that it can turn on the target cell responses within a few seconds. If running away from an immediate threat depended on producing new proteins, a process that can take several hours, many of us would have already perished. On the other hand, pregnancy maintenance is mediated by steroids, long-acting hormones, which is reflected by the fact that pregnancy is a long-term process. Thus, it is important that our bodies have hormones that can function over differing time scales.

Predict 2

A drug binds to a receptor and prevents target tissue from responding to a hormone. It is known that the drug is lipid-soluble and that it prevents the synthesis of messenger RNA. Explain how the hormone produces a response in its target tissue.

10.7 ENDOCRINE GLANDS AND THEIR HORMONES

Learning Outcomes *After reading this section, you should be able to*

A. State the location of each of the endocrine glands in the body.
B. Describe how the hypothalamus regulates hormone secretion from the pituitary.

C. Describe how the pituitary gland regulates the secretion of hormones from other endocrine glands.
D. Choose a hormone and use it to explain how negative feedback results in homeostasis.
E. For each of the major hormones in the chapter, describe the endocrine gland from which it is secreted, its target tissue, the response of the target tissue, and the means by which its secretion is regulated.
F. List the effects of hyper- and hyposecretion of the major hormones.

The endocrine system consists of ductless glands that secrete hormones into the interstitial fluid (see figure 10.1). The hormones then enter the blood. Not surprisingly, the organs in the body with the richest blood supply are endocrine glands, such as the adrenal gland and the thyroid gland.

Some glands of the endocrine system perform functions in addition to hormone secretion. For example, the endocrine part of the pancreas has cells that secrete hormones, whereas the much larger exocrine portion of the pancreas secretes digestive enzymes. Portions of the ovaries and testes secrete hormones, but other parts of the ovaries and testes produce oocytes (female sex cells) or sperm cells (male sex cells), respectively.

Pituitary and Hypothalamus

The **pituitary** (pi-too′i-tār-rē; *pituita*, phlegm or thick mucous secretion) **gland** is also called the *hypophysis* (hī-pof′i-sis; *hypo*, under + *physis*, growth). It is a small gland about the size of a pea (figure 10.12, *top*). It rests in a depression of the sphenoid bone inferior to the hypothalamus of the brain. The **hypothalamus** (hī′pō-thal′ă-mŭs; *hypo*, under + *thalamos*) is an important autonomic nervous system and endocrine control center of the brain located inferior to the thalamus. The pituitary gland lies posterior to the optic chiasm and is connected to the hypothalamus by a stalk called the **infundibulum** (in-fŭn-dib′ŭ-lŭm; a funnel). The pituitary gland is divided into two parts: The **anterior pituitary** is made up of epithelial cells derived from the embryonic oral cavity; the **posterior pituitary** is an extension of the brain and is composed of nerve cells. The hormones secreted from each lobe of the pituitary gland are listed in table 10.2.

Hormones from the pituitary gland control the functions of many other glands in the body, such as the ovaries, the testes, the thyroid gland, and the adrenal cortex (figure 10.12, *bottom*). The pituitary gland also secretes hormones that influence growth, kidney function, birth, and milk production by the mammary glands. Historically, the pituitary gland was known as the body's *master gland* because it controls the function of so many other glands. However, we now know that the hypothalamus controls the pituitary gland in two ways: hormonal control and direct innervation.

Hormonal Control of the Anterior Pituitary

Neurons of the hypothalamus produce and secrete neuropeptides that act on cells of the anterior pituitary gland (figure 10.13). They act as either releasing hormones or inhibiting hormones. Each releasing hormone stimulates the production and secretion of a specific hormone by the anterior pituitary, whereas each inhibiting hormone decreases the secretion of a specific anterior pituitary hormone. Releasing and inhibiting hormones enter a capillary bed

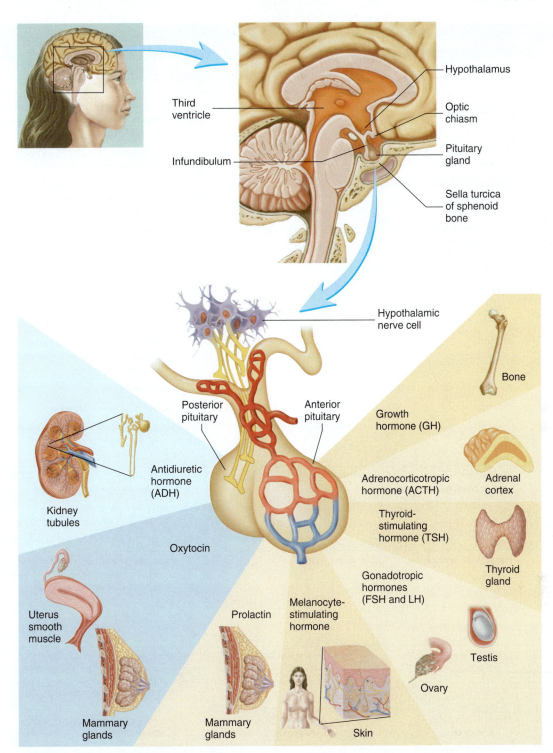

Figure 10.12 AP|R **Pituitary Gland, Its Hormones, and Their Target Tissues**

(*Top*) The pea-sized pituitary gland lies inferior to the hypothalamus of the brain. The infundibulum connects the hypothalamus and the pituitary gland. (*Bottom*) Hormones secreted by the posterior pituitary are indicated by *blue shading,* and hormones secreted by the anterior pituitary by *peach shading.*

in the hypothalamus and are transported through veins to a second capillary bed in the anterior pituitary. There they leave the blood and bind to membrane-bound receptors involved with regulating anterior pituitary hormone secretion. The capillary beds and veins that transport the releasing and inhibiting hormones are called the **hypothalamic-pituitary portal system.**

Direct Innervation of the Posterior Pituitary

Stimulation of neurons within the hypothalamus controls the secretion of hormones from the posterior pituitary (figure 10.14). The cell bodies of these neurons are in the hypothalamus, and their axons extend through the infundibulum to the posterior pituitary. Hormones are produced in the nerve cell bodies and transported through the

Gland	Hormone	Target Tissue	Response
TABLE 10.2	**Endocrine Glands, Hormones, and Their Target Tissues**		
Pituitary gland			
Anterior	Growth hormone	Most tissues	Increases gene expression, breakdown of lipids, and release of fatty acids from cells; increases blood glucose levels
	Thyroid-stimulating hormone (TSH)	Thyroid gland	Increases thyroid hormone secretion (thyroxine and triiodothyronine)
	Adrenocorticotropic hormone (ACTH)	Adrenal cortex	Increases secretion of glucocorticoid hormones, such as cortisol; increases skin pigmentation at high concentrations
	Melanocyte-stimulating hormone (MSH)	Melanocytes in skin	Increases melanin production in melanocytes to make skin darker in color
	Luteinizing hormone (LH) or interstitial cell-stimulating hormone (ICSH)	Ovary in females, testis in males	Promotes ovulation and progesterone production in ovary; promotes testosterone synthesis and support for sperm cell production in testis
	Follicle-stimulating hormone (FSH)	Follicles in ovary in females, seminiferous tubules in males	Promotes follicle maturation and estrogen secretion in ovary; promotes sperm cell production in testis
	Prolactin	Ovary and mammary gland in females, testis in males	Stimulates milk production and prolongs progesterone secretion following ovulation and during pregnancy in women; increases sensitivity to LH in males
Posterior	Antidiuretic hormone (ADH)	Kidney	Conserves water; constricts blood vessels
	Oxytocin	Uterus	Increases uterine contractions
		Mammary gland	Increases milk letdown from mammary glands
Thyroid gland	Thyroid hormones (thyroxine, triiodothyronine)	Most cells of the body	Increase metabolic rates, essential for normal process of growth and maturation
	Calcitonin	Primarily bone	Decreases rate of bone breakdown; prevents large increase in blood Ca^{2+} levels following a meal
Parathyroid glands	Parathyroid hormone	Bone, kidney	Increases rate of bone breakdown by osteoclasts; increases vitamin D synthesis, essential for maintenance of normal blood calcium levels
Adrenal medulla	Epinephrine mostly, some norepinephrine	Heart, blood vessels, liver, fat cells	Increases cardiac output; increases blood flow to skeletal muscles and heart; increases release of glucose and fatty acids into blood; in general, prepares body for physical activity
Adrenal cortex	Mineralocorticoids (aldosterone)	Kidneys; to lesser degree, intestine and sweat glands	Increase rate of sodium transport into body; increase rate of potassium excretion; secondarily favor water retention
	Glucocorticoids (cortisol)	Most tissues (e.g., liver, fat, skeletal muscle, immune tissues)	Increase fat and protein breakdown; increase glucose synthesis from amino acids; increase blood nutrient levels; inhibit inflammation and immune response
	Adrenal androgens	Most tissues	Insignificant in males; increase female sexual drive, growth of pubic and axillary hair
Pancreas	Insulin	Especially liver, skeletal muscle, adipose tissue	Increases uptake and use of glucose and amino acids
	Glucagon	Primarily liver	Increases breakdown of glycogen and release of glucose into the circulatory system

TABLE 10.2   *continued*

Gland	Hormone	Target Tissue	Response
Reproductive organs			
Testes 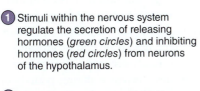	Testosterone	Most tissues	Aids in sperm cell production, maintenance of functional reproductive organs, secondary sexual characteristics, sexual behavior
Ovaries	Estrogens, progesterone	Most tissues	Aid in uterine and mammary gland development and function, external genitalia structure, secondary sexual characteristics, sexual behavior, menstrual cycle
Uterus, ovaries, inflamed tissues	Prostaglandins	Most tissues	Mediate inflammatory responses; increase uterine contractions and ovulation
Thymus	Thymosin	Immune tissues	Promotes immune system development and function
Pineal gland	Melatonin	Among others, hypothalamus	Inhibits secretion of gonadotropin-releasing hormone, thereby inhibiting reproduction

Pineal gland

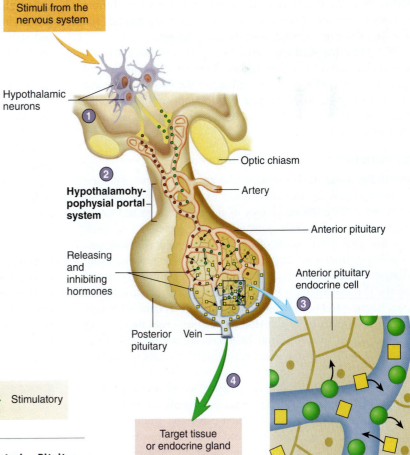

1. Stimuli within the nervous system regulate the secretion of releasing hormones (*green circles*) and inhibiting hormones (*red circles*) from neurons of the hypothalamus.

2. Releasing hormones and inhibiting hormones pass through the hypothalamohypophysial portal system to the anterior pituitary.

3. Releasing hormones and inhibiting hormones (*green and red circles*) leave capillaries and stimulate or inhibit the release of hormones (*yellow squares*) from anterior pituitary cells.

4. In response to releasing hormones, anterior pituitary hormones (*yellow squares*) travel in the blood to their target tissues (*green arrow*), which in some cases, are other endocrine glands.

Stimuli from the nervous system

Hypothalamic neurons

Optic chiasm

Hypothalamohy-pophysial portal system

Artery

Anterior pituitary

Releasing and inhibiting hormones

Anterior pituitary endocrine cell

Posterior pituitary

Vein

Target tissue or endocrine gland

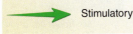 Stimulatory

PROCESS Figure 10.13 AP|R **Hypothalamus and Anterior Pituitary**

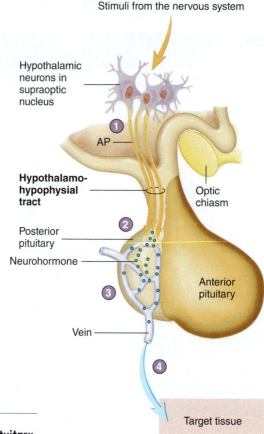

Stimuli from the nervous system

1. Stimuli within the nervous system cause hypothalamic neurons to either increase or decrease their action potential frequency.

2. Action potentials are conducted by axons of the hypothalamic neurons through the hypothalamohypophysial tract to the posterior pituitary. The axon endings of neurons store neurohormones in the posterior pituitary.

3. In the posterior pituitary gland, action potentials cause the release of neurohormones (*blue circles*) from axon terminals into the circulatory system.

4. The neurohormones pass through the circulatory system and influence the activity of their target tissues.

Hypothalamic neurons in supraoptic nucleus

AP

Hypothalamo-hypophysial tract

Optic chiasm

Posterior pituitary

Neurohormone

Anterior pituitary

Vein

Target tissue

PROCESS Figure 10.14 APR **Hypothalamus and Posterior Pituitary**

Endocrine

axons to the posterior pituitary, where they are stored in the axon endings. When these nerve cells are stimulated, action potentials from the hypothalamus travel along the axons to the posterior pituitary and cause the release of hormones from the axon endings.

Within the hypothalamus and pituitary, the nervous and endocrine systems are closely interrelated. Emotions such as joy and anger, as well as chronic stress, influence the endocrine system through the hypothalamus. Conversely, hormones of the endocrine system can influence the functions of the hypothalamus and other parts of the brain.

Hormones of the Anterior Pituitary

Growth hormone (GH) stimulates the growth of bones, muscles, and other organs by increasing gene expression. It also resists protein breakdown during periods of food deprivation and favors lipid breakdown. Too little growth hormone secretion can result from abnormal development of the pituitary gland. A young person suffering from a deficiency of growth hormone remains small, although normally proportioned, and is called a **pituitary dwarf** (dwŏrf). This condition can be treated by administering growth hormone. Because GH is a protein, it is difficult to produce artificially using conventional techniques. However, human genes for GH have been successfully introduced into bacteria using genetic engineering techniques. The gene in the bacteria causes GH synthesis, and the GH can be extracted from the medium in which the bacteria are grown. Thus, modern genetic engineering has provided a source of human GH for people who produce inadequate quantities.

Excess growth hormone secretion can result from hormone-secreting tumors of the pituitary gland. If excess growth hormone is present before bones finish growing in length, exaggerated bone

growth occurs. The person becomes abnormally tall, a condition called **giantism** (jī'an-tizm). If excess hormone is secreted after growth in bone length is complete, growth continues in bone diameter only. As a result, the facial features and hands become abnormally large, a condition called **acromegaly** (ak-rō-meg'ă-lē).

The secretion of growth hormone is controlled by two hormones from the hypothalamus. A releasing hormone stimulates growth hormone secretion, and an inhibiting hormone inhibits its secretion. Most people have a rhythm of growth hormone secretion, with daily peak levels occurring during deep sleep. Growth hormone secretion also increases during periods of fasting and exercise. Blood growth hormone levels do not become greatly elevated during periods of rapid growth, although children tend to have somewhat higher blood levels of growth hormone than do adults. In addition to growth hormone, genetics, nutrition, and sex hormones influence growth.

Part of the effect of growth hormone is influenced by a group of protein hormones called **insulin-like growth factors (IGFs),** or *somatomedins* (sō'mă-tō-mē'dinz). Growth hormone increases IGF secretion from tissues such as the liver, and the IGF molecules bind to receptors on the cells of tissues such as bone and cartilage, where they stimulate growth. The IGFs are similar in structure to insulin and can bind, to some degree, to insulin receptors. Also, insulin, at high concentrations, can bind to IGF receptors.

Predict 3

Mr. Hoops has a son who wants to be a basketball player. Mr. Hoops knows something about GH. He asked his son's doctor if he would prescribe some GH for his son, so that he can grow taller. What do you think the doctor tells Mr. Hoops? Why?

MICROBES IN YOUR BODY Do our bacteria make us fat?

Obesity has increased at an alarming rate over the last three decades. It is estimated that over 150 billion adults worldwide are overweight or obese. In the United States, 1/3 of adults are obese. As obesity rates have increased, so have the rates of obesity-related health conditions such as insulin resistance, diabetes, and cardiovascular disease. Why this dramatic increase? There are two main reasons for obesity: diet/lifestyle and gut bacteria; and it seems these two may be related.

The most familiar cause of obesity is diet and lifestyle. The "typical" Western diet consists of frequent large meals high in refined grains, red meat, saturated fats, and sugary drinks. This is in sharp contrast to healthier diets rich in whole grains, vegetables, fruits, and nuts that help with weight control and prevention of chronic disease. From an evolutionary perspective, our bodies are adapted to conserve energy because food sources were scarce for ancient humans. Many of us now have easy access to energy-rich foods. Combined with a reduction in physical activity and less sleep for many Americans, the Western diet and lifestyle can lead to obesity and poor health.

However, could humans' gut microbiota be just as responsible (or even more responsible) for obesity? Comparisons between the gut microbiota of lean versus obese individuals seem to suggest the possibility of an important link between gut microbiota and our weight. The human gut, like other animals, is densely populated with microbiota consisting of at least 100 trillion microbial cells divided into approximately 1000 different species. The majority (90%) of human gut bacteria fall into two groups: Firmicutes and Bacteroidetes. Lean people have more Bacteroidetes than Firmicutes, while the opposite is true for obese people.

We now know that gut microbiota affect nutrient processing and absorption, hormonal regulation of nutrient use by body cells, and even our hunger level. In addition, our diet can influence the type of bacteria in our GI system. Studies of humans on carbohydrate-restricted or fat-restricted diets demonstrated that after weight loss, the number of Bacteroidetes ("lean person" bacteria) increased, while the number of Firmicutes ("obese person" bacteria) decreased. This makes sense in light of the fact that Firmicutes bacteria break down ingested food more completely than Bacterioidetes, which makes the food's energy easier to absorb by the human gut. Obese individuals store the absorbed energy in adipose tissue, which contributes to weight gain.

Furthermore, experiments with germ-free mice—mice lacking normal gut microbiota—have demonstrated just how important normal gut bacteria are for homeostasis. In the absence of normal gut microbiota, malfunctions in germ-free mice are widespread and significant. For example, when germ-free mice received gut microbiota transplants from normal mice, their body fat increased significantly to normal levels within 2 weeks even though their diet and exercise level did not change. Studies have also shown that germ-free mice lack normal gastric immunity, but upon transplantation, their gastric immune system becomes functional. Germ-free mice also lack cell membrane proteins important for tight junction formation between the cells of the intestinal lining (see chapter 4). Without the normal microbiota, germ-free mice intestines are "leaky" meaning they could easily be penetrated by pathogens. Finally, germ-free mice display an enhanced stress response, which is substantially reduced upon implantation of gut microbiota. Overall, these experiments demonstrate that there is a much greater correlation between bacteria, gut health, obesity, and anxiety than ever before realized.

Changes in gut microbiota also alter the hormonal regulation of nutrient use. Inflammation-promoting effects of an imbalanced gut microbiota is thought to induce obesity via promoting insulin resistance, a known autoimmune malfunction. This observation is supported by the reduction in diabetes symptoms after gastric by-pass surgery when patients exhibit a major shift in gut microbiota populations. Finally, it is well documented that normal gut microbiota metabolism is critical for secretion of several anti-hunger hormones, and anti-depressive neurotransmitters and neurochemicals. Shifts in normal gut microbiota, as related to diet, may very well disrupt normal anti-hunger signals and gut permeability leading to the overeating and inflammation related to obesity.

These observations beg the question: can we manipulate gut microbiota in obese people to cause them to become lean? Several possibilities exist, including the distinct possibility that prescribing antibiotics against bacteria associated with obesity could shift the metabolism of an obese person to become leaner. Another possibility is the use of prebiotics—non-digestible sugars that enhance the growth of beneficial microbiota. Finally, probiotic use is another possible intervention for obesity. Probiotics are non-pathogenic live bacteria that confer a health benefit to the host. This is a rapidly expanding field that holds much promise, but it is still in its beginning stages of our understanding.

Thyroid-stimulating hormone (TSH) binds to membrane-bound receptors on cells of the thyroid gland and causes the cells to secrete thyroid hormone. When too much TSH is secreted, the thyroid gland enlarges and secretes too much thyroid hormone. When too little TSH is secreted, the thyroid gland decreases in size and secretes too little thyroid hormone. The rate of TSH secretion is increased by a releasing hormone from the hypothalamus.

Adrenocorticotropic (a-drē′nō-kōr′ti-kō-trō′pik) **hormone (ACTH)** binds to membrane-bound receptors on cells in the cortex of the adrenal glands. ACTH increases the secretion of a hormone from the adrenal cortex called **cortisol** (kōr′ti-sol), also called hydrocortisone. ACTH is required to keep the adrenal cortex from degenerating. ACTH molecules also bind to melanocytes in the skin and increase skin pigmentation (see chapter 5). One symptom of too much ACTH secretion is darkening of the skin. The rate of ACTH secretion is increased by a releasing hormone from the hypothalamus.

Gonadotropins (gō′nad-ō-trō′pinz) are hormones that bind to membrane-bound receptors on the cells of the gonads (ovaries and testes). They regulate the growth, development, and functions of the gonads. In females, **luteinizing** (loo′tē-ĭ-nīz-ing) **hormone**

(LH) causes the ovulation of oocytes and the secretion of the sex hormones estrogen and progesterone from the ovaries. In males, LH stimulates interstitial cells of the testes to secrete the sex hormone testosterone and thus is sometimes referred to as **interstitial** (in-ter-stish′ăl) **cell–stimulating hormone (ICSH). Follicle-stimulating hormone (FSH)** stimulates the development of follicles in the ovaries and sperm cells in the testes. Without LH and FSH, the ovaries and testes decrease in size, no longer produce oocytes or sperm cells, and no longer secrete hormones. A single releasing hormone from the hypothalamus increases the secretion of both LH and FSH.

Prolactin (prō-lak′tin; *pro,* precursor + *lact,* milk) binds to membrane-bound receptors in cells of the breast, where it helps promote development of the breast during pregnancy and stimulates the production of milk following pregnancy. The regulation of prolactin secretion is complex and may involve several substances released from the hypothalamus. There are two main regulatory hormones; one increases prolactin secretion and one decreases it.

Melanocyte-stimulating (mel′ă-nō-sīt) **hormone (MSH)** binds to membrane-bound receptors on melanocytes and causes them to synthesize melanin. The structure of MSH is similar to that of ACTH, and oversecretion of either hormone causes the skin to darken. Regulation of MSH is not well understood, but there appear to be two regulatory hormones from the hypothalamus— one that increases MSH secretion and one that decreases it.

Hormones of the Posterior Pituitary

Antidiuretic (an′tē-dī-ŭ-ret′ik; *anti,* against + *uresis,* urine volume) **hormone (ADH)** binds to membrane-bound receptors and increases water reabsorption by kidney tubules. This results in less water lost as urine. ADH can also cause blood vessels to constrict when released in large amounts. Consequently, it is sometimes also called *vasopressin* (vā-sō-pres′in). Reduced ADH release from the posterior pituitary results in large amounts of dilute urine.

A lack of ADH secretion causes diabetes insipidus, which is the production of a large amount of dilute urine. The consequences of diabetes insipidus are not obvious until the condition becomes severe, producing many liters of urine each day. The large urine volume causes an increase in the concentration of the body fluids and the loss of important electrolytes, such as Ca^{2+}, Na^+, and K^+. The lack of ADH secretion may be familiar to some who have ever had alcohol to drink. The diuretic actions of alcohol are due to its inhibition of ADH secretion.

Oxytocin (ok′sĭ-tō′sin; swift birth) binds to membrane-bound receptors, and causes contraction of the smooth muscle cells of the uterus as well as milk letdown from the breasts in lactating women. Commercial preparations of oxytocin, known as Pitocin, are given under certain conditions to assist in childbirth and to constrict uterine blood vessels following childbirth.

Thyroid Gland

The **thyroid** (thī′royd; shield-shaped) **gland** is made up of two lobes connected by a narrow band called the **isthmus** (is′mŭs; a constriction). The lobes are located on each side of the trachea, just inferior to the larynx (figure 10.15*a,b*). The thyroid gland is one of the largest endocrine glands. It appears more red than the surrounding tissues because it is highly vascular. It is surrounded by a connective tissue capsule. The main function of the thyroid gland is to secrete **thyroid hormones,** which bind to nuclear receptors in cells and regulate the rate of metabolism in the body (table 10.2). Thyroid hormones are synthesized and stored within the gland in numerous **thyroid follicles,** which are small spheres with walls composed of simple cuboidal epithelium (figure 10.15*c,d*). Each thyroid follicle is filled with the protein thyroglobulin (thī-rō-glob′ū-lin), to which thyroid hormones are attached. Between the follicles is a network of loose connective tissue that contains capillaries and scattered parafollicular cells, or C cells, which secrete the hormone calcitonin.

Thyroid hormone secretion is regulated by hormones from the hypothalamus and pituitary. The hypothalamus secretes TSH-releasing hormone, also known as TRH, which travels to the anterior pituitary and stimulates the secretion of thyroid-stimulating hormone (TSH) (figure 10.16). In turn, TSH stimulates the secretion of thyroid hormones from the thyroid gland. Small fluctuations in blood TSH levels occur on a daily basis, with a small increase at night. Increasing blood levels of TSH increase the synthesis and release of thyroid hormones from thyroglobulin. Decreasing blood levels of TSH decrease the synthesis and release of thyroid hormones.

The thyroid hormones have a negative-feedback effect on the hypothalamus and pituitary, so that increasing levels of thyroid hormones inhibit the secretion of TSH-releasing hormone from the hypothalamus and inhibit TSH secretion from the anterior pituitary gland. Decreasing thyroid hormone levels allow additional TSH-releasing hormone and TSH to be secreted. Because of the negative-feedback effect, the thyroid hormones fluctuate within a narrow concentration range in the blood. However, a loss of negative feedback will result in excess TSH. This causes the thyroid gland to enlarge, a condition called a **goiter** (goy′ter). One type of goiter develops if iodine in the diet is too low. As less thyroid hormone is synthesized and secreted, TSH-releasing hormone and TSH secretion increase above normal levels and cause dramatic enlargement of the thyroid gland.

Without a normal rate of thyroid hormone secretion, growth and development cannot proceed normally. A lack of thyroid hormones is called **hypothyroidism** (hī′pō-thī′royd-izm). In infants, hypothyroidism can result in **cretinism** (krē′tin-izm), characterized by mental retardation, short stature, and abnormally formed skeletal structures. In adults, the lack of thyroid hormones results in a decreased metabolic rate, sluggishness, a reduced ability to perform routine tasks, and **myxedema** (mik-se-dē′mă), which is the accumulation of fluid and other molecules in the subcutaneous tissue. An elevated rate of thyroid hormone secretion, known as **hyperthyroidism** (hī-per-thī′royd-izm), causes an increased metabolic rate, extreme nervousness, and chronic fatigue. **Graves disease** is a type of hyperthyroidism that results when the immune system produces abnormal proteins that are similar in structure and function to TSH. Graves disease is often accompanied by bulging of the eyes, a condition called **exophthalmia** (ek-sof-thal′mē-ă) (see Systems Pathology, "Graves Disease").

The thyroid gland requires iodine to synthesize thyroid hormones. Iodine is taken up by the thyroid follicles. One thyroid hormone, called **thyroxine** (thī-rok′sin) or *tetraiodothyronine* (tet′ră-ī′ō-dō-thī′rō-nēn), contains four iodine atoms and is abbreviated T_4. The other thyroid hormone, **triiodothyronine** (trī-ī′ō-dō-thī′rō-nēn), contains three iodine atoms and is abbreviated T_3. If insufficient iodine is present, production and secretion of the thyroid hormones decrease.

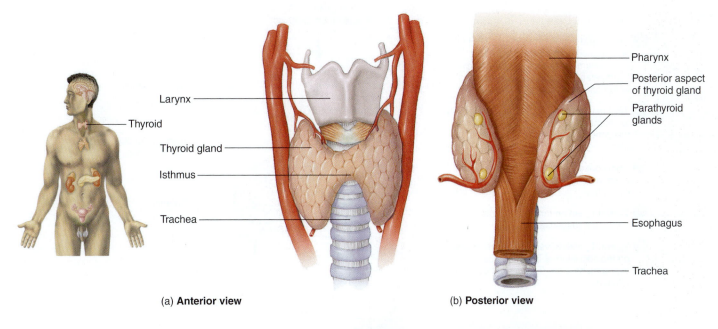

Larynx

Thyroid

Thyroid gland

Isthmus

Trachea

(a) **Anterior view**

Pharynx

Posterior aspect of thyroid gland

Parathyroid glands

Esophagus

Trachea

(b) **Posterior view**

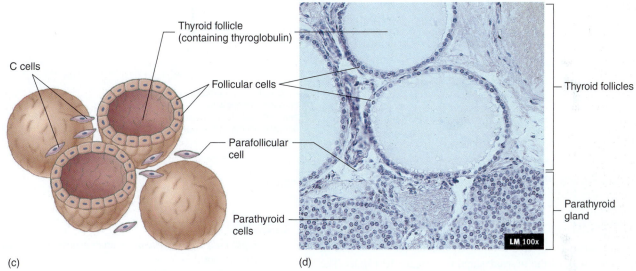

Thyroid follicle (containing thyroglobulin)

C cells

Follicular cells

Parafollicular cell

Parathyroid cells

Thyroid follicles

Parathyroid gland

LM 100x

(c)

(d)

Figure 10.15 AP|R **Thyroid and Parathyroid Glands**

(*a*) Anterior view of the thyroid gland. (*b*) The four small parathyroid glands are embedded in the posterior surface of the thyroid gland. (*c*) Three-dimensional interpretive drawing of thyroid follicles and parafollicular cells. (*d*) Light micrograph of thyroid and parathyroid tissue.

A lack of iodine in the diet results in reduced T_3 and T_4 synthesis. A deficiency of iodine is not as common in the United States as it once was. Table salt with iodine added to it (iodized salt) is available in grocery stores, and vegetables grown in soil rich in iodine can be shipped to most places.

Predict 4

In people with Graves disease (hyperthyroidism), the immune system produces a large amount of a protein that is so much like TSH that it binds to cells of the thyroid gland and acts like TSH. Unlike TSH, however, the secretion of this protein does not respond to negative feedback. Predict the effect of this abnormal protein on the structure and function of the thyroid gland and the release of hormones from the hypothalamus and anterior pituitary gland.

In addition to secreting thyroid hormones, the parafollicular cells of the thyroid gland secrete a hormone called **calcitonin** (kal-si-tō′nin) (see figure 10.15*c*). Calcitonin is secreted if the blood concentration of Ca^{2+} becomes too high, and it causes Ca^{2+} levels to decrease to their normal range (figure 10.17). Calcitonin binds to membrane-bound receptors of osteoclasts and reduces the rate of Ca^{2+} resaborption (breakdown) from bone by inhibiting the osteoclasts. Calcitonin may prevent blood Ca^{2+} levels from becoming overly elevated following a meal that contains a high concentration of Ca^{2+}.

Calcitonin helps prevent elevated blood Ca^{2+} levels, but a lack of calcitonin secretion does not result in a prolonged increase in those levels. Other mechanisms controlling blood Ca^{2+} levels compensate for the lack of calcitonin secretion.

Endocrine

① Stress and hypothermia cause TRH to be released from neurons within the hypothalamus. It passes through the hypothalamohypophysial portal system to the anterior pituitary.

② TRH causes cells of the anterior pituitary to secrete TSH, which passes through the general circulation to the thyroid gland.

③ TSH causes increased synthesis and release of T₃ and T₄ into the general circulation.

④ T₃ and T₄ act on target tissues to produce a response.

⑤ T₃ and T₄ also have an inhibitory effect on the secretion of TRH from the hypothalamus and TSH from the anterior pituitary.

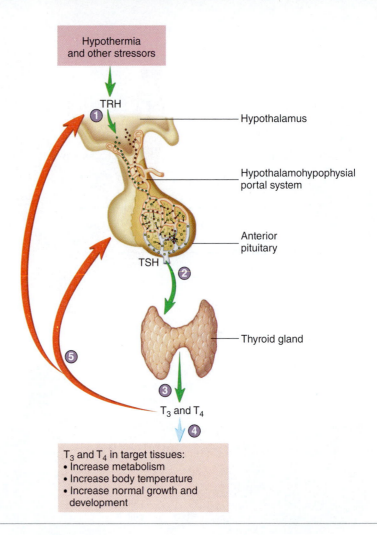

→ Stimulatory

→ Inhibitory

PROCESS Figure 10.16 Regulation of Thyroid Hormone (T₃ and T₄) Secretion

Parathyroid Glands

Four tiny **parathyroid** (par-ă-thī′royd) **glands** are embedded in the posterior wall of the thyroid gland (see figure 10.15b,d). The parathyroid glands secrete a hormone called **parathyroid hormone (PTH),** which is essential for the regulation of blood calcium levels (table 10.2). In fact, PTH is more important than calcitonin in regulating blood Ca^{2+} levels. PTH has many effects:

1. PTH binds to membrane-bound receptors of renal tubule cells, which increases active vitamin D formation. Vitamin D causes the epithelial cells of the intestine to increase Ca^{2+} absorption.
2. PTH binds to receptors on osteoblasts. Substances released by the osteoblasts increase osteoclast activity and cause reabsorption of bone tissue to release Ca^{2+} into the circulatory system.
3. PTH binds to receptors on cells of the renal tubules and decreases the rate at which Ca^{2+} is lost in the urine.
4. PTH acts on its target tissues to raise blood Ca^{2+} levels to normal.

Vitamin D is produced from precursors in the skin that are modified by the liver and kidneys. Ultraviolet light acting on the skin is required for the first stage of vitamin D synthesis, and the final stage of synthesis in the kidney is stimulated by PTH. Vitamin D can also be supplied in the diet.

Predict 5

Explain why a lack of vitamin D results in bones that are softer than normal.

Decreasing blood Ca^{2+} levels stimulate an increase in PTH secretion (figure 10.17). For example, if too little Ca^{2+} is consumed in the diet or if a person suffers from a prolonged lack of vitamin D, blood Ca^{2+} levels decrease, and PTH secretion increases. The increased PTH increases the rate of bone reabsorption. Blood Ca^{2+} levels can be maintained within a normal range, but prolonged reabsorption of bone results in reduced bone density, as manifested by soft, flexible bones that are easily deformed in young people and porous, fragile bones in older people.

Increasing blood Ca^{2+} levels cause a decrease in PTH secretion (figure 10.17). The decreased PTH secretion leads to reduced blood Ca^{2+} levels. In addition, increasing blood Ca^{2+} levels stimulate calcitonin secretion, which also causes blood Ca^{2+} levels to decline.

An abnormally high rate of PTH secretion is called **hyperparathyroidism.** One cause is a tumor in a parathyroid gland. The elevated blood levels of PTH increase bone reabsorption and elevate blood Ca^{2+} levels. As a result, bones can become soft, deformed, and easily fractured. In addition, the elevated blood Ca^{2+} levels make nerve and muscle cells less excitable, resulting in fatigue and muscle weakness. The excess Ca^{2+} can be deposited in soft tissues of the body, causing inflammation. In addition, kidney stones can result.

Endocrine

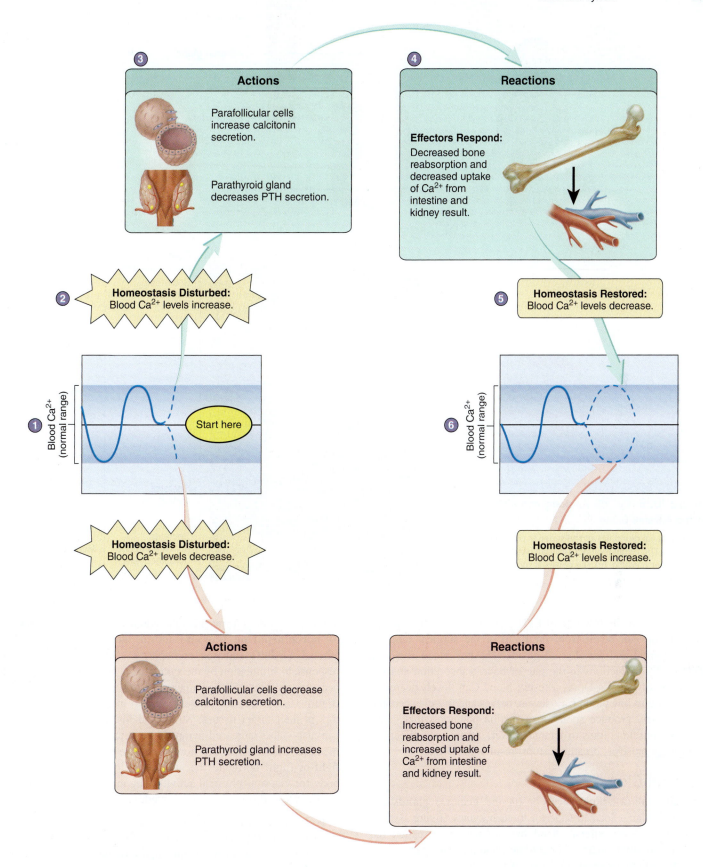

Homeostasis Figure 10.17 Regulation of Calcium Ion Blood Levels

(1) Blood Ca^{2+} is within its normal range. (2) Blood Ca^{2+} level increases outside the normal range. (3) The parafollicular cells and the parathyroid gland cells detect elevated blood Ca^{2+}. The parafollicular cells secrete calcitonin; the parathyroid gland cells decrease PTH secretion. (4) There is less bone reabsorption and less uptake of Ca^{2+} from both the kidney and the intestine. (5) Blood Ca^{2+} level drops back to its normal range. (6) Homeostasis is restored. Observe the response to a drop in blood Ca^{2+} by following the red arrows.

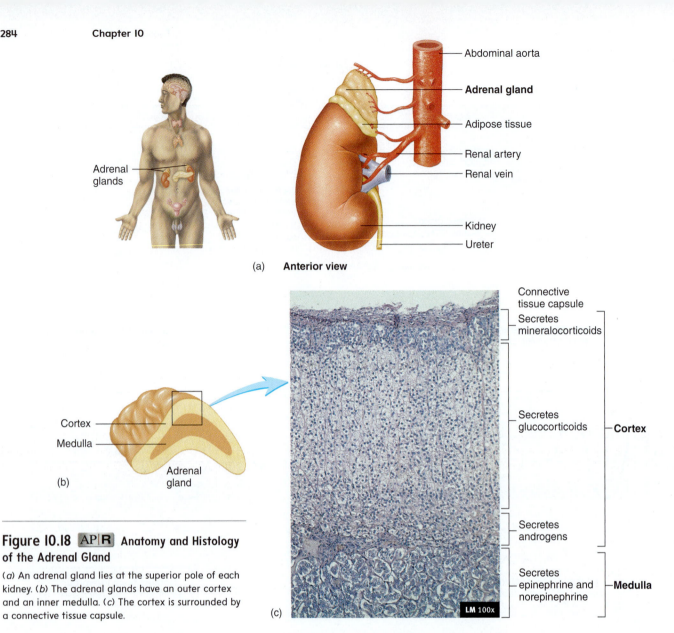

(a)   **Anterior view**

- Abdominal aorta
- **Adrenal gland**
- Adipose tissue
- Renal artery
- Renal vein
- Kidney
- Ureter

Adrenal glands

Cortex
Medulla

(b)   Adrenal gland

- Connective tissue capsule
- Secretes mineralocorticoids
- Secretes glucocorticoids — **Cortex**
- Secretes androgens
- Secretes epinephrine and norepinephrine — **Medulla**

LM 100x

Figure 10.18 AP|R **Anatomy and Histology of the Adrenal Gland**

(a) An adrenal gland lies at the superior pole of each kidney. (b) The adrenal glands have an outer cortex and an inner medulla. (c) The cortex is surrounded by a connective tissue capsule.

An abnormally low rate of PTH secretion, called **hypoparathyroidism,** can result from injury to or the surgical removal of the thyroid and parathyroid glands. The low blood levels of PTH lead to reductions in the rate of bone reabsorption and the formation of vitamin D. As a result, blood Ca^{2+} levels decrease. In response to low blood Ca^{2+} levels, nerves and muscles become excitable and produce spontaneous action potentials that cause frequent muscle cramps or tetanus. Severe tetanus can affect the respiratory muscles; breathing stops, resulting in death.

Adrenal Glands

The **adrenal** (ă-drē′năl; near or on the kidneys) **glands** are two small glands located superior to each kidney (figure 10.18a; table 10.2). Each adrenal gland has an inner part, called the **adrenal medulla** (marrow, or middle), and an outer part, called the **adrenal cortex** (bark, or outer). The adrenal medulla and the adrenal cortex function as separate endocrine glands.

Adrenal Medulla

The principal hormone released from the adrenal medulla is **epinephrine** (ep′i-nef′rin; *epi,* upon + *nephros,* kidney), or *adrenaline* (ă-dren′ă-lin; from the adrenal gland). Small amounts of **norepinephrine** (nōr′ep-i-nef′rin) are also released. The adrenal medulla releases epinephrine and norepinephrine in response to stimulation by the sympathetic nervous system, which becomes most active when a person is excited or physically active (figure 10.19). These hormones bind to membrane-bound receptors in their target tissues. Stress and low blood glucose levels can also cause increased sympathetic stimulation of the adrenal medulla. Epinephrine and norepinephrine are referred to as the **fight-or-flight** hormones because of their role in preparing the body for vigorous physical activity. The major effects of the hormones released from the adrenal medulla are

1. Increases in the breakdown of glycogen to glucose in the liver, the release of the glucose into the blood, and the release of fatty acids from adipose tissue. The glucose and fatty acids serve as energy sources to maintain the body's increased rate of metabolism.
2. Increased heart rate, which causes blood pressure to rise
3. Stimulation of smooth muscle in the walls of arteries supplying the internal organs and the skin, but not those supplying skeletal muscle. Blood flow to internal organs

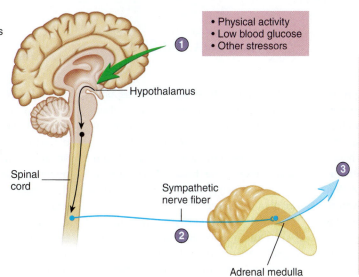

① Stress, physical activity, and low blood glucose levels act as stimuli to the hypothalamus, resulting in increased sympathetic nervous system activity.

② An increased frequency of action potentials conducted through the sympathetic division of the autonomic nervous system stimulates the adrenal medulla to secrete epinephrine and some norepinephrine into the circulatory system.

③ Epinephrine and norepinephrine act on their target tissues to produce responses.

Hypothalamus

Spinal cord

Sympathetic nerve fiber

Adrenal medulla

- Physical activity
- Low blood glucose
- Other stressors

Epinephrine and norepinephrine in the target tissues:
- Increase the release of glucose from the liver into the blood
- Increase the release of fatty acids from adipose tissue into the blood
- Increase heart rate
- Decrease blood flow through blood vessels of most internal organs
- Increase blood flow through blood vessels of skeletal muscle and the heart
- Increase blood pressure
- Decrease the function of visceral organs
- Increase the metabolic rate of skeletal muscles

PROCESS Figure 10.19 AP|R **Regulation of Adrenal Medullary Secretions**

Stimulation of the hypothalamus by stress, physical activity, or low blood glucose levels causes action potentials to travel through the sympathetic nervous system to the adrenal medulla. In response, the adrenal medulla releases epinephrine and smaller amounts of norepinephrine into the general circulation. These hormones have several effects that prepare the body for physical activity.

and the skin decreases, as do the functions of the internal organs. Blood flow through skeletal muscles increases.

4. Increased blood pressure due to smooth muscle contraction in the walls of blood vessels in the internal organs and the skin

5. Increased metabolic rate of several tissues, especially skeletal muscle, cardiac muscle, and nervous tissue

Responses to hormones from the adrenal medulla reinforce the effect of the sympathetic division of the autonomic nervous system. Thus, the adrenal medulla and the sympathetic division function together to prepare the body for physical activity and to produce the fight-or-flight response and many other responses to stress.

Adrenal Cortex

The adrenal cortex secretes three classes of steroid hormones: mineralocorticoids, glucocorticoids, and androgens. The molecules of all three classes of steroid hormones enter their target cells and bind to nuclear receptor molecules. However, the hormones and the receptors of each class have unique structural and functional characteristics.

The first class of hormones, secreted by the outer layer of the adrenal cortex, the **mineralocorticoids** (min′er-al-ō-kōr′ti-koydz), helps regulate blood volume and blood levels of K^+ and Na^+. **Aldosterone** (al-dos′ter-ōn) is the major hormone of this class (figure 10.20). Aldosterone primarily binds to receptor molecules in the kidney, but it also affects the intestine, sweat glands, and salivary glands. Aldosterone causes Na^+ and water to be retained in the body and increases the rate at which K^+ is eliminated.

Blood levels of K^+ and Na^+ directly affect the adrenal cortex to influence aldosterone secretion. The adrenal gland is much more sensitive to changes in blood K^+ levels than to changes in blood Na^+ levels. The rate of aldosterone secretion increases when blood K^+ levels increase or when blood Na^+ levels decrease.

Changes in blood pressure indirectly affect the rate of aldosterone secretion. Low blood pressure causes the release of a protein molecule called **renin** (rē′nin) from the kidney. Renin, which acts as an enzyme, causes a blood protein called **angiotensinogen**

(an′jē-ō-ten-sin′ō-jen) to be converted to **angiotensin I** (an-jē-ō-ten′sin). Then, a protein called **angiotensin-converting enzyme** causes angiotensin I to be converted to **angiotensin II.** Angiotensin II causes smooth muscle in blood vessels to constrict, and angiotensin II acts on the adrenal cortex to increase aldosterone secretion. Aldosterone causes retention of Na^+ and water, which leads to an increase in blood volume (figure 10.20). Both blood vessel constriction and increased blood volume help raise blood pressure.

> **Predict 6**
>
> *Predict the effects of reduced aldosterone secretion on blood levels of Na^+ and K^+, as well as on blood pressure.*

The second class of hormones, secreted by the middle layer of the adrenal cortex, the **glucocorticoids** (gloo-kō-kōr′ti-koydz), helps regulate blood nutrient levels. The major glucocorticoid hormone is **cortisol** (kōr′ti-sol), which increases the breakdown of proteins and lipids and increases their conversion to forms of energy the body can use. For example, cortisol causes the liver to convert amino acids to glucose, and it acts on adipose tissue, causing lipids to be broken down to fatty acids. The glucose and fatty acids are released into the blood, taken up by tissues, and used as a source of energy. Cortisol also causes proteins to be broken down to amino acids, which are then released into the blood.

Cortisol reduces the inflammatory and immune responses. A closely related steroid, **cortisone** (kōr′ti-sōn), or other similar drugs are often given to reduce inflammation caused by injuries. Cortisone can also reduce the immune and inflammatory responses that result from allergic reactions or abnormal immune responses, such as rheumatoid arthritis or asthma.

In response to stressful conditions, cortisol is secreted in larger than normal amounts; thus, it aids the body by providing energy sources for tissues. However, if stressful conditions are prolonged, the immune system can be suppressed enough to make the body susceptible to stress-related conditions (see Clinical Impact, "Hormones and Stress").

Endocrine

CLINICAL IMPACT   Hormones and Stress

Stress, in the form of disease, physical injury, or emotional anxiety, initiates a specific response that involves the nervous and endocrine systems. The stressful condition influences the hypothalamus, and through the hypothalamus, the sympathetic division of the autonomic nervous system is activated. The sympathetic division prepares the body for physical activity. It increases heart rate and blood pressure, shunts blood from the intestine and other visceral structures to skeletal muscles, and increases the metabolic rate in several tissues, especially skeletal muscle. Part of the response of the sympathetic division is due to the release of epinephrine from the adrenal medulla.

In addition to sympathetic responses, stress causes the release of ACTH from the pituitary. ACTH acts on the adrenal cortex to cause the release of glucocorticoids. These hormones increase blood glucose levels and break down protein and lipids, making nutrients more readily available to tissues.

Although the ability to respond to stress is adaptive for short periods of time, responses triggered by stressful conditions are harmful if they occur for long periods. Prolonged stress can lead to hypertension (elevated blood pressure), heart disease, ulcers, inhibited immune system function, changes in mood, and other conditions. Humans are frequently exposed to prolonged psychological stress from

high-pressure jobs, the inability to meet monetary obligations, or social expectations. Although responses to stress prepare a person for physical activity, increased physical activity is often not an appropriate response to the situation causing the stress. Long-term exposure to stress under conditions in which physical activity and emotions must be constrained may be harmful. Techniques that effectively reduce responses to stressful conditions, such as biofeedback, meditation, or other relaxation exercises can help people who are exposed to chronic stress. Adequate rest, relaxation, and regular physical exercise are also important in maintaining good health and reducing unhealthy responses to stressful situations.

 Stimulatory

1. Increased blood K⁺ levels or decreased blood Na⁺ levels cause the adrenal cortex to increase the secretion of aldosterone into the general circulation.

2. The kidneys detect a decrease in blood pressure. In response, they increase the secretion of renin into the general circulation. Renin converts angiotensinogen to angiotensin I. A converting enzyme changes angiotensin I to angiotensin II, which causes constriction of blood vessels, resulting in increased blood pressure.

3. Angiotensin II causes increased secretion of aldosterone, which primarily affects the kidneys.

4. Aldosterone stimulation of the kidneys causes Na⁺ retention, K⁺ excretion, and decreased water loss.

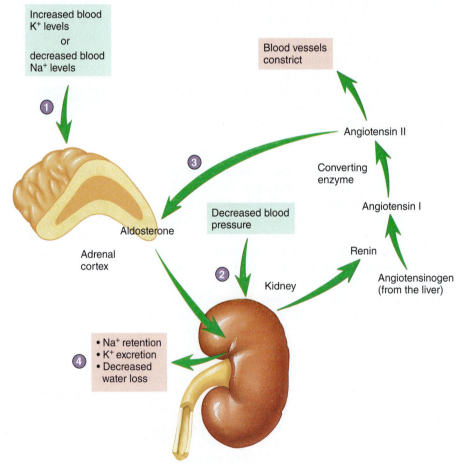

PROCESS Figure 10.20 Regulation of Aldosterone Secretion from the Adrenal Cortex

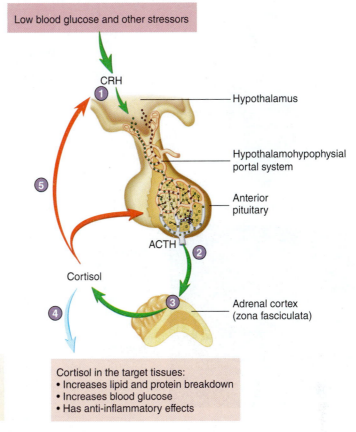

1. Corticotropin-releasing hormone (CRH) is released from hypothalamic neurons in response to stress or low blood glucose and passes, by way of the hypothalamohypophysial portal system, to the anterior pituitary.

2. In the anterior pituitary, CRH binds to and stimulates cells that secrete adrenocorticotropic hormone (ACTH).

3. ACTH binds to membrane-bound receptors on cells of the adrenal cortex and stimulates the secretion of glucocorticoids, primarily cortisol.

4. Cortisol acts on target tissues, resulting in increased lipid and protein breakdown, increased glucose levels, and anti-inflammatory effects.

5. Cortisol has a negative-feedback effect because it inhibits CRH release from the hypothalamus and ACTH secretion from the anterior pituitary.

Low blood glucose and other stressors

CRH

Hypothalamus

Hypothalamohypophysial portal system

Anterior pituitary

ACTH

Cortisol

Adrenal cortex (zona fasciculata)

→ Stimulatory

→ Inhibitory

Cortisol in the target tissues:
- Increases lipid and protein breakdown
- Increases blood glucose
- Has anti-inflammatory effects

Endocrine

PROCESS Figure 10.21 Regulation of Cortisol Secretion from the Adrenal Cortex

Adrenocorticotropic hormone (ACTH) molecules from the anterior pituitary bind to membrane-bound receptors and regulate the secretion of cortisol from the adrenal cortex in the process depicted in figure 10.21. When blood glucose levels decline, cortisol secretion increases. The low blood glucose acts on the hypothalamus to increase the secretion of the ACTH-releasing hormone, which, in turn, stimulates ACTH secretion from the anterior pituitary. ACTH then stimulates cortisol secretion. Without ACTH, the adrenal cortex atrophies and loses its ability to secrete cortisol.

Predict 7

Cortisone is sometimes given to people who have severe allergies or extensive inflammation or to people who suffer from autoimmune diseases. Taking this substance for long periods of time can damage the adrenal cortex. Explain how this damage can occur.

The third class of hormones, secreted by the inner layer of the adrenal cortex, is composed of the **androgens** (an′drō-jenz), which stimulate the development of male sexual characteristics. Small amounts of androgens are secreted from the adrenal cortex in both males and females. In adult males, most androgens are secreted by the testes. In adult females, the adrenal androgens influence the female sex drive. If the secretion of sex hormones from the adrenal cortex is abnormally high, exaggerated male characteristics develop in both males and females. This condition is most apparent in females and in males before puberty, when the effects are not masked by the secretion of androgens by the testes.

Pancreas, Insulin, and Diabetes

The endocrine part of the **pancreas** (pan′krē-as) consists of **pancreatic islets** (islets of Langerhans), which are dispersed throughout the exocrine portion of the pancreas (figure 10.22). The islets secrete three hormones—insulin, glucagon, and somatostatin—which help regulate the blood levels of nutrients, especially glucose (table 10.3). **Alpha cells** secrete glucagon, **beta cells** secrete insulin, and delta cells secrete somatostatin.

It is very important to maintain blood glucose levels within a normal range (figure 10.23). A below-normal blood glucose level causes the nervous system to malfunction because glucose is the nervous system's main source of energy. When blood glucose decreases, other tissues rapidly break down lipids and proteins to provide an alternative energy source. As lipids are broken down, the liver converts some of the fatty acids to acidic **ketones** (kē′tōnz), which are released into the blood. When blood glucose levels are very low, the breakdown of lipids can cause the release of enough fatty acids and ketones to reduce the pH of the body fluids below normal, a condition called **acidosis** (as-i-dō′sis). The amino acids of proteins are broken down and used by the liver to synthesize glucose.

If blood glucose levels are too high, the kidneys produce large volumes of urine containing substantial amounts of glucose. Because of the rapid loss of water in the form of urine, dehydration can result.

The hormone **insulin** (in′sŭ-lin) is released from the beta cells primarily in response to the elevated blood glucose levels

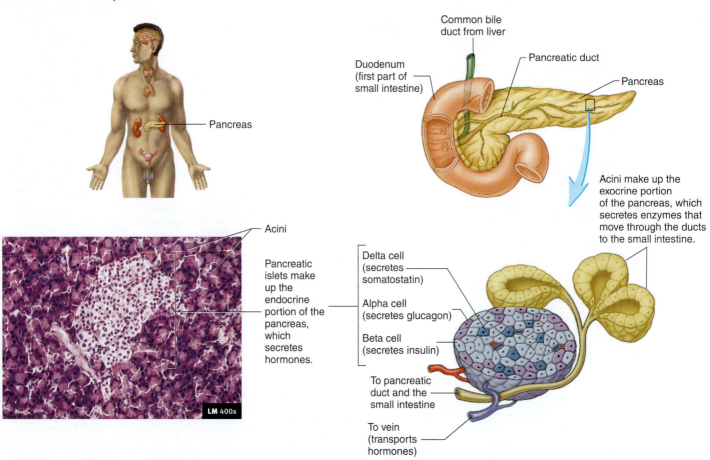

Figure 10.22 AP|R **Structure of the Pancreas**

The endocrine portion of the pancreas is made up of scattered pancreatic islets. Alpha cells secrete glucagon, beta cells secrete insulin, and delta cells secrete somatostatin. The exocrine portion of the pancreas surrounds the pancreatic islets and produces digestive enzymes that are carried through a system of ducts to the small intestine. The stain used for the light micrograph does not distinguish between alpha, beta, and delta cells.

TABLE 10.3	Effects of Insulin and Glucagon on Target Tissues	
Target Tissue	**Insulin Responses**	**Glucagon Responses**
Skeletal muscle, cardiac muscle, cartilage, bone fibroblasts, blood cells, mammary glands	Increases glucose uptake and glycogen synthesis; increases uptake of amino acids	Has little effect
Liver	Increases glycogen synthesis; increases use of glucose for energy	Causes rapid increase in the breakdown of glycogen to glucose and release of glucose into the blood; increases the formation of glucose from amino acids and, to some degree, from lipids; increases metabolism of fatty acids
Adipose cells	Increases glucose uptake, glycogen synthesis, lipid synthesis	High concentrations cause breakdown of lipids; probably unimportant under most conditions
Nervous system	Has little effect except to increase glucose uptake in the satiety center	Has no effect

and increased parasympathetic stimulation associated with digestion of a meal. Increased blood levels of certain amino acids also stimulate insulin secretion. Decreased insulin secretion results from decreasing blood glucose levels and from stimulation of the pancreas by the sympathetic division of the nervous system, which occurs during physical activity. Decreased insulin levels allow

blood glucose to be conserved to provide the brain with adequate glucose and to allow other tissues to metabolize fatty acids and glycogen stored in the cells.

The major target tissues for insulin are the liver, adipose tissue, muscles, and the area of the hypothalamus that controls appetite, called the **satiety** (sa-tī′-ĕ-tē; fulfillment of hunger)

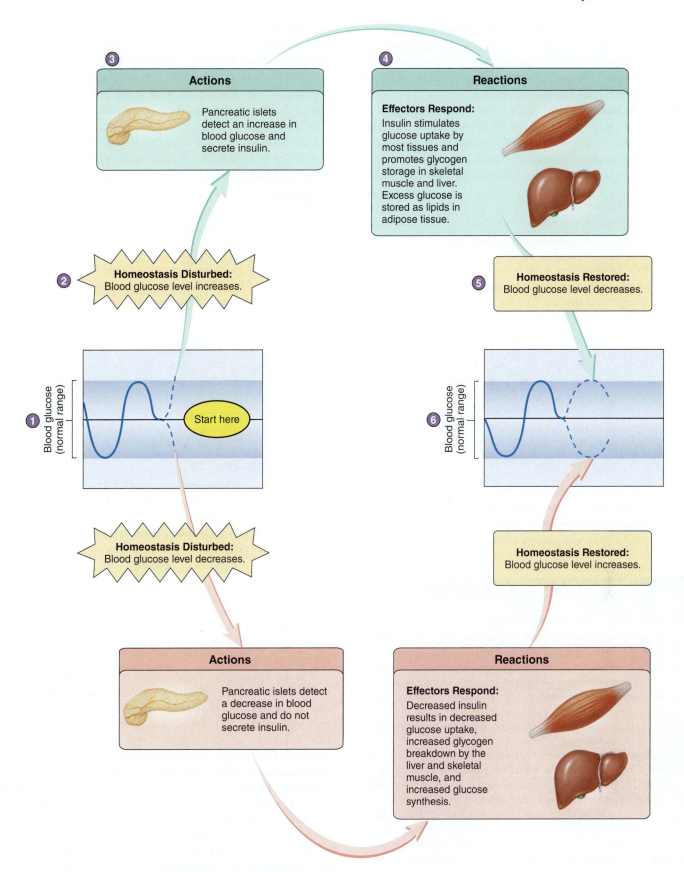

Homeostasis Figure 10.23 Regulation of Blood Glucose Levels

(1) Blood glucose is within its normal range. (2) Blood glucose level increases outside the normal range. (3) The pancreatic islets secrete insulin in direct response to elevated blood glucose. (4) Most tissues take up glucose when insulin binds to its receptor on the tissues. (5) Blood glucose level drops back to its normal range. (6) Homeostasis is restored. Observe the response to a drop in blood glucose by following the red arrows.

center. Insulin binds to membrane-bound receptors and, either directly or indirectly, increases the rate of glucose and amino acid uptake in these tissues. Glucose is converted to glycogen or lipids, and the amino acids are used to synthesize protein. The effects of insulin on target tissues are summarized in table 10.3.

Diabetes mellitus (dī-ă-bē′tēz′ me-lī′tŭs; much urine + honey or sweetened) has several causes. Type 1 diabetes mellitus occurs when too little insulin is secreted from the pancreas, and type 2 diabetes mellitus is caused by insufficient numbers of insulin receptors on target cells or by defective receptors that do not respond normally to insulin.

In type 1 diabetes mellitus, tissues cannot take up glucose effectively, causing blood glucose levels to become very high, a condition called **hyperglycemia** (hī′per-glī-sē′mē-ă; *hyper,* above + *glycemia,* blood glucose). Because glucose cannot enter the cells of the satiety center in the brain without insulin, the satiety center responds as if there were very little blood glucose, resulting in an exaggerated appetite. The excess glucose in the blood is excreted in the urine, making the urine volume much greater than normal. Because of excessive urine production, the person has a tendency to become dehydrated and thirsty. Even though blood glucose levels are high, lipids and proteins are broken down to provide an energy source for metabolism, resulting in the wasting away of body tissues, acidosis, and ketosis. People with this condition also exhibit a lack of energy. Insulin must be injected regularly to adequately control blood glucose levels. When too much insulin is present, as occurs when a diabetic is injected with too much insulin or has not eaten after an insulin injection, blood glucose levels become very low. The brain, which depends primarily on glucose for an energy source, malfunctions. This condition, called insulin shock, can cause disorientation and convulsions and may result in loss of consciousness. Fortunately genetic engineering has allowed synthetic insulin to become widely available to diabetics.

A CASE IN POINT

Type 2 Diabetes Mellitus

Kandy Barr is 60 years old. She is overweight, has been feeling lethargic and weak, and has had two urinary tract infections in the past 6 months. She visited her optometrist because she could not read the handwriting on checks at the bank where she works. Her optometrist recommended that Kandy see her physician, who ordered a blood test. The results indicated high blood glucose and blood lipid levels. After additional tests, Kandy's physician told her that she has type 2 diabetes mellitus, which results from reduced sensitivity of tissues to the effects of insulin because of abnormal insulin receptors or abnormal responses to the insulin receptors. Consequently, insulin is less able to facilitate the entry of glucose into the liver, skeletal muscles, and adipose tissues. Kandy's physician recommended that she try to control her diabetes by restricting food intake, especially carbohydrates and fats, and increasing exercise. Kandy's symptoms were reduced as a result of faithful adherence to her physician's recommendations. People who are obese are approximately 10 times more likely to develop type 2 diabetes than people of normal weight. Frequent infections, changes in vision, and fatigue are common signs of high blood glucose levels.

Glucagon (gloo′kă-gon) is released from the alpha cells when blood glucose levels are low. Glucagon binds to membrane-bound receptors primarily in the liver, causing the glycogen stored in the liver to be converted to glucose. The glucose is then released into the blood to increase blood glucose levels. After a meal, when blood glucose levels are elevated, glucagon secretion is reduced.

Somatostatin (sō′mă-tō′statĭn) is released by the delta cells in response to food intake. Somatostatin inhibits the secretion of insulin and glucagon and inhibits gastric tract activity.

Predict 8

How are the rates of insulin and glucagon secretion affected immediately following a large meal rich in carbohydrates? How are they affected after 12 hours without eating?

Insulin and glucagon together regulate blood glucose levels (see figure 10.23). When blood glucose levels increase, insulin secretion increases, and glucagon secretion decreases. When blood glucose levels decrease, the rate of insulin secretion declines, and the rate of glucagon secretion increases. Other hormones, such as epinephrine, cortisol, and growth hormone, also maintain blood levels of nutrients. When blood glucose levels decrease, these hormones are secreted at a greater rate. Epinephrine and cortisol cause the breakdown of protein and lipids and the synthesis of glucose to help increase blood levels of nutrients. Growth hormone slows protein breakdown and favors lipid breakdown.

Testes and Ovaries

The testes of the male and the ovaries of the female secrete sex hormones, in addition to producing sperm cells or oocytes, respectively. The hormones produced by these organs play important roles in the development of sexual characteristics. Structural and functional differences between males and females, as well as the ability to reproduce, depend on the sex hormones (see table 10.2).

The main sex hormone in the male is **testosterone** (tes′tos′tĕ-rōn), which is secreted by the testes. It is responsible for the growth and development of the male reproductive structures, muscle enlargement, the growth of body hair, voice changes, and the male sexual drive.

In the female, two main classes of sex hormones, secreted by the ovaries, affect sexual characteristics: **estrogen** (es′trō-jen) and **progesterone** (prō-jes′ter-ōn). Together, these hormones contribute to the development and function of female reproductive structures and other female sexual characteristics. Two such characteristics are enlargement of the breasts and the distribution of adipose tissue, which influences the shape of the hips, breasts, and thighs. In addition, the female menstrual cycle is controlled by the cyclical release of estrogen and progesterone from the ovaries.

LH and FSH stimulate the secretion of hormones from the ovaries and testes. Releasing hormone from the hypothalamus controls the rate of LH and FSH secretion in males and females. LH and FSH, in turn, control the secretion of hormones from the ovaries and testes. Hormones secreted by the ovaries and testes also have a negative-feedback effect on the hypothalamus and anterior pituitary. The control of hormones that regulate reproductive functions is discussed in greater detail in chapter 19.

Thymus

The **thymus** lies in the upper part of the thoracic cavity (see figure 10.1 and table 10.2). It is important in the function of the immune system. The thymus secretes a hormone called **thymosin** (thī′mō-sin), which aids the development of white blood cells called T cells. T cells help protect the body against infection by foreign organisms. The thymus is most important early in life; if an infant is born without a thymus, the immune system does not develop normally, and the body is less capable of fighting infections (see chapter 14).

Pineal Gland

The **pineal** (pin′ē-ăl; pinecone) **gland** is a small, pinecone-shaped structure located superior and posterior to the thalamus of the brain (see chapter 8). The pineal gland produces a hormone called **melatonin** (mel-ă-tōn′in), which is thought to decrease the secretion of LH and FSH by decreasing the release of hypothalamic-releasing hormones (see table 10.2). Thus, melatonin inhibits the functions of the reproductive system. Animal studies have demonstrated that the amount of available light controls the rate of melatonin secretion. In many animals, short day length causes an increase in melatonin secretion, whereas longer day length causes a decrease in melatonin secretion. Some evidence suggests that melatonin plays an important role in the onset of puberty in humans. Tumors may develop in the pineal gland, which increase pineal secretions in some cases but decrease them in others.

Predict 9

A tumor can destroy the pineal gland's ability to secrete melatonin. How would that affect the reproductive system of a young person?

10.8 OTHER HORMONES

Learning Outcome *After reading this section, you should be able to*

A. Describe the functions of hormones secreted by the stomach and small intestine, the functions of prostaglandins, and the functions of erythropoietin.

Cells in the lining of the stomach and small intestine secrete hormones that stimulate the production of digestive juices from the stomach, pancreas, and liver. This secretion occurs when food is present in the digestive system, but not at other times. Hormones secreted from the small intestine also help regulate the rate at which food passes from the stomach into the small intestine (see chapter 16).

Prostaglandins are widely distributed in tissues of the body, where they function as intercellular signals. Unlike most hormones, they are usually not transported long distances in the blood but function mainly as autocrine or paracrine chemical signals (see table 10.1). Thus, their effects occur in the tissues where they are produced. Some prostaglandins cause relaxation of smooth muscle, such as dilation of blood vessels. Others cause contraction of smooth muscle, such as contraction of the uterus during the delivery of a baby. Because of their action on the uterus, prostaglandins have been used medically to initiate abortion. Prostaglandins also play a role in inflammation. They are released by damaged tissues and cause blood vessel dilation, localized swelling, and pain. Prostaglandins produced by platelets appear to be necessary for normal blood clotting. The ability of aspirin and related substances to reduce pain and inflammation, to help prevent the painful cramping of uterine smooth muscle, and to treat headache is a result of their inhibitory effect on prostaglandin synthesis.

The kidneys secrete the hormone **erythropoietin** (ĕ-rith′rō-poy′ĕ-tin) in response to reduced oxygen levels in the kidney. Erythropoietin acts on bone marrow to increase the production of red blood cells (see chapter 11).

In pregnant women, the placenta is an important source of hormones that maintain pregnancy and stimulate breast development. These hormones are estrogen, progesterone, and **human chorionic gonadotropin** (gō′nad-ō-trō′pin), which is similar in structure and function to LH. Chapter 20 describes the function of these hormones in more detail.

10.9 EFFECTS OF AGING ON THE ENDOCRINE SYSTEM

Learning Outcome *After reading this section, you should be able to*

A. Describe the major age-related changes that occur in the endocrine system.

Age-related changes to the endocrine system include a gradual decrease in the secretion of some, but not all, endocrine glands. Some of the decreases in secretion may be due to the fact that older people commonly engage in less physical activity.

GH secretion decreases as people age, but the decrease is greatest in those who do not exercise, and it may not occur at all in older people who exercise regularly. Decreasing GH levels may explain the gradual decrease in bone and muscle mass and the increase in adipose tissue seen in many elderly people. So far, administering GH to slow or prevent the consequences of aging has not been found to be effective, and unwanted side effects are possible.

A decrease in melatonin secretion may influence age-related changes in sleep patterns, as well as the decreased secretion of some hormones, such as GH and testosterone.

The secretion of thyroid hormones decreases slightly with age. Age-related damage to the thyroid gland by the immune system can occur. Approximately 10% of elderly women experience some reduction in thyroid hormone secretion; this tendency is less common in men.

The kidneys of the elderly secrete less renin, reducing the ability to respond to decreases in blood pressure.

Reproductive hormone secretion gradually declines in elderly men, and women experience menopause (see chapter 19).

Secretion of thymosin from the thymus decreases with age. Fewer functional lymphocytes are produced, and the immune system becomes less effective in protecting the body against infections and cancer.

Parathyroid hormone secretion increases to maintain blood calcium levels if dietary Ca^{2+} and vitamin D levels decrease, as they often do in the elderly. Consequently, a substantial decrease in bone matrix may occur.

In most people, the ability to regulate blood glucose does not decrease with age. However, there is an age-related tendency to develop type 2 diabetes mellitus for those who have a familial tendency, and it is correlated with age-related increases in body weight.

Endocrine

Name: Grace

Gender: Female

Age: 36

Comments

Grace manages a business, has several employees, and works hard to make time for her husband and two children. Over several months, she slowly noticed that she was sweating excessively and appeared flushed. In addition, she often felt her heart pounding, was much more nervous than usual, and found it difficult to concentrate. Then Grace began to feel weak and lose weight, even though her appetite was greater than normal. Her family recognized some of these changes and noticed that her eyes seemed larger than usual. They encouraged her to see her physician. After an examination and some blood tests, Grace was diagnosed with Graves disease, a type of hyperthyroidism.

Graves Disease (Hyperthyroidism)

Background Information

Graves disease is caused by altered regulation of hormone secretion—specifically, the elevated secretion of thyroid hormones from the thyroid gland. In approximately 95% of Graves disease cases, the immune system produces an unusual antibody type, which binds to receptors on the cells of the thyroid follicle and stimulates them to secrete increased amounts of thyroid hormone. The secretion of the releasing hormone and thyroid-stimulating hormone is inhibited by elevated thyroid hormones. However, the antibody is produced in large amounts, and is not inhibited by thyroid hormones. A very elevated rate of thyroid hormone secretion is therefore maintained. In addition, the size of the thyroid gland increases, and connective tissue components are deposited behind the eyes, causing them to bulge (figure 10A). Enlargement of the thyroid gland is called a goiter.

Grace was treated with radioactive iodine (^{131}I) atoms that were actively transported into thyroid cells, where they destroyed a substantial portion of the thyroid gland. Data indicate that this treatment has few side effects and is effective in treating most cases of Graves disease. Other options include (1) drugs that inhibit the synthesis and secretion of thyroid hormones and (2) surgery to remove part of the thyroid gland.

Predict 10

Explain why removal of part of the thyroid gland is an effective treatment for Graves disease.

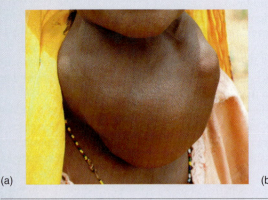
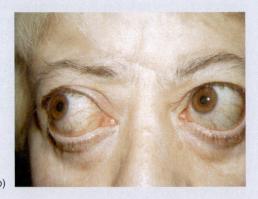

(a)   (b)

Figure 10A

(*a*) A goiter and (*b*) protruding eyes are symptoms of hyperthyroidism.

Endocrine

Graves Disease (Hyperthyroidism)

SKELETAL
Some increased bone reabsorption occurs, which can decrease bone density; increased blood Ca^{2+} levels can occur in severe cases.

INTEGUMENTARY
Excessive sweating, flushing, and warm skin result from the elevated body temperature caused by the increased rate of metabolism. The elevated metabolic rate makes amino acids unavailable for protein synthesis, resulting in fine, soft, straight hair, along with hair loss.

MUSCULAR
Muscle atrophy and muscle weakness result from increased metabolism, which causes the breakdown of muscle and the increased use of muscle proteins as energy sources.

REPRODUCTIVE
Reduced regularity of menstruation or lack of menstruation may occur in women because of the elevated metabolism. In men, the primary effect is loss of sex drive.

NERVOUS
Enlargement of the extrinsic eye muscles, edema in the area of the orbits, and the accumulation of fibrous connective tissue cause protrusion of the eyes in 50–70% of individuals with Graves disease. Damage to the retina and optic nerve and paralysis of the extraocular muscles can occur. Restlessness, short attention span, compulsive movement, tremor, insomnia, and increased emotional responses are consistent with hyperactivity of the nervous system.

DIGESTIVE
Weight loss occurs, with an associated increase in appetite. Increased peristalsis in the intestines leads to frequent stools or diarrhea. Nausea, vomiting, and abdominal pain also result. Hepatic glycogen stores and adipose and protein stores are increasingly used for energy, and serum lipid levels (including triglycerides, phospholipids, and cholesterol) decrease. The tendency to develop vitamin deficiencies increases.

Symptoms
- Hyperactivity
- Rapid weight loss
- Exopthalmos
- Excessive sweating

Treatment
- Exposure to radioactive iodine
- Treatment with drugs that inhibit thyroid hormone synthesis
- Removal of all or a part of the thyroid gland

CARDIOVASCULAR
An increased amount of blood pumped by the heart leads to increased blood flow through the tissues, including the skin. The heart rate is greater than normal, heart sounds are louder than normal, and the heartbeats may be out of rhythm periodically.

RESPIRATORY
Breathing may be labored, and the volume of air taken in with each breath may be decreased. Weak contractions of muscles of inspiration contribute to respiratory difficulties.

LYMPHATIC AND IMMUNE
Antibodies that bind to receptors for thyroid-stimulating hormone on the cells of the thyroid gland have been found in nearly all people with Graves disease. The condition, therefore, is classified as an autoimmune disease in which antibodies produced by the lymphatic system result in abnormal functions.

Endocrine

SUMMARY

10.1 Principles of Chemical Communication (p. 264)

1. The four classes of chemical messengers are autocrine, paracrine, neurotransmitter, and endocrine.
2. Endocrine chemical messengers are called hormones.

10.2 Functions of the Endocrine System (p. 265)

1. The endocrine system has 10 major functions that contribute to homeostasis.

10.3 Characteristics of the Endocrine System (p. 266)

1. The endocrine system includes glands and specialized endocrine cells that secrete hormones into the bloodstream.
2. A hormone is a chemical messenger that is secreted into the blood, travels to a distant target tissue, and binds to specific receptors to produce a coordinated set of events in that target tissue.

10.4 Hormones (p. 266)

Chemical Nature of Hormones

1. There are two chemical categories of hormones: lipid-soluble and water-soluble.
2. Lipid-soluble hormones include steroids, thyroid hormones, and some fatty acid derivatives.
3. Water-soluble hormones include proteins, peptides, and amino acid derivatives.
   - Water-soluble hormones circulate freely in the blood.
   - Proteases degrade protein and peptide hormones in the circulation; the breakdown products are then excreted in the urine. However, some water-soluble hormones have chemical modifications, such as the addition of a carbohydrate group, which prolongs their lifespan.

10.5 Control of Hormone Secretion (p. 267)

Stimulation of Hormone Release

Three types of stimuli result in hormone secretion: humoral, neural, and hormonal.

1. Humoral stimulation is exhibited by hormones that are sensitive to circulating blood levels of certain molecules, such as glucose or calcium.
2. Neural stimuli cause hormone secretion in direct response to action potentials in neurons, as occurs during stress or exercise. Hormones from the hypothalamus that cause the release of other hormones are called releasing hormones.
3. Hormonal stimulation of other hormone secretion is common in the endocrine system. Hormones from the anterior pituitary that stimulate hormones from other endocrine glands are called tropic hormones.

Inhibition of Hormone Release

Although the stimulus of hormone secretion is important, inhibition is equally important.

1. Humoral substances can inhibit the secretion of hormones.
2. Neural stimuli can prevent hormone secretion.
3. Inhibiting hormones prevent hormone release.

Regulation of Hormone Levels in the Blood

Two processes regulate the overall blood levels of hormones: negative feedback and positive feedback.

1. Negative feedback prevents further hormone secretion once a set point is achieved.
2. Positive feedback is a self-promoting system whereby the stimulation of hormone secretion increases over time.

10.6 Hormone Receptors and Mechanisms of Action (p. 269)

Classes of Receptors

Each of the two groups of hormones has its own class of receptors.

1. Lipid-soluble hormones bind to nuclear receptors located inside the nucleus of the target cell. Some lipid-soluble hormones have rapid actions that are most likely mediated via a membrane-bound receptor.
2. Water-soluble hormones bind to membrane-bound receptors, which are integral membrane proteins.

Action of Nuclear Receptors

1. Nuclear receptors have portions that allow them to bind to the DNA in the nucleus once the hormone is bound.
   - The hormone-receptor complex activates genes, which in turn activate the DNA to produce mRNA.
   - The mRNA increases the synthesis of certain proteins that produce the target cell's response.
2. Nuclear receptors cannot respond immediately because it takes time to produce the mRNA and the protein.

Membrane-Bound Receptors and Signal Amplification

1. Membrane-bound receptors activate a cascade of events once the hormone binds.
2. Some membrane-bound receptors are associated with membrane proteins called G proteins. When a hormone binds to a membrane-bound receptor, G proteins are activated. The α subunit of the G protein can bind to ion channels and cause them to open or change the rate of synthesis of intracellular mediators, such as cAMP.
3. Second-messenger systems act rapidly because they act on already existing enzymes and produce an amplification effect.

10.7 Endocrine Glands and Their Hormones (p. 274)

1. The endocrine system consists of ductless glands.
2. Some glands of the endocrine system perform more than one function.

Pituitary and Hypothalamus

1. The pituitary is connected to the hypothalamus in the brain by the infundibulum. It is divided into anterior and posterior portions.
2. Secretions from the anterior pituitary are controlled by hormones that pass through the hypothalamic-pituitary portal system from the hypothalamus.
3. Hormones secreted from the posterior pituitary are controlled by action potentials carried by axons that pass from the hypothalamus to the posterior pituitary.
4. The hormones released from the anterior pituitary are growth hormone (GH), thyroid-stimulating hormone (TSH), adrenocorticotropic hormone (ACTH), luteinizing hormone (LH), follicle-stimulating hormone (FSH), prolactin, and melanocyte-stimulating hormone (MSH).
5. Hormones released from the posterior pituitary include antidiuretic hormone (ADH) and oxytocin.

Thyroid Gland

The thyroid gland secretes thyroid hormones, which control the metabolic rate of tissues, and calcitonin, which helps regulate blood Ca^{2+} levels.

Parathyroid Glands

The parathyroid glands secrete parathyroid hormone, which helps regulate blood Ca^{2+} levels. Active vitamin D also helps regulate blood Ca^{2+} levels.

Adrenal Glands

1. The adrenal medulla secretes primarily epinephrine and some norepinephrine. These hormones help prepare the body for physical activity.
2. The adrenal cortex secretes three classes of hormones.
   a. Glucocorticoids (cortisol) reduce inflammation and break down proteins and lipids, making them available as energy sources to other tissues.
   b. Mineralocorticoids (aldosterone) help regulate blood Na^+ and K^+ levels and water volume. Renin, secreted by the kidneys, helps regulate blood pressure by increasing angiotensin II and aldosterone production. These hormones cause blood vessels to constrict and enhance Na^+ and water retention by the kidney.
   c. Adrenal androgens increase female sexual drive but normally have little effect in males.

Pancreas, Insulin, and Diabetes

1. The pancreas secretes insulin in response to elevated levels of blood glucose and amino acids. Insulin increases the rate at which many tissues, including adipose tissue, the liver, and skeletal muscles, take up glucose and amino acids.
2. The pancreas secretes glucagon in response to reduced blood glucose and increases the rate at which the liver releases glucose into the blood.
3. The pancreas secretes somatostatin in response to food intake. Somatostatin inhibits insulin and glucagon secretion and gastric tract activity.

Testes and Ovaries

1. The testes secrete testosterone, and the ovaries secrete estrogen and progesterone. These hormones help control reproductive processes.
2. LH and FSH from the pituitary gland control hormone secretion from the ovaries and testes.

Thymus

The thymus secretes thymosin, which enhances the function of the immune system.

Pineal Gland

The pineal gland secretes melatonin, which may help regulate the onset of puberty by acting on the hypothalamus.

10.8 Other Hormones (p. 291)

1. Hormones secreted by cells in the stomach and intestine help regulate stomach, pancreatic, and liver secretions.
2. The prostaglandins are hormones that have a local effect, produce numerous effects on the body, and play a role in inflammation.
3. Erythropoietin from the kidney stimulates red blood cell production.
4. The placenta secretes human chorionic gonadotropin, estrogen, and progesterone, which are essential to the maintenance of pregnancy.

10.9 Effects of Aging on the Endocrine System (p. 291)

1. Age-related changes include a gradual decrease in
   a. GH in people who do not exercise
   b. melatonin
   c. thyroid hormones (slight decrease)
   d. reproductive hormones
   e. thymosin
2. Parathyroid hormones increase if vitamin D and Ca^{2+} levels decrease.
3. There is an increase in type 2 diabetes in people with a familial tendency.

Endocrine

REVIEW AND COMPREHENSION

1. What are the major functional differences between the endocrine system and the nervous system?

2. List the functions of the endocrine system.

3. List the major differences between intracellular and intercellular chemical signals.

4. List the intercellular chemical signals that are classified on the basis of the cells from which they are secreted and their target cells.

5. Explain the relationship between a hormone and its receptor.

6. Describe the mechanisms by which membrane-bound receptors produce responses in their target tissues.

7. Describe the mechanisms by which intracellular receptors produce responses in their target tissues.

8. Compare the means by which hormones that can and cannot cross the cell membrane produce a response.

9. Define endocrine gland and hormone.

10. What makes one tissue a target tissue for a hormone and another not a target tissue?

11. Into what chemical categories can hormones be classified?

12. Name three ways that hormone secretion is regulated.

13. Describe how secretions of the anterior and posterior pituitary hormones are controlled.

14. What are the functions of growth hormone? What happens when too little or too much growth hormone is secreted?

15. Describe the effect of gonadotropins on the ovary and testis.

16. What are the functions of the thyroid hormones, and how is their secretion controlled? What happens when too large or too small an amount of the thyroid hormones is secreted?

17. Explain how calcitonin, parathyroid hormone, and vitamin D are involved in maintaining blood Ca^{2+} levels. What happens when too little or too much parathyroid hormone is secreted?

18. List the hormones secreted from the adrenal gland, give their functions, and compare the means by which the secretion rate of each is controlled.

19. What are the major functions of insulin and glucagon? How is their secretion regulated? What is the effect if too little insulin is secreted or the target tissues are not responsive to insulin?

20. List the effects of testosterone, estrogen, and progesterone.

21. What hormones are produced by the thymus and by the pineal gland? Name the effects of these hormones.

22. List the effects of prostaglandins. How is aspirin able to reduce the severity of the inflammatory response?

23. List the hormones secreted by the placenta.

24. List the major age-related changes that affect the endocrine system.

CRITICAL THINKING

1. A hormone is known to bind to a membrane-bound receptor. A drug that inhibits the breakdown of cyclic AMP causes an increased response to the hormone. A drug that inhibits the binding of GTP to proteins reduces the response to this drug. Based on these observations, describe the mechanism by which the membrane-bound receptor is most likely to produce a response to the hormone.

2. Aldosterone and antidiuretic hormone play important roles in regulating blood volume and concentration of blood. The response to one of these hormones is evident within minutes, and the response to the other requires several hours. Explain the difference in response time for these two hormones.

3. Biceps Benny figured that if a small amount of a vitamin was good, a lot should be better, so he began to take supplements that included a large amount of vitamin D. Predict the effect of vitamin D on his blood Ca^{2+} levels and on the secretion of hormones that regulate blood Ca^{2+} levels.

4. If a person's adrenal cortex degenerated and no longer secreted hormones, what would be the consequences?

5. Predict the effect of elevated aldosterone secretion from the adrenal cortex.

6. Explain how the blood levels of glucocorticoids, epinephrine, insulin, and glucagon change after a person has gone without food for 24 hours.

7. Stetha Scope wanted to go to medical school to become a physician. While attending college, she knew her grades had to be excellent. Stetha worked very hard and worried constantly. By the end of each school year, she had a cold and suffered from stomach pains. Explain why she might be susceptible to these symptoms.

Answers in Appendix D

11

Blood

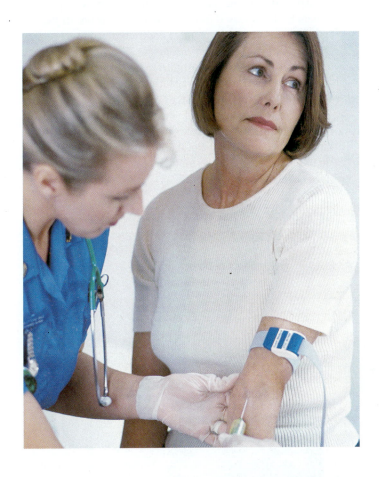

11.1 FUNCTIONS OF BLOOD

Learning Outcome *After reading this section, you should be able to*

A. State the functions of blood.

Blood has always fascinated humans, and throughout history they have speculated about its function. Some societies consider blood the "essence of life" because the uncontrolled loss of it can result in death. Many cultures around the world, both ancient and modern, believe blood has magical qualities. Blood has also been thought to define our character and emotions. For example, people of a noble bloodline are sometimes described as "blue bloods," whereas criminals are said to have "bad" blood. People commonly say that anger causes their blood to "boil," whereas fear makes it "curdle." The scientific study of blood reveals characteristics as fascinating as any of these fantasies. Blood performs many functions essential to life and can reveal much about our health. The heart pumps blood through blood vessels that extend throughout the body. Blood helps maintain homeostasis in several ways:

1. *Transport of gases, nutrients, and waste products.* Oxygen enters the blood in the lungs and is carried to cells. Carbon dioxide, produced by cells, is carried in the blood to the lungs, from which it is expelled. The blood transports ingested nutrients, ions, and water from the digestive tract to cells, and the blood transports the waste products of the cells to the kidneys for elimination.

2. *Transport of processed molecules.* Many substances are produced in one part of the body and transported in the blood to another part, where they are modified. For example, the precursor to vitamin D is produced in the skin (see chapter 5) and transported by the blood to the liver and then to the kidneys for processing into active vitamin D. Then the blood transports active vitamin D to the small intestine, where it promotes the uptake of calcium. Another example is lactate produced by skeletal muscles during anaerobic respiration (see chapter 7). The blood carries lactate to the liver, where it is converted into glucose.

Module 9 Cardiovascular System

3. *Transport of regulatory molecules.* The blood carries many of the hormones and enzymes that regulate body processes from one part of the body to another.

4. *Regulation of pH and osmosis.* Buffers (see chapter 2), which help keep the blood's pH within its normal limits of 7.35–7.45, are found in the blood. The osmotic composition of blood is also critical for maintaining normal fluid and ion balance.

5. *Maintenance of body temperature.* Warm blood is transported from the interior of the body to the surface, where heat is released from the blood. This is one of the mechanisms that helps regulate body temperature.

6. *Protection against foreign substances.* Certain cells and chemicals in the blood constitute an important part of the immune system, protecting against foreign substances, such as microorganisms and toxins.

7. *Clot formation.* When blood vessels are damaged, blood clotting protects against excessive blood loss. When tissues are damaged, the blood clot that forms is also the first step in tissue repair and the restoration of normal function (see chapter 4).

II.2 COMPOSITION OF BLOOD

Learning Outcome   *After reading this section, you should be able to*

A. List the components of blood.

Blood is a type of connective tissue that consists of a liquid matrix containing cells and cell fragments. The liquid matrix is the **plasma** (plaz′mă), and the cells and cell fragments are the **formed elements** (figure 11.1). The plasma accounts for slightly more than half of the total blood volume, and the formed elements account for slightly less than half. The total blood volume in the average adult is about 4–5 liters (L) in females and 5–6 L in males. Blood makes up about 8% of total body weight.

II.3 PLASMA

Learning Outcome   *After reading this section, you should be able to*

A. Name the components of plasma, and give their functions.

Plasma is a pale yellow fluid that consists of about 91% water, 7% proteins, and 2% other components, such as ions, nutrients, gases, waste products, and regulatory substances (figure 11.1 and table 11.1). Unlike the fibrous proteins found in other connective tissues, such as loose connective tissue, plasma contains dissolved proteins. Plasma proteins include albumin, globulins, and fibrinogen. **Albumin** (al-bū′min) makes up 58% of the plasma proteins. Although the osmotic pressure (see chapter 3) of blood

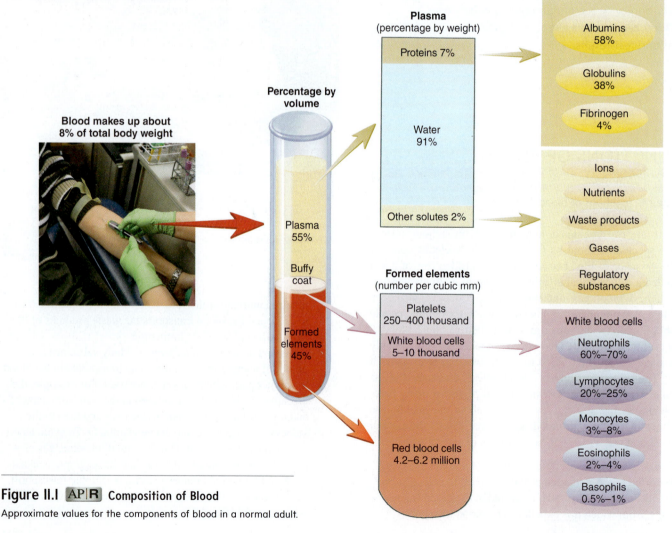

Figure II.I AP|R **Composition of Blood**
Approximate values for the components of blood in a normal adult.

results primarily from sodium chloride, albumin also makes an important contribution. The water balance between the blood and the tissues is determined by the movement of water into and out of the blood by osmosis. **Globulins** (glob′ū-linz; globule) account for 38% of the plasma proteins. Some globulins, such as antibodies and complement, are part of the immune system (see chapter 14). Other globulins and albumin function as transport molecules because they bind to molecules, such as hormones (see chapter 10), and carry them in the blood throughout the body. Some globulins are clotting factors, which are necessary for the formation of blood clots. **Fibrinogen** (fī-brin′ō-jen) is a clotting factor that constitutes 4% of plasma proteins. Activation of clotting factors results in the conversion of fibrinogen to **fibrin** (fi′brin), a threadlike protein that forms blood clots (see "Blood Clotting" later in this chapter). **Serum** (ser′um) is plasma without the clotting factors.

Plasma volume and composition remains relatively constant. Normally, water intake through the digestive tract closely matches water loss through the kidneys, lungs, digestive tract, and skin. Oxygen enters the blood in the lungs, and carbon dioxide enters the blood from tissues. Other suspended or dissolved substances in the blood come from the liver, kidneys, intestines, endocrine glands, and immune tissues, such as the lymph nodes and spleen. The concentration of these substances in the blood is also regulated and maintained within narrow limits.

II.4 FORMED ELEMENTS

Learning Outcomes *After reading this section, you should be able to*

A. Describe the origin and production of the formed elements.
B. Describe the structure, function, and life history of red blood cells.
C. Compare the structures and functions of the five types of white blood cells.
D. Describe the origin and structure of platelets.

About 95% of the volume of the formed elements consists of **red blood cells (RBCs),** or *erythrocytes* (ĕ-rith′rō-sītz; *erythro-,* red + *kytos,* cell). The remaining 5% of the volume of the formed elements consists of **white blood cells (WBCs),** or *leukocytes* (loo′kō-sītz; *leuko-,* white), and cell fragments called **platelets** (plāt′letz), or *thrombocytes* (throm′bō-sītz; *thrombo-,* clot). Red blood cells are 700 times more numerous than white blood cells and 17 times more numerous than platelets. Table 11.2 illustrates the formed elements of the blood.

Production of Formed Elements

The process of blood cell production is called **hematopoiesis** (hē′mă-tō-poy-ē′sis; *hemato-,* blood + *poiesis,* making). In the fetus, hematopoiesis occurs in several tissues, including the liver, thymus, spleen, lymph nodes, and red bone marrow. After birth, hematopoiesis is confined primarily to red bone marrow, but some white blood cells are produced in lymphatic tissues (see chapter 14).

All the formed elements of blood are derived from a single population of cells called **stem cells,** or *hemocytoblasts.* These stem cells differentiate to give rise to different cell lines, each of which ends with the formation of a particular type of formed element (figure 11.2). The development of each cell line is regulated by specific growth factors. That is, growth factors determine the types of formed elements derived from the stem cells and how many formed elements are produced.

Red Blood Cells

Normal red blood cells are disk-shaped, with edges that are thicker than the center of the cell (figure 11.3). The biconcave shape increases the cell's surface area compared to a flat disk of the same size. The greater surface area makes it easier for gases to move into and out of the red blood cell. In addition, the red blood cell can bend or fold around its thin center, decreasing its size and enabling it to pass more easily through smaller blood vessels.

During their development, red blood cells lose their nuclei and most of their organelles. Consequently, they are unable to divide. Red blood cells live for about 120 days in males and 110 days in females. One-third of a red blood cell's volume is the pigmented protein **hemoglobin** (hē-mō-glō′bin), which is responsible for the cell's red color.

Function

The primary functions of red blood cells are to transport oxygen from the lungs to the various tissues of the body and to help transport carbon dioxide from the tissues to the lungs. Oxygen transport is accomplished by hemoglobin, which consists of four protein chains and four heme groups. Each protein, called a **globin**

Cardiovascular

TABLE II.I	Composition of Plasma
Plasma Components	**Functions and Examples**
Water	Acts as a solvent and suspending medium for blood components
Proteins	Maintain osmotic pressure (albumin), destroy foreign substances (antibodies and complement), transport molecules (albumin and globulins), and form clots (fibrinogen)
Ions	Involved in osmotic pressure (sodium and chloride ions), membrane potentials (sodium and potassium ions), and acid-base balance (hydrogen, hydroxide, and bicarbonate ions)
Nutrients	Source of energy and "building blocks" of more complex molecules (glucose, amino acids, triglycerides)
Gases	Involved in aerobic respiration (oxygen and carbon dioxide)
Waste products	Breakdown products of protein metabolism (urea and ammonia salts) and red blood cells (bilirubin)
Regulatory substances	Catalyze chemical reactions (enzymes) and stimulate or inhibit many body functions (hormones)

CLINICAL IMPACT   Stem Cells and Cancer Therapy

Many cancer therapies attack the types of rapidly dividing cells found in tumors. An undesirable side effect, however, can be the destruction of nontumor cells that divide rapidly, such as the stem cells and their derivatives in red bone marrow. After treatment for cancer, some patients are prescribed growth factors to stimulate the rapid regeneration of the red bone marrow. Although not a treatment for the cancer itself, the growth factors can speed recovery from the side effects of cancer therapy.

Some types of leukemia and genetic immune deficiency diseases can be treated with a bone marrow transplant containing blood stem cells. To avoid tissue rejection, families with a history of these disorders can freeze the umbilical cord blood of their newborn children. The cord blood contains many stem cells and can be used instead of a bone marrow transplant.

TABLE II.2   Formed Elements of the Blood

Cell Type	Illustration	Description	Function
Red Blood Cell		Biconcave disk; no nucleus; contains hemoglobin, which colors the cell red; 6.5–8.5 μm in diameter	Transports oxygen and carbon dioxide
White Blood Cells		Spherical cells with a nucleus	Five types of white blood cells, each with specific functions
Granulocytes			
Neutrophil		Nucleus with two to four lobes connected by thin filaments; cytoplasmic granules stain a light pink or reddish purple; 10–12 μm in diameter	Phagocytizes microorganisms and other substances
Basophil		Nucleus with two indistinct lobes; cytoplasmic granules stain blue-purple; 10–12 μm in diameter	Releases histamine, which promotes inflammation, and heparin, which prevents clot formation
Eosinophil		Nucleus often bilobed; cytoplasmic granules stain orange-red or bright red; 11–14 μm in diameter	Participates in inflammatory response of allergic reactions and asthma; attacks certain worm parasites
Agranulocytes			
Lymphocyte		Round nucleus; cytoplasm forms a thin ring around the nucleus; 6–14 μm in diameter	Produces antibodies and other chemicals responsible for destroying microorganisms; contributes to allergic reactions, graft rejection, tumor control, and regulation of immune system
Monocyte		Nucleus round, kidney-shaped, or horseshoe-shaped; contains more cytoplasm than does lymphocyte; 12–20 μm in diameter	Phagocytic cell in the blood; leaves the blood and becomes a macrophage, which phagocytizes bacteria, dead cells, cell fragments, and other debris within tissues
Platelet		Cell fragment surrounded by a plasma membrane and containing granules; 2–4 μm in diameter	Forms platelet plugs; releases chemicals necessary for blood clotting

(glō′bin), is bound to one **heme** (hēm), a red-pigmented molecule. Each heme contains one iron atom, which is necessary for the normal function of hemoglobin. Each iron in a heme molecule can reversibly bind to an oxygen molecule. Hemoglobin picks up oxygen in the lungs and releases oxygen in other tissues (see chapter 15). Hemoglobin that is bound to oxygen is bright red, whereas hemoglobin without bound oxygen is a darker red. Hemoglobin is responsible for 98.5% of the oxygen transported in blood. The remaining 1.5% is transported dissolved in plasma.

Because iron is necessary for oxygen transport, it is not surprising that two-thirds of the body's iron is found in hemoglobin. Small amounts of iron are required in the diet to replace the small amounts

Stem cell (hemocytoblast)

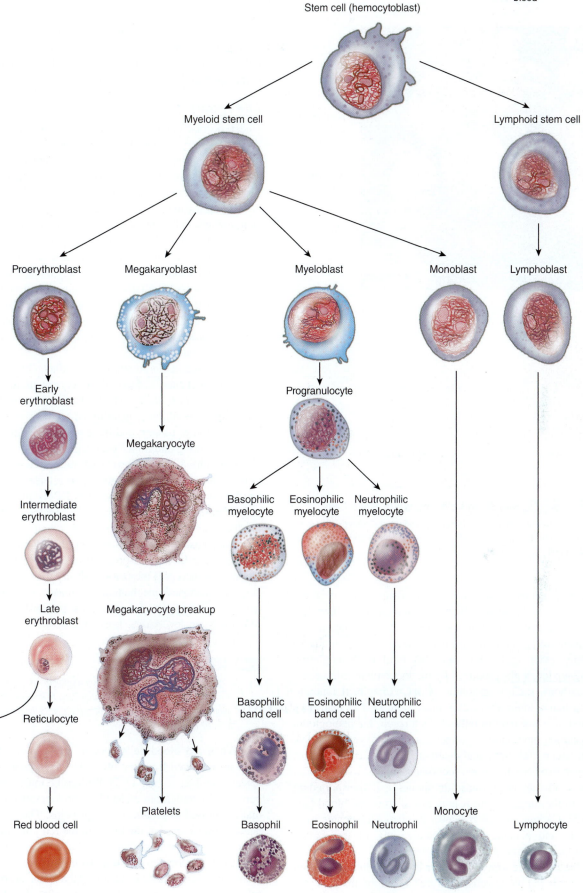

Myeloid stem cell

Lymphoid stem cell

Proerythroblast   Megakaryoblast   Myeloblast   Monoblast   Lymphoblast

Early
erythroblast

Progranulocyte

Megakaryocyte

Intermediate
erythroblast

Basophilic   Eosinophilic   Neutrophilic
myelocyte   myelocyte   myelocyte

Late
erythroblast   Megakaryocyte breakup

Nucleus
extruded

Reticulocyte   Basophilic   Eosinophilic   Neutrophilic
band cell   band cell   band cell

Platelets

Red blood cell   Platelets   Basophil   Eosinophil   Neutrophil   Monocyte   Lymphocyte

Figure II.2 Hematopoiesis

Stem cells give rise to the cell lines that produce the formed elements.

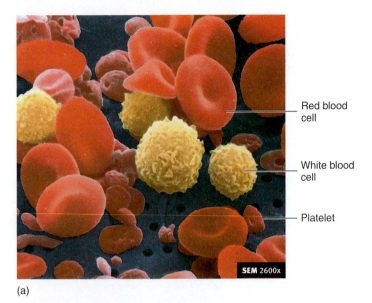

(a)

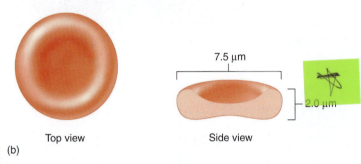

Top view   Side view

7.5 μm

2.0 μm

(b)

Figure II.3 AP|R Red Blood Cells and White Blood Cells

(*a*) Scanning electron micrograph of formed elements: red blood cells (*red doughnut shapes*) and white blood cells (*yellow*). (*b*) Shape and dimensions of a red blood cell.

lost in the urine and feces, but otherwise the existing iron is recycled, as described later in this section. Women need more dietary iron than men do because women lose iron as a result of menstruation.

Although oxygen is the primary molecule that binds to hemoglobin, other molecules can also bind to hemoglobin. Carbon monoxide, a gas produced by the incomplete combustion of hydrocarbons, such as gasoline, is one example. It binds to the iron in hemoglobin about 210 times more readily than does oxygen and does not tend to unbind. As a result, the hemoglobin bound to carbon monoxide no longer transports oxygen. Nausea, headache, unconsciousness, and death are possible consequences of prolonged exposure to carbon monoxide.

Carbon dioxide is produced in tissues and transported in the blood to the lungs, where it is removed from the blood (see chapter 15). Carbon dioxide transport involves bicarbonate ions, hemoglobin, and plasma. Approximately 70% of the carbon dioxide in blood is transported in the form of bicarbonate ions. The enzyme **carbonic anhydrase** (kar-bon′ik an-hī′drās), found primarily inside red blood cells, catalyzes a reaction that converts carbon dioxide (CO_2) and water (H_2O) into a hydrogen ion (H^+) and a bicarbonate ion (HCO_3^-):

$$CO_2 + H_2O \rightleftarrows H^+ + HCO_3^-$$

Carbon dioxide can bind reversibly to the globin part of hemoglobin. About 23% of the CO_2 in blood is transported bound to hemoglobin or other blood proteins. The remaining 7% of CO_2 is transported dissolved in plasma.

Life History of Red Blood Cells

Under normal conditions, about 2.5 million red blood cells are destroyed every second. Fortunately, new red blood cells are produced just as rapidly. Stem cells form **proerythroblasts** (prō-ĕ-rith′rō-blastz; *pro-*, before + *erythro-*, red + *blastos*, germ), which give rise to the red blood cell line (see figure 11.2). Red blood cell production involves a series of cell divisions. After each cell division, the new cells change and become more like mature red blood cells. In the later divisions, the newly formed cells manufacture large amounts of hemoglobin. After the final cell division, the cells lose their nuclei and become completely mature red blood cells.

The process of cell division requires the B vitamins folate and B_{12}, which are necessary for the synthesis of DNA (see chapter 3). Iron is required for the production of hemoglobin. Consequently, a lack of folate, vitamin B_{12}, or iron can interfere with normal red blood cell production.

Red blood cell production is stimulated by low blood oxygen levels. Typical causes of low blood oxygen are decreased numbers of red blood cells, decreased or defective hemoglobin, diseases of the lungs, high altitude, inability of the cardiovascular system to deliver blood to tissues, and increased tissue demand for oxygen, as occurs during endurance exercises.

Low blood oxygen levels stimulate red blood cell production by increasing the formation and release of the glycoprotein **erythropoietin** (ĕ-rith-rō-poy′ĕ-tin, **EPO**), primarily by the kidneys (figure 11.4). Erythropoietin stimulates red bone marrow to produce more red blood cells. Thus, when oxygen levels in the blood decrease, the production of erythropoietin increases, which increases red blood cell production. The greater number of red blood cells increases the blood's ability to transport oxygen. This negative-feedback mechanism increases the blood's capacity to transport oxygen and maintains homeostasis. Conversely, if blood oxygen levels rise, less erythropoietin is released, and red blood cell production decreases.

Predict 2

Cigarette smoke produces carbon monoxide. If a nonsmoker smoked a pack of cigarettes a day for a few weeks, what would happen to the number of red blood cells in the person's blood? Explain.

When red blood cells become old, abnormal, or damaged, they are removed from the blood by macrophages located in the spleen and liver (figure 11.5). Within the macrophage, the globin part of the hemoglobin molecule is broken down into amino acids that are reused to produce other proteins. The iron released from heme is transported in the blood to the red bone marrow and used to produce new hemoglobin. Thus, the iron is recycled. The heme molecules are converted to **bilirubin** (bil-i-roo′bin), a yellow pigment molecule. Bilirubin is normally taken up by the liver and released into the small intestine as part of the bile (see chapter 16). If the liver is not functioning normally, or if the flow of bile from the liver to the small intestine is hindered, bilirubin builds up in the circulation and produces **jaundice** (jawn′dis; *jaune*, yellow), a

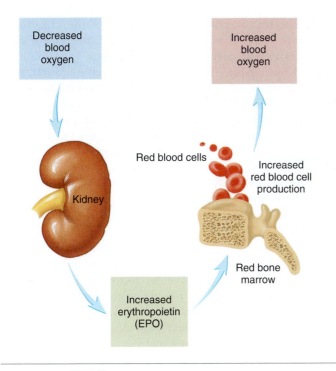

Figure 11.4 AP|R **Red Blood Cell Production**

In response to decreased blood oxygen, the kidneys release erythropoietin into the general circulation. The increased erythropoietin stimulates red blood cell production in the red bone marrow. This process increases blood oxygen levels.

yellowish color to the skin. After it enters the intestine, bilirubin is converted by bacteria into other pigments. Some of these pigments give feces their brown color, whereas others are absorbed from the intestine into the blood, modified by the kidneys, and excreted in the urine, contributing to the characteristic yellow color of urine.

White Blood Cells

White blood cells are spherical cells that lack hemoglobin. When the components of blood are separated from one another, white blood cells as well as platelets make up the **buffy coat,** a thin, white layer of cells between plasma and red blood cells (see figure 11.1). White blood cells are larger than red blood cells, and each has a nucleus (table 11.2). Although white blood cells are components of the blood, the blood serves primarily as a means of transporting these cells to other body tissues. White blood cells can leave the blood and travel by **ameboid** (ă-mē′boyd; like an ameba) **movement** through the tissues. In this process, the cell projects a cytoplasmic extension that attaches to an object. Then the rest of the cell's cytoplasm flows into the extension. Two functions of white blood cells are (1) to protect the body against invading microorganisms and other pathogens and (2) to remove dead cells and debris from the tissues by phagocytosis.

Cardiovascular

1. In macrophages, the globin part of hemoglobin is broken down to individual amino acids (*red arrow*) and metabolized or used to build new proteins.

2. The heme of hemoglobin releases iron. The heme is converted into bilirubin.

3. Blood transports iron to the red bone marrow, where it is used to produce new hemoglobin (*green arrows*).

4. Blood transports bilirubin (*blue arrows*) to the liver.

5. Bilirubin is excreted as part of the bile into the small intestine. Some bilirubin derivatives contribute to the color of feces.

6. Other bilirubin derivatives are reabsorbed from the intestine into the blood and excreted from the kidneys in the urine.

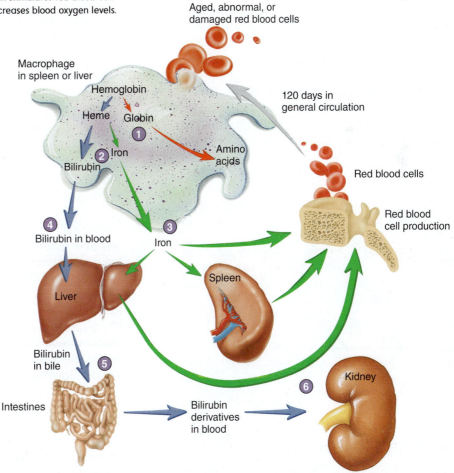

PROCESS Figure 11.5 AP|R **Hemoglobin Breakdown**

Macrophages break down hemoglobin, and the breakdown products are used or excreted.

Each white blood cell type is named according to its appearance in stained preparations. Those containing large cytoplasmic granules are *granulocytes* (gran'ū-lō-sītz; *granulo-,* granular), and those with very small granules that cannot be seen easily with the light microscope are **agranulocytes** (ă-gran'ū-lō-sītz; *a-,* without).

There are three kinds of granulocytes: neutrophils, basophils, and eosinophils. **Neutrophils** (noo'trō-filz; *neutro-,* neutral + *philos,* loving), the most common type of white blood cells, have small cytoplasmic granules that stain with both acidic and basic dyes (figure 11.6). Their nuclei are commonly lobed, with the number of lobes varying from two to four. Neutrophils usually remain in the blood for a short time (10–12 hours), move into other tissues, and phagocytize microorganisms and other foreign substances. Dead neutrophils, cell debris, and fluid can accumulate as **pus** at sites of infections.

Basophils (bā'sō-filz; *baso-,* base), the least common of all white blood cells, contain large cytoplasmic granules that stain blue or purple with basic dyes (table 11.2). Basophils release histamine and other chemicals that promote inflammation (see chapters 4 and 14). They also release heparin, which prevents the formation of clots.

Eosinophils (ē-ō-sin'o-filz) contain cytoplasmic granules that stain bright red with eosin, an acidic stain. They often have a two-lobed nucleus (table 11.2). Eosinophils are involved in inflammatory responses associated with allergies and asthma. In addition, chemicals from eosinophils are involved in destroying certain worm parasites.

There are two kinds of agranulocytes: lymphocytes and monocytes. **Lymphocytes** (lim'fō-sītz; *lympho-,* lymph) are the smallest of the white blood cells (table 11.2). The lymphocytic cytoplasm consists of only a thin, sometimes imperceptible ring around the nucleus. There are several types of lymphocytes, and they play an important role in the body's immune response. Their diverse activities involve the production of antibodies and other chemicals that destroy microorganisms, contribute to allergic reactions, reject grafts, control tumors, and regulate the immune system. Chapter 14 considers these cells in more detail.

Monocytes (mon'ō-sītz) are the largest of the white blood cells (table 11.2). After they leave the blood and enter tissues, monocytes enlarge and become **macrophages** (mak'rō-fā-jez; *macro-,* large + *phagō,* to eat), which phagocytize bacteria, dead cells, cell fragments, and any other debris within the tissues. In addition, macrophages can break down phagocytized foreign substances and present the processed substances to lymphocytes, causing activation of the lymphocytes (see chapter 14).

Predict 3

Based on their morphology, identify each of the white blood cells shown in figure 11.7.

Platelets

Platelets are minute fragments of cells, each consisting of a small amount of cytoplasm surrounded by a cell membrane (see figure 11.6). They are produced in the red bone marrow from **megakaryocytes** (meg-ă-kar'ē-ō-sītz; *mega-,* large + *karyon,* nucleus), which are large cells (see figure 11.2). Small fragments of these cells break off and enter the blood as platelets, which play an important role in preventing blood loss.

11.5 PREVENTING BLOOD LOSS

Learning Outcomes *After reading this section, you should be able to*

A. Explain the formation and function of platelet plugs and blood clots.

B. Describe the regulation of clot formation and how clots are removed.

When a blood vessel is damaged, blood can leak into other tissues and interfere with normal tissue function, or blood can be lost from the body. The body can tolerate a small amount of blood loss and can produce new blood to replace it. But a large amount of blood loss can lead to death. Fortunately, when a blood vessel is damaged, loss of blood is minimized by three processes: vascular spasm, platelet plug formation, and blood clotting.

Vascular Spasm

Vascular spasm is an immediate but temporary constriction of a blood vessel that results when smooth muscle within the wall of the vessel contracts. This constriction can close small vessels completely and stop the flow of blood through them. Damage to blood vessels can activate nervous system reflexes that cause vascular spasm. Chemicals also produce vascular spasm. For example, platelets release **thromboxanes** (throm'bok-zānz), which are derived from certain prostaglandins, and endothelial (epithelial) cells lining blood vessels release the peptide **endothelin** (en-dō-thē'lin).

Platelet Plug Formation

A **platelet plug** is an accumulation of platelets that can seal up a small break in a blood vessel. Platelet plug formation is very important in maintaining the integrity of the blood vessels of the cardiovascular system because small tears occur in the smaller vessels and capillaries many times each day. People who lack the normal number of platelets tend to develop numerous small hemorrhages in their skin and internal organs.

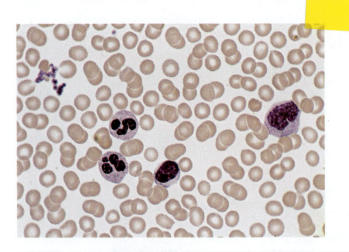

Figure 11.6 AP|R **Standard Blood Smear**

A thin film of blood is spread on a microscope slide and stained. The white blood cells have pink cytoplasm and purple nuclei. The red blood cells do not have nuclei. The center of a red blood cell appears whitish because light shines more readily through the thin center of the disk than through the thicker edges. The platelets are purple cell fragments.

CLINICAL IMPACT   Clinical Importance of Activating Platelets

Platelet activation results in platelet plug formation and the production of chemicals, such as phospholipids, that are important for blood clotting. Thus, inhibition of platelet activation reduces the formation of blood clots.

Thromboxanes, which activate platelets, are derived from certain prostaglandins. Aspirin inhibits prostaglandin synthesis and, therefore, thromboxane synthesis, which results in reduced platelet activation. If an expectant mother ingests aspirin near the end of pregnancy, thromboxane synthesis is inhibited. As a result, the mother experiences excessive bleeding after delivery, and the baby can exhibit numerous localized hemorrhages over the surface of its body,

both due to decreased platelet function. If the quantity of ingested aspirin is large, the infant, the mother, or both may die as a result of bleeding.

Platelet plugs and blood clots can block blood vessels, producing heart attacks and strokes. Suspected heart attack victims are routinely given aspirin en route to the emergency room as part of their treatment. The United States Preventive Services Task Force (USPSTF) and the American Heart Association recommend low-dose aspirin therapy (75–160 mg/day) for all men and women at high risk for cardiovascular disease. Determining high risk involves analyzing many factors and should be done in consultation with a physician.

The decreased risk of cardiovascular disease from aspirin therapy must be weighed against the increased risk of hemorrhagic stroke and gastrointestinal bleeding. Risk factors include age (men over 40 and postmenopausal women), high cholesterol, high blood pressure, a history of smoking, diabetes, a family history of cardiovascular disease, and a previous clotting event, such as a heart attack, a transient ischemic attack, or an occlusive stroke.

Plavix (clopidogrel bisulfate) is a drug that reduces the activation of platelets by blocking the ADP receptors on the surface of platelets. It is administered to prevent clotting and, along with other anticlotting drugs, to treat heart attacks.

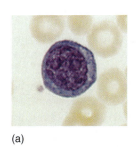 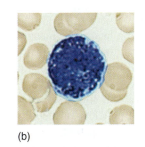 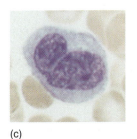 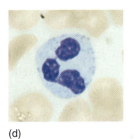 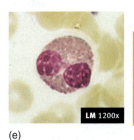

LM 1200x

(a)   (b)   (c)   (d)   (e)

Cardiovascular

Figure 11.7 AP|R Identification of White Blood Cells
See Predict 2.

The formation of a platelet plug can be described as a series of steps, but in actuality many of these steps occur at the same time (figure 11.8). First, platelets stick to the collagen exposed by blood vessel damage; this phenomenon is called **platelet adhesion.** Most platelet adhesion is mediated through **von Willebrand factor,** a protein produced and secreted by blood vessel endothelial cells. Von Willebrand factor forms a bridge between collagen and platelets by binding to platelet surface receptors and collagen. After platelets adhere to collagen, they become activated, change shape, and release chemicals.

In the **platelet release reaction,** platelets release chemicals, such as ADP and thromboxane, which bind to their respective receptors on the surfaces of other platelets, activating the platelets. These activated platelets also release ADP and thromboxane, which activates more platelets. Thus, a cascade of chemical release activates many platelets. This is an example of positive feedback. As platelets become activated, they express surface receptors called **fibrinogen receptors,** which can bind to fibrinogen, a plasma protein. In **platelet aggregation,** fibrinogen forms bridges between the fibrinogen receptors of numerous platelets, resulting in a platelet plug.

A CASE IN POINT

Idiopathic Thrombocytopenic Purpura

Following a weekend touch football game, Bruz Moore noticed that he had developed a number of small bruises. Although the bruising was unusual, he dismissed it. A few days later, Bruz suddenly developed a rash on his legs, which he also ignored. But when he noticed bleeding from his gums while brushing his teeth, he went to see his doctor. There, Bruz learned that the rash was caused by many pinhead-sized hemorrhages in the skin. This sign, together with the bruising and the bleeding from the gums, indicated a below-normal ability to stop bleeding. A blood test confirmed that Bruz's platelet count was lower than normal. In the absence of any evidence of toxic exposure to chemicals or diseases associated with a low platelet count, the doctor diagnosed idiopathic thrombocytopenic purpura (id′ē-ō-path′ik throm′bō-sī-tō-pē′nik pŭr′poo-ra) (ITP). In this disorder, the immune system makes antibodies (see chapter 14) that bind to platelets. As a result, the "tagged" platelets are removed from the blood by phagocytic cells in the spleen faster than they are produced in the red bone marrow. Consequently, platelet numbers decrease, causing bleeding problems. For more information on thrombocytopenia, see "Platelet Count" later in this chapter.

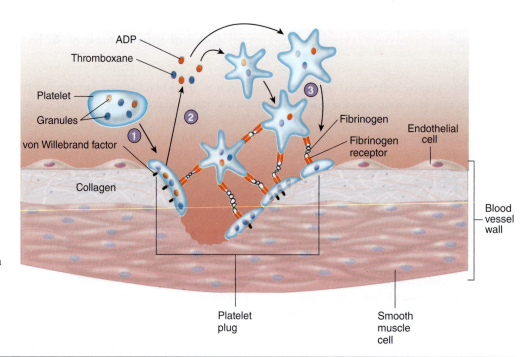

① Platelet adhesion occurs when von Willebrand factor connects exposed collagen to platelets.

② During the platelet release reaction, ADP, thromboxanes, and other chemicals are released and activate other platelets.

③ Platelet aggregation occurs when fibrinogen receptors on activated platelets bind to fibrinogen, connecting the platelets to one another. The accumulating mass of platelets forms a platelet plug.

PROCESS Figure II.8 Platelet Plug Formation

Blood Clotting

Blood vessel constriction and platelet plugs alone are not sufficient to close large tears or cuts in blood vessels. When a blood vessel is severely damaged, **blood clotting,** or *coagulation* (kō-ag-ū-lā′shŭn), results in the formation of a clot. A **clot** is a network of threadlike protein fibers, called **fibrin** (fī′brin), that traps blood cells, platelets, and fluid.

The formation of a blood clot depends on a number of proteins found within plasma, called **clotting factors.** Though normally present in the plasma, the clotting factors are inactive and do not cause clotting. Following injury, however, the clotting factors are activated. Clot formation is a complex process involving many chemical reactions, but it can be summarized in three stages (figure 11.9).

1. The chemical reactions can be started in two ways:
   (a) Inactive clotting factors come in contact with exposed connective tissue, resulting in their activation, or
   (b) chemicals, such as **thromboplastin,** are released from injured tissues, causing activation of clotting factors. After the initial clotting factors are activated, they in turn activate other clotting factors. A series of reactions results in which each clotting factor activates the next until the clotting factor **prothrombinase** (prō-throm′bi-nās), or *prothrombin activator*, is formed.
2. Prothrombinase converts an inactive clotting factor called **prothrombin** (prō-throm′bin) to its active form, **thrombin** (throm′bin).
3. Thrombin converts the plasma protein **fibrinogen** (fī-brin′ō-jen) to fibrin.

At each step of the clotting process, each clotting factor activates many additional clotting factors, resulting in the formation of a clot.

Most clotting factors are manufactured in the liver, and many of them require vitamin K for their synthesis. In addition, many of the chemical reactions of clot formation require Ca^{2+} and the chemicals released from platelets. The clotting process can be severely impaired by low levels of vitamin K, low levels of Ca^{2+}, low numbers of platelets, or reduced synthesis of clotting factors because of liver dysfunction.

Humans rely on two sources of vitamin K. About half comes from the diet, and the other half comes from bacteria within the large intestine. Antibiotics taken to fight bacterial infections sometimes kill these intestinal bacteria, reducing vitamin K levels and causing bleeding problems. Vitamin K supplements may be necessary for patients on prolonged antibiotic therapy. Newborns lack these intestinal bacteria and thus routinely receive a vitamin K injection at birth. Infants can also obtain vitamin K from food, such as milk. Because cow's milk contains more vitamin K than does human milk, breast-fed infants are more susceptible to bleeding than are bottle-fed infants. However, maternal supplementation with vitamin K, such as with oral vitamins, adequately elevates breast-fed infant vitamin K levels and decreases the risk of bleeding.

Control of Clot Formation

Without control, clotting would spread from the point of its initiation throughout the blood vessels. Fortunately, the blood contains several **anticoagulants** (an′tē-kō-ag′ū-lantz), which prevent clotting factors from forming clots under normal conditions. For example, **antithrombin** (an-tē-throm′bin) and **heparin** (hep′ă-rin) inactivate thrombin. Without thrombin, fibrinogen is not converted to fibrin, and no clot forms. At an injury site, however, the activation of clotting factors is very rapid. Enough clotting factors are activated that the anticoagulants can no longer

Cardiovascular

CLINICAL IMPACT   The Danger of Unwanted Clots

When platelets encounter damaged or diseased areas of blood vessels or heart walls, an attached clot, called a **thrombus** (throm'bus), can form. A thrombus that breaks loose and begins to float through the circulation is called an **embolus** (em'bō-lŭs). Both thrombi and emboli can cause death if they block vessels that supply blood to essential organs, such as the heart, brain, or lungs. Abnormal coagulation can be prevented or hindered by administering an anticoagulant, such as heparin, which acts rapidly. Warfarin (war'fă-rin), brand name Coumadin (koo'mă-din), acts more slowly than heparin. Warfarin prevents clot formation by suppressing the liver's production of vitamin K–dependent clotting factors.

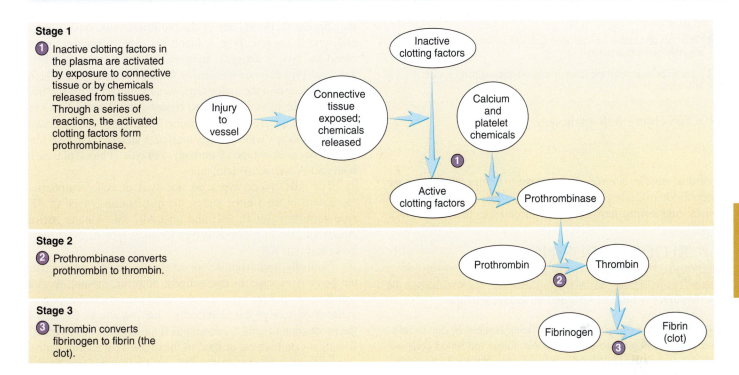

Stage 1

① Inactive clotting factors in the plasma are activated by exposure to connective tissue or by chemicals released from tissues. Through a series of reactions, the activated clotting factors form prothrombinase.

Stage 2

② Prothrombinase converts prothrombin to thrombin.

Stage 3

③ Thrombin converts fibrinogen to fibrin (the clot).

PROCESS Figure 11.9 Clot Formation

Clot formation has three stages.

prevent a clot from forming. Away from the injury site, there are enough anticoagulants to prevent clot formation from spreading.

Clot Retraction and Fibrinolysis

After a clot has formed, it begins to condense into a more compact structure through a process known as **clot retraction.** Platelets contain the contractile proteins actin and myosin, which operate in a fashion similar to that of the actin and myosin in muscle (see chapter 7). Platelets form small extensions that attach to fibrin through surface receptors. Contraction of the extensions pulls on the fibrin and leads to clot retraction. During clot retraction, serum, which is plasma without the clotting factors, is squeezed out of the clot.

Retraction of the clot pulls the edges of the damaged blood vessel together, helping stop the flow of blood, reducing the probability of infection, and enhancing healing. The vessel is repaired as fibroblasts move into the damaged area and new connective tissue forms. In addition, epithelial cells around the wound divide and fill in the torn area (see chapter 4).

Clots are dissolved by a process called **fibrinolysis** (fī-bri-nol'-i-sis) (figure 11.10). An inactive plasma protein called **plasminogen** (plaz-min'ō-jen) is converted to its active form, **plasmin** (plaz'min). Thrombin, other clotting factors activated during clot formation, and **tissue plasminogen activator (t-PA)** released from surrounding tissues can stimulate the conversion of plasminogen to plasmin. Over a few days, plasmin slowly breaks down the fibrin.

A heart attack can result when a clot blocks blood vessels that supply the heart. One treatment for a heart attack is to inject certain chemicals into the blood that activate plasmin. Unlike aspirin and anticoagulant therapies, which are used to prevent heart attacks, plasmin activators quickly dissolve the clot and restore blood flow

Cardiovascular

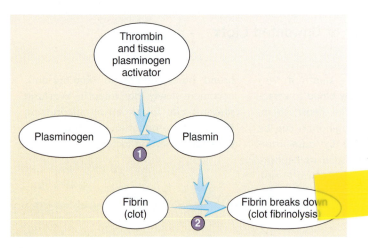

1 Thrombin and tissue plasminogen activator convert inactive plasminogen into plasmin.

2 Plasmin breaks down the fibrin in a blood clot, resulting in clot fibrinolysis.

PROCESS Figure II.I0 Fibrinolysis

to cardiac muscle, thus reducing damage to tissues. **Streptokinase** (strep-tō-kīn′ās), a bacterial enzyme, and t-PA, produced through genetic engineering, have been used successfully to dissolve clots.

II.6 BLOOD GROUPING

Learning Outcome *After reading this section, you should be able to*

A. Explain the basis of ABO and Rh incompatibilities.

If large quantities of blood are lost during surgery or due to injury, a patient can go into shock and die unless red blood cells are replaced to restore the blood's oxygen-carrying capacity. In this event, either a transfusion or an infusion is called for. A **transfusion** is the transfer of blood or blood components from one individual to another. An **infusion** is the introduction of a fluid other than blood, such as a saline or glucose solution, into the blood. In many cases, the return of blood volume to normal levels is all that is necessary to prevent shock. Eventually, the body produces enough red blood cells to replace those that were lost.

Early attempts to transfuse blood were often unsuccessful because they resulted in **transfusion reactions,** characterized by clumping or rupture of blood cells and clotting within blood vessels. We now know that transfusion reactions are caused by interactions between antigens and antibodies (see chapter 14). In brief, the surfaces of red blood cells have molecules called **antigens** (an′ti-jenz), and the plasma includes proteins called **antibodies** (an′te-bod-ēz; *anti-,* against). Antibodies are very specific, meaning that each antibody can bind only to a certain antigen. When the antibodies in the plasma bind to the antigens on the surface of the red blood cells, they form molecular bridges that connect the red blood cells together. As a result, **agglutination** (ă-gloo-ti-nā′shŭn; *ad,* to + *gluten,* glue), or clumping of the cells, occurs. The combination of the antibodies with the antigens can also initiate reactions that cause **hemolysis** (hē-mol′i-sis; *hemo-,* blood + *lysis,*

destruction), or rupture of the red blood cells. The debris formed from the ruptured red blood cells can cause severe tissue damage, particularly in the kidneys. Hemoglobin released from lysed red blood cells can damage kidney tissue, reducing its blood-filtering ability. If the damage is extensive, the lack of kidney function could result in death.

The antigens on the surface of red blood cells have been categorized into **blood groups.** Although many blood groups are recognized, the ABO and Rh blood groups are the most important when discussing transfusion reactions.

ABO Blood Group

The **ABO blood group** system is used to categorize human blood. In this blood group system, there are two types of antigens that may appear on the surface of the red blood cells, type A antigen and type B antigen. Type A blood has type A antigens, type B blood has type B antigens, and type AB blood has both types of antigens. Type O blood has neither A nor B antigens (figure 11.11). Antibodies against the antigens are usually present in the plasma of blood. Plasma from type A blood contains anti-B antibodies, which act against type B antigens; plasma from type B blood contains anti-A antibodies, which act against type A antigens. Type AB blood plasma has neither type of antibody, and type O blood plasma has both anti-A and anti-B antibodies.

The ABO blood types do not exist in equal numbers. In caucasians in the United States, the distribution is type O, 47%; type A, 41%; type B, 9%; and type AB, 3%. Among African-Americans, the distribution is type O, 46%; type A, 27%; type B, 20%; and type AB, 7%.

Normally, antibodies do not develop against an antigen unless the body is exposed to that antigen; however, the anti-A and/or anti-B antibodies are present in the blood even without exposure to antigens on foreign red blood cells. One possible explanation for the production of anti-A and/or anti-B antibodies is that type A or B antigens on bacteria or food in the digestive tract stimulate the formation of antibodies against antigens that are different from the body's own antigens. In support of this explanation, anti-A and anti-B antibodies are not found in the blood until about 2 months after birth. It is possible that an infant with type A blood would produce anti-B antibodies against the B antigens on bacteria or food. Meanwhile, an infant with A antigens would not produce antibodies against the A antigens on bacteria or food because mechanisms exist in the body to prevent the production of antibodies that would react with the body's own antigens (see chapter 14).

When a blood transfusion is performed, the **donor** is the person who gives blood, and the **recipient** is the person who receives it. Usually, a recipient can successfully receive blood from a donor as long as they both have the same blood type. For example, a person with type A blood can receive blood from a person with type A blood. No ABO transfusion reaction occurs because the recipient has no anti-A antibodies against the type A antigen. On the other hand, if type A blood were donated to a person with type B blood, a transfusion reaction would occur because the person with type B blood has anti-A antibodies against the type A antigen, causing agglutination (figure 11.12).

Historically, people with type O blood have been called universal donors because they can usually give blood to the other ABO blood types without causing an ABO transfusion reaction. Their red

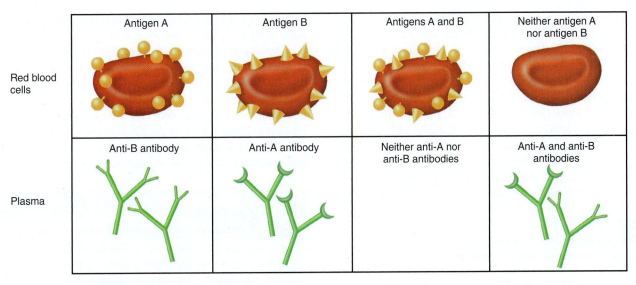

Type A	Type B	Type AB	Type O
Red blood cells with type A surface antigens and plasma with anti-B antibodies	Red blood cells with type B surface antigens and plasma with anti-A antibodies	Red blood cells with both type A and type B surface antigens and neither anti-A nor anti-B plasma antibodies	Red blood cells with neither type A nor type B surface antigens but both anti-A and anti-B plasma antibodies

Figure II.II ABO Blood Groups

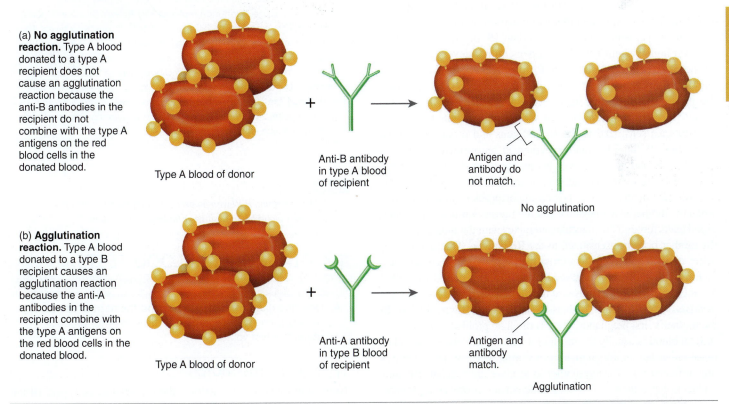

Figure II.I2 Agglutination Reaction

blood cells have no ABO surface antigens and therefore do not react with the recipient's anti-A or anti-B antibodies. For example, if a person with type A blood receives type O blood, the type O red blood cells do not react with the anti-B antibodies in the recipient's blood.

However, the term *universal* donor is misleading. Transfusion of type O blood can still produce a transfusion reaction in one of two ways: First, mismatching blood groups other than the ABO blood group can cause a transfusion reaction. To reduce the likelihood

Cardiovascular

of a transfusion reaction, all the blood groups must be correctly matched (see "Type and Crossmatch" in the next section of this chapter). Second, antibodies in the donor's blood can react with antigens on the recipient's red blood cells. For example, type O blood has anti-A and anti-B antibodies. If type O blood is transfused into a person with type A blood, the anti-A antibodies (in the type O donor blood) react against the A antigens (on the red blood cells in the type A recipient blood). Usually, such reactions are not serious because the antibodies in the donor's blood are diluted in the large volume of the recipient's blood. Even though such transfusion reactions seldom occur, type O blood is given to a person with another blood type only in life-or-death situations.

Predict 4

Historically, people with type AB blood were called universal recipients. What is the rationale for this term? Explain why the term is misleading.

Rh Blood Group

Another important blood group is the **Rh blood group,** so named because it was first studied in the rhesus monkey. People are Rh-positive if they have certain Rh antigens on the surface of their red blood cells, and they are Rh-negative if they do not have these Rh antigens. About 85% of caucasians and 95% of African-Americans are Rh-positive. The ABO blood type and the Rh blood type are usually expressed together. For example, a person designated as type A in the ABO blood group and Rh-positive is said to be A-positive. The rarest combination in the United States is AB-negative, which occurs in less than 1% of the population.

Antibodies against the Rh antigens do not develop unless an Rh-negative person is exposed to Rh-positive red blood cells. This can occur through a transfusion or by the transfer of blood across the placenta to a mother from her fetus. When an Rh-negative person receives a transfusion of Rh-positive blood, the recipient becomes sensitized to the Rh antigens and produces anti-Rh antibodies. If the Rh-negative person is unfortunate enough to receive a second transfusion of Rh-positive blood after becoming sensitized, a transfusion reaction results.

Rh incompatibility can pose a major problem in a pregnancy when the mother is Rh-negative and the fetus is Rh-positive. If fetal blood leaks through the placenta and mixes with the mother's blood, the mother becomes sensitized to the Rh antigen. The mother produces anti-Rh antibodies that cross the placenta and cause agglutination and hemolysis of fetal red blood cells. This disorder is called **hemolytic** (hē-mō-lit′ik) **disease of the newborn (HDN),** or *erythroblastosis fetalis* (ĕ-rith′rō-blas-tō′sis fē-ta′lis) (figure 11.13). In the mother's first pregnancy, there is often no problem. The leakage of fetal blood is usually the result of a tear in the placenta that takes place either late in the pregnancy or during delivery. Thus, there is not sufficient time for the mother to produce enough anti-Rh antibodies to harm the fetus. In later pregnancies, however, a problem can arise because the mother has been sensitized to the Rh antigen. Consequently, if the fetus is Rh-positive and if any fetal blood leaks into the mother's blood, she rapidly produces large amounts of anti-Rh antibodies, which can cross the placenta to the fetus, resulting in HDN. Because HDN can be fatal to the fetus, the levels of anti-Rh antibodies in the mother's blood should be monitored. If they increase to unacceptable levels, the fetus should be tested to determine the severity of the HDN. In severe cases, a transfusion to

replace lost red blood cells can be performed through the umbilical cord, or the baby can be delivered if mature enough.

Prevention of HDN is often possible if the Rh-negative mother is injected with a specific preparation called Rho(D) immune globulin (RhoGAM), which contains antibodies against Rh antigens. The injection can be given during the pregnancy, before delivery, or immediately after each delivery, miscarriage, or abortion. The injected antibodies bind to the Rh antigens of any fetal red blood cells that may have entered the mother's blood. This treatment inactivates the fetal Rh antigens and prevents sensitization of the mother.

A CASE IN POINT

Treatment of Hemolytic Disease of the Newborn

Billy Rubin was born with HDN. He was treated with phototherapy, exchange transfusion, and erythropoietin, all effective treatments.

During fetal development, the increased rate of red blood cell destruction caused by the mother's anti-Rh antibodies leads to increased production of bilirubin. Although high levels of bilirubin can damage the brain by killing nerve cells, this is not usually a problem in the fetus because the placenta removes the bilirubin. Following birth, bilirubin levels can increase because red blood cells continue to lyse and the newborn's liver is unable to handle the large bilirubin load. However, using phototherapy, blood passing through the skin is exposed to blue or white lights, which break down bilirubin to less toxic compounds, which the newborn's liver can then remove.

An exchange transfusion replaces Billy Rubin's blood with donor blood, thus decreasing the bilirubin and anti-Rh antibody levels. The presence of fewer anti-Rh antibodies decreases the agglutination and lysis of red blood cells. The exchange transfusion also helps alleviate the low number of red blood cells, a condition called anemia. The exchange transfusion replaces the newborn's blood with blood that has more red blood cells. Finally, administration of erythropoietin also treats the anemia by stimulating the newborn's body to produce more red blood cells.

Predict 5

When treating HDN with an exchange transfusion, should the donor's blood be Rh-positive or Rh-negative? Explain.

11.7 DIAGNOSTIC BLOOD TESTS

Learning Outcome   After reading this section, you should be able to

A. Describe diagnostic blood tests and the normal values for the tests, and give examples of disorders that produce abnormal test values.

Type and Crossmatch

To prevent transfusion reactions, the blood must be typed. **Blood typing** determines the ABO and Rh blood groups of a blood sample. Typically, the cells are separated from the serum and then tested with known antibodies to determine the type of antigen on the cell surface. For example, if a patient's blood cells agglutinate when mixed with anti-A antibodies but do not agglutinate when mixed with anti-B antibodies, the cells have type A antigen. In a similar fashion, the serum is mixed with known cell types (antigens) to determine the type of antibodies in the serum.

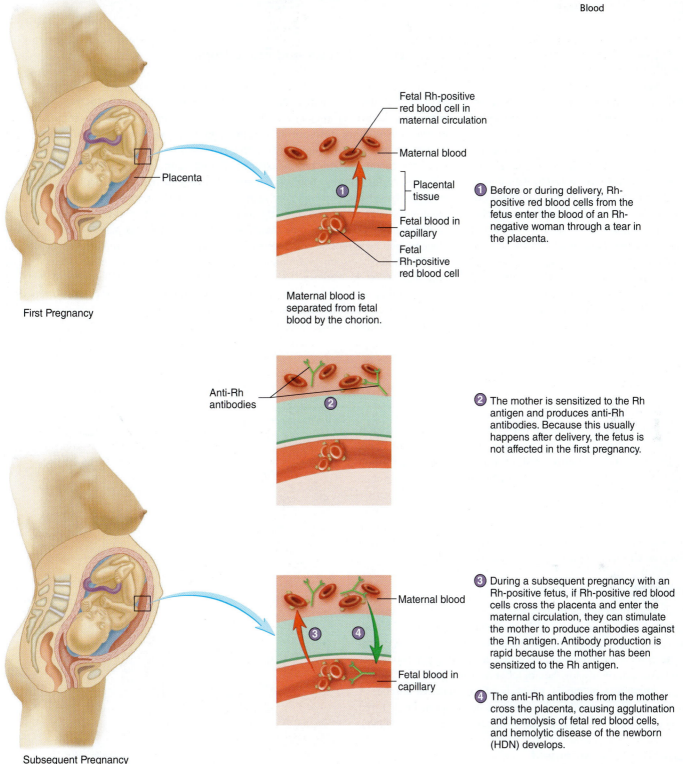

Fetal Rh-positive red blood cell in maternal circulation

Maternal blood

Placental tissue

Fetal blood in capillary

Fetal Rh-positive red blood cell

Maternal blood is separated from fetal blood by the chorion.

First Pregnancy

① Before or during delivery, Rh-positive red blood cells from the fetus enter the blood of an Rh-negative woman through a tear in the placenta.

Anti-Rh antibodies

② The mother is sensitized to the Rh antigen and produces anti-Rh antibodies. Because this usually happens after delivery, the fetus is not affected in the first pregnancy.

Placenta

Maternal blood

Fetal blood in capillary

Subsequent Pregnancy

③ During a subsequent pregnancy with an Rh-positive fetus, if Rh-positive red blood cells cross the placenta and enter the maternal circulation, they can stimulate the mother to produce antibodies against the Rh antigen. Antibody production is rapid because the mother has been sensitized to the Rh antigen.

④ The anti-Rh antibodies from the mother cross the placenta, causing agglutination and hemolysis of fetal red blood cells, and hemolytic disease of the newborn (HDN) develops.

Cardiovascular

PROCESS Figure 11.13 Hemolytic Disease of the Newborn (HDN)

Normally, donor blood must match the ABO and Rh type of the recipient. However, because other blood groups can cause a transfusion reaction, a crossmatch is performed. In a **crossmatch,** the donor's blood cells are mixed with the recipient's serum, and the donor's serum is mixed with the recipient's cells. The donor's blood is considered safe for transfusion only if no agglutination occurs in either match.

Complete Blood Count

A **complete blood count (CBC)** is an analysis of blood that provides much useful information. A CBC consists of a red blood cell count, hemoglobin and hematocrit measurements, and a white blood cell count.

CLINICAL IMPACT   Erythrocytosis and Blood Doping

Erythrocytosis is an overabundance of red blood cells, leading to increased blood viscosity, reduced flow rates, and, if severe, plugged capillaries. **Relative erythrocytosis** results from decreased plasma volume, such as that caused by dehydration, diuretics, and burns. **Primary erythrocytosis**, often called **polycythemia vera** (pol′ē-sī-thē′mē-ā ve′ra), is a stem cell defect of unknown cause that results in the overproduction of red blood cells, granulocytes, and platelets. Erythropoietin levels are low, and the spleen may be enlarged.

Secondary erythrocytosis results from a decreased oxygen supply, as occurs at high altitudes, in chronic obstructive pulmonary disease, or in congestive heart failure. The resulting decrease in oxygen delivery to the kidneys stimulates erythropoietin secretion and red blood cell production. In primary and secondary erythrocytosis, the greater number of red blood cells increases blood viscosity and blood volume. Capillaries may become clogged, and hypertension can develop.

Blood doping is an intentional process that serves to increase the number of circulating red blood cells. Having more red blood cells increases the blood's ability to transport oxygen. Recall from chapter 7 that muscles require significant amounts of ATP for contraction, so increasing oxygen supply to muscle tissue improves muscle performance, strength, and endurance. Athletes have been known to use blood doping to enhance their performance. Blood doping can be accomplished in several ways, including the use of blood transfusions or by taking drugs that stimulate the process of red blood cell production. Though blood doping may sound harmless at first, the overall result is erythrocytosis—a very dangerous side effect.

Red Blood Count

Blood cell counts are usually performed electronically with a machine, but they can also be done manually with a microscope. A normal **red blood count (RBC)** for a male is 4.6–6.2 million red blood cells per microliter (μL) of blood; for a female, a normal RBC count is 4.2–5.4 million per μL of blood. (A microliter is equivalent to 1 cubic millimeter [mm³] or 10^{-6} L, and one drop of blood is approximately 50 μL). The condition called **erythrocytosis** (ĕ-rith′rō-sī-tō′sis) is an overabundance of red blood cells (see Clinical Impact "Erythrocytosis and Blood Doping").

Hemoglobin Measurement

The amount of hemoglobin in a given volume of blood is usually expressed in terms of grams of hemoglobin per 100 mL of blood. The normal hemoglobin measurement for a male is 14–18 grams (g) per 100 mL of blood, and for a female 12–16 g per 100 mL of blood. An abnormally low hemoglobin measurement is an indication of **anemia** (ă-nē′mē-ă), which is either a reduced number of red blood cells or a reduced amount of hemoglobin in each red blood cell (see the Clinical Impact "Anemia" later in this chapter).

Hematocrit Measurement

The percentage of the total blood volume that is composed of red blood cells is the **hematocrit** (hē′mă-tō-krit, hem′a-tō-krit). One way to determine hematocrit is to place blood in a capillary tube and spin it in a centrifuge. The formed elements, which are heavier than the plasma, are forced to one end of the tube. Of these, the white blood cells and platelets form the buffy coat between the plasma and the red blood cells (figure 11.14). The red blood cells account for 40–52% of the total blood volume in males and 38–48% in females. The hematocrit measurement is affected by the number and size of red blood cells because it is based on volume. For example, a decreased hematocrit can result from a decreased number of normal-size red blood cells or a normal number of small red blood

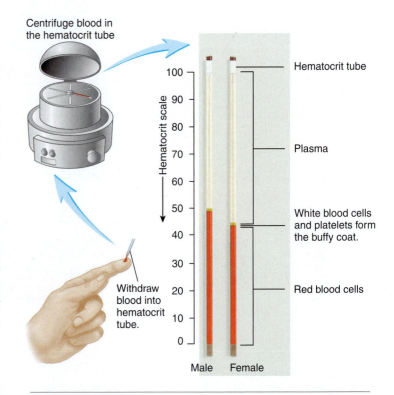

Centrifuge blood in the hematocrit tube

Hematocrit tube

Plasma

White blood cells and platelets form the buffy coat.

Red blood cells

Withdraw blood into hematocrit tube.

Male   Female

Figure II.I4 Hematocrit

Blood is withdrawn into a capillary tube and spun in a centrifuge. The blood is separated into plasma and red blood cells, with a narrow layer of white blood cells and platelets forming in between. The hematocrit is the percentage of the blood volume that is red blood cells. It does not measure the white blood cells and platelets. As shown here, the normal hematocrit for a male lies in the range of 40–52% of the total blood volume and for a female within the range of 38–48%.

CLINICAL IMPACT   Anemia

Anemia is a deficiency of normal hemoglobin in the blood, resulting from a decreased number of red blood cells, a decreased amount of hemoglobin in each red blood cell, or both. Anemia can also be caused by abnormal hemoglobin production.

Anemia reduces the blood's ability to transport oxygen. People with anemia lack energy and feel excessively tired and listless. They may appear pale and become short of breath after only slight exertion.

One general cause of anemia is insufficient production of red blood cells. **Aplastic** (ā-plas′tik) **anemia** is caused by an inability of the red bone marrow to produce red blood cells and, often, white blood cells and platelets. It is usually acquired as a result of damage to the stem cells in red marrow by chemicals, such as benzene, certain antibiotics and sedatives, or radiation.

Red blood cell production can also be lower than normal due to nutritional deficiencies. **Iron-deficiency anemia** results from insufficient intake or absorption of iron or from excessive iron loss. Consequently, not enough hemoglobin is produced, the number of red blood cells decreases, and the red blood cells that are manufactured are smaller than normal.

Folate deficiency can also cause anemia. The usual cause of folate deficiency is inadequate folate in the diet, with the disorder developing most often in the poor, in pregnant women, and in chronic alcohol-

ics. Because folate helps in the synthesis of DNA, a folate deficiency results in fewer cell divisions and therefore decreased red blood cell production. A deficiency in folate during pregnancy is also associated with birth disorders called neural tube defects, such as spina bifida.

Another type of nutritional anemia is **pernicious** (per-nish′ŭs) **anemia,** which is caused by inadequate vitamin B_{12}. Because vitamin B_{12} is important for folate synthesis, inadequate amounts of vitamin B_{12} can also result in decreased red blood cell production. Although inadequate levels of vitamin B_{12} in the diet can cause pernicious anemia, the usual cause is insufficient absorption of the vitamin. Normally, the stomach produces **intrinsic factor,** a protein that binds to vitamin B_{12}. The combined molecules pass into the lower intestine, where intrinsic factor facilitates the absorption of the vitamin. Without adequate levels of intrinsic factor, insufficient vitamin B_{12} is absorbed, and pernicious anemia develops. Most cases of pernicious anemia probably result from an autoimmune disease in which the body's immune system damages the cells in the stomach that produce intrinsic factor.

Another general cause of anemia is loss or destruction of red blood cells. **Hemorrhagic** (hem-ŏ-raj′ik) **anemia** results from a loss of blood due to trauma, ulcers, or excessive menstrual bleeding. Chronic blood loss, in which small amounts of blood are lost over a period of time, can

cause iron-deficiency anemia. **Hemolytic** (hē-mō-lit′ik) **anemia** occurs when red blood cells rupture or are destroyed at an excessive rate. It can be caused by inherited defects in the red blood cells. For example, one kind of inherited hemolytic anemia results from a defect in the cell membrane that causes red blood cells to rupture easily. Many kinds of hemolytic anemia result from unusual damage to the red blood cells by drugs, snake venom, artificial heart valves, autoimmune disease, or hemolytic disease of the newborn.

Anemia can also stem from a reduced rate of synthesis of the globin chains in hemoglobin. **Thalassemia** (thal-ă-sē′mē-ă) is a hereditary disease found in people of Mediterranean, Asian, and African ancestry. If hemoglobin production is severely depressed, death usually occurs before age 20. In less severe cases, thalassemia produces a mild anemia.

Some anemias are caused by defective hemoglobin production. **Sickle-cell anemia** is a hereditary disease found mostly in people of African descent. The red blood cells assume a rigid sickle shape and plug up small blood vessels. They are also more fragile than normal. In severe cases, so much abnormal hemoglobin is produced that the disease is usually fatal before age 30. In many cases, however, enough normal hemoglobin is produced to compensate for the abnormal hemoglobin, and the person exhibits no symptoms.

Cardiovascular

cells. The average size of a red blood cell is calculated by dividing the hematocrit by the red blood cell count. A number of disorders cause red blood cells to be smaller or larger than normal. For example, inadequate iron in the diet can impair hemoglobin production. Consequently, red blood cells do not fill up with hemoglobin during their formation, and they remain smaller than normal.

White Blood Count

A **white blood count (WBC)** measures the total number of white blood cells in the blood. There are normally 5000–9000 white blood cells per microliter of blood. **Leukopenia** (loo-kō-pē′nē-ă) is a lower than normal WBC resulting from decreased production or destruction of the red marrow. Radiation, drugs, tumors, viral infections, or a deficiency of the vitamins folate or B_{12} can cause leukopenia. **Leukocytosis** (loo′kō-sī-tō′sis) is an abnormally high

WBC. Bacterial infections often cause leukocytosis by stimulating neutrophils to increase in number. **Leukemia** (loo-kē′mē-ă), cancer of the red marrow characterized by abnormal production of one or more of the white blood cell types, can cause leukocytosis. However, the white blood cells do not function normally. Because these cells are usually immature or abnormal and lack normal immunological functions, people with leukemia are very susceptible to infections. The excess production of white blood cells in the red marrow can also interfere with the formation of red blood cells and platelets and thus lead to anemia and bleeding.

Differential White Blood Count

A **differential white blood count** determines the percentage of each of the five kinds of white blood cells. Normally, neutrophils account for 60–70%, lymphocytes 20–25%, monocytes 3–8%, eosinophils 2–4%, and basophils 0.5–1% of all white blood cells. Much insight into a patient's condition can be obtained from a differential white blood count. For example, if a bacterial infection is present, the neutrophil count is often greatly increased, whereas in allergic reactions, the eosinophil and basophil counts are elevated.

Clotting

The blood's ability to clot can be assessed by the platelet count and the prothrombin time measurement.

Platelet Count

A normal **platelet count** is 250,000–400,000 platelets per microliter of blood. In the condition called **thrombocytopenia** (throm′bō-sī-tō-pē′nē-ă), the platelet count is greatly reduced, resulting in chronic bleeding through small vessels and capillaries. It can be caused by decreased platelet production as a result of hereditary disorders, lack of vitamin B_{12} (pernicious anemia), drug therapy, or radiation therapy.

Prothrombin Time Measurement

Prothrombin time measurement calculates how long it takes for the blood to start clotting, which is normally 9–12 seconds. Prothrombin time is determined by adding thromboplastin to whole plasma. Thromboplastin is a chemical released from injured tissues that starts the process of clotting (see figure 11.9). Prothrombin time is officially reported as the International Normalized Ratio (INR), which standardizes the time it takes to clot on the basis of the slightly different thromboplastins used by different labs. Because many clotting factors have to be activated to form fibrin, a deficiency of any one of them can cause the prothrombin time to be abnormal. Vitamin K deficiency, certain liver diseases, and drug therapy can increase prothrombin time.

Blood Chemistry

The composition of materials dissolved or suspended in the plasma can be used to assess the functioning of many of the body's systems. For example, high blood glucose levels can indicate that the pancreas is not producing enough insulin; high blood urea nitrogen (BUN) is a sign of reduced kidney function; increased bilirubin can indicate liver dysfunction; and high cholesterol levels can signify an increased risk of cardiovascular disease. A number of blood chemistry tests are routinely done when a blood sample is taken, and additional tests are available.

Predict 6

When a patient complains of acute pain in the abdomen, the physician suspects appendicitis, a bacterial infection of the appendix. What blood test can provide supporting evidence for that diagnosis?

Cardiovascular

DISEASES AND DISORDERS: Blood

CONDITION	DESCRIPTION
Erythrocytosis	
Relative erythrocytosis	Overabundance of red blood cells due to decreased blood volume, as may result from dehydration, diuretics, or burns
Primary erythrocytosis (polycythemia vera)	Stem cell defect of unknown cause; results in overproduction of red blood cells, granulocytes, and platelets; signs include low erythropoietin levels and enlarged spleen; increased blood viscosity and blood volume can cause clogging of the capillaries and hypertension
Secondary erythrocytosis	Overabundance of red blood cells resulting from decreased oxygen supply, as occurs at high altitudes, in chronic obstructive pulmonary disease, or in congestive heart failure; decreased oxygen delivery to the kidney stimulates the secretion of erythropoietin, resulting in increased blood viscosity and blood volume that can cause clogging of the capillaries and hypertension
Clotting Disorders	
Disseminated intravascular coagulation (DIC)	Clotting throughout the vascular system, followed by bleeding; may develop when normal regulation of clotting by anticoagulants is overwhelmed, as occurs due to massive tissue damage; also caused by alteration of the lining of the blood vessels resulting from infections or snake bites
von Willebrand disease	Most common inherited bleeding disorder; platelet plug formation and the contribution of activated platelets to blood clotting are impaired; treatments are injection of von Willebrand factor or administration of drugs that increase von Willebrand factor levels in blood, which helps platelets adhere to collagen and become activated
Hemophilia	Genetic disorder in which clotting is abnormal or absent; each of the several types results from deficiency or dysfunction of a clotting factor; most often a sex-linked trait that occurs almost exclusively in males (see chapter 20)
Infectious Diseases of the Blood	
Septicemia (blood poisoning)	Spread of microorganisms and their toxins by the blood; often the result of a medical procedure, such as insertion of an intravenous tube; release of toxins by bacteria can cause septic shock, producing decreased blood pressure and possibly death
Malaria	Caused by a protozoan introduced into blood by *Anopheles* mosquito; symptoms include chills and fever produced by toxins released when the protozoan causes red blood cells to rupture
Infectious mononucleosis	Caused by Epstein-Barr virus, which infects salivary glands and lymphocytes; symptoms include fever, sore throat, and swollen lymph nodes, all probably produced by the immune system response to infected lymphocytes
Acquired immunodeficiency syndrome (AIDS)	Caused by human immunodeficiency virus (HIV), which infects lymphocytes and suppresses immune system

ANSWER TO LEARN TO PREDICT

Jessica's feeling of fatigue and her blood test results are consistent with anemia. A low red blood cell count with microcytic cells, low hemoglobin, and a low hematocrit are all indicators of iron-deficiency anemia.

Jessica's low red blood cell count caused less oxygen to be transported to her tissues. Recall that low oxygen levels stimulate the secretion of erythropoietin from the kidneys, which stimulates the production of red blood cells in the red bone marrow. Because

of Jessica's iron deficiency, which caused hemoglobin synthesis to slow, the newly synthesized red blood cells were smaller than normal, or microcytic. Remember, Jessica also complained of intense abdominal pain. The evidence of hemoglobin in her feces suggests that Jessica is losing blood from her digestive system, which, considering her abdominal pain, would be consistent with having an ulcer. Jessica's doctor would need to order additional tests to confirm the presence of ulcers before determining treatment.

Answers to the rest of this chapter's Predict questions are in Appendix E.

 SUMMARY

11.1 Functions of Blood   (p. 297)

1. Blood transports gases, nutrients, waste products, processed molecules, and regulatory molecules.
2. Blood regulates pH as well as fluid and ion balance.
3. Blood is involved with temperature regulation and protects against foreign substances, such as microorganisms and toxins.
4. Blood clotting prevents fluid and cell loss and is part of tissue repair.

11.2 Composition of Blood   (p. 298)

1. Blood is a connective tissue consisting of plasma and formed elements.
2. Total blood volume in an average adult is approximately 5 L.

11.3 Plasma   (p. 298)

1. Plasma is 91% water and 9% suspended or dissolved substances.
2. Plasma maintains osmotic pressure, is involved in immunity, prevents blood loss, and transports molecules.

11.4 Formed Elements   (p. 299)

The formed elements are cells (red blood cells and white blood cells) and cell fragments (platelets).

Production of Formed Elements

Formed elements arise (hematopoiesis) in red bone marrow from stem cells.

Red Blood Cells

1. Red blood cells are disk-shaped cells containing hemoglobin, which transports oxygen and carbon dioxide. Red blood cells also contain carbonic anhydrase, which is involved with carbon dioxide transport.
2. In response to low blood oxygen levels, the kidneys produce erythropoietin, which stimulates red blood cell production in red bone marrow.
3. Worn-out red blood cells are phagocytized by macrophages in the spleen or liver. Hemoglobin is broken down, iron and amino acids are reused, and heme becomes bilirubin that is secreted in bile.

White Blood Cells

1. White blood cells protect the body against microorganisms and remove dead cells and debris.
2. Granulocytes contain cytoplasmic granules. The three types of granulocytes are neutrophils, small phagocytic cells; basophils, which promote inflammation; and eosinophils, which defend against parasitic worms and influence inflammation.

3. Agranulocytes have very small granules and are of two types: Lymphocytes are involved in antibody production and other immune system responses; monocytes become macrophages that ingest microorganisms and cellular debris.

Platelets

Platelets are cell fragments involved with preventing blood loss.

11.5 Preventing Blood Loss   (p. 304)

Vascular Spasm

Blood vessels constrict in response to injury, resulting in decreased blood flow.

Platelet Plug Formation

1. Platelet plugs repair minor damage to blood vessels.
2. Platelet plugs form when platelets adhere to collagen, release chemicals (ADP and thromboxanes) that activate other platelets, and connect to one another with fibrinogen.

Blood Clotting

1. Blood clotting, or coagulation, is the formation of a clot (a network of protein fibers called fibrin).
2. There are three steps in the clotting process: activation of clotting factors by connective tissue and chemicals, resulting in the formation of prothrombinase; conversion of prothrombin to thrombin by prothrombinase; and conversion of fibrinogen to fibrin by thrombin.
3. Anticoagulants in the blood, such as antithrombin and heparin, prevent clot formation.
4. Clot retraction condenses the clot, pulling the edges of damaged tissue closer together.
5. Serum is plasma without clotting factors.
6. Fibrinolysis (clot breakdown) is accomplished by plasmin.

11.6 Blood Grouping   (p. 308)

1. Blood groups are determined by antigens on the surface of red blood cells.
2. In transfusion reactions, antibodies can bind to red blood cell antigens, resulting in agglutination or hemolysis of red blood cells.

ABO Blood Group

1. Type A blood has A antigens, type B blood has B antigens, type AB blood has A and B antigens, and type O blood has neither A nor B antigens.

2. Type A blood has anti-B antibodies, type B blood has anti-A antibodies, type AB blood has neither anti-A nor anti-B antibodies, and type O blood has both anti-A and anti-B antibodies.
3. Mismatching the ABO blood group can result in transfusion reactions.

Rh Blood Group

1. Rh-positive blood has Rh antigens, whereas Rh-negative blood does not.
2. Antibodies against the Rh antigen are produced when an Rh-negative person is exposed to Rh-positive blood.
3. The Rh blood group is responsible for hemolytic disease of the newborn, which can occur when the fetus is Rh-positive and the mother is Rh-negative.

11.7 Diagnostic Blood Tests   (p. 310)

Type and Crossmatch

1. Blood typing determines the ABO and Rh blood groups of a blood sample.
2. A crossmatch tests for agglutination reactions between donor and recipient blood.

Complete Blood Count

The complete blood count consists of the red blood count (million/μL), the hemoglobin measurement (grams of hemoglobin per 100 mL of blood), the hematocrit measurement (percent volume of red blood cells), and the white blood count (million/μL).

Differential White Blood Count

The differential white blood count determines the percentage of each type of white blood cell.

Clotting

Platelet count and prothrombin time measurement determine the blood's ability to clot.

Blood Chemistry

The composition of materials dissolved or suspended in plasma (e.g., glucose, urea nitrogen, bilirubin, and cholesterol) can be used to assess the functioning and status of the body's systems.

REVIEW AND COMPREHENSION

1. Describe the functions of blood.
2. Define plasma, and list its functions.
3. Define formed elements, and name the different types of formed elements. Explain how and where the formed elements arise through hematopoiesis.
4. Describe the two basic parts of a hemoglobin molecule. Which part is associated with iron? What gases are transported by each part?
5. What is the role of carbonic anhydrase in gas transport?
6. Why are the vitamins folate and B_{12} important in red blood cell production?
7. Explain how low blood oxygen levels result in increased red blood cell production.
8. Where are red blood cells broken down? What happens to the breakdown products?
9. Give two functions of white blood cells.
10. Name the five types of white blood cells, and state a function for each type.
11. What are platelets, and how are they formed?
12. Describe the role of blood vessel constriction and platelet plugs in preventing bleeding. Describe the three steps of platelet plug formation.
13. What are clotting factors? Describe the three steps of activation that result in the formation of a clot.
14. Explain the function of anticoagulants in the blood, and give an example of an anticoagulant.
15. What is clot retraction, and what does it accomplish?
16. Define fibrinolysis, and name the chemicals responsible for this process.
17. What are blood groups, and how do they cause transfusion reactions? List the four ABO blood types. Why is a person with type O blood considered a universal donor?
18. What is meant by the term *Rh-positive?* How can Rh incompatibility affect a pregnancy?
19. For each of the following tests, define the test and give an example of a disorder that would cause an abnormal test result:
    a. type and crossmatch
    b. red blood count
    c. hemoglobin measurement
    d. hematocrit measurement
    e. white blood count
    f. differential white blood count
    g. platelet count
    h. prothrombin time
    i. blood chemistry

CRITICAL THINKING

1. Red Packer, a physical education major, wanted to improve his performance in an upcoming marathon race. About 6 weeks before the race, he had 1 L of blood removed from his body and the formed elements separated from the plasma. The formed elements were frozen, and the plasma was reinfused into his body. Just before the race, the formed elements were thawed and injected into his circulatory system. Explain why this procedure, called blood doping or blood boosting, would help Red's performance. Can you suggest any possible bad effects?

2. Chemicals such as benzene can destroy red bone marrow, causing aplastic anemia. What symptoms would you expect as a result of the lack of (a) red blood cells, (b) platelets, and (c) white blood cells?

3. E. Z. Goen habitually used barbiturates to ease feelings of anxiety. Because barbiturates depress the respiratory centers in the brain, they cause hypoventilation (i.e., a slower than normal breathing rate). What happens to the red blood cell count of a habitual user of barbiturates? Explain.

4. What blood problems would you expect in a patient after a total gastrectomy (removal of the stomach)?

5. According to an old saying, "Good food makes good blood." Name three substances in the diet that are essential for "good blood." What blood disorders develop if these substances are absent from the diet?

6. Why do many anemic patients have gray feces? (*Hint:* The feces are lacking their normal coloration.)

7. Reddie Popper has a cell membrane defect in her red blood cells that makes them more susceptible to rupture. Her red blood cells are destroyed faster than they can be replaced. Would her RBC, hemoglobin, hematocrit, and bilirubin levels be below normal, normal, or above normal? Explain.

Answers in Appendix D

12 Heart

12.1 FUNCTIONS OF THE HEART

Learning Outcome *After reading this section, you should be able to*

A. List the major functions of the heart.

People often refer to the heart as the seat of strong emotions. For example, we may describe a very determined person as having "a lot of heart" or say that a person who has been disappointed romantically has a "broken heart." Emotions, however, are a product of brain function, not heart function. The heart is a muscular organ that is essential for life because it pumps blood through the body.

Fluids flow through a pipe only if they are forced to do so. The force is commonly produced by a pump, which increases the pressure of the liquid at the pump above the pressure in the pipe. Thus, the liquid flows from the pump through the pipe from an area of higher pressure to an area of lower pressure. If the pressure produced by the pump increases, flow of liquid through the pipe increases. If the pressure produced by the pump decreases, flow of liquid through the pipe decreases.

Like a pump that forces water through a pipe, the heart contracts forcefully to pump blood through the blood vessels of the body. Together, the heart, the blood vessels, and the blood make up the **cardiovascular system** (figure 12.1). The heart of a healthy adult, at rest, pumps approximately 5 liters (L) of blood per minute. For most people, the heart continues to pump at approximately that rate for more than 75 years. During short periods of vigorous exercise, the amount of blood pumped per minute increases

Module 9 Cardiovascular System

several-fold. But if the heart loses its pumping ability for even a few minutes, blood flow through the blood vessels stops, and the person's life is in danger.

The heart is actually two pumps in one. The right side of the heart pumps blood to the lungs and back to the left side of the heart through vessels of the **pulmonary circulation** (figure 12.2). The left side of the heart pumps blood to all other tissues of the body and back to the right side of the heart through vessels of the **systemic circulation.**

The functions of the heart are

1. *Generating blood pressure.* Contractions of the heart generate blood pressure, which is required to force blood through the blood vessels.

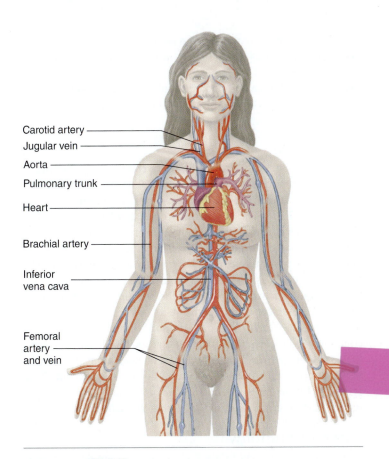

Figure 12.1 AP|R **Cardiovascular System**

The heart, the blood vessels, and the blood are the major components of the cardiovascular system.

Carotid artery
Jugular vein
Aorta
Pulmonary trunk
Heart
Brachial artery
Inferior
vena cava
Femoral
artery
and vein

2. *Routing blood.* The heart separates the pulmonary and systemic circulations, which ensures the flow of oxygen-rich blood to tissues.
3. *Ensuring one-way blood flow.* The valves of the heart ensure a one-way flow of blood through the heart and blood vessels.
4. *Regulating blood supply.* Changes in the rate and force of heart contraction match blood flow to the changing metabolic needs of the tissues during rest, exercise, and changes in body position.

12.2 SIZE, FORM, AND LOCATION OF THE HEART

Learning Outcome *After reading this section, you should be able to*

A. Describe the size, shape, and location of the heart, and explain why knowing its location is important.

The adult heart is shaped like a blunt cone and is approximately the size of a closed fist. It is larger in physically active adults than in less active but otherwise healthy adults. The heart generally decreases in size after approximately age 65, especially in people who are not physically active. The blunt, rounded point of the heart is the **apex** and the larger, flat part at the opposite end of the heart is the **base.**

The heart is located in the thoracic cavity between the two pleural cavities that surround the lungs. The heart, trachea, esophagus, and associated structures form a midline partition, the **mediastinum** (me′dē-as-tī′nŭm; see figure 1.13). The heart is surrounded by its own cavity, the **pericardial cavity** (*peri*, around + *cardio*, heart) (see chapter 1).

Cardiovascular

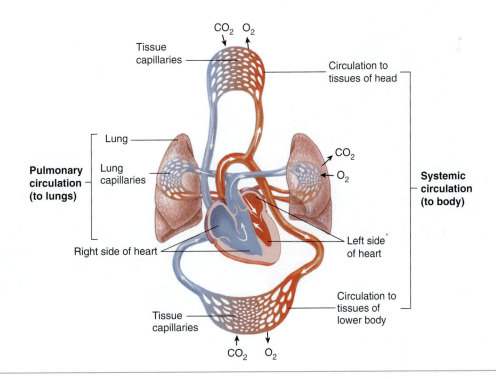

CO_2 O_2

Tissue capillaries

Circulation to tissues of head

Lung

Lung capillaries

Pulmonary circulation (to lungs)

CO_2
O_2

Systemic circulation (to body)

Right side of heart

Left side of heart

Tissue capillaries

Circulation to tissues of lower body

CO_2 O_2

Figure 12.2 AP|R **Overview of the Circulatory System**

The circulatory system consists of the pulmonary and systemic circulations. The right side of the heart pumps blood through vessels to the lungs and back to the left side of the heart through the pulmonary circulation. The left side of the heart pumps blood through vessels to the body tissues and back to the right side of the heart through the systemic circulation.

It is important for health professionals to know the location and shape of the heart in the thoracic cavity. This knowledge enables them to accurately place a stethoscope to hear the heart sounds, to place chest leads for an electrocardiogram (described later in this chapter), or to administer **cardiopulmonary resuscitation** (kar′dē-ō-pŭl′mo-nār-ē rē-sŭs′i-tā-shun; **CPR).** CPR is an emergency procedure that maintains blood flow in the body if a person's heart stops (see A Case In Point, "Hands Only Cardiopulmonary Resuscitation (CPR)," later in this chapter).

The heart lies obliquely in the mediastinum, with its base directed posteriorly and slightly superiorly and its apex directed anteriorly and slightly inferiorly. The apex is also directed to the left, so that approximately two-thirds of the heart's mass lies to the left of the midline of the sternum (figure 12.3*a*). The base of the heart is located deep to the sternum and extends to the level of the second intercostal space. The apex is deep to the left fifth intercostal space, approximately 7–9 centimeters (cm) to the left of the sternum near the midclavicular line, which is a perpendicular line that extends down from the middle of the clavicle (figure 12.3*b*).

12.3 ANATOMY OF THE HEART

Learning Outcomes   *After reading this section, you should be able to*

A. Describe the structure of the pericardium.
B. Give the location and function of the coronary arteries.

C. Describe the chambers of the heart.
D. Name the valves of the heart, and state their locations and functions.
E. Describe the flow of blood through the heart, and name each of the chambers and structures through which the blood passes.

Pericardium

The heart lies in the **pericardial cavity.** The pericardial cavity is formed by the **pericardium** (per-i-kar′dē-ŭm), or *pericardial sac,* tissues that surround the heart and anchor it within the mediastinum (figure 12.4). The pericardium consists of two layers. The tough, fibrous connective tissue outer layer is called the **fibrous pericardium,** and the inner layer of flat epithelial cells, with a thin layer of connective tissue, is called the **serous pericardium.** The portion of the serous pericardium lining the fibrous pericardium is the **parietal pericardium,** whereas the portion covering the heart surface is the **visceral pericardium,** or *epicardium* (ep-i-kar′dē-ŭm; upon the heart). The parietal and visceral pericardia are continuous with each other where the great vessels enter or leave the heart. The pericardial cavity, located between the visceral and parietal pericardia, is filled with a thin layer of **pericardial fluid** produced by the serous pericardium. The pericardial fluid helps reduce friction as the heart moves within the pericardium.

External Anatomy

The right and left **atria** (a′trē-ă; sing. atrium, entrance chamber) are located at the base of the heart, and the right and left **ventricles** (ven′tri-klz; cavities) extend from the base of the heart

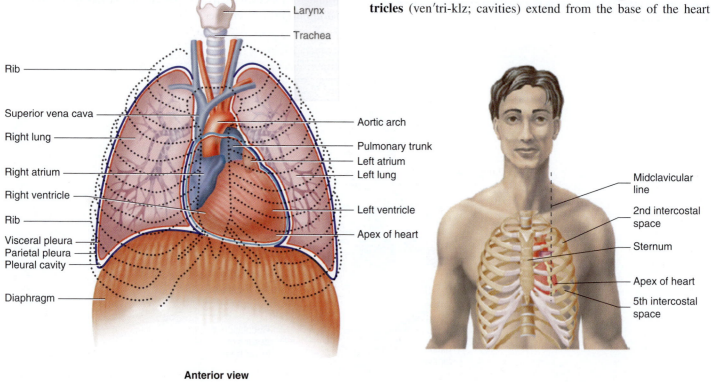

Larynx
Trachea
Rib
Superior vena cava
Right lung
Right atrium
Right ventricle
Rib
Visceral pleura
Parietal pleura
Pleural cavity
Diaphragm

Aortic arch
Pulmonary trunk
Left atrium
Left lung
Left ventricle
Apex of heart

Midclavicular line
2nd intercostal space
Sternum
Apex of heart
5th intercostal space

Anterior view

(a)   (b)

Figure 12.3 AP|R **Location of the Heart**

(*a*) The heart is in the thoracic cavity between the lungs, deep and slightly to the left of the sternum. (*b*) The base of the heart, located deep to the sternum, extends superiorly to the second intercostal space, and the apex of the heart is located deep to the fifth intercostal space, approximately 7–9 cm to the left of the sternum where the midclavicular line intersects with the fifth intercostal space.

Cardiovascular

CLINICAL IMPACT   Disorders of the Pericardium

Pericarditis (per'i-kar-dī'tis) is an inflammation of the serous pericardium. The cause is frequently unknown, but it can result from infection, diseases of connective tissue, or damage due to radiation treatment for cancer. The condition can cause extremely painful sensations that are referred to the back and to the chest and can be confused with a myocardial infarction (heart attack). Pericarditis can lead to a small amount of fluid accumulation within the pericardial sac.

Cardiac tamponade (tam-pŏ-nād'; a pack or plug) is a potentially fatal condition in which fluid or blood accumulates in the pericardial cavity and compresses the heart from the outside. The heart is a powerful muscle, but it relaxes passively. When it is compressed by fluid within the pericardial cavity, it cannot expand when the cardiac muscle relaxes. Consequently, the heart cannot fill with blood during relaxation, which makes pumping impossible. Cardiac tamponade can cause a

person to die quickly unless the fluid is removed. Causes of cardiac tamponade include rupture of the heart wall following a myocardial infarction, rupture of blood vessels in the pericardium after a malignant tumor invades the area, damage to the pericardium due to radiation therapy, and trauma, such as that resulting from a traffic accident.

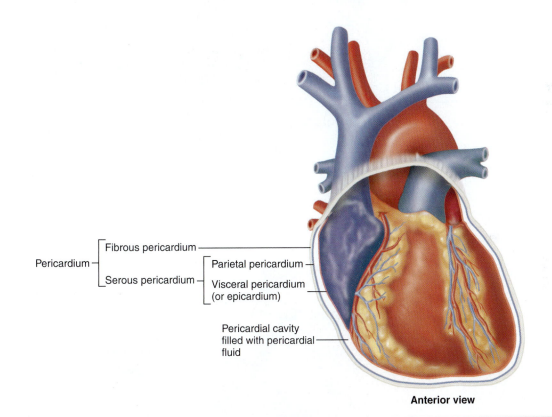

Pericardium
— Fibrous pericardium
— Serous pericardium

Parietal pericardium
Visceral pericardium (or epicardium)

Pericardial cavity filled with pericardial fluid

Anterior view

Figure 12.4 AP|R Heart in the Pericardium

The pericardium consists of an outer fibrous pericardium and an inner serous pericardium. The serous pericardium has two parts: The parietal pericardium lines the fibrous pericardium, and the visceral pericardium (epicardium) covers the surface of the heart. The pericardial cavity, between the parietal pericardium and visceral pericardium, is filled with a small amount of pericardial fluid.

toward the apex (figure 12.5). A **coronary** (kōr'o-nār-ē) **sulcus** (sool'kus) extends around the heart, separating the atria from the ventricles. In addition, two grooves, or sulci, which indicate the division between the right and left ventricles, extend inferiorly from the coronary sulcus. The **anterior interventricular sulcus** extends inferiorly from the coronary sulcus on the anterior surface of the heart, and the **posterior interventricular sulcus**

extends inferiorly from the coronary sulcus on the posterior surface of the heart.

Six large veins carry blood to the heart (figure 12.5a,c): The **superior vena cava** and **inferior vena cava** carry blood from the body to the right atrium, and four **pulmonary** (pŭl'mō-nār-ē; lung) **veins** carry blood from the lungs to the left atrium. Two arteries, the **pulmonary trunk** and the **aorta** (ā-ōr'tă), exit the

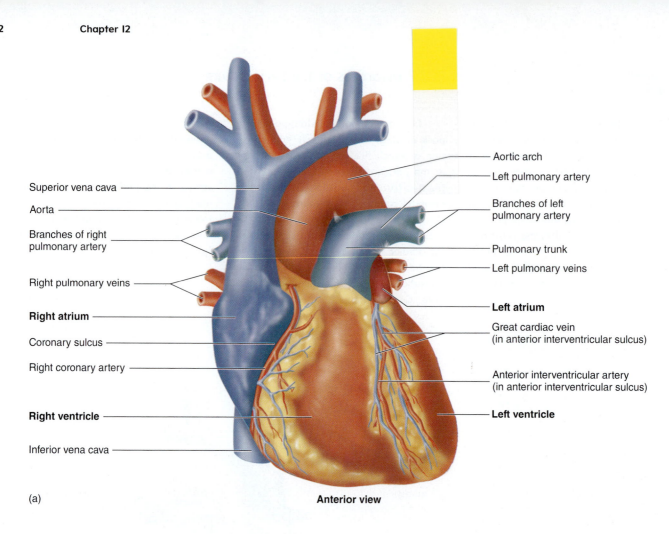

Aortic arch

Left pulmonary artery

Superior vena cava

Branches of left
pulmonary artery

Aorta

Branches of right
pulmonary artery

Pulmonary trunk

Left pulmonary veins

Right pulmonary veins

Left atrium

Right atrium

Great cardiac vein
(in anterior interventricular sulcus)

Coronary sulcus

Right coronary artery

Anterior interventricular artery
(in anterior interventricular sulcus)

Right ventricle

Left ventricle

Inferior vena cava

(a)

Anterior view

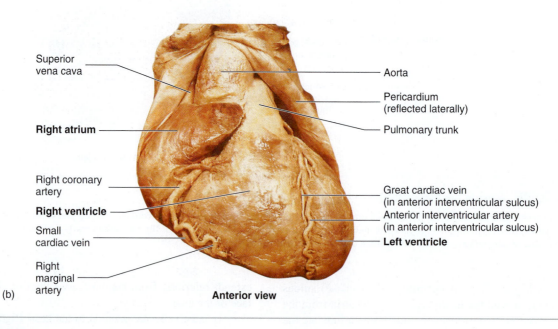

Superior
vena cava

Aorta

Pericardium
(reflected laterally)

Right atrium

Pulmonary trunk

Right coronary
artery

Great cardiac vein
(in anterior interventricular sulcus)

Right ventricle

Anterior interventricular artery
(in anterior interventricular sulcus)

Small
cardiac vein

Left ventricle

Right
marginal
artery

(b)

Anterior view

Figure 12.5 `AP|R` Surface of the Heart

(*a*) In this anterior view of the heart, the two atria (right and left) are located superiorly, and the two ventricles (right and left) are located inferiorly. The superior and inferior venae cavae enter the right atrium. The pulmonary veins enter the left atrium. The pulmonary trunk exits the right ventricle, and the aorta exits the left ventricle. (*b*) Photograph of the anterior surface of the heart.

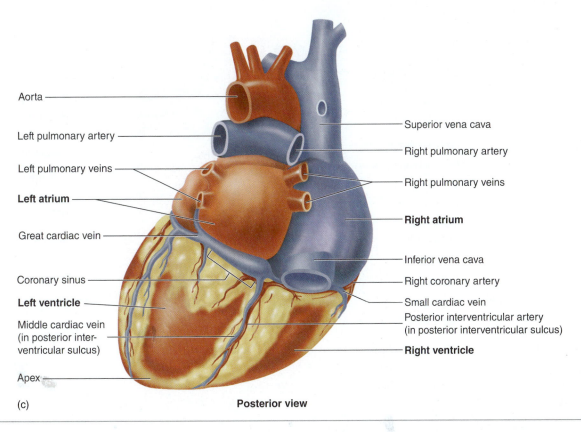

Aorta

Left pulmonary artery

Left pulmonary veins

Left atrium

Great cardiac vein

Coronary sinus

Left ventricle

Middle cardiac vein
(in posterior inter-
ventricular sulcus)

Apex

(c)

Superior vena cava

Right pulmonary artery

Right pulmonary veins

Right atrium

Inferior vena cava

Right coronary artery

Small cardiac vein

Posterior interventricular artery
(in posterior interventricular sulcus)

Right ventricle

Posterior view

Figure 12.5 Surface of the Heart (continued)

(c) In this posterior view of the heart, the two atria (right and left) are located superiorly, and the two ventricles (right and left) are located inferiorly. The superior and inferior venae cavae enter the right atrium, and the four pulmonary veins enter the left atrium.

heart. The pulmonary trunk, arising from the right ventricle, splits into the right and left **pulmonary arteries,** which carry blood to the lungs. The aorta, arising from the left ventricle, carries blood to the rest of the body.

Heart Chambers and Internal Anatomy

The heart is a muscular pump consisting of four chambers: the right and left atria and the right and left ventricles (figure 12.6).

Right and Left Atria

The atria of the heart receive blood from veins. The atria function primarily as reservoirs, where blood returning from veins collects before it enters the ventricles. Contraction of the atria forces blood into the ventricles to complete ventricular filling. The right atrium receives blood through three major openings. The superior vena cava and the inferior vena cava drain blood from most of the body (figure 12.6), and the smaller coronary sinus drains blood from most of the heart muscle. The left atrium receives blood through the four pulmonary veins, which drain blood from the lungs. The two atria are separated from each other by a partition called the **interatrial** (between the atria) **septum.**

Right and Left Ventricles

The ventricles of the heart are its major pumping chambers. They eject blood into the arteries and force it to flow through the circulatory system. The atria open into the ventricles, and each

ventricle has one large outflow route located superiorly near the midline of the heart. The right ventricle pumps blood into the pulmonary trunk, and the left ventricle pumps blood into the aorta. The two ventricles are separated from each other by the muscular **interventricular** (between the ventricles) **septum** (figure 12.6).

The wall of the left ventricle is thicker than the wall of the right ventricle, and the wall of the left ventricle contracts more forcefully and generates a greater blood pressure than the wall of the right ventricle. When the left ventricle contracts, the pressure increases to approximately 120 mm Hg. When the right ventricle contracts, the pressure increases to approximately one-fifth of the pressure in the left ventricle. However, the left and right ventricles pump nearly the same volume of blood. The higher pressure generated by the left ventricle moves blood through the larger systemic circulation, whereas the lower pressure generated by the right ventricle moves blood through the smaller pulmonary circulation (see figure 12.2).

Heart Valves

The **atrioventricular** (**AV**) **valves** are located between the right atrium and the right ventricle and between the left atrium and the left ventricle. The AV valve between the right atrium and the right ventricle has three cusps and is called the **tricuspid valve** (figure 12.7a). The AV valve between the left atrium and the left ventricle has two cusps and is called the **bicuspid valve** or *mitral* (resembling a bishop's miter, a two-pointed hat) *valve* (figure 12.7b). These valves allow blood to flow from the atria into the ventricles but prevent it from flowing back into the atria. When the ventricles

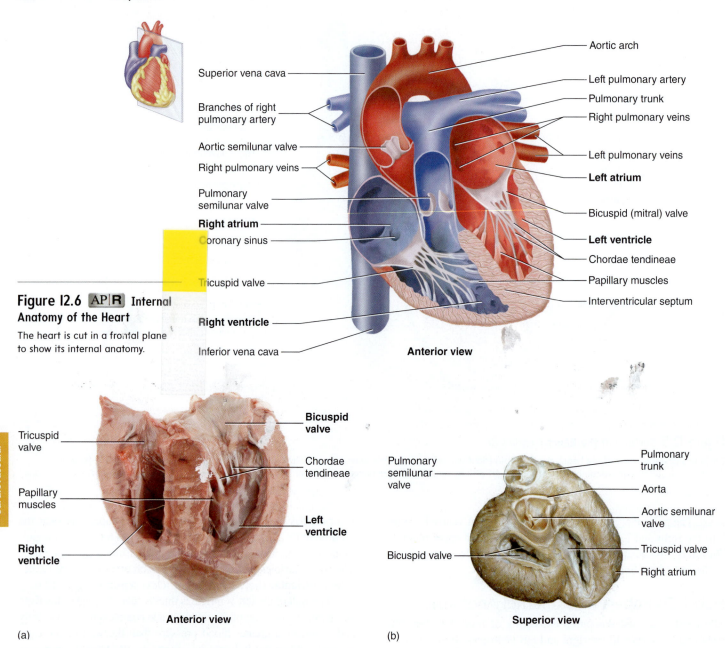

Figure I2.6 AP|R **Internal Anatomy of the Heart**

The heart is cut in a frontal plane to show its internal anatomy.

Labels (Anterior view):
- Aortic arch
- Superior vena cava
- Left pulmonary artery
- Pulmonary trunk
- Branches of right pulmonary artery
- Right pulmonary veins
- Aortic semilunar valve
- Left pulmonary veins
- Right pulmonary veins
- **Left atrium**
- Pulmonary semilunar valve
- Bicuspid (mitral) valve
- **Right atrium**
- **Left ventricle**
- Coronary sinus
- Chordae tendineae
- Papillary muscles
- Tricuspid valve
- Interventricular septum
- **Right ventricle**
- Inferior vena cava
- **Anterior view**

(a) Anterior view labels:
- Tricuspid valve
- **Bicuspid valve**
- Chordae tendineae
- Papillary muscles
- **Left ventricle**
- **Right ventricle**
- **Anterior view**

(b) Superior view labels:
- Pulmonary semilunar valve
- Pulmonary trunk
- Aorta
- Aortic semilunar valve
- Bicuspid valve
- Tricuspid valve
- Right atrium
- **Superior view**

Figure I2.7 AP|R **Heart Valves**

(*a*) Anterior view of the tricuspid valve, bicuspid valve, the chordae tendineae, and the papillary muscles. (*b*) In superior view, note that the three cusps of each semilunar valve meet to prevent the backflow of blood.

relax, the higher pressure in the atria forces the AV valves to open, and blood flows from the atria into the ventricles (figure 12.8*a*). In contrast, when the ventricles contract, blood flows toward the atria and causes the AV valves to close (figure 12.8*b*).

Each ventricle contains cone-shaped, muscular pillars called **papillary** (pap′ĭ-lār-ē) **muscles.** These muscles are attached by thin, strong, connective tissue strings called **chordae tendineae** (kōr′dē ten′di-nē-ē; heart strings) to the free margins of the cusps of the atrioventricular valves. When the ventricles contract, the papillary muscles contract and prevent the valves from opening into the atria by pulling on the chordae tendineae attached to the valve cusps (see figures 12.6 and 12.7*a*; figure 12.8).

The aorta and pulmonary trunk possess **aortic** and **pulmonary semilunar valves,** respectively (see figure 12.6). Each valve consists of three pocketlike semilunar (half-moon-shaped) cusps (see figure 12.7*b*; figure 12.8). When the ventricles relax, the pressures in the aorta and pulmonary trunk are higher than in the ventricles. Blood flows back from the aorta or pulmonary trunk toward the ventricles and enters the pockets of the cusps, causing them to bulge toward and meet in the center of the aorta or pulmonary trunk, thus closing the vessels and blocking blood flow back into the ventricles (figure 12.8*a*). When the ventricles contract, the increasing pressure within the ventricles forces the semilunar valves to open (figure 12.8*b*).

Cardiovascular

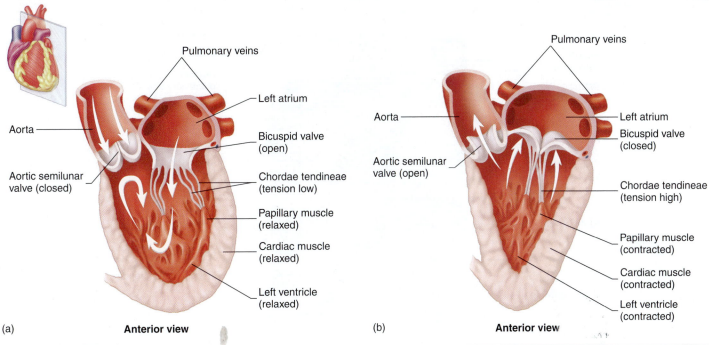

Pulmonary veins

Left atrium

Aorta

Bicuspid valve (open)

Aortic semilunar valve (closed)

Chordae tendineae (tension low)

Papillary muscle (relaxed)

Cardiac muscle (relaxed)

Left ventricle (relaxed)

(a)   **Anterior view**

Pulmonary veins

Aorta

Left atrium

Bicuspid valve (closed)

Aortic semilunar valve (open)

Chordae tendineae (tension high)

Papillary muscle (contracted)

Cardiac muscle (contracted)

Left ventricle (contracted)

(b)   **Anterior view**

Figure 12.8 AP|R **Function of the Heart Valves**

(a) As the ventricle relaxes, the pressure in the ventricle becomes lower than the pressure in the atrium. Blood flowing into the left atrium opens the bicuspid valve, and blood flows into the left ventricle. At the same time, in the aorta, blood flowing back toward the relaxed ventricle causes the aortic semilunar valve to close, and no blood can reenter the ventricle from the aorta. (b) When the ventricle contracts, blood flowing toward the atrium causes the bicuspid valve to close. The increased pressure in the left ventricle forces the aortic semilunar valve open.

A plate of connective tissue, sometimes called the **cardiac skeleton,** consists mainly of fibrous rings that surround the atrioventricular and semilunar valves and give them solid support (figure 12.9). This connective tissue plate also serves as electrical insulation between the atria and the ventricles and provides a rigid attachment site for cardiac muscle.

Route of Blood Flow Through the Heart

The route of blood flow through the heart is depicted in figure 12.10. Even though blood flow is described for the right and then the left side of the heart, it is important to understand that both atria contract at the same time, and both ventricles contract at the same time. This concept is most important when considering the electrical activity, pressure changes, and heart sounds.

Blood enters the right atrium from the systemic circulation through the superior and inferior venae cavae, and from heart muscle through the coronary sinus. Most of the blood flowing into the right atrium flows into the right ventricle while the right ventricle relaxes following the previous contraction. Before the end of ventricular relaxation, the right atrium contracts, and enough blood is pushed from the right atrium into the right ventricle to complete right ventricular filling.

Following right atrial contraction, the right ventricle begins to contract. This contraction pushes blood against the tricuspid valve, forcing it closed. After pressure within the right ventricle increases, the pulmonary semilunar valve is forced open, and blood flows into the pulmonary trunk. As the right ventricle relaxes, its pressure falls rapidly, and pressure in the pulmonary trunk becomes greater than in the right ventricle. The backflow of blood forces the pulmonary semilunar valve to close.

The pulmonary trunk branches to form the right and left pulmonary arteries, which carry blood to the lungs, where CO_2 is released and O_2 is picked up. Blood returning from the lungs enters

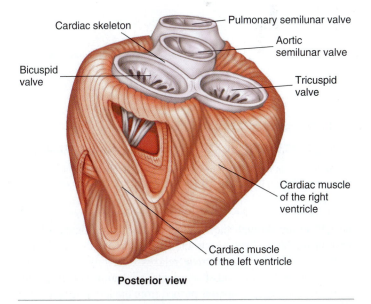

Cardiac skeleton

Pulmonary semilunar valve

Aortic semilunar valve

Bicuspid valve

Tricuspid valve

Cardiac muscle of the right ventricle

Cardiac muscle of the left ventricle

Posterior view

Figure 12.9 Cardiac Skeleton

The cardiac skeleton consists of fibrous connective tissue rings that surround the heart valves and separate the atria from the ventricles. Cardiac muscle attaches to the fibrous connective tissue. The muscle fibers are arranged so that, when the ventricles contract, a wringing motion is produced, and the distance between the apex and the base of the heart shortens.

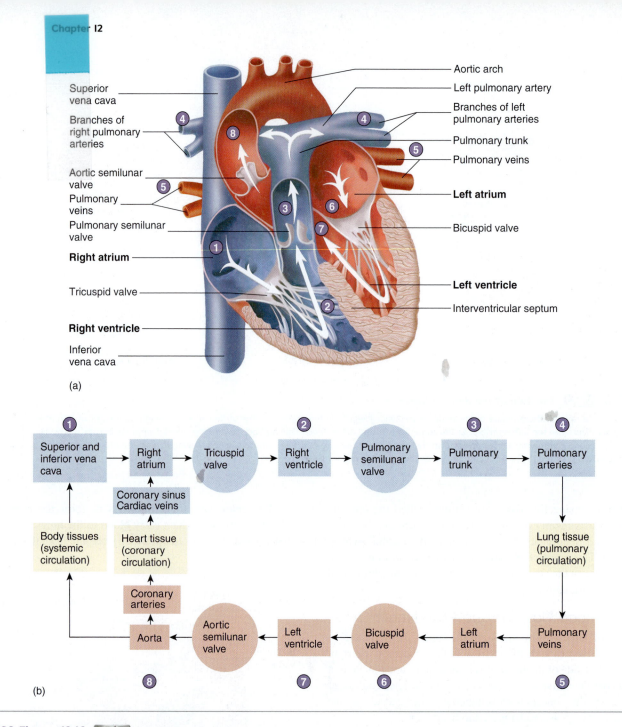

PROCESS Figure 12.10 **AP|R** Blood Flow Through the Heart

(*a*) Frontal section of the heart, revealing the four chambers and the direction of blood flow (*purple numbers*). (*b*) Diagram listing in order the structures through which blood flows in the systemic and pulmonary circulations. The heart valves are indicated by *circles*, deoxygenated blood by *blue*, and oxygenated blood by *red*.

the left atrium through the four pulmonary veins (figure 12.10). Most of the blood flowing into the left atrium passes into the left ventricle while the left ventricle relaxes following the previous contraction. Before the end of ventricular relaxation, the left atrium contracts, and enough blood is pushed from the left atrium into the left ventricle to complete left ventricular filling.

Following left atrial contraction, the left ventricle begins to contract. This contraction pushes blood against the bicuspid valve, forcing it closed. After pressure within the left ventricle increases, the aortic semilunar valve is forced open, and blood flows into the aorta. Blood flowing through the aorta is distributed to all parts of the

body, except to those parts of the lungs supplied by the pulmonary blood vessels. As the left ventricle relaxes, its pressure falls rapidly, and pressure in the aorta becomes greater than in the left ventricle. The backflow of blood forces the aortic semilunar valve to close.

Blood Supply to the Heart

Coronary Arteries

The cardiac muscle in the wall of the heart is thick and metabolically very active. Two **coronary arteries** supply blood to the wall of the heart (figure 12.11*a*). The coronary arteries originate from the base of the aorta, just above the aortic semilunar valves. The **left coronary artery** originates on the left side of the aorta. It has three major branches: The **anterior interventricular artery** lies in the anterior interventricular sulcus; the **circumflex artery** extends around the coronary sulcus on the left to the posterior surface of the heart; and the **left marginal artery** extends inferiorly along the lateral wall of the left ventricle from the circumflex artery. The branches of the left coronary artery supply much of the anterior wall of the heart and most of the left ventricle. The **right coronary artery** originates on the right side of the aorta. It extends around the coronary sulcus on the right to the posterior surface of the heart and gives rise to the **posterior interventricular artery,** which lies in the posterior interventricular sulcus. The **right marginal artery** extends inferiorly along the lateral wall of the right ventricle. The right coronary artery and its branches supply most of the wall of the right ventricle.

In a resting person, blood flowing through the coronary arteries gives up approximately 70% of its O_2. In comparison, blood flowing through arteries to skeletal muscle gives up only about 25% of its O_2. The percentage of O_2 the blood releases to skeletal muscle increases to 70% or more during exercise, but the percentage of O_2 the blood releases to cardiac muscle cannot increase substantially during exercise. Therefore, the rate of blood flow through the coronary arteries must increase above its resting level to provide cardiac muscle with adequate O_2 during exercise. Blood flow into the coronary circulation is greatest while the ventricles of the heart are relaxed and contraction of the cardiac muscle does not compress the coronary arteries. Blood flow into other arteries of the body is highest during contraction of the ventricles.

Predict 2

Predict the effect on the heart if blood flow through the anterior interventricular artery is restricted or completely blocked (Hint: See figure 12.11a).

Cardiac Veins

The **cardiac veins** drain blood from the cardiac muscle. Their pathways are nearly parallel to the coronary arteries, and most of them drain blood into the **coronary sinus,** a large vein located within the coronary sulcus on the posterior aspect of the heart. Blood flows from the coronary sinus into the right atrium (figure 12.11b). Some small cardiac veins drain directly into the right atrium.

12.4 HISTOLOGY OF THE HEART

Learning Outcomes *After reading this section, you should be able to*

A. List the components of the heart wall, and describe the structure and function of each.

B. Describe the structural and functional characteristics of cardiac muscle cells.

Heart Wall

The heart wall is composed of three layers of tissue: the epicardium, the myocardium, and the endocardium (figure 12.12). The **epicardium** (ep-i-kar′dē-ŭm), also called the *visceral pericardium,* is a thin, serous membrane forming the smooth outer surface of the heart. It consists of simple squamous epithelium overlying

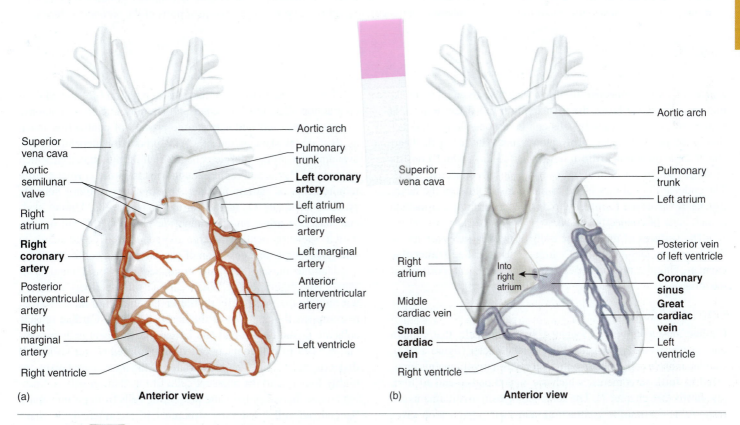

Figure 12.11 AP|R **Blood Supply to the Heart**

The vessels of the anterior surface of the heart depicted here are darker in color, whereas the vessels of the posterior surface are seen through the heart and are lighter in color. (*a*) Coronary arteries supply blood to the wall of the heart. (*b*) Cardiac veins carry blood from the wall of the heart back to the right atrium.

CLINICAL IMPACT   Heart Attack

When a blood clot, or **thrombus** (throm′bōs; a clot), suddenly blocks a coronary blood vessel, a **heart attack,** or *coronary thrombosis* (throm-bō′sis), occurs. The area that has been cut off from its blood supply suffers from lack of O₂ and nutrients and dies if the blood supply is not quickly reestablished. The region of dead heart tissue is called an **infarct** (in′farkt), or *myocardial infarction.* If the infarct is large, the heart may be unable to pump enough blood to keep the person alive. People can reduce their risk of having a heart attack by taking small amounts of aspirin daily, which inhibits thrombus formation (see chapter 11).

Aspirin is also administered to many people who are exhibiting clear symptoms of a heart attack. In some cases, it is possible to treat heart attacks with enzymes, such as **tissue plasminogen** (plaz-min′o-jen) **activator (t-PA),** which break down blood clots. One of the enzymes is injected into the circulatory system of a heart attack patient, where it reduces or removes the blockage in

the coronary artery. If the clot is broken down quickly, the blood supply to cardiac muscle is reestablished, and the heart may suffer little permanent damage.

Coronary arteries can become blocked more gradually by **atherosclerotic** (ath′er-ō-skler-ot′ik; *athero,* pasty material + *sclerosis,* hardness) **plaque.** These thickenings in the walls of arteries can contain deposits that are high in cholesterol and other lipids. The lesions narrow the lumen (opening) of the arteries, thus restricting blood flow. The ability of cardiac muscle to function is reduced when it is deprived of an adequate blood supply. The person suffers from fatigue and often experiences pain in the area of the chest and usually in the left arm with the slightest exertion. The pain is called **angina pectoris** (an-jī′nă, pain; pek′tō-ris, in the chest).

Angioplasty (an′jē-ō-plas-tē) is a procedure in which a surgeon threads a small balloon through the aorta and into a coronary artery. After entering a partially blocked coronary artery, the balloon is inflated, flattening the atherosclerotic

deposits against the vessel wall and opening the blocked blood vessel. This technique improves the function of cardiac muscle in patients experiencing inadequate blood flow to the cardiac muscle through the coronary arteries. Some controversy exists about its effectiveness, at least in some patients, because dilation of the coronary arteries can be reversed within a few weeks or months and because blood clots can form in coronary arteries following angioplasty. Small rotating blades and lasers are also used to remove lesions from coronary vessels, or a small coil device, called a **stent,** may be placed in a vessel to hold it open following angioplasty.

A **coronary bypass** is a surgical procedure that relieves the effects of obstructions in the coronary arteries. The technique involves taking healthy segments of blood vessels from other parts of the patient's body and using them to bypass, or create an alternative path around, obstructions in the coronary arteries. The technique is common in cases of severe blockage of parts of the coronary arteries.

a layer of loose connective tissue and adipose tissue. The thick, middle layer of the heart, the **myocardium** (mī-ō-kar′dē-ŭm), is composed of cardiac muscle cells and is responsible for contraction of the heart chambers. The smooth inner surface of the heart chambers is the **endocardium** (en-dō-kar′dē-ŭm), which consists of simple squamous epithelium over a layer of connective tissue. The endocardium allows blood to move easily through the heart. The heart valves are formed by folds of endocardium that include a thick layer of connective tissue.

The surfaces of the interior walls of the ventricles are modified by ridges and columns of cardiac muscle called **trabeculae carneae.** Smaller muscular ridges are also present in portions of the atria.

Cardiac Muscle

Cardiac muscle cells are elongated, branching cells that contain one, or occasionally two, centrally located nuclei (figure 12.13). Cardiac muscle cells contain actin and myosin myofilaments organized to form sarcomeres, which are joined end-to-end to form myofibrils (see chapter 7). The actin and myosin myofilaments are responsible for muscle contraction, and their organization gives cardiac muscle a striated (banded) appearance much like that of skeletal muscle. However, the striations are less regularly arranged and less numerous than in skeletal muscle.

Like skeletal muscle, cardiac muscle relies on Ca²⁺ and ATP for contraction. Calcium ions enter cardiac muscle cells in response to action potentials and activate the process of contraction much as they do in skeletal muscle. ATP production depends on O₂ availability. Cardiac muscle cells are rich in mitochondria, which produce ATP at a rate rapid enough to sustain the normal energy requirements of cardiac muscle. An extensive capillary network provides adequate O₂ to the cardiac muscle cells. Unlike skeletal muscle, cardiac muscle cannot develop a significant oxygen deficit. Development of a large oxygen deficit could result in muscular fatigue and cessation of cardiac muscle contraction.

Cardiac muscle cells are organized into spiral bundles or sheets (see figure 12.9). When cardiac muscle fibers contract, not only do the muscle fibers shorten but the spiral bundles twist to compress the contents of the heart chambers. Cardiac muscle cells are bound end-to-end and laterally to adjacent cells by specialized cell-to-cell contacts called **intercalated** (in-ter′kă-lā-ted) **disks** (figure 12.13). The membranes of the intercalated disks are highly folded, and the adjacent cells fit together, greatly increasing contact between them and preventing cells from pulling apart. Specialized cell membrane structures in the intercalated disks called **gap junctions** (see chapter 4) allow cytoplasm to flow freely between cells. This enables action potentials to pass quickly and easily from one cell to the next. The cardiac muscle cells

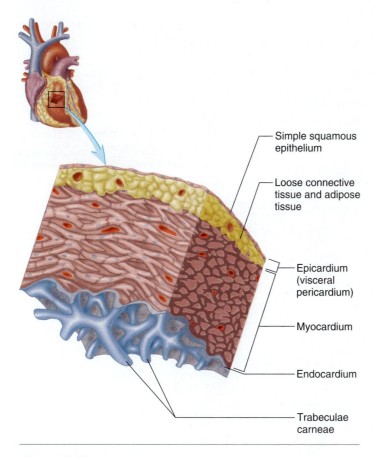

- Simple squamous epithelium
- Loose connective tissue and adipose tissue
- Epicardium (visceral pericardium)
- Myocardium
- Endocardium
- Trabeculae carneae

Figure 12.12 Heart Wall

This enlarged section of the heart wall illustrates the epicardium (visceral pericardium), the myocardium, and the endocardium.

of the atria or ventricles, therefore, contract at nearly the same time. The heart's highly coordinated pumping action depends on this characteristic.

12.5 ELECTRICAL ACTIVITY OF THE HEART

Learning Outcomes *After reading this section, you should be able to*

A. Describe the characteristics of action potentials in cardiac muscle.

B. Explain the structure and function of the conduction system of the heart.

C. Describe the waves of an electrocardiogram, and relate each of them to contractions of the heart.

Action Potentials in Cardiac Muscle

Like action potentials in skeletal muscle and neurons, those in cardiac muscle exhibit depolarization followed by repolarization. In cardiac muscle, however, a period of slow repolarization greatly prolongs the action potential (figure 12.14). In contrast to action potentials in skeletal muscle, which take less than 2 milliseconds (ms) to complete, action potentials in cardiac muscle take approximately 200 to 500 ms to complete. In addition, unlike in skeletal muscle, action potentials in cardiac muscle are conducted from cell to cell.

In cardiac muscle, each action potential consists of a **depolarization phase** followed by a period of slow repolarization called the **plateau phase.** At the end of the plateau phase, a rapid **repolarization phase** takes place. During the final repolarization phase, the membrane potential achieves its maximum degree of repolarization (figure 12.14b) and returns to the resting membrane potential.

The opening and closing of membrane channels is responsible for the changes in the permeability of the cell membrane that produce action potentials (see chapter 8). The initial, rapid depolarization phase of the action potential results from the opening of voltage-gated Na^+ channels, which increases the permeability of the cell membrane to Na^+. Sodium ions then diffuse into the cell, causing depolarization. This depolarization stimulates the opening of voltage-gated Ca^{2+} channels, and Ca^{2+} begins diffusing into the cell, contributing to the overall depolarization. At the peak of depolarization, the Na^+ channels close, and a small number of K^+ channels open. However, the Ca^{2+} channels remain open. Thus, the exit of K^+ from the cell is counteracted by the continued movement of Ca^{2+} into the cell. Consequently, the plateau phase is primarily the result of the opening of voltage-regulated Ca^{2+} channels. The slow diffusion of Ca^{2+} into the cell is the reason the cardiac muscle fiber action potential lasts longer than the action potentials in skeletal muscle fibers. The plateau phase ends, and the repolarization phase begins as the Ca^{2+} channels close and many K^+ channels open, allowing K^+ to move out of the cell.

Action potentials in cardiac muscle exhibit a **refractory period,** like that of action potentials in skeletal muscle and in neurons. The refractory period lasts about as long as the plateau phase of the action potential in cardiac muscle. The prolonged action potential and refractory period allow cardiac muscle to contract and relax almost completely before another action potential can be produced. Also, the long refractory period in cardiac muscle prevents tetanic contractions from occurring, thus ensuring a rhythm of contraction and relaxation for cardiac muscle. Therefore, action potentials in cardiac muscle are different from those in skeletal muscle because the plateau phase makes the action potential and its refractory period last longer.

Predict 3

Why is it important to prevent tetanic contractions in cardiac muscle but not in skeletal muscle?

Conduction System of the Heart

Unlike skeletal muscle that requires neural stimulation to contract, cardiac muscle can contract without neural stimulations. Contraction of the atria and ventricles is coordinated by specialized cardiac muscle cells in the heart wall that form the **conduction system of the heart** (figure 12.15).

All the cells of the conduction system can produce spontaneous action potentials. The sinoatrial node, atrioventricular node, atrioventricular bundle, right and left bundle branches, and Purkinje fibers constitute the conduction system of the heart. The **sinoatrial (SA) node,** which functions as the heart's pacemaker, is located in the superior wall of the right atrium and initiates the contraction of the heart. Action potentials originate in the SA node and spread over the right and left atria, causing them to contract. The SA node produces action potentials at a faster rate than other areas of the heart and has a larger number of Ca^{2+} channels than

Cardiovascular

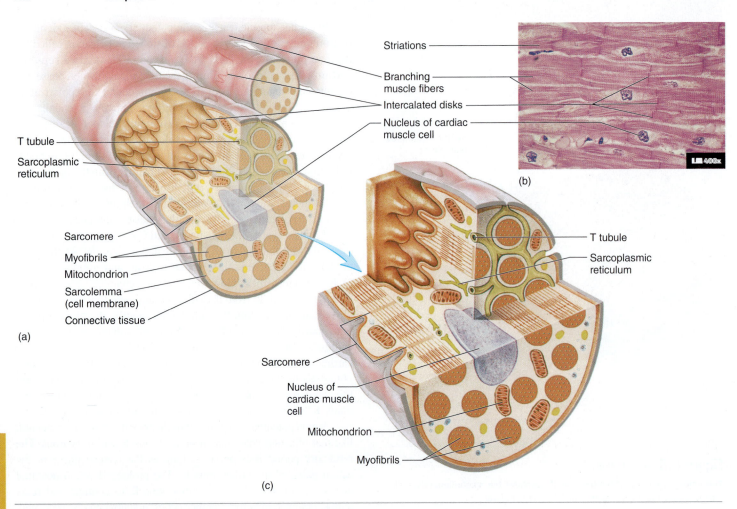

Figure I2.I3 AP|R **Cardiac Muscle Cells**

(a) Cardiac muscle cells are branching cells with centrally located nuclei. The cells are joined to one another by intercalated disks, which allow action potentials to pass from one cardiac muscle cell to the next. (b) In a light micrograph of cardiac muscle tissue, the muscle fibers appear striated because of the arrangement of the individual myofilaments. (c) As in skeletal muscle, sarcomeres join end-to-end to form myofibrils, and mitochondria provide ATP for contraction. Sarcoplasmic reticulum and T tubules are visible but are not as numerous as they are in skeletal muscle.

other cells in the heart. In addition, the Na⁺ and Ca²⁺ channels in the SA node spontaneously open and close at a rhythmic rate. The heart rate can be affected by certain drugs. Calcium channel blocking agents, for example, are drugs that slow the heart by decreasing the rate of action potential production in the SA node. Calcium channel blockers decrease the rate at which Ca^{2+} moves through Ca^{2+} channels. As a result, it takes longer for depolarization to reach threshold, and the interval between action potentials increases.

A second area of the heart, the **atrioventricular (AV)** (ā-trē-ō-ven′trik′-ū′lăr) **node,** is located in the lower portion of the right atrium. When action potentials reach the AV node, they spread slowly through it and then into a bundle of specialized cardiac muscle called the **atrioventricular (AV) bundle.** The slow rate of action potential conduction in the AV node allows the atria to complete their contraction before action potentials are delivered to the ventricles.

After action potentials pass through the AV node, they are rapidly transmitted through the AV bundle, which projects through the fibrous connective tissue plate that separates the atria from the ventricles (see figure 12.9). The AV bundle then divides into two branches of conducting tissue, called the **left** and **right bundle branches** (figure 12.15). At the tips of the left and right bundle branches, the conducting tissue forms many small bundles of **Purkinje** (pŭr-kĭn′jē) **fibers.** The Purkinje fibers pass to the apex of the heart and then extend to the cardiac muscle of the ventricle walls. The AV bundle, the bundle branches, and the Purkinje fibers are composed of specialized cardiac muscle fibers that conduct action potentials more rapidly than do other cardiac muscle fibers. Consequently, action potentials are rapidly delivered to all the cardiac muscle of the ventricles. The coordinated contraction of the ventricles depends on the conduction of action potentials by the conduction system.

Predict 4

If blood supply is reduced in a small area of the heart through which the left bundle branch passes, predict the effect on ventricular contraction.

Following their contraction, the ventricles begin to relax. After the ventricles have completely relaxed, another action potential originates in the SA node to begin the next cycle of contractions.

Skeletal Muscle

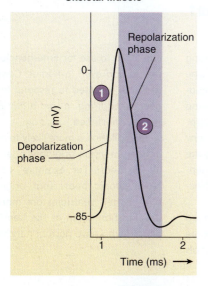

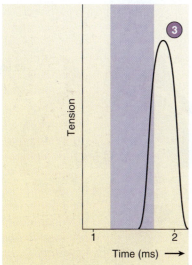

Cardiac Muscle

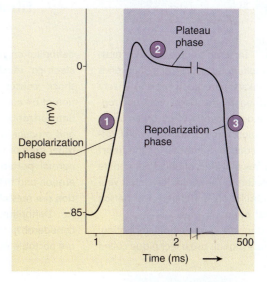

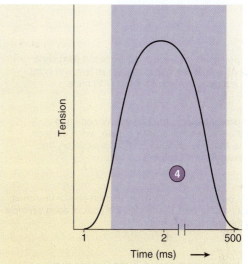

(a)

1 Depolarization phase
- Na⁺ channels open.
- K⁺ channels begin to open.

2 Repolarization phase
- Na⁺ channels close.
- K⁺ channels continue to open, causing repolarization.
- K⁺ channels close at the end of repolarization and return the membrane potential to its resting value.

3 Refractory period effect on tension
- Maximum tension is obtained after the refractory period (purple shaded area) is completed allowing for increased tension with additional stimulation.

(b)

1 Depolarization phase
- Na⁺ channels open.
- Ca²⁺ channels open.

2 Plateau phase
- Na⁺ channels close.
- Some K⁺ channels open, causing repolarization.
- Ca²⁺ channels are open, producing the plateau by slowing further repolarization.

3 Repolarization phase
- Ca²⁺ channels close.
- Many K⁺ channels open.

4 Refractory period effect on tension
- Cardiac muscle contracts and relaxes almost completely during the refractory period (purple shaded area).

PROCESS Figure 12.14 Comparison of Action Potentials in Skeletal and Cardiac Muscle

(a) An action potential in skeletal muscle consists of depolarization and repolarization phases. The refractory period is indicated by the purple shaded area. (b) An action potential in cardiac muscle consists of depolarization, plateau, and repolarization phases. Cardiac muscle does not repolarize as rapidly as skeletal muscle (*indicated by the break in the curve*) because of the plateau phase. Due to the prolonged action potential and refractory period (indicated by the purple shaded area), cardiac muscle contracts and relaxes almost completely before another action potential can be produced.

CLINICAL IMPACT   Fibrillation of the Heart

In some people, cardiac muscle can malfunction—it is as if thousands of pacemakers were each making a very small portion of the heart contract rapidly and independently of all other areas. This condition, called **fibrillation** (fĭ-bri-lā′shŭn), reduces the output of the heart to only a few milliliters of blood per minute when it occurs in the ventricles. Unless ventricular fibrillation is stopped, the person dies in just a few minutes.

To stop the process of fibrillation, health professionals use a technique called defibrillation, in which they apply a strong electrical shock to the chest region. The shock causes simultaneous depolarization of all cardiac muscle fibers. Following depolarization, the SA node can recover and produce action potentials before any other area of the heart. Consequently, the normal pattern of action potential generation and the normal rhythm of contraction are reestablished.

Defibrillator machines have changed considerably over the years. Now available are portable models that nonprofessionals can learn to use for emergencies in workplaces and even in homes. Consequently, the time required to respond to an emergency requiring a defibrillator can be greatly shortened.

Fibrillation is more likely to occur when action potentials originate at ectopic sites in the heart. For example, people who have ectopic beats that originate from one of their ventricles are more likely to develop fibrillation of the heart than are people who have normal heartbeats.

1. Action potentials originate in the sinoatrial (SA) node and travel across the wall of the atrium (*arrows*) from the SA node to the atrioventricular (AV) node.

2. Action potentials pass through the AV node and along the atrioventricular (AV) bundle, which extends from the AV node, through the fibrous skeleton, into the interventricular septum.

3. The AV bundle divides into right and left bundle branches, and action potentials descend to the apex of each ventricle along the bundle branches.

4. Action potentials are carried by the Purkinje fibers from the bundle branches to the ventricular walls.

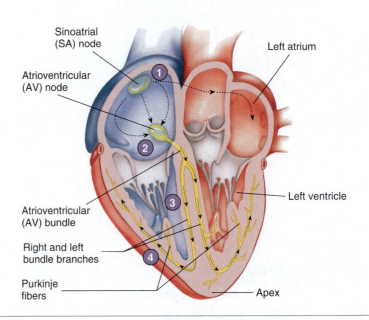

PROCESS Figure 12.15 AP|R **Conduction System of the Heart**

The SA node is the pacemaker of the heart, but other cells of the conduction system are also capable of producing action potentials spontaneously. For example, if the SA node is unable to function, another area, such as the AV node, becomes the pacemaker. The resulting heart rate is much slower than normal. When action potentials originate in an area of the heart other than the SA node, the result is called an **ectopic** (ek-top′ik) **beat.**

Electrocardiogram

Action potentials conducted through the heart during the cardiac cycle produce electrical currents that can be measured at the surface of the body. Electrodes placed on the body surface and attached to a recording device can detect the small electrical changes resulting from the action potentials in all of the cardiac muscle cells. The record of these electrical events is an **electrocardiogram** (ē-lek-trō-kar′dē-ō-gram; **ECG** or **EKG**) (figure 12.16).

The normal ECG consists of a P wave, a QRS complex, and a T wave. The **P wave** results from depolarization of the atrial myocardium, and the beginning of the P wave precedes the onset of atrial contraction. The **QRS complex** consists of three individual waves: the Q, R, and S waves. The QRS complex results from depolarization of the ventricles, and the beginning of the QRS complex precedes ventricular contraction. The **T wave** represents repolarization of the ventricles, and the beginning of the T wave precedes ventricular relaxation. A wave representing repolarization of the atria cannot be seen because it occurs during the QRS complex.

The time between the beginning of the P wave and the beginning of the QRS complex is the **PQ interval,** commonly called the **PR interval** because the Q wave is very small. During the PQ interval, the atria contract and begin to relax. At the end of the PQ interval, the ventricles begin to depolarize.

The **QT interval** extends from the beginning of the QRS complex to the end of the T wave and represents the length of time required for ventricular depolarization and repolarization. Table 12.1 describes several conditions associated with abnormal heart rhythms.

The ECG is not a direct measurement of mechanical events in the heart, and neither the force of contraction nor the blood pressure can be determined from it. However, each deflection in the ECG record indicates an electrical event within the heart and correlates with a subsequent mechanical event. Consequently, the ECG is an extremely valuable tool for diagnosing a number of cardiac abnormalities, particularly because it is painless, easy to record, and nonsurgical. Analysis of an ECG can reveal abnormal heart rates or rhythms; problems in conduction pathways, such as blockages; hypertrophy or atrophy of portions of the heart; and the approximate location of damaged cardiac muscle.

Predict 5

> Explain how the ECGs would appear for a person who has a damaged left bundle branch (see Predict 4) and for a person who has many ectopic beats originating from her atria.

A CASE IN POINT

Hands Only Cardiopulmonary Resuscitation (CPR)

Willie May Kitt is a 65-year-old bank manager. While walking up a short flight of stairs to his office, he experienced a crushing pain in his chest and exhibited substantial pallor. Willie fell to the floor, lost consciousness, and then stopped breathing. A coworker noticed the pallor, saw Willie fall, and ran to his aid. He could detect no pulse and decided to administer cardiopulmonary resuscitation (CPR). Another coworker called 911 and then assisted the first. Neither of Willie's coworkers had received training in CPR; however, they were both familiar with the Hands Only CPR recommended by the American Heart Association for teens and adults. One coworker pushed down firmly on Willie's sternum at a rate of 100 compressions per minute (about the tempo of the Bee Gee's song "Stayin' Alive"). Pushing down on the sternum compresses the ventricles of the heart and forces blood into the aorta and pulmonary trunk. Between compressions, blood flows into the ventricles from the atria. This maintains blood flow to the body tissues until emergency medical help arrives.

Fortunately, it took only about 5 minutes for emergency medical technicians to arrive. They confirmed the lack of a pulse and used portable equipment to record an electrocardiogram, which indicated that the heart was fibrillating (see Clinical Impact, "Fibrillation of the Heart"). They quickly used a portable defibrillator to apply a strong electrical shock to Willie's chest. Willie's heart responded by beginning to beat rhythmically.

Willie's heart may have first developed arrhythmia and then ventricular fibrillation. Willie was very fortunate. Most people whose heart suddenly stops pumping do not survive. In Willie's case, CPR was administered quickly and effectively, and emergency help arrived in a very short time.

Willie was transported to a hospital. His condition could be due to a myocardial infarction (see Systems Pathology, "Myocardial Infarction") or to some other cause that needs to be identified and treated.

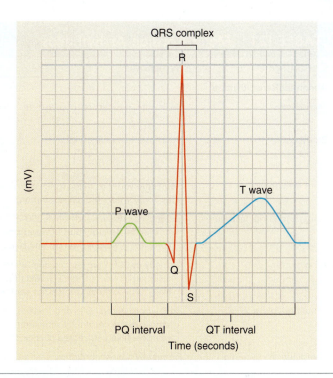

Figure 12.16 Electrocardiogram

The major waves and intervals of an electrocardiogram are labeled. Each thin horizontal line on the ECG recording represents 1 millivolt (mV), and each thin vertical line represents 0.04 second.

12.6 CARDIAC CYCLE

Learning Outcome *After reading this section, you should be able to*

A. Describe the cardiac cycle and the relationships among the contraction of each of the chambers, the pressure in each of the chambers, the phases of the electrocardiogram, and the heart sounds.

The right and left halves of the heart can be viewed as two separate pumps. Each pump consists of a primer pump—the atrium—and a power pump—the ventricle. The atria act as primer pumps because they complete the filling of the ventricles with blood, and the ventricles act as power pumps because they produce the major force that causes blood to flow through the pulmonary and systemic circulations. The term **cardiac cycle** refers to the repetitive pumping process that begins with the onset of cardiac muscle contraction and ends with the beginning of the next contraction (figure 12.17). Pressure changes produced within the heart chambers as a result of cardiac muscle contraction move blood from areas of higher pressure to areas of lower pressure.

Atrial systole (sis′tō-lē; a contracting) refers to contraction of the two atria. **Ventricular systole** refers to contraction of the two ventricles. **Atrial diastole** (dī-as′tō-lē; dilation) refers to relaxation of the two atria, and **ventricular diastole** refers to relaxation of the two ventricles. When the terms *systole* and *diastole* are used alone, they refer to ventricular contraction or relaxation. The ventricles contain more cardiac muscle than the atria and produce far greater pressures, which force blood to circulate throughout the vessels of the body.

TABLE 12.1   Major Cardiac Arrhythmias

Condition	Symptoms	Possible Causes
Abnormal Heart Rhythms		
Tachycardia	Heart rate in excess of 100 beats per minute (bpm)	Elevated body temperature, excessive sympathetic stimulation, toxic conditions
Bradycardia	Heart rate less than 60 bpm	Increased stroke volume in athletes, excessive vagus nerve stimulation, nonfunctional SA node, carotid sinus syndrome
Sinus arrhythmia	Heart rate varies as much as 5% during respiratory cycle and up to 30% during deep respiration.	Cause not always known; occasionally caused by ischemia, inflammation, or cardiac failure
Paroxysmal atrial tachycardia	Sudden increase in heart rate to 150–250 bpm for a few seconds or even for several hours; P waves precede every QRS complex; P wave is inverted and superimposed on T wave	Excessive sympathetic stimulation, abnormally elevated permeability of cardiac muscle to Ca^{2+}
Atrial flutter	As many as 300 P waves/min and 125 QRS complexes/min; resulting in two or three P waves (atrial contractions) for every QRS complex (ventricular contraction)	Ectopic beats in atria
Atrial fibrillation	No P waves, normal QRS and T waves, irregular timing; ventricles are constantly stimulated by atria; reduced ventricle filling; increased chance of fibrillation	Ectopic beats in atria
Ventricular tachycardia	Frequently causes fibrillation	Often associated with damage to AV node or ventricular muscle
Heart Blocks		
SA node block	No P waves; low heart rate resulting from AV node acting as the pacemaker; normal QRS complexes and T waves	Ischemia, tissue damage resulting from infarction; cause sometimes unknown
AV node blocks		
First-degree	PQ interval greater than 0.2 s	Inflammation of AV bundle
Second-degree	PQ interval 0.25–0.45 s; some P waves trigger QRS complexes and others do not; examples of 2:1, 3:1, and 3:2 P wave/QRS complex ratios	Excessive vagus nerve stimulation, AV node damage
Third-degree (complete heart block)	P wave dissociated from QRS complex; atrial rhythm about 100 bpm; ventricular rhythm less than 40 bpm	Ischemia of AV node or compression of AV bundle
Premature Contractions		
Premature atrial contractions	Occasional shortened intervals between one contraction and the succeeding contraction; frequently occurs in healthy people	Excessive smoking, lack of sleep, too much caffeine
Premature ventricular contractions (PVCs)	Prolonged QRS complex; exaggerated voltage because only one ventricle may depolarize; possible inverted T wave; increased probability of fibrillation	Ectopic beat in ventricles, lack of sleep, too much caffeine, irritability; occasionally occurs with coronary thrombosis

Cardiovascular

During the cardiac cycle, changes in chamber pressure and the opening and closing of the heart valves determine the direction of blood movement. As the cardiac cycle is described, it is important to focus on these pressure changes and heart valve movements. Before we start, it is also important to have a clear image of the state of the heart. At the beginning of the cardiac cycle, the atria and ventricles are relaxed, the AV valves are open, and the semilunar valves are closed. Blood returning to the heart first enters the atria. Since the AV valves are open, blood flows into the ventricles, filling them to approximately 70% of their volume (figure 12.17, step 1). The major events of the cardiac cycle are as follows:

- *Atrial systole*—The atria contract, forcing additional blood to flow into the ventricles to complete their filling (figure 12.17, step 2). The semilunar valves remain closed.

- *Ventricular systole*—At the beginning of ventricular systole, contraction of the ventricles pushes blood toward the atria, causing the AV valves to close as the pressure begins to increase (figure 12.17, step 3).

- As ventricular systole continues, the increasing pressure in the ventricles exceeds the pressure in the pulmonary trunk and aorta, the semilunar valves are forced open, and blood is ejected into the pulmonary trunk and aorta (figure 12.17, step 4).

- *Ventricular diastole*—At the beginning of ventricular diastole, the pressure in the ventricles decreases below the pressure in the aorta and pulmonary trunk. The semilunar valves close and prevent blood from flowing back into the ventricles (figure 12.17, step 5).

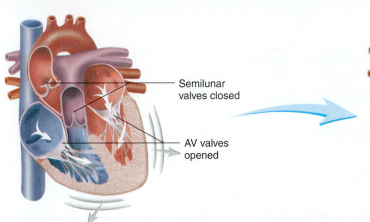

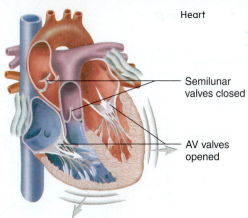

① The atria and ventricles are relaxed. AV valves open, and blood flows into the ventricles. The ventricles fill to approximately 70% of their volume.

Semilunar valves closed

AV valves opened

② The atria contract and complete ventricular filling.

Semilunar valves closed

AV valves opened

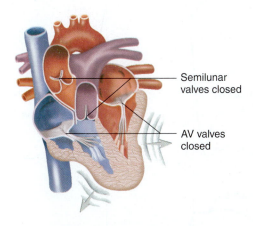

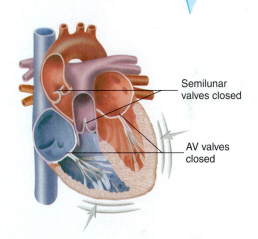

Semilunar valves closed

AV valves closed

⑤ At the beginning of ventricular diastole, the ventricles relax, and the semilunar valves close (the second heart sound).

Semilunar valves closed

AV valves closed

③ Contraction of the ventricles causes pressure in the ventricles to increase. Almost immediately, the AV valves close (the first heart sound). The pressure in the ventricles continues to increase.

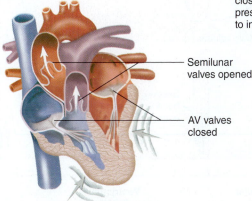

Semilunar valves opened

AV valves closed

④ Continued ventricular contraction causes the pressure in the ventricles to exceed the pressure in the pulmonary trunk and aorta. As a result, the semilunar valves are forced open, and blood is ejected into the pulmonary trunk and aorta.

Cardiovascular

PROCESS Figure 12.17 AP|R **Cardiac Cycle**

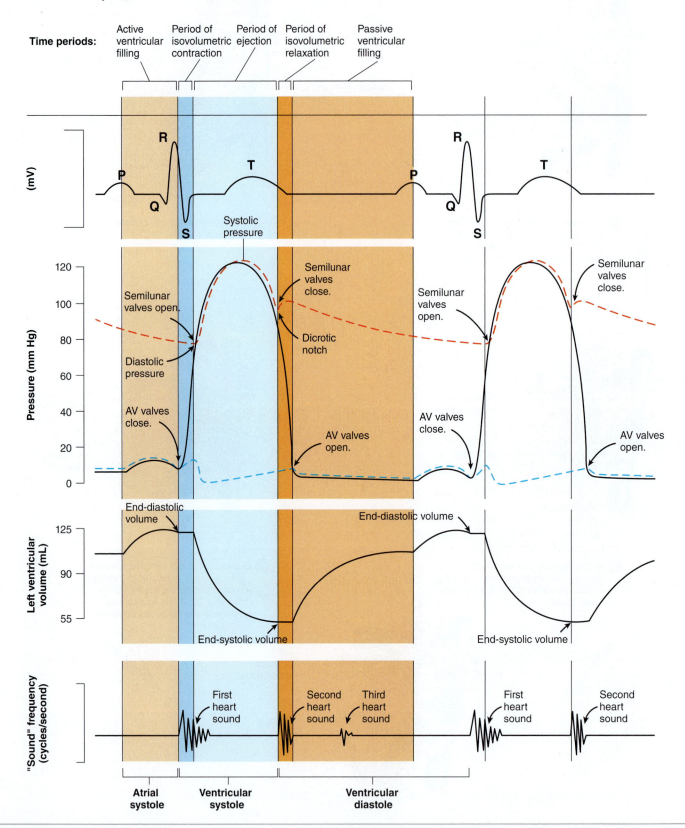

Figure 12.18 Events of the Cardiac Cycle

(*Top*) The cardiac cycle is divided into five periods. *From top to bottom,* these graphs show an electrocardiogram; pressure changes for the left atrium (*blue dashed line*), the left ventricle (*black line*), and the aorta (*red dashed line*); left ventricular volume curve; and the sequence of the heart sounds.

CLINICAL IMPACT   Consequences of an Incompetent Bicuspid Valve

In some individuals, a heart valve does not close completely and thus is called an **incompetent valve.** Such valves leak when they are supposed to be closed and allow blood to flow in the reverse direction. For example, an incompetent bicuspid valve allows blood to flow from the left ventricle to the left atrium during ventricular systole. This reduces the amount of blood pumped into the aorta. It also dramatically increases the blood pressure in the left atrium and in the pulmonary veins during ventricular systole. During diastole, the excess blood pumped into the atrium once again flows into the ventricle, along with the blood that normally flows from the lungs to the left atrium. Therefore, the volume of blood entering the left ventricle is greater than normal. The increased filling of the left ventricle gradually causes it to hypertrophy and can lead to heart failure. The increased pressure in the pulmonary veins can cause edema in the lungs.

• As diastole continues, the pressure continues to decline in the ventricles until atrial pressures are greater than ventricular pressures. Then the AV valves open, and blood flows directly from the atria into the relaxed ventricles. During the previous ventricular systole, the atria were relaxed and blood collected in them. When the ventricles relax and the AV valves open, blood flows into the ventricles (figure 12.17, step 1) and they begin to fill again.

Figure 12.18 displays the main events of the cardiac cycle and should be examined from top to bottom for each period. The ECG indicates the electrical events that cause the atria and ventricles to contract and relax. The pressure graph shows the pressure changes within the left atrium (blue dashed line), left ventricle (black line), and aorta (red dashed line) resulting from atrial and ventricular contraction and relaxation. The pressure changes in the right side of the heart are not shown here, but they are similar to those in the left side, only lower. The volume graph presents the changes in left ventricular volume as blood flows into and out of the left ventricle as a result of the pressure changes. The sound graph records the closing of valves caused by blood flow. See figure 12.17 for illustrations of the valves and blood flow.

Predict 6

Predict the effect of a leaky (incompetent) aortic semilunar valve on the volume of blood in the left ventricle just before ventricular contraction. Predict the effect of a severely narrowed opening through the aortic semilunar valves on the amount of work the heart must do to pump the normal volume of blood into the aorta during each beat of the heart.

12.7 HEART SOUNDS

Learning Outcome   *After reading this section, you should be able to*

A. Describe the heart sounds and their significance.

The **stethoscope** (steth′ō-skōp; *stetho*, the chest) was originally developed to listen to the sounds of the lungs and heart and is now used to listen to other sounds of the body as well. Figure 12.19 shows the sites in the thorax where the heart sounds can best be heard with a stethoscope. There are two main heart sounds. The first heart sound can be represented by the syllable *lubb*, and the

second heart sound can be represented by *dupp*. The first heart sound has a lower pitch than the second. The first heart sound occurs at the beginning of ventricular systole and results from closure of the AV valves (see figure 12.17, step 3, and figure 12.18). The second heart sound occurs at the beginning of ventricular diastole and results from closure of the semilunar valves (see figure 12.17, step 5, and figure 12.18). The valves usually do not make sounds when they open.

Clinically, ventricular systole occurs between the first and second heart sounds. Ventricular diastole occurs between the second heart sound and the first heart sound of the next beat. Because ventricular diastole lasts longer than ventricular systole, there is less time between the first and second heart sounds than between the second heart sound and the first heart sound of the next beat.

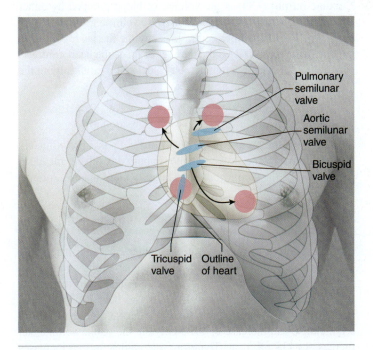

Figure 12.19 Location of the Heart Valves in the Thorax
Surface markings of the heart in an adult male. The positions of the four heart valves are indicated by *blue ellipses,* and the sites where the sounds of the valves are best heard with a stethoscope are indicated by *pink circles.*

Labels in figure: Pulmonary semilunar valve; Aortic semilunar valve; Bicuspid valve; Tricuspid valve; Outline of heart

Predict 7

Compare the rate of blood flow out of the ventricles between the first and second heart sounds of the same beat to the rate of blood flow out of the ventricles between the second heart sound of one beat and the first heart sound of the next beat.

Abnormal heart sounds called **murmurs** are usually a result of faulty valves. For example, an incompetent valve fails to close tightly, so that blood leaks through the valve when it is closed (see Clinical Impact, "Consequences of an Incompetent Bicuspid Valve"). A murmur caused by an incompetent valve makes a swishing sound immediately after the valve closes. For example, an incompetent bicuspid valve produces a swishing sound immediately after the first heart sound.

When the opening of a valve is narrowed, or **stenosed** (sten′ozd; a narrowing), a swishing sound precedes closure of the stenosed valve. For example, when the bicuspid valve is stenosed, a swishing sound precedes the first heart sound.

Predict 8

If normal heart sounds are represented by lubb-dupp, lubb-dupp, what does a heart sound represented by lubb-duppshhh, lubb-duppshhh represent? What does lubb-shhhdupp, lubb-shhhdupp represent (assuming that shhh represents a swishing sound)?

12.8 REGULATION OF HEART FUNCTION

Learning Outcomes *After reading this section, you should be able to*

A. Describe intrinsic and extrinsic regulation of the heart.

B. Give the conditions for which the major heart medications and treatments are administered.

Various measurements can be taken to assess the heart's function. **Cardiac output (CO)** is the volume of blood pumped by either ventricle of the heart each minute. **Stroke volume (SV)** is the volume of blood pumped per ventricle each time the heart contracts, and the **heart rate (HR)** is the number of times the heart contracts each minute. Cardiac output can be calculated by multiplying the stroke volume times the heart rate:

$$CO = SV \times HR$$
$$\text{(mL/min)} \quad \text{(mL/beat)} \quad \text{(beats/min)}$$

Under resting conditions, the heart rate is approximately 72 beats/min, and the stroke volume is approximately 70 mL/beat. Consequently, the cardiac output is slightly more than 5 L/min:

$$CO = SV \times HR$$
$$= 70 \text{ mL/beat} \times 72 \text{ bpm}$$
$$= 5040 \text{ mL/min (approximately 5 L/min)}$$

The heart rate and the stroke volume vary considerably among people. Athletes tend to have a higher stroke volume and lower heart rate at rest because exercise has increased the size of their hearts. Nonathletes are more likely to have a higher heart rate and lower stroke volume. During exercise, the heart rate in a nonathlete can increase to 190 bpm, and the stroke volume can increase to 115 mL/beat. Therefore, the cardiac output increases to approximately 22 L/min:

$$CO = SV \times HR$$
$$= 115 \text{ mL/beat} \times 190 \text{ bpm}$$
$$= 21,850 \text{ mL/min (approximately 22 L/min)}$$

This cardiac output is several times greater than the cardiac output under resting conditions. Athletes can increase their cardiac output to a greater degree than nonathletes. The control mechanisms that modify the stroke volume and the heart rate are classified as intrinsic and extrinsic.

Intrinsic Regulation of the Heart

Intrinsic regulation refers to mechanisms contained within the heart itself. The force of contraction produced by cardiac muscle is related to the degree of stretch of cardiac muscle fibers. The amount of blood in the ventricles at the end of ventricular diastole determines the degree to which cardiac muscle fibers are stretched. **Venous return** is the amount of blood that returns to the heart, and the degree to which the ventricular walls are stretched at the end of diastole is called **preload.** If venous return increases, the heart fills to a greater volume and stretches the cardiac muscle fibers, producing an increased preload. In response, cardiac muscle fibers contract with a greater force. The greater force of contraction causes an increased volume of blood to be ejected from the heart, resulting in an increased stroke volume. As venous return increases, resulting in an increased preload, cardiac output increases. Conversely, if venous return decreases, resulting in a decreased preload, the cardiac output decreases. The relationship between preload and stroke volume is called **Starling's law of the heart.**

A CASE IN POINT

Paroxysmal Atrial Tachycardia

Speedy Beat is a 70-year-old man. He and his daughter, Normal, were getting out of the car at a restaurant where they planned to have dinner. Before Speedy could get completely out of the car, he became dizzy. He exhibited substantial pallor and experienced chest pains. Normal checked her father's pulse, which was close to 180–200 bpm and irregular. She helped him back into the car and drove to the emergency room of a nearby hospital. There, physicians determined that Speedy's blood pressure was low, even though his heart rate was rapid. Having been previously diagnosed with paroxysmal atrial tachycardia, Speedy regularly takes a calcium channel blocking agent.

Tachycardia is characterized by rapid and ectopic beats that originate in the atria. The term *paroxysmal* means that the cause of the tachycardia is not known. The heart rate is irregular because the rate at which action potentials responsible for atrial contractions occur is greater than the rate at which the ventricles can contract. Consequently, not every atrial contraction is followed by a ventricular contraction. Speedy's blood pressure is low because his heart rate is so fast that there is little time for blood to fill the rapidly contracting chambers of the heart between the contractions. Consequently, Speedy's stroke volume and cardiac output are low. Chest pains result because the heart muscle is working hard but blood flow to cardiac muscle through the coronary vessels is reduced, so that the heart muscle suffers from an inadequate supply of O_2 (ischemia).

Once Speedy's heart rate and rhythm were stabilized, the amount of calcium channel blocking agent prescribed for him was adjusted, and he was released from the hospital the next day. Along with the calcium channel blocking agent, another drug used to treat this condition is digoxin, which has the overall effect of increasing the force and slowing the rate of cardiac muscle contraction.

CLINICAL IMPACT   Consequences of Heart Failure

Although heart failure can occur in young people, it usually results from progressive weakening of the heart muscle in elderly people. In heart failure, the heart is not capable of pumping all the blood that is returned to it because further stretching of the cardiac muscle fibers does not increase the stroke volume. Consequently, blood backs up in the veins. For example, heart failure that affects the right ventricle, called **right heart failure,** causes blood to back up in the veins that return blood from systemic vessels to the heart. Filling of the veins with blood causes edema, especially in the legs and feet, due to the accumulation of fluid in the tissues outside blood vessels. Heart failure that affects the left ventricle, called **left heart failure,** causes blood to back up in the veins that return blood from the lungs to the heart. Filling of these veins causes edema in the lungs, which makes breathing difficult.

Because venous return is influenced by many conditions, Starling's law of the heart has a major influence on cardiac output. For example, muscular activity during exercise causes increased venous return, resulting in increased preload, stroke volume, and cardiac output. This is beneficial because increased cardiac output is needed during exercise to supply O_2 to exercising skeletal muscles.

Afterload refers to the pressure against which the ventricles must pump blood. People suffering from hypertension have an increased afterload because their aortic pressure is elevated during contraction of the ventricles. The heart must do more work to pump blood from the left ventricle into the aorta, which increases the workload on the heart and can eventually lead to heart failure. A reduced afterload decreases the work the heart must do. People who have lower blood pressure have a reduced afterload and develop heart failure less often than people who have hypertension. However, the afterload influences cardiac output less than preload influences it. The afterload must increase substantially before it decreases the volume of blood pumped by a healthy heart.

Extrinsic Regulation of the Heart

Extrinsic regulation refers to mechanisms external to the heart, such as either nervous or chemical regulation.

Nervous Regulation: Baroreceptor Reflex

Nervous influences of heart activity are carried through the autonomic nervous system. Both sympathetic and parasympathetic nerve fibers innervate the heart and have a major effect on the SA node. Stimulation by sympathetic nerve fibers causes the heart rate and the stroke volume to increase, whereas stimulation by parasympathetic nerve fibers causes the heart rate to decrease.

The **baroreceptor** (bar′ō-rē-sep′ter; *baro,* pressure) **reflex** is a mechanism of the nervous system that plays an important role in regulating heart function. **Baroreceptors** are stretch receptors that monitor blood pressure in the aorta and in the wall of the internal carotid arteries, which carry blood to the brain. Changes in blood pressure result in changes in the stretch of the walls of these blood vessels—and changes in the frequency of action potentials produced by the baroreceptors. The action potentials are transmitted along nerve fibers from the stretch receptors to the medulla oblongata of the brain.

Within the medulla oblongata is a **cardioregulatory center,** which receives and integrates action potentials from the baroreceptors. The cardioregulatory center controls the action potential frequency in sympathetic and parasympathetic nerve fibers that extend from the brain and spinal cord to the heart. The cardioregulatory center also influences sympathetic stimulation of the adrenal gland. Epinephrine and norepinephrine, released from the adrenal gland, increase the stroke volume and heart rate.

Figure 12.20 shows how the baroreceptor reflex keeps the heart rate and stroke volume within normal ranges. When the blood pressure increases, the baroreceptors are stimulated. Action potentials are sent along the nerve fibers to the medulla oblongata at increased frequency. This prompts the cardioregulatory center to increase parasympathetic stimulation and decrease sympathetic stimulation of the heart. As a result, the heart rate and stroke volume decrease, causing blood pressure to decline.

When the blood pressure decreases, less stimulation of the baroreceptors occurs. A lower frequency of action potentials is sent to the medulla oblongata of the brain, and this triggers a response in the cardioregulatory center. The cardioregulatory center responds by increasing sympathetic stimulation of the heart and decreasing parasympathetic stimulation. Consequently, the heart rate and stroke volume increase, causing blood pressure to increase. If the decrease in blood pressure is large, sympathetic stimulation of the adrenal medulla also increases. The epinephrine and norepinephrine secreted by the adrenal medulla increase the heart rate and stroke volume, also causing the blood pressure to increase toward its normal value (figure 12.20).

Predict 9

In response to a severe hemorrhage, blood pressure lowers, the heart rate increases dramatically, and the stroke volume lowers. If low blood pressure activates a reflex that increases sympathetic stimulation of the heart, why is the stroke volume low?

Chemical Regulation: Chemoreceptor Reflex

Epinephrine and small amounts of norepinephrine released from the adrenal medulla in response to exercise, emotional excitement, or stress also influence the heart's function. Epinephrine and norepinephrine bind to receptor proteins on cardiac muscle and cause increased heart rate and stroke volume. Excitement, anxiety, or anger can affect the cardioregulatory center, resulting in increased sympathetic stimulation of the heart and increased cardiac output. Depression, on the other hand, can increase parasympathetic stimulation of the heart, causing a slight reduction in cardiac output.

The medulla oblongata of the brain also contains chemoreceptors that are sensitive to changes in pH and CO_2 levels. A decrease in pH, often caused by an increase in CO_2, results in sympathetic stimulation of the heart (figure 12.21).

Cardiovascular

③ **Actions**

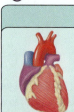

Baroreceptors in the carotid arteries and aorta detect an increase in blood pressure.

The cardioregulatory center in the brain decreases sympathetic stimulation of the heart and adrenal medulla and increases parasympathetic stimulation of the heart.

④ **Reactions**

Effectors Respond:
The SA node and cardiac muscle decrease activity and heart rate and stroke volume decrease.

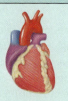

② **Homeostasis Disturbed:**
Blood pressure increases.

⑤ **Homeostasis Restored:**
Blood pressure decreases.

① Blood pressure (normal range)

Start here

⑥ Blood pressure (normal range)

Homeostasis Disturbed:
Blood pressure decreases.

Homeostasis Restored:
Blood pressure increases.

Actions

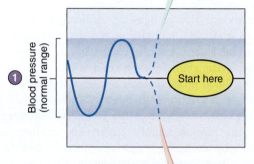

Baroreceptors in the carotid arteries and aorta detect a decrease in blood pressure.

The cardioregulatory center in the brain increases sympathetic stimulation of the heart and adrenal medulla and decreases parasympathetic stimulation of the heart.

Reactions

Effectors Respond:
The SA node and cardiac muscle increase activity and heart rate and stroke volume increase.

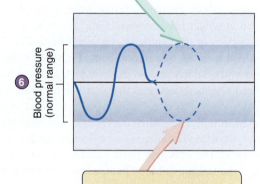

Cardiovascular

Homeostasis Figure 12.20 AP|R Baroreceptor Reflex

The baroreceptor reflex maintains homeostasis in response to changes in blood pressure. (1) Blood pressure is within its normal range. (2) Blood pressure increases outside the normal range, which causes homeostasis to be disturbed. (3) Baroreceptors in the carotid arteries and aorta detect the increase in blood pressure and the cardioregulatory center in the brain alters autonomic stimulation of the heart. (4) Heart rate and stroke volume decrease. (5) These changes cause blood pressure to decrease. (6) Blood pressure returns to its normal range, and homeostasis is restored.

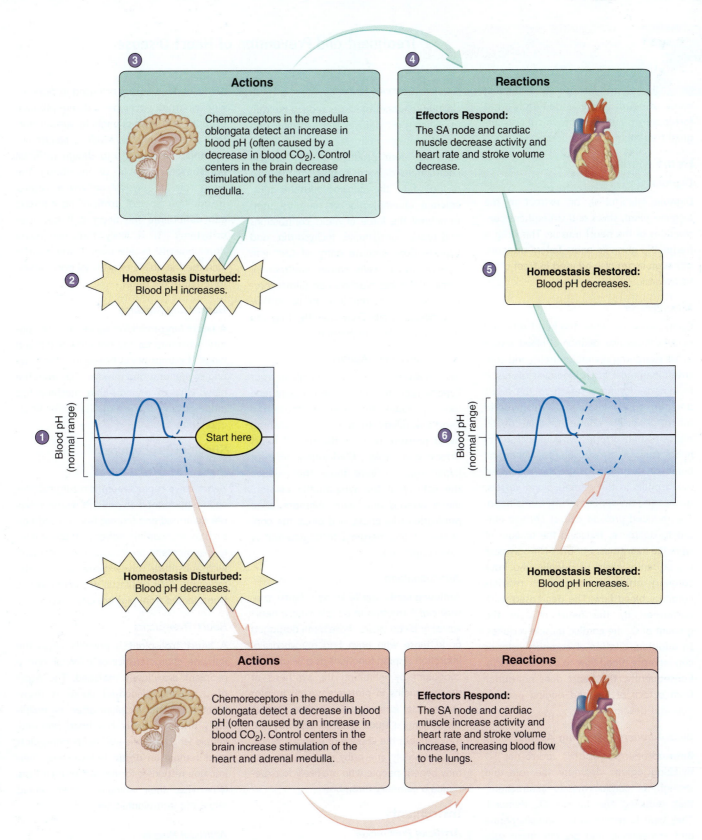

Cardiovascular

Homeostasis Figure 12.21 $\boxed{\text{AP}\,|\,\text{R}}$ **Chemoreceptor Reflex—pH**

The chemoreceptor reflex maintains homeostasis in response to changes in blood concentrations of CO_2 and H^+ (or pH). (1) Blood pH is within its normal range. (2) Blood pH increases outside the normal range. (3) Chemoreceptors in the medulla oblongata detect increased blood pH. Control centers in the brain decrease sympathetic stimulation of the heart and adrenal medulla. (4) Heart rate and stroke volume decrease, reducing blood flow to lungs. (5) These changes cause blood pH to decrease (as a result of increase in blood CO_2). (6) Blood pH returns to its normal range, and homeostasis is restored.

CLINICAL IMPACT   Treatment and Prevention of Heart Disease

Heart disease is a leading cause of death in the United States, but fortunately a number of medical and surgical treatments are available.

Heart Medications

Digitalis

Digitalis (dij′i-tal′is), an extract of the foxglove plant, slows and strengthens contractions of the heart muscle. This drug is frequently given to heart failure patients, although it can also be used to treat atrial tachycardia.

Nitroglycerin

During exercise, when the heart rate and stroke volume rise, dilation of blood vessels in the exercising skeletal muscles and constriction in most other blood vessels result in increased venous return to the heart and increased preload. **Nitroglycerin** (nī-trō-glis′er-in) causes dilation of all the veins and arteries without increasing heart rate or stroke volume. When all blood vessels dilate, a greater volume of blood pools in the dilated blood vessels, decreasing the venous return to the heart. The reduced preload causes cardiac output to decrease, reducing the amount of work the heart must perform. Nitroglycerin is frequently given to people who have coronary artery disease, which restricts coronary blood flow. The decreased work performed by the heart reduces the amount of O_2 the cardiac muscle requires. In addition, dilation of coronary arteries can increase blood flow to cardiac muscle. Consequently, the heart does not suffer from a lack of O_2, and angina pectoris does not develop.

Beta-Adrenergic Blocking Agents

Beta-adrenergic (bā-tă ad-rĕ-ner′jik) **blocking agents** decrease the rate and strength of cardiac muscle contractions, thus reducing the heart's O_2 demand. They bind to receptors for norepinephrine and epinephrine and prevent these substances from having their normal effects. Beta-adrenergic blocking agents are often used to treat people who have rapid heart rates, certain types of arrhythmias, and hypertension.

Calcium Channel Blockers

Calcium channel blockers reduce the rate at which Ca^{2+} diffuses into cardiac and smooth muscle cells. Because the action potentials that produce cardiac muscle contractions depend in part on the flow of Ca^{2+} into the cardiac muscle cells, calcium channel blockers can be used to control the force of heart contractions and reduce arrhythmia, tachycardia, and hypertension. Because entry of Ca^{2+} into smooth muscle cells causes contraction, calcium channel blockers can dilate coronary blood vessels and increase blood flow to cardiac muscle. Consequently, they are used to treat angina pectoris.

Antihypertensive Agents

Several drugs are used specifically to treat hypertension. Because these drugs reduce blood pressure, they also decrease the heart's workload. In addition, the lowered blood pressure reduces the risk of heart attacks and strokes. Medications to treat hypertension include those that reduce the activity of the sympathetic division, dilate arteries and veins, increase urine production (diuretics), and block the conversion of angiotensin I to angiotensin II (see chapter 13).

Anticoagulants

Anticoagulants (an′tē-kō-ag′ū-lants) prevent clot formation in people whose heart valves or blood vessels have been damaged or in those who have had a myocardial infarction. Aspirin functions as a weak anticoagulant by inhibiting the synthesis of prostaglandins in platelets, which in turn reduces clot formation. Some data suggest that taking a small dose of aspirin regularly reduces the chance of a heart attack. Approximately one baby aspirin each day may benefit people who are likely to experience a coronary thrombosis.

Instruments

Artificial Pacemaker

An **artificial pacemaker** is an instrument placed beneath the skin that is equipped with an electrode extending to the heart. The pacemaker provides an electrical stimulus to the heart at a set frequency.

Artificial pacemakers are used in patients whose natural pacemakers do not produce a heart rate high enough to sustain normal physical activity. Modern electronics has made it possible to design artificial pacemakers capable of increasing the heart rate as physical activity increases. In addition, special artificial pacemakers can defibrillate the heart if it becomes arrhythmic. It is likely that continuing developments in electronics will further increase the ability of artificial pacemakers to regulate the heart.

Heart-Lung Machine

A **heart-lung machine** serves as a temporary substitute for the patient's heart and lungs. It pumps blood throughout the body and oxygenates and removes CO_2 from the blood. Use of a heart-lung machine has made possible many surgeries on the heart and lungs.

Surgical Procedures

Heart Valve Replacement

Heart valve replacement is a surgical procedure performed on patients whose valves are deformed and scarred from a condition such as endocarditis. Valves that are severely incompetent or stenosed are replaced with substitute valves made of synthetic materials, such as plastic or Dacron; valves transplanted from pigs are also used.

Heart Transplants

A **heart transplant** is possible when the immune characteristics of a donor and a recipient are closely matched. The heart of a recently deceased donor is transplanted into the recipient after the recipient's diseased heart has been removed. People who have received heart transplants must remain on drugs that suppress their immune responses for the rest of their lives. Otherwise, their immune system would reject the transplanted heart.

Artificial Hearts

Artificial hearts have been used on an experimental basis to extend the life of an individual until an acceptable transplant can be found to replace the heart permanently. However, the technology

CLINICAL IMPACT *(continued)*

currently available has not yet produced an artificial heart capable of permanently sustaining a high quality of life.

Prevention of Heart Disease

Proper nutrition is important in reducing the risk of heart disease. A recommended diet is low in fats, especially saturated fats and cholesterol, and low in refined sugar. The diet should also be high in fiber, whole grains, fruits, and vegetables. Nutritionists advise people to reduce their intake of sodium chloride (salt) and to limit their total food intake to avoid obesity.

In addition, tobacco and excessive use of alcohol should be avoided. Smoking increases the risk of heart disease by at least tenfold, and excessive use of alcohol substantially increases the risk of heart disease.

Chronic stress, frequent emotional upsets, and lack of physical exercise can increase the risk of cardiovascular disease. Remedies include relaxation techniques and aerobic activities, such as swimming, walking, jogging, or dancing, with gradual increases in duration and difficulty.

Hypertension, an abnormally high systemic blood pressure, affects approximately one-fifth of the population. Regular blood pressure monitoring is important because hypertension does not produce obvious symptoms. If hypertension cannot be controlled by diet and exercise, drugs should be prescribed. The cause of hypertension is unknown in the majority of cases.

1. Sensory neurons (*green*) carry action potentials from baroreceptors to the cardioregulatory center. Chemoreceptors in the medulla oblongata influence the cardioregulatory center.

2. The cardioregulatory center controls the frequency of action potentials in the parasympathetic neurons (*red*) extending to the heart. The parasympathetic neurons decrease the heart rate.

3. The cardioregulatory center controls the frequency of action potentials in the sympathetic neurons (*blue*) extending to the heart. The sympathetic neurons increase the heart rate and the stroke volume.

4. The cardioregulatory center influences the frequency of action potentials in the sympathetic neurons (*blue*) extending to the adrenal medulla. The sympathetic neurons increase the secretion of epinephrine and some norepinephrine into the general circulation. Epinephrine and norepinephrine increase the heart rate and stroke volume.

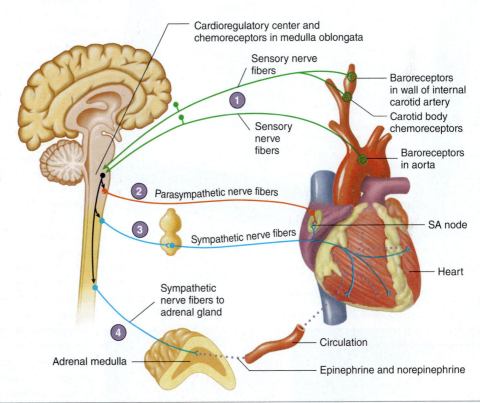

PROCESS Figure 12.22 Summary of Extrinsic Regulation

Sensory nerves (*green*) carry action potentials from sensory receptors to the medulla oblongata. Sympathetic nerves (*blue*) and parasympathetic nerves (*red*) exit the spinal cord or medulla oblongata and extend to the heart to regulate its function. Epinephrine and norepinephrine from the adrenal gland also help regulate the heart's action. (SA = sinoatrial)

Changes in the extracellular concentration of K^+, Ca^{2+}, and Na^+, which influence other electrically excitable tissues, also affect cardiac muscle function. An excess of extracellular K^+ causes the heart rate and stroke volume to decrease. If the extracellular K^+ concentration increases further, normal conduction of action potentials through cardiac muscle is blocked, and death can result. An excess of extracellular Ca^{2+} causes the heart to contract arrhythmically. Reduced extracellular Ca^{2+} cause both the heart rate and stroke volume to decrease.

Figure 12.22 summarizes how nervous and chemical factors interact to regulate the heart rate and stroke volume.

Body temperature affects the metabolism in the heart just as it affects other tissues. Elevated body temperature increases the heart rate, and reduced body temperature slows the heart rate. For

Name: Paul
Gender: Male
Age: 49

Comments

Paul, an overweight, out-of-shape executive, was a smoker and indulged in high-fat foods. Yesterday, Paul felt intense pain in his chest that radiated down his left arm. Out of breath and dizzy, Paul had to lie down on the floor. He became anxious and disoriented and eventually lost consciousness, but he did not stop breathing. Paul's coworker noticed him and called 911. Paramedics arrived and found that Paul's blood pressure was low and he exhibited arrhythmia and tachycardia. They conducted an ECG and administered O_2 and medication to control the arrhythmia.

Myocardial Infarction

Background Information

At the hospital, Paul was given tissue plasminogen activator (t-PA) to improve blood flow to the damaged area of the heart. Paul remained in the hospital. Over the next few days, blood levels of enzymes, such as creatine phosphokinase, increased in Paul's blood, which confirmed that cardiac muscle had been damaged by an infarction. Paul also began to experience shortness of breath because of pulmonary edema, and after a few days he developed pneumonia. He was treated for the pneumonia, and his condition gradually improved over the next few weeks. An **angiogram** (an'jē-ō-gram; *angeion*, a vessel + *gramma*, a writing) (figure I2A), which is an imaging technique used to visualize the coronary arteries, was performed several days after Paul's infarction. The angiogram showed some restricted blood flow in Paul's coronary arteries and a permanent reduction in a significant part of the lateral wall of Paul's left ventricle, but his physicians decided neither angioplasty nor bypass surgery was necessary.

Paul experienced a myocardial infarction in which a thrombosis in one of the branches of the left coronary artery reduced blood supply to the lateral wall of the left ventricle, resulting in ischemia of the left ventricle wall (figure I2A). The fact that t-PA, which activates plasminogen and thereby dissolves blood clots, was an effective treatment supports the conclusion that the infarction was due to a thrombosis. An ischemic area of the heart wall was not able to contract normally; therefore, the heart's pumping effectiveness was dramatically reduced. The reduced pumping capacity was responsible for the low blood pressure, which decreased the blood flow to Paul's brain, resulting in confusion, disorientation, and unconsciousness. Low blood pressure, increasing blood CO_2 levels, pain, and anxiousness increased sympathetic stimulation of the heart and adrenal glands. Increased sympathetic stimulation of the adrenal medulla caused the release of epinephrine. Increased parasympathetic stimulation of the heart resulted from pain sensations. The heart rate was periodically arrhythmic due to the combined effects of parasympathetic stimulation, sympathetic stimulation, and the release of epinephrine and norepinephrine from the adrenal gland. In addition, the ischemic areas of the left ventricle produced ectopic beats.

Pulmonary edema resulted from the increased pressure in the pulmonary veins because of the left ventricle's reduced ability to pump blood. The edema allowed bacteria to infect the lungs and cause pneumonia.

Figure I2A

An angiogram is a picture of a blood vessel that is usually obtained by injecting the blood vessel with dye that can be detected by x-rays. Note the partially occluded (blocked) coronary blood vessel in this angiogram on the left, which has been computer-enhanced to show colors. The angiogram on the right is of a normal heart.

Occluded coronary artery

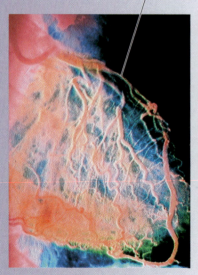
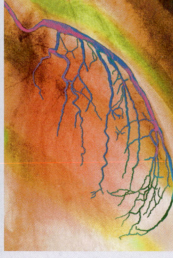

Cardiovascular

MUSCULAR

Skeletal muscle activity is reduced because of lack of blood flow to the brain and because blood is shunted from blood vessels that supply skeletal muscles to those that supply the heart and brain.

INTEGUMENTARY

Pallor of the skin results from intense vasoconstriction of peripheral blood vessels, including those in the skin.

NERVOUS

Decreased blood flow to the brain, decreased blood pressure, and pain due to ischemia of heart muscle result in increased sympathetic and parasympathetic stimulation of the heart. Loss of consciousness occurs when the blood flow to the brain decreases so that not enough O_2 is available to maintain normal brain function, especially in the reticular activating system.

URINARY

Blood flow to the kidney decreases dramatically in response to sympathetic stimulation. If the kidney becomes ischemic, the kidney tubules can be damaged, resulting in acute renal failure and reduced urine production. Increased blood urea nitrogen, increased blood levels of K^+, and edema are indications that the kidneys cannot eliminate waste products and excess water. If damage is not too great, the period of reduced urine production may last up to 3 weeks; then the rate of urine production slowly returns to normal as the kidney tubules heal.

Myocardial Infarctions

Symptoms
- Chest pain that radiates down left arm
- Tightness and pressure in chest
- Difficulty breathing
- Nausea and vomiting
- Dizziness and fatigue

Treatment
- Restore blood flow to cardiac muscle
- Medication to reduce blood clotting (aspirin) and increase blood flow (t-PA)
- Supplemental O_2 to restore normal O_2 to heart tissue
- Prevention and control of hypertension
- Angioplasty or bypass surgery

ENDOCRINE

When blood pressure becomes low, antidiuretic hormone (ADH) is released from the posterior pituitary gland, and renin, released from the kidney, activates the renin-angiotensin-aldosterone mechanism. ADH, secreted in large amounts, and angiotensin II cause vasoconstriction of peripheral blood vessels. ADH and aldosterone act on the kidneys to retain water and ions. Increased blood volume increases venous return, which results in increased stroke volume and blood pressure unless damage to the heart is very severe.

DIGESTIVE

Intense sympathetic stimulation reduces blood flow to the digestive system to very low levels, which often results in nausea and vomiting.

LYMPHATIC AND IMMUNE

White blood cells, including macrophages, move to the area of cardiac muscle damage and phagocytize any dead cardiac muscle cells.

RESPIRATORY

Decreased blood pressure results in decreased blood flow to the lungs. The decrease in gas exchange leads to increased blood CO_2 levels, acidosis, and decreased blood O_2 levels. Initially, respiration becomes deep and labored because of the elevated CO_2 levels, decreased blood pH, and depressed O_2 levels. If the blood O_2 levels decrease too much, the person loses consciousness. Pulmonary edema can result when the pumping effectiveness of the left ventricle is substantially reduced.

The heart began to beat rhythmically in response to medication because the infarction had not damaged the heart's conducting system, which is an indication that there were no permanent arrhythmias. (Permanent arrhythmias are indications of damage done to cardiac muscle cells specialized to conduct action potentials in the heart.)

Analysis of the electrocardiogram, blood pressure measurements, and the angiogram indicated that the infarction was located on the left side of Paul's heart.

Paul's physician made it very clear that he was lucky to have survived a myocardial infarction. The physician recommended a weight-loss program and a low-sodium and low-fat diet. He also advised Paul to take a small amount of aspirin regularly and to stop smoking. He explained that Paul would have to take medication for high blood pressure if his blood pressure did not decrease in response to the recommended changes. The physician recommended an aerobic exercise program after a period of recovery and suggested that Paul seek ways to reduce the stress associated with his job. Paul followed the doctor's recommendations, and after several months, his blood pressure was normal, and he felt better than he had in years.

Predict 10

Severe ischemia in the wall of a ventricle can cause the death of cardiac muscle cells. Inflammation develops around the necrotic (dead) tissue, and macrophages invade and phagocytize dead cells. At the same time, blood vessels and connective tissue grow into the necrotic area and begin to deposit connective tissue to replace the necrotic tissue. While Paul was still recovering in the hospital, another patient was admitted with a similar condition. After about a week, that person's blood pressure suddenly decreased to very low levels, and he died within a short time. An autopsy revealed a large amount of blood in the pericardial sac and a rupture in the wall of the left ventricle. Explain these effects.

example, the heart rate is usually elevated when a person has a fever. During heart surgery, the body temperature is sometimes intentionally lowered to slow the heart rate and metabolism.

12.9 EFFECTS OF AGING ON THE HEART

Learning Outcome *After reading this section, you should be able to*

A. List the major age-related changes that affect the heart.

Gradual changes in heart function are associated with aging. These changes are minor under resting conditions but become more obvious during exercise and in response to age-related diseases.

By age 70, a person's cardiac output has probably decreased by approximately one-third. Because of the decrease in the reserve strength of the heart, many elderly people have a limited ability to respond to emergencies, infections, blood loss, or stress.

Hypertrophy (enlargement) of the left ventricle is a common age-related change. This appears to result from a gradual increase in the pressure in the aorta (afterload), against which the left ventricle must pump. The increased aortic pressure results from a gradual decrease in the aorta's elasticity. Because the left ventricle is enlarged, its ability to pump out blood is reduced, which can cause an increase in left atrial pressure and lead to increased pulmonary edema. Consequently, older people tend to feel out of breath when they exercise strenuously.

Aging cardiac muscle requires a greater amount of time to contract and relax. Thus, the maximum heart rate decreases. Both the resting and maximum cardiac output slowly lower as people grow older, and by age 85, the cardiac output has decreased 30–60%.

Age-related changes also affect the connective tissue of the heart valves. The connective tissue becomes less flexible, and calcium deposits develop in the valves. As a result, the aortic semilunar valve may become stenosed or incompetent.

An age-related increase in cardiac arrhythmias occurs as a consequence of a decreased number of cardiac cells in the SA node and because of cell replacement in the AV bundle.

Coronary artery disease and heart failure are also age-related. Approximately 10% of elderly people over age 80 have heart failure, and a major contributing factor is coronary heart disease. Advanced age, malnutrition, chronic infections, toxins, severe anemias, hyperthyroidism, and hereditary factors can also lead to heart failure.

Exercise has many beneficial effects on the heart. Regular aerobic exercise improves the heart's functional capacity at all ages, providing the person has no other conditions that cause the increased workload on the heart to be harmful.

DISEASES AND DISORDERS: Heart

CONDITION	DESCRIPTION
Inflammation of Heart Tissue	
Endocarditis	Inflammation of the endocardium; affects the valves more severely than other areas of the endocardium; may lead to scarring, causing stenosed or incompetent valves
Pericarditis	Inflammation of the pericardium; see Clinical Impact, "Disorders of the Pericardium"
Cardiomyopathy	Disease of the myocardium of unknown cause or occurring secondarily to other disease; results in weakened cardiac muscle, causing all chambers of the heart to enlarge; may eventually lead to congestive heart failure
Rheumatic heart disease	Results from a streptococcal infection in young people; toxin produced by the bacteria can cause rheumatic fever several weeks after the infection that can result in rheumatic endocarditis
Reduced Blood Flow to Cardiac Muscle	
Coronary heart disease	Reduces the amount of blood the coronary arteries can deliver to the myocardium; see Clinical Impact, "Heart Attack"
Coronary thrombosis	Formation of blood clot in a coronary artery
Myocardial infarction	Damaged cardiac muscle tissue resulting from lack of blood flow to the myocardium; often referred to as a heart attack; see Systems Pathology
Congenital Heart Diseases (Occur at Birth)	
Septal defect	Hole in the septum between the left and right sides of the heart, allowing blood to flow from one side of the heart to the other and greatly reducing the heart's pumping effectiveness
Patent ductus arteriosus	Ductus arteriosus fails to close after birth, allowing blood to flow from the aorta to the pulmonary trunk under a higher pressure, which damages the lungs; also, the left ventricle must work harder to maintain adequate systemic pressure
Stenosis of the heart valves	Narrowed opening through one or more of the heart valves; aortic or pulmonary semilunar stenosis increases the heart's workload; bicuspid valve stenosis causes blood to back up in the left atria and lungs, resulting in edema of the lungs; tricuspid valve stenosis results in similar blood flow problems and edema in the peripheral tissues
Cyanosis (sī-ă-nō′sis; *cyan*, blue + *osis*, condition of)	Symptom of inadequate heart function in babies with congenital heart disease; the infant's skin appears blue because of low O_2 levels in the blood in peripheral blood vessels

Answers to the rest of this chapter's Predict questions are in Appendix E.

ANSWER TO LEARN TO PREDICT

We learned in this chapter that the heart valves maintain a one-way flow of blood through the heart—from the atria to the ventricles. We also learned that an incompetent valve is one that leaks, or allows some blood to flow in the opposite direction—from the ventricles to the atria. An irregular swooshing noise following the first heart sound, as noted by Stan's regular physician, is a typical sign of an incompetent valve. The first heart sound is produced when the bicuspid and tricuspid valves close. The swooshing sound is the regurgitation of blood into atria. The cardiologist determined that the bicuspid valve was incompetent, resulting in abnormal blood flow on the left side of the heart.

Stan's difficulty breathing results from the abnormal blood flow caused by his incompetent valve. After reviewing the blood flow through the heart in this chapter, we are aware that blood entering the left atrium is returning from the lungs through the pulmonary veins. As a result of the incompetent bicuspid valve, the pressure in the left atrium, which is normally low, increases substantially during ventricular contraction. The increased left atrial pressure causes the pressure in the pulmonary veins and pulmonary capillaries to increase. As a result, fluid leaks from the pulmonary capillaries and accumulates in the lungs, causing pulmonary edema, making it difficult for Stan to breathe.

Answers to the rest of this chapter's Predict questions are in Appendix E.

SUMMARY

12.1 Functions of the Heart   (p. 318)

1. The heart generates blood pressure.
2. The heart routes blood through the systemic and pulmonary circulations.
3. The heart's pumping action and its valves ensure a one-way flow of blood through the heart and blood vessels.
4. The heart helps regulate blood supply to tissues.

12.2 Size, Form, and Location of the Heart (p. 319)

The heart is approximately the size of a closed fist and is located in the pericardial cavity.

12.3 Anatomy of the Heart   (p. 320)

Pericardium

1. The pericardium is a sac consisting of fibrous and serous pericardia. The fibrous pericardium is lined by the parietal pericardium.
2. The outer surface of the heart is lined by the visceral pericardium (epicardium).
3. Between the visceral and parietal pericardia is the pericardial cavity, which is filled with pericardial fluid.

External Anatomy

1. The atria are separated externally from the ventricles by the coronary sulcus. The right and left ventricles are separated externally by the interventricular sulci.
2. The inferior and superior venae cavae enter the right atrium. The four pulmonary veins enter the left atrium.
3. The pulmonary trunk exits the right ventricle, and the aorta exits the left ventricle.

Heart Chambers and Internal Anatomy

1. There are four chambers in the heart. The left and right atria receive blood from veins and function mainly as reservoirs. Contraction of the atria completes ventricular filling.
2. The atria are separated internally from each other by the interatrial septum.

3. The ventricles are the main pumping chambers of the heart. The right ventricle pumps blood into the pulmonary trunk, and the left ventricle, which has a thicker wall, pumps blood into the aorta.
4. The ventricles are separated internally by the interventricular septum.

Heart Valves

1. The heart valves ensure one-way flow of blood.
2. The tricuspid valve (three cusps) separates the right atrium and the right ventricle, and the bicuspid valve (two cusps) separates the left atrium and the left ventricle.
3. The papillary muscles attach by the chordae tendineae to the cusps of the tricuspid and bicuspid valves and adjust tension on the valves.
4. The aorta and pulmonary trunk are separated from the ventricles by the semilunar valves.
5. The skeleton of the heart is a plate of fibrous connective tissue that separates the atria from the ventricles, acts as an electrical barrier between the atria and ventricles, and supports the heart valves.

Route of Blood Flow Through the Heart

1. The left and right sides of the heart can be considered separate pumps.
2. Blood flows from the systemic vessels to the right atrium and from the right atrium to the right ventricle. From the right ventricle, blood flows to the pulmonary trunk and from the pulmonary trunk to the lungs. From the lungs, blood flows through the pulmonary veins to the left atrium, and from the left atrium, blood flows to the left ventricle. From the left ventricle, blood flows into the aorta and then through the systemic vessels.

Blood Supply to the Heart

1. The left and right coronary arteries originate from the base of the aorta and supply the heart.
2. The left coronary artery has three major branches: the anterior interventricular, the circumflex, and the left marginal arteries.
3. The right coronary artery has two major branches: the posterior interventricular and the right marginal arteries.
4. Blood returns from heart tissue through cardiac veins to the coronary sinus and into the right atrium. Small cardiac veins also return blood directly to the right atrium.

12.4 Histology of the Heart   (p. 327)

Heart Wall

The heart wall consists of the outer epicardium, the middle myocardium, and the inner endocardium.

Cardiac Muscle

1. Cardiac muscle is striated; it depends on ATP for energy and on aerobic metabolism.
2. Cardiac muscle cells are joined by intercalated disks that allow action potentials to be propagated throughout the heart.

12.5 Electrical Activity of the Heart   (p. 329)

Action Potentials in Cardiac Muscle

1. Action potentials in cardiac muscle are prolonged compared to those in skeletal muscle and have a depolarization phase, a plateau phase, and a repolarization phase.
2. The depolarization is due mainly to opening of the voltage-gated Na^+ channels, and the plateau phase is due to opened voltage-gated Ca^{2+} channels. Repolarization at the end of the plateau phase is due to the opening of K^+ channels for a brief period.
3. The prolonged action potential in cardiac muscle ensures that contraction and relaxation occur and prevents tetany.
4. The SA node located in the upper wall of the right atrium is the normal pacemaker of the heart, and cells of the SA node have more voltage-gated Ca^{2+} channels than do other areas of the heart.

Conduction System of the Heart

1. The conduction system of the heart is made up of specialized cardiac muscle cells.
2. The SA node produces action potentials that are propagated over the atria to the AV node.
3. The AV node and the atrioventricular bundle conduct action potentials to the ventricles.
4. The right and left bundle branches conduct action potentials from the atrioventricular bundle through Purkinje fibers to the ventricular muscle.
5. An ectopic beat results from an action potential that originates in an area of the heart other than the SA node.

Electrocardiogram

1. An ECG is a record of electrical events within the heart.
2. An ECG can be used to detect abnormal heart rates or rhythms, abnormal conduction pathways, hypertrophy or atrophy of the heart, and the approximate location of damaged cardiac muscle.
3. A normal ECG consists of a P wave (atrial depolarization), a QRS complex (ventricular depolarization), and a T wave (ventricular repolarization).
4. Atrial contraction occurs during the PQ interval, and the ventricles contract and relax during the QT interval.

12.6 Cardiac Cycle   (p. 333)

1. Atrial systole is contraction of the atria, and ventricular systole is contraction of the ventricles. Atrial diastole is relaxation of the atria, and ventricular diastole is relaxation of the ventricles.
2. During atrial systole, the atria contract and complete filling of the ventricles.

3. During ventricular systole, the AV valves close, pressure increases in the ventricles, the semilunar valves are forced open, and blood flows into the aorta and pulmonary trunk.
4. At the beginning of ventricular diastole, pressure in the ventricles decreases. The semilunar valves close to prevent backflow of blood from the aorta and pulmonary trunk into the ventricles.
5. When the pressure in the ventricles is low enough, the AV valves open, and blood flows from the atria into the ventricles.

12.7 Heart Sounds   (p. 337)

1. The first heart sound results from closure of the AV valves. The second heart sound results from closure of the semilunar valves.
2. Abnormal heart sounds, called murmurs, can result from incompetent (leaky) valves or stenosed (narrowed) valves.

12.8 Regulation of Heart Function   (p. 338)

Cardiac output (volume of blood pumped per ventricle per minute) is equal to the stroke volume (volume of blood ejected per beat) times the heart rate (beats per minute).

Intrinsic Regulation of the Heart

1. *Intrinsic regulation* refers to regulation mechanisms contained within the heart.
2. As venous return to the heart increases, the heart wall is stretched, and the increased stretch of the ventricular walls is called preload.
3. A greater preload causes the cardiac output to increase because stroke volume increases (Starling's law of the heart).
4. Afterload is the pressure against which the ventricles must pump blood.

Extrinsic Regulation of the Heart

1. *Extrinsic regulation* refers to nervous and chemical mechanisms.
2. Sympathetic stimulation increases stroke volume and heart rate; parasympathetic stimulation decreases heart rate.
3. The baroreceptor reflex detects changes in blood pressure. If blood pressure increases suddenly, the reflex causes a decrease in heart rate and stroke volume; if blood pressure decreases suddenly, the reflex causes an increase in heart rate and stroke volume.
4. Emotions influence heart function by increasing sympathetic stimulation of the heart in response to exercise, excitement, anxiety, or anger and by increasing parasympathetic stimulation in response to depression.
5. Alterations in body fluid levels of CO_2, pH, and ion concentrations, as well as changes in body temperature, influence heart function.

12.9 Effects of Aging on the Heart   (p. 346)

The following age-related changes are common:
1. By age 70, cardiac output has often decreased by one-third.
2. Hypertrophy of the left ventricle can cause pulmonary edema.
3. Decrease in the maximum heart rate by 30–60% by age 85 leads to decreased cardiac output.
4. The aortic semilunar valve can become stenotic or incompetent.
5. Coronary artery disease and congestive heart failure can develop.
6. Aerobic exercise improves the functional capacity of the heart at all ages.

REVIEW AND COMPREHENSION

1. Describe the size and location of the heart, including its base and apex.

2. Describe the structure and function of the pericardium.

3. What chambers make up the left and right sides of the heart? What are their functions?

4. Describe the structure and location of the tricuspid, bicuspid, and semilunar valves. What is the function of these valves?

5. What are the functions of the atria and ventricles?

6. Starting in the right atrium, describe the flow of blood through the heart.

7. Describe the vessels that supply blood to the cardiac muscle.

8. Define heart attack and infarct. How does atherosclerotic plaque affect the heart?

9. Describe the three layers of the heart. Which of the three layers is most important in causing contractions of the heart?

10. Describe the structure of cardiac muscle cells, including the structure and function of intercalated disks.

11. Describe the events that result in an action potential in cardiac muscle.

12. Explain how cardiac muscle cells in the SA node produce action potentials spontaneously and why the SA node is the heart's pacemaker.

13. What is the function of the conduction system of the heart? Starting with the SA node, describe the route of an action potential as it goes through the conduction system of the heart.

14. Explain the electrical events that generate each portion of the electrocardiogram. How do they relate to contraction events?

15. What contraction and relaxation events occur during the PQ interval and the QT interval of the electrocardiogram?

16. Define cardiac cycle, systole, and diastole.

17. Describe blood flow and the opening and closing of heart valves during the cardiac cycle.

18. Describe the pressure changes that occur in the left atrium, left ventricle, and aorta during ventricular systole and diastole (see figure 12.18).

19. What events cause the first and second heart sounds?

20. Define murmur. Describe how either an incompetent or a stenosed valve can cause a murmur.

21. Define cardiac output, stroke volume, and heart rate.

22. What is Starling's law of the heart? What effect does an increase or a decrease in venous return have on cardiac output?

23. Describe the effect of parasympathetic and sympathetic stimulation on heart rate and stroke volume.

24. How does the nervous system detect and respond to the following?
    a. a decrease in blood pressure
    b. an increase in blood pressure

25. What is the effect of epinephrine on the heart rate and stroke volume?

26. Explain how emotions affect heart function.

27. What effects do the following have on cardiac output?
    a. a decrease in blood pH
    b. an increase in blood CO_2

28. How do changes in body temperature influence the heart rate?

29. List the common age-related heart diseases that develop in elderly people.

CRITICAL THINKING

1. A friend tells you that an ECG revealed that her son has a slight heart murmur. Should you be convinced that he has a heart murmur? Explain.

2. Predict the effect on Starling's law of the heart if the parasympathetic (vagus) nerves to the heart are cut.

3. Predict the effect on heart rate if the sensory nerve fibers from the baroreceptors are cut.

4. An experiment is performed on a dog in which the arterial blood pressure in the aorta is monitored before and after the common carotid arteries are clamped. Explain the change in arterial blood pressure that would occur. (*Hint:* Baroreceptors are located in the internal carotid arteries, which are superior to the site of clamping of the common carotid arteries.)

5. Predict the consequences on the heart if a person took a large dose of a drug that blocks calcium channels.

6. What happens to cardiac output following the ingestion of a large amount of fluid?

7. At rest, the cardiac output of athletes and nonathletes can be equal, but the heart rate of athletes is lower than that of nonathletes. At maximum exertion, the maximum heart rate of athletes and nonathletes can be equal, but the cardiac output of athletes is greater than that of nonathletes. Explain these differences.

8. Explain why it is useful that the walls of the ventricles are thicker than those of the atria.

9. Predict the effect of an incompetent aortic semilunar valve on ventricular and aortic pressure during ventricular systole and diastole.

Answers in Appendix D

aprevealed.com

13

Blood Vessels and Circulation

LEARN TO PREDICT

T. J. and Tyler were building a treehouse. While searching for a board in a pile of lumber, T. J. stepped on a rusty nail that penetrated deep into his foot, causing it to bleed. Neither T. J. nor Tyler wanted to tell their parents about the accident, but after three days, T. J.'s foot became infected and the infection spread into his bloodstream. The infection quickly became much worse and T. J. went into septic shock.

After reading this chapter and recalling information about the structure and function of the heart described in chapter 12, explain how T. J.'s blood volume, blood pressure, heart rate, and stroke volume changed due to septic shock. Also, explain how blood flow in the periphery changed and how it affected T. J.'s appearance. Finally, explain the consequences if T. J.'s blood pressure remained abnormally low for a prolonged period of time.

13.1 FUNCTIONS OF THE CIRCULATORY SYSTEM

Learning Outcome *After reading this section, you should be able to*

A. List the functions of the circulatory system.

The blood vessels of the body form a network more complex than an interstate highway system. The blood vessels carry blood to within two or three cell diameters of nearly all the trillions of cells that make up the body. Blood flow through them is regulated, so that cells receive adequate nutrients and so that waste products are removed. Blood vessels remain functional, in most cases, in excess of 70 years, and when they are damaged, they repair themselves.

Blood vessels outside the heart are divided into two classes: (1) the **pulmonary vessels,** which transport blood from the right ventricle of the heart through the lungs and back to the left atrium, and (2) the **systemic vessels,** which transport blood through all parts of the body, from the left ventricle of the heart and back to the right atrium (see chapter 12 and figure 12.2). Together, the pulmonary vessels and the systemic vessels constitute the **circulatory system.**

Module 9 Cardiovascular System

The heart provides the major force that causes blood to circulate, and the circulatory system has five functions:

1. *Carries blood.* Blood vessels carry blood from the heart to all the tissues of the body and back to the heart.
2. *Exchanges nutrients, waste products, and gases with tissues.* Nutrients and oxygen diffuse from blood vessels to cells in essentially all areas of the body. Waste products and carbon dioxide diffuse from the cells, where they are produced, to blood vessels.
3. *Transports substances.* Blood transports hormones, components of the immune system, molecules required for coagulation, enzymes, nutrients, gases, waste products, and other substances to all areas of the body.

4. *Helps regulate blood pressure.* The circulatory system and the heart work together to regulate blood pressure within a normal range.

5. *Directs blood flow to the tissues.* The circulatory system directs blood to tissues when increased blood flow is required to maintain homeostasis.

13.2 GENERAL FEATURES OF BLOOD VESSEL STRUCTURE

Learning Outcome   *After reading this section, you should be able to*

A. Describe the structure and function of arteries, capillaries, and veins.

The three main types of blood vessels are arteries, capillaries, and veins. **Arteries** (ar′ter-ēz) carry blood away from the heart; usually, the blood is oxygen-rich. Blood is pumped from the ventricles of the heart into large, elastic arteries, which branch repeatedly to form progressively smaller arteries. As they become smaller, the artery walls undergo a gradual transition from having more elastic tissue than smooth muscle to having more smooth muscle than elastic tissue (figure 13.1*a–c*). The arteries are normally classified as elastic arteries, muscular arteries, or arterioles, although they form a continuum from the largest to the smallest branches.

Blood flows from arterioles into **capillaries** (kap′i-lār-ēz), where exchange occurs between the blood and the tissue fluid. Capillaries have thinner walls than do arteries (figure 13.1*d*). Blood flows through them more slowly, and there are far more of them than of any other blood vessel type.

From the capillaries, blood flows into veins. **Veins** (vānz) carry blood toward the heart; usually, the blood is oxygen-poor. Compared to arteries, the walls of veins are thinner and contain less elastic tissue and fewer smooth muscle cells (figure 13.1*e–g*). Starting at capillaries and proceeding toward the heart, small-diameter veins come together to form larger-diameter veins, which are fewer in number. Veins increase in diameter and decrease in number as they progress toward the heart, and their walls increase in thickness. Veins may be classified as venules, small veins, medium-sized veins, or large veins.

Except in capillaries and venules, blood vessel walls consist of three layers, or **tunics** (too′niks). From the inner to the outer wall, the tunics are (1) the tunica intima, (2) the tunica media, and (3) the tunica adventitia, or tunica externa (figure 13.2; see figure 13.1).

The **tunica intima** (too′ni-kă in′ti-mă), or innermost layer, consists of an endothelium composed of simple squamous epithelial cells, a basement membrane, and a small amount of connective tissue. In muscular arteries, the tunica intima also contains a layer of thin elastic connective tissue. The **tunica media,** or middle layer, consists of smooth muscle cells arranged circularly around the blood vessel. It also contains variable amounts of elastic and collagen fibers, depending on the size and type of the vessel. In muscular arteries, a layer of elastic connective tissue forms the outer margin of the tunica media. The **tunica adventitia** (ad-ven-tish′ă) is composed of dense connective tissue adjacent to the tunica media; the tissue becomes loose connective tissue toward the outer portion of the blood vessel wall.

Arteries

Elastic arteries are the largest-diameter arteries and have the thickest walls (see figure 13.1*a*). Compared to other arteries, a greater proportion of their walls is composed of elastic tissue, and a smaller proportion is smooth muscle. The aorta and pulmonary trunk are examples of elastic arteries. Elastic arteries stretch when the ventricles of the heart pump blood into them. The elastic recoil of these arteries prevents blood pressure from falling rapidly and maintains blood flow while the ventricles are relaxed.

The **muscular arteries** include medium-sized and small arteries. The walls of medium-sized arteries are relatively thick compared to their diameter. Most of the wall's thickness results from smooth muscle cells of the tunica media (see figure 13.1*b*). Medium-sized arteries are frequently called **distributing arteries** because the smooth muscle tissue enables these vessels to control blood flow to different body regions. Contraction of the smooth muscle in blood vessels, called **vasoconstriction** (vā′sō-kon-strik′shŭn), decreases blood vessel diameter and blood flow. Relaxation of the smooth muscle in blood vessels, called **vasodilation** (vā′sō-dī-lā′shŭn), increases blood vessel diameter and blood flow.

Medium-sized arteries supply blood to small arteries. Small arteries have about the same structure as the medium-sized arteries, except for a smaller diameter and thinner walls. The smallest of the small arteries have only three or four layers of smooth muscle in their walls.

Arterioles (ar-tēr′ē-ōlz) transport blood from small arteries to capillaries. Arterioles (see figure 13.1*c*) are the smallest arteries in which the three tunics can be identified; the tunica media consists of only one or two layers of circular smooth muscle cells. Small arteries and arterioles are adapted for vasodilation and vasoconstriction.

Capillaries

Blood flows from arterioles into capillaries, which branch to form networks (figure 13.3; see figure 13.1*d*). Blood flow through capillaries is regulated by smooth muscle cells called **precapillary sphincters** located at the origin of the branches. Capillary walls consist of **endothelium** (en-dō-thē′lē-ŭm), which is a layer of simple squamous epithelium surrounded by delicate loose connective tissue. The thin walls of capillaries facilitate diffusion between the capillaries and surrounding cells. Each capillary is 0.5–1 millimeter (mm) long. Capillaries branch without changing their diameter, which is approximately the same as the diameter of a red blood cell (7.5 μm).

Red blood cells flow through most capillaries in single file and are frequently folded as they pass through the smaller-diameter capillaries. As blood flows through capillaries, blood gives up O_2 and nutrients to the tissue spaces and takes up CO_2 and other by-products of metabolism. Capillary networks are more numerous and more extensive in the lungs and in highly metabolic tissues, such as the liver, kidneys, skeletal muscle, and cardiac muscle, than in other tissue types.

Veins

Blood flows from capillaries into venules and from venules into small veins. **Venules** (ven′oolz) have a diameter slightly larger than that of capillaries and are composed of endothelium resting on a delicate connective tissue layer (see figure 13.1*e*). The

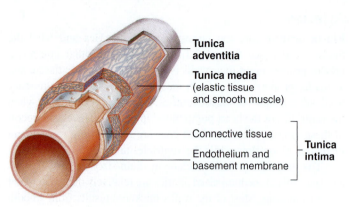

(a) **Elastic arteries.** The tunica media is mostly elastic connective tissue. Elastic arteries recoil when stretched, which prevents blood pressure from falling rapidly.

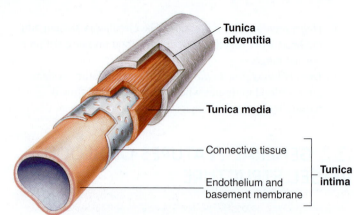

(g) **Large veins.** All three tunics are present. The tunica media is thin but can regulate vessel diameter because blood pressure in the venous system is low. The predominant layer is the tunica adventitia.

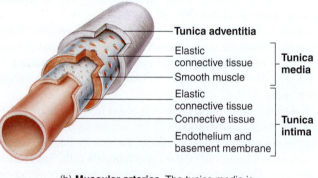

(b) **Muscular arteries.** The tunica media is a thick layer of smooth muscle. Muscular arteries regulate blood flow to different regions of the body.

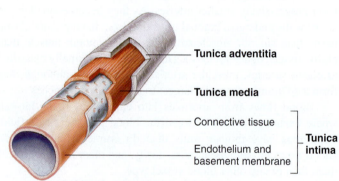

(f) **Small and medium veins.** All three tunics are present.

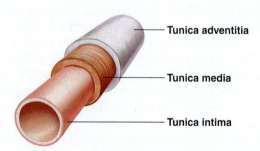

(c) **Arterioles.** All three tunics are present; the tunica media consists of only one or two layers of circular smooth muscle cells.

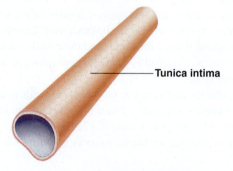

(e) **Venules.** Only the tunica intima resting on a delicate layer of dense connective tissue is present.

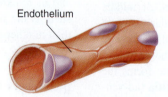

(d) **Capillaries.** Walls consist of only a simple endothelium surrounded by delicate loose connective tissue.

Figure 13.1 AP|R **Blood Vessel Structure**

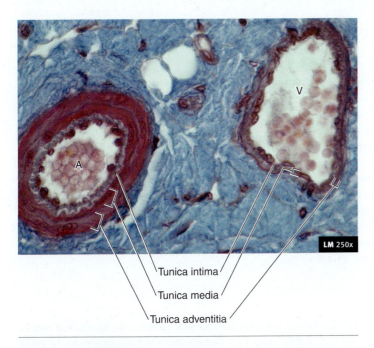

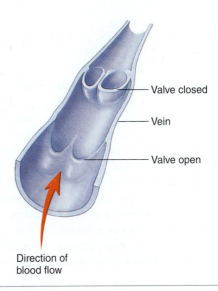

Figure 13.4 **Valves**

Folds in the tunica intima form the valves of veins, which allow blood to flow toward the heart but not in the opposite direction.

Figure 13.2 AP|R **Photomicrograph of an Artery and a Vein**

The typical structure of a medium-sized artery (A) and vein (V). Note that the artery has a thicker wall than the vein. The predominant layer in the wall of the artery is the tunica media with its circular layers of smooth muscle. The predominant layer in the wall of the vein is the tunica adventitia, and the tunica media is thinner.

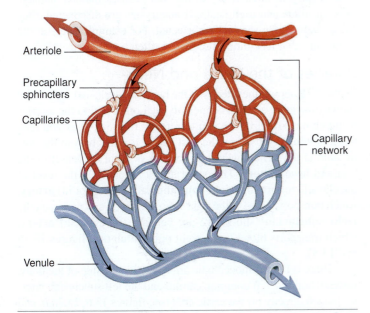

Figure 13.3 **Capillary Network**

A capillary network stems from an arteriole. The network forms numerous branches. Blood flows through capillaries into venules. Smooth muscle cells, called precapillary sphincters, regulate blood flow through the capillaries. Blood flow decreases when the precapillary sphincters constrict and increases when they dilate.

media contains a continuous layer of smooth muscle cells, and the connective tissue of the tunica adventitia surrounds the tunica media (see figure 13.1*f*).

Medium-sized veins collect blood from small veins and deliver it to large veins. The three thin but distinctive tunics make up the wall of the medium-sized and large veins. The tunica media contains some circular smooth muscle and sparsely scattered elastic fibers. The predominant layer is the outer tunica adventitia, which consists primarily of dense collagen fibers (see figures 13.1*g* and 13.2). Consequently, veins are more distensible than arteries. The connective tissue of the tunica adventitia determines the degree to which they can distend.

Veins having diameters greater than 2 mm contain **valves,** which allow blood to flow toward the heart but not in the opposite direction (figure 13.4). Each valve consists of folds in the tunica intima that form two flaps, which are similar in shape and function to the semilunar valves of the heart. There are many valves in medium-sized veins and more valves in veins of the lower limbs than in veins of the upper limbs. This prevents blood from flowing toward the feet in response to the pull of gravity.

13.3 BLOOD VESSELS OF THE PULMONARY CIRCULATION

Learning Outcome   *After reading this section, you should be able to*

A. Describe the blood vessels of the pulmonary circulation.

The **pulmonary circulation** is the system of blood vessels that carries blood from the right ventricle of the heart to the lungs and back to the left atrium of the heart. Blood from the right ventricle is pumped into a short vessel called the **pulmonary** (pŭl′mō-nār-ē) **trunk** (figure 13.5). The pulmonary trunk then branches into the **right** and **left pulmonary arteries,** which extend to the right and left lungs, respectively. These arteries carry oxygen-poor blood to

structure of venules, except for their diameter, is very similar to that of capillaries. **Small veins** are slightly larger in diameter than venules. All three tunics are present in small veins. The tunica

CLINICAL IMPACT   Varicose Veins

Varicose (văr′ĭ-kōs) **veins** result when the veins of the lower limbs become so dilated that the cusps of the valves no longer overlap to prevent the backflow of blood. As a consequence, venous pressure is greater than normal in the veins of the lower limbs and can result in edema (swelling) of the limb. Some people have a genetic tendency to develop varicose veins. The condition is further encouraged by activities that increase the pressure in veins. For example, stand-ing in place for prolonged periods allows the pressure of the blood to stretch the veins, and the enlarged uterus in preg-nancy compresses the veins in the pelvis, increasing pressure in the veins that drain the lower limbs.

The blood in varicose veins can become so stagnant that it clots. The clots, called **thromboses** (throm-bō′sēz), can cause inflammation of the veins, a condition called **phlebitis** (fle-bī′tis). If the condition becomes severe enough, the blocked veins can prevent blood from flowing through the capillaries that are drained by the veins. The lack of blood flow can lead to tissue death and infection with anaerobic bacteria, a condition called **gangrene** (gang′grēn). In addition, frag-ments of the clots may dislodge and travel through the veins to the lungs, where they can cause severe damage. Fragments of thromboses that dislodge and float in the blood are called **emboli** (em′bō-lī).

the pulmonary capillaries in the lungs, where the blood takes up O_2 and releases CO_2. Blood rich in O_2 flows from the lungs to the left atrium. Four **pulmonary veins** (two from each lung) exit the lungs and carry the oxygen-rich blood to the left atrium. Figure 13.5 traces the path of the pulmonary circulation and shows it in rela-tion to the systemic circulation, described next.

13.4 BLOOD VESSELS OF THE SYSTEMIC CIRCULATION: ARTERIES

Learning Outcome   *After reading this section, you should be able to*

A. List the major arteries that supply each of the body areas, and describe their functions.

The **systemic circulation** is the system of blood vessels that carries blood from the left ventricle of the heart to the tissues of the body and back to the right atrium. Oxygen-rich blood from the pulmo-nary veins passes from the left atrium into the left ventricle and from the left ventricle into the aorta. Arteries distribute blood from the aorta to all portions of the body (figure 13.6).

Aorta

All arteries of the systemic circulation branch directly or indi-rectly from the **aorta** (ā-ōr′tă). The aorta is usually considered in three parts—the ascending aorta, the aortic arch, and the descend-ing aorta; the last segment is further divided into the thoracic aorta and the abdominal aorta (figure 13.7a).

The **ascending aorta** is the part of the aorta that passes supe-riorly from the left ventricle. The right and left **coronary arteries** arise from the base of the ascending aorta and supply blood to the heart (see chapter 12).

The aorta arches posteriorly and to the left as the **aortic arch.** Three major arteries, which carry blood to the head and upper limbs, originate from the aortic arch: the brachiocephalic artery, the left common carotid artery, and the left subclavian artery (figure 13.7b).

The **descending aorta** is the longest part of the aorta. It extends through the thorax and abdomen to the upper margin of the pelvis. The part of the descending aorta that extends through the thorax to the diaphragm is called the **thoracic** (thō-ras′ik) **aorta.** The part of the descending aorta that extends from the diaphragm to the point at which it divides into the two **common iliac** (il′ē-ak) **arteries** is called the **abdominal** (ab-dom′i-năl) **aorta** (figure 13.7a,c).

An **arterial aneurysm** (an′ū-rizm; a dilation) is a localized dilation of an artery that usually develops in response to trauma or a congenital (existing at birth) weakness of the artery wall. Rupture of a large aneurysm in the aorta is almost always fatal, and rupture of an aneurysm in an artery of the brain causes massive damage to brain tissue and even death. If aneurysms are discovered, they can sometimes be surgically corrected. For example, large aortic aneurysms that leak blood slowly can often be repaired.

Arteries of the Head and Neck

Figure 13.8 depicts the arteries of the head and neck, and figure 13.9 shows how these arteries connect to each other and to the arteries of the thorax. It may help you to refer to figure 13.9 as you read the following discussion. The first vessel to branch from the aor-tic arch is the **brachiocephalic** (brā′kē-ō-se-fal′ik; vessel to the arm and head) **artery.** This short artery branches at the level of the clavicle to form the **right common carotid** (ka-rot′id) **artery,** which transports blood to the right side of the head and neck, and the **right subclavian** (sŭb-klā′vē-an; beneath the clavicle) **artery,** which transports blood to the right upper limb (see figures 13.7b and 13.8).

There is no brachiocephalic artery on the left side of the body. Instead, both the left common carotid and the left subclavian arter-ies branch directly off the aortic arch (see figures 13.6, 13.7a,b, and 13.9). They are the second and third branches of the aortic arch. The **left common carotid artery** transports blood to the left side of the head and neck, and the **left subclavian artery** transports blood to the left upper limb.

The common carotid arteries extend superiorly along each side of the neck to the angle of the mandible, where they branch into **internal** and **external carotid arteries.** The base of each internal carotid artery is slightly dilated to form a **carotid sinus,** which contains structures important in monitoring blood pres-sure (baroreceptors). The external carotid arteries have several branches that supply the structures of the neck, face, nose, and

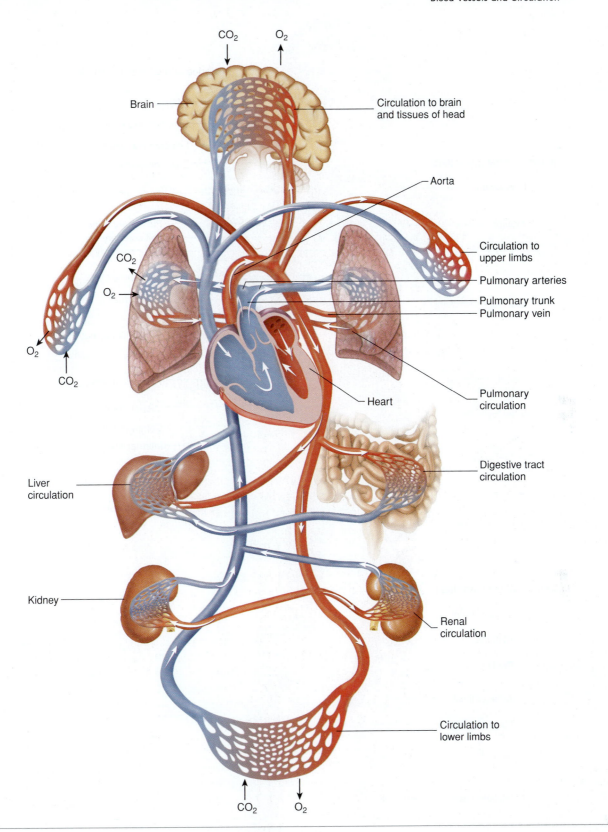

Figure 13.5 AP|R Blood Flow Through the Circulatory System

Veins and venules return blood from the body to the right atrium. After passing from the right atrium to the right ventricle, blood is pumped into the pulmonary trunk. The pulmonary trunk divides into the right and left pulmonary arteries, which carry oxygen-poor blood to the lungs. In the lung capillaries, CO_2 is given off, and the blood picks up O_2. Blood, now rich in O_2, flows from each lung to the left atrium. Blood passes from the left atrium to the left ventricle. The left ventricle then pumps the blood into the aorta, which distributes the blood through its branches to all of the body. Blood returns to the heart through the venous system, and the cycle continues.

Cardiovascular

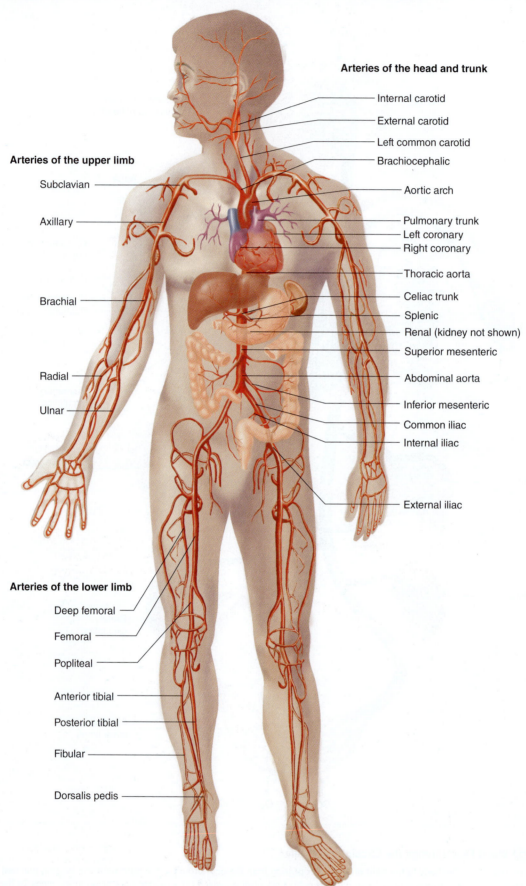

Arteries of the head and trunk

Internal carotid

External carotid

Left common carotid

Brachiocephalic

Aortic arch

Pulmonary trunk

Left coronary

Right coronary

Thoracic aorta

Celiac trunk

Splenic

Renal (kidney not shown)

Superior mesenteric

Abdominal aorta

Inferior mesenteric

Common iliac

Internal iliac

External iliac

Arteries of the upper limb

Subclavian

Axillary

Brachial

Radial

Ulnar

Arteries of the lower limb

Deep femoral

Femoral

Popliteal

Anterior tibial

Posterior tibial

Fibular

Dorsalis pedis

Figure I3.6 AP|R **Major Arteries**

These major arteries carry blood from the left ventricle of the heart to the body tissues.

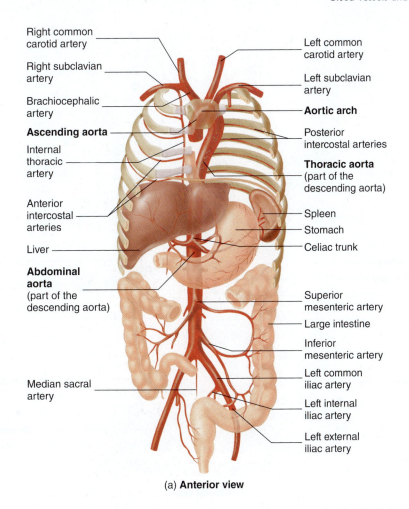

Right common carotid artery

Right subclavian artery

Brachiocephalic artery

Ascending aorta

Internal thoracic artery

Anterior intercostal arteries

Liver

Abdominal aorta (part of the descending aorta)

Median sacral artery

Left common carotid artery

Left subclavian artery

Aortic arch

Posterior intercostal arteries

Thoracic aorta (part of the descending aorta)

Spleen

Stomach

Celiac trunk

Superior mesenteric artery

Large intestine

Inferior mesenteric artery

Left common iliac artery

Left internal iliac artery

Left external iliac artery

(a) **Anterior view**

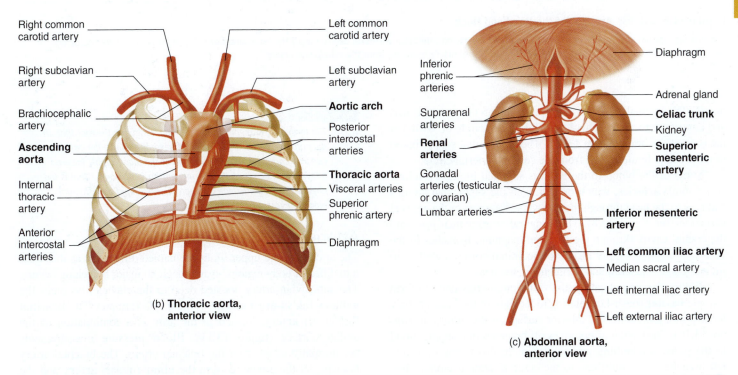

Right common carotid artery

Right subclavian artery

Brachiocephalic artery

Ascending aorta

Internal thoracic artery

Anterior intercostal arteries

Left common carotid artery

Left subclavian artery

Aortic arch

Posterior intercostal arteries

Thoracic aorta

Visceral arteries

Superior phrenic artery

Diaphragm

(b) **Thoracic aorta, anterior view**

Inferior phrenic arteries

Suprarenal arteries

Renal arteries

Gonadal arteries (testicular or ovarian)

Lumbar arteries

Diaphragm

Adrenal gland

Celiac trunk

Kidney

Superior mesenteric artery

Inferior mesenteric artery

Left common iliac artery

Median sacral artery

Left internal iliac artery

Left external iliac artery

(c) **Abdominal aorta, anterior view**

Figure 13.7 AP|R Branches of the Aorta

(*a*) The aorta is considered in three portions: the ascending aorta, the aortic arch, and the descending aorta. The descending aorta consists of the thoracic and abdominal aortas. (*b*) The thoracic aorta. (*c*) The abdominal aorta.

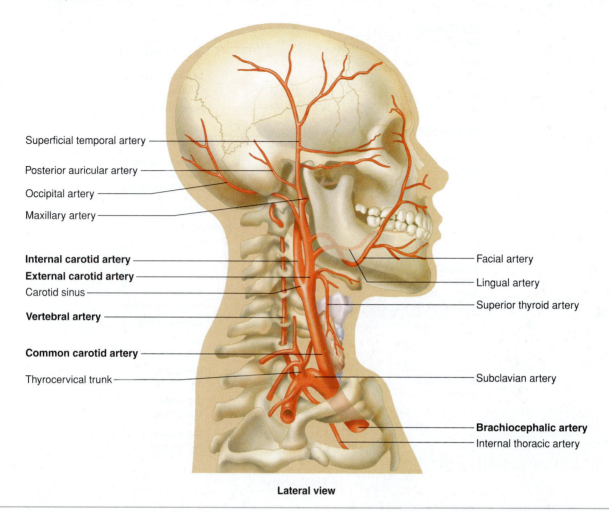

Superficial temporal artery

Posterior auricular artery

Occipital artery

Maxillary artery

Internal carotid artery

External carotid artery

Carotid sinus

Vertebral artery

Common carotid artery

Thyrocervical trunk

Facial artery

Lingual artery

Superior thyroid artery

Subclavian artery

Brachiocephalic artery

Internal thoracic artery

Lateral view

Figure 13.8 AP|R **Arteries of the Head and Neck**

The brachiocephalic artery, the right common carotid artery, and the right vertebral artery supply the head and neck. The right common carotid artery branches from the brachiocephalic artery, and the vertebral artery branches from the subclavian artery.

mouth. The internal carotid arteries pass through the carotid canals and contribute to the **cerebral arterial circle** (circle of Willis) at the base of the brain (figure 13.10). The vessels that supply blood to most of the brain branch from the cerebral arterial circle.

Some of the blood to the brain is supplied by the **vertebral** (ver′tĕ-brăl) **arteries,** which branch from the subclavian arteries and pass to the head through the transverse foramina of the cervical vertebrae (see figure 13.8). The vertebral arteries then pass into the cranial cavity through the foramen magnum. Branches of the vertebral arteries supply blood to the spinal cord, as well as to the vertebrae, muscles, and ligaments in the neck.

Within the cranial cavity, the vertebral arteries unite to form a single **basilar** (bas′i-lăr; relating to the base of the brain) **artery** located along the anterior, inferior surface of the brainstem (figure 13.10). The basilar artery gives off branches that supply blood to the pons, cerebellum, and midbrain. It also forms right and left branches that contribute to the cerebral arterial circle. Most of the blood supply to the brain is through the internal carotid arteries; however, not enough blood is supplied to the brain to maintain life if either the vertebral arteries or the carotid arteries are blocked.

Predict 2

The term carotid *means to put to sleep, reflecting the fact that, if the carotid arteries are blocked for several seconds, the patient can lose consciousness. Interruption of the blood supply for even a few minutes can result in permanent brain damage. What is the physiological significance of atherosclerosis (lipid deposits that block the vessels) in the carotid arteries?*

Arteries of the Upper Limbs

The arteries of the upper limbs are named differently as they pass into different body regions, even though no major branching occurs. The subclavian artery, located deep to the clavicle, becomes the **axillary** (ak′sil-ār-ē) **artery** in the axilla (armpit). The **brachial** (brā′kē-ăl) **artery,** located in the arm, is a continuation of the axillary artery (figure 13.11). Blood pressure measurements are normally taken from the brachial artery. The brachial artery branches at the elbow to form the **ulnar** (ul′năr) **artery** and the **radial** (rā′dē-ăl) **artery,** which supply blood to the forearm and hand. The radial artery is the one most commonly used for taking a pulse. The pulse can be detected easily on the thumb (radial) side of the anterior surface of the wrist.

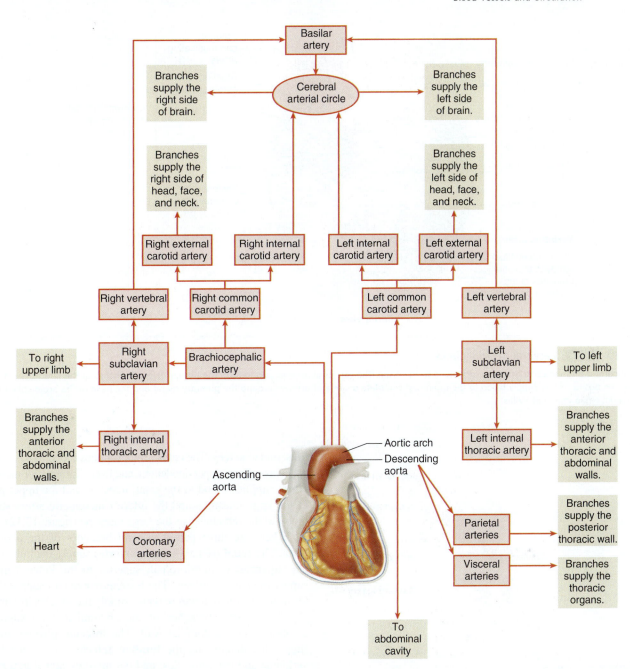

Figure 13.9 Major Arteries of the Head and Thorax

Thoracic Aorta and Its Branches

The branches of the thoracic aorta can be divided into two groups: The **visceral** (vis′er-ăl) **arteries** supply the thoracic organs, and the **parietal** (pă-rī′ĕ-tăl) **arteries** supply the thoracic wall. The visceral branches of the thoracic aorta supply the esophagus, the trachea, the parietal pericardium, and part of the lung. The major parietal arteries are the **posterior intercostal** (in-ter-kos′tăl) **arteries,** which arise from the thoracic aorta and extend between the ribs (see figure 13.7a,b). They supply the intercostal muscles, the vertebrae, the spinal cord, and the deep muscles of the back. The **superior phrenic** (fren′ik; diaphragm) **arteries** supply the diaphragm.

The **internal thoracic arteries** are branches of the subclavian arteries. They descend along the internal surface of the anterior thoracic wall and give rise to branches called the **anterior intercostal arteries,** which extend between the ribs to supply the anterior chest wall (see figure 13.7a,b).

Abdominal Aorta and Its Branches

The branches of the abdominal aorta, like those of the thoracic aorta, can be divided into visceral and parietal groups. The visceral arteries are divided into paired and unpaired branches. There are three major unpaired branches: the **celiac** (sē′lē-ak; belly) **trunk,** the **superior mesenteric** (mez-en-ter′ik) **artery,** and the **inferior**

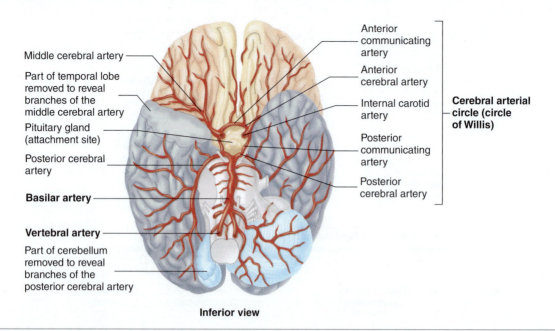

Inferior view

Figure 13.10 AP|R Cerebral Arterial Circle

The internal carotid and vertebral arteries carry blood to the brain. The vertebral arteries join to form the basilar artery. Branches of the internal carotid arteries and the basilar artery supply blood to the brain and complete a circle of arteries around the pituitary gland and the base of the brain called the cerebral arterial circle (circle of Willis).

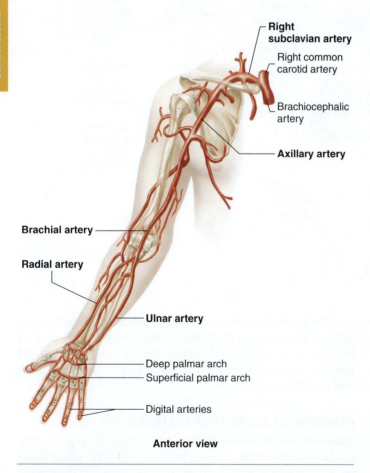

Anterior view

Figure 13.11 AP|R Arteries of the Upper Limb

Anterior view of the arteries of the right upper limb and their branches.

mesenteric artery. The celiac trunk supplies blood to the stomach, pancreas, spleen, upper duodenum, and liver. The superior mesenteric artery supplies blood to the small intestine and the upper portion of the large intestine, and the inferior mesenteric artery supplies blood to the remainder of the large intestine (figure 13.12).

There are three paired visceral branches of the abdominal aorta. The **renal** (rē′nal; kidney) **arteries** supply the kidneys, and the **suprarenal** (sū′pră-rē′năl; superior to the kidney) **arteries** supply the adrenal glands. The **testicular arteries** supply the testes in males; the **ovarian arteries** supply the ovaries in females.

The parietal branches of the abdominal aorta supply the diaphragm and abdominal wall. The **inferior phrenic arteries** supply the diaphragm; the **lumbar arteries** supply the lumbar vertebrae and back muscles; and the **median sacral artery** supplies the inferior vertebrae.

Arteries of the Pelvis

The abdominal aorta divides at the level of the fifth lumbar vertebra into two **common iliac arteries.** Each common iliac artery divides to form an **external iliac artery,** which enters a lower limb, and an **internal iliac artery,** which supplies the pelvic area (see figure 13.7c). Visceral branches of the internal iliac artery supply organs such as the urinary bladder, rectum, uterus, and vagina. Parietal branches supply blood to the walls and floor of the pelvis; the lumbar, gluteal, and proximal thigh muscles; and the external genitalia.

Arteries of the Lower Limbs

Like the arteries of the upper limbs, the arteries of the lower limbs are named differently as they pass into different body regions, even though there are no major branches. The external iliac artery in the

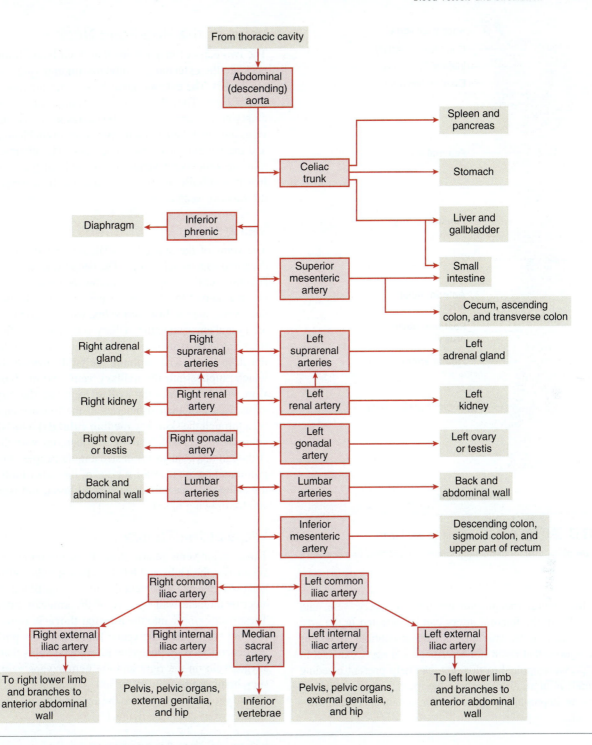

Figure 13.12 Major Arteries of the Abdomen and Pelvis
Visceral branches include those that are unpaired (celiac trunk, superior mesenteric, inferior mesenteric) and those that are paired (renal, suprarenal, testicular or ovarian). Parietal branches include inferior phrenic, lumbar, and median sacral.

pelvis becomes the **femoral** (fem′ŏ-răl) **artery** in the thigh, and it becomes the **popliteal** (pop-lit′ē-ăl) **artery** in the popliteal space, which is the posterior region of the knee. The popliteal artery branches slightly inferior to the knee to give off the **anterior tibial artery** and the **posterior tibial artery,** both of which give rise to arteries that supply blood to the leg and foot (figure 13.13). The anterior tibial artery becomes the **dorsalis pedis** (dōr-sāl′lis pē′dis;

pes, foot) **artery** at the ankle. The posterior tibial artery gives rise to the **fibular artery,** or *peroneal artery,* which supplies the lateral leg and foot.

The femoral triangle is located in the superior and medial area of the thigh. Its margins are formed by the inguinal ligament, the medial margin of the sartorius muscle, and the lateral margin of the adductor longus muscle (see figures 7.14a and 7.26a). Passing

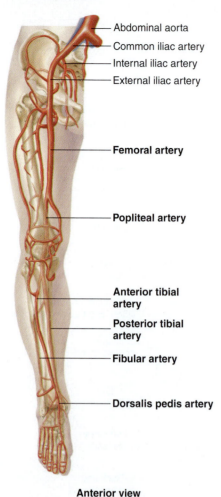

Abdominal aorta
Common iliac artery
Internal iliac artery
External iliac artery

Femoral artery

Popliteal artery

Anterior tibial artery

Posterior tibial artery

Fibular artery

Dorsalis pedis artery

Anterior view

Figure 13.13 AP|R **Arteries of the Lower Limb**
Anterior view of the arteries of the right lower limb and their branches.

through the femoral triangle are the femoral artery, vein, and nerve. A pulse in the femoral artery can be detected in the area of the femoral triangle. This area is also susceptible to serious traumatic injuries that result in hemorrhage and nerve damage. In addition, pressure applied to this area can help prevent bleeding from wounds in more distal areas of the lower limb. The femoral triangle is an important access point for certain medical procedures as well.

13.5 BLOOD VESSELS OF THE SYSTEMIC CIRCULATION: VEINS

Learning Outcome *After reading this section, you should be able to*

A. List the major veins that carry blood from each of the body areas, and describe their functions.

The oxygen-poor blood from the tissues of the body returns to the heart through veins. The **superior vena cava** (vē′nă kā′vă) returns blood from the head, neck, thorax, and upper limbs to the right atrium of the heart, and the **inferior vena cava** returns blood from the abdomen, pelvis, and lower limbs to the right atrium (figure 13.14).

Veins of the Head and Neck

The two pairs of major veins that drain blood from the head and neck are the **external** and **internal jugular** (jŭg′ū-lar) **veins** (figure 13.15). The external jugular veins are the more superficial of the two sets. They drain blood from the posterior head and neck, emptying primarily into the subclavian veins. The internal jugular veins are much larger and deeper. They drain blood from the brain and the anterior head, face, and neck. The internal jugular veins join the **subclavian veins** on each side of the body to form the **brachiocephalic veins.** The brachiocephalic veins join to form the superior vena cava.

Veins of the Upper Limbs

The veins of the upper limbs (figure 13.16) can be divided into deep and superficial groups. The deep veins, which drain the deep structures of the upper limbs, follow the same course as the arteries and are named for their respective arteries. The only noteworthy deep veins are the **brachial veins,** which accompany the brachial artery and empty into the axillary vein. The superficial veins drain the superficial structures of the upper limbs and then empty into the deep veins. The **cephalic** (sĕ′fal′ik; toward the head) **vein,** which empties into the **axillary vein,** and the **basilic** (ba-sil′ik) **vein,** which becomes the axillary vein, are the major superficial veins. Many of their tributaries in the forearm and hand can be seen through the skin. The **median cubital** (kū′bi-tal) **vein** usually connects the cephalic vein or its tributaries with the basilic vein. Although this vein varies in size among people, it is usually quite prominent on the anterior surface of the upper limb at the level of the elbow, an area called the **cubital fossa,** and is often used as a site for drawing blood.

Veins of the Thorax

Three major veins return blood from the thorax to the superior vena cava: the **right** and **left brachiocephalic veins** and the **azygos** (az-ĭ′gos, az′i-gos) **vein** (figure 13.17). Blood drains from the anterior thoracic wall by way of the **anterior intercostal veins.** These veins empty into the **internal thoracic veins,** which empty into the brachiocephalic veins. Blood from the posterior thoracic wall is collected by **posterior intercostal veins** that drain into the azygos vein on the right and the **hemiazygos vein** or the **accessory hemiazygos vein** on the left. The hemiazygos and accessory hemiazygos veins empty into the azygos vein, which drains into the superior vena cava (see figure 13.15).

Veins of the Abdomen and Pelvis

Figure 13.18 provides an overview of the major veins of the abdomen and pelvis. It may help to refer to this figure as you read the following discussion. Blood from the posterior abdominal wall drains through **ascending lumbar veins** into the azygos vein. Blood from the rest of the abdomen and from the pelvis and lower limbs returns to the heart through the inferior vena cava. The gonads (testes or ovaries), kidneys, adrenal glands, and liver are the only abdominal organs outside the pelvis from which blood drains directly into the inferior vena cava. The **internal iliac veins** drain the pelvis and join the **external iliac veins** from the lower limbs to form the **common iliac veins.** The common iliac veins combine to form the inferior vena cava (see figure 13.14).

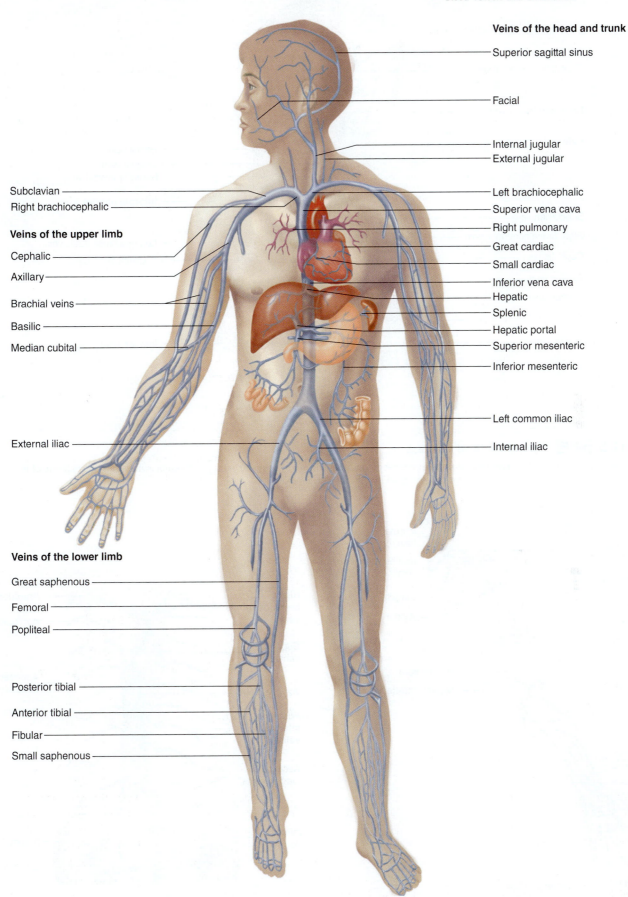

Veins of the head and trunk

Superior sagittal sinus

Facial

Internal jugular
External jugular

Left brachiocephalic
Superior vena cava
Right pulmonary
Great cardiac
Small cardiac
Inferior vena cava
Hepatic
Splenic
Hepatic portal
Superior mesenteric
Inferior mesenteric

Left common iliac

Internal iliac

Subclavian
Right brachiocephalic

Veins of the upper limb

Cephalic

Axillary

Brachial veins

Basilic

Median cubital

External iliac

Veins of the lower limb

Great saphenous

Femoral

Popliteal

Posterior tibial

Anterior tibial

Fibular

Small saphenous

Anterior view

Cardiovascular

Figure I3.I4 Major Veins

These major veins return blood from the body tissues to the right atrium.

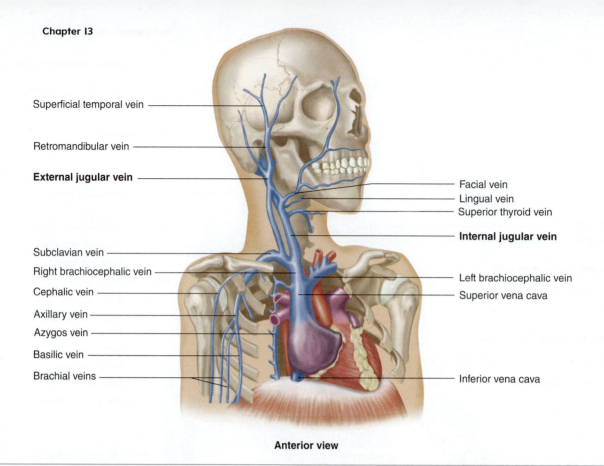

Superficial temporal vein

Retromandibular vein

External jugular vein

Facial vein
Lingual vein
Superior thyroid vein
Internal jugular vein

Subclavian vein
Right brachiocephalic vein
Cephalic vein
Axillary vein
Azygos vein
Basilic vein
Brachial veins

Left brachiocephalic vein
Superior vena cava

Inferior vena cava

Anterior view

Figure 13.15 AP|R Veins of the Head and Neck

The external and internal jugular veins drain blood from the head and neck. The internal jugular veins join the subclavian veins on each side of the body to form the brachiocephalic veins. The external jugular veins drain into the subclavian veins.

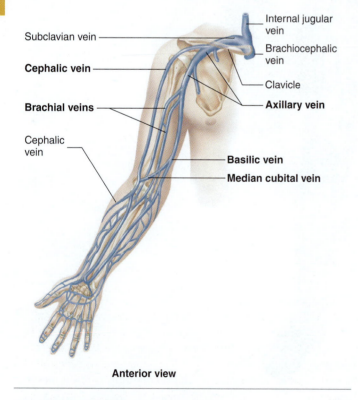

Subclavian vein
Internal jugular vein
Brachiocephalic vein
Cephalic vein
Clavicle
Brachial veins
Axillary vein
Cephalic vein
Basilic vein
Median cubital vein

Anterior view

Figure 13.16 AP|R Veins of the Upper Limb

Anterior view of the major veins of the right upper limb and their branches.

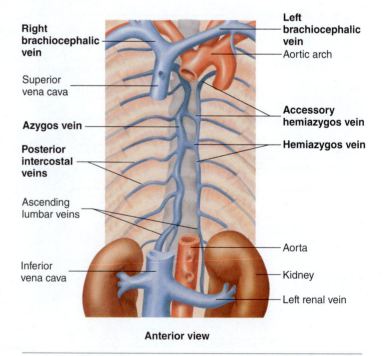

Right brachiocephalic vein
Left brachiocephalic vein
Aortic arch
Superior vena cava
Accessory hemiazygos vein
Azygos vein
Hemiazygos vein
Posterior intercostal veins
Ascending lumbar veins
Aorta
Kidney
Inferior vena cava
Left renal vein

Anterior view

Figure 13.17 AP|R Veins of the Thorax

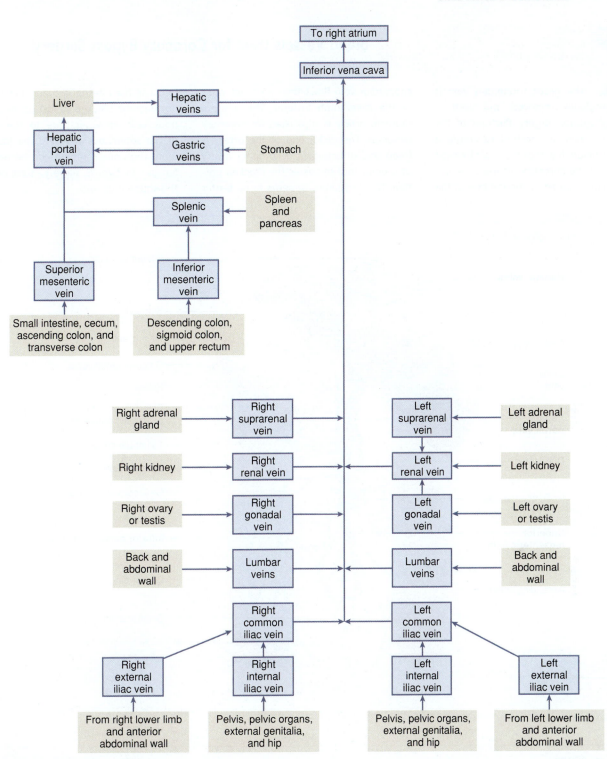

Figure 13.18 Major Veins of the Abdomen and Pelvis

Blood from the capillaries within most of the abdominal viscera, such as the stomach, intestines, pancreas, and spleen, drains through a specialized system of blood vessels to the liver. The liver is a major processing center for substances absorbed by the intestinal tract. A **portal** (pŏr′tăl) **system** is a vascular system that begins and ends with capillary beds and has no pumping mechanism, such as the heart, in between. The **hepatic** (he-pa′tik) **portal system** (figure 13.19) begins with capillaries

in the viscera and ends with capillaries in the liver. The major tributaries of the hepatic portal system are the **splenic** (splen′ik) **vein** and the **superior mesenteric vein.** The **inferior mesenteric vein** empties into the splenic vein. The splenic vein carries blood from the spleen and pancreas. The superior and inferior mesenteric veins carry blood from the intestines. The splenic vein and the superior mesenteric vein join to form the **hepatic portal vein,** which enters the liver.

CLINICAL IMPACT   Blood Vessels Used for Coronary Bypass Surgery

The great saphenous vein is often surgically removed and used in coronary bypass surgery. Portions of the saphenous vein are grafted to create a route for blood flow that bypasses blocked portions of the coronary arteries. The circulation interrupted by the removal of the saphenous vein flows through other veins of the lower limb instead. The internal thoracic artery is also used for coronary bypasses. The distal end of the artery is freed and attached to a coronary artery at a point that bypasses the blocked portion. This technique appears to be better because the internal thoracic artery does not become blocked as fast as the saphenous vein. However, due to the length of the saphenous vein and the fact that more than one segment can be removed for use in bypass surgery, surgeons use the saphenous vein.

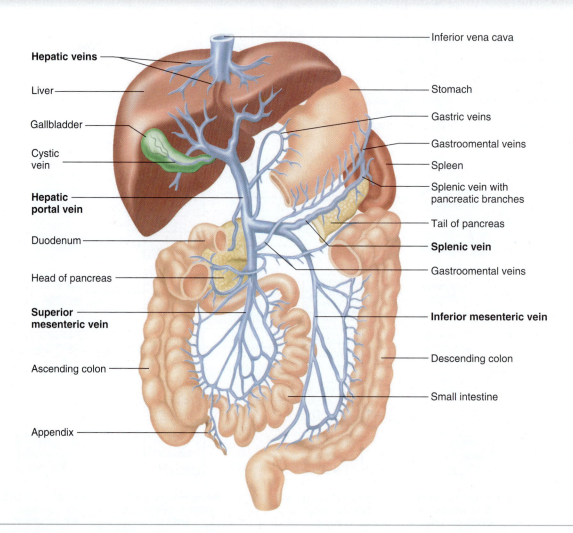

Figure 13.19 AP|R Veins of the Hepatic Portal System

The hepatic portal system begins as capillary beds in the stomach, pancreas, spleen, small intestine, and large intestine. The veins of the hepatic portal system converge on the hepatic portal vein, which carries blood to a series of capillaries in the liver. Hepatic veins carry blood from capillaries in the liver to the hepatic veins, which then enter the inferior vena cava.

Blood from the liver flows into **hepatic veins,** which join the inferior vena cava. Blood entering the liver through the hepatic portal vein is rich in nutrients collected from the intestines, but it may also contain a number of toxic substances that are potentially harmful to body tissues. Within the liver, nutrients are taken up and stored or modified, so that they can be used by other cells of the body. Also within the liver, toxic substances are converted to nontoxic substances. These substances can be removed from the blood or carried by the blood to the kidneys for excretion. The liver is discussed more fully in chapter 16.

Other veins of the abdomen and pelvis include the renal veins, the suprarenal veins, and the gonadal veins. The **renal veins** drain

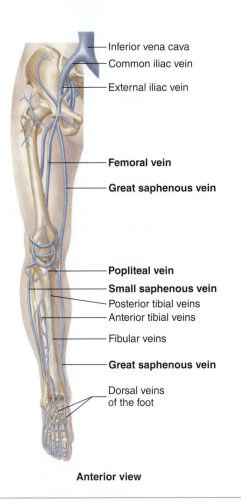

- Inferior vena cava
- Common iliac vein
- External iliac vein
- **Femoral vein**
- **Great saphenous vein**
- **Popliteal vein**
- **Small saphenous vein**
- Posterior tibial veins
- Anterior tibial veins
- Fibular veins
- **Great saphenous vein**
- Dorsal veins of the foot

Anterior view

Figure 13.20 AP|R **Veins of the Lower Limb**

Anterior view of the major veins of the right lower limb and their branches.

the kidneys, and the **suprarenal veins** drain the adrenal glands. The **testicular veins** drain the testes in males; the **ovarian veins** drain the ovaries in females.

Veins of the Lower Limbs

The veins of the lower limbs (figure 13.20), like those of the upper limbs, consist of deep and superficial groups. The deep veins follow the same path as the arteries and are named for the arteries they accompany. The superficial veins consist of the great and small **saphenous** (să-fē′nŭs, sa′fĕ-nŭs) **veins.** The **great saphenous vein** originates over the dorsal and medial side of the foot and ascends along the medial side of the leg and thigh to empty into the femoral vein. The **small saphenous vein** begins over the lateral side of the foot and joins the **popliteal vein,** which becomes the femoral vein. The femoral vein empties into the external iliac vein.

Predict 3

If a thrombus in the posterior tibial vein gave rise to an embolus, name in order the parts of the circulatory system the embolus would pass through before lodging in a blood vessel in the lungs. Why are the lungs the most likely place for the embolus to lodge?

13.6 PHYSIOLOGY OF CIRCULATION

Learning Outcomes   *After reading this section, you should be able to*

A. Explain how blood pressure and resistance to flow change as blood flows through the blood vessels.

B. Describe the exchange of material across a capillary wall.

The function of the circulatory system is to maintain adequate blood flow to all body tissues. Adequate blood flow is required to provide nutrients and O_2 to the tissues and to remove the waste products of metabolism from the tissues. Blood flows through the arterial system primarily as a result of the pressure produced by contractions of the heart ventricles.

Blood Pressure

Blood pressure is a measure of the force blood exerts against the blood vessel walls. In arteries, blood pressure values go through a cycle that depends on the rhythmic contractions of the heart. When the ventricles contract, blood is forced into the arteries, and the pressure reaches a maximum value called the **systolic pressure.**

When the ventricles relax, blood pressure in the arteries falls to a minimum value called the **diastolic pressure.** The standard unit for measuring blood pressure is millimeters of mercury (mm Hg). For example, if the blood pressure is 100 mm Hg, the pressure is great enough to lift a column of mercury 100 mm.

Health professionals most often use the **auscultatory** (aws-kŭl′tă-tō-rē; to listen) method to determine blood pressure (figure 13.21). A blood pressure cuff connected to a **sphygmomanometer** (sfig′mō-mă-nom′ĕ-ter) is wrapped around the patient's arm, and a **stethoscope** (steth′ō-skōp) is placed over the brachial artery. The blood pressure cuff is then inflated until the brachial artery is completely blocked. Because no blood flows through the constricted area at this point, no sounds can be heard through the stethoscope (figure 13.21, step 1). The pressure in the cuff is then gradually lowered. As soon as the pressure in the cuff declines below the systolic pressure, blood flows through the constricted area each time the left ventricle contracts. The blood flow is turbulent immediately downstream from the constricted area. This turbulence produces vibrations in the blood and surrounding tissues that can be heard through the stethoscope. These sounds are called **Korotkoff** (Kō-rot′kof) **sounds,** and the pressure at which the first Korotkoff sound is heard is the systolic pressure (figure 13.21, step 2).

As the pressure in the blood pressure cuff is lowered still more, the Korotkoff sounds change tone and loudness (figure 13.21, step 3). When the pressure has dropped until the brachial artery is no longer constricted and blood flow is no longer turbulent, the sound disappears completely. The pressure at which the Korotkoff sounds disappear is the diastolic pressure (figure 13.21, step 4). The brachial artery remains open during systole and diastole, and continuous blood flow is reestablished.

The systolic pressure is the maximum pressure produced in the large arteries. It is also a good measure of the maximum pressure within the left ventricle. The diastolic pressure is close to the lowest pressure within the large arteries. During relaxation of the left ventricle, the aortic semilunar valve closes, trapping the blood that was ejected during ventricular contraction in the aorta. The pressure in the ventricles falls to 0 mm Hg during ventricular relaxation. However, the blood trapped in the elastic

CLINICAL IMPACT Hypertension

Hypertension, or *high blood pressure,* affects at least 20% of all people at some time in their lives. The following guidelines* categorize blood pressure for adults:

- *Normal:* less than 120 mm Hg systolic and 80 mm Hg diastolic
- *Prehypertension:* from 120 mm Hg systolic and 80 mm Hg diastolic to 139 mm Hg systolic and 89 mm Hg diastolic
- *Stage 1 hypertension:* from 140 mm Hg systolic and 90 mm Hg diastolic to 159 mm Hg systolic and 99 mm Hg diastolic
- *Stage 2 hypertension:* greater than 160 mm Hg systolic and 100 mm Hg diastolic

Individuals with prehypertension pressure should monitor their blood pressure regularly and consider lifestyle changes that can reduce blood pressure.

Hypertension requires the heart to perform a greater-than-normal amount of work because of the increased afterload (see chapter 12). The extra work leads to hypertrophy of cardiac muscle, especially in the left ventricle, and can result in heart failure. Hypertension also increases the rate of arteriosclerosis development. Arteriosclerosis, in turn, increases the chance that blood clots will form and that blood vessels will rupture. Common conditions associated with hypertension

are cerebral hemorrhage, heart attack, hemorrhage of renal blood vessels, and poor vision resulting from burst blood vessels in the retina.

The most common treatments for hypertension are those that dilate blood vessels, increase the rate of urine production, or decrease cardiac output. A low-salt diet is also recommended to reduce the amount of sodium chloride and water absorbed from the intestine into the bloodstream.

*Joint National Committee on Prevention, Detection, Evaluation, and Treatment of High Blood Pressure.

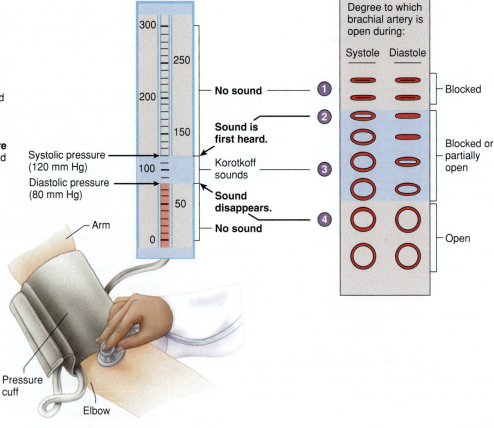

1. When the cuff pressure is high enough to keep the brachial artery closed, no blood flows through it, and no sound is heard.

2. When cuff pressure decreases and is no longer able to keep the brachial artery closed, blood is pushed through the partially opened brachial artery, producing turbulent blood flow and a sound. **Systolic pressure** is the pressure at which a sound is first heard.

3. As cuff pressure continues to decrease, the brachial artery opens even more during systole. At first, the artery is closed during diastole, but as cuff pressure continues to decrease, the brachial artery partially opens during diastole. Turbulent blood flow during systole produces Korotkoff sounds, although the pitch of the sounds changes as the artery becomes more open.

4. Eventually, cuff pressure decreases below the pressure in the brachial artery, and it remains open during systole and diastole. Nonturbulent flow is reestablished, and no sounds are heard. **Diastolic pressure** is the pressure at which the sound disappears.

PROCESS Figure 13.21 Measuring Blood Pressure

arteries is compressed by the recoil of the elastic arteries, and the pressure falls more slowly, reaching the diastolic pressure (see figure 12.18).

Pressure and Resistance

The values for systolic and diastolic pressure vary among healthy people, making the range of normal values quite broad. In addition, other factors, such as physical activity and emotions, affect blood pressure values in a normal person. A standard blood pressure for a resting young adult male is 120 mm Hg for the systolic pressure and 80 mm Hg for the diastolic pressure, commonly expressed as 120/80.

As blood flows from arteries through the capillaries and veins, blood pressure falls progressively to about 0 mm Hg or even slightly lower by the time blood is returned to the right atrium. In addition, the fluctuations in blood pressure are damped, meaning that the difference between the systolic and diastolic pressures is decreased in the small-diameter vessels. The decrease in fluctuations in pressure is the result of increased resistance to blood flow in smaller and smaller vessels. By the time blood reaches the capillaries, the smallest of the vessels, there is no variation in blood pressure, and only a steady pressure of about 30 mm Hg remains (figure 13.22).

Resistance to blood flow is related to the diameter of the blood vessel. The smaller the diameter of the blood vessel, the greater the resistance to flow, and the more rapidly the pressure decreases as blood flows through it. The most rapid decline in blood pressure occurs in the arterioles and capillaries because their small diameters increase the resistance to blood flow. Blood pressure declines slowly as blood flows from large to medium-sized arteries because their diameters are larger and the resistance to blood flow is not as great.

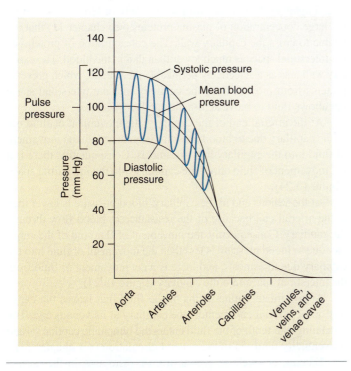

Figure 13.22 Blood Pressure in the Various Blood Vessel Types

In small arteries and arterioles, blood pressure fluctuations between systole and diastole are damped, so the fluctuations become smaller. No fluctuations in blood pressure occur in capillaries and veins.

A CASE IN POINT

Atherosclerosis

Mori Payne and Leslie Payne were preparing for their fortieth wedding anniversary celebration. While they were at the shopping mall, Mori's legs began to ache, especially his left leg. He had noticed before that when he walked briskly or very far, his legs hurt. Mori sat down to rest while Leslie shopped at a nearby store. After about 5 minutes, Mori experienced excruciating pain in his left leg that was not relieved by massage or rest. Leslie asked a passerby to help her get Mori to their car, and she drove him to the hospital.

In the emergency room, the examining physician observed that Mori's left leg was mottled and cyanotic (bluish) distal to the knee and was cool to the touch. His right leg was pink and warm. A Doppler test performed on the left leg revealed decreased pulses with faint, abnormal sounds called bruits (broo-ez') in the distal left popliteal artery. Normal pulses and no obvious bruits were observed in other arteries of Mori's thigh and leg, including the right popliteal artery. His temperature was slightly elevated, and his blood pressure was 165/95 mm Hg. His pulse was 96 beats per minute and regular. His respiratory rate was 20 respirations per minute.

The physician explained to Leslie that reduced pulses indicate reduced blood flow and that bruits indicate turbulent blood flow. Both of these observations and the sudden onset of the pain are consistent with the formation of an arterial thrombus partially blocking Mori's distal left popliteal artery. He was quickly treated with an enzyme that breaks down the fibrin of blood clots, and this treatment successfully dissolved the thrombus. As blood flow to the leg increased, the pain and cyanosis decreased.

An angiogram on Mori's left leg revealed severe stenosis (narrowing) in his left distal popliteal artery due to an atherosclerotic plaque. When an artery is stenosed, blood delivery may be adequate to maintain tissue homeostasis as long as the person is at rest, but not during exercise. In Mori's case, other, smaller atherosclerotic lesions were likely in both legs, because both of his legs hurt in response to brisk walking. The lack of sufficient O_2 resulted in anaerobic respiration and acidic by-products.

The sudden, dramatic increase in pain that Mori experienced resulted from the formation of a thrombus on a large atherosclerotic plaque that nearly blocked the blood flow through the left popliteal artery. After the thrombus was dissolved, angioplasty was performed to further increase blood flow through the left popliteal artery.

It was later determined that Mori has high blood cholesterol and blood glucose levels, conditions that are associated with the increased development of atherosclerotic plaque in arteries. The high blood glucose may be a sign of type 2 diabetes mellitus (see chapter 10). From now on, it is important for Mori to reduce his blood pressure, blood cholesterol, and blood glucose levels.

Because of veins' large diameters, resistance to blood flow in them is low. The low resistance results in low blood pressure in the veins. However, though pressure is low, blood continues to flow through the veins toward the heart. Blood flow rate is ensured and amplified by valves that prevent the backflow of blood in the veins, as well as skeletal muscle movements that periodically compress veins, forcing blood to flow toward the heart.

The muscular arteries, arterioles, and precapillary sphincters are capable of constricting (vasoconstriction) and dilating (vasodilation). If vessels constrict, resistance to blood flow increases, and the volume of blood flowing through the vessels declines. Because muscular arteries are able to constrict and dilate, they help control the amount of blood flowing to each region of the body. In contrast, arterioles and precapillary sphincters regulate blood flow through local tissues.

Pulse Pressure

The difference between the systolic and diastolic pressures is called the **pulse pressure.** For example, if a person has a systolic pressure of 120 mm Hg and a diastolic pressure of 80 mm Hg, the pulse pressure is 40 mm Hg. Two factors affect pulse pressure: stroke volume and vascular compliance. When the stroke volume increases, the systolic pressure increases more than the diastolic pressure, causing the pulse pressure to increase. During periods of exercise, the stroke volume and pulse pressure increase substantially.

Vascular compliance is related to the elasticity of the blood vessel wall. In people who have **arteriosclerosis** (ar-tēr′ē-ō-skler-ō′sis; hardening of the arteries), the arteries are less elastic than normal. Arterial pressure increases rapidly and falls rapidly in these less elastic arteries. The effect that this change has on blood pressure is that the systolic pressure increases substantially, and the diastolic pressure may be somewhat lower than normal or slightly increased. The same amount of blood ejected into a less elastic artery results in a higher systolic pressure than it would have in a more elastic artery. Therefore, the pulse pressure is greater than normal, even though the same amount of blood is ejected into the aorta as in a normal person. Arteriosclerosis increases the amount of work the heart performs because the left ventricle must produce a greater pressure to eject the same amount of blood into a less elastic artery. In severe cases, the heart's increased workload leads to heart failure.

Ejection of blood from the left ventricle into the aorta produces a pressure wave, or **pulse,** which travels rapidly along the arteries. A pulse can be felt at locations where large arteries are close to the surface of the body (figure 13.23). Health professionals should know the major locations of pulse detection because monitoring the pulse can yield important information about the heart rate, the heart rhythm, and other characteristics. For example, a weak pulse usually indicates a decreased stroke volume or increased constriction of the arteries.

Predict 4

A weak pulse occurs in response to premature heartbeats and during cardiovascular shock due to hemorrhage. Stronger-than-normal pulses occur in a healthy person during exercise. Explain the causes for the changes in the pulse under these conditions.

Capillary Exchange

There are about 10 billion capillaries in the body. Nutrients diffuse across the capillary walls into the interstitial spaces, and waste products diffuse in the opposite direction. In addition, a small amount of fluid is forced out of the capillaries into the interstitial space at their arterial ends. Most of that fluid, but not all, reenters the capillaries at their venous ends.

The major forces responsible for moving fluid through the capillary wall are blood pressure and osmosis (figure 13.24). Blood pressure forces fluid out of the capillary, and osmosis moves fluid into the capillary. Fluid moves by osmosis from the interstitial space into the capillary because blood has a greater osmotic pressure than does interstitial fluid. The greater the concentration of molecules

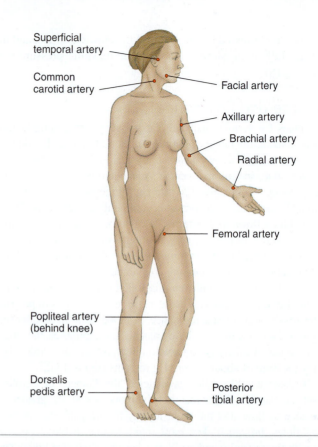

Figure 13.23 Major Points at Which the Pulse Can Be Monitored
Each pulse point is named after the artery on which it occurs.

Labels: Superficial temporal artery; Common carotid artery; Facial artery; Axillary artery; Brachial artery; Radial artery; Femoral artery; Popliteal artery (behind knee); Dorsalis pedis artery; Posterior tibial artery

dissolved in a fluid, the greater the osmotic pressure of the fluid (see chapter 3). The greater osmotic pressure of blood is caused by the large concentration of plasma proteins (see chapter 11) that are unable to cross the capillary wall. The concentration of proteins in the interstitial space is much lower than that in the blood. The capillary wall acts as a selectively permeable membrane, which prevents proteins from moving from the capillary into the interstitial space but allows fluid to move across the capillary wall.

At the arterial end of the capillary, the movement of fluid out of the capillary due to blood pressure is greater than the movement of fluid into the capillary due to osmosis. Consequently, there is a net movement of fluid out of the capillary into the interstitial space (figure 13.24).

At the venous end of the capillary, blood pressure is lower than at the arterial end because of the resistance to blood flow through the capillary. Consequently, the movement of fluid out of the capillary due to blood pressure is less than the movement of fluid into the capillary due to osmosis, and there is a net movement of fluid from the interstitial space into the capillary (figure 13.24).

Approximately nine-tenths of the fluid that leaves the capillary at the arterial end reenters the capillary at its venous end. The remaining one-tenth of the fluid enters the lymphatic capillaries and is eventually returned to the general circulation (see chapter 14).

Edema, or swelling, results from a disruption in the normal inwardly and outwardly directed pressures across the capillary walls. For example, inflammation increases the permeability of capillaries. Proteins, mainly albumin, leak out of the capillaries into the interstitial spaces. The proteins increase the osmotic pressure in the

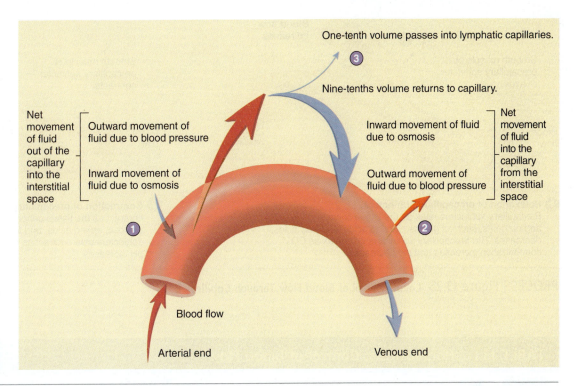

1. At the arterial end of the capillary, the movement of fluid out of the capillary due to blood pressure is greater than the movement of fluid into the capillary due to osmosis.

2. At the venous end of the capillary, the movement of fluid into the capillary due to osmosis is greater than the movement of fluid out of the capillary due to blood pressure.

3. Approximately nine-tenths of the fluid that leaves the capillary at its arterial end reenters the capillary at its venous end. About one-tenth of the fluid passes into the lymphatic capillaries.

One-tenth volume passes into lymphatic capillaries.

Nine-tenths volume returns to capillary.

Net movement of fluid out of the capillary into the interstitial space
- Outward movement of fluid due to blood pressure
- Inward movement of fluid due to osmosis

Inward movement of fluid due to osmosis

Outward movement of fluid due to blood pressure

Net movement of fluid into the capillary from the interstitial space

Blood flow

Arterial end

Venous end

PROCESS Figure 13.24 **AP|R** **Capillary Exchange**

interstitial fluid. Consequently, fluid passes from the arterial end of capillaries into the interstitial spaces at a greater rate, and fluid passes from the interstitial spaces into the venous ends of capillaries at a slower rate. The lymphatic capillaries cannot carry all the fluid away. Thus, fluid accumulates in the interstitial spaces, resulting in edema.

Predict 5

Explain edema (a) in response to a decrease in plasma protein concentration and (b) as a result of increased blood pressure within a capillary.

13.7 CONTROL OF BLOOD FLOW IN TISSUES

Learning Outcomes *After reading this section, you should be able to*

A. Explain how local control mechanisms regulate blood flow.
B. Explain how nervous mechanisms control blood flow.

Blood flow provided to the tissues by the circulatory system is highly controlled and matched closely to the metabolic needs of tissues. Mechanisms that control blood flow through tissues are classified as (1) local control or (2) nervous and hormonal control.

Local Control of Blood Flow

Local control of blood flow is achieved by the periodic relaxation and contraction of the precapillary sphincters. When the sphincters relax, blood flow through the capillaries increases. When the sphincters contract, blood flow through the capillaries decreases. The precapillary sphincters are controlled by the metabolic needs of the tissues. For example, blood flow increases when by-products of

metabolism buildup in tissue spaces. During exercise, the metabolic needs of skeletal muscle increase dramatically, and the by-products of metabolism are produced more rapidly. The precapillary sphincters relax, increasing blood flow through the capillaries.

Other factors that control blood flow through the capillaries are the tissue concentrations of O_2 and nutrients, such as glucose, amino acids, and fatty acids (figure 13.25 and table 13.1). Blood flow increases when O_2 levels decrease or, to a lesser degree, when glucose, amino acids, fatty acids, and other nutrients decrease. An increase in CO_2 or a decrease in pH also causes the precapillary sphincters to relax, thereby increasing blood flow.

Predict 6

A student has been sitting for a short time with her legs crossed. After getting up to walk out of class, she notices a red blotch on the back of one of her legs. On the basis of what you know about local control of blood flow, explain why this happens.

In addition to the control of blood flow through existing capillaries, if the metabolic activity of a tissue increases often, additional capillaries gradually grow into the area. The additional capillaries allow local blood flow to increase to a level that matches the metabolic needs of the tissue. For example, the density of capillaries in the well-trained skeletal muscles of athletes is greater than that in skeletal muscles on a typical nonathlete (table 13.1).

Nervous and Hormonal Control of Blood Flow

Nervous control of blood flow is carried out primarily through the sympathetic division of the autonomic nervous system. Sympathetic nerve fibers innervate most blood vessels of the body, except the capillaries and precapillary sphincters, which have no nerve supply (figure 13.26).

Cardiovascular

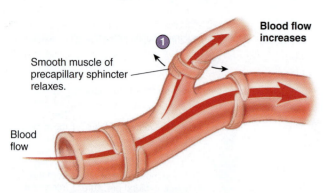

Blood flow increases

Smooth muscle of precapillary sphincter relaxes.

Blood flow

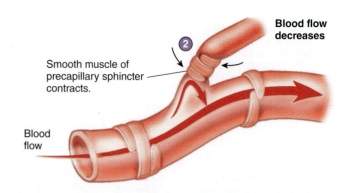

Blood flow decreases

Smooth muscle of precapillary sphincter contracts.

Blood flow

① **Relaxation of precapillary sphincters.** Precapillary sphincters relax as the tissue concentration of O_2 and nutrients, such as glucose, amino acids, and fatty acids, decreases. The precapillary sphincters also relax as CO_2 concentration increases and pH decreases.

② **Contraction of precapillary sphincters.** Precapillary sphincters contract as the tissue concentration of O_2 and nutrients, such as glucose, amino acids, and fatty acids, increases. The precapillary sphincters also contract as the CO_2 concentration decreases and pH increases.

PROCESS Figure 13.25 Local Control of Blood Flow Through Capillary Beds

TABLE 13.1	Homeostasis: Control of Blood Flow*
Stimulus	**Response**
Regulation by Metabolic Needs of Tissues	
Increased CO_2 and decreased pH or decreased O_2 and nutrients, such as glucose, amino acids, and fatty acids, due to increased metabolism	Relaxation of precapillary sphincters and subsequent increase in blood flow through capillaries
Decreased CO_2 and increased pH or increased O_2 and nutrients, such as glucose, amino acids, and fatty acids	Contraction of precapillary sphincters and subsequent decrease in blood flow through capillaries
Regulation by Nervous Mechanisms	
Increased physical activity or increased sympathetic activity	Constriction of blood vessels in skin and viscera
Increased body temperature detected by neurons of hypothalamus	Dilation of blood vessels in skin (see chapter 5)
Decreased body temperature detected by neurons of hypothalamus	Constriction of blood vessels in skin (see chapter 5)
Decrease in skin temperature below a critical value	Dilation of blood vessels in skin (protects skin from extreme cold)
Anger or embarrassment	Dilation of blood vessels in skin of face and upper thorax
Regulation by Hormonal Mechanisms (Reinforces Increased Activity of Sympathetic Division)	
Increased physical activity and increased sympathetic activity, causing release of epinephrine and small amounts of norepinephrine from adrenal medulla	Constriction of blood vessels in skin and viscera; dilation of blood vessels in skeletal and cardiac muscle
Long-Term Local Blood Flow	
Increased metabolic activity of tissues over a long period, as occurs in athletes who train regularly	Increased number of capillaries
Decreased metabolic activity of tissues over a long period, as occurs during periods of reduced physical activity	Decreased number of capillaries
*The mechanisms operate when the systemic blood pressure is maintained within a normal range of values.	

An area of the lower pons and upper medulla oblongata, called the **vasomotor center,** continually transmits a low frequency of action potentials to the sympathetic nerve fibers. As a consequence, the peripheral blood vessels are continually in a partially constricted state, a condition called **vasomotor** (vā-sō-mō′ter) **tone.** An increase in vasomotor tone causes blood vessels to constrict further and blood pressure to increase. A decrease in vasomotor tone causes blood vessels to dilate and blood pressure to decrease. Nervous control of blood vessel diameter is an important way that blood pressure is regulated.

Nervous control of blood vessels also causes blood to be shunted from one large area of the body to another. For example, nervous control of blood vessels during exercise increases vasomotor tone in the viscera and skin and reduces vasomotor tone in exercising skeletal muscles. As a result, blood flow to the viscera and skin decreases, and blood flow to skeletal muscle increases. Nervous control of blood vessels during exercise and dilation of precapillary sphincters as muscle activity increases together increase blood flow through exercising skeletal muscle severalfold (table 13.1).

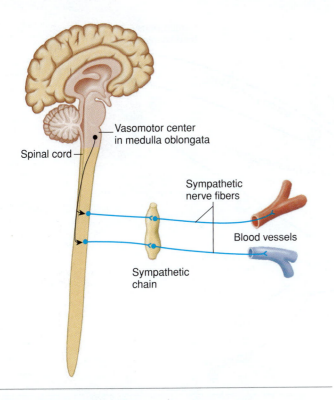

Figure I3.26 Nervous Regulation of Blood Vessels

Most arteries and veins are innervated by sympathetic nerve fibers. The vasomotor center within the medulla oblongata regulates the frequency of action potentials in nerve fibers that innervate blood vessels. In most blood vessels, increased action potential frequencies cause vasoconstriction, and decreased action potential frequencies cause vasodilation.

The sympathetic division also regulates hormonal control of blood flow through the release of epinephrine and norepinephrine from the adrenal medulla. These hormones are transported in the blood to all parts of the body. In most blood vessels, these hormones cause constriction, which reduces blood flow. But in some tissues, such as skeletal muscle and cardiac muscle, these hormones cause the blood vessels to dilate, increasing blood flow.

Predict 7

Raynaud syndrome is a treatable condition in which blood vessels, primarily in the fingers and toes, undergo exaggerated vasoconstriction in response to emotions or exposure to cold. Predict the effects on the fingers and toes of a person who has severe, untreated Raynaud syndrome. Explain why these consequences occur.

I3.8 REGULATION OF ARTERIAL PRESSURE

Learning Outcome *After reading this section, you should be able to*

A. Describe the short-term and long-term mechanisms that regulate arterial pressure.

Adequate blood pressure is required to maintain blood flow through the blood vessels of the body, and several regulatory mechanisms ensure that blood pressure remains adequate for this task. The **mean arterial blood pressure (MAP)** is slightly less than the average of the systolic and diastolic pressures in the aorta because diastole lasts longer than systole. The mean arterial pressure is about 70 mm Hg at birth, is maintained at about 95 mm Hg from adolescence to middle age, and may reach 110 mm Hg in a healthy older person.

The body's MAP is equal to the **cardiac output (CO)** times the **peripheral resistance (PR),** which is the resistance to blood flow in all the blood vessels:

$$MAP = CO \times PR$$

Because the cardiac output is equal to the **heart rate (HR)** times the **stroke volume (SV),** the mean arterial pressure is equal to the heart rate times the stroke volume times the peripheral resistance (PR):

$$MAP = HR \times SV \times PR$$

Thus, the MAP increases in response to increases in HR, SV, or PR, and the MAP decreases in response to decreases in HR, SV, or PR. The MAP is controlled on a minute-to-minute basis by changes in these variables. For example, when blood pressure suddenly drops because of hemorrhage or some other cause, control systems attempt to reestablish blood pressure by increasing HR, SV, and PR, so that blood pressure is maintained at a value consistent with life. Mechanisms are also activated to increase the blood volume to its normal value.

Baroreceptor Reflexes

Baroreceptor reflexes activate responses that keep the blood pressure within its normal range. **Baroreceptors** respond to stretch in arteries caused by increased pressure. They are scattered along the walls of most of the large arteries of the neck and thorax, and many are located in the carotid sinus at the base of the internal carotid artery and in the walls of the aortic arch. Action potentials travel from the baroreceptors to the medulla oblongata along sensory nerve fibers (figure 13.27).

A sudden increase in blood pressure stretches the artery walls and increases action potential frequency in the baroreceptors. The increased action potential frequency delivered to the vasomotor and cardioregulatory centers in the medulla oblongata causes responses that lower the blood pressure. One major response is a decrease in vasomotor tone, resulting in dilation of blood vessels and decreased peripheral resistance. Other responses, controlled by the cardioregulatory center, are an increase in the parasympathetic stimulation of the heart, which decreases the heart rate, and a decrease in the sympathetic stimulation of the heart, which reduces the stroke volume. The decreased heart rate, stroke volume, and peripheral resistance lower the blood pressure toward its normal value (figure 13.28).

A sudden decrease in blood pressure results in a decreased action potential frequency in the baroreceptors. The decreased frequency of action potentials delivered to the vasomotor and cardioregulatory centers in the medulla oblongata produces responses that raise blood pressure. Sympathetic stimulation of the heart increases, which increases the heart rate and stroke volume. In addition, vasomotor tone increases, resulting in constriction of blood vessels and increased peripheral resistance. The increased heart rate, stroke volume, and peripheral resistance raise the blood pressure toward its normal value (figure 13.28).

① Baroreceptors in the carotid sinus and aortic arch monitor blood pressure.

② Sensory nerves conduct action potentials to the cardioregulatory and vasomotor centers in the medulla oblongata.

③ Increased parasympathetic stimulation of the heart decreases the heart rate.

④ Increased sympathetic stimulation of the heart increases the heart rate and stroke volume.

⑤ Increased sympathetic stimulation of blood vessels increases vasoconstriction.

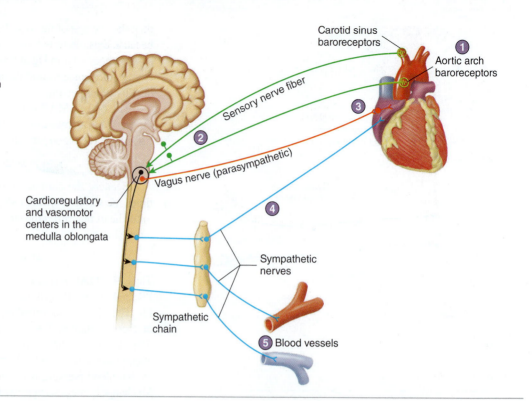

Carotid sinus baroreceptors

Aortic arch baroreceptors

Sensory nerve fiber

Vagus nerve (parasympathetic)

Cardioregulatory and vasomotor centers in the medulla oblongata

Sympathetic nerves

Sympathetic chain

Blood vessels

PROCESS Figure I3.27 AP|R **Baroreceptor Reflex Mechanisms**
The baroreceptor reflex helps control blood pressure.

Cardiovascular

These **baroreceptor reflexes** regulate blood pressure on a moment-to-moment basis. When a person rises rapidly from a sitting or lying position, blood pressure in the neck and thoracic regions drops dramatically due to the pull of gravity on the blood. This reduction in blood pressure can be so great that it reduces blood flow to the brain enough to cause dizziness or even loss of consciousness. The falling blood pressure activates the baroreceptor reflexes, which reestablish normal blood pressure within a few seconds. A healthy person usually experiences only a temporary sensation of dizziness.

Chemoreceptor Reflexes

Carotid bodies are small structures that lie near the carotid sinuses, and **aortic bodies** lie near the aortic arch. These structures contain sensory receptors that respond to changes in blood O_2 concentration, CO_2 concentration, and pH. Because they are sensitive to chemical changes in the blood, they are called **chemoreceptors.** They send action potentials along sensory nerve fibers to the medulla oblongata. There are also chemoreceptors in the medulla oblongata.

When O_2 or pH levels decrease or when CO_2 levels increase, the chemoreceptors respond with an increased frequency of action potentials and activate the **chemoreceptor reflexes** (figure 13.29). In response, the vasomotor and cardiovascular centers decrease parasympathetic stimulation of the heart, which increases the heart rate. The vasomotor and cardioregulatory centers also increase sympathetic stimulation of the heart, which further increases heart rate, stroke volume, and vasomotor tone. All these changes result in increased blood pressure. This increased blood pressure causes a greater rate of blood flow to the lungs, which helps raise blood O_2 levels and reduce blood CO_2 levels. The chemoreceptor

reflexes function under emergency conditions and usually do not play an important role in regulating the cardiovascular system. They respond strongly only when the O_2 levels in the blood fall to very low levels or when CO_2 levels become substantially elevated.

Hormonal Mechanisms

In addition to the rapidly acting baroreceptor and chemoreceptor reflexes, four important hormonal mechanisms help control blood pressure.

Adrenal Medullary Mechanism

Stimuli that lead to increased sympathetic stimulation of the heart and blood vessels also cause increased stimulation of the adrenal medulla. The adrenal medulla responds by releasing epinephrine and small amounts of norepinephrine into the blood (figure 13.30). Epinephrine increases heart rate and stroke volume and causes vasoconstriction, especially of blood vessels in the skin and viscera. Epinephrine also causes vasodilation of blood vessels in skeletal muscle and cardiac muscle, thereby increasing the supply of blood flowing to those muscles and preparing the body for physical activity.

Renin-Angiotensin-Aldosterone Mechanism

In response to reduced blood flow, the kidneys release an enzyme called **renin** (rē′nin) into the circulatory system (figure 13.31). Renin acts on the blood protein **angiotensinogen** (an′jē-ō-ten-sin′ō-jen) to produce **angiotensin I** (an-jē-ō-ten′sin). Another enzyme, called **angiotensin-converting enzyme (ACE),** found in

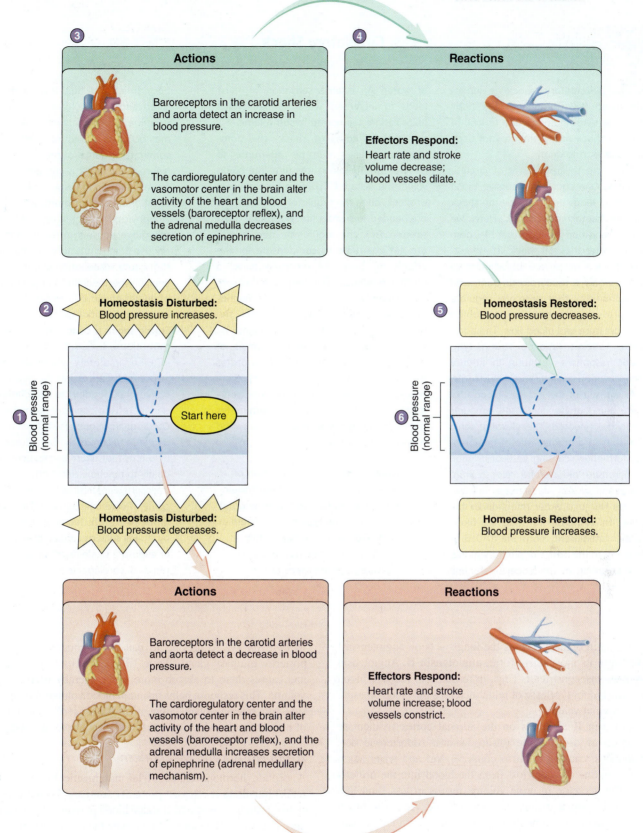

③ Actions

Baroreceptors in the carotid arteries and aorta detect an increase in blood pressure.

The cardioregulatory center and the vasomotor center in the brain alter activity of the heart and blood vessels (baroreceptor reflex), and the adrenal medulla decreases secretion of epinephrine.

④ Reactions

Effectors Respond:
Heart rate and stroke volume decrease; blood vessels dilate.

② **Homeostasis Disturbed:**
Blood pressure increases.

⑤ **Homeostasis Restored:**
Blood pressure decreases.

Blood pressure (normal range)

① Start here

Blood pressure (normal range)

⑥

Homeostasis Disturbed:
Blood pressure decreases.

Homeostasis Restored:
Blood pressure increases.

Actions

Baroreceptors in the carotid arteries and aorta detect a decrease in blood pressure.

The cardioregulatory center and the vasomotor center in the brain alter activity of the heart and blood vessels (baroreceptor reflex), and the adrenal medulla increases secretion of epinephrine (adrenal medullary mechanism).

Reactions

Effectors Respond:
Heart rate and stroke volume increase; blood vessels constrict.

Cardiovascular

Homeostasis Figure 13.28 Baroreceptor Effects on Blood Pressure

(1) Blood pressure is within its normal range. (2) Blood pressure increases outside the normal range, which causes homeostasis to be disturbed. (3) Baroreceptors detect the increase in blood pressure. The blood pressure control centers in the brain respond to changes in blood pressure. (4) Neural and hormonal changes alter the activity of cardiac muscle of the heart and the smooth muscle of the blood vessels (effectors) causing heart rate and stroke volume to decrease and blood vessels to dilate. (5) These changes cause blood pressure to decrease. (6) Blood pressure returns to its normal range, and homeostasis is restored.

CLINICAL IMPACT   Circulatory Shock

Circulatory shock is inadequate blood flow throughout the body that causes tissue damage due to lack of O_2. Severe shock may damage vital body tissues and lead to death.

There are several causes of circulatory shock. One cause is excessive blood loss, which leads to the type of shock known as hemorrhagic shock. Here, we use this condition to illustrate the general characteristics of shock. If hemorrhagic shock is not severe, blood pressure decreases only a moderate amount, and the mechanisms that normally regulate blood pressure are able to reestablish normal pressure and blood flow. The baroreceptor reflexes produce strong sympathetic responses, resulting in intense vasoconstriction and increased heart rate.

As a result of the reduced blood flow through the kidneys, increased amounts of renin are released. The elevated renin level results in a greater rate of angiotensin II formation, causing vasoconstriction and increased aldosterone release from the adrenal cortex. Aldosterone, in turn, promotes Na^+ and water retention by the kidneys. In response to reduced blood pressure, antidiuretic hormone (ADH) is released from the posterior pituitary gland; ADH also enhances the kidneys' retention of water. An intense sensation of thirst leads to increased water intake, which helps restore the normal blood volume.

In mild cases of shock, the baroreceptor reflexes can be adequate to compensate for blood loss until the blood volume is restored. In more severe cases of shock, all of the regulatory mechanisms are needed to sustain life. But in the most severe cases, the regulatory mechanisms are not adequate to compensate for the effects of shock. As a consequence, a positive-feedback cycle begins to develop: The blood pressure regulatory mechanisms lose their ability to control the blood pressure, and shock worsens. As shock becomes worse, the effectiveness of the regulatory mechanisms deteriorates even further. The positive-feedback cycle proceeds until death occurs or until treatment, such as a transfusion, terminates the cycle. Five types of shock are classified based on their cause:

1. **Hypovolemic shock** is the result of reduced blood volume. **Hemorrhagic shock,** caused by internal or external bleeding, is one type of hypovolemic shock. **Plasma loss shock** results from loss of plasma, as may occur in severely burned areas of the body. **Interstitial fluid loss shock** is reduced blood volume resulting from the loss of interstitial fluid, as may occur as a result of diarrhea, vomiting, or dehydration.

2. **Neurogenic shock** is caused by vasodilation in response to emotional upset or anesthesia.

3. **Anaphylactic shock** is caused by an allergic response that results in the release of inflammatory substances that cause vasodilation and increased capillary permeability. Large amounts of fluid then move from capillaries into the interstitial spaces.

4. **Septic shock** is caused by infections that release toxic substances into the circulatory system (*blood poisoning*), depressing the heart's activity and leading to vasodilation and increased capillary permeability.

5. **Cardiogenic shock** results from a decrease in cardiac output caused by events that decrease the heart's ability to function. Heart attack (myocardial infarction) is a common cause of cardiogenic shock. Fibrillation of the heart, which can be initiated by stimuli such as cardiac arrhythmias or exposure to electrical shocks, also results in cardiogenic shock.

---

large amounts in organs, such as the lungs, acts on angiotensin I to convert it to its most active form, **angiotensin II.** Angiotensin II is a potent vasoconstrictor. Thus, in response to reduced blood pressure, the kidneys' release of renin increases the blood pressure toward its normal value.

Angiotensin II also acts on the adrenal cortex to increase the secretion of **aldosterone** (al-dos′ter-ōn). Aldosterone acts on the kidneys, causing them to conserve Na^+ and water. As a result, the volume of water lost from the blood into the urine is reduced. The decrease in urine volume results in less fluid loss from the body, which maintains blood volume. Adequate blood volume is essential to maintain normal venous return to the heart and thereby maintain blood pressure (see chapter 12).

Antidiuretic Hormone Mechanism

When the concentration of solutes in the plasma increases or when blood pressure decreases substantially, nerve cells in the hypothalamus respond by causing the release of **antidiuretic** (an′tē-dī-ū-ret′ik; to decrease urine production) **hormone (ADH),** also called *vasopressin* (vā-sō-pres′in; to cause vasoconstriction), from the posterior pituitary gland (figure 13.32). ADH acts on the kidneys and causes them to absorb more water, thereby decreasing urine volume. This response helps maintain blood volume and blood pressure. The release of large amounts of ADH causes vasoconstriction of blood vessels, which causes blood pressure to increase.

Atrial Natriuretic Mechanism

A peptide hormone called **atrial natriuretic** (ā′trē-ăl nā′trē-ū-ret′ik) **hormone** is released primarily from specialized cells of the right atrium in response to elevated blood pressure. Atrial natriuretic hormone causes the kidneys to promote the loss of Na^+ and water in the urine, increasing urine volume. Loss of water in the urine causes blood volume to decrease, thus decreasing the blood pressure.

Summary of Regulatory Mechanisms

Blood pressure regulation involves both short-term and long-term mechanisms. Baroreceptor mechanisms are most important in controlling blood pressure on a short-term basis (see figures 13.27 and

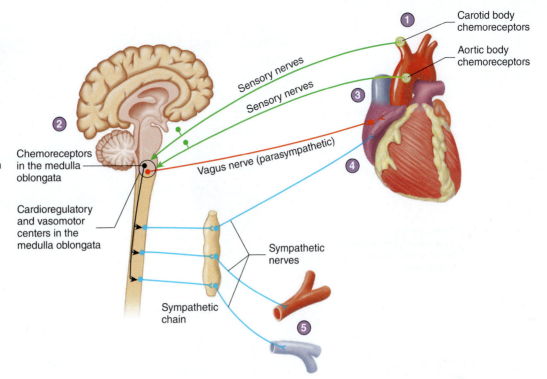

① Chemoreceptors in the carotid and aortic bodies monitor blood O_2, CO_2, and pH.

② Chemoreceptors in the medulla oblongata monitor blood CO_2 and pH.

③ Decreased blood O_2, increased CO_2, and decreased pH decrease parasympathetic stimulation of the heart, which increases the heart rate.

④ Decreased blood O_2, increased CO_2, and decreased pH increase sympathetic stimulation of the heart, which increases the heart rate and stroke volume.

⑤ Decreased blood O_2, increased CO_2, and decreased pH increase sympathetic stimulation of blood vessels, which increases vasoconstriction.

PROCESS Figure 13.29 Chemoreceptor Reflex Mechanisms

The chemoreceptor reflex helps control blood pressure.

① The same stimuli that increase sympathetic stimulation of the heart and blood vessels cause action potentials to be carried to the medulla oblongata.

② Descending pathways from the medulla oblongata to the spinal cord increase sympathetic stimulation of the adrenal medulla, resulting in secretion of epinephrine and some norepinephrine.

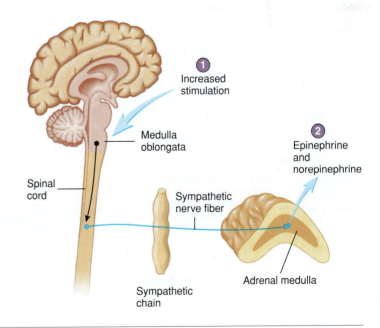

PROCESS Figure 13.30 Hormonal Regulation: Adrenal Medullary Mechanism

Stimuli that increase sympathetic stimulation of the heart and blood vessels also increase sympathetic stimulation of the adrenal medulla and lead to secretion of epinephrine and some norepinephrine.

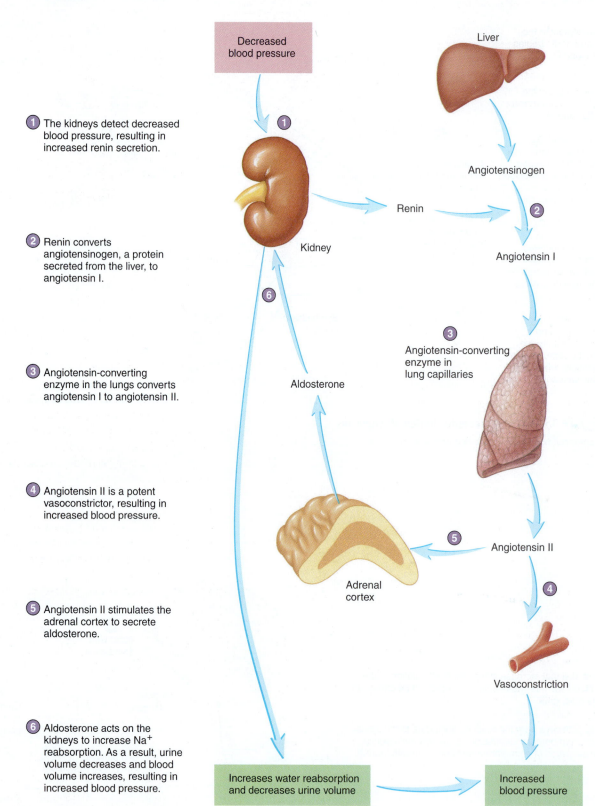

1 The kidneys detect decreased blood pressure, resulting in increased renin secretion.

2 Renin converts angiotensinogen, a protein secreted from the liver, to angiotensin I.

3 Angiotensin-converting enzyme in the lungs converts angiotensin I to angiotensin II.

4 Angiotensin II is a potent vasoconstrictor, resulting in increased blood pressure.

5 Angiotensin II stimulates the adrenal cortex to secrete aldosterone.

6 Aldosterone acts on the kidneys to increase Na⁺ reabsorption. As a result, urine volume decreases and blood volume increases, resulting in increased blood pressure.

Decreased blood pressure

Liver

Angiotensinogen

Renin

Kidney

Angiotensin I

Angiotensin-converting enzyme in lung capillaries

Aldosterone

Adrenal cortex

Angiotensin II

Vasoconstriction

Increases water reabsorption and decreases urine volume

Increased blood pressure

PROCESS Figure 13.31   Hormonal Regulation: Renin-Angiotensin-Aldosterone Mechanism

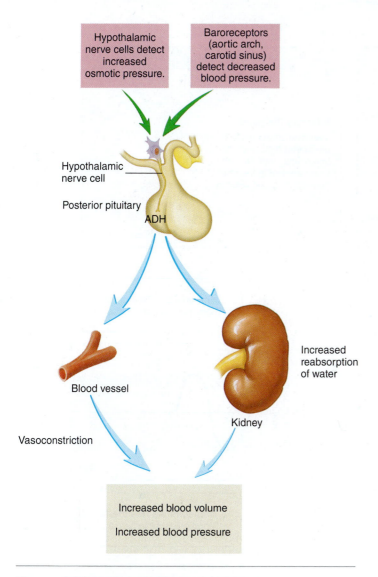

Figure 13.32 Hormonal Regulation: Antidiuretic Hormone Mechanism

Increases in the osmolality of blood or decreases in blood pressure result in antidiuretic hormone (ADH) secretion. ADH increases water reabsorption by the kidneys, and large amounts of ADH result in vasoconstriction. These changes maintain blood pressure.

13.28). They are sensitive to sudden changes in blood pressure, and they respond quickly. The chemoreceptor and adrenal medullary reflexes are also sensitive to sudden changes in blood pressure and respond quickly, but they respond to large changes in blood pressure. The renin-angiotensin-aldosterone, antidiuretic hormone, and atrial natriuretic mechanisms are more important in maintaining blood pressure on a long-term basis. They are influenced by small changes in blood pressure or concentration and respond by gradually bringing the blood pressure back into its normal range (figure 13.33).

Predict 8

Suppose that a hemorrhage results in rapid loss of a large volume of blood. What blood pressure regulation mechanisms will come into play? If the hemorrhage results in the loss of the same volume of blood over a period of several hours, will the same mechanisms respond? Explain.

13.9 EFFECTS OF AGING ON THE BLOOD VESSELS

Learning Outcome *After reading this section, you should be able to*

A. Describe the effects of aging on blood vessels.

The walls of all arteries undergo changes as people age. Some arteries change more rapidly than others, and some individuals are more susceptible to change than others. The most significant effects of aging occur in the large, elastic arteries, such as the aorta; in large arteries carrying blood to the brain; and in the coronary arteries.

Changes in arteries that make them less elastic, referred to as arteriosclerosis, occur in nearly every individual, and become more severe with advancing age. A type of arteriosclerosis called **atherosclerosis** (ath′er-ō-skler-ō′sis) results from the deposition of material in the walls of arteries that forms plaques (figure 13.34). The material is composed of a fatlike substance containing cholesterol. The fatty material can eventually be dominated by the deposition of dense connective tissue and calcium salts.

Several factors influence the development of atherosclerosis. Lack of exercise, smoking, obesity, and a diet high in cholesterol and fats appear to increase the severity of atherosclerosis and the rate at which it develops. Severe atherosclerosis is more prevalent in some families than in others, which suggests a genetic influence. Some evidence suggests that a low-fat diet, mild exercise, and relaxation activities slow the progression of atherosclerosis and may even reverse it to some degree.

Atherosclerosis greatly increases resistance to blood flow because the deposits narrow the inside diameter of the arteries. The added resistance hampers normal circulation to tissues and greatly increases the work the heart must perform. Furthermore, the rough atherosclerotic plaques attract platelets, which adhere to them and increase the chance of thrombus formation.

Capillaries narrow and become more irregular in shape with age. Their walls become thicker, and consequently the efficiency of capillary exchange decreases.

Veins tend to develop patchy thickenings in their walls, resulting in narrowing in these areas. The tendency to develop varicose veins increases with age (see Clinical Impact earlier in this chapter), as does the tendency to develop hemorrhoids (varicose veins of the rectum or anus) because some veins increase in diameter due to weakening of the connective tissue in their walls. There is a related increase in the development of thromboses and emboli, especially in veins that are dilated or those in which blood flow is sluggish.

Cardiovascular

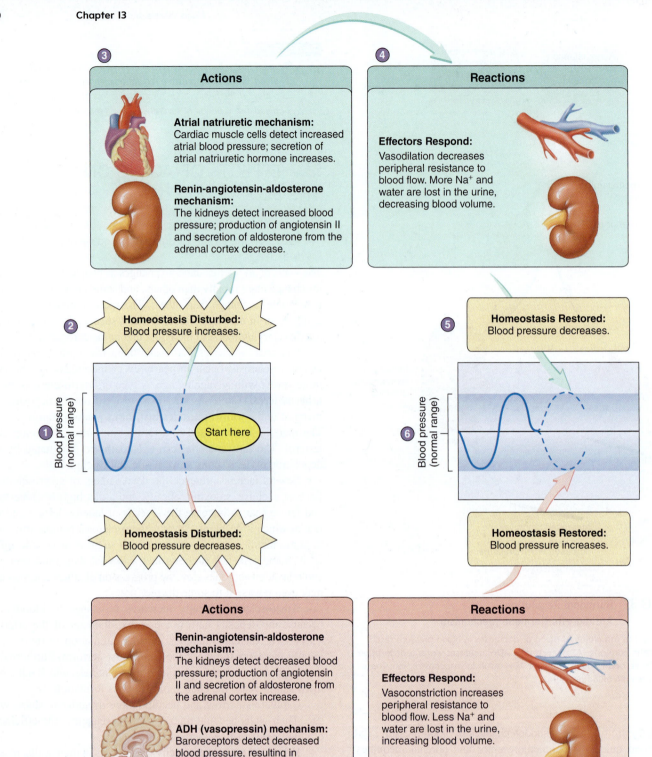

Homeostasis Figure 13.33 Long-Term Control of Blood Pressure

(1) Blood pressure is within its normal range. (2) Blood pressure increases outside the normal range, which causes homeostasis to be disturbed. (3) Increased blood pressure is detected by cardiac muscle cells and the kidneys (receptors). The heart and kidneys (control center) respond to increased blood pressure by the secretions of hormones. (4) Blood vessels of the body and the kidneys (effectors) respond to the hormones by dilating or adjusting blood volume through urine formation. (5) These changes cause blood pressure to decrease. (6) Blood pressure returns to its normal range, and homeostasis is restored.

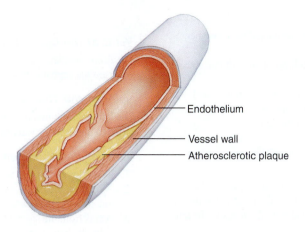

Endothelium

Vessel wall

Atherosclerotic plaque

Figure 13.34 Atherosclerotic Plaque
Atherosclerotic plaque develops within the tissue of the artery wall.

ANSWER TO LEARN TO PREDICT

After reading the Clinical Impact "Circulatory Shock" in this chapter, we learned that circulatory shock is inadequate blood flow throughout the body and that septic shock is one type of circulatory shock. Septic shock results from infections that cause the release of toxic substances into the circulatory system that depress heart activity, cause vasodilation, and increase capillary permeability. After T. J. developed septic shock, we would expect his blood volume to decrease as fluid moved from the more permeable capillaries to the interstitial spaces. The reduc-

tion in blood volume would lead to a drop in his blood pressure, stimulating the baroreceptor reflex mechanism and subsequently an increased heart rate. We would also expect T. J.'s stroke volume to decrease with a drop in blood volume. Increased sympathetic stimulation would cause vasoconstriction of blood vessels as T. J.'s body tried to maintain normal blood pressure. With the reduction in blood flow through the skin, T. J. would appear very pale. If T. J.'s blood pressure were not maintained, his blood pressure could continue to decrease to lethal levels.

Answers to the rest of this chapter's Predict questions are in Appendix E.

 SUMMARY

13.1 Functions of the Circulatory System (p. 350)

The circulatory system can be divided into the pulmonary vessels and the systemic vessels. The circulatory system and the heart maintain sufficient blood flow to tissues. The circulatory system carries blood; exchanges nutrients, waste products, and gases; transports hormones; regulates blood pressure; and directs blood flow.

13.2 General Features of Blood Vessel Structure (p. 351)

1. The heart pumps blood through elastic arteries, muscular arteries, and arterioles to the capillaries.
2. Blood returns to the heart from the capillaries through venules, small veins, medium-sized veins, and large veins.
3. Except for capillaries and venules, blood vessels have three layers:
   a. The tunica intima, the innermost layer, consists of endothelium, a basement membrane, and connective tissue.
   b. The tunica media, the middle layer, contains circular smooth muscle and elastic fibers.
   c. The outer tunica adventitia is composed of connective tissue.

Arteries

1. Large, elastic arteries have many elastic fibers but little smooth muscle in their walls. They carry blood from the heart to smaller arteries with little decrease in pressure.
2. The walls of muscular arteries have much smooth muscle and some elastic fibers. They undergo vasodilation and vasoconstriction to control blood flow to different regions of the body.
3. Arterioles, the smallest arteries, have smooth muscle cells and a few elastic fibers. They undergo vasodilation and vasoconstriction to control blood flow to local areas.

Capillaries

1. Capillaries consist of only endothelium and are surrounded by a basement membrane and loose connective tissue.
2. Nutrient and waste exchange is the principal function of capillaries.
3. Blood is supplied to capillaries by arterioles. Precapillary sphincters regulate blood flow through capillary networks.

Veins

1. Venules are composed of endothelium surrounded by a basement membrane.
2. Small veins are venules covered with a layer of smooth muscle and a layer of connective tissue.
3. Medium-sized and large veins contain less smooth muscle and fewer elastic fibers than arteries of the same size.
4. Valves prevent the backflow of blood in the veins.

13.3 Blood Vessels of the Pulmonary Circulation   (p. 353)

The pulmonary circulation moves blood to and from the lungs. The pulmonary trunk carries oxygen-poor blood from the heart to the lungs, and pulmonary veins carry oxygen-rich blood from the lungs to the left atrium of the heart.

13.4 Blood Vessels of the Systemic Circulation: Arteries   (p. 354)

Aorta

The aorta leaves the left ventricle to form the ascending aorta, the aortic arch, and the descending aorta, which consists of the thoracic aorta and the abdominal aorta.

Arteries of the Head and Neck

1. The brachiocephalic, left common carotid, and left subclavian arteries branch from the aortic arch to supply the head and the upper limbs.
2. The common carotid arteries and the vertebral arteries supply the head. The common carotid arteries divide to form the external carotids (which supply the face and mouth) and the internal carotids (which supply the brain).

Arteries of the Upper Limbs

The subclavian artery continues as the axillary artery and then as the brachial artery, which branches to form the radial and ulnar arteries.

Thoracic Aorta and Its Branches

The thoracic aorta has visceral branches, which supply the thoracic organs, and parietal branches, which supply the thoracic wall.

Abdominal Aorta and Its Branches

The abdominal aorta has visceral branches, which supply the abdominal organs, and parietal branches, which supply the abdominal wall.

Arteries of the Pelvis

Branches of the internal iliac arteries supply the pelvis.

Arteries of the Lower Limbs

The common iliac arteries give rise to the external iliac arteries, and the external iliac artery continues as the femoral artery and then as the popliteal artery in the leg. The popliteal artery divides to form the anterior and posterior tibial arteries.

13.5 Blood Vessels of the Systemic Circulation: Veins   (p. 362)

The superior vena cava drains the head, neck, thorax, and upper limbs. The inferior vena cava drains the abdomen, pelvis, and lower limbs.

Veins of the Head and Neck

1. The internal jugular veins drain the brain, anterior head, and anterior neck.
2. The external jugular veins drain the posterior head and posterior neck.

Veins of the Upper Limbs

The deep veins are the brachial, axillary, and subclavian; the superficial veins are the cephalic, basilic, and median cubital.

Veins of the Thorax

The left and right brachiocephalic veins and the azygos veins return blood to the superior vena cava.

Veins of the Abdomen and Pelvis

1. Posterior abdominal wall veins join the azygos veins.
2. Veins from the kidneys, adrenal glands, and gonads directly enter the inferior vena cava.
3. Veins from the stomach, intestines, spleen, and pancreas connect with the hepatic portal vein, which transports blood to the liver for processing. The hepatic veins from the liver join the inferior vena cava.

Veins of the Lower Limbs

1. The deep veins course with the deep arteries and have similar names.
2. The superficial veins are the great and small saphenous veins.

13.6 Physiology of Circulation   (p. 367)

Blood Pressure

1. Blood pressure is a measure of the force exerted by blood against the blood vessel walls.
2. Blood pressure moves blood through vessels.
3. Blood pressure can be measured by listening for Korotkoff sounds produced as blood flows through arteries partially constricted by a blood pressure cuff.

Pressure and Resistance

In a normal adult, blood pressure fluctuates between 120 mm Hg (systolic) and 80 mm Hg (diastolic) in the aorta. If blood vessels constrict, resistance to blood flow increases, and blood flow decreases.

Pulse Pressure

1. Pulse pressure is the difference between systolic and diastolic pressures. Pulse pressure increases when stroke volume increases.
2. A pulse can be detected when large arteries are near the body surface.

Capillary Exchange

1. Most exchange across the wall of the capillary occurs by diffusion.
2. Blood pressure, capillary permeability, and osmosis affect movement of fluid across the wall of the capillaries. There is a net movement of fluid from the blood into the tissues. The fluid gained by the tissues is removed by the lymphatic system.

13.7 Control of Blood Flow in Tissues   (p. 371)

Local Control of Blood Flow

Blood flow through a tissue is usually proportional to the metabolic needs of the tissue and is controlled by the precapillary sphincters.

Nervous and Hormonal Control of Blood Flow

1. The vasomotor center (sympathetic division) controls blood vessel diameter. Other brain areas can excite or inhibit the vasomotor center.
2. Vasomotor tone is the state of partial constriction of blood vessels.
3. The nervous system is responsible for routing the flow of blood, except in the capillaries, and for maintaining blood pressure.
4. Epinephrine and norepinephrine released by the adrenal medulla alter blood vessel diameter.

Cardiovascular

13.8 Regulation of Arterial Pressure   (p. 373)

Mean arterial pressure (MAP) is proportional to cardiac output times the peripheral resistance.

Baroreceptor Reflexes

1. Baroreceptors are sensitive to stretch.
2. Baroreceptors are located in the carotid sinuses and the aortic arch.
3. The baroreceptor reflex changes peripheral resistance, heart rate, and stroke volume in response to changes in blood pressure.

Chemoreceptor Reflexes

1. Chemoreceptors are sensitive to changes in blood O_2, CO_2, and pH.
2. Chemoreceptors are located in the carotid bodies and the aortic bodies.
3. The chemoreceptor reflex increases peripheral resistance in response to low O_2 levels, high CO_2 levels, and reduced blood pH.

Hormonal Mechanisms

1. Epinephrine released from the adrenal medulla as a result of sympathetic stimulation increases heart rate, stroke volume, and vasoconstriction.
2. The kidneys release renin in response to low blood pressure. Renin promotes the production of angiotensin II, which causes vasoconstriction and increased secretion of aldosterone. Aldosterone reduces urine output.
3. ADH released from the posterior pituitary causes vasoconstriction and reduces urine output.
4. The heart releases atrial natriuretic hormone when atrial blood pressure increases. Atrial natriuretic hormone stimulates an increase in urine production, causing a decrease in blood volume and blood pressure.

Summary of Regulatory Mechanisms

1. The baroreceptor, chemoreceptor, and adrenal medullary reflex mechanisms are most important in short-term regulation of blood pressure.
2. Hormonal mechanisms, such as the renin-angiotensin-aldosterone system, antidiuretic hormone, and atrial natriuretic hormone, are more important in long-term regulation of blood pressure.

13.9 Effects of Aging on the Blood Vessels   (p. 379)

1. Reduced elasticity and thickening of arterial walls result in hypertension and decreased ability to respond to changes in blood pressure.
2. Atherosclerosis is an age-related condition.
3. The efficiency of capillary exchange decreases with age.
4. Walls of veins thicken in some areas and dilate in others. Thromboses, emboli, varicose veins, and hemorrhoids are age-related conditions.

 # REVIEW AND COMPREHENSION

1. Name, in order, all the types of blood vessels, starting at the heart, going to the tissues, and returning to the heart.
2. Name the three layers of a blood vessel. What kinds of tissue are in each layer?
3. Relate the structures of the different types of arteries to their functions.
4. Describe a capillary network. Name the structure that regulates blood flow through the capillary network.
5. Describe the structure of capillaries, and explain their major function.
6. Describe the structure of veins.
7. What is the function of valves in blood vessels, and which blood vessels have valves?
8. List the parts of the aorta. Name the major arteries that branch from the aorta and deliver blood to the vessels that supply the heart, the head and upper limbs, and the lower limbs.
9. Name the arteries that supply the major areas of the head, upper limbs, thorax, abdomen, and lower limbs. Describe the area each artery supplies.
10. Name the major vessels that return blood to the heart. What area of the body does each drain?
11. List the veins that drain blood from the thorax, abdomen, and pelvis. What specific area of the body does each drain? Describe the hepatic portal system.
12. List the major veins that drain the upper and lower limbs.
13. Define blood pressure, and describe how it is normally measured.
14. Describe the changes in blood pressure, starting in the aorta, moving through the vascular system, and returning to the right atrium.
15. Define pulse pressure, and explain what information can be determined by monitoring the pulse.
16. Explain how blood pressure and osmosis affect the movement of fluid between capillaries and tissues. What happens to excess fluid that enters the tissues?
17. Explain what is meant by local control of blood flow through tissues, and describe what carries out local control.
18. Describe nervous control of blood vessels. Define vasomotor tone.
19. Define mean arterial pressure. How is it related to heart rate, stroke volume, and peripheral resistance?
20. Where are baroreceptors located? Describe the baroreceptor reflex when blood pressure increases and when it decreases.
21. Where are the chemoreceptors for CO_2 and pH located? Describe what happens when O_2 levels in the blood decrease.
22. For each of the following hormones—epinephrine, renin, angiotensin II, aldosterone, ADH, and atrial natriuretic hormone—state where each is produced, what stimulus causes increased hormone production, and the hormone's effect on the circulatory system.
23. Describe the changes that occur in arteries as people age.

CRITICAL THINKING

1. For each of the following destinations, name all the arteries a red blood cell encounters if it starts its journey in the left ventricle:
   a. brain
   b. external part of the skull
   c. left hand
   d. anterior portion of the right leg

2. For each of the following starting places, name all the veins a red blood cell encounters on its way back to the right atrium:
   a. left side of the brain
   b. external part of the right side of the skull
   c. left hand
   d. medial portion of the right leg
   e. kidney
   f. small intestine

3. In angioplasty, a surgeon threads a catheter through blood vessels to a blocked coronary artery. The tip of the catheter can expand, stretching the coronary artery and unblocking it, or the tip of the catheter can be equipped with tiny blades capable of removing the blockage. Typically, the catheter is first inserted into a large blood vessel in the superior, medial part of the thigh. Starting with this blood vessel, name all the blood vessels the catheter passes through to reach the anterior interventricular artery.

4. A 55-year-old man has a colonoscopy that reveals a large tumor in his colon. His doctor orders a liver scan to determine if the cancer has spread from the colon to the liver. Based on your knowledge of blood vessels, explain how cancer cells from the colon can end up in the liver.

5. High blood pressure can be caused by advanced atherosclerosis of the renal arteries, even though blood flow appears sufficient to allow a normal volume of urine to be produced. Explain how atherosclerosis of the renal arteries can result in high blood pressure.

6. Hugo Faster ran a race during which his stroke volume and heart rate increased. Vasoconstriction occurred in his viscera, and his blood pressure rose, but not dramatically. Explain these changes in his circulatory system.

7. Nitroglycerin is often given to people who experience angina pains. This drug causes vasodilation of arteries and veins, which reduces the amount of work the heart performs and increases blood flow through the coronary arteries. Explain why dilation of arteries and veins reduces the heart's amount of work.

Answers in Appendix D

14 Lymphatic System and Immunity

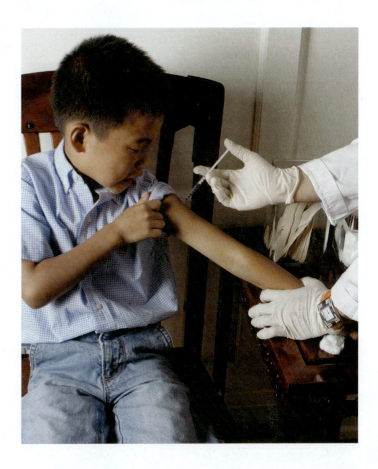

14.1 FUNCTIONS OF THE LYMPHATIC SYSTEM

Learning Outcomes *After reading this section, you should be able to*

A. Describe the functions of the lymphatic system.

B. Explain how lymph is formed.

One of the basic tenets of life is that many organisms consume or use other organisms in order to survive. Some microorganisms, such as certain bacteria or viruses, use humans as a source of nutrients and as an environment where they can survive and reproduce. As a result, some of these microorganisms can damage the body, causing disease or even death. Any substance or microorganism that causes disease or damage to the tissues of the body is considered a **pathogen.** Not surprisingly, the body has ways to resist or destroy pathogens. This chapter considers how the lymphatic system and the components of other systems, such as white blood cells and phagocytes, continually provide protection against pathogens.

The **lymphatic** (lim-fat′ik) **system** functions are

1. *Fluid balance.* About 30 liters (L) of fluid pass from the blood capillaries into the interstitial spaces each day, whereas only 27 L pass from the interstitial spaces back into the blood capillaries (see chapter 13). If the extra 3 L of interstitial fluid remained in the interstitial spaces, edema would result, causing tissue damage and eventually death. Instead, the 3 L of fluid enters the lymphatic capillaries,

Module 10 Lymphatic System

where it is called **lymph** (limf), and it passes through the lymphatic vessels to return to the blood. In addition to water, lymph contains solutes derived from two sources: (a) Substances in plasma, such as ions, nutrients, gases, and some proteins, pass from blood capillaries into the interstitial spaces and become part of the lymph; (b) substances such as hormones, enzymes, and waste products, derived from cells within the tissues, are also part of the lymph.

2. *Lipid absorption.* The lymphatic system absorbs lipids and other substances from the digestive tract (see figure 16.14) through lymphatic vessels called **lacteals** (lak′tē-ălz) located in the lining of the small intestine. Lipids enter the lacteals and pass through the lymphatic vessels to the venous circulation. The lymph passing through these lymphatic vessels appears white because of its lipid content and is called **chyle** (kīl).

3. *Defense.* Pathogens, such as microorganisms and other foreign substances, are filtered from lymph by lymph nodes and from blood by the spleen. In addition, lymphocytes and other cells are capable of destroying pathogens. Because the lymphatic system is involved with fighting infections, as well as filtering blood and lymph to remove pathogens, many infectious diseases produce symptoms associated with the lymphatic system (see the Diseases and Disorders table at the end of this chapter).

14.2 ANATOMY OF THE LYMPHATIC SYSTEM

Learning Outcomes *After reading this section, you should be able to*

A. Describe how lymph is transported.
B. Describe the structure and function of tonsils, lymph nodes, the spleen, and the thymus.

Lymphatic Capillaries and Vessels

The lymphatic system includes lymph, lymphocytes, lymphatic vessels, lymph nodes, the tonsils, the spleen, and the thymus (figure 14.1*a*). The lymphatic system, unlike the circulatory system, does not circulate fluid to and from tissues. Instead, the lymphatic system carries fluid in one direction, from tissues to the circulatory system. Fluid moves from blood capillaries into tissue spaces (see figure 13.24). Most of the fluid returns to the blood, but some of the fluid moves from the tissue spaces into lymphatic capillaries to become lymph (figure 14.2*a*). The **lymphatic capillaries** are tiny, closed-ended vessels consisting of simple squamous epithelium. The lymphatic capillaries are more permeable than blood capillaries because they lack a basement membrane, and fluid moves easily into them. Overlapping squamous cells of the lymphatic capillary walls act as valves that prevent the backflow of fluid (figure 14.2*b*). After fluid enters lymphatic capillaries, it flows through them.

Lymphatic capillaries are present in most tissues of the body. Exceptions are the central nervous system, bone marrow, and tissues lacking blood vessels, such as the epidermis and cartilage. A superficial group of lymphatic capillaries drains the dermis and subcutaneous tissue, and a deep group drains muscle, the viscera, and other deep structures.

The lymphatic capillaries join to form larger **lymphatic vessels,** which resemble small veins (figure 14.2*b*). Small lymphatic vessels have a beaded appearance because they have one-way valves that are similar to the valves of veins (see chapter 13). When a lymphatic vessel is compressed, the valves prevent backward movement of lymph. Consequently, compression of the lymphatic vessels causes lymph to move forward through them. Three factors cause compression of the lymphatic vessels: (1) contraction of

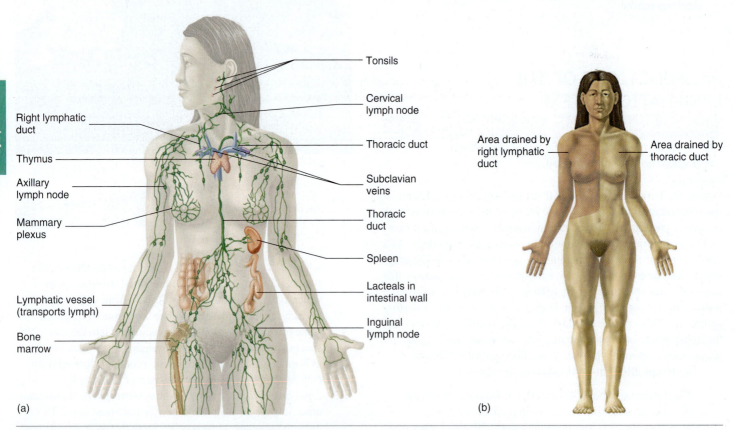

(a)

(b)

Right lymphatic duct

Thymus

Axillary lymph node

Mammary plexus

Lymphatic vessel (transports lymph)

Bone marrow

Tonsils

Cervical lymph node

Thoracic duct

Subclavian veins

Thoracic duct

Spleen

Lacteals in intestinal wall

Inguinal lymph node

Area drained by right lymphatic duct

Area drained by thoracic duct

Figure 14.1 Lymphatic System and Lymph Drainage

(*a*) The major lymphatic organs are the tonsils, the lymph nodes, the spleen, and the thymus. Lymph nodes are located along lymphatic vessels throughout the body, but aggregations of them are found in the cervical, axillary, and inguinal areas. (*b*) Lymph from the uncolored areas drains through the thoracic duct. Lymph from the darkened area drains through the right lymphatic duct.

Lymphatic

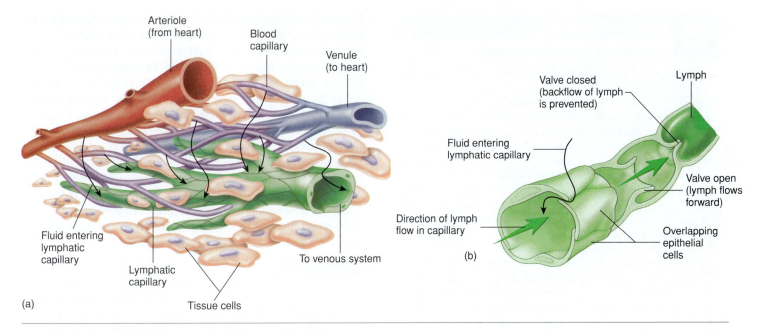

Figure 14.2 Lymph Formation and Movement

(*a*) Fluid moves from blood capillaries into tissues and from tissues into lymphatic capillaries to form lymph. (*b*) The overlap of epithelial cells of the lymphatic capillary allows fluid to enter easily but prevents it from moving back into the tissue. Valves, located farther along in lymphatic vessels, also ensure one-way flow of lymph.

surrounding skeletal muscle during activity, (2) periodic contraction of smooth muscle in the lymphatic vessel wall, and (3) pressure changes in the thorax during breathing.

The lymphatic vessels converge and eventually empty into the blood at two locations in the body. Lymphatic vessels from the right upper limb and the right half of the head, neck, and chest form the **right lymphatic duct,** which empties into the right subclavian vein. Lymphatic vessels from the rest of the body enter the **thoracic duct,** which empties into the left subclavian vein (see figure 14.1*b*).

Lymphatic Organs

The **lymphatic organs** include the tonsils, the lymph nodes, the spleen, and the thymus. **Lymphatic tissue,** which consists of many lymphocytes and other cells, such as macrophages, is found within lymphatic organs. The lymphocytes originate from red bone marrow (see chapter 11) and are carried by the blood to lymphatic organs. These lymphocytes divide and increase in number when the body is exposed to pathogens. The increased number of lymphocytes is part of the immune response that causes the destruction of pathogens. In addition to cells, lymphatic tissue has very fine reticular fibers (see chapter 4). These fibers form an interlaced network that holds the lymphocytes and other cells in place. When lymph or blood filters through lymphatic organs, the fiber network also traps microorganisms and other items in the fluid.

Tonsils

There are three groups of **tonsils** (figure 14.3; see figure 15.2). The **palatine** (pal′ă-tīn; palate) **tonsils** are located on each side of the posterior opening of the oral cavity; these are the ones usually referred to as "the tonsils." The **pharyngeal** (fă-rin′jē-ăl) **tonsil** is located near the internal opening of the nasal cavity. When the

pharyngeal tonsil is enlarged, it is commonly called the **adenoid** (ad′ĕ-noid), or adenoids. An enlarged pharyngeal tonsil can interfere with normal breathing. The **lingual** (ling′gwăl; tongue) **tonsil** is on the posterior surface of the tongue.

The tonsils form a protective ring of lymphatic tissue around the openings between the nasal and oral cavities and the pharynx. They protect against pathogens and other potentially harmful material entering from the nose and mouth. Sometimes the palatine or pharyngeal tonsils become chronically infected and must be removed. The lingual tonsil becomes infected less often than the other tonsils and is more difficult to remove. In adults, the tonsils decrease in size and may eventually disappear.

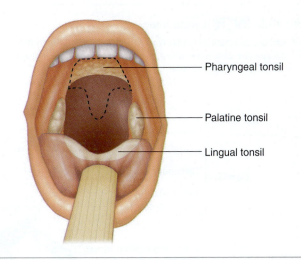

Figure 14.3 AP|R

Anterior view of the oral cavity, showing the tonsils. Part of the palate is removed (*dotted line*) to show the pharyngeal tonsil.

A CASE IN POINT

Tonsillectomy and Adenoidectomy

Audie Tory is a 4-year-old girl who has some hearing loss and an associated delay in speech development. She has a history of frequent sore throats and middle ear infections, which have been treated with antibiotics. Recently, she has experienced difficulty in swallowing; she also snores and sleeps with her mouth open. Audie's physician tells her parents that her palatine tonsils and adenoid are enlarged and chronically infected. Her enlarged tonsils make swallowing difficult, and the infection stimulates inflammation that causes her throat to hurt. Her enlarged adenoid is restricting airflow, causing her to snore and sleep with her mouth open—and it is also the probable cause of her middle ear infections because the openings to the auditory tubes are located next to the adenoid. Chronic middle ear infections are associated with loss of hearing, which affects speech development. The doctor recommends a procedure called a **T&A**—that is, a **tonsillectomy** (ton′si-lek′tō-mē), which is removal of the palatine tonsils, and an **adenoidectomy** (ad′ĕ-noy-dek′tō-mē), which is removal of the adenoid.

Lymph Nodes

Lymph nodes are rounded structures, varying from the size of a small seed to that of a shelled almond. Lymph nodes are distributed along the various lymphatic vessels (see figure 14.1a), and most lymph passes through at least one lymph node before entering the blood. Although lymph nodes are found throughout the body, there are three superficial aggregations of lymph nodes on each side of the body: inguinal nodes in the groin, axillary nodes in the axilla (armpit), and cervical nodes in the neck.

A dense connective tissue **capsule** surrounds each lymph node (figure 14.4). Extensions of the capsule, called **trabeculae,** subdivide a lymph node into compartments containing lymphatic tissue and lymphatic sinuses. The lymphatic tissue consists of lymphocytes and other cells that can form dense aggregations of tissue called

lymphatic nodules. Lymphatic sinuses are spaces between the lymphatic tissue that contain macrophages on a network of fibers. Lymph enters the lymph node through afferent vessels, passes through the lymphatic tissue and sinuses, and exits through efferent vessels.

As lymph moves through the lymph nodes, two functions are performed. One function is to activate the immune system. Pathogens in the lymph can stimulate lymphocytes in the lymphatic tissue to divide. The lymphatic nodules containing the rapidly dividing lymphocytes are called **germinal centers.** The newly produced lymphocytes are released into the lymph and eventually reach the blood, where they circulate and enter other lymphatic tissues. The lymphocytes are part of the adaptive immune response (see "Adaptive Immunity" later in this chapter) that destroys pathogens. The second function of the lymph nodes is to remove pathogens from the lymph through the action of macrophages.

Predict 2

Cancer cells can spread from a tumor site to other areas of the body through the lymphatic system. At first, however, as the cancer cells pass through the lymphatic system, they are trapped in the lymph nodes, which filter the lymph. During radical cancer surgery, malignant (cancerous) lymph nodes are removed, and their vessels are cut and tied off to prevent the cancer from spreading. Predict the consequences of tying off the lymphatic vessels.

Spleen

The **spleen** (splēn) is roughly the size of a clenched fist and is located in the left, superior corner of the abdominal cavity (figure 14.5). The spleen has an outer **capsule** of dense connective tissue and a small amount of smooth muscle. **Trabeculae** from the capsule divide the spleen into small, interconnected compartments containing two specialized types of lymphatic tissue. **White pulp** is lymphatic tissue surrounding the arteries within the spleen. **Red pulp** is associated with the veins. It consists of a fibrous network, filled with macrophages and red blood cells, and enlarged capillaries that connect to the veins.

Figure 14.4 AP|R Lymph Node

Arrows indicate direction of lymph flow. As lymph moves through the sinuses, phagocytic cells remove foreign substances. The germinal centers are sites of lymphocyte production.

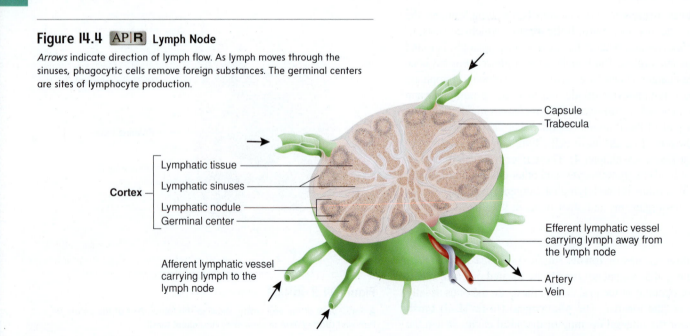

Cortex
- Lymphatic tissue
- Lymphatic sinuses
- Lymphatic nodule
- Germinal center

Afferent lymphatic vessel carrying lymph to the lymph node

Capsule
Trabecula

Efferent lymphatic vessel carrying lymph away from the lymph node

Artery
Vein

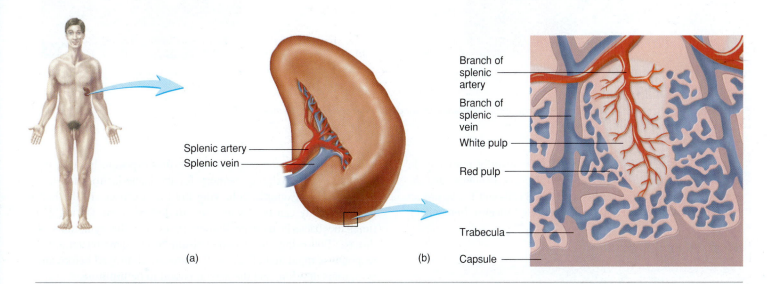

(a)   (b)

Figure 14.5 AP|R Spleen

(*a*) Inferior view of the spleen. (*b*) Section showing the arrangement of arteries, veins, white pulp, and red pulp. White pulp is associated with arteries, and red pulp is associated with veins.

The spleen filters blood instead of lymph. Cells within the spleen detect and respond to foreign substances in the blood and destroy worn-out red blood cells. Lymphocytes in the white pulp can be stimulated in the same manner as in lymph nodes. Before blood leaves the spleen through veins, it passes through the red pulp. Macrophages in the red pulp remove foreign substances and worn-out red blood cells through phagocytosis.

The spleen also functions as a blood reservoir, holding a small volume of blood. In emergency situations, such as hemorrhage, smooth muscle in splenic blood vessels and in the splenic capsule can contract, allowing a small amount of blood to move out of the spleen into the general circulation.

Thymus

The **thymus** (thi′mŭs) is a bilobed gland roughly triangular in shape (figure 14.6*a*). It is located in the superior mediastinum, the partition dividing the thoracic cavity into left and right parts. Each lobe of the thymus is surrounded by a thin connective tissue **capsule. Trabeculae** from the capsule divide each lobe into **lobules** (figure 14.6*b*). Near the capsule and trabeculae, the lymphocytes are numerous and form dark-staining areas called the **cortex.** A lighter-staining, central portion of the lobules, called the **medulla,** has fewer lymphocytes.

The thymus is the site for the maturation of a class of lymphocytes called T cells (described in "Adaptive Immunity" later in this chapter). Large numbers of T cells are produced in the thymus, but most degenerate. The T cells that survive the maturation process are capable of reacting to foreign substances. The mature T cells migrate to the medulla, enter the blood, and travel to other lymphatic tissues, where they help protect against pathogens. Production of T cells declines later in life due to decreased function of the thymus (see "Effects of Aging on the Lymphatic System and Immunity" later in this chapter).

Overview of the Lymphatic System

Figure 14.7 summarizes the processes performed by the lymphatic system. Lymphatic capillaries and vessels remove fluid from tissues and absorb lipids from the small intestine. Lymph nodes filter lymph, and the spleen filters blood.

Figure 14.7 also illustrates two types of lymphocytes, called B cells and T cells. B cells originate and mature in red bone marrow. Pre-T cells are produced in red bone marrow and migrate to the thymus, where they mature to become T cells. B cells from red bone marrow and T cells from the thymus circulate to, and populate, other lymphatic tissues.

Lymphatic

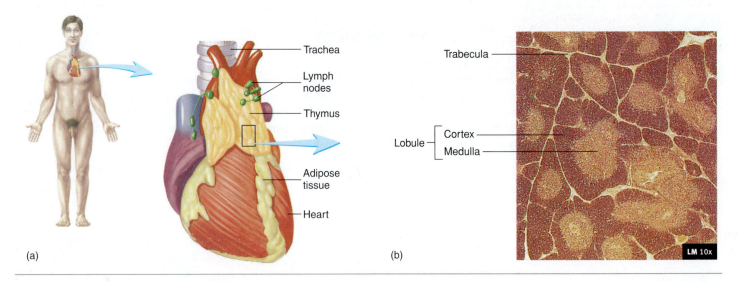

Figure 14.6 AP|R

(*a*) Location and shape of the thymus. (*b*) Histology of thymic lobules, showing the outer cortex and the inner medulla.

B cells and T cells are responsible for much of immunity. In response to infections, B cells and T cells increase in number and circulate to lymphatic and other tissues. How B cells and T cells protect the body is discussed later in this chapter (see "Adaptive Immunity").

14.3 IMMUNITY

Learning Outcome   *After reading this section, you should be able to*

A. Define the concepts of specificity and memory as they apply to immunity.

Immunity (i-mū′ni-tē) is the ability to resist damage from pathogens, such as microorganisms; harmful chemicals, such as toxins released by microorganisms; and internal threats, such as cancer cells. Immunity is categorized as **innate** (i′nāt, i-nāt′) **immunity** (also called *nonspecific resistance*) and **adaptive immunity** (also called *specific immunity*), although the two systems are fully integrated in the body. In innate immunity, the body recognizes and destroys certain pathogens, but the response to them is the same each time the body is exposed. In adaptive immunity, the body recognizes and destroys pathogens, but the response to them improves each time the pathogen is encountered.

Specificity and memory are characteristics of adaptive immunity, but not innate immunity. **Specificity** is the ability of adaptive immunity to recognize a particular substance. For example, innate immunity can act against bacteria in general, whereas adaptive immunity can distinguish among various kinds of bacteria. **Memory** is the ability of adaptive immunity to "remember" previous encounters with a particular substance. As a result, future responses are faster, stronger, and longer-lasting.

In innate immunity, each time the body is exposed to a substance, the response is the same because specificity and memory of previous encounters are not present. For example, each time a bacterial cell is introduced into the body, it is phagocytized with the same speed and efficiency. In adaptive immunity, the response during the second exposure to the same bacteria is faster and stronger than the response to the first exposure because the immune system exhibits memory for the bacteria from the first exposure. For example, following the first exposure to the bacteria, the body can take many days to destroy them. During this time, the bacteria damage tissues, producing the symptoms of disease. Following the second exposure to the same bacteria, the response is rapid and effective. Bacteria are destroyed before any symptoms develop, and the person is said to be **immune.**

Innate and adaptive immunity are intimately linked. Most important, mediators of innate immunity are required for the initiation and regulation of the adaptive response.

14.4 INNATE IMMUNITY

Learning Outcomes   *After reading this section, you should be able to*

A. Define innate immunity, and describe the cells and chemical mediators involved.
B. List the events of an inflammatory response, and explain their significance.

Innate immunity is accomplished by physical barriers, chemical mediators, white blood cells, and the inflammatory response.

Physical Barriers

Physical barriers prevent pathogens and chemicals from entering the body in two ways: (1) The skin and mucous membranes form barriers that prevent their entry, and (2) tears, saliva, and urine wash these substances from body surfaces. Pathogens cannot cause a disease if they cannot get into the body.

Chemical Mediators

Chemical mediators are molecules responsible for many aspects of innate immunity. Some chemicals on the surface of cells destroy pathogens or prevent their entry into the cells. For example, lysozyme in tears and saliva kills certain bacteria, and mucus on the mucous membranes prevents the entry of some pathogens. Other chemical

1. Lymphatic capillaries remove fluid from tissues. The fluid becomes lymph (see figure 14.2a).

2. Lymph flows through lymphatic vessels, which have valves that prevent the backflow of lymph (see figure 14.2b).

3. Lymph nodes filter lymph (see figure 14.4) and are sites where lymphocytes respond to infections.

4. Lymph enters the thoracic duct or the right lymphatic duct.

5. Lymph enters the blood.

6. Lacteals in the small intestine (see figure 16.14) absorb lipids, which enter the thoracic duct.

7. Chyle, which is lymph containing lipids, enters the blood.

8. The spleen (see figure 14.5) filters blood and is a site where lymphocytes respond to infections.

9. Lymphocytes (pre-B and pre-T cells) originate from stem cells in the red bone marrow (see figure 14.9). The pre-B cells become mature B cells in the red bone marrow and are released into the blood. The pre-T cells enter the blood and migrate to the thymus.

10. The thymus (see figure 14.6) is where pre-T cells derived from red bone marrow increase in number and become mature T cells that are released into the blood (see figure 14.9).

11. B cells and T cells from the blood enter and populate all lymphatic tissues. These lymphocytes can remain in tissues or pass through them and return to the blood. B cells and T cells can also respond to infections by dividing and increasing in number. Some of the newly formed cells enter the blood and circulate to other tissues.

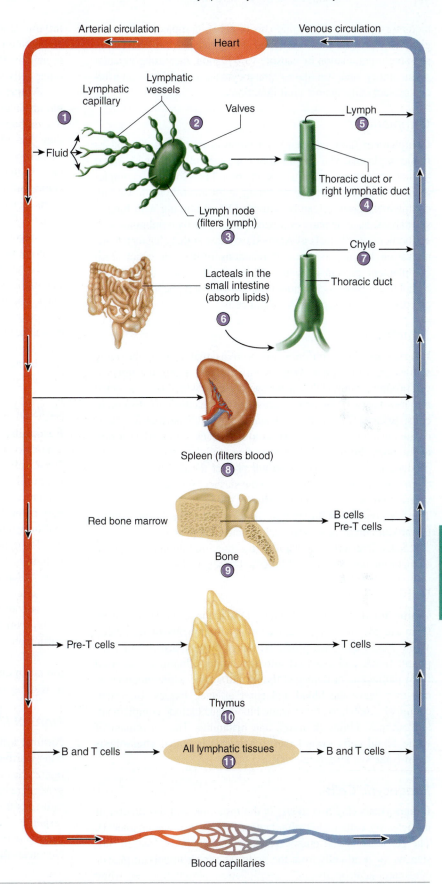

PROCESS Figure 14.7 AP|R Overview of the Lymphatic System

mediators, such as histamine (his′tă-mēn), complement, prosta-glandins (pros-tă-glan′dinz), and leukotrienes (loo-kō-trī′ēnz), promote inflammation by causing vasodilation, increasing vascular permeability, and stimulating phagocytosis. In addition, interferons protect cells against viral infections.

Complement

Complement (kom′plĕ-ment) is a group of more than 20 proteins found in plasma. The operation of complement proteins is similar to that of clotting proteins (see chapter 11). Normally, complement proteins circulate in the blood in an inactive form. Certain complement proteins can be activated by combining with foreign substances, such as parts of a bacterial cell, or by combining with antibodies (see "Effects of Antibodies" later in this chapter). Once activation begins, a series of reactions results, in which each complement protein activates the next. Once activated, certain complement proteins promote inflammation and phagocytosis and can directly lyse (rupture) bacterial cells.

Interferons

Interferons (in-ter-fēr′onz) are proteins that protect the body against viral infections. When a virus infects a cell, the infected cell produces viral nucleic acids and proteins, which are assembled into new viruses. The new viruses are then released to infect other cells. Because infected cells usually stop their normal functions or die during viral replication, viral infections are clearly harmful to the body. Fortunately, viruses often stimulate infected cells to produce interferons, which do not protect the cell that produces them. Instead, interferons bind to the surface of neighboring cells, where they stimulate those cells to produce antiviral proteins. These antiviral proteins inhibit viral reproduction by preventing the production of new viral nucleic acids and proteins.

Some interferons play a role in activating immune cells, such as macrophages and natural killer cells (see next section).

White Blood Cells

White blood cells (see chapter 11) and the cells derived from them are the most important cellular components of immunity. White blood cells are produced in red bone marrow and lymphatic tissue and released into the blood. Chemicals released from pathogens or damaged tissues attract the white blood cells, and they leave the blood and enter affected tissues. Important chemicals known to attract white blood cells include complement, leukotrienes, kinins (kī′ninz), and histamine. The movement of white blood cells toward these chemicals is called **chemotaxis** (kem-ō-tak′sis, kē-mō-tak′sis).

Phagocytic Cells

Phagocytosis (fag′ō-sī-tō′sis) is the ingestion and destruction of particles by cells called **phagocytes** (fag′ō-sītz) (see chapter 3). The particles can be microorganisms or their parts, foreign substances, or dead cells from the body. The most important phagocytes are neutrophils and macrophages, although other white blood cells also have limited phagocytic ability.

Neutrophils (noo′trō-filz) are small phagocytic cells that are usually the first cells to enter infected tissues from the blood in large numbers. They release chemical signals that increase the inflammatory response by recruiting and activating other immune cells. Neutrophils often die after phagocytizing a single microorganism. **Pus** is an accumulation of fluid, dead neutrophils, and other cells at a site of infection.

Macrophages (mak′rō-fā′jes) are monocytes that leave the blood, enter tissues, and enlarge about fivefold. Monocytes and macrophages form the **mononuclear phagocytic system** because they are phagocytes with a single (mono), unlobed nucleus. Sometimes macrophages are given specific names, such as *dust cells* in the lungs, *Kupffer cells* in the liver, and *microglia* in the central nervous system. Macrophages can ingest more and larger items than can neutrophils. Macrophages usually appear in infected tissues after neutrophils do, and they are responsible for most of the phagocytic activity in the late stages of an infection, including cleaning up dead neutrophils and other cellular debris.

In addition to leaving the blood in response to an infection, macrophages are also found in uninfected tissues. If pathogens enter uninfected tissue, the macrophages may phagocytize the microorganisms before they can replicate or cause damage. For example, macrophages are located at potential points where pathogens may enter the body, such as beneath the skin and mucous membranes, and around blood and lymphatic vessels. They also protect lymph in lymph nodes and blood in the spleen and liver.

Cells of Inflammation

Basophils, which are derived from red bone marrow, are motile white blood cells that can leave the blood and enter infected tissues. **Mast cells,** which are also derived from red bone marrow, are nonmotile cells in connective tissue, especially near capillaries. Like macrophages, mast cells are located at points where pathogens may enter the body, such as the skin, lungs, gastrointestinal tract, and urogenital tract.

Basophils and mast cells can be activated through innate immunity (e.g., by complement) or through adaptive immunity (see "Adaptive Immunity" later in this chapter). When activated, they release chemicals, such as histamine and leukotrienes, that produce an inflammatory response or activate other mechanisms, such as smooth muscle contraction in the lungs. **Eosinophils** also participate in inflammation associated with allergies and asthma.

Inflammation is beneficial in the fight against pathogens, but too much inflammation can be harmful, destroying healthy tissues as well as the microorganisms.

Natural Killer Cells

Natural killer (NK) cells, a type of lymphocyte produced in red bone marrow, account for up to 15% of lymphocytes. NK cells recognize classes of cells, such as tumor cells or virus-infected cells, in general, rather than specific tumor cells or cells infected by a specific virus. For this reason, and because NK cells do not exhibit a memory response, they are classified as part of innate immunity. NK cells use a variety of methods to kill their target cells, including releasing chemicals that damage cell membranes and cause the cells to lyse.

Inflammatory Response

The **inflammatory response** to injury involves many of the chemicals and cells previously discussed. Most inflammatory responses are very similar, although some details vary, depending on the intensity of the response and the type of injury. In figure 14.8, we use a bacterial infection to illustrate an inflammatory response. Bacteria enter the tissue, causing damage that stimulates the release or activation of chemical mediators, such as histamine, prostaglandins, leukotrienes, complement, and kinins. These chemicals produce several effects: (1) Vasodilation increases blood flow and brings phagocytes and other white blood cells to the area; (2) phagocytes leave the blood and enter the tissue; and (3) increased vascular permeability allows fibrinogen and complement to enter the tissue from the blood. Fibrinogen is con- verted to fibrin (see chapter 11), which isolates the infection by walling off the infected area. Complement further enhances the inflammatory response and attracts additional phagocytes. This process of releasing chemical mediators and attracting phago- cytes and other white blood cells continues until the bacteria are destroyed. Phagocytes remove microorganisms and dead tissue, and the damaged tissues are repaired.

Inflammation can be local or systemic. **Local inflammation** is an inflammatory response confined to a specific area of the body. Symptoms of local inflammation include redness, heat, and swelling due to increased blood flow and increased vascular permeability, as well as pain caused by swelling and by chemical mediators acting on pain receptors. The tissue destruction, swelling, and pain lead to loss of function (see chapter 4).

Systemic inflammation is an inflammatory response that is generally distributed throughout the body. In addition to the local symptoms at the sites of inflammation, three additional features can be present:

1. Red bone marrow produces and releases large numbers of neutrophils, which promote phagocytosis.
2. **Pyrogens** (pī′rō-jenz; fever-producing), chemicals released by microorganisms, neutrophils, and other cells, stimulate fever production. Pyrogens affect the body's temperature-regulating mechanism in the hypothalamus in the brain. As a consequence, heat production and conservation increase, raising body temperature. Fever promotes the activities of the immune system, such as phagocytosis, and inhibits the growth of some microorganisms.
3. In severe cases of systemic inflammation, vascular permeability can increase so much that large amounts of fluid are lost from the blood into the tissues. The decreased blood volume can cause shock and death (see Clinical Impact "Circulatory Shock" in chapter 13).

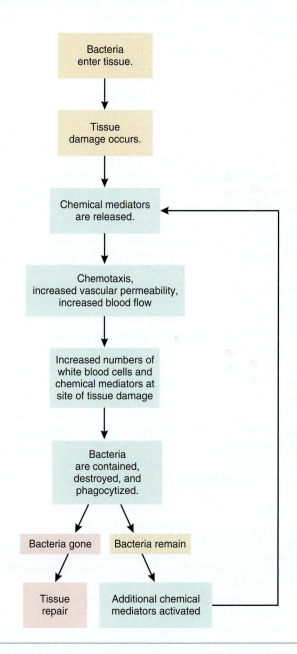

Figure 14.8 Inflammatory Response

Bacteria cause tissue damage and the release of chemical mediators that initiate inflammation and phagocytosis, resulting in the destruction of the bacteria. If any bacteria remain, additional chemical mediators are activated. After all the bacteria are destroyed, the tissue is repaired.

Lymphatic

MICROBES IN YOUR BODY — Do Our Gut Bacteria Drive Immune Development and Function?

"All disease begins in the gut." This quote from Hippocrates (460-377 B.C.), the father of western medicine, is still relevant today. Over the last four decades, there have been increasing numbers of people suffering from allergies and autoimmune disorders. Researchers hypothesize that the increase in these conditions stems from inadequate development of immune function. In turn, they hypothesize that underdeveloped immune function is due to deficiencies in our gut microbiota. This has led to the Hygiene Hypothesis, which states that increased use of antibiotics and antimicrobial chemicals damages normal gut and other microbiota critical for immune system development and function.

Could the Hygiene Hypothesis explain the observed increases in allergies and autoimmune disorders? Much of the evidence for the importance of gut microbiota for immune function is derived from studies with germ-free mice. These lab-raised mice lack the natural microorganisms in their gut and in their body. As a result, the mice have multiple defects with their lymphatic tissues, such as fewer and smaller Peyer patches in the gut and fewer B and T lymphocytes. However, if scientists place intestinal or fecal microbiota from normal mice into the gut of germ-free mice, the germ-free mice immune tissues begin developing and functioning normally.

The importance of the gut in immune development is further supported by the fact that it contains the largest concentration of lymphatic tissue and microbiota in the human body. In humans, the gut microbiota begin to appear just before birth, and as the baby passes through the birth canal, more microorganisms are transferred from the mother to the baby. The makeup of a baby's microbiota is influenced by many factors including genetics, mode of delivery (vaginal or C-section), antibiotic use, stress and the mother's diet during late pregnancy. The first year of life is the most critical for accumulation of gut bacteria, but this process continues through childhood. At about 10 years of age, a person's gut microbiota is established and remains similar throughout life. Humans and their gut microbiota have a symbiotic relationship in that the gut provides space and nutrients for the microbiota which, in turn, provide their host with specialized nutrition, physiological regulators and protection against pathogens. Because of these ever-present microbiota ("good" bacteria), human gut epithelial and immune cells must maintain tolerance to them, but still protect against invading gut pathogens ("bad" bacteria).

How do our cells distinguish between "good" and "bad" bacteria? As it turns out, gut microbiota help stimulate development of immune cells by triggering production of different receptors. These receptors are found in cell membranes of white blood cells such as macrophages, neutrophils, and also in cell membranes of intestinal epithelial cells. The surface of all bacterial cells has bacteria-specific molecules that can be recognized by the receptors of defense cells and is what allows for distinction between "good" and "bad" microorganisms. Activation of the receptors triggers a cascade of events that results in immune responses such as T-lymphocyte activation and production of immunity chemicals. In addition, the "good" microbiota also attack invading "bad" bacteria by secreting antimicrobial substances against them and competing with them for nutrients and space. Thus, without appropriate amounts and/or types of gut microbiota, the body's immune system may not have all of the essential messages for production of specific immune cells and chemicals that kill pathogenic intestinal microorganisms.

Medical professionals are interested in manipulating gut microbiota to reduce allergies and other diseases, and to promote healing. First, and perhaps most importantly, is to get the desired population of gut microbiota started immediately through breastfeeding. Human breast milk contains certain carbohydrates that stimulate growth of specific intestinal microbiota but also prevent infection by some pathogens. If problems should arise later in life, the use of prebiotics (non-digestible carbohydrates that promote the growth of healthy microbiota) and probiotics (live normal gut microbiota) is being actively explored. However, there is still much work to be done before we fully understand the extent to which gut microbiota are involved in human immune function.

14.5 ADAPTIVE IMMUNITY

Learning Outcomes *After reading this section, you should be able to*

A. Define antigen.

B. Describe the origin, development, activation, and proliferation of lymphocytes.

C. Define antibody-mediated immunity and cell-mediated immunity, and name the cells responsible for each.

D. Diagram the structure of an antibody, and describe the effects produced by antibodies.

E. Discuss the primary and secondary responses to an antigen. Explain the basis for long-lasting immunity.

F. Describe the functions of T cells.

Adaptive immunity exhibits specificity and memory. As explained earlier in this chapter, specificity is the ability to recognize a particular substance, and memory is the ability to respond with increasing effectiveness to successive exposures to the antigen. **Antigens** (an'ti-jenz; *anti* (body) + -*gen*, producing) are substances that stimulate adaptive immune responses. Antigens can be divided into two groups: foreign antigens and self-antigens.

Foreign antigens are introduced from outside the body. Microorganisms, such as bacteria and viruses, and chemicals released by microorganisms are examples of foreign antigens. Pollen, animal hairs, foods, and drugs can cause an **allergic reaction** because they are foreign antigens that produce an overreaction of the immune system (see the Diseases and Disorders table at the end of this chapter). Transplanted tissues and organs contain foreign antigens, and the response to these antigens can cause rejection of the transplant.

Self-antigens are molecules the body produces to stimulate an immune system response. The response to self-antigens can be beneficial. For example, the recognition of tumor antigens can result in destruction of the tumor. But the response to self-antigens can also be harmful. **Autoimmune disease** results when self-antigens stimulate unwanted destruction of normal tissue. An example is rheumatoid arthritis, which destroys the tissue within joints.

Adaptive immunity can be divided into antibody-mediated immunity and cell-mediated immunity. **Antibody-mediated immunity** involves a group of lymphocytes called **B cells** and proteins called **antibodies** (an'tĕ-bod-ēz), which are found in the plasma. Antibodies are produced by **plasma cells,** which are derived from the B cells. **Cell-mediated immunity** involves the actions of a second type of lymphocyte, called **T cells.** Several subpopulations of T cells exist. For example, **cytotoxic** (sī-tō-tok'sik; destructive to cells) **T cells** produce the effects of cell-mediated immunity, and **helper T cells** can promote or inhibit the activities of both antibody-mediated immunity and cell-mediated immunity.

Table 14.1 summarizes and contrasts the main features of innate immunity and the two categories of adaptive immunity (i.e., antibody-mediated immunity and cell-mediated immunity).

Origin and Development of Lymphocytes

To understand how lymphocytes are responsible for antibody-mediated and cell-mediated immunity, it is important to know how lymphocytes originate and become specialized immune cells. **Stem cells** in red bone marrow are capable of giving rise to all the blood cells (see figure 11.2). Some stem cells give rise to pre-T cells, which migrate through the blood to the thymus, where they

divide and are processed into T cells (figure 14.9). Other stem cells produce pre-B cells, which are processed in the red bone marrow into B cells.

B cells are released from red bone marrow, and T cells are released from the thymus. Both types of cells move through the blood to lymphatic tissues (see figure 14.7). These lymphocytes live for a few months to many years and continually circulate between the blood and the lymphatic tissues. Normally, there are about five T cells for every B cell in the blood. When stimulated by an antigen, B cells and T cells divide, producing cells that are responsible for the destruction of antigens.

Small groups of identical B cells or T cells, called **clones,** form during embryonic development. Each clone is derived from a single, unique B cell or T cell. Each clone can respond only to a particular antigen. However, there is such a large variety of clones that the immune system can react to most antigens. Among the antigens to which the clones can respond are self-antigens. Because this response could destroy the body's own cells, clones acting against self-antigens are normally eliminated or suppressed. Most of this process occurs during prenatal development, but it also continues after birth and throughout a person's lifetime.

Activation and Multiplication of Lymphocytes

The specialized B-cell or T-cell clones can respond to antigens and produce an adaptive immune response. For the adaptive immune response to be effective, two events must occur: (1) antigen recognition by lymphocytes and (2) proliferation of the lymphocytes recognizing the antigen.

Antigen Recognition

Lymphocytes have proteins, called **antigen receptors,** on their surfaces. The antigen receptors on B cells are called **B-cell receptors,** and those on T cells are called **T-cell receptors.** Each receptor

TABLE 14.1	Comparison of Innate and Adaptive Immunity					
Primary Cells	**Origin of Cells**	**Site of Maturation**	**Location of Mature Cells**	**Primary Secretory Product**	**Primary Actions**	**Allergic Reactions**
Innate Immunity						
Neutrophils, eosinophils, basophils, mast cells, monocytes, and macrophages	Red bone marrow	Red bone marrow (neutrophils, eosinophils, basophils, and monocytes) and tissues (mast cells and macrophages)	Blood, connective tissue, and lymphatic tissue	Histamine, complement, prostaglandins, leukotrienes, kinins, and interferons	Inflammatory response and phagocytosis	None
Adaptive Immunity						
Antibody-mediated immunity (B cells)	Red bone marrow	Red bone marrow	Blood and lymphatic tissue	Antibodies	Protection against extracellular antigens (bacteria, toxins, and viruses outside of cells)	Immediate hypersensitivity
Cell-mediated immunity (T cells)	Red bone marrow	Thymus	Blood and lymphatic tissue	Cytokines	Protection against intracellular antigens (viruses and intracellular bacteria) and tumors; responsible for graft rejection	Delayed hypersensitivity

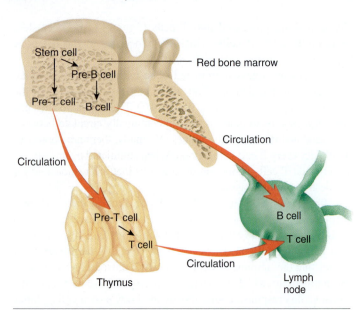

Figure 14.9 Origin and Processing of B Cells and T Cells

Both B cells and T cells originate from stem cells in red bone marrow. B cells are processed from pre-B cells in the red marrow, whereas T cells are processed from pre-T cells in the thymus. Both B cells and T cells circulate to other lymphatic tissues, such as lymph nodes.

binds with only a specific antigen. Each clone consists of lymphocytes that have identical antigen receptors on their surfaces. When antigens combine with the antigen receptors of a clone, the lymphocytes in that clone can be activated, and the adaptive immune response begins.

B cells and T cells typically recognize antigens after large molecules have been processed or broken down into smaller fragments. Antigen-presenting cells, such as macrophages, present antigens to B and T cells. The antigens are taken into macrophages by phagocytosis and are broken down into smaller antigen fragments. The processed antigen fragments are bound to major histocompatibility complex molecules, transported to the surface of the macrophages, and presented to B cells and T cells (figure 14.10, step 1).

Major histocompatibility complex (MHC) molecules are glycoproteins that have binding sites for antigens. Different MHC molecules have different binding sites—that is, they are specific for certain antigens. It is important not to confuse the MHC molecules with the antigen receptors described above. Though both types of receptors interact with antigens, MHC molecules are a different group of receptors found on the membrane of many types of cells. There are two classes of MHC molecules. MHC class I molecules are found on the membranes of most nucleated cells and MHC class II molecules are found on the membranes of antigen-presenting cells, B lymphocytes, and other defense cells. The MHC molecules function as "serving trays" that hold and present a processed antigen on the outer surface of the cell membrane. The combined MHC molecule and processed antigen can then bind to the antigen receptor on a B cell or T cell and stimulate it. For example, figure 14.10, step 2, illustrates how helper T cells are stimulated when combined with MHC class II molecules.

The MHC molecule/antigen combination is usually only the first signal necessary to produce a response from a B cell or T cell.

In many cases, **costimulation** by a second signal is also required. Costimulation can be achieved by **cytokines** (sī′tō-kīnz), which are proteins or peptides secreted by one cell as a regulator of neighboring cells. For example, **interleukin-1** (in-ter-loo′kin) is a cytokine released by macrophages that can stimulate helper T cells (figure 14.10, step 3).

Lymphocytes have other surface molecules besides MHC molecules that help bind cells together and stimulate a response. For example, helper T cells have a glycoprotein called CD4, which helps connect helper T cells to the macrophage by binding to MHC class II molecules. The CD4 protein is also bound by the virus that causes AIDS (see the Diseases and Disorders table at the end of this chapter). As a result, the virus preferentially infects helper T cells. Cytotoxic T cells have a glycoprotein called CD8, which helps connect cytotoxic T cells to cells displaying MHC class I molecules.

Lymphocyte Proliferation

Before exposure to a particular antigen, the number of helper T cells that can respond to that antigen is too small to produce an effective response against it. After the antigen is processed and presented to a helper T cell by a macrophage, the helper T cell responds by producing **interleukin-2** and **interleukin-2 receptors** (figure 14.10, step 4). Interleukin-2 binds to the receptors and stimulates the helper T cell to divide (figure 14.10, step 5). The "daughter" helper T cells produced by this division can again be presented with the antigen by macrophages and again be stimulated to divide. Thus, the number of helper T cells is greatly increased (figure 14.10, step 6).

It is important for the number of helper T cells to increase because helper T cells are necessary for the activation of most B cells or T cells (figure 14.10, step 7). For example, B cells have receptors that can recognize antigens. Most B cells, however, do not respond to antigens without stimulation from helper T cells. Without functional helper T cells, the immune response of B cells would not be effective to prevent disease.

B-cell proliferation begins when a B cell takes in the same kind of antigen that stimulated the helper T cell (figure 14.11, step 1). The antigen is processed by the B cell and presented on the B-cell surface by an MHC class II molecule (figure 14.11, step 2). A helper T cell is stimulated when it binds to the MHC class II/antigen complex (figure 14.11, step 3). There is also costimulation involving CD4 and interleukins (figure 14.11, steps 4 and 5). As a result, the B cell divides into two "daughter" cells (figure 14.11, step 6). These daughter cells may differentiate into **plasma cells,** which produce antibodies (figure 14.11, step 7). The division process continues, eventually producing many cells capable of producing antibodies and resulting in sufficient antibodies to destroy all the antigen.

Predict 3

How does elimination of the antigen stop the production of antibodies?

Antibody-Mediated Immunity

Exposure of the body to an antigen can lead to the activation of B cells and the production of antibodies. The antibodies bind to the antigens, which can be destroyed through several different mechanisms. Because antibodies are in body fluids, antibody-

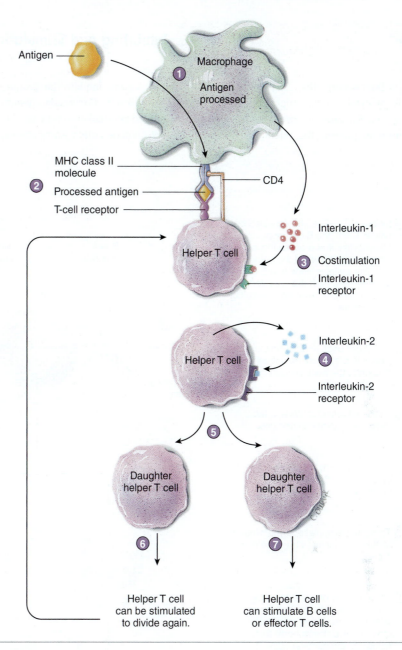

1. Antigen-presenting cells, such as macrophages, phagocytize, process, and display antigens on the cell's surface.

2. The antigens are bound to MHC class II molecules, which present the processed antigen to the T-cell receptor of the helper T cell.

3. Costimulation results from interleukin-1, secreted by the macrophage, and the CD4 glycoprotein of the helper T cell.

4. Interleukin-1 stimulates the helper T cell to secrete interleukin-2 and to produce interleukin-2 receptors.

5. The helper T cell stimulates itself to divide when interleukin-2 binds to interleukin-2 receptors.

6. The "daughter" helper T cells resulting from this division can be stimulated to divide again if they are exposed to the same antigen that stimulated the "parent" helper T cell. This greatly increases the number of helper T cells.

7. The increased number of helper T cells can facilitate the activation of B cells or effector T cells.

PROCESS Figure 14.10 **AP|R** **Proliferation of Helper T Cells**

An antigen-presenting cell (macrophage) stimulates helper T cells to divide.

mediated immunity is effective against extracellular antigens, such as bacteria, viruses (when they are outside cells), and toxins. Antibody-mediated immunity is also involved in certain allergic reactions.

Structure of Antibodies

Antibodies are proteins produced in response to an antigen. They are Y-shaped molecules consisting of four polypeptide chains: two identical heavy chains and two identical light chains (figure 14.12). The end of each "arm" of the antibody is the **variable region,** the part of the antibody that combines with the antigen. The variable region of a particular antibody can join only with a particular antigen; this is similar to the lock-and-key model of enzymes

(see chapter 2). The rest of the antibody is the **constant region,** and it has several functions. For example, the constant region can activate complement, or it can attach the antibody to cells, such as macrophages, basophils, and mast cells.

Antibodies make up a large portion of the proteins in plasma. Most plasma proteins can be separated into albumin and alpha, beta, and gamma globulin portions (see chapter 11). Antibodies are sometimes called **gamma globulins** (glob′ū-linz), because they are found mostly in the gamma globulin part of plasma, or **immunoglobulins (Ig),** because they are globulin proteins involved in immunity. The five general classes of antibodies are denoted IgG, IgM, IgA, IgE, and IgD (table 14.2).

CLINICAL IMPACT   Inhibiting and Stimulating Immunity

Decreasing the production or activity of cytokines can suppress the immune system. For example, cyclosporine, a drug used to prevent the rejection of transplanted organs, inhibits the production of interleukin-2. Conversely, genetically engineered interleukins can be used to stimulate the immune system. Administering interleukin-2 has promoted the destruction of cancer cells in some cases by increasing the activities of T cells.

1. Before a B cell can be activated by a helper T cell, the B cell must phagocytize and process the same antigen that activated the helper T cell. The antigen binds to a B-cell receptor, and both the receptor and the antigen are taken into the cell by endocytosis.

2. The B cell uses an MHC class II molecule to present the processed antigen to the helper T cell.

3. The T-cell receptor binds to the MHC class II/antigen complex.

4. There is costimulation of the B cell by CD4 and other surface molecules.

5. There is costimulation by interleukins (cytokines) released from the helper T cell.

6. The B cell divides, the resulting daughter cells divide, and so on, eventually producing many cells (only two are shown here) that recognize the same antigen.

7. Many of the daughter cells differentiate to become plasma cells, which produce antibodies. Antibodies are part of the immune response that eliminates the antigen.

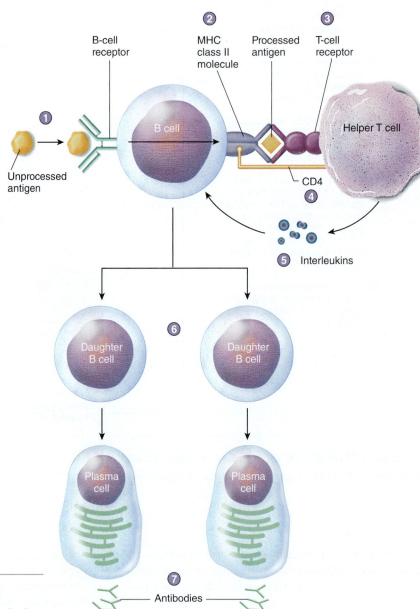

PROCESS Figure 14.11 Proliferation of B Cells

A helper T cell stimulates a B cell to divide and produce antibodies.

Effects of Antibodies

Antibodies can affect antigens either directly or indirectly. Direct effects occur when a single antibody binds to an antigen and inactivates the antigen, or when many antigens are bound together and are inactivated by many antibodies (figure 14.13a,b). The ability of antibodies to join antigens together is the basis for many clinical tests, such as blood typing, because when enough antigens are bound together, they form visible clumps.

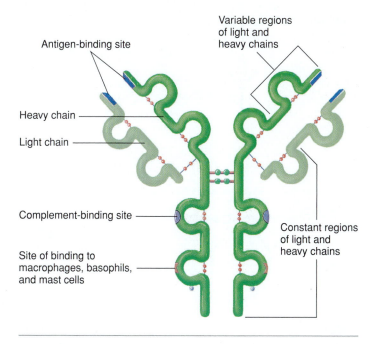

Figure 14.12 Structure of an Antibody

The Y-shaped antibody has two "arms." Each arm has a variable region that functions as an antigen-binding site. The constant region can activate complement or bind to other immune system cells, such as macrophages, basophils, or mast cells.

Most of the effectiveness of antibodies results from indirect effects (figure 14.13*c–e*). After an antibody has attached by its variable region to an antigen, the constant region of the antibody can activate other mechanisms that destroy the antigen. For example, the constant region of antibodies can activate complement, which stimulates inflammation, attracts white blood cells through chemotaxis, and lyses bacteria. When an antigen combines with the antibody, the constant region triggers the release of inflammatory chemicals from mast cells and basophils. For example, people who have hay fever inhale the antigens (usually plant pollens), which are then absorbed through the respiratory mucous membrane. The combination of the antigen with antibodies stimulates mast cells to release inflammatory chemicals, such as histamine. The resulting localized inflammatory response produces swelling and excess mucus production in the respiratory tract. Finally, macrophages can attach to the constant region of the antibody and phagocytize both the antibody and the antigen.

Antibody Production

The production of antibodies after the first exposure to an antigen is different from that following a second or subsequent exposure. The **primary response** results from the first exposure of a B cell to an antigen (figure 14.14, step 1). When the antigen binds to the antigen-binding receptor on the B cell and the B cell has been activated by a helper T cell, the B cell undergoes several divisions to form plasma cells and memory B cells. Plasma cells produce antibodies. The primary response normally takes 3–14 days to produce enough antibodies to be effective against the antigen. In the meantime, the individual usually develops disease symptoms because the antigen has had time to cause tissue damage.

    Memory B cells are responsible for the **secondary response,** or *memory response,* which occurs when the immune system is exposed to an antigen against which it has already produced a primary response (figure 14.14, step 2). When exposed to the antigen, the memory B cells quickly divide to form plasma cells, which rapidly produce antibodies. The secondary response provides better protection than the primary response for two reasons: (1) The time required to start producing antibodies is less (hours to a few days), and (2) more plasma cells and antibodies are produced. As a consequence, the antigen is quickly destroyed, no disease symptoms develop, and the person is immune.

Lymphatic

TABLE 14.2	Classes of Antibodies and Their Functions		
Antibody	**Total Serum Antibody (%)**	**Structure**	**Description**
IgG	80–85	IgG	Activates complement and increases phagocytosis; can cross the placenta and provide immune protection to the fetus and newborn; responsible for Rh reactions, such as hemolytic disease of the newborn
IgM	5–10	IgM IgA	Activates complement and acts as an antigen-binding receptor on the surface of B cells; responsible for transfusion reactions in the ABO blood system; often the first antibody produced in response to an antigen
IgA	15	Heavy chain	Secreted into saliva, into tears, and onto mucous membranes to protect body surfaces; found in colostrum and milk to provide immune protection to the newborn
IgE	0.002	IgE IgD Light chain	Binds to mast cells and basophils and stimulates the inflammatory response
IgD	0.2		Functions as an antigen-binding receptor on B cells

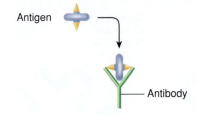

(a) **Inactivate the antigen.** An antibody binds to an antigen and inactivates it.

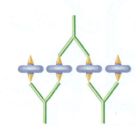

(b) **Bind antigens together.** Antibodies bind several antigens together.

(c) **Activate the complement cascade.** An antigen binds to an antibody. As a result, the antibody can activate complement proteins, which can produce inflammation, chemotaxis, and lysis.

(d) **Initiate the release of inflammatory chemicals.** An antibody binds to a mast cell or a basophil. When an antigen binds to the antibody, it triggers the release of chemicals that cause inflammation.

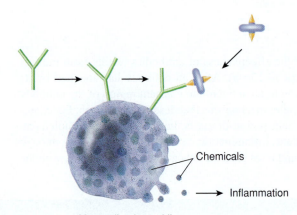

(e) **Facilitate phagocytosis.** An antibody binds to an antigen and then to a macrophage, which phagocytizes the antibody and antigen.

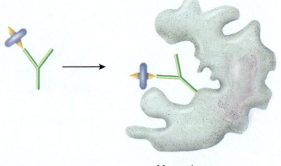

Figure 14.13 Effects of Antibodies

Antibodies directly affect antigens by inactivating the antigens or by binding the antigens together. Antibodies indirectly affect antigens by activating other mechanisms through the constant region of the antibody. Indirect mechanisms include activation of complement, increased inflammation resulting from the release of inflammatory chemicals from mast cells or basophils, and increased phagocytosis resulting from antibody attachment to macrophages.

Lymphatic

CLINICAL IMPACT Use of Monoclonal Antibodies

A **monoclonal antibody** is a pure antibody preparation that is specific for only one antigen. Monoclonal antibodies are grown in laboratories and have many clinical uses. Monoclonal antibodies are produced by injecting an antigen into a laboratory animal to activate a B-cell clone against the antigen. These B cells are removed from the animal and fused with tumor cells. The resulting cells have two ideal characteristics: (1) They produce

only one (mono) specific antibody because they are derived from one B-cell clone, and (2) they divide rapidly because they are derived from tumor cells. The result is many cells producing a specific antibody.

Monoclonal antibodies are used for determining pregnancy and for diagnosing diseases such as gonorrhea, syphilis, hepatitis, rabies, and cancer. These tests are specific and rapid because the monoclonal antibodies bind only to the antigen

being tested. Monoclonal antibodies were initially hailed as "magic bullets" because of their potential for delivering drugs to cancer cells (see "Immunotherapy" later in this chapter). This potential is now being realized, as over 20 monoclonal antibodies are being used to treat diseases, including some types of cancer, and hundreds more monoclonal antibodies are in clinical trials.

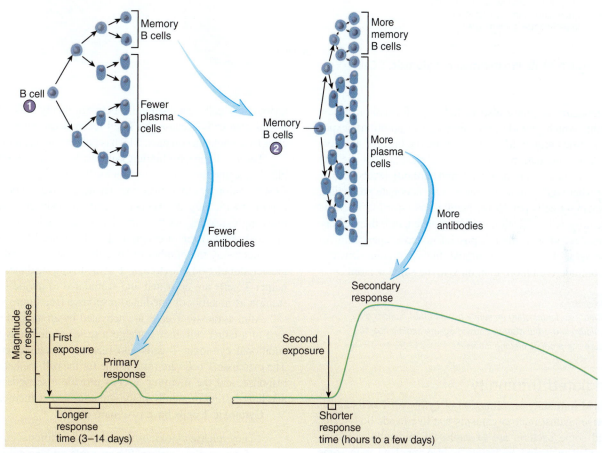

1 **Primary response.** The primary response occurs when a B cell is first activated by an antigen. The B cell proliferates to form plasma cells and memory cells. The plasma cells produce antibodies.

2 **Secondary response.** The secondary response occurs when another exposure to the same antigen causes the memory cells to rapidly form plasma cells and additional memory cells. The secondary response is faster and produces more antibodies than the primary response.

PROCESS Figure 14.14 AP|R Antibody Production

1. An MHC class I molecule displays an antigen, such as a viral protein, on the surface of a target cell.

2. The activation of a cytotoxic T cell begins when the T-cell receptor binds to the MHC class I/antigen complex.

3. There is costimulation of the cytotoxic T cell by CD8 and other surface molecules.

4. There is costimulation by cytokines, such as interleukin-2, released from helper T cells.

5. The activated cytotoxic T cell divides, the resulting daughter cells divide, and so on, eventually producing many cytotoxic T cells (only two are shown here).

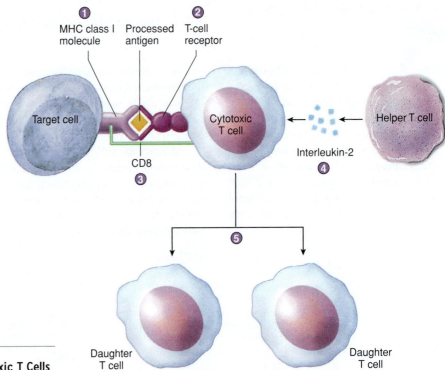

PROCESS Figure 14.15 Proliferation of Cytotoxic T Cells

The secondary response also includes the formation of new memory cells, which provide protection against additional exposures to a specific antigen. Memory cells are the basis of adaptive immunity. After destruction of the antigen, plasma cells die, the antibodies they released are degraded, and antibody levels decline to the point where they can no longer provide adequate protection. However, memory cells persist for many years—for life, in some cases. If memory cell production is not stimulated, or if the memory cells produced are short-lived, it is possible to have repeated infections of the same disease. For example, the same cold virus can cause the common cold more than once in the same person.

Predict 4

One theory for long-lasting immunity assumes that humans are continually exposed to disease-causing agents. Explain how this exposure can produce lifelong immunity.

Cell-Mediated Immunity

Cell-mediated immunity is a function of cytotoxic T cells and is most effective against microorganisms that live inside body cells. Viruses and some bacteria are examples of intracellular microorganisms. Cell-mediated immunity is also involved with some allergic reactions, the control of tumors, and graft rejection.

Cell-mediated immunity is essential for fighting viral infections. When viruses infect cells, they direct the cells to make new viruses, which are then released to infect other cells. Thus, cells are turned into virus-manufacturing plants. While inside the cell, viruses have a safe haven from antibody-mediated immunity because antibodies cannot cross the cell membrane. Cell-mediated immunity fights viral infections by destroying virally infected cells. When viruses infect cells, some viral proteins are broken down and become processed antigens that are combined with MHC class I molecules and displayed on the surface of the infected cell (figure 14.15, step 1).

Cytotoxic T cells can distinguish between virally infected cells and noninfected cells because the T-cell receptor can bind to the MHC class I/viral antigen complex, which is not present on uninfected cells.

The T-cell receptor binding with the MHC class I/antigen complex is a signal for activating cytotoxic T cells (figure 14.15, step 2). Costimulation by other surface molecules, such as CD8, also occurs (figure 14.15, step 3). Helper T cells provide costimulation by releasing cytokines, such as interleukin-2, which stimulate activation and cell division of cytotoxic T cells (figure 14.15, step 4).

Increasing the number of "daughter" helper T cells results in greater stimulation of cytotoxic T cells. In cell-mediated responses, helper T cells are activated and stimulated to divide in the same fashion as in antibody-mediated responses (see figure 14.10).

After cytotoxic T cells are activated by an antigen on the surface of a target cell, they undergo a series of divisions to produce additional cytotoxic T cells and memory T cells (figure 14.16). The cytotoxic T cells are responsible for the cell-mediated immune response, and the **memory T cells** provide a secondary response and long-lasting immunity in the same fashion as memory B cells.

Cytotoxic T cells have two main effects:

1. They release cytokines that activate additional components of the immune system. For example, some cytokines attract innate immune cells, especially macrophages. These cells are then responsible for phagocytosis of the antigen and the production of an inflammatory response. Cytokines also activate additional cytotoxic T cells, which increases the effectiveness of the cell-mediated response.

2. Cytotoxic T cells can come in contact with other cells and kill them. Virally infected cells have viral antigens, tumor cells have tumor antigens, and tissue transplants have foreign antigens that can stimulate cytotoxic T-cell activity. The cytotoxic T cells bind to the antigens on the surfaces of these cells and cause the cells to lyse.

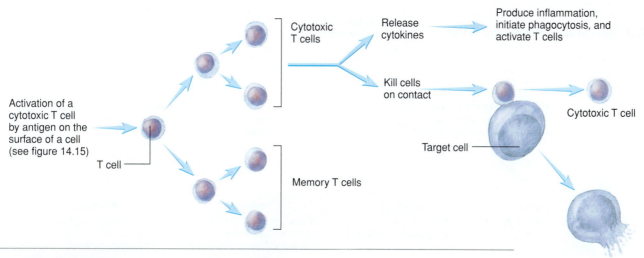

Figure 14.16 `AP|R` **Stimulation and Effects of T Cells**

When activated, cytotoxic T cells form many additional cytotoxic T cells, as well as memory T cells. The cytotoxic T cells release cytokines that promote the destruction of the antigen or cause the lysis of target cells, such as virally infected cells, tumor cells, or transplanted cells. The memory T cells are responsible for the secondary response.

A CASE IN POINT

Sjogren Syndrome

Ima Akin has a toothache, so she goes to her dentist, Dr. Dekay, for help. Dr. Dekay finds an abscessed tooth, several dental cavities, and enlarged parotid salivary glands, but little saliva production. On questioning, Ima tells the dentist she has been experiencing a dry mouth and dry eyes and she feels fatigued all the time. Dr. Dekay refers Ima to her family physician, Dr. Hurtt, for further testing. Eventually, the physician determines that Ima Akin has Sjogren (show′grin) syndrome, a systemic, autoimmune inflammatory disorder affecting glands and mucous membranes.

Innate, antibody-mediated, and cell-mediated immunity normally work together to protect us against foreign antigens. In autoimmune disorders, self-antigens activate immune responses, resulting in the destruction of healthy tissues. In Sjogren syndrome, damage to the salivary glands leads to decreased saliva production and a dry mouth, which increases the likelihood of developing cavities. Damage to the lacrimal glands results in decreased tear production and dry eyes, which damage the conjunctiva. Sjogren syndrome is one of the most common autoimmune disorders. About 50% of the time, it occurs alone, but the remainder of cases occur in conjunction with other autoimmune diseases, such as rheumatoid arthritis, systemic lupus erythematosus, and scleroderma. Nine out of 10 people with Sjogren syndrome are women.

14.6 ACQUIRED IMMUNITY

Learning Outcome *After reading this section, you should be able to*

A. Explain the four ways that adaptive immunity can be acquired.

There are four ways to acquire adaptive immunity: active natural, active artificial, passive natural, and passive artificial (figure 14.17). **Active immunity** results when an individual is exposed to an antigen (either naturally or artificially) and the response of the individual's own immune system is the cause of the immunity. **Passive immunity** occurs when another person or an animal develops immunity and the immunity is transferred to a nonimmune individual. *Natural* and *artificial* refer to the method of exposure or antibody transfer. *Natural* implies that contact with the antigen or transfer of antibodies occurs as part of everyday living and is not deliberate. *Artificial* implies that deliberate introduction of an antigen or antibody into the body has occurred.

Active Natural Immunity

Active natural immunity results from natural exposure to an antigen, such as a disease-causing microorganism, that stimulates the immune system to respond against the antigen. Because the individual is not immune during the first exposure, he or she usually develops the symptoms of the disease.

Active Artificial Immunity

In **active artificial immunity,** an antigen is deliberately introduced into an individual to stimulate the immune system. This process is called **vaccination** (vak′si-nā′shŭn), and the introduced antigen is a **vaccine** (vak′sēn, vak-sēn′). A vaccine is usually administered by injection. Examples of vaccinations are the DTP injection against diphtheria, tetanus, and pertussis (whooping cough) and the MMR injection against mumps, measles, and rubella (German measles).

The vaccine usually consists of part of a pathogen, either a dead microorganism or a live, altered one. The antigen has been changed so that it will stimulate an immune response but will not cause the disease symptoms. Because active artificial immunity produces long-lasting immunity without disease symptoms, it is the preferred method of acquiring adaptive immunity.

Passive Natural Immunity

Passive natural immunity results when antibodies are transferred from a mother to her child across the placenta before birth. During her life, the mother has been exposed to many antigens, either naturally or artificially, and she has antibodies against many of these antigens, which protect her and the developing fetus against disease. Some of the antibodies (IgG) can cross the placenta and enter the fetal blood. Following birth, the antibodies protect the

Figure 14.17 Ways to Acquire Adaptive Immunity

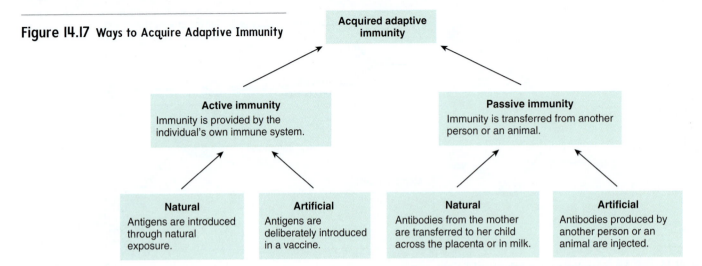

baby for the first few months. Eventually, the antibodies break down, and the baby must rely on its own immune system. If the mother breastfeeds her baby, antibodies (IgA) in the mother's milk may also provide some protection for the baby.

Passive Artificial Immunity

Achieving **passive artificial immunity** begins with vaccinating an animal, such as a horse. After the animal's immune system responds to the antigen, antibodies are removed from the animal and injected into the human requiring immunity. Alternatively, a human who has developed immunity through natural exposure or vaccination can serve as a source of antibodies. Passive artificial immunity provides immediate protection because the antibodies either directly or indirectly destroy the antigen. Passive artificial immunity is therefore the preferred treatment when not enough time is available for the individual to develop his or her own active immunity. However, the technique provides only temporary immunity because the antibodies are used or eliminated by the recipient.

Antibodies that provide passive artificial immunity are referred to by the general term **antiserum** because the antibodies are found in serum, which is plasma minus the clotting factors. Antisera are available against microorganisms that cause disease, such as rabies, hepatitis, and measles; bacterial toxins, such as those that cause tetanus, diphtheria, and botulism; and venoms from poisonous snakes and spiders.

14.7 OVERVIEW OF IMMUNE INTERACTIONS

Learning Outcome *After reading this section, you should be able to*

A. Explain how innate, antibody-mediated, and cell-mediated immunity can function together to eliminate an antigen.

Although the immune system can be described in terms of innate, antibody-mediated, and cell-mediated immunity, these categories are artificial divisions used to emphasize particular aspects of immunity. In actuality, there is only one immune system, but its responses often involve components of more than one type of immunity. For example, although adaptive immunity can recognize and remember specific antigens, once recognition has

occurred the antigen is destroyed with the help of many innate immunity activities, including inflammation and phagocytosis (figure 14.18; see table 14.1).

14.8 IMMUNOTHERAPY

Learning Outcome *After reading this section, you should be able to*

A. Define and give examples of immunotherapy.

Knowledge of how the immune system operates has produced two fundamental benefits: (1) an understanding of the cause and progression of many diseases and (2) the development or proposed development of methods to prevent, stop, or even reverse diseases.

Immunotherapy treats disease by altering immune system function or by directly attacking harmful cells. Some approaches attempt to boost immune system function in general. For example, administering cytokines or other agents can promote inflammation and activate immune cells, which can help destroy tumor cells. On the other hand, sometimes inhibiting the immune system is helpful. For example, multiple sclerosis is an autoimmune disease in which the immune system treats self-antigens as foreign antigens, destroying the myelin that covers axons. A type of cytokine called interferon beta blocks the expression of MHC molecules that display self-antigens and is used to treat multiple sclerosis.

Some immunotherapy takes a more specific approach. For example, vaccination can prevent many diseases (see "Acquired Immunity"). The ability to produce monoclonal antibodies can result in effective therapies for treating tumors. If an antigen unique to tumor cells can be found, then monoclonal antibodies can be used to deliver radioactive isotopes, drugs, toxins, enzymes, or cytokines that kill the tumor cell directly or activate the immune system to kill the cell. Unfortunately, so far researchers have found no antigen on tumor cells that is not also present on normal cells. Nonetheless, this approach can be useful if damage to normal cells is minimal.

Some uses of monoclonal antibodies to treat tumors are yielding promising results. For example, monoclonal antibodies with radioactive iodine (^{131}I) have been found to cause the regression of B-cell lymphomas while producing few side effects. Herceptin, a monoclonal antibody, binds to a growth factor receptor that is overexpressed in 25–30% of primary breast cancers. The antibody

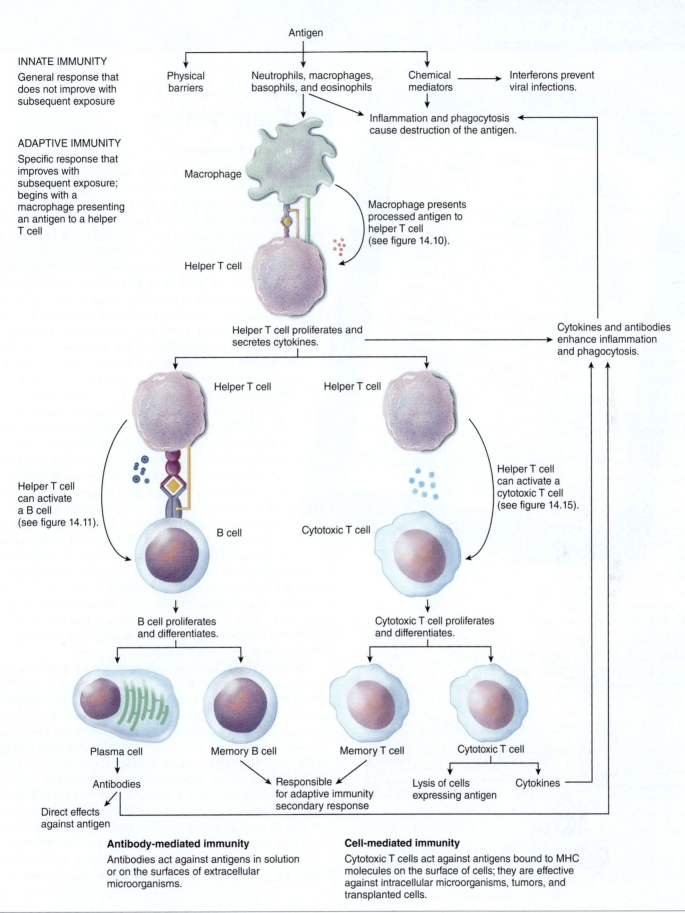

Antigen

INNATE IMMUNITY
General response that does not improve with subsequent exposure

Physical barriers

Neutrophils, macrophages, basophils, and eosinophils

Chemical mediators → Interferons prevent viral infections.

Inflammation and phagocytosis cause destruction of the antigen.

ADAPTIVE IMMUNITY
Specific response that improves with subsequent exposure; begins with a macrophage presenting an antigen to a helper T cell

Macrophage

Macrophage presents processed antigen to helper T cell (see figure 14.10).

Helper T cell

Helper T cell proliferates and secretes cytokines.

Cytokines and antibodies enhance inflammation and phagocytosis.

Helper T cell

Helper T cell

Helper T cell can activate a B cell (see figure 14.11).

Helper T cell can activate a cytotoxic T cell (see figure 14.15).

B cell

Cytotoxic T cell

B cell proliferates and differentiates.

Cytotoxic T cell proliferates and differentiates.

Plasma cell

Memory B cell

Memory T cell

Cytotoxic T cell

Antibodies

Responsible for adaptive immunity secondary response

Lysis of cells expressing antigen

Cytokines

Direct effects against antigen

Antibody-mediated immunity
Antibodies act against antigens in solution or on the surfaces of extracellular microorganisms.

Cell-mediated immunity
Cytotoxic T cells act against antigens bound to MHC molecules on the surface of cells; they are effective against intracellular microorganisms, tumors, and transplanted cells.

Figure 14.18 AP|R **Immune Interactions**
The major interactions and responses of innate and adaptive immunity to an antigen.

Lymphatic

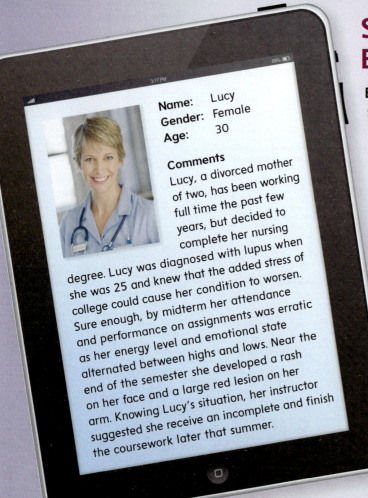

Name: Lucy
Gender: Female
Age: 30

Comments

Lucy, a divorced mother of two, has been working full time the past few years, but decided to complete her nursing degree. Lucy was diagnosed with lupus when she was 25 and knew that the added stress of college could cause her condition to worsen. Sure enough, by midterm her attendance and performance on assignments was erratic as her energy level and emotional state alternated between highs and lows. Near the end of the semester she developed a rash on her face and a large red lesion on her arm. Knowing Lucy's situation, her instructor suggested she receive an incomplete and finish the coursework later that summer.

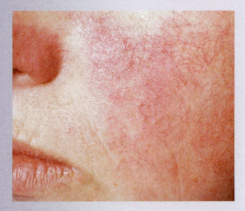

Figure 14A

Inflammation in the skin caused by systemic lupus erythematosus results in the characteristic "butterfly rash" on the face.

Systemic Lupus Erythematosus

Background Information

Systemic lupus erythematosus (SLE) (lū′pŭs er-i-thē-mă-tō′sŭs) is a disease in which tissues and cells are damaged by the immune system. The name describes the skin rash that is characteristic of the disease (figure 14A). The term *lupus* literally means "wolf" and originally referred to eroded (as if gnawed by a wolf) lesions of the skin. *Erythematosus* refers to redness of the skin resulting from inflammation. Skin lesions take three forms: (1) an inflammatory redness that can be in the shape of a butterfly, extending from the bridge of the nose to the cheeks; (2) small, localized, pimplelike eruptions accompanied by scaling of the skin; (3) areas of atrophied, depigmented skin with borders of increased pigmentation. Unfortunately, as the term *systemic* implies, the disorder is not confined to the skin but can affect tissues and cells throughout the body. Another systemic effect is the presence of low-grade fever in most cases of active SLE.

SLE is an autoimmune disorder in which a large variety of antibodies are produced that recognize self-antigens, such as nucleic acids, phospholipids, coagulation factors, red blood cells, and platelets. The combination of the antibodies with self-antigens forms immune complexes that circulate throughout the body and are deposited in various tissues, where they stimulate inflammation and tissue destruction. Thus, SLE can affect many body systems. For example, the most common antibodies act against DNA released from damaged cells. Normally, the liver removes the DNA, but when DNA and antibodies form immune complexes, they tend to be deposited in the kidneys and other tissues. Approximately 40–50% of individuals with SLE develop renal disease. In some cases, the antibodies can bind to antigens on cells, causing the cells to lyse. For example, antibodies binding to red blood cells cause hemolysis and anemia.

The cause of SLE is unknown. The most popular hypothesis suggests that a viral infection disrupts the function of the T cells that normally prevent an immune response to self-antigens. Genetic factors probably contribute to the development of the disease. The likelihood of developing SLE is much higher if a family member also has it.

Approximately 1 out of 2000 individuals in the United States has SLE. The first symptoms usually appear between 15 and 25 years of age, affecting women approximately nine times as often as men. The progress of the disease is unpredictable, with flare-ups followed by periods of remission. The survival after diagnosis is greater than 90% after 10 years. The most frequent causes of death are kidney failure, CNS dysfunction, infections, and cardiovascular disease.

There is no cure for SLE, nor is there one standard of treatment because the course of the disease is highly variable, and patient his-

Lymphatic

Systemic Lupus Erythematosus

Symptoms
(Highly variable)
- Skin lesions, particularly on face
- Fever
- Fatigue
- Arthritis
- Anemia

Treatment
- Anti-inflammatory drugs
- Antimalarial drugs

SKELETAL
Arthritis, tendinitis, and death of bone tissue can develop.

INTEGUMENTARY
Skin lesions frequently occur and are made worse by exposure to the sun. Hair loss results in diffuse thinning of the hair.

MUSCULAR
Destruction of muscle tissue and muscular weakness occur.

URINARY
Renal lesions and glomerulonephritis can result in progressive failure of kidney functions. Excess proteins are lost in the urine, resulting in lower than normal blood proteins, which can produce edema.

NERVOUS
Memory loss, intellectual deterioration, disorientation, psychosis, reactive depression, headache, seizures, nausea, and loss of appetite can occur. Stroke is a major cause of dysfunction and death. Cranial nerve involvement results in facial muscle weakness, drooping of the eyelid, and double vision. Central nervous system lesions can cause paralysis.

DIGESTIVE
Ulcers develop in the oral cavity and pharynx. Abdominal pain and vomiting are common, but no cause can be found. Inflammation of the pancreas and occasionally enlargement of the liver and minor abnormalities in liver function tests occur.

ENDOCRINE
Sex hormones may play a role in SLE because 90% of the cases occur in females, and females with SLE have reduced levels of androgens.

RESPIRATORY
Chest pain may be caused by inflammation of the pleural membranes; fever, shortness of breath, and hypoxemia may occur due to inflammation of the lungs; alveolar hemorrhage can develop.

CARDIOVASCULAR
Inflammation of the pericardium (pericarditis) with chest pain can develop. Damage to heart valves, inflammation of cardiac tissue, tachycardia, arrhythmias, angina, and myocardial infarction also occur. Hemolytic anemia and leukopenia can be present (see chapter 11). Antiphospholipid antibody syndrome, through an unknown mechanism, increases coagulation and thrombus formation, which increases the risk of stroke and heart attack.

---

tories differ widely. Treatment usually begins with mild medications and proceeds to increasingly potent therapies as conditions warrant. Aspirin and nonsteroidal anti-inflammatory drugs are prescribed to suppress inflammation. Antimalarial drugs are used to treat skin rash and arthritis in SLE, but the mechanism of action is unknown. Patients who do not respond to these drugs or who have severe SLE are helped by glucocorticoids. Although these steroids effectively suppress inflammation, they can produce undesirable side effects, including suppression of normal adrenal gland functions. In patients with life-threatening SLE, very high doses of glucocorticoids are used.

Predict 5

The red lesion Lucy developed on her arm is called purpura (pŭr′pū-ră), and it is caused by bleeding into the skin. The lesions gradually change color and disappear in 2 or 3 weeks. Explain how SLE produces purpura.

"tags" cancer cells, which are then lysed by natural killer cells. Herceptin slows disease progression and increases survival time, but it is not a cure for breast cancer.

Many other immunotherapy approaches are being studied, and more treatments that use the immune system are sure to develop in the future.

DISEASES AND DISORDERS: Lymphatic System

CONDITION	DESCRIPTION
LYMPHATIC SYSTEM	
Infections	
Lymphadenitis (lim-fad′ĕ-nī′tis)	Inflammation of the lymph nodes; nodes become enlarged and tender as pathogens are trapped and destroyed
Lymphangitis (lim-fan-jī′tis)	Inflammation of the lymphatic vessels; often results in visible red streaks in the skin that extend from the site of infection
Bubonic (bū-bon′ik) plague	Enlarged lymph nodes caused by bacterial infection (transferred by flea bites from rats); without treatment, bacteria enter blood and death occurs rapidly due to septicemia (see chapter 11); known as the Black Death in the Middle Ages
Lymphedema (limf′e-de′ma)	
Abnormal accumulation of lymph in tissues, often the limbs; 70%–90% cases in women; can be caused by developmental defects, disease, or damage to the lymphatic system	
Elephantiasis (el-ĕ-fan-tī′ă-sis)	Caused by long, slender roundworms transferred to humans by mosquito bites; adult worms lodge in lymphatic vessels and block lymph flow, so that a limb can become permanently swollen and enlarged; major cause of lymphedema worldwide
Lymphedema following cancer treatment	Caused by removal of lymph nodes near a tumor by surgery or radiation therapy; sentinel lymph nodes (those nearest the tumor) are first examined for cancer cells
Lymphoma (lim-fō′mă)	
Cancer of lymphocytes that often begins in lymph nodes; immune system becomes depressed, with increased susceptibility to infections	
IMMUNE SYSTEM	
Immediate Allergic Reactions	
Symptoms occur within a few minutes of exposure to an antigen because antibodies are already present from prior exposure.	
Hay fever	Often caused by inhalation of plant pollen antigens
Asthma (az′mă)	Antigen combines with antibodies on mast cells or basophils in the lungs, which then release inflammatory chemicals that cause constriction of the air tubes, so that the patient has trouble breathing
Urticaria (er′ti-kar′i-á)	Skin rash or localized swelling; can be caused by an ingested antigen; also called hives
Anaphylaxis (an′ă -fī-lak′sis, an′ă-fĭ-lak′sis)	Systemic allergic reaction, often resulting from insect stings or drugs such as penicillin; chemicals released from mast cells and basophils cause systemic vasodilation, increased vascular permeability, drop in blood pressure, and possibly death
Delayed Allergic Reactions	
Symptoms occur in hours to days following exposure to the antigen because these types of reactions involve migration of T cells to the antigen, followed by release of cytokines.	
Poison ivy and poison oak	Antigen absorbed by epithelial cells, which are then destroyed by T cells, causing inflammation and tissue destruction; itching can be intense
Autoimmune Diseases	
Similar to allergic reactions, except that the immune system incorrectly treats self-antigens as foreign antigens. Many types of autoimmune diseases exist, including type I diabetes, celiac disease, rheumatoid arthritis, multiple sclerosis, systemic lupus erythematosus, and Graves disease.	
Congenital Immunodeficiencies	
Usually involve failure of the fetus to form adequate numbers of B cells, T cells, or both	
Severe combined immunodeficiency (SCID)	Both B cells and T cells fail to form; unless patient kept in a sterile environment or provided with a compatible bone marrow transplant, death from infection results
Acquired Immunodeficiencies	
Many causes—for example, diseases, stress, or drugs	
Acquired immunodeficiency syndrome (AIDS)	Life-threatening disease caused by the human immunodeficiency virus (HIV); HIV is transmitted in body fluids; infection begins when the virus binds to the CD4 protein found primarily on helper T cells; without helper T cells, cytotoxic T-cell and B-cell activation is impaired, and adaptive immunity is suppressed; course of HIV infection varies, with most people surviving 10 or more years
Transplanted tissue rejection	Caused by a normal immune response to foreign antigens encoded by the major histocompatibility complex genes, also called human leukocyte antigen (HLA) genes; drugs that suppress the immune system must be administered for life to prevent graft rejection

Lymphatic

14.9 EFFECTS OF AGING ON THE LYMPHATIC SYSTEM AND IMMUNITY

Learning Outcome   *After reading this section, you should be able to*

A. Describe how aging affects the lymphatic system and immunity.

Aging appears to have little effect on the lymphatic system's ability to remove fluid from tissues, absorb lipids from the digestive tract, or remove defective red blood cells from the blood. However, aging has a severe impact on the immune system.

With age, people eventually lose the ability to produce new, mature T cells in the thymus. By age 40, much of the thymus has been replaced with adipose tissue, and after age 60, the thymus decreases in size to the point that it can be difficult to detect. While the number of T cells remains stable in most individuals due to the replication (not maturation) of T cells in lymphatic tissues, the T cells are less functional. In many individuals, the ability of helper T cells to proliferate in response to antigens decreases. Thus, antigen exposure produces fewer helper T cells, which results in less stimulation of B cells and cytotoxic T cells. Consequently, both antibody-mediated immunity and cell-mediated immunity responses to antigens decrease with age.

Both primary and secondary antibody responses decrease with age. More antigen is required to produce a response, the response is slower, less antibody is produced, and fewer memory cells result. Thus, a person's ability to resist infections and develop immunity decreases.

The ability of cell-mediated immunity to resist intracellular pathogens also decreases with age. For example, the elderly are more susceptible to influenza (flu) and should be vaccinated every year. Some pathogens cause disease but are not eliminated from the body; with age, decreased immunity can lead to reactivation of the pathogen. An example is the virus that causes chickenpox in children, which can remain latent within nerve cells, even though the disease seems to have disappeared. Later in life, the virus can leave the nerve cells and infect skin cells, causing painful lesions known as herpes zoster, or shingles.

Autoimmune disease occurs when immune responses destroy otherwise healthy tissue. There is very little increase in the number of new-onset autoimmune diseases in the elderly. However, the chronic inflammation and immune responses that began earlier in life have a cumulative, damaging effect. Likewise, the increased incidence of cancer in the elderly is likely to be caused by a combination of repeated exposure to and damage from cancer-causing agents and decreased immunity.

ANSWER TO LEARN TO PREDICT

First, recall that vaccines are a form of artificial active immunity caused by deliberately introducing an antigen to the body, stimulating a primary response by the immune system. The immune system responds to the vaccine by increasing the number of specific memory cells and antibodies for the particular disease. This provides long-lasting immunity without disease symptoms. The "booster" shot stimulates a memory (secondary) response, resulting in the formation of even more memory cells and antibodies. Since Shay will be exposed to more people at school, there is an increased chance that he will also be exposed to the particular pathogen. His "booster" shot improves the effectiveness of his immune system to fight the types of infections for which he was vaccinated.

Answers to the rest of this chapter's Predict questions are in Appendix E.

SUMMARY

14.1 Functions of the Lymphatic System   (p. 385)

The lymphatic system maintains fluid balance in tissues, absorbs lipids from the small intestine, and defends against pathogens.

14.2 Anatomy of the Lymphatic System   (p. 386)

The lymphatic system consists of lymph, lymphocytes, lymphatic vessels, lymph nodes, tonsils, the spleen, and the thymus.

Lymphatic Capillaries and Vessels

1. Lymphatic vessels carry lymph away from tissues. Valves in the vessels ensure the one-way flow of lymph.
2. Skeletal muscle contraction, contraction of lymphatic vessel smooth muscle, and thoracic pressure changes move the lymph through the vessels.
3. The thoracic duct and right lymphatic duct empty lymph into the blood.

Lymphatic Organs

1. Lymphatic tissue produces lymphocytes when exposed to foreign substances, and it filters lymph and blood.
2. The tonsils protect the openings between the nasal and oral cavities and the pharynx.

3. Lymph nodes, located along lymphatic vessels, filter lymph.
4. The white pulp of the spleen responds to foreign substances in the blood, whereas the red pulp phagocytizes foreign substances and worn-out red blood cells. The spleen also functions as a reservoir for blood.
5. The thymus processes lymphocytes that move to other lymphatic tissue to respond to foreign substances.

Overview of the Lymphatic System

The lymphatic system removes fluid from tissues, absorbs lipids from the small intestine, and produces B cells and T cells, which are responsible for much of immunity.

14.3 Immunity   (p. 390)

1. Immunity is the ability to resist the harmful effects of pathogens.
2. Immunity is classified as innate or adaptive.

14.4 Innate Immunity   (p. 390)

Physical Barriers

1. The skin and mucous membranes are barriers that prevent microorganisms from entering the body.

2. Tears, saliva, and urine wash away microorganisms.

Chemical Mediators

1. Chemical mediators kill pathogens, promote phagocytosis, and increase inflammation.
2. Lysozyme in tears and complement in plasma are examples of chemicals involved in innate immunity.
3. Interferons prevent the replication of viruses.

White Blood Cells

1. Chemotaxis is the ability of cells to move toward pathogens or sites of tissue damage.
2. Neutrophils are the first phagocytic cells to respond to pathogens.
3. Macrophages are large phagocytic cells that are active in the latter part of an infection. Macrophages are positioned at sites where pathogens may enter tissues.
4. Basophils and mast cells promote inflammation. Eosinophils also play a role in inflammation associated with allergic reactions.
5. Natural killer cells lyse tumor cells and virus-infected cells.

Inflammatory Response

1. Chemical mediators cause vasodilation and increase vascular permeability, allowing chemicals to enter damaged tissues. Chemicals also attract phagocytes.
2. The amount of chemical mediators and phagocytes increases until the cause of the inflammation is destroyed. Then the tissues undergo repair.
3. Local inflammation produces redness, heat, swelling, pain, and loss of function. Symptoms of systemic inflammation include an increase in neutrophil numbers, fever, and shock.

14.5 Adaptive Immunity   (p. 394)

1. Antigens are molecules that stimulate adaptive immunity.
2. B cells are responsible for antibody-mediated immunity; T cells are involved with cell-mediated immunity.

Origin and Development of Lymphocytes

1. B cells and T cells originate in red bone marrow. T cells are processed in the thymus, and B cells are processed in red bone marrow.
2. B cells and T cells move to lymphatic tissue from their processing sites. They continually circulate from one lymphatic tissue to another.

Activation and Multiplication of Lymphocytes

1. B cells and T cells have antigen receptors on their surfaces. Clones are lymphocytes with the same antigen receptor.
2. Major histocompatibility complex (MHC) molecules present processed antigens to B or T cells.
3. Costimulation by cytokines, such as interleukins, and surface molecules, such as CD4, are required in addition to MHC molecules.
4. Macrophages present processed antigens to helper T cells, which divide and increase in number.
5. Helper T cells stimulate B cells to divide and differentiate into plasma cells that produce antibodies.

Antibody-Mediated Immunity

1. Antibodies are proteins. The variable region combines with antigens and is responsible for antibody specificity. The constant region activates complement or attaches the antibody to cells. The five classes of antibodies are IgG, IgM, IgA, IgE, and IgD.
2. Antibodies directly inactivate antigens or cause them to clump together. Antibodies indirectly destroy antigens by promoting phagocytosis and inflammation.
3. The primary response results from the first exposure to an antigen. B cells form plasma cells, which produce antibodies, and memory B cells.
4. The secondary (memory) response results from exposure to an antigen after a primary response. Memory B cells quickly form plasma cells and new memory B cells.

Cell-Mediated Immunity

1. Exposure to an antigen activates cytotoxic T cells and produces memory T cells.
2. Cytotoxic T cells lyse virally infected cells, tumor cells, and tissue transplants. Cytotoxic T cells produce cytokines, which promote inflammation and phagocytosis.

14.6 Acquired Immunity   (p. 403)

1. Active natural immunity results from everyday exposure to an antigen against which the person's own immune system mounts a response.
2. Active artificial immunity results from deliberate exposure to an antigen (vaccine) to which the person's own immune system responds.
3. Passive natural immunity is the transfer of antibodies from a mother to her fetus during gestation or to her baby during breastfeeding.
4. Passive artificial immunity is the transfer of antibodies from an animal or another person to a person requiring immunity.

14.7 Overview of Immune Interactions   (p. 404)

Innate immunity, antibody-mediated immunity, and cell-mediated immunity can function together to eliminate an antigen.

14.8 Immunotherapy   (p. 404)

Immunotherapy stimulates or inhibits the immune system to treat diseases.

14.9 Effects of Aging on the Lymphatic System and Immunity   (p. 409)

1. Aging has little effect on the lymphatic system's ability to remove fluid from tissues, absorb lipids from the digestive tract, or remove defective red blood cells from the blood.
2. Decreased helper T-cell proliferation results in decreased antibody-mediated and cell-mediated immune responses.
3. The primary and secondary antibody responses decrease with age.
4. The ability to resist intracellular pathogens decreases with age.

⎯⎯⋀⎯⎯ REVIEW AND COMPREHENSION

1. List the parts of the lymphatic system, and describe the three main functions of the lymphatic system.
2. What is the function of the valves in lymphatic vessels? What causes lymph to move through lymphatic vessels?
3. Which parts of the body are drained by the right lymphatic duct and which by the thoracic duct?
4. Describe the cells and fibers of lymphatic tissue, and explain the functions of lymphatic tissue.
5. Name the three groups of tonsils. What is their function?
6. Where are lymph nodes found? What is the function of the germinal centers within lymph nodes?
7. Where is the spleen located? What is the function of white pulp and red pulp within the spleen? What other function does the spleen perform?
8. Where is the thymus located, and what function does it perform?
9. What is the difference between innate immunity and adaptive immunity?

10. How do physical barriers and chemical mediators provide protection against pathogens? Describe the effects of complement and interferons.

11. Describe the functions of the two major phagocytic cell types of the body. What is the mononuclear phagocytic system?

12. Name the cells involved in promoting inflammation.

13. What protective function do natural killer cells perform?

14. Describe the effects that take place during an inflammatory response. What are the symptoms of local and systemic inflammation?

15. Define antigen. What is the difference between a self-antigen and a foreign antigen?

16. Which cells are responsible for antibody-mediated immunity and for cell-mediated immunity?

17. Describe the origin and development of B cells and T cells.

18. What is the function of antigen receptors and major histocompatibility proteins?

19. What is costimulation? Give an example.

20. Describe the process by which an antigen can cause an increase in helper T-cell numbers.

21. Describe the process by which helper T cells can stimulate B cells to divide, differentiate, and produce antibodies.

22. What are the functions of the variable and constant regions of an antibody?

23. Describe the direct and indirect ways that antibodies destroy antigens.

24. What are the functions of plasma cells and memory B cells?

25. Define primary and secondary responses. How do they differ from each other in regard to speed of response and amount of antibody produced?

26. Explain how cytotoxic T cells are activated.

27. What are the functions of cytotoxic T cells and memory T cells?

28. Define active natural, active artificial, passive natural, and passive artificial immunity. Give an example of each.

29. What effect does aging have on the major functions of the lymphatic system?

30. Describe the effects of aging on B cells and T cells. Give examples of how this affects antibody-mediated and cell-mediated immunity responses.

CRITICAL THINKING

1. A patient is suffering from edema in the right lower limb. Explain why elevation and massage of the limb help remove the excess fluid.

2. If the thymus of an experimental animal is removed immediately following birth, the animal exhibits the following characteristics:
   a. increased susceptibility to infections
   b. decreased numbers of lymphocytes
   c. greatly decreased ability to reject grafts

   Explain these observations.

3. Adjuvants are substances that slow, but do not stop, the release of an antigen from an injection site into the blood. Suppose injection A of a given amount of antigen is given without an adjuvant and injection B of the same amount of antigen is given with an adjuvant that caused the release of antigen over a period of 2 to 3 weeks. Does injection A or injection B result in the greater amount of antibody production? Explain.

4. Compare how long active immunity and passive immunity last. Explain the difference between the two types of immunity. In what situations is one type preferred over the other?

5. Tetanus is caused by bacteria (*Clostridium tetani*) that enter the body through wounds in the skin. The bacteria produce a toxin that causes spastic muscle contractions. Death often results from failure of the respiratory muscles. A patient goes to the emergency room after stepping on a nail. If the patient has been vaccinated against tetanus, he is given a tetanus booster shot, which consists of the toxin altered so that it is harmless. If the patient has never been vaccinated against tetanus, he is given an antiserum shot against tetanus. Explain the rationale for this treatment strategy. Sometimes both a booster and an antiserum shot are given, but at different locations on the body. Explain why this is done and why the shots are given in different locations.

6. A child appears healthy until approximately 9 months of age. Then the child develops severe bacterial infections, one after another. Fortunately, the infections are successfully treated with antibiotics. When infected with measles and other viral diseases, the child recovers without difficulty. Explain.

7. A patient had many allergic reactions. As part of the treatment scheme, physicians decided to try to identify the allergens that stimulated the allergic reactions. A series of solutions, each containing a common allergen, was composed. Each solution was then injected into the skin at different locations on the patient's back. The following results were obtained:
   a. Within a few minutes, one injection site became red and swollen.
   b. At another injection site, swelling and redness did not appear until 2 days later.
   c. No redness or swelling developed at the other sites.

   Explain what happened for each observation and what caused the redness and swelling.

8. Ivy Hurtt developed a poison ivy rash after a camping trip. Her doctor prescribed a cortisone ointment to relieve the inflammation. A few weeks later, Ivy scraped her elbow, which became inflamed. Because she had some leftover cortisone ointment, she applied it to the scrape. Was the ointment a good idea for the poison ivy? Was it an appropriate treatment for the scrape?

9. Suzy Withitt has just had her ears pierced. To her dismay, she finds that wearing inexpensive (but tasteful) jewelry causes an inflammatory (allergic) reaction to the metal in the jewelry by the end of the day. Is this because of antibodies or cytokines?

Answers in Appendix D

aprevealed.com

Lymphatic

15 Respiratory System

LEARN to PREDICT

Flashing lights at 2 a.m. alerted the neighbors that something was wrong at the Theron home. Mr. Theron, who has moderate emphysema, could not stop coughing, so his wife called 911. In the emergency room, a physician listened to Mr. Theron's respiratory sounds and concluded that his left lung had collapsed.

By using some information about the peripheral circulation from chapter 13 and reading chapter 15, explain how emphysema affected Mr. Theron's breathing, what caused his lung to collapse, and how the physician was able to detect the collapsed lung. How would a ventilator help?

15.1 FUNCTIONS OF THE RESPIRATORY SYSTEM

Learning Outcome *After reading this section, you should be able to*

A. Describe the functions of the respiratory system.

Module 11 Respiratory System

Studying, sleeping, talking, eating, and exercising all involve breathing. From our first breath at birth, the rate and depth of our breathing are unconsciously matched to our activities. Although we can voluntarily stop breathing, within a few minutes we must breathe again. Breathing is so characteristic of life that, along with the pulse, it is one of the first things health professionals check to determine if an unconscious person is alive.

Respiration includes the following processes: (1) ventilation, or breathing, which is the movement of air into and out of the lungs; (2) the exchange of oxygen (O_2) and carbon dioxide (CO_2) between the air in the lungs and the blood; (3) the transport of O_2 and CO_2 in the blood; and (4) the exchange of O_2 and CO_2 between the blood and the tissues. It can be confusing to hear the term *respiration* alone because sometimes it also refers to cellular metabolism, or **cellular respiration** (discussed in chapter 17); in fact, the two processes are directly related. Breathing provides the O_2 needed in cellular respiration to make ATP from glucose.

Breathing also rids the body of potentially toxic CO_2, the waste produced during cellular respiration. In addition to respiration, the respiratory system performs the following functions:

1. *Regulation of blood pH.* The respiratory system can alter blood pH by changing blood CO_2 levels.
2. *Voice production.* Air movement past the vocal cords makes sound and speech possible.
3. *Olfaction.* The sensation of smell occurs when airborne molecules are drawn into the nasal cavity.
4. *Innate immunity* (see chapter 14). The respiratory system protects against some microorganisms and other pathogens, such as viruses, by preventing them from entering the body and by removing them from respiratory surfaces.

15.2 ANATOMY OF THE RESPIRATORY SYSTEM

Learning Outcomes *After reading this section, you should be able to*

A. Describe the anatomy of the respiratory passages, beginning at the nose and ending with the alveoli.

B. Describe the structure of the lungs, and define respiratory membrane.

The **respiratory system** is divided into the upper respiratory tract and the lower respiratory tract (figure 15.1). The **upper respiratory tract** includes the external nose, the nasal cavity, the pharynx, and associated structures; the **lower respiratory tract** includes the larynx, the trachea, the bronchi, and the lungs. Keep in mind, however, that *upper* and *lower respiratory tract* are not official anatomical terms. Rather, they are arbitrary divisions for the purposes of discussion, and some anatomists define them differently. For example, one alternative places the larynx in the upper respiratory tract. Also, the oral cavity is considered part of the digestive system, not the respiratory system, even though air frequently passes through the oral cavity.

Nose

The **nose** consists of the external nose and the nasal cavity. The **external nose** is the visible structure that forms a prominent feature of the face. Most of the external nose is composed of hyaline cartilage, although the bridge of the external nose consists of bone (see figure 6.12). The bone and cartilage are covered by connective tissue and skin.

The **nares** (nā'rēs; sing. nā'ris), or *nostrils,* are the external openings of the nose, and the **choanae** (kō'an-ē; funnels) are the openings into the pharynx. The **nasal** (nā'zăl) **cavity** extends from the nares to the choanae (figure 15.2). The **nasal septum** is a partition dividing the nasal cavity into right and left parts. A **deviated nasal septum** occurs when the septum bulges to one side. The **hard palate** (pal'ăt) forms the floor of the nasal cavity, separating the nasal cavity from the oral cavity. Air can flow through the nasal cavity when the oral cavity is closed or full of food.

Three prominent bony ridges called **conchae** (kon'kē) are present on the lateral walls on each side of the nasal cavity. The conchae increase the surface area of the nasal cavity and cause air to churn, so that it can be cleansed, humidified, and warmed.

The **paranasal** (par-ă-nā'săl) **sinuses** are air-filled spaces within bone. They include the maxillary, frontal, ethmoidal, and sphenoidal sinuses, each named for the bones in which they are located. The paranasal sinuses open into the nasal cavity and are lined with a mucous membrane. They reduce the weight of the skull, produce mucus, and influence the quality of the voice by acting as resonating chambers.

The **nasolacrimal** (nā-zō-lak'ri-măl) **ducts,** which carry tears from the eyes, also open into the nasal cavity. Sensory receptors for the sense of smell are in the superior part of the nasal cavity (see chapter 9).

Air enters the nasal cavity through the nares. Just inside the nares, the lining of the cavity is composed of stratified squamous epithelium containing coarse hairs. The hairs trap some of the large particles of dust suspended in the air. The rest of the nasal cavity is lined with pseudostratified columnar epithelial cells containing cilia and many mucus-producing goblet cells (see chapter 4). Mucus produced by the goblet cells also traps debris in the air. The cilia sweep the mucus posteriorly to the pharynx, where it is swallowed. As air flows through the nasal cavities, it is humidified by moisture from the mucous epithelium and warmed by blood flowing through the superficial capillary networks underlying the mucous epithelium.

The **sneeze reflex** dislodges foreign substances from the nasal cavity. Sensory receptors detect the foreign substances, and action potentials are conducted along the trigeminal nerves to the medulla oblongata, where the reflex is triggered. During the sneeze reflex, the uvula and the soft palate are depressed, so that rapidly flowing air from the lungs is directed primarily through the nasal passages, although a considerable amount passes through the oral cavity.

About 17–25% of people have a photic sneeze reflex, which is stimulated by exposure to bright light, such as the sun. Another reflex, the pupillary reflex, causes the pupils to constrict in response to bright light. Physiologists speculate that the complicated "wiring" of the pupillary and sneeze reflexes are intermixed in some people, so that, when bright light activates a pupillary reflex, it also activates a sneeze reflex. Sometimes the photic sneeze reflex is fancifully called ACHOO, which stands for *a*utosomal-*d*ominant-*c*ompelling-*h*elio-*o*phthalmic-*o*utburst.

Predict 2

Explain what happens to your throat when you sleep with your mouth open, especially when your nasal passages are plugged as a result of a cold. Explain what may happen to your lungs when you run a long way in very cold weather while breathing rapidly through your mouth.

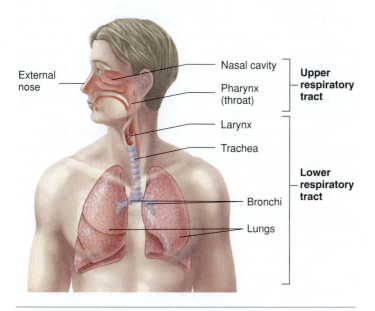

External nose — Nasal cavity ⎫
Pharynx (throat) ⎬ **Upper respiratory tract**
Larynx ⎫
Trachea ⎬
Bronchi ⎬ **Lower respiratory tract**
Lungs ⎭

Figure 15.1 AP|R Respiratory System

The upper respiratory tract consists of the external nose, nasal cavity, and pharynx (throat). The lower respiratory tract consists of the larynx, trachea, bronchi, and lungs.

A CASE IN POINT

Sinusitis

Cy Ness has a cold; his symptoms include sneezing, achiness, and fatigue. Although Cy recovers from the cold, a stuffy nose and recurrent headaches linger for several weeks. Cy's doctor listens to his history of symptoms, examines his nose and throat, and orders a sinus x-ray. He concludes that Cy Ness has **sinusitis** (sī-nŭ-sī′tis), inflammation of the mucous membrane of a sinus, especially one or more of the paranasal sinuses. Viral infections, such as the common cold, can cause mucous membranes to become inflamed and swollen and to produce excess mucus. As a result, the sinus opening into the nasal cavity can be partially or completely blocked. In addition, mucus accumulation within the sinus can promote the development of a bacterial infection. The combination of built-up mucus and inflamed and infected mucous membranes produces pain.

Treatment of sinusitis consists of taking antibiotics to kill the bacteria and decongestants to promote sinus drainage; drinking fluids to maintain hydration; and inhaling steam. Decongestants reduce tissue swelling, or edema. When mucous membranes become less swollen, breathing is easier, and more mucus can move out of the paranasal sinuses and nasal cavity. Decongestants, such as pseudoephedrine hydrochloride (e.g., Sudafed), reduce swelling by causing the release of norepinephrine from sympathetic neurons supplying blood vessels. Increased vasoconstriction of blood vessels in mucous membranes reduces blood flow and the movement of fluid from the blood into tissues. Sinusitis can also result from inflammation caused by allergies or from benign growths, called polyps, that obstruct a sinus opening into the nasal cavity.

Pharynx

The **pharynx** (far′ingks; throat) is the common passageway for both the respiratory and the digestive systems. Air from the nasal cavity and air, food, and water from the mouth pass through the pharynx. Inferiorly, the pharynx leads to the rest of the respiratory system through the opening into the larynx and to the digestive system through the esophagus. The pharynx can be divided into three regions: the nasopharynx, the oropharynx, and the laryngopharynx (figure 15.2a).

The **nasopharynx** (nā′zō-far′ingks) is the superior part of the pharynx. It is located posterior to the choanae and superior to the **soft palate,** which is an incomplete muscle and connective tissue partition separating the nasopharynx from the oropharynx. The **uvula** (ū′vū-lă; a little grape) is the posterior extension of the soft palate. The soft palate forms the floor of the nasopharynx. The nasopharynx is lined with pseudostratified ciliated columnar epithelium that is continuous with the nasal cavity. The auditory tubes extend from the middle ears and open into the nasopharynx. The posterior part of the nasopharynx contains the **pharyngeal** (fă-rin′jē-ăl) **tonsil,** which helps defend the body against infection (see chapter 14). The soft palate is elevated during swallowing; this movement closes the nasopharynx and prevents food from passing from the oral cavity into the nasopharynx.

The **oropharynx** (ōr′ō-far′ingks) extends from the uvula to the epiglottis, and the oral cavity opens into the oropharynx. Thus, food, drink, and air all pass through the oropharynx. The oropharynx is lined with stratified squamous epithelium, which protects against

abrasion. Two sets of tonsils, the palatine tonsils and the lingual tonsil, are located near the opening between the mouth and the oropharynx. The **palatine** (pal′ă-tīn) **tonsils** are located in the lateral walls near the border of the oral cavity and the oropharynx. The **lingual tonsil** is located on the surface of the posterior part of the tongue.

The **laryngopharynx** (lă-ring′gō-far-ingks) passes posterior to the larynx and extends from the tip of the epiglottis to the esophagus. Food and drink pass through the laryngopharynx to the esophagus. A small amount of air is usually swallowed with the food and drink. Swallowing too much air can cause excess gas in the stomach and may result in belching. The laryngopharynx is lined with stratified squamous epithelium and ciliated columnar epithelium.

Larynx

The **larynx** (lar′ingks) is located in the anterior throat and extends from the base of the tongue to the trachea (figure 15.2a). It is a passageway for air between the pharynx and the trachea. The larynx consists of an outer casing of nine cartilages connected to one another by muscles and ligaments (figure 15.3). Three of the nine cartilages are unpaired, and six of them form three pairs. The largest cartilage is the unpaired **thyroid** (thī′royd; shield-shaped) **cartilage,** or *Adam's apple.* The thyroid cartilage is attached superiorly to the hyoid bone. The most inferior cartilage of the larynx is the unpaired **cricoid** (krī′koyd; ring-shaped) **cartilage,** which forms the base of the larynx on which the other cartilages rest. The thyroid and cricoid cartilages maintain an open passageway for air movement.

The third unpaired cartilage is the **epiglottis** (ep-i-glot′is; on the glottis). It differs from the other cartilages in that it consists of elastic cartilage rather than hyaline cartilage. Its inferior margin is attached to the thyroid cartilage anteriorly, and the superior part of the epiglottis projects superiorly as a free flap toward the tongue. The epiglottis helps prevent swallowed materials from entering the larynx. As the larynx elevates during swallowing, the epiglottis tips posteriorly to cover the opening of the larynx.

The six paired cartilages consist of three cartilages on each side of the posterior part of the larynx (figure 15.3b). The top cartilage on each side is the **cuneiform** (kū′nē-i-fōrm; wedge-shaped) **cartilage,** the middle cartilage is the **corniculate** (kōr-nik′ū-lāt; horn-shaped) **cartilage,** and the bottom cartilage is the **arytenoid** (ar-i-tē′noyd; ladle-shaped) **cartilage.** The arytenoid cartilages articulate with the cricoid cartilage inferiorly. The paired cartilages form an attachment site for the vocal folds.

Two pairs of ligaments extend from the posterior surface of the thyroid cartilage to the paired cartilages. The superior pair forms the **vestibular** (ves-tib′ū-lăr) **folds,** or *false vocal cords,* and the inferior pair composes the **vocal folds,** or *true vocal cords* (figure 15.4). When the vestibular folds come together, they prevent air from leaving the lungs, as when a person holds his or her breath. Along with the epiglottis, the vestibular folds also prevent food and liquids from entering the larynx.

The vocal folds are the primary source of voice production. Air moving past the vocal folds causes them to vibrate, producing sound. Muscles control the length and tension of the vocal folds. The force of air moving past the vocal folds controls the loudness, and the tension of the vocal folds controls the pitch of the voice. An inflammation of the mucous epithelium of the vocal folds is called **laryngitis** (lar-in-jī′tis). Swelling of the vocal folds during laryngitis inhibits voice production.

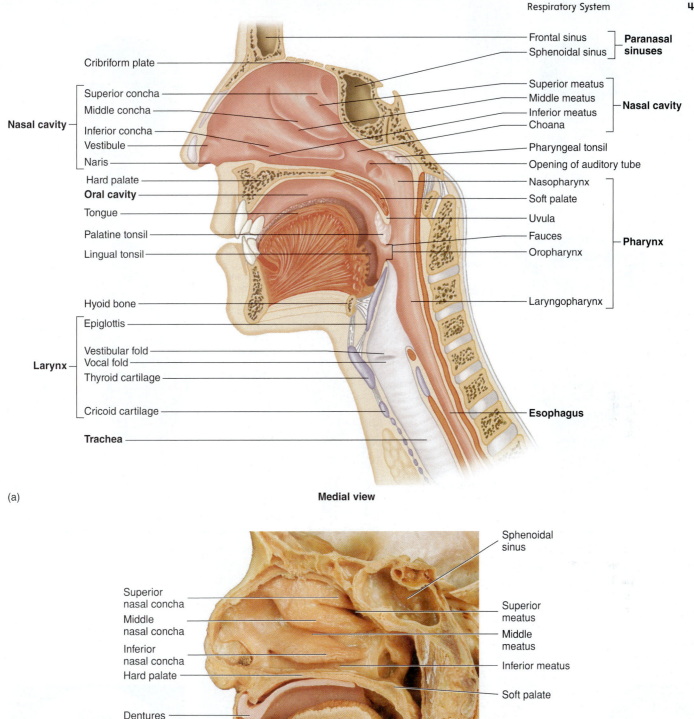

(a)

Medial view

(b)   **Medial view**

Figure 15.2 AP|R **Nasal Cavity and Pharynx**

(*a*) Sagittal section through the nasal cavity and pharynx. (*b*) Photograph of sagittal section of the nasal cavity. Note: This cadaver wore dentures during life.

Trachea

The **trachea** (trā′kē-ă), or *windpipe,* is a membranous tube attached to the larynx. It consists of connective tissue and smooth muscle, reinforced with 16–20 C-shaped pieces of hyaline cartilage (see figure 15.3). The adult trachea is about 1.4–1.6 centimeters (cm) in diameter and about 10–11 cm long. It begins immediately inferior to the cricoid cartilage, which is the most inferior cartilage of the larynx. The trachea projects through the mediastinum and

divides into the right and left primary bronchi at the level of the fifth thoracic vertebra (figure 15.5). The esophagus lies immediately posterior to the trachea (see figure 15.2*a*).

C-shaped cartilages form the anterior and lateral sides of the trachea. The cartilages protect the trachea and maintain an open passageway for air. The posterior wall of the trachea has no cartilage and consists of a ligamentous membrane and smooth muscle (see figure 15.3*b*). The smooth muscle can alter the diameter of

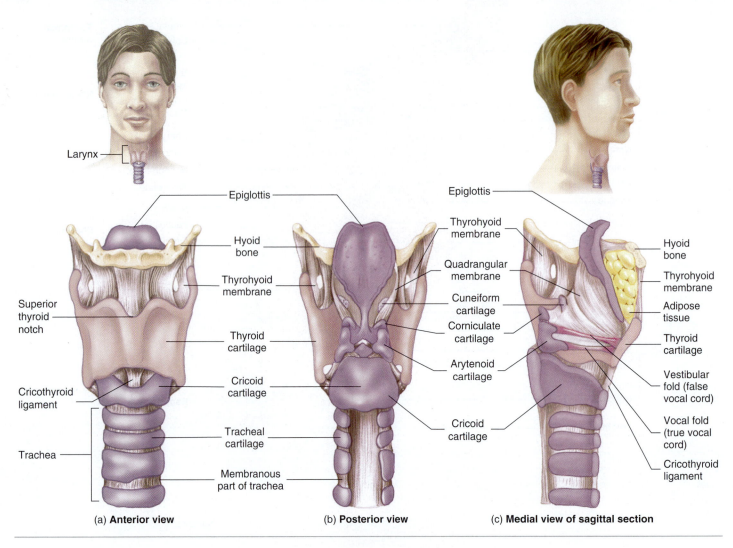

Larynx

Epiglottis

Hyoid
bone

Thyrohyoid
membrane

Superior
thyroid
notch

Thyroid
cartilage

Cricoid
cartilage

Cricothyroid
ligament

Trachea

Tracheal
cartilage

Membranous
part of trachea

(a) **Anterior view**

Epiglottis

Thyrohyoid
membrane

Quadrangular
membrane

Cuneiform
cartilage

Corniculate
cartilage

Arytenoid
cartilage

Cricoid
cartilage

(b) **Posterior view**

Epiglottis

Hyoid
bone

Thyrohyoid
membrane

Adipose
tissue

Thyroid
cartilage

Vestibular
fold (false
vocal cord)

Vocal fold
(true vocal
cord)

Cricothyroid
ligament

(c) **Medial view of sagittal section**

Figure 15.3 Anatomy of the Larynx

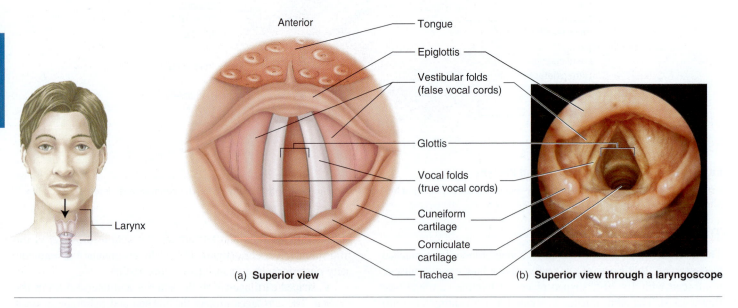

Anterior

Tongue

Epiglottis

Vestibular folds
(false vocal cords)

Glottis

Vocal folds
(true vocal cords)

Cuneiform
cartilage

Corniculate
cartilage

Larynx

Trachea

(a) **Superior view**

(b) **Superior view through a laryngoscope**

Figure 15.4 AP|R Vestibular and Vocal Folds

(*Far left*) The arrow shows the direction of viewing the vestibular and vocal folds. (*a*) The relationship of the vestibular folds to the vocal folds and the laryngeal cartilages. (*b*) Superior view of the vestibular and vocal folds as seen through a laryngoscope.

Respiratory

CLINICAL IMPACT   Establishing Airflow

In cases of extreme emergency, when the upper air passageway is blocked by a foreign object to the extent that the victim cannot breathe, quick action is required to save the person's life. In abdominal thrusts, also known as the **Heimlich maneuver,** the sudden application of pressure to the abdomen can force air up the trachea to dislodge the obstruction. The person who performs the maneuver stands behind the victim with his or her arms under the victim's arms and his or her hands over the victim's abdomen between the navel and the rib cage. With one hand formed into a fist, the other hand suddenly pulls the fist toward the abdomen with an accompanying upward motion. The pressure pushes up on the diaphragm and

therefore increases air pressure in the lungs. If this maneuver is done properly, it causes air to flow from the lungs with sufficient force to dislodge most foreign objects.

There are other ways to establish airflow, but they should be performed only by trained medical personnel. **Intubation** is the insertion of a tube into an opening, a canal, or a hollow organ. A tube can be passed through the mouth or nose into the pharynx and then through the larynx to the trachea. Sometimes it is necessary to make an opening for the tube. The preferred point of entry in an emergency is through the membrane between the cricoid and thyroid cartilages (see figure 15.3a), a procedure referred to as a **cricothyrotomy** (krī′kō-thī-rot′ō-mē). A

tube is then inserted into the opening to facilitate the passage of air.

A **tracheostomy** (trā′kē-os′tō-mē; tracheo- + stoma, mouth) is an operation to make an opening into the trachea. Usually, the opening is intended to be permanent, and a tube is inserted into the trachea to allow airflow and provide a way to remove secretions. The term **tracheotomy** (trā-kē-ot′ō-mē; tracheo- + tome, incision) refers to the actual cutting into the trachea (although sometimes the terms tracheostomy and tracheotomy are used interchangeably). It is not advisable to enter the air passageway through the trachea in emergency cases because arteries, nerves, and the thyroid gland overlie the anterior surface of the trachea.

the trachea. This occurs, for example, when the **cough reflex** dislodges foreign substances from the trachea. Sensory receptors detect the foreign substance, and action potentials travel along the vagus nerves to the medulla oblongata, where the cough reflex is triggered. During coughing, the smooth muscle of the trachea contracts, decreasing the trachea's diameter. As a result, air moves rapidly through the trachea, which helps expel mucus and foreign substances. Also, the uvula and soft palate are elevated, so that air passes primarily through the oral cavity.

Predict 3

Explain what happens to the shape of the trachea when a person swallows a large mouthful of food. Why is this process advantageous?

The trachea is lined with pseudostratified columnar epithelium, which contains numerous cilia and goblet cells. The cilia propel mucus produced by the goblet cells, as well as foreign particles embedded in the mucus, out of the trachea, through the larynx, and into the pharynx, from which they are swallowed.

Constant, long-term irritation of the trachea by cigarette smoke can cause the tracheal epithelium to change to stratified squamous epithelium. The stratified squamous epithelium has no cilia and therefore cannot clear the airway of mucus and debris. The accumulations of mucus provide a place for microorganisms to grow, resulting in respiratory infections. Constant irritation and inflammation of the respiratory passages stimulate the cough reflex, resulting in "smoker's cough."

Bronchi

The trachea divides into the left and right **main bronchi** (brong′kī; sing. bronchus, brong′kŭs; windpipe), or *primary bronchi*, each of which connects to a lung. The left main bronchus is more

horizontal than the right main bronchus because it is displaced by the heart (figure 15.5). Foreign objects that enter the trachea usually lodge in the right main bronchus, because it is wider, shorter, and more vertical than the left main bronchus and is more in direct line with the trachea. The main bronchi extend from the trachea to the lungs. Like the trachea, the main bronchi are lined with pseudostratified ciliated columnar epithelium and are supported by C-shaped pieces of cartilage.

Lungs

The **lungs** are the principal organs of respiration. Each lung is cone-shaped, with its base resting on the diaphragm and its apex extending superiorly to a point about 2.5 cm above the clavicle (figure 15.5). The right lung has three **lobes,** called the superior, middle, and inferior lobes. The left lung has two lobes, called the superior and inferior lobes (figure 15.6). The lobes of the lungs are separated by deep, prominent fissures on the lung surface. Each lobe is divided into **bronchopulmonary segments** separated from one another by connective tissue septa, but these separations are not visible as surface fissures. Individual diseased bronchopulmonary segments can be surgically removed, leaving the rest of the lung relatively intact, because major blood vessels and bronchi do not cross the septa. There are nine bronchopulmonary segments in the left lung and ten in the right lung.

The main bronchi branch many times to form the **tracheobronchial tree** (see figure 15.5). Each main bronchus divides into lobar bronchi as they enter their respective lungs (figure 15.6). The **lobar bronchi** (or *secondary bronchi*), two in the left lung and three in the right lung, conduct air to each lobe. The lobar bronchi in turn give rise to **segmental bronchi** (or *tertiary bronchi*), which extend to the bronchopulmonary segments of the lungs. The bronchi continue to branch many times, finally giving rise to **bronchioles**

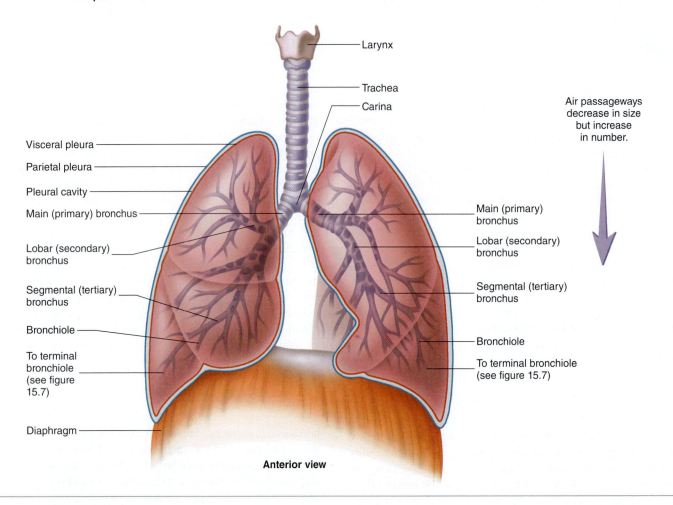

Larynx

Trachea

Carina

Air passageways
decrease in size
but increase
in number.

Visceral pleura

Parietal pleura

Pleural cavity

Main (primary) bronchus

Main (primary)
bronchus

Lobar (secondary)
bronchus

Lobar (secondary)
bronchus

Segmental (tertiary)
bronchus

Segmental (tertiary)
bronchus

Bronchiole

Bronchiole

To terminal
bronchiole
(see figure
15.7)

To terminal bronchiole
(see figure 15.7)

Diaphragm

Anterior view

Figure 15.5 AP|R **Anatomy of the Trachea and Lungs**
The trachea and lungs and the branching of the bronchi are shown. Each lung is surrounded by a pleural cavity, formed by the visceral and parietal pleurae.

(brong′kē-ōlz). The bronchioles also subdivide numerous times to give rise to **terminal bronchioles,** which then subdivide into **respiratory bronchioles** (figure 15.7). Each respiratory bronchiole subdivides to form **alveolar** (al-vē′ō-lăr) **ducts,** which are like long, branching hallways with many open doorways. The doorways open into **alveoli** (al-vē′ō-lī; hollow sacs), which are small air sacs. The alveoli become so numerous that the alveolar duct wall is little more than a succession of alveoli. The alveolar ducts end as two or three **alveolar sacs,** which are chambers connected to two or more alveoli. There are about 300 million alveoli in the lungs.

As the air passageways of the lungs become smaller, the structure of their walls changes. The amount of cartilage decreases and the amount of smooth muscle increases until, at the terminal bronchioles, the walls have a prominent smooth muscle layer but no cartilage. Relaxation and contraction of the smooth muscle within the bronchi and bronchioles can change the diameter of the air passageways. For example, during exercise the diameter can increase, thus increasing the volume of air moved. During an **asthma attack,** however, contraction of the smooth muscle in the terminal bronchioles can result in greatly reduced airflow (see Systems Pathology at the end of this chapter). In severe cases, air movement can be so restricted that death results. Fortunately, medications, such as albuterol (al-bū′ter-ol), help counteract the effects of an asthma attack by

promoting smooth muscle relaxation in the walls of terminal bronchioles, so that air can flow more freely.

As the air passageways of the lungs become smaller, the lining of their walls also changes. The trachea and bronchi have pseudostratified ciliated columnar epithelium, the bronchioles have ciliated simple columnar epithelium, and the terminal bronchioles have ciliated simple cuboidal epithelium. The ciliated epithelium of the air passageways functions as a mucus-cilia escalator, which traps debris from the inhaled air and removes it from the respiratory system.

As the air passageways beyond the terminal bronchioles become smaller, their walls become thinner. The walls of the respiratory bronchioles are composed of cuboidal epithelium, and those of the alveolar ducts and alveoli are simple squamous epithelium. The **respiratory membrane** of the lungs is where gas exchange between the air and blood takes place. It is formed mainly by the walls of the alveoli and the surrounding capillaries (figure 15.8), but the alveolar ducts and respiratory bronchioles also contribute. The respiratory membrane is very thin to facilitate the diffusion of gases. It consists of six layers:

1. a thin layer of fluid lining the alveolus
2. the alveolar epithelium, composed of simple squamous epithelium
3. the basement membrane of the alveolar epithelium

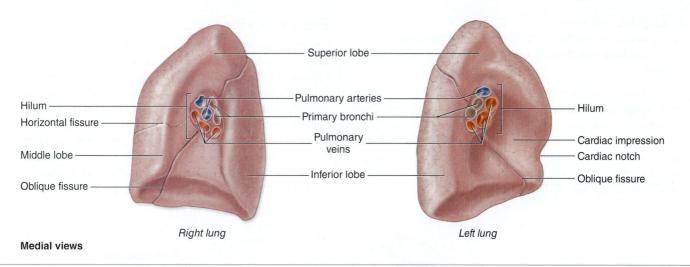

Superior lobe

Hilum

Horizontal fissure

Middle lobe

Oblique fissure

Pulmonary arteries

Primary bronchi

Pulmonary veins

Inferior lobe

Hilum

Cardiac impression

Cardiac notch

Oblique fissure

Right lung

Left lung

Medial views

Figure 15.6 Lungs, Lung Lobes, and Bronchi

The right lung is divided into three lobes by the horizontal and oblique fissures. The left lung is divided into two lobes by the oblique fissure. A main bronchus supplies each lung, a lobar bronchus supplies each lung lobe, and segmental bronchi supply the bronchopulmonary segments (not visible).

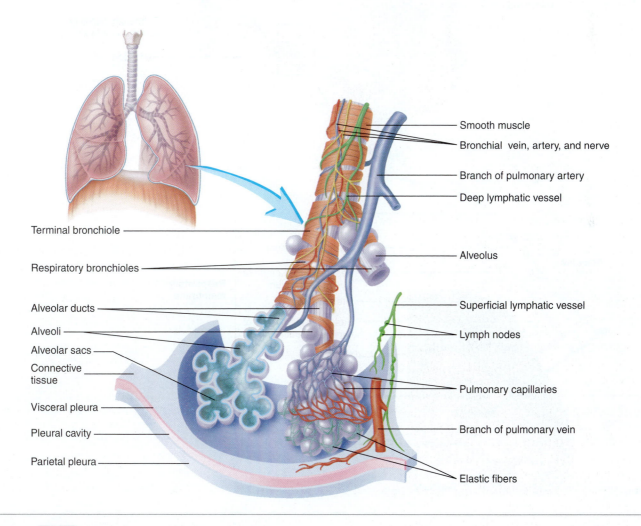

Smooth muscle

Bronchial vein, artery, and nerve

Branch of pulmonary artery

Deep lymphatic vessel

Terminal bronchiole

Respiratory bronchioles

Alveolus

Alveolar ducts

Alveoli

Alveolar sacs

Connective tissue

Visceral pleura

Pleural cavity

Parietal pleura

Superficial lymphatic vessel

Lymph nodes

Pulmonary capillaries

Branch of pulmonary vein

Elastic fibers

Respiratory

Figure 15.7 AP|R Bronchioles and Alveoli

A terminal bronchiole branches to form respiratory bronchioles, which give rise to alveolar ducts. Alveoli connect to the alveolar ducts and respiratory bronchioles. The alveolar ducts end as two or three alveolar sacs.

4. a thin interstitial space
5. the basement membrane of the capillary endothelium
6. the capillary endothelium, composed of simple squamous epithelium

The elastic fibers surrounding the alveoli (figure 15.8) allow them to expand during inspiration and recoil during expiration. The lungs are very elastic and, when inflated, are capable of expelling the air and returning to their original, uninflated state. Specialized secretory cells within the walls of the alveoli (figure 15.8) secrete a chemical, called surfactant, that reduces the tendency of alveoli to recoil (see "Lung Recoil" later in this chapter).

Pleural Cavities

The lungs are contained within the thoracic cavity. In addition, each lung is surrounded by a separate **pleural** (ploor′ăl; relating to the ribs) **cavity.** Each pleural cavity is lined with a serous membrane called the **pleura.** The pleura consists of a parietal and a visceral part. The **parietal pleura,** which lines the walls of the thorax, diaphragm, and mediastinum, is continuous with the **visceral pleura,** which covers the surface of the lung (figure 15.9; see figure 15.5).

The pleural cavity, between the parietal and visceral pleurae, is filled with a small volume of pleural fluid produced by the pleural membranes. The pleural fluid performs two functions:

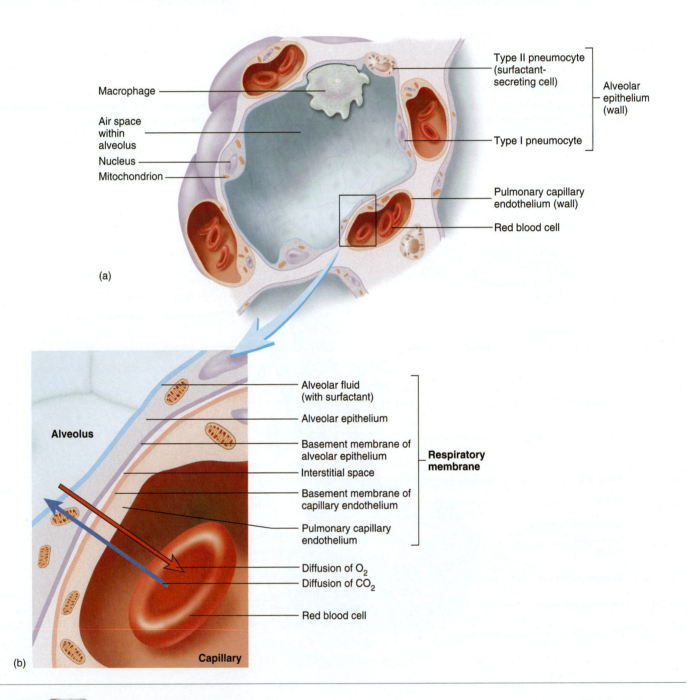

Respiratory

(a)

Macrophage

Air space within alveolus

Nucleus

Mitochondrion

Type II pneumocyte (surfactant-secreting cell)

Alveolar epithelium (wall)

Type I pneumocyte

Pulmonary capillary endothelium (wall)

Red blood cell

(b)

Alveolus

Alveolar fluid (with surfactant)

Alveolar epithelium

Basement membrane of alveolar epithelium

Interstitial space

Basement membrane of capillary endothelium

Pulmonary capillary endothelium

Respiratory membrane

Diffusion of O_2

Diffusion of CO_2

Red blood cell

Capillary

Figure I5.8   AP|R   Alveolus and the Respiratory Membrane

(a) Section of an alveolus, showing the air-filled interior and thin walls composed of simple squamous epithelium. The alveolus is surrounded by elastic connective tissue and blood capillaries. (b) O_2 and CO_2 diffuse across the six thin layers of the respiratory membrane.

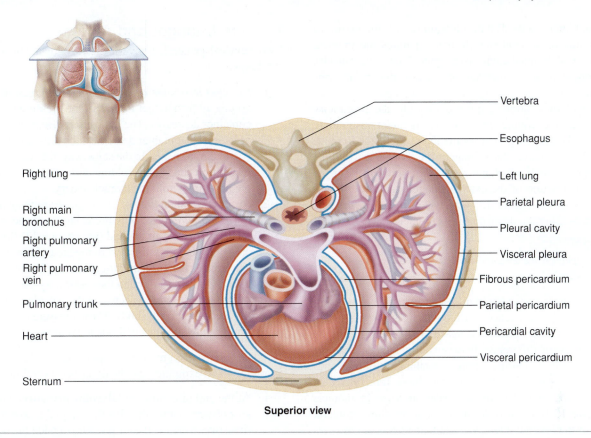

- Vertebra
- Esophagus
- Left lung
- Parietal pleura
- Pleural cavity
- Visceral pleura
- Fibrous pericardium
- Parietal pericardium
- Pericardial cavity
- Visceral pericardium

Right lung
Right main bronchus
Right pulmonary artery
Right pulmonary vein
Pulmonary trunk
Heart
Sternum

Superior view

Figure 15.9 AP|R **Pleural Cavities and Membranes**

Transverse section of the thorax, showing the relationship of the pleural cavities to the thoracic organs. Each lung is surrounded by a pleural cavity. The parietal pleura lines the wall of each pleural cavity, and the visceral pleura covers the surface of the lungs. The space between the parietal and visceral pleurae is small and filled with pleural fluid.

(1) It acts as a lubricant, allowing the visceral and parietal pleurae to slide past each other as the lungs and thorax change shape during respiration, and (2) it helps hold the pleural membranes together. The pleural fluid acts similarly to a thin film of water between two sheets of glass (the visceral and parietal pleurae); the glass sheets can slide over each other easily, but it is difficult to separate them.

Predict 4

> Pleurisy is an inflammation of the pleural membranes. Explain why this condition is so painful, especially when a person takes deep breaths.

Lymphatic Supply

The lungs have two lymphatic supplies (see figure 15.7). The **superficial lymphatic vessels** are deep to the visceral pleura; they drain lymph from the superficial lung tissue and the visceral pleura. The **deep lymphatic vessels** follow the bronchi; they drain lymph from the bronchi and associated connective tissues. No lymphatic vessels are located in the walls of the alveoli. Both the superficial and deep lymphatic vessels exit the lungs at the main bronchi.

Phagocytic cells within the lungs phagocytize most carbon particles and other debris from inspired air and move them to the lymphatic vessels. In older people, the surface of the lungs can appear gray to black because of the accumulation of these particles, especially if the person smoked or lived primarily in a city with air pollution. Other materials, such as cancer cells from the lungs, can also spread to other parts of the body through the lymphatic vessels.

15.3 VENTILATION AND RESPIRATORY VOLUMES

Learning Outcomes   *After reading this section, you should be able to*

A. Explain how contraction of the muscles of respiration causes changes in thoracic volume during quiet breathing and during labored breathing.

B. Describe the changes in alveolar pressure that are responsible for moving air into and out of the lungs.

C. Explain how surfactant and pleural pressure prevent the lungs from collapsing and how changes in pleural pressure cause alveolar volume to change.

D. List the respiratory volumes and capacities, and define each of them.

Ventilation, or *breathing,* is the process of moving air into and out of the lungs. There are two phases of ventilation: (1) **Inspiration,** or *inhalation,* is the movement of air into the lungs; (2) **expiration,** or *exhalation,* is the movement of air out of the lungs. Ventilation is regulated by changes in thoracic volume, which produce changes in air pressure within the lungs.

Changing Thoracic Volume

The muscles associated with the ribs are responsible for ventilation (figure 15.10). The **muscles of inspiration** include the diaphragm and the muscles that elevate the ribs and sternum, such

Respiratory

as the external intercostals. The **diaphragm** (dī′a-fram; partition) is a large dome of skeletal muscle that separates the thoracic cavity from the abdominal cavity (see figure 7.19). The **muscles of expiration,** such as the internal intercostals, depress the ribs and sternum.

At the end of a normal, quiet expiration, the respiratory muscles are relaxed (figure 15.10a). During quiet inspiration, contraction of the diaphragm causes the top of the dome to move inferiorly, which increases the volume of the thoracic cavity. The largest change in thoracic volume results from movement of the diaphragm. Contraction of the external intercostals also elevates the ribs and sternum (figure 15.10b), which increases thoracic volume by increasing the diameter of the thoracic cage.

Predict 5

> During inspiration, the abdominal muscles relax. How is this advantageous?

Expiration during quiet breathing occurs when the diaphragm and external intercostals relax and the elastic properties of the thorax and lungs cause a passive decrease in thoracic volume.

There are several differences between normal, quiet breathing and labored breathing. During labored breathing, all the inspiratory muscles are active, and they contract more forcefully than during quiet breathing, causing a greater increase in thoracic volume (figure 15.10b). Also during labored breathing, forceful contraction of the internal intercostals and the abdominal muscles produces a faster and greater decrease in thoracic volume than would be produced by the passive recoil of the thorax and lungs.

Pressure Changes and Airflow

Two physical principles govern the flow of air into and out of the lungs:

1. *Changes in volume result in changes in pressure.* As the volume of a container increases, the pressure within the container decreases. The opposite is also true. As the volume of a container decreases, the pressure within the container increases. In the same way, the muscles of respiration change the volume of the thorax and therefore the pressure within the thoracic cavity.

2. *Air flows from an area of higher pressure to an area of lower pressure.* If the pressure is higher at one end of a tube than at the other, air or fluid (see chapter 13) flows from the area of higher pressure toward the area of lower pressure. The greater the pressure difference, the greater the rate of airflow. Air flows through the respiratory passages because of pressure differences between the outside of the body and the alveoli inside the body. These pressure differences are produced by changes in thoracic volume.

The volume and pressure changes responsible for one cycle of inspiration and expiration can be described as follows:

1. At the end of expiration, **alveolar pressure,** which is the air pressure within the alveoli, is equal to **atmospheric pressure,** which is the air pressure outside the body. No air moves into or out of the lungs because alveolar pressure and atmospheric pressure are equal (figure 15.11, step 1).

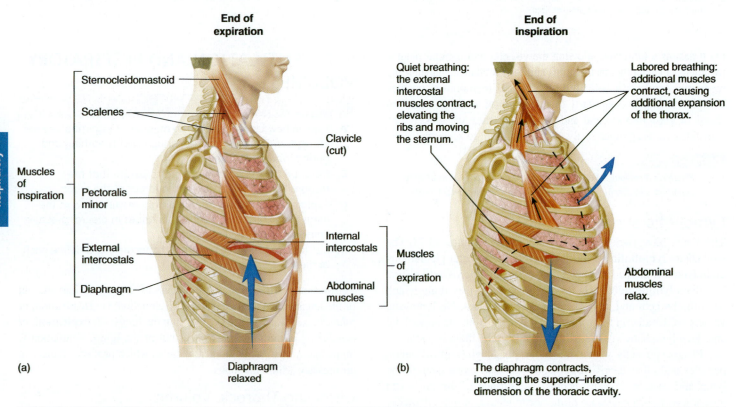

End of expiration

Sternocleidomastoid
Scalenes
Clavicle (cut)

Muscles of inspiration

Pectoralis minor
External intercostals
Internal intercostals
Diaphragm
Abdominal muscles

(a)

Diaphragm relaxed

End of inspiration

Quiet breathing: the external intercostal muscles contract, elevating the ribs and moving the sternum.

Labored breathing: additional muscles contract, causing additional expansion of the thorax.

Muscles of expiration

Abdominal muscles relax.

(b)

The diaphragm contracts, increasing the superior–inferior dimension of the thoracic cavity.

Figure 15.10 AP|R Effect of the Muscles of Respiration on Thoracic Volume

(a) Muscles of respiration at the end of expiration. (b) Muscles of respiration at the end of inspiration.

2. During inspiration, contraction of the muscles of inspiration increases the volume of the thoracic cavity. The increased thoracic volume causes the lungs to expand, resulting in an increase in alveolar volume (see "Changing Alveolar Volume" later in this section). As the alveolar volume increases, alveolar pressure becomes less than atmospheric pressure, and air flows from outside the body through the respiratory passages to the alveoli (figure 15.11, step 2).

3. At the end of inspiration, the thorax and alveoli stop expanding. When the alveolar pressure and atmospheric pressure become equal, airflow stops (figure 15.11, step 3).

4. During expiration, the thoracic volume decreases, producing a corresponding decrease in alveolar volume. Consequently, alveolar pressure increases above atmospheric pressure, and air flows from the alveoli through the respiratory passages to the outside (figure 15.11, step 4).

As expiration ends, the decrease in thoracic volume stops, and the process repeats, beginning at step 1.

Lung Recoil

During quiet expiration, thoracic volume and lung volume decrease because of **lung recoil,** the tendency for an expanded lung to decrease in size. The thoracic wall also recoils due to the elastic properties of its tissues. Lung recoil is able to occur because the connective tissue of the lungs contains elastic fibers and because the film of fluid lining the alveoli has surface tension. **Surface tension** exists because the oppositely charged ends of water molecules are attracted to each other (see chapter 2). As the water molecules pull together, they also pull on the alveolar walls, causing the alveoli to recoil and become smaller.

Two factors keep the lungs from collapsing: (1) surfactant and (2) pressure in the pleural cavity.

Surfactant

Surfactant (ser-fak′tănt; *surf*ace *act*ing *agent*) is a mixture of lipoprotein molecules produced by secretory cells of the alveolar epithelium. The surfactant molecules form a single layer on the surface of the thin fluid layer lining the alveoli, reducing surface tension. Without surfactant, the surface tension causing the alveoli to recoil can be ten times greater than when surfactant is present. Thus, surfactant greatly reduces the tendency of the lungs to collapse.

A CASE IN POINT

Infant Respiratory Distress Syndrome

Tu Soun was born 3 months prematurely. She presented with a respiration rate of 68 breaths per minute; blue lips, tongue, and nail beds; nasal flaring during inspiration; inward movement of the thoracic cage and outward movement of the abdomen during inspiration; an expiratory grunt; and a negative shake test. Tu Soun has infant respiratory distress syndrome (IRDS), caused by too little surfactant, a substance that covers the lining of the lung alveoli, where it helps reduce surface tension. IRDS, also called *hyaline membrane disease,* is common in premature infants because surfactant is not produced in adequate quantities until about the seventh month of gestation. Thereafter, the amount produced increases as the fetus matures. Pregnant women who are likely to deliver prematurely can be given cortisol, which crosses the placenta into the fetus and stimulates surfactant synthesis.

The normal respiration rate for a newborn is around 40 breaths per minute. Tu's high respiratory rate is stimulated by high blood CO_2 levels and low blood O_2 levels (see "Chemical Control of Breathing" later in this chapter). Her blue lips, tongue, and nail beds are signs of cyanosis caused by deoxygenated blood. The nasal flaring occurs to maximize air intake.

If too little surfactant has been produced by the time of birth, the lungs tend to collapse, and the muscles of respiration must exert a great deal of energy to keep the lungs inflated; even then, ventilation is inadequate. The thoracic cage in newborns is very pliable. During labored inspiration, the increased inferior movement of the diaphragm causes such a decrease in thoracic cavity pressure that the thoracic cage is pulled inward as the abdomen expands. The more labored the breathing, the more exaggerated the expansion of the abdomen and the inward movement of the thoracic cage.

An expiratory grunt is a gruff, throaty sound made during expiration. It is caused by partial closure of the vestibular and vocal folds. Expiratory grunting increases airway pressure and helps prevent alveolar collapse.

The shake test determines the presence of surfactant in lung fluid. Fetal lung fluid is either swallowed by the fetus or passed through the mouth into the amniotic fluid, so 30 minutes after delivery the fetus's gastric fluid still contains swallowed lung fluid and amniotic fluid. To perform the shake test, a sample of gastric fluid is collected and then mixed with saline and alcohol, placed in a tube, and shaken. A positive shake test produces bubbles, but a negative shake test does not, which indicates that very little surfactant is present because the bubbles collapse.

Without specialized treatment, most babies with IRDS die soon after birth due to inadequate ventilation of the lungs and respiratory muscle fatigue.

Treatment strategies include forcing enough oxygen-rich air into the lungs to inflate them and administering purified, natural surfactant via an intubation tube directly into the lungs.

Pleural Pressure

When **pleural pressure,** the pressure in the pleural cavity, is less than alveolar pressure, the alveoli tend to expand. This principle can be understood by considering a balloon. The balloon expands when the pressure outside it is less than the pressure inside. This pressure difference is normally achieved by increasing the pressure inside the balloon by blowing into it. This pressure difference, however, can also be achieved by decreasing the pressure outside the balloon. For example, if the balloon is placed in a chamber from which air is removed, the pressure around the balloon becomes lower than atmospheric pressure, and the balloon expands. The lower the pressure outside the balloon, the greater the tendency for the higher pressure inside the balloon to cause it to expand. In a similar fashion, decreasing pleural pressure can result in expansion of the alveoli.

Normally, the alveoli are in the expanded state because pleural pressure is lower than alveolar pressure. Pleural pressure is lower than alveolar pressure because of a suction effect caused by fluid removal by the lymphatic system (see chapter 14) and by lung recoil. As the lungs recoil, the visceral and parietal pleurae tend to be pulled apart. Normally, the lungs do not pull away from the thoracic wall because pleural fluid holds the visceral and parietal pleurae together. Nonetheless, this pull decreases pressure in the pleural cavity. You can appreciate this effect by putting water on the palms of your hands and then placing them together. As you gently pull your hands apart, you will feel a sensation of negative pressure.

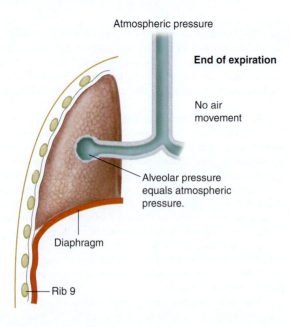

Atmospheric pressure

End of expiration

No air movement

Alveolar pressure equals atmospheric pressure.

Diaphragm

Rib 9

(1) At the end of expiration, alveolar pressure is equal to atmospheric pressure, and there is no air movement.

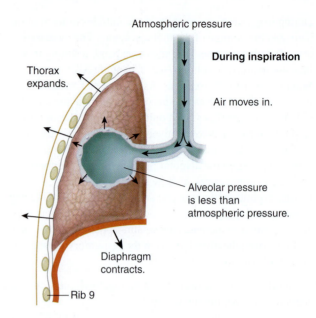

Atmospheric pressure

During inspiration

Thorax expands.

Air moves in.

Alveolar pressure is less than atmospheric pressure.

Diaphragm contracts.

Rib 9

(2) During inspiration, increased thoracic volume results in increased alveolar volume and decreased alveolar pressure. Atmospheric pressure is greater than alveolar pressure, and air moves into the lungs.

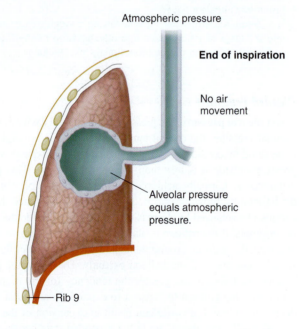

Atmospheric pressure

End of inspiration

No air movement

Alveolar pressure equals atmospheric pressure.

Rib 9

(3) At the end of inspiration, alveolar pressure is equal to atmospheric pressure, and there is no air movement.

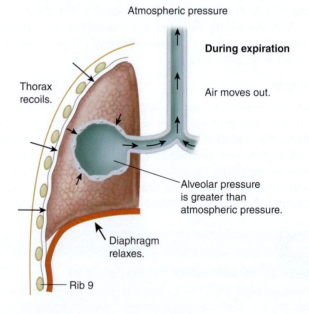

Atmospheric pressure

During expiration

Thorax recoils.

Air moves out.

Alveolar pressure is greater than atmospheric pressure.

Diaphragm relaxes.

Rib 9

(4) During expiration, decreased thoracic volume results in decreased alveolar volume and increased alveolar pressure. Alveolar pressure is greater than atmospheric pressure, and air moves out of the lungs.

PROCESS Figure 15.11 AP|R **Alveolar Pressure Changes During Inspiration and Expiration**

The combined space of all the alveoli is represented by a large "bubble" (*blue*). The alveoli are actually microscopic and cannot be seen in the illustration.

Respiratory

A **pneumothorax** (noo-mō-thōr′aks) is the introduction of air into the pleural cavity, the space between the parietal and visceral pleurae that normally contains only pleural fluid. Air can enter by an external route, as when a sharp object, such as a bullet or a broken rib, penetrates the thoracic wall, or air can enter the pleural cavity by an internal route if alveoli at the lung surface rupture, as can occur in a patient with emphysema. When the pleural cavity is connected to the outside by such openings, the pressure in the pleural cavity increases and becomes equal to the air pressure outside the body. Thus, pleural pressure is also equal to alveolar pressure because pressure in the alveoli at the end of expiration is equal to air pressure outside the body. When pleural pressure and alveolar pressure are equal, there is no tendency for the alveoli to expand, lung recoil is unopposed, and the lungs collapse. A pneumothorax can occur in one lung while the other remains inflated because the two pleural cavities are separated by the mediastinum.

When pleural pressure is lower than alveolar pressure, the alveoli tend to expand. This expansion is opposed by the tendency of the lungs to recoil. Therefore, the alveoli expand when the pleural pressure is low enough that lung recoil is overcome. If the pleural pressure is not low enough to overcome lung recoil, the alveoli collapse, as is the case with a pneumothorax (see the Clinical Impact: "Pneumothorax").

Predict 6

Treatment of a pneumothorax involves closing the opening into the pleural cavity that caused the pneumothorax, then placing a tube into the pleural cavity. In order to inflate the lung, should this tube pump in air under pressure (as in blowing up a balloon), or should the tube apply suction? Explain.

Changing Alveolar Volume

Changes in alveolar volume cause the changes in alveolar pressure that are responsible for moving air into and out of the lungs (see figure 15.11). Alveolar volume changes result from changes in pleural pressure. For example, during inspiration, pleural pressure decreases, and the alveoli expand. The decrease in pleural pressure occurs for two reasons:

1. Increasing the volume of the thoracic cavity results in a decrease in pleural pressure because a change in volume affects pressure.
2. As the lungs expand, lung recoil increases, increasing the suction effect and lowering the pleural pressure. The increased lung recoil of the stretched lung is similar to the increased force generated in a stretched rubber band.

The events of inspiration and expiration can be summarized as follows:

1. During inspiration, pleural pressure decreases because of increased thoracic volume and increased lung recoil. As pleural pressure decreases, alveolar volume increases, alveolar pressure decreases, and air flows into the lungs.
2. During expiration, pleural pressure increases because of decreased thoracic volume and decreased lung recoil. As pleural pressure increases, alveolar volume decreases, alveolar pressure increases, and air flows out of the lungs.

Respiratory Volumes and Capacities

Spirometry (spī-rom′ĕ-trē) is the process of measuring volumes of air that move into and out of the respiratory system, and the **spirometer** (spī-rom′ĕ-ter) is the device that measures these respiratory volumes. Measurements of the respiratory volumes can provide information about the health of the lungs. **Respiratory volumes** are measures of the amount of air movement during different portions of ventilation, whereas **respiratory capacities** are sums of two or more respiratory volumes. The four respiratory volumes and their normal values for a young adult male are shown in figure 15.12:

1. **Tidal volume** is the volume of air inspired or expired with each breath. At rest, quiet breathing results in a tidal volume of about 500 milliliters (mL).
2. **Inspiratory reserve volume** is the amount of air that can be inspired forcefully beyond the resting tidal volume (about 3000 mL).
3. **Expiratory reserve volume** is the amount of air that can be expired forcefully beyond the resting tidal volume (about 1100 mL).
4. **Residual volume** is the volume of air still remaining in the respiratory passages and lungs after maximum expiration (about 1200 mL).

The tidal volume increases when a person is more active. Because the maximum volume of the respiratory system does not change from moment to moment, an increase in the tidal volume causes a decrease in the inspiratory and expiratory reserve volumes.

Predict 7

*The **minute ventilation** is the total amount of air moved into and out of the respiratory system each minute, and it is equal to the tidal volume times the respiratory rate. The **respiratory rate** is the number of breaths taken per minute. Calculate the minute ventilation of a resting person who has a tidal volume of 500 mL and a respiratory rate of 12 respirations/min and the minute ventilation of an exercising person who has a tidal volume of 4000 mL and a respiratory rate of 24 respirations/min.*

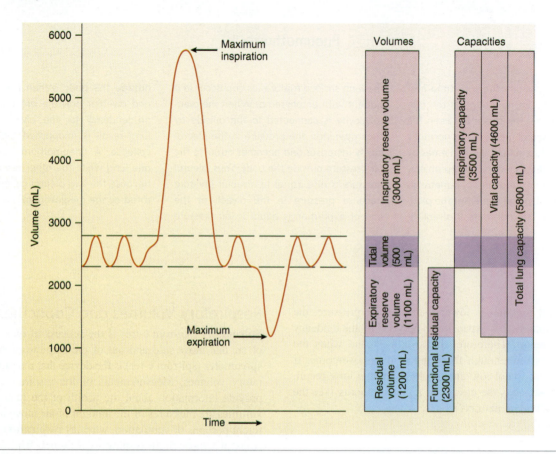

Figure 15.12 Respiratory Volumes and Respiratory Capacities
The tidal volume shown here is during resting conditions. Respiratory volumes are measurements of the volume of air moved into and out of the lungs during breathing. Respiratory capacities are the sum of two or more respiratory volumes.

Values of respiratory capacities, the sum of two or more pulmonary volumes, are shown in figure 15.12:

1. **Functional residual capacity** is the expiratory reserve volume plus the residual volume. This is the amount of air remaining in the lungs at the end of a normal expiration (about 2300 mL at rest).
2. **Inspiratory capacity** is the tidal volume plus the inspiratory reserve volume. This is the amount of air a person can inspire maximally after a normal expiration (about 3500 mL at rest).
3. **Vital capacity** is the sum of the inspiratory reserve volume, the tidal volume, and the expiratory reserve volume. It is the maximum volume of air that a person can expel from the respiratory tract after a maximum inspiration (about 4600 mL).
4. **Total lung capacity** is the sum of the inspiratory and expiratory reserves and the tidal and residual volumes (about 5800 mL). The total lung capacity is also equal to the vital capacity plus the residual volume.

Factors such as sex, age, and body size influence the respiratory volumes and capacities. For example, the vital capacity of adult females is usually 20–25% less than that of adult males. The vital capacity reaches its maximum amount in young adults and gradually decreases in the elderly. Tall people usually have

a greater vital capacity than short people, and thin people have a greater vital capacity than obese people. Well-trained athletes can have a vital capacity 30–40% above that of untrained people. In patients whose respiratory muscles are paralyzed by spinal cord injury or diseases such as poliomyelitis or muscular dystrophy, the vital capacity can be reduced to values not consistent with survival (less than 500–1000 mL).

The **forced expiratory vital capacity** is the rate at which lung volume changes during direct measurement of the vital capacity. It is a simple and clinically important pulmonary test. The individual inspires maximally and then exhales maximally as rapidly as possible into a spirometer. The spirometer records the volume of air expired per second. This test can help identify conditions in which the vital capacity might not be affected but the expiratory flow rate is reduced. Abnormalities that increase the resistance to airflow slow the rate at which air can be forced out of the lungs. For example, in people who have asthma, contraction of the smooth muscle in the bronchioles increases the resistance to airflow. In people who have emphysema, changes in the lung tissue result in the destruction of the alveolar walls, collapse of the bronchioles, and decreased elasticity of the lung tissue. The collapsed bronchioles increase the resistance to airflow. In people who have chronic bronchitis, the air passages are inflamed. The swelling, increased mucus secretion, and gradual loss of cilia result in narrowed bronchioles and increased resistance to airflow.

15.4 GAS EXCHANGE

Learning Outcomes *After reading this section, you should be able to*

A. Explain the factors that affect gas movement through the respiratory membrane.

B. Describe the partial pressure gradients for O_2 and CO_2.

Ventilation supplies atmospheric air to the alveoli. The next step in the process of respiration is the diffusion of gases between the alveoli and the blood in the pulmonary capillaries. As previously stated, gas exchange between air and blood occurs in the respiratory membrane of the lungs (see figure 15.8). The major area of gas exchange is in the alveoli, although some takes place in the respiratory bronchioles and alveolar ducts. Gas exchange between blood and air does not occur in other areas of the respiratory passageways, such as the bronchioles, bronchi, and trachea. The volume of these passageways is therefore called **anatomical dead space.**

The exchange of gases across the respiratory membrane is influenced by the thickness of the membrane, the total surface area of the respiratory membrane, and the partial pressure of gases across the membrane.

Respiratory Membrane Thickness

The thickness of the respiratory membrane increases during certain respiratory diseases. For example, in patients with pulmonary edema, fluid accumulates in the alveoli, and gases must diffuse through a thicker than normal layer of fluid. If the thickness of the respiratory membrane is doubled or tripled, the rate of gas exchange is markedly decreased. Oxygen exchange is affected before CO_2 exchange because O_2 diffuses through the respiratory membrane about 20 times less easily than does CO_2.

Surface Area

The total surface area of the respiratory membrane is about 70 square meters (m^2) in the normal adult, which is approximately the floor area of a 25- × 30-ft room, or roughly the size of a racquetball court (20 × 40 ft). Under resting conditions, a decrease in the surface area of the respiratory membrane to one-third or one-fourth of normal can significantly restrict gas exchange. During strenuous exercise, even small decreases in the surface area of the respiratory membrane can adversely affect gas exchange. Possible reasons for having a decreased surface area include the surgical removal of lung tissue, the destruction of lung tissue by cancer, and the degeneration of the alveolar walls by emphysema. Collapse of the lung—as occurs in pneumothorax—dramatically reduces the volume of the alveoli, as well as the surface area for gas exchange.

Partial Pressure

Gas molecules move randomly from higher concentration to lower concentration until an equilibrium is achieved. One measurement of the concentration of gases is partial pressure. The **partial pressure** of a gas is the pressure exerted by a specific gas in a mixture of gases, such as air. For example, if the total pressure of all the gases in a mixture of gases is 760 millimeters of mercury (mm Hg), which is the atmospheric pressure at sea level, and 21% of the mixture is made up of O_2, the partial pressure for O_2 is 160 mm Hg (0.21 × 760 mm Hg = 160 mm Hg). If the composition of air is 0.04% CO_2 at sea level, the partial pressure for CO_2 is 0.3 mm Hg (0.0004 × 760 = 0.3 mm Hg) (table 15.1). It is traditional to designate the partial pressure of individual gases in a mixture with a capital P followed by the symbol for the gas. Thus, the partial pressure of O_2 is P_{O_2}, and that of CO_2 is P_{CO_2}.

When air is in contact with a liquid, gases in the air dissolve in the liquid. The gases dissolve until the partial pressure of each gas in the liquid is equal to the partial pressure of that gas in the air. Gases in a liquid, like gases in air, diffuse from areas of higher partial pressure toward areas of lower partial pressure, until the partial pressures of the gases are equal throughout the liquid. In other words, gases diffuse *down* their pressure gradient; when atmospheric air comes into contact with the water-based fluid in the lungs, CO_2 and O_2 dissolve in the fluid and each diffuses down its pressure gradient.

Diffusion of Gases in the Lungs

The cells of the body use O_2 and produce CO_2. Thus, blood returning from tissues and entering the lungs has a decreased P_{O_2} and an increased P_{CO_2} compared to alveolar air (figure 15.13). Oxygen diffuses from the alveoli into the pulmonary capillaries because the P_{O_2} in the alveoli is greater than that in the pulmonary capillaries. In contrast, CO_2 diffuses from the pulmonary capillaries into the alveoli because the P_{CO_2} is greater in the pulmonary capillaries than in the alveoli (figure 15.13, step 1) .

When blood enters a pulmonary capillary, the P_{O_2} and P_{CO_2} in the capillary are different from the P_{O_2} and P_{CO_2} in the alveolus. By the time blood flows through the first third of the pulmonary capillary, an equilibrium is achieved, and the P_{O_2} and P_{CO_2} in the capillary are the same as in the alveolus. Thus, in the lungs, the blood gains O_2 and loses CO_2 (figure 15.13, step 2).

During breathing, atmospheric air mixes with alveolar air. The air entering and leaving the alveoli keeps the P_{O_2} higher in the alveoli than in the pulmonary capillaries. Increasing the breathing rate makes the P_{O_2} even higher in the alveoli than it is during slow breathing. During labored breathing, the rate of O_2 diffusion

TABLE 15.1	Partial Pressures of Gases at Sea Level						
	Dry Air		**Humidified Air**		**Alveolar Air**		**Expired Air**	
Gases	**mm HG**	**%**	**mm HG**	**%**	**mm HG**	**%**	**mm HG**	**%**
Nitrogen	600.2	78.98	563.4	74.09	569.0	74.9	566.0	74.5
Oxygen	159.5	20.98	149.3	19.67	104.0	13.6	120.0	15.7
Carbon dioxide	0.3	0.04	0.3	0.04	40.0	5.3	27.0	3.6
Water vapor	0.0	0.0	47.0	6.20	47.0	6.2	47.0	6.2

into the pulmonary capillaries increases because the difference in partial pressure between the alveoli and the pulmonary capillaries has increased. There is a slight decrease in Po_2 in the pulmonary veins due to mixing with deoxygenated blood from veins draining the bronchi and bronchioles; however, the Po_2 in the blood is still higher than that in the tissues (figure 15.13, step 3).

Increasing the rate of breathing also makes the Pco_2 lower in the alveoli than it is during normal, quiet breathing. Because the alveolar Pco_2 decreases, the difference in partial pressure between the alveoli and the pulmonary capillaries increases, which increases the rate of CO_2 diffusion from the pulmonary capillaries into the alveoli.

PROCESS Figure 15.13 AP|R Gas Exchange

Differences in partial pressure are responsible for the exchange of O_2 and CO_2 that occurs between the alveoli and the pulmonary capillaries and between the tissues and the tissue capillaries.

1. Oxygen diffuses into the arterial ends of pulmonary capillaries, and CO_2 diffuses into the alveoli because of differences in partial pressures.
2. As a result of diffusion at the venous ends of pulmonary capillaries, the Po_2 in the blood is equal to the Po_2 in the alveoli, and the Pco_2 in the blood is equal to the Pco_2 in the alveoli.
3. The Po_2 of blood in the pulmonary veins is less than in the pulmonary capillaries because of mixing with deoxygenated blood from veins draining the bronchi and bronchioles.
4. Oxygen diffuses out of the arterial ends of tissue capillaries, and CO_2 diffuses out of the tissue because of differences in partial pressures.
5. As a result of diffusion at the venous ends of tissue capillaries, the Po_2 in the blood is equal to the Po_2 in the tissue, and the Pco_2 in the blood is equal to the Pco_2 in the tissue. Go back to *step 1*.

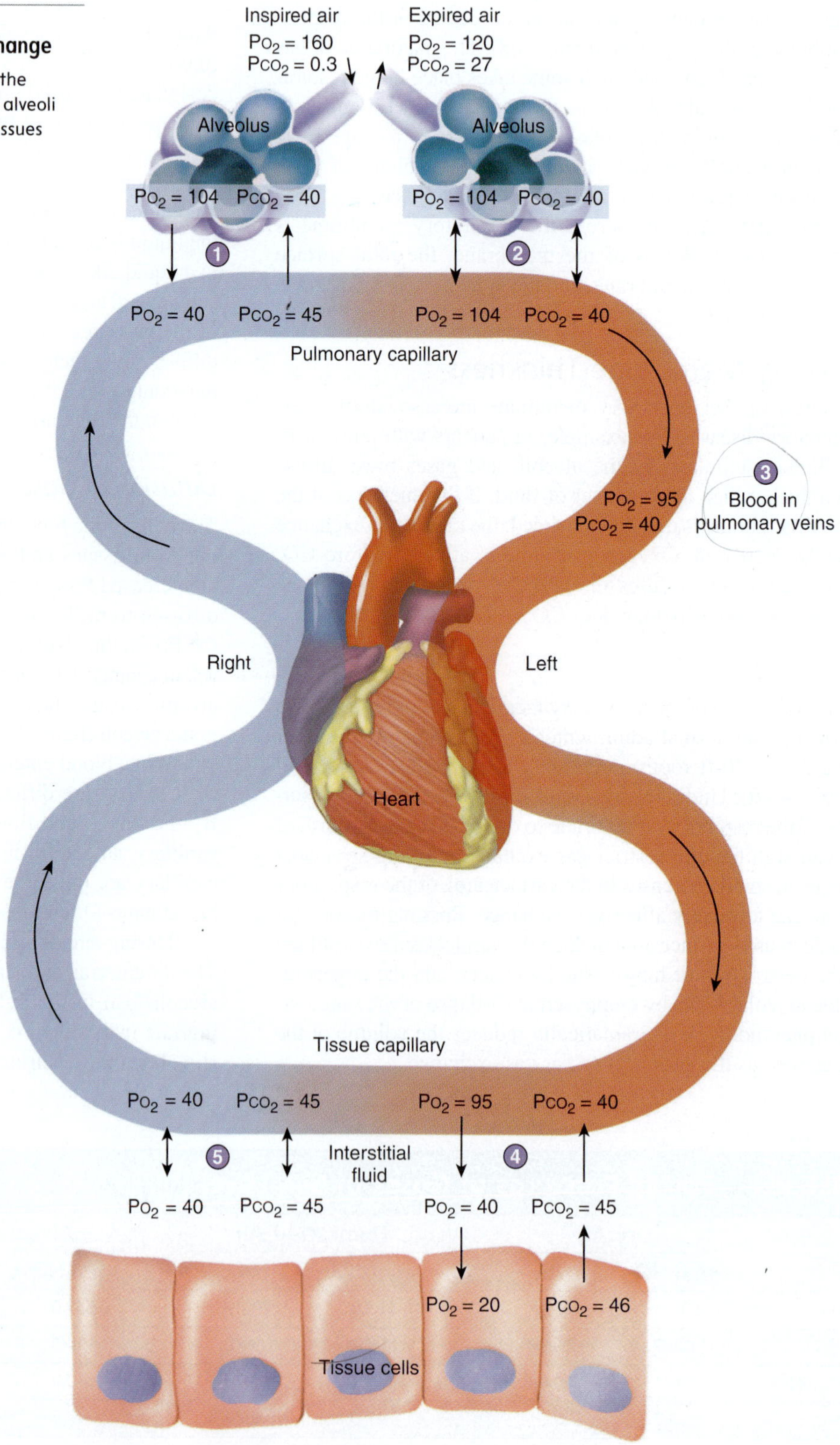

Diffusion of Gases in the Tissues

Blood flows from the lungs through the left side of the heart to the tissue capillaries. Figure 15.13 illustrates the partial pressure differences for O_2 and CO_2 across the wall of a tissue capillary. Oxygen diffuses from the capillary into the interstitial fluid because the P_{O_2} is lower in the interstitial fluid than in the capillary. Oxygen diffuses from the interstitial fluid into cells, in which the P_{O_2} is less than in the interstitial fluid (figure 15.13, step 4). Within the cells, O_2 is used in cellular respiration. There is a constant difference in P_{O_2} between the tissue capillaries and the cells because the cells continuously use O_2. There is also a constant diffusion gradient for CO_2 from the cells. Carbon dioxide therefore diffuses from cells into the interstitial fluid and from the interstitial fluid into the tissue capillaries, and an equilibrium between the blood and tissues is achieved (figure 15.13, step 5).

15.5 GAS TRANSPORT IN THE BLOOD

Learning Outcome *After reading this section, you should be able to*

A. Explain how O_2 and CO_2 are transported in the blood.

Oxygen Transport

After O_2 diffuses through the respiratory membrane into the blood, about 98.5% of the O_2 transported in the blood combines reversibly with the iron-containing heme groups of hemoglobin (see chapter 11). About 1.5% of the O_2 remains dissolved in the plasma. Hemoglobin with O_2 bound to its heme groups is called **oxyhemoglobin** (ok'sē-hē-mō-glō'bin).

The ability of hemoglobin to bind to O_2 depends on the P_{O_2}. At high P_{O_2}, hemoglobin binds to O_2, and at low P_{O_2}, hemoglobin releases O_2. In the lungs, P_{O_2} normally is sufficiently high so that hemoglobin holds as much O_2 as it can. In the tissues, P_{O_2} is lower because the tissues are using O_2. Consequently, hemoglobin releases O_2 in the tissues. Oxygen then diffuses into the cells, which use it in cellular respiration. At rest, approximately 23% of the O_2 picked up by hemoglobin in the lungs is released to the tissues.

The amount of O_2 released from oxyhemoglobin is influenced by four factors. More O_2 is released from hemoglobin if (1) the P_{O_2} is low, (2) the P_{CO_2} is high, (3) the pH is low, and (4) the temperature is high. Increased muscular activity results in a decreased P_{O_2}, an increased P_{CO_2}, a reduced pH, and an increased temperature. Consequently, during physical exercise, as much as 73% of the O_2 picked up by hemoglobin in the lungs is released into skeletal muscles.

Carbon Dioxide Transport and Blood pH

Carbon dioxide diffuses from cells, where it is produced, into the tissue capillaries. After CO_2 enters the blood, it is transported

in three ways: (1) About 7% is transported as CO_2 dissolved in the plasma; (2) 23% is transported in combination with blood proteins, primarily hemoglobin; and (3) 70% is transported in the form of bicarbonate ions.

Carbon dioxide (CO_2) reacts with water to form carbonic acid (H_2CO_3), which then dissociates to form H^+ and bicarbonate ions (HCO_3^-):

$$\text{CO}_2 + \text{H}_2\text{O} \underset{}{\overset{\text{Carbonic anhydrase}}{\rightleftharpoons}} \text{H}_2\text{CO}_3 \rightleftharpoons \text{H}^+ + \text{HCO}_3^-$$

| Carbon dioxide | Water | Carbonic acid | Hydrogen ion | Bicarbonate ion |

An enzyme called **carbonic anhydrase** (kar-bon'ik an-hī'drās) is located inside red blood cells and on the surface of capillary epithelial cells. Carbonic anhydrase increases the rate at which CO_2 reacts with water to form H^+ and HCO_3^- in the tissue capillaries (figure 15.14a). Thus, carbonic anhydrase promotes the uptake of CO_2 by red blood cells.

In the capillaries of the lungs, the process is reversed, so that the HCO_3^- and H^+ combine to produce H_2CO_3, which then forms CO_2 and H_2O (figure 15.14b). The CO_2 diffuses into the alveoli and is expired.

Carbon dioxide has an important effect on the pH of blood. As CO_2 levels increase, the blood pH decreases (becomes more acidic) because CO_2 reacts with H_2O to form H_2CO_3. The H^+ that results from the dissociation of H_2CO_3 is responsible for the decrease in pH. Conversely, as blood levels of CO_2 decline, the blood pH increases (becomes less acidic, or more basic).

15.6 RHYTHMIC BREATHING

Learning Outcomes *After reading this section, you should be able to*

A. Describe the respiratory areas of the brainstem and how they produce a rhythmic pattern of ventilation.
B. Name the neural mechanisms that can modify the normal rhythmic pattern of ventilation.
C. Explain how blood pH, CO_2, and O_2 levels affect ventilation.

The normal rate of breathing in adults is between 12 and 20 breaths per minute. In children, the rates are higher and may vary from 20 to 40 per minute. The rate of breathing is determined by the number of times respiratory muscles are stimulated. The basic rhythm of breathing is controlled by neurons within the medulla oblongata that stimulate the muscles of respiration. An increased depth of breathing results from stronger contractions of the respiratory muscles caused by recruitment of muscle fibers and increased frequency of stimulation of muscle fibers.

Respiratory Areas in the Brainstem

Neurons involved with respiration are located in the brainstem. The neurons that are active during inspiration and those active during expiration are intermingled in these areas.

Respiratory

1. In the tissues, carbon dioxide (CO_2) diffuses into the plasma and into red blood cells. Some of the carbon dioxide remains in the plasma.

2. In red blood cells, carbon dioxide reacts with water (H_2O) to form carbonic acid (H_2CO_3) in a reaction catalyzed by the enzyme carbonic anhydrase (CA).

3. Carbonic acid dissociates to form bicarbonate ions (HCO_3^-) and hydrogen ions (H^+).

4. In the chloride shift, an antiporter allows HCO_3^- to diffuse out of the red blood cells and chloride ions (Cl^-) to diffuse into them, which maintains their electrical neutrality.

5. Oxygen (O_2) is released from hemoglobin (Hb). Oxygen diffuses out of red blood cells and plasma into the tissue.

6. Hydrogen ions combine with hemoglobin, which promotes the release of oxygen from hemoglobin (Bohr effect).

7. Carbon dioxide combines with hemoglobin. Hemoglobin that has released oxygen readily combines with carbon dioxide (Haldane effect).

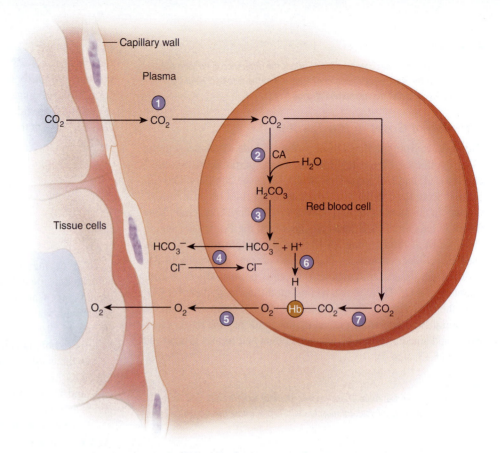

(a) **Gas exchange in the tissues**

1. In the lungs, carbon dioxide (CO_2) diffuses from red blood cells and plasma into the alveoli.

2. Carbonic anhydrase catalyzes the formation of CO_2 and H_2O from H_2CO_3.

3. Bicarbonate ions and H^+ combine to replace H_2CO_3.

4. In the chloride shift, an antiporter allows HCO_3^- to diffuse into the red blood cells and chloride ions (Cl^-) to diffuse out of them, which maintains their electrical neutrality.

5. Oxygen diffuses into the plasma and into red blood cells. Some of the oxygen remains in the plasma. Oxygen binds to hemoglobin.

6. Hydrogen ions are released from hemoglobin, which promotes the uptake of oxygen by hemoglobin (Bohr effect).

7. Carbon dioxide is released from hemoglobin. Hemoglobin that is bound to oxygen readily releases carbon dioxide (Haldane effect).

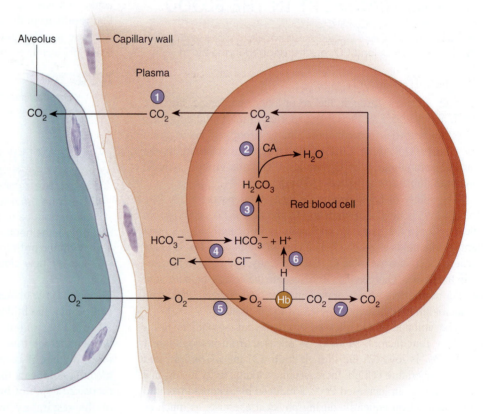

(b) **Gas exchange in the lungs**

PROCESS Figure 15.14   AP|R   Gas Exchange in the Tissues and in the Lungs

(a) In the tissues, CO_2 diffuses into red blood cells, where the enzyme carbonic anhydrase (CA) is located. CA catalyzes the reaction of CO_2 with H_2O to form carbonic acid (H_2CO_3). H_2CO_3 dissociates to form bicarbonate ions (HCO_3^-) and hydrogen ions (H^+). Oxygen is released from hemoglobin (Hb) and diffuses into tissue cells. (b) In the lungs, CO_2 diffuses from red blood cells into the alveoli. CA catalyzes the formation of CO_2 and H_2O from H_2CO_3. H^+ and HCO_3^- combine to replace H_2CO_3. Oxygen diffuses into red blood cells and binds to hemoglobin.

Respiratory

The **medullary respiratory center** consists of two **dorsal respiratory groups,** each forming a longitudinal column of cells located bilaterally in the dorsal part of the medulla oblongata, and two **ventral respiratory groups,** each forming a longitudinal column of cells located bilaterally in the ventral part of the medulla oblongata (figure 15.15). The dorsal respiratory group is primarily responsible for stimulating contraction of the diaphragm. The ventral respiratory group is primarily responsible for stimulating the external intercostal, internal intercostal, and abdominal muscles. A part of the ventral respiratory group, the **pre-Bötzinger complex,** is now known to establish the basic rhythm of breathing.

The **pontine respiratory group** is a collection of neurons in the pons (figure 15.15). It has connections with the medullary respiratory center and appears to play a role in switching between inspiration and expiration.

Generation of Rhythmic Breathing

The medullary respiratory center generates the basic pattern of spontaneous, rhythmic breathing. Although the precise mechanism is not well understood, the generation of rhythmic breathing involves the integration of stimuli that start and stop inspiration.

1. *Starting inspiration.* The neurons in the medullary respiratory center that promote inspiration are continuously active. The medullary respiratory center constantly receives stimulation from many sources, such as receptors that monitor blood gas levels and the movements of muscles and joints. In addition, stimulation can come from parts of the brain concerned with voluntary respiratory movements and emotions. When the inputs from all these sources reach a threshold level, somatic nervous system neurons stimulate respiratory muscles via action potentials, and inspiration starts.

2. *Increasing inspiration.* Once inspiration begins, more and more neurons are activated. The result is progressively stronger stimulation of the respiratory muscles, which lasts for approximately 2 seconds (s).

3. *Stopping inspiration.* The neurons stimulating the muscles of respiration also stimulate the neurons in the medullary respiratory center that are responsible for stopping inspiration. The neurons responsible for stopping inspiration also receive input from the pontine respiratory neurons, stretch receptors in the lungs, and probably other sources. When the inputs to these neurons exceed a threshold level, they cause the neurons stimulating respiratory muscles to be inhibited. Relaxation of respiratory muscles results in expiration, which lasts approximately 3 s. The next inspiration begins with step 1.

Although the medullary neurons establish the basic rate and depth of breathing, their activities can be influenced by input from other parts of the brain and from peripherally located receptors.

Nervous Control of Breathing

Higher brain centers can modify the activity of the respiratory center (figure 15.16*a*). For example, controlling air movements out of the lungs makes speech possible, and emotions can make us sob or gasp. In addition, breathing can be consciously controlled—that is, it is possible to breathe or to stop breathing voluntarily. Some people can hold their breath until they lose consciousness due to

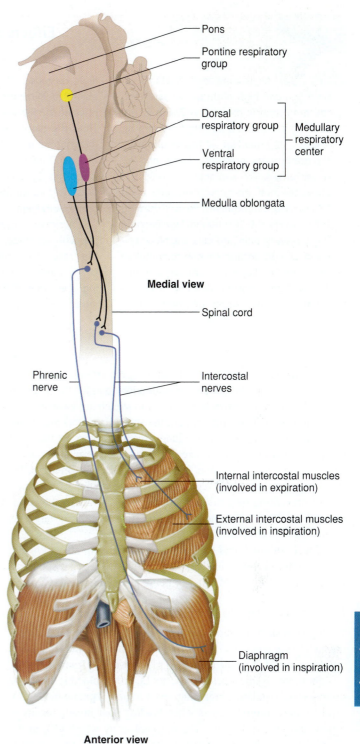

Figure 15.15 Respiratory Structures in the Brainstem
Specific structures in the brainstem correlate with the nerves that innervate the muscles of respiration.

lack of O_2 in the brain. Children have used this strategy to encourage parents to give them what they want. However, as soon as conscious control of respiration is lost, automatic control resumes, and the person starts to breathe again.

Several reflexes, such as sneeze and cough reflexes, can modify breathing. The **Hering-Breuer** (her′ing broy′er) **reflex** supports

CLINICAL IMPACT   Effects of High Altitude and Emphysema

Air is composed of 21% O_2 at both low and high altitudes. At sea level, the atmospheric pressure is 760 mm Hg, and Po_2 is about 160 mm Hg (760 mm Hg × 0.21 = 160 mm Hg). At higher altitudes, the atmospheric pressure is lower, and Po_2 is decreased. For example, at 10,000 ft above sea level, the atmospheric pressure is 523 mm Hg. Consequently, Po_2 is 110 mm Hg (523 mm Hg × 0.21 = 110 mm Hg). Because Po_2 is lower at high altitudes, the blood levels of O_2 can decline enough to stimulate the carotid and aortic bodies. Oxygen then becomes an important stimulus for elevating the rate and depth of breathing. At high altitudes, the respiratory system's ability to eliminate CO_2 is not adversely affected by the low atmospheric pressure. Thus, the blood CO_2 levels become lower than normal because of the increased rate and depth of breathing stimulated by the low blood O_2 levels. The decreased blood CO_2 levels cause the blood pH to become abnormally high.

A similar situation can exist in people with emphysema; the destruction of the respiratory membrane allows less O_2 to move into the blood. The resulting low arterial Po_2 levels stimulate an increased rate and depth of respiration. At first, arterial Pco_2 levels may be unaffected by the reduced surface area of the respiratory membrane, because CO_2 diffuses across the respiratory membrane 20 times more readily than does O_2. However, if alveolar ventilation increases to the point that CO_2 exchange increases above normal, arterial CO_2 becomes lower than normal. More severe emphysema, in which the respiratory membrane surface area is reduced to a minimum, can decrease CO_2 exchange to the point that arterial CO_2 becomes elevated.

---

rhythmic respiratory movements by limiting the extent of inspiration (figure 15.16d). As the muscles of inspiration contract, the lungs fill with air. Sensory receptors that respond to stretch are located in the lungs, and as the lungs fill with air, the stretch receptors are stimulated. Action potentials from the lung stretch receptors are then sent to the medulla oblongata, where they inhibit the respiratory center neurons and cause expiration. In infants, the Hering-Breuer reflex plays an important role in regulating the basic rhythm of breathing and in preventing overinflation of the lungs. In adults, however, the reflex is important only when the tidal volume is large, as occurs during heavy exercise.

Touch, thermal, and pain receptors in the skin also stimulate the respiratory center, which explains why we gasp in response to being splashed with cold water or being pinched (figure 15.16e,f).

Chemical Control of Breathing

During cellular respiration, the body's cells consume O_2 and produce CO_2 (see chapter 17). The primary function of the respiratory system is to add O_2 to the blood and to remove CO_2 from the blood.

Surprisingly, the level of CO_2, not O_2, in the blood is the major driving force regulating breathing. Even a small increase in the CO_2 level (**hypercapnia**), such as when holding your breath, results in a powerful urge to breathe. The mechanism by which CO_2 in the blood stimulates breathing involves the change in pH that accompanies an increase in CO_2 levels. Receptors in the medulla oblongata called **chemoreceptors** are sensitive to small changes in H^+ concentration. Recall from the section "Carbon Dioxide Transport and Blood pH" that blood CO_2 combines with water, which increases H^+ concentration. Thus, it is the H^+ that is detected by the medullary chemoreceptors (figure 15.16b).

Predict 11

Explain why a person who breathes rapidly and deeply (hyperventilates) for several seconds experiences a short period of time during which breathing does not occur (apnea) before normal breathing resumes.

Although O_2 levels are not the major driving force of breathing, there are O_2-sensitive chemoreceptors in the carotid and aortic bodies (figure 15.16c). When blood O_2 levels decline to a low level (**hypoxia**) such as during exposure to high altitude, emphysema, shock, and asphyxiation, the aortic and carotid bodies are strongly stimulated. They send action potentials to the respiratory center and produce an increase in the rate and depth of breathing, which increases O_2 diffusion from the alveoli into the blood.

Because CO_2 levels affect blood pH, the medullary chemoreceptors are important for more than just regulating breathing rate; the medullary chemoreceptors play a crucial role in maintaining blood pH. Figure 15.17 depicts the role breathing rate has on blood pH. If blood CO_2 levels decrease, such as during more rapid breathing, blood pH will increase (become more basic). Thus, the homeostatic mechanism is that the medullary chemoreceptors signal a decreased breathing rate, which retains CO_2 in the blood. More CO_2 in the blood causes H^+ levels to increase, which causes blood pH to decrease to normal levels. Alternatively, if blood CO_2 levels increase, such as during increased physical activity when the body's cells are producing more CO_2 as waste, blood pH will decrease (become more acidic). The medullary chemoreceptors will detect the elevated H^+ and signal a faster breathing rate. As breathing rate goes up, more CO_2 will diffuse out of the blood and blood pH will return to normal. Thus, CO_2 levels are very influential on breathing rate. The opposite is also true, which is why hyperventilation without accompanying increases in CO_2 levels due to physical exercise can cause someone to pass out.

Effect of Exercise on Breathing

The mechanisms by which breathing is regulated during exercise are controversial, and no single factor can account for all the observed responses. Breathing during exercise can be divided into two phases:

1. *Breathing increases abruptly.* At the onset of exercise, the rate of breathing immediately increases. This initial increase can be as much as 50% of the total increase that

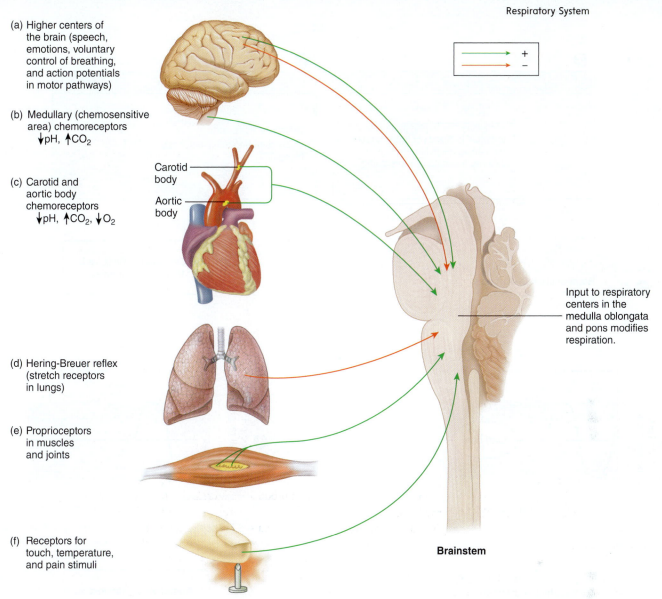

(a) Higher centers of the brain (speech, emotions, voluntary control of breathing, and action potentials in motor pathways)

(b) Medullary (chemosensitive area) chemoreceptors ↓pH, ↑CO_2

(c) Carotid and aortic body chemoreceptors ↓pH, ↑CO_2, ↓O_2

Carotid body

Aortic body

(d) Hering-Breuer reflex (stretch receptors in lungs)

(e) Proprioceptors in muscles and joints

(f) Receptors for touch, temperature, and pain stimuli

+
−

Input to respiratory centers in the medulla oblongata and pons modifies respiration.

Brainstem

Figure 15.16 Nervous and Chemical Mechanisms of Breathing

Several regulatory mechanisms affect the rate and depth of breathing. A plus sign indicates that the mechanism increases breathing and a minus sign indicates that it results in a decrease in breathing.

will occur. The immediate increase occurs too quickly to be explained by changes in metabolism or blood gases. As axons pass from the motor cortex of the cerebrum through the motor pathways, numerous collateral fibers project to the respiratory center. During exercise, action potentials in the motor pathways stimulate skeletal muscle contractions, and action potentials in the collateral fibers stimulate the respiratory center (see figure 15.16).

Furthermore, during exercise, body movements stimulate proprioceptors in the joints of the limbs. Nerve fibers from these proprioceptors extend to the spinal cord to connect with sensory nerve tracts ascending to the brain. Collateral fibers from these nerve tracts connect to the respiratory center; therefore, movement of the limbs has a strong stimulatory influence on the respiratory center (see figure 15.16e).

There may also be a learned component in the breathing response during exercise. After a period of training, the brain "learns" to match breathing with the intensity of the exercise. Well-trained athletes match their respiratory movements more efficiently with their level of physical activity than do untrained individuals. Thus, centers in the brain involved in learning have an indirect influence on the respiratory center, but the exact mechanism is unclear.

2. *Breathing increases gradually.* After the immediate increase in breathing, breathing continues to increase gradually and then levels off within 4–6 minutes after the onset of exercise. Factors responsible for the immediate increase in breathing may play a role in the gradual increase as well.

Despite large changes in O_2 consumption and CO_2 production during exercise, the *average* arterial O_2, CO_2, and pH levels remain constant and close to resting levels as long as the exercise is aerobic (see chapter 7). This suggests that changes in blood gases and pH do not play an important role in regulating breathing

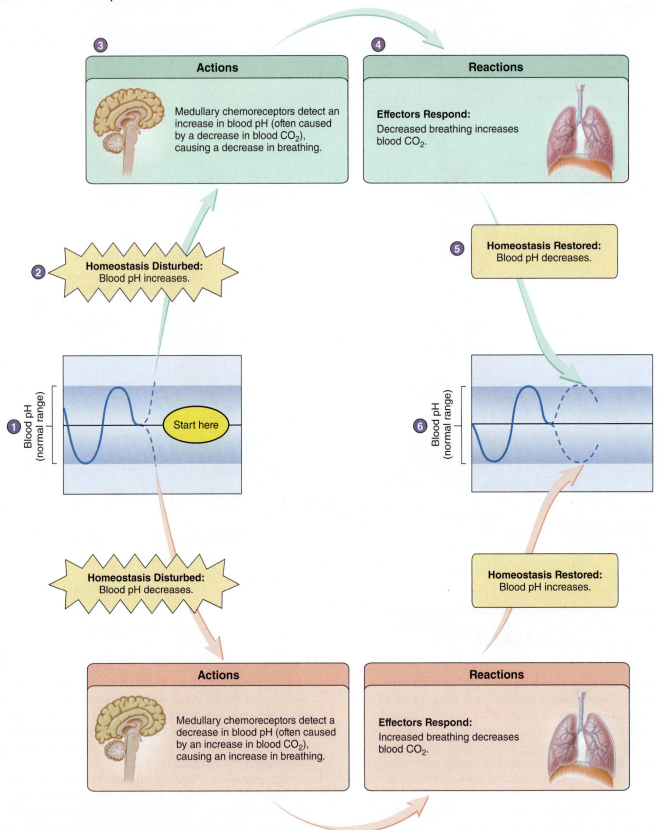

Homeostasis Figure 15.17 Regulation of Blood pH

(1) Blood pH is in its normal range. (2) Blood pH increases outside its normal range, which disturbs homeostasis. (3) The control centers for blood pH, the medullary chemoreceptors, detect an increase in blood pH (blood becomes more basic) and respond to the increased pH by signaling a decreased breathing rate. (4) The effectors, the diaphragm and other respiratory muscles, respond by slowing their contraction rate, which lowers the rate of breathing. (5) As a result, more CO_2 is retained, which causes pH to drop (blood becomes more acidic). (6) Blood pH returns to its normal range and homeostasis is maintained. Observe the responses to a decrease in blood pH by following the red arrows.

DISEASES AND DISORDERS: Respiratory System

CONDITION	DESCRIPTION
	RESPIRATORY DISORDERS
Bronchi and Lungs	
Bronchitis (brong-kī′tis)	Inflammation of the bronchi caused by irritants, such as cigarette smoke or infections; swelling impairs breathing; bronchitis can progress to emphysema
Emphysema (em-fi-sē′mă)	Destruction of alveolar walls; increased coughing increases pressure on the alveoli, causing rupture and destruction; loss of alveoli decreases surface area for gas exchange and decreases the lungs' ability to expel air; progression can be slowed, but there is no cure; alone or in combination with bronchitis, the condition is known as **chronic obstructive pulmonary disease (COPD)**
Adult respiratory distress syndrome (ARDS)	Caused by damage to the respiratory membrane, which promotes inflammation; amount of surfactant is reduced, and fluid fills the alveoli, lessening gas exchange; ARDS usually develops after an injurious event, such as inhaling smoke from a fire or breathing toxic fumes
Cystic fibrosis (fi-brō′sis)	Genetic disorder that affects mucus secretions throughout the body due to an abnormal transport protein; mucus is much more viscous and accumulates in ducts and tubes, such as the bronchioles; airflow is restricted, and infections are more likely
Pulmonary fibrosis	Replacement of lung tissue with fibrous connective tissue, making the lungs less elastic; exposure to asbestos or coal dust are common causes
Lung cancer	Occurs in the epithelium of the respiratory tract; can easily spread to other parts of the body because of the rich blood and lymphatic supply to the lungs
Circulatory System	
Thrombosis of the pulmonary arteries	Blood clot in lung blood vessels; inadequate blood flow through the pulmonary capillaries, affecting respiratory function
Anemia	Reduced hemoglobin lowers oxygen-carrying capacity of blood
Carbon monoxide poisoning	Carbon monoxide binds more strongly to hemoglobin than does O_2 and prevents already-bound O_2 from entering tissues
Nervous System	
Sudden infant death syndrome (SIDS)	Most frequent cause of death of infants between 2 weeks and I year of age; cause is still unknown, but at-risk babies can be placed on monitors that warn if breathing stops
Paralysis of the respiratory muscles	Damage to the spinal cord in the cervical or thoracic region interrupts nervous signals to the muscles of respiration
Thoracic Wall	Decreased elasticity of the thoracic wall prevents it from expanding to full capacity and reduces air movement; two spine curvature conditions that reduce elasticity of the thoracic wall are scoliosis (skō-lē-ō′sis) and kyphosis (kī-fō′sis)
	INFECTIOUS DISEASES OF THE RESPIRATORY SYSTEM
Upper Respiratory Tract	
Strep throat	Caused by streptococcal bacteria (*Streptococcus pyogenes*); characterized by inflammation of the pharynx and fever
Diphtheria (dif-thēr′ē-ă)	Caused by the bacterium *Corynebacterium diphtheriae*; a grayish membrane forms in the throat and can completely block respiratory passages; DPT immunization for children partially targets diphtheria
Common cold	Results from a viral infection
Lower Respiratory Tract	
Whooping cough (pertussis; per-tŭs′is)	Caused by the bacterium *Bordetella pertussis*, which destroys cilia lining the respiratory epithelium, allowing mucus to accumulate; leads to a very severe cough; DPT immunization for children partially targets pertussis
Tuberculosis (tū-ber′kyū-lō′sis)	Caused by the bacterium *Clostridium tuberculosis*, which forms small, lumplike lesions called tubercles; immune system targets tubercles and causes larger lesions; certain strains of tuberculosis are resistant to antibiotics
Pneumonia (noo-mō′nē-ă)	Many bacterial or viral infections of the lungs that cause fever, difficulty in breathing, and chest pain; edema in the lungs reduces their inflation ability and reduces gas exchange
Flu (influenza; in-flū-en′ză)	Viral infection of the respiratory system; does not affect the digestive system, as is commonly misunderstood; causes chills, fever, headache, and muscle aches
Fungal diseases	Fungal spores enter the respiratory tract attached to dust particles, usually resulting in minor respiratory infections that in some cases can spread to other parts of the body

during aerobic exercise. However, during exercise, the values of arterial O_2, CO_2, and pH levels rise and fall more than they do at rest. Thus, even though their average values do not change, their oscillations may be a signal for helping control breathing.

The highest level of exercise that can be performed without causing a significant change in blood pH is the **anaerobic threshold.** If the exercise intensity becomes high enough to exceed the anaerobic threshold, skeletal muscles produce lactate through the

Asthma

Background Information

Asthma (az′mă; difficult breathing) is characterized by abnormally increased constriction of the trachea and bronchi in response to various stimuli, which results in narrowed air passageways and decreased ventilation efficiency. Symptoms include rapid and shallow breathing, wheezing, coughing, and shortness of breath (figure 15A). In contrast to many other respiratory disorders, the symptoms of asthma typically reverse either spontaneously or with therapy.

There is no definitive pathological feature or diagnostic test for asthma, but three important characteristics of the disease are chronic airway inflammation, airway hyperreactivity, and airflow obstruction. The inflammation results in tissue damage, edema, and mucus buildup, which can block airflow through the bronchi. Airway hyperreactivity means that the smooth muscle in the trachea and bronchi contracts greatly in response to a stimulus, thus decreasing the diameter of the airway and increasing resistance to airflow. The effects of inflammation and airway hyperreactivity combine to cause airflow obstruction (figure 15B).

Many cases of asthma appear to be associated with a chronic inflammatory response by the immune system. The number of immune cells in the bronchi, including mast cells, eosinophils, neutrophils, macrophages, and lymphocytes, increases. Inflammation appears to be linked to airway hyperreactivity by some chemical mediators released by immune cells (e.g., leukotrienes, prostaglandins, and interleukins), which increase the airway's sensitivity to stimulation and cause smooth muscle contraction.

The stimuli that prompt airflow obstruction in asthma vary from one individual to another. Some asthmatics react to particular allergens, which are foreign substances that evoke an inappropriate immune system response (see chapter 14). Examples include inhaled pollen, animal dander, and dust mites. Many cases of asthma are

Name: Will
Gender: Male
Age: 18

Comments

Will is an 18-year-old track athlete in seemingly good health. Despite suffering from a slight cold, Will went jogging one morning with his running buddy, Al. After a few minutes of exercise, Will felt that he could hardly get enough air. Even though he stopped jogging, he continued to breathe rapidly and wheeze forcefully. Because his condition was not improving, Al took him to the emergency room of a nearby hospital.

The emergency room doctor used a stethoscope to listen to Will's lungs and noted that air movement was poor. Will inhaled a bronchodilator drug, which rapidly improved his condition.

Respiratory

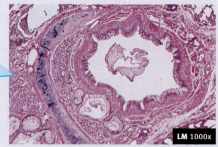

Asthmatic bronchiole:
Note how constricted and mucus-filled it is.

`LM 1000x`

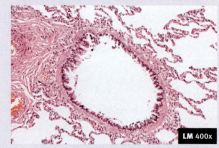

Normal bronchiole:
Note how clear it is.

`LM 400x`

Figure 15A

Strenuous exercise is one of the many factors that can bring on an asthma attack.

Figure 15B

SKELETAL
Many of the immune cells responsible for the inflammatory response of asthma are produced in the red bone marrow. The thoracic cage is necessary for respiration.

INTEGUMENTARY
Cyanosis, a bluish skin color, results from decreased blood O_2 content.

MUSCULAR
Skeletal muscles are necessary for respiratory movements and the cough reflex. Increased muscular work during a severe asthma attack can cause metabolic acidosis because of anaerobic respiration and excessive lactate production.

URINARY
Modifying hydrogen ion secretion into the urine helps compensate for acid-base imbalances caused by asthma.

Asthma

Symptoms
- Rapid and shallow breathing
- Wheezing
- Coughing
- Shortness of breath

Treatment
- Avoiding the causative agent
- Taking anti-inflammatory medication
- Using bronchodilators

NERVOUS
Emotional upset or stress can provoke an asthma attack. Peripheral and central chemoreceptor reflexes affect ventilation. The cough reflex helps remove mucus from respiratory passages. Pain, anxiety, and death from asphyxiation can result from the altered gas exchange caused by asthma. One theory of the cause of asthma is an imbalance in the autonomic nervous system (ANS) control of bronchiolar smooth muscle, and drugs that enhance the sympathetic effects or block the parasympathetic effects are used to treat asthma.

DIGESTIVE
Ingested substances, such as aspirin, sulfiting agents (preservatives), tartrazine, certain foods, and reflux of stomach acid into the esophagus can provoke an asthma attack.

LYMPHATIC AND IMMUNE
Immune cells release chemical mediators that promote inflammation, increase mucus production, and cause bronchiolar constriction, which is believed to be a major factor in asthma. Ingested allergens, such as aspirin or sulfites in food, can provoke an asthma attack.

ENDOCRINE
Steroids from the adrenal gland help regulate inflammation and are used in asthma therapy.

CARDIOVASCULAR
Increased vascular permeability of lung blood vessels results in edema. Blood carries ingested substances that provoke an asthma attack to the lungs. Blood also carries immune cells from red bone marrow to the lungs. Tachycardia commonly occurs during an asthma attack, and the normal effects of respiration on venous return are exaggerated, resulting in large fluctuations in blood pressure.

---

caused by an allergic reaction to substances in the droppings and carcasses of cockroaches, which may explain the higher rate of asthma in poor, urban areas. However, other inhaled substances, such as chemicals in the workplace or cigarette smoke, can provoke an asthma attack without stimulating an allergic reaction. Over 200 substances have been associated with occupational asthma. An asthma attack can also be stimulated by ingested substances, such as aspirin; nonsteroidal anti-inflammatory compounds, such as ibuprofen (ī-bū′prō-fen); sulfites in food preservatives; and tartrazine (tar′tră-zēn) in food colorings. Asthmatics can substitute acetaminophen (as-et-ă-mē′nō-fen, a-set-ă-min′ō-fen) (e.g., Tylenol) for aspirin.

Other stimuli, such as strenuous exercise (especially in cold weather) can precipitate an asthma attack. Such episodes can often be avoided by using a bronchodilator prior to exercise. Viral infections, emotional upset, stress, air pollution, and even reflux of stomach acid into the esophagus are known to elicit asthma attacks.

Treatment of asthma involves avoiding the causative stimulus and taking medications. Steroids and mast cell–stabilizing agents, which prevent the release of chemical mediators from mast cells, can reduce airway inflammation. Bronchodilators are used to increase airflow.

Predict 12

It is not usually necessary to assess arterial blood gases when diagnosing and treating asthma. However, this information can sometimes be useful in severe asthma attacks. Suppose that Will had a P_{O_2} of 60 mm Hg and a P_{CO_2} of 30 mm Hg when he first went to the emergency room. Explain how that could happen.

anaerobic process of anaerobic respiration (see figure 17.5). Lactate released into the blood contributes to a decrease in blood pH, which stimulates the carotid bodies, resulting in increased breathing. In fact, ventilation can increase so much that arterial CO_2 levels fall below resting levels, and arterial O_2 levels rise above resting levels.

15.7 RESPIRATORY ADAPTATIONS TO EXERCISE

Learning Outcome   *After reading this section, you should be able to*

A. Describe the regulation of ventilation during exercise and the changes in the respiratory system that result from exercise training.

In response to training, athletic performance increases because the cardiovascular and respiratory systems become more efficient at delivering O_2 and picking up CO_2. In most individuals, breathing does not limit performance because breathing can increase to a greater extent than can cardiovascular function.

After training, vital capacity increases slightly, and residual volume decreases slightly. Tidal volume at rest and during standardized, submaximal exercise (activities normally encountered in everyday life) does not change. At maximal exercise, however, the tidal volume increases. Increased efficiency of the respiratory system in response to training is evident because the respiratory rate at rest or during standardized submaximal exercise in trained individuals is slightly lower; however, at maximal exercise, their respiratory rate is usually increased.

Minute ventilation is affected by changes in tidal volume and breathing rate. After training, minute ventilation is essentially unchanged or slightly reduced at rest, slightly reduced during standardized submaximal exercise, and greatly increased at maximal exercise. For example, an untrained person with a minute ventilation of 120 liters per minute (L/min) can increase his or her minute ventilation to 150 L/min after training. Increases to 180 L/min are typical of highly trained athletes.

15.8 EFFECTS OF AGING ON THE RESPIRATORY SYSTEM

Learning Outcome   *After reading this section, you should be able to*

A. Describe the effects of aging on the respiratory system.

Aging affects most aspects of the respiratory system. But even though vital capacity, maximum ventilation rates, and gas exchange decrease with age, the elderly can engage in light to moderate exercise because the respiratory system has a large reserve capacity.

With age, mucus accumulates within the respiratory passageways. The mucus-cilia escalator is less efficient because the mucus becomes more viscous and the number of cilia and their rate of movement decrease. As a consequence, the elderly are more susceptible to respiratory infections and bronchitis.

Vital capacity decreases with age because of reduced ability to fill the lungs (decreased inspiratory reserve volume) and to empty the lungs (decreased expiratory reserve volume). As a result, maximum minute ventilation rates decrease, which in turn decreases the ability to perform intense exercise. These changes are related to the weakening of respiratory muscles and the stiffening of cartilage and ribs.

Residual volume increases with age as the alveolar ducts and many of the larger bronchioles increase in diameter. This increases the dead space, which decreases the amount of air available for gas exchange. In addition, gas exchange across the respiratory membrane declines because parts of the alveolar walls are lost, which decreases the surface area available for gas exchange, and the remaining walls thicken, which decreases the diffusion of gases. A gradual increase in resting tidal volume with age compensates for these changes.

ANSWER TO LEARN TO PREDICT

Mr. Theron suffers from emphysema, a respiratory disorder that results in the destruction of alveoli. This chapter explained that the alveoli form the respiratory membrane, the site of gas exchange between the atmosphere and the blood. Alveolar destruction would directly reduce the respiratory membrane surface and, therefore, gas exchange. As a consequence, Mr. Theron has exaggerated respiratory movements to compensate for the reduction in surface area. Blood P_{O_2} is an important stimulus for the respiratory center, and the increased respiratory movements keep the ventilation just adequate to maintain blood P_{O_2} in the low normal range. Because CO_2 diffuses across the respiratory membrane at a faster rate than O_2, the elevated respiration required to maintain blood P_{O_2} causes too much CO_2 to be expired, with the result that his blood P_{CO_2} level has dropped below normal.

This chapter explained that a pneumothorax, or the introduction of air into the pleural cavity through an opening in the thoracic wall or lung, can cause a lung to collapse. Due to his emphysema, Mr. Theron's lung collapsed when alveoli near the surface of the lung ruptured, allowing air to enter the pleural space. The physician was able to diagnose Mr. Theron's collapsed lung by listening for respiratory sounds with a stethoscope. He detected respiratory sounds in the right lung but not in the left lung. We would expect Mr. Theron's respiratory movements to be even more exaggerated, since the respiratory membrane was reduced by one-half when his left lung collapsed. The reduction in the respiratory membrane would cause a drop in P_{O_2} and an increase in P_{CO_2}, both of which would stimulate respiratory centers to increase ventilation. While in the hospital, Mr. Theron's breathing was assisted by a ventilator. If his emphysema continues to severely worsen, he may require a ventilator for the remainder of his life.

Answers to the rest of this chapter's Predict questions are in Appendix E.

SUMMARY

Respiration includes ventilation, the movement of air into and out of the lungs; the exchange of gases between the air and the blood; the transport of gases in the blood; and the exchange of gases between the blood and the tissues.

15.1 Functions of the Respiratory System (p. 412)

The respiratory system exchanges O_2 and CO_2 between the air and the blood, regulates blood pH, helps produce sounds, moves air over the sensory receptors that detect smell, and protects against some pathogens.

15.2 Anatomy of the Respiratory System (p. 413)

Nose

1. The nose consists of the external nose and the nasal cavity.
2. The bridge of the nose is bone, and most of the external nose is cartilage.
3. The nasal cavity warms, humidifies, and cleans the air. The nares open to the outside, and the choanae lead to the pharynx. The nasal cavity is divided by the nasal septum into right and left parts. The paranasal sinuses and the nasolacrimal duct open into the nasal cavity. Hairs just inside the nares trap debris. The nasal cavity is lined with pseudostratified epithelium containing cilia that trap debris and move it to the pharynx.

Pharynx

1. The nasopharynx joins the nasal cavity through the choanae and contains the opening to the auditory tube and the pharyngeal tonsils.
2. The oropharynx joins the oral cavity and contains the palatine and lingual tonsils.
3. The laryngopharynx opens into the larynx and the esophagus.

Larynx

1. The larynx consists of three unpaired cartilages and six paired ones. The thyroid cartilage and cricoid cartilage form most of the larynx. The epiglottis covers the opening of the larynx during swallowing.
2. The vestibular folds can prevent air, food, and liquids from passing into the larynx.
3. The vocal folds (true vocal cords) vibrate and produce sounds when air passes through the larynx. The force of air movement controls loudness, and changes in the length and tension of the vocal folds determine pitch.

Trachea

The trachea connects the larynx to the main bronchi.

Bronchi

The main bronchi extend from the trachea to each lung.

Lungs

1. There are two lungs.
2. The airway passages of the lungs branch and decrease in size. The main bronchi form the lobar bronchi, which go to each lobe of the lungs. The lobar bronchi form the segmental bronchi, which extend to each bronchopulmonary segment of the lungs. The segmental bronchi branch many times to form the bronchioles. The bronchioles branch to form the terminal bronchioles, which give rise to the respiratory bronchioles, from which alveolar ducts branch. Alveoli are air sacs connected to the alveolar ducts and respiratory bronchioles.

3. The epithelium from the trachea to the terminal bronchioles is ciliated to facilitate removal of debris. Cartilage helps hold the tube system open (from the trachea to the bronchioles). Smooth muscle controls the diameter of the tubes (especially the bronchioles). The alveoli are formed by simple squamous epithelium, and they facilitate diffusion of gases.
4. The respiratory membrane has six layers, including a film of water, the walls of the alveolus and the capillary, and an interstitial space. The respiratory membrane is thin and has a large surface area that facilitates gas exchange.

Pleural Cavities

The pleural membranes surround the lungs and provide protection against friction.

Lymphatic Supply

Superficial and deep lymphatic vessels drain the lungs.

15.3 Ventilation and Respiratory Volumes (p. 421)

Changing Thoracic Volume

1. Inspiration occurs when the diaphragm contracts and the external intercostal muscles lift the rib cage, thus increasing the volume of the thoracic cavity. During labored breathing, additional muscles of inspiration increase rib movement.
2. Expiration can be passive or active. Passive expiration during quiet breathing occurs when the muscles of inspiration relax. Active expiration during labored breathing occurs when the diaphragm relaxes and the internal intercostal and abdominal muscles depress the rib cage to forcefully decrease the volume of the thoracic cavity.

Pressure Changes and Airflow

1. Respiratory muscles cause changes in thoracic volume, which in turn cause changes in alveolar volume and pressure.
2. During inspiration, air flows into the alveoli because atmospheric pressure is greater than alveolar pressure.
3. During expiration, air flows out of the alveoli because alveolar pressure is greater than atmospheric pressure.

Lung Recoil

1. The lungs tend to collapse because of the elastic recoil of the connective tissue and the surface tension of the fluid lining the alveoli.
2. The lungs normally do not collapse because surfactant reduces the surface tension of the fluid lining the alveoli, and pleural pressure is lower than alveolar pressure.

Changing Alveolar Volume

1. Increasing thoracic volume results in decreased pleural pressure, increased alveolar volume, decreased alveolar pressure, and air movement into the lungs.
2. Decreasing thoracic volume results in increased pleural pressure, decreased alveolar volume, increased alveolar pressure, and air movement out of the lungs.

Respiratory Volumes and Capacities

1. There are four measurements of respiratory volume: tidal, inspiratory reserve, expiratory reserve, and residual.
2. Respiratory capacities are the sum of two or more respiratory volumes; they include vital capacity and total lung capacity.

3. The forced expiratory vital capacity measures the rate at which air can be expelled from the lungs.

15.4 Gas Exchange   (p. 427)

1. Gas exchange between air and blood occurs in the respiratory membrane.
2. The parts of the respiratory passageways where gas exchange between air and blood does not occur constitute the dead space.

Respiratory Membrane Thickness

Increases in the thickness of the respiratory membrane result in decreased gas exchange.

Surface Area

Small decreases in surface area adversely affect gas exchange during strenuous exercise. When the surface area is decreased to one-third to one-fourth of normal, gas exchange is inadequate under resting conditions.

Partial Pressure

1. The pressure exerted by a specific gas in a mixture of gases is reported as the partial pressure of that gas.
2. Oxygen diffuses from a higher partial pressure in the alveoli to a lower partial pressure in the pulmonary capillaries. Oxygen diffuses from a higher partial pressure in the tissue capillaries to a lower partial pressure in the tissue spaces.
3. Carbon dioxide diffuses from a higher partial pressure in the tissues to a lower partial pressure in the tissue capillaries. Carbon dioxide diffuses from a higher partial pressure in the pulmonary capillaries to a lower partial pressure in the alveoli.

15.5 Gas Transport in the Blood   (p. 429)

Oxygen Transport

1. Most (98.5%) O_2 is transported bound to hemoglobin. Some (1.5%) O_2 is transported dissolved in plasma.
2. Oxygen is released from hemoglobin in tissues when the partial pressure for O_2 is low, the partial pressure for CO_2 is high, pH is low, and temperature is high.

Carbon Dioxide Transport and Blood pH

1. Carbon dioxide is transported in solution as plasma (7%), in combination with blood proteins (23%), and as bicarbonate ions (70%).
2. In tissue capillaries, CO_2 combines with water inside the red blood cells to form carbonic acid that dissociates to form HCO_3^- and H^+. This reaction promotes the transport of CO_2.
3. In lung capillaries, HCO_3^- combines with H^+ to form carbonic acid. The carbonic acid dissociates to form CO_2, which diffuses out of the red blood cells.
4. As blood CO_2 levels increase, blood pH decreases; as blood CO_2 levels decrease, blood pH increases. Changes in breathing change blood CO_2 levels and pH.

15.6 Rhythmic Breathing   (p. 429)

Respiratory Areas in the Brainstem

1. The medullary respiratory center, specifically the pre-Bötzinger complex, establishes rhythmic breathing.
2. The pontine respiratory group appears to be involved with the switch between inspiration and expiration.

Generation of Rhythmic Breathing

1. Inspiration begins when stimuli from many sources, such as receptors that monitor blood gases, reach a threshold.
2. Expiration begins when the neurons causing inspiration are inhibited.

Nervous Control of Breathing

1. Higher brain centers allow voluntary control of breathing. Emotions and speech production affect breathing.
2. The Hering-Breuer reflex inhibits the respiratory center when the lungs are stretched during inspiration.
3. Touch, thermal, and pain receptors can stimulate breathing.

Chemical Control of Breathing

1. Carbon dioxide is the major chemical regulator of breathing. An increase in blood CO_2 causes a decrease in blood pH, resulting in increased breathing.
2. Low blood levels of O_2 can stimulate chemoreceptors in the carotid and aortic bodies, also resulting in increased breathing.
3. Chemoreceptors in the medulla oblongata respond to changes in blood pH. Usually, changes in blood pH are produced by changes in blood CO_2.

Effect of Exercise on Breathing

Input from higher brain centers and from proprioceptors stimulates the respiratory center during exercise.

15.7 Respiratory Adaptations to Exercise   (p. 438)

Training results in increased minute volume at maximal exercise because of increased tidal volume and respiratory rate.

15.8 Effects of Aging on the Respiratory System   (p. 438)

1. Vital capacity and maximum minute ventilation decrease with age because of weakening of the respiratory muscles and stiffening of the thoracic cage.
2. Residual volume and dead space increase because the diameter of respiratory passageways increases.
3. An increase in resting tidal volume compensates for increased dead space, loss of alveolar walls (surface area), and thickening of alveolar walls.
4. The ability to remove mucus from the respiratory passageways decreases with age.

REVIEW AND COMPREHENSION

1. Define respiration.
2. What are the functions of the respiratory system?
3. Describe the structures of the nasal cavity and their functions.
4. Name the three parts of the pharynx. With what structures does each part communicate?
5. Name and give the functions of the three unpaired cartilages of the larynx.
6. What are the functions of the vestibular and vocal folds? How are sounds of different loudness and pitch produced?
7. Starting at the larynx, name in order all the tubes air passes through to reach an alveolus.
8. What is the function of the C-shaped cartilages in the trachea? What happens to the amount of cartilage in the tube system of the respiratory system as the tubes become smaller? Explain why breathing becomes more difficult during an asthma attack.

9. What is the function of the ciliated epithelium in the tracheobronchial tree?

10. Distinguish among the lungs, a lobe of the lung, and a bronchopulmonary segment.

11. List the components of the respiratory membrane.

12. Describe the pleurae of the lungs. What is their function?

13. Describe the lymphatic supply of the lungs. What is its function?

14. Explain how the muscles of inspiration and expiration change thoracic volume.

15. Describe the pressure changes that cause air to move into and out of the lungs. What causes these pressure changes?

16. Give two reasons the lungs tend to recoil. What two factors keep the lungs from collapsing?

17. Explain how changes in thoracic volume result in changes in pleural pressure, alveolar volume, alveolar pressure, and airflow during inspiration and expiration.

18. Define tidal volume, inspiratory reserve volume, expiratory reserve volume, and residual volume. Define vital capacity, total lung capacity, and forced expiratory vital capacity.

19. Describe the factors that affect the diffusion of gases across the respiratory membrane. Give some examples of diseases that decrease diffusion by altering these factors.

20. What is the partial pressure of a gas? Describe the diffusion of O_2 and CO_2 between the alveoli and the pulmonary capillaries and between the tissue capillaries and the tissues in terms of partial pressures.

21. List the ways in which blood transports O_2. What factors promote the release of O_2 in tissues?

22. List the ways in which blood transports CO_2.

23. How does the level of blood CO_2 affect blood pH? How can changes in ventilation affect blood pH?

24. Name the respiratory areas of the brainstem, and explain how rhythmic breathing is generated.

25. Describe how higher brain centers and the Hering-Breuer reflex can modify breathing.

26. Explain the role of blood pH, CO_2, and O_2 in modifying breathing.

27. How is breathing regulated during exercise?

28. What effect does exercise training have on the respiratory system?

29. Why do vital capacity, alveolar ventilation, and diffusion of gases across the respiratory membrane decrease with age? Why are the elderly more likely to develop respiratory infections and bronchitis?

CRITICAL THINKING

1. Cardiopulmonary resuscitation (CPR) has replaced former methods of sustaining ventilation. The back pressure/arm lift method is one such technique that is no longer used. This procedure must be performed with the victim lying face down. The rescuer presses firmly on the base of the scapulae for several seconds, then grasps the arms and lifts them. The sequence is then repeated. Although this procedure is less efficient than CPR, it does result in ventilation of the lungs. Explain why.

2. Another technique for artificial respiration is mouth-to-mouth resuscitation. The rescuer takes a deep breath, blows air into the victim's mouth, and then lets air flow out of the victim's lungs. The process is repeated. Explain the following:
   a. Why do the victim's lungs expand?
   b. Why does air move out of the victim's lungs?

3. A person's vital capacity was measured while she was standing and while she was lying down. What difference, if any, in the measurement would you predict, and why?

4. If water vapor forms 10% of the gases in air at sea level, what is the partial pressure of water?

5. A patient has pneumonia, and fluids accumulate within the alveoli. Explain why this results in an increased rate of respiration. How can O_2 therapy return this rate to normal?

6. A patient has severe emphysema that has extensively damaged the alveoli and reduced the surface area of the respiratory membrane. Although the patient is receiving O_2 therapy, he still has a tremendous urge to take a breath (i.e., he does not feel as if he is getting enough air). Why does this occur?

7. Patients with diabetes mellitus who are not being treated with insulin therapy rapidly metabolize lipids, and acidic by-products of lipid metabolism accumulate in the bloodstream. How does this affect ventilation? Why is the change in ventilation beneficial?

8. Ima Anxious was hysterical and hyperventilating. The doctor made her breathe into a paper bag. Because you are an especially astute student, you say to the doctor, "When Ima was hyperventilating, she was reducing blood CO_2 levels; when she breathed into the paper bag, CO_2 was trapped in the bag, and she was rebreathing it, thus causing her blood CO_2 levels to increase. As blood CO_2 levels increased, the urge to breathe should have increased. Instead, she began to breathe more slowly. Please explain." How do you think the doctor would respond? (*Hint:* Recall that the effect of decreased blood CO_2 on the vasomotor center results in vasodilation and a sudden decrease in blood pressure.)

9. Hyperventilating before swimming underwater can increase the time spent under water. Explain how that can happen. Sometimes a person who has hyperventilated before swimming under water passes out while still under water and drowns. Explain.

10. The blood pH of a runner was monitored during a race. Shortly after the beginning of the race, her blood pH increased for a short time. Propose an explanation to account for the increased pH values following the start of the race.

Answers in Appendix D

Respiratory

16 Digestive System

16.1 FUNCTIONS OF THE DIGESTIVE SYSTEM

Learning Outcome *After reading this section, you should be able to*

A. List the major functions of the digestive system.

Every cell of the body requires nutrients, yet most cells cannot leave their position in the body and travel to a food source. Therefore, the digestive system must help deliver food to them. The **digestive system** (figure 16.1), with the assistance of the circulatory system, is like a gigantic "meals on wheels," serving over 100 trillion customers the nutrients they need. It also has its own quality control and waste disposal methods. Food is taken into the digestive system, where it is broken down into smaller and smaller particles. Enzymes in the digestive system break the particles down into very small molecules, which are absorbed into the circulation and transported all over the body. There, those molecules are broken down by other enzymes to release energy or are assembled into new molecules to build tissues and organs. This chapter describes the structure and function of the digestive organs and their accessory glands.

The functions of the digestive system include the following:

1. *Ingestion of food.* Food and water enter the body through the mouth.

Module 12 Digestive System

2. *Digestion of food.* During the process of digestion, food is broken down from complex particles to smaller molecules that can be absorbed.
3. *Absorption of nutrients.* The epithelial cells that line the lumen of the small intestine absorb the small molecules of nutrients (amino acids, monosaccharides, fatty acids, vitamins, minerals, and water) that result from the digestive process.
4. *Elimination of wastes.* Undigested material, such as fiber from food, plus waste products excreted into the digestive tract are eliminated in the feces.

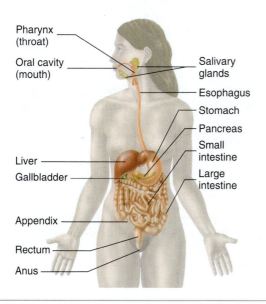

Pharynx
(throat)

Oral cavity
(mouth)

Salivary
glands

Esophagus

Stomach

Pancreas

Small
intestine

Liver

Gallbladder

Large
intestine

Appendix

Rectum

Anus

Figure 16.1 AP|R **Digestive System**

16.2 ANATOMY AND HISTOLOGY OF THE DIGESTIVE SYSTEM

Learning Outcome   *After reading this section, you should be able to*

A. Describe the general histology of the digestive tract.

The digestive system consists of the **digestive tract,** or *gastrointestinal (GI;* gas′trō-in-tes′tin-ăl) *tract,* plus specific associated organs. Because the digestive tract is open at the mouth and anus, the inside of the tract is continuous with the outside environment, and food entering the digestive tract may contain not only useful nutrients but also indigestible components such as fiber, or harmful materials such as bacteria. Therefore, the inner lining of the digestive tract serves as a protective barrier to those indigestible and harmful materials and nutrients must be specifically transported across the wall of the digestive tract. Once across the wall of the digestive tract, the nutrients enter the circulation to access tissues of the body.

The digestive tract consists of the oral cavity, pharynx, esophagus, stomach, small intestine, large intestine, and anus. Accessory glands are associated with the digestive tract (figure 16.1). The salivary glands empty into the oral cavity, and the liver and pancreas are connected to the small intestine.

Various parts of the digestive tract are specialized for different functions. Nearly all segments of the digestive tract consist of four layers, called tunics. These are the mucosa, the submucosa, the muscularis, and a serosa or an adventitia (figure 16.2):

1. The innermost tunic, the **mucosa** (mū-kō′să), consists of **mucous epithelium,** a loose connective tissue called the **lamina propria,** and a thin smooth muscle layer, the **muscularis mucosae.** The epithelium in the mouth, esophagus, and anus resists abrasion, and the epithelium in the stomach and intestine absorbs and secretes.

2. The **submucosa** lies just outside the mucosa. It is a thick layer of loose connective tissue containing nerves, blood vessels, and small glands. An extensive network of nerve cell processes forms a **plexus** (network). Autonomic nerves innervate this plexus.

3. The next tunic is the **muscularis.** In most parts of the digestive tract it consists of an inner layer of **circular smooth muscle** and an outer layer of **longitudinal smooth muscle.** Another nerve plexus, also innervated by autonomic nerves, lies between the two muscle layers. Together, the nerve plexuses of the submucosa and muscularis compose the **enteric** (en-těr′ik) **nervous system.** This nervous system, which is a division of the autonomic nervous system, is extremely important in controlling movement and secretion within the tract (see chapter 8).

4. The fourth, or outermost, layer of the digestive tract is either a serosa or an adventitia. The **serosa** consists of the peritoneum, which is a smooth epithelial layer, and its underlying connective tissue. Regions of the digestive tract not covered by peritoneum are covered by a connective tissue layer called the **adventitia** (ad′ven-tish′ă; foreign, coming from outside), which is continuous with the surrounding connective tissue.

Peritoneum

The body wall of the abdominal cavity and the abdominal organs is covered with serous membranes (figure 16.3). The serous membrane that covers the organs is the serosa, or **visceral peritoneum** (per′i-tō-nē′ŭm; to stretch over). The serous membrane that lines the wall of the abdominal cavity is the **parietal peritoneum.**

Many of the organs of the abdominal cavity are held in place by connective tissue sheets called **mesenteries** (mes′en-ter-ēz). *Mesentery* is a general term referring to the serous membranes attached to the abdominal organs. The mesenteries consist of two layers of serous membranes with a thin layer of loose connective tissue between them. The mesentery connecting the lesser curvature of the stomach to the liver and diaphragm is called the **lesser omentum** (ō-men′tŭm), and the mesentery connecting the greater curvature of the stomach to the transverse colon and posterior body wall is called the **greater omentum.** The greater omentum is unusual in that it is a long, double fold of mesentery that extends inferiorly from the stomach before looping back to the transverse colon to create a cavity, or pocket, called the **omental bursa** (ber′să). Adipose tissue accumulates in the greater omentum, giving it the appearance of a fat-filled apron that covers the anterior surface of the abdominal viscera. The mesentery that attaches the small intestine to the posterior abdominal wall is called the **mesentery proper.**

Predict 2

If you placed a pin completely through both folds of the greater omentum, through how many layers of simple squamous epithelium would the pin pass?

Other abdominal organs lie against the abdominal wall, have no mesenteries, and are described as **retroperitoneal** (re′trō-per′i-tō-nē′ăl; behind the peritoneum). The retroperitoneal organs include the duodenum, pancreas, ascending colon, descending colon, rectum, kidneys, adrenal glands, and urinary bladder.

Digestive

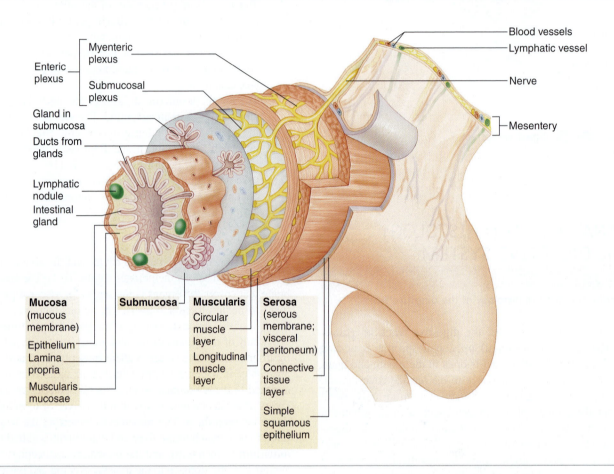

Figure 16.2 AP|R **Digestive Tract Histology**

The four tunics are the mucosa, the submucosa, the muscularis, and a serosa or an adventitia. Glands may exist along the digestive tract as part of the epithelium, within the submucosa, or as large glands outside the digestive tract.

16.3 ORAL CAVITY, PHARYNX, AND ESOPHAGUS

Learning Outcomes *After reading this section, you should be able to*

A. Describe the structure of a tooth.
B. Describe the major salivary glands. Compare their structures and functions.
C. Describe mastication and swallowing.

Anatomy of the Oral Cavity

The **oral cavity** (figure 16.4), or mouth, is the first part of the digestive tract. It is bounded by the lips and cheeks and contains the teeth and tongue. The **lips** are muscular structures, formed mostly by the **orbicularis oris** (ōr-bik´ū-lā´ris ōr´is) **muscle** (see figure 7.16). The outer surfaces of the lips are covered by skin. The keratinized stratified epithelium of the skin becomes thin at the margin of the lips. The color from the underlying blood vessels can be seen through the thin, transparent epithelium, giving the lips a reddish-pink appearance. At the internal margin of the lips, the epithelium is continuous with the moist stratified squamous epithelium of the mucosa in the oral cavity.

The cheeks form the lateral walls of the oral cavity. The **buccinator** (bŭk´si-nā-tōr) **muscles** (see figure 7.16), located within the cheeks, flatten the cheeks against the teeth. The lips and cheeks are important in the process of **mastication** (mas-ti-kā´shŭn), or

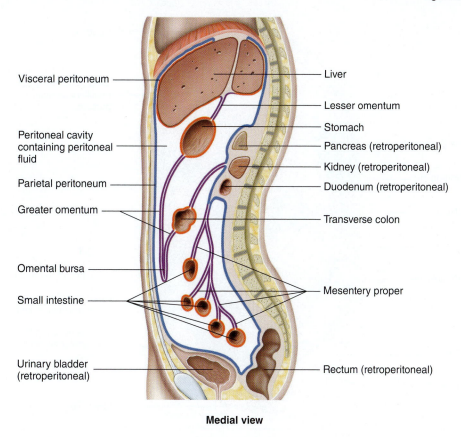

Medial view

Figure 16.3 AP|R Peritoneum and Mesenteries

The parietal peritoneum lines the abdominal cavity (*blue*), and the visceral peritoneum covers abdominal organs (*red*). Retroperitoneal organs are behind the parietal peritoneum. The mesenteries are membranes that connect abdominal organs to each other and to the body wall.

chewing. They help manipulate the food within the oral cavity and hold the food in place while the teeth crush or tear it. Mastication begins the process of mechanical digestion, which breaks down large food particles into smaller ones. The cheeks also help form words during the speech process.

The **tongue** is a large, muscular organ that occupies most of the oral cavity. The major attachment of the tongue is in the posterior part of the oral cavity. The anterior part of the tongue is relatively free, except for an anterior attachment to the floor of the mouth by a thin fold of tissue called the **frenulum** (fren′ū-lŭm) (figure 16.4). The muscles associated with the tongue are described in chapter 7. The anterior two-thirds of the tongue is covered by papillae, some of which contain taste buds (see chapter 9). The posterior one-third of the tongue is devoid of papillae and has only a few scattered taste buds. In addition, the posterior portion does contain a large amount of lymphatic tissue, which helps form the lingual tonsil (see chapter 14).

The tongue moves food in the mouth and, in cooperation with the lips and cheeks, holds the food in place during mastication. It also plays a major role in the process of swallowing. In addition, the tongue is a major sensory organ for taste, as well as one of the major organs of speech.

Teeth

There are 32 **teeth** in the normal adult mouth, located in the mandible and maxillae. The teeth can be divided into quadrants:

right upper, left upper, right lower, and left lower. In adults, each quadrant contains one central and one lateral **incisor** (in-sī′zŏr; to cut); one **canine** (kā′nīn; dog); first and second **premolars** (prē-mō′lărz; *molaris,* a millstone); and first, second, and third **molars** (mō′lărz). The third molars are called **wisdom teeth** because they usually appear in the late teens or early twenties, when the person is old enough to have acquired some degree of wisdom.

The teeth of adults are called **permanent teeth,** or *secondary teeth* (figure 16.5*a*). Most of them are replacements for the 20 **primary teeth,** or *deciduous* (dē-sid′ū-ŭs) *teeth,* also called milk or baby teeth, which are lost during childhood (figure 16.5*b*).

Each tooth consists of a **crown** with one or more **cusps** (points), a **neck,** and a **root** (figure 16.6). The center of the tooth is a **pulp cavity,** which is filled with blood vessels, nerves, and connective tissue, called **pulp.** The pulp cavity is surrounded by a living, cellular, bonelike tissue called **dentin** (den′tin; *dens,* tooth). The dentin of the tooth crown is covered by an extremely hard, acellular substance called **enamel,** which protects the tooth against abrasion and acids produced by bacteria in the mouth. The surface of the dentin in the root is covered with **cementum** (se-men′tŭm), which helps anchor the tooth in the jaw.

The teeth are rooted within **alveoli** (al-vē′ō-lī; sockets) along the alveolar processes of the mandible and maxillae. The alveolar processes are covered by dense fibrous connective tissue and moist stratified squamous epithelium, referred to as the **gingiva** (jin′ji-vă), or gums. The teeth are held in place by **periodontal**

Digestive

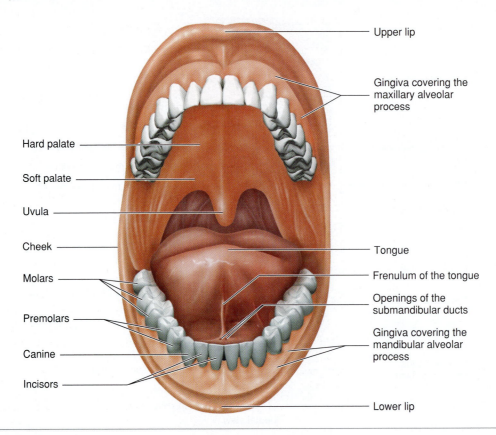

Upper lip

Gingiva covering the maxillary alveolar process

Hard palate

Soft palate

Uvula

Cheek

Tongue

Frenulum of the tongue

Molars

Openings of the submandibular ducts

Premolars

Gingiva covering the mandibular alveolar process

Canine

Incisors

Lower lip

Figure 16.4 AP|R **Oral Cavity**

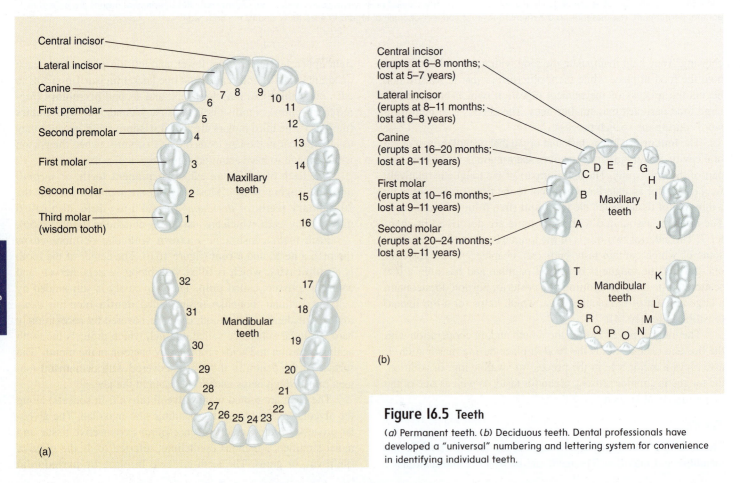

Central incisor

Lateral incisor

Canine

First premolar

Second premolar

First molar

Second molar

Third molar (wisdom tooth)

7 8 9 10
6   11
5   12
4   13
            14
3
            15
2
1   16

Maxillary teeth

Central incisor (erupts at 6–8 months; lost at 5–7 years)

Lateral incisor (erupts at 8–11 months; lost at 6–8 years)

Canine (erupts at 16–20 months; lost at 8–11 years)

First molar (erupts at 10–16 months; lost at 9–11 years)

Second molar (erupts at 20–24 months; lost at 9–11 years)

C D E F G
B   H
A   I
        J

Maxillary teeth

32   17
31   18
30   19
29   20
28   21
27   22
26 25 24 23

Mandibular teeth

T   K
S   L
R   M
Q P O N

Mandibular teeth

(b)

(a)

Figure 16.5 **Teeth**

(*a*) Permanent teeth. (*b*) Deciduous teeth. Dental professionals have developed a "universal" numbering and lettering system for convenience in identifying individual teeth.

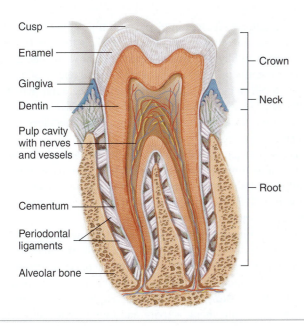

- Cusp
- Enamel
- Gingiva
- Dentin
- Pulp cavity with nerves and vessels
- Cementum
- Periodontal ligaments
- Alveolar bone

- Crown
- Neck
- Root

Figure 16.6 Molar Tooth in Place in the Alveolar Bone

A tooth consists of a crown, a neck, and a root. The root is covered with cementum, and the tooth is held in the socket by periodontal ligaments. Nerves and vessels enter and exit the tooth through a foramen in the part of the root deepest in the alveolus.

(per′ē-ō-don′tăl; around the teeth) **ligaments,** which are connective tissue fibers that extend from the alveolar walls and are embedded into the cementum.

Formation of **dental caries** (kār′ēz), or tooth decay, is the result of the breakdown of enamel by acids produced by bacteria on the tooth surface. Enamel is nonliving and cannot repair itself. Consequently, a dental filling is necessary to prevent further damage. **Periodontal disease** is inflammation and degeneration of the periodontal ligaments, gingiva, and alveolar bone. This disease is the most common cause of tooth loss in adults.

A CASE IN POINT

Lymph Nodes and a Toothache

Tu Thake went to his dentist for a checkup because he had experienced pain in his right mandible for several days. During the routine exam, the dentist felt along the anterior and posterior edges of his sternocleidomastoid muscle on each side and along the inferior edges of the mandible on each side. Just inferior to the angle of the right mandible, the dentist noted a lump, suggesting an enlargement of the superior cervical lymph nodes. Enlargement of the superior cervical lymph nodes indicates a problem in the face, because all lymphatic drainage from the face goes through these nodes. Additional swellings along the inferior edge of the anterior or central part of the body of the mandible suggest a problem in the mandible. The "problem" could be an infection or a cancerous growth, or it could be idiopathic (of unknown origin). An infection may occur in a tooth, in the bone, or in the soft tissues of the area. In Tu's case, further examination revealed a small abscess near his right, first mandibular molar. The dentist opened the abscess and treated it with topical and systemic antibiotics. The infection disappeared, as did the swelling in the superior cervical lymph nodes.

Palate and Tonsils

The **palate** (pal′ăt), or roof of the oral cavity, separates the oral cavity from the nasal cavity and prevents food from passing into the nasal cavity during chewing and swallowing. The palate consists of two parts. The anterior part contains bone and is called the **hard palate,** whereas the posterior portion consists of skeletal muscle and connective tissue and is called the **soft palate** (see figure 16.4). The **uvula** (ū′vū-lă; a grape) is a posterior extension of the soft palate.

The **tonsils** (ton′silz) are located in the lateral posterior walls of the oral cavity, in the nasopharynx, and in the posterior surface of the tongue. The tonsils are described in chapter 14.

Salivary Glands

There are three major pairs of **salivary** (sal′i-văr-ē) **glands:** the parotid, submandibular, and sublingual glands (figure 16.7). A considerable number of other salivary glands are scattered throughout the oral cavity, including on the tongue. Salivary glands produce saliva, which is a mixture of **serous** (watery) and **mucous** fluids. The salivary glands are compound alveolar glands. They have branching ducts with clusters of alveoli, resembling grapes, at the ends of the ducts (see chapter 4).

The largest of the salivary glands, the **parotid** (pă-rot′id; beside the ear) **glands,** are serous glands located just anterior to each ear. Parotid ducts enter the oral cavity adjacent to the second upper molars.

Mumps (mŭmpz) is an inflammation of the parotid gland caused by a viral infection. The inflamed parotid glands become swollen, often making the cheeks quite large. The virus causing mumps can also infect other structures. Mumps in an adult male may involve the testes and can result in sterility.

The **submandibular** (sŭb-man-dib′ū-lăr; below the mandible) **glands** produce more serous than mucous secretions. Each gland can be felt as a soft lump along the inferior border of the mandible. The submandibular ducts open into the oral cavity on each side of the frenulum of the tongue (see figure 16.4).

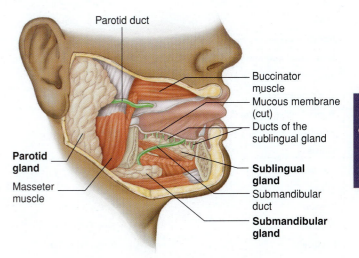

- Parotid duct
- Buccinator muscle
- Mucous membrane (cut)
- Ducts of the sublingual gland
- **Sublingual gland**
- Submandibular duct
- **Submandibular gland**
- **Parotid gland**
- Masseter muscle

Figure 16.7 AP|R Salivary Glands

The large salivary glands are the parotid glands, the submandibular glands, and the sublingual glands.

Digestive

The **sublingual** (sŭb-ling′gwăl; below the tongue) **glands,** the smallest of the three paired salivary glands, produce primarily mucous secretions. They lie immediately below the mucous membrane in the floor of the oral cavity. Each sublingual gland has 10–12 small ducts opening onto the floor of the oral cavity.

Saliva

Saliva (să-lī′vă) helps keep the oral cavity moist and contains enzymes that begin the process of digestion. Saliva is secreted at the rate of approximately 1 liter (L) per day. The serous part of saliva, produced mainly by the parotid and submandibular glands, contains a digestive enzyme called **salivary amylase** (am′il-ās) (table 16.1), which breaks the covalent bonds between glucose molecules in starch and other polysaccharides to produce the disaccharides maltose and isomaltose. Maltose and isomaltose have a sweet taste; thus, the digestion of polysaccharides by salivary amylase enhances the sweet taste of food.

Food spends very little time in the mouth. Consequently, only about 5% of the total carbohydrates humans absorb are digested in the mouth. Also, most starches are contained in plant cells, which are surrounded by cell walls composed primarily of the polysaccharide **cellulose** (sel′ū-lōs). Humans lack the necessary enzymes to digest cellulose. Cooking and thorough chewing of food disrupt the cellulose covering and increase the efficiency of the digestive process.

In addition to its role in digestion, saliva protects the mouth from bacterial infection by washing the oral cavity with **lysozyme** (lī′sō-zīm), a mildly antibacterial enzyme. Saliva also neutralizes the pH in the mouth, which reduces the harmful effects of bacterial acids on tooth enamel. Lack of salivary gland secretion (which can result from radiation therapy) increases the chance of ulceration and infection of the oral mucosa and caries (cavities) formation in the teeth.

The serous part of saliva dissolves molecules, which must be in solution to stimulate taste receptors. The mucous secretions of the submandibular and sublingual glands contain a large amount of **mucin** (mū′sin), a proteoglycan that gives a lubricating quality to the secretions of the salivary glands.

Salivary gland secretion is regulated primarily by the autonomic nervous system, with parasympathetic stimulation being the most important. Salivary secretions increase in response to a variety of stimuli, such as tactile stimulation in the oral cavity and certain tastes, especially sour. Higher brain centers can stimulate parasympathetic activity and thus increase the activity of the salivary glands in response to the thought of food, to odors, or to the sensation of hunger. Sympathetic stimulation increases the mucous content of saliva. When a person becomes frightened and the sympathetic division of the autonomic nervous system is stimulated, the person may have a dry mouth with thick mucus.

TABLE 16.1   Functions of Digestive Secretions

Fluid or Enzyme	Source	Function
Mouth		
Saliva (water, bicarbonate ions, mucus)	Salivary glands	Moistens and lubricates food, neutralizes bacterial acids, flushes bacteria from oral cavity
Salivary amylase	Salivary glands	Digests starch
Lysozyme	Salivary glands	Has weak antibacterial action
Stomach		
Hydrochloric acid	Gastric glands	Kills bacteria, converts pepsinogen to pepsin
Pepsin*	Gastric glands	Digests protein
Mucus	Mucous cells	Protects stomach lining
Intrinsic factor	Gastric glands	Binds to vitamin B_{12}, aids in its absorption
Small Intestine and Associated Glands		
Bile salts	Liver	Emulsify fats
Bicarbonate ions	Pancreas	Neutralize stomach acid
Trypsin*, chymotrypsin*, carboxypeptidase*	Pancreas	Digest protein
Pancreatic amylase	Pancreas	Digests starch
Lipase	Pancreas	Digests lipid (triglycerides)
Nucleases	Pancreas	Digest nucleic acid (DNA or RNA)
Mucus	Duodenal glands and goblet cells	Protects duodenum from stomach acid and digestive enzymes
Peptidases**	Small intestine	Digest polypeptide
Sucrase**	Small intestine	Digests sucrose
Lactase**	Small intestine	Digests lactose
Maltase**	Small intestine	Digests maltose

*These enzymes are secreted as inactive forms, then activated.
**These enzymes remain in the microvilli.

CLINICAL IMPACT   Dietary Fiber

Even though humans cannot digest cellulose, it is important to normal digestive function. Cellulose provides bulk, or fiber, in the diet. The presence of this bulk facilitates movement of material through the digestive tract by providing mass against which the muscular wall of the digestive tract can push. In the 1950s, some nutritionists speculated that

eventually all the nutrients we need could be reduced into a single tablet and that we no longer would have to eat food. It is now known that indigestible bulk is very important to the normal function of the digestive tract. Adults should consume fiber in the form of cellulose, hemicellulose, pectin, vegetable gums, muscilage, lignin, and beta-glucan, among others.

Dietary fiber is classified as either soluble or insoluble fiber. Soluble fiber binds fat and cholesterol in the intestines, lowering low-density lipoprotein concentrations in the blood, and slows the absorption of glucose, preventing a glucose surge and insulin spike. Insoluble fiber pushes food through the intestinal tract, preventing constipation, and dilutes carcinogens.

Mastication

Food taken into the mouth is chewed, or masticated, by the teeth. The incisors and canines primarily cut and tear food, whereas the premolars and molars primarily crush and grind it. Mastication breaks large food particles into many small ones, which have a much larger total surface area than a few large particles would have. Because digestive enzymes act on molecules only at the surface of the food particles, mastication increases the efficiency of digestion.

Pharynx

The **pharynx** (far′ingks), or throat, which connects the mouth with the esophagus, consists of three parts: the nasopharynx, the oropharynx, and the laryngopharynx (see chapter 15). Normally, only the oropharynx and laryngopharynx transmit food. The posterior walls of the oropharynx and laryngopharynx are formed by the superior, middle, and inferior **pharyngeal constrictor muscles.**

Esophagus

The **esophagus** (ē-sof′ă-gŭs; gullet) is a muscular tube, lined with moist stratified squamous epithelium, that extends from the pharynx to the stomach. It is about 25 centimeters (cm) long and lies anterior to the vertebrae and posterior to the trachea within the mediastinum. The upper two-thirds of the esophagus has skeletal muscle in its wall, while the lower one-third has smooth muscle in its wall. It passes through the diaphragm and ends at the stomach. The esophagus transports food from the pharynx to the stomach. Upper and lower **esophageal sphincters,** located at the upper and lower ends of the esophagus, respectively, regulate the movement of food into and out of the esophagus. The lower esophageal sphincter is sometimes called the **cardiac sphincter.** Numerous mucous glands produce a thick, lubricating mucus that coats the inner surface of the esophagus.

Swallowing

Swallowing, or *deglutition* (dē-gloo-tish′ŭn), can be divided into three phases: the voluntary phase, the pharyngeal phase, and the esophageal phase (figure 16.8). During the **voluntary phase,** a bolus, or mass of food, is formed in the mouth. The bolus is pushed by the tongue against the hard palate, forcing the bolus toward the posterior part of the mouth and into the oropharynx.

The **pharyngeal phase** of swallowing is a reflex that is initiated when a bolus of food stimulates receptors in the oropharynx. This phase of swallowing begins with the elevation of the soft palate, which closes the passage between the nasopharynx and oropharynx. The pharynx elevates to receive the bolus of food from the mouth. The three **pharyngeal constrictor muscles** then contract in succession, forcing the food through the pharynx. At the same time, the upper esophageal sphincter relaxes, and food is pushed into the esophagus. As food passes through the pharynx, the vestibular and vocal folds close, and the **epiglottis** (ep-i-glot′is; upon the glottis, opening of the larynx) is tipped posteriorly, so that the opening into the larynx is covered. These movements prevent food from passing into the larynx.

Predict 3

What would happen if a person had a cleft of the soft palate, so that the soft palate did not completely close the passage between the nasopharynx and the oropharynx during swallowing? What happens if a person has an explosive burst of laughter while trying to swallow a liquid? What happens if a person tries to swallow and speak at the same time?

The **esophageal phase** of swallowing is responsible for moving food from the pharynx to the stomach. Muscular contractions of the esophagus occur in **peristaltic** (per-i-stal′tik; *peri,* around + *stalsis,* constriction) **waves** (figure 16.9). A wave of relaxation of the esophageal muscles precedes the bolus of food down the esophagus, and a wave of strong contraction of the circular muscles follows and propels the bolus through the esophagus. Gravity assists the movement of material, especially liquids, through the esophagus. However, the peristaltic contractions that move material through the esophagus are sufficiently forceful to allow a person to swallow even while doing a headstand or floating in the zero-gravity environment of space. The peristaltic contractions cause relaxation of the lower esophageal sphincter in the esophagus as the peristaltic waves approach the stomach.

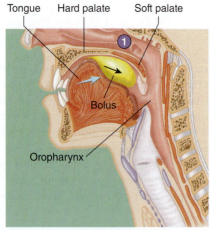

Tongue   Hard palate   Soft palate

Bolus

Oropharynx

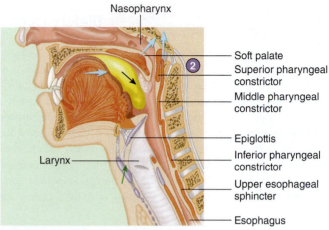

Nasopharynx

Soft palate
Superior pharyngeal constrictor
Middle pharyngeal constrictor
Epiglottis
Inferior pharyngeal constrictor
Upper esophageal sphincter
Esophagus

Larynx

1 During the **voluntary phase**, a bolus of food (*yellow*) is pushed by the tongue against the hard and soft palates and posteriorly toward the oropharynx (*blue arrow* indicates tongue movement; *black arrow* indicates movement of the bolus). *Tan:* bone; *purple:* cartilage; *red:* muscle.

2 During the **pharyngeal phase**, the soft palate is elevated, closing off the nasopharynx. The pharynx and larynx are elevated (*blue arrows* indicate muscle movement; *green arrow* indicates elevation of the larynx).

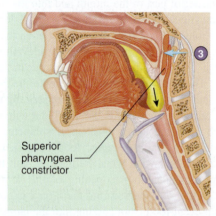

Superior pharyngeal constrictor

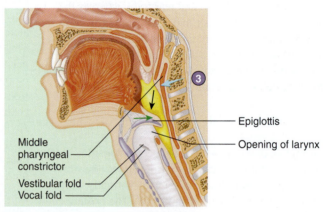

Middle pharyngeal constrictor

Vestibular fold
Vocal fold

Epiglottis

Opening of larynx

3 Successive constriction of the pharyngeal constrictors from superior to inferior (*blue arrows*) forces the bolus through the pharynx and into the esophagus. As this occurs, the vestibular and vocal folds expand medially to close the passage of the larynx. The epiglottis (*green arrow*) is bent down over the opening of the larynx largely by the force of the bolus pressing against it.

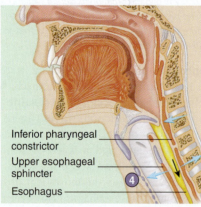

Inferior pharyngeal constrictor

Upper esophageal sphincter

Esophagus

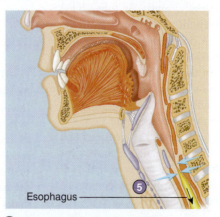

Esophagus

4 As the inferior pharyngeal constrictor contracts, the upper esophageal sphincter relaxes (outwardly directed *blue arrows*), allowing the bolus to enter the esophagus.

5 During the **esophageal phase**, the bolus is moved by peristaltic contractions of the esophagus toward the stomach (inwardly directed *blue arrows*).

PROCESS Figure 16.8 AP|R Events During the Three Phases of Swallowing

Digestive

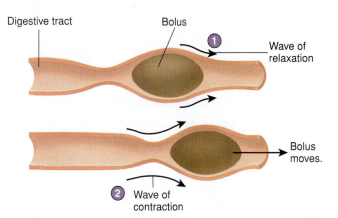

① A wave of smooth muscle relaxation moves ahead of the bolus, allowing the digestive tract to expand.

② A wave of contraction of the smooth muscle behind the bolus propels it through the digestive tract.

PROCESS Figure 16.9 Peristalsis

16.4 STOMACH

> **Learning Outcomes** *After reading this section, you should be able to*
>
> **A.** Outline the anatomical and physiological characteristics of the stomach.
> **B.** Describe the stomach secretions, their functions, and their regulation.
> **C.** Describe gastric movements and stomach emptying and how they are regulated.

Anatomy of the Stomach

The **stomach** (figure 16.10) is an enlarged segment of the digestive tract in the left superior part of the abdomen. The opening from the esophagus into the stomach is called the **gastroesophageal opening**. The region of the stomach around the gastroesophageal opening is called the **cardiac region** because it is near the heart. The most superior part of the stomach is the **fundus** (fŭn′dŭs). The largest part of the stomach is the **body,** which turns to the right, forming a **greater curvature** on the left and a **lesser curvature** on the right. The opening from the stomach into the small intestine is the **pyloric** (pī-lōr′ik; gatekeeper) **opening,** which is surrounded by a relatively thick ring of smooth muscle called the **pyloric sphincter.** The region of the stomach near the pyloric opening is the **pyloric region.**

The muscular layer of the stomach is different from other regions of the digestive tract in that it consists of three layers: an outer longitudinal layer, a middle circular layer, and an inner oblique layer. These muscular layers produce a churning action in the stomach, important in the digestive process. The submucosa and mucosa of the stomach are thrown into large folds called **rugae** (roo′gē; wrinkles) (figure 16.10a) when the stomach is empty. These folds allow the mucosa and submucosa to stretch, and the folds disappear as the stomach is filled.

The stomach is lined with simple columnar epithelium. The mucosal surface forms numerous tubelike **gastric pits** (figure 16.10b), which are the openings for the **gastric glands.** The epithelial cells of the stomach can be divided into five groups. The first group consists of **surface mucous cells** on the inner surface of the stomach and lining the gastric pits. Those cells produce mucus, which coats and protects the stomach lining. The remaining four cell types are in the gastric glands. They are

1. **mucous neck cells,** which produce mucus;
2. **parietal cells,** which produce hydrochloric acid and intrinsic factor;

3. **endocrine cells,** which produce regulatory chemicals; and
4. **chief cells,** which produce **pepsinogen** (pep-sin′ō-jen), a precursor of the protein-digesting enzyme **pepsin** (pep′sin; *pepsis,* digestion).

Secretions of the Stomach

The stomach functions primarily as a storage and mixing chamber for ingested food. As food enters the stomach, it is mixed with stomach secretions to become a semifluid mixture called **chyme** (kīm; juice). Although some digestion occurs in the stomach, that is not its principal function.

Stomach secretions from the gastric glands include hydrochloric acid, pepsin, mucus, and intrinsic factor (table 16.1). **Hydrochloric acid** produces a pH of about 2.0 in the stomach. The acid kills microorganisms and activates **pepsin** from its inactive form, called pepsinogen. Pepsin breaks covalent bonds of proteins to form smaller peptide chains. Pepsin exhibits optimum enzymatic activity at a pH of about 2.0. A thick layer of **mucus** lubricates the epithelial cells of the stomach wall and protects them from the damaging effect of the acidic chyme and pepsin. Irritation of the stomach mucosa stimulates the secretion of a greater volume of mucus. **Intrinsic** (in-trin′sik) **factor** binds with vitamin B_{12} and makes it more readily absorbed in the small intestine. Vitamin B_{12} is important in deoxyribonucleic acid (DNA) synthesis and in red blood cell production.

Regulation of Stomach Secretions

Approximately 2 L of gastric secretions (gastric juice) are produced each day. Both nervous and hormonal mechanisms regulate gastric secretions. The neural mechanisms involve central nervous system (CNS) reflexes integrated within the medulla oblongata. Higher brain centers can influence these reflexes. Local reflexes are integrated within the enteric plexus in the wall of the digestive tract and do not involve the CNS. Hormones produced by the stomach and intestine help regulate stomach secretions.

Regulation of stomach secretions can be divided into three phases: the cephalic, gastric, and intestinal phases. The cephalic phase can be viewed as the "get started" phase, when the stomach secretions are increased in anticipation of incoming food. This is followed by the gastric, "go for it," phase, when most of the stimulation of secretion occurs. Finally, the intestinal phase is the "slow down" phase, during which stomach secretion decreases.

In the **cephalic** (se-fal′ik; *kephale,* head) **phase** (figure 16.11a), sensations of taste, the smell of food, stimulation of tactile

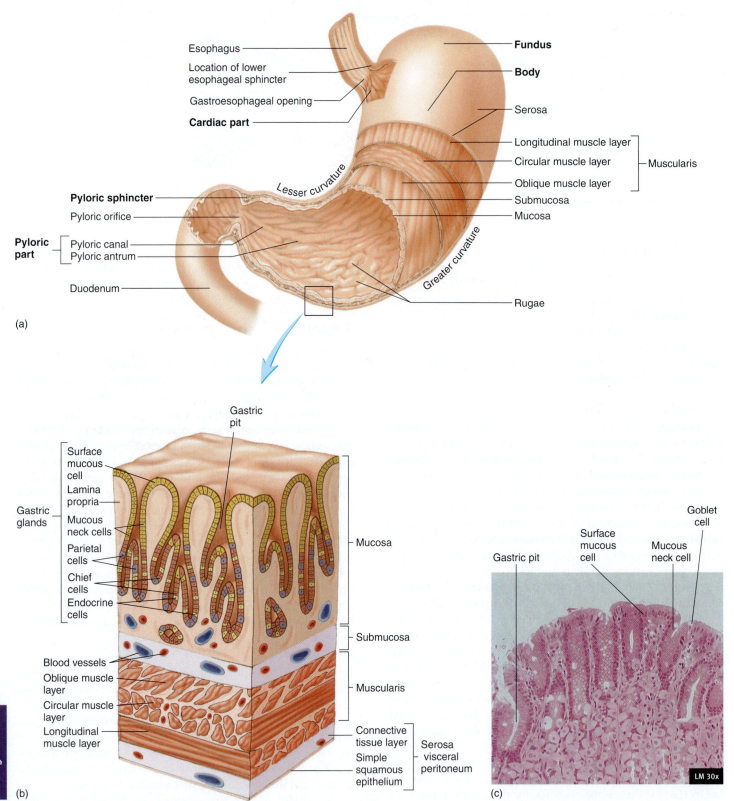

(a)

Esophagus
Location of lower esophageal sphincter
Gastroesophageal opening
Cardiac part
Pyloric sphincter
Pyloric orifice
Pyloric part
Pyloric canal
Pyloric antrum
Duodenum
Lesser curvature
Fundus
Body
Serosa
Longitudinal muscle layer
Circular muscle layer
Oblique muscle layer
Muscularis
Submucosa
Mucosa
Greater curvature
Rugae

(b)

Gastric pit
Gastric glands
Surface mucous cell
Lamina propria
Mucous neck cells
Parietal cells
Chief cells
Endocrine cells
Blood vessels
Oblique muscle layer
Circular muscle layer
Longitudinal muscle layer
Mucosa
Submucosa
Muscularis
Connective tissue layer
Simple squamous epithelium
Serosa visceral peritoneum

(c)

Gastric pit
Surface mucous cell
Goblet cell
Mucous neck cell
LM 30x

Figure 16.10 Anatomy and Histology of the Stomach

(*a*) Cutaway section reveals muscular layers and internal anatomy. (*b*) A section of the stomach wall illustrates its histology, including several gastric pits and glands. (*c*) Photomicrograph of gastric glands.

CLINICAL IMPACT   Hypertrophic Pyloric Stenosis

Hypertrophic pyloric stenosis is a common defect of the stomach in infants. It occurs in 1 in 150 males and 1 in 750 females. The pyloric sphincter is greatly thickened and thus interferes with normal stomach emptying. Infants with this defect develop symptoms 3 to 10 weeks after birth. The infants exhibit projectile (forceful) vomiting. Because the pyloric opening is blocked, little food enters the intestine, and the infant fails to gain weight. The defect can be corrected with surgery to enlarge the pyloric opening.

receptors during the process of chewing and swallowing, and pleasant thoughts of food stimulate centers within the medulla oblongata that influence gastric secretions. Action potentials are sent from the medulla oblongata along parasympathetic axons within the vagus nerves to the stomach. Within the stomach wall, the preganglionic neurons stimulate postganglionic neurons in the enteric plexus. The postganglionic neurons stimulate secretory activity in the cells of the stomach mucosa, causing the release of hydrochloric acid, pepsin, mucus, and intrinsic factor. The neurons also stimulate the release of gastrin and histamine from endocrine cells. **Gastrin** (gas′trin) is a hormone that enters the circulation and is carried back to the stomach, where it stimulates additional secretory activity (table 16.2). Histamine is both a paracrine chemical signal that acts locally and a hormone that enters the blood to stimulate gastric gland secretory activity. Histamine is the most potent stimulator of hydrochloric acid secretion. Drugs that block the actions of histamine can lower acid levels.

The **gastric phase** is the period during which the greatest volume of gastric secretion occurs (figure 16.11*b*). The gastric phase is activated by the presence of food in the stomach. During the gastric phase, the food in the stomach is mixed with gastric secretions. Distention of the stomach stimulates stretch receptors. Action potentials generated by these receptors activate CNS reflexes and local reflexes, resulting in the cascade of events that increases secretion, as in the cephalic phase. Peptides, produced by the action of pepsin on proteins, stimulate the secretion of gastrin, which in turn stimulates additional hydrochloric acid secretion.

The **intestinal phase** of gastric secretion primarily inhibits gastric secretions (figure 16.11*c*). It is controlled by the entrance of acidic chyme into the duodenum, which initiates both neural and hormonal mechanisms. When the pH of the chyme entering the duodenum drops to 2.0 or below, the inhibitory influence of the intestinal phase is greatest. The hormone **secretin** (se-krē′tin), which inhibits gastric secretions, is released from the duodenum in response to low pH (table 16.2). Fatty acids and peptides in the duodenum initiate the release of the hormone **cholecystokinin** (kō′lē-sis-tō-kī′nin), which also inhibits gastric secretions (table 16.2). Acidic chyme (pH < 2.0) in the duodenum also inhibits CNS stimulation and initiates local reflexes that inhibit gastric secretion.

A CASE IN POINT

Heartburn

Hart Burne had a large meat-lover's pizza delivered to his apartment. He had been eating chips and drinking beer before the pizza came. Hart consumed the whole pizza and two more bottles of beer as he watched the last half of a game on TV. By the end of the game, he was feeling uncomfortably full, so he lay down on the couch. Within half an hour, Hart felt a severe pain in his chest and headed for the bathroom to find an antacid.

Heartburn, or *gastritis,* is a painful or burning sensation in the chest usually associated with an increase in gastric acid secretion and/or a backflush of acidic chyme into the esophagus. Overeating, eating fatty foods, lying down immediately after a meal, consuming too much alcohol or caffeine, smoking, and wearing extremely tight clothing can all cause heartburn. Medications can relieve the symptoms of heartburn by blocking gastric acid secretion. Cimetidine (si-met′i-dēn; Tagamet) and ranitidine (ră-nī′ti-dēn; Zantac) block histamine stimulation of acid release from parietal cells, and esomeprazole (eh-sō-meh′pra-zōl; Nexium) blocks the proton pump in parietal cells that generates gastric acid. Antacids (e.g., Tums) neutralize acids already secreted into the stomach.

TABLE 16.2   Functions of the Major Digestive System Hormones

Hormone	Source	Function
Gastrin	Gastric glands	Increases gastric secretions
Secretin	Duodenum	Decreases gastric secretions
		Increases pancreatic and bile secretions high in bicarbonate ions
		Decreases gastric motility
Cholecystokinin	Duodenum	Decreases gastric secretions
		Strongly decreases gastric motility
		Increases gallbladder contraction
		Increases pancreatic enzyme secretion

Cephalic Phase

① The taste, smell, or thought of food or tactile sensations of food in the mouth stimulate the medulla oblongata (*green arrows*).

② Vagus nerves carry parasympathetic action potentials to the stomach (*pink arrow*), where enteric plexus neurons are activated.

③ Postganglionic neurons stimulate secretion by parietal and chief cells and stimulate gastrin and histamine secretion by endocrine cells.

④ Gastrin is carried through the circulation back to the stomach (*purple arrow*), where, along with histamine, it stimulates secretion.

(a)

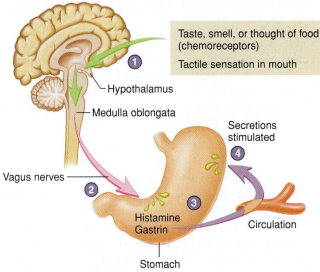

Gastric Phase

① Distention of the stomach stimulates mechanoreceptors (stretch receptors) and activates a parasympathetic reflex. Action potentials generated by the mechanoreceptors are carried by the vagus nerves to the medulla oblongata (*green arrow*).

② The medulla oblongata increases action potentials in the vagus nerves that stimulate secretions by parietal and chief cells and stimulate gastrin and histamine secretion by endocrine cells (*pink arrow*).

③ Distention of the stomach also activates local reflexes that increase stomach secretions (*orange arrow*).

④ Gastrin is carried through the circulation back to the stomach (*purple arrow*), where, along with histamine, it stimulates secretion.

(b)

Intestinal Phase

① Chyme in the duodenum with a pH less than 2 or containing fat digestion products (lipids) inhibits gastric secretions by three mechanisms (2–4).

② Chemoreceptors in the duodenum are stimulated by H⁺ (low pH) or lipids. Action potentials generated by the chemoreceptors are carried by the vagus nerves to the medulla oblongata (*green arrow*), where they inhibit parasympathetic action potentials (*pink arrow*), thereby decreasing gastric secretions.

③ Local reflexes activated by H⁺ or lipids also inhibit gastric secretion (*orange arrows*).

④ Secretin and cholecystokinin produced by the duodenum (*brown arrows*) decrease gastric secretions in the stomach.

(c)

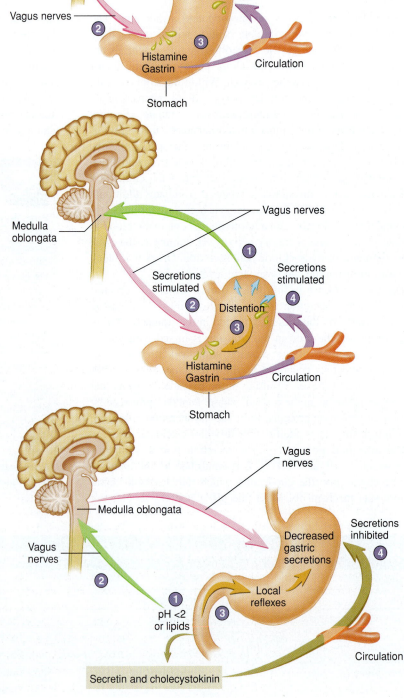

PROCESS Figure 16.11 **AP|R** Regulation of Stomach Secretions

CLINICAL IMPACT   Peptic Ulcers

Approximately 10% of people in the United States will develop a **peptic ulcer** during their lifetime. Peptic ulcers are caused when the gastric juices (acid and pepsin) digest the mucosal lining of the digestive tract. Peptic ulcers can occur in the duodenum, stomach, or esophagus.

Nearly all peptic ulcers are due to infection by a specific bacterium,

Helicobacter pylori, which is also linked to gastritis and gastric cancer. Because diet, stress, smoking and alcohol cause excess acid secretion in the stomach, these lifestyle patterns were deemed responsible for ulcers for many years. Although these factors can contribute to ulcers, it is now clear that the root cause is *H. pylori*.

Antibiotic treatment to eradicate *H. pylori* is the best therapy for ulcers. Other treatments involve drugs that prevent histamine-stimulated acid secretion or that directly inhibit the proton pumps that secrete the acid, but these are only temporary.

To summarize, once gastric acid secretion begins, further secretion is controlled by negative-feedback loops involving nerves and hormones. First, during the gastric phase, high acid levels in the stomach trigger a decrease in additional acid secretion. Second, during the intestinal phase, acidic chyme entering the duodenum triggers a decrease in gastric acid secretion. These negative-feedback loops ensure that the acidic chyme entering the duodenum is neutralized, which is required for the digestion of food by pancreatic enzymes and for the prevention of peptic ulcer formation.

Movement in the Stomach

Two types of stomach movement aid digestion and help move chyme through the digestive tract: mixing waves and peristaltic waves (figure 16.12). Both types of movement result from smooth muscle contractions in the stomach wall. The contractions occur about every 20 seconds and proceed from the body of the stomach toward the pyloric sphincter. Relatively weak contractions result in **mixing waves,** which thoroughly mix ingested food with stomach secretions to form chyme. The more fluid part of the chyme is pushed toward the pyloric sphincter, whereas the more solid center moves back toward the body of the stomach. Stronger contractions result in **peristaltic waves,** which force the chyme toward and through the pyloric sphincter. The pyloric sphincter usually remains closed because of mild tonic contraction. Each peristaltic contraction is sufficiently strong to cause partial relaxation of the pyloric sphincter and to pump a few milliliters of chyme through the pyloric opening and into the duodenum. Increased motility leads to increased emptying.

If the stomach empties too fast, the efficiency of digestion and absorption in the small intestine is reduced. However, if the rate of emptying is too slow, the highly acidic contents of the stomach may damage the stomach wall. To prevent these two extremes, stomach emptying is regulated. The hormonal and neural mechanisms that stimulate stomach secretions also increase stomach motility, so that the increased secretions are effectively mixed with the stomach contents. The major stimulus of gastric motility and emptying is distension of the stomach wall. Inhibition of gastric motility and emptying is accomplished by the same negative-feedback loops associated with the intestinal phase of gastric secretion. In particular, cholecystokinin is a major inhibitor of motility and emptying (table 16.2). Hence, stomach emptying is slower after a fatty meal due to the release of cholecystokinin.

In an empty stomach, peristaltic contractions that approach tetanic contractions can occur for about 2 to 3 minutes. The contractions are increased by low blood glucose levels and are sufficiently strong to create an uncomfortable sensation called a "hunger pang." Hunger pangs usually begin 12 to 24 hours after the previous meal; in less time for some people. If nothing is ingested, hunger pangs reach their maximum intensity within 3 or 4 days and then become progressively weaker.

16.5 SMALL INTESTINE

Learning Outcomes   *After reading this section, you should be able to*

A. List the characteristics of the small intestine that account for its large surface area.

B. Describe the secretions and movements that occur in the small intestine.

Anatomy of the Small Intestine

The **small intestine** is about 6 meters (m) long and consists of three parts: the duodenum, the jejunum, and the ileum (figure 16.13). The **duodenum** (doo-od′ĕ-nŭm, doo-ō-dĕ′nŭm) is about 25 cm long (the term *duodenum* means 12, suggesting that it is 12 in. long). The **jejunum** (jĕ-joo′nŭm) is about 2.5 m long and makes up two-fifths of the total length of the small intestine. The **ileum** (il′ē-ŭm) is about 3.5 m long and makes up three-fifths of the small intestine.

The duodenum nearly completes a 180-degree arc as it curves within the abdominal cavity. Part of the pancreas lies within this arc. The **common bile duct** from the liver and the **pancreatic duct** from the pancreas join and empty into the duodenum (see figure 16.17).

The small intestine is the major site of digestion and absorption of food, which are accomplished due to the presence of a large surface area. The small intestine has three modifications that increase its surface area about 600-fold: circular folds, villi, and microvilli. The mucosa and submucosa form a series of **circular folds** that run perpendicular to the long axis of the digestive tract (figure 16.14a). Tiny, fingerlike projections of the mucosa form numerous **villi** (vil′i; sing. villus), which are 0.5–1.5 mm long (figure 16.14b). Most of the cells composing the surface of the villi have numerous cytoplasmic extensions, called **microvilli**

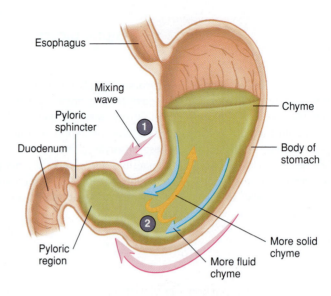

1. A mixing wave initiated in the body of the stomach progresses toward the pyloric sphincter (*pink arrows directed inward*).

2. The more fluid part of the chyme is pushed toward the pyloric sphincter (*blue arrows*), whereas the more solid center of the chyme squeezes past the peristaltic constriction back toward the body of the stomach (*orange arrow*).

3. Peristaltic waves (*purple arrows*) move in the same direction and in the same way as the mixing waves but are stronger.

4. Again, the more fluid part of the chyme is pushed toward the pyloric region (*blue arrows*), whereas the more solid center of the chyme squeezes past the peristaltic constriction back toward the body of the stomach (*orange arrow*).

5. Peristaltic contractions force a few milliliters of the most fluid chyme through the pyloric opening into the duodenum (*small red arrows*). Most of the chyme, including the more solid portion, is forced back toward the body of the stomach for further mixing (*yellow arrow*).

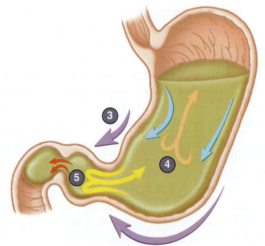

PROCESS Figure 16.12 Movements in the Stomach

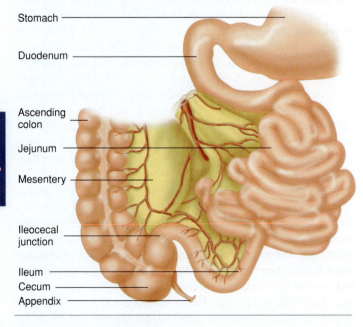

Figure 16.13 Small Intestine

(mī′krō-vil′ī) (figure 16.14c,d). Each villus is covered by simple columnar epithelium. Within the loose connective tissue core of each villus are a blood capillary network and a lymphatic capillary called a **lacteal** (lak′tē-ăl; resembling milk) (figure 16.14c). The blood capillary network and the lacteal are very important in transporting absorbed nutrients.

The mucosa of the small intestine is simple columnar epithelium with four major cell types: (1) **absorptive cells,** which have microvilli, produce digestive enzymes, and absorb digested food; (2) **goblet cells,** which produce a protective mucus; (3) **granular cells,** which may help protect the intestinal epithelium from bacteria; and (4) **endocrine cells,** which produce regulatory hormones.

The epithelial cells are located within tubular glands of the mucosa, called **intestinal glands** or *crypts of Lieberkühn,* at the base of the villi. Granular and endocrine cells are located in the bottom of the glands. The submucosa of the duodenum contains mucous glands, called **duodenal glands,** which open into the base of the intestinal glands.

The duodenum, jejunum, and ileum are similar in structure. However, progressing from the duodenum through the ileum,

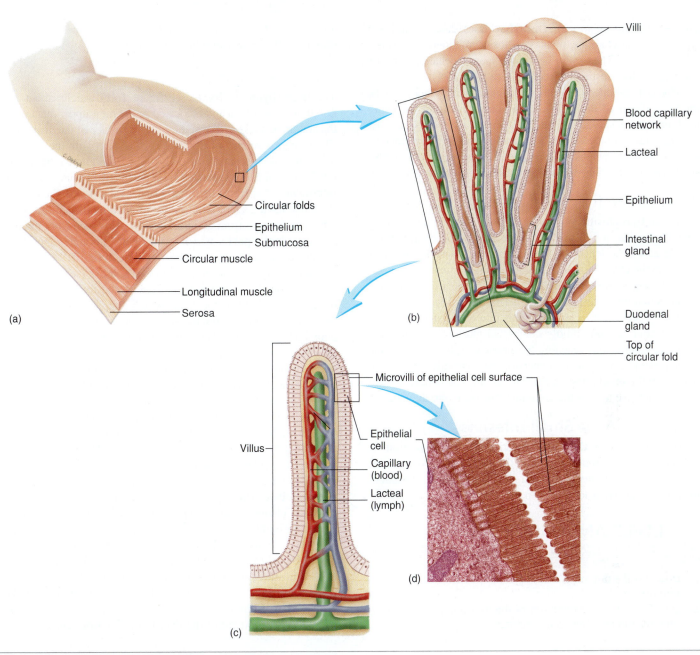

Figure 16.14 [AP|R] **Anatomy and Histology of the Duodenum**

(*a*) Wall of the duodenum, showing the circular folds. (*b*) Villi on a circular fold. (*c*) A single villus, showing the lacteal and capillary network. (*d*) Transmission electron micrograph of microvilli on the surface of a villus.

there are gradual decreases in the diameter of the small intestine, in the thickness of the intestinal wall, in the number of circular folds, and in the number of villi. Lymphatic nodules are common along the entire length of the digestive tract, and clusters of lymphatic nodules, called **Peyer patches,** are numerous in the ileum. These lymphatic tissues help protect the intestinal tract from harmful pathogens.

The site where the ileum connects to the large intestine is called the **ileocecal** (il′ē-ō-sē′kăl) **junction.** It has a ring of smooth muscle, the **ileocecal sphincter,** and an **ileocecal valve** (see figure 16.21*a*), which allow the intestinal contents to move from the ileum to the large intestine, but not in the opposite direction.

Secretions of the Small Intestine

Secretions from the mucosa of the small intestine contain mainly mucus, ions, and water. Intestinal secretions lubricate and protect the intestinal wall from the acidic chyme and the action of digestive enzymes. They also keep the chyme in the small intestine in a liquid form to facilitate the digestive process. Most of the secretions entering the small intestine are produced by the intestinal mucosa, but the secretions of the liver and the pancreas also enter the small intestine and play important roles in digestion.

The epithelial cells in the walls of the small intestine have enzymes, bound to their free surfaces, that are significant in the final steps of digestion. **Peptidases** (pep′ti-dās-ez) break the peptide bonds in proteins to form amino acids. **Disaccharidases** (dī-sak′ă-rid-ās-ez) break down disaccharides, such as maltose, into monosaccharides, such as glucose. The amino acids and monosaccharides can be absorbed by the intestinal epithelium (see table 16.1).

Mucus is produced by duodenal glands and by goblet cells, which are dispersed throughout the epithelial lining of the entire small intestine and within intestinal glands. Hormones released from the intestinal mucosa stimulate liver and pancreatic secretions. Secretion by duodenal glands is stimulated by the vagus nerve, secretin release, and chemical or tactile irritation of the duodenal mucosa.

Movement in the Small Intestine

Mixing and propulsion of chyme are the primary mechanical events that occur in the small intestine. **Peristaltic contractions** proceed along the length of the intestine for variable distances and cause the chyme to move along the small intestine (see figure 16.9). **Segmental contractions** are propagated for only short distances and mix intestinal contents (figure 16.15).

The ileocecal sphincter at the juncture of the ileum and the large intestine remains mildly contracted most of the time, but peristaltic contractions reaching the ileocecal sphincter from the small intestine cause the sphincter to relax and allow chyme to move from the small intestine into the cecum. The ileocecal valve prevents movement from the large intestine back into the ileum.

Absorption in the Small Intestine

A major function of the small intestine is the **absorption** of nutrients. Most absorption occurs in the duodenum and jejunum, although some absorption also occurs in the ileum (see "Digestion, Absorption, and Transport" later in this chapter).

16.6 LIVER AND PANCREAS

Learning Outcomes   *After reading this section, you should be able to*

A. Describe the anatomy, histology, and ducts of the liver and pancreas.
B. Describe the major functions of the liver and pancreas, and explain how they are regulated.

Two large accessory glands, the liver and the pancreas, produce secretions that empty into the duodenum.

Anatomy of the Liver

The **liver** (figure 16.16*a,b*; see figure 16.1) weighs about 1.36 kilograms (kg) (3 lb) and is located in the right upper quadrant of the abdomen, tucked against the inferior surface of the diaphragm. The posterior surface of the liver is in contact with the right ribs 5–12. It is divided into two major lobes, the **right lobe** and the **left lobe,** which are separated by a connective tissue septum, the **falciform** (fal′si-fōrm) **ligament.** Two smaller lobes, the **caudate** (kaw′dāt; having a tail) **lobe** and the **quadrate** (kwah′drāt; square) **lobe,** can be seen from an inferior view. Also seen from the inferior view is the **porta,** which is the "gate" through which blood vessels, ducts, and nerves enter or exit the liver.

The liver receives blood from two sources (see chapter 13). The **hepatic** (he-pa′tik; associated with the liver) **artery** takes oxygen-rich blood to the liver, which supplies liver cells with oxygen. The **hepatic portal vein** carries blood that is oxygen-poor but rich in absorbed nutrients and other substances from the digestive tract to the liver. Liver cells process nutrients and detoxify harmful substances from the blood. Blood exits the liver through **hepatic veins,** which empty into the inferior vena cava.

Many delicate connective tissue septa divide the liver into **lobules** with portal triads at their corners. The **portal triads** contain three structures: the hepatic artery, the hepatic portal vein, and the hepatic duct (figure 16.16*c*). **Hepatic** (he-pa′tik) **cords,** formed by platelike groups of cells called **hepatocytes** (hep′ă-tō-sīts), are located between the center and the margins of each lobule. The hepatic cords are separated from one another by blood channels called **hepatic sinusoids** (si′nŭ-soydz, sī′nū-soydz; resembling cavities). The sinusoid epithelium contains phagocytic cells that help remove foreign particles from the blood. Blood from the hepatic portal vein and the hepatic artery flows into the sinusoids and mixes together. The mixed blood flows toward the center of each lobule into a **central vein.** The central veins from all the lobes unite to form the hepatic veins, which carry blood out of the liver to the inferior vena cava.

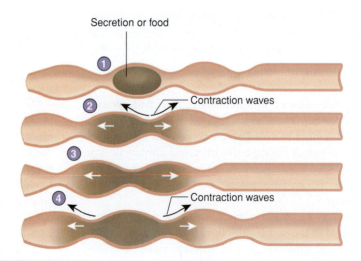

1. A secretion introduced into the digestive tract or into food within the tract begins in one location.

2. Segments of the digestive tract alternate between contraction and relaxation.

3. Material (*brown*) in the intestine is spread out in both directions from the site of introduction.

4. The secretion or food is spread out in the digestive tract and becomes more diffuse (*lighter color*) through time.

Secretion or food

Contraction waves

Contraction waves

PROCESS Figure 16.15 Segmental Contractions in the Small Intestine

Digestive

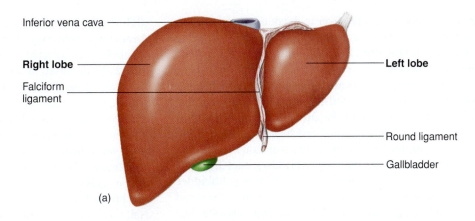

Inferior vena cava

Right lobe

Falciform
ligament

Left lobe

Round ligament

Gallbladder

Liver

(a)

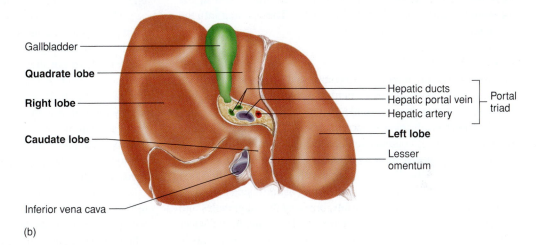

Gallbladder

Quadrate lobe

Right lobe

Caudate lobe

Hepatic ducts

Hepatic portal vein

Hepatic artery

Portal
triad

Left lobe

Lesser
omentum

Inferior vena cava

(b)

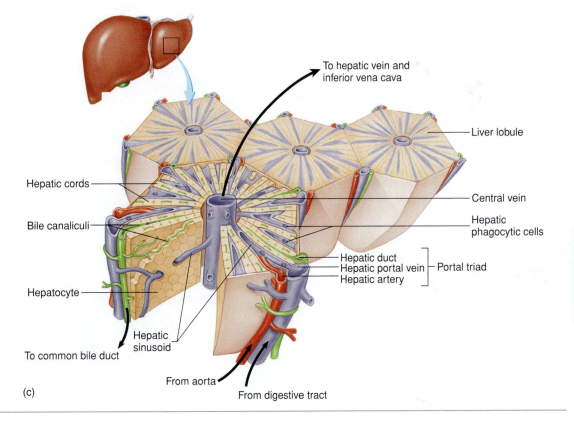

To hepatic vein and
inferior vena cava

Liver lobule

Hepatic cords

Central vein

Bile canaliculi

Hepatic
phagocytic cells

Hepatic duct

Hepatic portal vein

Hepatic artery

Portal triad

Hepatocyte

Hepatic
sinusoid

To common bile duct

From aorta

From digestive tract

(c)

Figure 16.16 AP|R **Liver**

(*a*) Anterior view. (*b*) Inferior view. (*c*) Histology.

Digestive

A system of ducts from the liver to the duodenum serves as a pathway for bile and other secretions (figure 16.17). A cleftlike lumen, the **bile canaliculus** (kan′ă-lik′ū-lŭs; pl. kan′ă lik′ū-lī, little canals), is between the cells of each hepatic cord. Bile, produced by the hepatocytes, flows through the bile canaliculi to the hepatic ducts in the portal triads. The hepatic ducts converge and empty into the right and left **hepatic ducts,** which transport bile out of the liver. The right and left hepatic ducts unite to form a single **common hepatic duct.** The common hepatic duct is joined by the **cystic** (sis′tik; *kystis,* bladder) **duct** from the gallbladder to form the **common bile duct.** The **gallbladder** is a small sac on the inferior surface of the liver that stores and concentrates bile (see figure 16.16a,b). The common bile duct joins the pancreatic duct and opens into the duodenum at the **duodenal papilla** (pă-pil′ă) (figure 16.17). The opening into the duodenum is regulated by a sphincter.

Functions of the Liver

The liver performs important digestive and excretory functions, stores and processes nutrients, detoxifies harmful chemicals, and synthesizes new molecules (table 16.3).

The liver produces and secretes about 600–1000 mL of bile each day. **Bile** (bīl) contains no digestive enzymes, but it plays an important role in digestion by diluting and neutralizing stomach acid and by dramatically increasing the efficiency of fat digestion and absorption. Digestive enzymes cannot act efficiently on large fat globules. **Bile salts** emulsify fats, breaking the fat globules into smaller droplets, much like the action of detergents in dishwater (see tables 16.1 and 16.3). The small droplets are more easily digested by digestive enzymes. Bile also contains excretory products, such as cholesterol, fats, and bile pigments, including **bilirubin**

(bil-i-roo′bin), a bile pigment that results from the breakdown of hemoglobin (see chapter 11). **Gallstones** may form if the amount of cholesterol secreted by the liver becomes excessive and is not able to be dissolved by the bile salts.

Neural and hormonal stimuli regulate the secretion and release of bile (figure 16.18). Bile secretion by the liver is stimulated by parasympathetic stimulation through the vagus nerve. Secretin, which is released from the duodenum, also stimulates bile secretion and release. Cholecystokinin stimulates the gallbladder to contract and release bile into the duodenum. In addition, most (90% of) bile salts are reabsorbed in the ileum. The blood carries the bile salts back to the liver, where they stimulate additional bile salt secretion and are once again secreted into the bile. The loss of bile salts in the feces is reduced by this recycling process.

The liver can remove sugar from the blood and store it in the form of glycogen (table 16.3). It can also store fat, vitamins, copper, and iron. This storage function is usually short-term.

Foods are not always ingested in the proportion needed by the tissues. If this is the case, the liver can convert some nutrients into others (table 16.3). For example, if a person eats a meal that is very high in protein, a large amount of amino acids and only a small amount of lipids and carbohydrates are delivered to the liver. The liver can break down the amino acids and cycle many of them through metabolic pathways to produce ATP and to synthesize lipids and glucose (see chapter 17).

The liver also transforms some nutrients into more readily usable substances. Ingested fats, for example, can be combined with choline and phosphorus in the liver to produce phospholipids, which are essential components of cell membranes.

Many ingested substances are harmful to body cells. In addition, the body itself produces many by-products of metabolism

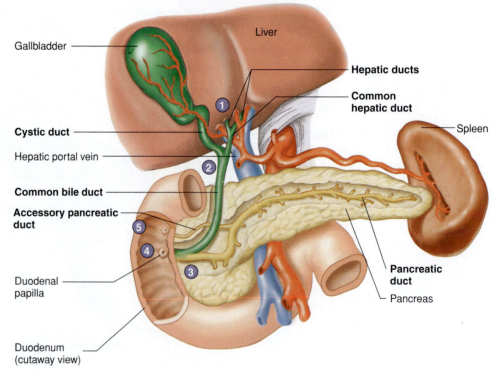

1 The hepatic ducts from the liver lobes combine to form the common hepatic duct.

2 The common hepatic duct combines with the cystic duct from the gallbladder to form the common bile duct.

3 The common bile duct joins the pancreatic duct.

4 The combined duct empties into the duodenum at the duodenal papilla.

5 Pancreatic secretions may also enter the duodenum through an accessory pancreatic duct, which also empties into the duodenum.

Gallbladder

Liver

Hepatic ducts

Common hepatic duct

Spleen

Cystic duct

Hepatic portal vein

Common bile duct

Accessory pancreatic duct

Pancreatic duct

Pancreas

Duodenal papilla

Duodenum (cutaway view)

PROCESS Figure 16.17 **AP|R** **Flow of Bile and Pancreatic Secretions Through the Duct System of the Liver, Gallbladder, and Pancreas**

Digestive

TABLE 16.3   Functions of the Liver

Function	Explanation
Digestion	Bile neutralizes stomach acid and emulsifies fats, which facilitates fat digestion.
Excretion	Bile contains excretory products, such as cholesterol, fats, and bile pigments (e.g., bilirubin), that result from hemoglobin breakdown.
Nutrient storage	Liver cells remove sugar from the blood and store it in the form of glycogen; they also store fat, vitamins (A, B_{12}, D, E, and K), copper, and iron.
Nutrient conversion	Liver cells convert some nutrients into others; for example, amino acids can be converted to lipids or glucose, fats can be converted to phospholipids, and vitamin D is converted to its active form.
Detoxification of harmful chemicals	Liver cells remove ammonia from the circulation and convert it to urea, which is eliminated in the urine; other substances are detoxified and secreted in the bile or excreted in the urine.
Synthesis of new molecules	The liver synthesizes blood proteins, such as albumin, fibrinogen, globulins, and clotting factors.

that, if accumulated, are toxic. The liver is an important line of defense against many of those harmful substances. It detoxifies them by altering their structure, which makes their excretion easier (table 16.3). For example, the liver removes ammonia, a toxic by-product of amino acid metabolism, from the circulation and converts it to urea, which is then secreted into the circulation and eliminated by the kidneys in the urine. The liver also removes other substances from the circulation and excretes them into the bile.

The liver can also produce unique new compounds (table 16.3). Many of the blood proteins, such as albumins, fibrinogen, globulins, and clotting factors, are synthesized in the liver and released into the circulation.

Anatomy of the Pancreas

The **pancreas** is located retroperitoneal, posterior to the stomach in the inferior part of the left upper quadrant (see figure 16.1). It has a **head** near the midline of the body and a **tail** that extends to the left, where it touches the spleen (figure 16.19; see figure 16.17). It is a complex organ composed of both endocrine and exocrine tissues that perform several functions. The endocrine part of the pancreas consists of **pancreatic islets,** or *islets of Langerhans.* The islet cells produce the hormones insulin and glucagon, which enter the blood. These hormones are very important in controlling blood levels of nutrients, such as glucose and amino acids (see chapter 10).

1. Vagus nerve stimulation (*red arrow*) causes the gallbladder to contract, thereby releasing bile into the duodenum.

2. Secretin, produced by the duodenum (*purple arrows*) and carried through the circulation to the liver, stimulates bile secretion by the liver (*green arrows inside the liver*).

3. Cholecystokinin, produced by the duodenum (*pink arrows*) and carried through the circulation to the gallbladder, stimulates the gallbladder to contract and the sphincters to relax, thereby releasing bile into the duodenum (*green arrow outside the liver*).

4. Bile salts also stimulate bile secretion. Over 90% of bile salts are reabsorbed in the ileum and returned to the liver (*green arrows*), where they stimulate additional secretion of bile salts.

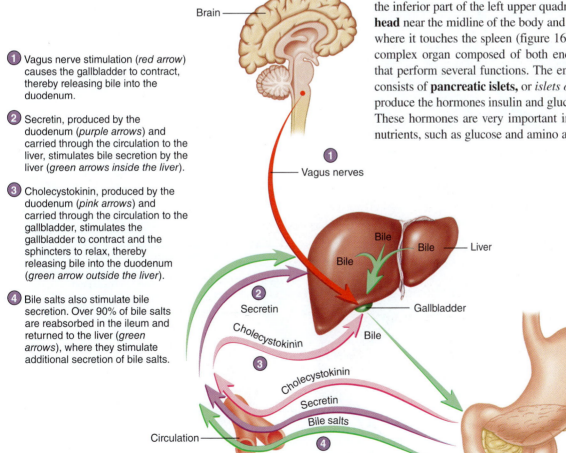

PROCESS Figure 16.18 **Control of Bile Secretion and Release**

The exocrine part of the pancreas is a compound acinar gland (see chapter 4). The **acini** (as′i-nī; grapes) produce digestive enzymes. Clusters of acini are connected by small ducts, which join to form larger ducts, and the larger ducts join to form the **pancreatic duct.** The pancreatic duct joins the common bile duct and empties into the duodenum.

Functions of the Pancreas

The exocrine secretions of the pancreas include bicarbonate ions (HCO_3^-), which neutralize the acidic chyme that enters the small intestine from the stomach. The increased pH resulting from the secretion of HCO_3^- stops pepsin digestion but provides the proper environment for the function of pancreatic enzymes. Pancreatic enzymes are also present in the exocrine secretions and are important in digesting all major classes of food (see table 16.1). Without the enzymes produced by the pancreas, lipids, proteins, and carbohydrates cannot be adequately digested.

The major proteolytic (protein-digesting) enzymes are **trypsin** (trip′sin), **chymotrypsin** (kī-mō-trip′sin), and **carboxypeptidase** (kar-box′ē-pep′ti-dās). These enzymes continue the protein digestion that started in the stomach, and **pancreatic amylase** (am′il-ās) continues the polysaccharide digestion that began in the oral cavity. The pancreatic enzymes also include **lipase** (lip′ās), a lipid-digesting enzyme, and **nucleases** (noo′klē-ās-ez), which are enzymes that degrade DNA and RNA to their component nucleotides.

The exocrine secretory activity of the pancreas is controlled by both hormonal and neural mechanisms (figure 16.20; see table 16.2). Secretin initiates the release of a watery pancreatic solution that contains a large amount of HCO_3^-. The primary stimulus for secretin release is the presence of acidic chyme in the duodenum. Cholecystokinin stimulates the pancreas to release an enzyme-rich solution. The primary stimulus for cholecystokinin release is the presence of fatty acids and amino acids in the duodenum. In turn, enzymes secreted by the pancreas act to digest these fatty acids and amino acids. Parasympathetic stimulation through the vagus nerves also stimulates the secretion of pancreatic juices rich in pancreatic enzymes. Sympathetic action potentials inhibit pancreatic secretion.

Predict 4

Explain how secretin production in response to acidic chyme and bicarbonate ion secretion in response to secretin constitute a negative-feedback mechanism.

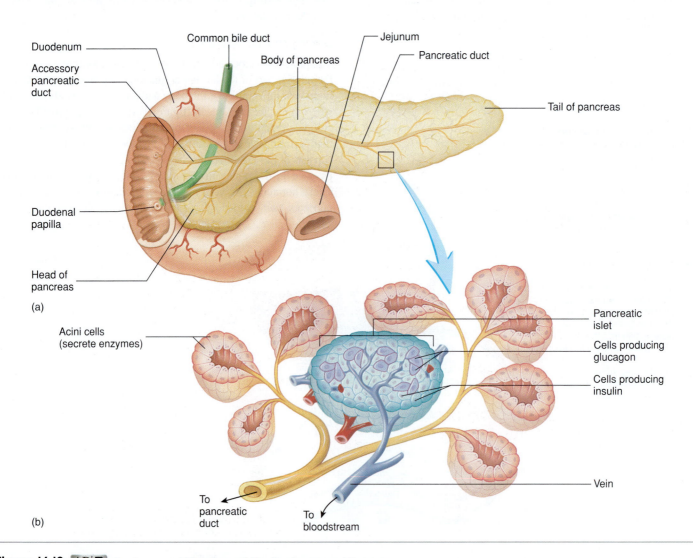

(a) Duodenum, Common bile duct, Jejunum, Accessory pancreatic duct, Body of pancreas, Pancreatic duct, Tail of pancreas, Duodenal papilla, Head of pancreas

(b) Acini cells (secrete enzymes), Pancreatic islet, Cells producing glucagon, Cells producing insulin, Vein, To pancreatic duct, To bloodstream

Figure 16.19 AP|R Anatomy and Histology of the Duodenum and Pancreas

(*a*) The head of the pancreas lies within the duodenal curvature, with the pancreatic duct emptying into the duodenum. (*b*) Histology of the pancreas, showing both the acini and the pancreatic duct system.

Digestive

16.7 LARGE INTESTINE

Learning Outcomes *After reading this section, you should be able to*

A. List the parts of the large intestine, and describe its anatomy and histology.
B. Describe the major functions of the large intestine, and explain how movement is regulated.

Anatomy of the Large Intestine

The **large intestine** consists of the cecum, colon, rectum, and anal canal (figure 16.21; see figure 16.1).

Cecum

The **cecum** (sē′kŭm) is the proximal end of the large intestine where it joins with the small intestine at the ileocecal junction. The cecum is located in the right lower quadrant of the abdomen near the iliac fossa. The cecum is a sac that extends inferiorly about 6 cm past the ileocecal junction. Attached to the cecum is a tube about 9 cm long called the **appendix.**

Colon

The **colon** (kō′lon) is about 1.5–1.8 m long and consists of four parts: the ascending colon, the transverse colon, the descending colon, and the sigmoid colon (figure 16.21). The **ascending colon** extends superiorly from the cecum to the right colic flexure, near the liver, where it turns to the left. The **transverse colon** extends from the right colic flexure to the left colic flexure near the spleen, where the colon turns inferiorly; and the **descending colon** extends from the left colic flexure to the pelvis, where it becomes the **sigmoid colon.** The sigmoid colon forms an S-shaped tube that extends medially and then inferiorly into the pelvic cavity and ends at the rectum.

A CASE IN POINT

Appendicitis

Lowe Payne had been feeling nauseated for a couple of days and had lost his appetite. Suddenly, he felt a sharp pain in his lower right abdomen. The pain was so intense that his mother, Lotta, took Lowe to the hospital. There, he was diagnosed as having **appendicitis.** Part of that diagnosis includes applying slight pressure, such as by pushing with the fingertips, to a specific point in the right lower quadrant of the abdomen. That point, called the McBurney point, is midway between the umbilicus and the right anterior superior iliac spine of the coxal bone.

Appendicitis is an inflammation of the appendix that usually occurs because of obstruction. Secretions from the appendix cannot pass the obstruction; therefore, they accumulate, causing enlargement and pain. Bacteria in the area can cause infection. Symptoms include sudden abdominal pain, particularly in the right lower quadrant, along with a slight fever, loss of appetite, constipation or diarrhea, nausea, and vomiting. If the appendix bursts, the infection can spread throughout the peritoneal cavity, causing **peritonitis** (see Clinical Impact earlier in this chapter), with life-threatening results. Each year, 500,000 people in the United States suffer from appendicitis. The usual treatment is an **appendectomy,** surgical removal of the appendix.

The mucosal lining of the colon contains numerous straight, tubular glands called **crypts,** which contain many mucus-producing goblet cells. The longitudinal smooth muscle layer of the colon does not completely envelop the intestinal wall but forms three bands called **teniae coli** (tē′nē-ē kō′lī).

Rectum

The **rectum** is a straight, muscular tube that begins at the termination of the sigmoid colon and ends at the anal canal (figure 16.21). The muscular tunic is composed of smooth muscle and is relatively thick in the rectum compared to the rest of the digestive tract.

Anal Canal

The last 2–3 cm of the digestive tract is the **anal canal.** It begins at the inferior end of the rectum and ends at the **anus** (external digestive tract opening). The smooth muscle layer of the anal canal is

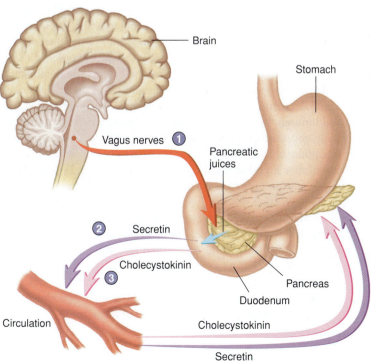

1. Parasympathetic stimulation from the vagus nerve (*red arrow*) causes the pancreas to release a secretion rich in digestive enzymes.

2. Secretin (*purple arrows*), released from the duodenum, stimulates the pancreas to release a watery secretion, rich in bicarbonate ions.

3. Cholecystokinin (*pink arrows*), released from the duodenum, causes the pancreas to release a secretion rich in digestive enzymes.

PROCESS Figure 16.20 Control of Pancreatic Secretion

Digestive

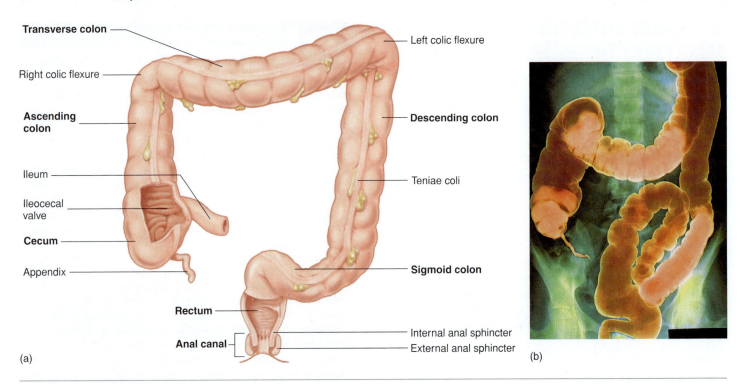

Figure 16.21 AP|R **Large Intestine**

(*a*) The large intestine consists of the cecum, colon, rectum, and anal canal. The teniae coli are bands of smooth muscle along the length of the colon. (*b*) Radiograph of the large intestine following a barium enema.

even thicker than that of the rectum and forms the **internal anal sphincter** at its superior end. The **external anal sphincter** at the inferior end of the anal canal is formed by skeletal muscle.

Hemorrhoids are enlarged or inflamed rectal, or hemorrhoidal, veins that supply the anal canal. Hemorrhoids may cause pain, itching, and/or bleeding around the anus. Treatments include increasing bulk (indigestible fiber) in the diet, taking sitz baths, and using hydrocortisone suppositories. Surgery may be necessary if the condition is extreme and does not respond to other treatments.

Functions of the Large Intestine

Normally, 18–24 hours are required for material to pass through the large intestine, in contrast to the 3–5 hours required for chyme to move through the small intestine. While in the colon, chyme is converted to **feces** (fē'sēz). The formation of feces involves the absorption of water and salts, the secretion of mucus, and extensive action of microorganisms. The colon stores the feces until they are eliminated by the process of **defecation** (def-ĕ-kā'shŭn).

Numerous microorganisms inhabit the colon. They reproduce rapidly and ultimately constitute about 30% of the dry weight of the feces. Some bacteria in the intestine synthesize vitamin K and other vitamins, which are passively absorbed in the colon.

Every 8–12 hours, large parts of the colon undergo several strong contractions, called **mass movements,** which propel the colon contents a considerable distance toward the anus. Each mass movement contraction extends over 20 or more centimeters of the large intestine, which is a much longer part of the digestive tract than that covered by a peristaltic contraction. These mass movements are very common following some meals, especially breakfast.

Feces distend the rectal wall and stimulate the **defecation reflex,** which involves local and parasympathetic reflexes. Local reflexes cause weak contractions, whereas parasympathetic reflexes cause strong contractions and are normally responsible for most of the defecation reflex. Action potentials produced in response to the distention travel along sensory nerve fibers to the sacral region of the spinal cord, where motor action potentials are initiated that reinforce peristaltic contractions in the lower colon and the rectum. Action potentials from the spinal cord also cause the internal anal sphincter to relax. The external anal sphincter, which is composed of skeletal muscle and is under conscious cerebral control, prevents feces from moving out of the rectum and through the anal opening. If this sphincter is relaxed voluntarily, feces are expelled. The defecation reflex persists for only a few minutes and quickly subsides. Generally, the reflex is reinitiated after a period that may be as long as several hours. Mass movements in the colon are usually the reason for the reinitiation of the defecation reflex.

Defecation can be initiated by voluntary actions that stimulate a defecation reflex. These actions include a large inspiration of air, followed by closure of the larynx and forceful contraction of the abdominal muscles. As a consequence, the pressure in the abdominal cavity increases and forces feces into the rectum. Stretch of the rectum initiates a defecation reflex. The increased abdominal pressure also helps push feces through the rectum.

Predict 5

Explain how an enema stimulates defecation.

16.8 DIGESTION, ABSORPTION, AND TRANSPORT

A. Describe the digestion, absorption, and transport of carbohydrates, proteins, vitamins, and minerals.

B. Describe the digestion, absorption, and transport of fats and lipids.

C. Discuss water movement into and out of the digestive tract.

Digestion is the breakdown of food to molecules that are small enough to be absorbed into the circulation. **Mechanical digestion** breaks large food particles into smaller ones. **Chemical digestion** involves the breaking of covalent chemical bonds in organic molecules by digestive enzymes. Carbohydrates break down into monosaccharides, lipids break down into fatty acids and monoglycerides, and proteins break down into amino acids (figure 16.22).

Absorption begins in the stomach, where some small, lipid-soluble molecules, such as alcohol and aspirin, can diffuse through the stomach epithelium into the circulation. Most absorption occurs in the duodenum and jejunum, although some occurs in the ileum. Some molecules can diffuse through the intestinal wall, whereas others must be transported across the intestinal wall. **Transport** requires carrier molecules and includes facilitated diffusion, cotransport, and active transport. Cotransport and active transport require energy to move the transported molecules across the intestinal wall.

Carbohydrates

Ingested **carbohydrates** (kar-bō-hī′drāts) consist primarily of starches, cellulose, sucrose (table sugar), and small amounts of fructose (fruit sugar) and lactose (milk sugar). Starches, cellulose, sucrose, and fructose are derived from plants, and lactose is derived from animals. **Polysaccharides** (pol-ē-sak′ă-rīdz) are large carbohydrates, such as starches, cellulose, and glycogen, that consist of many sugars linked by chemical bonds. Starch is

an energy-storage molecule in plants. Cellulose forms the walls of plant cells. Glycogen is an energy-storage molecule in animals and is contained in muscle and in the liver. When uncooked meats are processed or stored, the glycogen is broken down to glucose, which is further broken down, so that little, if any, glycogen remains. Therefore, almost all dietary carbohydrates come from plants. Starch is broken down by enzymes. Cellulose is a polysaccharide that is not digested but is important for providing fiber in the diet.

Salivary amylase begins the digestion of carbohydrates in the mouth (figure 16.23). The carbohydrates then pass to the stomach, where digestion continues until the food is well mixed with acid, which inactivates salivary amylase. In the duodenum, **pancreatic amylase** continues the digestion of carbohydrates, and absorption begins. The amylases break down polysaccharides to **disaccharides** (dī-sak′ă-rīdz; two sugars chemically linked; see chapter 2). A group of enzymes called **disaccharidases** that are bound to the microvilli of the intestinal epithelium break down the disaccharides to monosaccharides. One disaccharidase is lactase, which breaks down lactose (milk sugar); many people have heard of lactase, because lack of it leads to lactose intolerance. Lactase is made at birth; however, in 5–15% of the European-American population and 80–90% of the African-American and Asian-American populations, lactase synthesis sharply declines 3–4 years after weaning. In the absence of lactase, ingestion of dairy products causes intestinal cramping, bloating, and diarrhea. An increasing selection of foods for lactose-intolerant people can be found in the supermarket.

The **monosaccharides** (mon-ō-sak′ă-rīdz; single sugars) glucose, galactose, and fructose are taken up through the intestinal epithelial cells (figure 16.24; see chapter 3). Cotransport of the major monosaccharide, glucose, and Na^+ is driven by a Na^+ concentration gradient that is established by the sodium-potassium pump. Diffusion of Na^+ down its concentration gradient provides the energy to transport glucose across the cell membrane. This mechanism is also used for galactose transport, while fructose is taken up by facilitated diffusion (see chapter 3). Once inside the intestinal epithelial cell, monosaccharides are transported into the capillaries of the intestinal villi and are carried by the hepatic portal system to the liver. Liver cells convert different types of monosaccharides to **glucose,** which then leaves the liver via the circulation to be distributed throughout the body. Glucose enters the cells by facilitated diffusion. The rate of glucose transport into most types of cells is greatly influenced by **insulin** and can increase tenfold in the presence of insulin. Without insulin, glucose enters most cells very slowly.

Lipids

Lipid molecules are insoluble or only slightly soluble in water (see chapter 2). They include triglycerides, phospholipids, steroids, and fat-soluble vitamins. **Triglycerides** (trī-glis′er-īdz) are the most common type of lipid. They consist of three fatty acids bound to glycerol. Triglycerides are often referred to as fats. Fats are **saturated** if their fatty acids have only single bonds between carbons and **unsaturated** if they have one (monounsaturated) or more (polyunsaturated) double bonds between carbons (see chapter 2). Saturated fats are solid at room temperature, whereas polyunsaturated fats are liquid at room temperature. Saturated fats are found in meat, dairy products, eggs, nuts, coconut oil, and palm oil. Unsaturated fats are found in fish and most plant oils.

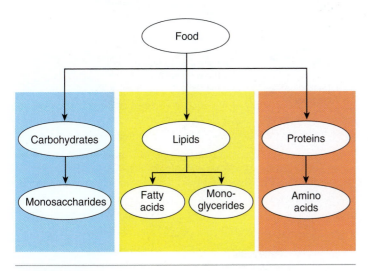

Figure 16.22 Digestion

Food consists primarily of carbohydrates, lipids, and proteins. Carbohydrates are broken down into monosaccharides, lipids into fatty acids and monoglycerides, and proteins into amino acids.

Digestive

CLINICAL IMPACT   High- and Low-Density Lipids

Cholesterol is a steroid molecule that plays important roles in the body (see chapter 2). However, cholesterol levels in the blood are of great concern to many adults because people with high blood cholesterol run a much greater risk of heart disease and stroke than do people with low cholesterol. A total cholesterol level of less than 180 milligrams (mg)/dL is considered low, which is usually good, although an extremely low cholesterol level can be harmful. A cholesterol level above 200 mg/dL is considered too high. People with high levels should seek advice from their physician, reduce their intake of foods rich in cholesterol and other fats, and increase their level of exercise. Some people with very high cholesterol levels may have to take medication to lower their cholesterol.

Fats, including cholesterol, are not soluble in water; thus, they are transported in the blood as lipid-protein complexes, or lipoproteins (see figure 16.26). Cholesterol is found primarily in two types of lipoproteins. Low-density lipoproteins (LDLs) carry cholesterol to the tissues for use by the cells. High-density lipoproteins (HDLs) transport cholesterol from the tissues to the liver. In addition to assessing total cholesterol levels, blood tests can reveal the levels of HDLs and LDLs in a person's blood. LDL is commonly considered "bad" because it deposits cholesterol in arterial walls, which leads to atherosclerosis. HDL is considered "good" because it transports cholesterol to the liver for removal from the body by excretion in the bile. A high HDL/LDL ratio in the bloodstream is related to a lower risk of heart disease. Aerobic exercise is one way to elevate HDL and decrease LDL levels.

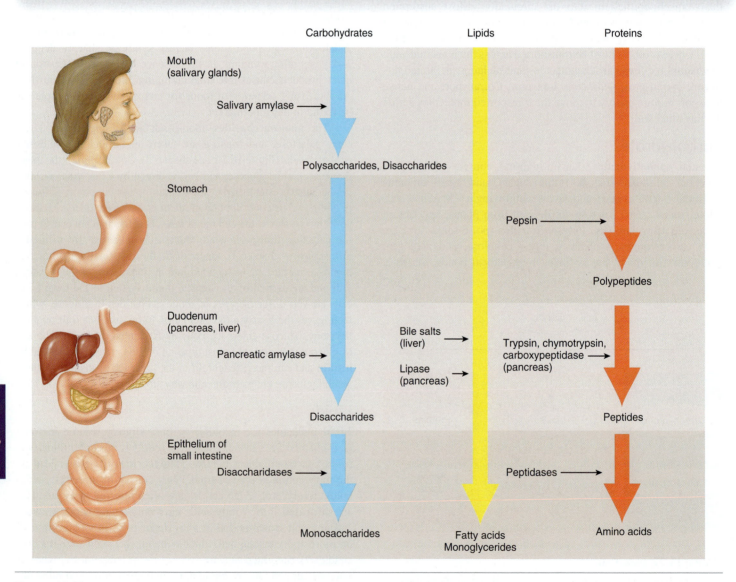

Figure 16.23 Digestion of Carbohydrates, Lipids, and Proteins

The enzymes involved in digesting carbohydrates, lipids, and proteins are depicted in relation to the region of the digestive tract where each functions.

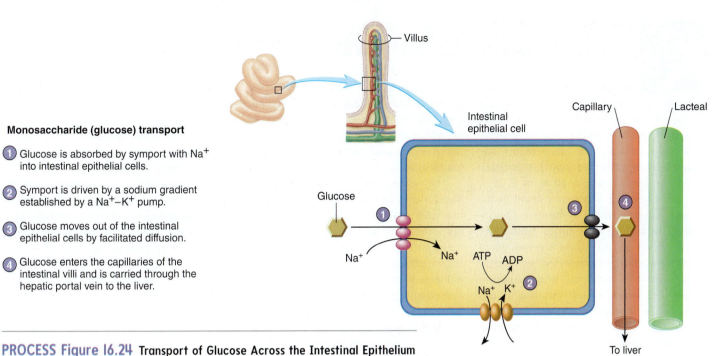

Monosaccharide (glucose) transport

1 Glucose is absorbed by symport with Na⁺ into intestinal epithelial cells.

2 Symport is driven by a sodium gradient established by a Na⁺–K⁺ pump.

3 Glucose moves out of the intestinal epithelial cells by facilitated diffusion.

4 Glucose enters the capillaries of the intestinal villi and is carried through the hepatic portal vein to the liver.

PROCESS Figure 16.24 Transport of Glucose Across the Intestinal Epithelium

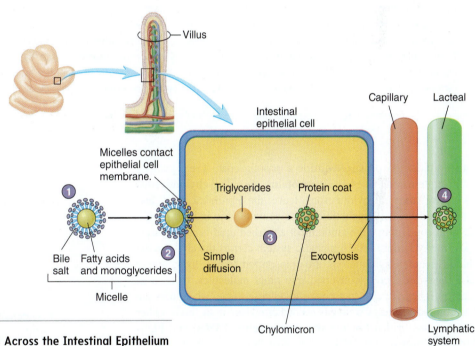

Lipid transport

1 Bile salts surround fatty acids and monoglycerides to form micelles.

2 Micelles attach to the plasma membranes of intestinal epithelial cells, and the fatty acids and monoglycerides pass by simple diffusion into the intestinal epithelial cells.

3 Within the intestinal epithelial cell, the fatty acids and monoglycerides are converted to triglycerides; proteins coat the triglycerides to form chylomicrons, which move out of the intestinal epithelial cells by exocytosis.

4 The chylomicrons enter the lacteals of the intestinal villi and are carried through the lymphatic system to the general circulation.

PROCESS Figure 16.25 Transport of Lipids Across the Intestinal Epithelium

The first step in lipid digestion is **emulsification** (ē-mŭl′si-fi-kā′shŭn), by which large lipid droplets are transformed into much smaller droplets. Emulsification is accomplished by bile salts secreted by the liver. The enzymes that digest lipids are soluble in water and can digest the lipids only by acting at the surface of the droplets. The emulsification process increases the surface area of the lipid droplets exposed to the digestive enzymes by increasing the number of lipid droplets and decreasing the size of each droplet.

Lipase, secreted by the pancreas, digests lipid molecules (see figure 16.23). The primary products of this digestive process are fatty acids and monoglycerides.

In the intestine, bile salts aggregate around small droplets of digested lipids to form **micelles** (mi-selz′, mī-selz′; small morsels) (figure 16.25). The hydrophobic (water-fearing) ends of the bile salts are directed toward the lipid particles, and the hydrophilic (water-loving) ends are directed outward, toward the water environment. When a micelle comes in contact with the epithelial cells of the small intestine, the lipids, fatty acids, and monoglyceride molecules pass, by simple diffusion, from the micelles through the cell membranes of the epithelial cells.

Once inside the intestinal epithelial cells, the fatty acids and monoglycerides are recombined to form triglycerides. These, and

MICROBES IN YOUR BODY Fecal Transplants

Would you be shocked if your doctor said that the one thing that could save your life was feces? Unfortunately, we are in the midst of a global, hospital-acquired diarrhea epidemic. The cause of this epidemic is a bacterium called *Clostridium difficile* (commonly referred to as *C. diff*), a pathogen that is normally found in the colon, but is controlled by the normal microbiota. One of the most critical functions of the normal gut microbiota is prevention of infections through competition with pathogens. As a consequence, when a patient takes antibiotics, *C. diff* can flourish and cause life-threatening diarrhea. Treatment of *C. diff* infections with specific antibiotics will often stop the diarrhea initially. However, *C. diff* are spore-forming bacteria. Spores are very stable structures that allow bacteria to withstand harsh conditions until favorable conditions return and the bacteria can regrow. Thus, antibiotics kill only the *C. diff* cells, not the spores. Hence, it is very common for patients to suffer multiple recurrences of diarrhea for months, which can

lead to death in some patients. Additionally, a more virulent, resistant strain of *C. diff* has emerged. This strain is resistant to certain antibiotics and makes a greater number of spores and more of the toxins responsible for the diarrhea.

So, where do feces come into play? Because antibiotic treatments are not effective (65% infection recurrence), physicians are considering an old treatment: fecal transplants. The first documented case of transplanting feces from a healthy donor into a diseased recipient was in 1958. Fecal transplantation has since been used successfully in veterinary medicine for decades. However, due to the unappealing nature of this treatment, it has only recently been considered an option in humans. Now more commonly known as intestinal fecal transplantation (IFT), it has been shown to effectively treat diarrhea in over 90% of *C. diff* infections. The idea is that a healthy donor—usually a close family household member such as a spouse or significant partner—donates their feces. The feces are mixed with physiological

saline, filtered, and then introduced into the recipient's gastrointestinal (GI) tract by one of two ways: the upper GI tract route or the lower GI tract route. The upper GI tract route uses either a gastroscope or nasogastric tube to transfer the material to the recipient's intestine. Of the two, this one is easier and costs less. However, there is the possibility the donor microbiota may not reach the end of the colon or that the patient may vomit the fecal material. The lower GI tract route uses a colonoscope or enema and is the preferred approach, but does run the risk of perforating the colon. Thus, as of yet, there is no standardized method for transferring the donor feces. However, research is showing that more and more patients may overcome their initial reluctance when presented with a predictable success rate and greater reliability than other protocols. In addition, the recent "RePOOPulating" study shows promise that doctors may soon be able to treat *C. diff* infections simply by prescribing a pill that contains normal microbiota.

---

other lipids, are packaged inside a protein coat. The packaged lipid-protein complexes, or **lipoproteins,** are called **chylomicrons** (kī-lō-mi′kronz). Chylomicrons leave the epithelial cells and enter the lacteals, lymphatic capillaries within the intestinal villi. Lymph containing large amounts of absorbed lipid is called **chyle** (kīl; milky lymph). The lymphatic system carries the chyle to the bloodstream. Chylomicrons are transported to the liver, where the lipids are stored, converted into other molecules, or used as energy. They are also transported to adipose tissue, where they are stored until an energy source is needed elsewhere in the body. Other lipoproteins, called **low-density lipoproteins (LDLs)** and **high-density lipoproteins (HDLs),** transport cholesterol and fats in the blood (figure 16.26).

Proteins

Proteins are chains of **amino acids.** They are found in most of the plant and animal products we eat. **Pepsin** is an enzyme secreted by the stomach that breaks down proteins, producing shorter amino acid chains called **polypeptides** (see figure 16.23). Only about 10–20% of the total ingested protein is digested by pepsin. After the remaining proteins and polypeptide chains leave the stomach and enter the small intestine, the enzymes **trypsin, chymotrypsin,** and **carboxypeptidase,** produced by the pancreas in their inactive forms and activated in the intestine, continue the digestive process. These enzymes produce small peptides, which are further broken down

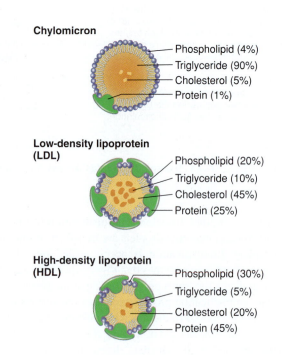

Chylomicron
- Phospholipid (4%)
- Triglyceride (90%)
- Cholesterol (5%)
- Protein (1%)

Low-density lipoprotein (LDL)
- Phospholipid (20%)
- Triglyceride (10%)
- Cholesterol (45%)
- Protein (25%)

High-density lipoprotein (HDL)
- Phospholipid (30%)
- Triglyceride (5%)
- Cholesterol (20%)
- Protein (45%)

Figure 16.26 Lipoproteins

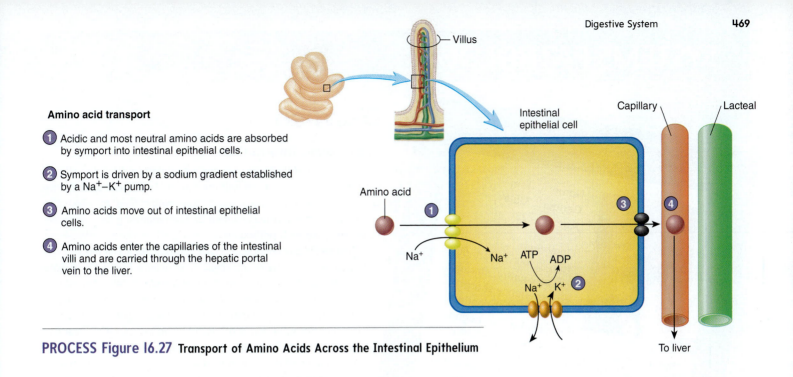

Amino acid transport

1. Acidic and most neutral amino acids are absorbed by symport into intestinal epithelial cells.

2. Symport is driven by a sodium gradient established by a Na^+–K^+ pump.

3. Amino acids move out of intestinal epithelial cells.

4. Amino acids enter the capillaries of the intestinal villi and are carried through the hepatic portal vein to the liver.

PROCESS Figure 16.27 Transport of Amino Acids Across the Intestinal Epithelium

into tripeptides (three amino acids), dipeptides (two amino acids), or single amino acids by digestive enzymes called **peptidases.** The peptidases are bound to the microvilli of the small intestine.

Absorption of tripeptides, dipeptides, or individual amino acids occurs through the intestinal epithelial cells by various cotransport mechanisms. Many, but not all, amino acids are taken up by cotransport with Na^+ similar to glucose cotransport (figure 16.27). Within the intestinal epithelial cells, tripeptides and dipeptides are broken down into amino acids. The amino acids then enter blood capillaries in the villi and are carried by the hepatic portal vein to the liver. The amino acids may be modified in the liver, or they may be released into the bloodstream and distributed throughout the body.

Amino acids are actively transported into the various cells of the body. This transport is stimulated by growth hormone and insulin. Most amino acids are used as building blocks to form new proteins, but some may be metabolized, with a portion of the released energy used to produce ATP. The body cannot store excess amino acids. Instead, they are partially broken down and used to synthesize glycogen or lipids, which can be stored. The body can store only small amounts of glycogen, so most of the excess amino acids are converted to lipids.

Water and Minerals

Approximately 9 L of water enter the digestive tract each day (figure 16.28). We ingest about 2 L in food and drink, and the remaining 7 L are from digestive secretions. Approximately 92% of that water is absorbed in the small intestine, about 7% is absorbed in the large intestine, and about 1% leaves the body in the feces. Water can move in either direction by osmosis across the wall of the digestive tract. The direction of its movement is determined by osmotic gradients across the epithelium. When the chyme is diluted, water moves out of the intestine into the blood. If the chyme is concentrated and contains little water, water moves into the lumen of the small intestine.

Sodium, potassium, calcium, magnesium, and phosphate ions are actively transported from the small intestine. Vitamin D is

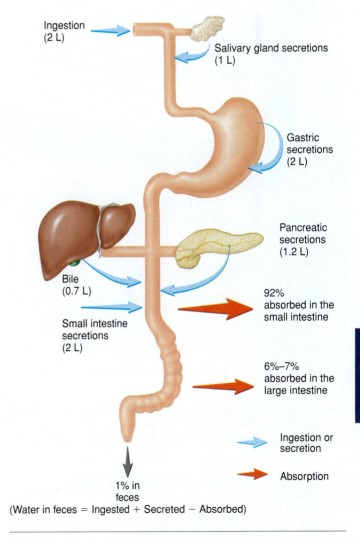

Figure 16.28 Fluid Volumes in the Digestive Tract

Name: Tom
Gender: Male
Age: 38

Comments
Tom, a tourist from out of the country, started to experience sharp pains in his abdominal region. He also began to feel hot and sweaty and felt an extreme urge to defecate. Tom anxiously inquired about the nearest facility. Once the immediate crisis was taken care of, Tom and his wife went back to their hotel room, where they remained while Tom recovered. During the next two days, his stools were frequent and watery. He also vomited a couple of times. Tom was encouraged to rest and drink plenty of fluids. He was feeling much better, although a little weak, in a couple of days.

Figure 16A
Many tourists develop diarrhea.

Diarrhea

Background Information

Diarrhea is one of the most common complaints in clinical medicine. Diarrhea affects more than half the tourists in developing countries (figure 16A), where it may result from eating food to which the digestive tract is not accustomed or from ingesting food or water contaminated with pathogens.

Diarrhea is any change in bowel habits involving increased stool frequency or volume or increased stool fluidity. It is not a disease in itself, but it can be a symptom of a wide variety of disorders. Diarrhea that lasts less than 2–3 weeks is acute diarrhea; diarrhea lasting longer is considered chronic. Acute diarrhea is usually self-limiting, but some forms of diarrhea can be fatal if not treated. Diarrhea results from either a decrease in fluid absorption in the intestine or an increase in fluid secretion. It can also be caused by increased bowel motor activity that moves chyme rapidly through the small intestine, so that more water enters the colon. Normally, about 600 mL of fluid enter the colon each day, and all but 150 mL are reabsorbed. The loss of more than 200 mL of fluid per day in the stool is considered abnormal.

Mucus secretion by the colon increases dramatically in response to diarrhea. This mucus contains large quantities of bicarbonate ions, which come from the dissociation of carbonic acid into bicarbonate ions and hydrogen ions within the blood supply to the colon. The bicarbonate ions enter the mucus secreted by the colon, whereas the hydrogen ions remain in the circulation; as a result, the blood pH decreases. Thus, a condition called metabolic acidosis can develop (see chapter 18).

Diarrhea is usually caused by bacteria, viruses, amoebic parasites, or chemical toxins. Symptoms can begin from as little as 1–2 hours after bacterial toxins are ingested to as long as 24 hours or more for some strains of bacteria. Nearly any bacterial species is capable of causing diarrhea. Some types of bacterial diarrhea are associated with severe vomiting, whereas others are not. Some bacterial toxins also induce fever. Identifying the causal organism usually requires laboratory analysis of the food or stool, but in cases of acute diarrhea, the infectious agent is seldom identified.

required for the transport of Ca^{2+}. Negatively charged Cl^- moves passively through the wall of the duodenum and jejunum with the positively charged Na^+, but Cl^- is actively transported from the ileum.

16.9 EFFECTS OF AGING ON THE DIGESTIVE SYSTEM

Learning Outcome *After reading this section, you should be able to*

A. Describe the effects of aging on the digestive system.

As a person ages, the connective tissue layers of the digestive tract—the submucosa and serosa—tend to thin. The blood supply to the digestive tract decreases. There is also a decrease in the number of smooth muscle cells in the muscularis, resulting in decreased motility in the digestive tract. In addition, goblet cells within the mucosa secrete less mucus. Glands along the digestive tract, such as the gastric glands, the liver, and the pancreas, also tend to secrete less with age. However, these changes by themselves don't appreciably decrease the function of the digestive system.

Digestive

Diarrhea

MUSCULAR
Muscular weakness may result due to ion loss, metabolic acidosis, fever, and general malaise. The stimulus to defecate may become so strong that it overcomes the voluntary control mechanisms.

INTEGUMENTARY
Pallor is due to vasoconstriction of blood vessels in the skin, resulting from a decrease in blood volume. Pallor and sweating increase in response to abdominal pain and anxiety.

NERVOUS
Local reflexes in the colon respond to increased colon fluid volume by stimulating mass movements and the defecation reflex. Abdominal pain, much of which is felt as referred pain, can occur as the result of inflammation and distention of the colon. Nervous system function decreases due to ion loss. Reduced blood volume stimulates a sensation of thirst in the CNS.

URINARY
In the event of fluid loss from the intestine, the kidney is activated to compensate for metabolic acidosis by increasing hydrogen ion secretion and bicarbonate ion reabsorption.

ENDOCRINE
A decrease in extracellular fluid volume, due to the loss of fluid in the feces, stimulates the release of hormones (antidiuretic hormone from the posterior pituitary and aldosterone from the adrenal cortex) that increase water retention and sodium reabsorption in the kidney. In addition, decreased extracellular fluid volume and anxiety result in increased release of epinephrine and norepinephrine from the adrenal medulla.

RESPIRATORY
Increased HCO_3^- secretion and H^+ absorption reduce blood pH. As the result of reduced blood pH, the rate of respiration increases to eliminate carbon dioxide, which helps eliminate excess H^+.

Diarrhea

Symptoms
- Increased stool frequency
- Increased stool volume
- Increased stool fluidity

Treatments
- Replacing lost fluid and ions
- Eating a bland diet without dairy

LYMPHATIC AND IMMUNE
White blood cells migrate to the colon in response to infection and inflammation. In the case of bacterial diarrhea, the immune response is initiated to begin production of antibodies against bacteria and bacterial toxins.

CARDIOVASCULAR
Movement of extracellular fluid into the colon results in decreased blood volume. The reduced blood volume activates the baroreceptor reflex, which maintains blood volume and blood pressure.

Treatment of diarrhea involves replacing lost fluids and ions. The diet should be limited to clear fluids during at least the first day or so. Medicines that may help combat diarrhea include bismuth subsalicylate (sŭb-să-lis′i-lāt), which increases mucus and HCO_3^- secretion and decreases pepsin activity, and loperamide (lō-per′ă-mīd), which slows intestinal motility. Patients should avoid milk and milk products. Breads, rice, and baked fish or chicken can be added to the diet as the person's condition improves. A normal diet can be resumed after 2–3 days.

Predict 6

Predict the effects of prolonged diarrhea on the cardiovascular system.

Through the years, the digestive tract, like the skin and lungs, is directly exposed to materials from the outside environment. Some of those substances can cause mechanical damage to the digestive tract, and others may be toxic to the tissues. Because the connective tissue of the digestive tract becomes thin with age and because the protective mucous covering is reduced, the digestive tract of elderly people becomes less and less protected from these outside influences. In addition, the mucosa of elderly people tends to heal more slowly following injury. The liver's ability to detoxify certain chemicals tends to decline; the ability of the hepatic phagocytic cells to remove particulate contaminants decreases; and the liver's ability to store glycogen decreases.

This overall decline in the defenses of the digestive tract leaves elderly people more susceptible to infections and to the effects of toxic agents. Elderly people are more likely to develop ulcerations and cancers of the digestive tract. Colorectal cancers, for example, are the second leading cause of cancer deaths in the United States, with an estimated 135,000 new cases and 57,000 deaths each year.

Gastroesophageal reflux disorder (GERD) increases with advancing age. It is probably the main reason that elderly people take antacids and inhibitors of hydrochloric acid secretion. Disorders that are not necessarily age-induced, such as hiatal hernia and irregular or inadequate esophageal motility, may be worsened by the effects of aging, because of general decreased motility in the digestive tract.

The enamel on the surface of elderly people's teeth becomes thinner with age and may expose the underlying dentin. In addition, the gingiva covering the tooth root recedes, exposing additional dentin. Exposed dentin may become painful and change the person's eating habits. Many elderly people also lose teeth, which can have a marked effect on eating habits unless artificial teeth are provided. The muscles of mastication tend to become weaker; as a result, older people tend to chew their food less before swallowing.

DISEASES AND DISORDERS: Digestive System

CONDITION	DESCRIPTION
Stomach	
Vomiting	Contraction of the diaphragm and abdominal muscles and relaxation of the esophageal sphincters to forcefully expel gastric contents; vomiting reflex is initiated by irritation of the stomach or small intestine
Peptic ulcer	Lesions in the lining of the stomach or duodenum, usually due to infection by the bacterium *Helicobacter pylori*; stress, diet, smoking, or alcohol may be predisposing factors; antibiotic therapy is the accepted treatment
Liver	
Cirrhosis (sir-ō′sis)	Characterized by damage to and death of hepatic cells and replacement by connective tissue; results in loss of normal liver function and interference with blood flow through the liver; a common consequence of alcoholism
Hepatitis (hep-ă-tī′tis)	Inflammation of the liver that causes liver cell death and replacement by scar tissue; if not corrected, results in loss of liver function and eventually death; symptoms include nausea, abdominal pain, fever, chills, malaise, and jaundice; caused by any of seven distinct viruses
Hepatitis A	Infectious hepatitis; usually transmitted by poor sanitation practices or from mollusks living in contaminated waters
Hepatitis B	Serum hepatitis; usually transmitted through blood or other body fluids through either sexual contact or contaminated hypodermic needles
Hepatitis C	Often a chronic disease leading to cirrhosis and possibly cancer of the liver
Gallstones	Most often due to excess cholesterol in the bile; gallstones can occasionally enter the cystic duct, where they block the release of bile and/or pancreatic enzymes, which interferes with digestion
Intestine	
Inflammatory bowel disease (IBD)	Localized inflammatory degeneration that may occur anywhere along the digestive tract but most commonly involves the distal ileum and proximal colon; the intestinal wall often becomes thickened, constricting the lumen, with ulcers and fissures in the damaged areas; symptoms include diarrhea, abdominal pain, fever, fatigue, and weight loss; cause is unknown; treatments involve anti-inflammatory drugs, avoidance of foods that produce symptoms, and surgery in some cases; also called Crohn's disease or ulcerative colitis
Irritable bowel syndrome (IBS)	Disorder of unknown cause marked by alternating bouts of constipation and diarrhea; may be linked to stress or depression; high familial incidence
Gluten enteropathy (celiac disease)	Malabsorption in the small intestine due to the effects of gluten, a protein in certain grains, especially wheat; the reaction can destroy newly formed epithelial cells, causing the intestinal villi to become blunted and decreasing the intestinal surface, which reduces absorption of nutrients
Constipation (kon-sti-pā′ shŭn)	Slow movement of feces through the large intestine, causing the feces to become dry and hard because of increased fluid absorption while being retained; often results from inhibiting normal defecation reflexes; spasms of the sigmoid colon resulting from irritation can also result in slow feces movement and constipation; high-fiber diet can be preventative
	INFECTIONS OF THE DIGESTIVE TRACT
Food poisoning	Caused by ingesting bacteria or toxins, such as *Staphylococcus aureus*, *Salmonella*, or *Escherichia coli*; symptoms include nausea, abdominal pain, vomiting, and diarrhea; in severe cases, death can occur
Typhoid (tī′foyd) fever	Caused by a virulent strain of the bacterium *Salmonella typhi*, which can cross the intestinal wall and invade other tissues; symptoms include severe fever, headaches, and diarrhea; usually transmitted through poor sanitation practices; leading cause of death in many underdeveloped countries
Cholera (kol′er-a)	Caused by a bacterium, *Vibrio cholerae*, in contaminated water; bacteria produce a toxin that stimulates the secretion of chloride, bicarbonate, and water into the intestine, resulting in severe diarrhea; the loss of as much as 12–20 L of fluid per day causes shock, circulatory collapse, and even death; still a major health problem in parts of Asia
Giardiasis (jē-ar-dī′a-sis)	Caused by a protozoan, *Giardia lamblia*, that invades the intestine; symptoms include nausea, abdominal cramps, weakness, weight loss, and malaise; bacteria are transmitted in the feces of humans and other animals, often by drinking from contaminated wilderness streams
Intestinal parasites	Common under conditions of poor sanitation; parasites include tapeworms, pinworms, hookworms, and roundworms
Diarrhea (dī-ă-rē′ă)	Intestinal mucosa secretes large amounts of water and ions due to irritation, inflammation, or infection; diarrhea moves feces out of the intestine more rapidly, which speeds recovery
Dysentery (dis′ en-tār-ē)	Severe form of diarrhea with blood or mucus in the feces; can be caused by bacteria, protozoa, or amoebae

ANSWER TO LEARN TO PREDICT

The first thing we learn is that Demondre is African-American and becomes uncomfortable after eating food with lactose. Next, we discover that digestive enzymes are bound to the epithelial cells of the small intestine and that these enzymes are specific for certain nutrients, including lactose. Demondre's doctor tells him to avoid lactose because it gives him the symptoms that make him sick. We can conclude that Demondre has a problem with lactase, the enzyme that digests lactose. This isn't too surprising, since 80–90% of African-Americans synthesize lower quantities of lactase after childhood and cannot fully digest lactose. This condition is called **lactose intolerance.**

Answers to the rest of this chapter's Predict questions are in Appendix E.

SUMMARY

16.1 Functions of the Digestive System   (p. 442)

The functions of the digestive system are to take in food, break down the food, absorb the digested molecules, provide nutrients to the body, and eliminate wastes.

16.2 Anatomy and Histology of the Digestive System   (p. 443)

The digestive tract is composed of four tunics: the mucosa, the submucosa, the muscularis, and a serosa or an adventitia.

Peritoneum

1. The peritoneum is a serous membrane that lines the abdominal cavity and covers the organs.
2. Mesenteries are double layers of peritoneum that extend from the body wall to many of the abdominal organs.
3. Retroperitoneal organs are located behind the parietal peritoneum.

16.3 Oral Cavity, Pharynx, and Esophagus   (p. 444)

Anatomy of the Oral Cavity

1. The lips and cheeks are involved in mastication and speech.
2. The tongue is involved in speech, taste, mastication, and swallowing.
3. There are 32 permanent teeth, including incisors, canines, premolars, and molars. Each tooth consists of a crown, a neck, and a root.
4. The roof of the oral cavity is divided into the hard and soft palates.
5. Salivary glands produce serous and mucous secretions. The three major pairs of large salivary glands are the parotid, submandibular, and sublingual glands.

Saliva

Saliva helps protect the mouth from oral bacteria, starts starch digestion, and provides lubrication.

Mastication

Mastication is accomplished by the teeth, which cut, tear, and crush the food.

Pharynx

The pharynx consists of the nasopharynx, oropharynx, and laryngopharynx.

Esophagus

The esophagus connects the pharynx to the stomach. The upper and lower esophageal sphincters regulate movement.

Swallowing

1. During the voluntary phase of swallowing, a bolus of food is moved by the tongue from the oral cavity to the pharynx.
2. During the pharyngeal phase of swallowing, the soft palate closes the nasopharynx, and the epiglottis closes the opening into the larynx. Pharyngeal muscles elevate the pharynx and larynx and then move the bolus to the esophagus.
3. During the esophageal phase of swallowing, a wave of constriction (peristalsis) moves the food down the esophagus to the stomach.

16.4 Stomach   (p. 451)

Anatomy of the Stomach

1. The stomach connects to the esophagus at the gastroesophageal opening and to the duodenum at the pyloric opening.
2. The wall of the stomach consists of three muscle layers: longitudinal, circular, and oblique.
3. Gastric glands contain mucous neck cells, parietal cells, endocrine cells, and chief cells.

Secretions of the Stomach

1. Mucus protects the stomach lining.
2. Hydrochloric acid kills microorganisms and activates pepsin.
3. Pepsin starts protein digestion.
4. Intrinsic factor aids in vitamin B_{12} absorption.

Regulation of Stomach Secretions

1. Parasympathetic stimulation, gastrin, and histamine increase stomach secretions.
2. During the cephalic phase, the stomach secretions are initiated by the sight, smell, taste, or thought of food.
3. During the gastric phase, partially digested proteins and distention of the stomach promote secretion.
4. During the intestinal phase, acidic chyme in the duodenum stimulates neuronal reflexes and the secretion of hormones that inhibit gastric secretions by negative-feedback loops. Secretin and cholecystokinin inhibit gastric secretion.

Movement in the Stomach

1. Mixing waves mix the stomach contents with the stomach secretions to form chyme.
2. Peristaltic waves move the chyme into the duodenum. Increased motility increases emptying.
3. Distention of the stomach increases gastric motility. Neural and hormonal feedback loops from the duodenum inhibit gastric motility. Cholecystokinin is a major inhibitor of gastric motility.

16.5 Small Intestine   (p. 455)

Anatomy of the Small Intestine

1. The small intestine is divided into the duodenum, jejunum, and ileum.

2. Circular folds, villi, and microvilli greatly increase the surface area of the intestinal lining.
3. Goblet cells and duodenal glands produce mucus.

Secretions of the Small Intestine

1. Mucus protects against digestive enzymes and stomach acids.
2. Chemical or tactile irritation, vagal stimulation, and secretin stimulate intestinal secretion.

Movement in the Small Intestine

1. Peristaltic contractions occur over the length of the intestine and propel chyme through the intestine.
2. Segmental contractions occur over short distances and mix the intestinal contents.

Absorption in the Small Intestine

Most absorption occurs in the duodenum and jejunum.

16.6 Liver and Pancreas   (p. 458)

Anatomy of the Liver

1. The liver consists of four lobes. It receives blood from the hepatic artery and the hepatic portal vein.
2. Branches of the hepatic artery and hepatic portal vein empty into hepatic sinusoids, which empty into a central vein in the center of each lobe. The central veins empty into hepatic veins, which exit the liver.
3. The liver is divided into lobules with portal triads at the corners. Portal triads contain branches of the hepatic portal vein, hepatic artery, and hepatic duct.
4. Hepatic cords, formed by hepatocytes, form the substance of each lobule. A bile canaliculus, between the cells of each cord, joins the hepatic duct system.
5. Bile leaves the liver through the hepatic duct system. The right and left hepatic ducts join to form the common hepatic duct. The gallbladder stores bile. The cystic duct joins the common hepatic duct to form the common bile duct. The common bile duct joins the pancreatic duct and empties into the duodenum.

Functions of the Liver

1. The liver produces bile, which contains bile salts that emulsify fats, and excretory products.
2. The liver stores and processes nutrients, produces new molecules, and detoxifies molecules.

Anatomy of the Pancreas

The pancreas is both an endocrine and an exocrine gland. Its endocrine function is to control blood nutrient levels. Its exocrine function is to produce bicarbonate ions and digestive enzymes.

Functions of the Pancreas

1. The pancreas produces HCO_3^- and digestive enzymes.
2. Acidic chyme stimulates the release of a watery bicarbonate solution that neutralizes acidic chyme. Fatty acids and amino acids in the duodenum stimulate the release of pancreatic enzymes.

16.7 Large Intestine   (p. 463)

Anatomy of the Large Intestine

1. The cecum forms a blind sac at the junction of the small and large intestines. The appendix is a blind sac off the cecum.
2. The colon consists of ascending, transverse, descending, and sigmoid portions.
3. The large intestine contains mucus-producing crypts.
4. The rectum is a straight tube that ends at the anal canal.
5. The anal canal is surrounded by an internal anal sphincter (smooth muscle) and an external anal sphincter (skeletal muscle).

Functions of the Large Intestine

1. The functions of the large intestine are feces production and water absorption.
2. Mass movements occur three or four times a day.
3. Defecation is the elimination of feces. Reflex activity moves feces through the internal anal sphincter. Voluntary activity regulates movement through the external anal sphincter.

16.8 Digestion, Absorption, and Transport
(p. 465)

Digestion is the chemical breakdown of organic molecules into their component parts. After the molecules are digested, some diffuse through the intestinal wall; others must be transported across the intestinal wall.

Carbohydrates

1. Polysaccharides are split into disaccharides by salivary and pancreatic amylases.
2. Disaccharides are broken down to monosaccharides by disaccharidases on the surface of the intestinal epithelium.
3. The major monosaccharide, glucose, is absorbed by cotransport with Na^+ into intestinal epithelial cells. The transport is driven by the Na^+ gradient generated by the sodium-potassium pump.
4. Glucose is carried by the hepatic portal vein to the liver and enters most cells in the body by facilitated diffusion. Insulin increases the rate of glucose transport into most cells.

Lipids

1. Bile salts emulsify lipids.
2. Lipase breaks down lipids. The breakdown products aggregate with bile salts to form micelles.
3. Micelles come in contact with the intestinal epithelium, and their contents diffuse into the cells, where they are packaged and released into the lacteals.
4. Lipids are stored in adipose tissue and in the liver, both of which release the lipids into the blood when energy sources are needed elsewhere in the body.

Proteins

1. Proteins are split into small polypeptides by enzymes secreted by the stomach and pancreas, and peptidases on the surface of intestinal epithelial cells.
2. Tripeptides, dipeptides, and amino acids are absorbed into intestinal epithelial cells.
3. Amino acids are actively transported into cells under the influence of growth hormone and insulin.
4. Amino acids are used to build new proteins or to serve as a source of energy.

Water and Minerals

Water can move across the intestinal wall in either direction, depending on osmotic conditions. Approximately 99% of the water entering the intestine is absorbed. Most minerals are actively transported across the wall of the small intestine.

16.9 Effects of Aging on the Digestive System
(p. 470)

1. With advancing age, the layers of the digestive tract thin, and the blood supply decreases.
2. Mucus secretion and motility also decrease in the digestive tract.
3. The defenses of the digestive tract decline, leaving it more sensitive to infection and the effects of toxic agents.
4. Tooth enamel becomes thinner, and the gingiva recede, exposing dentin, which may become painful and affect eating habits.

Digestive

REVIEW AND COMPREHENSION

1. What are the functions of the digestive system?
2. What are the major layers, or tunics, of the digestive tract?
3. What are the peritoneum, mesenteries, and retroperitoneal organs?
4. List the functions of the lips, cheeks, and tongue.
5. Distinguish between the deciduous teeth and the permanent teeth. Name the different kinds of teeth.
6. Describe the three parts of a tooth. What are dentin, enamel, and pulp?
7. What is the function of the palate? What are the hard and soft palates?
8. Name and give the locations of the three pairs of salivary glands.
9. What are the functions of saliva?
10. Where is the esophagus located?
11. Describe the three phases of swallowing.
12. Describe the parts of the stomach. How are the muscles in the stomach different from those in the esophagus?
13. What are gastric pits and gastric glands? Name the secretions they produce.
14. List the stomach secretions, and give their functions.
15. Describe the three phases of the regulation of stomach secretion.
16. What are the two kinds of stomach movements? What do they accomplish?
17. Name and describe the three parts of the small intestine.
18. What are circular folds, villi, and microvilli in the small intestine? What are their functions?
19. List the secretions of the small intestine, and give their functions.
20. Describe the kinds of movements in the small intestine, and explain what they accomplish.
21. Describe the anatomy and location of the liver and pancreas, including their duct systems.
22. What are the functions of the liver?
23. Name the exocrine secretions of the pancreas. What are their functions?
24. Describe the parts of the large intestine.
25. How is chyme converted to feces?
26. Describe the defecation reflex.
27. Describe carbohydrate digestion, absorption, and transport.
28. Describe the role of bile salts in lipid digestion and absorption.
29. Describe protein digestion and amino acid absorption. What enzymes are responsible for the digestion?
30. Describe the movement of water into and out of the digestive tract.
31. Describe the effects of aging on the digestive system.

CRITICAL THINKING

1. While anesthetized, patients sometimes vomit. Given that the anesthetic eliminates the swallowing reflex, explain why it is dangerous for an anesthetized patient to vomit.
2. Achlorhydria (ā-klōr-hī′drē-ă) is a condition in which the stomach stops producing hydrochloric acid and other secretions. What effect does achlorhydria have on the digestive process?
3. Victoria Worrystudent developed a duodenal ulcer during final examination week. Describe the possible reasons. Explain what habits could have contributed to the ulcer, and recommend a reasonable remedy.
4. Gallstones sometimes obstruct the common bile duct. What would be the consequences of such a blockage?
5. Many people have a bowel movement shortly after a meal, especially breakfast. Why does this occur?

Answers in Appendix D

17 Nutrition, Metabolism, and Body Temperature Regulation

17.1 NUTRITION

A. Define nutrition, essential nutrient, and kilocalorie.
B. For carbohydrates, lipids, and proteins, describe their dietary sources, their uses in the body, and the daily recommended amounts of each in the diet.
C. List the common vitamins and minerals, and give a function for each.

When choosing from a menu or selecting food to prepare, we are often more concerned with the food's taste than with its nutritional value. However, knowing about nutrition is important because the food we eat provides us with the energy and the building blocks necessary to synthesize new molecules. What happens if we don't obtain enough vitamins, or if we eat too much sugar and fat? Health claims about foods and food supplements bombard us every day. Which ones are ridiculous, and which ones have merit? A basic understanding of nutrition can help us answer these and other questions so that we can develop a healthful diet. It can also allow us to know which questions currently do not have good answers.

Nutrition (noo-trish′ŭn; to nourish) is the process by which food is taken into and used by the body; it includes digestion, absorption, transport, and metabolism. Nutrition is also the study of food and drink requirements for normal body function.

Module 12 Digestive System

Nutrients

Nutrients are the chemicals taken into the body that provide energy and building blocks for new molecules. Some substances in food are not nutrients but provide bulk (fiber) in the diet. Nutrients can be divided into six major classes: carbohydrates, lipids, proteins, vitamins, minerals, and water. Carbohydrates, lipids, and proteins are the major organic nutrients and are broken down by enzymes into their individual subunits during digestion. Subsequently, many of these subunits are broken down further to supply energy, whereas other subunits are used as building blocks for making new carbohydrates, lipids, and proteins. Vitamins, minerals, and water are taken into the body without being broken down. They are essential participants in the chemical reactions necessary to maintain life. The body requires some nutrients in fairly substantial quantities and others, called **trace elements,** in only minute amounts.

Essential nutrients are nutrients that must be ingested because the body cannot manufacture them—or it cannot manufacture them in adequate amounts. The essential nutrients include certain amino acids, certain fatty acids, most vitamins, minerals, water, and some carbohydrates. The term *essential* does not mean that the body requires only the essential nutrients. Other nutrients are necessary, but if they are not ingested, they can be synthesized from the essential nutrients. Most of this synthesis takes place in the liver, which has a remarkable ability to transform and manufacture molecules. A balanced diet consists of enough nutrients in the correct proportions to support normal body functions.

Every 5 years, the U.S. Department of Health and Human Services (HHS) and the U.S. Department of Agriculture (USDA) jointly make recommendations on what Americans should eat to be healthy. The latest recommendations, the Dietary Guidelines for Americans, 2010, were published in January 2011. In light of the increasing problem of obesity in the United States, the latest recommendations focus on two concepts: (1) balancing calorie intake to obtain and maintain a healthy weight, and (2) increasing consumption of healthy, nutrient-rich foods. In June 2011, the USDA also introduced MyPlate, a new food icon to replace the former food guide icon, called MyPyramid. MyPlate (figure 17.1) is a simple visual reminder of how to build a healthy meal. The MyPlate icon shows a plate and glass with portions representing foods from the fruits, vegetables, grains, proteins, and dairy food groups. To emphasize the importance of making healthy food choices, half the plate is fruits and vegetables. In addition to the MyPlate icon, the USDA also launched ChooseMyPlate.gov, a website that includes information on how to make healthy dietary choices.

The importance of a healthy diet has been proposed as a key to a healthy life. Two studies completed in 2000 compared the eating habits of 51,529 men and 67,272 women to the government's Healthy Eating Index, a measure of how well a diet conforms to recommended dietary guidelines. Results from these studies demonstrate that eating a healthy diet does provide certain health benefits. People who ate the best, according to the index, were compared to those who ate the worst. Men who ate best had a 28% reduction in heart disease and an 11% decrease in chronic diseases compared to men who ate worst. Women who ate best had a 14% reduction in heart disease but no significant decrease in chronic diseases compared to women who ate worst. There was no significant difference in cancer rates between the men and women who ate best compared to those who ate worst.

Kilocalories

The energy the body uses is stored within the chemical bonds of certain nutrients. A **calorie** (kal′ō-rē; heat) **(cal)** is the amount of energy (heat) necessary to raise the temperature of 1 gram (g) of water 1°C. A **kilocalorie** (kil′ō-kal-ō-rē) **(kcal)** is 1000 cal and is used to express the larger amounts of energy supplied by foods and released through metabolism. For example, one slice of white bread contains about 75 kcal, 1 cup of whole milk contains 150 kcal, a banana contains 100 kcal, a hot dog contains 170 kcal (not counting the bun and dressings), a McDonald's Big Mac has 540 kcal, and a soft drink adds another 145 kcal. For each gram of carbohydrate or protein metabolized by the body, about 4 kcal of energy are released. Fats contain more energy per unit of weight than carbohydrates and proteins, and they yield about 9 kcal/g. Table 17.1 lists the kilocalories supplied by some typical foods. A typical diet in the United States consists of 50–60% carbohydrates, 35–45% fats, and 10–15% proteins. Table 17.1 also lists the carbohydrate, fat, and protein composition of some foods.

A kilocalorie is often called a *Calorie* (with a capital *C*). Unfortunately, this usage has resulted in confusion of the term *calorie* (with a lowercase *c*) with *Calorie* (with a capital *C*). It is common practice on food labels to use the term *calorie* when *Calorie* (*kilocalorie*) is the proper term.

Carbohydrates

Sources in the Diet

Carbohydrates (kar-bō-hī′drātz) include monosaccharides, disaccharides, and polysaccharides (see chapter 2). Although most of the carbohydrates we ingest are derived from plants, lactose is derived from animals. The most common monosaccharides in the diet are glucose and fructose. Plants capture energy from sunlight and use that energy to produce glucose, which can be found in vegetables, fruits, molasses, honey, and syrup. Fructose is most often derived from fruits and berries.

The disaccharide sucrose (table sugar) is what most people think of when they use the term *sugar*. Sucrose consists of one glucose and one fructose molecule joined together, and its principal sources are sugarcane and sugar beets. Maltose (malt sugar), derived from germinating cereals, is a combination of two glucose molecules, and lactose (milk sugar) consists of one glucose molecule and one galactose molecule.

Complex carbohydrates are large polysaccharides, which are composed of long chains of glucose (see figure 2.11). Examples are starch, glycogen, and cellulose, which differ from one another in the arrangement of the glucose molecules and the structure of the chemical bonds holding them together.

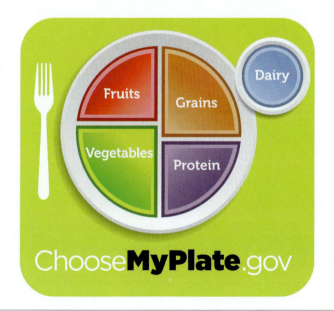

Figure 17.1 MyPlate

The MyPlate icon provides a visual reminder for making choices at mealtime, by selecting healthy foods from five food groups. Half the meal should be fruits and vegetables.

Source: U.S. Department of Agriculture.

TABLE 17.1   Food Composition

Food	Quantity	Food Energy (kcal)	Carbohydrate (g)	Fat (g)	Protein (g)
Dairy Products					
Whole milk (3.3% fat)	1 cup	150	11	8	8
Low-fat milk (2% fat)	1 cup	120	12	5	8
Butter	1 T	100	—	12	—
Grain					
Bread, white enriched	1 slice	75	24	1	2
Bread, whole-wheat	1 slice	65	14	1	3
Fruit					
Apple	1	80	20	1	—
Banana	1	100	25	—	1
Orange	1	65	16	—	1
Vegetables					
Corn, canned	1 cup	140	33	1	4
Peas, canned	1 cup	150	29	1	8
Lettuce	1 cup	5	2	—	—
Celery	1 cup	20	5	—	1
Potato, baked	1 large	145	33	—	4
Meat, Fish, and Poultry					
Lean ground beef (10% fat)	3 oz	185	—	10	23
Shrimp, french fried	3 oz	190	9	9	17
Tuna, canned	3 oz	170	—	7	24
Chicken breast, fried	3 oz	160	1	5	26
Bacon	2 slices	85	—	8	4
Hot dog	1	170	1	15	7
Fast Foods					
McDonald's Egg McMuffin	1	300	30	12	18
McDonald's Big Mac	1	540	45	29	25
Taco Bell Beef Burrito Supreme	1	420	53	15	17
Arby's roast beef, regular	1	360	37	14	22
Pizza Hut Super Supreme	1 slice	245	29	17	14
Long John Silver's fish	2 pieces	520	32	34	24
Dairy Queen Oreo Cookie Blizzard, medium	1	680	100	25	14
Desserts					
Chocolate chip cookie	1	50	7	2	1
Apple pie	1 piece	135	49	14	3
Soft ice cream	1 cup	377	38	23	7
Beverage					
Cola soft drink	12 oz	145	37	—	—
Beer	12 oz	144	13	—	—
Wine	$3\frac{1}{2}$ oz	73	2	—	—
Hard liquor (86 proof)	$1\frac{1}{2}$ oz	105	—	—	—
Miscellaneous					
Egg	1	80	1	6	6
Mayonnaise	1 T	100	—	11	—
Sugar	1 T	45	12	—	—

Starch is an energy-storage molecule in plants and is found primarily in vegetables, fruits, and grains. Glycogen is an energy-storage molecule in animals and is located primarily in muscle and in the liver. Cellulose forms plant cell walls.

Uses of Carbohydrates in the Body

During digestion, polysaccharides and disaccharides are split into monosaccharides, which are absorbed into the blood (see chapter 16). Humans have enzymes that can break the bonds between the glucose molecules of starch and glycogen, but they do not have the enzymes necessary to digest cellulose. Therefore, it is important to thoroughly cook or chew plant matter. Cooking and chewing break down the plant cell walls and expose the starches contained inside the cells to digestive enzymes. The undigested cellulose provides fiber, or "roughage," which increases the bulk of feces, making it easier to defecate.

Fructose and other monosaccharides absorbed into the blood are converted into glucose by the liver. Glucose, whether absorbed directly from the digestive tract or synthesized by the liver, is an energy source used to produce ATP. Because the brain relies almost entirely on glucose for its energy, the body carefully regulates blood glucose levels.

If an excess amount of glucose is present in the diet, the glucose is used to make glycogen, which is stored in muscle and in the liver. Glycogen can be rapidly converted back to glucose when energy is needed. Because cells can store only a limited amount of glycogen, additional glucose that is ingested is converted into lipids for long-term storage in adipose tissue.

In addition to being used as an energy source, sugars form part of deoxyribonucleic acid (DNA), ribonucleic acid (RNA), and ATP molecules. Sugars also combine with proteins to form glycoproteins, some of which function as receptor molecules on the outer surface of the cell membrane.

Recommended Consumption of Carbohydrates

According to the Dietary Guidelines Advisory Committee, the **Acceptable Macronutrient Distribution Range (AMDR)** for carbohydrates is 45–65% of total kilocalories. Although a minimum level of carbohydrates is not known, it is assumed that amounts of 100 g or less per day result in overuse of the body's proteins and lipids for energy sources. Because muscles are primarily protein, the use of proteins for energy can result in the breakdown of muscle tissue. The extensive use of lipids as an energy source can result in acidosis (see chapter 18).

Complex carbohydrates are recommended in the diet because many starchy foods contain other valuable nutrients, such as vitamins and minerals, and because the slower rate of digestion and absorption of complex carbohydrates does not result in large increases and decreases in blood glucose level, as the consumption of large amounts of simple sugars does. Eating whole grains also helps a person meet the recommended fiber intake (25 g/day for females and 38 g/day for males). Foods containing large amounts of simple sugars, such as soft drinks and candy, are rich in carbohydrates but have few other nutrients. For example, a typical soft drink is mostly sucrose, containing 9 teaspoons of sugar per 12-oz container. In excess, the consumption of these kinds of foods usually results in obesity and tooth decay.

Lipids

Sources in the Diet

Lipids (lip′idz) include triglycerides, steroids, phospholipids, and fat-soluble vitamins. **Triglycerides** (trī-glis′er-īdz), also called *triacylglycerols* (trī-as′il-glis′er-olz), are the most common type of lipid in the diet, accounting for about 95% of the total lipid intake. Triglyceride molecules consist of three fatty acids bound to one glycerol molecule (see figure 2.12). Triglycerides are often referred to as fats. If the fat is a liquid at room temperature, it is referred to as an oil. Fats are **saturated** if their fatty acids have only single covalent bonds between carbon atoms and **unsaturated** if they have one or more double bonds (see figure 2.13). **Monounsaturated fats** have one double bond, and **polyunsaturated fats** have two or more double bonds. Saturated fats are found in meat, dairy products, eggs, nuts, coconut oil, and palm oil (table 17.1). Monounsaturated fats include olive and peanut oils; polyunsaturated fats are found in fish, safflower, sunflower, and corn oils.

Solid fats, mainly shortening and margarine, work better than liquid oils in preparing some foods, such as pastries. Polyunsaturated vegetable oils can be changed from a liquid to a solid by making them more saturated—that is, by decreasing the number of double covalent bonds in their polyunsaturated fatty acids. To saturate an unsaturated oil, the oil can be *hydrogenated,* which means that hydrogen gas is bubbled through the oil, producing a change in molecular shape that solidifies the oil. The more saturated the product, the harder it becomes at room temperature. These processed fats are usually referred to as *trans* **fats.**

Processed foods and oils account for most of the *trans* fats in the American diet, although some *trans* fats occur naturally in food from animal sources. *Trans* fatty acids raise the concentration of low-density lipoproteins and lower the concentration of high-density lipoproteins in the blood (see chapter 16). These changes are associated with a greater risk of cardiovascular disease. In 2006, the Food and Drug Administration (FDA) required that food labels include a detailed list of the amounts of saturated and *trans* fats, allowing the consumer to make better dietary choices.

The remaining 5% of ingested lipids include steroids and phospholipids. **Cholesterol** (kō-les′ter-ol) is a steroid (see chapter 2) found in high concentrations in the brain, the liver, and egg yolks, but it is also present in whole milk, cheese, butter, and meats. Cholesterol is not found in plants. Phospholipids, such as **lecithin** (les′i-thin; *lekithos,* egg yolk), are major components of cell membranes and are found in a variety of foods. A good source of lecithin is egg yolks.

Uses of Lipids in the Body

Triglycerides are an important source of energy that can be used to produce ATP. A gram of triglyceride delivers over twice as many calories as does a gram of carbohydrate or protein. Some cells, such as skeletal muscle cells, derive most of their energy from triglycerides.

Ingested triglyceride molecules not immediately used are stored in adipose tissue or in the liver. When energy is required, the stored triglycerides are broken down, and the fatty acids are released into the blood. The fatty acids can be taken up and used by various tissues. In addition to storing energy, adipose tissue surrounds, pads, and protects organs. Adipose tissue located under the skin is an insulator, which helps reduce heat loss.

CLINICAL IMPACT   Fatty Acids and Blood Clotting

The essential fatty acids are used to synthesize molecules that affect blood clotting. Linoleic acid can be converted to **arachidonic** (ă-rak-i-don′ik; *arakis*) **acid,** which is used to produce prostaglandins that *increase* blood clotting. Alpha-linolenic acid can be converted to **eicosapentaenoic** (ī′kō-să-pen-tă-nō′ik) **acid (EPA),** which is used to produce prostaglandins that *decrease* blood clotting. Normally, most prostaglandins are synthesized from linoleic acid because it is more plentiful in the diet. However, individuals who consume foods rich in EPA, such as herring, salmon, tuna, and sardines, increase the synthesis of prostaglandins from EPA. People who eat these fish two or more times per week have a lower risk of heart attack than those who don't, probably because of reduced blood clotting. Although EPA can be obtained using fish oil supplements, this is not currently recommended because fish oil supplements contain high amounts of cholesterol, vitamins A and D, and uncommon fatty acids, all of which can cause health problems when taken in large amounts.

Cholesterol is an important molecule with many functions in the body. It is obtained in food, or it can be manufactured by the liver and most other tissues. Cholesterol is a component of the cell membrane, and it can be modified to form other useful molecules, such as bile salts and steroid hormones. Bile salts emulsify fats, which is important for fat digestion and absorption (see chapter 16). Steroid hormones include the sex hormones estrogen, progesterone, and testosterone, which regulate the reproductive system. Eicosanoids, which are derived from fatty acids, are involved in inflammation, tissue repair, smooth muscle contraction, and other functions.

Phospholipids (see chapter 2) are part of the cell membrane and are used to construct myelin sheaths around the axons of nerve cells. Lecithin is found in bile and helps emulsify fats.

Recommended Consumption of Lipids

The AMDR for fats is 20–35% for adults, 25–35% for children and adolescents 4 to 18 years of age, and 30–35% for children 2 to 3 years of age. Saturated fats should be no more than 10% of total kilocalories, or as low as possible. Most dietary fat should come from sources of polyunsaturated and monounsaturated fats. Cholesterol should be limited to 300 mg (the amount in one egg yolk) or less per day, and *trans* fat consumption should be as low as possible. These guidelines reflect the belief that excess amounts of fats, especially saturated fats, *trans* fats, and cholesterol, contribute to cardiovascular disease. The typical American diet derives 35–45% of its kilocalories from fats, indicating that most Americans need to reduce their fat consumption. See table 17.1 for a sampling of the fat composition in foods.

If a person does not consume enough fats, the body can synthesize fats from carbohydrates and proteins. **Linoleic** (lin-ō-lē′ik; *linum,* flax + *oleum,* oil) **acid** and **alpha linolenic** (lin-ō-len′ik) **acid** are called **essential fatty acids** because the body cannot synthesize them, so they must be ingested. They are found in plant oils, such as canola or soybean oil.

Proteins

Sources in the Diet

Proteins (prō′tēnz) are chains of amino acids (see figure 2.16). They are found in most of the plant and animal products we eat (see table 17.1). Proteins in the body are constructed of 20 different kinds of amino acids, which are divided into two groups: essential and nonessential amino acids. The body cannot synthesize **essential amino acids,** so they must be obtained in the diet. The nine essential amino acids are histidine, isoleucine, leucine, lysine, methionine, phenylalanine, threonine, tryptophan, and valine. **Nonessential amino acids** are necessary to construct our proteins but do not necessarily need to be ingested, since they can be synthesized from the essential amino acids. A **complete protein** food contains all nine essential amino acids in the needed proportions, whereas an incomplete protein food does not. Animal proteins tend to be complete proteins, whereas plant proteins tend to be incomplete. Examples of complete proteins are red meat, fish, poultry, milk, cheese, and eggs. Examples of incomplete proteins are leafy green vegetables, grains, and legumes (peas and beans). If two incomplete proteins, such as rice and beans, are ingested, each can provide amino acids lacking in the other. Thus, a vegetarian diet, if balanced correctly, provides all the essential amino acids.

Uses of Proteins in the Body

Proteins perform numerous functions in the human body, as the following examples illustrate. Collagen provides structural strength in connective tissue, as does keratin in the skin. The combination of actin and myosin makes muscle contraction possible. Enzymes are responsible for regulating the rate of chemical reactions in the body, and protein hormones regulate many physiological processes (see chapter 10). Proteins in the blood act as clotting factors, transport molecules, and buffers (which limit changes in pH). Proteins also function as ion channels, carrier molecules, and receptor molecules in the cell membrane. Antibodies, lymphokines, and complement are all proteins that function in the immune system.

Proteins can also be used as a source of energy, yielding approximately the same amount of energy as that derived from carbohydrates. If excess proteins are ingested, the body can use the energy from the proteins to produce glycogen and lipid molecules, which can be stored. When protein intake is adequate in a healthy adult, the synthesis and breakdown of proteins occur at the same rate.

Recommended Consumption of Proteins

The AMDR for protein is 10–35% of total kilocalories. See table 17.1 for a sampling of the protein composition in foods.

CLINICAL IMPACT   Free Radicals and Antioxidants

Free radicals are molecules, produced as part of normal metabolism, that are missing an electron. Free radicals can replace the missing electron by taking an electron from cell molecules, such as lipids, proteins, or DNA, resulting in damage to the cell. Damage from free radicals may contribute to aging and certain diseases, such as atherosclerosis and cancer.

The loss of an electron from a molecule is called oxidation. **Antioxidants** are substances that prevent oxidation of cell components by donating an electron to free radicals. Examples of antioxidants are beta carotene (provitamin A), vitamin C, and vitamin E.

Many studies have attempted to determine whether taking large doses of antioxidants is beneficial. Among scientists establishing the RDAs, the best evidence presently available does not support the claim that large doses of antioxidants can prevent chronic disease or otherwise improve health. On the other hand, the amount of antioxidants normally found in a balanced diet (including fruits and vegetables rich in antioxidants), combined with the complex mix of other chemicals found in food, can be beneficial.

Vitamins

Vitamins (vīt′ă-minz; life-giving chemicals) are organic molecules that exist in minute quantities in food and are essential to normal metabolism (table 17.2). **Essential vitamins** cannot be produced by the body and must be obtained through the diet. Because no single food item or nutrient class provides all the essential vitamins, it is necessary to maintain a balanced diet by eating a variety of foods. The absence of an essential vitamin in the diet can result in a specific deficiency disease. A few vitamins, such as vitamin K, are produced by intestinal bacteria, and a few others can be formed by the body from substances called provitamins. A **provitamin** is a part of a vitamin that the body can assemble or modify into a functional vitamin. Beta carotene is an example of a provitamin that can be modified by the body to form vitamin A. The other provitamins are **7-dehydrocholesterol** (dē-hī′drō-kō-les′ter-ol), which can be converted to vitamin D, and **tryptophan** (trip′tō-fan), which can be converted to niacin.

Vitamins are not broken down by catabolism but are used by the body in their original or slightly modified forms. After the chemical structure of a vitamin is destroyed, its function is lost. The chemical structure of many vitamins is destroyed by heat, as when food is overcooked.

Most vitamins function as **coenzymes,** which combine with enzymes to make the enzymes functional (see chapter 2). Vitamins B_2 and B_3, biotin (bī′ō-tin), and pantothenic (pan-tō-then′ik) acid are critical for some of the chemical reactions involved in the production of ATP. Folate (fō′lāt) and vitamin B_{12} are involved in nucleic acid synthesis. Vitamins A, B_1, B_6, B_{12}, C, and D are necessary for growth. Vitamin K is necessary for the synthesis of proteins involved in blood clotting (table 17.2).

Predict 2
Predict what would happen if vitamins were broken down during the process of digestion rather than being absorbed intact into the circulation.

Vitamins are either fat-soluble or water-soluble. **Fat-soluble vitamins,** such as vitamins A, D, E, and K, are absorbed from the intestine along with lipids. Some of them can be stored in the body for a long time. Because they can be stored, these vitamins can accumulate in the body to the point of toxicity. **Water-soluble vitamins,** such as the B vitamins and vitamin C, are absorbed with water from the intestinal tract and typically remain in the body only a short time before being excreted in the urine.

Vitamins were first identified at the beginning of the twentieth century. They were found to be associated with certain foods that were known to protect people from diseases such as rickets and beriberi. In 1941, the first Food and Nutrition Board established the **Recommended Dietary Allowances (RDAs),** which are the nutrient intakes that are sufficient to meet the needs of nearly all people in certain age and gender groups. RDAs were established for different-aged males and females, starting with infants and continuing on to adults. RDAs were also set for pregnant and lactating women. The RDAs have been reevaluated every 4–5 years and updated, when necessary, on the basis of new information.

The RDAs establish a minimum intake of vitamins and minerals that should protect almost everyone (97%) in a given group from diseases caused by vitamin or mineral deficiencies. Although personal requirements can vary, the RDAs are a good benchmark. The farther below the RDAs an individual's dietary intake is, the more likely that person is to develop a nutritional deficiency. On the other hand, consuming too much of some nutrients can have harmful effects. For example, the long-term ingestion of 3–10 times the RDA for vitamin A can cause bone and muscle pain, skin disorders, hair loss, and an enlarged liver. Consuming 5–10 times the RDA of vitamin D over the long term can result in calcium deposits in the kidneys, heart, and blood vessels, and regularly consuming more than 2 g (which is more than 20 times the RDA) of vitamin C daily can cause stomach inflammation and diarrhea.

Minerals

Minerals (min′er-ălz) are inorganic nutrients that are essential for normal metabolic functions. People ingest minerals alone or in combination with organic molecules. They constitute about 4–5% of total body weight and are involved in a number of important functions, such as establishing resting membrane potentials and generating action potentials; adding mechanical strength to bones and teeth; combining with organic molecules; and acting as coenzymes, buffers, or regulators of osmotic pressure. A balanced diet can provide all the necessary minerals, with a few possible

TABLE 17.2   The Principal Vitamins

Vitamin	Fat- (F) or Water- (W) Soluble	Source	Function	Symptoms of Deficiency	Reference Daily Intake (RDI)s[a]
A (retinol)	F	From provitamin carotene found in yellow and green vegetables; preformed in liver, egg yolk, butter, and milk	Necessary for rhodopsin synthesis, normal health of epithelial cells, and bone and tooth growth	Rhodopsin deficiency, night blindness, retarded growth, skin disorders, and increased infection risk	900 RE[b]
B₁ (thiamine)	W	Yeast, grains, and milk	Involved in carbohydrate and amino acid metabolism; necessary for growth	Beriberi—muscle weakness (including cardiac muscle), neuritis, and paralysis	1.2 mg
B₂ (riboflavin)	W	Green vegetables, liver, wheat germ, milk, and eggs	Component of flavin adenine dinucleotide; involved in citric acid cycle	Eye disorders and skin cracking, especially at corners of the mouth	1.3 mg
B₃ (niacin)	W	Fish, liver, red meat, yeast, grains, peas, beans, and nuts	Component of nicotinamide adenine dinucleotide; involved in glycolysis and citric acid cycle	Pellagra—diarrhea, dermatitis, and nervous system disorder	16 mg
Pantothenic acid	W	Liver, yeast, green vegetables, grains, and intestinal bacteria	Constituent of coenzyme-A; glucose production from lipids and amino acids; steroid hormone synthesis	Neuromuscular dysfunction and fatigue	5 mg
Biotin	W	Liver, yeast, eggs, and intestinal bacteria	Fatty acid and nucleic acid synthesis; movement of pyruvic acid into citric acid cycle	Mental and muscle dysfunction, fatigue, and nausea	30 µg
B₆ (pyridoxine)	W	Fish, liver, yeast, tomatoes, and intestinal bacteria	Involved in amino acid metabolism	Dermatitis, retarded growth, and nausea	1.7 mg
Folate	W	Liver, green leafy vegetables, and intestinal bacteria	Nucleic acid synthesis; hematopoiesis; prevents birth defects	Macrocytic anemia (enlarged red blood cells) and spina bifida	0.4 mg
B₁₂ (cobalamins)	W	Liver, red meat, milk, and eggs	Necessary for red blood cell production, some nucleic acid and amino acid metabolism	Pernicious anemia and nervous system disorders	2.4 µg
C (ascorbic acid)	W	Citrus fruit, tomatoes, and green vegetables	Collagen synthesis; general protein metabolism	Scurvy—defective bone formation and poor wound healing	90 mg
D (cholecalciferol, ergosterol)	F	Fish liver oil, enriched milk, and eggs; provitamin D converted by sunlight to cholecalciferol in the skin	Promotes calcium and phosphorus use; normal growth and bone and tooth formation	Rickets—poorly developed, weak bones, osteomalacia; bone reabsorption	10 µg[c]
E (tocopherol, tocotrienols)	F	Wheat germ; cottonseed, palm, and rice oils; grain; liver; and lettuce	Prevents oxidation of cell membranes and DNA	Hemolysis of red blood cells	15 mg
K (phylloquinone)	F	Alfalfa, liver, spinach, vegetable oils, cabbage, and intestinal bacteria	Required for synthesis of a number of clotting factors	Excessive bleeding due to retarded blood clotting	120 µg

[a] RDIs for people over 4 years of age; IU = international units.
[b] Retinol equivalents (RE). 1 retinol equivalent = 1 µg retinol or 6 µg β-carotene.
[c] As cholecalciferol. 1 µg cholecalciferol = 40 IU (international units) vitamin D.

TABLE 17.3 Important Minerals

Mineral	Function	Symptoms of Deficiency	Reference Daily Intake (RDI)s[a]
Calcium	Bone and teeth formation; blood clotting; muscle activity; and nerve function	Spontaneous action potential generation in neurons and tetany	1300 mg
Chlorine	Blood acid-base balance; hydrochloric acid production in stomach	Acid-base imbalance	2.3 g[b]
Chromium	Associated with enzymes in glucose metabolism	Unknown	35 µg
Cobalt	Component of vitamin B_{12}; red blood cell production	Anemia	Unknown
Copper	Hemoglobin and melanin production; electron-transport system	Anemia and loss of energy	0.9 mg
Fluorine	Provides extra strength in teeth; prevents dental caries	No real pathology	4 mg
Iodine	Thyroid hormone production; maintenance of normal metabolic rate	Goiter and decrease in normal metabolism	150 µg
Iron	Component of hemoglobin; ATP production in electron-transport system	Anemia, decreased oxygen transport, and energy loss	18 mg
Magnesium	Coenzyme constituent; bone formation; muscle and nerve function	Increased nervous system irritability, vasodilation, and arrhythmias	420 mg
Manganese	Hemoglobin synthesis; growth; activation of several enzymes	Tremors and convulsions	2.3 mg
Molybdenum	Enzyme component	Unknown	45 µg
Phosphorus	Bone and teeth formation; important in energy transfer (ATP); component of nucleic acids	Loss of energy and cellular function	1250 mg
Potassium	Muscle and nerve function	Muscle weakness, abnormal electrocardiogram, and alkaline urine	4.7 g
Selenium	Component of many enzymes	Unknown	55 µg
Sodium	Osmotic pressure regulation; nerve and muscle function	Nausea, vomiting, exhaustion, and dizziness	1.5 g[b]
Sulfur	Component of hormones, several vitamins, and proteins	Unknown	Unknown
Zinc	Component of several enzymes; carbon dioxide transport and metabolism; necessary for protein metabolism	Deficient carbon dioxide transport and deficient protein metabolism	11 mg

[a] RDIs for people over 4 years of age, except for sodium.
[b] 3.8 g sodium chloride (table salt)

exceptions. For example, individuals suffering from chronic bleeding or women who experience excessive menstrual bleeding may need an iron supplement. Table 17.3 lists some minerals and their functions.

Daily Values

Daily Values appear on food labels to help consumers plan a healthful diet and to minimize confusion. However, not all possible Daily Values are required to be listed on food labels. Daily Values are based on two other sets of reference values—Reference Daily Intakes and Daily Reference Values:

- **Reference Daily Intakes (RDIs)** are based on the 1968 RDAs for certain vitamins and minerals. RDIs have been set for four categories of people: infants, toddlers, people over 4 years of age, and pregnant or lactating women. Generally, the RDIs are set to the highest 1968 RDA value of an age category. For example, the highest RDA for iron in males over 4 years of age is 10 mg/day and for females over 4 years of age is 18 mg/day. Thus, the RDI for iron is set at 18 mg/day.

- **Daily Reference Values (DRVs)** are set for total fat, saturated fat, cholesterol, total carbohydrate, dietary fiber, sodium, potassium, and protein.

The Daily Values appearing on food labels are based on a 2000-kcal reference diet, which approximates the weight maintenance requirements of postmenopausal women, women who exercise moderately, teenage girls, and sedentary men (figure 17.2). On large food labels, additional information is listed based on a daily intake of 2500 kcal, which is adequate for young men.

The Daily Values for energy-producing nutrients are determined as a percentage of daily kilocaloric intake: 60% for carbohydrates, 30% for total fats, 10% for saturated fats, and 10% for proteins. The Daily Value for fiber is 14 g for each 1000 kcal of intake. The Daily Values for a nutrient in a 2000 kcal/day diet can be calculated on the basis of the recommended daily percentage of the nutrient and the kilocalories in a gram of the nutrient. For example, carbohydrates should constitute 60% of a 2000 kcal/day diet, or 1200 kcal/day (0.60 × 2000). Because there are 4 kilocalories in a gram of carbohydrate, the Daily Value for carbohydrate is 300 g/day (1200/4).

Nutrition Facts

Serving Size 1 donut (about 52g)
Servings Per Container 12

Amount Per Serving

Calories 200   **Calories From Fat** 100

	%Daily Value*
Total Fat 12g	18%
Saturated Fat 3g	15%
Trans Fat 4g	
Cholesterol 5mg	1%
Sodium 95mg	4%
Total Carbohydrate 22g	7%
Dietary Fiber <1g	1%
Sugars 10g	
Protein 2g	

Vitamin A	0%	•	Vitamin C	2%
Calcium	6%	•	Iron	4%

*Percent of Daily Values (DV) are based on a 2,000 calorie diet.

(a)

Figure 17.2 Food Label

(*a*) Current food label. (*b*) Revised food label proposed in early 2014.

Source: U.S. Food and Drug Administration.

Nutrition Facts

8 servings per container

Serving size   2/3 cup (55g)

Amount per 2/3 cup

Calories   **230**

% DV*

12%	**Total Fat** 8g
5%	Saturated Fat 1g
	Trans Fat 0g
0%	**Cholesterol** 0mg
7%	**Sodium** 160mg
12%	**Total Carbs** 37g
14%	Dietary Fiber 4g
	Sugars 1g
	Added Sugars 0g
	Protein 3g
10%	Vitamin D 2mcg
20%	Calcium 260mg
45%	Iron 8mg
5%	Potassium 235mg

* Footnote on Daily Values (DV) and calories reference to be inserted here.

(b)

The Daily Values for some nutrients are the uppermost limit considered desirable because of the link between these nutrients and certain diseases. Thus, the Daily Values for total fats are less than 65 g; saturated fats, less than 20 g; and cholesterol, less than 300 mg because of their association with increased risk of heart disease. The Daily Value for sodium is less than 2400 mg because of its association with high blood pressure in some people.

For a particular food, the Daily Values are used to calculate the **Percent Daily Value (% Daily Value)** for some of the nutrients in one serving of the food (figure 17.2). For example, if a serving of food has 3 g of fat and the Daily Value for total fat is 65 g, then the % Daily Value is 5% (3/65 = 0.05, or 5%). The Food and Drug Administration (FDA) requires % Daily Values to be on food labels, so that the public has useful and accurate dietary information.

In early 2014, the FDA proposed a major revision of food nutrition labels. The newly proposed nutrition labels will have calorie counts displayed more prominently, representing more reasonable serving sizes (Figure 17.2*b*). Manufacturers will also be required to list the amount of added sugar on the nutrition label, to assist the consumer in making better choices about total sugar intake, which has been linked to the current increase in obesity in the United States. Saturated fat and *trans* fat information will also be included, but the "Calories from Fat" on the old label will no longer be included. This reflects the modern view of nutritionists that the types of fats we ingest are more important than the amount. Once the new food label has been approved, manufacturers will have to include the revisions on most packaged foods within two years.

Predict 3

One serving of a food has 30 g of carbohydrate. What % Daily Value for carbohydrate is on the food label for this food?

The % Daily Values for nutrients related to energy consumption are based on a 2000 kcal/day diet. For people who maintain their weight on a 2000 kcal/day diet, the total of the % Daily Values for each of these nutrients should be no more than 100%. However, for individuals consuming more or fewer than 2000 kcal per day, the total of the % Daily Values can be more or fewer than 100%. For example, for a person consuming 2200 kcal/day, the total of the % Daily Values for each of these nutrients should be no more than 110% because 2200/2000 = 1.10, or 110%.

Predict 4

Suppose a person consumes 1800 kcal/day. What total % Daily Values for energy-producing nutrients is recommended?

When using the % Daily Values of a food to determine how the amounts of certain nutrients in the food fit into the overall diet, the number of servings in a container or package needs to be considered. For example, suppose a small (2.25-oz) bag of corn chips has a % Daily Value of 16% for total fat. A person might suppose that eating the bag of chips accounts for 16% of total fat for the day. The bag, however, contains 2.5 servings. All the chips in the bag account for 40% (16% × 2.5) of the maximum recommended total fat.

A CASE IN POINT

Vegetarian Diet

Hazel Nutt decides to switch from a typical American diet to a vegetarian diet because she believes a vegetarian diet is better for her health, and she no longer wants to eat animals. A strict **vegetarian** (or vegan) diet includes only plant foods.

Plants alone can provide all the protein required for good health. But to get adequate amounts of the essential amino acids, a person following a vegan diet should consume a variety of protein sources, such as grains and legumes. In addition, the *Dietary Guidelines for Americans* recommends that vegan diets be supplemented with vitamin B$_{12}$, vitamin D, calcium, iron, and zinc. This is especially important for children and pregnant and lactating women. Plant sources do not supply vitamin B$_{12}$ or sufficient amounts of vitamin D, although the body can produce vitamin D with adequate exposure to sunlight (see chapter 5). Calcium is found in green leafy vegetables and nuts. Iron and zinc are in whole grains, nuts, and legumes. However, these minerals are either present in low amounts or not easily absorbed.

17.2 METABOLISM

Learning Outcomes *After reading this section, you should be able to*

A. Define metabolism, anabolism, and catabolism.
B. List three ways in which enzyme activity is controlled.
C. Describe glycolysis, and name its products.
D. Describe the citric acid cycle and its products.
E. Describe the electron-transport chain and how ATP is produced in the process.
F. Explain how the breakdown of glucose yields 2 ATP molecules in anaerobic respiration and 38 ATP molecules in aerobic respiration.
G. Describe the basic steps involved in using lipids and amino acids as energy sources.
H. Differentiate between the absorptive and postabsorptive metabolic states.
I. Define metabolic rate.

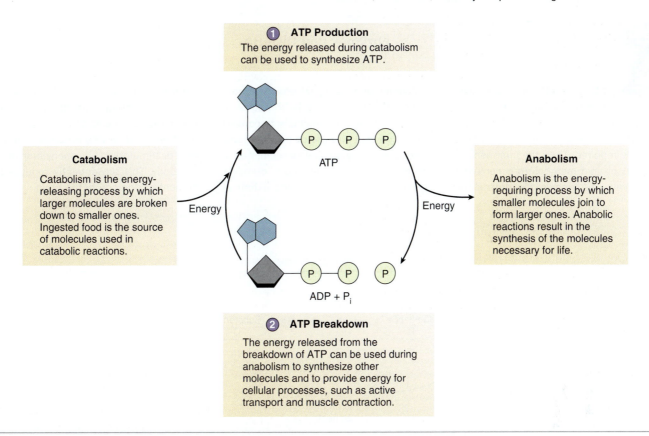

1 ATP Production
The energy released during catabolism can be used to synthesize ATP.

Catabolism

Catabolism is the energy-releasing process by which larger molecules are broken down to smaller ones. Ingested food is the source of molecules used in catabolic reactions.

Energy

ATP

Energy

Anabolism

Anabolism is the energy-requiring process by which smaller molecules join to form larger ones. Anabolic reactions result in the synthesis of the molecules necessary for life.

$ADP + P_i$

2 ATP Breakdown
The energy released from the breakdown of ATP can be used during anabolism to synthesize other molecules and to provide energy for cellular processes, such as active transport and muscle contraction.

PROCESS Figure 17.3 AP|R **ATP Derived from Catabolic Reactions Drives Anabolic Reactions**

Metabolism (mĕ-tab′ō-lizm; change) is the total of all the chemical reactions that occur in the body. It consists of **catabolism** (kă-tab′ō-lizm), the energy-releasing process by which large molecules are broken down into smaller ones, and **anabolism** (ă-nab′ō-lizm), the energy-requiring process by which small molecules are joined to form larger ones. Catabolism begins during the process of digestion and is concluded within individual cells. Anabolism occurs in all cells of the body as they divide to form new cells, maintain their own intracellular structure, and produce molecules such as hormones, neurotransmitters, or extracellular matrix molecules for export. The energy derived from catabolism is used to drive anabolic reactions.

Metabolism can be divided into the chemical reactions that occur during digestion and the chemical reactions that occur after the products of digestion are taken up by cells. The chemical reactions that occur within cells are often referred to as **cellular metabolism.** The digestive products of carbohydrates, proteins, and lipids can be further broken down inside cells. The energy released during this breakdown can be used to combine **ADP** and an inorganic phosphate group (P_i) to form **ATP** (figure 17.3).

ATP is often called the energy currency of the cell. When ATP is broken down to ADP, cells can use the released energy for active transport, muscle contraction, and molecule synthesis. Because the body has high energy demands, it uses ATP rapidly.

Regulation of Metabolism

The products of digestion, such as glucose, fatty acids, and amino acids, are molecules containing energy within their chemical bonds. A series of chemical reactions, called a **biochemical pathway,**

controls the energy release from these molecules. At some of the steps, small amounts of energy are released; some of this energy is used to synthesize ATP (figure 17.4). About 40% of the energy in foods is incorporated into ATP; the rest is lost as heat.

There are several different biochemical pathways inside cells. Which pathways function and how much each pathway is used is determined by enzymes because each step in the pathway requires a specific enzyme (see chapter 2). In turn, enzymes are regulated in several ways:

1. *Enzyme synthesis.* Enzymes are proteins, and their synthesis depends on DNA (see chapter 3). Thus, the types and amounts of enzymes present in cells are under genetic control.
2. *Receptor-mediated enzyme activity.* The combination of a chemical signal, such as a neurotransmitter or hormone, with a membrane-bound or intracellular receptor can activate or inhibit enzyme activity (see chapter 10).
3. *Product control of enzyme activity.* The end product of a biochemical pathway can inhibit the enzyme responsible for the first reaction in the pathway. This negative-feedback regulation prevents accumulation of the intermediate products and the end product of the pathway (figure 17.4).

Carbohydrate Metabolism

Monosaccharides are the breakdown products of carbohydrate digestion. Of these, glucose is the most important in terms of cellular metabolism. Glucose is transported in the circulation to all tissues of the body, where it is used as a source of energy. Any excess glucose in the blood following a meal can be used to form

CLINICAL IMPACT   Enzymes and Disease

Many metabolic disorders result from missing or dysfunctional enzymes. For example, in Tay-Sachs disease, the breakdown of lipids within lysosomes is impaired. Thus, the intermediate products of lipid metabolism accumulate abnormally, resulting in the destruction of neurons and death by age 3–4 years. Phenylketonuria (fen′il-kē′tō-noo′rē-a) (PKU) results from the inability to convert the amino acid phenylalanine to tyrosine. Therefore, phenylalanine accumulates in nerve cells and causes brain damage. Unlike Tay-Sachs disease, the extent of damage resulting from PKU can be limited by controlling the diet. Restricting the intake of phenylalanine is an effective treatment for PKU because it reduces the accumulation of phenylalanine in the body. This subsequently reduces the amount of damage to the brain.

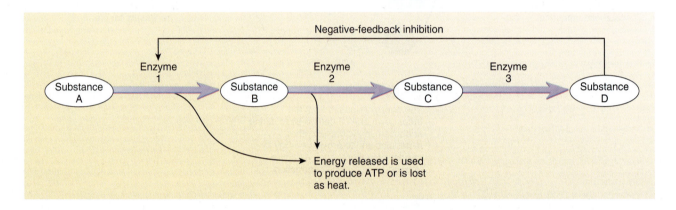

Figure 17.4 Biochemical Pathway

Each step in the pathway is regulated by a specific enzyme. Substance D can inhibit enzyme I, thus regulating its own production. Some of the energy released by reactions in the pathway can be used to synthesize ATP.

---

glycogen (glī′kō-jen; *glyks,* sweet), or it can be partially broken down and the components used to form lipids. Glycogen is a short-term energy-storage molecule that the body can store only in limited amounts, whereas lipids are long-term energy-storage molecules that the body can store in large amounts. Most of the body's glycogen is in skeletal muscle and in the liver.

Glycolysis

Glycolysis (glī-kol′i-sis) is a series of chemical reactions that occurs in the fluid part of cytoplasm surrounding the organelles. It results in the breakdown of glucose to two **pyruvic** (pī-roo′vik) **acid** molecules (figure 17.5). When glucose is converted to pyruvic acid, two ATP molecules are used and four ATP molecules are produced, for a net gain of two ATP molecules.

Glucose consists of 6 carbon atoms, 12 hydrogen atoms, and 6 oxygen atoms covalently bonded together. During the breakdown of glucose, a hydrogen ion (H⁺) and two electrons (e⁻) are released and can attach to a **carrier molecule,** which moves the H⁺ and electrons to other parts of the cell. A very common carrier molecule in cells is **nicotinamide adenine dinucleotide** (nik-ō-tin′a-mīd ad′ĕ-nēn dī-noo′klē-ō-tīd) **(NADH):**

$$NAD^+ + 2\,e^- + H^+ \rightarrow NADH$$

The H⁺ and high-energy electrons in the NADH molecules can be used in other chemical reactions or in the production of ATP molecules in the electron-transport chain (described later in this section).

The pyruvic acid and NADH produced in glycolysis can be used in two different biochemical pathways, depending on the availability of O_2. When the amounts of O_2 are limited, anaerobic respiration can take place. Anaerobic respiration does not require O_2 and can quickly produce a few ATP molecules for a short time. If the cell has adequate amounts of O_2, the pyruvic acid and NADH produced in glycolysis are used in aerobic respiration to produce many more ATP.

Anaerobic Respiration

Lactic acid fermentation, a form of anaerobic respiration, is the breakdown of glucose in the absence of O_2 to produce two molecules of **lactic** (lak′tik) **acid** and two molecules of ATP (figure 17.5). The ATP thus produced is a source of energy during activities, such as intense exercise, when insufficient O_2 is delivered to tissues. Lactic acid fermentation can be divided into two phases:

1. *Glycolysis.* Glucose undergoes several reactions to produce two pyruvic acid molecules, two ATP, and two NADH.
2. *Lactic acid formation.* Pyruvic acid is converted to lactic acid, a reaction that requires the input of energy from the NADH produced in phase 1 of lactic acid fermentation.

Lactic acid is released from the cells that produce it, and blood transports it to the liver. When O_2 becomes available, the lactic acid in the liver can be converted through a series of chemical reactions into glucose. The glucose then can be released from the

① Glycolysis converts glucose to two pyruvic acid molecules. There is a net gain of two ATP and two NADH from glycolysis.

② Anaerobic respiration, which does not require O_2, includes glycolysis and converts the two pyruvic acid molecules produced by glycolysis to two lactic acid molecules. This conversion requires energy, which is derived from the NADH generated in glycolysis.

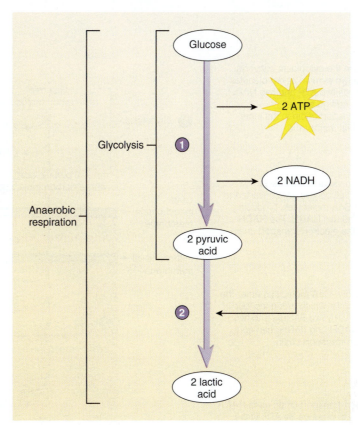

PROCESS Figure 17.5 Glycolysis and Anaerobic Respiration

liver and transported in the blood to cells that use glucose as an energy source. Some of the reactions that convert lactic acid to glucose require the input of energy derived from ATP that is produced by aerobic respiration. The O_2 necessary to make enough ATP for the synthesis of glucose from lactic acid contributes to the **oxygen deficit** (see chapter 7).

Aerobic Respiration

Aerobic (ār-ō′bik) **respiration** is the breakdown of glucose in the presence of O_2 to produce CO_2, water, and 38 molecules of ATP (figure 17.6). Aerobic respiration can be divided into four phases:

1. *Glycolysis.* The six-carbon glucose molecule is broken down to form two molecules of pyruvic acid, each consisting of three carbon atoms. Two ATP and two NADH molecules are also produced.
2. *Acetyl-CoA formation.* Each pyruvic acid moves from the cytoplasm into a mitochondrion, where enzymes remove a carbon atom from the three-carbon pyruvic acid molecule to form CO_2 and a two-carbon acetyl (as′e-til, a-set′il) group. Hydrogen ions and electrons are released in the reaction and can be used to produce NADH. Each acetyl group combines with coenzyme A (CoA), derived from vitamin B_2, to form **acetyl-CoA.** Because two pyruvic acid molecules are produced in phase 1, phase 2 results in two acetyl-CoAs, two CO_2 molecules, and two NADH molecules for each glucose molecule.

3. *Citric acid cycle.* Each acetyl-CoA combines with a four-carbon molecule to form a six-carbon citric acid molecule, which enters the citric acid cycle. The **citric** (sit′rik) **acid cycle** is also called the *tricarboxylic* (trī-kar-bok′sil-ik) *acid (TCA) cycle* (citric acid has three carboxylic acid groups) or the *Krebs cycle,* after its discoverer, British biochemist Sir Hans Krebs (1900–1981). The citric acid cycle is a series of reactions wherein the six-carbon citric acid molecule is converted, in a number of steps, into a four-carbon molecule (figure 17.6). The four-carbon molecule can then combine with another acetyl-CoA molecule to form another citric acid molecule and reinitiate the cycle. During the cycle, two carbon atoms are used to form CO_2, and energy, H^+, and electrons are released. Some of the energy can be used to produce ATP. Most of the energy, H^+, and electrons are used to form NADH molecules and another carrier molecule, called **flavin** (flā′vin) **adenine dinucleotide (FADH$_2$).** These molecules are used in the electron-transport chain to generate additional ATP. Carbon dioxide diffuses out of the cell and into the blood. It is transported by the circulatory system to the lungs, where it is expired. Thus, the carbon atoms that constitute food molecules, such as glucose, are eventually eliminated from the body as CO_2. We literally breathe out part of the food we eat!
4. *Electron-transport chain.* The **electron-transport chain** is a series of electron-transport molecules attached to the inner mitochondrial membrane (figure 17.7). This membrane

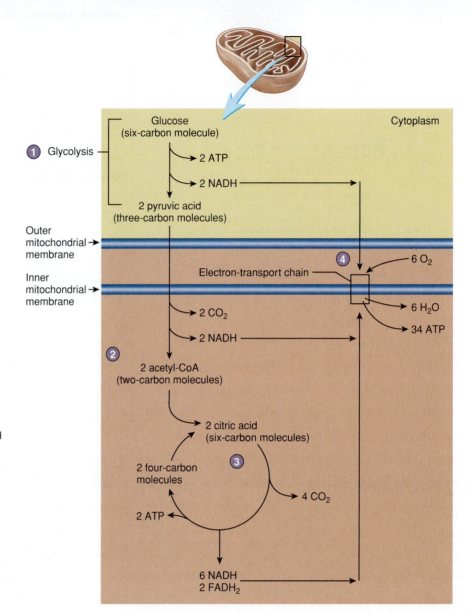

1. Glycolysis in the cytoplasm converts glucose to two pyruvic acid molecules and produces two ATP and two NADH. The NADH can go to the electron-transport chain in the inner mitochondrial membrane.

2. The two pyruvic acid molecules produced in glycolysis are converted to two acetyl-CoA molecules, producing two CO_2 and two NADH. The NADH can go to the electron-transport chain.

3. The two acetyl-CoA molecules enter the citric acid cycle, which produces four CO_2, six NADH, two $FADH_2$, and two ATP. The NADH and $FADH_2$ can go to the electron-transport chain.

4. The electron-transport chain uses NADH and $FADH_2$ to produce 34 ATP. This process requires O_2, which combines with H^+ to form H_2O.

PROCESS Figure 17.6 Aerobic Respiration

Aerobic respiration involves four phases: (1) glycolysis, (2) acetyl-CoA formation, (3) the citric acid cycle, and (4) the electron-transport chain. The number of carbon atoms in a molecule is indicated after the molecule's name. As glucose is broken down, the carbon atoms from glucose are incorporated into carbon dioxide.

divides the interior of the mitochondrion into an inner compartment and an outer compartment. Electrons are transferred from NADH and $FADH_2$ to the electron-transport carriers, and H^+ is released into the inner mitochondrial compartment. Some of the electron-transport carriers are also H^+ pumps, which use some of the energy from the transported electrons to pump H^+ from the inner to the outer mitochondrial compartment. Because of an increased H^+ concentration in the outer compartment, the H^+ passes by diffusion back into the inner compartment. The H^+ passes through channels in the inner mitochondrial membrane that couple the movement of the H^+ to ATP production. In the last step of the electron-transport chain, two H^+ and two electrons combine with an O_2 atom to form H_2O:

$$2\,H^+ + 2\,e^- + \tfrac{1}{2}O_2 \rightarrow H_2O$$

Without O_2 to accept the H^+ and electrons, the citric acid cycle and the electron-transport chain cannot function. Note that the O_2 we breathe in is eventually bound to two hydrogen atoms to become water, which has many uses in the body (see chapter 2).

Predict 5

> *Many poisons function by blocking certain steps in the metabolic pathways. For example, cyanide blocks the last step in the electron-transport chain. Explain why this blockage causes death.*

Summary of Anaerobic Respiration and Aerobic Respiration

In anaerobic respiration, each glucose molecule ($C_6H_{12}O_6$) yields 2 ATP and 2 lactic acid molecules ($C_3H_6O_3$) through glycolysis:

$$C_6H_{12}O_6 + 2\,ADP + P_i \rightarrow 2\,C_3H_6O_3 + 2\,ATP$$

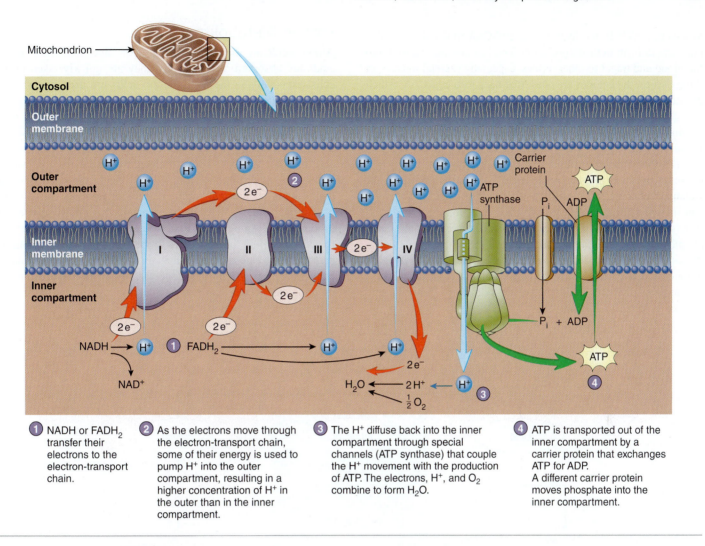

Mitochondrion

Cytosol

Outer membrane

Outer compartment

Inner membrane

Inner compartment

I   II   III   IV

$2e^-$ $2e^-$ $2e^-$ $2e^-$ $2e^-$

NADH → H$^+$

NAD$^+$

FADH$_2$ → H$^+$

H$^+$

H_2O ← $2 H^+$
         $\frac{1}{2} O_2$

ATP synthase

Carrier protein

ATP

P$_i$ ADP

P$_i$ + ADP

ATP

① NADH or FADH$_2$ transfer their electrons to the electron-transport chain.

② As the electrons move through the electron-transport chain, some of their energy is used to pump H$^+$ into the outer compartment, resulting in a higher concentration of H$^+$ in the outer than in the inner compartment.

③ The H$^+$ diffuse back into the inner compartment through special channels (ATP synthase) that couple the H$^+$ movement with the production of ATP. The electrons, H$^+$, and O$_2$ combine to form H$_2$O.

④ ATP is transported out of the inner compartment by a carrier protein that exchanges ATP for ADP. A different carrier protein moves phosphate into the inner compartment.

PROCESS Figure 17.7 AP|R **Electron-Transport Chain**

The electron-transport chain in the inner membrane consists of four protein complexes (*purple;* numbered I to IV) with carrier molecules.

In contrast, in aerobic respiration, each glucose molecule yields 38 ATP:

$$C_6H_{12}O_6 + 6\,O_2 + 38\,ADP + 38\,P_i \rightarrow 6\,CO_2 + 6\,H_2O + 38\,ATP$$

Of the 38 ATP molecules, 2 result from glycolysis, 2 result from the citric acid cycle, and 34 form through the electron-transport chain. Thus, aerobic respiration is much more efficient at producing ATP than is anaerobic respiration. In addition, many of the chemical reactions of aerobic respiration can be used to produce energy from other food molecules, such as lipids and proteins (see "Lipid Metabolism" and "Protein Metabolism" next).

The number of ATP molecules produced during aerobic respiration can also be reported as 36. The 2 NADH molecules produced by glycolysis cannot cross the inner mitochondrial membrane; thus, their electrons are donated to a shuttle molecule that carries the electrons to the electron-transport chain. Depending on the shuttle molecule, each glycolytic NADH molecule can produce 2 or 3 ATP molecules. In skeletal muscle and in the brain, only 2 ATP molecules are produced for each NADH mol-

ecule formed during glycolysis, resulting in a total number of 36 ATP molecules; however, in the liver, kidneys, and heart, 3 ATP molecules are produced for each NADH molecule, and the total number of ATP molecules formed is 38.

Lipid Metabolism

Triglycerides, or fat, are the body's main energy-storage molecules. In a normal person, fat is responsible for about 99% of the body's energy storage, and glycogen accounts for the remaining 1%.

Between meals, triglycerides in adipose tissue are broken down into fatty acids and glycerol. Some of the fatty acids produced are released into the blood. Other tissues, especially skeletal muscle and the liver, use the fatty acids as a source of energy.

The metabolism of fatty acids takes place in the mitochondria. It occurs by a series of reactions wherein two carbon atoms are removed from the end of a fatty acid chain to form acetyl-CoA (figure 17.8). As the process continues, carbon atoms are removed two at a time until the entire fatty acid chain is converted into acetyl-CoA. Acetyl-CoA can enter the citric acid cycle and be used to generate ATP. In the liver, two acetyl-CoA molecules can also combine to form **ketones** (kē′tōnz). The ketones are released into the blood and travel to other tissues, especially skeletal muscle. In these tissues, the ketones are converted back to acetyl-CoA, which can enter the citric acid cycle to produce ATP.

The presence of small amounts of ketones in the blood is normal and beneficial, but excessive production of ketones is called **ketosis** (ke-to′sis). If the increased number of ketones exceeds the capacity of the body's buffering systems, acidosis, a decrease in blood pH, can occur (see "Regulation of Acid-Base Balance," chapter 18). Conditions that increase lipid metabolism can increase the rate of ketone formation. Examples are starvation (see the Clinical Impact "Starvation and Obesity" later in this chapter), diets consisting of proteins and fats with few carbohydrates, and untreated diabetes mellitus (see chapter 10). Because ketones are excreted by the kidneys and lungs, characteristics of untreated diabetes mellitus include ketones in the urine and "acetone breath."

Protein Metabolism

Amino acids are the products of protein digestion. Once amino acids are absorbed into the body, they are quickly taken up by cells, especially in the liver. Amino acids are used primarily to synthesize needed proteins and only secondarily as a source of energy. If serving as a source of energy, amino acids can be used in two ways (figure 17.8): (1) The amino acids can be converted into the molecules of carbohydrate metabolism, such as pyruvic acid and acetyl-CoA. These molecules can be metabolized to yield ATP. (2) The amine group ($-NH_2$) can be removed from the amino acid, leaving ammonia and an α-keto acid. This process produces NADH, which can enter the electron-transport chain to produce ATP. Ammonia is toxic to cells, so the liver converts it to urea, which the blood carries to the kidneys, where it is eliminated. The α-keto acid can enter the citric acid cycle or can be converted into pyruvic acid, acetyl-CoA, or glucose. Although proteins can serve as an energy source, they are not considered major storage molecules.

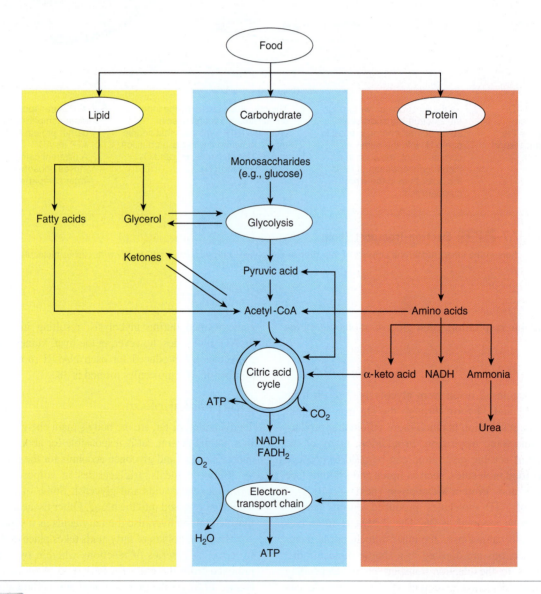

Figure 17.8 **AP|R** **Overall Pathways for the Metabolism of Food**
Carbohydrates, lipids, and proteins enter biochemical pathways to produce energy in the cell.

Metabolic States

The body experiences two major metabolic states. The first is the **absorptive state,** the period immediately after a meal, when nutrients are being absorbed through the intestinal wall into the circulatory and lymphatic systems (figure 17.9). The absorptive state usually lasts about 4 hours after each meal, and most of the glucose that enters the circulation is used by cells for the energy they require. The remainder of the glucose is converted into glycogen or lipids. Most of the absorbed fats are deposited in adipose tissue. Many of the absorbed amino acids are used by cells in gene expression, and some are used for energy; still others enter the liver and are converted to lipids or carbohydrates.

The second state, the **postabsorptive state,** occurs late in the morning, late in the afternoon, or during the night after each absorptive state is concluded (figure 17.10). Normal blood glucose levels range between 70 and 110 mg/dL, and it is vital to the body's homeostasis that this range be maintained. During the postabsorptive state, blood glucose levels are maintained by the conversion of other molecules to glucose. The first source of blood glucose during the postabsorptive state is the glycogen stored in the liver. However, this glycogen supply can provide glucose for only about four hours. The glycogen stored in skeletal muscles can also be used during vigorous exercise. As the glycogen stores are depleted, the body uses lipids as an energy source. The glycerol from triglycerides can be converted to glucose. The fatty acids can be converted to acetyl-CoA, moved into the citric acid cycle, and used as a source of energy to produce ATP. In the liver, acetyl-CoA can be used to produce ketone bodies that other tissues can use for energy. The use of fatty acids as an energy source can partly eliminate the need to use glucose for energy, so that less glucose is removed from the blood and homeostasis is maintained. The amino acids of proteins can be converted to glucose or can be used to produce energy, again sparing blood glucose.

Metabolic Rate

Metabolic rate is the total amount of energy produced and used by the body per unit of time. Metabolic rate is usually estimated by measuring the amount of oxygen used per minute. One liter of oxygen consumed by the body is estimated to produce 4.825 kcal of energy.

Metabolic energy can be used in three ways: for basal metabolism, for muscle contraction, and for the assimilation of food, which involves processes such as the production of digestive enzymes and the active transport of digested molecules. The **basal metabolic rate (BMR)** is the energy needed to keep the resting body functional. It is the metabolic rate calculated in expended kilocalories per square meter of body surface area per hour. BMR is measured when a person is awake but restful and has not eaten for 12 hours. A typical BMR for a 70-kg (154-lb) male is 38 kcal/m²/hour.

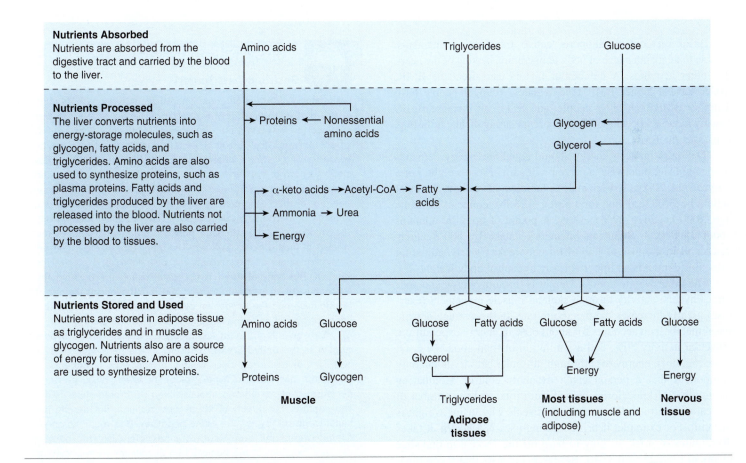

Figure 17.9 Events of the Absorptive State

Absorbed molecules, especially glucose, are used as sources of energy. Molecules not immediately needed for energy are stored: Glucose is converted to glycogen or triglycerides, triglycerides are deposited in adipose tissue, and amino acids are converted to triglycerides or carbohydrates.

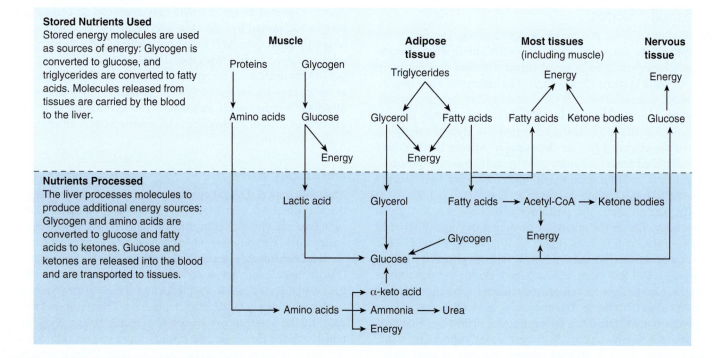

Stored Nutrients Used
Stored energy molecules are used as sources of energy: Glycogen is converted to glucose, and triglycerides are converted to fatty acids. Molecules released from tissues are carried by the blood to the liver.

Nutrients Processed
The liver processes molecules to produce additional energy sources: Glycogen and amino acids are converted to glucose and fatty acids to ketones. Glucose and ketones are released into the blood and are transported to tissues.

Figure 17.10 Events of the Postabsorptive State
Stored energy molecules are used as sources of energy: Glycogen is converted to glucose; triglycerides are broken down to fatty acids, some of which are converted to ketones; and proteins are converted to glucose.

Basal metabolism supports active transport mechanisms, muscle tone, maintenance of body temperature, beating of the heart, and other activities. A number of factors can affect the BMR. Males have a higher BMR than females, younger people have a higher BMR than older people, and fever can increase BMR. Greatly reduced kilocaloric input, as occurs during dieting or fasting, depresses the BMR.

The daily input of energy should equal the energy demands of metabolism; otherwise, a person will gain or lose weight. For a 23-year-old, 70-kg (154-lb) male to maintain his weight, the input should be 2700 kcal/day; for a 58-kg (128-lb) female of the same age, 2000 kcal/day are necessary. A pound of body fat contains about 3500 kcal. Reducing kilocaloric intake by 500 kcal/day can result in the loss of 1 lb of fat per week. Clearly, adjusting kilocaloric input is an important way to control body weight.

The other way to control weight is through energy expenditure. Physical activity through skeletal muscle movement (exercise) greatly increases the metabolic rate. In the average person, basal metabolism accounts for about 60% of energy expenditure, muscular activity 30%, and assimilation of food about 10%. Of these amounts, energy loss through muscular activity is the only component that a person can reasonably control. Comparing the number of kilocalories gained from food and the number of kilocalories burned during exercise reveals why losing weight is difficult. For example, if brisk walking uses 225 kcal/h, it takes 20 minutes of brisk walking to burn off the 75 kcal in one slice of bread (75/225 = 0.33 h). Research suggests that a combination of appropriate physical activity and appropriate kilocaloric intake is the best approach to maintaining a healthy body composition and weight.

A CASE IN POINT

Gastric Bypass Surgery

Les Moore is a 35-year-old male who weighs 420 pounds. Although at one time he weighed less as a result of dieting, he has not only regained the lost weight but also gained additional weight. Les has type 2 diabetes, high blood cholesterol, and high blood pressure. Because of his extreme obesity and related health problems, Les elects to have a Roux-en-Y gastric bypass. In this procedure, the digestive tract is surgically rearranged to form a Y-shaped structure. One arm of the Y is formed by separating the superior part of the stomach to form a small pouch that is connected to the jejunum. The small size of the stomach pouch dramatically reduces the amount of food a person can eat. The other arm of the Y consists of the remainder of the stomach, the duodenum, and part of the jejunum. The body of the Y is the distal jejunum. Food from the esophagus enters the small stomach pouch and bypasses the rest of the stomach and the duodenum by entering the jejunum. Bile and pancreatic juices enter the duodenum and pass into the jejunum.

Roux-en-Y is the most commonly performed gastric bypass surgery in the United States. It results in the permanent loss of over 50% of presurgery body weight for over 90% of patients. It also cures most cases of type 2 diabetes, lowers blood pressure, decreases cholesterol levels, reduces the risk of premature death, and improves the quality of life in most patients. But despite the benefits, gastric bypass has some negatives. It is major surgery—complications and even death are possible. Also, the normal digestive functions of the stomach and duodenum are bypassed as food moves quickly through the stomach pouch into the jejunum. As a result, vitamin and mineral deficiencies can develop—particularly, deficiencies of vitamin B$_{12}$, calcium, and iron.

CLINICAL IMPACT   Starvation and Obesity

Starvation

Starvation is the inadequate intake of nutrients or the inability to metabolize or absorb nutrients. Starvation can be caused by a number of factors, such as prolonged fasting, anorexia, deprivation, or disease. No matter what the cause, starvation takes approximately the same course and consists of three phases. The events of the first two phases occur even during relatively short periods of fasting or dieting. The third phase occurs in prolonged starvation and ends in death.

During the first phase of starvation, blood glucose levels are maintained through the production of glucose from glycogen, proteins, and fats. At first, glycogen is broken down into glucose. However, enough glycogen is stored in the liver to last only a few hours. Thereafter, blood glucose levels are maintained by the breakdown of proteins and fats. Fats are decomposed into fatty acids and glycerol. Fatty acids can be used as a source of energy, especially by skeletal muscle, thus decreasing the use of glucose by tissues other than the brain. The brain cannot use fatty acids as an energy source, so the conservation of glucose is critical to normal brain function. Glycerol can be used to make a small amount of glucose, but most of the glucose is formed from the amino acids of proteins. In addition, some amino acids can be used directly for energy.

In the second stage, which can last for several weeks, fats are the primary energy source. The liver metabolizes fatty acids into ketone bodies that can serve as a source of energy. After about a week of fasting, the brain begins to use ketone bodies, as well as glucose, for energy. This usage decreases the demand for glucose, and the rate of protein breakdown diminishes but does not stop. In addition, the use of proteins is selective; that is, proteins not essential for survival are used first.

The third stage of starvation begins when the fat reserves are nearly depleted and the body switches to proteins as the major energy source. Muscles, the largest source of protein in the body, are rapidly depleted. At the end of this stage, proteins essential for cellular functions are broken down, and cell function degenerates. Death can occur very rapidly.

Symptoms of starvation, in addition to weight loss, include apathy, listlessness, withdrawal, and increased susceptibility to infectious diseases. Few people die directly from starvation because they usually die of an infectious disease first. Other signs of starvation include changes in hair color, flaky skin, and massive edema in the abdomen and lower limbs, causing the abdomen to appear bloated.

During starvation, the body's ability to consume normal volumes of food also decreases. Foods high in bulk but low in protein content often cannot reverse the process of starvation. Intervention involves feeding the starving person low-bulk foods containing ample protein, calories, vitamins, and minerals. Starvation also results in dehydration; thus, rehydration is an important part of intervention. Even with intervention, a victim may be so affected by disease or weakness that he or she cannot recover.

Obesity

Obesity is the storage of excess fat, and it results from the ingestion of more food than is necessary for the body's energy needs. Obesity can be defined on the basis of body weight, body mass index, or body fat. "Desirable body weight" is listed in a table produced by the Metropolitan Life Insurance Company and indicates, for any height, the weight associated with a maximum lifespan. Overweight is defined as 10% more than the desirable weight, and obesity is 20% more than the desirable weight. **Body mass index (BMI)** can be calculated by dividing a person's weight (Wt) in kilograms by the square of his or her height (Ht) in meters:

$$BMI = Wt/Ht^2$$

A BMI greater than 25–27 is overweight, and a value greater than 30 is defined as obese. About 26% of Americans have a BMI of 30 or greater. In terms of the percent of total body weight contributed by fat, 15% body fat in men and 25% body fat in women is associated with reduced health risks. Obesity is defined as more than 25% body fat in men and 30–35% in women.

The distribution of fat in obese individuals can vary. Fat can accumulate mainly in the upper body, such as in the abdominal region, or it can be associated with the hips and buttocks. These distribution differ-

ences can be clinically significant because upper body obesity is associated with an increased likelihood of diabetes mellitus, cardiovascular disease, stroke, and death.

In some cases, obesity is caused by a medical condition. For example, a tumor in the hypothalamus can stimulate overeating. In most cases, however, no specific cause can be recognized. In fact, obesity can occur for many reasons and can have more than one cause in the same individual. There seems to be a genetic component for obesity; if one or both parents are obese, their children are more likely to be obese. Environmental factors, such as eating habits, can also play an important role. For example, adopted children can exhibit obesity similar to that of their adoptive parents. In addition, psychological factors, such as overeating as a way to deal with stress, can contribute to obesity.

In **hypertrophic** (hī-per-trof′ik; *hyper-*, above normal + *trophē*, nourishment) **obesity,** the number of adipose cells is usually normal, but the amount of fat contained in each adipose cell is increased. This type of obesity is characteristic of adult-onset obesity. People who were thin or of average weight and quite active when young become less active as they age. Although they no longer use as many kilocalories, they still consume the same amount of food as when they were younger. The excess kilocalories (see "Metabolic Rate" earlier in this section) are used to synthesize fat. In this type of obesity, the amount of fat in each adipose cell increases, and if the amount of stored fat continues to increase, the total number of adipose cells may also increase. It is estimated that the average U.S. resident gains 1.25–1.5 lb of fat per year after age 25 and, at the same time, loses 0.25–0.5 lb of lean body weight (muscle mass) per year.

In **hyperplastic** (hī-per-plas′tik; *hyper-* + *plasis,* a molding) **obesity,** which is characteristic of juvenile-onset obesity, the number of adipose cells is increased. This condition may also be accompanied by an increase in cell size (hypertrophic obesity). Hyperplastic obesity has a very strong hereditary component, but family eating habits can also have a great influence. People with hyperplastic obesity are obese as children and become more obese with age. This type of obesity is a major health problem in school-age children.

Predict 6

If watching TV uses 95 kcal/h, how long does it take to burn off the kilocalories in one cola or beer (see table 17.1)? If jogging at a pace of 6 mph uses 580 kcal/h, how long does it take to use the kilocalories in one cola or beer?

17.3 BODY TEMPERATURE REGULATION

Learning Outcome   *After reading this section, you should be able to*

A. Describe heat production and regulation in the body.

Humans can maintain a relatively constant internal body temperature despite changes in the temperature of the surrounding environment. A constant body temperature is very important to homeostasis. For example, environmental temperatures are too low for normal enzyme function. The heat produced by metabolism and muscle contraction helps maintain the body temperature at a steady, elevated level that is high enough for normal enzyme function. Excessively high temperatures can alter enzyme structure, resulting in loss of the enzyme's function.

Free energy is the total amount of energy that can be liberated by the complete catabolism of food. About 40% of the total energy released by catabolism is used to accomplish biological work, such as anabolism, muscular contraction, and other cellular activities. The remaining energy is lost as **heat.**

Predict 7

Explain why we become warm during exercise. Why is shivering useful when it is cold?

Normal body temperature is regulated in the same way as other homeostatic conditions in the body. The average normal temperature is usually considered to be 37°C (98.6°F) when measured orally and 37.6°C (99.7°F) when measured rectally. Rectal temperature comes closer to the true core body temperature, but an oral temperature is more easily obtained in older children and adults and therefore is the preferred measure. The normal oral temperature may vary from person to person, with a range of approximately 36.1–37.2°C (97–99°F).

Body temperature is maintained by balancing heat input with heat loss. The body and the environment exchange heat in a number of ways (figure 17.11). **Radiation** is the gain or loss of heat as infrared energy between two objects that are not in physical contact with each other. For example, heat can be gained by radiation from the sun, a hot coal, or the hot sand of a beach. On the other hand, heat can be lost as radiation to cool vegetation, water in the ocean, or snow on the ground. **Conduction** is the exchange of heat between objects that are in direct contact with each other, such as the bottom of the feet and the ground. **Convection** is a transfer of heat between the body and the air or water. A cool breeze moves air over the body and causes loss of heat from the body. **Evaporation** is the conversion of water from a liquid to a gaseous form. As water evaporates from body surfaces, heat is lost.

The amount of heat exchanged between the environment and the body is determined by the difference in temperature between the two. The greater the temperature difference, the greater the rate of heat exchange. Control of the temperature difference can be used to regulate body temperature. For example, if the environmental

Figure 17.11 Heat Exchange

Heat exchange between a person and the environment occurs by convection, radiation, evaporation, and conduction. *Arrows* show the direction of net heat gain or loss in this environment.

temperature is very cold, as occurs on a winter day, there is a large temperature difference between the body and the environment, and therefore a large loss of heat. Behaviorally, we can reduce the loss of heat by selecting a warmer environment, such as by going inside a heated house or by putting on extra clothes. Physiologically, the body controls the temperature difference through dilation and constriction of blood vessels in the skin. When the blood vessels dilate, they bring warm blood to the surface of the body, raising skin temperature; when the blood vessels constrict, blood flow and skin temperature decrease (see figure 5.8).

Predict 8

Explain why constriction of skin blood vessels on a cold winter day is beneficial.

When environmental temperature is greater than body temperature, dilation of blood vessels in the skin brings blood to the skin, causing an increase in skin temperature that decreases the gain of heat from the environment. At the same time, evaporation carries away excess heat to prevent heat gain and overheating.

Body temperature regulation is an example of a negative-feedback system (figure 17.12). Maintenance of a specific body temperature is accomplished by neurons in the hypothalamus, which regulate body temperature around a set point. A small area in the anterior part of the hypothalamus can detect slight increases in body temperature through changes in blood temperature. As a

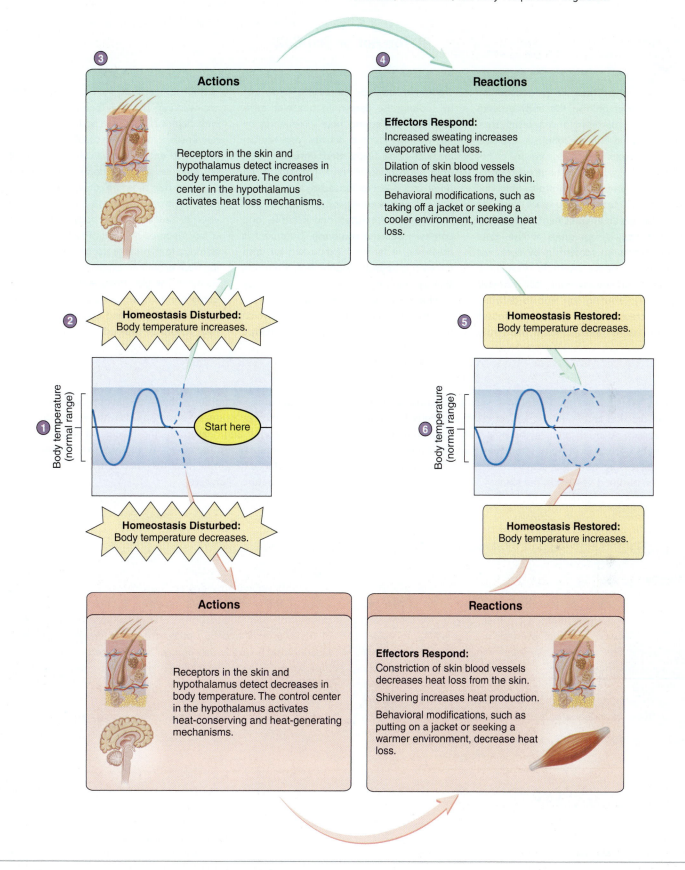

Homeostasis Figure 17.12 Temperature Regulation

(1) Body temperature is within normal range. (2) Body temperature increases outside the normal range, which causes homeostasis to be disturbed. (3) Receptors in the skin and hypothalamus detect the increase in body temperature and the control center in the hypothalamus responds to the change in body temperature. (4) The effectors are activated. Blood vessels in the skin dilate and sweating increases to promote heat loss and evaporative cooling. (5) Body temperature decreases. (6) Body temperature returns to its normal range, and homeostasis is restored.

CLINICAL IMPACT   Too Hot or Too Cold

Hyperthermia is a condition in which heat gain in the body exceeds heat loss. Hyperthermia can result from exposure to a hot environment, exercise, or fever. Prolonged exposure to a hot environment can lead to **heat exhaustion,** in which normal temperature reduction mechanisms cannot keep pace with the excessive environmental heat, thus allowing the body temperature to rise. Heat exhaustion is characterized by cool, wet skin due to heavy sweating. Weakness, dizziness, and nausea usually occur as well. The heavy sweating can lead to dehydration, decreased blood volume, decreased blood pressure, and increased heart rate.

Treatment involves increasing heat loss by moving the person to a cooler environment, decreasing heat production by decreasing muscular activity, and replacing lost body fluids. **Heat stroke** results from an increase in the hypothalamic set point and is characterized by dry, flushed skin because sweating is inhibited. The person becomes confused and irritable, and can even become comatose. Treatment is the same as for heat exhaustion but also involves increasing evaporation from the skin by applying water to the skin or by placing the person in cool water.

Hypothermia is a condition in which heat loss exceeds heat gain. The body's normal temperature increase mechanisms are working, but they cannot keep pace with heat loss, and the body temperature decreases. Hypothermia usually results from prolonged exposure to a cold environment or even to a cool, damp environment because the moisture draws heat away from the body. Treatment for hypothermia calls for rewarming the body at a rate of a few degrees per hour. **Frostbite** is local damage to the skin or deeper tissues resulting from prolonged exposure to a cold environment. The best treatment for frostbite is immersion in a warm-water bath. Rubbing the affected area and applying local, dry heat should be avoided.

result, mechanisms that cause heat loss, such as dilation of blood vessels in the skin and sweating, are activated, and body temperature decreases. A small area in the posterior hypothalamus can detect slight decreases in body temperature and can initiate heat gain by increasing muscular activity (shivering) and by initiating constriction of blood vessels in the skin.

Under some conditions, the hypothalamus set point is actually changed. For example, during a fever, the set point is raised. Heat-conserving and heat-producing mechanisms are stimulated, and body temperature increases. To recover from a fever, the set point is reduced to normal, heat loss mechanisms are initiated, and body temperature decreases.

ANSWER TO LEARN TO PREDICT

Although Sadie's suggested snack did contain a lot of calories, which she and David needed for their day at the park, the food choices were not ideal. Most of the calories in the cookies and grape soda were actually from simple sugars. Eating large amounts of simple sugars, such as Sadie's suggested snack, could result in large fluctuations in blood glucose levels. Though the children may initially have an increase in energy, they will most likely experience a drastic decrease in energy as well.

Foods that include complex carbohydrates, such as those in the snack Sadie's mom suggested, have other nutrients, such as vitamins, many of which are necessary for normal metabolism. In addition, complex carbohydrates are digested and absorbed at a slower rate and do not contribute to drastic changes in blood glucose levels. Essentially, Sadie's mom selected food that would provide the children with energy and additional beneficial nutrients.

Answers to the rest of this chapter's Predict questions are in Appendix E.

SUMMARY

17.1 Nutrition   (p. 476)

Nutrition is the ingestion and use of food.

Nutrients

1. Nutrients, the chemicals used by the body, consist of carbohydrates, lipids, proteins, vitamins, minerals, and water.
2. Essential nutrients either cannot be produced by the body or cannot be produced in adequate amounts.

3. The MyPlate icon provides a visual reminder for making choices at mealtime, by selecting healthy foods from five food groups. Half the meal should be fruits and vegetables.

Kilocalories

A kilocalorie is the energy required to raise the temperature of 1000 g of water 1°C. A kilocalorie (Calorie) is the unit of measurement used to express the energy content of food.

Carbohydrates

1. Carbohydrates include monosaccharides, disaccharides, and polysaccharides.
2. Most of the carbohydrates we ingest are from plants.
3. Carbohydrates are used as an energy source and for making DNA, RNA, and ATP.
4. The Acceptable Macronutrient Distribution Range (AMDR) for carbohydrates is 45–65% of total kilocalories.

Lipids

1. Lipids include triglycerides, steroids, phospholipids, and fat-soluble vitamins. Triglycerides are a major source of energy. Eicosanoids are involved in inflammation, tissue repair, and smooth muscle contraction. Cholesterol and phospholipids are part of the cell membrane. Some steroid hormones regulate the reproductive system.
2. The AMDR for lipids is 20–35%.

Proteins

1. Proteins are chains of amino acids.
2. Animal proteins tend to be complete proteins, whereas plant proteins tend to be incomplete.
3. Proteins are involved in structural strength, muscle contraction, regulation, buffering, clotting, transport, ion channels, receptors, and the immune system.
4. The AMDR for protein is 10–35% of total kilocalories.

Vitamins

1. Most vitamins are not produced by the body and must be obtained in the diet. Some vitamins can be formed from provitamins.
2. Vitamins are important in energy production, nucleic acid synthesis, growth, and blood clotting.
3. Vitamins are classified as either fat-soluble or water-soluble.
4. Recommended dietary allowances (RDAs) are a guide for estimating the nutritional needs of groups of people on the basis of their age, their gender, and other factors.

Minerals

Minerals are essential for normal metabolic functions. They are involved with establishing the resting membrane potential; generating action potentials; adding mechanical strength to bones and teeth; combining with organic molecules; and acting as coenzymes, buffers, or regulators of osmotic pressure.

Daily Values

1. Daily Values are dietary references that can be used to plan a healthful diet.
2. Daily Values for vitamins and minerals are based on Reference Daily Intakes (RDIs), which are generally the highest 1968 RDA value of an age category.
3. Daily Values are based on Daily Reference Values. The Daily Reference Values for energy-producing nutrients (carbohydrates, total fat, saturated fat, and proteins) and dietary fiber are recommended percentages of the total kilocalories ingested daily for each nutrient. The Daily Reference Values for total fats, saturated fats, cholesterol, and sodium are the uppermost limits considered desirable because of their link to diseases.
4. The % Daily Value is the percentage of the recommended Daily Value of a nutrient found in one serving of a particular food.

17.2 Metabolism   (p. 484)

1. Metabolism consists of catabolism and anabolism. Catabolism, the breakdown of molecules, gives off energy. Anabolism, the synthesis of molecules, requires energy.

2. The energy in carbohydrates, lipids, and proteins is used to produce ATP.
3. The energy from ATP can be used for active transport, muscle contraction, and the synthesis of molecules.

Regulation of Metabolism

1. A biochemical pathway is a series of chemical reactions, some of which release energy that can be used to synthesize ATP.
2. Each step in a biochemical pathway requires enzymes.
3. Enzyme synthesis is determined by DNA. Enzyme activity is modified by receptor-mediated and end-product processes.

Carbohydrate Metabolism

1. Glycolysis is the breakdown of glucose to two pyruvic acid molecules. Two ATP molecules are also produced.
2. Anaerobic respiration is the breakdown of glucose in the absence of oxygen to two lactic acid molecules and two ATP molecules.
3. Lactic acid can be converted to glucose using aerobically produced ATP; the necessary oxygen contributes to the oxygen deficit.
4. Aerobic respiration is the breakdown of glucose in the presence of oxygen to produce carbon dioxide, water, and 38 molecules of ATP. The first phase of aerobic respiration is glycolysis; the second phase is the conversion of pyruvic acid to acetyl-CoA; the third phase is the citric acid cycle; and the fourth phase is the electron-transport chain, which uses carrier molecules, such as NADH, to synthesize ATP.

Lipid Metabolism

1. Lipids are broken down in adipose tissue, and fatty acids are released into the blood.
2. Fatty acids are taken up by cells and broken down into acetyl-CoA, which can enter the citric acid cycle. Acetyl-CoA can also be converted into ketones by the liver. Ketones released from the liver into the blood are used as energy sources by other cells.

Protein Metabolism

1. Amino acids are used to synthesize proteins.
2. Amino acids can be used for energy, yielding ammonia as a by-product. Ammonia is converted to urea and excreted by the kidneys.

Metabolic States

1. In the absorptive state, nutrients are used as energy, with the remainder being stored.
2. In the postabsorptive state, stored nutrients are used for energy.

Metabolic Rate

1. The metabolic rate is the total energy expenditure per unit of time.
2. Metabolic energy is used for basal metabolism, muscular activity, and the assimilation of food.

17.3 Body Temperature Regulation   (p. 494)

1. Body temperature is a balance between heat gain and heat loss.
2. Heat is produced through metabolism.
3. Heat is exchanged through radiation, conduction, convection, and evaporation.
4. The greater the temperature difference, the greater the rate of heat exchange.
5. Body temperature is maintained around a set point by neural circuits in the hypothalamus.
6. Dilation of blood vessels in the skin and sweating increase heat loss from the body.
7. Constriction of blood vessels in the skin and shivering promote heat gain by the body.

REVIEW AND COMPREHENSION

1. Define a nutrient, and list the six major classes of nutrients. What is an essential nutrient?

2. What is a kilocalorie (Calorie)? Distinguish between a Calorie and a calorie.

3. List some sources of carbohydrates, lipids, and proteins in the diet.

4. List the recommended consumption amounts of carbohydrates, lipids, and proteins.

5. What are vitamins and provitamins? Name the water-soluble vitamins and the fat-soluble vitamins. List some of the functions of vitamins.

6. What are the Recommended Dietary Allowances (RDAs)?

7. List some of the minerals, and give their functions.

8. What are the Daily Values? How are the Daily Values related to total daily kilocaloric intake? Why are some Daily Values considered the uppermost amounts that should be consumed?

9. Define a % Daily Value.

10. Define a biochemical pathway. How are the steps in a biochemical pathway controlled? What are three ways in which enzymes are regulated?

11. Describe glycolysis. What molecule is the end product of glycolysis? How many ATP and NADH molecules are produced?

12. What determines whether the pyruvic acid produced in glycolysis becomes lactic acid or acetyl-CoA?

13. Describe the two phases of anaerobic respiration. How many ATP molecules are produced? What happens to the lactic acid produced when oxygen becomes available?

14. Define aerobic respiration, and state how many ATP molecules are produced.

15. Describe the citric acid cycle.

16. What is the function of the electron-transport chain?

17. What happens to the carbon atoms in ingested food during metabolism? What happens to the oxygen we breathe in during metabolism?

18. Describe the events occurring during the absorptive and postabsorptive metabolic states.

19. What is meant by metabolic rate? Name three ways that the body uses metabolic energy.

20. Describe how heat is produced by and lost from the body. How is body temperature regulated?

CRITICAL THINKING

1. One serving of a food contains 2 g of saturated fat. What % Daily Value for saturated fat would appear on the food label? (See the bottom of figure 17.2a for information needed to answer this question.)

2. An active teenage boy consumes 3000 kcal/day. What is the maximum amount (weight) of total fats he should consume, according to the % Daily Values?

3. If the teenager in question 2 eats a food that has a total fat content of 10 g/serving, what is his total fat % Daily Value?

4. Suppose the food in question 3 is in a package that lists a serving size of $\frac{1}{2}$ cup, with four servings in the package. If the teenager eats half the contents of the package (1 cup), how much of his % Daily Value does he consume?

5. Why can some people lose weight on a 1200 kcal/day diet, whereas other people cannot?

6. Lotta Bulk, a bodybuilder, wanted to increase her muscle mass. Knowing that proteins are the main components of muscle, she consumed large amounts of protein daily (high-protein diet), along with small amounts of lipids and carbohydrates. Explain why this strategy will or will not work.

7. After consuming a high-protein diet for several days, does Lotta Bulk's urine contain less, the same amount of, or more urea than before she consumed the proteins? Explain.

8. Thyroid hormone is known to increase the activity of the sodium-potassium exchange pump, which is an active-transport mechanism, thereby increasing the breakdown of ATP. If a person produced excess amounts of thyroid hormone, what effect would this have on basal metabolic rate, body weight, and body temperature? How might the body attempt to compensate for the changes in body weight and temperature?

9. On learning that sweat evaporation results in the loss of calories, an anatomy and physiology student decides that sweating is an easier way to lose weight than dieting. He knows that a liter (about a quart) of water weighs 1000 g, which is equivalent to 580,000 cal (or 580 kcal) of heat when lost as sweat. He believes that instead of reducing his caloric intake by 580 kcal/day, he can lose about a pound of fat a week by losing a liter of sweat every day in the sauna. Will this approach work? Explain.

10. It is recommended that a person on a diet drink six to eight glasses of cool water per day. How could this practice aid weight loss?

11. In some diseases, an infection results in a high fever. The patient is on the way to recovery when the crisis is over and body temperature begins to return to normal. If you were looking for symptoms in a patient who had just passed through the crisis state, would you look for a dry, pale skin or a wet, flushed skin? Explain.

Answers in Appendix D

18 Urinary System and Fluid Balance

LEARN TO PREDICT

Baby Sadie was born early one morning. Not long after this picture was taken on her first day home from the hospital, Sadie's parents noticed that her diapers were excessively wet hour after hour throughout the day and night. In addition, Sadie was irritable, had a slight fever, and had vomited even though she had not eaten for several feedings. Her parents took her to the pediatrician, who ordered blood tests. The tests indicated that Sadie had normal levels of antidiuretic hormone (ADH). After reading this chapter, predict Sadie's disorder.

18.1 FUNCTIONS OF THE URINARY SYSTEM

Learning Outcomes *After reading this section, you should be able to*

A. List the structures that make up the urinary system.

B. List the major functions of the urinary system.

Although you probably know that each person has two kidneys and you most likely can identify their general location, you may not be aware of the many functions the kidneys perform. On the other hand, you probably have a much better understanding of the function of the urinary bladder and a great appreciation for the attention required when it is filled with the urine produced by the kidneys.

The **urinary** (ūr′i-nār-ē) **system** consists of two kidneys, two ureters, the urinary bladder, and the urethra (figure 18.1). A large volume of blood flows through the kidneys, which remove substances from the blood to form urine. The urine contains excess water and ions, metabolic wastes (such as urea), and toxic substances. The urine produced by the kidneys flows through the ureters to the urinary bladder, where it is stored until it is eliminated through the urethra.

The kidneys can suffer extensive damage and still maintain their extremely important role in the maintenance of homeostasis. As long as about one-third of one kidney remains functional, survival is possible. However, if the functional ability of the kidneys fails completely, death will result unless the person receives medical treatment.

Module 13 Urinary System

The major function of the urinary system is to control the composition and volume of body fluids. The kidneys perform this function through multiple processes:

1. *Excretion.* The kidneys are the major excretory organs of the body. They remove waste products from the blood. Many waste products are toxic, but most are metabolic by-products of cells and substances absorbed from the intestine. The skin, liver, lungs, and intestines eliminate some of these waste products, but they cannot compensate if the kidneys fail to function.

2. *Regulation of blood volume and pressure.* The kidneys play a major role in controlling the extracellular fluid volume in the body. They can produce either a large volume of dilute urine or a small volume of concentrated urine. Thereby, the kidneys regulate blood volume and blood pressure.

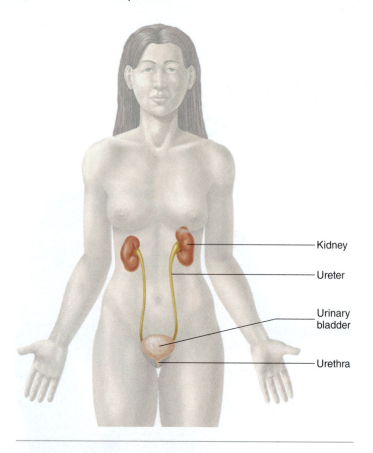

Figure 18.1 AP|R **Urinary System**

The urinary system consists of two kidneys, two ureters, the urinary bladder, and the urethra.

(labels: Kidney, Ureter, Urinary bladder, Urethra)

3. *Regulation of the concentration of solutes in the blood.* The kidneys help regulate the concentration of the major molecules and ions, such as glucose, Na^+, Cl^-, K^+, Ca^{2+}, HCO_3^-, and HPO_4^{2-}.

4. *Regulation of extracellular fluid pH.* The kidneys excrete variable amounts of H^+ to help regulate extracellular fluid pH.

5. *Regulation of red blood cell synthesis.* The kidneys secrete a hormone, erythropoietin, which regulates the synthesis of red blood cells in bone marrow (see chapter 11).

6. *Regulation of vitamin D synthesis.* The kidneys play an important role in controlling blood levels of Ca^{2+} by regulating the synthesis of vitamin D (see chapter 6).

18.2 ANATOMY OF THE KIDNEYS

Learning Outcomes *After reading this section, you should be able to*

A. Describe the location and anatomy of the kidneys.

B. Describe the structure of the nephron and the location of the parts of the nephron in the kidney.

The **kidneys** are bean-shaped organs, each about the size of a tightly clenched fist. They lie on the posterior abdominal wall, behind the peritoneum, with one kidney on each side of the vertebral column (figure 18.2). Structures that are behind the peritoneum are said to be **retroperitoneal** (re′trō-per′i-tō-nē′ăl). A layer of connective tissue

called the **renal** (derived from the Latin word for kidney) **capsule** surrounds each kidney. Around the renal capsule is a thick layer of adipose tissue, which protects the kidney from mechanical shock. On the medial side of each kidney is the **hilum** (hī′lŭm), where the renal artery and nerves enter and where the renal vein, ureter, and lymphatic vessels exit the kidney (figure 18.3). The hilum opens into a cavity called the **renal sinus,** which contains blood vessels, part of the system for collecting urine, and adipose tissue.

The kidney is divided into an outer **cortex** and an inner **medulla,** which surround the renal sinus. The bases of several cone-shaped **renal pyramids** are located at the boundary between the cortex and the medulla, and the tips of the renal pyramids project toward the center of the kidney. A funnel-shaped structure called a **calyx** (kā′liks; pl. calyces kal′i-sēz) surrounds the tip of each renal pyramid. The calyces from all the renal pyramids join to form a larger funnel called the **renal pelvis.** The renal pelvis then narrows to form a small tube, the **ureter** (ū-rē′ter, ū′re-ter), which exits the kidney and connects to the urinary bladder. Urine passes from the tips of the renal pyramids into the calyces. From the calyces, urine collects in the renal pelvis and exits the kidney through the ureter (figure 18.3).

The functional unit of the kidney is the **nephron** (nef′ron; Greek for kidney), and there are approximately 1.3 million of them in each kidney. Each nephron consists of a **renal corpuscle,** a **proximal convoluted tubule,** a **loop of Henle,** and a **distal convoluted tubule** (figure 18.4). Fluid is forced into the renal corpuscle and then flows into the proximal convoluted tubule. From there, it flows into the loop of Henle. Each loop of Henle consists of a descending limb and an ascending limb. The limbs are further categorized into segments: the thin segment of the descending limb, the thin segment of the ascending limb, and the thick segment of the ascending limb. The descending limb extends toward the renal sinus, where it makes a hairpin turn, and the ascending limb extends back toward the cortex. The fluid flows through the ascending limb of the loop of Henle to the distal convoluted tubule. Several distal convoluted tubules empty into a **collecting duct,** which carries the fluid from the cortex, through the medulla. Multiple collecting ducts empty into a single **papillary duct,** and the papillary ducts empty their contents into a calyx.

The renal corpuscle and both convoluted tubules are in the renal cortex (figure 18.4). The collecting duct and loop of Henle enter the medulla. Approximately 15% of the nephrons, called **juxtamedullary** (next to the medulla) **nephrons,** have loops of Henle that extend deep into the medulla of the kidney. The other nephrons (85%), called **cortical nephrons,** have loops of Henle that do not extend deep into the medulla.

The renal corpuscle of the nephron consists of the Bowman capsule and the glomerulus (figure 18.5; see figure 18.4). The **Bowman capsule** consists of the enlarged end of the nephron, which is indented to form a double-walled chamber. The **glomerulus** (glō-mār′ū-lŭs) is a tuft of capillaries that resembles a ball of yarn and lies within the indentation of the Bowman capsule. The cavity of the Bowman capsule opens into the proximal convoluted tubule, which carries fluid away from the capsule. The inner layer of the Bowman capsule consists of specialized cells called **podocytes** (pod′ō-sīts), which wrap around the glomerular capillaries. The outer layer of the Bowman capsule consists of simple squamous epithelial cells.

The glomerular capillaries have pores in their walls, and the podocytes have numerous cell processes with gaps between them.

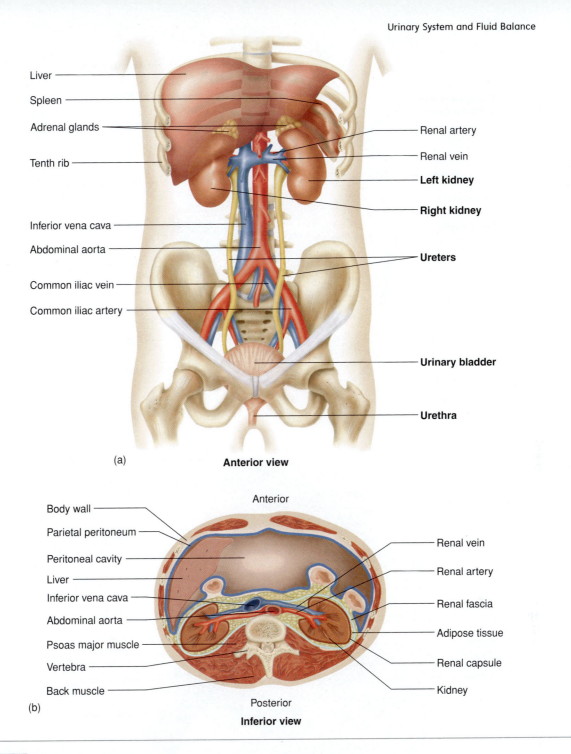

Liver
Spleen
Adrenal glands
Tenth rib
Inferior vena cava
Abdominal aorta
Common iliac vein
Common iliac artery

Renal artery
Renal vein
Left kidney
Right kidney
Ureters
Urinary bladder
Urethra

(a)   **Anterior view**

Anterior

Body wall
Parietal peritoneum
Peritoneal cavity
Liver
Inferior vena cava
Abdominal aorta
Psoas major muscle
Vertebra
Back muscle

Renal vein
Renal artery
Renal fascia
Adipose tissue
Renal capsule
Kidney

(b)

Posterior

Inferior view

Figure 18.2 AP|R **Anatomy of the Urinary System**

(*a*) The kidneys are located in the abdominal cavity, with the right kidney just below the liver and the left kidney below the spleen. A ureter extends from each kidney to the urinary bladder within the pelvic cavity. An adrenal gland is located at the superior pole of each kidney. (*b*) The kidneys are located behind the parietal peritoneum, surrounded by adipose tissue. A connective tissue layer, the renal fascia, anchors the kidney to the abdominal wall. The renal arteries extend from the abdominal aorta to each kidney, and the renal veins extend from the kidneys to the inferior vena cava.

The endothelium of the glomerular capillaries, the podocytes, and the basement membrane together form a **filtration membrane** (figure 18.5*d*). In the first step of urine formation, fluid, consisting of water and solutes smaller than proteins, passes from the blood in the glomerular capillaries through the filtration membrane into the Bowman capsule. The fluid that passes across the filtration membrane is called **filtrate.**

The proximal convoluted tubules, the thick segments of the loops of Henle, the distal convoluted tubules, and the collecting ducts consist of simple cuboidal epithelium. The cuboidal epithelial cells have microvilli and many mitochondria. These portions of the nephron actively transport molecules and ions across the wall of the nephron. The thin segments of the descending and ascending limbs of the loops of Henle have very thin walls made up of simple

Urinary

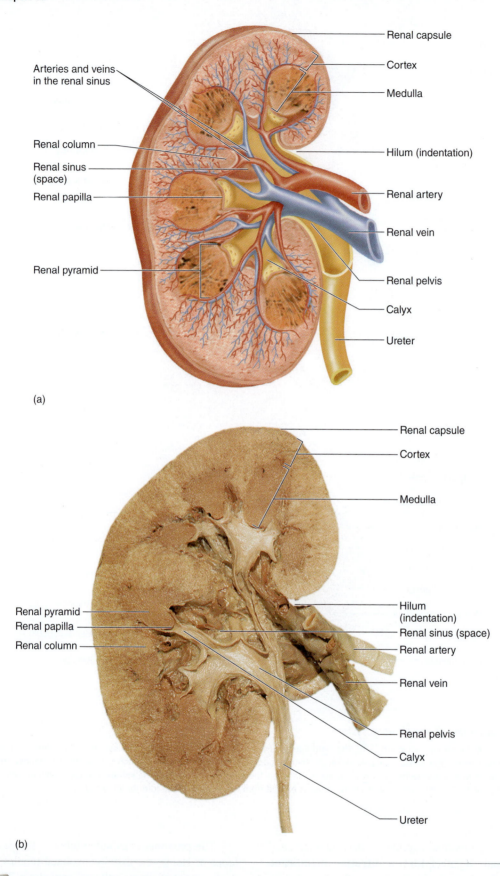

Arteries and veins
in the renal sinus

Renal capsule

Cortex

Medulla

Renal column

Hilum (indentation)

Renal sinus
(space)

Renal artery

Renal papilla

Renal vein

Renal pelvis

Renal pyramid

Calyx

Ureter

(a)

Renal capsule

Cortex

Medulla

Renal pyramid

Hilum
(indentation)

Renal papilla

Renal sinus (space)

Renal column

Renal artery

Renal vein

Renal pelvis

Calyx

Ureter

(b)

Figure 18.3 AP|R **Longitudinal Section of the Kidney**

(a) The cortex and the medulla of the kidney surround the renal sinus. The renal sinus is a space containing the renal pelvis, calyces, blood vessels, adipose tissue, and other connective tissues. The renal pyramids extend from the cortex of the kidney to the renal sinus. The tip of each renal pyramid is surrounded by a calyx. The calyces connect to the renal pelvis. Urine flows from the tip of the renal pyramid through the calyx and renal pelvis into the ureter. (b) Photograph of a longitudinal section through a human kidney.

Urinary

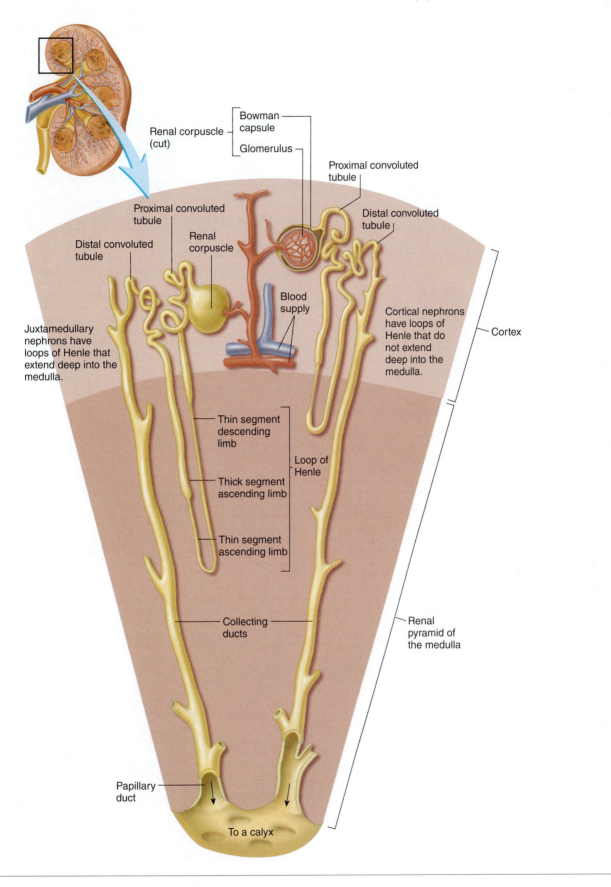

Figure l8.4 **Functional Unit of the Kidney—the Nephron**

A nephron consists of a renal corpuscle, a proximal convoluted tubule, a loop of Henle, and a distal convoluted tubule. The distal convoluted tubule empties into a collecting duct. Juxtamedullary nephrons (those near the medulla of the kidney) have loops of Henle that extend deep into the medulla of the kidney, whereas cortical nephrons do not. Collecting ducts merge into larger papillary ducts, which empty into a calyx.

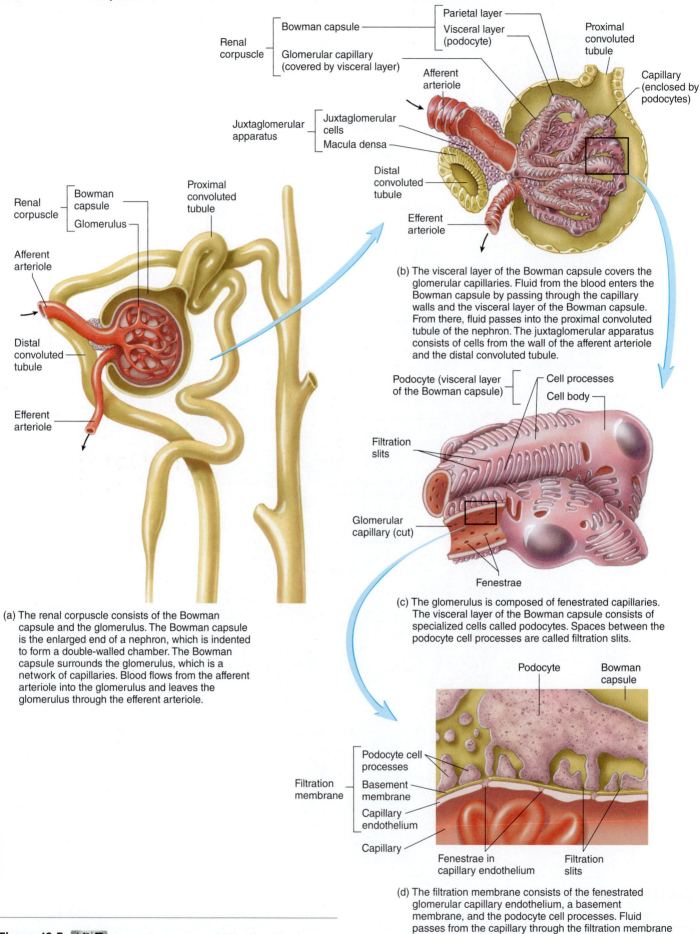

Renal corpuscle
- Bowman capsule
- Glomerular capillary (covered by visceral layer)

- Parietal layer
- Visceral layer (podocyte)

Bowman capsule

Proximal convoluted tubule

Afferent arteriole

Capillary (enclosed by podocytes)

Juxtaglomerular apparatus
- Juxtaglomerular cells
- Macula densa

Distal convoluted tubule

Efferent arteriole

(b) The visceral layer of the Bowman capsule covers the glomerular capillaries. Fluid from the blood enters the Bowman capsule by passing through the capillary walls and the visceral layer of the Bowman capsule. From there, fluid passes into the proximal convoluted tubule of the nephron. The juxtaglomerular apparatus consists of cells from the wall of the afferent arteriole and the distal convoluted tubule.

Renal corpuscle
- Bowman capsule
- Glomerulus

Proximal convoluted tubule

Afferent arteriole

Distal convoluted tubule

Efferent arteriole

(a) The renal corpuscle consists of the Bowman capsule and the glomerulus. The Bowman capsule is the enlarged end of a nephron, which is indented to form a double-walled chamber. The Bowman capsule surrounds the glomerulus, which is a network of capillaries. Blood flows from the afferent arteriole into the glomerulus and leaves the glomerulus through the efferent arteriole.

Podocyte (visceral layer of the Bowman capsule)
- Cell processes
- Cell body

Filtration slits

Glomerular capillary (cut)

Fenestrae

(c) The glomerulus is composed of fenestrated capillaries. The visceral layer of the Bowman capsule consists of specialized cells called podocytes. Spaces between the podocyte cell processes are called filtration slits.

Podocyte

Bowman capsule

Filtration membrane
- Podocyte cell processes
- Basement membrane
- Capillary endothelium

Capillary

Fenestrae in capillary endothelium

Filtration slits

(d) The filtration membrane consists of the fenestrated glomerular capillary endothelium, a basement membrane, and the podocyte cell processes. Fluid passes from the capillary through the filtration membrane into the Bowman capsule.

Figure 18.5 AP|R **Renal Corpuscle and Filtration Membrane**

squamous epithelium. Water and solutes pass through the walls of these portions of the nephron by diffusion. The thin segment of the descending limb of the loop of Henle is permeable to water and, to a lesser degree, solutes, and the thin segment of the ascending limb is permeable to solutes, but not to water.

Arteries and Veins

A system of blood vessels allows the exchange of materials that occurs in the kidneys. The **renal arteries** branch off the abdominal aorta and enter the kidneys (figure 18.6a). They give rise to several branches. The **interlobar** (in-ter-lō′bar; between the lobes) **arteries** pass between the renal pyramids and give rise to the **arcuate** (ar′kū-āt; arched) **arteries,** which arch between the cortex and the medulla. **Interlobular arteries** branch off the arcuate arteries and project into the cortex. The **afferent arterioles** arise from branches of the interlobular arteries and extend to the glomerular capillaries (figure 18.6b). **Efferent arterioles** extend from the glomerular capillaries to the **peritubular** (around the tubes) **capillaries,** which surround the proximal convoluted and distal convoluted tubules and the loops of Henle. The **vasa recta** (vā′să rek′tă; straight vessels) are specialized portions of the peritubular capillaries that extend deep into the medulla of the kidney and surround the loops of Henle and collecting ducts. Blood from the peritubular capillaries, including the vasa recta, enters the interlobular veins. The veins of the kidney run parallel to the arteries and have similar names (figure 18.6).

A structure called the **juxtaglomerular** (jŭks′-tă-glŏ-mer′ū-lăr) **apparatus** (pl. apparatuses) is formed where the distal convoluted tubule comes in contact with the afferent arteriole next to the Bowman capsule (see figure 18.5b). The juxtaglomerular apparatus consists of specialized cells of the walls of the afferent arteriole and the distal convoluted tubules. Certain cells of the juxtaglomerular apparatus secrete the enzyme, renin, and play an important role in blood pressure regulation (see "Hormonal Mechanisms" later in the chapter).

18.3 URINE PRODUCTION

Learning Outcomes *After reading this section, you should be able to*

A. Identify the principal factors that influence filtration pressure, and explain how they affect the rate of filtrate production.

B. Give the function of the proximal convoluted tubule, descending and ascending limbs of the loop of Henle, distal convoluted tubule, and collecting duct.

C. Explain how the movement of substances across the wall of the nephron and collecting duct influences the composition of the filtrate.

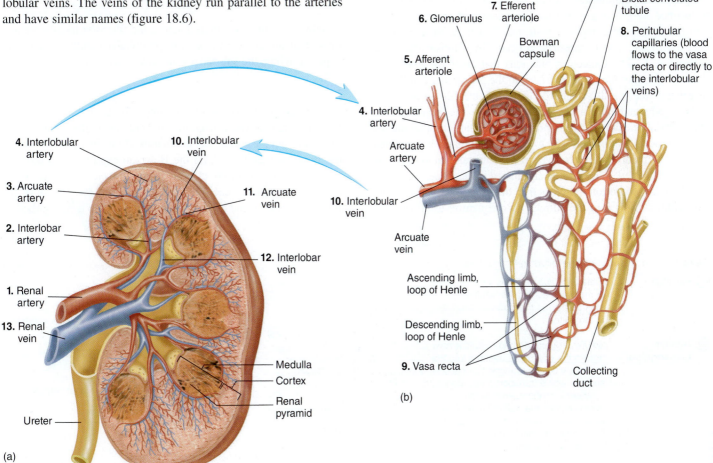

Figure 18.6 **AP|R** Blood Flow Through the Kidney

(a) Blood flow through the larger arteries and veins of the kidney. (b) Blood flow through the arteries, capillaries, and veins that provide circulation to the nephrons.

The primary function of the kidney is regulation of body fluid composition. The kidney is the organ that sorts the substances from the blood for either removal in the urine or return to the blood. Substances that are waste products, toxins, and excess materials are permanently removed from the body, whereas other substances need to be conserved to maintain homeostasis. The structural components that perform this sorting are the nephrons, the functional units of the kidney. If you have ever decided to organize your "junk" drawer in your desk or kitchen, you may realize just how difficult it is to quickly sort through all its contents. In fact, you may have found yourself simply emptying the drawer onto a table and then sorting the contents one by one as you place objects into a "save" group or a "throw away" group. In a sense, the kidney uses the same approach when regulating blood composition. The "throw away" items end up in the urine, and the "save" items go back into the blood. Urine is mostly water and some organic waste products, as well as excess ions (table 18.1).

Scientists usually categorize urine formation into three major processes: filtration, tubular reabsorption, and tubular secretion (figure 18.7).

Filtration occurs when blood pressure nonselectively forces water and other small molecules out of glomerular capillaries and into the Bowman capsule, forming a fluid called filtrate.

Tubular reabsorption is the movement of substances from the filtrate across the wall of the nephron back into the blood of the peritubular capillaries. Certain solute molecules and ions are reabsorbed by processes such as active transport and cotransport into the cells of the nephron wall and then from there into the interstitial fluid. Water reabsorption occurs by osmosis across the nephron wall. The molecules and ions that enter the interstitial fluid surrounding the nephron pass into the peritubular capillaries. In general, the useful substances that enter the filtrate are reabsorbed, and metabolic waste products remain in the filtrate and are eliminated. For example, when proteins are metabolized, ammonia is a by-product. Ammonia, which is toxic to humans, is converted into urea by the liver. Urea forms part of the filtrate; although some of it is reabsorbed, much of it is eliminated in the urine.

Tubular secretion is the active transport of solutes across the nephron walls into the filtrate. Consequently, urine consists of substances that are filtered across the filtration membrane and those that are secreted from the peritubular capillaries into the nephron, minus the substances that are reabsorbed.

Filtration

An average of 21% of the blood pumped by the heart each minute flows through the kidneys. Of the total volume of blood plasma that flows through the glomerular capillaries, about 19% passes through the filtration membrane into the Bowman capsule to become filtrate. In all the nephrons of both kidneys, about 180 liters (L) of filtrate are produced each day, but only about 1% or less of the filtrate becomes urine because most of the filtrate is reabsorbed.

TABLE 18.1	The Average Concentration of Major Urine Substances	
Substance	**Plasma**	**Urine**
Water (L/day)		1.4
Organic molecules (mg/dL)		
Protein	3900–5000	0*
Glucose	100	0
Urea	26	1820
Uric acid	3	42
Creatinine	1	196
Ions (mEq/L)		
Na$^+$	142	128
K$^+$	5	60
Cl$^-$	103	134
HCO$_3^-$	28	14
Specific gravity (g/ml)†	1.019–1.022	1.005–1.030
pH	7.35–7.45	4.5–8.0

*Trace amounts of protein can be found in the urine.
†The specific gravity increases as the concentration of solutes in urine increases.

Urine formation results from the following three processes:

① Filtration   Filtration (*blue arrow*) is the movement of materials across the filtration membrane into the Bowman capsule to form filtrate.

② Tubular reabsorption   Solutes are reabsorbed (*purple arrow*) across the wall of the nephron into the interstitial fluid by transport processes, such as active transport and cotransport.
    Water is reabsorbed (*orange arrow*) across the wall of the nephron by osmosis. Water and solutes pass from the interstitial fluid into the peritubular capillaries.

③ Tubular secretion   Solutes are secreted (*green arrow*) across the wall of the nephron into the filtrate.

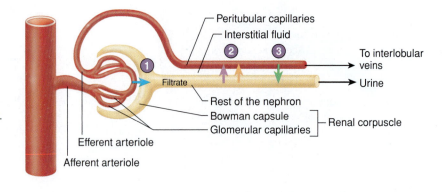

PROCESS Figure 18.7 Urine Formation

Filtration is a nonspecific process whereby materials are separated based on size or charge. A simple example of size filtration is demonstrated by a drip coffeemaker. In this case, the driving force of filtration is gravity. The kidneys also use size filtration to remove substances from the blood by filtering it, but in this case, the driving force of this filtration is blood pressure.

The filtration membrane allows some substances, but not others, to pass from the blood into the Bowman capsule. Water and small solutes readily pass through the openings of the filtration membrane, but blood cells and most proteins, which are too large, do not enter the Bowman capsule. One example of a small blood protein that can enter the filtrate in very small amounts is albumin. Consequently, the filtrate contains no cells and little protein.

The formation of filtrate depends on a pressure gradient, called the **filtration pressure,** which forces fluid from the glomerular capillary across the filtration membrane into the Bowman capsule (figure 18.8). The filtration pressure results from forces that move fluid out of the glomerular capillary into the Bowman capsule minus the forces that move fluid out of the Bowman capsule into the glomerular capillary. The **glomerular capillary pressure** is the blood pressure in the glomerular capillary. It is the major force causing fluid to move from the glomerular capillary across the filtration membrane into the Bowman capsule. There are two major forces opposing the movement of fluid into the lumen of the Bowman capsule: capsular pressure and colloid osmotic pressure. The **capsular pressure** is caused by the pressure of filtrate already inside the Bowman capsule, and the **colloid osmotic pressure** is within the glomerular capillary. Because most plasma proteins do not pass through the filtration membrane, they produce an osmotic pressure that favors fluid movement into the glomerular capillary from the Bowman capsule. Therefore, the filtration pressure is calculated as follows:

> Glomerular capillary pressure
> − Capsular pressure
> − Colloid osmotic pressure
> _____
> Filtration pressure

The filtration pressure forces fluid from the glomerulus into the Bowman capsule because the glomerular capillary pressure is greater than both the capsular and the colloid osmotic pressures. Under most conditions, the filtration pressure remains within a narrow range of values. However, when the filtration pressure increases, both the filtrate volume and the urine volume increase, and when the filtration pressure decreases, both the filtrate volume and the urine volume decrease.

The filtration pressure is influenced by the blood pressure in the glomerular capillaries, the blood protein concentration, and the pressure in the Bowman capsule. The blood pressure is normally higher in the glomerular capillaries than it is in most capillaries. The filtration pressure increases if the blood pressure in the glomerular capillaries increases further. The filtration pressure decreases if the blood pressure in the glomerular capillaries decreases.

The filtration pressure is also influenced by the concentration of proteins in the blood. An increase in blood protein concentration encourages the movement of water by osmosis back into the glomerular capillaries and therefore reduces the overall filtration pressure (see chapter 3). On the other hand, a decrease in blood protein concentration inhibits the movement of water by osmosis back into the glomerular capillaries, which increases the overall filtration pressure.

Regulation of Filtration

The blood pressure within the glomerular capillaries is fairly constant because the afferent and efferent arterioles either dilate or constrict to regulate the blood pressure there, even though the systemic blood pressure may fluctuate substantially. Also, the concentration of blood proteins and the pressure inside the Bowman capsule are fairly constant. As a consequence, the filtration pressure and the rate of filtrate formation are maintained within a narrow range of values most of the time.

Predict 2

Predict the change in filtration pressure if the afferent arteriole is constricted. Would urine production tend to increase or decrease? Explain.

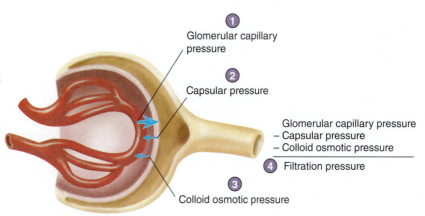

① Glomerular capillary pressure, the blood pressure within the glomerulus, moves fluid from the blood into the Bowman capsule.

② Capsular pressure, the pressure inside the Bowman capsule, moves fluid from the capsule into the blood.

③ Colloid osmotic pressure, produced by the concentration of blood proteins, moves fluid from the Bowman capsule into the blood by osmosis.

④ Filtration pressure is equal to the glomerular capillary pressure minus the capsular and colloid osmotic pressures.

① Glomerular capillary pressure

② Capsular pressure

Glomerular capillary pressure
− Capsular pressure
− Colloid osmotic pressure
_____
④ Filtration pressure

③ Colloid osmotic pressure

PROCESS Figure 18.8 Filtration Pressure

However, the filtration pressure does change dramatically under some conditions. Sympathetic neurons innervate the blood vessels of the kidney. Sympathetic stimulation constricts the arteries, causing a decrease in renal blood flow and filtrate formation in proportion to the intensity of the stimulation. It is possible to decrease filtrate formation to only a few milliliters per minute. Consequently, only a small volume of urine is produced (see figure 18.17). For instance, during **cardiovascular shock,** the filtration pressure and filtrate formation fall dramatically. One of the dangers of cardiovascular shock is that the renal blood flow can be so low that the kidneys suffer from lack of O_2. If the O_2 level remains too low for a long enough time, permanent kidney damage or complete kidney failure results. One important reason for treating cardiovascular shock quickly is to avoid damage to the kidneys. Other conditions, such as intense physical activity or trauma, also increase sympathetic stimulation of renal arteries and decrease urine production to very low levels. On the other hand, increased blood pressure decreases sympathetic stimulation of renal blood arteries, and urine volume increases (see figure 18.17).

In addition to the effects of blood pressure on filtration pressure, decreases in the concentration of plasma proteins, caused by conditions such as inflammation of the liver, where most of the blood proteins are produced, increase the filtration pressure. The increased filtration pressure causes the filtrate and urine volume to increase.

Tubular Reabsorption

As the filtrate flows from the Bowman capsule through the proximal convoluted tubule, loop of Henle, distal convoluted tubule, and collecting duct, many of the solutes in the filtrate are reabsorbed. About 99% of the original filtrate volume is reabsorbed and enters the peritubular capillaries. The reabsorbed filtrate flows through the renal veins to enter the general circulation. Only 1% of the original filtrate volume becomes urine (figure 18.9). Because excess ions and metabolic waste products are not readily reabsorbed, the small volume of urine produced contains a high concentration of ions and metabolic waste products.

The proximal convoluted tubule is the primary site for the reabsorption of solutes and water. The cuboidal cells of the proximal convoluted tubule have numerous microvilli and mitochondria, and they are well adapted to transport molecules and ions across the nephron wall by active transport and cotransport. Substances transported from the proximal convoluted tubule include proteins, amino acids, glucose, and fructose molecules, as well as Na^+, K^+, Ca^{2+}, HCO_3^-, and Cl^-. The proximal convoluted tubule is permeable to water. As solute molecules are transported out of the proximal convoluted tubule into the interstitial fluid, water moves by osmosis in the same direction. The solutes and water then enter the peritubular capillaries. Consequently, 65% of the filtrate volume is reabsorbed from the proximal convoluted tubule (figures 18.9, step 1, and 18.10).

The descending limb of the loop of Henle further concentrates the filtrate. The renal medulla contains very concentrated interstitial fluid that has large amounts of Na^+, Cl^-, and urea. The wall of the thin segment of the descending limb is permeable to water and moderately permeable to solutes. As the filtrate passes through the descending limb of the loop of Henle into the medulla of the kidney, water moves out of the nephron by osmosis, and some solutes move into the nephron by diffusion. By the time the filtrate has passed through the descending limb, another 15% of the filtrate

volume has been reabsorbed, and the filtrate is as concentrated as the interstitial fluid of the medulla. The reabsorbed water and solutes enter the vasa recta (figure 18.11; see figure 18.9, step 2).

The ascending limb of the loop of Henle dilutes the filtrate by removing solutes. The thin segment of the ascending limb is not permeable to water, but it is permeable to solutes. Consequently, solutes diffuse out of the nephron (figure 18.11; see figure 18.9, step 3).

The cuboidal epithelial cells of the thick segment of the ascending limb actively transport Na^+ out of the nephron, and K^+ and Cl^- are cotransported with Na^+. The thick segment of the ascending limb is not permeable to water. As a result, Na^+, K^+, and Cl^-, but little water, are removed from the filtrate (see figure 18.9, step 4). Because of the efficient removal of these solutes, the highly concentrated filtrate that enters the ascending limb of the loop of Henle is converted to a dilute solution by the time it reaches the distal convoluted tubule (figure 18.12; see figure 18.9, step 5). As the filtrate enters the distal convoluted tubule, it is more dilute than the interstitial fluid of the renal cortex. Also, because of the volume of filtrate reabsorbed in the proximal convoluted tubule and the descending limb of the loop of Henle, only about 20% of the original filtrate volume remains. The solutes transported from the ascending limb of the loop of Henle enter the interstitial fluid of the medulla and help keep the concentration of solutes in the medulla high. Excess solutes enter the vasa recta.

The cuboidal cells of the distal convoluted tubule and collecting duct remove water and additional solutes. Na^+ and Cl^- are reabsorbed. Sodium ions are actively transported, and chloride ions are cotransported. Also, 19% of the original filtrate volume is reabsorbed by osmosis, leaving about 1% of the original filtrate as urine (figure 18.13; see figure 18.9, steps 6 and 7). The reabsorbed water and solutes from the distal convoluted tubule enter the peritubular capillaries and the vasa recta from the collecting ducts.

The reabsorption of water and solutes from the distal convoluted tubule and collecting duct is controlled by hormones, which have a great influence on urine concentration and volume (see "Regulation of Urine Concentration and Volume").

In summary, most of the useful solutes that pass through the filtration membrane into the Bowman capsule are reabsorbed in the proximal convoluted tubule. Filtrate volume is reduced by 65% in the proximal convoluted tubule and by 15% in the descending limb of the loop of Henle. In the ascending limb of the loop of Henle, Na^+, K^+, and Cl^-, but little water, are removed from the filtrate. Consequently, the filtrate becomes dilute. In the distal convoluted tubule and the collecting duct, additional Na^+ and Cl^- are removed, water moves out by osmosis, and the filtrate volume is reduced by another 19%, leaving 1% of the original filtrate volume as urine.

Predict 3

People who suffer from untreated diabetes mellitus can experience very high levels of glucose in the blood. The glucose can easily cross the filtration membrane into the Bowman capsule. Normally, all the glucose is reabsorbed from the nephron. However, if the concentration of glucose in the nephron becomes too high, not all the glucose can be reabsorbed because the number of transport molecules in the cells of the proximal convoluted tubule is limited. How does the volume of urine produced by a person with untreated diabetes mellitus differ from that of a healthy person, and how does the concentration of the urine differ from that of a healthy person?

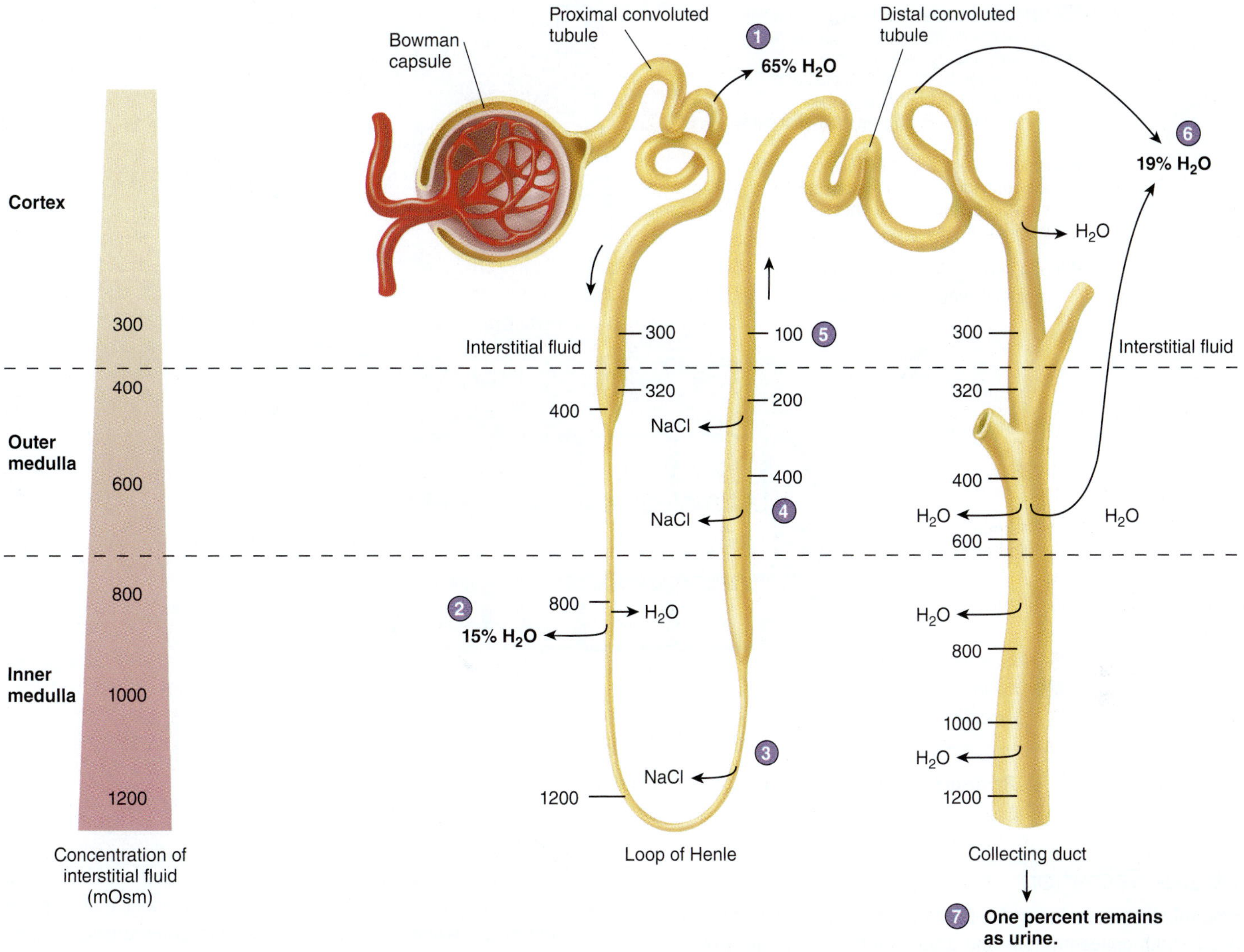

PROCESS Figure 18.9 Urine-Concentrating Mechanism

The concentration gradient from the cortex to the inner medulla is shown on the left. Interstitial fluid increases in concentration from 300 mOsm/L in the cortex to 1200 mOsm/L in the medulla. The concentrations of the filtrate in different parts of the nephron are also shown.

1. Approximately 180 L of filtrate enters the nephrons each day; of that volume, 65% is reabsorbed in the proximal convoluted tubule. In the proximal convoluted tubule, solute molecules move by active transport and cotransport from the lumen of the tubule into the interstitial fluid. Water moves by osmosis because the cells of the tubule wall are permeable to water (see figure 18.10).

2. Approximately 15% of the filtrate volume is reabsorbed in this segment of the descending limb of the loop of Henle. The descending limb passes through the concentrated interstitial fluid of the medulla. Because the wall of the descending limb is permeable to water, water moves by osmosis from the tubule into the more concentrated interstitial fluid (see figure 18.11, step 1). By the time the filtrate reaches the tip of the renal pyramid, the concentration of the filtrate is equal to the concentration of the interstitial fluid.

3. The ascending limb of the loop of Henle is not permeable to water. Solutes diffuse out of the thin segment (see figure 18.11, step 2).

4. Na^+ are actively transported, and K^+ and Cl^- are cotransported, from the filtrate of the thick segment into the interstitial fluid (see figure 18.12).

5. The volume of the filtrate doesn't change as it passes through the ascending limb, but the concentration is greatly reduced (see figure 18.12). By the time the filtrate reaches the cortex of the kidney, the concentration is approximately 100 mOsm/L, which is less concentrated than the interstitial fluid of the cortex (300 mOsm/L).

6. The distal convoluted tubule and collecting duct are permeable to water if ADH is present. If ADH is present, water moves by osmosis from the less concentrated filtrate into the more concentrated interstitial fluid (see figure 18.13). By the time the filtrate reaches the tip of the renal pyramid, an additional 19% of the filtrate is reabsorbed.

7. One percent or less remains as urine, when ADH is present.

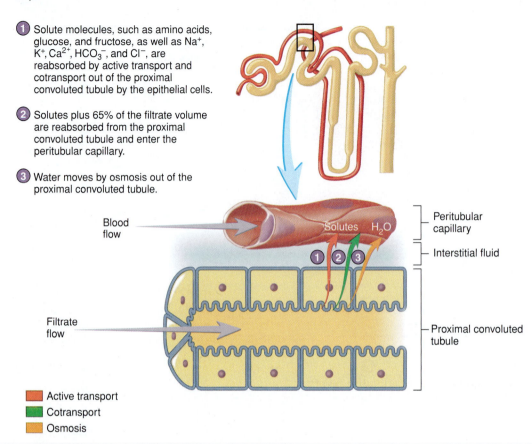

① Solute molecules, such as amino acids, glucose, and fructose, as well as Na^+, K^+, Ca^{2+}, HCO_3^-, and Cl^-, are reabsorbed by active transport and cotransport out of the proximal convoluted tubule by the epithelial cells.

② Solutes plus 65% of the filtrate volume are reabsorbed from the proximal convoluted tubule and enter the peritubular capillary.

③ Water moves by osmosis out of the proximal convoluted tubule.

Blood flow

Solutes H_2O

Peritubular capillary

Interstitial fluid

Filtrate flow

Proximal convoluted tubule

- ■ Active transport
- ■ Cotransport
- ■ Osmosis

PROCESS Figure 18.10 Reabsorption in the Proximal Convoluted Tubule

Tubular Secretion

Some substances, including by-products of metabolism that become toxic in high concentrations and drugs or other molecules not normally produced by the body, are secreted into the nephron from the peritubular capillaries. As with tubular reabsorption, tubular secretion can be either active or passive. For example, ammonia diffuses into the lumen of the nephron, whereas H^+, K^+, creatinine, histamine, and penicillin are actively transported into the nephron.

Hydrogen ions are actively transported into the proximal convoluted tubule. The epithelial cells actively transport large quantities of H^+ across the nephron wall into the filtrate. The secretion of H^+ plays an important role in regulating the body fluid pH.

In the proximal convoluted tubule, K^+ is reabsorbed. However, in the distal convoluted tubule and collecting duct, K^+ is secreted, resulting in a net loss of K^+ in the urine.

18.4 REGULATION OF URINE CONCENTRATION AND VOLUME

Learning Outcome *After reading this section, you should be able to*

A. Explain how antidiuretic hormone, aldosterone, and atrial natriuretic hormone influence the volume and concentration of urine.

Given a solution in a container, such as a pan on a stove, it is possible to change its concentration by adding water to it or by boiling

it, thereby removing water. Similarly, the kidneys maintain the concentration of the body fluids by increasing water reabsorption from the filtrate when the body fluid concentration increases and by reducing water reabsorption from the filtrate when the body fluid concentration decreases. The volume and composition of urine therefore change, depending on conditions in the body. If body fluid concentration increases above normal levels, the kidneys produce a small volume of concentrated urine. This eliminates solutes and conserves water, both of which help lower the body fluid concentration back to normal. On the other hand, if the body fluid concentration decreases, the kidneys produce a large volume of dilute urine. As a result, water is lost, solutes are conserved, and the body fluid concentration increases.

Urine production also maintains blood volume and therefore blood pressure. An increase in blood volume can increase blood pressure, and a decrease in blood volume can decrease blood pressure. When blood volume increases above normal, the kidneys produce a large volume of urine. The loss of water in the urine lowers blood volume. Conversely, if blood volume decreases below normal, the kidneys produce a small volume of urine to conserve water and maintain blood volume.

Hormonal Mechanisms

Three major hormonal mechanisms are involved in regulating urine concentration and volume: the renin-angiotensin-aldosterone mechanism, the antidiuretic hormone (ADH) mechanism, and the atrial natriuretic hormone (ANH) mechanism. Each mechanism

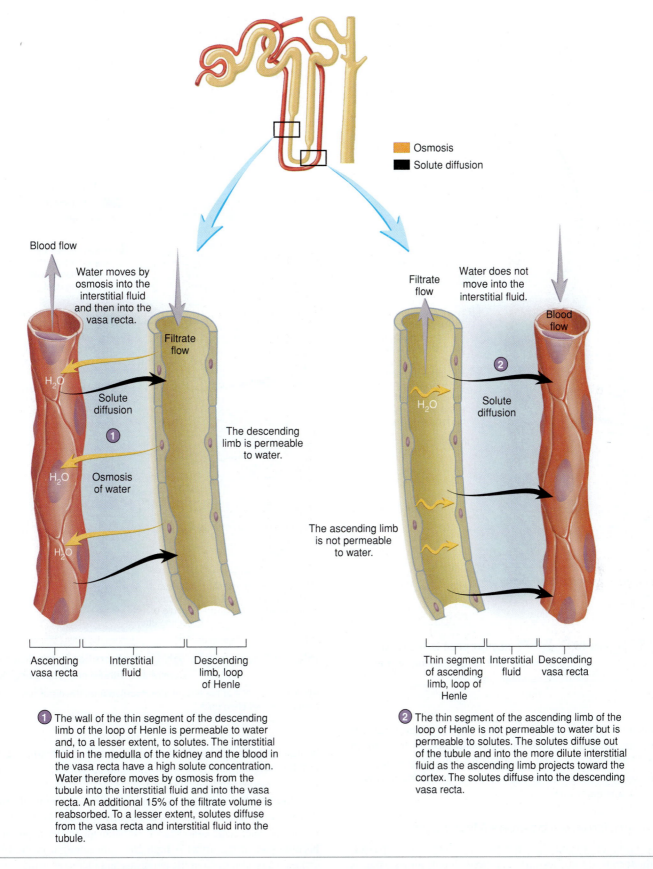

Blood flow

Water moves by osmosis into the interstitial fluid and then into the vasa recta.

Filtrate flow

H₂O

Solute diffusion

1

Osmosis of water

H₂O

H₂O

The descending limb is permeable to water.

The ascending limb is not permeable to water.

Filtrate flow

Water does not move into the interstitial fluid.

Blood flow

2

Solute diffusion

H₂O

| Ascending vasa recta | Interstitial fluid | Descending limb, loop of Henle |

| Thin segment of ascending limb, loop of Henle | Interstitial fluid | Descending vasa recta |

Osmosis
Solute diffusion

1 The wall of the thin segment of the descending limb of the loop of Henle is permeable to water and, to a lesser extent, to solutes. The interstitial fluid in the medulla of the kidney and the blood in the vasa recta have a high solute concentration. Water therefore moves by osmosis from the tubule into the interstitial fluid and into the vasa recta. An additional 15% of the filtrate volume is reabsorbed. To a lesser extent, solutes diffuse from the vasa recta and interstitial fluid into the tubule.

2 The thin segment of the ascending limb of the loop of Henle is not permeable to water but is permeable to solutes. The solutes diffuse out of the tubule and into the more dilute interstitial fluid as the ascending limb projects toward the cortex. The solutes diffuse into the descending vasa recta.

PROCESS Figure 18.11 **AP|R** Reabsorption in the Loop of Henle: The Descending Limb and the Thin Segment of the Ascending Limb

Urinary

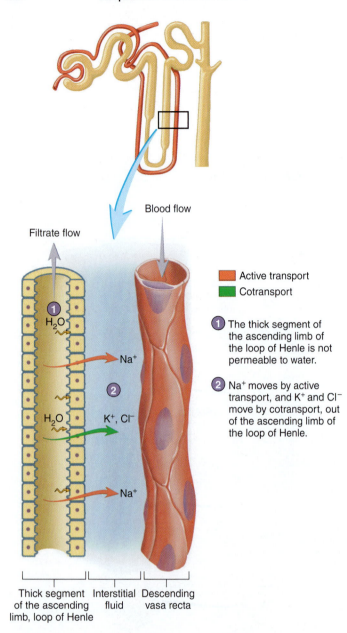

Blood flow

Filtrate flow

Active transport
Cotransport

① The thick segment of the ascending limb of the loop of Henle is not permeable to water.

② Na⁺ moves by active transport, and K⁺ and Cl⁻ move by cotransport, out of the ascending limb of the loop of Henle.

Thick segment of the ascending limb, loop of Henle | Interstitial fluid | Descending vasa recta

PROCESS Figure 18.12 AP|R Reabsorption in the Loop of Henle: The Thick Segment of the Ascending Limb

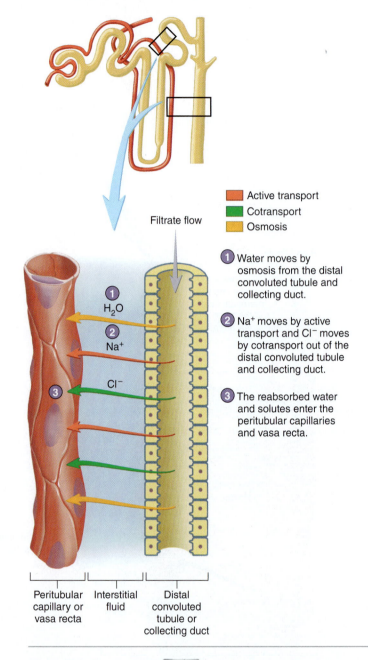

Filtrate flow

Active transport
Cotransport
Osmosis

① Water moves by osmosis from the distal convoluted tubule and collecting duct.

② Na⁺ moves by active transport and Cl⁻ moves by cotransport out of the distal convoluted tubule and collecting duct.

③ The reabsorbed water and solutes enter the peritubular capillaries and vasa recta.

Peritubular capillary or vasa recta | Interstitial fluid | Distal convoluted tubule or collecting duct

PROCESS Figure 18.13 AP|R Reabsorption in the Distal Convoluted Tubule and Collecting Duct

Approximately 19% of the filtrate is reabsorbed from the distal convoluted tubule and collecting duct.

is activated by different stimuli, but they work together to achieve homeostasis. Both the renin-angiotensin-aldosterone mechanism and the ANH mechanism are more sensitive to changes in blood pressure, whereas the ADH mechanism is more sensitive to changes in blood concentration.

Renin-Angiotensin-Aldosterone Mechanism

Renin (rē′nin, ren′in) and **angiotensin** (an′jē-ō-ten′sin) help regulate **aldosterone** (al-dos′ter-ōn) secretion. Renin, an enzyme, is secreted by cells of the juxtaglomerular apparatuses in the kidneys (see figure 18.5b). Renin acts on **angiotensinogen** (an′jē-ō-ten-sin′ō-jen), a plasma protein produced by the liver, and converts it to **angiotensin I**. Angiotensin I is rapidly converted to a smaller peptide called **angiotensin II** by **angiotensin-converting enzyme (ACE)**. Angiotensin II acts on the adrenal cortex, causing it to secrete aldosterone (see chapter 13).

Aldosterone increases the rate of active transport of Na⁺ in the distal convoluted tubules and collecting ducts. In the absence of aldosterone, large amounts of Na⁺ remain in the nephron and become part of the urine. A high Na⁺ concentration in the filtrate causes water to remain in the nephrons and increases urine volume. Therefore, when the rate of active transport of Na⁺ is slow, urine volume increases, and the urine contains a high concentration of Na⁺. Because Cl⁻ is attracted by the positive charge on Na⁺, Cl⁻ is cotransported with Na⁺.

Urinary

Diuretics (di-u-ret′iks) are chemicals that increase the rate of urine formation. Although the definition is simple, a number of physiological mechanisms are involved.

Diuretics are used to treat hypertension, as well as several types of edema caused by congestive heart failure, cirrhosis of the liver, and other disorders. However, treatment with diuretics can lead to complications, including dehydration and electrolyte imbalances.

The varying degrees of diuretic chemical types are outlined in the following descriptions, along with their physiological mechanisms.

Sodium ion reabsorption inhibitors include thiazide-type diuretics. They promote the loss of Na^+, Cl^-, and water in the urine. These diuretics are sometimes given to people who have hypertension. The increased loss of water in the urine lowers blood volume, and thus blood pressure.

Osmotic diuretics freely pass into the filtrate and undergo limited reabsorption by the nephron. These diuretics increase urine volume by elevating the osmotic concentration of the filtrate, thus reducing the amount of water moving by osmosis out of the nephron. Urea, mannitol, and glycerine

have been used as osmotic diuretics and can be effective in treating patients who have cerebral edema and edema in acute renal failure.

Caffeine and related substances act as diuretics partly because they increase renal blood flow and the rate of glomerular filtrate formation. They also influence the nephron by decreasing Na^+ and Cl^- reabsorption.

Alcohol acts as a diuretic, although it is not used clinically for that purpose. It inhibits ADH secretion from the posterior pituitary and results in increased urine volume.

When blood pressure suddenly decreases (figure 18.14; see figure 18.17) or when the concentration of Na^+ in the filtrate becomes too low, the kidney releases renin. The resultant increase in aldosterone causes an increase in Na^+ and Cl^- reabsorption from the nephrons. Water follows the Na^+ and Cl^-. Thus, the volume of water lost in the form of urine declines. This method of conserving water helps prevent a further decline in blood pressure (figure 18.14; see figure 18.17).

Predict 4

Drugs that increase the urine volume are called diuretics. Some diuretics inhibit the active transport of Na^+ in the nephron. Explain how these diuretic drugs can cause an increase in urine volume.

Antidiuretic Hormone Mechanism

Antidiuretic (an′tē-dī-ū-ret′ik) **hormone (ADH),** secreted by the posterior pituitary gland, passes through the circulatory system to the kidneys. ADH regulates the amount of water reabsorbed by the distal convoluted tubules and collecting ducts. When ADH levels increase, the permeability of the distal convoluted tubules and collecting ducts to water increases, and more water is reabsorbed from the filtrate. Consequently, an increase in ADH results in the production of a small volume of concentrated urine. On the other hand, when ADH levels decrease, the distal convoluted tubules and collecting ducts become less permeable to water. As a result, less water is reabsorbed, and a large volume of dilute urine is produced (figure 18.15; see figure 18.17).

The release of ADH from the posterior pituitary is regulated by the hypothalamus. Certain cells of the hypothalamus are sensitive to changes in solute concentration. When the solute concentration in the blood increases, action potentials are sent along the axons of the ADH-secreting neurons of the hypothalamus to the posterior pituitary, and ADH is released from the ends of the axons (figure 18.15; see chapter 10). A reduced solute concentration in the blood causes inhibition of ADH release.

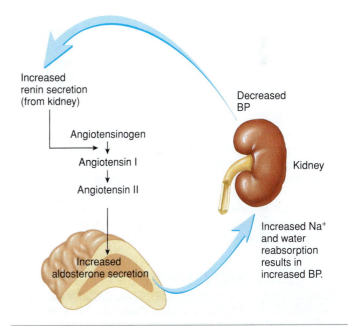

Figure 18.14   Aldosterone and the Regulation of Na^+ and Water in Extracellular Fluid

Low blood pressure (BP) stimulates renin secretion from the kidney. Renin stimulates the production of angiotensin I, which is converted to angiotensin II, which in turn stimulates aldosterone secretion from the adrenal cortex. Aldosterone increases Na^+ and water reabsorption in the kidney.

Baroreceptors that monitor blood pressure also influence ADH secretion. A large decrease in blood pressure causes an increase in ADH secretion (figure 18.15; see figure 18.17), and a large increase in blood pressure decreases ADH secretion.

Atrial Natriuretic Hormone

Atrial natriuretic (nā′trē-ū-ret′ik) **hormone (ANH)** is secreted from cardiac muscle cells in the right atrium of the heart when

Increased blood solute concentration or
large decrease in BP

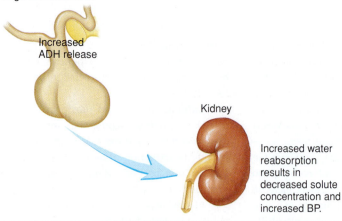

Increased
ADH release

Kidney

Increased water
reabsorption
results in
decreased solute
concentration and
increased BP.

Figure 18.15   ADH and the Regulation of Extracellular Fluid Concentration

Increased blood solute concentration affects hypothalamic neurons, and decreased blood pressure affects baroreceptors. As a result of these stimuli, the posterior pituitary secretes ADH, which increases water reabsorption by the kidney.

Increased blood
pressure in right atrium

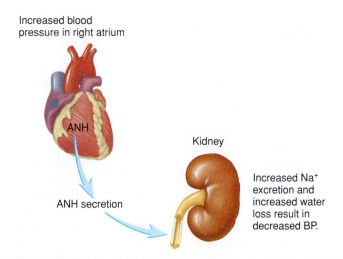

ANH

Kidney

ANH secretion

Increased Na^+
excretion and
increased water
loss result in
decreased BP.

Figure 18.16   ANH and the Regulation of Na⁺ and Water in Extracellular Fluid

Increased blood pressure in the right atrium of the heart causes increased secretion of ANH, which increases Na^+ excretion and water loss in the form of urine.

blood pressure in the right atrium increases above normal (figures 18.16 and 18.17). ANH acts on the kidney to decrease Na^+ reabsorption. Therefore, Na^+ and water remain in the nephron to become urine. The increased loss of Na^+ and water as urine reduces the blood volume and the blood pressure.

Predict 5

Ivy Saline worked as a nurse in a hospital. Because she was very observant, she recognized that one of her patients had received a much larger volume of an intravenous (IV) saline solution than had been ordered. A saline solution consists of NaCl but also contains other solutes, such as small amounts of KCl. Saline solutions have the same concentration as body fluids. Predict the effect of the large volume of IV saline solution on the patient's rate of urine production, and describe the role of ADH, ANH, and aldosterone in controlling the change in urine production.

18.5 URINE MOVEMENT

Learning Outcomes   *After reading this section, you should be able to*

   A. Describe the ureters, urinary bladder, and urethra.
   B. Describe the micturition reflex.

Anatomy and Histology of the Ureters, Urinary Bladder, and Urethra

The **ureters** are small tubes that carry urine from the renal pelvis of the kidney to the posterior inferior portion of the urinary bladder (figure 18.18). The **urinary bladder** is a hollow, muscular container that lies in the pelvic cavity just posterior to the pubic symphysis. It stores urine; thus, its size depends on the quantity of urine present. The urinary bladder can hold from a few milliliters (mL) to a maximum of about 1000 mL of urine. When the urinary bladder reaches a volume of a few hundred mL, its wall is stretched enough to activate a reflex that causes the smooth muscle of the urinary bladder to contract, and most of the urine flows out of the urinary bladder through the urethra.

The **urethra** is the tube that carries urine from the urinary bladder to the outside of the body. The triangle-shaped portion of the urinary bladder located between the opening of the ureters and the opening of the urethra is called the **trigone** (trī′gōn; triangle).

The ureters and the urinary bladder are lined with transitional epithelium, which is specialized to stretch (see chapter 4). As the volume of the urinary bladder increases, the epithelial cells change in shape from columnar to flat, and the number of epithelial cell layers decreases. As the volume of the urinary bladder decreases, transitional epithelial cells assume their columnar shape and form a greater number of cell layers.

The walls of the ureter and urinary bladder are composed of layers of smooth muscle and connective tissue. Regular waves of smooth muscle contraction in the ureters produce the force that causes urine to flow from the kidneys to the urinary bladder. Contractions of smooth muscle in the urinary bladder force urine to flow from the bladder through the urethra.

At the junction of the urinary bladder and the urethra, the smooth muscle of the bladder wall forms the **internal urinary sphincter** in males. (There is no functional internal urinary sphincter in females.) In males, the internal urinary sphincter contracts to keep semen from entering the urinary bladder during sexual intercourse (see chapter 19). In both males and females, the **external urinary sphincter** is formed of skeletal muscle that surrounds the urethra as the urethra extends through the pelvic floor. The external urinary sphincter is under voluntary control, allowing a person to start or stop the flow of urine through the urethra.

In males, the urethra extends to the end of the penis, where it opens to the outside. The female urethra is much shorter (approximately 4 cm) than the male urethra (approximately 20 cm) and opens into the vestibule anterior to the vaginal opening.

Predict 6

Cystitis (sis-tī′tis) is an inflammation of the urinary bladder often caused by a bacterial infection. Typically, bacteria from outside the body enter the bladder. Are males or females more prone to cystitis? Explain.

③ **Actions**

Pituitary:
Baroreceptors inhibit posterior pituitary ADH secretion when blood volume increases.

Kidney:
Juxtaglomerular apparati inhibit renin release when blood volume increases, which decreases aldosterone secretion.

Heart:
Atrial cardiac muscle cells secrete ANH when blood volume increases.

Blood vessels:
Sympathetic division baroreceptors detect increased blood volume, which causes vasodilation of renal arteries.

④ **Reactions**

Effectors Respond:
Decreased ADH decreases water reabsorption by the distal convoluted tubules and collecting ducts. Less water returns to the blood and more water is lost in the urine, which decreases blood volume.

Decreased aldosterone and increased ANH decrease Na⁺ reabsorption from the distal convoluted tubule and collecting duct. More Na⁺ and water are lost in the urine, which decreases blood volume.

Increased renal blood flow increases the rate of filtrate formation, and more water is lost in the urine.

② **Homeostasis Disturbed:**
High blood volume induces elevated blood pressure.

⑤ **Homeostasis Restored:**
Reduced blood volume due to loss of water and Na⁺ in the urine lowers blood pressure.

① Blood volume (normal range)

Start here

⑥ Blood volume (normal range)

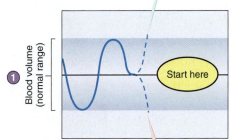
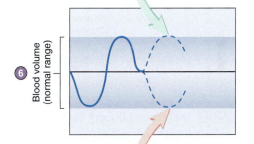

Homeostasis Disturbed:
Low blood volume induces lowered blood pressure.

Homeostasis Restored:
Increased blood volume due to decreased Na⁺ and water loss in the urine raises blood pressure.

Actions

Blood vessels:
Sympathetic division baroreceptors detect decreased blood volume, which causes vasoconstriction of renal arteries.

Heart:
Atrial cardiac muscle cells do not secrete ANH when blood volume decreases.

Kidney:
Juxtaglomerular apparati stimulate renin release when blood volume decreases, which increases aldosterone secretion.

Pituitary:
Baroreceptors stimulate posterior pituitary ADH secretion when blood volume decreases.

Reactions

Effectors Respond:
Decreased renal blood flow decreases filtrate formation, and less water is lost in urine, which increases blood volume.

Increased aldosterone and decreased ANH increase Na⁺ reabsorption in the distal convoluted tubule and the collecting duct. Less Na⁺ and water are lost in the urine, which increases blood volume.

Increased ADH increases the permeability of the distal convoluted tubule and the collecting duct to water. Increased ADH also increases the sensation of thirst. Less water is lost in the urine.

Urinary

Homeostasis Figure 18.17 **Hormonal Regulation of Blood Volume and Its Effect on Urine Volume and Concentration**

(1) Blood volume is in its normal range. (2) Blood volume increases outside the normal range, which causes homeostasis to be disturbed. (3) The control centers respond to the change in blood volume. (4) The control centers cause ADH and aldosterone secretion to decrease, which reduces water reabsorption. The control centers also cause dilation of renal arteries, which increases urine production. (5) These changes cause blood volume and thus blood pressure to decrease. (6) Blood volume returns to its normal range and homeostasis is restored. Observe the responses to a decrease in blood volume outside its normal range by following the *red arrows.*

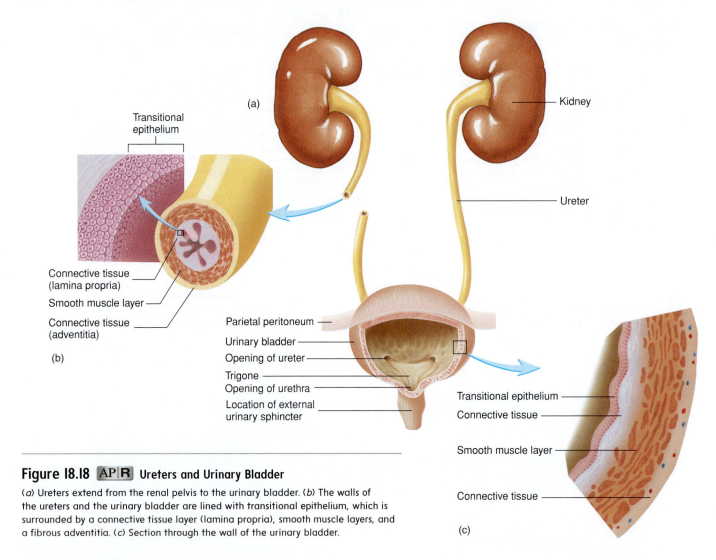

Figure 18.18 AP|R **Ureters and Urinary Bladder**

(*a*) Ureters extend from the renal pelvis to the urinary bladder. (*b*) The walls of the ureters and the urinary bladder are lined with transitional epithelium, which is surrounded by a connective tissue layer (lamina propria), smooth muscle layers, and a fibrous adventitia. (*c*) Section through the wall of the urinary bladder.

A CASE IN POINT

Cystitis

Ima Burning was attending a 3-day business meeting out of town. On the last day of the meeting, she noticed a frequent urge to urinate, even though her urine volume was small. She also felt a burning sensation when urinating. By the time Ima returned home, she was experiencing lower abdominal pain. Her urine appeared cloudy and had an unpleasant odor. Ima made an appointment with her physician, Dr. Blatter, who requested a urine sample. Urine is normally sterile, but Ima's urine contained abundant bacteria. Her physician diagnosed cystitis, an inflammation of the urinary bladder usually resulting from a bacterial infection. Infection by the bacterium *E. coli* is the most common cause of cystitis. In Ima's case, Dr. Blatter was unable to identify a specific cause of the infection, and she explained to Ima that 30% of women experience cystitis during their lifetime. She prescribed an antibiotic. Within 3 days, Ima was feeling normal again. A urine sample taken several days later showed no sign of infection. It is important to recognize cystitis early and treat it, because the infection can migrate up the ureters to affect the kidneys.

Micturition Reflex

The **micturition** (mik-choo-rish′ŭn) **reflex** is activated by stretch of the urinary bladder wall. As the urinary bladder fills with urine, pressure increases, stimulating stretch receptors in the wall of the urinary bladder. Action potentials are conducted from the urinary bladder to the spinal cord through the pelvic nerves. Integration of the reflex occurs in the spinal cord, and action potentials are conducted along parasympathetic nerve fibers to the urinary bladder. Parasympathetic action potentials cause the urinary bladder to contract (figure 18.19).

The external urinary sphincter is normally contracted as a result of stimulation from the somatic motor nervous system. Because of the micturition reflex, action potentials conducted along somatic motor nerves to the external urinary sphincter decrease, which causes the sphincter to relax. The micturition reflex is an automatic reflex, but it can be inhibited or stimulated by higher centers in the brain. The higher brain centers prevent micturition by sending action potentials through the spinal cord to decrease the intensity of the autonomic reflex that stimulates urinary bladder contractions and to stimulate nerve fibers that keep the external urinary sphincter contracted. The ability to voluntarily inhibit micturition develops at the age of 2–3 years.

Control of the micturition reflex by higher brain centers

(A) Ascending pathways carry an increased frequency of action potentials up the spinal cord to the pons and cerebrum when the urinary bladder becomes stretched. This increases the conscious urge to urinate.

(B) Descending pathways carry action potentials to the sacral region of the spinal cord to tonically inhibit the micturition reflex, preventing automatic urination when the urinary bladder is full. Descending pathways carry action potentials from the cerebrum to the sacral region of the spinal cord to facilitate the reflex when stretch of the urinary bladder produces the conscious urge to urinate and when a person voluntarily chooses to urinate. This reinforces the micturition reflex.

Micturition reflex

(1) Urine in the urinary bladder stretches the urinary bladder wall.

(2) Action potentials produced by stretch receptors are carried along pelvic nerves (*green line*) to the sacral region of the spinal cord.

(3) Action potentials are carried by parasympathetic nerves (*red line*) to contract the smooth muscles of the urinary bladder. Decreased action potentials carried by somatic motor nerves (*purple line*) cause the external urinary sphincter to relax.

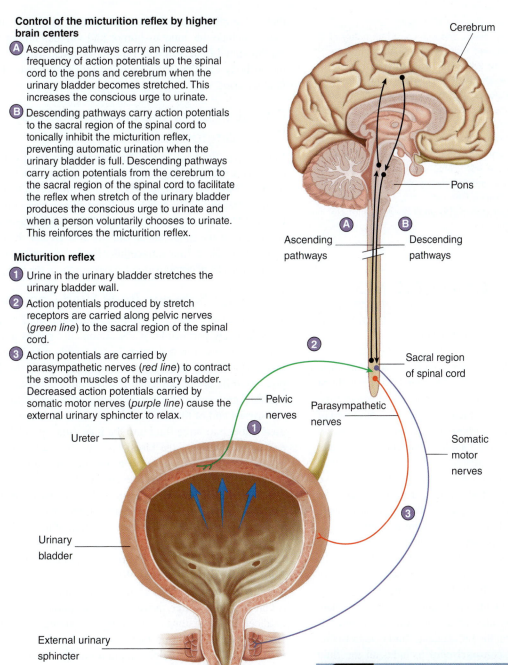

PROCESS Figure I8.I9 AP|R **Micturition Reflex**

When a person feels the urge to urinate, the higher brain centers alter action potentials sent to the spinal cord to facilitate the micturition reflex and relax the external urinary sphincter. Awareness of the need to urinate occurs because stretch of the urinary bladder stimulates sensory nerve fibers that increase action potentials carried to the brain by ascending tracts in the spinal cord. Irritation of the urinary bladder or the urethra by a bacterial infection or some other condition can also initiate the urge to urinate, even though the urinary bladder is nearly empty.

A CASE IN POINT

Kidney Stones (Renal Calculi)

After a long day at work, Harry Payne had just sat down to watch the news when he noticed rapidly developing discomfort in his left lateral abdominal region. Before long, the pain had radiated to the inguinal region on his left side and had become excruciating and debilitating. Harry suspected a kidney stone, or renal calculus, because he had passed a kidney stone about 2 years before. His wife helped him to the car and drove him to the emergency room. X-rays confirmed the presence of a kidney stone in the upper portion of Harry's left ureter. Peristaltic contractions of the ureter were forcing the irregularly shaped kidney stone through the narrow ureter, causing inflammation and pain. Harry's physician, Dr. Stone, prescribed strong analgesics to reduce the intense pain. Over the next 24 hours, urine was collected, and finally a kidney stone was recovered. Once the kidney stone had passed through the ureter, the pain decreased dramatically.

Urinary

Like the majority of kidney stones, this one consisted mainly of calcium oxalate. Dr. Stone explained that Harry needed to increase his fluid intake to produce at least 2 liters of urine each day. Keeping the urine dilute reduces the likelihood of calcium salts precipitating and forming additional kidney stones. Also, as a precaution, Dr. Stone recommended that Harry reduce his intake of animal protein.

In addition, Dr. Stone ordered a CT scan, which revealed a large kidney stone in Harry's left renal pelvis. This large kidney stone could give rise to additional kidney stones, obstruct the flow of urine from the left kidney through the left ureter, or become a chronic source of irritation, which could lead to kidney infections. Therefore, Dr. Stone recommended lithotripsy (lith′ō-trip-sē), an ultrasound technique that pulverizes kidney stones into small particles that can pass easily through the ureter.

18.6 BODY FLUID COMPARTMENTS

Learning Outcome *After reading this section, you should be able to*

A. List the major body fluid compartments.

Approximately 60% of the total body weight of an adult male consists of water. Approximately 50% of the total body weight of an adult female is water. Because the water content of adipose tissue is relatively low, the fraction of the body's weight composed of water decreases as the amount of adipose tissue increases. A smaller percentage of the body weight of an adult female consists of water because females generally have a greater percentage of body fat than do males. Water and the ions dissolved in it are distributed in two major compartments: the intracellular fluid compartment and the extracellular fluid compartment (table 18.2). Water and ions move between these compartments, but their movement is regulated.

The **intracellular fluid compartment** includes the fluid inside all the cells of the body. The cell membranes of the individual cells enclose the intracellular compartment, which actually consists of trillions of small compartments. Both the composition of the fluid in all these compartments and the regulation of fluid movement across all these cell membranes are similar. Approximately two-thirds of all the water in the body is in the intracellular fluid compartment.

The **extracellular fluid compartment** includes all the fluid outside the cells. It constitutes approximately one-third of the total body water. The extracellular fluid compartment includes the interstitial fluid, the plasma within blood vessels, and the fluid in the lymphatic vessels. A small portion of the extracellular fluid volume is separated by membranes into subcompartments. These special subcompartments contain fluid with a composition different from that of the other extracellular fluid. Fluids within the subcompartments include the aqueous humor and vitreous humor of the eye, cerebrospinal fluid, synovial fluid in the joint cavities, serous fluid in the body cavities, fluid secreted by glands, renal filtrate, and bladder urine.

Composition of the Fluid in the Body Fluid Compartments

Intracellular fluid has a similar composition from cell to cell. It contains a relatively high concentration of ions, such as K^+, magnesium (Mg^{2+}), phosphate (PO_4^{3-}), and sulfate (SO_4^{2-}), compared to the extracellular fluid. It has a lower concentration of Na^+, Ca^{2+}, Cl^-, and HCO_3^- than does the extracellular fluid. The concentration of protein in the intracellular fluid is also greater than that in the extracellular fluid. Like intracellular fluid, the extracellular fluid has a fairly consistent composition from one area of the body to another.

Exchange Between Body Fluid Compartments

The cell membranes that separate the body fluid compartments are selectively permeable. Water continually passes through them, but ions dissolved in the water do not readily pass through the cell membrane. Water movement is regulated mainly by hydrostatic pressure differences and osmotic differences between the compartments. For example, water moves across the wall of the capillary at the arterial end of the capillary because the blood pressure there is great enough to force fluid into the interstitial space. At the venous end of the capillary, the blood pressure is much lower, and fluid returns to the capillary because the osmotic pressure is higher inside the capillary than outside it (see chapter 13).

The major influence controlling the movement of water between the intracellular and extracellular spaces is osmosis. For example, if the extracellular concentration of ions increases, water moves by osmosis from cells into the extracellular fluid.

The intracellular fluid can help maintain the extracellular fluid volume if it is depleted. When a person becomes dehydrated, the concentration of ions in the extracellular fluid increases. As a consequence, water moves from the intracellular fluid to the extracellular fluid, thus maintaining the extracellular fluid volume. Because blood is an important component of the extracellular fluid volume, this process helps maintain blood volume. Movement of water from the intracellular fluid compartment to the extracellular fluid compartment can help prolong the time a person can survive a condition such as dehydration or cardiovascular shock.

If the concentration of ions in the extracellular fluid decreases, water moves by osmosis from the extracellular fluid into the cells. This water movement can cause the cells to swell. Under most

TABLE 18.2	Approximate Volumes of Body Fluid Compartments*				
Age of Person	Total Body Water	Intracellular Fluid	Plasma	Extracellular Fluid Interstitial	Total
Infant	75	45	4	26	30
Adult male	60	40	5	15	20
Adult female	50	35	5	10	15

*Expressed as percentages of body weight.

conditions, the movement of water between the intracellular and extracellular fluid compartments is maintained within limits that are consistent with survival of the individual.

18.7 REGULATION OF EXTRACELLULAR FLUID COMPOSITION

Learning Outcome *After reading this section, you should be able to*

A. Describe the mechanisms by which Na⁺, K⁺, and Ca²⁺ are regulated in the extracellular fluid.

Homeostasis requires that the intake of substances equals their elimination. Needed water and ions enter the body by ingestion; excess water and ions exit the body by excretion. The amounts of water and ions entering and leaving the body can vary over the short term. For example, greater quantities of water and ions are lost in the form of perspiration on warm days than on cool days, and varying amounts of water and ions may be lost in the form of feces. However, over a long period, the total amount of water and ions in the body does not change unless the individual is growing, gaining weight, or losing weight. Regulating the amounts of water and ions in the body involves the coordinated participation of several organ systems, but the kidneys are the most important, with the skin, liver, and digestive tract playing supporting roles. Two mechanisms help regulate the levels of ions in the extracellular fluid: thirst regulation and ion concentration regulation.

Thirst Regulation

Water intake is controlled by neurons in the hypothalamus, collectively called the **thirst center.** When blood becomes more concentrated, the thirst center responds by initiating the sensation of thirst (figure 18.20). When water or another dilute solution is consumed, the blood becomes less concentrated and the sensation of thirst decreases. Similarly, when blood pressure drops, as occurs during shock, the thirst center is activated, and the sensation of thirst

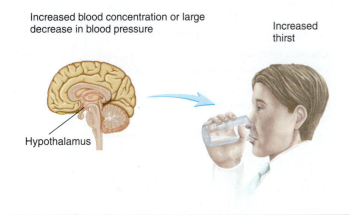

Increased blood concentration or large decrease in blood pressure

Increased thirst

Hypothalamus

Figure 18.20   Thirst and the Regulation of Extracellular Fluid Concentration

Increased blood concentration affects hypothalamic neurons, and large decreases in blood pressure affect baroreceptors in the aortic arch, carotid sinuses, and right atrium. These stimuli cause an increase in thirst, which increases water intake. Increased water intake reduces blood concentration and increases blood volume.

is triggered. Consumption of water increases the blood volume and allows the blood pressure to return to its normal value. Other stimuli can also trigger the sensation of thirst. For example, if the mucosa of the mouth becomes dry, the thirst center is activated. Thirst is one of the important means of regulating extracellular fluid volume and concentration.

Ion Concentration Regulation

If the water content or concentration of ions in the extracellular fluid deviates from its normal range, cells cannot control the movement of substances across their cell membranes or the composition of their intracellular fluid. The consequence is abnormal cell function or even cell death. Keeping the extracellular fluid composition within a normal range is therefore required to sustain life.

Regulating the concentrations of positively charged ions, such as Na⁺, K⁺, and Ca²⁺, in the body fluids is particularly important. Action potentials, muscle contraction, and normal cell membrane permeability depend on the maintenance of a narrow range of concentrations for these ions. Important mechanisms control the concentrations of these ions in the body. Negatively charged ions, such as Cl⁻, are secondarily regulated by the mechanisms that control the positively charged ions. The negatively charged ions are attracted to the positively charged ions; when the positively charged ions are transported, the negatively charged ions move with them.

Sodium Ions

Sodium ions (Na⁺) are the dominant ions in the extracellular fluid. About 90–95% of the osmotic pressure of the extracellular fluid results from sodium ions and from the negative ions associated with them.

The recommended intake of Na⁺ is 2.4 grams per day (g/day), because of its association with high blood pressure in some people. Most people in the United States consume two to three times the recommended amount of Na⁺. The kidneys provide the major route by which the excess Na⁺ is excreted.

Stimuli that control aldosterone secretion influence the reabsorption of Na⁺ from nephrons of the kidneys and the total amount of Na⁺ in the body fluids. Reabsorption of Na⁺ from the distal convoluted tubules and collecting ducts is very efficient, and little Na⁺ is lost in the urine when aldosterone is present. When aldosterone is absent, reabsorption of Na⁺ in the nephron is greatly reduced, and the amount of Na⁺ lost in the urine increases. Aldosterone also plays an essential role in regulating the extracellular K⁺ concentration (figure 18.21).

Sodium ions are also excreted from the body in perspiration, or **sweat.** Normally, only a small quantity of Na⁺ is lost each day in the form of sweat, but the amount increases during heavy exercise in a warm environment.

Because Na⁺ has such a large effect on the osmotic pressure of the extracellular fluid, mechanisms that influence Na⁺ concentrations in the extracellular fluid also influence the extracellular fluid volume. The mechanisms that play important roles in controlling these levels are the renin-angiotensin-aldosterone mechanism, the atrial natriuretic (ANH) mechanism, and antidiuretic hormone (ADH) (see figures 18.14, 18.15, 18.16, and 18.17). For example, low blood pressure increases renin and ADH secretion. The result is an increase in Na⁺ and water reabsorption in the kidney to bring blood pressure and the Na⁺ concentration back to their normal

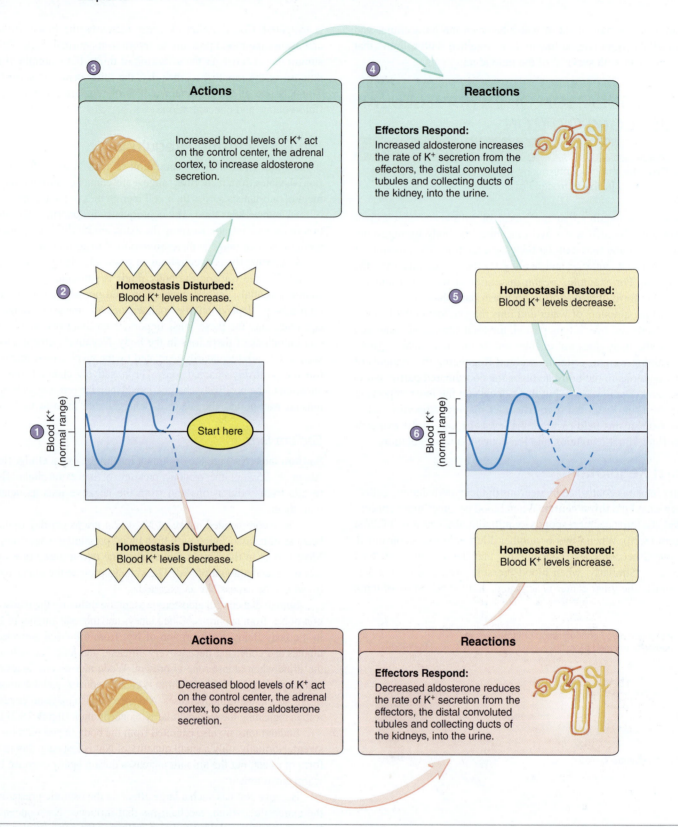

Homeostasis Figure 18.21 Regulation of Blood K⁺ Levels

(1) Blood K⁺ is in its normal range. (2) Blood K⁺ increases outside its normal range, which causes homeostasis to be disturbed. (3) The control center in the adrenal cortex responds to the change in blood K⁺. (4) The control center causes aldosterone secretion. (5) Aldosterone stimulates K⁺ secretion from the distal convoluted tubule and collecting duct, reducing blood K⁺ levels. (6) Blood K⁺ returns to its normal range, and homeostasis is restored. Observe the responses to a decrease in blood K⁺ outside its normal range by following the *red arrows*.

Urinary

ranges. Increased blood pressure inhibits renin and ADH secretion, and stimulates ANH secretion. The result is a decrease in Na^+ reabsorption and an increase in urine production to bring blood pressure and the blood's Na^+ concentration into their normal ranges.

Potassium Ions

Electrically excitable tissues, such as muscles and nerves, are highly sensitive to slight changes in the extracellular K^+ concentration. The extracellular concentration of K^+ must be maintained within a narrow range for these tissues to function normally.

Aldosterone plays a major role in regulating the concentration of K^+ in the extracellular fluid. Dehydration, circulatory system shock resulting from plasma loss, and tissue damage due to injuries such as severe burns all cause extracellular K^+ concentrations to increase above normal. In response, aldosterone secretion from the adrenal cortex increases and causes K^+ secretion to increase (figure 18.21).

If the K^+ concentration in the extracellular fluid decreases, aldosterone secretion from the adrenal cortex decreases. In response, the rate of K^+ secretion by the kidneys is reduced (figure 18.21).

Calcium Ions

The extracellular concentration of Ca^{2+}, like that of other ions, is maintained within a narrow range. Increases and decreases in the extracellular concentration of Ca^{2+} have dramatic effects on the electrical properties of excitable tissues. For example, decreased extracellular Ca^{2+} concentrations make cell membranes more permeable to Na^+, thus making them more electrically excitable. Decreased extracellular concentrations of Ca^{2+} cause spontaneous action potentials in nerve and muscle cells, resulting in hyperexcitability and muscle tetany. Increased extracellular Ca^{2+} concentrations make cell membranes less permeable to Na^+, thus making them less electrically excitable. Increased extracellular concentrations of Ca^{2+} inhibit action potentials in nerve and muscle cells, resulting in reduced excitability and either muscle weakness or paralysis.

Parathyroid hormone (PTH), secreted by the parathyroid glands, increases extracellular Ca^{2+} concentrations. The rate of PTH secretion is regulated by the extracellular Ca^{2+} concentration (see figure 10.17). An elevated Ca^{2+} concentration inhibits the secretion of PTH and a reduced Ca^{2+} concentration stimulates the secretion of PTH. PTH causes osteoclasts to degrade bone and release Ca^{2+} into the body fluids. PTH also increases the rate of Ca^{2+} reabsorption from kidney nephrons.

Vitamin D increases Ca^{2+} concentration in the blood by increasing the rate of Ca^{2+} absorption by the intestine. Some vitamin D is consumed in food, and the body produces the rest (see chapter 5). PTH affects the intestinal uptake of Ca^{2+} because PTH increases the rate of vitamin D production in the body.

Calcitonin (kal-si-tō′nin) is secreted by the thyroid gland. Calcitonin reduces the blood Ca^{2+} concentration when it is too high. An elevated blood Ca^{2+} concentration causes the thyroid gland to secrete calcitonin, and a low blood Ca^{2+} concentration inhibits calcitonin secretion. Calcitonin reduces the rate at which bone is broken down and decreases the release of Ca^{2+} from bone (see figure 10.17).

Phosphate and Sulfate Ions

Some ions, such as **phosphate ions** (PO_4^{3-}) and **sulfate ions** (SO_4^{2-}), are reabsorbed by active transport in the kidneys. The

rate of reabsorption is slow, so that if the concentration of these ions in the filtrate exceeds the nephron's ability to reabsorb them, the excess is excreted into the urine. As long as the concentration of these ions is low, nearly all of them are reabsorbed by active transport. This mechanism plays a major role in regulating the concentration of PO_4^{3-} and SO_4^{2-} in the body fluids.

18.8 REGULATION OF ACID-BASE BALANCE

Learning Outcome   *After reading this section, you should be able to*

A. Illustrate how the mechanisms that regulate the body fluid pH function, by explaining how they respond to decreasing and increasing pH in the body fluids.

The concentration of H^+ in the body fluids is reported as the pH. The body fluid pH is maintained between 7.35 and 7.45; any deviation from that range is life-threatening. Consequently, the mechanisms that regulate body fluid pH are critical for survival. The pH of body fluids is controlled by three factors: buffers, the respiratory system, and the kidneys. When the pH of body fluids is not properly maintained, the result is acidosis or alkalosis.

Buffers

Buffers are chemicals that resist a change in the pH of a solution when either acids or bases are added to the solution. The buffers in the body fluids contain salts of either weak acids or weak bases that combine with H^+ when H^+ increases in those fluids, or release H^+ when H^+ decreases in those fluids. Buffers tend to keep the H^+ concentration, and thus the pH, within a narrow range of values (figure 18.22) because of these characteristics. The three major buffers in the body fluids are the proteins, the PO_4^{3-} buffer system, and the HCO_3^- buffer system.

Proteins and PO_4^{3-} in the body fluids are able to combine reversibly with hydrogen ions. When the H^+ concentration increases, proteins and PO_4^{3-} prevent a decrease in pH by combining with the hydrogen ions. Conversely, when the H^+ concentration decreases, proteins and PO_4^{3-} release H^+, preventing an increase in pH.

The following reaction illustrates how PO_4^{3-} buffers work:

$$\begin{array}{ccccc} HPO_4^{2-} & + & H^+ & \rightleftharpoons & H_2PO_4^- \\ \text{Monohydrogen} & & \text{Hydrogen} & & \text{Dihydrogen} \\ \text{phosphate ion} & & \text{ion} & & \text{phosphate ion} \end{array}$$

Monohydrogen phosphate ions (HPO_4^{2-}) combine with H^+ to form dihydrogen phosphate ions ($H_2PO_4^-$) when excess H^+ is present. When H^+ concentration declines, some of the H^+ separate from the $H_2PO_4^-$.

Proteins are able to function as buffers because their amino acids have side chains that function as weak acids and weak bases. Many side chains contain carboxyl groups (–COOH) or amine groups (–NH_2). Both of these groups are able to function as buffers because of the following reactions:

$$\begin{array}{ccccc} -COO^- & + & H^+ & \rightleftharpoons & -COOH \\ \text{Carboxyl group} & & \text{Hydrogen} & & \text{Carboxyl} \\ \text{(ionized)} & & \text{ion} & & \text{group} \end{array}$$

$$-NH_2 \quad + \quad H^+ \quad \rightleftarrows \quad -NH_3$$
Amine group   Hydrogen ion   Ammonium group

The bicarbonate (HCO_3^-) buffer system is unable to combine with as many hydrogen ions as can proteins and PO_4^{3-} buffers, but the HCO_3^- buffer system is critical because it can be regulated by the respiratory and urinary systems. Carbon dioxide (CO_2) combines with water (H_2O) to form carbonic acid (H_2CO_3), which in turn forms H^+ and HCO_3^- as follows:

$$H_2O \quad + \quad CO_2 \quad \rightleftarrows \quad H_2CO_3 \quad \rightleftarrows \quad H^+ \quad + \quad HCO_3^-$$
Water   Carbon   Carbonic   Hydrogen   Bicarbonate
dioxide   acid   ion   ion

The reaction between CO_2 and H_2O is catalyzed by an enzyme, called carbonic anhydrase, which is found in red blood cells and on the surface of capillary epithelial cells (see chapter 15). The enzyme accelerates the rate at which the reaction proceeds in either direction.

The higher the concentration of CO_2, the greater the amount of H_2CO_3 formed, and the greater the amount of H^+ and HCO_3^- formed. This results in a decreased pH. However, the reaction is reversible. If CO_2 levels decline, the equilibrium shifts in the opposite direction. That is, H^+ and HCO_3^- combine to form H_2CO_3, which then forms CO_2 and H_2O, and the pH increases.

Respiratory System

The **respiratory system** responds rapidly to a change in pH and helps bring the pH of body fluids back toward normal. Increasing CO_2 levels and decreasing body fluid pH stimulate neurons in the respiratory center of the brain and cause elevated rate and depth of ventilation. As a result, CO_2 is eliminated from the body through the lungs at a greater rate, and the concentration of CO_2 in the body fluids decreases. As CO_2 levels decline, the concentration of H^+ also declines. The pH therefore rises back toward its normal range (figure 18.22).

Predict 7

Under stressful conditions, some people hyperventilate. Predict the effect of the rapid rate of ventilation on the pH of body fluids. In addition, explain why a person who is hyperventilating may benefit from breathing into a paper bag.

If CO_2 levels become too low or the pH of the body fluids is elevated, the rate and depth of respiration decline. As a consequence, the rate at which CO_2 is eliminated from the body is reduced. Carbon dioxide then accumulates in the body fluids because it is continually produced as a by-product of metabolism. As CO_2 accumulates in the body fluids, so does H^+, resulting in a decreased pH.

Kidneys

The nephrons of the kidneys secrete H^+ into the urine and therefore can directly regulate the pH of the body fluids. The kidney is a powerful regulator of pH, but it responds more slowly than does the respiratory system. Cells in the walls of the distal convoluted tubule are primarily responsible for the secretion of H^+. As the pH of the body fluids drops below normal, the rate at which the distal convoluted tubules secrete H^+ increases (figure 18.22). At the same time, reabsorption of HCO_3^- increases. The increased rate of H^+ secretion and the increased rate of HCO_3^- reabsorption both cause the blood pH to rise toward its normal value. On the other hand, as the body fluid pH elevates above normal, the rate of H^+ secretion by the distal convoluted tubules declines, and the amount of HCO_3^- lost in the urine increases. Consequently, the blood pH drops toward its normal value.

Acidosis and Alkalosis

Failure of the buffer systems, the respiratory system, or the urinary system to maintain normal pH levels can result in acidosis or alkalosis.

Acidosis

Acidosis (as-i-dō′sis) occurs when the blood pH falls below 7.35. The central nervous system malfunctions, and the individual becomes disoriented and, as the condition worsens, may become comatose. Acidosis is separated into two categories. **Respiratory acidosis** results when the respiratory system is unable to eliminate adequate amounts of CO_2. Carbon dioxide accumulates in the circulatory system, causing the pH of the body fluids to decline. **Metabolic acidosis** results from excess production of acidic substances, such as lactic acid and ketone bodies, because of increased metabolism or decreased ability of the kidneys to eliminate H^+ in the urine.

Predict 8

Adam suffers from severe emphysema. Gas exchange in his lungs is not adequate, and he requires supplementary O_2. But even though his blood CO_2 levels are elevated, his blood pH is close to normal. Explain.

Alkalosis

Alkalosis (al-kă-lō′sis) occurs when the blood pH increases above 7.45. A major effect of alkalosis is hyperexcitability of the nervous system. Peripheral nerves are affected first, resulting in spontaneous nervous stimulation of muscles. Spasms and tetanic contractions result, as can extreme nervousness or convulsions. Tetany of respiratory muscles can cause death. **Respiratory alkalosis** results from hyperventilation, as can occur in response to stress. **Metabolic alkalosis** usually results from the rapid elimination of H^+ from the body, as occurs during severe vomiting or when excess aldosterone is secreted by the adrenal cortex.

③ Actions

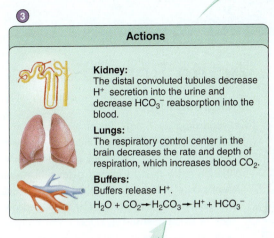

Kidney:
The distal convoluted tubules decrease H^+ secretion into the urine and decrease HCO_3^- reabsorption into the blood.

Lungs:
The respiratory control center in the brain decreases the rate and depth of respiration, which increases blood CO_2.

Buffers:
Buffers release H^+.
$$H_2O + CO_2 \rightarrow H_2CO_3 \rightarrow H^+ + HCO_3^-$$

④ Reactions

Effectors Respond:
Fewer H^+ are removed from the blood, and fewer HCO_3^- are available to bind H^+.

Increased blood CO_2 reacts with water to produce carbonic acid, which dissociates to increase H^+.
$$H_2O + CO_2 \rightarrow H_2CO_3 \rightarrow H^+ + HCO_3^-$$

The number of H^+ in the blood increases.

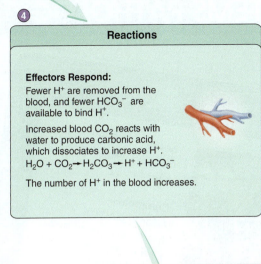

② Homeostasis Disturbed:
Blood pH increases (H^+ decreases).

⑤ Homeostasis Restored:
Blood pH decreases (H^+ increases).

① Blood pH (normal range) — Start here

⑥ Blood pH (normal range)

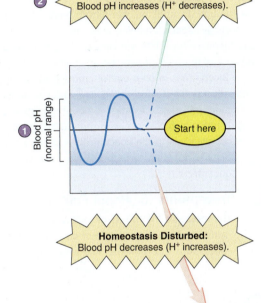

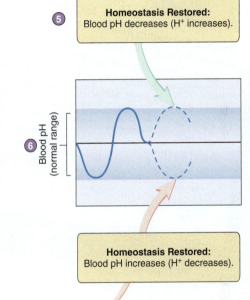

Homeostasis Disturbed:
Blood pH decreases (H^+ increases).

Homeostasis Restored:
Blood pH increases (H^+ decreases).

Actions

Buffers:
Buffers bind H^+.
$$H_2O + CO_2 \leftarrow H_2CO_3 \leftarrow H^+ + HCO_3^-$$

Lungs:
The respiratory control center in the brain increases the rate and depth of respiration, which decreases blood CO_2.

Kidney:
The distal convoluted tubules increase H^+ secretion into the urine, and increase HCO_3^- reabsorption into the blood.

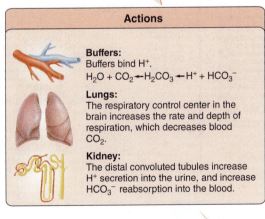

Reactions

Effectors Respond:
The number of H^+ in the blood decreases.

Decreased blood CO_2 causes H^+ to react with HCO_3^- to form carbonic acid, which decreases H^+ in the blood.
$$H_2O + CO_2 \leftarrow H_2CO_3 \leftarrow H^+ + HCO_3^-$$

More H^+ are removed from the blood, and more HCO_3^- are available to bind H^+.

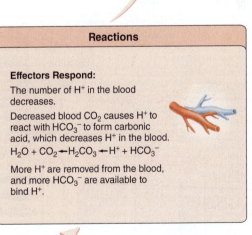

Homeostasis Figure 18.22 Regulation of Acid-Base Balance

(1) Blood pH is in its normal range. (2) Blood pH increases outside the normal range, which causes homeostasis to be disturbed. (3) The blood pH control centers respond to the change in blood pH. (4) The control centers cause decreased H^+ secretion from the blood and increased carbonic acid production, which increases blood H^+ concentration. (5) These changes cause blood pH to decrease. (6) Blood pH returns to its normal range, and homeostasis is restored. Observe the responses to a decrease in blood pH outside its normal range by following the *red arrows*.

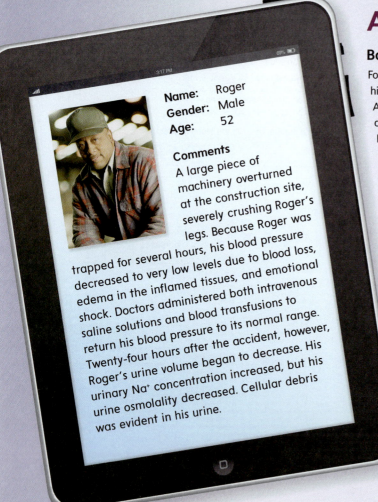

Name: Roger
Gender: Male
Age: 52

Comments
A large piece of machinery overturned at the construction site, severely crushing Roger's legs. Because Roger was trapped for several hours, his blood pressure decreased to very low levels due to blood loss, edema in the inflamed tissues, and emotional shock. Doctors administered both intravenous saline solutions and blood transfusions to return his blood pressure to its normal range. Twenty-four hours after the accident, however, Roger's urine volume began to decrease. His urinary Na^+ concentration increased, but his urine osmolality decreased. Cellular debris was evident in his urine.

Acute Renal Failure

Background Information

For approximately 7 days, Roger required renal dialysis to maintain his blood volume and ion concentrations within normal ranges. After about 3 weeks, his kidney functions slowly began to improve, although many months passed before they returned to normal. In Roger's case, the events 24 hours after his accident are consistent with acute renal failure caused by prolonged low blood pressure and lack of blood flow to the kidneys. The reduced blood flow to the kidneys was severe enough to cause damage to the epithelial lining of the kidney tubules. The period of reduced urine volume resulted from tubular damage. Dead and damaged tubular cells sloughed off into the tubules and blocked them, so that filtrate could not flow through. In addition, the filtrate leaked from the blocked or partially blocked tubules back into the interstitial spaces and therefore back into the circulatory system. As a result, the amount of filtrate that became urine was markedly reduced.

Blood levels of urea and of creatine usually increase due to reduced filtrate formation and reduced function of the tubular epithelium. A small amount of urine is produced that has a high Na^+ concentration, although the osmolality is usually close to the concentration of the body fluids. The kidney is not able to reabsorb Na^+, nor can it effectively concentrate urine.

Treatments for Renal Failure

Hemodialysis (hē′mō-dī-al′i-sis) is used when a person is suffering from severe acute or chronic kidney failure. The procedure substitutes for the excretory functions of the

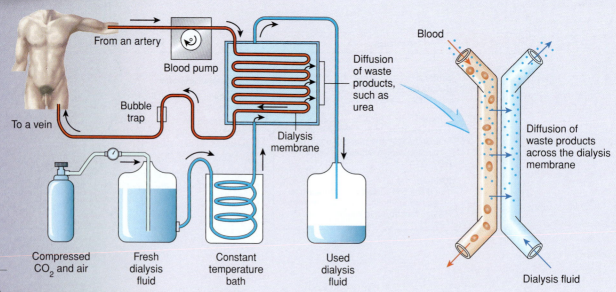

Figure 18A

During hemodialysis, blood flows through a system of tubes composed of a selectively permeable membrane. Dialysis fluid, which has a composition similar to that of normal blood (except that the concentration of waste products is very low), flows in the opposite direction on the outside of the dialysis tubes. Waste products, such as urea, diffuse from the blood into the dialysis fluid. Other substances, such as Na^+, K^+, and glucose, can diffuse from the blood into the dialysis fluid if they are present in higher than normal concentrations, because these substances are present in the dialysis fluid at the same concentrations found in normal blood.

Acute Renal Failure

Symptoms
- Decreased urine volume
- Increased Na$^+$ in urine
- Decreased urine osmolality

Treatment
- Hemodialysis
- Kidney transplant

SKELETAL
Bone resorption can result because of excessive and chronic loss of Ca^{2+} in the urine. Vitamin D levels may be reduced.

INTEGUMENTARY
Anemia causes pallor, and bruising results from clotting proteins lacking in the blood because they are lost in the urine. Accumulation of urinary pigments changes skin tone. High urea gives a yellow cast to light-skinned people, and white crystals of urea, called uremic frost, may appear on areas of the skin where there is heavy perspiration.

MUSCULAR
Neuromuscular irritability results from the toxic effect of metabolic wastes on the central nervous system and ionic imbalances, such as elevated blood K$^+$ levels. Involuntary jerking and twitching can occur as neuromuscular irritability develops.

DIGESTIVE
Decreased appetite, mouth infections, nausea, and vomiting result from altered digestive tract functions due to the effects of ionic imbalances on the nervous system.

NERVOUS
Elevated blood K$^+$ levels and the toxic effects of metabolic wastes result in depolarization of neurons. Slowing of action potential conduction, burning sensations, pain, numbness, or tingling results. Also, decreased mental acuity, reduced ability to concentrate, apathy, and lethargy occur. Or in severe cases, confusion and coma occur.

RESPIRATORY
Early during acute renal failure, the depth of breathing increases and becomes labored as acidosis develops because the kidneys are not able to secrete H$^+$. Pulmonary edema often develops because of water and Na$^+$ retention as a result of reduced urine production. The likelihood of infection increases as a result of pulmonary edema.

ENDOCRINE
Major hormone deficiencies include vitamin D deficiency. In addition, the secretion of reproductive hormones decreases due to the effects of metabolic wastes and ionic imbalances on the hypothalamus.

LYMPHATIC
There are no major direct effects on the lymphatic system, except that increased lymph flow occurs as a result of edema.

CARDIOVASCULAR
Water and Na$^+$ retention cause edema in peripheral tissues and in the lungs, leading to increased blood pressure and congestive heart failure. Elevated blood K$^+$ levels result in dysrhythmias and can cause cardiac arrest. Anemia due to decreased erythropoietin production by the damaged kidney exists.

---

kidney. Hemodialysis is based on blood flow through tubes composed of a selectively permeable membrane. Blood is usually taken from an artery, passed through the tubes of the dialysis machine, and then returned to a vein (figure 18A). On the outside of the dialysis tubes is a fluid, called dialysis fluid, which contains the same concentration of solutes as normal plasma, except for the metabolic waste products. As a consequence, the metabolic wastes diffuse from the blood to the dialysis fluid. The dialysis membrane has pores that are too small to allow plasma proteins to pass through them, and because the dialysis fluid contains the same beneficial solutes as the plasma, the net movement of these substances is zero.

Peritoneal (per′i-tō-nē′ăl) **dialysis** is sometimes used to treat kidney failure. The principles by which peritoneal dialysis works are the same as for hemodialysis, but the dialysis fluid flows through a tube inserted into the peritoneal cavity. The visceral and parietal peritonea act as the dialysis membrane. Waste products diffuse from the blood vessels beneath the peritoneum, across the peritoneum, and into the dialysis fluid.

Kidney transplants are sometimes performed on people who have severe renal failure. Often, the donor suffered an accidental death and had granted permission to have his or her kidneys used for transplantation. The major cause of kidney transplant failure is rejection by the recipient's immune system. Physicians therefore attempt to match the immune characteristics of the donor and recipient to reduce the tendency for rejection. Even with careful matching, recipients have to take medication for the rest of their lives to suppress their immune reactions. But in most cases, the transplanted kidney functions well, and the tendency of the recipient's immune system to reject the transplanted kidney can be controlled.

Predict 9

Nine days after the accident, Roger began to appear pale, became dizzy on standing, and was very weak and lethargic. His hematocrit was elevated, and his heart was arrhythmic. Explain these manifestations.

DISEASES AND DISORDERS: Urinary System

CONDITION	DESCRIPTION
Inflammation of the Kidneys	
Glomerulonephritis (glō-mār′ū-lō-nef-rī′tis)	Inflammation of the filtration membrane within the renal corpuscle, causing increased membrane permeability; plasma proteins and blood cells enter the filtrate, causing increased urine volume due to increased osmotic concentration of the filtrate
Acute glomerulonephritis	Often occurs 1–3 weeks after a severe bacterial infection, such as strep throat; normally subsides after several days
Chronic glomerulonephritis	Long-term and progressive process whereby the filtration membrane thickens and is eventually replaced by connective tissue and the kidneys become nonfunctional
Renal Failure	Can result from any condition that interferes with kidney function
Acute renal failure	Occurs when damage to the kidney is rapid and extensive; leads to accumulation of wastes in the blood; can lead to death in 1–2 weeks if renal failure is complete
Chronic renal failure	Results from permanent damage to so many nephrons that the remaining nephrons are inadequate for normal kidney function; can be caused by chronic glomerulonephritis, trauma to the kidneys, tumors, or kidney stones

ANSWER TO LEARN TO PREDICT

First we learn that Sadie is unable to retain water and is urinating excessively. This helps us to narrow our focus to kidney function. Next we learn that Sadie has normal levels of ADH. Excessive urination falls under the category of diabetes (derived from the Greek for "siphon"); however, diabetes insipidus is a condition involving ADH abnormality. There are two main causes of diabetes insipidus: The posterior pituitary fails to secrete ADH, or the kidney tubules have abnormal receptors for ADH and do not respond to the presence of ADH. In people suffering from diabetes insipidus, much of the filtrate entering both the proximal convoluted and the distal convoluted tubules becomes urine. People with this condition can produce as much as 20–30 L of urine each day. Because they lose so much water, they are continually in danger of severe dehydration. Even though their urine is dilute, producing such a large volume of urine leads to the loss of Na⁺, Ca²⁺, and other ions. The resulting ionic imbalances cause the nervous system and cardiac muscle to function abnormally. Because Sadie's levels of ADH are normal, we can conclude that she has the type of diabetes insipidus involving abnormal receptors. Treatments for Sadie include making sure she drinks plenty of water and giving her a sodium-sparing diuretic so that her kidneys retain sodium.

Answers to the rest of this chapter's Predict questions are in Appendix E.

SUMMARY

The urinary system consists of two kidneys, two ureters, the urinary bladder, and the urethra.

18.1 Functions of the Urinary System   (p. 499)

1. The kidneys excrete waste products.
2. The kidneys control blood volume and blood pressure by regulating the volume of urine produced.
3. The kidneys help regulate the concentration of major ions in the body fluids.
4. The kidneys help regulate the pH of the extracellular fluid.
5. The kidneys regulate the concentration of red blood cells in the blood.
6. The kidneys participate, with the skin and liver, in regulating vitamin D synthesis.

18.2 Anatomy of the Kidneys   (p. 500)

1. Each kidney is behind the peritoneum and surrounded by a renal capsule and adipose tissue.
2. The kidney is divided into an outer cortex and an inner medulla.
3. Each renal pyramid has a base located at the boundary between the cortex and the medulla. The tip of the renal pyramid extends toward the center of the kidney and is surrounded by a calyx.
4. Calyces are extensions of the renal pelvis, which is the expanded end of the ureter within the renal sinus.
5. The functional unit of the kidney is the nephron. The parts of a nephron are the renal corpuscle, the proximal convoluted tubule, the loop of Henle, and the distal convoluted tubule.
6. The filtration membrane is formed by the glomerular capillaries, the basement membrane, and the podocytes of the Bowman capsule.

Arteries and Veins

1. Renal arteries give rise to branches that lead to afferent arterioles.
2. Afferent arterioles supply the glomeruli.
3. Efferent arterioles carry blood from the glomeruli to the peritubular capillaries.
4. Blood from the peritubular capillaries flows to the renal veins.

18.3 Urine Production   (p. 505)

Urine is produced by filtration, tubular reabsorption, and tubular secretion.

Filtration

1. The renal filtrate passes from the glomerulus into the Bowman capsule and contains no blood cells and few blood proteins.
2. Filtration pressure is responsible for filtrate formation.
3. Increased sympathetic activity decreases blood flow to the kidney, decreases filtrate formation, and decreases urine production. Decreased sympathetic activity has the opposite effect.

Tubular Reabsorption

1. About 99% of the filtrate volume is reabsorbed; 1% becomes urine.
2. Among the substances reabsorbed are proteins, amino acids, glucose, fructose, Na^+, K^+, Ca^{2+}, HCO_3^-, and Cl^-.
3. About 65% of the filtrate volume is reabsorbed in the proximal convoluted tubule, 15% is reabsorbed in the descending limb of the loop of Henle, and another 19% is reabsorbed in the distal convoluted tubule and collecting duct.

Tubular Secretion

Hydrogen ions, some by-products of metabolism, and some drugs are actively secreted into the nephron.

18.4 Regulation of Urine Concentration and Volume   (p. 510)

Hormonal Mechanisms

1. Renin is secreted from the kidney when the blood pressure decreases. Renin converts angiotensinogen to angiotensin I, which is then converted to angiotensin II by angiotensin-converting enzyme. Angiotensin II stimulates aldosterone secretion, and aldosterone increases the rate of Na^+ and Cl^- reabsorption from the nephron.
2. ADH is secreted from the posterior pituitary when the concentration of blood increases or when blood pressure decreases. ADH increases the permeability to water of the distal convoluted tubule and collecting duct. It increases water reabsorption by the kidney.
3. Atrial natriuretic hormone, secreted from the right atrium in response to increases in blood pressure, acts on the kidney to increase Na^+ and water loss in the urine.

18.5 Urine Movement   (p. 514)

Anatomy and Histology of the Ureters, Urinary Bladder, and Urethra

1. Each ureter carries urine from a renal pelvis to the urinary bladder.
2. The urethra carries urine from the urinary bladder to the outside of the body.
3. The ureters and urinary bladder are lined with transitional epithelium and have smooth muscle in their walls.
4. The external urinary sphincter regulates the flow of urine through the urethra.

Micturition Reflex

1. Increased volume in the urinary bladder stretches its wall and activates the micturition reflex.
2. Parasympathetic action potentials cause the urinary bladder to contract. Reduced somatic motor action potentials cause the external urinary sphincter to relax.
3. Higher brain centers control the micturition reflex. Stretching of the urinary bladder stimulates sensory neurons that carry impulses to the brain and inform the brain of the need to urinate.

18.6 Body Fluid Compartments   (p. 518)

1. Water and the ions dissolved in the water are distributed in the intracellular and extracellular fluid compartments.
2. Approximately two-thirds of the total body water is found within cells.
3. Approximately one-third of the total body water is found outside cells, mainly in interstitial fluid, blood plasma, and lymph.

Composition of the Fluid in the Body Fluid Compartments

1. Intracellular fluid contains more K^+, Mg^{2+}, PO_4^{3-}, SO_4^{2-}, and protein than does extracellular fluid.
2. Extracellular fluid contains more Na^+, Ca^{2+}, Cl^-, and HCO_3^- than does intracellular fluid.

Exchange Between Body Fluid Compartments

Water moves continuously between compartments in response to hydrostatic pressure differences and osmotic differences between the compartments.

18.7 Regulation of Extracellular Fluid Composition   (p. 519)

The total amount of water and electrolytes in the body does not change unless the person is growing, gaining weight, or losing weight.

Thirst Regulation

The sensation of thirst increases if extracellular fluid becomes more concentrated or if blood pressure decreases.

Ion Concentration Regulation

1. Sodium ions are the dominant extracellular ions. Aldosterone increases Na^+ reabsorption from the filtrate, ADH increases water reabsorption from the nephron, and ANH increases Na^+ loss in the urine.
2. Aldosterone increases K^+ secretion in the urine. Increased blood levels of K^+ stimulate, and decreased blood levels of K^+ inhibit, aldosterone secretion.
3. Parathyroid hormone secreted from the parathyroid glands increases extracellular Ca^{2+} levels by causing bone resorption and increased Ca^{2+} uptake in the kidney. Parathyroid hormone increases vitamin D synthesis. Calcitonin, secreted by the thyroid gland, inhibits bone resorption and lowers blood Ca^{2+} levels when they are too high.
4. When PO_4^{3-} and SO_4^{2-} levels in the filtrate are low, nearly all PO_4^{3-} and SO_4^{2-} are reabsorbed. When levels are high, excess is lost in the urine.

18.8 Regulation of Acid-Base Balance   (p. 521)

Buffers

Three principal classes of buffers in the body fluids resist changes in the pH: proteins, the phosphate buffer system, and the bicarbonate buffer system.

Respiratory System

The respiratory system rapidly regulates pH. An increased respiratory rate raises the pH because the rate of CO_2 elimination is increased, and a reduced respiratory rate reduces the pH because the rate of CO_2 elimination is reduced.

Kidneys

The kidneys excrete H^+ in response to a decreasing blood pH, and they reabsorb H^+ in response to an increasing blood pH.

Acidosis and Alkalosis

1. Acidosis occurs when the pH of the blood falls below 7.35. The two major types are respiratory acidosis and metabolic acidosis.
2. Alkalosis occurs when the pH of the blood increases above 7.45. The two major types are respiratory alkalosis and metabolic alkalosis.

Urinary

REVIEW AND COMPREHENSION

1. Name the structures that make up the urinary system. List the functions of the urinary system.

2. What structures surround the kidney?

3. Describe the relationships of the renal pyramids, calyces, renal pelvis, and ureter.

4. What is the functional unit of the kidney? Name its parts.

5. Describe the blood supply to the kidney.

6. Name the three general processes involved in the production of urine.

7. Describe the filtration membrane. What substances do not pass through it?

8. How do changes in blood pressure in the glomerulus affect the volume of filtrate produced?

9. What effect does sympathetic stimulation have on the kidneys?

10. What substances are reabsorbed in the nephron? What happens to most of the filtrate volume that enters the nephron?

11. In what parts of the nephron are large volumes of filtrate reabsorbed? In what part of the nephron is no filtrate reabsorbed?

12. In general, what substances are secreted into the nephron?

13. What effect does ADH have on urine volume? Name the factors that cause an increase in ADH secretion.

14. Where is renin produced, and what stimulates its secretion?

15. Explain how renin controls the synthesis of angiotensin I. What enzyme regulates the conversion of angiotensin I to angiotensin II?

16. Describe the effect of angiotensin II on aldosterone secretion.

17. Where is aldosterone produced, and what effect does it have on urine volume? What factors stimulate aldosterone secretion?

18. Where is atrial natriuretic hormone produced, and what effect does it have on urine production?

19. What are the functions of the ureters, urinary bladder, and urethra? Describe their structure.

20. Describe the micturition reflex. How is voluntary control of micturition accomplished?

21. What stimuli result in an increased sensation of thirst?

22. Describe how Na^+ levels are regulated in the body fluids.

23. Describe how K^+ levels are regulated in the body fluids.

24. Describe how Ca^{2+} levels are regulated in the body fluids.

25. Explain how buffers respond to changes in the pH of body fluids.

26. Explain how the respiratory system and the kidneys respond to changes in the pH of body fluids.

27. Define respiratory acidosis, metabolic acidosis, respiratory alkalosis, and metabolic alkalosis.

CRITICAL THINKING QUESTIONS

1. Mucho McPhee decided to do an experiment after reading the urinary system chapter in his favorite anatomy and physiology textbook. He drank 2 L of water in 15 min and then monitored his rate of urine production and urine concentration over the next 2 hours. What did he observe? Explain the major mechanism involved.

2. A student ate a full bag of salty (NaCl) potato chips but drank no liquids. What effect did this have on urine concentration and the rate of urine production? Explain the mechanisms involved.

3. During severe exertion in a hot environment, a person can lose up to 4 L of sweat per hour (sweat is less concentrated than extracellular fluid in the body). What effect would this loss have on urine concentration and rate of production? Explain the mechanisms involved.

4. Which of the following symptoms are consistent with reduced secretion of aldosterone: excessive urine production, low blood pressure, high plasma potassium levels, high plasma sodium levels? Explain.

5. Propose as many ways as you can to decrease the rate at which filtrate enters the Bowman capsule.

6. Swifty Trotts has an enteropathogenic *Escherichia coli* infection that causes severe diarrhea. Diarrhea produces a large volume of mucus that contains high concentrations of HCO_3^-. How would this diarrhea affect this patient's blood pH, urine pH, and respiration rate?

7. Spanky and his mother went to a grocery store, where Spanky eyed some candy he wanted. His mother refused to buy it, so Spanky became angry. He held his breath for 2 min. What effect did this have on his body fluid pH? After the 2 min, what mechanisms were most important in reestablishing the normal body fluid pH?

8. Martha suffered from severe nausea for 2 days. She vomited frequently and was so nauseated that she could not tolerate to eat or drink anything. Explain how each of the following levels had changed in her body by the second day: blood pH, blood ADH, blood aldosterone, and urine pH.

Answers in Appendix D

19

Reproductive System

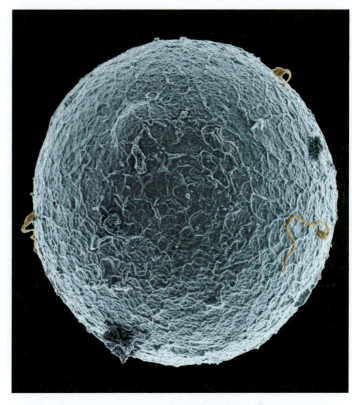

▲ Color-enhanced scanning electron micrograph of a human oocyte.

19.1 FUNCTIONS OF THE REPRODUCTIVE SYSTEM

Learning Outcome   *After reading this section, you should be able to*

A. List the functions of the male and female reproductive systems.

The human species could not survive without functional male and female reproductive systems. The reproductive systems play essential roles in the development of the structural and functional differences between males and females, influence human behavior, and produce offspring. However, a reproductive system, unlike other organ systems, is not necessary for the survival of an individual human.

Most of the body's organ systems show little difference between males and females, but the male and female reproductive systems exhibit striking differences. On the other hand, they have a number of similarities. For example, many reproductive organs of males and females are derived from the same embryological structures (see chapter 20), and some hormones are the same in males and females, even though they produce different responses. The reproductive system performs the following functions:

1. *Production of gametes.* The reproductive system produces gametes: sperm cells in the testes of males and oocytes (eggs) in the ovaries of females.
2. *Fertilization.* The reproductive system enhances fertilization of the oocyte by the sperm. The duct system in males

Module 14 Reproductive System

nourishes sperm until they are mature and are deposited in the female reproductive tract by the penis. The female reproductive system receives the male's sperm and transports them to the fertilization site.
3. *Development and nourishment of a new individual.* The female reproductive system nurtures the development of a new individual in the uterus until birth and provides nourishment (milk) after birth.
4. *Production of reproductive hormones.* Hormones produced by the reproductive system control its development and the development of the gender-specific body form. These hormones are also essential for the normal function of the reproductive system and for reproductive behavior.

19.2 FORMATION OF GAMETES

Learning Outcome *After reading this section, you should be able to*

A. Describe the function of meiosis in the formation of sperm cells and oocytes.

The testes in males and the ovaries in females (figure 19.1) produce **gametes** (gam′ētz), or *sex cells*. The formation of gametes in males and females occurs by a type of cell division called **meiosis** (mī-ō′sis; a lessening) (see chapter 3).

Meiosis occurs only in the testis and ovary. During meiosis, one cell undergoes two consecutive cell divisions to produce four daughter cells, each having half as many chromosomes as the parent cell.

The two divisions of meiosis are called meiosis I and meiosis II. Like mitosis, each division of meiosis has prophase, metaphase, anaphase, and telophase. However, there are distinct differences between meiosis and mitosis.

Before meiosis begins, all the chromosomes are duplicated. At the beginning of meiosis, each of the 46 chromosomes consists of 2 chromatids connected by a centromere (figure 19.2, step 1). The chromosomes align as pairs in a process called **synapsis** (si-nap′sis; a connection) (figure 19.2, step 2). Because each chromosome consists of 2 chromatids, the pairing of the chromosomes brings 2 chromatids of each chromosome close together. Occasionally, part of a chromatid of 1 chromosome breaks off and is exchanged with part of another chromatid from the other chromosome. This event, called **crossing over,** allows the exchange of genetic material between chromosomes.

The chromosomes align along the center of the cell (figure 19.2, step 3), and then the pairs of chromosomes are separated to each side of the cell (figure 19.2, step 4). As a consequence, when meiosis I is complete, each daughter cell has 1 chromosome from each of the pairs (figure 19.2, step 5), or 23 chromosomes. Each of the 23 chromosomes in each daughter cell consists of 2 chromatids joined by a centromere.

It is during the first meiotic division that the chromosome number is reduced from 46 (23 pairs) to 23 chromosomes. The first meiotic division is therefore called a reduction division.

The second meiotic division is similar to mitosis. The chromosomes, each consisting of 2 chromatids (figure 19.2, steps 6 and 7), align along the center of the cell. Then the chromatids separate at the centromere, and each daughter cell receives 1 of the chromatids from each chromosome (figure 19.2, steps 8 and 9). When the centromere separates, each of the chromatids is called a chromosome. Consequently, each of the 4 daughter cells produced by meiosis contains 23 chromosomes.

During fertilization, the zygote receives 1 set of chromosomes (23) from each parent. Although half the genetic material of a zygote comes from each parent, the genetic makeup of each zygote is unique.

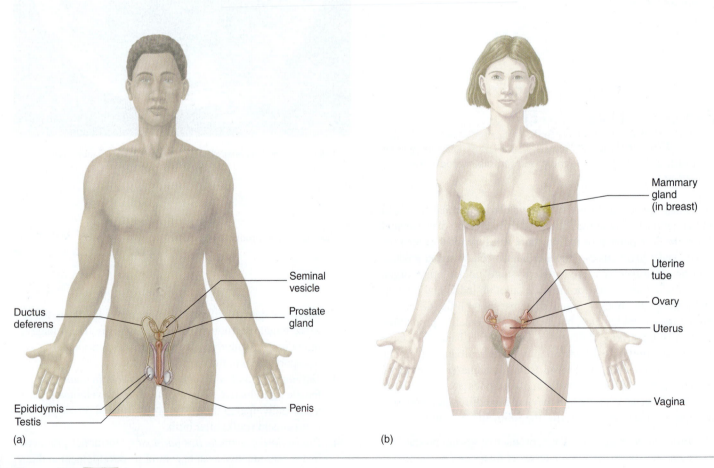

(a)

(b)

Mammary gland (in breast)

Uterine tube

Ovary

Uterus

Vagina

Seminal vesicle

Prostate gland

Ductus deferens

Epididymis

Testis

Penis

Figure 19.1 APR Major Organs of the Reproductive System

(*a*) The male reproductive system: testes, epididymides, ducta deferentia, seminal vesicles, prostate gland, and penis. (*b*) The female reproductive system: ovaries, uterine tubes, uterus, vagina, and mammary glands.

First Meiotic Division (Meiosis I)

Second Meiotic Division (Meiosis II)
(continued from the bottom of previous column)

1 Early prophase I
The duplicated chromosomes become visible chromatids.

Centromere

Chromosome

Nucleus

Chromatids

Centrioles

6 Prophase II
Each chromosome consists of two chromatids.

2 Middle prophase I
Pairs of chromosomes synapse. Crossing over may occur at this stage.

Pair of chromosomes

Spindle fibers

3 Metaphase I
Pairs of chromosomes align along the center of the cell. Random assortment of chromosomes occurs.

Equatorial plane

7 Metaphase II
Chromosomes align along the center of the cell.

4 Anaphase I
Chromosomes move apart to opposite sides of the cell.

8 Anaphase II
Chromatids separate, and each is now called a chromosome.

Cleavage furrow

5 Telophase I
New nuclei form, and the cell divides. Each cell now has two sets of half the chromosomes.

9 Telophase II
New nuclei form around the chromosomes. The cells divide to form four daughter cells with half as many chromosomes as the parent cell.

Prophase II (top of next column)

PROCESS Figure 19.2 AP|R **Meiosis**

19.3 MALE REPRODUCTIVE SYSTEM

Learning Outcomes *After reading this section, you should be able to*

A. Describe the scrotum and its role in regulating the temperature of the testes.

B. Describe the structure of the testes, the specialized cells of the testes, and the process of spermatogenesis.

C. Describe the ducts of the male reproductive system and their functions.

D. Describe the structure of the penis, seminal vesicles, bulbourethral glands, and prostate gland, and explain their functions.

The male reproductive system consists of the testes (sing. testis), a series of ducts, accessory glands, and supporting structures. The ducts include the epididymides (sing. epididymis), the ducta deferentia (sing. ductus deferens; also, vas deferens), and the urethra. Accessory glands include the seminal vesicles, the prostate gland, and the bulbourethral glands. Supporting structures include the scrotum and the penis (figure 19.3). The sperm cells are very heat-sensitive and must develop at a temperature slightly less than normal body temperature. The testes, in which the sperm cells develop, are located outside the body cavity in the scrotum, where the temperature is lower. Sperm cells travel from each testis to the prostate gland and then empty into the urethra within the prostate gland. The urethra exits the pelvis, passes through the penis, and opens to the outside of the body.

Scrotum

The **scrotum** (skrō′tum) is a saclike structure containing the testes. It is divided into right and left internal compartments by an incomplete connective tissue septum. Externally, the scrotum consists of skin. Beneath the skin are a layer of loose connective tissue and a layer of smooth muscle called the **dartos** (dar′tōs) **muscle.**

In cold temperatures, the dartos muscle contracts, causing the skin of the scrotum to become firm and wrinkled and reducing the overall size of the scrotum. At the same time, extensions of abdominal muscles into the scrotum, called **cremaster** (krē-mas′ter) **muscles,** contract (see figure 19.6). Consequently, the testes are pulled nearer the body, and their temperature is elevated. During warm weather or exercise, the dartos and cremaster muscles relax, the skin of the scrotum becomes loose and thin, and the testes descend away from the body, which lowers their temperature. The response of the dartos and cremaster muscles is important in regulating the temperature in the testes. If the testes become too warm or too cold, normal sperm cell development does not occur.

Testes

The **testes** (tes′tēz), or male gonads (gō′nădz; *gonē*, seed), are oval organs, each about 4–5 cm long, within the scrotum (see figure 19.3). The outer part of each testis consists of a thick, white connective tissue capsule. Extensions of the capsule project into the interior of the testis and divide each testis into about 250 cone-shaped lobules (figure 19.4a). The lobules contain **seminiferous**

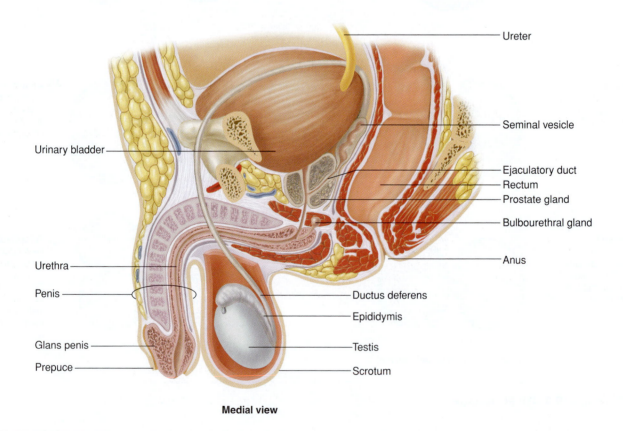

Medial view

Figure 19.3 AP|R Male Reproductive Structures

Medial view of the male pelvis, showing the male reproductive structures.

CLINICAL IMPACT   Descent of the Testes

The testes develop in the abdominopelvic cavity. They move from the abdominopelvic cavity through the **inguinal** (ing′gwi-năl; *inguen,* groin) **canal** to the scrotum. The inguinal canals and the internal layers of the scrotum originate as outpocketings of the abdominal cavity along the lateral, superior margin of the pubis. The descent of the testes occurs during the seventh or eighth month of fetal development or, in some cases, shortly after birth.

Failure of the testes to descend into the scrotal sac is called **cryptorchidism** (kriptōr′ki-dizm; *crypto,* concealed + *orchis,* testis). It results in sterility because of the inhibiting effect of normal body temperature on sperm cell development. In addition, about 10% of testicular cancer cases occur in men with a history of cryptorchidism. Fortunately, if the testis has not descended into the scrotum by the age of 4 months, it can be surgically repaired.

After the testes descend, the inguinal canals narrow permanently, but they remain as weak spots in the abdominal wall. An inguinal canal that enlarges or ruptures can result in an **inguinal hernia** (her′nē-ă; rupture), through which a loop of intestine can protrude. This herniation can be quite painful and even very dangerous, especially if the inguinal canal compresses the intestine and cuts off its blood supply. Fortunately, inguinal hernias can be repaired surgically.

(sem′ĭ-nif′er-ŭs) **tubules,** in which sperm cells develop. Delicate connective tissue surrounding the seminiferous tubules contains clusters of endocrine cells called **interstitial** (in-ter-stish′ăl) **cells,** or *Leydig* (lī′dig) *cells,* which secrete testosterone. The seminiferous tubules contain **germ cells** and **sustentacular cells,** or *Sertoli* (ser-tō′lē) *cells* (figure 19.4*b*). Sustentacular cells are large and extend from the periphery to the lumen of the seminiferous tubule. They nourish the germ cells and produce a number of hormones.

Spermatogenesis

Spermatogenesis (sper′mă-tō-jen′ĕ-sis) is the formation of sperm cells. Before puberty, the testes remain relatively simple and unchanged from the time of their initial development. The interstitial

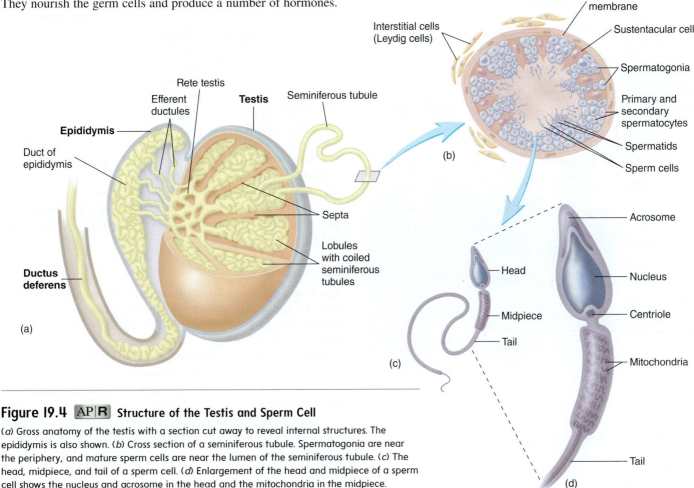

Figure 19.4 AP|R Structure of the Testis and Sperm Cell

(*a*) Gross anatomy of the testis with a section cut away to reveal internal structures. The epididymis is also shown. (*b*) Cross section of a seminiferous tubule. Spermatogonia are near the periphery, and mature sperm cells are near the lumen of the seminiferous tubule. (*c*) The head, midpiece, and tail of a sperm cell. (*d*) Enlargement of the head and midpiece of a sperm cell shows the nucleus and acrosome in the head and the mitochondria in the midpiece.

Reproductive

cells are not prominent, and the seminiferous tubules are small and not yet functional. At the time of puberty, the interstitial cells increase in number and size, the seminiferous tubules enlarge, and spermatogenesis begins.

Germ cells are partially embedded in the sustentacular cells. The most peripheral germ cells are **spermatogonia** (sper′mă-tō-gō′nē-ă; undeveloped sperm cells), which divide through mitosis (figure 19.5). Some daughter cells produced from these mitotic divisions remain as spermatogonia and continue to divide by mitosis. Other daughter cells form **primary spermatocytes** (sper′mă-tō-sītz; sperm cells), which divide by meiosis and become sperm cells.

A primary spermatocyte contains 46 chromosomes, each consisting of 2 chromatids. Each primary spermatocyte passes through the first meiotic division to produce 2 **secondary spermatocytes.** Each secondary spermatocyte undergoes the second meiotic division to produce 2 smaller cells called **spermatids** (sper′mă-tidz), each having 23 chromosomes. After the second meiotic division, the spermatids undergo major structural changes to form sperm cells (figure 19.5; see figure 19.4b,c). Much of the cytoplasm of the spermatids is eliminated, and each spermatid develops a head, midpiece, and flagellum (tail) to become a **sperm cell,** or **spermatozoon** (figure 19.5; see figure 19.4b–d). The nucleus of the sperm cell is located in the head of the sperm cell. Just anterior to the nucleus is a vesicle called the **acrosome** (ak′rō-sōm), which contains enzymes that are released during the process of fertilization and are necessary for the sperm cell to penetrate the oocyte, or egg cell.

At the end of spermatogenesis, the developing sperm cells are located around the lumen of the seminiferous tubules, with their heads directed toward the surrounding sustentacular cells and their tails directed toward the center of the lumen (figure 19.5; see figure 19.4b). Finally, sperm cells are released into the lumen of the seminiferous tubules.

Ducts

After their production, sperm cells are transported through the seminiferous tubules and a series of ducts to the exterior of the body.

Epididymis

The seminiferous tubules of each testis empty into a tubular network called the **rete** (rē′tē; net) **testis** (see figure 19.4a). The rete testis empties into 15–20 tubules called the **efferent ductules** (ef′er-ent dŭk′toolz). The efferent ductules carry sperm cells from the testis to a tightly coiled series of threadlike tubules that form a comma-shaped structure on the posterior side of the testis called the **epididymis** (ep-i-did′i-mis) (figure 19.6; see figures 19.3 and 19.4). The sperm cells continue to mature within the epididymis, developing the capacity to swim and the ability to bind to the oocyte. Sperm cells taken directly from the testes are not capable of fertilizing oocytes, but after maturing for several days in the epididymis, the sperm cells develop the capacity to function as gametes. Final changes in sperm cells, called **capacitation** (kă-pas′i-tă′shun), occur after ejaculation of semen into the vagina and prior to fertilization.

Ductus Deferens

The **ductus deferens** (dŭk′tŭs def′er-enz), or *vas deferens,* emerges from the epididymis and ascends along the posterior side of the testis to become associated with the blood vessels and nerves that supply the testis. These structures form the **spermatic cord** (figure 19.6a). Each spermatic cord consists of the ductus deferens, testicular artery and veins, lymphatic vessels, and testicular nerve. It is surrounded by the cremaster muscle and two connective tissue sheaths.

Each ductus deferens extends, in the spermatic cord, through the abdominal wall by way of the inguinal canal. Each ductus deferens then crosses the lateral wall of the pelvic cavity and loops behind the posterior surface of the urinary bladder to approach the prostate gland (figure 19.6a; see figure 19.3). The total length of

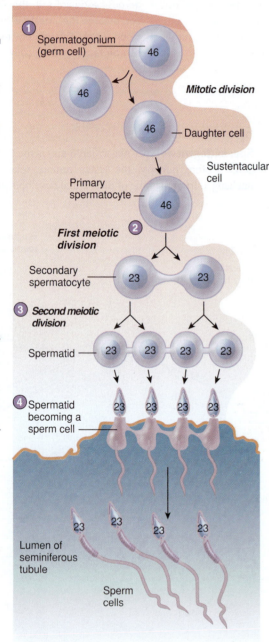

1. Spermatogonia are the cells from which sperm cells arise. The spermatogonia divide by mitosis. One daughter cell remains a spermatogonium that can divide again by mitosis. One daughter cell becomes a primary spermatocyte.

2. The primary spermatocyte divides by meiosis to form secondary spermatocytes.

3. The secondary spermatocytes divide by meiosis to form spermatids.

4. The spermatids differentiate to form sperm cells.

1. Spermatogonium (germ cell)   46

Mitotic division

46   Daughter cell

Sustentacular cell

Primary spermatocyte   46

First meiotic division

2. Secondary spermatocyte   23   23

3. **Second meiotic division**

Spermatid —   23   23   23   23

4. Spermatid becoming a sperm cell —   23   23   23   23

23   23   23   23

Lumen of seminiferous tubule

Sperm cells

PROCESS Figure 19.5 AP|R Spermatogenesis

A section of the seminiferous tubule illustrates the process of meiosis and sperm cell formation. The number in each cell indicates the total chromosome number.

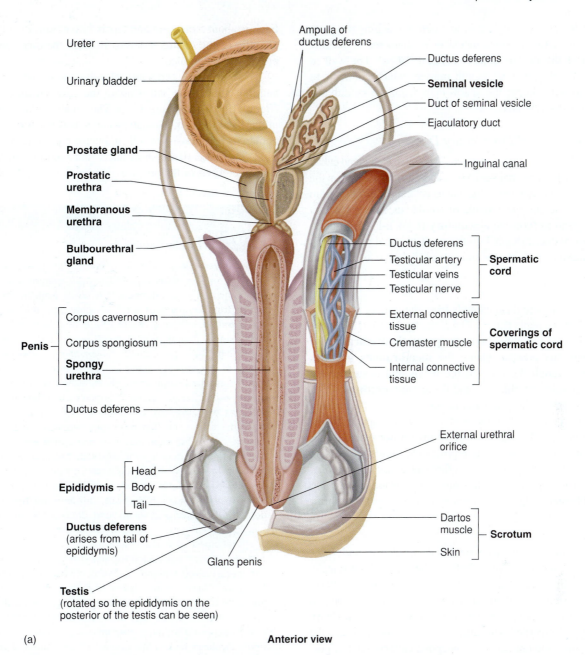

Ureter

Urinary bladder

Ampulla of
ductus deferens

Ductus deferens

Seminal vesicle

Duct of seminal vesicle

Ejaculatory duct

Inguinal canal

Prostate gland

**Prostatic
urethra**

**Membranous
urethra**

**Bulbourethral
gland**

Ductus deferens
Testicular artery   **Spermatic
Testicular veins   cord**
Testicular nerve

Penis

Corpus cavernosum

Corpus spongiosum

**Spongy
urethra**

External connective
tissue

Cremaster muscle   **Coverings of
spermatic cord**

Internal connective
tissue

Ductus deferens

External urethral
orifice

Epididymis Head
Body
Tail

Ductus deferens
(arises from tail of
epididymis)

Dartos
muscle   **Scrotum**

Skin

Glans penis

Testis
(rotated so the epididymis on the
posterior of the testis can be seen)

(a)   **Anterior view**

Figure 19.6 AP|R Male Reproductive Organs

(*a*) Frontal view of a testis, an epididymis, and a ductus deferens, along with
the penis and glands of the male reproductive system. The testis is viewed
in the scrotal sac with the smooth muscle and cremaster muscles on one
side. The ductus deferens extends from the epididymis in the scrotal sac and
passes through the inguinal canal and pelvic cavity to the prostate gland.
Note that the ductus deferens and the testicular artery, vein, and nerves
are surrounded by the cremaster muscle and connective tissue to form the
spermatic cord. (*b*) A cross section of the penis illustrates the two dorsal
corpora cavernosa and the ventral corpus spongiosum. Connective tissue
sheaths and skin cover the three erectile bodies. Blood vessels, including
the dorsal artery and vein, and the dorsal nerve of the penis are visible.

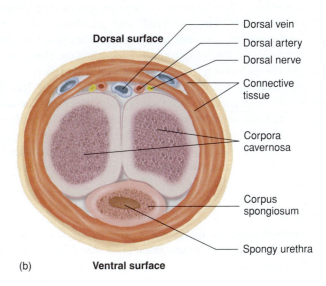

Dorsal surface

Dorsal vein

Dorsal artery

Dorsal nerve

Connective
tissue

Corpora
cavernosa

Corpus
spongiosum

Spongy urethra

(b)   **Ventral surface**

Reproductive

the ductus deferens is about 45 cm. Just before reaching the prostate gland, the ductus deferens increases in diameter to become the **ampulla of the ductus deferens** (figure 19.6*a*). The wall of the ductus deferens contains smooth muscle, which contracts in peristaltic waves to propel the sperm cells from the epididymis through the ductus deferens.

Seminal Vesicle and Ejaculatory Duct

Near the ampulla of each ductus deferens is a sac-shaped gland called the **seminal vesicle** (sem′i-năl ves′i-kl). A short duct extends from the seminal vesicle to the ampulla of the ductus deferens. The ducts from the seminal vesicle and the ampulla of the ductus deferens join at the prostate gland to form the **ejaculatory** (ē-jak′ū-lă-tōr-ē) **duct.** Each ejaculatory duct extends into the prostate gland and ends by joining the urethra within the prostate gland (figure 19.6*a*).

Urethra

The male **urethra** (ū-rē′thră) extends from the urinary bladder to the distal end of the penis (figure 19.6*a*; see figure 19.3). The urethra can be divided into three parts: the **prostatic urethra,** which passes through the prostate gland; the **membranous urethra,** which passes through the floor of the pelvis and is surrounded by the external urinary sphincter; and the **spongy urethra,** which extends the length of the penis and opens at its end. The urethra is a passageway for both urine and male reproductive fluids. However, urine and the reproductive fluids do not exit the urethra at the same time. While male reproductive fluids are passing through the urethra, a sympathetic reflex causes the internal urinary sphincter to contract, which keeps semen from passing into the urinary bladder and prevents urine from entering the urethra.

Penis

The **penis** (pe′nis) is the male organ of copulation and functions in the transfer of sperm cells from the male to the female. The penis contains three columns of erectile tissue (figure 19.6). Engorgement of this erectile tissue with blood causes the penis to enlarge and become firm, a process called **erection** (ē-rek′shŭn). Two columns of erectile tissue form the dorsal portion and the sides of the penis and are called the **corpora cavernosa** (kōr′pōr-ă kav-er-nōs′ă). The third, smaller erectile column occupies the ventral portion of the penis and is called the **corpus spongiosum** (kōr′pŭs spŭn-gē-ō′sŭm). It expands over the distal end of the penis to form a cap, the **glans** (glanz) **penis.** The spongy urethra passes through the corpus spongiosum, including the glans penis, and opens to the exterior as the **external urethral orifice.**

The shaft of the penis is covered by skin that is loosely attached to the connective tissue surrounding the penis. The skin is firmly attached at the base of the glans penis, and a thinner layer of skin tightly covers the glans penis. The skin of the penis, especially the glans penis, is well supplied with sensory receptors. A loose fold of skin, called the **prepuce** (prē′pūs), or *foreskin,* covers the glans penis (see figure 19.3).

Glands

The seminal vesicles are glands consisting of many saclike structures located next to the ampulla of the ductus deferens (figure 19.6*a*; see figure 19.3). There are two seminal vesicles.

Each is about 5 cm long and tapers into a short duct that joins the ampulla of the ductus deferens to form the ejaculatory duct, as previously mentioned.

The **prostate** (pros′tāt) **gland** consists of both glandular and muscular tissue and is about the size and shape of a walnut (figure 19.6*a*; see figure 19.3). The prostate gland surrounds the urethra and the two ejaculatory ducts. It consists of a capsule and numerous partitions. The cells lining the partitions secrete prostatic fluid. There are 10–20 short ducts (not seen in figure 19.6*a*) that carry secretions of the prostate gland to the prostatic urethra.

A CASE IN POINT

Prostate Cancer

Sixty-five-year-old Chance O'Prostancer has a wellness checkup every year. Ten years ago, a prostate-specific antigen (PSA) test indicated that Chance's PSA levels were higher than the results from his previous tests. His physician reported moderate enlargement of the prostate gland but detected no obvious tumorlike structures in a physical examination. Because of the increasing PSA levels, Chance's physician recommended a needle biopsy of the prostate gland through the rectum. The pathology report described suspicious cells consistent with prostate cancer in one of the tissue samples. Chance's physician had the biopsy samples examined by another pathology laboratory, which did not confirm the first pathology report. As a consequence, Chance's physician explained that one option was to do nothing and continue having regular checkups because prostate cancer typically develops slowly. This is what Chance chose to do.

Eight years later, a PSA test showed another substantial increase in Chance's PSA levels, although no tumor could be detected by a physical exam, and Chance had no complaints, such as difficulty in urinating. This time, a needle biopsy of the prostate gland revealed cancer cells in two of the six biopsy samples. Chance's physician explained that so far the cancer had not metastasized (spread) to areas outside the prostate gland. Therefore, Chance could choose to do nothing, have his prostate gland surgically removed, or treat the cancer with radiation therapy, hormonal therapy, or chemotherapy. Statistics indicate that surgery and radiation therapy have similar success rates for small, localized tumors like Chance's. The trauma of surgery and the higher probability of erectile dysfunction following surgery convinced Chance that radiation therapy, which focuses radiation on the prostate gland to kill the cancer cells, was preferable. Chance's physician indicated that doing nothing is a reasonable option for men who are significantly older than Chance because older men diagnosed with prostate cancer often die of other conditions before they succumb to prostate cancer. Chance's physician explained that, for patients like him, approximately 85% are cancer free after 5 years. Chance was grateful that he had had annual checkups. Prostate cancer represents 29% of cancers in males in the United States and 14% of the deaths due to cancer. Only lung cancer results in more cancer deaths in men.

Predict 2

Changes in the size and texture of the prostate gland can be an indication of developing prostate cancer. Suggest a way that the size and texture of the prostate gland can be examined by palpation without surgical techniques (see figure 19.3).

CLINICAL IMPACT   Circumcision

Circumcision (ser-kŭm-sizh′kŭn) is the surgical removal of the prepuce, usually shortly after birth. There are few compelling medical reasons for circumci-sion. Uncircumcised males have a higher incidence of penile cancer, but the under-lying causes seem related to chronic infec-tions and poor hygiene. In those few cases in which the prepuce is "too tight" to be moved over the glans penis, circumcision can be necessary to avoid chronic infec-tions and maintain normal circulation.

The **bulbourethral** (bul′bō-ū-rē′thrăl) **glands,** or *Cowper glands,* are a pair of small, mucus-secreting glands located near the base of the penis (figure 19.6*a*; see figure 19.3). In young adults, each is about the size of a pea, but they decrease in size with age. A single duct from each gland enters the urethra.

Secretions

Semen (sē′men) is a mixture of sperm cells and secretions from the male reproductive glands. The seminal vesicles produce about 60% of the fluid, the prostate gland contributes approximately 30%, the testes contribute 5%, and the bulbourethral glands contribute 5%.

The bulbourethral glands and the mucous glands in the urethra produce a mucous secretion, which lubricates the urethra, helps neutralize the contents of the normally acidic urethra, provides a small amount of lubrication during intercourse, and helps reduce acidity in the vagina.

Testicular secretions include sperm cells and a small amount of fluid. The thick, mucuslike secretion of the seminal vesicles contains the sugar fructose and other nutrients that nourish sperm cells. The seminal vesicle secretions also contain proteins that weakly coagulate after ejaculation and enzymes that are thought to help destroy abnormal sperm cells. Prostaglandins, which stimulate smooth muscle contractions, are present in high concentrations in the secretions of the seminal vesicles and can cause contractions of the female reproductive tract, which help transport sperm cells through the tract.

The thin, milky secretions of the prostate have an alkaline pH and help neutralize the acidic urethra, as well as the acidic secretions of the testes, the seminal vesicles, and the vagina. The increased pH is important for normal sperm cell function. The movement of sperm cells is not optimal until the pH is increased to between 6.0 and 6.5. In contrast, the pH of vaginal secretions is between 3.5 and 4.0. Prostatic secretions also contain proteolytic enzymes that break down the coagulated proteins of the seminal vesicles and make the semen more liquid. The normal volume of semen is 2–5 milliliters (mL), with each milliliter of semen typically containing about 100 million sperm cells.

19.4 PHYSIOLOGY OF MALE REPRODUCTION

Learning Outcomes   *After reading this section, you should be able to*

- **A.** List the hormones that influence the male reproductive system, and describe their functions.
- **B.** Describe the changes that occur in males during puberty.
- **C.** Explain the events that occur during the male sexual act.

The male reproductive system depends on both hormonal and neural mechanisms to function normally. Hormones control the development of reproductive structures, the development of secondary sexual characteristics, spermatogenesis, and some aspects of sexual behavior. The mature neural mechanisms are primarily involved in controlling the sexual act and in the expression of sexual behavior.

Regulation of Reproductive Hormone Secretion

The hypothalamus of the brain, the anterior pituitary gland, and the testes (figure 19.7) produce hormones that influence the male reproductive system. **Gonadotropin-releasing** (gō′nad-ō-trō′pin) **hormone (GnRH)** is released from neurons in the hypothalamus and passes to the anterior pituitary gland (table 19.1). GnRH causes cells in the anterior pituitary gland to secrete two hormones, **luteinizing** (loo′tē-ĭ-nīz-ing) **hormone (LH)** and **follicle-stimulating hormone (FSH),** into the blood. LH and FSH are named for their functions in females, but they are also essential reproductive hormones in males.

LH binds to the interstitial cells in the testes and causes them to secrete testosterone. FSH binds primarily to sustentacular cells in the seminiferous tubules and promotes sperm cell development. It also increases the secretion of a hormone called **inhibin** (in-hib′in; to inhibit).

The blood levels of reproductive hormones are under negative-feedback control. Testosterone has a negative-feedback effect on the secretion of GnRH from the hypothalamus and the secretion of LH and FSH from the anterior pituitary gland. Inhibin has a negative-feedback effect on the secretion of FSH from the anterior pituitary gland.

For GnRH to stimulate LH and FSH release, the pituitary gland must be exposed to a series of brief increases and decreases in GnRH. If GnRH is maintained at a high level in the blood for days or weeks, the anterior pituitary cells become insensitive to it. GnRH can be produced synthetically and is useful in treating some people who are infertile. Synthetic GnRH must be administered in small amounts in frequent pulses or surges. GnRH can also inhibit reproduction because long-term administration of GnRH can sufficiently reduce LH and FSH levels to prevent sperm cell production in males or ovulation in females.

Puberty in Males

Puberty (pū′ber-tē) is the sequence of events by which a child is transformed into a young adult. The reproductive system matures and assumes its adult functions, and the structural differences

1. Gonadotropin-releasing hormone (GnRH) from the hypothalamus stimulates the secretion of luteinizing hormone (LH) and follicle-stimulating hormone (FSH) from the anterior pituitary.

2. LH stimulates testosterone secretion from the interstitial cells.

3. FSH stimulates sustentacular cells of the seminiferous tubules to increase spermatogenesis and to secrete inhibin.

4. Testosterone has a stimulatory effect on the sustentacular cells of the seminiferous tubules, as well as on the development of reproductive organs and secondary sexual characteristics.

5. Testosterone has a negative-feedback effect on the hypothalamus and pituitary to reduce GnRH, LH, and FSH secretion.

6. Inhibin has a negative-feedback effect on the anterior pituitary to reduce FSH secretion.

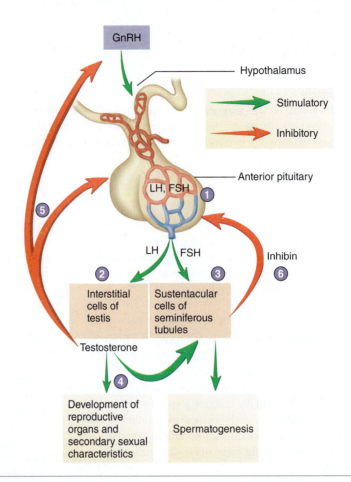

PROCESS Figure 19.7 Regulation of Reproductive Hormone Secretion in Males

between adult males and females become more apparent. In boys, puberty commonly begins between the ages of 12 and 14 and is largely completed by age 18. Before puberty, small amounts of testosterone, secreted by the testes and the adrenal cortex, inhibit GnRH, LH, and FSH secretion. Beginning just before puberty and continuing throughout puberty, developmental changes in the hypothalamus cause the hypothalamus and the anterior pituitary gland to become much less sensitive to the inhibitory effect of testosterone, and the rate of GnRH, LH, and FSH secretion increases. Consequently, elevated FSH levels promote spermatogenesis, and elevated LH levels cause the interstitial cells to secrete larger amounts of testosterone. Testosterone still has a negative-feedback effect on the hypothalamus and anterior pituitary gland, but GnRH, LH, and FSH secretion occurs at substantially higher levels.

Effects of Testosterone

Testosterone (tes′tos′tĕ-rōn) is the major male hormone secreted by the testes. Testosterone influences reproductive organs and nonreproductive structures (table 19.2). During puberty, testosterone causes the enlargement and differentiation of the male genitals and the reproductive duct system. It is necessary for spermatogenesis and for the development of male secondary sexual characteristics. The **secondary sexual characteristics** are those structural and behavioral changes, other than in the reproductive organs, that develop at puberty and distinguish males from females.

Secondary sexual characteristics in males include hair distribution and growth, skin texture, fat distribution, skeletal muscle growth, and changes in the larynx. After puberty, testosterone maintains the adult structure of the male genitals, reproductive ducts, and secondary sexual characteristics.

Male Sexual Behavior and the Male Sex Act

Testosterone is required for normal sexual behavior. Testosterone enters certain cells within the brain, especially within the hypothalamus, and influences their functions. The blood levels of testosterone remain relatively constant throughout the lifetime of a male, from puberty until about 40 years of age. Thereafter, the levels slowly decline to approximately 20% of this value by 80 years of age, causing a slow decrease in sex drive and fertility.

Predict 3

Predict the effect on secondary sexual characteristics, external genitalia, and sexual behavior if the testes fail to produce normal amounts of testosterone at puberty.

The male sex act is a complex series of reflexes that result in erection of the penis, secretion of mucus into the urethra, emission, and ejaculation. **Emission** (ē-mish′ŭn) is the movement of sperm cells, mucus, prostatic secretions, and seminal vesicle secretions into the prostatic, membranous, and spongy urethra. **Ejaculation** (ē-jak′ū-lā′shŭn) is the forceful expulsion of the secretions that

TABLE 19.1 Major Reproductive Hormones in Males and Females

Hormone	Source	Target Tissue	Response
Male Reproductive System			
Gonadotropin-releasing hormone (GnRH)	Hypothalamus	Anterior pituitary	Stimulates secretion of LH and FSH
Luteinizing hormone (LH)	Anterior pituitary	Interstitial cells of the testes	Stimulates synthesis and secretion of testosterone
Follicle-stimulating hormone (FSH)	Anterior pituitary	Seminiferous tubules (sustentacular cells)	Supports spermatogenesis and inhibin secretion
Testosterone	Interstitial cells of testes	Testes; body tissues	Development and maintenance of reproductive organs; supports spermatogenesis and causes the development and maintenance of secondary sexual characteristics
		Anterior pituitary and hypothalamus	Inhibits GnRH, LH, and FSH secretion through negative feedback
Inhibin	Sustentacular cells	Anterior pituitary	Inhibits FSH secretion through negative feedback
Female Reproductive System			
Gonadotropin-releasing hormone (GnRH)	Hypothalamus	Anterior pituitary	Stimulates secretion of LH and FSH
Luteinizing hormone (LH)	Anterior pituitary	Ovaries	Causes follicles to complete maturation and undergo ovulation; causes ovulation; causes the ovulated follicle to become the corpus luteum
Follicle-stimulating hormone (FSH)	Anterior pituitary	Ovaries	Causes follicles to begin development
Estrogen	Follicles of ovaries and corpus luteum	Uterus	Proliferation of endometrial cells
		Breasts	Development of mammary glands (especially duct systems)
		Anterior pituitary and hypothalamus	Positive feedback before ovulation, resulting in increased LH and FSH secretion; negative feedback with progesterone on the hypothalamus and anterior pituitary after ovulation, resulting in decreased LH and FSH secretion
		Other tissues	Development and maintenance of secondary sexual characteristics
Progesterone	Corpus luteum of ovaries	Uterus	Enlargement of endometrial cells and secretion of fluid from uterine glands; maintenance of pregnant state
		Breasts	Development of mammary glands (especially alveoli)
		Anterior pituitary	Negative feedback, with estrogen, on the hypothalamus and anterior pituitary after ovulation, resulting in decreased LH and FSH secretion
		Other tissues	Secondary sexual characteristics
Oxytocin	Posterior pituitary	Uterus and mammary glands	Contraction of uterine smooth muscle and contraction of cells in the breast, resulting in milk letdown in lactating women
Human chorionic gonadotropin	Placenta	Corpus luteum of ovaries	Maintains corpus luteum and increases its rate of progesterone secretion during the first one-third (first trimester) of pregnancy; increases testosterone production in testes of male fetuses

CLINICAL IMPACT   Anabolic Steroids

Some athletes, especially those who depend on muscle strength, may either ingest or inject synthetic **androgens** (an'drō-jenz; *andros,* male), which are hormones that have testosterone-like effects, such as stimulating the development of male sexual characteristics. The synthetic androgens are commonly called **anabolic steroids,** or simply **steroids,** and they are used in an attempt to increase muscle mass. Many of the synthetic androgens are structurally different from testosterone. Their effect on muscle is greater than their effect on the reproductive organs. However, when taken in large amounts, they can influence the reproductive system. Large doses of synthetic androgens have a negative-feedback effect on the hypothalamus and pituitary, reducing GnRH, LH, and FSH levels. As a result, the testes can atrophy, and sterility can develop. Other side effects of large doses of synthetic androgens include kidney and liver damage, heart attack, and stroke. In addition, synthetic androgens cause abrupt mood swings, usually toward intense anger and rage. Taking synthetic androgens is highly discouraged by the medical profession, violates the rules of most athletic organizations, and is illegal without a prescription.

TABLE 19.2	Effects of Testosterone on Target Tissues
Target Tissue	**Response**
Penis and scrotum	Enlargement and differentiation
Hair follicles	Hair growth and coarser hair in the pubic area, legs, chest, axillary region, face, and occasionally back; male pattern baldness on the head if the person has the appropriate genetic makeup
Skin	Coarser texture of skin; increased rate of secretion of sebaceous glands, frequently resulting in acne at the time of puberty; increased secretion of sweat glands in axillary regions
Larynx	Enlargement of larynx and deeper masculine voice
Most tissues	Increased rate of metabolism
Red blood cells	Increased rate of red blood cell production; a red blood cell count increase by about 20% as a result of increased erythropoietin secretion
Kidneys	Retention of sodium and water to a small degree, resulting in increased extracellular fluid volume
Skeletal muscle	A skeletal muscle mass increase at puberty; average increase is greater in males than in females
Bone	Rapid bone growth, resulting in increased rate of growth and early cessation of bone growth; males who mature sexually at a later age do not exhibit a rapid period of growth, but they grow for a longer time and can become taller than men who mature earlier

have accumulated in the urethra to the exterior. Sensations, normally interpreted as pleasurable, occur during the male sex act and result in an intense sensation called an **orgasm** (ōr'gazm), or *climax.* In males, orgasm is closely associated with ejaculation, although they are separate functions and do not always occur simultaneously. A phase called **resolution** occurs after ejaculation. During resolution, the penis becomes flaccid, an overall feeling of satisfaction exists, and the male is unable to achieve erection and a second ejaculation.

Sensory Impulses and Integration

Sensory action potentials from the genitals are carried to the sacral region of the spinal cord, where reflexes that result in the male sex act are integrated. Action potentials also travel from the spinal cord to the cerebrum to produce conscious sexual sensations.

Rhythmic massage of the penis, especially the glans, and surrounding tissues, such as the scrotal, anal, and pubic regions, provide important sources of sensory action potentials. Engorgement of the prostate gland and seminal vesicles with secretions or irritation of the urethra, urinary bladder, ducta deferentia, and testes can also cause sexual sensations.

Psychological stimuli, such as sight, sound, odor, or thoughts, have a major effect on male sexual reflexes. Ejaculation while sleeping (nocturnal emission) is a relatively common event in young males and is thought to be triggered by psychological stimuli associated with dreaming.

Erection, Emission, and Ejaculation

Erection is the first major component of the male sex act. Parasympathetic action potentials from the sacral region of the spinal cord cause the arteries that supply blood to the erectile tissues to dilate. Blood then fills small venous sinuses called **sinusoids** in the erectile tissue and compresses the veins, which reduces blood flow from the penis. The increased blood pressure in the sinusoids causes the erectile tissue to become inflated and rigid. Parasympathetic action potentials also cause the mucous glands within the urethra and the bulbourethral glands to secrete mucus.

Failure to achieve erections, or **erectile dysfunction (ED),** sometimes called *impotence,* can be a major source of frustration. The inability to achieve erections can be due to reduced testosterone secretion resulting from hypothalamic, pituitary, or testicular complications. In other cases, ED can be due to defective stimulation of the erectile tissue by nerve fibers or reduced response of the blood vessels to neural stimulation. Some men can achieve erections by taking oral medications, such as sildenafil (Viagra), tadalafil (Cialis), or verdenafil (Livitra), or by having specific drugs injected into the base of the penis. These drugs increase blood flow into the erectile tissue of the penis, resulting in erection for many minutes.

Before ejaculation, the ductus deferens begins to contract rhythmically, propelling sperm cells and testicular fluid from the epididymis through the ductus deferens. Contractions of the ductus deferens, seminal vesicles, and ejaculatory ducts cause the sperm cells, testicular secretions, and seminal fluid to move into the urethra, where they mix with prostatic secretions released by contraction of the prostate.

Emission is stimulated by sympathetic action potentials that originate in the lumbar region of the spinal cord. Action potentials cause the reproductive ducts to contract and stimulate the seminal vesicles and the prostate gland to release secretions. Consequently, semen accumulates in the urethra.

Ejaculation results from the contraction of smooth muscle in the wall of the urethra and skeletal muscles surrounding the base of the penis. Just before ejaculation, action potentials are sent to the skeletal muscles that surround the base of the penis. Rhythmic contractions are produced that force the semen out of the urethra, resulting in ejaculation. In addition, muscle tension increases throughout the body.

Infertility in Males

Infertility (in-fer-til′i-tē) is reduced or diminished fertility. The most common cause of infertility in males is a low sperm cell count. If the sperm cell count drops to below 20 million sperm cells per milliliter, the male is usually sterile.

The sperm cell count can decrease because of damage to the testes as a result of trauma, radiation, cryptorchidism (See Clinical Impact "Descent of the Testes"), or infections, such as mumps, which block the ducts in the epididymis. Reduced sperm cell counts can also result from inadequate secretion of luteinizing hormone and follicle-stimulating hormone, which can be caused by hypothyroidism, trauma to the hypothalamus, infarctions of the hypothalamus or anterior pituitary gland, or tumors. Decreased testosterone secretion reduces the sperm cell count as well.

Even when the sperm cell count is normal, fertility can be reduced if sperm cell structure is abnormal, as occurs due to chromosomal abnormalities caused by genetic factors. Reduced sperm cell

motility also results in infertility. A major cause of reduced sperm cell motility is the presence of antisperm antibodies, which are produced by the immune system and bind to sperm cells.

In cases of infertility due to low sperm cell count or reduced motility, fertility can sometimes be achieved by collecting several ejaculations, concentrating the sperm cells, and inserting them into the female's reproductive tract, a process called **artificial insemination** (in-sem-i-nā′shŭn).

19.5 FEMALE REPRODUCTIVE SYSTEM

Learning Outcomes *After reading this section, you should be able to*

A. Name the organs of the female reproductive system, and describe their structure.
B. Describe the anatomy and histology of the ovaries.
C. Discuss the development of the oocyte and the follicle, and describe ovulation and fertilization.
D. Describe the structure of the uterine tubes, uterus, vagina, external genitalia, and mammary glands.

The female reproductive organs consist of the ovaries, the uterine tubes (or fallopian tubes), the uterus, the vagina, the external genitalia, and the mammary glands (see figure 19.1b). The internal reproductive organs of the female are located within the pelvis, between the urinary bladder and the rectum (figure 19.8). The uterus and the vagina are in the midline, with an ovary to each side of the uterus (figure 19.9). The internal reproductive organs are held in place within the pelvis by a group of ligaments. The most conspicuous is the **broad ligament,** which spreads out on both sides of the uterus and attaches to the ovaries and uterine tubes.

Ovaries

The two **ovaries** (ō′vă-rēz; *ovum,* egg) are small organs suspended in the pelvic cavity by ligaments. The **suspensory ligament** extends from each ovary to the lateral body wall, and the **ovarian ligament** attaches the ovary to the superior margin of the uterus (figure 19.9). In addition, the ovaries are attached to the posterior surface of the broad ligament by folds of peritoneum called the **mesovarium** (mez′ō-vā′rē-ŭm). The ovarian arteries, veins, and nerves traverse the suspensory ligament and enter the ovary through the mesovarium.

A layer of visceral peritoneum covers the surface of the ovary. The outer part of the ovary is composed of dense connective tissue and contains **ovarian follicles** (figure 19.10). Each of the ovarian

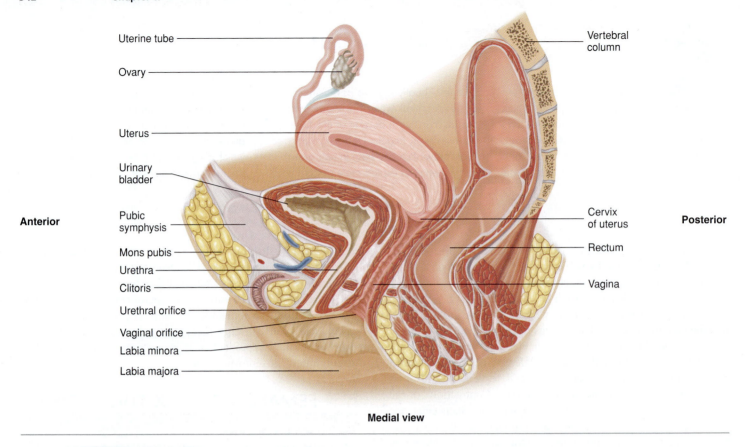

Medial view

Figure 19.8 `AP|R` Female Pelvis

The female reproductive tract, including the uterus, the vagina, and the surrounding pelvic structures, is shown in a medial view of the female pelvis. Note that the female reproductive and urinary tracts open separately to the exterior.

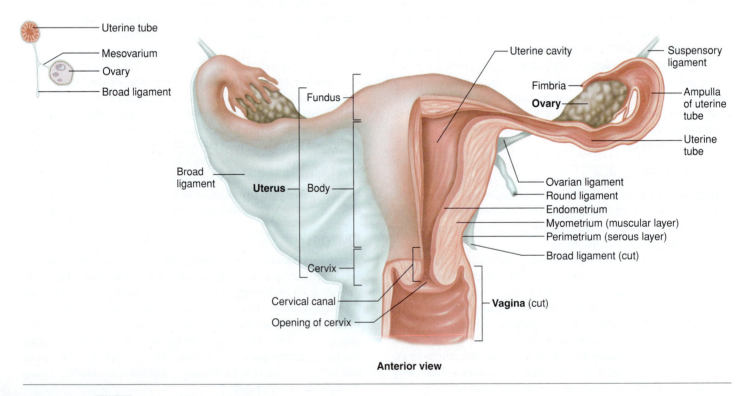

Anterior view

Figure 19.9 `AP|R` Female Reproductive Organs

Anterior view of the uterus, uterine tubes, and associated ligaments. The uterus and uterine tubes are cut in section (on the left side), and the vagina is cut to show the internal anatomy. The inset shows the relationships among the ovary, the uterine tube, and the ligaments that suspend them in the pelvic cavity.

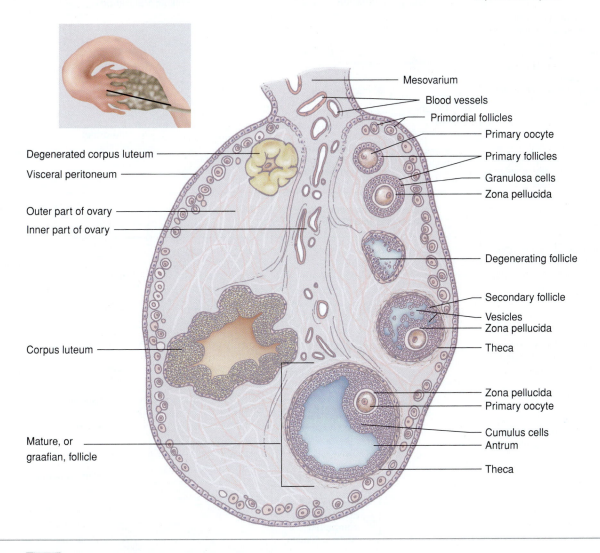

Figure 19.10 **AP|R** **Structure of the Ovary and Ovarian Follicles**

The ovary is sectioned to illustrate its internal structure (inset shows plane of section). Represented are ovarian follicles from each major stage of development, as well as a corpus luteum.

follicles contains an **oocyte** (ōʹō-sīt; *ōon,* egg), the female sex cell. Loose connective tissue makes up the inner part of the ovary, where blood vessels, lymphatic vessels, and nerves are located.

Oogenesis and Fertilization

The formation of female gametes begins during fetal development, even before the female is born. By the fourth month of development, the ovaries contain 5 million **oogonia** (ō-ō-gōʹnē-ă), the cells from which oocytes develop (figure 19.11). By the time of birth, many of the oogonia have degenerated, and the remaining ones have begun meiosis. Also, some data indicate that oogonia can form after birth from stem cells, but the extent to which this occurs, and how long it occurs, is not clear. As in meiosis in males, the genetic material is duplicated, and two cell divisions occur (see "Formation of Gametes" earlier in this chapter). Meiosis stops, however, during the first meiotic division at prophase I. The cell at this stage is called a **primary oocyte,** and at birth there are about 2 million of them. From birth to puberty, many primary oocytes degenerate. The number of primary oocytes decreases to

around 300,000 to 400,000; of these, only about 400 will complete development and be released from the ovaries. Nearly all others degenerate after partial development.

Ovulation is the release of an oocyte from an ovary (figure 19.11, step 8). Just before ovulation, the primary oocyte completes the first meiotic division to produce a **secondary oocyte** and a **polar body.** Unlike meiosis in males, cytoplasm is not split evenly between the two cells. Most of the cytoplasm of the primary oocyte remains with the secondary oocyte. The polar body either degenerates or divides to form two polar bodies. The secondary oocyte begins the second meiotic division but stops in metaphase II.

After ovulation, the secondary oocyte may be fertilized by a sperm cell (figure 19.11, step 9). **Fertilization** (ferʹtil-i-zāʹshŭn) begins when a sperm cell penetrates the cytoplasm of a secondary oocyte. Subsequently, the secondary oocyte completes the second meiotic division to form 2 cells, each containing 23 chromosomes. One of these cells has very little cytoplasm and is another polar body that degenerates. In the other, larger cell, the 23 chromosomes from the sperm join with the 23 from the female gamete to

1 Oogonia give rise to oocytes. Before birth, oogonia multiply by mitosis. During development of the fetus, many oogonia begin meiosis, but stop in prophase I and are now called primary oocytes. They remain in this state until puberty.

2 Before birth, the primary oocytes become surrounded by a single layer of granulosa cells, creating a primordial follicle. These are present until puberty.

3 After puberty, primordial follicles develop into primary follicles when the granulosa cells enlarge and increase in number.

4 Secondary follicles form when fluid-filled vesicles develop and theca cells arise on the outside of the follicle.

5 Mature follicles form when the vesicles create a single antrum.

6 Just before ovulation, the primary oocyte completes meiosis I, creating a secondary oocyte and a nonviable polar body.

7 The secondary oocyte begins meiosis II, but stops at metaphase II.

8 During ovulation, the secondary oocyte is released from the ovary.

9 The secondary oocyte only completes meiosis II if it is fertilized by a sperm cell. The completion of meiosis II forms an oocyte and a second polar body. Fertilization is complete when the oocyte nucleus and the sperm cell nucleus unite, creating a zygote.

10 Following ovulation, the granulosa cells divide rapidly and enlarge to form the corpus luteum.

11 The corpus luteum degenerates to form a scar, or corpus albicans.

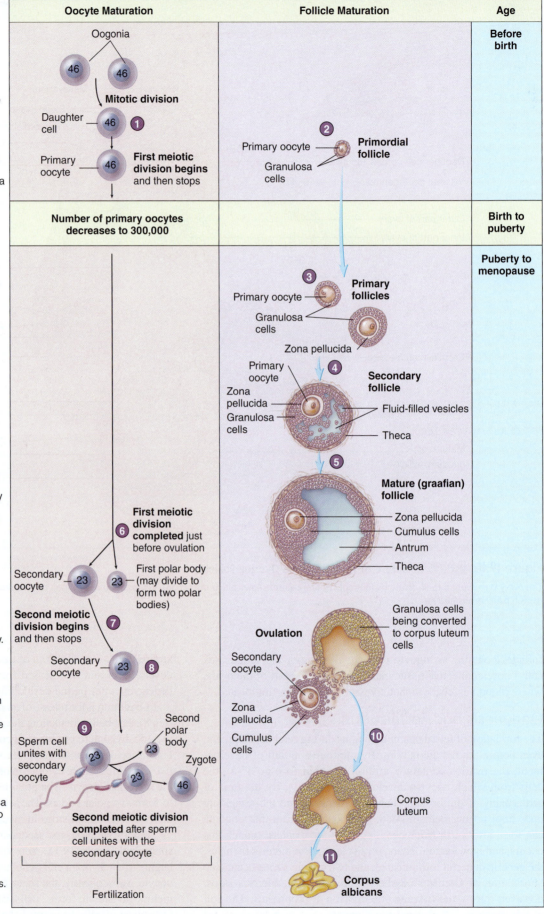

PROCESS Figure 19.11 **AP|R** Maturation of the Oocyte and Follicle

The events leading to oocyte and follicle maturation are closely linked. Once these structures are mature, fertilization can result. The numbers written in the cells are total numbers of chromosomes.

Cancer of the cervix is relatively common in women. In the United States, approximately 11,000 women are diagnosed with cervical cancer annually, and about 3700 women die from it each year. It has been estimated that 70% of all cases of cervical cancer can be linked to infection with 2 of the over 100 types of the human papillomavirus (HPV, types 16 and 18). An immunization (Gardisil) has been developed that targets types 16 and 18, as well as 2 types (6 and 11) that are linked to 90% of genital warts cases. Fortunately, cervical cancer can be detected and treated. A **Pap smear** is a diagnostic test to determine if a woman has cancer of the cervix. By inserting a swab through the vagina, a physician obtains a sample of epithelial cells from the area of the cervix and the wall of the vagina. These cells are smeared on a glass slide and later stained and examined microscopically for signs of cancer. Early in the development of cervical cancer, the cells of the cervix change in a characteristic way. Cells that are cancerous appear less mature than the characteristic epithelial cells of the cervix or vaginal wall.

form a **zygote** (zī′gōt) and complete fertilization. The zygote has 23 pairs of chromosomes (a total of 46 chromosomes). All cells of the human body contain 23 pairs of chromosomes, except for the male and female gametes. The zygote divides by mitosis to form 2 cells, which divide to form 4 cells, and so on. The mass of cells formed may eventually implant in, or attach to, the uterine wall and develop into a new individual (see chapter 20).

Follicle Development

As we discussed, when a female is in her mother's uterus, her ovaries have already begun oocyte formation. The primary oocytes present at birth are surrounded by a primordial follicle. A **primordial follicle** is a primary oocyte surrounded by a single layer of flat cells, called **granulosa cells** (figure 19.11). Once puberty begins, some of the primordial follicles are converted to **primary follicles** when the oocyte enlarges and the single layer of granulosa cells becomes enlarged and cuboidal. Subsequently, several layers of granulosa cells form, and a layer of clear material called the **zona pellucida** (zō′nă pel-lū′sid-dă) is deposited around the primary oocyte.

Approximately every 28 days, hormonal changes stimulate some of the primary follicles to continue to develop (figure 19.11). The primary follicle becomes a **secondary follicle** as fluid-filled spaces called **vesicles** form among the granulosa cells, and a capsule called the **theca** (thē′kă; a box) forms around the follicle.

The secondary follicle continues to enlarge, and when the fluid-filled vesicles fuse to form a single, fluid-filled chamber called the **antrum** (an′trŭm), the follicle is called the **mature follicle,** or *graaf-ian* (graf′ē-ăn) *follicle.* The primary oocyte is pushed off to one side and lies in a mass of granulosa cells called the **cumulus cells.**

The mature follicle forms a lump on the surface of the ovary. During ovulation, the mature follicle ruptures, forcing a small amount of blood, follicular fluid, and the secondary oocyte, surrounded by the cumulus cells, into the peritoneal cavity. In most cases, only one of the follicles that begin to develop forms a mature follicle and undergoes ovulation. The other follicles degenerate. After ovulation, the remaining cells of the ruptured follicle are transformed into a glandular structure called the **corpus luteum** (kōr′pŭs, body; loo′tē-ŭm, yellow). If pregnancy occurs, the corpus luteum enlarges in response to a hormone secreted by the placenta called **human chorionic gonadotropin hormone (hCG)** (kō-rē-on′ik gō′nad-o-trō′pin) (see table 19.1). If pregnancy does not occur, the corpus luteum lasts for 10–12 days and then begins to degenerate.

Uterine Tubes

A **uterine tube,** also called a *fallopian* (fa-lō′pē-an) *tube* or *oviduct* (ō′vi-dŭct), is associated with each ovary. The uterine tubes extend from the area of the ovaries to the uterus. They open directly into the peritoneal cavity near each ovary and receive the secondary oocyte. The opening of each uterine tube is surrounded by long, thin processes called **fimbriae** (fim′brē-ē; fringes) (see figure 19.9).

The fimbriae nearly surround the surface of the ovary. As a result, as soon as the secondary oocyte is ovulated, it comes into contact with the surface of the fimbriae. Cilia on the fimbriae surface sweep the oocyte into the uterine tube. Fertilization usually occurs in the part of the uterine tube near the ovary, called the **ampulla** (am-pul′lă). The fertilized oocyte then travels to the uterus, where it embeds in the uterine wall in a process called **implantation.**

Uterus

The **uterus** (ū′ter-ŭs; womb) is as big as a medium-sized pear (see figures 19.8 and 19.9). It is oriented in the pelvic cavity with the larger, rounded part directed superiorly. The part of the uterus superior to the entrance of the uterine tubes is called the **fundus** (fŭn′dŭs). The main part of the uterus is called the **body,** and the narrower part, the **cervix** (ser′viks; neck), is directed inferiorly. Internally, the **uterine cavity** in the fundus and uterine body continues through the cervix as the **cervical canal,** which opens into the vagina. The cervical canal is lined by mucous glands.

The uterine wall is composed of three layers: a serous layer, a muscular layer, and a layer of endometrium (see figure 19.9). The outer layer, called the **perimetrium** (per-i-mē′trē-ŭm), or *serous layer,* of the uterus is formed from visceral peritoneum. The middle layer, called the **myometrium** (mī′ō-mē′trē-ŭm), or *muscular layer,* consists of smooth muscle, is quite thick, and accounts for the bulk of the uterine wall. The innermost layer of the uterus is the **endometrium** (en′dō-mē′trē-ŭm), which consists of simple columnar epithelial cells with an underlying connective tissue layer. Simple tubular glands, called spiral glands, are formed by folds of the endometrium. The superficial part of the endometrium is sloughed off during menstruation.

The uterus is supported by the broad ligament and the **round ligament.** In addition to these ligaments, much support is provided inferiorly to the uterus by skeletal muscles of the pelvic floor. If ligaments that support the uterus or muscles of the pelvic floor are weakened, as may occur due to childbirth, the uterus can extend inferiorly into the vagina, a condition called a **prolapsed uterus.** Severe cases require surgical correction.

Vagina

The **vagina** (vă-jī′nă) is the female organ of copulation; it receives the penis during intercourse. It also allows menstrual flow and childbirth. The vagina extends from the uterus to the outside of the body (see figures 19.8 and 19.9). The superior portion of the vagina is attached to the sides of the cervix, so that a part of the cervix extends into the vagina.

The wall of the vagina consists of an outer muscular layer and an inner mucous membrane. The muscular layer is smooth muscle and contains many elastic fibers. Thus, the vagina can increase in size to accommodate the penis during intercourse, and it can stretch greatly during childbirth. The mucous membrane is moist stratified squamous epithelium that forms a protective surface layer. Lubricating fluid passes through the vaginal epithelium into the vagina.

In young females, the vaginal opening is covered by a thin mucous membrane called the **hymen** (hī′men; membrane). In rare cases, the hymen may completely close the vaginal orifice and it must be removed to allow menstrual flow. More commonly, the hymen is perforated by one or several holes. The openings in the hymen are usually greatly enlarged during the first sexual intercourse. The hymen can also be perforated or torn earlier in a young female's life during a variety of activities, including strenuous exercise. The condition of the hymen is therefore an unreliable indicator of virginity.

External Genitalia

The external female genitalia, also called the **vulva** (vŭl′vă) or *pudendum* (pū-den′dŭm), consist of the vestibule and its surround-ing structures (figure 19.12). The **vestibule** (ves′ti-bool) is the space into which the vagina and urethra open. The urethra opens just anterior to the vagina. The vestibule is bordered by a pair of thin, longitudinal skin folds called the **labia minora** (lā′bē-ă, lips; mī-nō′ră, small). A small, erectile structure called the **clitoris** (klit′ŏ-ris, klī′tŏ-ris) is located in the anterior margin of the vesti-bule. The two labia minora unite over the clitoris to form a fold of skin called the **prepuce.**

The clitoris (see figure 19.8) consists of a shaft and a dis-tal glans. Like the glans penis, the clitoris is well supplied with sensory receptors, and it is made up of erectile tissue. Additional erectile tissue is located on each side of the vaginal opening.

On each side of the vestibule, between the vaginal opening and the labia minora, are openings of the **greater vestibular glands.** These glands produce a lubricating fluid that helps main-tain the moistness of the vestibule.

Lateral to the labia minora are two prominent, rounded folds of skin called the **labia majora** (mă-jō′ră; large). The two labia majora unite anteriorly at an elevation of tissue over the pubic symphysis called the **mons pubis** (monz pū′bis) (figure 19.12). The lateral surfaces of the labia majora and the surface of the mons pubis are covered with coarse hair. The medial surfaces of the labia majora are covered with numerous sebaceous and sweat glands. The space between the labia majora is called the **pudendal cleft.** Most of the time, the labia majora are in contact with each other across the midline, closing the pudendal cleft and covering the deeper structures within the vestibule.

The region between the vagina and the anus is the **clinical perineum** (per′i-nē′um; area between the thighs). The skin and muscle of this region can tear during childbirth. To prevent such tearing, an incision called an **episiotomy** (e-piz-ē-ot′ŏ-mē) is sometimes made in the clinical perineum. Traditionally, this clean, straight incision has been thought to result in less injury, less trouble in healing, and less pain. However, many studies report less injury and pain when no episiotomy is performed.

Mammary Glands

The **mammary** (mam′ă-rē; relating to breasts) **glands** are the organs of milk production and are located in the **breasts** (fig-ure 19.13). The mammary glands are modified sweat glands. Externally, each of the breasts of both males and females has a raised **nipple** surrounded by a circular, pigmented area called the **areola** (ă-rē′ō-lă).

In prepubescent children, the general structure of the male and female breasts is similar, and both males and females possess a rudimentary duct system. The female breasts begin to enlarge during puberty, under the influence of estrogen and progesterone. Some males also experience a minor and temporary enlargement of the breasts at puberty. Occasionally, the breasts of a male can become permanently enlarged, a condition called **gynecomastia** (gī′nĕ-kō-mas′tē-ă). Causes of gynecomastia include hormonal imbalances and the abuse of anabolic steroids.

Each adult female breast contains mammary glands con-sisting of usually 15–20 glandular **lobes** covered by adipose tissue (figure 19.13a,b). It is primarily this superficial adipose tissue that gives the breast its form. Each lobe possesses a single **lactiferous duct** that opens independently to the surface of the

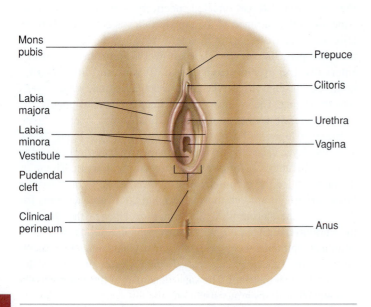

Mons pubis
Prepuce
Clitoris
Labia majora
Labia minora
Urethra
Vestibule
Vagina
Pudendal cleft
Clinical perineum
Anus

Figure 19.12 AP|R **Female External Genitalia**

CLINICAL IMPACT   Cancer of the Breast

Cancer of the breast is a serious, often fatal disease that most often occurs in women. Regular self-examination of the breast can lead to early detection of breast cancer and effective treatment. In addition, **mammography** (ma-mog′ră fe) often allows tumors to be identified even before they can be detected by palpation. Mammography uses low-intensity x-rays to detect tumors in the soft tissue of the breast. Once a tumor is identified, a biopsy is normally performed to determine whether the tumor is benign or malignant. Most tumors of the mammary glands are benign, but those that are malignant can spread to other areas of the body and ultimately lead to death.

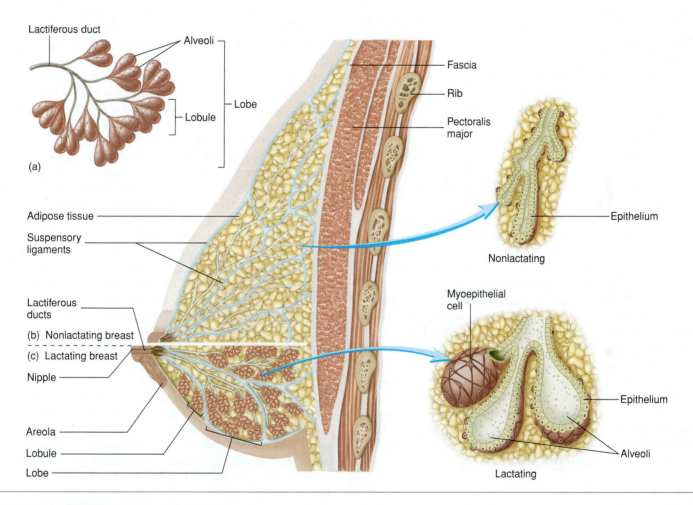

Figure 19.13   AP|R   Anatomy of the Breast

(*a*) Each lactiferous duct of the mammary gland branches. At the end of each branch is one or more alveoli. (*b*) The nonlactating breast has a duct system that is not extensively developed. The branches of the lactiferous ducts end as small, tubelike structures. (*c*) The lactating breast has a well-developed duct system with many branches. The branches of the lactiferous duct end with well-developed alveoli. Adipose tissue is abundant in both the nonlactating and the lactating breast.

nipple. The duct of each lobe is formed as several smaller ducts, which originate from **lobules,** converge. Within a lobule, the ducts branch and become even smaller. In the milk-producing, or lactating, mammary gland, the ends of these small ducts expand to form secretory sacs called **alveoli. Myoepithelial cells** surround the alveoli and contract to expel milk from the alveoli (figure 19.13*c*).

The breasts are supported by suspensory ligaments that extend from the fascia over the pectoralis major muscles to the skin over the breasts (figure 19.13*b*).

The nipples are very sensitive to tactile stimulation and contain smooth muscle. When the smooth muscle contracts in response to stimuli, such as touch, cold, and sexual arousal, the nipple becomes erect.

Reproductive

19.6 PHYSIOLOGY OF FEMALE REPRODUCTION

A. Describe the changes that occur in females during puberty and the changes in the ovary and uterus that occur during the menstrual cycle.

B. List the hormones of the female reproductive system, and explain how their secretion is regulated.

C. Explain the events that occur during the female sexual act.

D. Define menopause, and describe the changes that result from it.

As in the male, female reproduction is controlled by hormonal and nervous system mechanisms.

Puberty in Females

The initial change that results in puberty is most likely maturation of the hypothalamus. In girls, puberty, which typically begins between ages 11 and 13 and is largely completed by age 16, is marked by the first episode of menstrual bleeding, which is called **menarche** (me-nar′kē; *mēn,* month + *archē,* beginning). During puberty, the vagina, uterus, uterine tubes, and external genitalia begin to enlarge. Adipose tissue is deposited in the breasts and around the hips, causing them to enlarge and assume an adult form. In addition, pubic and axillary hair grows. The development of sexual drive is also associated with puberty.

The changes associated with puberty primarily result from the increasing rate of estrogen and progesterone secretion by the ovaries. Before puberty, estrogen and progesterone are secreted in very small amounts. At puberty, the cyclical adult pattern of hormone secretion is gradually established.

Before puberty, the rate of GnRH secretion from the hypothalamus and the rate of LH and FSH secretion from the anterior pituitary are very low. Estrogen and progesterone from the ovaries have a strong negative-feedback effect on the hypothalamus and pituitary. After the onset of puberty, the hypothalamus and anterior pituitary secrete larger amounts of GnRH, LH, and FSH. Estrogen and progesterone have less of a negative-feedback effect on the hypothalamus and pituitary, and a sustained increase in estrogen concentration has a positive-feedback effect. The normal cyclical pattern of reproductive hormone secretion that occurs during the menstrual cycle becomes established.

Menstrual Cycle

The term **menstrual** (men′stroo-ăl) **cycle** refers to the series of changes that occur in sexually mature, nonpregnant females and that culminate in menses. **Menses** (men′sēz; month) is a period of mild hemorrhage, during which part of the endometrium is sloughed and expelled from the uterus. Typically, the menstrual cycle is about 28 days long, although it can be as short as 18 days or as long as 40 days (figure 19.14 and table 19.3). The menstrual cycle results from the cyclical changes that occur in the endometrium of the uterus. These changes, in turn, result from the cyclical changes that occur in the ovary and are controlled by the secretions of FSH and LH from the anterior pituitary gland.

The first day of menstrual bleeding (menses), when the endometrium sloughs off, is considered day 1 of the menstrual cycle. Sloughing of the endometrium is inhibited by progesterone but

stimulated by estrogen. Menses typically lasts 4 or 5 days and can be accompanied by strong uterine contractions, called menstrual cramps, that are extremely uncomfortable in some women. Menstrual cramps are the result of forceful myometrial contractions that occur before and during menstruation. The cramps can result from excessive secretion of prostaglandins. As the endometrium of the uterus sloughs off, it becomes inflamed, and prostaglandins are produced as part of the inflammation. Many women can alleviate painful cramps by taking medications, such as aspirin-like drugs, that inhibit prostaglandin biosynthesis just before the onset of menstruation. These medications, however, are not effective in treating all painful menstruation, especially when the cause of the pain, such as that experienced by women who have tumors of the myometrium, is not due to the inflammatory response.

Ovulation occurs on about day 14 of the menstrual cycle, although the timing of ovulation varies from individual to individual and can vary within an individual from one menstrual cycle to the next. To avoid or optimize contraception, it is critical to predict ovulation; however, there is no single reliable method that can predict its exact timing. The simplest method of predicting ovulation is looking for a drop in basal body temperature preceding ovulation, but it is the least reliable method.

Between the end of menses and ovulation is the **proliferative phase,** which refers to proliferation of the endometrium. During the proliferative phase, the secondary follicles in the ovary mature; as they do so, they secrete increasing amounts of estrogen. Estrogen acts on the uterus and causes the epithelial cells of the endometrium to divide rapidly. The endometrium thickens, and spiral glands form.

The sustained increase of estrogen secreted by the developing follicles stimulates GnRH secretion from the hypothalamus. GnRH, in turn, triggers FSH and LH secretion from the anterior pituitary gland. FSH stimulates estrogen secretion at an increasing rate from the developing follicles. This positive-feedback loop produces a series of larger and larger surges of FSH and LH secretion. Ovulation occurs in response to the large increases in LH levels that normally occur on about day 14 of the menstrual cycle. This large increase in LH is also responsible for the development of the corpus luteum.

Following ovulation, the corpus luteum begins to secrete progesterone and smaller amounts of estrogen. Progesterone acts on the uterus, causing the cells of the endometrium to become larger and to secrete a small amount of fluid. Together, progesterone and estrogen act on the hypothalamus and anterior pituitary gland to inhibit GnRH, LH, and FSH secretion. Thus, LH and FSH levels decline after ovulation.

Between ovulation and the next menses is the **secretory phase** of the menstrual cycle, called this because of the small amount of fluid secreted by the cells of the endometrium. During the secretory phase, the lining of the uterus reaches its greatest degree of development.

If fertilization occurs, the zygote undergoes several cell divisions to produce a collection of cells called the **blastocyst** (blas′tō-sist). The blastocyst passes through the uterine tube and arrives in the uterus by 7 or 8 days after ovulation. The endometrium is prepared to receive the blastocyst, which becomes implanted in the endometrium, where it continues to develop. If the secondary oocyte is not fertilized, the endometrium sloughs away as a result of declining blood progesterone levels. Unless the secondary oocyte is fertilized, the corpus luteum begins to produce less progesterone by

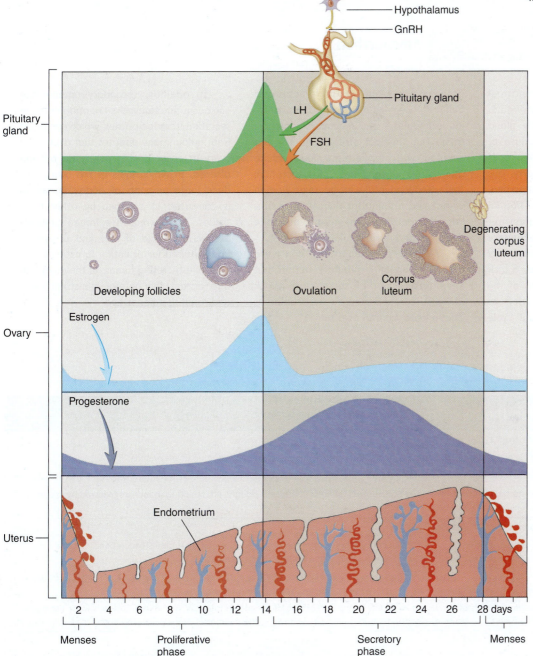

Figure 19.14 AP|R
Menstrual Cycle
Over approximately 30 days, fluctuations occur in the levels of follicle-stimulating hormone (FSH) and luteinizing hormone (LH) secretion from the anterior pituitary gland and in the levels of estrogen and progesterone secretion from the ovary. In addition, changes in the ovary and changes in the endometrium of the uterus are correlated with the changes in hormone secretion throughout the menstrual cycle. Ovulation occurs on about day 14.

day 24 or 25 of the menstrual cycle. By day 28, the declining progesterone causes the endometrium to slough away to begin menses and the next menstrual cycle. The declining progesterone secretion results in a small increase in FSH secretion at the beginning of the next menses, which triggers more follicles to mature.

Predict 4

Predict the effect of administering a relatively large amount of progesterone and estrogen just before the increase in LH that precedes ovulation.

An **ectopic pregnancy** results if implantation occurs anywhere other than in the uterine cavity. The most common site of ectopic pregnancy is the uterine tube. Implantation in the uterine tube is eventually fatal to the fetus and can cause the tube to rupture. In some rare cases, implantation occurs in the mesenteries of the abdominal cavity; the fetus can develop normally but must be delivered by cesarean

section. However, maternal mortality rates for abdominal pregnancies are significantly higher than for fallopian tube ectopic pregnancies.

Menopause

When a woman is 40–50 years old, the menstrual cycles become less regular, and ovulation does not occur consistently during each cycle. Eventually, the cycles stop completely. The cessation of menstrual cycles is called **menopause** (men′ō-pawz; *mēn,* month + *pausis,* cessation), and the whole time period from the onset of irregular cycles to their complete cessation is called the female **climacteric** (klī-mak′ter-ik).

The major cause of menopause is age-related changes in the ovaries. The number of follicles remaining in the ovaries of menopausal women is small. In addition, the follicles that remain become less sensitive to stimulation by FSH and LH, and therefore fewer mature follicles and corpora lutea are produced.

Reproductive

CLINICAL IMPACT   Amenorrhea

The absence of a menstrual cycle is called **amenorrhea** (ă-men-ō-rē′ă; without menses). If the pituitary gland does not function properly because of abnormal development, a woman does not begin to menstruate at puberty. This condition is called **primary amenorrhea.** In contrast, if a woman has had normal menstrual cycles and later stops menstruating, the condition is called **secondary amenorrhea.** One cause of secondary amenorrhea is anorexia, in which a lack of food causes the hypothalamus of the brain to decrease GnRH secretion to levels so low that the menstrual cycle

cannot occur. Many female athletes and ballet dancers who have rigorous training schedules have secondary amenorrhea. The physical stress that can be coupled with inadequate food intake also results in very low GnRH secretion. Increased food intake for anorexic women or reduced training for dancers and athletes generally restores normal hormone secretion and normal menstrual cycles.

Secondary amenorrhea can also be the result of a pituitary tumor that decreases FSH and LH secretion or a lack of GnRH secretion from the hypothalamus due to head trauma or a tumor.

In addition, secondary amenorrhea can occur due to a lack of normal hormone secretion from the ovaries, which can result from autoimmune diseases that attack the ovary or occur due to **polycystic ovarian disease,** in which cysts in the ovary produce large amounts of androgen, which is converted to estrogen by other body tissues. The increased estrogen prevents the normal cycle of FSH and LH secretion required for ovulation to occur. Other hormone-secreting tumors of the ovary can also disrupt the normal menstrual cycle and result in amenorrhea.

TABLE 19.3   Events During the Menstrual Cycle

Menses (day 1 to day 4 or 5 of the menstrual cycle)

Pituitary gland	The rate of FSH and LH secretion is low, but the rate of FSH secretion increases as progesterone levels decline.
Ovary	The rate of estrogen and progesterone secretion is low after degeneration of the corpus luteum produced during the previous menstrual cycle.
Uterus	In response to declining progesterone levels, the endometrial lining of the uterus sloughs off, resulting in menses followed by repair of the endometrium.

Proliferative Phase (from day 4 or 5 until ovulation on about day 14)

Pituitary gland	The rate of FSH and LH secretion is only slightly elevated during most of the proliferative phase; FSH and LH secretions increase near the end of the proliferative phase in response to increasing estrogen secretion from the ovaries.
Ovary	Developing follicles secrete increasing amounts of estrogen, especially near the end of the proliferative phase; increasing FSH and LH cause additional estrogen secretion from the ovaries near the end of the proliferative phase.
Uterus	Estrogen causes endometrial cells of the uterus to divide. The endometrium of the uterus thickens, and tubelike glands form. Estrogen causes the cells of the uterus to be more sensitive to progesterone by increasing the number of progesterone receptors in uterine tissues.

Ovulation (about day 14)

Pituitary gland	The rate of FSH and LH secretion increases rapidly just before ovulation in response to increasing estrogen levels. Increasing FSH and LH levels stimulate estrogen secretion, resulting in a positive-feedback cycle.
Ovary	LH causes final maturation of a mature follicle and initiates the process of ovulation. FSH acts on immature follicles and causes several of them to begin to enlarge.
Uterus	The endometrium continues to divide in response to estrogen.

Secretory Phase (from about day 14 to day 28)

Pituitary gland	Estrogen and progesterone reach levels high enough to inhibit FSH and LH secretion from the pituitary gland.
Ovary	After ovulation, the follicle is converted to the corpus luteum; the corpus luteum secretes large amounts of progesterone and smaller amounts of estrogen from shortly after ovulation until about day 24 or 25. If fertilization does not occur, the corpus luteum degenerates after about day 25, and the rate of progesterone secretion rapidly declines to low levels.
Uterus	In response to progesterone, the endometrial cells enlarge, the endometrial layer thickens, and the glands of the endometrium reach their greatest degree of development; the endometrial cells secrete a small amount of fluid. After progesterone levels decline, the endometrium begins to degenerate.

Menses (day 1 to day 4 or 5 of the next menstrual cycle)

Pituitary gland	The rate of LH remains low, and the rate of FSH secretion increases as progesterone levels decline.
Ovary	The rate of estrogen and progesterone secretion is low.
Uterus	In response to declining progesterone levels, the endometrial lining of the uterus sloughs off, resulting in menses followed by repair of the endometrium.

Older women experience gradual changes in response to the reduced amount of estrogen and progesterone produced by the ovaries (table 19.4). For example, during the climacteric, some women experience sudden episodes of uncomfortable sweating (hot flashes), fatigue, anxiety, temporary decrease in libido, and occasionally emotional disturbances. Many of these symptoms can be treated successfully with hormone replacement therapy (HRT), which usually consists of small amounts of estrogen and progesterone. HRT has been linked to a slightly increased risk of developing breast cancer, uterine cancer, heart attack, stroke, or blood clots. On the positive side, HRT slows the decrease in bone density that can become severe in some women after menopause, and it decreases the risk of developing colorectal cancer.

Female Sexual Behavior and the Female Sex Act

Sexual drive in females, like sexual drive in males, is dependent on hormones. Testosterone-like hormones, and possibly estrogen, affect brain cells (especially in the area of the hypothalamus) and influence sexual behavior. Testosterone-like hormones are produced primarily in the adrenal cortex. Psychological factors also play a role in sexual behavior. The sensory and motor neural pathways involved in controlling female sexual responses are similar to those found in the male.

During sexual excitement, erectile tissue within the clitoris and around the vaginal opening becomes engorged with blood. The mucous glands within the vestibule, especially the greater vestibular glands, secrete small amounts of mucus. Larger amounts of mucus-like fluid are also extruded into the vagina through its wall. These secretions provide lubrication to allow easy entry and movement of the penis in the vagina during intercourse. Tactile stimulation of the female's genitals during sexual intercourse and psychological stimuli normally trigger an **orgasm,** or *climax.* The vaginal and uterine smooth muscle, as well as the surrounding skeletal muscles, contract rhythmically, and muscle tension increases throughout much of the body. After the sex act, there is a period of **resolution,** which is characterized by an overall sense of satisfaction and relaxation. Females are sometimes receptive to further immediate stimulation, however, and can experience successive orgasms. Orgasm is not necessary for fertilization to occur. Ovulation results from hormonal stimuli and is not dependent on the female sex act.

A CASE IN POINT

Endometriosis

Helen Hurtz is in her mid-twenties and has been married for 4 years. She has become very frustrated because she experiences pain during and after sexual intercourse, and the pain has become worse over the past 2 years. She also has an increasing, persistent pain in her pelvic region that is especially uncomfortable before and during menstruation and becomes more intense during urination and bowel movements. In addition, she recently developed periodic bouts of diarrhea. She reported all these symptoms to her physician.

Helen's physician suspects **endometriosis** (en'dō-mē-trē-ō'sis), a condition in which endometrial tissue migrates from the lining of the uterus into the peritoneal cavity, where it attaches to the surface of organs. Common sites of attachment are the ovaries and the pelvic peritoneum. Other possible sites are the intestines, uterus, urinary bladder, and vagina. If the attached endometrial tissue has an adequate blood supply, it proliferates, breaks down, and bleeds in response to the hormones produced during the menstrual cycle. Unlike the normal endometrium, which is shed each month during menstruation, endometrial tissue attached outside of the uterus causes lesions or tumors to develop, resulting in internal bleeding, scar tissue formation, inflammation, and pain. Other major complications of endometriosis are infertility and ovarian cyst formation. Between 40% and 50% of infertile women have endometriosis.

Helen's physician explained that the most accurate method of confirming the diagnosis is by laparoscopy, a procedure that allows the physician to visually observe the abdominopelvic cavity. After a few weeks of thinking about the procedure, Helen agreed to the laparoscopic examination. During the procedure, the physician observed several lesions characteristic of endometriosis and removed them with a laser instrument that vaporizes them.

Helen's physician explained to her that there is no cure for endometriosis, but the condition can be managed by administering medication and removing the endometrial lesions periodically.

TABLE 19.4	Possible Changes in Postmenopausal Women Caused by Decreased Ovarian Hormone Secretion
	Changes
Menstrual cycle	Five to seven years before menopause, the cycle becomes irregular; the number of cycles in which ovulation does not occur and in which corpora lutea do not develop increases.
Uterus	Gradual increase in irregular menstruations is followed by no menstruation; the endometrium finally atrophies, and the uterus becomes smaller.
Vagina and external genitalia	The epithelial lining becomes thinner; the external genitalia become thinner and less elastic; the labia majora become smaller; the pubic hair decreases; reduced secretion leads to dryness; the vagina is more easily inflamed and infected.
Skin	The epidermis becomes thinner.
Cardiovascular system	Hypertension and atherosclerosis occur more frequently.
Vasomotor instability	Hot flashes and increased sweating are correlated with vasodilation of cutaneous blood vessels; the hot flashes are related to decreased estrogen levels.
Libido	Temporary changes, usually a decrease in libido, are associated with the onset of menopause.
Fertility	Fertility begins to decline about 10 years before the onset of menopause; by age 50, almost all the oocytes and follicles have degenerated.
Pituitary function	Low levels of estrogen and progesterone produced by the ovaries cause the pituitary gland to secrete larger than normal amounts of LH and FSH; increased levels of these hormones have little effect on the postmenopausal ovaries.

CLINICAL IMPACT   Control of Pregnancy

Many methods are used to prevent pregnancy, either by preventing fertilization (contraception) or by preventing implantation of the developing embryo. Many of these techniques are quite effective when used *perfectly* and consistently. But most of these methods also have disadvantages, and the use of some of them is controversial.

Behavioral Methods

Abstinence, or refraining from sexual intercourse, is 100% effective in preventing pregnancy when it is practiced consistently. It is not an effective method when used only occasionally.

Coitus interruptus (kō'i-tŭs int-ĕ-rŭp'tŭs), or **withdrawal,** is removal of the penis from the vagina just before ejaculation. This is a very unreliable method of preventing pregnancy because it requires perfect awareness and willingness to withdraw the penis at the correct time. Statistically, about 23 women out of 100 become pregnant while relying on this method. The withdrawal method also ignores the fact that some sperm cells are present in preejaculatory emissions.

The **calendar method** requires abstaining from sexual intercourse near the time of ovulation. A major factor in the success of this method is the ability to predict accurately the time of ovulation. Although the calendar method provides some protection against becoming pregnant, it has a relatively high failure rate because of both the inability to predict the time of ovulation and the failure to abstain from intercourse around that time. About 9 women out of 100 become pregnant while using the calendar method.

Continuous breastfeeding, or **lactation** (lak-tā'shŭn) (also known as lactation amenorrhea, or LAM), often stops the menstrual cycle for up to the first 6 months after childbirth, as long as the baby is exclusively breastfed and the mother does not resume menstruation while lactating. This method is 99% effective. Continuous breastfeeding works because action potentials sent to the hypothalamus in response to infant suckling inhibit GnRH release from the hypothalamus. Reduced GnRH reduces LH, which prevents ovulation. Eventually, the menstrual cycle resumes. Because ovulation normally precedes menstruation, relying on lactation to prevent pregnancy after the first 6 months postdelivery is not effective.

Barrier Methods

A male **condom** (kon'dom) is a sheath made of animal membrane, rubber, or plastic (figure 19A*a*). When placed over the erect penis, a condom is a barrier device because it collects the semen instead of allowing it to be released into the vagina. Condoms also provide some protection against sexually transmitted diseases. Condoms alone are 98% effective when used correctly, 99% effective when used with spermicide.

A **vaginal condom** (or female condom) also acts as a barrier. A woman can place the vaginal condom into the vagina before sexual intercourse. Female condoms are 95% effective. Using spermicide further increases their effectiveness.

Methods to prevent sperm cells from reaching the oocyte once they are in the vagina include a diaphragm, spermicidal agents, and a vaginal sponge. The **diaphragm** and the **cervical cap** (figure 19A*b*) are flexible latex domes that are placed over the cervix within the vagina, where they prevent sperm cells from passing from the vagina through the cervical canal of the uterus. The diaphragm is a larger, shallow latex cup, and the cervical cap is a smaller, thimble-shaped latex cup. Diaphragms are 94% effective, whereas cervical cap effectiveness ranges from 71% in a woman who has previously been pregnant to 86% in a woman who has never been pregnant. The most commonly used **spermicidal agents** are foams or creams that kill sperm cells (figure 19A*c*). They are inserted into the vagina before sexual intercourse, often in conjunction with diaphragm or condom use. Alone, spermicidal agents are only about 85% effective.

Intrauterine (in'trā-yū'ter-in) **devices (IUDs)** (figure 19A*d*) are inserted into the uterus through the cervix. The two types of IUDs now available in the United States are the copper-containing ParaGard and the progestin hormone–coated Mirena. The ParaGard may be left in place for 12 years, whereas the Mirena may be left in place for 5 years. Both types of IUDs thicken cervical mucus, which bars sperm from entering the uterus. Some women stop ovulating when they have an IUD implanted. IUDs also alter the endometrium, which may prevent implantation of an embryo. IUDs are 99.99% effective in preventing pregnancy.

Chemical Methods

Synthetic estrogen and progesterone in **oral contraceptives** (birth control pills) (figure 19A*e*) are among the most effective contraceptives, providing 99.9% effectiveness. The synthetic steroids can have more than one action, but they reduce LH and FSH release from the anterior pituitary. Estrogen and progesterone are present in high enough concentrations to have a negative-feedback effect on the pituitary, which prevents the large increase in LH and FSH secretion that triggers ovulation. Over the years, the dose of estrogen and progesterone in birth control pills has been reduced. The current lower dose of birth control pills has fewer side effects than earlier dosages. However, the risk of heart attack or stroke increases in female users of oral contraceptives who smoke or who have a history of hypertension or coagulation disorders. For most females, the pill is effective and has a minimum frequency of complications, until at least age 35. The **mini-pill** is an oral contraceptive that contains only synthetic progesterone. It reduces and thickens the mucus of the cervix, which prevents sperm cells from reaching the oocyte. It also prevents blastocysts from implanting in the uterus.

Progesterone-like chemicals, such as medroxyprogesterone (med-rok'-sē-prō-jes'ter-ōn) (Depo-Provera), which are injected intramuscularly and slowly released into the circulatory system, can act as effective contraceptives. Injected progesterone-like chemicals can protect against pregnancy for approximately 1 month, depending on the amount injected, and are 99.9% effective. The **patch** (Ortho Evra) is an adhesive skin patch containing synthetic estrogen and progesterone. It is worn on the lower abdomen, buttocks, or upper body and is 99.9% effective. The **vaginal contraceptive ring** (Nuva Ring) is inserted into the vagina, where it releases synthetic estrogen and progesterone; it is 99.9% effective.

A drug called **RU486,** or *mifepristone* (mif'pris-tōn), blocks the action of progesterone, causing the endometrium of the uterus

CLINICAL IMPACT *(continued)*

to slough off as it does at the time of menstruation. It can therefore be used to induce menstruation and reduce the possibility of implantation when sexual intercourse has occurred near the time of ovulation. It can also be used to terminate pregnancies.

Morning-after pills, similar in composition to birth control pills, are available. Alternatively, doubling the number of birth control pills after sexual intercourse within 3 days and again after 12 more hours is sometimes recommended. These techniques can be used after intercourse, but they are only about 75% effective. The elevated blood levels of estrogen and progesterone may inhibit the increase in LH that causes ovulation, alter the rate at which the fertilized oocyte is transported through the uterine tube to the uterus, or inhibit implantation.

Surgical Methods

Vasectomy (va-sek′tō-mē) is a common method used to render males permanently infertile without affecting the performance of the sex act. Vasectomy is a surgical procedure in which the ductus deferens from each testis is cut and tied off within the scrotal sac (figure 19A*f*). This procedure prevents sperm cells from passing through the ductus deferens and becoming part of the ejaculate. Because such a small volume of ejaculate comes from the testis and epididymis, vasectomy has little effect on the volume of the ejaculated semen. The sequestered sperm cells are reabsorbed in the epididymis. Only 1–4 in 1000 surgeries of this type fail.

A common method of permanent birth control in females is **tubal ligation** (lī-gā′shŭn), in which the uterine tubes are tied and cut or clamped by means of an incision through the wall of the abdomen (figure 19A*g*). This procedure closes off the path between the sperm cells and the oocyte. Commonly, a technique called **laparoscopy** (lap-ă-ros′kŏ-pē) is used, in which an instrument is inserted into the abdomen through a small incision, so that only small openings need to be made to perform the operation.

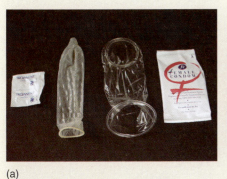
(a)

(b)

(c)

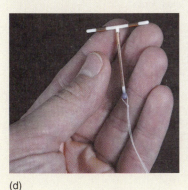
(d)

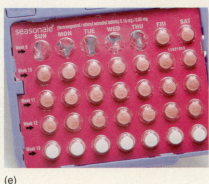
(e)

Ductus deferens within spermatic cord

Ductus deferens (vas deferens) cut and tied

(f)

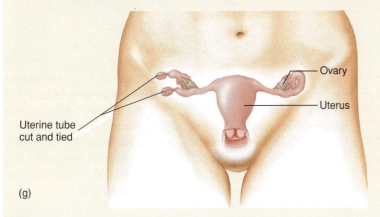
Ovary

Uterus

Uterine tube cut and tied

(g)

Figure 19A

Contraceptive devices and techniques include (*a*) condom, (*b*) cervical cap and diaphragm used with spermicidal jelly, (*c*) spermicidal foam, (*d*) intrauterine device (IUD), (*e*) oral contraceptives, (*f*) vasectomy, and (*g*) tubal ligation.

Name: Molly
Gender: Female
Age: 43

Comments

Molly had four children and was 43. She noticed that menstruation was becoming gradually more severe and lasting up to several days longer each time. After menstruating almost continuously for 2 months, Molly made an appointment with her physician. The physician performed a pelvic examination, including tests for conditions such as cervical and uterine cancer. Palpation of the uterus indicated the presence of enlarged masses in Molly's uterus. The doctor performed a D&C, which is dilation of the cervix and scraping (curettage) of the endometrium to remove growths or other abnormal tissues. The results of the D&C indicated that Molly had developed leiomyomas.

Benign Uterine Tumors

Background Information

Leiomyomas (lī′-ō-mī-ō′măz), also called uterine fibroids, are fibroid tumors of the uterus (figure 19B). They are one of the most common disorders of the uterus, and the most frequent tumor in women, affecting one of every four. However, three-fourths of the women with this condition experience no symptoms. The enlarged masses that originate from smooth muscle tissue compress the uterine lining (endometrium), resulting in ischemia and inflammation. The increased inflammation, which shares some characteristics with menstruation, results in frequent and severe menses, with associated abdominal cramping due to strong uterine contractions. Constant menstruation is a frequent manifestation of these tumors, and it is one of the most common reasons women elect to have the uterus removed, a procedure called a **hysterectomy** (his-ter-ek′tō-mē).

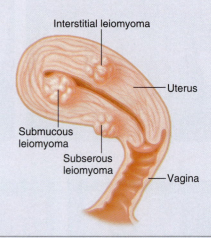

Figure 19B

Leiomyomas, or fibroid tumors, are enlarged masses of smooth muscle. They are located near the mucosa (submucous), within the myometrium (interstitial), or near the serosa (subserous).

Infertility in Females

Causes of infertility in females include malfunctions of the uterine tubes, reduced hormone secretion from the pituitary gland or the ovaries, and interruption of implantation. Uterine tube malfunction can occur when infections result in pelvic inflammatory disease (PID), which causes adhesions to form in one or both of the uterine tubes. Inadequate secretion of LH and FSH can occur for a variety of reasons, including hypothyroidism or a tumor in or trauma to the hypothalamus or anterior pituitary. Decreased secretion of LH and FSH interrupts ovulation.

Interruption of implantation can result from uterine tumors or conditions causing abnormal ovarian hormone secretion. For example, premature degeneration of the corpus luteum causes progesterone levels to decline and menses to occur. If the corpus luteum degenerates before the placenta begins to secrete progesterone, the endometrium and the developing embryonic mass will degenerate and be eliminated from the uterus. The conditions that result in secondary amenorrhea also reduce fertility.

Benign Uterine Tumors

Symptoms
- None in 75% of cases
- Frequent and severe menses
- Strong menstrual uterine cramping

Treatment
- Hysterectomy

SKELETAL
The rate of red blood cell synthesis in the red bone marrow increases.

INTEGUMENTARY
If anemia develops, the skin can appear pale because of the reduced hemoglobin in red blood cells.

MUSCULAR
If anemia develops and is severe, muscle weakness results because of the reduced ability of the cardiovascular system to deliver adequate oxygen to muscles.

URINARY
The kidney increases erythropoietin secretion in response to the reduced oxygen-carrying capacity of the blood. The erythropoietin increases red blood cell synthesis in red bone marrow. An enlarged uterine tumor can put pressure on the urinary bladder, resulting in frequent and painful urination.

CARDIOVASCULAR
A chronic loss of blood, as occurs in prolonged menstruation over many months to years, frequently results in iron-deficiency anemia. Manifestations of anemia include pale skin, reduced hematocrit, reduced hemoglobin concentration, smaller than normal red blood cells (microcytic anemia), and increased heart rate.

RESPIRATORY
Because of anemia, the oxygen-carrying capacity of the blood is reduced. Increased respiration during physical exertion and rapid fatigue are likely to occur if anemia develops.

Predict 5

When discussing her condition with her mother, Molly discovered that her mother had experienced frequent menses that were irregular and prolonged when she was in her late 40s. Molly's mother did not have a hysterectomy, and in a few years, the frequency gradually began to subside. Explain.

19.7 EFFECTS OF AGING ON THE REPRODUCTIVE SYSTEM

Learning Outcome *After reading this section, you should be able to*

A. Describe the major age-related changes in the reproductive system.

Aging affects the reproductive system in both men and women in several ways. Sexual activity is often maintained in men and women as they age, but the frequency of sexual intercourse usually decreases gradually.

In men, benign prostatic enlargement is common after 50 years of age. A major consequence of prostatic enlargement is blockage of the prostatic urethra. Although benign prostatic enlargement is not preventable, treatments are available to reverse its negative effects. The frequency of prostate cancer also increases as men age and is a significant cause of death in men. In addition, the tendency for erectile dysfunction increases as men age. However, less than 15% of men age 60 or under experience abnormal erectile dysfunction. Keeping physically healthy can minimize some factors leading to abnormal erectile dysfunction, and medical treatments are available.

In women, the most significant age-related change is menopause. By age 50, the amount of estrogen and progesterone produced by the ovaries has decreased. The uterus decreases in size, and the endometrium decreases in thickness. The times between menses become irregular and longer until menstruation stops. The vaginal wall becomes thinner and less elastic, and there is less lubricant in the vagina, resulting in an increased tendency for vaginal yeast infections. However, wearing cotton underwear and loose clothing reduces this possibility. In the event of infections, very effective medical treatments are available.

Approximately 10% of all women will develop breast cancer. The incidence of breast cancer is greatest between 45 and 65 years of age and is greater for women who have a family history of breast cancer. The single most important measure to guard against death from breast cancer is early detection through breast self-exams and yearly mammograms after age 40. The incidences of uterine cancer, ovarian cancer, and cervical cancer all increase between 50 and 65 years of age. Annual medical checkups, including Pap smears for cervical cancer, are important in order to detect cancer at early stages, when it can be easily treated.

DISEASES AND DISORDERS: Reproductive System

CONDITION	DESCRIPTION
Infectious Diseases	
Pelvic inflammatory disease (PID)	Bacterial infection of the female pelvic organs; commonly caused by a vaginal or uterine infection with the bacteria gonorrhea or chlamydia; early symptoms include increased vaginal discharge and pelvic pain; antibiotics are effective; if untreated, can lead to sterility or be life-threatening
Sexually Transmitted Diseases	Commonly known as STDs; spread by intimate sexual contact
Nongonococcal urethritis (non-gon′ō-kok′ăl u-rē-thrī′tis)	Inflammation of the urethra that is not caused by gonorrhea; can be caused by trauma, insertion of a nonsterile catheter, or sexual contact; usually due to infection with the bacterium *Chlamydia trachomatis* (kla-mid′ē-ă tra-kō′mă-tis); may go unnoticed and result in pelvic inflammatory disease or sterility; antibiotics are effective treatment
Trichomoniasis (trik-ō-mō-nī′ă-sis)	Caused by *Trichomonas* (trik′ō-mō′nas), a protozoan commonly found in the vagina of women and in the urethra of men; results in a greenish-yellow discharge with a foul odor; more common in women than in men
Gonorrhea (gon-ō-rē′ă)	Caused by the bacterium *Neisseria gonorrhoeae* (nī-sē′rē-ă gon-ō-rē′ă), which attaches to the epithelial cells of the vagina or male urethra and causes pus to form; pain and discharge from the penis occur in men; asymptomatic in women in the early stages; can lead to sterility in men and pelvic inflammatory disease in women
Genital herpes (her′pēz)	Caused by herpes simplex 2 virus; characterized by lesions on the genitals that progress into blisterlike areas, making urination, sitting, and walking painful; antiviral drugs can be effective
Genital warts	Caused by a viral infection; very contagious; warts vary from separate, small growths to large, cauliflower-like clusters; lesions are not painful, but sexual intercourse with lesions is; treatments include topical medicines and surgery to remove the lesions
Syphilis (sif′i-lis)	Caused by the bacterium *Treponema pallidum* (trep-ō-nē′mă pal′i-dŭm); can be spread by sexual contact; multiple disease stages occur; children born to infected mothers may be developmentally delayed; antibiotics are effective
Acquired immunodeficiency syndrome (AIDS)	Caused by the human immunodeficiency virus (HIV), which ultimately destroys the immune system (see chapter 14); transmitted through intimate sexual contact or by allowing infected body fluids into the interior of another person

ANSWER TO LEARN TO PREDICT

In this chapter, we learned that meiosis is cell division that produces haploid cells. When comparing meiosis in males and females, we find that the processes differ in several ways: the stage in an individual's life when meiosis begins, the types of cells produced, the number of functional cells produced with each cell division, and the stage of life when meiosis ceases to occur.

In males, meiosis begins at puberty in the seminiferous tubules. Spermatogonia give rise to primary spermatocytes, which will undergo the process of meiosis. During this process, each primary spermatocyte eventually gives rise to four equal-sized mature sperm cells.

Meiosis in females is more complex. The process actually begins before a female is even born. During fetal development, many of the oogonia in the ovaries degenerate. The remaining oogonia actually begin meiosis I and are called primary oocytes. At birth, the existing primary oocytes stop meiosis. After puberty and just before the ovulation of each oocyte, the primary oocyte that will be ovulated completes the first meiotic division to produce two different-sized cells: one large secondary oocyte and one small polar body. The secondary oocyte begins the second meiotic division but will complete the process only if fertilized by a sperm cell. In the case of fertilization, the secondary oocyte divides unevenly to form two cells. The smaller cell is another polar body and degenerates. In the larger cell, the haploid sperm nucleus combines with the haploid oocyte nucleus to form a zygote. So, in females, each primary oocyte produces only one large functional cell. The advantage of the size difference lies in the fact that sperm cells contribute only their DNA to the zygote; it is the oocyte that contributes cytoplasm and all the organelles to the zygote.

In addition, one final difference between male and female meiosis is that in males, meiosis can continue until death and for females, the process of meiosis stops at menopause.

Answers to the rest of this chapter's Predict questions are in Appendix E.

SUMMARY

19.1 Functions of the Reproductive System
(p. 529)

The reproductive system produces male and female gametes, enhances fertilization of an oocyte by a sperm, nurtures the new individual until birth (in the female), and produces reproductive hormones.

19.2 Formation of Gametes   (p. 530)

The reproductive organs in males and females produce gametes by meiosis.
1. Two consecutive cell divisions halve the chromosome number from 46 total chromosomes to 23 total chromosomes.
2. Meiosis forms male and female gametes.

19.3 Male Reproductive System   (p. 532)

Scrotum
1. The scrotum is a sac containing the testes.
2. The dartos and cremaster muscles help regulate testes temperature.

Testes

The testes are divided into lobules containing the seminiferous tubules and interstitial cells.

Spermatogenesis
1. Spermatogenesis begins in the seminiferous tubules at the time of puberty.
2. Sustentacular cells nourish the sperm cells and produce small amounts of hormones.
3. Spermatogonia divide (mitosis) to form primary spermatocytes.
4. Primary spermatocytes divide by meiosis to produce first secondary spermatocytes and then spermatids. The spermatids then mature to form sperm cells.
5. A spermatid develops a head, midpiece, and flagellum to become a sperm cell. The head contains the acrosome and the nucleus.

Ducts
1. The epididymis, a coiled tube system, is located on the testis and is the site of sperm maturation. Final changes, called capacitation of sperm cells, occur after ejaculation.
2. The seminiferous tubules lead to the rete testis, which opens into the efferent ductules that extend to the epididymis.
3. The ductus deferens passes from the epididymis into the abdominal cavity.
4. The ejaculatory duct is formed by the joining of the ductus deferens and the duct from the seminal vesicle.
5. The ejaculatory ducts join the prostatic urethra within the prostate gland.
6. The urethra extends from the urinary bladder through the penis to the outside of the body.

Penis
1. The penis consists of erectile tissue.
2. The two corpora cavernosa form the dorsum and the sides.
3. The corpus spongiosum forms the ventral portion and the glans penis, and it encloses the spongy urethra. The prepuce covers the glans penis.

Glands
1. The seminal vesicles empty into the ejaculatory duct.
2. The prostate gland consists of glandular and muscular tissue and empties into the urethra.
3. The bulbourethral glands empty into the urethra.

Secretions
1. Semen is a mixture of sperm cells and gland secretions.
2. The bulbourethral glands and the urethral mucous glands produce mucus that neutralizes the acidic pH of the urethra.
3. The testicular secretions contain sperm cells.
4. The seminal vesicle fluid contains nutrients, prostaglandins, and proteins that coagulate.
5. The prostate fluid contains nutrients and proteolytic enzymes, and it neutralizes the pH of the vagina.

19.4 Physiology of Male Reproduction   (p. 537)

Regulation of Reproductive Hormone Secretion
1. GnRH is produced in the hypothalamus and released in surges.
2. GnRH stimulates release of LH and FSH from the anterior pituitary.
3. LH stimulates the interstitial cells to produce testosterone.
4. FSH binds to sustentacular cells and stimulates spermatogenesis and secretion of inhibin.
5. Testosterone has a negative-feedback effect on GnRH, LH, and FSH secretion.
6. Inhibin has a negative-feedback effect on FSH secretion.

Puberty in Males
1. Before puberty, small amounts of testosterone inhibit GnRH release.
2. During puberty, testosterone does not completely suppress GnRH release, resulting in increased production of FSH, LH, and testosterone.

Effects of Testosterone
1. Testosterone causes enlargement of the genitals and is necessary for spermatogenesis.
2. Testosterone is responsible for the development of secondary sexual characteristics.

Male Sexual Behavior and the Male Sex Act
1. Testosterone is required for normal sex drive.
2. Stimulation of the sex act can be tactile or psychological.
3. Sensory impulses pass to the sacral region of the spinal cord.
4. Motor stimulation causes erection, mucus production, emission, and ejaculation.

Infertility in Males

The most common cause of male infertility is a low sperm cell count.

19.5 Female Reproductive System   (p. 541)

Ovaries, Oogenesis, and Fertilization
1. By the fourth month of development, the ovaries contain 5 million oogonia.
2. By birth, many oogonia have degenerated, and for the remaining oogonia meiosis has stopped in prophase I, causing them to become primary oocytes.
3. By puberty, 300,000 to 400,000 primary oocytes remain, of which about 400 will be released from the ovaries.
4. Ovulation is the release of an oocyte from an ovary. The first meiotic division is completed, and a secondary oocyte is released.
5. A sperm cell penetrates the secondary oocyte, the second meiotic division is completed, and the nuclei of the oocyte and sperm cell are united to complete fertilization.
6. A primordial follicle is a primary oocyte surrounded by a single layer of flat granulosa cells.

7. In primary follicles, the oocyte enlarges, and granulosa cells become cuboidal and form more than one layer. A zona pellucida is present.

8. In a secondary follicle, fluid-filled vesicles appear, and a theca forms around the follicle.

9. In a mature follicle, vesicles fuse to form an antrum, and the primary oocyte is surrounded by cumulus cells.

10. During ovulation, the mature follicle ruptures, releasing the secondary oocyte, surrounded by cumulus cells, into the peritoneal cavity.

11. The remaining granulosa cells in the follicle develop into the corpus luteum.

12. If fertilization occurs, the corpus luteum persists. If there is no fertilization, it degenerates.

Uterine Tubes

1. The ovarian end of the uterine tube is surrounded by fimbriae.
2. Cilia on the fimbriae move the oocyte into the uterine tube.
3. Fertilization usually occurs in the ampulla of the uterine tube, which is near the ovary.

Uterus

1. The uterus is a pear-shaped organ. The uterine cavity and the cervical canal are the spaces formed by the uterus.
2. The wall of the uterus consists of the perimetrium, or serous layer; the myometrium (smooth muscle); and the endometrium.

Vagina

1. The vagina connects the uterus (cervix) to the vestibule.
2. The vagina consists of a layer of smooth muscle and an inner lining of moist stratified squamous epithelium.
3. The wall of the vagina produces lubricating fluid.
4. The hymen covers the vestibular opening of the vagina in young females.

External Genitalia

1. The vestibule is a space into which the vagina and the urethra open.
2. The clitoris is composed of erectile tissue and contains many sensory receptors important in detecting sexual stimuli.
3. The labia minora are folds that cover the vestibule and form the prepuce.
4. The greater vestibular glands produce a mucous fluid.
5. The labia majora cover the labia minora, and the pudendal cleft is a space between the labia majora.
6. The mons pubis is an elevated area superior to the labia majora.

Mammary Glands

1. Mammary glands are the organs of milk production.
2. The mammary glands are modified sweat glands that consist of glandular lobes and adipose tissue.
3. The lobes connect to the nipple through ducts. The nipple is surrounded by the areola.
4. The female breast enlarges during puberty under the influence of estrogen and progesterone.

19.6 Physiology of Female Reproduction   (p. 548)

Puberty in Females

1. Puberty begins with the first menstrual bleeding (menarche).
2. Puberty begins when GnRH, LH, and FSH levels increase.

Menstrual Cycle

1. The cyclical changes in the uterus are controlled by estrogen and progesterone produced by the ovary.
2. Menses (from day 1 to day 4 or 5): Menstrual fluid is produced by degeneration of the endometrium.
3. Proliferative phase (from day 5 to day of ovulation): Epithelial cells multiply and form glands.
4. Secretory phase (from day of ovulation to day 28): The endometrium becomes thicker, and endometrial glands secrete. The uterus is prepared for implantation of the developing blastocyst by day 21.
5. Estrogen stimulates proliferation of the endometrium, and progesterone causes thickening of the endometrium. Decreased progesterone causes menses.
6. FSH initiates the development of the follicles.
7. Estrogen produced by the follicles stimulates GnRH, FSH, and LH secretion, and FSH and LH stimulate more estrogen secretion. This positive-feedback mechanism causes FSH and LH levels to increase near the time of ovulation.
8. LH stimulates ovulation and formation of the corpus luteum.
9. Estrogen and progesterone inhibit LH and FSH secretion following ovulation.
10. If fertilization does not occur, progesterone secretion by the corpus luteum decreases and menses begins.
11. If fertilization does occur, the corpus luteum continues to secrete progesterone and menses does not occur.

Menopause

The cessation of the menstrual cycle is called menopause.

Female Sexual Behavior and the Female Sex Act

1. Female sex drive is partially influenced by testosterone-like hormones produced by the adrenal cortex and estrogen produced by the ovary.
2. Autonomic nerves cause erectile tissue to become engorged with blood, the vestibular glands to secrete mucus, and the vagina to produce a lubricating fluid.

Infertility in Females

Causes of infertility in females include malfunctions of the uterine tubes, reduced hormone secretion from the pituitary or ovary, and interruption of implantation.

19.7 Effects of Aging on the Reproductive System   (p. 555)

1. Benign prostatic enlargement affects men as they age, and it blocks urine flow through the prostatic urethra.
2. Prostate cancer is more common in elderly men.
3. Menopause is the most common age-related change in females.
4. Cancers of the breast, the cervix, and the ovaries increase in elderly women.
5. Early detection is key to the successful treatment of most cancers.

REVIEW AND COMPREHENSION

1. List the functions of the male and female reproductive systems.

2. What is the scrotum? Explain the function of the dartos and cremaster muscles.

3. Where, specifically, are sperm cells produced in the testes? Describe the process of spermatogenesis.

4. Name the ducts the sperm cells traverse from their site of production to the outside of the body.

5. In which duct do sperm cells develop their ability to fertilize?

6. Name the erectile tissues of the penis, and describe how erectile tissue becomes erect.

7. State where the seminal vesicles, prostate gland, and bulbourethral glands empty into the male reproductive duct system.

8. Define emission and ejaculation.

9. Define semen. What structures give rise to secretions that make up the semen? Describe the composition of the secretions of each gland.

10. Describe where GnRH, FSH, LH, and testosterone are produced, and explain how their secretion is regulated.

11. Describe the effects of testosterone on males during puberty and in adulthood.

12. Describe the regulation of the male sex act.

13. Describe the process of follicle development and ovulation.

14. What is the corpus luteum? What happens to the corpus luteum if fertilization occurs? If fertilization does not occur?

15. Describe the normal pathway the oocyte follows after ovulation. Where does fertilization usually take place?

16. Describe the relationship among the uterus, vagina, vestibule, and external genitalia.

17. Describe the labia minora, the prepuce, the labia majora, the pudendal cleft, and the mons pubis.

18. What are the effects of estrogen and progesterone on the uterus?

19. Describe the hormonal changes that result in ovulation. Explain the sequence of events during each phase of the menstrual cycle.

20. Define menopause and female climacteric. What causes these changes?

21. List the major age-related diseases and conditions that occur in the male reproductive system.

22. List the major age-related changes that occur in the female reproductive system.

23. List the major age-related types of cancer that occur in the female reproductive system.

 CRITICAL THINKING

1. If an adult male were castrated by having his testes removed, what would happen to the levels of GnRH, FSH, LH, and testosterone in his blood?

2. Birth control pills for women contain estrogen and progesterone compounds. Explain how these hormones can prevent pregnancy.

3. During the secretory phase of the menstrual cycle, you would normally expect
   a. the highest levels of progesterone that occur during the menstrual cycle.
   b. a follicle present in the ovary that is ready to undergo ovulation.
   c. the endometrium to reach its greatest degree of development.
   d. a and b
   e. a and c

4. During approximately days 12–14 of the menstrual cycle, you would normally expect
   a. increasing blood levels of estrogen.
   b. increasing blood levels of LH.

   c. blood levels of progesterone to be near their maximum.
   d. a and b
   e. a, b, and c

5. On day 15 of the menstrual cycle, you would normally expect
   a. decreasing blood levels of LH.
   b. decreasing blood levels of estrogen.
   c. increasing blood levels of progesterone.
   d. a and b
   e. a, b, and c

6. Predict the consequences if a drug that blocks the effect of progesterone is taken by a woman 2 or 3 days following ovulation or by a woman who is pregnant.

7. During menopause, which reproductive hormones are reduced in the blood and which are increased?

Answers in Appendix D

20 Development, Heredity, and Aging

20.1 PRENATAL DEVELOPMENT

Learning Outcomes *After reading this section, you should be able to*

A. List the prenatal periods, and state the major developmental events associated with each.

B. Describe the process of fertilization.

C. Describe the blastocyst, the process of implantation, and placental formation.

D. Describe the maternal hormonal changes that occur during pregnancy.

E. List the three germ layers, describe their formation, and list the adult derivatives of each layer.

F. Describe the formation of the neural tube and the neural crest cells.

G. Describe the formation of the limbs, the face, and the digestive tract.

H. Explain how the single heart tube is divided into four chambers.

The human lifespan is usually considered the period between birth and death; however, the 9 months before birth are a critical part of existence. What happens in these 9 months profoundly affects the rest of a person's life. Although most people develop normally and are born without defects, approximately 3 of every 100 people are born with a birth defect so severe that it requires medical attention during the first year of life. Later in life, many more people discover previously unknown problems, such as the tendency to develop asthma, certain brain disorders, or cancer.

Module 14 Reproductive System

The **prenatal** (prē-nā′tăl; before birth) **period,** the period from conception to birth, can be divided into three parts: (1) During approximately the first 2 weeks of development, the primitive germ layers are formed; (2) from about the second to the eighth week of development, the major organ systems come into existence; and (3) during the last 7 months of the prenatal period, the organ systems grow and become more mature. Between the time of fertilization and 8 weeks of development, the developing human is called an **embryo** (em′brē-ō). From 8 weeks to birth, the developing human is called a **fetus** (fē′tus; offspring).

To calculate the **clinical age** of an unborn child, the medical community uses the mother's **last menstrual** (men′stroo-ăl) **period (LMP).** An embryo or a fetus is therefore considered to be a certain number of days post-LMP. Most embryologists, on the other hand, use **developmental age,** which begins with fertilization, to describe the timing of developmental events. Because fertilization is assumed to occur approximately 14 days after LMP, the developmental age is 14 days less than the clinical age. The times presented in this chapter are based on developmental age.

Fertilization

Fertilization is the union of a sperm cell and an oocyte, along with their genetic material (chromosomes), to produce a new individual. After sperm cells are ejaculated into the vagina, they are transported through the cervix and the body of the uterus to the uterine tubes, where fertilization occurs. The swimming ability of the sperm cells and the muscular contractions of the uterus and uterine tubes are responsible for the movement of sperm cells through the female reproductive tract. Both oxytocin released by the female posterior pituitary and prostaglandins within the semen stimulate contractions in the uterus and uterine tubes.

While passing through the uterus and the uterine tubes, the sperm cells undergo capacitation. **Capacitation** (kă-pas′i-tā′shŭn) makes the sperm cells capable of releasing the concentrated enzymes contained in the acrosome, a region of the sperm cell head. The enzymes digest a pathway through the cumulus cells and the zona pellucida of the **secondary oocyte** (ō′ō-sīt; egg cell). One sperm cell attaches to the oocyte cell membrane and enters the oocyte (figure 20.1, step 1).

The secondary oocyte is capable of being fertilized for up to about 1 day after ovulation, and some sperm cells remain viable in the female reproductive tract for up to 6 days, although most of them degenerate before that time.

Predict 2

During what days of the menstrual cycle is sexual intercourse most likely to result in pregnancy?

Hundreds of sperm cells reach the secondary oocyte, but normally a change in the oocyte cell membrane prevents more than one sperm cell from entering the secondary oocyte. The secondary oocyte undergoes the second meiotic division only after a sperm cell enters it (figure 20.1, step 2). After the second meiotic division, the oocyte nucleus moves to the center of the cell, where it meets the nucleus of the sperm cell. Each of these nuclei has 23 chromosomes. Their fusion, which completes the process of fertilization, produces a **zygote** (zī′gōt; having a yoke) that has 46 chromosomes (figure 20.1, step 3). The zygote develops into the embryo.

Early Cell Division

About 18–36 hours after fertilization, the zygote divides to form two cells (figure 20.2). Those two cells divide to form four cells, which divide to form eight, and so on. Even though the number of cells increases, the size of each cell decreases, so that the total mass of cells remains about the same size as the zygote. These cells have the ability to develop into a wide range of tissues. As a result, the total number of cells can be decreased, increased, or reorganized during this period without affecting normal development.

A CASE IN POINT

Twins

The parents of twin boys named the babies Juan and Hamall because, at birth, they looked so much alike that, if you've seen Juan, you've seen Hamall. As the twins grew older, they continued to look very much alike but not exactly the same. Their parents wanted to know if they were identical or fraternal twins, so they had an expensive genetic analysis performed. The results indicated that the twins were identical, despite their slight differences in appearance.

In rare cases, following early cell divisions, the cells separate and form two individuals, called "identical twins," or **monozygotic** (mon-ō-zī-got′ik) **twins.** Identical twins therefore have identical genetic information in their cells. Identical twins can also occur by other mechanisms a little later in development.

Occasionally, a woman can ovulate two or more secondary oocytes at the same time. Fertilization of multiple oocytes by different sperm cells results in "fraternal twins," or **dizygotic** (dī′zī-got′ik) **twins.**

Even though who we are and what we look like are determined in large part by genes, identical twins—although they have the same genes—may not look exactly alike. In fact, identical twins often look more like mirror images of each other. Genes interact with minute environmental cues in the embryo and in the child to determine the final form of the individual.

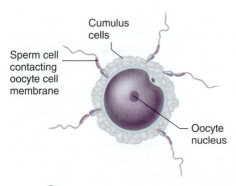

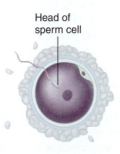

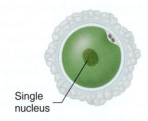

① Many sperm cells come in contact with the cumulus cells of the secondary oocyte, but only one sperm cell will penetrate the oocyte's cell membrane.

② The head of the sperm, carrying the genetic material, enters the oocyte, which completes the second meiotic division.

③ The two nuclei fuse to form a single nucleus. Fertilization is complete, and a zygote results.

Labels (figure 20.1): Cumulus cells; Sperm cell contacting oocyte cell membrane; Oocyte nucleus; Head of sperm cell; Single nucleus

PROCESS Figure 20.1 AP|R **Fertilization**

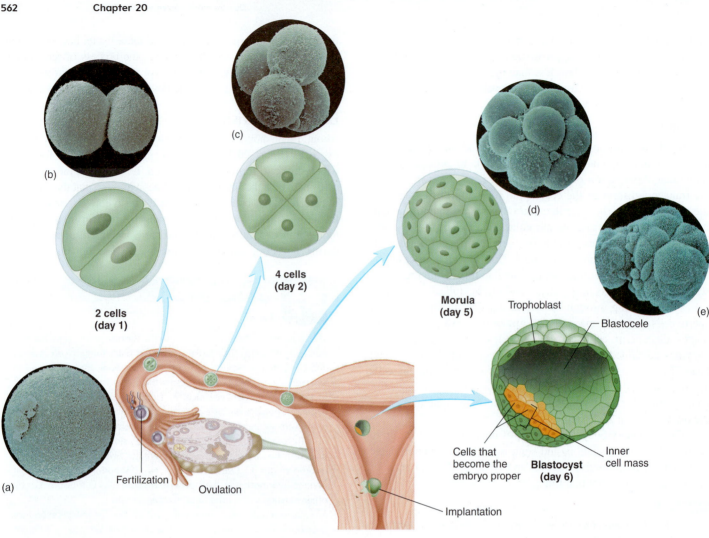

Figure 20.2 AP|R **Development of the Blastocyst and Implantation**

Successive cell divisions produce a multicellular morula by day 5, which becomes a hollow blastocyst on day 6. In the figure of the blastocyst, *green* cells are trophoblastic and *orange* cells are embryonic. (*a*) Zygote (120 µM in diameter with two polar bodies attached). (*b-e*) During the early cell divisions, the embryo divides into more and more cells, but the total size of the embryo remains relatively constant.

Blastocyst

After fertilization, multiple cell divisions have produced an embryonic mass of about 12–16 cells, called a **morula** (mōr′ū-lă; mulberry) (figure 20.2). Most of the cells of the morula will not form the embryo proper but will form support structures, such as the placenta.

When a cavity begins to appear within the mass of cells, the whole structure is called a **blastocyst** (blas′tō-sist) (figure 20.2). The fluid-filled cavity is called the **blastocele** (blas′tō-sēl). Most of the blastocele is surrounded by a single layer of cells, but at one end of the blastocyst, the cells are several layers thick. This thickened area is called the **inner cell mass.** Not all cells of the blastocyst give rise to the new individual. The embryo proper, which will become the new individual, will develop from only a few cells of the inner cell mass. These cells are commonly referred to as **stem cells** because they give rise to all the cell types within the body. The remaining cells of the blastocyst are called the **trophoblast** (trof′ō-blast, trō′fō-blast), which forms the embryonic part of the placenta and the membranes (chorion and amnion) surrounding the embryo.

Implantation of the Blastocyst and Development of the Placenta

All these early events, from the first cell division to formation of the blastocele, occur as the embryonic mass moves from the site of fertilization in the uterine tube to the site of **implantation** (im-plan-tā′shun) in the uterus. By 7 or 8 days after ovulation, the endometrium of the uterus is prepared for implantation. About 7 days after fertilization, the blastocyst attaches itself to the uterine wall and begins the process of implantation. The trophoblast cells of the blastocyst digest the uterine tissues as the blastocyst burrows into the uterine wall. Before implantation and for a short time afterward, the embryo is insensitive to environmental toxins. During the first few days of development, each cell has enough yolk to supply its own energy needs and requires few external nutrients. Furthermore, during the first couple of weeks of development, large numbers of cells can die, yet the embryo can fully recover.

As the blastocyst burrows into the uterine wall, trophoblast cells, called the **chorion** (kō′rē-on), form the embryonic portion of the **placenta** (plă-sen′tă), the organ of nutrient and waste product exchange between the embryo and the mother. Fingerlike projections, called **chorionic villi,** protrude into cavities formed within the maternal endometrium. Those cavities, called **lacunae** (lă-koo′nē), are filled with maternal blood (figure 20.3). In the mature placenta, the embryonic blood supply is separated from the maternal blood supply by the embryonic capillary wall, a basement membrane, and a thin layer of chorion. As a result, the embryonic blood and maternal blood do not mix. Nutrients and waste products must cross this semipermeable barrier between the two circulations.

Initially, the embryo is attached to the placenta by a connecting stalk. As the embryo matures, the connecting stalk elongates and becomes the **umbilical** (ŭm-bil′i-kăl; navel) **cord** (figure 20.3b). Within the umbilical cord, blood vessels carry blood from the embryo to the placenta and from the placenta to the embryo.

Maternal Hormonal Changes

The chorion secretes **human chorionic gonadotropin** (gō′nad-ō-trō′pin) **(hCG),** which travels in the blood to the maternal ovary and causes the corpus luteum to remain functional. The secretion of hCG begins shortly after implantation, increases rapidly, and reaches a peak about 8 or 9 weeks after fertilization. Subsequently, hCG levels decline to a lower level, where they are maintained throughout the remainder of the pregnancy (figure 20.4). Most pregnancy tests are designed to detect hCG in either urine or blood.

The estrogen and progesterone secreted by the corpus luteum (see chapter 19) are essential for maintaining the endometrium for the first 3 months of pregnancy. After the placenta forms, it also begins to secrete estrogen and progesterone. By the third month of pregnancy, the placenta has become an endocrine gland that secretes sufficient quantities of estrogen and progesterone to maintain pregnancy, and the corpus luteum is no longer needed. Estrogen and progesterone levels increase in the mother's blood throughout pregnancy.

Formation of the Germ Layers

After implantation, a new cavity, called the **amniotic** (am-nē-ot′ik) **cavity,** forms inside the inner cell mass and causes the part of the inner cell mass nearest the blastocele to separate as a flat disk of tissue called the **embryonic disk** (figure 20.5). The amniotic cavity is bounded by a membrane called the **amnion** and is filled with **amniotic fluid.** The embryo will grow in the amniotic cavity, where the amniotic fluid forms a protective cushion. The embryonic disk at first is composed of two layers of cells: an **epiblast** adjacent to the amniotic cavity and a **hypoblast** on the side of the disk opposite the amniotic cavity. A third cavity, the **yolk sac,** forms inside the blastocele from the hypoblast.

At about 14 days after fertilization, the embryonic disk has become a slightly elongated, oval structure. Some of the epiblast cells migrate toward the center of the disk, forming a thickened

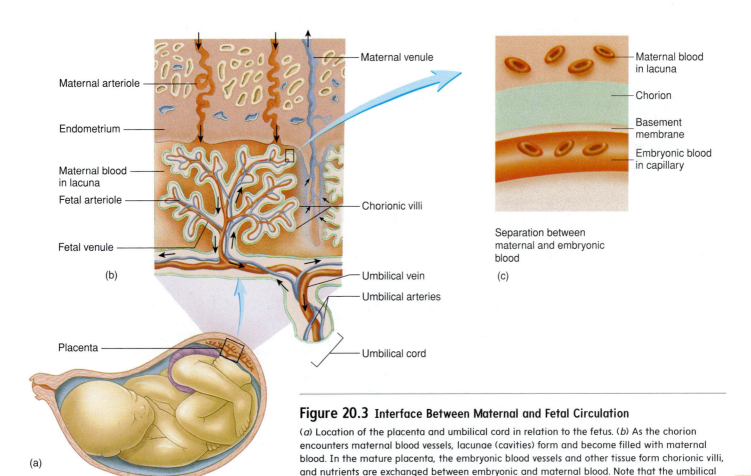

Maternal venule

Maternal arteriole

Endometrium

Maternal blood in lacuna

Fetal arteriole

Fetal venule

(b)

Chorionic villi

Umbilical vein

Umbilical arteries

Placenta

Umbilical cord

(a)

Maternal blood in lacuna

Chorion

Basement membrane

Embryonic blood in capillary

Separation between maternal and embryonic blood

(c)

Figure 20.3 Interface Between Maternal and Fetal Circulation

(*a*) Location of the placenta and umbilical cord in relation to the fetus. (*b*) As the chorion encounters maternal blood vessels, lacunae (cavities) form and become filled with maternal blood. In the mature placenta, the embryonic blood vessels and other tissue form chorionic villi, and nutrients are exchanged between embryonic and maternal blood. Note that the umbilical arteries and fetal arterioles contain deoxygenated blood and that the umbilical vein and fetal venules contain oxygenated blood. (*c*) Under normal conditions, the maternal and fetal blood are separated by the chorion and a basement membrane and do not mix.

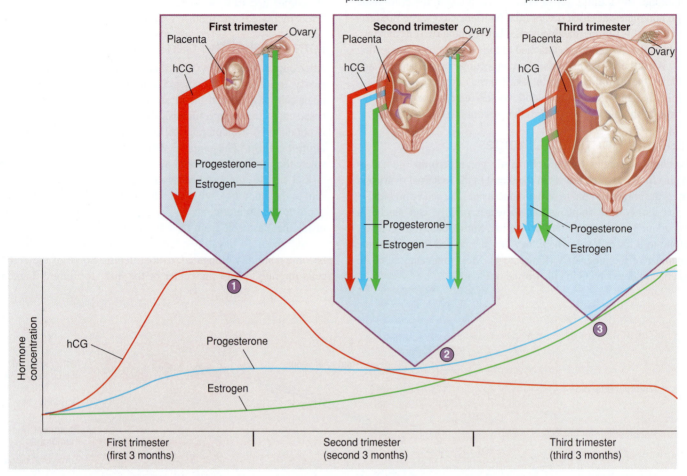

1 Human chorionic gonadotropin (*red line*) (hCG) increases until it reaches a maximum concentration near the end of the first 3 months of pregnancy and then decreases to a low level thereafter.

2 Progesterone (*blue line*) continues to increase until it levels off near the end of pregnancy. Early in pregnancy, progesterone is produced by the corpus luteum in the ovary; by the second trimester, production shifts to the placenta.

3 Estrogen (*green line*) increases slowly throughout pregnancy but increases more rapidly as the end of pregnancy approaches. Early in pregnancy, estrogen is produced only in the ovary; by the second trimester, production shifts to the placenta.

PROCESS Figure 20.4 Changes in Hormone Concentration During Pregnancy

During pregnancy, hCG, progesterone, and estrogen are secreted. The placenta secretes hCG. Early in pregnancy, the ovary secretes estrogen and progesterone. During midpregnancy, there is a shift toward estrogen and progesterone secretion by the placenta.

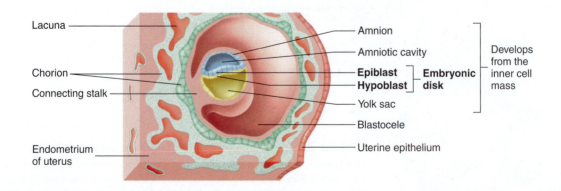

Figure 20.5 Early Embryo and Surrounding Structures in the Placenta

The embryonic disk consists of epiblast and hypoblast surrounded by the amniotic cavity and yolk sac. The connecting stalk, which attaches the embryo to the uterus, becomes part of the umbilical cord.

line called the **primitive streak** (figure 20.6). The formation of the primitive streak establishes the central axis of the embryo. Some of the epiblast cells migrate through the primitive streak. Of these migrating cells, some displace the hypoblast to form the **endoderm** (en'dō-derm; inside layer) while others emerge between the epiblast and endoderm as a new germ layer, called the **mesoderm** (mez'ō-derm; middle layer). Those epiblast cells that do not migrate form the **ectoderm** (ek'tō-derm; outside layer). This process of cell migration and the formation of three distinct germ layers

is called **gastrulation.** The embryo is now three-layered, having ectoderm, mesoderm, and endoderm (figure 20.6). All the tissues of an adult can be traced to these three germ layers (table 20.1).

From about day 14 until about day 35, the embryo is at maximum risk from environmental toxins and drugs that can cause birth defects. The causes of birth defects are a major unsolved issue in biology at present. However, it appears that oxidative damage to key molecules in certain developing cells and/or changes in the DNA function within those cells may be involved.

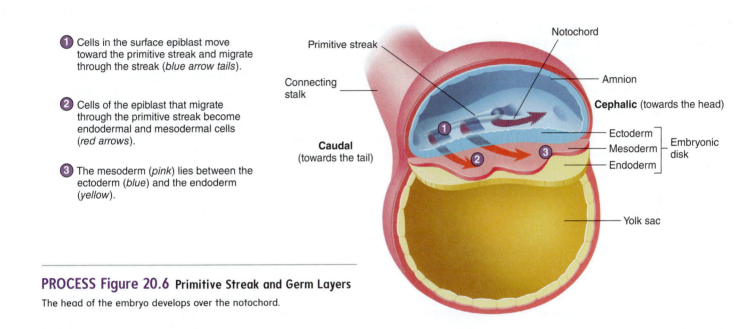

1. Cells in the surface epiblast move toward the primitive streak and migrate through the streak (*blue arrow tails*).

2. Cells of the epiblast that migrate through the primitive streak become endodermal and mesodermal cells (*red arrows*).

3. The mesoderm (*pink*) lies between the ectoderm (*blue*) and the endoderm (*yellow*).

PROCESS Figure 20.6 Primitive Streak and Germ Layers
The head of the embryo develops over the notochord.

TABLE 20.I	Tissues Derived from Each Germ Layer	
Endoderm	**Ectoderm**	**Mesoderm**
Lining of digestive tract	Epidermis of skin	Dermis of skin
Lining of lungs	Tooth enamel	Heart and blood vessels
Lining of hepatic, pancreatic, and other exocrine ducts	Lens and cornea of eye	Parenchyma (substance) of glands
Kidney ducts and urinary bladder	Nasal cavity	Kidneys
Thymus	Anterior pituitary	Gonads
Thyroid gland	Neuroectoderm	Muscle
Parathyroid glands	Brain and spinal cord	Bones (except facial)
Tonsils	Somatic motor neurons	Microglia
	Preganglionic autonomic neurons	
	Posterior pituitary	
	Neuroglial cells (except microglia)	
	Neural crest cells	
	Melanocytes	
	Sensory neurons	
	Postganglionic autonomic neurons	
	Adrenal medulla	
	Facial bones	
	Teeth (dentin, pulp, and cementum) and gingiva	
	A few skeletal muscles in head	

CLINICAL IMPACT   In Vitro Fertilization and Embryo Transfer

In a small number of women, normal pregnancy is not possible because of an anatomical or physiological condition. In 87% of these cases, the uterine tubes are incapable of transporting the zygote to the uterus or allowing sperm cells to reach the oocyte. Since 1978, in vitro fertilization and embryo transfer have made pregnancy possible in hundreds of such women. **In vitro fertilization (IVF)** involves removing secondary oocytes from a woman, placing the oocytes into a petri dish, and adding sperm cells to the dish, where fertilization and early development occur in vitro, which means "in glass." **Embryo transfer** involves removing the developing embryo from the petri dish and introducing the embryo into the uterus of a recipient woman.

For in vitro fertilization and embryo transfer to be accomplished, a woman is first injected with a substance similar to luteinizing hormone (LH), which causes more than one follicle to ovulate at a time. Just before the follicles rupture, the secondary oocytes are surgically removed from the ovary. The oocytes are then incubated in a dish and maintained at body temperature for 6 hours. Then, sperm cells are added to the dish. Different techniques may then be utilized that enhance sperm entry into the oocyte.

After 24–48 hours, several of the embryos are transferred to the uterus. Several embryos are introduced into the uterus to increase the success rate as much as possible, because only a few of them survive. The woman is usually required to lie perfectly still for

several hours after the embryos have been introduced into the uterus to prevent possible expulsion before implantation can occur. It is not fully understood why such expulsion does not occur in natural fertilization and implantation. Implantation and subsequent development then proceed in the uterus as they would for natural implantation.

The success rate of embryo transfer varies from clinic to clinic, with the age of the embryo at the time of transfer, and with the age of the patient. The current success rate for achieving pregnancy following IVF is 31%. Of these pregnancies, 83% result in live births. Multiple births have occurred frequently following embryo transfer because of the practice of introducing more than one embryo into the uterus.

A specialized group of cells at the cephalic (towards the head) end of the primitive streak moves from one end of the primitive streak to the other and, in some yet unknown way, organizes the embryo. A cordlike structure called the **notochord** (nō'tō-kōrd) is formed by these cells as they move down the primitive streak. The notochord marks the central axis of the developing embryo (figure 20.6).

Predict 3

Predict the result if two primitive streaks form in one embryonic disk. What if the two primitive streaks are touching each other?

Neural Tube and Neural Crest Formation

At about 18 days after fertilization, the ectoderm overlying the notochord thickens to form the **neural plate**. The lateral edges of the plate begin to rise like two ocean waves coming together. These edges are called the **neural folds,** and between them lies a **neural groove.** The neural folds begin to meet in the midline and fuse into a **neural tube** (figure 20.7). The cells of the neural tube are called **neuroectoderm** (noor-ō-ek'tō-derm) (table 20.1). Neuroectoderm becomes the brain, the spinal cord, and parts of the peripheral nervous system. The neural tube becomes completely closed by day 26. If the neural tube fails to close, major defects of the central nervous system can result.

Anencephaly (an'en-sef'ă-lē; no brain) is a birth defect wherein much of the brain fails to form because the neural tube did not close in the region of the head. A baby born with anencephaly cannot survive. **Spina bifida** (spī'nă bif'i-dă; split spine) is a general term describing defects of the spinal cord or vertebral column. Spina bifida can range from a simple defect with one or more vertebral spinous processes split or missing but no clinical manifestation to a more severe defect that can result in paralysis

of the limbs or the bowels and bladder, depending on where the defect occurs. More severe forms of spina bifida result from failure of the neural tube in the area of the spinal cord to close. It has been demonstrated that adequate amounts of the B vitamin folate, more commonly referred to as folic acid, in the diet during pregnancy can reduce the risk of such defects.

As the neural folds come together and fuse, a population of cells breaks away from the neuroectoderm all along the crests of the folds. Most of these **neural crest cells** become part of the peripheral nervous system or become melanocytes in the skin. In the head, neural crest cells also contribute to the skull, the dentin of teeth, blood vessels, and general connective tissue.

Formation of the General Body Structure

Arms and legs first appear at about 28 days after fertilization as **limb buds** (figure 20.8) and quickly begin to elongate. At about 35 days, expansions called hand and foot plates form at the ends of the limb buds. Zones of cell death between the future fingers and toes of the hand and foot plates help sculpt the fingers and toes.

The face develops by fusion of five growing masses of tissue, called **processes** (figure 20.9). One, the **frontonasal process,** forms the forehead, nose, and center of the upper jaw and lip. Two **maxillary processes** form the maxillae (upper lip and jaw), and two **mandibular processes** form the mandible (lower lip and jaw).

The nose begins as two structures, one on each side of the forehead mass. As the brain enlarges and the face matures, the two parts of the nose approach each other in the midline and fuse (figure 20.9). The two masses forming the upper jaw expand toward the midline and fuse with part of the nose to form the upper jaw and lip. A **cleft lip** results from failure of these structures to fuse. Cleft lips usually do not occur in the midline, but

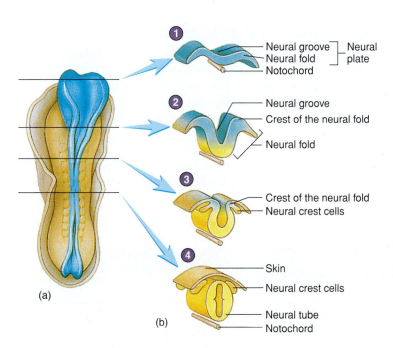

(a)

(b)

Neural groove ⎤ Neural
Neural fold ⎦ plate
Notochord

Neural groove
Crest of the neural fold
Neural fold

Crest of the neural fold
Neural crest cells

Skin
Neural crest cells
Neural tube
Notochord

1 The neural plate forms from ectoderm.

2 Neural folds form as parallel ridges along the embryo.

3 Neural crest cells begin to form from the crest of the neural folds.

4 The neural folds meet at the midline to form the neural tube, and neural crest cells separate from the neural folds.

PROCESS Figure 20.7 Formation of the Neural Tube

The neural folds, which consist of neuroectoderm, come together at the midline and fuse to form a neural tube. This fusion begins in the center and moves both cranially and caudally. (*a*) The embryo shown is about 21 days after fertilization. (*b*) These cross sections represent progressive closure of the neural tube.

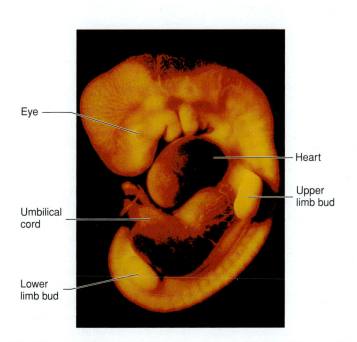

Eye

Umbilical cord

Lower limb bud

Heart

Upper limb bud

Figure 20.8 Human Embryo 35 Days After Fertilization

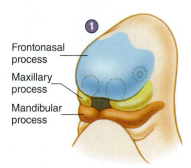

1 **28 days after fertilization**
The face develops from five processes: frontonasal (*blue*), two maxillary (*yellow*), and two mandibular (*orange;* already fused).

Frontonasal process
Maxillary process
Mandibular process

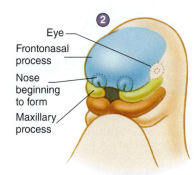

2 **33 days after fertilization**
Nasal placodes, areas of thickening, appear in the frontonasal process.

Eye
Frontonasal process
Nose beginning to form
Maxillary process

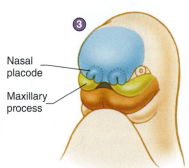

3 **40 days after fertilization**
Maxillary processes extend toward the midline. The nasal placodes also move toward the midline and fuse with the maxillary processes to form the jaw and lip.

Nasal placode
Maxillary process

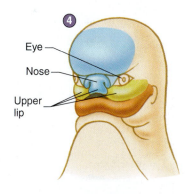

4 **48 days after fertilization**
Continued growth brings structures more toward the midline.

Eye
Nose
Upper lip

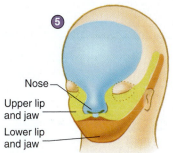

5 **14 weeks after fertilization**
Colors show the contributions of each process to the adult face.

Nose
Upper lip and jaw
Lower lip and jaw

PROCESS Figure 20.9 Development of the Face

to one side (or both sides). The cleft can vary in severity from a slight indentation in the lip to a fissure that extends from the mouth to the nares (nostril).

The roof of the mouth, or palate, begins to form as vertical shelves of tissue that grow on the inside of the maxillary masses. These shelves swing to a horizontal position and begin to fuse with each other at about 56 days of development. If the palate does not fuse, a midline cleft in the roof of the mouth called a **cleft palate** results. A cleft palate can range in severity from a slight cleft of the uvula to a fissure extending the entire length of the palate. A cleft lip and cleft palate can occur together, forming a continuous fissure.

Development of the Organ Systems

The major organ systems appear and begin to develop during the embryonic period (second to eighth week of development). This period is therefore also called the period of **organogenesis** (ōr′gă-nō-jen′ĕ-sis). The individual organ systems are listed in table 20.2; only general comments about a few select systems are presented in the text.

While the neural tube is forming (18–26 days), the remainder of the embryo is folding to form a tube along the upper part of the yolk sac (figure 20.10). The developing digestive tract pinches off from the yolk sac as a tube but remains attached in the center to the yolk sac by a yolk stalk.

TABLE 20.2	Development of the Organ Systems					
	Age (Days Since Fertilization)					
	1–5	6–10	11–15	16–20	21–25	26–30
General Features	Fertilization, blastocyst	Blastocyst implants.	Primitive streak, three germ layers	Neural plate	Neural tube closed	Limb buds and other "buds" appear.
Integumentary System			Ectoderm, mesoderm		Melanocytes form from neural crest.	
Skeletal System			Mesoderm		Neural crest cells (form facial bones)	Limb buds
Muscular System			Mesoderm	Somites (body segments) begin to form.		Somites are all formed.
Nervous System			Ectoderm	Neural plate	Neural tube complete; neural crest forms; eyes and ears begin to form.	Lens begins to form.
Endocrine System			Ectoderm, mesoderm, endoderm	Thyroid begins to develop.		Parathyroid glands and pancreas appear.
Cardiovascular System			Mesoderm	Blood islands form; two-tubed heart forms.	Single-tubed heart begins to beat.	Interatrial septum forms.
Lymphatic System			Mesoderm			Thymus appears.
Respiratory System			Mesoderm, endoderm		Diaphragm begins to form.	Trachea, lung buds appear.
Digestive System			Mesoderm, endoderm		Tooth dentin forms; foregut and hindgut form.	Liver and pancreas appear as buds; tongue bud appears.
Urinary System			Mesoderm, endoderm		Embryonic kidneys appear.	Embryonic kidneys elongate.
Reproductive System			Mesoderm, endoderm		Primordial germ cells form on yolk sac.	Male reproductive ducts appear; external genital structures begin to form.

A considerable number of outpocketings appear at about 28 days after fertilization along the entire length of the digestive tract (figure 20.11). A surprisingly large number of important internal organs develop from those outpocketings, including the auditory tubes, tonsils, thymus, anterior pituitary gland, thyroid gland, parathyroid glands, lungs, liver, pancreas, and urinary bladder.

The heart develops from two blood vessels, which lie side by side in the early embryo and fuse about 21 days after fertilization into a single, midline heart (figure 20.12, steps 1 and 2). At about this time, the primitive heart begins to beat. Blood vessels form from "blood islands" on the surface of the yolk sac and inside the embryo. These islands expand and fuse to form the circulatory system.

The major chambers of the heart, the atrium and ventricle, expand rapidly. The single ventricle is subdivided into two chambers by the development of an **interventricular** (in-ter-ven-trik′ū-lăr) **septum** (figure 20.12, steps 3 and 4). If the interventricular septum does not grow enough to completely separate the ventricles, a ventricular septal defect results.

An **interatrial** (in-ter-ā′trē-ăl) **septum** forms to separate the two atria (figure 20.12, steps 3–5). An opening in the interatrial septum called the **foramen ovale** (ō-val′ē) connects the two atria and allows blood to flow from the right to the left atrium in the fetus. Because of the foramen ovale, some blood in the fetus passes from the right atrium to the left atrium and bypasses the

Age (Days Since Fertilization)					
31–35	**36–40**	**41–45**	**46–50**	**51–55**	**56–60**
Hand and foot plates on limbs	Fingers and toes appear; lips form; embryo 15 mm	External ear forming; embryo 20 mm	Embryo 25 mm	Limbs elongate to adult proportions; embryo 35 mm	Face is distinctly human in appearance.
Sensory receptors appear in skin.		Collagen fibers are clearly present in skin.		Extensive sensory nerve endings in skin	
Mesoderm condensation in areas of future bone	Cartilage in site of future humerus	Cartilage in site of future ulna and radius	Cartilage in site of future hand and fingers		Ossification begins in clavicle and then in other bones.
Muscle precursor cells enter limb buds.			Functional muscle		Nearly all muscles appearing in an adult form
Nerve processes enter limb buds.		External ear forming; olfactory nerves begin to form.		Semicircular canals in inner ear are complete.	Eyelids form; cochlea in inner ear is complete.
Pituitary gland appears as evaginations from brain and mouth.	Gonads begin to form; adrenal glands form.		Pineal body appears.	Thyroid gland in adult position	Anterior pituitary gland loses its connection to mouth.
Interventricular septum begins to form.		Interventricular septum is complete.	Interatrial septum is complete but foramen ovale remains until birth.		
Large lymphatic vessels form in neck.	Spleen appears.			Adult lymph pattern is formed.	
Secondary bronchi to lobes form.	Tertiary bronchi to lobules form.		Tracheal cartilage begins to form.		
Mouth opens to outside.		Palate begins to form; tooth buds begin to form.			Palate begins to fuse (fusion complete by 90 days); anus opens.
Adult kidneys begin to develop.				Embryonic kidneys degenerate.	
	Gonads begin to form.	Primordial germ cells enter gonads.	Female reproductive ducts appear.		Uterus is forming; external genitalia begin to differentiate in male and female.

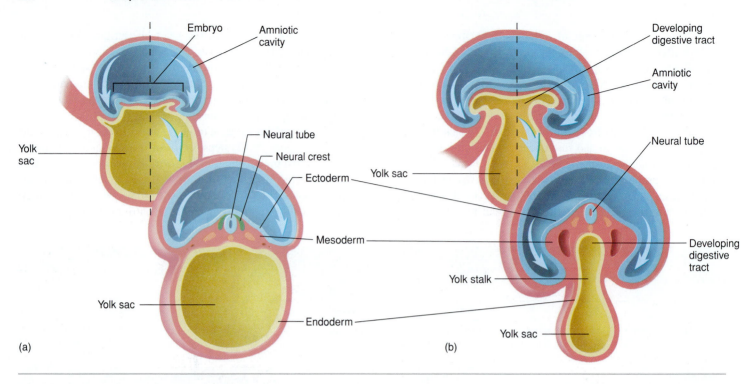

Figure 20.10 Development of the Digestive Tract

The digestive tract develops along the dorsal side of the yolk sac (*yellow*) as the body folds into a tube (*blue arrows*). The figures in back are shown in sagittal section. The figures in front are shown in cross section. The *dashed line* on the figures in back shows the plane of section in the figures in front. (*a*) An early embryo (about 24 days). (*b*) A slightly older embryo (about 28 days).

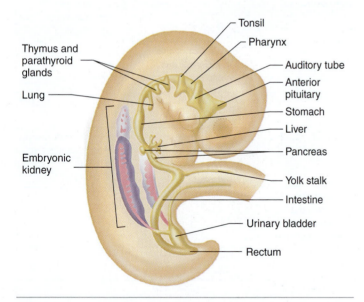

Figure 20.11 Embryonic Digestive and Urinary Systems

Outpocketings of the digestive tract (*yellow*), form many adult structures, such as the lungs and glands. The embryonic kidney is also shown (*purple*).

right ventricle and the lungs. The foramen ovale normally closes off at the time of birth, and blood then circulates through the right ventricle and the lungs. If this does not occur, an interatrial septal defect occurs. An interatrial septal defect or a ventricular septal defect usually results in a heart murmur.

The kidneys develop from mesoderm located along the lateral wall of the body cavity (see figure 20.11). The embryonic kidneys are much more extensive than the adult kidneys, extending the entire length of the body cavity. They are closely associated with internal reproductive organs, such as the ovaries or testes, and reproductive ducts, such as the uterine tubes or ductus deferens. Most of the embryonic kidneys degenerate, with only a very small part forming the adult kidney.

Growth of the Fetus

The embryo becomes a fetus about 8 weeks after fertilization (figure 20.13). The beginning of the fetal period is marked by the beginning of bone ossification. In the embryo, most of the organ systems are developing, whereas in the fetus the organs are present. During the fetal period, the organ systems enlarge and mature. The fetus grows from about 3 cm and 2.5 g (0.09 oz) at 8 weeks to 50 cm and 3300 g (7 lb, 4 oz) at the end of pregnancy. The growth during the fetal period represents more than a 15-fold increase in length and a 1400-fold increase in weight.

The amniotic fluid contains toxic waste products from the fetus's digestive tract and kidneys. Fine, soft hair called **lanugo** (lă-noo′gō) covers the fetus, and a waxy coat of loose epithelial cells called **vernix caseosa** (ver′niks kā′sē-ō′să) forms a protective layer between the fetus and the amniotic fluid.

Subcutaneous adipose tissue that accumulates in the fetus provides a nutrient reserve, helps insulate, and aids the newborn in sucking by strengthening and supporting the cheeks, so that a small vacuum can be developed in the oral cavity.

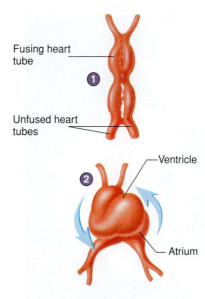

1 20 days after fertilization
At this age, the heart consists of two parallel tubes that will fuse into a single, midline heart.

Fusing heart tube

Unfused heart tubes

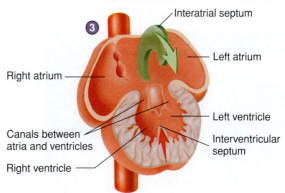

2 22 days after fertilization
The two parallel tubes have fused to form one tube. This tube bends as it elongates (*blue arrows* suggest the direction of bending) within the confined space of the pericardium.

Ventricle

Atrium

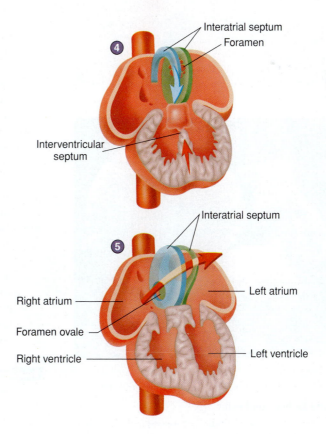

3 31 days after fertilization
The interatrial septum (*green*) and the interventricular septum grow toward the center of the heart.

Interatrial septum

Left atrium

Right atrium

Left ventricle

Canals between atria and ventricles

Interventricular septum

Right ventricle

4 35 days after fertilization
The interventricular septum is nearly complete. A foramen, which will become the left side of the foramen ovale, opens in the left side of the interatrial septum (*green*) as the right side of the interatrial septum begins to form (*blue*).

Interatrial septum

Foramen

Interventricular septum

5 Final embryonic condition of the interatrial septum
A foramen remains in the right side of the interatrial septum (*blue*), which forms the right part of the foramen ovale. Blood from the right atrium can flow through the foramen ovale into the left atrium. After birth, as blood begins to flow in the other direction, the left side of the interatrial septum is forced against the right side, closing the foramen ovale.

Interatrial septum

Right atrium

Left atrium

Foramen ovale

Right ventricle

Left ventricle

PROCESS Figure 20.12 Formation of the Heart

Reproductive

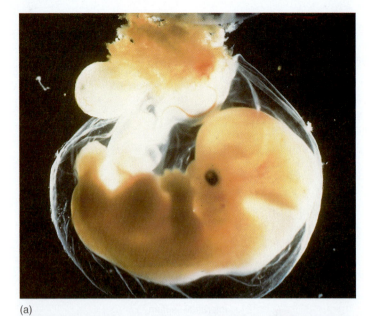

(a)

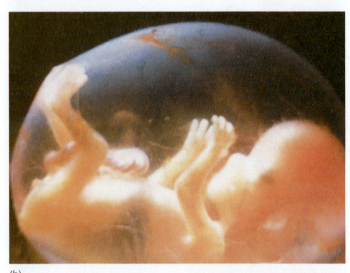

(b)

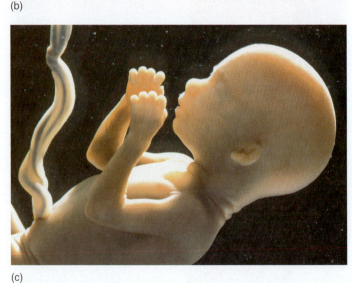

(c)

Figure 20.13 Late Embryo and Fetus

(*a*) Embryo at 50 days of development. (*b*) Fetus at 3 months of development. (*c*) Fetus at 4 months after fertilization.

Peak body growth occurs late in gestation, but as the placenta reaches its maximum size, the oxygen and nutrient supply to the fetus becomes limited. Growth of the placenta essentially stops at about 35 weeks, limiting fetal growth.

At approximately 38 weeks of development, the fetus is ready to be delivered. The average weight at this point is 3250 g (7 lb, 2 oz) for a female fetus and 3300 g (7 lb, 4 oz) for a male fetus.

20.2 PARTURITION

Learning Outcome   *After reading this section, you should be able to*

A. Explain the events that occur during parturition.

Physicians usually calculate the **gestation** (jes-tā′shŭn) **period** (length of pregnancy) as 280 days (40 weeks, or 10 lunar months) from the LMP to the date of delivery of the fetus. **Parturition** (par-toor-ish′ŭn) is the process by which the baby is born (figure 20.14). Near the end of pregnancy, the uterus becomes progressively more excitable and usually exhibits occasional contractions that become stronger and more frequent until parturition is initiated. The cervix gradually dilates, and strong uterine contractions help expel the fetus from the uterus through the vagina.

Labor is the period during which uterine contractions occur that result in expulsion of the fetus. Although labor may differ greatly from woman to woman and from one pregnancy to another for the same woman, it can usually be divided into three stages.

1. The **first stage** of labor, often called the *dilation phase,* begins with the onset of regular uterine contractions and extends until the cervix dilates to a diameter about the size of the fetus's head (10 cm) (figure 20.14). This stage takes approximately 24 hours, but it may be as short as a few minutes in some women who have had more than one child. During this phase, the amnion surrounding the fetus ruptures, and amniotic fluid flows through the vagina to the exterior. This event is commonly referred to as the "water breaking" and usually occurs naturally, but the amnion may need to be ruptured artificially.
2. The **second stage** of labor, often called the *expulsion phase,* lasts from the time of maximum cervical dilation until the time the baby exits the vagina. This stage may last from 1 minute to 1 hour or more. During this stage, contraction of the woman's abdominal muscles assists the uterine contractions.
3. The **third stage** of labor, often called the *placental stage,* involves the expulsion of the placenta from the uterus. Contractions of the uterus cause the placenta to tear away from the wall of the uterus. Some bleeding from the uterine wall occurs because of the intimate contact between the placenta and the uterus. However, bleeding is normally limited because uterine smooth muscle contractions compress the blood vessels.

Predict 4

Compare and contrast clinical age and developmental age for fertilization, implantation, the beginning of the fetal period, and parturition.

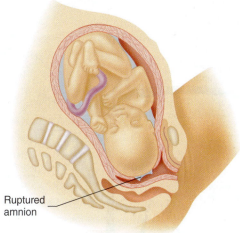

1 **First stage.** The cervix dilates, and the amnion ruptures.

Ruptured amnion

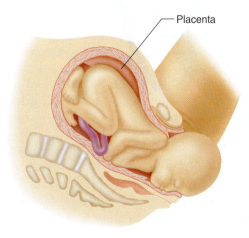

Placenta

2 **Second stage.** The fetus is expelled from the uterus.

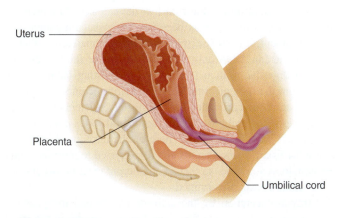

Uterus

Placenta

Umbilical cord

3 **Third stage.** The placenta is expelled.

PROCESS Figure 20.14 **Parturition**

During the 4 or 5 weeks following parturition, the uterus becomes much smaller, but it remains somewhat larger than it was before pregnancy. A vaginal discharge composed of small amounts of blood and degenerating endometrium can persist for several weeks after parturition.

The precise signal that triggers parturition is unknown, but many factors that support it have been identified (figure 20.15).

Before parturition, the progesterone concentration in the mother's blood has reached its highest level. Progesterone has an inhibitory effect on uterine smooth muscle cells. However, estrogen levels continually increase in the maternal circulation, exciting uterine smooth muscle. Thus, the inhibitory influence of progesterone on smooth muscle is overcome by the stimulatory effect of estrogen near the end of pregnancy.

The fetus also plays a role in stimulating parturition. Stress on the fetus triggers the secretion of a releasing hormone from the fetal hypothalamus, which in turn causes adrenocorticotropic hormone (ACTH) to be released from the fetal anterior pituitary. ACTH stimulates the fetal adrenal gland to secrete hormones from the adrenal cortex that reduce progesterone secretion, increase estrogen secretion, and increase prostaglandin production by the placenta. Prostaglandins strongly stimulate uterine contractions.

During parturition, stretching of the uterine cervix initiates nervous reflexes that cause the mother's posterior pituitary gland to release oxytocin. Oxytocin stimulates uterine contractions, which move the fetus farther into the cervix, causing further stretch. A positive-feedback mechanism is established. This positive-feedback mechanism stops after delivery, when the cervix is no longer stretched.

20.3 THE NEWBORN

Learning Outcome *After reading this section, you should be able to*

A. Discuss the respiratory, circulatory, and digestive changes that occur in the newborn at the time of birth.

The newborn, or **neonate** (nē′ō-nāt; newborn), experiences several dramatic changes at the time of birth. The major and earliest changes are the separation of the infant from the maternal circulation and the transfer from a fluid to a gaseous environment.

Respiratory and Circulatory Changes

The large, forced gasps of air that occur when an infant cries at the time of delivery help inflate the lungs. The fetal lungs produce a substance called **surfactant** (ser-fak′tănt), which coats the inner surface of the alveoli, reduces surface tension in the lungs, and allows the newborn lungs to inflate (see chapter 15).

Surfactant is not manufactured in the fetal lungs before about 6 months after fertilization. If a fetus is born before the lungs can produce surfactant, the surface tension inside the lungs is too great for the lungs to inflate. Under these conditions, the newborn may die of respiratory distress. Therefore, premature newborns are treated with bovine or synthetic surfactant.

The initial inflation of the lungs causes important changes in the cardiovascular system (figure 20.16). Expansion of the lungs reduces the resistance to blood flow through the lungs, resulting in increased blood flow from the right ventricle of the heart through the pulmonary arteries. Consequently, an increased amount of blood flows from the right atrium to the right ventricle and into the pulmonary arteries, and less blood flows from the right atrium through the foramen ovale to the left atrium. The reduced resistance to blood flow through the lungs and the increasing volume of blood returning from the lungs through the pulmonary veins to the left atrium make the pressure in the left atrium greater than that in the right atrium. This pressure difference forces blood against the interatrial septum, closing a flap of tissue that develops in that region over the foramen

1. The fetal hypothalamus secretes a releasing hormone that stimulates adrenocorticotropic hormone (ACTH) secretion from the pituitary. The fetal pituitary secretes ACTH in greater amounts near parturition.

2. ACTH causes the fetal adrenal gland to secrete greater quantities of adrenal cortical steroids.

3. Adrenal cortical steroids travel in the umbilical blood to the placenta.

4. In the placenta, the adrenal cortical steroids cause progesterone synthesis to level off and estrogen and prostaglandin synthesis to increase, making the uterus more excitable.

5. The stretching of the uterus produces action potentials that are transmitted to the brain through ascending pathways.

6. Action potentials stimulate the secretion of oxytocin from the mother's posterior pituitary.

7. Oxytocin causes the uterine smooth muscle to contract.

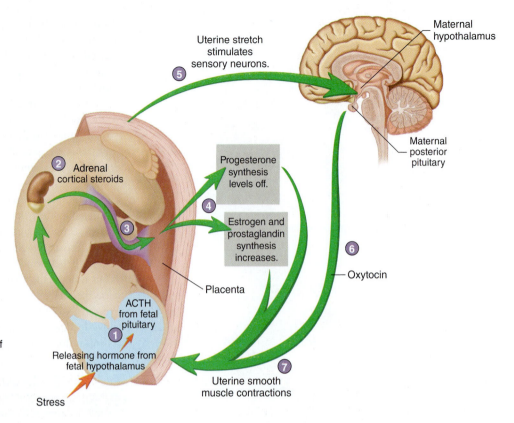

PROCESS Figure 20.15 Factors That Influence Parturition

Although the precise control of parturition in humans is unknown, these changes appear to play a role.

---

ovale. This action completes the separation of the heart into two pumps: the right side and the left side of the heart.

A short artery, the **ductus arteriosus** (ar-tēr′ē-ō-sŭs), connects the pulmonary trunk to the aorta. Before birth, the ductus arteriosus carries blood from the pulmonary trunk to the aorta, bypassing the fetal lungs. This artery closes off shortly after birth, forcing blood to flow through the lungs.

Also before birth, the deoxygenated fetal blood passes to the placenta through **umbilical** (ŭm-bil′i-kăl) **arteries,** which originate in the internal iliac arteries. As the blood passes through the placenta, nutrient and waste exchange occurs between the fetal blood and the maternal blood. Oxygenated fetal blood then returns to the fetus through an **umbilical vein.** The umbilical vein passes through the liver but bypasses the sinusoids of the liver by way of the **ductus venosus** (vē-nō′sŭs) and joins the inferior vena cava. When the umbilical cord is tied and cut, no more blood flows through the umbilical vein and arteries, and they degenerate. The remnant of the umbilical vein becomes the round ligament of the liver.

Digestive Changes

Late in gestation, the fetus swallows amniotic fluid from time to time. Shortly after birth, this swallowed fluid plus cells sloughed from the mucosal lining of the digestive tract, mucus produced by intestinal mucous glands, and bile from the liver are eliminated as a greenish anal discharge called **meconium** (mē-kō′nē-ŭm).

After birth, the neonate is suddenly separated from its source of nutrients, the maternal circulation. Because of this separation

and the shock of birth, the neonate usually loses 5–10% of its total body weight during the first few days of life. Although the digestive system of the fetus becomes somewhat functional late in development, it is still very immature in comparison to that of the adult and can digest only a limited number of food types. The newborn digestive system is capable of digesting lactose (milk sugar) from the time of birth. The pancreatic secretions are sufficiently mature for a milk diet, but the digestive system only gradually develops the ability to digest more solid foods over the first year or two. New foods should therefore be introduced gradually during the first 2 years. Parents are also advised to introduce only one new food at a time, so that, if an allergic reaction occurs, the cause is more easily determined.

Amylase secretion by the salivary glands and the pancreas remains low until after the first year. Lactase activity in the small intestine is high at birth but declines during infancy, although the levels still exceed those in adults. In many adults, lactase activity is lost, and an intolerance to milk develops.

20.4 LACTATION

Learning Outcome *After reading this section, you should be able to*

A. Describe the events of lactation.

Lactation (lak-tā′shŭn) is the production of milk by the mammary glands (figure 20.17). It normally occurs in women following parturition and may continue for up to 2 or 3 years.

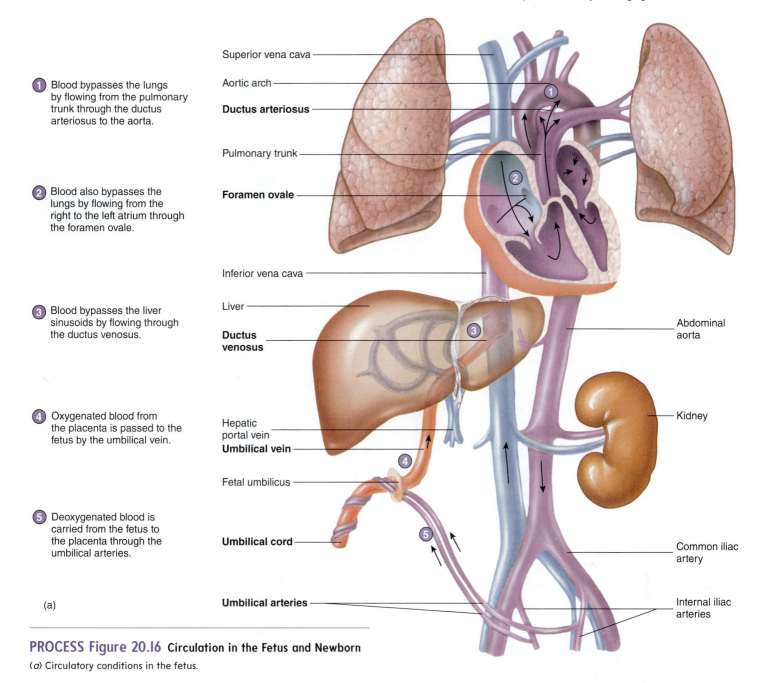

1 Blood bypasses the lungs by flowing from the pulmonary trunk through the ductus arteriosus to the aorta.

2 Blood also bypasses the lungs by flowing from the right to the left atrium through the foramen ovale.

3 Blood bypasses the liver sinusoids by flowing through the ductus venosus.

4 Oxygenated blood from the placenta is passed to the fetus by the umbilical vein.

5 Deoxygenated blood is carried from the fetus to the placenta through the umbilical arteries.

Superior vena cava

Aortic arch

Ductus arteriosus

Pulmonary trunk

Foramen ovale

Inferior vena cava

Liver

Ductus venosus

Hepatic portal vein

Umbilical vein

Fetal umbilicus

Umbilical cord

(a)

Umbilical arteries

Abdominal aorta

Kidney

Common iliac artery

Internal iliac arteries

PROCESS Figure 20.16 Circulation in the Fetus and Newborn

(*a*) Circulatory conditions in the fetus.

During pregnancy, the high concentration and continuous presence of estrogen and progesterone cause expansion of the duct system and the secretory units within the breast. Other hormones, including a prolactin-like hormone produced by the placenta, help support the development of the breasts. Also, additional adipose tissue is deposited; thus, the size of the breasts increases throughout pregnancy. Estrogen and progesterone prevent the secretory part of the breast from producing milk during pregnancy.

Blood levels of estrogen and progesterone fall dramatically after parturition. Once the placenta has been dislodged from the uterus, the source of these hormones is gone. After parturition, in the absence of estrogen and progesterone, prolactin produced by the anterior pituitary stimulates milk production. During suckling, sensory action potentials are sent from the nipple to the brain, stimulating the release of prolactin from the anterior pituitary

(figure 20.17). For the first few days following parturition, the mammary glands secrete **colostrum** (kō-los′trŭm), a high-protein material that contains many antibodies. Although colostrum is high in proteins, it contains little fat and less lactose than milk. Eventually, milk with a higher fat and lactose content is produced. Colostrum and milk provide nutrition and antibodies that help protect the baby from infections.

At the time of breastfeeding, milk contained in the alveoli and ducts of the breast is forced out of the breast by contractions of cells surrounding the alveoli. Suckling produces action potentials that are carried to the hypothalamus, where they cause the release of oxytocin from the posterior pituitary (figure 20.17). Oxytocin stimulates cells surrounding the alveoli to contract. As a result, milk flows from the breasts, a process called **milk letdown.** Higher brain centers can also cause the release of oxytocin as a

① When air enters the lungs, blood is forced through the pulmonary arteries to the lungs. The ductus arteriosus closes (*gray*).

② The foramen ovale closes and becomes the fossa ovalis. Blood can no longer flow from the right to the left atrium.

③ The ductus venosus degenerates and becomes the ligamentum venosum (*gray*).

④ The umbilical arteries and vein are cut. The umbilical vein becomes the round ligament of the liver (*gray*).

⑤ The umbilical arteries also degenerate (*gray*).

(b)

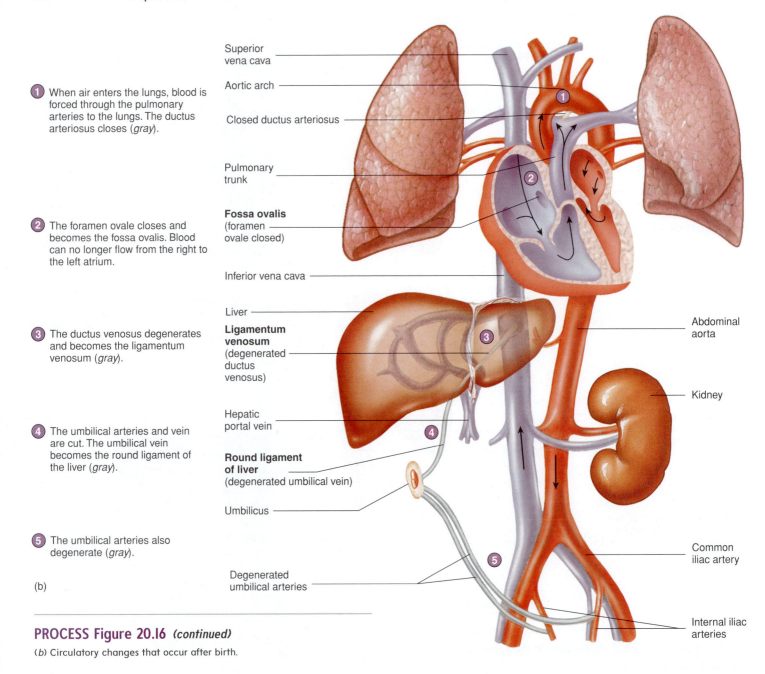

PROCESS Figure 20.16 *(continued)*

(*b*) Circulatory changes that occur after birth.

result of a conditioned reflex in response to such stimuli as hearing an infant cry or thinking about breastfeeding.

Repeated stimulation of prolactin release makes breastfeeding possible for several years. If breastfeeding is stopped, prolactin release stops, and within a few days the breasts' ability to respond to prolactin is lost, and milk production ceases.

Predict 5

While breastfeeding her baby, a woman felt cramps in her abdomen. Explain what was happening.

20.5 FIRST YEAR FOLLOWING BIRTH

Learning Outcome   *After reading this section, you should be able to*

A. Describe the changes that occur during the first year after birth.

A great number of changes occur in the infant from the time of birth until 1 year of age. The time when these changes occur may vary considerably from child to child, and the dates given are only rough estimates. The brain is still developing at this time, and much of what the infant can accomplish depends on the amount of brain development achieved. It is estimated that the newborn's central nervous system contains nearly all the adult number of neurons, but subsequent growth and maturation of the brain add new neuroglial cells, new myelin sheaths, and new connections between neurons, which may continue throughout life.

By 6 weeks, the baby is usually able to hold up her head when placed in a prone position and begins to smile in response to people or objects. At 3 months of age, the infant exercises his limbs aimlessly. However, he can control his arms and hands enough that voluntary thumb sucking can occur. The infant can follow a

PROCESS Figure 20.17 AP|R Hormonal Control of Lactation

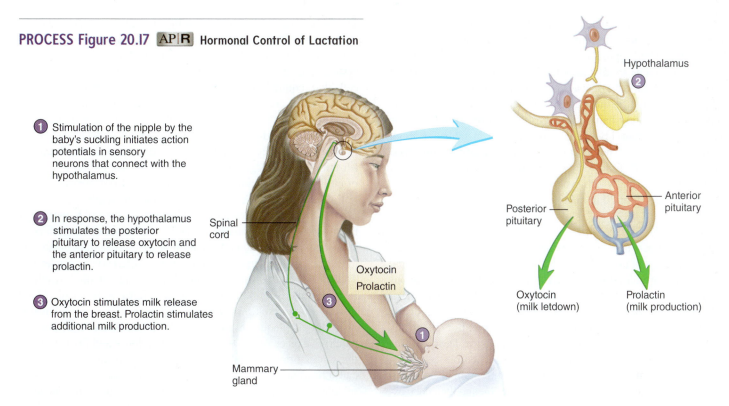

1. Stimulation of the nipple by the baby's suckling initiates action potentials in sensory neurons that connect with the hypothalamus.

2. In response, the hypothalamus stimulates the posterior pituitary to release oxytocin and the anterior pituitary to release prolactin.

3. Oxytocin stimulates milk release from the breast. Prolactin stimulates additional milk production.

Spinal cord

Oxytocin
Prolactin

Mammary gland

Hypothalamus

Posterior pituitary

Anterior pituitary

Oxytocin (milk letdown)

Prolactin (milk production)

moving person with his eyes. At 4 months, the infant begins to do push-ups—that is, raise himself by the arms. The infant can begin to grasp objects placed in his hands, coo and gurgle, roll from back to side, listen quietly when hearing a person's voice or music, hold the head erect, and play with his hands. At 5 months, the infant can usually laugh out loud, reach for objects, turn her head to follow an object, lift her head and shoulders, sit with support, and roll over. At 8 months, the infant can recognize familiar people, sit up without support, and reach for specific objects. At 12 months, the infant may pull herself to a standing position and may be able to walk without support. The child can pick up objects in her hands and examine them carefully. A 12-month-old child can understand much of what is said and may say several words.

20.6 LIFE STAGES

Learning Outcomes *After reading this section, you should be able to*

A. List the stages of life, and describe the major events associated with each stage.
B. Describe the process of aging.
C. Describe the events that occur at the time of death.

The stages of life and the activities associated with those stages are issues of great interest in today's society. We view life stages very differently today than we did in the past. For example, in 1960, 20% of males and 12% of females graduating from high school attended college. Today, over half of all people 25 and older have attended college for some period of time. In addition, there are many more nontraditional college students than there were just a few years ago.

In 1900, only 5% of the U.S. population was over age 65. Today, about 16% of the population is over age 65, and by 2030

more than 20% will be older than 65. The average life expectancy in 1900 was about 47 years, in 1940 it was about 63 years, and today it is about 78 years. In 1900, nearly 70% of all males over age 65 were still working; today, only about 20% are still working past age 65. Though older people are more likely to be retired today, they are healthier and more active than in past generations. Instead of working, older people may be enjoying the later stages of life by participating in recreational activities.

The stages of life from fertilization to death are divided into three prenatal stages and five postnatal stages as follows:

1. **Germinal** (jer′mi-năl) **period,** fertilization to 14 days
2. **Embryo,** 14–56 days after fertilization
3. **Fetus,** 56 days after fertilization to birth
4. **Neonate,** birth to 1 month after birth
5. **Infant** (in′fănt), 1 month to 1 or 2 years after birth (the end of infancy is sometimes set at the time that the child begins to walk)
6. **Child,** 1 or 2 years old to puberty (about 11–14 years)
7. **Adolescent** (ad-ō-les′ent), teenage years, from puberty to 20 years old
8. **Adult,** 20 years old to death

Adulthood is sometimes divided into three periods: **young adult,** 20–40 years old; **middle age,** 40–65 years old; and **older adult,** or *senior citizen,* 65 years old to death. Much of this designation is associated more with social norms than with physiology.

During childhood, the individual grows in size and develops considerably. Many of the emotional characteristics a person possesses throughout life are formed during early childhood.

Major physical and physiological changes occur during adolescence, and many of these changes also affect the emotions and behavior of the individual. Other emotional changes occur as the adolescent attempts to fit into an adult world. **Puberty** (pū′ber-tē) is the time when reproductive cells begin to mature and when gonadal hormones are first secreted in substantial amounts. These hormones stimulate the development and maturation of secondary sexual characteristics, such as enlargement of the female breasts and growth of body hair in both sexes. Puberty usually begins in females at about 11–13 years of age and in males at about 12–14 years of age. The onset of puberty is usually accompanied by a growth spurt, followed by a period of slower growth. Full adult stature is usually achieved before age 17 or 18 in females and before age 19 or 20 in males.

The Aging Process

Development of a new human being begins at fertilization, and so does the process of aging. Cell division occurs at an extremely rapid rate during early development and then begins to slow as various cells become committed to specific functions within the body.

Many cells of the body, such as skin cells, continue to divide throughout life, replacing dead or damaged tissue. But other cells, such as mature neurons in the brain, cease to divide. Dead neurons tend not to be replaced. After the number of neurons reaches a peak (at approximately the time of birth), the number begins to decline. Neuronal loss is most rapid early in life and then decreases to a slower, steady rate.

Young embryonic tissue has relatively small amounts of collagen, and the collagen that is present is not highly cross-linked, making it very flexible and elastic. However, many of the collagen fibers produced during development are permanent components of the individual. As the individual ages, more and more cross-links form between the collagen molecules, rendering the tissues more rigid and less flexible.

The tissues with the highest collagen content and the greatest dependency on collagen for their function are the most severely affected by the collagen cross-linking and tissue rigidity associated with aging. One of the first structures to exhibit age-related changes as a result of this increased rigidity is the lens of the eye. Seeing close objects becomes more difficult with advancing age until most middle-aged people require reading glasses (see chapter 9). Loss of elasticity also affects other tissues, including the joints, kidneys, lungs, and heart, and greatly reduces their functional ability.

As with nervous tissue, the number of skeletal muscle fibers declines with age. The strength of skeletal muscle reaches a peak between 20 and 35 years of life and usually declines steadily thereafter. Skeletal muscle strength depends primarily on the size of the muscle fibers, but the total number of fibers is probably also important to muscle strength. As most people age, both the number of fibers and the size of each tend to decline. The decline in muscle fiber size may be more related to a general decrease in activity than to any specific age-related changes. Like the collagen of connective tissue, however, the macromolecules of skeletal muscle cells undergo biochemical changes during aging, rendering the muscle tissue less functional. A good exercise program can slow or even partially reverse the process of muscular aging.

Cardiac muscle cells also do not normally divide after birth. Age-related changes in cardiac muscle cell function probably contribute to a decline in cardiac function with advancing age.

The heart loses elastic recoil ability and muscular contractility. As a result, total cardiac output declines, and less oxygen and fewer nutrients reach the body cells supplied by the cardiovascular system. This decline in blood flow can be particularly harmful to cells that require high oxygen levels, such as neurons of the brain, and cells that are already compromised, such as cartilage cells of the joints, contributing to the general decline in these tissues.

Reduced cardiac function also can result in decreased blood flow to the kidneys, contributing to decreases in the kidneys' filtration ability. Degeneration of the connective tissues as a result of collagen cross-linking and other factors also decreases the filtration efficiency of the glomerular basement membrane.

Arteriosclerosis (ar-tēr′ē-ō-skler-ō′sis) is general hardening of the arteries affecting mainly arterioles. **Atherosclerosis** (ath′er-ō-skler-ō′sis) is the gradual formation of lipid-containing plaques in the arterial wall of large and medium-sized arteries (see chapter 13). These plaques then may become fibrotic and calcified, resulting in arteriosclerosis. Atherosclerosis interferes with normal blood flow and may result in a **thrombosis** (throm-bō′sis), the formation of a clot or plaque inside a vessel. An **embolus** (em′bō-lŭs; a patch) is a piece of a clot that has broken loose and floats through the circulation. An embolus can lodge in smaller arteries and cause heart attacks or strokes. Although atherosclerosis affects all middle-aged and elderly people to some extent and can even occur in young people, some people appear more at risk because of high blood cholesterol levels. This condition seems to have a hereditary component, and blood tests are available to screen people for high blood cholesterol levels.

Many other organs, including the liver, pancreas, stomach, and colon, undergo degenerative changes with age. The ingestion of harmful agents can accelerate such changes. Examples of these types of accelerated changes are the degenerative changes induced in the lungs (aside from lung cancer) by cigarette smoke and sclerotic changes in the liver resulting from excessive alcohol consumption.

In addition to the previously described changes, cellular wear and tear, or cellular aging, contributes to aging. Progressive damage from many sources, such as radiation and toxic substances, can result in irreversible cellular insults and may be one of the major factors leading to aging. Although the data are mixed and their interpretation is controversial, some studies suggest that ingesting moderate amounts of vitamins E and C in combination may help slow aging by stimulating cell repair. Vitamin C also stimulates collagen production and may slow the loss of tissue elasticity associated with aging collagen.

According to the **free radical theory of aging,** free radicals, which are atoms or molecules with an unpaired electron, can react with and alter the structure of molecules that are critical for normal cell function. Alteration of these molecules can result in cell dysfunction, cancer, or other types of cellular damage. Free radicals are produced as a normal part of metabolism and are introduced into the body from the environment through the air we breathe and the food we eat. The damage caused by free radicals may accumulate with age. Antioxidants, such as beta carotene (provitamin A), vitamin C, and vitamin E, can donate electrons to free radicals without themselves becoming harmful. Thus, antioxidants may prevent the damage caused by the free radicals and may ward off age-related disorders ranging from wrinkles to cancer. Again, the data are mixed and their interpretation is controversial.

One characteristic of aging is an overall decrease in ATP production. This decline is associated with a decrease in oxidative phosphorylation, which has been shown in many cases to be associated with **mitochondrial DNA mutations.**

Immune system changes may also be a major contributing factor to aging. The aging immune system loses its ability to respond to outside antigens and begins to be more sensitive to the body's own antigens. Immune responses to one's own tissues can result in the degeneration of the tissues and may be responsible for such conditions as arthritic joint disorders, chronic glomerulonephritis, and hyperthyroidism. In addition, T lymphocytes tend to lose their functional capacity with aging and cannot destroy abnormal cells as efficiently. This change may be one reason that certain types of cancer occur more frequently in older people.

One of the greatest disadvantages of aging is the increasing lack of ability to adjust to stress. Homeostasis is far more precarious in elderly people, and eventually the body encounters some stressor so great that the body's ability to recover is surpassed and death results.

Death

Death is usually not attributed to old age. Some other problem, such as heart failure, renal failure, or stroke, is usually listed as the cause of death.

Death was once defined as the loss of heartbeat and respiration. In recent years, however, more precise definitions of death have been developed because both the heart and the lungs can be kept working artificially, as occurs during cardiopulmonary resuscitation, and the heart can even be temporarily replaced by an artificial device. Modern definitions of death are based on the permanent cessation of life functions and the cessation of integrated tissue and organ function. The most widely accepted indication of death in humans is whole brain death, which is manifested clinically by the absence of (1) response to stimulation, (2) natural respiration and heart function, and (3) brainstem reflexes, in addition to an electroencephalogram that remains isoelectric ("flat") for at least 30 minutes.

When determining death, certain conditions also need to be ruled out. For example, some central nervous system poisons can cause a flat electroencephalogram, but the patient can be revived if the effects of the poison are eliminated. Hypothermia slows all chemical reactions, including those involved in degenerative changes that begin at the time of death. As a result, a person suffering from hypothermia can exhibit no response to stimulation, exhibit no respiration or heartbeat, and have a flat electroencephalogram for more than 30 minutes and still be revived.

Neocortical (nē-ō-kōr′ti-kăl) **death** is a condition in which major portions of the cerebrum are no longer functioning. The patient is comatose and incapable of responding to stimuli. However, heartbeat and respiration still continue because of some relatively unimpaired brainstem functions. Also, because some brainstem function remains, the electroencephalogram is not flat but exhibits some level of activity. Under these conditions, some state laws require that the patient be kept alive by intravenous feeding and by other support equipment. The patient may have previously stated in a living will that, if neocortical death occurs and he or she cannot be returned to a reasonably normal level of function, no artificial support should be applied in an attempt to keep the body alive.

20.7 GENETICS

Learning Outcomes *After reading this section, you should be able to*

A. Define genetics, and explain how chromosomes are related to inheritance.

B. Describe the major types of inheritance.

Genetics is the study of heredity—that is, the characteristics children inherit from their parents. Many of a person's characteristics, including specific abilities, susceptibility to certain diseases, and even lifespan, are influenced by heredity. The functional units of heredity are **genes,** which are carried on chromosomes.

Chromosomes

Deoxyribonucleic (dē-oks′ē-rī′bō-noo-klē′ic) **acid (DNA)** molecules and their associated proteins become visible as densely stained bodies, called **mitotic chromosomes** (krō′mō-sōmz), during cell division (see chapter 3). **Somatic** (sō-mat′ik) **cells,** all the cells of the body except the sex cells, contain 23 pairs of chromosomes, or 46 total chromosomes. The sex cells, or **gametes** (gam′ētz), contain 23 unpaired chromosomes.

A **karyotype** (kar′ē-ō-tīp) is a display of the chromosomes in a somatic cell (figure 20.18; see chapter 3). There are 22 pairs of **autosomal** (aw-tō-sō′măl) **chromosomes,** which are all the chromosomes except the sex chromosomes, and there is one pair of **sex chromosomes.** A normal female has two **X chromosomes** (XX) in each somatic cell, whereas a normal male has one X and one **Y chromosome** (XY).

Gametes are produced by **meiosis** (mī-ō′sis) (see chapter 19). Meiosis is called a reduction division because it produces gametes

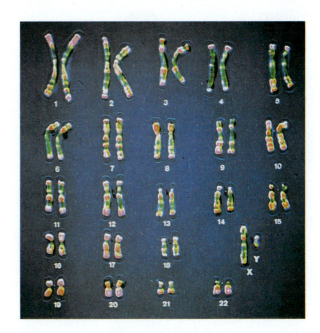

Figure 20.18 Human Karyotype

The 23 pairs of chromosomes in humans consist of 22 pairs of autosomal chromosomes (numbered 1–22) and 1 pair of sex chromosomes. The autosomal chromosome pairs are numbered in order from largest to smallest. This is the karyotype of a male, as evidenced by the presence of an X and a Y sex chromosome. A female karyotype would have 2 X chromosomes.

that have half the number of chromosomes that somatic cells have. When a sperm cell and a secondary oocyte fuse during fertilization, each contributes one-half of the chromosomes necessary to produce new somatic cells. Therefore, half of an individual's genetic makeup comes from the father, and half comes from the mother.

During meiosis, the chromosomes are distributed in such a way that each gamete receives only one chromosome from each **homologous** (hŏ-mol′ō-gŭs) pair of chromosomes (see chapter 19). Homologous chromosomes contain the same complement of genetic information. The determination of sex illustrates, in part, how chromosomes are distributed during gamete formation and fertilization. During meiosis and gamete formation, the pair of sex chromosomes separates, so that each secondary oocyte receives one X chromosome, whereas each sperm cell receives either an X chromosome or a Y chromosome (figure 20.19). When a sperm cell fertilizes an oocyte to form a single cell, the sex of the individual is determined randomly. If the oocyte is fertilized by a sperm cell with a Y chromosome, a male results; if the oocyte is fertilized by a sperm cell with an X chromosome, a female results. Estimating the probability that any given zygote will be male or female is much like flipping a coin. For any given fertilization event, there is a 50% probability that the individual will be female and 50% probability that the individual will be male.

Genes

Each chromosome contains thousands of genes. Each **gene** consists of a certain portion of a DNA molecule, but not necessarily a continuous stretch of DNA. Genes determine the proteins in a cell. Both chromosomes of a given pair contain similar but not necessarily identical genes. Similar genes on homologous chromosomes are called **alleles** (ă-lēlz′). If the two allelic genes are identical, the person is **homozygous** (hō-mō-zī′gŭs) for the trait specified by that gene. If the two alleles are slightly different, the person is **heterozygous** (het′er-ō-zī′gŭs) for the trait. All the genes in one homologous set of 23 chromosomes in one individual constitute that person's **genome** (jē′nōm).

Through the processes of meiosis, gamete formation, and fertilization, the distribution of genes received from each parent is essentially random. This random distribution is influenced by several factors, however. First, all the genes on a given chromosome are **linked;** that is, they tend to be inherited as a set rather than as individual genes because chromosomes, not individual genes, segregate during meiosis. Second, homologous chromosomes may exchange genetic information during meiosis by **crossing over.**

Segregation errors may occur during meiosis. As the chromosomes separate, the two members of a homologous pair may not segregate. As a result, one of the daughter cells receives both chromosomes of a given pair, and the other daughter cell receives none. When the gametes are fertilized, the resulting zygote has either 47 chromosomes or 45 chromosomes, rather than the normal 46. When this condition results in an abnormal autosomal chromosome number, it is usually, but not always, lethal and is one reason for a high rate of early embryo loss. **Down syndrome,** or *trisomy 21,* which results when there are three #21 chromosomes, is one of the few autosomal trisomies that is not lethal. In contrast, sex chromosome abnormalities are not usually lethal. For example, **Turner Syndrome** or *monosomy X,* in which only one X chromosome is present, results in sterility and abnormal sexual development but does not affect the mental development of the female.

Dominant and Recessive Genes

Most human genetic traits are recognized because defective alleles for those traits exist in the population. For example, on chromosome 11 is a gene that produces an enzyme necessary for the synthesis of melanin, the pigment responsible for skin, hair, and eye color (see chapter 5). An abnormal allele, however, produces a defective enzyme not capable of catalyzing one of the steps in melanin synthesis. If a given person inherits two defective alleles, a homozygous condition, the person is unable to produce melanin and therefore lacks normal pigment. This condition is referred to as **albinism** (al′bi-nizm).

For many genetic traits, the effects of one allele for that trait can mask the effect of another allele for the same trait. For example, a person who is heterozygous for the melanin-producing enzyme gene has one normal gene for melanin production and one defective gene for melanin production. One copy of the gene and its resulting enzymes are enough to make normal melanin. As a result, the person who is heterozygous produces melanin and appears normal. In this case, the allele that produces the normal enzyme is said to be **dominant,** whereas the allele producing the abnormal enzyme is **recessive.** By convention, dominant traits are indicated by uppercase letters, and recessive traits are indicated by lowercase letters. In this example, the letter *A* designates the dominant normal, pigmented allele, the letter *a* the recessive abnormal allele. Not all dominant traits are the normal condition, and not all recessive traits are abnormal. Many examples exist of abnormal

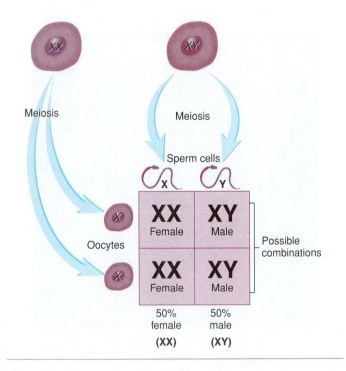

Figure 20.19 Inheritance of Sex

The female produces oocytes containing one X chromosome, whereas the male produces sperm cells with either an X or a Y chromosome. There are four possible combinations of an oocyte with a sperm cell, half of which produce females and half of which produce males.

dominant traits. For example, **Huntington disease,** a degenerative neurological disease, is caused by an abnormal dominant gene.

The possible combinations of dominant and recessive alleles for normal melanin production versus albinism are *AA* (homozygous dominant, normal), *Aa* (heterozygous, normal), and *aa* (homozygous recessive, albino). The combination of alleles a person has for a given trait is called the **genotype** (jen′ō-tīp). The person's appearance is called the **phenotype** (fē′nō-tīp). A person with the genotype *AA* or *Aa* has the phenotype of normal pigmentation, whereas a person with the genotype *aa* has the phenotype of albinism. The recessive trait is expressed only when no allele for the dominant trait is present. A **carrier** is a heterozygous person with an abnormal recessive gene but a normal phenotype because the normal dominant allele for that gene is also present.

Predict 6

> *Polydactyly* (pol-ē-dak′ti-lē) *is the condition of having extra fingers or toes. Given that polydactyly is a dominant trait, list all the possible genotypes and phenotypes for polydactyly. Use the letters D and d for the alleles.*

The inheritance of dominant and recessive traits can be determined if the genotypes of the parents are known. For example, if an albino person (*aa*) mates with a heterozygous normal person (*Aa*), the probability that the child will be albino (*aa*) is one-half, and the probability that the child will be normal (*Aa*) is one-half. If two carriers (*Aa*) mate, the probability that the child will be homozygous dominant (*AA*) or homozygous recessive (*aa*) is one-fourth for either genotype. The probability that the child will be heterozygous (*Aa*) is one-half. Such a probability can be determined easily by using a table called a **Punnett square** (figure 20.20).

Sex-Linked Traits

Traits affected by genes on the sex chromosomes are called **sex-linked traits.** Most sex-linked traits are **X-linked;** that is, they are on the X chromosome. Only a few **Y-linked** traits exist, largely because the Y chromosome is very small. An example of an X-linked trait is **hemophilia A** (classic hemophilia), in which the person is unable to produce one of the clotting factors (see chapter 11). Consequently, clotting is impaired, and persistent bleeding can occur, either spontaneously or as a result of an injury. Hemophilia A is a recessive trait, and the allele for the trait is located on the X chromosome. The possible genotypes and phenotypes are therefore

$X^H X^H$ (normal homozygous female)
$X^H X^h$ (normal heterozygous female)
$X^h X^h$ (hemophiliac homozygous female)
$X^H Y$ (normal male)
$X^h Y$ (hemophiliac male)

Note that a female must have both recessive alleles to exhibit hemophilia, whereas a male has hemophilia if he has one recessive allele, because he has only one X chromosome. A Punnett square representing the inheritance of hemophilia is illustrated in figure 20.21. If a woman who is a carrier for hemophilia mates with a man who does not have hemophilia, none of their daughters will have hemophilia; however, the probability that a son will have hemophilia is one-half. On the other hand, the probability that a daughter will be a carrier is one-half.

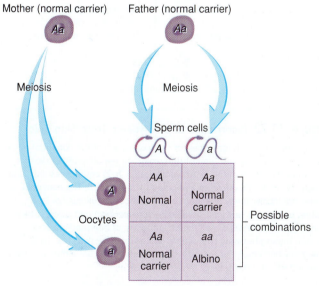

Genotype and phenotype probabilities:
$\frac{1}{4}$ *AA* (normal) : $\frac{1}{2}$ *Aa* (normal carrier) : $\frac{1}{4}$ *aa* (albino)

Figure 20.20 Inheritance of a Recessive Trait: Albinism

A represents the normal pigmented condition, and *a* represents the recessive unpigmented condition. This Punnett square represents a mating between two normal carriers.

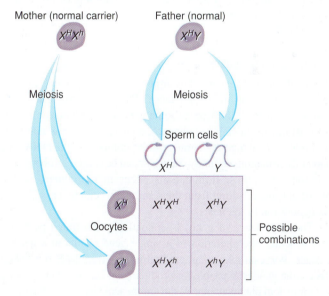

Genotype and phenotype probabilities:
$\frac{1}{4} X^H X^H$ (normal female) : $\frac{1}{4} X^H X^h$ (carrier female) :
$\frac{1}{4} X^H Y$ (normal male) : $\frac{1}{4} X^h Y$ (male with hemophilia)

Figure 20.21 Inheritance of an X-Linked Trait: Hemophilia

X^H represents the normal X chromosome condition with all clotting factors, and X^h represents the X chromosome lacking an allele for one clotting factor. This Punnett square represents a mating between a normal male and a normal carrier female.

Predict 7

Predict the probability of a girl with Turner Syndrome having hemophilia if her mother is a carrier for hemophilia.

Other Types of Gene Expression

In some cases, the dominant allele does not completely mask the effects of the recessive allele. This is called **incomplete dominance.** An example of incomplete dominance is **sickle-cell anemia,** in which the hemoglobin produced by the gene is abnormal. The result is sickle-shaped red blood cells, which are likely to stick in capillaries and tend to rupture more easily than normal red blood cells. The normal hemoglobin allele (S) is dominant over the sickle-cell allele (s). A normal person (SS) has normal hemoglobin, and a person with sickle-cell anemia (ss) has abnormal hemoglobin. A person who is heterozygous (Ss) has half normal hemoglobin and half abnormal hemoglobin and usually has only a few sickle-shaped red blood cells. This condition is called **sickle-cell trait.**

In another type of gene expression, called **codominance** (kō-dom′i-năns), two alleles can combine to produce an effect without either of them being dominant or recessive. For example, a person with type AB blood has A antigens and B antigens on the surface of his or her red blood cells (see chapter 11). The antigens result from a gene that causes the production of the A antigen and a different gene that causes the production of the B antigen, and neither gene is dominant or recessive in relation to the other.

Many traits, called **polygenic** (pol-ē-jen′ik) **traits,** are determined by the expression of multiple genes on different chromosomes. Examples are height, intelligence, eye color, and skin color. Polygenic traits typically have a great amount of variability. For example, there are many shades of eye color and skin color (figure 20.22).

Genetic Disorders

Genetic disorders are caused by abnormalities in a person's genetic makeup—that is, in his or her DNA. They may involve a single gene or an entire chromosome. Some genetic disorders result from a **mutation** (mū-tā′shŭn; to change), a change in a gene that usually involves a change in the nucleotides composing the DNA (see chapter 2). Mutations occur by chance or can be caused by chemicals, radiation, or viruses. If mutations occur in reproductive cells, abnormal traits resulting from these mutations can be passed from one generation to the next.

The importance of genes is dramatically illustrated by situations in which the alteration of a single gene results in a genetic disorder. For example, in **phenylketonuria** (fen′il-kē′tō-nū′rē-ă) **(PKU),** the gene responsible for producing an enzyme that converts the amino acid phenylalanine to the amino acid tyrosine is defective. Therefore, phenylalanine accumulates in the blood and is eventually converted to harmful substances that can cause mental retardation.

Genetic Counseling

Genetic counseling includes predicting the possible results of matings involving carriers of harmful genes and talking to parents or prospective parents about the possible outcomes and treatments of a genetic disorder. With this knowledge, prospective parents can make informed decisions about having children.

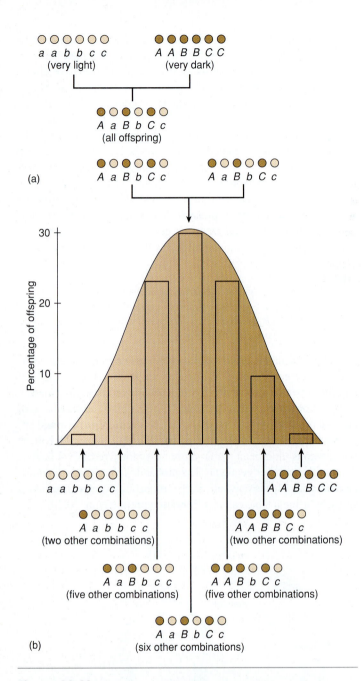

Figure 20.22 Inheritance of a Polygenic Trait: Skin Color

This example shows three genes for skin color. Each of the dominant alleles (A, B, C) contributes one "unit of dark color" to the offspring (indicated by a *dark dot*). Each of the recessive alleles (a, b, c) contributes one "unit of light color" to the offspring (indicated by a *light dot*). The dominant alleles are incompletely dominant over the recessive alleles. (a) In a mating between a very light-skinned person (aabbcc) and a very dark-skinned person (AABBCC), all the offspring are of intermediate color. (b) In a mating between two people of intermediate skin color (AaBbCc), a very low percentage of the offspring (less than 2%) are either very light or very dark, and most are of intermediate color.

The first step in genetic counseling is to attempt to determine the genotype of the individuals involved. A family tree, or **pedigree,** provides historical information about family members. Figure 20.23 shows the pedigree for a simple dominant trait, such as Huntington disease, a neurological disorder. Sometimes,

CLINICAL IMPACT   Human Genome Project

The **human genome** comprises all the genes in one homologous set of human chromosomes. Researchers estimate that humans have 20,000–25,000 genes.

A **genomic** (jĕ-nom′ik, je-nōm′ik) **map** is a description of the DNA nucleotide sequences of the genes and their locations on the chromosomes (figure 20A). The

Human Genome Project, a large-scale project to characterize the entire human genome, was completed in February 2003.

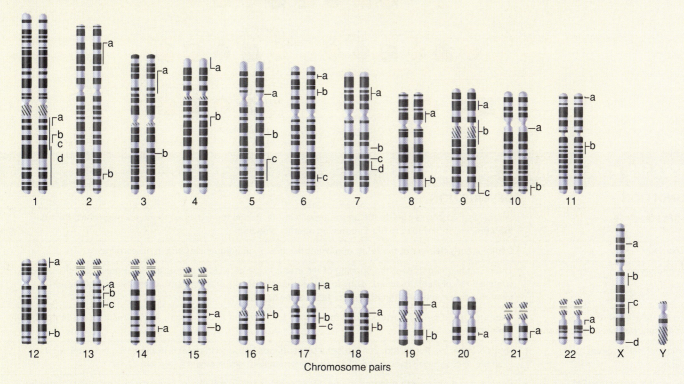

Chromosome pairs

1. a. Gaucher disease
   b. Prostate cancer
   c. Glaucoma
   d. Alzheimer disease*
2. a. Familial colon cancer*
   b. Waardenburg syndrome
3. a. Lung cancer
   b. Retinitis pigmentosa*
4. a. Huntington disease
   b. Parkinson disease
5. a. Cockayne syndrome
   b. Familial polyposis of the colon
   c. Asthma
6. a. Spinocerebellar ataxia
   b. Diabetes*
   c. Epilepsy*

7. a. Diabetes*
   b. Osteogenesis imperfecta
   c. Cystic fibrosis
   d. Obesity*
8. a. Werner syndrome
   b. Burkitt lymphoma
9. a. Malignant melanoma
   b. Friedreich ataxia
   c. Tuberous sclerosis
10. a. Multiple endocrine neoplasia, type 2
    b. Gyrate atrophy
11. a. Sickle-cell anemia
    b. Multiple endocrine neoplasia
12. a. Zellweger syndrome
    b. Phenylketonuria (PKU)

13. a. Breast cancer*
    b. Retinoblastoma
    c. Wilson disease
14. a. Alzheimer disease*
15. a. Marfan syndrome
    b. Tay-Sachs disease
16. a. Polycystic kidney disease
    b. Crohn disease*
17. a. Tumor suppressor protein
    b. Breast cancer*
    c. Osteogenesis imperfecta
18. a. Amyloidosis
    b. Pancreatic cancer*
19. a. Familial hypercholesterolemia
    b. Myotonic dystrophy

20. a. Severe combined immunodeficiency
21. a. Amyotrophic lateral sclerosis*
22. a. DiGeorge syndrome
    b. Neurofibromatosis, type 2
X a. Duchenne muscular dystrophy
    b. Menkes syndrome
    c. X-linked severe combined immunodeficiency
    d. Factor VIII deficiency (hemophilia A)

*Gene responsible for only some cases.

Figure 20A

Representative genetic defects mapped to date. The bars and lines indicate the locations of the genes listed for each chromosome.

knowing the phenotypes of relatives makes it possible to determine a person's genotype. As part of the process of collecting information, a karyotype can be prepared. For some genetic disorders, the amount of a given substance, such as an enzyme, produced by a carrier can be tested. For example, carriers for cystic fibrosis produce more salt in their sweat than is normal.

Figure 20.23 Pedigree of a Simple Dominant Trait
Males are indicated by *squares,* females by *circles.*
People who express the trait are indicated in *purple.*
The *horizontal line* between symbols represents a mating.
The symbols connected to the mating line by *vertical* and
horizontal lines represent the children resulting from the
mating in order of birth from left to right. Matings not
related to the pedigree are not shown.

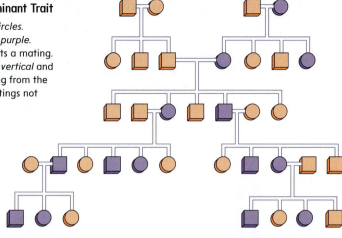

DISEASES AND DISORDERS: Pregnancy

CONDITION	DESCRIPTION
Ectopic pregnancy	Pregnancy that occurs outside the uterus, usually in the uterine tube; as the pregnancy progresses, the tube ruptures, which may lead to life-threatening internal bleeding
Miscarriage	Death or early delivery of the fetus; approximately 15% of all pregnancies end in miscarriage
Placenta previa	The placenta implants near the opening of the uterus in the cervical canal; as the fetus grows and the uterus stretches, the placenta may tear away from the uterine wall, resulting in death of the fetus; associated hemorrhaging can be life-threatening to the mother as well.
Toxemia (tok-sē′mē-ă) (pregnancy-induced hypertension)	Increased blood pressure associated with pregnancy; may result in convulsions, kidney failure, and death of both the mother and fetus
Teratogens (ter′ă-tō-jenz)	Drugs or other chemicals that can cross the placenta and cause birth defects in the developing embryo; for example, thalidomide (tha-lid′o-mīd) causes abnormal limb development
Fetal alcohol syndrome (FAS)	Brain dysfunction, growth retardation, and facial peculiarities in children of women who consumed substantial amounts of alcohol while pregnant
Infections	
Rubella (German measles)	If rubella is contracted by a pregnant female, the fetus may be severely affected; results include visual and hearing defects as well as mental retardation.
Neonatal gonorrheal ophthalmia	Severe form of conjunctivitis contracted by an infant passing through the birth canal of a mother infected with gonorrhea; carries a high risk of blindness; can be prevented by treating a newborn's eye with silver nitrate and antibiotics
Chlamydial conjunctivitis	Contracted by an infant passing through the birth canal of a mother infected with chlamydia
Human immunodeficiency virus (HIV)	Virus that causes acquired immunodeficiency syndrome (AIDS); can infect the fetus in utero, during parturition, or during breastfeeding; the drug azidothymidine (AZT) inhibits HIV replication and can reduce the number of infants who contract AIDS if given to HIV-infected pregnant women and to their newborn infants

ANSWER TO LEARN TO PREDICT

In this chapter we learned that the production of breast milk depends on a number of hormones. Estrogen and progesterone cause expansion of the duct system and the secretory units of the breast. Estrogen and progesterone levels during pregnancy prevent the breast from producing milk, but after parturition, these hormone levels drop dramatically. Prolactin produced by the anterior pituitary stimulates milk production. We also learned that suckling stimulates the release of prolactin and oxytocin, which stimulate continued milk production and milk letdown. If Ming stops breastfeeding for a prolonged period of time, prolactin secretion will stop, which will cause milk production to decline. Ming can retain the capacity to breastfeed her baby by using a breast pump while she takes the prescribed medication. After finishing the antibiotic, Ming can then continue to breastfeed her baby.

Answers to the rest of this chapter's Predict questions are in Appendix E.

SUMMARY

20.1 Prenatal Development (p. 560)

1. Prenatal development is an important part of an individual's life. About 3 of every 100 people are born with a birth defect requiring early medical attention.
2. The developing human is called an embryo during the first 8 weeks of the prenatal period and a fetus from 8 weeks until birth.
3. Developmental age is 14 days less than clinical age.

Fertilization

Fertilization, the union of the secondary oocyte and sperm cell, results in a zygote.

Early Cell Division

The zygote undergoes divisions until it becomes a mass of cells.

Blastocyst

1. The embryonic mass develops a cavity and is known as the blastocyst.
2. The blastocyst consists of a trophoblast and an inner cell mass, where the embryo forms.

Implantation of the Blastocyst and Development of the Placenta

1. The blastocyst implants in the uterus about 7 days after fertilization.
2. The embryonic portion of the placenta is derived from the trophoblast of the blastocyst.

Maternal Hormonal Changes

1. Human chorionic gonadotropin levels are high in early pregnancy but then decline to low levels.
2. Progesterone and estrogen levels are low in early pregnancy but increase to high levels late in pregnancy.

Formation of the Germ Layers

1. The embryo forms around the primitive streak, which forms about 14 days after fertilization.
2. All tissues of the body are derived from the three primary germ layers: ectoderm, mesoderm, and endoderm.

Neural Tube and Neural Crest Formation

The nervous system develops from a neural tube that forms in the ectodermal surface of the embryo and from neural crest cells derived from the developing neural tube.

Formation of the General Body Structure

1. The limbs develop as outgrowths called limb buds.
2. The face develops by fusion of five tissue masses.

Development of the Organ Systems

1. The digestive tract develops as the developing embryo closes off part of the yolk sac.
2. The heart develops as two blood vessels fuse into a single tube that develops septa to form four chambers.
3. The kidneys and the reproductive system are closely related in their development.

Growth of the Fetus

During the fetal period, the fetus increases 15-fold in length and 1400-fold in weight.

20.2 Parturition (p. 572)

1. Uterine contractions force the baby out of the uterus during labor.
2. Increased estrogen, decreased progesterone, and secretions from the fetal adrenal cortex initiate parturition.
3. Stretching of the uterus stimulates oxytocin secretion, which stimulates uterine contractions.

20.3 The Newborn (p. 573)

Respiratory and Circulatory Changes

1. Inflation of the lungs at birth results in closure of the foramen ovale and the ductus arteriosus.
2. When the umbilical cord is cut, blood no longer flows through the umbilical vessels.

Digestive Changes

The digestive system only gradually develops the ability to digest a variety of foods.

20.4 Lactation (p. 574)

1. Estrogen and progesterone help stimulate the growth of the breasts during pregnancy.
2. Suckling stimulates prolactin and oxytocin synthesis. Prolactin stimulates milk production, and oxytocin stimulates milk letdown.

20.5 First Year Following Birth (p. 576)

Many of the important changes that occur during the first year after birth are linked to continued development of the brain.

20.6 Life Stages (p. 577)

The eight stages of life are germinal period (fertilization to 14 days); embryo (14–56 days after fertilization); fetus (56 days after fertilization to birth); neonate (birth–1 month); infant (1 month–1 or 2 years); child (1 or 2 years–puberty); adolescent (puberty–20 years); adult (20 years–death).

The Aging Process

1. Aging occurs as irreplaceable cells wear out and the tissue becomes more brittle and less able to repair damage.
2. Atherosclerosis is the deposit of lipids in the arteries. Arteriosclerosis is hardening of the arteries.

Death

Death is defined as the absence of brain response to stimulation, the absence of natural respiration and heart function, and a flat electroencephalogram for 30 minutes.

20.7 Genetics (p. 579)

Chromosomes

1. Humans have 46 chromosomes in 23 pairs.
2. Males have the sex chromosomes XY, and females have XX.
3. During gamete formation, the chromosomes of each pair of chromosomes separate; therefore, half of a person's genetic makeup comes from the father and half from the mother.

Genes

1. A gene is a portion of a DNA molecule. Genes determine the proteins in a cell.
2. Genes are paired (located on the paired chromosomes).
3. Dominant genes mask the effects of recessive genes.
4. Sex-linked traits result from genes on the sex chromosomes.
5. In incomplete dominance, the heterozygote expresses a trait that is intermediate between the two homozygous traits.
6. In codominance, neither gene is dominant or recessive, but both are fully expressed.

7. Polygenic traits result from the expression of multiple genes.

Genetic Disorders

1. A mutation is a change in the DNA.
2. Some genetic disorders result from an abnormal distribution of chromosomes during gamete formation.

Genetic Counseling

1. A pedigree (family history) can be used to determine the risk of having children with a genetic disorder.
2. Specific chemical tests or an examination of a person's karyotype can be used to determine a person's genotype.

REVIEW AND COMPREHENSION

1. Define clinical age and developmental age, and state the difference between the two in number of days. Define embryo and fetus.
2. What are the events during the first week after fertilization? Define zygote, morula, and blastocyst.
3. How does the placenta develop?
4. Describe the formation of the germ layers and the role of the primitive streak.
5. How are the neural tube and neural crest cells formed? What do they become?
6. Describe the formation of the limbs and face.
7. Describe the formation of the digestive tract.
8. How does the single heart tube become four-chambered?
9. What major events distinguish embryonic and fetal development?
10. Describe the hormonal changes that take place before and during parturition.
11. What changes take place in the newborn's circulatory system and digestive system shortly after birth?
12. What hormones are involved in preparing the breasts for lactation? What hormones are involved in milk production and milk letdown?
13. Describe the changes in motor and language skills that take place during the first year of life.
14. List the different life stages.

15. How does the loss of cells that are not replaced affect the aging process? Give examples.
16. How does the loss of tissue elasticity affect the aging process? Give examples.
17. How does aging affect the immune system?
18. Define death.
19. Give the number and type of chromosomes in the karyotype of a human somatic cell. How do the chromosomes of a male and a female differ from each other?
20. How do the chromosomes in somatic cells and gametes differ from each other?
21. What is a gene, and how are genes responsible for the structure and function of cells?
22. Define homozygous dominant, heterozygous, and homozygous recessive.
23. What is a sex-linked trait? Give an example.
24. How do sickle-cell anemia, type AB blood, and a person's height result from the expression of genes?
25. What is a mutation?
26. What is the cause of the genetic disorder Down syndrome?
27. How are pedigrees, karyotypes, and chemical tests used in genetic counseling?

CRITICAL THINKING

1. A physician tells a woman that her pregnancy has progressed 44 days since her last menstrual period (LMP). How many days has the embryo been developing, and what developmental events are occurring?
2. A high fever can prevent neural tube closure. If a woman has a high fever about 35 to 45 days post-LMP, what kinds of birth defects may be seen in the developing embryo?
3. A drug that stops the production of milk in the breast after a few days probably has which effect?
   a. inhibits prolactin secretion
   b. inhibits oxytocin secretion
   c. increases estrogen secretion
   d. increases progesterone secretion
   e. increases prolactin secretion

4. Dimpled cheeks are inherited as a dominant trait. Is it possible for two parents, each of whom has dimpled cheeks, to have a child without dimpled cheeks? Explain.
5. The ability to roll the tongue to form a "tube" results from a dominant gene. Suppose that a woman and her son can both roll their tongues, but her husband cannot. Is it possible to determine if the husband is the father of her son based on this trait?
6. A woman who does not have hemophilia marries a man who has the disorder. Determine the genotypes of both parents if they have a daughter with hemophilia.

Answers in Appendix D

Appendix A

Table of Measurements

TABLE A.1	Table of Measurements		
Unit	**Metric Equivalent**	**Symbol**	**U.S. Equivalent**
Measures of Length			
1 kilometer	= 1000 meters	km	0.62137 mile
1 meter	= 10 decimeters or 100 centimeters	m	39.37 inches
1 decimeter	= 10 centimeters	dm	3.937 inches
1 centimeter	= 10 millimeters	cm	0.3937 inch
1 millimeter	= 1000 micrometers	mm	
1 micrometer[a]	= 1/1000 millimeter or 1000 nanometers	µm	
1 nanometer[a]	= 10 angstroms or 1000 picometers	nm	No U.S. equivalent
1 angstrom	= 1/10,000,000 millimeter	Å	
1 picometer	= 1/1,000,000,000 millimeter	pm	
Measures of Volume			
1 cubic meter	= 1000 cubic decimeters	m^3	1.308 cubic yards
1 cubic decimeter	= 1000 cubic centimeters	dm^3	0.03531 cubic foot
1 cubic centimeter	= 1000 cubic millimeters or 1 milliliter	cm^3 (cc)	0.06102 cubic inch
Measures of Capacity			
1 kiloliter	= 1000 liters	kL	264.18 gallons
1 liter	= 10 deciliters	L	1.0567 quarts
1 deciliter	= 100 milliliters	dL	0.4227 cup
1 milliliter	= volume of 1 gram of water at standard temperature and pressure	mL	0.3381 ounce
1 microliter	= 1 mm^3 (cubic millimeter) or 10^{-6} L	µL	
Measures of Mass			
1 kilogram	= 1000 grams	kg	2.2046 pounds
1 gram	= 100 centigrams or 1000 milligrams	g	0.0353 ounce
1 centigram	= 10 milligrams	cg	0.1543 grain
1 milligram	= 1/1000 gram	mg	

[a]A micrometer was formerly called a micron (µ), and a nanometer was formerly called a millimicron (mµ).

TABLE A.2	Comparative Temperature Scales

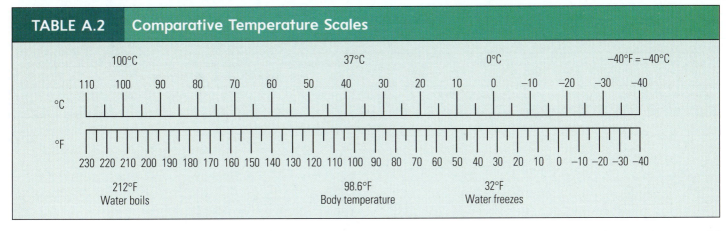

Appendix B

Some Reference Laboratory Values

TABLE B.I	Blood, Plasma, or Serum Values	
Test	**Normal Values**	**Clinical Significance**
Acetoacetate plus acetone	0.32–2 mg/100 mL	Values increase in diabetic acidosis, fasting, high-fat diet, and toxemia of pregnancy.
Ammonia	9–33 μmol/L (micromol/L)	Values decrease with proteinuria and as a result of severe burns; values increase in multiple myeloma.
Amylase	4–25 U/mL[a]	Values increase in acute pancreatitis, intestinal obstruction, and mumps; values decrease in cirrhosis of the liver, toxemia of pregnancy, and chronic pancreatitis.
Barbiturate	0	Coma level: phenobarbital, approximately 10 mg/100 mL; most other drugs, 1–3 mg/100 mL.
Bilirubin	0.4 mg/100 mL	Values increase in conditions causing red blood cell destruction, biliary obstruction, or liver inflammation.
Blood volume	8.5–9% of body weight in kilograms	
Calcium	8.5–10.5 mg/100 mL	Values increase in hyperparathyroidism, vitamin D hypervitaminosis; values decrease in hypoparathyroidism, malnutrition, and severe diarrhea.
Carbon dioxide content	24–30 mEq/L[b] 20–26 mEq/L in infants (as HCO_3^-)	Values increase in respiratory diseases, vomiting, and intestinal obstruction; values decrease in acidosis, nephritis, and diarrhea.
Carbon monoxide	0	Symptoms with over 20% saturation
Chloride	100–106 mEq/L	Values increase in Cushing syndrome, nephritis, and hyperventilation; values decrease in diabetic acidosis, Addison disease, and diarrhea, and also after severe burns.
Creatine phosphokinase (CPK)	Female: 5–35 mU/mL Male: 5–55 mU/mL	Values increase in myocardial infarction and skeletal muscle diseases, such as muscular dystrophy.
Creatinine	0.6–1.5 mg/100 mL	Values increase in certain kidney diseases.
Ethanol	0	0.3–0.4%, marked intoxication 0.4–0.5%, alcoholic stupor Over 0.5%, alcoholic coma
Glucose	Fasting 70–110 mg/100 mL	Values increase in diabetes mellitus, liver diseases, nephritis, hyperthyroidism, and pregnancy; values decrease in hyperinsulinism, hypothyroidism, and Addison disease.
Iron	50–150 μg/100 mL	Values increase in various anemias and liver disease; values decrease in iron-deficiency anemia.
Lactic acid	0.6–1.8 mEq/L	Values increase with muscular activity and in congestive heart failure, severe hemorrhage, shock, and anaerobic exercise.
Lactic dehydrogenase	60–120 U/mL	Values increase in pernicious anemia, myocardial infarction, liver diseases, acute leukemia, and widespread carcinoma.

[a]A unit (U) is the quantity of a substance that has a physiological effect.
[b]A milliequivalent (mEq) is defined in Appendix C.

TABLE B.I	Blood, Plasma, or Serum Values (continued)

Test	Normal Values	Clinical Significance
Lipids	Cholesterol 120–220 mg/100 mL; cholesterol esters 60–75% of cholesterol; phospholipids 9–16 mg/100 mL as lipid phosphorus; total fatty acids 190–420 mg/100 mL Total lipids 450–1000 mg/100 mL Triglycerides 40–150 mg/100 mL	Increased values for cholesterol and triglycerides are connected with increased risk of cardiovascular disease, such as heart attack and stroke.
Lithium	Toxic levels 2 mEq/L	
Osmolality	285–295 mOsm/kg water	
Oxygen saturation (arterial) (see P_{O_2})	96–100%	
P_{CO_2}	35–43 mm Hg	Values decrease in acidosis, nephritis, and diarrhea; values increase in respiratory diseases, intestinal obstruction, and vomiting.
pH	7.35–7.45	Values decrease as a result of hypoventilation, severe diarrhea, Addison disease, and diabetic acidosis; values increase due to hyperventilation, Cushing syndrome, and vomiting.
P_{O_2}	75–100 mm Hg (breathing room air)	Values increase in polycythemia and decrease in anemia and obstructive pulmonary diseases.
Phosphatase (acid)	Male: total 0.13–0.63 U/mL Female: total 0.01–0.56 U/mL	Values increase in cancer of the prostate gland, hyperparathyroidism, some liver diseases, myocardial infarction, and pulmonary embolism.
Phosphatase (alkaline)	13–39 IU/L[c] (infants and adolescents up to 104 IU/L)	Values increase in hyperparathyroidism, some liver diseases, and pregnancy.
Phosphorus (inorganic)	3–4.5 mg/100 mL (infants in first year up to 6 mg/100 mL)	Values increase in hypoparathyroidism, acromegaly, vitamin D hypervitaminosis, and kidney diseases; values decrease in hyperparathyroidism.
Potassium	3.5–5 mEq/100 mL	
Protein	Total 6–8.4 g/100 mL Albumin 3.5–5 g/100 mL Globulin 2.3–3.5 g/100 mL	Total protein values increase in severe dehydration and shock; values decrease in severe malnutrition and hemorrhage.

Salicylate
  Therapeutic
  Toxic
	0	
		20–25 mg/100 mL
		Over 30 mg/100 mL
		Over 20 mg/100 mL after age 60

Sodium	135–145 mEq/L	Values increase in nephritis and severe dehydration; values decrease in Addison disease, myxedema, kidney disease, and diarrhea.
Sulfonamide Therapeutic	0	 5–15 mg/100 mL
Urea nitrogen	8–25 mg/100 mL	Values increase in response to increased dietary protein intake; values decrease in impaired renal function.
Uric acid	3–7 mg/100 mL	Values increase in gout and toxemia of pregnancy and as a result of tissue damage.

[c]A unit (U) is the quantity of a substance that has a physiological effect.

TABLE B.2   Blood Count Values

Test	Normal Values	Clinical Significance
Clotting (coagulation) time	5–10 min	Values increase in afibrinogenemia and hyperheparinemia, severe liver damage.
Fetal hemoglobin	Newborns: 60–90% Before age 2: 0–4% Adults: 0–2%	Values increase in thalassemia, sickle-cell anemia, and leakage of fetal blood into maternal bloodstream during pregnancy.
Hemoglobin	Male: 14–16.5 g/100 mL Female: 12–15 g/100 mL Newborn: 14–20 g/100 mL	Values decrease in anemia, hyperthyroidism, cirrhosis of the liver, and severe hemorrhage; values increase in polycythemia, congestive heart failure, and obstructive pulmonary disease and at high altitude.
Hematocrit	Male: 40–54% Female: 38–47%	Values increase in polycythemia, severe dehydration, and shock; values decrease in anemia, leukemia, cirrhosis, and hyperthyroidism.
Ketone bodies	0.3–2 mg/100 mL Toxic level: 20 mg/100 mL	Values increase in ketoacidosis, fever, anorexia, fasting, starvation, and a high-fat diet.
Platelet count	250,000–400,000/μL	Values decrease in anemias and allergic conditions and during cancer chemotherapy; values increase in cancer, trauma, heart disease, and cirrhosis.
Prothrombin time	11–15 s	Values increase in prothrombin and vitamin deficiency, liver disease, and hypervitaminosis A.
Red blood cell count	Males: 4.6–6.2 million/μL Females: 4.2–5.4 million/μL	Values decrease in systemic lupus erythematosus, anemias, and Addison disease; values increase in polycythemia and dehydration and following hemorrhage.
Reticulocyte count	1–3%	Values decrease in iron-deficiency and pernicious anemias and radiation therapy; values increase in hemolytic anemia, leukemia, and metastatic carcinoma.
White blood cell count, differential	Neutrophils 60–70% Eosinophils 2–4% Basophils 0.5–1% Lymphocytes 20–25% Monocytes 3–8%	Neutrophils increase in acute infections; eosinophils and basophils increase in allergic reactions; monocytes increase in chronic infections; lymphocytes increase during antigen-antibody reactions.
White blood cell count, total	5000–9000/μL	Values decrease in diabetes mellitus, anemias and following cancer chemotherapy; values increase in acute infections, trauma, some malignant diseases, and some cardiovascular diseases.

TABLE B.3   Urine Values

Test	Normal Values	Clinical Significance
Acetone and acetoacetate	0	Values increase in diabetic acidosis and during fasting.
Albumin	0 to trace	Values increase in glomerular nephritis and hypertension.
Ammonia	20–70 mEq/L	Values increase in diabetes mellitus and liver disease.
Bacterial count	Under 10,000/mL	Values increase in urinary tract infection.
Bile and bilirubin	0	Values increase in biliary tract obstruction.
Calcium	Under 250 mg/24 h	Values increase in hyperparathyroidism and decrease in hypoparathyroidism.
Chloride	110–254 mEq/24 h	Values decrease in pyloric obstruction and diarrhea; values increase in Addison disease and dehydration.
Potassium	25–100 mEq/L	Values decrease in diarrhea, malabsorption syndrome, and adrenal cortical insufficiency; values increase in chronic renal failure, dehydration, and Cushing syndrome.
Sodium	75–200 mg/24 h	Values decrease in diarrhea, acute renal failure, and Cushing syndrome; values increase in dehydration, starvation, and diabetic acidosis.
Creatinine clearance	100–140 mL/min	Values increase in renal diseases.
Creatinine	1–2 g/24 h	Values increase in infections and decrease in muscular atrophy, anemia, and certain kidney diseases.
Glucose	0	Values increase in diabetes mellitus and certain pituitary gland disorders.
Urea clearance	Over 40 mL of blood cleared of urea per min	Values increase in certain kidney diseases.
Urea	25–35 g/24 h	Values decrease in complete biliary obstruction and severe diarrhea; values increase in liver diseases and hemolytic anemia.
Uric acid	0.6–1 g/24 h	Values increase in gout and decrease in certain kidney diseases.
Casts		
Epithelial	Occasional	Values increase in nephrosis and heavy metal poisoning.
Granular	Occasional	Values increase in nephritis and pyelonephritis.
Hyaline	Occasional	Values increase in glomerular membrane damage and fever.
Red blood cell	Occasional	Values increase in pyelonephritis; blood cells appear in urine in response to kidney stones and cystitis.
White blood cell	Occasional	Values increase in kidney infections.
Color	Amber, straw, transparent yellow	Color varies with hydration, diet, and disease states.
Odor	Aromatic	Odor becomes acetone-like in diabetic ketosis.
Osmolality	500–800 mOsm/kg water	Values decrease in aldosteronism and diabetes insipidus; values increase in high-protein diets, heart failure, and dehydration.
pH	4.6–8	Values decrease in acidosis, emphysema, starvation, and dehydration; values increase in urinary tract infections and severe alkalosis.

TABLE B.4   Hormone Levels

Test	Normal Values
Steroid hormones	
Aldosterone	Excretion: 5–19 µg/24 h[a]
Fasting, at rest, 210 mEq sodium diet	Supine: 48 ± 29 pg/mL[b]
	Upright: 65 ± 23 pg/mL
Fasting, at rest, 10 mEq sodium diet	Supine: 175 ± 75 pg/mL
	Upright: 532 ± 228 pg/mL
Cortisol	
Fasting	8 AM: 5–25 µg/100 mL
At rest	8 PM: Below 10 µg/100 mL
Testosterone	Adult male: 300–1100 ng/100 mL[c]
	Adolescent male: over 100 ng/100 mL
	Female: 25–90 ng/100 mL
Peptide hormones	
Adrenocorticotropin (ACTH)	15–170 pg/mL
Calcitonin	Undetectable in normal individuals
Growth hormone (GH)	
Fasting, at rest	Below 5 ng/mL
After exercise	Children: over 10 ng/mL
	Male: below 5 ng/mL
	Female: up to 30 ng/mL
Insulin	
Fasting	6–26 µU/mL
During hypoglycemia	Below 20 µU/mL
After glucose	Up to 150 µU/mL
Luteinizing hormone (LH)	Male: 6–18 mU/mL
	Preovulatory or postovulatory female: 5–22 mU/mL
	Midcycle peak 30–250 mU/mL
Parathyroid hormone	Less than 10 µEq/L[d]
Prolactin	2–15 ng/mL
Renin activity	
Normal diet	
Supine	1.1 ± 0.8 ng/mL/h
Upright	1.9 ± 1.7 ng/mL/h
Low-sodium diet	
Supine	2.7 ± 1.8 ng/mL/h
Upright	6.6 ± 2.5 ng/mL/h
Thyroid-stimulating hormone (TSH)	0.5–3.5 µU/mL
Thyroxine-binding globulin	15.25 µg T_4/100 mL
Total thyroxine	4–12 µg/100 mL

[a] 1 microgram (1 µg) is equal to 10^{-6} g.
[b] 1 picogram (1 pg) is equal to 10^{-12} g.
[c] 1 nanogram (1 ng) is equal to 10^{-9} g.
[d] µEq microequivalent, 1/1000 mEq, which is defined in Appendix C.

Appendix C

Solution Concentrations

Physiologists often express solution concentration in terms of percent, molarity, molality, and equivalents.

Percent

The weight-volume method of expressing percent concentrations states the weight of a solute in a given volume of solvent. For example, to prepare a 10% solution of sodium chloride, 10 g of sodium chloride is dissolved in a small amount of water (solvent) to form a salt solution. Then additional water is added to the salt solution to form 100 mL of salt solution. Note that the sodium chloride was dissolved in water and then diluted to the required volume. The sodium chloride was not dissolved directly in 100 mL of water.

Molarity

Molarity determines the number of moles of solute dissolved in a given volume of solvent. A **mole** (abbreviated mol) of a substance contains 6.023×10^{23} number (Avogadro number) of particles, such as atoms, ions, compounds, or molecules. A 1-molar (1-M) solution is made by dissolving 1 mol of a substance in enough water to make 1 L of solution. For example, 1 mol of sodium chloride solution is made by dissolving 58.44 g of sodium chloride in enough water to make 1 L of solution. One mole of glucose solution is made by dissolving 180.2 g of glucose in enough water to make 1 L of solution. Both solutions have the same number (Avogadro number) of sodium chloride or glucose compounds in solution.

Molality

Although 1-M solutions have the same number of solute particles, they do not have the same number of solvent (water) molecules. Because 58.5 g of sodium chloride occupies less volume than 180 g of glucose, the sodium chloride solution has more water molecules. **Molality** is a method of calculating concentrations that takes into account the number of solute and solvent molecules. A 1-molal solution (1-*m*) is 1 mol of a substance dissolved in 1 kg of water. Thus, a 1-*m* solution of sodium chloride and a 1-*m* solution of glucose contain the same number of sodium chloride and glucose compounds dissolved in the same amount of water.

When sodium chloride (NaCl) is dissolved in water, it dissociates, or separates, to form two ions: a Na^+ and a Cl^-. Glucose, however, does not dissociate when dissolved in water. Although 1-*m* solutions of sodium chloride and of glucose have the same number of sodium chloride and glucose compounds, because of dissociation the sodium chloride solution contains twice as many particles as the glucose solution (one Na^+ and one Cl^- for each glucose molecule). To report the concentration of these substances in a way that reflects the number of particles in a given mass of solvent, the concept of **osmolality** is used. The osmolality of a solution is the molality of the solution times the number of particles into which the solute dissociates in 1 kg of solvent. Thus, 1 mol of sodium chloride in 1 kg of water is a 2-osmolal (2-osm) solution because sodium chloride dissociates to form two ions.

The osmolality of a solution is a reflection of the number, not the type, of particles in a solution. Thus, a 1-osm solution contains 1 osm of particles per kilogram of solvent, but the particles may be all one type or a complex mixture of different types.

The concentration of particles in body fluids is so low that the measurement milliosmole (mOsm), 1/1000 of an osmole, is used. Most body fluids have an osmotic concentration of approximately 300 mOsm and consist of many different ions and molecules. The osmotic concentration of body fluids is important because it influences the movement of water into or out of cells (see chapter 3).

Equivalents

Equivalents are a measure of the concentrations of ionized substances. One equivalent (Eq) is 1 mol of an ionized substance multiplied by the absolute value of its charge. For example, 1 mol of NaCl dissociates into 1 mol of Na^+ and 1 mol of Cl^-. Thus, there are 1 Eq of Na^+ (1 mol $\times$ 1) and 1 Eq of Cl^- (1 mol $\times$ 1). One mole of $CaCl_2$ dissociates into 1 mol of Ca^{2+} and 2 mol of Cl^-. Thus, there are 2 Eq of Ca^{2+} (1 mol $\times$ 2) and 2 Eq of Cl^- (2 mol $\times$ 1). In an electrically neutral solution, the equivalent concentration of positively charged ions is equal to the equivalent concentration of the negatively charged ions. One milliequivalent (mEq) is 1/1000 of an equivalent.

Appendix D

Answers to Critical Thinking Questions

Chapter 1

1. D is correct. Positive-feedback mechanisms result in movement away from homeostasis and are usually harmful. The continually decreasing blood pressure is an example. Negative-feedback mechanisms result in a return to homeostasis. The elevated heart rate is a negative-feedback mechanism that attempts to return blood pressure to a normal value. In this case, the negative-feedback mechanism was inadequate to restore homeostasis, and blood pressure continued to decrease. Medical intervention (a transfusion) increased blood volume and blood pressure. With the increase in blood pressure, the positive-feedback mechanism is interrupted and the negative-feedback mechanism is able to maintain blood pressure.

2. Student B is correct. As the muscles become more active, they use more O_2. The increased rate of respiration maintains blood and muscle tissue O_2 levels (homeostasis) and is a negative-feedback mechanism.

3. a. Inferior
   b. Posterior (dorsal) or deep
   c. Distal or inferior
   d. Lateral

4. The heel is distal, inferior, and posterior to the kneecap. The kneecap is proximal, superior, and anterior to the heel.

5. The wedding band should be worn proximal to the engagement ring.

6. The pancreas is located in the right-upper and left-upper quadrants; it is located in the epigastric and left hypochondriac regions. The top of the urinary bladder is located in the right-lower quadrant and the left-lower quadrant; it is located in the hypogastric region. The rest of the urinary bladder is located in the pelvic cavity.

7. The pelvic cavity contains the uterus, which increases greatly in size during pregnancy as the fetus within the uterus grows. However, the pelvic cavity is surrounded by bones, which are not expandable. Therefore, the uterus moves superiorly into the abdominal cavity, crowding abdominal organs and dramatically increasing the size of the abdominal cavity.

8. After passing through the left thoracic wall, the first membrane to be encountered is the parietal pleura. Continuing through the pleural cavity, the visceral peritoneum and the left lung are pierced. Leaving the lung, the bullet penetrates the visceral pleura, the pleural cavity, and the parietal pleura (remember that the lung is surrounded by a double membrane sac). Next the parietal pericardium, the pericardial cavity, the visceral pericardium, and the heart are encountered.

9. The kidneys are located in the abdominal cavity but are retroperitoneal. When a person is lying prone, it is possible to cut through the posterior abdominal wall and remove a kidney without cutting through parietal peritoneum.

Chapter 2

1. Because atoms are electrically neutral, the iodine atom has the same number of protons and electrons. The gain of an electron means the iodine ion has one more electron than protons and therefore a charge of minus one. The correct symbol is I^{-1}.

2. a. Dissociation
   b. Synthesis
   c. Decomposition

3. Muscle contains proteins and, to increase muscle mass, proteins must be synthesized. The synthesis of molecules in living organisms results from chemical reactions that require the input of energy. The energy needed to drive these synthesis reactions comes from chemical reactions that release energy. The energy-releasing reactions occur during the decomposition of food molecules.

4. The sodium bicarbonate dissociates to form sodium ions (Na^-) and bicarbonate ions (HCO_3^-). Because this is a reversible reaction, the HCO_3^- added to the solution bind with H^+ to form CO_2 and water. The decrease in H^+ causes the pH to increase (become more basic).

5. The slight amount of heat functioned as activation energy and started a chemical reaction. The reaction released energy, especially heat, which caused the solution to become very hot.

6. There was most likely at least one enzyme present that was required for one of the initial reactions. Boiling denatured any enzymes present in the solution. Without enzymes, the reaction(s) will not occur.

Chapter 3

1. B is the most logical conclusion. Swollen lung tissue suggests that the tissues had been submerged in a hypotonic solution, such as fresh water. Because the bay contains salt water, which is slightly hypertonic to blood, it is unlikely he drowned in the bay. It is more likely he drowned in fresh water and was later placed in the bay. Although this is the most logical conclusion of those given, there are possibilities other than murder. For example, he may have accidentally drowned in a freshwater stream and then been washed into the bay.

2. B is correct. A solution that is isotonic causes cells to neither shrink nor swell. Therefore, there is no net movement of water between the blood and the dialysis fluid. Also, if the solution has the same composition as blood except for having no urea, only urea diffuses from a higher concentration in the blood to the lower concentration in the dialysis fluid. A solution that is isotonic and contains only protein will cause many different small molecules (not just urea) that are found in blood to diffuse into the dialysis fluid. Distilled water is hypotonic and causes blood cells to swell and undergo lysis. Also, molecules that are able to pass through the dialysis membrane diffuse from the blood into the distilled water. Blood is not a good solution to use because it would have to be blood that has already had urea removed in order to be useful as dialysis fluid.

3. Because the plasma membrane stays the same size, even though small pieces of membrane from secretory vesicles are continually added to the plasma membrane, one must conclude that some process is removing small pieces of membrane at the same rate that they are added. The cell membrane is constantly being recycled.

4. The main function of this cell is to secrete proteins. Well-developed rough endoplasmic reticulum for protein production, well-developed Golgi apparatuses for the packaging of proteins for secretion, and many vesicles for exocytosis are typical of many cells that secrete proteins. Many mitochondria produce large amounts of ATP. The ATP serves as an energy source for forming the bonds between the amino acids, for the active transport of amino acids into the cell, and for exocytosis of the completed proteins. Many microvilli, which greatly increase surface area for transport, can also be found in cells that produce and secrete proteins. However, large numbers of mitochondria or large amounts of microvilli can be found in other types of cells as well.

5. The sickle-cell gene for producing the protein has a different nucleotide than the normal gene for producing the protein. Because the nucleotides are the set of instructions for making the protein, the change means the instructions contain an error. When mRNA copies the instructions from the faulty DNA, the error is also copied. Consequently, when the amino acids are joined together to form the protein at the ribosome, the instructions are wrong and the protein is incorrectly assembled. A substitution of a single nucleotide results in a different amino acid in the chain of amino acids that make up the protein. This change alters the shape of the protein. Just as the function of enzymes depends on their shape, the functions of other proteins depend on their shapes. The incorrectly manufactured proteins in the red blood cells have the wrong shape and do not stack correctly. As a result of the abnormal stacking, the red blood cells have an abnormal, sickle shape.

Chapter 4

1. Pseudostratified columnar ciliated epithelium is found in the trachea. It produces mucus that traps dust and debris in air. The cilia move the mucus with entrapped dust and debris to the throat, where it is swallowed. In heavy smokers, the pseudostratified columnar epithelium is replaced by stratified squamous epithelium, which serves a protective function against the irritating materials in the smoke. Unfortunately, this type of epithelium is not ciliated, so removal of foreign materials from the trachea is more difficult. After 2 years without smoking, the original pseudostratified columnar ciliated epithelium replaces the stratified squamous epithelium and debris removal resumes.

2. Tight junctions prevent the passage of materials between the epithelial cells.

3. A secretory epithelium is generally a simple epithelium. To manufacture large amounts of enzymes, a cuboidal or columnar cell with the appropriate organelles, such as rough endoplasmic reticulum and Golgi apparatuses, would be expected. The pancreas is formed of simple columnar epithelium. The epithelium has microvilli on its free surface, which increase the surface area and facilitate secretion. The tight junctions that connect the epithelial cells of the pancreatic glands and duct system to each other prevent damage to the underlying tissues (by the action of pancreatic digestive enzymes).

4. a. The stratified squamous epithelium that lines the mouth provides protection. Replacement of it with simple columnar epithelium makes the lining of the mouth much more susceptible to damage because the single layer of epithelial cells is easier to damage.
   b. Tendons attach muscles to bone. When muscles contract, muscles pull on the tendons and, because the tendons are attached to bone, the bones move. If the tendons contained elastic tissue, the tendons would be more like elastic bands. The muscles would contract and stretch the tendons. Not all of the force of muscle contraction would be transferred to the bones to cause them to move.
   c. If bones were made of elastic cartilage, they would be much more flexible and could bend and then return to their original shape. They would not be rigid structures, like bone, that support our weight and result in efficient movement.

5. The fibers are organized along the direction of pull on the ligaments—that is, parallel to the length of the ligament. When the back is bent, the ligaments are stretched. The elastic recoil of the stretched ligaments helps straighten the back.

6. When blood is ejected into the aorta, the aortic wall expands and the diameter of the aorta increases. The elastic fibers are arranged in a circular fashion, so that, when the aortic wall expands, the elastic fibers are stretched. Recoil of the fibers causes the aorta to resume its original shape, which helps force blood to flow through the aorta.

7. Chemical mediators of inflammation normally produce beneficial responses, such as dilation of small-diameter blood vessels and increased blood vessel permeability. Blocking these effects could reduce the body's ability to deal with harmful agents, such as bacteria. On the other hand, antihistamines also reduce many of the unpleasant symptoms of inflammation, making the patient more comfortable, and this can be considered beneficial. Antihistamines are commonly taken to prevent allergy symptoms that are often an overreaction of the inflammatory response to foreign substances, such as pollen.

8. Collagen synthesis is required for granulation tissue and scar formation. If collagen synthesis does not occur because of a lack of vitamin C, or if collagen synthesis is slowed, wound healing does not occur or is slower than normal. One might expect that the density of collagen fibers in a scar is reduced and the scar is not as durable as a normal scar.

Chapter 5

1. Yes, the skin (dermis) can be overstretched as a result of obesity.

2. Carotene, a yellow pigment from ingested plants, accumulates in lipids. The stratum corneum of a callus has more layers of cells than other noncalloused parts of the skin and the cells in each layer are surrounded by lipids. The carotene in the lipids make the callus appear yellow.

3. The vermillion border is covered by keratinized epithelium that is a transition between the nonkeratinized stratified epithelium of the mucous membrane and the keratinized stratified epithelium of the facial skin. The mucous membrane has mucous glands, which secrete mucus (see chapter 4). The mucus helps keep the inner surface of the lips moist. In addition, the inner surface of the lips is "sealed off" most of the time and is moistened by saliva. In contrast to facial skin, the skin of the vermillion border does not have sebaceous or sweat glands. Without sebaceous glands, the surface of the vermillion border is not protected against drying by sebum. Also, the vermillion border is not as heavily keratinized as facial skin; that is, the number of cell layers with surrounding lipids is less. Consequently, the vermillion border dries out more easily.

4. The hair follicle, but not the hair, is surrounded with nerve endings that can detect movement or pulling of the hair. The hair is dead, keratinized epithelium, so cutting the hair is not painful.

5. Several methods have some degree of success in treating acne: (1) Kill the bacteria. One effective agent is benzoyl

peroxide in some acne medications. Antibiotics have also been used to treat severe cases of acne. (2) Prevent sebum production. Isotretinoin, or Accutane (a derivative of vitamin A), has proven effective in preventing sebum production. However, Accutane produces birth defects and should be used with caution. Estrogen, a female sex hormone, has also been used to treat severe cases of acne. (3) Unplug the follicle. Some sulfur compounds speed up peeling of the skin and unplug the follicle.

6. Rickets is a disease of children resulting from inadequate vitamin D. With inadequate vitamin D, there is insufficient absorption of calcium from the intestine, resulting in soft bones. If adequate vitamin D is ingested, rickets is prevented, whether one is dark- or fair-skinned. However, if dietary vitamin D is inadequate, when the skin is exposed to ultraviolet light, it can produce a precursor molecule that can be converted to vitamin D. Dark-skinned children are more susceptible to rickets because the additional melanin in their skin screens out the ultraviolet light and they produce less vitamin D.

7. When first exposed to the cold temperature just before starting the run, the blood vessels in the skin constrict to conserve heat. This produces a pale skin color. Dilation of skin blood vessels does not occur at this time because the skin has not been exposed to the cold long enough to cause skin temperature to fall below 15°C (27°F). After running awhile, as a result of the excess heat generated by the exercise, the blood vessels in the skin dilate. This results in heat loss and helps prevent overheating. Increased blood flow through the skin causes it to turn red. After the run, the body still has excess heat to eliminate, so for awhile the skin remains red.

Chapter 6

1. The injury separated the head (epiphysis) from the shaft (diaphysis) at the epiphyseal plate, which is cartilage. Because the epiphyseal plate is the site of bone elongation, damage to the epiphyseal plate can interfere with bone elongation, resulting in a shortened limb.

2. If Justin landed on his heels, he could have broken one or both of his heel bones, the calcaneus. When Hefty Stomper stepped on Justin's foot, the most likely bones broken would be the metatarsal bones, which make up the main structure of the foot.

3. The humerus articulating with the ulna and radius forms the elbow joint. The ulna fits tightly over the end of the humerus. The radius attaches to both the humerus and the ulna, but not as tightly to the humerus as does the ulna. Pulling forcefully on the forearm would, therefore, be more likely to pull the radius away from the humerus than the ulna.

4. Because the female pelvis is relatively wider than the male pelvis, the heads of the femora (pl. of *femur*) are placed wider apart. For the lower limbs to be placed directly below the center of gravity in the body, the femora must be angled in more sharply in females than in males. This angle may bring the knees closer together, resulting in the condition called knock-kneed.

5. All skeletons, including those of humans, have features unique to each species. A human skeleton can be identified based on its unique anatomical features. In addition, female skeletons have several anatomical features that distinguish them from male skeletons. For example, the female pelvis differs in several features from the male pelvis (see figure 6.32).

Chapter 7

1. Botulism toxin decreases acetylcholine release in the neuromuscular junction. This prevents action potentials in skeletal muscle cells. Thus, the respiratory muscles (e.g., the diaphragm) relax and do not contract. Other explanations that you could have proposed because they would also lead to respiratory failure are that the toxin prevents acetylcholine from binding to its receptor on skeletal muscle cells and that the toxin inhibits the production of acetylcholine.

2. Harvey's hands became fatigued as a result of ATP depletion. Without ATP, the muscles moving the fingers could not contract and, so, the fingers could not flex or extend. In addition, his inability to release his grip is an indication of physiological contracture, in which too little ATP is present to allow cross-bridge release.

3. Both the 100-m dash and weight lifting (to a lesser extent) involve short bursts of anaerobic activity, so in both cases the researcher should expect to find more fast-twitch muscle in the gastrocnemius of these athletes. The 10,000-m run involves aerobic metabolism, and the outstanding athlete should have more slow-twitch fibers.

4. The exercise that best builds the anterior arm muscles is flexion of the elbow against force, such as in pull-ups. A pull-up with the hand supinated builds both the biceps and the brachialis, but a curl with the hand pronated builds the brachialis to a greater extent. The posterior arm is best built by extension of the elbow, as in push-ups or dips. The anterior forearm muscles are best built up by forcefully flexing the wrist and fingers. The anterior thigh is best built by extending the knee, as in partial knee bends. The posterior leg muscles are best built by forcefully plantar flexing the ankle and toes, as in standing on one's toes. The abdominal muscles are best built by flexing the vertebral column against resistance, as in leg lifts or partial sit-ups. Of course, each exercise has to be repeated many times to increase the size of the muscles.

5. This is a typical hamstring injury (pulled hamstring), in which one or more of the hamstring muscles are either pulled away from their attachment to bone or torn. The slight flexion of the knee and the bulge in the muscle occur because muscle fibers contract in the hamstrings, but the tear does not allow antagonist muscles in the anterior thigh to pull on the muscle fibers to lengthen them. The pain results from the damaged tissue in the muscle (flexion and pain are also due to spasmodic contractions of the muscle in response to the injury). Because of the tissue damage and cramps, the muscles could not respond to voluntary nervous control and could not contract.

6. a. First, we are told that Javier suffered interruptions in functions under control of the muscular system (blurred vision, nausea, and spastic contractions). The fact that his muscles contracted without relaxing tells us that the poison has interferred with the body's ability to turn off the stimulation of the muscle cell. Organophosphate poisons inhibit the activity of acetylcholinesterase. Acetylcholinesterase breaks down acetylcholine at the neuromuscular junction and limits the length of time the acetylcholine stimulates the postsynaptic terminal of the muscle fiber. Under organophosphate poison influence, acetylcholine accumulates in the synaptic cleft and continuously stimulates the muscle fiber. As a result, the muscle remains contracted until it fatigues. Death is caused by the victim's inability to breathe. Either the respiratory muscles are in spastic paralysis or they are so depleted of ATP that they cannot contract at all.

  b. First, we learned from the Clinical Impact "Acetylcholine Antagonists" that certain chemicals can affect the metabolism of neurotransmitters such as acetylcholine. Medical practitioners sometimes use such chemicals to help counteract exposure to poisons. For example, we learned that the chemical curare binds to acetylcholine receptors and thus prevents acetylcholine from binding to them. Because curare blocks the acetylcholine receptors, muscles do not respond as normal to nervous stimulation. In the case of organophosphate poisoning, this is actually desirable, because the body has too much acetylcholine at the neuromuscular junction, since acetylcholinesterase is inhibited. Thus, by blocking the acetylcholine receptors with curare, we can prevent spastic paralysis of the respiratory muscles.

Chapter 8

1. If one series of neurons had more neurons, it would have more synapses, which should slow down the rate of action potential propagation. Also, if one series were unmyelinated and the other myelinated, the unmyelinated series would be slower.
2. The phrenic nerves are the nerves to the diaphragm. The left phrenic nerve would be cut to paralyze the left side of the diaphragm.
3. a. Radial nerve. This nerve supplies posterior (extensor) muscles. When these muscles are paralyzed, the flexor muscles are unopposed and result in the flexed limb described.
   b. Femoral nerve
4. The cerebellum controls motor functions, such as balance, muscle tone, and fine motor movements. Cerebellar dysfunction is characterized as a loss of balance, a loss of motor tone, and an inability to control fine motor movements, such as touching one's nose with the eyes closed. Such patients appear to be drunk (alcohol apparently most directly affects the cerebellum).
5. The blow to the head erased Louis's short-term memory, which would have extended to about 10 minutes before the blow, and that information was never transferred to long-term memory.
6. a. Hypoglossal nerve (cranial nerve XII)
   b. Optic nerve (cranial nerve II)
   c. Trigeminal nerve (cranial nerve V)
   d. Facial nerve (cranial nerve VII)
   e. Oculomotor nerve (cranial nerve III)
7. Nervous system stimulation of the digestive system occurs primarily through the vagus nerve (cranial nerve X), which arises from the brain. Therefore, an injury to the spinal cord at the level of C6 does not affect the functioning of this nerve.

Chapter 9

1. The lenses of the eye are biconvex structures and are most important in close vision as they are thickened (made more convex) by relaxation of the suspensory ligaments (accommodation). Therefore, if they were removed, close vision would be

most affected, and convex glasses would do the best job of replacing the lenses for near vision, such as when reading.
2. The fovea centralis, where a person focuses on an object, contains almost entirely cones. Cones do not detect faint light, so dim stars, which can be detected by the rods, are lost from vision when the person looks directly at them. When the person looks slightly to the side, the more peripherally located rods can again detect the faint light from the stars.
3. Under normal conditions, pressure changes in the middle ear occur through the auditory tube. The auditory tube connects the middle ear to the pharynx, which opens to the outside of the body through the oral and nasal cavities. This allows air pressure to equalize on both sides of the eardrum. When a person is under water, the increased water pressure on the outside of the eardrum forces the eardrum toward the middle ear because the external water pressure exceeds the air pressure in the middle ear. As a result, the eardrum cannot vibrate as freely as when the pressure is equalized, and sound transmission is dampened. If the pressure difference becomes too great, it can rupture the eardrum.
4. Sound is normally transmitted through the middle ear by vibration of the auditory ossicles. A vibrating tuning fork touching the mastoid process sets up vibrations in the temporal bone, causing the perilymph and endolymph of the inner ear to vibrate. These vibrations cause the basilar membrane to vibrate and are therefore perceived as sound.
5. When a contaminated hand rubs the eyes, the virus can be introduced into tears on the conjunctiva. From there the virus can spread into the lacrimal canaliculi and pass through the nasolacrimal duct into the nasal cavity.

Chapter 10

1. The hormone binds to a membrane-bound receptor and activates a G protein mechanism. Because a drug inhibits the binding of GTP to a protein, GTP cannot bind to the α subunit of the G protein complex. Also, the fact that a drug that inhibits the breakdown of cyclic-AMP causes an increased response suggests that the α subunit of the G protein complex, with GTP bound to it, activates the enzymes that are responsible for the synthesis of cyclic-AMP.
2. The effect of ADH on cells is mediated through membrane-bound receptors, whereas the effect of aldosterone on cells is mediated through intracellular receptors. In addition, ADH levels

increase more rapidly than aldosterone levels in response to a decrease in blood volume. ADH is secreted within minutes by the posterior pituitary gland in response to dehydration of the body. The hypothalamus detects a decrease in blood volume and an increase in the concentration of the blood, which stimulates neurons in the hypothalamus to secrete ADH. ADH binds to membrane-bound receptors in the kidney. The cells of the kidney respond within minutes. As a result, the kidney produces a small volume of very concentrated urine. Aldosterone is secreted in response to decreased blood pressure (caused by decreased blood volume), but the process involves other compounds. The kidneys respond to a decrease in blood pressure by secreting renin. Renin acts as an enzyme to break down angiotensinogen into angiotensin I. Angiotensin-converting enzyme converts angiotensin I to angiotensin II. Finally, angiotensin II stimulates the adrenal cortex to increase aldosterone secretion. Aldosterone binds to intracellular receptors in cells of the kidney. The response of these cells is to increase specific mRNA and protein synthesis. The new mRNA and protein synthesis requires a significant amount of time. The response to the synthesis of new proteins is to increase the rate of solute reabsorption in the kidney, including Na^+. This enhances water reabsorption by the kidney.
3. Vitamin D increases the rate of transport of Ca^{2+} across the wall of the intestine. Therefore, in response to high concentrations of vitamin D, blood Ca^{2+} levels can become abnormally high. Because the blood Ca^{2+} levels are elevated, the rate of PTH secretion decreases to low levels. Also, the elevated blood Ca^{2+} levels stimulate calcitonin secretion.
4. If the adrenal cortex degenerated, glucocorticoids, mineralocorticoids, and the androgens normally secreted by the adrenal cortex would no longer be secreted. A lack of glucocorticoid hormone secretion results in a reduced ability to maintain blood nutrient levels, such as glucose, between meals or during periods when no food is available. The ability to inhibit inflammation is also lost. Reduced levels of mineralocorticoids result in the loss of Na^+ in the urine and the loss of water in the urine. Thus, the ability to regulate blood pressure is reduced. Reduced secretion of androgens also occurs. Because the testes secrete androgens, the reduced androgen secretion in males is not significant. The effect of

reduced androgens in females is unclear, although adrenal androgens may influence sexual behavior, to some degree, in females.

5. Aldosterone increases Na^+ reabsorption from the distal nephron and it increases K^+ secretion into the nephron. Elevated aldosterone secretion results in the increased retention of Na^+ and H_2O in the kidney and an increased rate of K^+ elimination in the urine. As a result, blood pressure increases and blood K^+ levels can get very low.

6. After 24 hours without food, blood glucose levels will be decreasing. This will activate mechanisms that cause blood glucose levels to increase. The decreasing blood glucose levels will result in an increased rate of secretion of cortisol, a glucocorticoid, from the adrenal cortex, epinephrine from the adrenal medulla, and glucagon from pancreatic islets. The decreasing blood glucose levels result in increased ACTH secretion from the anterior pituitary gland; increased sympathetic nervous system stimulation of the adrenal medulla, resulting in increased epinephrine secretion; and increased secretion of glucagon from the pancreatic islets. Cortisol increases lipid and protein breakdown, so that these substances can be used as energy sources. It also causes increased glucose synthesis, primarily from amino acids. Epinephrine and glucagon both bind to receptors on the liver and increase the release of glucose from the liver. The decreasing blood glucose levels inhibit insulin secretion. The decreased insulin secretion slows the uptake of glucose by tissues. Thus, blood glucose levels are maintained within normal levels.

7. Stressful conditions cause increased ACTH secretion. Increased ACTH stimulates the secretion of glucocorticoids, such as cortisol. One of the effects of increased cortisol secretion is inhibition of the immune response. This could lead to an increased likelihood of a variety of infections, including colds and other diseases. Stomach pains may be due to inflammation of the stomach. Many stomach ulcers result from inflammation caused by bacteria.

Chapter II

1. Blood doping increases the number of red blood cells in the blood, increasing its oxygen-carrying capacity. However, the increased number of red blood cells also makes it more difficult for the blood to flow through the blood vessels, increasing the workload on the heart. The same effect is seen in erythrocytosis

(polycythemia). In addition, if the process is not done properly, there is a risk of infection.

2. Symptoms resulting from decreased red blood cells are associated with a decreased ability of the blood to carry O_2: shortness of breath, weakness, fatigue, and pallor. Symptoms resulting from decreased platelets are associated with a decreased ability to form platelet plugs and clots: small areas of hemorrhage in the skin, bruises, and decreased ability to stop bleeding. A symptom resulting from decreased white blood cells is an increased susceptibility to infections and diseases.

3. The hypoventilation results in decreased blood O_2, which stimulates the release of erythropoietin from the kidneys. The erythropoietin stimulates red blood cell production in the red bone marrow. Consequently, red blood cell numbers increase.

4. Removal of the stomach results in removal of intrinsic factor, which is necessary for adequate vitamin B_{12} absorption in the small intestine. Therefore, the patient would develop pernicious anemia. Vitamin B_{12} injections can be used to prevent pernicious anemia.

5. Vitamin B_{12} and folate are necessary for blood cell division. Lack of these vitamins results in anemia. Iron is necessary for the production of hemoglobin. Lack of iron results in iron-deficiency anemia. Vitamin K is necessary for the production of many blood-clotting factors. Lack of vitamin K can greatly increase blood-clotting time.

6. The anemia results from too little hemoglobin. Because there is less hemoglobin, less hemoglobin is broken down into bilirubin. Consequently, less bilirubin is excreted as part of the bile into the small intestine. With decreased bilirubin in the small intestine, bacteria produce fewer of the pigments that normally color the feces.

7. Reddie Popper has hemolytic anemia. The RBC is lower than normal because the red blood cells are being destroyed faster than they are replaced. With fewer red blood cells, hemoglobin and hematocrit are lower than normal. Bilirubin levels are above normal because of the breakdown of the hemoglobin released from the ruptured red blood cells.

Chapter I2

1. A heart murmur is an abnormal heart sound. It is unlikely that an ECG will reveal that he has a heart murmur because an ECG monitors the electrical activity of the heart, not heart sounds. A stethoscope

is a more likely instrument for detecting a heart murmur. A condition that causes a heart murmur could also cause an obvious abnormality in the ECG, but that is not always the case.

2. Starling's law of the heart is an intrinsic regulatory mechanism, whereas parasympathetic innervation of the heart is a component of the extrinsic regulation of the heart. Cutting the vagus nerve does not significantly affect the ability of Starling's law of the heart to operate.

3. Cutting sensory nerve fibers from baroreceptors would reduce the frequency of action potentials delivered to the medulla oblongata from the baroreceptors. This results because the normal blood pressure stimulates the baroreceptors. An increase in blood pressure increases the action potential frequency, and a decrease in blood pressure decreases the action potential frequency. Because cutting the sensory nerve fibers decreases the action potential frequency, this acts as a signal to the medulla oblongata that a decrease in blood pressure has occurred, even though the blood pressure has not decreased. The medulla oblongata responds by increasing sympathetic action potentials and reducing parasympathetic action potentials delivered to the heart. Consequently, the heart rate increases.

4. When the internal carotid arteries are clamped, blood flow and blood pressure in the clamped internal carotid arteries decrease dramatically. Thus, blood pressure in the area of the baroreceptors in the internal carotid arteries is low. Sensory neurons carry a lower frequency of action potentials from the baroreceptors of the internal carotid arteries to the medulla oblongata. The cardioregulatory center within the medulla oblongata responds as if the blood pressure has decreased by causing an increase in sympathetic stimulation and a decrease in parasympathetic stimulation of the heart. In addition, epinephrine and norepinephrine are released from the adrenal medulla. Consequently, there is an increased heart rate and stroke volume. Therefore, blood pressure in the aorta increases.

5. A drug that blocks Ca^{2+} channels decreases the heart rate and the force of contraction of the heart. This occurs because Ca^{2+} is involved in the depolarization of the cardiac muscle cells. If less Ca^{2+} flows into the cardiac muscle cells, the rate and degree of depolarization decrease. The result is that action potentials develop more slowly and depolarization does not occur to the

normal degree. Slower development of action potentials slows the heart rate. The degree of depolarization causes less Ca²⁺ to enter the cell; consequently, the force of contraction decreases.

6. Consuming a large amount of fluid increases the total volume of the blood, at least until the mechanisms that regulate blood volume decrease the blood volume to normal values. Increased blood volume causes an increase in venous return to the heart. Because of Starling's law of the heart, the increased venous return results in an increased stroke volume. A slight increase in the heart rate also occurs. Mechanisms that regulate blood pressure, such as the baroreceptor reflex, would prevent a large increase in blood pressure. Therefore, there may be an increased stroke volume but the blood pressure would not increase dramatically.

7. Cardiac output is influenced by the heart rate and stroke volume ($CO = SV \times HR$). An athlete's cardiac output can be equal to a nonathlete's cardiac output while they are both at rest, even though the athlete's heart rate is lower than the nonathlete's heart rate, because the stroke volume of the athlete's heart is greater than the stroke volume of the nonathlete's heart. Athletic training causes a gradual hypertrophy (enlargement) of the heart. Therefore, an athlete can maintain a cardiac output that is equal to a nonathlete's because the athlete has an increased stroke volume but a decreased heart rate. During exercise, an athlete's heart rate and stroke volume increase. The stroke volume of the athlete's heart is much greater than the stroke volume of the nonathlete's. Therefore, the cardiac output of the athlete's heart is greater than the cardiac output of the nonathlete's during exercise.

8. The walls of the ventricles are thicker than the walls of the atria because the force that the ventricles must produce is greater than the force the atria must produce. The pressure in the ventricles during ventricular systole is substantially higher than the pressure in the atria during atrial systole. In addition, most of the ventricular filling (approximately 70%) occurs before contraction of the atria. Contraction of the atria is responsible for only about 30% of ventricular filling. In contrast, all of the blood ejected from the ventricles during ventricular systole is ejected because of contraction of the ventricles.

9. An incompetent aortic semilunar valve is leaky. Thus, when it is closed, blood is able to leak through it back into the left ventricle from the aorta. The aortic semilunar valve is normally closed from the beginning of ventricular diastole until just after the beginning of ventricular systole. During ventricular diastole, blood flows from the left atrium into the left ventricle. If the aortic semilunar valve is leaky, blood will also flow through the aortic semilunar valve into the left ventricle. Consequently, the volume of the left ventricle and the pressure in the left ventricle become greater than normal during ventricular diastole. Because blood is leaking from the aorta, the aortic pressure, during ventricular diastole, is lower than normal in the aorta. Because of Starling's law of the heart during ventricular systole, the ventricular muscle contracts with a greater force and forces a greater volume of blood into the aorta. Consequently, the pressure in the left ventricle is greater than normal during ventricular systole. Also, pressure in the aorta is greater than normal during ventricular systole.

Chapter 13

1. a. For the right side of the brain: the aorta, the brachiocephalic artery, the right common carotid artery, the right internal carotid artery, the cerebral arterial circle, and then the brain tissue. For the left side of the brain: the aorta, the left common carotid artery, the left internal carotid artery, the cerebral arterial circle, and then the brain tissue

   b. For the left external portion of the skull: the aorta, the brachiocephalic artery, the right common carotid artery, the right external carotid artery, and then the external portion of the skull. For the left side of the external portion of the skull: the aorta, the left common carotid artery, the left external carotid artery, and then the external portion of the skull

   c. The aorta, the left subclavian artery, the left axillary artery, the left brachial artery, either the left radial or the ulnar artery, and the left hand

   d. The aorta, the right common iliac artery, the external iliac artery, the femoral artery, the popliteal artery, the anterior tibial artery, and the anterior portion of the right leg

2. a. The left internal jugular vein, the left brachiocephalic vein, the superior vena cava, and the right atrium of the heart

   b. The right external jugular vein, the right subclavian vein, the right brachiocephalic vein, the superior vena cava, and the right atrium of the heart

   c. The superficial veins of the left hand and forearm; either the left cephalic or the left basilic vein; the left cephalic vein and the left basilic vein empty into the axillary vein; the left axillary vein, the left subclavian vein, the left brachiocephalic vein, the superior vena cava, and the right atrium of the heart

   d. The right great saphenous vein, the right femoral vein, the right external iliac vein, the right common iliac vein, the inferior vena cava, and the right atrium of the heart

   e. The renal vein, the inferior vena cava, and the right atrium of the heart

   f. The superior mesenteric vein, the hepatic portal vein, the liver, the hepatic veins, the inferior vena cava, and right atrium of the heart

3. The femoral artery and vein are close to the surface in the femoral triangle, which is in the superior and medial part of the thigh. The femoral artery is the vessel into which the catheter will be placed. The femoral vein cannot be used because extending a catheter superiorly in the femoral vein will deliver the catheter to the right side of the heart. The anterior interventricular artery is a branch of the left coronary artery, which originates from the ascending aorta, just superior to the aortic semilunar valve. The catheter is inserted into the femoral artery. From there, it passes through the external iliac artery, through the common iliac artery, and through the aorta. It passes through the abdominal aorta, the thoracic aorta, and the aortic arch to the beginning of the left coronary artery and then to the anterior interventricular artery.

4. Cells from the tumor can spread from the colon to the liver through the hepatic portal vein. Cells from the tumor can enter the superior mesenteric or the inferior mesenteric vein. The cells can pass from the superior mesenteric vein to the hepatic portal vein. The cells can also pass from the inferior mesenteric vein to the splenic vein and then the hepatic portal vein. The cells can pass through the hepatic portal vein to the liver.

5. Reduced blood flow to the kidney stimulates renin secretion. Renin acts on angiotensinogen to produce angiotensin I. Angiotensin I is converted to angiotensin II by the action of angiotensin-converting enzyme. Angiotensin II causes vasoconstriction, which increases blood pressure. In addition, angiotensin II stimulates the secretion of aldosterone from the adrenal cortex. Aldosterone acts on the kidney, causing the reabsorption of Na⁺ and water, thus increasing the blood volume. The increased blood volume results in increased blood pressure.

6. During exercise, vasoconstriction occurs in the viscera, but vasodilation occurs in the exercising muscles. Even though the

cardiac output is increased because of the increase in stroke volume and heart rate, the blood pressure does not go up as much as it would if vasoconstriction occurred in the viscera without vasodilation of the blood vessels in the exercising skeletal muscles.

7. Dilation of arteries and veins allows blood to accumulate in the dilated blood vessels. Consequently, less blood is returned to the heart (decreased venous return). Because the venous return is reduced, the stroke volume of the heart decreases (see Starling's law of the heart—chapter 12). Consequently, the heart does less work and less O_2 is required to support the contraction of the heart. Therefore, angina pains, which are caused by inadequate O_2 delivery to the heart muscle, are reduced.

Chapter 14

1. Elevation of the limb allows gravity to assist the movement of lymph toward the heart. Massage moves lymph through the lymphatic vessels in the same fashion as does contraction of skeletal muscle. The application of pressure periodically to lymphatic vessels forces lymph to flow toward the trunk of the body, but valves prevent the flow of lymph in the reverse direction. The removal of lymph from the tissue helps relieve edema.

2. Normally, T cells are processed in the thymus and then migrate to other lymphatic tissues. Without the thymus, this processing is prevented. Because there are normally five T cells for every B cell, the number of lymphocytes is greatly reduced. The loss of T cells results in an increased susceptibility to infection and an inability to reject grafts because of the loss of cell-mediated immunity. In addition, because T cells are involved with the activation of B cells, antibody-mediated immunity is also depressed.

3. Injection B results in the greatest amount of antibody production. At first, the antigen causes a primary response. A few weeks later, the slowly released antigen causes a secondary response, resulting in a greatly increased production of antibodies. Injection A does not cause a secondary response because all of the antigen is eliminated by the primary response.

4. Active immunity varies from a few weeks (e.g., the common cold) to lifelong (e.g., measles). Immunity can be long-lasting if enough memory cells (B or T) are produced and persist to respond to later antigen exposure. Passive immunity is not long-lasting because the individual does not produce his or her own memory

cells. Because active immunity can last longer than passive immunity, it is the preferred method in most cases. However, passive immunity is preferred in situations in which immediate protection is needed, because it takes time for active immunity to develop.

5. If the patient has already been vaccinated, the booster shot stimulates a memory (secondary) response and rapid production of antibodies against the toxin. If the patient has never been vaccinated, vaccinating him is not effective because there is not enough time for the patient to develop his own primary response. Therefore, an antiserum is given to provide immediate, but temporary, protection. Sometimes both are given: The antiserum provides short-term protection and the tetanus vaccine stimulates the patient's immune system to provide long-term protection. If the shots are given at the same location in the body, the antiserum (antibodies against the tetanus toxin) could cancel the effects of the tetanus vaccine (tetanus toxin altered to be nonharmful).

6. Resistance to extracellular bacterial infections is accomplished by antibody-mediated immunity, which is not functioning properly here. Maternal antibodies (IGg) that crossed the placenta provided protection following birth but eventually were degraded. Resistance to intracellular viral infections is accomplished by cell-mediated immunity, which appears to be working normally.

7. At the first location, an antibody-mediated response results in an inflammatory response (immediate hypersensitivity). The combination of antibodies with the allergen triggers the release and activation of inflammatory chemicals. At the second location, a cell-mediated response also results in an inflammatory response (delayed hypersensitivity). This probably involves the release of lymphokines and the lysis of cells. At the other locations, there is neither an antibody-mediated nor a cell-mediated response.

8. The ointment was a good idea for the poison ivy, which caused a delayed hypersensitivity reaction—that is, too much inflammation. For the scrape, it is a bad idea because a normal amount of inflammation is beneficial and helps fight infection in the scrape.

9. Because both antibodies and cytokines produce inflammation, the fact that the metal in the jewelry resulted in inflammation is not enough information to answer the question. However, the fact that it took most of the day (many hours)

to develop the reaction indicates a delayed hypersensitivity reaction and, therefore, cytokines.

Chapter 15

1. Pressing firmly on the base of the scapulae decreases the volume of the thoracic cavity, pleural pressure increases, and alveolar pressure increases to a value greater than atmospheric pressure. Thus, air flows from the lungs. When the arms are lifted, gravity causes the thoracic cavity to sag downward and expand. The volume of the thoracic cavity increases, pleural pressure decreases, and alveolar pressure decreases to less than atmospheric pressure, causing air to flow into the lungs.

2. a. The victim's lungs expand because the rescuer is blowing air into the lungs at a pressure higher than alveolar pressure. When alveolar pressure exceeds pleural pressure, the lungs tend to expand.

   b. Air flows out of the lungs because of the natural recoil of the lungs. Elastic fibers in the lungs and surface tension of water in the alveoli cause the recoil of the lungs.

3. The best prediction would be that her vital capacity would be greatest when she is standing up. In the upright position, gravity tends to pull the abdominal organs downward. As a result, movement of the diaphragm is not as restricted and thoracic volume is increased. Lying down allows abdominal organs to exert pressure on the diaphragm, decreasing the thoracic volume.

4. At sea level, all the gases in the atmosphere exert a pressure of 760 mm Hg. If water vapor is 10% of that gas mixture, then water vapor must have a partial pressure of 760×0.10, or 76 mm Hg.

5. As fluid accumulates in the alveoli, the layer through which O_2 and CO_2 must diffuse in the alveoli becomes thicker. As the layer thickens, the rate at which gases diffuse slows. Consequently, the amount of O_2 that diffuses into the pulmonary capillaries is reduced and the amount of CO_2 that diffuses out of the pulmonary capillaries decreases. The decreased blood O_2 levels and the increased CO_2 levels stimulate the respiratory center and cause the rate and depth of respiration to increase. Oxygen therapy increases the P_{O_2} in the alveoli and reduces the P_{CO_2} in the alveoli. Therefore, O_2 diffuses more rapidly from the alveoli into the pulmonary capillaries and CO_2 diffuses more rapidly from the pulmonary capillaries into the alveoli. This establishes more normal blood levels of O_2 and CO_2. Because of increased

levels of O_2 and decreased levels of CO_2 in the blood, the rate of respiratory movements is decreased.

6. In severe emphysema, the surface area for gas exchange is reduced, so that not enough O_2 can diffuse from the alveoli into the pulmonary capillaries and not enough CO_2 can diffuse from the pulmonary capillaries into the alveoli even when O_2 therapy is given. The elevated CO_2 and the reduced O_2 levels in the blood both stimulate the respiratory center to produce an urge to take a breath.

7. The pH of the body fluids in patients who suffer from untreated diabetes mellitus decreases (becomes more acidic). A decreasing pH acts as a strong stimulus to the respiratory center. Consequently, the rate and depth of respiration increase. The increased rate and depth of respiration cause CO_2 to be released from the circulation at a greater rate. The lower blood level of CO_2 opposes the reduced pH because of the following reaction:

$$CO_2 + H_2O \rightleftharpoons H_2CO_3 \rightleftharpoons H^+ + HCO_3^-$$

As more CO_2 is removed from the blood, more H^+ and HCO_3^- combine to form H_2CO_3. Removal of the H^+ from the circulation resists a further reduction in the body fluid pH.

8. When Ima is hyperventilating, the stimulus for the hyperventilation is anxiousness, and the anxiousness is more important than the CO_2 in controlling respiratory movements. As the blood levels of CO_2 decrease during hyperventilation, vasodilation occurs in the periphery. As a result, the systemic blood pressure decreases. The systemic blood pressure can decrease enough that the blood flow to the brain decreases. Decreased blood flow to the brain results in a reduced O_2 level in the brain tissue, causing dizziness. Breathing into a paper bag increases Ima's blood levels of CO_2 toward normal. Because the CO_2 does not increase above normal, it does not increase the urge to breathe. The more normal level of CO_2 prevents the peripheral vasodilation. As Ima breathes into the paper bag, the anxiousness is likely to subside. As the anxiousness subsides, the normal regulation of respiration resumes.

9. Hyperventilating before swimming decreases the amount of CO_2 in the blood. The decreased amount of CO_2 allows a longer than normal period of time before the swimmer has a strong urge to breathe. This can be dangerous, however, because the reduced level of CO_2 causes vasodilation in the periphery and reduced blood flow to the brain and could cause a person to pass out. Passing out under water could lead to drowning.

10. Immediately after the beginning of a race, runners increase their rate and depth of respiration before blood CO_2 levels have a chance to increase. The rate and depth of respiration increase in anticipation of increased muscular activity. As a result, for a short time the increased rate and depth of respiration can actually cause blood levels of CO_2 to decrease. The lower levels of blood CO_2 result in a slightly increased blood pH. After the race has progressed, the increased metabolic activity of the muscles can produce enough extra CO_2 to cause blood CO_2 levels to increase above resting levels, even with an increased rate and depth of respiration.

Chapter 16

1. With the loss of the swallowing reflex, vomit can enter the larynx and trachea, block the respiratory tract, and keep the patient from breathing. Acidic stomach secretions cause severe inflammation and swelling of the respiratory passages.

2. Without adequate amounts of hydrochloric acid, the pH in the stomach is not low enough for the activation of pepsin. This loss of pepsin function results in inadequate protein digestion in the stomach. However, if the food is well chewed, proteolytic enzymes in the small intestine (e.g., trypsin, chymotrypsin) can still digest the protein. If the stomach secretion of intrinsic factor decreases, the absorption of vitamin B_{12} is hindered.

3. Even though bacteria apparently cause most ulcers, overproduction of hydrochloric acid because of stress is a possible contributing factor. Antibiotic therapy is an effective treatment, but reducing hydrochloric acid production is recommended. Possible solutions also include drugs that reduce stomach acid secretion, antacids to neutralize the stomach acid, smaller meals (distention of the stomach stimulates acid production), and proper diet. The person should avoid alcohol, caffeine, and large amounts of protein because they stimulate acid production. She should ingest fatty acids because they inhibit acid production, causing the release of gastric inhibitory polypeptide and cholecystokinin. Stress also stimulates the sympathetic nervous system, which inhibits duodenal gland secretion. As a result, the duodenum has less of a mucous coating and is more susceptible to gastric acid and enzymes.

Relaxing after a meal helps decrease sympathetic activities and increase parasympathetic activities.

4. Obstruction of the common bile duct blocks the flow of bile from the gallbladder to the small intestine. As a result, stomach acids are not diluted and neutralized to as great an extent as if bile were present. Lipids are not emulsified by the bile, resulting in decreased lipid digestion and absorption. Excretory products, such as bile pigments, cholesterol, and lipids, are not as readily removed from the body.

5. Introduction of food into the stomach increases the frequency of mass movements. The mass movements move feces into the rectum. Also, during the night, much of the material in the intestine has moved to the lower part of the large intestine. Mass movements following breakfast, therefore, are more likely to move feces into the rectum. Stretch of the rectum triggers the defecation reflex.

Chapter 17

1. The recommended daily consumption of saturated fats is less than 20 g. Therefore, the % Daily Value for 2 g of fat is 10% (2 / 20 = 0.10, or 10%).

2. It is recommended that no more than 30% of the daily total kilocaloric intake be fats. For a 3000 kcal/day diet, that is 900 kcal (3000 × 0.30). There are 9 kcal/g of fat. Therefore, the maximum amount (weight) of total fat consumed per day is 100 g (900 / 9).

3. From question 2, the teenager's daily maximum amount (weight) of fat consumed per day is 100 g. If he consumes a 10-g serving of fat, his % Daily Value is 10% (10 / 100 = 0.10, or 10%).

4. From question 3, one serving equals 10% Daily Value. If he eats half of the contents of the package, he consumes two servings, or 20% Daily Value.

5. An active person uses many more kilocalories in a day than a person who is not active, and so can lose weight on a higher-kilocalorie diet. In addition, some people have a higher basal metabolic rate and tend to burn more kilocalories.

6. Ingested proteins are digested to amino acids in the stomach and intestine and are then absorbed into the circulation. The amino acids can be used by cells as building blocks for proteins, which is very important if a person is attempting to build up muscle mass. However, beyond a certain point, the excess amino acids are broken down in the liver to make glucose. These amino acids are, therefore, no more helpful than any other source of energy.

If the amount of amino acids is excessive, ammonia or keto acids, the breakdown products of amino acid metabolism, can accumulate to toxic levels.

7. Lotta Bulk's urine would have more urea than before she began the high-protein diet. Excess protein is metabolized and used as a source of energy. One of the breakdown products of protein is ammonia. Ammonia is toxic, so it is converted in the liver to urea, which is carried by the blood to the kidneys and eliminated.

8. As ATP breakdown increases, more ATP is produced to replace what is used. Over an extended time, the ATP must be produced through aerobic respiration. Therefore, O_2 consumption and basal metabolic rate increase. The production of ATP requires the metabolism of carbohydrates, lipids, or proteins. As these molecules are used at a faster than normal rate, body weight decreases. Increased appetite and increased food consumption resist the loss in body weight. As ATP is produced and used, heat is released as a by-product. The heat raises body temperature, which is resisted by the dilation of blood vessels in the skin and by sweating.

9. No, this approach does not work because he is not losing stored energy from adipose tissue. In the sauna, he gains heat, primarily by convection from the hot air and by radiation from the hot walls. The evaporating sweat is removing heat gained from the sauna. The loss of water will make him thirsty, and he will regain the lost weight from drinking fluids and eating foods containing water.

10. Drinking cool water could help in two ways. Because the water is cool, raising the water to body temperature requires the expenditure of energy. Also, stretch of the stomach decreases appetite (see chapter 16).

11. During fever production, the body produces heat by shivering. The body also conserves heat by constricting blood vessels in the skin (producing pale skin) and by reducing sweat loss (producing dry skin). When the fever breaks—that is, "the crisis is over"—heat is lost from the body to lower body temperature to normal. This is accomplished by dilation of blood vessels in the skin (producing flushed skin) and increased sweat loss (producing wet skin).

Chapter 18

1. The rate of urine production increased over the next 2 hours, and the urine was dilute. The water he consumed increased his blood volume and decreased the concentration (osmolality) of his blood. Both of these changes inhibit ADH secretion from the posterior pituitary gland. The reduced ADH levels in the blood cause the kidney to produce a large volume of dilute urine.

2. The sodium chloride increased the concentration (osmolality) of his blood without affecting the volume of blood. The increased concentration (osmolality) of his blood stimulated ADH secretion from the posterior pituitary gland. The ADH caused the kidney to produce a small volume of concentrated urine.

3. The excessive sweating results in a reduced blood volume and the blood has a greater than normal concentration (osmolality). Both of these changes stimulate ADH secretion from the posterior pituitary gland. The ADH acts on the kidney to produce a small volume of concentrated urine.

4. Excessive urine production, low blood pressure, high plasma K^+ levels, but not high plasma Na^+ levels, result. Low aldosterone levels result in excessive loss of Na^+ in the urine. The Na^+ attract water molecules from the peritubular capillaries into the nephron, which results in a greater than normal urine volume. The loss of Na^+ in the urine reduces the plasma Na^+ concentration. The loss of water in the urine reduces the blood volume and therefore reduces the blood pressure. Because aldosterone causes Na^+ reabsorption from the nephron and K^+ secretion into the nephron, K^+ accumulate in the plasma.

5. The following all decrease the blood pressure in the glomerular capillaries and therefore decrease the glomerular filtration rate: decreased blood pressure and increased sympathetic stimulation that constricts the renal arteries and the afferent arterioles. Blocking the nephron causes the capsule pressure in the nephron to increase, so that it is equal to the pressure in the glomerular capillaries, which reduces the filtration rate. An increase in plasma proteins also decreases the filtration rate. The increased plasma protein concentration attracts water by osmosis and therefore reduces the tendency for water to pass from the glomerulus to Bowman capsule.

6. Removal of a large amount of bicarbonate ion (HCO_3^-) from the body fluids causes a decrease in the body fluid pH because there are fewer HCO_3^- to combine with hydrogen ions (H^+) according to the following formula:

$$CO_2 + H_2O \rightleftharpoons H_2CO_3 \rightleftharpoons H^+ + HCO_3^-$$

Carbon   Water   Carbonic   Hydro-   Bicarbon-
dioxide   acid   gen ion   ate ion

Consequently, the H^+ accumulate in the body fluids and decrease the body fluid pH. The rate and depth of respiration increase because a decreased body fluid pH stimulates the respiratory center. The decrease in the body fluid pH causes the kidneys to secrete additional H^+ into the nephron. Therefore, the pH of the urine decreases.

7. While Spanky held his breath, CO_2 accumulated in his body fluids. The increased CO_2 combined with water to form carbonic acid that, in turn, produced H^+ and HCO_3^-. The increased concentration of H^+ in the body fluids caused the pH of the body fluids to decrease. After 2 minutes, the reduced pH and increased body fluid CO_2 levels strongly stimulated the respiratory center to increase the rate and depth of respiration. The increased rate of respiration reduced the body fluid CO_2 levels and increased the body fluid pH back toward its normal value. The kidneys did not respond to the change in pH within 2 minutes. The kidneys are more powerful regulators of body fluid pH, but they require several hours to become maximally active.

8. Martha's blood pH increased. Her frequent vomiting resulted in the loss of H^+ from her stomach and an increase in HCO_3^- in her blood. Since Martha lost fluid from her body and she consumed no fluid, the concentration of her body fluids increased and her blood volume decreased. Consequently, ADH secretion increased. Because blood volume decreases, the blood flow to the kidneys decreases. This stimulates renin secretion from the kidney. This, in turn, converts angiotensinogen to angiotensin I. Angiotensin I is converted to angiotensin II, which stimulates aldosterone secretion from the adrenal cortex. In response to an increase in blood pH, the kidneys decrease their rate of H^+ secretion and HCO_3^- reabsorption. Consequently, this causes an increase in blood H^+ concentration and a decrease in blood HCO_3^-. The kidney's response to changes in pH takes many hours, but it has a large capacity to maintain pH homeostasis.

Chapter 19

1. Removing the testes eliminates the source of testosterone. Therefore, blood levels of testosterone would decrease. Because testosterone has a negative-feedback effect on the hypothalamus and pituitary gland, GnRH, FSH, and LH secretion increases and the blood levels of these hormones increase.

2. The estrogen and progesterone in birth control pills inhibit the large increase in LH secretion from the anterior pituitary, which is responsible for ovulation. Without the large increase in LH secretion, ovulation cannot occur.

3. E is the correct answer. The secretory phase of the menstrual cycle occurs after ovulation. It is following ovulation that the corpus luteum forms and produces progesterone. In addition, the progesterone acts on the endometrium of the uterus to cause its maximum development. Therefore, progesterone secretion reaches its maximum levels and the endometrium reaches its greatest degree of development during the secretory phase of the menstrual cycle.

4. D is the correct answer. Between approximately 12 and 14 days of the menstrual cycle, you would normally expect increasing blood levels of estrogen and LH. In the average menstrual cycle, ovulation occurs on day 14. Prior to that time, estrogen levels are increasing. The increasing estrogen stimulates LH secretion. The increasing LH, in turn, causes increased estrogen secretion from the developing follicle. Blood progesterone levels are low at this time. Progesterone is not secreted in large amounts until the corpus luteum is formed after ovulation.

5. E is the correct answer. On day 15 of the menstrual cycle, one would expect decreasing blood levels of LH, decreasing blood levels of estrogen, and increasing blood levels of progesterone. In the average menstrual cycle, ovulation occurs on day 14. After ovulation, the ovulated follicle develops into the corpus luteum. Estrogen levels decrease as the follicle is converted to the corpus luteum. The corpus luteum secretes progesterone in increasing quantities and small amounts of estrogen as the corpus luteum develops. The increasing blood levels of progesterone inhibit LH secretion from the pituitary gland. Consequently, blood levels of LH fall to low levels.

6. A drug that blocks the effect of progesterone on its target tissues will cause the tissue to respond as if no progesterone were present. Progesterone is secreted by the corpus luteum for about 7 days after ovulation. The progesterone affects the endometrium of the uterus and, in response, the endometrium becomes prepared for implantation. A decline in progesterone causes menstruation to occur. If a drug is taken 3–4 days following ovulation that blocks the effect of progesterone, the endometrium will not become fully prepared for implantation and menstruation will begin early. If a drug that blocks progesterone is taken by a woman who is pregnant, the effect will be to cause the endometrium to slough. This is similar to the events that occur during menstruation. If this occurs, it will terminate the pregnancy. Progesterone is necessary to maintain pregnancy. The corpus luteum of the ovary secretes progesterone until the end of the first 3 months of pregnancy; prior to the third month of pregnancy, the placenta begins to secrete progesterone. The placenta becomes the primary source of progesterone after the third month of pregnancy.

7. During menopause, estrogen and progesterone are produced by the ovary in only small amounts. Consequently, estrogen and progesterone levels in the blood are low. These hormones have a negative-feedback effect on the secretion of FSH and LH from the anterior pituitary gland. Therefore, in the absence of estrogen and progesterone, FSH and LH are secreted in greater amounts and the blood levels of FSH and LH increase. However, there is no large increase in LH, like the increases that occur prior to ovulation. The average concentration of LH and FSH is greater than the levels that occur either before or after ovulation.

Chapter 20

1. Fertilization occurs approximately 14 days after the LMP, so the embryo in this case is 30 days old. It has just completed the "budding" period; limb buds and other "buds" have just formed (see table 20.2).

2. At 35–45 days post-LMP, the embryo would be 21–31 days old, which overlaps the period of neural tube closure (18–26 days). A high fever during this period could prevent neural tube closure in the embryo, so that the newborn baby may have neural tube defects, such as anencephaly (open neural tube in the area of the brain, resulting in absence of the upper brain) or spina bifida (open spinal cord resulting from failure of the neural tube to close in that area).

3. The correct answer is A. Prolactin is responsible for the production of milk by the breast. Surges of prolactin are stimulated by the process of suckling and are required to maintain milk production. Drugs that inhibit prolactin release will cause the breast to cease milk production after a few days.

4. Yes. Because dimpled cheeks are dominant, a person with dimpled cheeks may have a genotype of Dd (D being the dominant gene for dimples and d being the normal gene). If two Dd, dimpled people have a child, there is a 25% probability that the child would not have dimples (dd).

5. No, not without additional information. To be able to roll their tongue into a tube, both mother and son need to have only one dominant gene. The son could inherit this dominant gene from his mother, so it is possible that his father is a non–tongue roller. It is also possible that his father is a tongue roller, but there is no proof for either hypothesis.

6. Hemophilia is an X-linked recessive gene. A female must have the genotype X^hX^h to have hemophilia, whereas because males have only one X chromosome, a single hemophilia gene will cause them to have the disorder (X^hY). For their daughter to have hemophilia, she would have the genotype X^hX^h. That would mean she inherited the X^h gene from both parents. Since the mother does not have hemophilia, she is a carrier and must be XX^h. The father's genotype must be X^hY (he has hemophilia).

Appendix E

Answers to Predict Questions

Chapter 1

2. To answer this question, you must first realize that regulation of our body's environment is due to homeostatic mechanisms. These mechanisms work to keep variables near their set point. In this chapter, you learned that in order to keep body temperature near its set point, the control center (the hypothalamus), receives input from thermoreceptors in the skin. The hypothalamus would then instruct the effectors (the sweat glands), to produce sweat if body temperature had risen too high. So, for this question, you can predict that swimming in cool water would prevent the hypothalamus from stimulating the sweat glands to produce sweat. The next part of the question asks what would happen if this mechanism was not sufficient to maintain normal body temperature. In other words, if you were swimming in cool water, the thermoreceptors would cause the hypothalamus to initiate shivering. Simply preventing sweating was insufficient to keep body temperature from dropping out of its normal range. If shivering continued to be ineffective, the body would conserve heat through loss of consciousness and severe hypothermia could set in.

3. To answer this question, you must first recall the major difference between negative-feedback mechanisms and positive-feedback mechanisms. Negative feedback returns a variable to its set point while positive feedback keeps a variable different from its set point. Also, you read that the body cells' internal fluid and solute environment is maintained within a narrow range. When you are thirsty your impulse is to drink something, thereby adding fluid to your body. This is a negative-feedback mechanism because drinking fluids restores blood volume and pressure. When these variables return to normal you no longer feel thirsty.

4. In order to recognize which correct term to use here, you must first realize that directional terms are relative terms to the body. Therefore, it doesn't matter what position your body is in compared to the earth, body parts always have the same relationship to each other. Thus the nose is always referred to as being superior to the mouth.

5. To determine the proper quadrants of these organs, it is important to know that the positions of organs in the body are consistent from person to person. Referring to figures 1.2 and 1.10*a*, it is simple to answer this question. Just remember to use the organs in figure 1.10*a* as reference points when comparing to organ positions in figure 1.2. Thus, spleen: left-upper; gall bladder: right-upper; left kidney: left-upper; right kidney: right-upper; stomach: mostly left-upper; liver: mostly right-upper.

6. The first step is to define the abdominopelvic and peritoneal cavities. The abdominopelvic cavity is located inferiorly to the diaphragm and superiorly to the pubic symphysis. The peritoneal cavity is located between the visceral peritoneum, which covers organs in the abdominopelvic cavity, and the parietal peritoneum, which lines the wall of the abdominopelvic cavity. So, look at figure 1.15*c* and take notice of the bright white area. This is the peritoneal cavity containing only peritoneal fluid. You can see that, although the peritoneal cavity is around these organs, they are not within the peritoneal cavity. Secondly, looking at the right side of figure 1.15*c* again, notice there are organs behind the parietal peritoneum, but inside the abdominopelvic cavity. These organs are also not within the peritoneal cavity and are considered retroperitoneal (e.g., the kidneys).

Chapter 2

2. The question asks you to differentiate between mass and weight. First, consider the definitions. Weight, in particular, is dependent on the force of gravity. Therefore, if an astronaut is in outer space, where the force of gravity from earth is nearly nonexistent, the astronaut is "weightless." However, the definition of mass is the amount of matter present in the object itself. Thus, no matter the location of an object, the mass remains constant.

3. The question asks you to predict the atomic structure of fluorine. By definition, the atomic number (9) is the number of protons. Therefore, there are 9 protons in fluorine. Since the number of electrons in an atom is the same as the number of protons, there are 9 electrons. To find the number of neutrons of any element, subtract the atomic number from the mass number ($19 - 9 = 10$). So there are 10 neutrons.

4. A fun, easy way to remember how loss of electrons affects the charge of an atom is to ask: "Atom, are you sure you lost an electron?" Atom: "I'm positive." Therefore, we know that the charge on an iron ion would be positive. The charge is also equal to the number of electrons lost, in this case, 3. Therefore, the iron's charge would be $+3$ and the symbol is Fe^{3+}.

5. To answer this question, you must recall the relationship between CO_2, H_2O, and H^+ in solution. Carbon dioxide readily combines with water, resulting in the production of free H^+. Therefore, as the amount of CO_2 decreases, the reversible reaction will shift in the other direction to form CO_2. Similar to the trough of water example, if CO_2 levels decrease, it's like raising the right side of the trough causing water to flow to the left. The reaction "flows" to the left: $CO_2 + H_2O \leftarrow H^+ + HCO_3^-$. In order for this to happen, free H^+ combines with HCO_3^- decreasing its level in the blood. Section 2.3 explains that this decrease in H^+ levels changes the acidity of the blood so that it becomes more basic.

6. During exercise, your body is doing work by muscular contractions. Work involves converting one form of energy into another, and as you read in the previous section, this conversion is not 100% efficient. As a result, heat energy is released. When you contract your muscles, potential energy is converted to kinetic energy and heat energy. Thus, more heat is produced than when at rest, and your body temperature increases.

7. To answer this question, first consider the pH scale. It's important to realize that it is an inverse scale. In other words, a high concentration of H^+ (acid) is represented by a low pH value. For example, a pH of 2 means there are many free H^+ in the solution. The question asks what will happen to the pH value if a base is added

to a solution. Because bases combine with H^+, there will be less free H^+ and the pH value will increase. To evaluate the effect of a buffer, it's necessary to know that a buffer is a chemical that resists changes in pH. In an acidic solution, a buffer will act as a H^+ reservoir, releasing H^+ when a base is added. The released H^+ combines with the base to inhibit or even prevent the solution from becoming more basic.

Chapter 3

2. First, we need to consider the normal process and identify the intracellular and extracellular areas involved. We are told that urea diffuses from liver cells, which is the intracellular region, to the blood, which is the extracellular region. This also defines the direction of urea diffusion, from the area of higher urea concentration inside the cells to the area of lower urea concentration in the blood. The kidneys remove the urea from the blood; therefore, if the kidneys stopped functioning, the concentration of urea in the blood would increase. Eventually, this would eliminate the urea concentration gradient or even reverse it. Urea would remain in the cells and increase to toxic levels that could damage or even kill the cells.

3. Remember that diffusion, whether simple or facilitated, is the movement of a substance down its concentration gradient. That means that the glucose concentration gradient between the extracellular fluid and the cytoplasm depends on the amount of glucose molecules only and is not affected by other molecules, even if they are similar. If glucose is converted to other molecules inside the cell, the concentration gradient is maintained and the cell can continue to take up more glucose.

4. In the description of cystic fibrosis we learn that it is the result of defective chloride ion channels that fail to transport Cl^- out of cells. This description allows us to easily predict the effect on Cl^- concentrations inside and outside of the cell, because ion channels allow ions to diffuse down their concentration gradients. The concentration of Cl^- would remain higher inside the cell and lower outside the cell.

5. To answer this question, we must first consider the functions involved in each cell described and identify the organelles that carry out these functions. Referring to table 3.1 would help to quickly identify the organelles and functions. (a) Cells that synthesize and secrete proteins would require organelles involved in the manufacturing, packaging, and releasing of proteins from the cell. The rough endoplasmic reticulum, with the attached ribosomes, carries out the synthesis of proteins that will be released from the cell. The Golgi apparatus is also involved in the packaging of cellular materials that are secreted by packaging the proteins into secretory vesicles that move to the plasma membrane. (b) Active transport requires ATP to move materials across the cell membrane, so we would expect the cell to have many mitochondria to produce ample ATP. (c) Cells that ingest foreign substances by endocytosis would require the enzymes needed to break down those substances. We learned that lysosomes are vesicles of digestive enzymes that break down materials brought into the cell.

6. Recall from "Gene Expression" that a gene is a sequence of DNA nucleotides that provides the instructions for making a specific protein. We also learned that the sequence of DNA nucleotides determines the sequence of RNA nucleotides of an mRNA, which then determines the sequence of amino acids of a protein in the processes of transcription and translation, respectively. We also learned that the information in mRNA is carried in three nucleotide groups called codons, which specify a particular amino acid or signal the end of translation. So, if one nucleotide in a DNA molecule is changed, that one codon in the mRNA molecule will also be changed. This can lead to several different outcomes. Since an amino acid can often be specified by more than one codon, the changed nucleotide may not change the amino acid and hence the protein structure will not be affected. Alternatively, the changed nucleotide may change the codon to specify a different amino acid. This single amino acid change could in turn change the protein structure. Depending on the changed amino acid, the structural change could be subtle or severe. Finally, the changed nucleotide could create a new stop codon or eliminate an existing stop codon. These changes would very likely cause dramatic changes in the protein structure since a premature stop codon would cut the protein short and removal of a stop codon would add extra amino acids not normally in the protein.

7. Recall that after cell division, the new cells undergo the process of differentiation, which is when the cell develops the specialized structures and functions of a mature cell. As a result, cells of different tissues do not all look the same. If cancer cells are continuously dividing, they do not undergo this differentiation process and therefore do not appear the same as a mature cell of the specific tissue.

Chapter 4

2. a. The question asks about the relationship between form and function of tissues. First, consider the name of the tissue type: nonkeratinized stratified epithelium. The term stratified means more than one layer of cells, whereas, the term simple means a single layer of cells. In the digestive tract, a principal function is absorption, a process that would be hindered by the many layers of stratified epithelium. Stratified epithelium is more suited to areas where the layers would protect underlying tissues from abrasion. Columnar cells are specialized for secretion and absorption. These cells contain a large number of organelles that produce the secretions and transporters needed to support absorption.

   b. In this scenario, both tissue types are stratified, but one type lacks keratin. The protein keratin provides a tough layer that retards water movement. If keratin were absent from the epidermis, the body could not retain water effectively and would be more prone to damage from abrasion.

3. The question asks how the structure of a tissue's components contributes to its function. When a muscle contracts, the pull it exerts is transmitted along the length of its tendons. The tendons need to be very strong in that direction but not as strong in others. The collagen fibers, which are like microscopic ropes, are therefore all arranged in the same direction to maximize their strength. In the skin, collagen fibers are oriented in many directions because the skin can be pulled in many directions. The collagen fibers can be somewhat randomly oriented, or they can be organized into alternating layers. The fibers within a layer run in the same direction, but the fibers of different layers run in different directions, similar to steel rods of rebar in a concrete structure.

4. The question tells you that: (1) vitamin C is required for collagen formation; (2) scars are formed from collagen; and (3) there is a lack of vitamin C. Therefore, if there is a lack of vitamin C, the density of collagen fibers in a scar may be reduced and the scar may not be as durable as a normal scar.

5. There is more than one way to organize a table that summarizes the characteristics of the major muscle types. See the following table.

6. First, define inflammation. Inflammation produces five main symptoms: redness, heat, swelling, pain, and disturbance of function. These symptoms occur because of an increased blood flow to the area. However,

TABLE E.I	Major Characteristics of the Three Muscle Types					
Muscle Type	Nuclei in Each Cell	Location of Nuclei	Control	Cell Shape	Striated	Branching Fibers
Skeletal	Many	Peripheral	Voluntary	Long and cylindrical	Yes	No
Cardiac	One	Central	Involuntary	Branching cylinders joined by intercalated disks	Yes	Yes
Smooth	One	Central	Involuntary	Spindle-shaped cells	No	No

if the area is so badly damaged that blood vessels are destroyed, no inflammation occurs in the damaged site. However, in the surviving tissues surrounding the dead tissue, inflammation occurs.

7. The question asks why stitches are helpful for healing a wound with a large space between the edges. Sutures bring the edges of a wound closer together so that there is (1) less space for regenerating cells to fill in; (2) a lower risk of infection; (3) a reduced amount of granulation tissue formation; and (4) a faster healing time. Additionally, the smaller amount of scar tissue and wound contracture lessens the visibility of the wound site later on.

Chapter 5

2. In the description of the epidermis, the superficial layer of the skin, we learned that the keratinized cells are coated with lipids to prevent fluid loss. Recall from chapter 3 that substances that are lipid-soluble easily diffuse through lipid layers but water-soluble substances do not. By applying the same principles of diffusion across cell membranes to diffusion across the skin, we can predict that lipid-soluble substances diffuse easily but water-soluble substances do not.

3. Recall that skin color is determined by three factors: skin pigments, blood flow in the skin, and the thickness of the stratum corneum. We should consider these three factors for each of the comparisons. (a) The lips are pinker or redder than the palms of the hands. The skin of the palms of the hands is much thicker than the skin of the lips. As a result, the redness due to blood flow is more visible in the thinner skin of the lips than in the thicker skin of the palms of the hand. (b) The palms of the hands of a person who does heavy manual labor are more likely to be calloused and therefore have a thicker stratum corneum. The thicker stratum corneum masks the redness due to blood flow and will also have more carotene pigmentation, so the palms will appear paler with a yellowish tint. The palms of the hands of the person who does not do heavy manual labor

will have a pinker appearance due to the lack of calluses and a thinner stratum corneum. (c) The posterior surface of the forearm is usually darker than the anterior surface of the forearm. Recall that the pigment melanin contributes to the darker, tanned appearance of skin. Also, melanin production increases with sun exposure. The posterior surface of the arm is usually exposed to more sunlight than the anterior surface, so we would expect more melanin production and a darker appearance on the posterior surface.

4. Recall that hair color is determined by the production and distribution of melanin by melanocytes in the hair bulb. Also, remember that hair growth begins at the hair bulb and that the pigmented cells are pushed farther and farther away from the hair bulb, as the hair grows longer. This would make it impossible for Marie Antoinette's hair to suddenly turn white; only new hair growth would show the loss of pigment. Since it takes quite a while for hair to grow, it is impossible that her hair could turn white in one day.

5. Recall that the redness of skin is due to increased blood flow through the skin. Also, we learned that blood flow determines the distribution of heat to different regions of the body. The red appearance of the nose and ears indicates increased blood flow in these areas, and we can assume that this leads to warming of this tissue that is probably very cold due to the weather.

6. We learned that one of the functions of the skin is to reduce water loss. Sam's burns resulted in severe damage to his skin, which most likely led to increased water loss at the injury site, causing dehydration and reduced urine production. We learned that Sam was administered a large volume of fluid to counteract his increased fluid loss. But how much fluid should be given? The amount of fluid given should match the amount that is lost, plus enough to keep the kidneys functioning properly. An adult receiving intravenous fluids should produce 30–50 mL of urine per hour, and children should produce 1 mL/kg of body weight per hour. By monitoring Sam's

urine output, the nurse can determine if he is getting enough fluids. If his urine output is too low, more fluids can be given.

Chapter 6

2. In the previous section, you learned that the minerals of bone give it weight-bearing strength much as concrete provides strength to columns of a bridge. Without the minerals, the bone matrix would consist of only collagen and would be overly flexible—it could be tied into a knot. On the other hand, you learned that collagen springs back to its original shape after it has been bent. Therefore, without collagen, bone would be too brittle and would easily break. Collagen loss is one reason the bones of many older people break easily.

3. Without cartilage growth, the long bones will be shorter than normal. Bone growth occurs in the epiphyseal plate. As you learned, the epiphyseal plate is a section of hyaline cartilage between each epiphysis and the diaphysis. In order for bone to grow in length, the hyaline cartilage in the epiphyseal plate must first increase in thickness and can then be replaced by bone with the action of osteoblasts. However, bone thickness does not require cartilage growth to occur first. Therefore, an adult that lacked cartilage growth in the long bones during childhood would have much shorter bones than normal, but their bones would be the same diameter, or even thicker than normal. One type of dwarfism exemplifies this. The person's head and trunk are normal in size, but the long bones of the limbs are very short.

4. Tears run over the surface of our eyes and enter the nasolacrimal duct, which empties into the nasal cavity. Therefore, when our eyes are producing more tears than normal, the increased liquid in our nasal cavity causes a "runny nose."

5. Just before the swimmer begins the power stroke, the arm is flexed (close to the body) and medially rotated (towards the center of the body) and the forearm is extended (away from the body) and pronated (palm down). During the power stroke, the arm is powerfully extended (pushed away from

the body), slightly abducted (moved away from center), and medially rotated (towards the center of the body). During the recovery stroke, the arm is circumducted (circled around the shoulder towards the back), laterally rotated (moved away from the center of the body), and flexed (brought closer to the body) in preparation for the next stroke. The forearm is flexed (close to the body) during the first part of the recovery stroke and extended (away from the body) during the last part.

6. Hopefully, Betty will discourage her granddaughter from smoking as it reduces estrogen levels. Lowered estrogen levels, as you learned while reading the background information, are a major contributor to osteoporosis. Additionally, Betty should encourage her granddaughter to get adequate calcium and vitamin D in her diet and to get some sunscreen-free sun exposure for vitamin D production (no more than 20 minutes a day). Betty and her granddaughter might consider taking a yoga class together as well. Exercise is an important source of increased bone density because it causes the muscles to put stress on the bones, which thickens the bones. Because the greater the density of bone before the onset of osteoporosis, the more tolerant it is to bone loss, most people (especially women in their 20s and 30s) need to make sure to get adequate calcium and exercise.

Chapter 7

2. To predict a mechanism for anesthetic activity, you must first understand how action potentials are initiated and propagated in electrically excitable cells. In this chapter, you learned that in both nerve cells and muscle fibers, it is Na^+ entry that depolarizes the cell membrane to threshold. Now, recall that at threshold, it is the opening of voltage-gated Na^+ channels that creates the action potential. Thus, you can predict that in order for an anesthetic to prevent action potential formation, it prevents Na^+ channels from opening. Without Na^+ entry, there is no action potential and no transmission of information, such as pain.

3. We learned that aerobic respiration is the principal energy source for resting muscles or muscles undergoing long-term exercise; therefore, a 10-mile run requires aerobic respiration, which can provide a consistent source of ATP. However, during the sprint, the intense contraction of the muscles uses ATP faster than aerobic respiration can replace it, and thus anaerobic respiration is beneficial because it can

produce ATP faster than can aerobic. However, anaerobic respiration can only produce ATP for 2–3 minutes and can't be maintained. Because muscles replenish O_2 and glucose stores after physical exercise, aerobic respiration and breathing rate remain elevated. This can be thought of as repaying the O_2 deficit.

4. Referring to figure 7.16 will help you answer this question. Raising eyebrows—occipitofrontalis; winking—orbicularis oculi; whistling—orbicularis oris and buccinators; smiling—zygomaticus; frowning—depressor anguli oris; sneering—levator labii superioris.

5. Shortening the right sternocleidomastoid muscle rotates the head to the left and slightly elevates the chin.

6. In the sprinter's stance and the bicyclist's racing posture, the thigh is flexed at a 45-degree angle because at that angle the gluteus maximus functions at its maximum in extending the thigh, thus providing maximum force.

7. DMD affects the muscles of respiration and causes deformity of the thoracic cavity. The reduced capacity of muscle tissue to contract is one factor that reduces the ability to breathe deeply or cough effectively. In addition, the thoracic cavity can become severely deformed because of the replacement of skeletal muscle with connective tissue. The deformity can reduce the ability to breathe deeply. DMD can also affect the muscle of the heart and cause heart failure.

Chapter 8

2. To answer this question, let's first describe the Na^+ concentration and its role in a normal excitable cell. We learned that in a normal cell, the concentration of Na^+ is much higher outside the cell than inside the cell. As a result, when Na^+ channels open Na^+ diffuses into the cell quickly causing the changes in the membrane potential that result in an action potential. The movement of Na^+ into the cell is the result of its steep concentration gradient. Remember, also, that enough Na^+ must enter a cell for the membrane potential to reach threshold, opening the voltage-regulated Na^+ channels and causing an action potential. If the extracellular concentration of Na^+ is reduced, then the concentration gradient would also be reduced. The effect would be that if the cell is stimulated, less Na^+ would enter the cell. The cell would not reach threshold and an action potential would not occur.

3. First, refer to figure 8.2 to see the direction of action potential propagation in sensory axons and motor axons. Sensory

axons carry action potentials from peripheral tissues to the central nervous system (CNS), which includes the brain and spinal cord. Motor axons carry action potentials from the CNS to peripheral tissues. Now, identify the types of axons in each structure listed. Spinal nerves, recall, have both sensory and motor axons, so action potentials are propagated both to the spinal cord and away from the spinal cord. Dorsal roots contain only sensory axons, so action potentials are conducted to the spinal cord only. Finally, ventral roots contain only motor axons, so action potentials are conducted away from the spinal cord.

4. Recall that the phrenic nerve innervates the diaphragm allowing for the contraction necessary for breathing. If the right phrenic nerve was damaged, then we would expect lack of muscle contraction in the right half of the diaphragm affecting breathing. To answer the second part of the question, we need to consider the location of spinal cord injury to predict the effect it would have on the diaphragm. Remember that the phrenic nerve is part of the cervical plexus, which includes spinal nerves C1–C4. If the spinal cord is completely severed in the thoracic region, the phrenic nerve would not be damaged and the diaphragm would not be affected. On the other hand, if the spinal cord is severed in the upper cervical region, the phrenic nerve would be damaged and the contractions of the diaphragm would not occur, eliminating the person's ability to breathe. Death would likely occur if medical assistance was not administered quickly.

5. We need to consider the functions of the medulla oblongata first to answer this question. Recall that the medulla oblongata contain nuclei that regulate heart rate, blood vessel diameter, breathing, swallowing, vomiting, coughing, sneezing, balance, and coordination. Compression of the medulla could cause damage to these nuclei and lead to other problems in the body. Changes in the regulation of heart rate, blood flow, and respiration are the most life threatening. The cells of the body require adequate levels of O_2 and nutrients to maintain homeostatic conditions. Blood flow and respiration are vital functions and interference with either may inhibit normal homeostatic functions. If not corrected, the loss of homeostasis results in death. Interestingly, the heart can continue to beat, even if the medulla oblongata is damaged. However, blood pressure would most likely drop, resulting in shock, which can ultimately result in death.

6. To answer this question, we need to first identify the type of sensation, the area of the cerebrum that detects this sensation, and the association area that allows for recognition of the object based on past experience. A person that is blindfolded would use the sense of touch to gather information about the object in order to describe it. Sensations from the skin of the hand are sent via sensory neurons to the dorsal horn of the spinal cord. The signals are then relayed to spinal cord neurons and transmitted to the brain. We learned earlier that in the brain, the sense of touch is perceived in the primary somatic sensory cortex of the parietal lobe of the cerebrum and that the somatic sensory association area allows for recognition of the stimulation. The information is passed from the primary somatic sensory cortex to the somatic sensory association area, where she will recognize the characteristics of the object. Now, that she has recognized the object, we need to describe the pathway involved in determining a word and initiating the muscle contractions necessary to say the word. This occurs when action potentials travel to the sensory speech area, where the object is given a name. From there, action potentials travel to motor speech area, where the spoken word is initiated. Action potentials from the motor speech area travel to the premotor cortex and then to the primary motor cortex, where action potentials are initiated that stimulate the muscles necessary to formulate the word.

7. The autonomic nervous system maintains homeostasis by adjusting body activity to different levels of stress and activity. Recall that the sympathetic division of the autonomic nervous system, often called the fight-or-flight system, prepares the body for physical activity but also prepares the body for stressful situations. The parasympathetic division, often called the rest-and-digest system, is consistent with resting conditions. Keeping these descriptions in mind and referring to table 8.8, we can easily answer this question. (a) A person who is extremely angry is experiencing a stressful situation and we would expect changes in the body associated with sympathetic activity to occur. Examples include increased heart rate, increased breathing rate, dilation of pupils, and decreased digestive and urinary functions. (b) A person who has just finished eating and is now relaxing would exhibit changes associated with the parasympathetic division. Examples include decreased heart rate, decreased breathing rate, constriction of pupils, and increased digestive and urinary functions.

8. In this chapter, we learned that the right side of the body is monitored and controlled by the left side of the brain. The crossing over of axons from one side of the brainstem or spinal cord to the other side of the brainstem or spinal cord results in this pattern of regulation. Since Scott exhibited loss of skeletal muscle movement and sensations on his right side, we can conclude that the left side of the brainstem was more severely affected by the stroke.

Chapter 9

2. To answer the first part of the question, we must recall the types of pain and the types of structures associated with each type of pain. Constipation is causing pain associated with the colon, a deep structure. Remember that the deeper structures lack tactile receptors, so the type of pain experienced is diffuse. We also learned after reading "Referred Pain" that the pain associated with deeper structures is usually felt in more superficial structures. Using figure 9.2, we can predict that the man feels the pain around his navel, the common area for referred pain associated with the colon.

3. We have all probably had a cold and know that a common symptom is a stuffy nose, making it hard to inhale through the nose. This also affects our ability to smell. Recall that many taste sensations are strongly influenced by olfactory sensations. If our sense of smell is decreased, our sense of taste will also decrease.

4. After reviewing the section "Lacrimal Apparatus," we understand that tears from the surface of the eyeball drain through the lacrimal caniculi into the lacrimal sac, which empties into the nasal cavity through the nasolacrimal duct. Our sense of smell is due to the presence of olfactory receptors in the nasal cavity. If medications are placed into the eyes, some may drain into the nasal cavity, which may stimulate the olfactory receptors of the olfactory organ. Our sense of taste is due to the presence of taste receptors in the mouth and pharynx. The ability to "taste" the medication is due to the fluid draining from the nasal cavity into the pharynx stimulating taste receptors.

5. Recall that there are two types of photoreceptors that allow for the sense of vision. We learned that cones allow for color vision and require more light for stimulation, whereas rods allow for black-and-white vision and can function in dim light. Hence, in dim light primarily rods, and not cones, are stimulated, so that we see objects in shades of gray, not in color.

6. This question addresses changes that occur in the eye as a person looks at things far away and close up. Let's rephrase the question to include whether or not the person is looking at a distant object or a near object. As the person is driving a car, she is looking at distant objects down the road. She then looks down to see the near speedometer, and then back to the distant objects down the road. Recall that the lenses change shape to focus on objects that are either distant or near. Now consider the changes that occur to the eye, particularly the lens, in the series of events. Referring to figure 9.15 will also help in answering this question. As she is looking at the road (distant), the ciliary muscles of the ciliary body are relaxed, the tension in the suspensory ligaments is high, and the lenses of the eyes are flattened. When she looks at the speedometer (close), the ciliary muscles contract, reducing the tension in the suspensory ligaments, causing the lenses to thicken. When she looks back to the road (distant), the ciliary muscles relax again, increasing the tension in the suspensory ligaments, causing the lenses to flatten again.

7. When you hear a faint sound, you turn your head toward it because sound waves are collected by the auricle and conducted through the external auditory canal to the tympanic membrane. The shape of the auricle is such that sounds coming from the side and front of the head are most efficiently conducted into the external auditory canal. Turning the head toward the sound helps maximize accumulation of the sound waves by the ear. In addition, because the sound reaches each ear at a slightly different time, you can localize the origin of the sound. Remember also that reflexes integrated in the superior colliculi cause the head and therefore the eyes to turn toward a sound, so that you can see what is making the sound as well.

8. We learned that the brain compares sensory input from the semicircular canals, eyes, and proprioceptors in the back and lower limbs and that conflicting information can cause motion sickness. If you close your eyes, one of these sources of information is eliminated and the brain has less conflicting input to compare, reducing the probability of motion sickness. We can perceive more motion in close objects than in distant objects, so looking at the horizon would also reduce the visual input of perceived motion to the brain and reduce the probability of motion sickness.

Chapter 10

2. The description of the drug is very important to determining the function of the hormone that it affects. First, the drug is lipid-soluble. Recall that many lipid-

soluble hormones bind to intracellular receptors. Next, we learn that the drug prevents the synthesis of messenger RNA. We can therefore predict that the hormone functions by diffusing across the cell membrane and binding to an intracellular receptor. The response of the intracellular receptor is to produce new messenger RNA, which leads to synthesis of new proteins (see chapter 3 for a description of the role of messenger RNA in protein synthesis). The new proteins produce the cell's response to the hormone.

3. Recall that GH targets cartilage in the epiphyseal plate of long bones and stimulates cell division of the chondrocytes. While Mr. Hoops' son is actively growing, GH administration would cause him to grow taller. However, there could also be unwanted changes consistent with acromegaly (oversecretion of GH). Other side effects such as abnormal joint formation and diabetes mellitus are also possible.

4. Remember that hormones of the hypothalamus and anterior pituitary regulate thyroid hormones. TRH from the hypothalamus stimulates the secretion of TSH from the anterior pituitary. TSH travels to the thyroid to promote secretion of thyroid hormones. In response to the large amount of this protein that is similar to TSH, thyroid hormones are oversecreted by the thyroid gland (hyperthyroidism). Because production of the protein cannot be inhibited by thyroid hormones as TSH is, oversecretion of thyroid gland is prolonged, and the symptoms associated with hyperthyroidism become obvious. Because TSH stimulates gland growth as well as hormone secretion, the thyroid gland enlarges. In addition, the increased thyroid hormones have a negative-feedback effect on the hypothalamus and pituitary gland. TSH-releasing hormone secretion from the hypothalamus and TSH secretion from the anterior pituitary gland are inhibited.

5. Insufficient vitamin D results in insufficient Ca^{2+} absorption by the intestine. As a result, blood Ca^{2+} levels begin to fall. In response to the low blood Ca^{2+} levels, PTH is secreted from the parathyroid glands. PTH acts primarily on bone, causing bone to be broken down and Ca^{2+} to be released into the blood to maintain blood Ca^{2+} levels within the normal range. So much Ca^{2+} can eventually be removed from bones that they become soft, fragile, bent or deformed, and easily broken. In adults the condition is called osteomalacia, and in children the condition is called rickets.

6. Aldosterone is secreted by the adrenal cortex under low blood pressure conditions and results in increased Na^+

reabsorption and water conservation. Therefore, if aldosterone levels are reduced, less Na^+ is reabsorbed and less water is conserved. This results in a lower blood volume and thus low blood pressure. Aldosterone targets the Na^+–K^+ pump, which if you recall from chapter 3, transports Na^+ and K^+ in opposite directions. So as Na^+ reabsorption decreases due to reduced aldosterone, we would expect elevated blood K^+ levels due to the reduced secretion of K^+. High blood K^+ levels lead to altered nerve and muscle function, which could be life threatening.

7. To answer this question, you must first realize that cortisone mimics the normal adrenal cortex secretion, cortisol, and can, therefore, provide negative feedback at the anterior pituitary. Cortisone inhibits ACTH secretion from the anterior pituitary. Second, it's important to understand that ACTH is required to prevent atrophy of the adrenal cortex. In the absence of ACTH, the adrenal cortex shrinks and may never recover to produce its normal secretions even if ACTH secretion increases again.

8. It's important to understand that a large meal, high in carbohydrates, will increase blood glucose as the nutrients are transported from the small intestine into the blood. The body cells need to take up the glucose across their cell membranes either to utilize it immediately or to store it for later use. Cells take up glucose only in the presence of insulin. Glucose serves as the stimulus for insulin secretion and the inhibitor of glucagon secretion. Thus, after eating, insulin levels increase and glucagon levels decrease. Long before 12 hours has passed after a meal, blood glucose levels drop. Again, it is glucose that serves as the signal for hormone secretion. Now the reduction in blood glucose stimulates glucagon secretion and inhibits insulin secretion. Glucagon then promotes release of stored glucose into the blood to be utilized by cells until the next meal.

9. The pineal gland secretes melatonin, which inhibits the release of reproductive hormones by acting on the hypothalamus of the brain. If the pineal gland secretes less melatonin, it should no longer have an inhibitory effect on the hypothalamus. As a result, reproductive hormones could be secreted in greater amounts. If this condition occurred in a child before puberty, we would predict exaggerated development of the reproductive system. The evidence for this mechanism is not as clear in humans as it is in other animals.

10. Removal of part of the thyroid gland reduces the amount of thyroid hormone secreted by the gland. Usually, enough

thyroid tissue can be removed to reduce secretion of thyroid hormone to within the normal range. In addition, the remaining thyroid tissue normally does not enlarge enough to produce more thyroid hormone, although there are exceptions. The removal of the thyroid tissue does not remove the influence of the abnormal antibodies on the tissues behind the eyes. Thus, in many cases the condition's effect on the eyes is not improved.

Chapter II

2. To answer this question, we need to first determine the effect carbon monoxide has on red blood cells. Recall from our reading of the function of red blood cells and the description of hemoglobin that carbon monoxide binds very readily to hemoglobin and does not tend to unbind. Therefore, carbon monoxide–bound hemoglobin in red blood cells can no longer transport O_2. This essentially leads to a decrease in blood O_2 levels. We also learned that low blood O_2 levels stimulate red blood cell production by causing the release of erythropoietin, primarily by the kidneys. Erythropoietin stimulates red blood cell production in the red bone marrow. So, we can predict that the number of red blood cells in the person's blood will increase.

3. Referring to table 11.2 will help identify each of the white blood cells. It is also helpful to compare the size of each cell to red blood cells as well. In each figure, red blood cells are visible for easy comparison. (a) The cell in this figure is slightly larger than red blood cells and has a large round nucleus with a small amount of cytoplasm. This cell is a lymphocyte. (b) The cell in this figure is larger than a red blood cell and has many bluish-purple granules. This cell is a basophil. (c) The cell in this figure is very large and has a kidney-shaped nucleus. It is a monocyte. (d) The cell in this figure has a three-lobed nucleus. It is most likely a neutrophil. (e) The cell in this figure has a bi-lobed nucleus and many bright red–stained granules. It is an eosinophil.

4. People with type AB blood were called universal recipients because they could receive types A, B, AB, or O blood with little likelihood of a transfusion reaction. Looking at figure 11.11, you can see that type AB blood does not have antibodies against A or B antigens. Transfusion of these antigens in types A, B, AB, or O blood therefore does not cause a transfusion reaction in a person with type AB blood. The term is misleading, however, for two reasons. First, other blood groups can cause a transfusion

reaction. Second, antibodies in the donor's blood can cause a transfusion reaction. For example, type O blood contains anti-A and anti-B antibodies that can react against the A and B antigens in type AB blood.

5. We learned that HDN is caused by an Rh incompatibility between a pregnant mother with Rh-negative blood and her Rh-positive child. If the mother is sensitized to the Rh antigen, she can produce anti-Rh antibodies that cross the placenta and cause agglutination and hemolysis of fetal red blood cells. To treat HDN with an exchange transfusion, the donor's blood should be Rh-negative, even though the newborn is Rh-positive. Rh-negative red blood cells do not have Rh antigens. Therefore, any anti-Rh antibodies in the newborn's blood do not react with the transfused Rh-negative red blood cells. Eventually, all of the Rh-negative red blood cells die, and only Rh-positive red blood cells are produced by the newborn.

6. After reading the question, we know to focus on blood tests that are associated with identification of bacterial infections. White blood cells defend the body against pathogens. A white blood differential cell count would be useful and would show an abnormally high neutrophil percentage, as those cells are associated with bacterial infections.

Chapter 12

2. After reviewing figure 12.11a, we can see that the anterior interventricular artery supplies blood to the anterior wall of the heart and much of the left ventricle. A blocked anterior interventricular artery reduces the O_2 supply to the tissue and the cardiac muscle in that area of the heart is not able to contract effectively.

3. Before answering the question, we need to define tetanic contraction. Recall from chapter 7 that a tetanic contraction is a sustained contraction in which the frequency of stimulation of the muscle is so rapid that no relaxation occurs. The purpose of cardiac muscle contractions is to pump blood through the circulation by contracting and relaxing in a repeated cycle. Tetanic contractions in cardiac muscle would interrupt the pumping action produced by the cycle of contraction and relaxation and the blood flow would cease. Tetanic skeletal muscle contractions are important to maintain posture or to hold a limb in a specific position.

4. If the normal blood supply is reduced in a small area of the heart through which the left bundle branch passes, O_2 supply

to the tissue, including the conducting system tissue, in the area is reduced and tissue function is reduced. The conduction of action potentials through that side of the heart is slowed or blocked. As a consequence, the left side of the heart contracts more slowly and its pumping effectiveness is reduced. Since the right bundle branch is not affected, the right side of the heart contracts more normally.

5. There are two different scenarios that we need to consider to answer this question. First, a person who has a damaged left bundle branch will exhibit the consequences outlined for Predict question 4, but we need to describe how the electrocardiogram will also be altered. We learned that the QRS complex results from depolarization of the ventricles after action potentials pass through the bundle branches to the Purkinje fibers. In this first scenario, action potentials pass through the right bundle branch normally but conduction of action potentials through the left bundle branch is slowed because of the damage. As a result, the QRS complex has an abnormal shape and it is prolonged.

   In the second scenario, many ectopic action potentials arise in the atria. Ectopic action potentials originate in areas other than the SA node (the pacemaker of the heart). Each ectopic action potential initiates a new heartbeat, so we can predict that many ectopic action potentials cause an increase in heart rate. It is possible for some ectopic action potentials arising in the atria to occur while the ventricles are depolarized, but these action potentials do not initiate ventricular contraction. Therefore, there can be more P waves than QRS complexes in the electrocardiogram. If ectopic action potentials do not occur in a regular fashion, they can cause the heart to beat at an irregular rate, or arrhythmically.

6. Before addressing the effects of a leaky aortic semilunar valve on left ventricle blood volume, first review the blood flow under normal conditions. In a normal person, the aortic semilunar valve closes during ventricular relaxation. At the same time, blood flows out of the left atrium and into the left ventricle. When the aortic semilunar valve is leaky (incompetent), some blood leaks back into the left ventricles from the aorta during ventricular relaxation. When this blood is added to the blood that normally enters the left ventricle from the left atrium, there is a greater than normal volume of blood in the left ventricle just before ventricular contraction.

A severely narrowed opening through the aortic semilunar valve increases the amount of work the heart must do to pump the normal volume of blood into the aorta. A greater pressure is required in the ventricle to force the same amount of blood through the narrowed opening during ventricular contraction.

7. Recall that the first heart sound occurs at the beginning of ventricular systole (contraction) and results from closure of the AV valves, and the second heart sound occurs at the beginning of ventricular diastole (relaxation) and results from closure of the semilunar valves. Most of the ventricular contraction occurs between the first and second heart sounds of the same beat. Between the first and second heart sounds, blood is ejected from the ventricles into the pulmonary trunk and the aorta. Between the second heart sound of one beat and the first heart sound of the next beat, the ventricles are relaxing and the semilunar valves are closed. No blood passes from the ventricles into the aorta or pulmonary trunk during that period.

8. The shhh sound made after a heart sound is created by the backward flow of blood after closure of a leaky or incompetent valve. A swishing sound immediately after the second heart sound (lubb-duppshhh) represents a leaky aortic semilunar or pulmonary semilunar valve. The shhh sound before a heart sound is created by blood being forced through a narrowed, or stenosed, valve just before the valve closes. The lubb-shhhdupp suggests that there is a swishing sound immediately before the second heart sound, indicating a stenosed aortic or pulmonary semilunar valve.

9. In response to severe hemorrhage, blood pressure decreased, which is detected by baroreceptors. The decreased blood pressure leads to a reduced frequency of action potentials sent from the baroreceptors to the medulla oblongata. This causes the cardioregulatory center to increase sympathetic stimulation of the heart and increase the heart rate. Normally, sympathetic stimulation of the heart also increases stroke volume, as long as the volume of blood returned to the heart is optimizing ventricular stretching, or preload. Following hemorrhage, however, the blood volume in the body is reduced, and the venous return to the heart from the body is reduced. This leads to a reduction in preload. According to Starling's law of the heart, as venous return and preload decrease, so does stroke volume. So, the heart rate is increased due to sympathetic stimulation, but the volume of blood returning to the heart is decreased;

thus, the ventricle does not fill with blood. As a consequence, the stroke volume is low and the heart rate is high.

10. Rupture of the left ventricle can occur several days after a myocardial infarction. As the necrotic tissues are being removed by macrophages, the wall of the ventricle becomes thinner and may bulge during systole. If the wall of the ventricle becomes very thin before new connective tissue is deposited, it can rupture. If the left ventricle ruptures, blood flows from the left ventricle into the pericardial sac, resulting in cardiac tamponade. As blood fills the pericardial sac, it compresses the ventricle from the outside. As a consequence, the ventricle is not able to fill with blood and its pumping ability is rapidly eliminated. Rupture of the left ventricular wall, therefore, quickly results in death.

Chapter 13

2. As stated in the question, atherosclerosis of vessels occurs when lipid deposits block the vessels, which results in reduced blood flow through the vessel. The tissues to which the blocked vessels supply blood will therefore have reduced O_2 and nutrients. Since the carotid artery supplies blood to the brain, we would expect atherosclerosis of these vessels to lead to reduced brain function, which may include confusion and loss of memory.

3. To answer this question, let's view figures 13.14 and 13.5 to see the direction of blood flow and the arrangement of vessels and parts of the heart. Also, remember that veins carry blood towards the heart. An embolus that formed in the posterior tibial vein would pass through the popliteal vein, femoral vein, external iliac vein, common iliac vein, and inferior vena cava. From the inferior vena cava, the embolus would then pass through the right atrium and right ventricle of the heart before moving into the pulmonary trunk. The embolus would then pass from the pulmonary trunk into either the right or left pulmonary arteries into the smaller vessels of the lungs. The embolus would most likely lodge in the lungs because as it moved from the posterior tibial vein towards the heart, it was passing through larger and larger vessels. The vessels of the lungs would be too small for the embolus to pass through.

4. The first condition to consider is why premature heartbeats result in a weak pulse. Premature beats of the heart result in the heart contracting before it has time to fill to its normal capacity. The volume of blood ejected from the heart will therefore be less causing a weak pulse. The second condition to consider is why would cardiovascular shock due to hemorrhage cause a weak pulse. Cardiovascular shock due to hemorrhage, or loss of large volumes of blood, will reduce the volume of blood returning to the heart. If this occurs, the stroke volume will also be greatly reduced since there is not as much blood to be pumped from the heart. Again, the low stroke volume leads to a weak pulse. The third condition to consider is why exercise causes a stronger pulse. Exercise causes an increase in heart rate and stroke volume. The increased stroke volume causes a stronger pulse.

5. Recall from reading "Capillary Exchange" and reviewing figure 13.25 that blood pressure in the capillaries causes outward movement of fluid out of the capillaries and osmotic pressure causes inward movement of fluid into the capillaries. (a) If the plasma protein concentration in the blood is decreased, the osmotic pressure will decrease and there is less movement of fluid from the tissues to the capillaries, causing edema. (b) If blood pressure within a capillary increases, more fluid is forced out of the capillary into the peripheral tissue, causing edema.

6. While the student was sitting with her legs crossed, the blood vessels, particularly in the skin, were blocked. O_2 and nutrient levels decrease and waste products accumulate in tissues supplied by the blocked blood vessels. Recall that the precapillary sphincters that regulate blood flow through capillaries are regulated by metabolic needs, such as O_2 and nutrients. When these levels decrease, the precapillary sphincters relax allowing more blood flow through the tissues, and causing the appearance of the red blotch in the area that previously lacked blood flow.

7. Raynaud syndrome causes severe vasoconstriction in the fingers and toes. This means that blood flow to the fingers and toes is severely restricted. The restricted blood flow will alter the appearance and the health of the tissue of the fingers and toes. Due to the lack of blood flow, the fingers and toes appear white. If blood flow is not sufficient to provide O_2 and nutrients to and remove waste products from the tissues of the fingers and toes, cell damage and death could occur. Necrotic (dead) tissue and gangrene may develop.

8. The rapid loss of a large volume of blood activates the mechanisms responsible for maintaining blood pressure. Recall that blood pressure regulatory mechanisms include the baroreceptor reflexes, chemoreceptor reflexes, and hormonal mechanisms. In response to a dramatic drop in blood pressure due to the rapid loss of a large volume of blood, the baroreceptor mechanism, adrenal medullary mechanism, and chemoreceptor mechanism increase the heart rate and result in vasoconstriction of blood vessels, especially in the skin and viscera. Angiotensin II is produced quickly; it causes vasoconstriction and stimulates aldosterone secretion. Aldosterone, which requires up to 24 hours to become maximally active, increases water reabsorption from the kidneys and reduces the loss of water in the form of urine. Blood volume is therefore increased. All of these mechanisms increase blood pressure back to its normal value.

If blood is lost over several hours, the decrease in blood pressure is not as dramatic as when blood loss occurs quickly. Consequently, mechanisms that respond to a rapid and large decrease in blood pressure are stimulated to a lesser degree. These include the chemoreceptor reflex, the vasopressin mechanism, and the adrenal medullary mechanism. The baroreceptor reflexes are most sensitive to sudden decreases in blood pressure, but the baroreceptor reflexes are still sensitive to decreases in blood pressure that occur over a period of several hours. The baroreceptor reflexes that trigger vasoconstriction in response to the blood loss are substantial. The kidneys detect even small decreases in blood volume. Consequently, the renin-angiotensin-aldosterone mechanism is activated and remains active until the blood pressure is returned to its normal range of values. Aldosterone secretion increases and, though it requires several hours to become maximally active, it continues to stimulate water reabsorption by the kidneys, increasing blood volume until the blood pressure returns to its normal range of values.

Chapter 14

2. The function of lymphatic vessels is to return fluid from the peripheral tissues to the circulation. If lymphatic vessels are cut and tied, the movement of the fluid would be interrupted and edema will occur.

3. Review figure 14.11 to answer this question. We learned that a B cell phagocytizes and processes antigens that combine with MHC molecules on the surface of the cell. Helper T cells interact with the MHC-antigen complex to stimulate the B cell to divide. The daughter cells then produce antibodies. If the antigens

are eliminated, the stimulus for B cell proliferation and antibody production is then removed.

4. The first exposure to the disease-causing agent (antigen) evokes a primary immune response, which destroys the existing pathogens but also produces memory cells that can respond to future infections. As time passes, the antibodies produced during the primary immune response will degrade and memory cells will die. If, before all of the memory cells are eliminated, a second exposure to the antigen occurs, a secondary response results, increasing the number of antibodies and memory cells again. The newly produced memory cells could provide immunity until the next exposure to the antigen.

5. After reading the Systems Pathology, we learned that SLE is an autoimmune disorder in which self-antigens activate immune responses. Often, this results in the formation of immune complexes and inflammation. Sometimes antibodies bind to antigens on cell membranes, resulting in the rupture of the cell membranes. Purpura results from bleeding into the skin. One cause of purpura is thrombocytopenia, a condition in which the number of platelets is greatly reduced, resulting in decreased platelet plug formation and blood clotting (see chapter 11). Considering that SLE is an autoimmune disorder, we can predict that the purpura is the result of the production of antibodies that destroy platelets causing thrombocytopenia.

Chapter 15

2. Another way to word this question would be, "What is the function of the nasal passageway?" We learned that the nasal passageway warms and humidifies the air. Therefore, breathing with your mouth open brings in drier than normal air, which irritates the throat and trachea. Running in cold weather with your mouth open results in the same scenario. Cold air is dry, and breathing through the mouth doesn't allow for humidification of the air.

3. We learned that the cartilage rings of the trachea are incomplete, C-shaped rings, and that the esophagus lies in the groove of those rings along the posterior side of the trachea. When a large mouthful of food is swallowed, it will stretch the esophagus in the area through which the food passes. Because the posterior region of the trachea is flexible, as the region of the esophagus expands, the adjacent area of the trachea momentarily collapses as the food passes. This is advantageous because

the structure of the trachea ensures that the air passageway remains open under most conditions due to the reinforcement by the cartilage, but the trachea has the flexibility to allow the esophagus to expand and food to move towards the stomach.

4. During respiratory movements, the parietal and visceral pleurae slide over each other. Normally, the pleural fluid in the pleural cavities lubricates the surfaces of these membranes. When the pleural membranes are inflamed, their surfaces become roughened. The rough surfaces rub against each other and create an intense pain. The pain is increased when a person takes a deep breath because the movement of the membranes is greater than during normal breaths.

5. In order for us to inhale, our thoracic cavity must be able to expand, which increases its volume. When the volume increases, the pressure is lowered allowing air to flow into the lungs down its pressure gradient. The diaphragm contracts downward onto the liver. By relaxing the abdominal muscles, the liver and other abdominal organs also move downward (inferiorly). This makes it easier for the diaphragm to move downward while expanding the thoracic cavity.

6. The tube should apply suction. In order for the lung to expand, the pressure in the alveoli must be greater than the pressure in the pleural cavity. This can be accomplished by lowering the pressure in the pleural cavity through suction. Applying air under pressure would make the pressure in the pleural cavity greater than the pressure in the alveoli, which would keep the alveoli collapsed.

7. The resting person with a tidal volume of 500 mL and breathing rate of 12 respirations/min has a minute ventilation of 6000 mL (500 mL $\times$ 12 respirations/min). The exercising person with a tidal volume of 4000 mL and a breathing rate of 24 respirations/min has a minute ventilation of 96,000 mL (4000 mL $\times$ 24 respirations/min). The difference between the two is 90,000 mL more air per minute than the person at rest.

8. The main principle allowing for acquisition of O_2 and removal of CO_2 are pressure differences. Therefore, inadequate ventilation causes a smaller difference in the P_{O_2} and P_{CO_2} across the respiratory membrane. Therefore, the rate of O_2 and CO_2 diffusion across the membrane decreases, causing O_2 levels in the blood to decrease and CO_2 levels to increase.

9. During exercise, skeletal muscle cells increase O_2 use in order to produce the ATP molecules required for muscle contraction.

Therefore, the P_{O_2} inside the cells declines, which increases the partial pressure difference for O_2 across the cell membrane. This results in increased movement of O_2 into the cells. The aerobic production of ATP also produces CO_2 (see chapter 17). Therefore the P_{CO_2} inside the cell goes up. This increases the partial pressure difference for CO_2 across the membrane causing increased movement of CO_2 out of cells.

10. A rapid rate of breathing increases the blood pH because CO_2 is eliminated from the blood more quickly during rapid respiration. As CO_2 is lost, H^+ and HCO_3^- combine to form H_2CO_3, which in turn dissociates to form CO_2 and H_2O. The lowered H^+ levels cause an increase in blood pH. Holding your breath results in a decrease in pH because CO_2 accumulates in the blood. The CO_2 combines with H_2O to form H_2CO_3, which dissociates to form H^+ and HCO_3^-. The elevation in H^+ levels causes a decrease in blood pH.

11. When a person breathes rapidly and deeply for several seconds, the CO_2 levels decrease and blood pH increases. Carbon dioxide is an important regulator of respiratory movements. A decrease in blood CO_2 and an increase in blood pH reduce the stimulus to the respiratory center. As a consequence, respiratory movements stop until blood CO_2 levels build up again in the body fluid. This normally takes only a short time.

12. By viewing figure 15.13 and Appendix B we see that normal blood P_{O_2} levels are 75–100 mm Hg and normal blood P_{CO_2} levels are 35–43 mm Hg. A P_{O_2} of 60 mm Hg and a P_{CO_2} of 30 mm Hg are both below normal. Let's first focus on how Will's P_{O_2} could be low. The movement of air into and out of the lungs is restricted because of the asthma, and there is a mismatch between ventilation of the alveoli and blood flow to the alveoli. Consequently, because of the ineffective ventilation, blood O_2 levels decrease. Will's hyperventilating also explains why his P_{CO_2} is low. As Will hyperventilates to help maintain blood O_2 levels it also results in lower than normal blood CO_2 levels. (If there were no hyperventilation, we would expect decreased blood O_2 but increased blood CO_2.)

Chapter 16

2. Four. Recall that the greater omentum is a localized mesentery, which consists of serous membranes. Each single layer of the mesentery has two layers of simple squamous epithelium. Since the greater omentum is folded back on itself, that results in 4 layers of simple squamous epithelium.

3. First, it is important to define the function of a normal palate. Normally, during swallowing, the soft palate is elevated, closing off the nasopharynx so that liquids and food bypass the nasal cavity and enter the esophagus. A cleft in the soft palate would result in food and liquid entering the nasal cavity during swallowing. If a person laughs suddenly while drinking a liquid, the liquid may be explosively expelled from the mouth and even the nose. If a person tries to swallow and speak at the same time, choking is most likely to occur. Speaking requires that the epiglottis be elevated, so that air can pass out of the larynx. When the epiglottis is raised, food or liquid could pass into the larynx and choke the person.

4. First, let's consider that acidic chyme in the small intestine is the stimulus. Stimuli are detected by receptors and then control centers send a signal to initiate a response that will regulate homeostasis. In this case, the control center is the pancreas. In response to the acidic chyme, pancreatic secretin stimulates bicarbonate ion secretion from the pancreas (the effector), which neutralizes the acidic chyme. Thus, secretin prevents the acid levels in the chyme from becoming too high, and keeps them in the normal range. The neutralization of the acidic chyme removes the stimulus for more secretin release and bicarbonate ion is no longer secreted. Because the response was inhibited, this is an example of a negative-feedback system.

5. An enema is introduction of fluid into the rectum, which causes it to distend. Recall that the defecation reflex is initiated by the movement of feces into the rectum and the subsequent stretch of the rectal wall. Therefore, because an enema stretches the rectal wall, it initiates the defecation reflex.

6. Recall that diarrhea is either increased stool frequency or increased stool volume, which can result in the abnormal loss of fluid and ions from the colon. This fluid loss from the colon affects the cardiovascular system in the same way that blood loss does. In either case, the result is hypovolemia, which causes a drop in blood pressure in a positive-feedback cycle. Eventually heart failure results from insufficient blood flow to the heart itself.

Chapter 17

2. Most of the vitamins, with the exception of A, D, and niacin, are essential vitamins, meaning they cannot be produced by the body but must be obtained from the diet. Recall that after a vitamin is destroyed, its function is lost. If the vitamins were broken down by digestion before being absorbed, they would not be functional and vitamin deficiencies would occur.

3. To answer this question we need to first identify the Daily Value for carbohydrates, which is 300 g/day. The % Daily Value is then determined by dividing the amount in the serving of food (30 g) by the Daily Value (300 g). The % Daily Value for carbohydrates from one serving of this food product is 10% (30/300 = 0.10, or 10%).

4. Recall that the % Daily Values for energy-producing nutrients are based on a 2000 kcal/day diet. We can use the % Daily Values of a food on the Nutrition Facts food label to determine how the amounts of certain nutrients in the food fit into the overall diet. If a person consumes 1800 kcal/day, the % Daily Values will be reduced proportionally. To calculate the adjusted % Daily Values, the actual caloric intake (1800 kcal/day) should be divided by 2000 kcal/day. On a 1800 kcal/day diet, the total percentage of Daily Values for energy-producing nutrients should add up to no more than 90% because 1800/2000 = 0.9, or 90%.

5. The last step in the electron-transport chain is when the electrons are passed to O_2 to form water. If this step is blocked, the citric acid and electron-transport chain cannot function, so ATP will not be produced aerobically. Lactic acid fermentation alone cannot produce sufficient levels of ATP to maintain normal cellular activity and death will occur.

6. We can see from table 17.1 that one cola or beer has about 145 kcal/serving. To determine the time it takes to burn these kilocalories we divide the kcal/serving by the number of kilocalories used per hour. Watching TV uses 95 kcal/h, so 145/95 is equal to about 1.5 hours, or 1 hour and 30 minutes. Jogging at a pace of 6 mph uses 580 kcal/h, so 145/580 is equal to 0.25 hours, or 15 minutes.

7. Recall that catabolism of food releases energy that can be used by the body for normal biological work, such as muscle contraction. However, about 40% of the total energy released is actually used for biological work. The remaining energy is lost as heat. Exercise increases the amount of biological work and therefore requires more energy in the form of ATP. As more ATP is produced to fuel the exercise, more heat is also generated as lost energy, thereby increasing body temperature. Shivering consists of small, rapid muscle contractions that produce heat in an effort to prevent a decrease in body temperature in the cold.

8. When blood vessels constrict, the flow of warm blood to the skin is reduced and the temperature of the skin is also reduced. The benefit is that less heat is lost through the skin to the environment and the internal body temperature is maintained. As the difference in temperature between the skin and the environment decreases, less heat is lost. If the skin temperature decreases too much, however, dilation of blood vessels to the skin occurs, which prevents the skin from becoming so cold that it is damaged.

Chapter 18

2. First, recall that the afferent arteriole supplies the glomerulus with blood to be filtered. The afferent arteriole has a larger diameter than the efferent arteriole. This means that blood enters the glomerulus at a faster rate than it exits, which causes blood in the glomerular capillaries to be under higher pressure than other capillary beds in the body. The high pressure in the glomerular capillaries is the driving force of filtration in the renal corpuscle. Thus, by changing the glomerular capillary pressure, the rate of filtration can be changed. Second, remember that arteriolar walls have a layer of smooth muscle and can vasoconstrict. If the afferent arteriole were to constrict, its diameter would be reduced, which would reduce the volume of blood entering the glomerulus. Reducing the volume of blood in the glomerulus reduces the pressure. Since more pressure equals more filtration, it follows that less pressure equals less filtration and therefore reduced urine production.

3. First, remember that urine formation is greatly influenced by osmosis. The ability of the kidney to produce concentrated urine by reabsorbing water depends on the standing salt gradient. If the concentration of salt or other compounds, such as glucose, in the urine exceeds that of the medullary interstitial fluid, the kidney won't be able to reabsorb the water from the urine. The glucose molecules attract water and, because the glucose molecules are trapped in the nephron, the amount of water that remains in the nephron is increased. Urine volume will therefore be much higher. In addition, because glucose cannot be reabsorbed by the kidney, the urine concentration will be greater than a healthy person's.

4. Reabsorption of water from the nephron is based on osmosis. If Na^+ and Cl^- are not actively transported out of the nephron, the concentration of ions inside the nephron stays elevated. The normal osmotic gradient that moves water out

of the nephron is greatly reduced and the water stays in the nephron, which increases urine volume.

5. Because the solution was a saline solution, it had the same concentration of solutes as the body fluids. Therefore, the excess IV solution did not change the concentration of the body fluids, but it did increase the volume of body fluids. An increased volume of saline solution increases the blood volume and blood pressure. Increased blood pressure stimulates baroreceptors, which results in inhibition of ADH secretion. Remember that ADH normally conserves water. The reduced ADH secretion causes the kidneys to produce a large volume of urine. At the same time, the increased blood volume stretches the walls of the atria, especially the right atrium, and causes the release of atrial natriuretic hormone. Atrial natriuretic hormone acts on the kidneys to reduce Na$^+$ reabsorption. Because Na$^+$ reabsorption is decreased, both Na$^+$ and water are lost in the urine. The increased pressure also results in less renin secretion from the kidney. The reduced renin causes less angiotensinogen to be converted to angiotensin. Consequently, less angiotensin II is formed, which reduces aldosterone secretion from the adrenal cortex. The decreased aldosterone slows Na$^+$ and water reabsorption, causing more Na$^+$ and water to be lost in the urine. Consequently, the urine volume and the amount of NaCl in the urine increase until the excess saline solution is eliminated.

6. Recall that the female urethra is much shorter than the male urethra and is more accessible to bacteria from the external environment. For this reason, females are more susceptible to bladder infections than males.

7. Remember that acidic pH is due to an elevated concentration of H$^+$. Hydrogen ions in the blood are derived from the combination of CO$_2$ and water. When someone is hyperventilating, the faster breathing rate results in a greater than normal rate of CO$_2$ loss from the circulatory system. Because CO$_2$ levels decrease, fewer H$^+$ are formed and the pH becomes more basic. Breathing into a paper bag corrects for the effects of hyperventilation because the person rebreathes air that has a higher concentration of CO$_2$. Carbon dioxide levels increase in the body, more H$^+$ are formed and pH levels drop back into the normal range.

8. Elevated blood CO$_2$ levels cause an increase in H$^+$ and a decrease in blood pH due to the following reaction:

$$CO_2 + H_2O \rightleftarrows H_2CO_3 \rightleftarrows H^+ + HCO_3^-$$

However, the kidney also plays an important role in the regulation of blood pH. The kidney's rate of H$^+$ secretion into the urine and reabsorption of HCO$_3^-$ increase. This helps prevent high blood H$^+$ levels and low blood pH in Adam.

9. After 7 days Roger's kidneys began to produce a large volume of urine with larger than normal Na$^+$ and K$^+$ concentrations. As a result, Roger became dehydrated by day 9. Dehydration results in reduced blood volume and blood pressure. His hematocrit was increased because the volume of his blood was decreased, but there was no decrease in the number of red blood cells. The percentage of the blood made of red blood cells therefore increased. The pale skin was the result of vasoconstriction, which was triggered by the reduced blood pressure. Dizziness resulted from reduced blood flow to the brain when Roger tried to stand and walk. He was lethargic in part because of reduced blood volume, but also because of low blood levels of K$^+$ and Na$^+$, caused by the loss of these ions in the urine. Low blood levels of Na$^+$ and K$^+$ alter the electrical activity of nerve and muscle cells and result in muscular weakness. The arrhythmia of his heart was due to low blood levels of K$^+$ and increased sympathetic stimulation, which was also triggered by low blood pressure.

Chapter 19

2. Recall that the prostate is inferior to the bladder and anterior to the rectum. Physicians can manually palpate the prostate through the wall of the rectum. The patient is fully awake and has relatively minor discomfort.

3. To answer the question, you must first remember that the testes are the major source of the hormone testosterone. Secondary sexual characteristics, external genitalia, and sexual behavior development are all driven by testosterone. Therefore, an inability of the testes to produce normal amounts of the hormone would result in the failure to develop into a sexually mature male. It is most likely that this individual's external genitalia would retain a juvenile appearance and normal adult sexual behavior would not develop.

4. The question has addressed the time period in the menstrual cycle just before the LH surge, which promotes ovulation. Referring to figure 19.14, it is evident that estrogen and progesterone are normally at their lowest levels before the LH surge. In contrast, progesterone is at its highest

level after ovulation and prevents further development of follicles. Therefore, administration of a large amount of progesterone and estrogen just before the preovulatory LH surge inhibits the release of GnRH, LH, and FSH. Consequently, ovulation does not occur. However, progesterone is the more potent hormone when it comes to inhibiting ovulation. Injections of a small amount of estrogen just before ovulation could stimulate GnRH, LH, and FSH secretion with little negative effect on ovulation.

5. Molly's mother could have had leiomyomas also, although, without direct data from medical examinations, one cannot be certain. If that was the cause of her irregular menstruations, they may have become less frequent as Molly's mother experienced menopause. During menopause, the uterus gradually becomes smaller, and eventually the cyclical changes in the endometrial lining cease. If the condition was relatively mild, the onset of menopause could explain the gradual disappearance of the irregular and prolonged menstruations. (Note: If the tumors are large, constant and severe menstruations are likely even if regular menstrual cycles stop due to menopause.)

Chapter 20

2. To determine the days of the menstrual cycle when fertilization is most likely to occur, we need to remember the timing of ovulation, when the secondary oocyte is released from the ovary and available for fertilization. Recall that ovulation usually occurs around day 14 post-LMP. Also recall that sperm cells remain viable in the female reproductive tract for up to 6 days and that the secondary oocyte is capable of being fertilized for up to 1 day after ovulation. Considering all of these factors, we can conclude that fertilization would occur if sexual intercourse occurred between 5 days before ovulation and 1 day following ovulation. That will be between 10 days and 15 days post-LMP. You may find it interesting that data indicate that the most fertile period during the menstrual cycle is between 2 days just before ovulation and the day of ovulation.

3. The primitive streak essentially forms the central axis of an embryo. If two primitive streaks formed in one embryonic disk, we would expect two different embryos, or essentially twins, to develop. If the two primitive streaks were touching each other, conjoined twins would develop. The degree to which the two primitive streaks are touching would determine the severity of the attachment.

4. Recall that clinical age is dependent on LMP (last menstrual period) of the mother and developmental age begins at fertilization, which is assumed to occur on day 14 after LMP. Most of the times reported in the text are developmental age. To determine the clinical age, add 14 to the developmental age. The one exception is parturition, which is reported as clinical age. To determine the developmental age, subtract 14 from the clinical age. We can easily construct a table to compare the ages:

	Clinical Age	Developmental Age
Fertilization	14 days	0 days
Implantation	21 days	7 days
Fetal period begins	70 days	56 days
Parturition	280 days	266 days

5. Suckling causes a reflex release of oxytocin from the mother's posterior pituitary. Oxytocin causes expulsion of milk from the breast, but it also causes contraction of the uterus. Contraction of the uterus is responsible for the sensation of cramps in her abdomen.

6. Genotypes are the alleles a person has for a given trait, and the phenotype is the person's appearance. In the case of polydactyly, there are three possible genotypes: DD (homozygous dominant), Dd (heterozygous), and dd (homozygous recessive). Since polydactyly is a dominant trait, we would expect the individuals with genotypes DD or Dd to exhibit polydactyly and the individuals with the genotype dd not to exhibit polydactyly.

7. Since we assume that the nondisjunction causing her Turner syndrome occurred in the father, the girl would have inherited her single X chromosome from her mother. Hemophilia is an X-linked gene. If her mother is a carrier ($X^H X^h$) for hemophilia, then there is a 50% chance that she will pass the X chromosome with the recessive gene (X^h) to this daughter and a 50% chance that she will pass the X chromosome with the normal gene (X^H) to this daughter. So, we can conclude that there is a 50% chance that a female with Turner syndrome, who inherited her only X chromosome from her mother (a carrier of hemophilia), would have hemophilia.

Glossary

Many of the words in this glossary and throughout the text are followed by a simplified phonetic spelling showing pronunciation. The pronunciation key reflects standard clinical usage, with minor modifications, as presented in *Stedman's Medical Dictionary* (27th edition), which has long been a leading reference volume in the health sciences.

ā as in day, ate, way
a as in mat, hat, act
ă as in alone, abortion, media
ah as in father
ar as in far
aw as in fall (fawl)
ē as in be, bee, meet
ĕ as in taken, genesis
er as in term, earn, learn
ī as in pie, pine, side
i as in pit, tip, fit
ĭ as in pencil
ō as in no, note, toe
o as in not, box, cot
ŏ as in occult, lemon, son
oo as in food, to, tool
ow as in cow, brow, plow, now
oy as in boy, toy, oil
u as in wood, foot, took
ŭ as in but, sun, bud, cup, up
ū as in pure, unit, union, future

abdomen (ab-dō′men, ab′dō-men) Belly, between the thorax and the pelvis.

abdominal cavity (ab-dom′i-năl) Space bounded by the diaphragm, the abdominal wall, and the pelvis.

abdominopelvic cavity (ab-dom′i-nō-pel′vik) Abdominal and pelvic cavities considered together.

abduction (ab-dŭk′shun) [abductio] Movement away from the midline.

absorption (ab-sōrp′shŭn) The taking in or reception of gases, liquids, light, heat, or solutes, such as the movement of digested molecules across the intestinal wall and into the bloodstream; the movement of substances through the skin; and the movement of fluid into the lymphatics from the interstitial fluid.

accommodation (ă-kom′ŏ-dā′shŭn) The act or state of adjustment or adaptation, such as the increase in the thickness and convexity of the lens of the eye in order to focus an object on the retina as the object moves closer to the eyes; decreasing sensitivity of a nerve cell to a stimulus of constant strength.

acetabulum (as-ĕ-tab′ū-lŭm) [L., shallow vinegar vessel or cup] Cup-shaped depression on the lateral surface of the coxal bone, where the head of the femur articulates.

acetylcholine (as-e-til-kō′lēn) Neurotransmitter substance released from motor neurons that innervate skeletal muscle fibers, all autonomic eganglionic neurons, all postganglionic parasympathetic neurons, some postganglionic sympathetic neurons, and some central nervous system neurons.

acetylcholinesterase (as′e-til-kō-lin-es′ter-ās) Enzyme that breaks down acetylcholine to acetic acid and choline.

acetyl-CoA (as′e-til) Acetyl-coenzyme A; formed by the combination of the two-carbon acetyl group with coenzyme A; the molecule that combines with a four-carbon molecule to enter the citric acid cycle.

Achilles (ă-kil′ēz) **tendon** Common tendon of the calf muscles that attaches to the heel (calcaneus); named after a mythical Greek warrior who was vulnerable only in the heel.

acid (as′id) Any substance that is a proton donor; or any substance that releases hydrogen ions.

acidic solution Solution with more hydrogen ions than hydroxide ions; has a pH of less than 7.

acidosis (as-i-dō′sis) Condition characterized by a lower than normal blood pH (pH of 7.35 or lower).

acinus (as′i-nŭs), pl. **acini** (as′i-nī) [L., berry, grape] Grape-shaped secretory portion of a gland.

acromegaly (ak-rō-meg′ă-lē) [G. *acro; megas,* large] Disorder marked by progressive enlargement of the bones of the head, face, hands, feet, and thorax as a result of excessive secretion of growth hormone by the anterior pituitary gland.

acromion (ă-krō′mē-on) [Fr. *akron,* tip + *omos,* shoulder] Lateral end of the spine of the scapula that projects as a broad, flattened process overhanging the glenoid fossa; articulates with the clavicle.

acrosome (ak′rō-sōm) [*acro,* tip + G. *soma,* body]. A caplike organelle surrounding the anterior portion of a sperm cell, containing enzymes that facilitate entry of the sperm cell through the zona pellucida.

actin myofilament (ak′tin mī-ō-fil′ă-ment) One of the two major kinds of protein fibers that make up a sarcomere; thin filaments; resemble two minute strands of pearls twisted together.

action potential All-or-none change in membrane potential in an excitable tissue that is propagated as an electrical signal.

activation energy Energy that must be added to atoms or molecules to start a chemical reaction.

active transport Carrier-mediated process that requires ATP and can move substances into or out of cells from a lower to a higher concentration.

adaptive immunity Immune system response in which there is an ability to recognize, remember, and destroy a specific antigen.

adduction (ă-dŭk′shŭn) [L. *adductus,* to bring toward] Movement toward the midline.

adductor (a-dŭk′ter, -tōr) [L. *adductus,* to bring toward] Muscle causing movement toward the midline.

adenoid (ad′ĕ-noyd) Enlarged pharyngeal tonsil.

adenosine triphosphate (ă-den′ō-sēn trī-fos′făt) (ATP) Adenosine, an organic base, with three phosphate groups attached to it; energy stored in adenosine triphosphate is used in nearly all the energy-requiring reactions in the body.

ADH See *antidiuretic hormone.*

adipose (ad′i-pōs) [L. *adeps,* fat] Fat; relating to fat tissue.

adrenal cortex (ă-drē′năl kōr′teks) The outer part of the adrenal gland, which secretes the following steroid hormones: glucocorticoids, mainly cortisol; mineralocorticoids, mainly aldosterone; and androgens.

adrenal gland (ă-drē′năl) [L. *ad,* to; *ren,* kidney, near or on the kidneys] One of two endocrine glands located on the superior pole of each kidney; secretes the hormones epinephrine, norepinephrine, aldosterone, cortisol, and androgens.

adrenal medulla (ă-drē′năl me-dool′ă) Inner part of the adrenal gland, which secretes mainly epinephrine but also small amounts of norepinephrine.

adrenaline (ă-dren′ă-lin) [from the adrenal gland] Synonym for *epinephrine.*

adrenocorticotropic hormone (ă-drē′nō-kōr′ti-kō-trō′pik) [L. *ad,* near + *ren,* kidney + *cortico,* cortex + *trophre,* nurture] (ACTH) Hormone of the anterior pituitary gland that stimulates the adrenal cortex to secrete cortisol.

adventitia (ad-ven-tish′ă) [L. *adventicius,* coming from abroad or outside; foreign] Outermost covering of an organ that is continuous with the surrounding connective tissue.

aerobic respiration (ār-ō′bik) Breakdown of glucose in the presence of oxygen to produce carbon dioxide, water, and approximately 38 ATP molecules; includes glycolysis, the citric acid cycle, and the electron-transport chain.

afferent (af′er-ent) [L. *afferens,* to bring to] Inflowing; conducting toward a center, denoting certain arteries, veins, lymphatics, and sensory nerves. Opposite of *efferent.*

afferent arteriole (ar-tēr′ē-ōl) Small artery in the renal cortex that supplies blood to the glomerulus.

afferent fiber Sensory nerve fiber going from the peripheral to the central nervous system; sensory or afferent fiber.

afterload Resistance against which the ventricles must pump blood; it is increased in people who have hypertension.

agglutination (ă-gloo′ti-nā′shŭn) [L. *ad,* to + *gluten,* glue] Process by which cells stick together to form clumps.

agonist (ăg′ōn-ist) [G. *agon,* a contest] Denoting a muscle in a state of contraction, with reference to its opposing muscle, or antagonist.

agranulocyte (ă-gran′ū-lō-sīt) [G. *a-,* without + *granular* + *kytos,* cell] White blood cell with very small cytoplasmic granules that cannot be easily seen with the light microscope; lymphocytes and monocytes.

aldosterone (al-dos′ter-ōn) Steroid hormone produced by the adrenal cortex that facilitates potassium exchange for sodium in the distal convoluted tubule and collecting duct, causing sodium ion reabsorption and potassium and hydrogen ion secretion.

alkaline solution (al′kă-līn) See *basic solution.*

alkalosis (al-kă-lō′sis) Condition characterized by a higher than normal blood pH (pH of 7.45 or above).

alveolar duct (al-vē′ō-lăr) Part of the respiratory passages beyond a respiratory bronchiole; from it arise alveolar sacs and alveoli.

alveolar sac Two or more alveoli that share a common opening.

alveolus (al-vē′ō-lŭs), pl. alveoli (al-vē′ō-lī) [L., small cavity or hollow sac] Cavity; examples include the sockets into which the teeth fit and the ends of the respiratory system.

amino acid (ă-mē′nō) Class of organic acids containing an amine group (NH_2) that makes up the building blocks of proteins.

amniotic cavity (am-nē-ot′ik) [G. *amnios,* lamb] Fluid-filled cavity surrounding and protecting the developing embryo.

amylase (am′il-ās) One of a group of starch-splitting enzymes that cleave starch, glycogen, and related polysaccharides.

anabolism (ă-nab′ō-lizm) [G. *anabole,* a raising up] All the synthesis reactions that occur within the body; requires energy.

anaerobic respiration (an-ār-ō′bik) Breakdown of glucose in the absence of oxygen to produce lactic acid and two ATP molecules; consists of glycolysis and the reduction of pyruvic acid to lactic acid.

anaphase (an′ă-fāz) [G. *ana,* up + *phases,* appearance] Stage of mitosis or meiosis in which the chromosomes move from the center area of the cell, the equatorial plane, toward the poles of the cell.

anatomical position (an′ă-tom′i-kăl) Position in which a person is standing erect with the feet facing forward, the arms hanging to the sides, and the palms of the hands facing forward.

anatomy (ă-nat′ō-mē) [G. *ana,* apart + *tome,* a cutting] Scientific discipline that investigates the structure of the body.

androgen (an′drō-jen) [G. *andros,* male] Hormone that stimulates the development of male sexual characteristics; includes testosterone.

anemia (ă-nē′mē-ă) [G. *an,* without + *haima,* blood] Condition that results in less than normal hemoglobin in the blood or a lower than normal number of red blood cells.

anencephaly (an′en-sef′ă-lē) Defective development of the brain with absence of the cerebral and cerebellar hemispheres and with only a rudimentary brainstem.

angina pectoris (an′ji-nă pek′tō-ris, an-jī′nă) Pain resulting from a reduced blood supply to cardiac muscle.

angioplasty (an′jē-ō-plas-tē) [G. *angio,* blood vessel] Technique used to dilate the coronary arteries by threading a small, balloonlike device into a partially blocked coronary artery and then inflating the balloon to enlarge the diameter of the vessel.

angiotensin (an-jē-ō-ten′sin) [*angio,* blood vessel + *tensus,* to stretch] Angiotensin I is a peptide derived when renin acts on angiotensinogen; angiotensin II is formed from angiotensin I when angiotensin-converting enzyme acts on angiotensin I; angiotensin II is a potent vasoconstrictor, and it stimulates the secretion of aldosterone from the adrenal cortex.

angiotensinogen (an′jē-ō-ten-sin′ō-jen) Protein found in the blood that gives rise to angiotensin I after renin, an enzyme secreted from the kidney, acts on it.

ANH See *atrial natriuretic hormone.*

antagonist (an-tag′ŏ-nist) Muscle that works in opposition to another muscle.

anterior (an-tēr′ē-ōr) [L., to go before] That which goes first; in humans, toward the belly or front.

anterior horn Part of the spinal cord gray matter containing motor neurons; also called the ventral horn or motor horn.

anterior pituitary gland Portion of the pituitary gland derived from the oral epithelium.

antibody (an′tē-bod-ē) [G. *anti,* against + *body,* a thing] Protein in the plasma that is responsible for antibody-mediated (humoral) immunity; binds specifically to an antigen.

antibody-mediated immunity Immunity resulting from B cells and the production of antibodies.

anticoagulant (an′tē-kō-ag′ū-lant) Chemical that prevents coagulation or blood clotting; an example is antithrombin.

antidiuretic hormone (an′tē-dī-ū-ret′ik) [G. *anti,* against + *uresis,* urine volume] (ADH) Hormone secreted from the posterior pituitary gland that acts on the kidney to reduce the output of urine; also called vasopressin.

antigen (an′ti-jen) [G. *anti* (*body*) + *-gen,* a thing] Substance that induces a state of sensitivity or resistance to microorganisms or toxic substances after a latent period; substance that stimulates the adaptive immune system; self-antigens are produced by the body, and foreign antigens are introduced into the body.

antigen receptor Molecule on the surface of lymphocytes that specifically binds antigens.

aorta (ā-ōr′tă) [G. *aorte,* from + *aeiro,* to lift up] Large, elastic artery that is the main trunk of the systemic arterial system, which carries blood from the left ventricle of the heart and passes through the thorax and abdomen.

aortic semilunar valve Semilunar valve consisting of three cusps of tissue located at the base of the aorta where it arises from the left ventricle; the cusps overlap during ventricular diastole to prevent leakage of blood from the aorta into the left ventricle.

apex (ā′peks) [L., tip] Extremity of a conical or pyramidal structure; the apex of the heart is the rounded tip directed anteriorly and slightly inferiorly.

aphasia (ă-fā′zē-ă) [G. *a-,* without + *phasis,* speech or speechlessness] Impaired or absent communication by speech, writing, or signs because of dysfunction of brain centers in the dominant cerebral hemisphere.

apocrine (ap′ō-krin) [G. *apo,* away from + *krino,* to separate] Type of gland whose cells contribute cytoplasm to its secretion; sweat glands that produce organic secretions traditionally are called apocrine; however, these sweat glands are now known to be merocrine glands; see also *merocrine* and *holocrine.*

aponeurosis (ap'ō-noo-rō'sis) [G. *neuron*, sinew; end of a muscle where it becomes a tendon] Sheet of fibrous connective tissue, or an expanded tendon, serving as the origin or insertion of a flat muscle.

appendicular (ap'en-dik'ū-lăr) [L. *appendo*, to hang something on] Relating to an appendage, such as the limbs and their associated girdles.

appendix (ă-pen'diks), pl. **appendices** (ă-pen'di-sēs) [L. *appendo*, to hang something on] Smaller structure usually attached by one end to a larger structure; a small blind extension of the colon attached to the cecum.

appositional growth (ap-ō-zish'ŭn-ăl) [L. *ap* + *pono*, to place at or to] To place one layer of bone, cartilage, or other connective tissue against an existing layer; increases the width or diameter of bones.

aqueous humor (ak'wē-ŭs, ā'kwē-ŭs) Watery, clear fluid that fills the anterior chamber and posterior chamber of the eye.

arachnoid mater (ă-rak'noyd ma'ter) [G. *arachne*, spiderlike, cobweb] Thin, cobweblike meningeal layer surrounding the brain and spinal cord; the middle of three layers.

areola (ă-rē-ō-lă), pl. **areolae** (ă-rē'ō-lē) [small areas] Pigmented area surrounding the nipple of a mammary gland.

areolar (ă-rē'ō-lăr) Relating to connective tissue with small spaces within it; loose connective tissue.

arrector pili (ă-rek'tōr pī'lī) [L., that which raises hair] Smooth muscle attached to the hair follicle and dermis that raises the hair when it contracts.

arteriosclerosis (ar-tēr'ē-ō-skler-ō'sis) [L. *arterio-* + G. *sklerosis*, hardness] Hardness of the arteries.

arteriosclerotic lesion (ar-tēr'ē-ō-skler-ot'ik) Lesion or growth in arteries that narrows the lumen, or passage, and makes the walls of the arteries less elastic.

artery (ar'ter-ē) [G. *arteria*, the windpipe] Blood vessel that carries blood away from the heart.

articulation (ar-tik-ū-lā'shŭn) [L. *articulatio*, a forming of vines] Place where two bones come together; a joint.

artificial heart Mechanical pump used to replace a diseased heart.

artificial pacemaker Electronic device implanted beneath the skin with an electrode that extends to the heart; provides periodic electrical stimuli to the heart and substitutes for a faulty SA node.

astrocyte (as'trō-sīt) [G. *astron*, star + *kytos*, a hollow cell] Star-shaped neuroglial cell that helps regulate the composition of fluid around the neurons of the central nervous system.

atherosclerosis (ath'er-ō-skler-ō'sis) [G. *athere*, gruel or soft, pasty material + *sklerosis*, hardness] Lipid deposits (plaques) in the tunica intima of large and medium-sized arteries.

atom (at'ŏm) [G. *atomos*, indivisible, uncut] Smallest particle into which an element can be divided using chemical methods; composed of neutrons, protons, and electrons.

atomic number (ă-tom'ik) Number of protons in an element.

ATP See *adenosine triphosphate.*

atrial natriuretic (ā'trē-ăl nā'tre-yū-ret'ik) **hormone (ANH)** Hormone released from cells in the atrial wall of the heart when atrial blood pressure is increased; lowers blood pressure by increasing the rate of urine production.

atrioventricular (AV) bundle (ā-trē-ō-ven-trik'ū-lar) Bundle of modified cardiac muscle fibers that projects from the AV node through the interventricular septum; conducts action potentials from the AV node rapidly through the interventricular septum; also called the bundle of His.

atrioventricular (AV) node Small collection of specialized cardiac muscle fibers located in the inferior part of the right atrium; delays action potential transmission to the atrioventricular bundle.

atrioventricular valve Valve between the atrium and the ventricle of the heart, the tricuspid valve between the right atrium and right ventricle, and the bicuspid (or mitral valve) between the left atrium and left ventricle.

atrium (ā'trē-ŭm), pl. **atria** (ā'trē-ă) [L., entrance chamber] One of the two chambers of the heart that collect blood during ventricular contraction and pump blood into the ventricles to complete ventricular filling at the end of ventricular relaxation; the right atrium receives blood from the inferior and superior venae cavae and from the coronary sinus and delivers blood to the right ventricle; the left atrium receives blood from the pulmonary veins and delivers blood to the left ventricle.

auditory (aw'di-tōr-ē) Relating to hearing.

auditory ossicles (os'i-klz) Bones of the middle ear; the malleus, incus, and stapes.

auditory tube Air-filled passageway between the middle ear and pharynx.

auricle (aw'ri-kl) [L. *auris*, ear] Fleshy part of the external ear on the outside of the head; a small, conical pouch projecting from the upper anterior part of each atrium of the heart.

auscultatory (aws-kŭl'tă-tō-rē) [L. *ausculto*, to listen] To listen to the sounds made by the various body structures, especially to Korotkoff sounds when determining blood pressure.

autocrine (aw'tō-krin) [G. *autos*, self + *krino*, to separate] Denoting self-stimulation through cellular production of a factor and a specific receptor for it.

autoimmune disease (aw-tō-i-mūn') Disorder resulting from a specific immune system reaction against self-antigens.

autonomic nervous system (ANS) (aw-tō-nom'ik) Part of the peripheral nervous system composed of efferent fibers that reach from the central nervous system to smooth muscle, cardiac muscle, and glands.

autosome (aw'tō-sōm) [G. *auto-*, self + *soma*, body] Any chromosome other than a sex chromosome; normally occurs in pairs in somatic cells and singly in gametes.

AV See *atrioventricular.*

axial (ak'sē-ăl) [L. *axle*, axis] Head, neck, and trunk as distinguished from the extremities.

axon (ak'son) [G., axis] Main process of a neuron; usually conducts action potentials away from the neuron cell body.

baroreceptor (bar'ō-rē-sep'ter) [G. *baro*, weight or pressure] Sensory nerve endings in the walls of the atria of the heart, aortic arch, and carotid sinuses; sensitive to stretching of the wall caused by increased blood pressure; also called a pressoreceptor.

baroreceptor reflex Process in which baroreceptors detect changes in blood pressure and produce changes in heart rate, force of heart contraction, and blood vessel diameter that return blood pressure toward normal levels.

basal nuclei Nuclei at the base of the cerebrum, diencephalon, and midbrain involved in controlling motor functions.

base Any substance that is a proton acceptor or any substance that binds to hydrogen ions; lower part or bottom of a structure; the base of the heart is the flat portion directed posteriorly and superiorly; veins and arteries project into and out of the base, respectively.

basement membrane Structure that attaches most epithelia (exceptions include lymphatic vessels and the liver sinusoids) to underlying tissue; consists of carbohydrates and proteins secreted by the epithelia and the underlying connective tissue.

basic solution (bā'sik) Solution with fewer hydrogen ions than hydroxide ions; has a pH greater than 7.

basilar membrane (bas'i-lăr) One of two membranes forming the cochlear duct; supports the spiral organ.

basophil (bā'sō-fil) [G. *basis*, base + *phileo*, to love] White blood cell with granules that stain purple with basic dyes; promotes inflammation and prevents clot formation.

belly Largest part of a muscle between the origin and insertion.

benign (bē-nīn′) [L. *benignus*, kind] Mild in character or nonmalignant; does not spread to distant sites.

beta-adrenergic (bā′tă ad-rĕ-ner′jik) **blocking agent** Drug that binds to and prevents adrenergic receptors from responding to adrenergic compounds that normally bind to beta-adrenergic receptors and cause them to function; beta-adrenergic blocking agents are used to treat certain arrhythmias in the heart and to treat tachycardia (rapid heart rate).

biceps brachii (bī′seps brā′kē-ī) Muscle in the anterior arm with two heads, or origins, on the scapula and an insertion onto the radius; flexes and supinates the forearm.

bicuspid valve (bī-kŭs′pid) Valve closing the opening between the left atrium and left ventricle of the heart; has two cusps; also called the mitral valve.

bile (bīl) Fluid secreted from the liver, stored in the gallbladder, and released into the duodenum; consists of bile salts, bile pigments, bicarbonate ions, fats, and other materials.

bile salt Organic salt secreted by the liver that emulsifies lipids.

bilirubin (bil-i-roo′bin) [L., bile + *ruber*, red] Bile pigment formed from the heme in hemoglobin during the destruction of red blood cells by macrophages.

biopsy (bī′op-sē) Process of removing tissue from living patients for diagnostic examination, or a specimen obtained by biopsy.

blastocele (blas′tō-sēl) [G. *blastos*, germ + *koilos*, hollow] Cavity in the blastocyst.

blastocyst (blas′tō-sist) [G. *blastos*, germ + *kystis*, bladder] Early stage of mammalian embryo development consisting of a hollow ball of cells with an inner cell mass and an outer trophoblast layer.

blood-brain barrier Cellular and matrix barrier made up primarily of blood vessel endothelium, with some help from the surrounding astrocytes; it allows some (usually small) substances to pass from the circulation into the brain but does not allow other (larger) substances to pass.

blood group Category of red blood cells based on the type of antigen on the surface of the red blood cell; for example, the ABO blood group is involved with transfusion reactions.

blood pressure [L. *pressus*, to press] The force blood exerts against the blood vessel walls; expressed relative to atmospheric pressure and reported in the form of millimeters of mercury (mm Hg) of pressure.

bony labyrinth (lab′i-rinth) Interconnecting tunnels and chambers within the temporal bone in which the inner ear is located.

Bowman's capsule Enlarged end of the nephron; Bowman's capsule and the glomerulus make up the renal corpuscle.

brachialis (brā′kē-ăl-is) Muscle of the anterior arm that originates on the humerus and inserts onto the ulna; flexes the forearm.

brachial plexus (brā′kē-ăl) [L. *brachium*, arm] Nerve plexus to the upper limb; originates from spinal nerves C5 to T1.

brainstem Portion of the brain consisting of the midbrain, pons, and medulla oblongata.

breathing (brēthing) Movement of air into and out of the lung; see *ventilation.*

bronchiole (brong′kē-ōl) One of the finer subdivisions of the bronchial tubes, less than 1 mm in diameter, that has no cartilage in its wall but has relatively more smooth muscle and elastic fibers than do larger bronchial tubes.

bronchus (brong′kŭs), pl. **bronchi** (brong′kī) [G. *bronchos*, windpipe] Any one of the air ducts conducting air from the trachea to the bronchioles.

buccinator (buk′sĭ-nā′tōr) Muscle making up the lateral sides of the oral cavity; flattens the cheeks.

buffer (bŭf′er) Chemical that resists changes in pH when either an acid or a base is added to a solution containing the buffer.

bundle of His See *atrioventricular bundle.*

burn Lesion caused by heat, acid, or other agents; a partial-thickness burn of the skin damages only the epidermis (first-degree burn) or the epidermis and part of the dermis (second-degree burn); a full-thickness (third-degree) burn destroys the epidermis and the dermis and sometimes the underlying tissue.

bursa (ber′să) [L., purse or pocket] Closed sac or pocket containing synovial fluid; usually found in areas where friction occurs.

calcaneus (kal-kā′nē-ŭs) [L., the heel] Largest tarsal bone forming the heel.

calcitonin (kal-si-tō′nin) Hormone, released from cells of the thyroid gland, that acts on tissues, especially bone, to cause a decrease in blood levels of calcium ions.

calcium channel blocker (kal′sē-ŭm) Class of drugs that specifically block channels in cell membranes through which calcium ions pass; calcium channel blockers are used to treat some kinds of cardiac arrhythmias.

callus (kal′ŭs) [L., hard skin] Thickening of the stratum corneum of skin in response to friction; the zone of tissue repair between fragments of a broken bone.

calorie (kal′ō-rē) [L. *calor*, heat] Unit of heat or energy content; the quantity of energy required to raise the temperature of 1 gram of water 1°C. A Calorie (Cal), or kilocalorie (kcal), is the amount of heat or energy

required to raise the temperature of 1000 grams of water from 14°C to 15°C.

calyx (kā′liks), pl. **calyces** (kal′i-sēz) [G., flower petal or cup of a flower] Small container into which urine flows as it leaves the collecting ducts at the tip of the renal pyramids; the calyces come together to form the renal pelvis.

canaliculus (kan-ă-lik′ū-lŭs) Tiny canal in bone between osteocytes containing osteocyte cell processes; a cleftlike lumen between the cells of each hepatic cord, connects medial corner of the eye to the lacrimal sac.

cancer (kan′ser) [L., a crab, suggesting crablike movement] Malignant neoplasm, or tumor.

capacitation (kă-pas′i-tā′shŭn) Process whereby the sperm cells develop the ability to fertilize oocytes.

capillary (kap′i-lār-ē) [L. *capillaris*, relating to hair, resembling a fine hair] Minute blood vessel consisting only of simple squamous epithelium and a basement membrane; major site for the exchange of substances between the blood and tissues.

carbohydrate (kar-bō-hī′drāt) Organic molecule made up of one or more monosaccharides chemically bound together; sugars and starches.

carbonic anhydrase (kar-bon′ik an-hī′drās) Enzyme that increases the rate at which carbon dioxide reacts with water to form hydrogen ions and bicarbonate ions.

carcinoma (kar-si-nō′mă) [G. *karkinoma*, cancer + *oma*, tumor] Malignant tumor derived from epithelial tissue.

cardiac cycle (kar′dē-ak) One complete sequence of cardiac systole and diastole.

cardiac output Volume of blood pumped by either ventricle of the heart per minute; about 5 L/min for the heart of a healthy adult at rest.

cardioregulatory center Specialized area within the medulla oblongata of the brain that receives sensory input and controls parasympathetic and sympathetic stimulation of the heart.

carotene (kar′ō-tēn) Yellow pigment in plants such as squash and carrots; accumulates in the lipids of the stratum corneum and in the fat cells of the dermis and hypodermis and is used as a source of vitamin A.

carotid bodies (ka-rot′id) Small organs near the carotid sinuses that detect changes in blood oxygen, carbon dioxide, and pH.

carotid sinus Enlargement of the internal carotid artery near the point where the internal carotid artery branches from the common carotid artery; contains baroreceptors.

carpal (kar′păl) [G. *karpos*, wrist] Associated with the wrist; bones of the wrist.

carrier molecule Protein that extends from one side of the cell membrane to the other; binds to molecules to be transported and moves them from one side of the membrane to the other.

cartilage (kar'ti-lij) [L. *cartilage*, gristle] Firm, smooth, resilient, nonvascular connective tissue.

cascade (kas-kād') [Fr. *cascare*, to fall] Series of sequential interactions, which once initiated continues to the final one; each interaction is activated by the preceding one, with cumulative effect.

catabolism (kă-tab'ō-lizm) [G. *katabole*, a casting down] All the decomposition reactions that occur in the body; releases energy.

catalyst (kat'ă-list) Substance that increases the rate of a chemical reaction; in the process, the catalyst is not permanently changed or used up.

cecum (sē'kŭm) [L. *caecus*, blind] Blind sac forming the beginning of the large intestine.

cell (sel) [L. *cella*, chamber] Basic living unit of all plants and animals.

cell-mediated immunity Immunity resulting from the actions of T cells.

cell membrane Plasma membrane; outermost component of the cell, surrounding and binding the rest of the cell contents.

central canal Small canal containing blood vessels, nerves, and loose connective tissue and running parallel to the long axis of a bone; also called a haversian canal.

central nervous system (CNS) Brain and spinal cord.

centriole (sen'trē-ōl) Small organelle that divides and migrates to each pole of the nucleus; spindle fibers extend from the centromeres to the centrioles during mitosis.

centromere (sen'trō-mēr) [G. *kentron*, center + *meros*, part] Specialized region where chromatids are linked together in a chromosome.

cerebellum (ser-e-bel'ŭm) [L., little brain] Part of the brain attached to the brainstem; important in maintaining muscle tone, balance, and coordination of movements.

cerebral aqueduct (ser'ē-brăl, sĕ-rē'brăl) Small connecting tube through the midbrain between the third and fourth ventricles.

cerebrospinal fluid (ser'ĕ-brō-spī'năl, sĕ-rē'brō-spī-năl) **(CSF)** Fluid filling the ventricles and surrounding the brain and spinal cord.

cerebrum (ser'ĕ-brŭm, sĕ-rē'brŭm) [L., brain] Largest part of the brain, consisting of two hemispheres and including the cortex, nerve tracts, and basal nuclei.

cerumen (sĕ-roo'men) [L. *cera*, wax] Specific type of sebum produced in the external auditory canal; earwax.

cervical (ser'vĭ-kal) Neck.

cervical plexus Nerve plexus of the neck; originates from spinal nerves C1–C4.

cervix (ser'viks) [L., neck] Lower part of the uterus extending to the vagina.

chemical (kem'i-kăl) Relating to chemistry, especially to the characteristics of atoms and molecules and to their interactions.

chemical bond Association between two atoms formed when the outermost electrons are transferred or shared between atoms.

chemical mediator of inflammation Chemical released or activated by injured tissues and adjacent blood vessels; produces vasodilation, increases vascular permeability, and attracts blood cells; includes histamine, kinins, prostaglandins, and leukotrienes.

chemical reaction Process by which atoms or molecules interact to form or break chemical bonds.

chemistry (kem'is-trē) [G. *chemeia*, alchemy] Science dealing with the atomic composition of substances and the reactions they undergo.

chemoreceptor reflex (kem'ō-rē-sep'tŏr) Process in which chemoreceptors detect changes in oxygen levels, carbon dioxide levels, and pH in the blood and produce changes in heart rate, force of heart contraction, and blood vessel diameter that return these values toward their normal levels.

cholecystokinin (kō'lē-sis-tō-kī'nin) [G. *chole*, bile + *kysis*, bladder + *kineo*, to move] Hormone released from the duodenum; inhibits gastric acid secretion and stimulates contraction of the gallbladder.

chondrocyte (kon'drō-sīt) [G. *chondrion*, gristle + *cyte*] Cartilage cell.

chordae tendineae (kōr'dē ten'di-nē-ē) [L., cord] Tendinous strands running from the papillary muscles to the free margin of the cusps that make up the tricuspid and bicuspid valves; prevent the cusps of these valves from extending up into the atria during ventricular contraction.

choroid (kō'royd) [G. *chorioeides*, membranelike or lacy] Portion of the vascular tunic associated with the sclera of the eye; prevents scattering of light.

choroid plexus Specialized group of ependymal cells in the ventricles; secretes cerebrospinal fluid.

chromatid (krō'mă-tid) [G. *chroma*, color] One of a pair of duplicated chromosomes, joined by the centromere, which separates from its partner during cell division.

chromatin (krō'ma-tin) [G. *chroma*, color] Genetic material of the nucleus consisting of deoxyribonucleic acid (DNA) associated with proteins.

chromosome (krō'mō-sōm) [G. *chroma*, color + *soma*, body] One of the bodies (normally 46 in humans) in the cell nucleus that carry the cell's genetic information.

chyle (kīl) [G. *chylos*, juice] Milky colored lymph with a high fat content.

chylomicron (kī-lō-mi'kron) [*chylo-* + G. *micros*, small] Lipid droplet synthesized in the epithelial cells of the small intestine containing triglycerides, cholesterol, and lipoproteins.

chyme (kīm) [G. *chymos*, juice] Semifluid mass of partly digested food passed from the stomach into the duodenum.

ciliary body (sil'ē-ar-ē) [like an eyelash] Structure continuous with the choroid layer of the eye at its anterior margin that contains smooth muscle cells and is attached to the lens by suspensory ligaments; regulates the thickness of the lens and produces aqueous humor.

cilium (sil'ē-ŭm), pl. **cilia** (sil'ē-ă) [L., eyelid] Mobile extension of a cell surface; varies from one to thousands per cell and contains specialized microtubules enclosed by the cell membrane.

citric acid cycle (sit'rik) Series of chemical reactions in which citric acid (six-carbon molecule) is converted into a four-carbon molecule, carbon dioxide is formed, and energy is released; the released energy is used to form ATP; the four-carbon molecule can combine with acetyl-CoA (two-carbon) to form citric acid and start the cycle again.

clavicle (klav'i-kl) [L., a small key] Bone between the sternum and shoulder; the collarbone.

climacteric (klī-mak'ter-ik, klī-mak-ter'ik) [G., the rung of a ladder] Period of endocrine, somatic, and transitory psychological changes occurring in the transition to menopause.

clitoris (klit'ō-ris) Small, erectile structure located in the anterior margin of the vestibule.

clot (klot) To coagulate; a soft, insoluble mass formed when blood coagulates.

clot retraction Condensation of the clot into a denser, more compact structure.

clotting factor One of many proteins found in the blood in an inactive state; activated in a series of chemical reactions that result in the formation of a blood clot.

coagulation (kō-ag-ū-lā'shŭn) Process of changing from a liquid to a solid, especially blood.

cochlea (kok'lē-ă) Portion of the inner ear involved in hearing; shaped like a snail shell.

codon (kō'don) Sequence of three nucleotides in mRNA that codes for a specific amino acid in a protein.

coenzyme (kō-en'zīm) Substance that enhances or is necessary for the function of an enzyme.

collagen (kol'lă-jen) [G. *koila*, glue + *gen*, producing] Ropelike protein of the extracellular matrix.

collecting duct Straight tubule that extends from the cortex of the kidney to the tip of the renal pyramid; filtrate from the distal convoluted tubules enters the collecting duct and is carried to the calyces.

colliculus (ko-lik′ū-lŭs) [L. *collis,* hill] One of four small mounds on the dorsal side of the midbrain; the superior two are involved in visual reflexes, and the inferior two are involved in hearing.

colon (kō′lon) Division of the large intestine that extends from the cecum to the rectum.

commissure (kom′ĭ-shūr) [L., a joining together] Bundle of nerve fibers passing from one side to the other in the brain or spinal cord.

common bile duct Duct formed by the union of the common hepatic and cystic ducts; it joins the pancreatic duct and empties into the duodenum.

common hepatic duct Duct formed by union of the right and left hepatic ducts; it joins the cystic duct to form the common bile duct.

compact bone Bone that is denser and has fewer spaces than spongy bone.

complement (kom′plĕ-ment) Group of serum proteins that stimulate phagocytosis, inflammation, and lysis of cells.

compound (kom′pound) [to place together] Substance containing two or more different kinds of atoms that are chemically combined.

concha (kon′kă) [L., shell] Structure resembling a shell in shape; the three bony ridges on the lateral wall of the nasal cavity.

condyle (kon′dīl) [G. *kondyles,* knuckle] Rounded, articulating surface of a joint.

cone Photoreceptor cell in the retina of the eye with cone-shaped photoreceptive process; important in color vision and visual acuity.

conjunctiva (kon-junk-tī′vă) [L. *conjungo,* to bind together] Mucous membrane covering the anterior surface of the eye and the inner lining of the eyelids.

connective tissue One of the four major tissue types; consists of cells usually surrounded by large amounts of extracellular material; holds other tissues together and provides a supporting framework for the body.

constant region Part of an antibody that does not combine with an antigen and is the same in different antibodies; responsible for activating complement and binding the antibody to cells such as macrophages, basophils, and mast cells.

corn [L. *cornu,* horn] Thickening of the stratum corneum of the skin over a bony projection in response to friction or pressure.

cornea (kōr′nē-ă) [hornlike] Transparent, anterior part of the fibrous tunic of the eye, through which light enters the eye.

corneum (kōr′nē-ŭm) See *stratum corneum.*

coronal plane (kor′ō-năl) [G. *korone,* crown] Plane separating the body into anterior and posterior portions; also called a frontal plane.

coronary artery (kōr′o-nār-ē) [circling like a crown] Artery that carries blood to the muscles of the heart; the left and right coronary arteries arise from the aorta.

coronary bypass Surgery in which a vein from another part of the body is grafted to a coronary artery in such a way as to allow blood flow past a blockage in the coronary artery.

coronary vein Vein that carries blood from the heart muscle, primarily to the right atrium.

corpus callosum (kōr′pus kă-lō′sŭm) [L., body; callous] Large, thick nerve fiber tract connecting the two cerebral hemispheres.

corpus luteum (loo-tē′ŭm) Yellow endocrine body formed in the ovary in the site of a ruptured follicle immediately after ovulation; secretes progesterone and estrogen.

cortex (kōr′teks), pl. cortices (kōr′ti-sēz) [L., bark] Outer part of an organ such as the brain, kidney, adrenal gland, or hair.

cortisol (kōr′ti-sol) Steroid hormone released by the adrenal cortex; increases blood glucose and inhibits inflammation; it is a glucocorticoid.

cotransport (kō-trans′pōrt) Transport of one substance across a cell membrane, coupled with the simultaneous transport of another substance across the same membrane in the same direction.

covalent bond (kō-vāl′ent) Chemical bond formed when two atoms share one or more pairs of electrons.

coxal (kok′săl) bone [L., hip] Bone of the hip.

cranial nerve (krā′nē-ăl) Peripheral nerve originating in the brain.

cranial vault Eight skull bones that surround and protect the brain; braincase.

cremaster muscle (krē-mas′ter) Extension of abdominal muscles; in the male, it raises the testis.

crenation (krē-nā′shŭn) [L. *crena,* a notch] Cell shrinkage that occurs when water moves by osmosis from a cell into a hypertonic solution.

cretinism (krē′tin-izm) Hypothyroidism in an infant; appears during the first years of life and results in stunting of body growth and mental development; hypothyroid dwarfism.

cricoid cartilage (krī′koyd) Most inferior laryngeal cartilage.

cricothyrotomy (krī′kō-thī-rot′ō-mē) Formation of an artificial opening in a victim's air passageway through the membrane between the cricoid and thyroid cartilage.

crown Part of the tooth formed of and covered by enamel.

crypt (kript) [G. *kryptos,* hidden] Pitlike depression.

cryptorchidism (krip-tōr′ki-dizm) Failure of the testes to descend into the scrotal sac.

cupula (koo′poo-lă) [L. *cupa,* a tub] Gelatinous mass that overlies the hair cells of the cristae ampullaris of the semicircular canals; responds to fluid movement.

cutaneous (kū-tā′nē-ŭs) [L. *cutis,* skin] Relating to the skin.

cuticle (kū′ti-kl) [L. *cutis,* skin] Outer thin layer, usually horny; for example, the outer covering of hair or the growth of the stratum corneum onto the nail.

cyanosis (sī-ă-nō′sis) [G., dark blue color] Blue coloration of the skin and mucous membranes caused by insufficient oxygenation of blood.

cystic duct (sis′tik) Duct from the gallbladder; it joins the common hepatic duct to form the common bile duct.

cytoplasm (sī′tō-plazm) [G. *cyto,* cell + *plasma,* a thing formed] Cellular material surrounding the nucleus.

cytoskeleton (sī-tō-skel′ĕ-ton) Collection of microtubules, microfilaments, and intermediate filaments that supports the cytoplasm and organelles; also involved with cell movements.

dartos muscle (dar′tōs) [Fr. *derō,* to skin] Layer of smooth muscle beneath the skin of the scrotum.

deciduous teeth (dē-sid′ū-ŭs) [L. *deciduus,* falling off] Primary teeth, which fall out to be replaced by the permanent teeth.

decomposition reaction (dē′kom-pō-zish′ŭn) Breakdown of a larger molecule into smaller molecules, ions, or atoms.

deep [O.E. *deop,* deep] Away from the surface, internal.

defecation (def-ĕ-kā′shŭn) [L. *defaeco,* to purify] Discharge of feces from the rectum.

deglutition (dē-gloo-tish′ŭn) [L. *de-,* from, away + *glutio,* to swallow] Swallowing.

deltoid (del′toyd) [triangular] Triangular muscle over the shoulder; inserts onto the humerus; abducts the arm.

denaturation (dē-na-tū-rā′shŭn) Change in shape of a protein caused by breaking hydrogen bonds; agents that cause denaturation include heat and changes in pH.

dendrite (den′drīt) [G. *dendrite,* tree] Short, treelike cell process of a neuron; receives stimuli.

dentin (den′tin) Bonelike material forming the mass of a tooth.

deoxyribonucleic (dē-oks′ē-rī′bō-noo-klē′ic) acid (DNA) Type of nucleic acid containing the sugar deoxyribose; the genetic material of cells.

depolarize Decrease in the difference in potential (charge) between two points, as between the inside and outside of a cell membrane.

dermis (der′mis) [G. *derma*, skin] Dense connective tissue that forms the deep layer of the skin; responsible for the structural strength of the skin.

desmosome (dez′mō-sōm) [G. *desmos*, a band + *soma*, body] Point of adhesion between two cells.

diabetes mellitus (dī-ă-bē′tēz me-lī′tŭs) Condition resulting from too little insulin secreted from the pancreatic islets, insufficient numbers of insulin receptors on target cells, or defective receptors that do not respond to insulin.

diaphragm (dī′ă-fram) [a partition wall] Muscular separation between the thoracic and abdominal cavities; its contraction results in inspiration.

diaphysis (dī-af′i-sis) [G., growing between] Shaft of a long bone.

diastole (dī-as′tō-lē) [G. *diastole*, dilation] Relaxation of the heart chambers, during which they fill with blood; usually refers to ventricular relaxation.

diastolic pressure Minimum arterial blood pressure achieved during ventricular diastole.

diencephalon (dī-en-sef′ă-lon) [G. *dia*, through + *enkephalos*, brain] Part of the brain inferior to and nearly surrounded by the cerebrum and connecting posteriorly and inferiorly to the brainstem.

diffusion (di-fū′zhŭn) [L. *diffundo*, to pour in different directions] Tendency for solute molecules to move from an area of higher concentration to an area of lower concentration in a solution; the product of the constant random motion of all atoms, ions, or molecules in a solution.

digestion (di-jes′chŭn, dī-jes′chŭn) Breakdown of carbohydrates, lipids, proteins, and other large molecules to their component parts.

digestive tract (di-jes′tiv, dī-jes′tiv) Tract from the mouth to the anus, including the stomach and intestines, where food is taken in, broken down, and absorbed.

digitalis (dij′i-tal′is) [L., relating to fingerlike flowers] Steroid used in the treatment of heart diseases, such as heart failure; increases the force of contraction of the heart; extracted from the foxglove plant (*Digitalis purpura*).

diploid (dip′loyd) Condition in which there are 2 copies of each autosome and 2 sex chromosomes (46 total chromosomes in humans).

disaccharidase (dī-sak′ă-rid-ās) Enzyme that breaks disaccharides down to monosaccharides; commonly found in the microvilli of the intestinal epithelium.

disaccharide (dī-sak′ă-rīd) [two sugars] Two monosaccharides chemically bound together; glucose and fructose chemically join to form sucrose.

dissociate (di-sō-sē-āt′) [L. *dis-* + *socio*, to disjoin, separate] Separation of positive and negative ions when they dissolve in water and are surrounded by water molecules.

distal (dis′tăl) [L. *di-* + *sto*, to be distant] Farther from the point of attachment to the body than another structure.

distal convoluted tubule Convoluted tubule of the nephron that extends from the ascending limb of the loop of Henle and ends in a collecting duct.

DNA See *deoxyribonucleic acid.*

dominant (dom′i-nant) [L. *dominus*, a master] In genetics, a gene that is expressed phenotypically to the exclusion of a contrasting recessive trait.

dorsal (dōr′săl) [L. *dorsum*, back] Back surface of the body; in humans, synonymous with *posterior.*

dorsal root Sensory root of a spinal nerve.

ductus arteriosus (dŭk′tŭs ar-tēr′ē-ō-sŭs) Short artery that extends from the pulmonary trunk to the aorta; in the fetus, blood flows through the ductus arteriosus from the pulmonary trunk into the aorta and bypasses the lungs.

ductus deferens (dŭk′tŭs def′er-enz) Duct of the testis, running from the epididymis to the ejaculatory duct; also called the vas deferens.

duodenum (doo-ō-dē′nŭm, doo-od′ě-nŭm) [L. *duodeni*, 12] First division of the small intestine; connects to the stomach.

dura mater (doo′ră mā′ter) [L., tough mother] Tough, fibrous membrane forming the outermost meningeal covering of the brain and spinal cord.

eardrum See *tympanic membrane.*

eccrine (ek′rin) [G. *ek*, out + *krino*, to separate] Exocrine; refers to water-producing sweat glands; see *merocrine.*

ECG See *electrocardiogram.*

ectoderm (ek′tō-derm) Outermost of the three germ layers of the embryo.

ectopic beat (ek-top′ik) Heartbeat that originates from an area of the heart other than the SA node.

edema (e-dē′mă) [G. *oidema*, a swelling] Excessive accumulation of fluid, usually causing swelling.

efferent (ef′er-ent) [L. *efferens*, to bring out] Conducting outward from a given organ or part, denoting certain arteries, veins, lymphatics, and motor nerves. Opposite of *afferent.*

efferent arteriole (ar-tēr′ē-ōl) Vessel that carries blood from the glomerulus to the peritubular capillaries.

efferent ductule (dŭk′tool) Small duct that leads from the testis to the epididymis.

efferent fiber Nerve fiber going from the central nervous system toward the peripheral nervous system; motor fiber.

ejaculation (ē-jak′ū-lā′shŭn) [to shoot out] Reflexive expulsion of semen from the penis.

ejaculatory duct (ē-jak′ū-lă-tōr-ē) Duct formed by the union of the ductus deferens and the excretory duct of the seminal vesicle, which opens into the urethra.

electrocardiogram (ē-lek-trō-kar′dē-ō-gram) (ECG) Graphic record of the heart's electrical currents obtained with an electronic recording instrument.

electrolyte (ē-lek′trō-līt) [G. *electro,* + *lytos,* soluble] Positive and negative ions that conduct electricity in solution.

electron (ē-lek′tron) Negatively charged particle around the nucleus of an atom.

electron-transport chain Series of energy-transfer molecules in the inner mitochondrial membrane; they receive energy and use it in the formation of ATP and water.

element (el′ě-ment) [L. *elementum,* a rudiment] Simplest type of matter with unique chemical properties.

embolus (em′bō-lŭs) [G. *embolos,* a plug] Detached clot or other foreign body that occludes a blood vessel.

embryo (em′brē-ō) [Fr. *en,* in + *bryō,* to swell] In prenatal development, the developing human from the time of fertilization to approximately the end of the second month.

emission (ē-mish′ŭn) [L. *emissio,* to send out] Discharge; formation and accumulation of semen prior to ejaculation.

emulsification (ē-mŭl′si-fi-kā-shŭn) Dispersal of one liquid, or very small globules of the liquid, within another liquid.

emulsify (ē-mŭl′si-fī) To form an emulsion, which is one liquid dispersed within another liquid.

enamel (ē-nam′ĕl) Hard substance covering the exposed portion of a tooth.

endocardium (en-dō-kar′dē-ŭm) [G. *endon,* within + *kardia,* heart] Innermost layer of the heart, including endothelium and connective tissue.

endochondral (en-dō-kon′drăl) [*endo* + G. *chondrion,* gristle] Growth of cartilage, which is then replaced by bone.

endochondral ossification (en-dō-kon′drăl os′i-fi-kā′shŭn) Bone formation within cartilage.

endocrine (en′dō-krin) [*endo* + G. *krino,* to separate] Ductless gland that secretes internally, usually into the circulatory system.

endocytosis (en′dō-sī-tō′sis) [*endo* + G. *kytos,* cell + *-osis,* condition] Bulk uptake of material through the cell membrane by taking it into a vesicle.

endoderm (en′dō-derm) [*endo* + G. *derma,* skin] Innermost of the three germ layers of the embryo.

endolymph (en′dō-limf) [endo + G. lympha, clear fluid or springwater] Fluid inside the membranous labyrinth of the inner ear.

endometrium (en′dō-mē′trē-ŭm) [endo + G. mētra, uterus] Mucous membrane that constitutes the inner layer of the uterine wall; consists of a simple columnar epithelium and a lamina propria that contains simple tubular uterine glands.

endomysium (en′dō-mis′ē-ŭm, en′dō-miz′ē-ŭm,) [endo + G. mys, muscle] Fine connective tissue sheath surrounding a muscle fiber.

endoplasmic reticulum (en′dō-plas′mik re-tik′ū-lŭm) [endo + G. plastos, formed a network] Membranous network inside the cytoplasm; rough endoplasmic reticulum has ribosomes attached to the surface; smooth endoplasmic reticulum does not have ribosomes attached.

endosteum (en-dos′tē-ŭm) [endo + G. osteon, bone] Membranous lining of the medullary cavity and the cavities of spongy bone.

endothelium, pl. endothelia (en-dō-thē′lē-ŭm) [G. endo + thēlē, nipple] Layer of flat cells lining especially blood and lymphatic vessels and the heart.

enzyme (en′zīm) [G. en, in + zyme, leaven] Protein molecule that increases the rate of a chemical reaction without being permanently altered; an organic catalyst.

eosinophil (ē-ō-sin′ō-fil) [eosin, an acidic dye + G. phileo, to love] White blood cell with granules that stain red with acidic dyes; inhibits inflammation.

ependymal (ep-en′di-măl) Neuroglial cell layer lining the ventricles of the brain.

epicardium (ep-i-kar′dē-ŭm) [G. epi, upon + kardia, heart] Serous membrane covering the surface of the heart; also called the visceral pericardium.

epicondyle (ep′i-kon′dīl) [epi + G. kondyle, knuckle] Projection on (usually to the side of) a condyle.

epidermis (ep-i-derm′is) [epi + G. derma, skin] Outer portion of the skin formed of epithelial tissue that rests on the dermis; resists abrasion and forms a permeability barrier.

epididymis (ep-i-did′i-mis) [epi + G. didymos, twin] Elongated structure connected to the posterior surface of the testis; site of storage and maturation of the sperm cells.

epiglottis (ep-i-glot′is) [epi + G. glottis, the mouth of the windpipe] Plate of elastic cartilage, covered with mucous membrane, that serves as a valve over the opening of the larynx during swallowing to prevent materials from entering the larynx.

epimysium (ep-i-mis′ē-ŭm, ep-i-miz′ē-ŭm) [epi + G. mys, muscle] Fibrous connective tissue layer surrounding a skeletal muscle.

epinephrine (ep′i-nef′rin) [epi + G. nephros, kidney] Hormone similar in structure to the neurotransmitter norepinephrine; major hormone released from the adrenal medulla; increases cardiac output and blood glucose levels.

epiphyseal line (ep-i-fiz′ē-ăl) Dense plate of bone in a bone that is no longer growing, indicating the former site of the epiphyseal plate.

epiphyseal plate Site at which bone growth in length occurs; located between the epiphysis and diaphysis of a long bone; area of cartilage where cartilage growth is followed by ossification; also called the growth plate.

epiphysis (e-pif′-i-sis) [epi, upon + G. physis, growth] End of a bone; separated from the remainder of the bone by the epiphyseal plate or epiphyseal line.

epiploic appendage (ep′i-plō′ik) One of a number of little, fat-filled processes of peritoneum projecting from the serous coat of the large intestine.

episiotomy (e-piz-ē-ot′ō-mē, e-pis-e-ot′ō-mē) [pubic region + G. tōme, incision] Incision in the clinical perineum, sometimes performed during childbirth.

epithalamus (ep′i-thal′ă-mŭs) [G. epi, upon + thalamus] Small dorsomedial area of the thalamus corresponding to the habenula and its associated structures, the stria medullaris of the thalamus, pineal gland, and habenular commissure.

epithelial tissue (ep-i-thē′lē-ăl) One of the four major tissue types consisting of cells with a basement membrane (exceptions are lymphatic vessels and liver sinusoids), little extracellular material, and no blood vessels; covers the surfaces of the body and forms glands.

epithelium (ep-i-thē′lē-ŭm) [G. epi, upon + thele, covering or lining] pl. epithelia (ep-i-thē′lē-ă) See epithelial tissue.

eponychium (ep-ō-nik′ē-ŭm) [G. epi + onyx, nail] Thin skin that attaches to the proximal part of the nail.

equilibrium (ē-kwi-lib′rē-ŭm) [G. aequus, equal + libra, a balance] State created by a chemical reaction proceeding in opposite directions (e.g., from reactants to products and from products to reactants) at equal speed.

erection (ē-rek′shŭn) Engorgement of erectile tissue with blood, such as in the erectile tissues of the penis, causing the penis to enlarge and become firm.

erector spinae (ē-rek′tōr spī′nē) Common name of the muscle group of the back; holds the back erect.

erythroblastosis fetalis (ĕ-rith′rō-blas-tō′sis fē-tă′lis) [erythroblast + G. -osis, condition] See hemolytic disease of the newborn.

erythrocyte (ĕ-rith′rō-sīt) [G. erythro, red + kytos, cell] See red blood cell.

erythropoietin (ĕ-rith′rō-poy′ĕ-tin) [erythrocyte + G. poiesis, a making] Protein hormone that stimulates red blood cell formation in red bone marrow.

esophagus (ē-sof′ă-gŭs) [G. oisophagos, gullet] Part of the digestive tract between the pharynx and stomach.

estrogen (es′trō-jen) Steroid hormone secreted primarily by the ovaries; involved in the maintenance and development of female reproductive organs, secondary sexual characteristics, and the menstrual cycle.

eustachian tube (ū-stā′shŭn) See auditory tube.

exchange reaction Combination of a decomposition reaction, in which molecules are broken down, and a synthesis reaction, in which the products of the decomposition reaction are combined to form new molecules.

exocrine (ek′sō-krin) [G. exo-, outside + krino, to separate] Gland that secretes to a surface or outward through a duct.

exocytosis (ek′sō-sī-to′sis) Elimination of material from a cell through the formation of vesicles.

exophthalmia (ek-sof-thal′mē-ă) [G. ex, cut + ophthalmos, eye] Bulging of the eyes that frequently accompanies Graves disease, due to accumulation of a type of connective tissue behind the eye.

expiration (eks-pi-rā′shŭn) To breathe out; to move air out of the lungs.

extension [L. extensio] To stretch out, usually to straighten out a joint.

extracellular (eks-tră-sel′ū-lăr) Outside of a cell.

extracellular matrix (mā′triks) Nonliving chemical substances located between cells; often consists of protein fibers, ground substance, and fluid.

extrinsic muscle (eks-trin′sik) Muscle located outside of the structure on which it acts.

extrinsic regulation Regulation of the heart that involves mechanisms outside the heart, including nervous and hormonal regulation.

facet (fas′et) [Fr., little face] Small, smooth articular surface.

facilitated diffusion (fă-sil′i-tā-tĭd di-fū′zhŭn) Carrier-mediated process that does not require ATP and moves substances into or out of cells from a higher to a lower concentration.

fascia (fash′ē-ă) [L., band or fillet] Loose areolar connective tissue beneath the skin (hypodermis), or dense connective tissue that encloses and separates muscles.

fasciculus (fă-sik′ū-lus) [L. fascis, bundle] Band or bundle of nerve or muscle fibers bound together by connective tissue.

fat Greasy, soft-solid lipid found in animal tissues and many plants; composed of glycerol and fatty acids.

fatty acid Straight chain of carbon atoms with a carboxyl group (−COOH) attached at one end; a building block of fats.

feces (fē′sēz) Matter discharged from the digestive tract during defecation, consisting of the undigested residue of food, epithelial cells, intestinal mucus, bacteria, and waste material.

fertilization (fer′til-i-zā′shŭn) Union of the sperm cell and oocyte to form a zygote.

fetus (fē′tŭs) In prenatal development, the developing human between approximately 56 days and birth.

fibrillation (fī-bri-lā′shŭn, fib-ri-lā′shŭn) Very rapid contraction of cardiac muscle fibers, but not of the muscle as a whole; results in dramatically reduced pumping action of the heart.

fibrin (fī′brin) [L. *fibra*, fiber] Threadlike protein fiber derived from fibrinogen by the action of thrombin; forms a clot— that is, a network of fibers that traps blood cells, platelets, and fluid—which stops bleeding.

fibrinogen (fī-brin′ō-jen) [L. *fibra*, fiber + *gen*, produce] Protein in plasma that gives rise to fibrin when acted on by thrombin to form a clot.

fibrinolysis (fī-bri-nol′i-sis) [L. *fibra*, fiber + G. *lysis*, dissolution] Breakdown of a clot by plasmin.

fibroblast (fī′brō-blast) Cell in connective tissue responsible for the production of collagen.

filtration (fil-trā′shŭn) Movement, resulting from a pressure difference, of a liquid through a filter, which prevents some or all of the substances in the liquid from passing.

filtration membrane Membrane formed by the glomerular capillary endothelium, the basement membrane, and the podocytes of Bowman's capsule.

fimbria (fim′brē-ă), pl. **fimbriae** (fim′brē-ē) Long, thin process that surrounds the opening of the uterine tube.

first heart sound Heart sound that results from the simultaneous closure of the tricuspid and bicuspid valves.

flagellum (flă-jel′ŭm), pl. **flagella** (flă-jel′ă) [L., whip] Whiplike locomotor organelle similar to cilia except longer, and there is usually one per sperm cell.

flexion (flek′shŭn) [L. *flectus*] To bend.

focal point Point at which light rays cross after passing through a concave lens.

follicle-stimulating hormone (fol′i-kl) **(FSH)** Hormone of the anterior pituitary gland that, in the female, stimulates the follicles of the ovary, assists in maturation of the follicle, and causes secretion of estrogen from the follicle; in the male, it stimulates the epithelium of the seminiferous tubules and is partially responsible for inducing spermatogenesis.

fontanel (fon′tă-nel′) [Fr., fountain] One of several membranous gaps between bones of the skull.

foramen (fō-rā′men) Hole; referring to a hole or opening in a bone.

foramen ovale (ō-val′ē) In the fetal heart, the oval opening in the interatrial septum with a valve that allows blood to flow from the right to the left atrium, but not in the opposite direction; becomes the fossa ovalis after birth.

formed element Cell, such as a red blood cell or white blood cell, or cell fragments, such as a platelet, in blood.

fossa (fos′ă) Depression below the level of the surface of a bone; usually longitudinal in shape.

fovea centralis (fō′vē-ă) Depression in the center of the macula of the eye; has the greatest visual acuity and only cones.

free energy Total amount of energy that can be liberated by the complete catabolism of food.

frenulum (fren′ū-lŭm) [L. *frenum*, bridle] Fold extending from the floor of the mouth to the middle of the undersurface of the tongue.

frontal plane Plane separating the body into anterior and posterior portions; also called a coronal plane.

FSH See *follicle-stimulating hormone.*

full-thickness burn Burn that destroys the epidermis and the dermis and sometimes the underlying tissue; sometimes called a third-degree burn.

fundus (fŭn′dŭs) [L., bottom] Bottom, or area farthest from the opening, of a hollow organ, such as the stomach or the uterus.

gamete (gam′ēt) [G. *gametēs*, husband; *gametē*, wife] Germ cell, such as an oocyte or a sperm cell.

gamma globulin (gam′ă glob′ū-lin) Family of proteins found in plasma.

ganglion (gang′glē-on), pl. **ganglia** (gang′glē-ă) [G., knot] Group of neuron cell bodies in the peripheral nervous system.

gap junction Small channel that allows materials to pass from one cell to an adjacent cell; provides a means of intercellular communication.

gastric gland (gas′trik) Gland within the stomach.

gastric inhibitory polypeptide Hormone released from the duodenum; inhibits gastric acid secretion.

gastrin (gas′trin) Hormone secreted in the mucosa of the stomach and duodenum that stimulates secretion of hydrochloric acid by the gastric glands.

gastrointestinal tract (gas′trō-in-tes′tin-ăl) Technically, only the stomach and intestines. Often used as a synonym for *digestive tract,* which extends from the mouth to the anus.

gene Sequence of nucleotides in DNA that is a chemical set of instructions for making a specific protein.

genetics (jĕ-net′iks) Branch of science that deals with heredity.

genotype (jen′ō-tīp) Genetic makeup of an individual.

GH See *growth hormone.*

giantism (jī′an-tizm) Abnormal growth in young people because of hypersecretion of growth hormone by the pituitary gland.

gingiva (jin′ji-vă) Dense fibrous tissue, covered by mucous membrane, that covers the alveolar processes of the upper and lower jaws and surrounds the necks of the teeth.

girdle (ger′dl) Bony ring or belt that attaches a limb to the body, such as the pectoral (shoulder) and pelvic girdles.

gland Single cell or multicellular structure that secretes substances into the blood, into a cavity, or onto a surface.

glia (glī′ă) See *neuroglia.*

glomerulus (glō-măr′ū-lŭs) [L. *glomus,* ball of yarn] Mass of capillary loops at the beginning of each nephron, nearly surrounded by Bowman's capsule.

glucagon (gloo′kă-gon) [*glucose* + *agō,* to lead] Hormone secreted from the pancreatic islets of the pancreas that acts primarily on the liver to release glucose into the circulatory system.

glucocorticoid (gloo-kō-kōr′ti-koyd) Hormone from the adrenal cortex capable of increasing the rate at which lipids are broken down to fatty acids and proteins are broken down to amino acids; elevates blood glucose levels and acts as an anti-inflammatory substance.

glycerol (glis′er-ol) Three-carbon molecule with a hydroxyl group attached to each carbon; a building block of fats.

glycogen (glī′kō-jen) Animal starch; composed of many glucose molecules bound together in chains that are highly branched; functions as a carbohydrate reserve; stores glucose molecules; in animal cells.

glycolysis (glī-kol′i-sis) [G. *glykys,* sweet + *lysis,* a loosening] Anaerobic process during which one glucose molecule is converted to two pyruvic acid molecules; a net of two ATP molecules is produced during glycolysis.

glycoprotein (glī-kō-prō′tēn) Organic molecule composed of a protein and a carbohydrate.

GnRH See *gonadotropin-releasing hormone.*

goblet cell Epithelial cell that has the end of the cell at the free surface distended with mucin.

goiter (goy′ter) [L. *guttur,* throat] Enlargement of the thyroid gland, not due to a neoplasm, usually caused by a lack of iodine in the diet.

Golgi apparatus (gol′jē) Named for Camillo Golgi, Italian histologist and Nobel laureate, 1843–1926. Stacks of flattened sacks, formed by membranes, that collect, modify, package, and distribute proteins and lipids.

gonad (gō′nad)[L. *gonē*, seed] Organ that produces gametes; a testis or an ovary.

gonadotropin (gō′nad-ō-trō′pin) [*gonē*, seed + *trope*, a turning] Hormone capable of promoting gonadal growth and function; two major gonadotropins are luteinizing hormone (LH) and follicle-stimulating hormone (FSH).

gonadotropin-releasing hormone (GnRH) Hypothalamic hormone that stimulates the secretion of LH and FSH from the anterior pituitary gland.

granulation tissue (gran′ū-lā′shŭn) Vascular connective tissue formed in wounds.

granulocyte (gran′ū-lō-sīt) [*granular* + G. *kytos*, cell] White blood cell named according to the appearance, in stained preparations, of large cytoplasmic granules; neutrophils, basophils, and eosinophils.

Graves disease Type of hyperthyroidism resulting from abnormal proteins produced by the immune system that are similar in structure and function to thyroid-stimulating hormone; often accompanied by exophthalmia.

growth hormone (GH) Protein hormone of the anterior pituitary gland; it promotes body growth, increases fat mobilization, and increases blood glucose levels because it inhibits glucose utilization.

gynecomastia (gī′nĕ-kō-mas′tē-ă) Enlarged breasts in males.

gyrus (jī′rŭs) [L. *gyros*, circle] Rounded elevation or fold on the surface of the brain.

hair Threadlike outgrowth of the skin, consisting of columns of dead, keratinized epithelial cells.

hair cell Cell of the inner ear containing hairlike processes (microvilli) that respond to bending of the hairs by depolarization.

hamstring muscle One of the three major muscles of the posterior thigh.

haploid (hap′loyd) Condition in which a cell has 1 copy of each autosome and 1 sex chromosome (23 total chromosomes in humans); characteristic of gametes.

haustra (haw′stră) Sacs of the colon, formed by the teniae coli, which are slightly shorter than the gut, so that the gut forms pouches.

haversian canal (ha-ver′shan) Named for seventeenth-century English anatomist Clopton Havers (1650–1702). See *central canal.*

haversian system See *osteon.*

hCG See *human chorionic gonadotropin.*

heart-lung machine Machine that pumps blood and carries out the process of gas exchange; it substitutes for the heart and lungs during heart surgery.

heart rate Number of complete cardiac cycles (heartbeats) per minute.

heart transplant Process of taking a healthy heart from a recently deceased donor and transplanting it into a recipient who has a diseased heart.

hematocrit (hē′mă-tō-krit, hem′a-tō-krit) [G. *hemato*, blood + *krino*, to separate] Percentage of total blood volume composed of red blood cells.

hematopoiesis (hē′mă-tō-poy-ē′sis) [G. *hemato*, blood + *poiesis*, a making] Production of blood cells.

hemidesmosome (hem-ē-des′mō-sōm) [G. *hemi*, one half] Half desmosome that occurs on the basal surface epithelial cells that rest on the basement membrane.

hemoglobin (hē-mō-glō′bin) [G. *hemato*, blood + *glob*, a ball] Red protein of red blood cells consisting of four globin proteins with an iron-containing red pigment, heme, bound to each globin protein; transports oxygen and carbon dioxide.

hemolysis (hē-mol′i-sis) [G. *hemo*, blood + *lysis*, destruction] Rupture of red blood cells.

hemolytic (hē-mō-lit′ik) **disease of the newborn (HDN)** Destruction of red blood cells in the fetus or newborn, caused by antibodies produced in the Rh-negative mother acting on the Rh-positive blood of the fetus or newborn.

hemorrhage (hem′ŏ-rij) [G. *haima*, blood + *rhegnymi*, to burst forth] Rupture or leaking of blood from vessels.

hepatic (he-pat′ik) [G. *hepar*, liver] Associated with the liver.

hepatic portal system [L. *porta*, gate] Blood flow through the veins that begin as capillary beds in the small intestine, spleen, pancreas, and stomach and carry blood to the liver, where they end as a capillary bed.

hepatic portal vein Vein that carries blood from the intestines, stomach, spleen, and pancreas to the liver.

Hering-Breuer reflex Named for German physiologist Heinrich E. Hering (1866–1948) and Austrian internist Josef Breuer (1842–1925). Process in which action potentials from stretch receptors in the lungs arrest inspiration; expiration then occurs.

heterozygous (het′er-ō-zī′gŭs) Having two different genes for a given trait.

hilum (hī′lŭm) [L., a small amount or trifle] Part of an organ where the nerves and vessels enter and leave.

histology (his-tol′ō-jē) [G. *histo*, web (tissue) + *logos*, study] Science that deals with the structure of cells, tissues, and organs in relation to their function.

holocrine (hol′ō-krin) [G. *holo*, whale + *krino*, to separate] Gland whose secretion consists of disintegrated cells of the gland; an example is a sebaceous gland.

homeostasis (hō′mē-ō-stā′sis) [G. *homoio*, like + *stasis*, a standing] Existence and maintenance of a relatively constant environment within the body with respect to functions and the composition of fluids and tissues.

homeotherm (hō′mē-ō-therm) [G. *homoiois*, like + *thermos*, warm] Any animal, including mammals and birds, that tends to maintain a constant body temperature; also referred to as warm-blooded.

homozygous (hō-mō-zī′gŭs) Having two identical genes for a given trait.

hormone (hōr′mōn) [G. *hormon*, to set into motion] Substance secreted by endocrine tissues into the blood that acts on a target tissue to produce a specific response.

human chorionic gonadotropin (hCG) (kō-rē-on′ik gō′nad-ō-trō′pin) Hormone, similar to LH, secreted from the placenta and essential for the maintenance of pregnancy for the first 3 months; prevents the corpus luteum from degenerating.

humerus (hū′mer-ŭs) [L., shoulder] Bone of the arm.

humoral immunity (hū′-mŏr-ăl i-mū′ni-tē) See *antibody-mediated immunity.*

hydrogen bond (hī′drō-jen) Weak attraction between the oppositely charged ends of two polar covalent molecules; the weak attraction between the end of a polar covalent molecule and an ion.

hydrophilic (hī-drō-fil′ik) [G. *hydro*, water + *philos*, fond or loving] Attracting or associating with water molecules; tending to dissolve or associate with water molecules; tending to dissolve in water; polar.

hydrophobic (hī-drō-fōb′ik) [G. *hydro*, water + *phobos*, fear] Lacking affinity for water molecules; tending not to dissolve in water; nonpolar.

hydroxyapatite (hī-drok′sē-ap-ă-tīt) Complex crystal structure that makes up the mineral portion of bones and teeth.

hymen (hī′men) [G. *hymen*, membrane] Thin, membranous fold highly variable in appearance; partly occludes the opening of the vagina prior to its rupture; may occur for a variety of reasons and is frequently absent.

hyoid (hī′oyd) [G., shaped like the letter epsilon, ε] U-shaped bone in the throat.

hypercapnia (hī-per-kap′nē-ă) [*hyper* + G. *kapnos*, smoke, vapor] Abnormally increased arterial carbon dioxide tension or levels.

hyperpolarize (hī′per-pō′lăr-īz) [G. *hyper*, above + *polaris*, polar] To increase polarization of membranes of nerve or muscle cells.

hypertension (hī′per-ten′shŭn) [G. *hyper*, above + *tensio*, tension] High blood pressure; generally, blood pressure greater than 140/90 is considered too high.

hyperthyroidism (hī-per-thī′royd-izm) Abnormality of the thyroid gland in which thyroid hormone secretion is increased.

hypertonic (hī-per-ton′ik) [G. *hyper,* above + *tonos,* tension] Solution that causes cells to shrink.

hypodermis (hī-pō-der′mis) [G. *hypo,* under + *dermis,* skin] Loose connective tissue under the dermis that attaches the skin to muscle and bone.

hypophysis (hī-pof′i-sis) [*hypo* + G., an undergrowth or growth] Endocrine gland attached to the hypothalamus by the infundibulum; the pituitary gland.

hypothalamic-pituitary portal system (hī′pō-thal′ă-mīk-pi-too′i-tār-ē) Series of blood vessels that carry blood from the area of the hypothalamus to the anterior pituitary gland; originates from capillary beds in the hypothalamus and terminates as a capillary bed in the anterior pituitary gland.

hypothalamus (hī′pō-thal′ă-mŭs) [*hypo* + G. *thalamos,* bedroom] Important autonomical and endocrine control center in the brain located beneath the thalamus.

hypothyroidism (hī′pō-thī′royd-izm) Reduced secretion of thyroid hormones from the thyroid gland, leading to cretinism in infants and symptoms of inadequate thyroid hormone secretion in adults.

hypotonic (hī-pō-ton′ik) [*hypo,* under + G. *tonos,* tension or tone] Solution that causes cells to swell.

hypoxia (hī-pok′sē-ă) [*hypo* + oxygen] Below-normal levels of oxygen in arterial blood.

ICSH See *interstitial cell–stimulating hormone.*

Ig See *immunoglobulin.*

ileocecal junction (il′ē-ō-se′kăl) Junction of the ileum of the small intestine and the cecum of the large intestine.

ileum (il′ē-ŭm) [L. *eileo,* to roll up, twist] Third portion of the small intestine, about 3.5 meters in length; extends from the jejunum to the ileocecal opening.

ilium (il′ē-ŭm) Broad, flaring portion of the hipbone; becomes fused with the ischium and pubis.

immunity (i-mū′ni-tē) Ability to resist damage from foreign substances, such as microorganisms, and harmful chemicals, such as toxins released by microorganisms.

immunoglobulin (im′ū-nō-glob-ū-lin) (Ig) Antibodies.

implantation (im-plan-tā′shŭn) Attachment of the blastocyst to the endometrium of the uterus; occurring 6 or 7 days after fertilization of the oocyte.

impotence (im′pŏ-tens) Inability of the male to achieve or maintain an erection and thus engage in sexual intercourse.

incompetent valve Leaky valve; usually refers to a heart valve that allows blood to flow through it when it is closed.

incus (ing′kus) [L., anvil] Middle bone of the middle ear; the anvil.

infarct (in′farkt) Area of necrosis resulting from a sudden insufficiency of arterial blood supply.

inferior (in-fē′rē-ōr) [L., lower] Down, or lower, with reference to the anatomical position.

inferior vena cava (vē′nă kā′vă) Receives blood from the lower limbs, pelvis, and abdominal organs and empties into the right atrium of the heart.

inflammatory (in-flam′ă-tōr-ē) **response** Complex sequence of events, involving chemicals and immune system cells, that results in the isolation and destruction of foreign substances, such as bacteria; symptoms include redness, heat, swelling, pain, and disturbance of function.

infundibulum (in-fŭn-dib′ū-lŭm) [L., a funnel] Funnel-shaped structure or passage—for example, the infundibulum that attaches the pituitary gland to the hypothalamus; funnel-like expansion of the uterine tube near the ovary.

inguinal canal (ing′gwi-năl) Passageway through which a testis passes as it descends from the abdominopelvic cavity to the scrotum.

inguinal hernia (her′nē-ă) Rupture that allows the potential protrusion of abdominal organs, such as the small intestine, through the inguinal canal.

innate immunity (i′nāt, i-nāt′) Immune system response that is the same on each exposure to an antigen; there is no ability to remember a previous exposure to a specific antigen.

inner cell mass Group of cells at one end of the blastocyst from which the embryo develops.

inorganic (in-ōr-gan′ik) Molecules that do not contain carbon atoms; originally defined as molecules that came from nonliving sources; the original definition is no longer valid because carbon dioxide produced by living organisms is considered an inorganic molecule.

insertion (in-ser′shŭn) The more movable attachment point of a muscle.

inspiration (in-spi-rā′shŭn) To breathe in; to move air into the lungs, or inhale.

insulin (in′sŭ-lin) Protein hormone, secreted from the pancreas, that increases the uptake of glucose and amino acids by most tissues.

interatrial septum (in-ter-ā′trē-ăl) Cardiac muscle partition separating the right and left atria.

intercalated disk (in-ter′kă-lā-ted) [inserted between] Connection between cardiac muscle cells; important in coordinating the contractions of cardiac muscle cells; contains gap junctions that allow action potentials to pass from one cardiac muscle cell to adjacent cardiac muscle cells.

intercostal muscle (in-ter-kos′tăl) Muscle located between ribs.

interferon (in-ter-fēr′on) Protein released by virally infected cells that binds to other cells and stimulates them to produce antiviral proteins that inhibit viral replication.

interkinesis (in′ter-ki-nē′sis) Short time period between the formation of the daughter cells of the first meiotic division and the second meiotic division.

interstitial cell (in-ter-stish′ăl) Cell between the seminiferous tubules of the testes; secretes testosterone; also called Leydig cell.

interstitial cell-stimulating hormone (ICSH) Luteinizing hormone in males. Hormone of the anterior pituitary gland that stimulates the secretion of testosterone in the testes. See *luteinizing hormone.*

interventricular septum (in-ter-ven-trik′ū-lăr) Cardiac muscle partition separating the right and left ventricles.

intestinal glands (in-tes′ti-năl) Tubular glands in the mucous membrane of the small intestine.

intracellular (in-tră-sel′ū-lăr) Inside a cell.

intramembranous ossification (in′tră-mem′brā-nŭs os′i-fi-kā′shŭn) Bone formation within connective tissue membranes.

intramural plexus (in′tră-mū′răl plek′sŭs) [L., within the wall] Nerve plexus within the walls of the digestive tract; involved in local and autonomic control of digestion.

intrinsic factor (in-trin′sik) Factor secreted by the gastric glands and required for adequate absorption of vitamin B_{12}.

intrinsic muscle Muscle located within the structure on which it acts.

ion (ī′on) Atom or group of atoms carrying an electrical charge because of a loss or gain of one or more electrons.

ionic bond (ī-on′ik) Chemical bond resulting from the attraction between ions of opposite charge.

iris (ī′ris) Specialized part of the vascular tunic of the eye; the "colored" part of the eye that can be seen through the cornea; consists of smooth muscles that regulate the amount of light entering the eye.

isometric contraction (ī-sō-met′rik) Muscle contraction in which the length of the muscle does not change but the amount of tension increases.

isotonic (ī′sō-ton′ik) [G. *iso,* equal + *tonos,* tension] Solution that causes cells to neither shrink nor swell.

isotonic contraction Muscle contraction in which the amount of tension is constant and the muscle shortens.

isotope (ī′sō-tōp) [G. *isos,* equal + *topos,* part] One of two or more elements that have the same number of protons and electrons but a different number of neutrons.

jaundice (jawn'dis) [Fr. *jaune*, yellow] Yellowish staining of the skin, sclerae, and deeper tissues and excretions with bile pigments.

jejunum (jĕ-joo'nŭm) [L. *jejunus*, empty] Portion of the small intestine, about 2.5 meters in length, between the duodenum and the ileum.

juxtaglomerular apparatus (jŭks'tă-glŏ-mer'ū-lăr) [L. *juxta*, close to + *glomerulus*] Specialized wall of the distal convoluted tubule and afferent arteriole that secretes renin.

keratin (ker'ă-tin) Protein that accumulates in cells of nails, hair, and the superficial layers of the epidermis of the skin.

keratinization (ker'ă-tin-i-zā'shŭn) Production of keratin and changes in the structure and shape of epithelial cells as they move to the skin surface.

kinetic (ki-net'ik) [G. *kinetikos*, of motion] Relating to motion or movement.

Korotkoff sound (kō-rot'kof) Named for Russian physician Nikolai Korotkoff (1874–1920). Sound heard over an artery when blood pressure is determined by the auscultatory method.

kyphosis (kī-fō'sis) [G., humpback] Abnormal posterior curvature, or flexion, of the spine.

labia majora (lā'bē-a) Two rounded folds of skin surrounding the labia minora and vestibule.

labia minora Two narrow, longitudinal folds of mucous membrane enclosed by the labia majora; they unite anteriorly to form the prepuce.

labyrinth (lab'i-rinth) Series of membranous and bony tunnels in the temporal bone; the part of the inner ear involved in hearing and balance.

lacrimal (lak'ri-măl) [L., a tear] Relating to tears or tear production.

lactation (lak'tā'shŭn) [L. *lactatio*, suckle] Period following childbirth during which milk is formed in the breasts.

lacteal (lak'tē-ăl) [relating to milk] Lymphatic vessel in the wall of the small intestine that carries chyle from the intestine and absorbs fat.

lactic acid (lak'tik) Three-carbon molecule derived from pyruvic acid as a product of anaerobic respiration.

lacuna (lă-koo'nă, pl. lacunae (lă-koo'nē) [L., a pit] Small space, cavity, or depression; a space in cartilage in which a chondrocyte is located; a space in bone matrix in which an osteocyte is located; a cavity containing maternal blood in the placenta.

lamella (lă-mel'ă, pl. lamellae (lă-mel'ē) [L. *lamina*, plate, leaf] Thin sheet or layer of bone.

lamina (lam'i-nă, pl. laminae (lam'i-nē) [L. *lamina*, plate, leaf] Layer; a portion of the vertebra that extends from the transverse process to the spinous process.

lamina propria (prō'prē-ă) Layer of connective tissue underlying the epithelium of a mucous membrane.

lanugo (lă-noo'gō) [L. *lana*, wool] Fine, soft fetal or embryonic hair.

laryngitis (lar-in-jī'tis) Inflammation of the mucous membrane of the larynx.

laryngopharynx (lă-ring'gō-far-ingks) Part of the pharynx lying below the tip of the epiglottis extending to the level of the cricoid cartilage of the larynx.

larynx (lar'ingks) Organ of voice production located between the pharynx and the trachea; it consists of a framework of cartilages and elastic membranes housing the vocal folds (true vocal cords) and the muscles that control the position and tension of these elements.

lateral [L. *latus*, side] Away from the middle or midline of the body.

lateral horn Small, lateral extension of spinal cord gray matter; located only in spinal cord regions T1–L2; containing preganglionic sympathetic neuron cell bodies.

lens Biconvex structure in the anterior part of the eye capable of being flattened or thickened to adjust the focus of light entering the eye.

leukemia (loo-kē'mē-ă) [G. *leukos*, white + *haima*, blood] Tumor of the red bone marrow that results in the production of large numbers of abnormal white blood cells; often accompanied by decreased production of red blood cells and platelets.

leukocyte (loo'kō-sīt) [G. *leukos*, white + *kytos*, cell] See *white blood cell*.

leukocytosis (loo'kō-sī-tō'sis) [*leukocyte* + G. *-osis*, a condition] Higher than normal number of white blood cells.

leukopenia (loo-kō-pē'nē-ă) [*leukocyte* + G. *penia*, poverty] Lower than normal number of white blood cells.

Leydig cell (lī'dig) Named for German anatomist Franz von Leydig (1821–1908). See *interstitial cell*.

LH See *luteinizing hormone*.

ligament (lig'ă-ment) Tough connective tissue band usually connecting bone to bone.

ligand (lig'and, lī'gand) [L. *ligo*, to bind] Molecule that binds to a macromolecule— for example, a ligand binding to a receptor.

limbic system (lim'bik) [L. *limbus*, a border or boundary] Primitive part of the brain involved in visceral and emotional response and in the response to odor.

linea alba (lin'ē-ă al'bă) White line in the center of the abdomen where muscles of the abdominal wall insert.

lipase (lip'ās) Enzyme that breaks down lipids.

lipid (lip'ĭd) [G. *lipos*, fat] Substance composed principally of carbon, oxygen, and hydrogen; generally soluble in nonpolar solvents; fats and cholesterol.

local inflammation Inflammation confined to a specific area of the body; symptoms include redness, heat, swelling, pain, and loss of function.

longitudinal section Cut made through the long axis of an organ.

loop of Henle U-shaped part of the nephron extending from the proximal to the distal convoluted tubule and consisting of descending and ascending limbs; many of the loops of Henle extend into the renal pyramids.

lordosis (lōr-dō'sis) [G., a bending backward; swayback] Abnormal anterior curvature of the spine, usually in the lumbar region; saddle back or swayback.

lower motor neuron Motor neuron located in the brainstem or spinal cord, as opposed to the cerebral cortex.

lumbosacral plexus (lŭm'bō-sā'krăl) [L. *lumbus*, loin + *sacrum*, sacred] Nerve plexus that innervates the lower limbs; originates from spinal nerves L1–S4.

lunula (loo'noo-lă) [L. *luna*, moon] White, crescent-shaped portion of the nail matrix visible through the proximal end of the nail.

luteinizing hormone (LH) (loo'tē-ĭ-nīz-ing) Hormone of the anterior pituitary gland that, in the female, initiates final maturation of the follicles, their rupture to release the oocyte, the conversion of the ruptured follicle into the corpus luteum, and the secretion of progesterone; in the male, it stimulates the secretion of testosterone in the testes and is sometimes referred to as interstitial cell–stimulating hormone (ICSH).

lymph (limf) [L. *lympha*, clear spring water] Clear or yellowish fluid derived from interstitial fluid and found in lymphatic vessels.

lymph node Encapsulated mass of lymphatic tissue found along lymphatic vessels; filters lymph and produces lymphocytes.

lymphocyte (lim'fō-sīt) [L. *lympho*, lymph + G. *kytos*, cell] Nongranulocytic white blood cell involved in the immune system; there are several types of lymphocytes with diverse functions, including antibody production, allergic reactions, graft rejections, tumor control, and regulation of the immune system.

lymphokine (lim'fō-kīn) Class of chemicals, produced by T cells, that activate macrophages and other immune cells; promote phagocytosis and inflammation.

lymphoma (lim-fō'mă) Neoplasm (tumor) of lymphatic tissue that is almost always malignant.

lysis (lī'sis) [G., dissolution or loosening] Rupturing or breaking of a cell membrane.

lysosome (lī'sō-sōm) [G. *lysis*, a loosening + *soma*, body] Membrane-bound vesicle containing intracellular digestive enzymes.

macrophage (mak′rō-fāj) [G. *makros*, large + *phago*, to eat] Large mononuclear, phagocytic cell.

macula (mak′ū-lă) [L., a spot] One of the sensory structures in the vestibule, consisting of hair cells and a gelatinous mass embedded with otoliths; responds to gravity. Also a small yellow spot in the posterior retina of the eye where the cones are concentrated; has no red tint because it is devoid of blood vessels.

malignant (mă-lig′nănt) [L. *maligno*, to do anything malicious, with malice or intent to do harm] In reference to a neoplasm, the property of spreading locally and to distant sites.

malleus (mal′ē-ŭs) [L., hammer] Most lateral of the middle ear bones, attached to the tympanic membrane; the hammer.

mamma (mam′ă) pl. **mammae** (mam′ē) See *mammary gland*.

mammary gland (mam′ă-rē) Organ of milk secretion, located in the breast, or mamma.

mastication (mas-ti-kā′shŭn) [L. *mastico*, to chew] Chewing.

mastoid (mas′toyd) [*mastos*, breast + *eidos*, resemblance] Resembling a breast—for example, the mastoid process of the temporal bone.

matrix (mā′triks) Substance between the cells of a tissue.

matter Anything that occupies space.

mean arterial blood pressure Average of the arterial blood pressure; it is slightly less than the average of the systolic and diastolic blood pressure because diastole lasts longer than systole.

meatus (mē-ā′tŭs) [L., to go, pass] Passageway or tunnel.

meconium (mē-kō′nē-ŭm) [L. *mēkōn*, poppy] Greenish anal discharge from the fetus; consists of fluid swallowed, epithelial cells from the mucosa of the gut, mucus from the intestinal glands, and bile from the liver.

medial (mē′dē-ăl) [L. *medialis*, middle] Toward the middle or midline of the body.

mediastinum (me′dē-as-tī′nŭm) [L., middle septum or wall] Middle wall of the thorax, consisting of the trachea, esophagus, thymus, heart, and other structures.

medulla (me-dool′ă) [L. *medius*, middle, marrow] Center, or core, of an organ, such as the adrenal gland, kidney, or hair.

medulla oblongata (ob-long-gah′tă) Inferior portion of the brainstem that connects the spinal cord with the brain; contains nuclei of cranial nerves plus autonomic control centers for heart rate, respiration, and so forth.

medullary cavity (med′ŭl-er-ē) Large, marrow-filled cavity in the diaphysis of a long bone.

medullary respiratory center (res′pi-ră-tōr-ē, rĕ-spīr′ă-tōr-ē) Nerve cells in the medulla oblongata and pons of the brain that control inspiration and expiration.

megakaryocyte (meg-ă-kar′ē-ō-sīt) Large cell in red bone marrow that gives rise to platelets.

meiosis (mī-ō′sis) [G., a lessening] Process of cell division that results in gametes. Consists of two cell divisions that result in four cells, each of which contains half the number of chromosomes as the parent cell; occur in the testes and ovaries.

melanin (mel′ă-nin) [G. *melas*, black] Brown to black pigment responsible for skin and hair color.

melanocyte (mel′ă-nō-sīt) [G. *melas*, black + *kytos*, cell] Cells, found mainly in the stratum basale of skin, that produce the brown or black pigment melanin.

melanocyte-stimulating hormone (MSH) (mel′ă-nō-sīt) [G. *melas*, black + *kytos*, cell] Peptide hormone secreted by the anterior pituitary gland; increases melanin production by melanocytes, making the skin darker in color.

melanoma (mel′ă-nō′mă) [G. *melas*, black + *oma*, tumor] Malignant tumor derived from melanocytes.

melanosome (mel′ă-nō′sōm) [G. *melano*, black + *soma*, body] Pigment granule produced by melanocytes.

melatonin (mel-ă-tōn′in) Hormone secreted by the pineal gland; may inhibit gonadotropin-releasing hormone secretion from the hypothalamus.

membranous labyrinth (mem′bră-nŭs lab′i-rinth) Membrane-bound set of tunnels and chambers of the inner ear.

memory cell Lymphocyte derived from a B cell or T cell that has been exposed to an antigen; when exposed to the same antigen a second time, the memory cell rapidly responds to provide immunity.

menarche (me-nar′kē) [G. *mēn*, month + *archē*, beginning] First menstrual period or flow.

meninges (mĕ-nin′jēz) [G. *meninx*, membrane] Series of three connective tissue membranes: the dura mater, arachnoid mater, and pia mater; they surround and protect the brain and spinal cord.

menopause (men′ō-pawz) [L. *mēn*, month + *pausis*, cessation] Permanent cessation of the menstrual cycle.

menses (men′sēz) [L. *mensis*, month] Loss of blood and tissue as the endometrium of the uterus sloughs away at the end of the menstrual cycle; occurring at about 28-day intervals in the nonpregnant female of reproductive age.

menstrual cycle (men′stroo-ăl) Series of changes that occur in sexually mature, nonpregnant females and result in menses; specifically includes the cyclical changes that occur in the uterus and ovary.

merocrine (mer′ō-krin) [G. *meros*, part + *krino*, to separate] Gland that secretes products with no loss of cellular material; an example is water-producing sweat glands; see *apocrine* and *holocrine*.

mesentery (mes′en-ter-ē) [G. *mesos*, middle + *enteron*, intestine] Double layer of peritoneum extending from the abdominal wall to the abdominopelvic organs; conveys blood vessels and nerves to abdominopelvic organs; holds and supports abdominopelvic organs.

mesoderm (mez′ō-derm) Middle of the three germ layers of the embryo.

mesovarium (mez′ō-vā′rē-ŭm) Mesentery of the ovary; mesentery that attaches the ovary to the posterior surface of the broad ligament.

metabolic rate Total amount of energy produced and used by the body per unit of time.

metabolism (mĕ-tab′ō-lizm) [G. *metabole*, change] Sum of the chemical changes that occur in tissues, consisting of the breakdown of molecules (catabolism) to produce energy and the buildup of molecules (anabolism), which requires energy.

metaphase (met′ă-fāz) [G. *meta*, after + *phasis*, an appearance] Stage of mitosis or meiosis in which the chromosomes become aligned near the center of the cell, at the equatorial plane, separating the centromeres, or chromosome pairs (see also *meiosis*). The centromeres of each chromosome divide, and the two daughter chromosomes are directed toward opposite poles of the cell.

metastasis (mĕ-tas′tă-sis) Shifting of a disease or a neoplasm from one part of the body to another remote from the original location.

micelle (mi-sel′, mī-sel′) [L. *micella*, small morsel] Droplet of digested lipid surrounded by bile salts in the small intestine.

microglia (mī-krog′lē-ă) [G. *micro*, small + *glia*, glue] Small neuroglial cells that become phagocytic and mobile in response to inflammation; considered to be macrophages of the central nervous system.

microtubule (mī-krō-too′būl) Hollow tube composed of tubulin; microtubules help support the cytoplasm of the cell and are components of certain cell organelles, such as cilia and flagella.

microvillus (mī′krō-vil′ŭs), pl. **microvilli** (mī′krō-vil′ī) One of the minute projections of the cell membrane that greatly increase the surface area of the cell membrane.

micturition reflex (mik-choo-rish′ŭn) Contraction of the urinary bladder stimulated by stretching of the urinary bladder wall; it results in emptying of the urinary bladder.

midbrain Superior end of the brainstem; located between the pons and diencephalon; contains fibers crossing from the brain to the spinal cord and vice versa, as well as nuclei and visual reflex centers.

midsagittal (mid′saj′i-tăl) Plane running vertically through the body and dividing it into equal right and left parts.

mineral (min′er-ăl) Inorganic nutrient necessary for normal metabolic functions.

mineralocorticoid (min′er-al-ō-kōr′ti-koyd) Steroid hormone released from the adrenal cortex; acts on the kidney to increase the rate of sodium ion reabsorption from the nephron and potassium and hydrogen ion secretion into the nephron of the kidney; an example is aldosterone.

mitochondrion (mī-tō-kon′drē-on), pl. mitochondria (mī-tō-kon′drē-ă) [G. mitos, thread + chandros, granule] Small, spherical, rod-shaped or thin filamentous structure in the cytoplasm that is a major site of ATP production.

mitosis (mī-tō′sis) [G., thread] Division of the nucleus. Process of cell division that results in two daughter cells with exactly the same number and type of chromosomes as the parent cell.

mitral valve (mī′trăl) [resembling a bishop's miter, a two-pointed hat] See bicuspid valve.

molecule (mol′ĕ-kūl) Two or more atoms chemically combined to form a structure that behaves as an independent unit.

monocyte (mon′ō-sīt) [G. mono, one or single + kytos, a cell] Type of white blood cell that transforms to become a macrophage.

mononuclear phagocytic system (mon-ō-noo′klē-ăr fag-ō-sit′ik) Phagocytic cells with a single nucleus, derived from monocytes; the cells either enter a tissue by chemotaxis in response to infection or tissue damage or are positioned to intercept microorganisms entering tissues.

monosaccharide (mon-ō-sak′ă-rīd) Basic building block from which more complex carbohydrates are constructed—for example, glucose and fructose.

mons pubis (monz pū′bis) [L., mountain] Prominence formed by a pad of fatty tissue over the pubic symphysis in the female.

morula (mōr′oo-lă) [L. morus, mulberry] Solid mass of cells resulting from the early cleavage divisions of the zygote.

motor neuron Neuron in the brain or spinal cord that innervates skeletal, smooth, or cardiac muscle cells or glands. Somatic motor neurons directly innervate skeletal muscle cells. Two autonomic motor neurons in series extend from the central nervous system to smooth or cardiac muscle cells or glands.

motor unit Single motor neuron and all the skeletal muscle fibers it innervates.

MSH See melanocyte-stimulating hormone.

mucin (mū′sin) Secretion containing mucopolysaccharides (proteoglycans), produced by mucous gland cells.

mucosa (mū-kō′să) [mucus-producing membrane] Mucous membrane consisting of the epithelium and connective tissue; in the digestive tract, there is also a layer of smooth muscle.

mucous membrane (mū′kŭs) Thin sheet consisting of epithelium and connective tissue that lines cavities opening to the outside of the body; many contain mucous glands, which secrete mucus.

mucus (mū′kŭs) Viscous secretion produced by and covering mucous membranes; lubricates and protects the mucous membrane and traps foreign substances.

murmur (mer′mer) Abnormal sound produced within the heart.

muscle fiber (mŭs′ĕl) Muscle cell.

muscle tissue One of the four major tissue types; consists of cells with the ability to contract; includes skeletal, cardiac, and smooth muscle.

muscle twitch Contraction of an entire muscle in response to a stimulus that causes an action potential in one or more muscle fibers.

muscularis (mŭs-kū-lā′ris) Outermost, smooth muscle coat of a hollow organ.

muscularis mucosa Inner, thin layer of smooth muscle found in most parts of the digestive tract outside the lamina propria.

myelinated (mī′ĕ-li-nāt-ed) [G. myelos, marrow] Nerve fibers having a myelin sheath.

myelin sheath (mī′ĕ-lin) Lipoprotein envelope made by wrapping the cell membrane of a Schwann cell or an oligodendrocyte around an axon.

myocardium (mī-ō-kar′dē-ŭm) [myo- + G. kordin, heart] Middle layer of the heart, consisting of cardiac muscle.

myofibril (mī-ō-fī′bril) Fine, longitudinal fibril within a skeletal muscle fiber; consisting of sarcomeres composed of thick (myosin) and thin (actin) myofilaments, placed end to end.

myofilament (mī-ō-fil′ă-ment) Ultramicroscopic protein thread helping form myofibrils in skeletal muscle; thin myofilaments are composed of actin, and thick myofilaments are composed of myosin.

myometrium (mī′ō-mē′trē-ŭm) Muscular wall of the uterus, composed of smooth muscle.

myosin myofilament (mī′ō-sin) One of the two major kinds of protein fibers of a sarcomere; thick filament resembling a bundle of golf clubs.

myxedema (mik-se-dē′mă) Hypothyroidism characterized by edema beneath the skin due to a change in the structure of the subcutaneous connective tissue.

NADH See nicotinamide adenine dinucleotide.

nail (nāl) Thin, horny plate at the ends of the fingers and toes, consisting of several layers of dead epithelial cells containing a hard keratin.

naris (nā′ris), pl. nares (nā′res) Nostril, the opening into the nasal cavity.

nasal cavity (nā′zăl) Cavity divided by the nasal septum and extending from the external nares anteriorly to the nasopharynx posteriorly; bounded inferiorly by the hard palate.

nasolacrimal duct (nā-zō-lak′ri-măl) [L. nasus, nose + lacrima, tear] Duct that leads from the lacrimal sac to the nasal cavity.

nasopharynx (nā′zō-far′ingks) Part of the pharynx that lies above the soft palate; anteriorly, it opens into the nasal cavity.

negative feedback Mechanism by which any deviation from an ideal normal value or set point is resisted or negated; returns a parameter to its normal range and thereby maintains homeostasis.

neonate (nē′ō-nāt) [G. neos, new + L. natalis, relating to birth] Newborn, from birth to 1 month.

neoplasm (nē′ō-plazm) [neo + G. plasma, thing formed] New growth; an abnormal tissue that grows by cellular proliferation; may be benign or malignant.

nephron (nef′ron) [G. nephros, kidney] Functional unit of the kidney, consisting of the renal corpuscle, the proximal convoluted tubule, the loop of Henle, and the distal convoluted tubule.

nerve (nerv) Collection of axons in the peripheral nervous system; conducts action potentials to and from the central nervous system.

nerve cell Cell capable of receiving a stimulus and propagating an action potential; a neuron.

nerve tract Bundle of axons, their sheaths, and accompanying connective tissues located in the central nervous system.

nervous tissue (ner′vŭs) One of the four major tissue types; consists of neurons, which have the ability to conduct action potentials, and neuroglia, which are support cells.

neural crest cell (noor′ăl) Cell derived during embryonic development from the crests of the neural folds; gives rise to facial structures, pigment cells, and peripheral nerve ganglia.

neural tube Tube formed from the neuroectoderm in the embryo by closure of the neural groove; develops into the brain and spinal cord.

neuroectoderm (noor-ō-ek′tō-derm) Part of the ectoderm that forms the neural tube and neural crest.

neuroglia (noo-rog′lē-ă) [G. neuro, nerve + glia, glue] Cells of the nervous system other than neurons; play a support role in the nervous system; include astrocytes, ependymal cells, microglia, oligodendrocytes, and Schwann cells; also called glia.

neurolemmocyte (noor-ō-lem′ō′sīt) [G. neuro, nerve + lemma, husk + kytos, cell] See Schwann cell.

neuromuscular junction (noor-rō-mŭs′kū-lăr) Synaptic junction between a nerve axon and a muscle fiber.

neuron (noor′on) [G., nerve] Nerve cell.

neurotransmitter (noor′ō-trans-mit′er) [G. neuro, nerve + L. transmitto, to send across] Chemical that is released by a presynaptic cell into the synaptic cleft and that acts on the postsynaptic cell to cause a response.

neutral solution (noo′trăl) Solution with equal numbers of hydrogen and hydroxide ions; has a pH of 7.0.

neutron (noo′tron) [L. *neuter,* neither] Electrically neutral particle in the nucleus of atoms.

neutrophil (noo′trō-fil) [L. *neuter,* neither + G. *phileo,* to love] White blood cell with granules that stains equally with either basic or acidic dyes; phagocytic white blood cell.

nevus (nē′vŭs), pl. **nevi** (nē′vī) Benign, localized overgrowth of the melanin-forming cells of the skin present at birth or appearing early in life; a mole.

nicotinamide adenine dinucleotide (NADH) (nik-ō-tin′ă-mīd ad′ĕ-nēn dī-noo′klē-ō-tīd) Base-containing organic molecule capable of accepting hydrogen atoms and transferring energy from glycolysis and the citric acid cycle to the electron-transport chain.

nitroglycerin (nī-trō-glis′er-in) Glyceryl trinitrate used as a vasodilator, especially in angina pectoris.

nociceptor (nō′si-sep′tŏr) [L. *oceo,* hurt, pain, injury + *capio,* to take] Peripheral sensory receptor or mechanism for the reception and transmission of painful or injurious stimuli.

node of Ranvier (ron′vē-ā) Unmyelinated area of an axon, every 0.1–1.0 mm, between adjacent oligodendrocytes of an axon in the central nervous system and between individual Schwann cells of the peripheral nervous system.

norepinephrine (nōr′ep-i-nef′rin) Neurotransmitter substance released from most of the postganglionic neurons of the sympathetic division; hormone released from the adrenal cortex that increases cardiac output and blood glucose levels.

notochord (nō′tō-kōrd) [G. *notor,* back + *chorde,* cord or string] Small rod of tissue lying ventral to the neural tube; characteristic of all vertebrates; in humans, it becomes the nucleus pulposus of the intervertebral disks.

nuclear pore (noo′klē-er) Point where the inner and outer membranes of the nuclear envelope come together to form a hole.

nuclease (noo′klē-ās) Enzyme that breaks down nucleic acids.

nucleic acid (noo-klē′ik, -klā′ik) Molecule consisting of many nucleotides chemically bound together; deoxyribonucleic acid and ribonucleic acid.

nucleolus (noo-klē′ō-lŭs), pl. **nucleoli** (noo-klē′ō-lī) Rounded, dense, well-defined nuclear body with no surrounding membrane; subunits of ribosomes are manufactured within the nucleolus.

nucleotide (noo′klē-ō-tīd) Basic building block of nucleic acids, consisting of a sugar molecule (either ribose or deoxyribose), one of several types of organic bases, and a phosphate group.

nucleus (noo′klē-ŭs), pl. **nuclei** (noo′klē-ī) [L., a little nut, stone of a fruit] Cell organelle containing most of the genetic material of the cell; center of an atom consisting of protons and neutrons; collection of neuron cell bodies in the central nervous system.

nutrient (noo′trē-ent) [L. *nutriens,* to nourish] Chemical taken into the body and used to produce energy, provide building blocks for new molecules, or function in other chemical reactions.

nutrition (noo-trish′ŭn) Process by which nutrients are obtained and used in the body.

oblique section (ob-lēk′) Cut made at other than a right angle to the long axis of an organ.

obturator (ob′toor-ā-tōr) [L., to occlude or stop up] Any occluding structure or a foramen so occluded, as with the obturator foramen of the hip.

occipital (ok-sip′i-tăl) Back of the head.

odorant (ō′-dōr-ănt) Substance with an odor.

olecranon (ō-lek′ră-non) Point of the elbow.

olfaction (ol-fak′shŭn) [L., to smell] Sense of smell.

olfactory (ol-fak′tō-rē) Relating to the sense of smell.

oligodendrocyte (ol′i-gō-den′drō-sīt) Neuroglial cell with multiple cell processes that form myelin sheaths around axons in the central nervous system.

omental bursa (ō-men′tăl ber′să) Pocketlike sac inside the fold of the greater omentum.

omentum (ō-men′tŭm) [L., membrane of the bowels] Fold of peritoneum extending from the stomach to another organ.

oncology (ong-kol′ō-jē) [G. *onco,* a tumor + *logos,* to study] Study of cancer and its associated problems.

oocyte (ō′ō-sīt) [G. *oon,* egg + *kytos,* cell] Female gamete, or sex cell; a secondary oocyte and a polar body result from the first meiotic division, which occurs prior to ovulation; a zygote and a polar body result from the second meiotic division, which occurs following union of the sperm cell with the secondary oocyte.

oogonium (ō-ō-gō′nē-ŭm), pl. **oogonia** (ō-ō-gō′nē-ă) [G. *oon,* egg + *gone,* generation] Cell that gives rise to oocytes; has a diploid number of chromosomes.

opsin Protein portion of the rhodopsin molecule; at least three different opsins are located in cone cells.

optic (op′tik) Relating to vision.

optic disc Region in the posterior wall of the eye where the optic nerve exits the eye; the blind spot.

optic nerve Nerve that leaves the eye and exits the orbit through the optic foramen to enter the cranial vault.

oral cavity (ōr′ăl) Mouth; the first portion of the digestive tract.

orbit (ōr′bit) Seven skull bones that surround and protect the eye; eye socket.

organ (ōr′găn) [G. *organon,* tool] Part of the body composed of two or more tissue types and performing one or more specific functions.

organ of Corti Named for Italian anatomist Alfonso Corti (1822–1888). Specialized region of the cochlear duct consisting of hair cells; produces action potentials in response to sound waves.

organ system Group of organs classified as a unit because of a common function or set of functions.

organelle (or′gă-nel) [G. *organon,* a tool + L. *-elle,* small, a little organ] Specialized part of a cell performing one or more specific functions.

organic (ōr-gan′ik) Molecules that contain a carbon atom (carbon dioxide is an exception); originally defined as molecules extracted from living organisms; the original definition became obsolete when it became possible to manufacture these molecules in the laboratory.

organism (ōr′gă-nizm) Any living thing considered as a whole, whether composed of one cell or many.

organogenesis (ōr′gă-nō-jen′ĕ-sis) Formation of organs during embryonic development.

orgasm (ōr′gazm) [G. *orgao,* to swell, be excited] Climax of the sexual act, often associated with a pleasurable sensation.

origin (ōr′i-jin) Less movable attachment point of a muscle.

oropharynx (ōr′ō-far′ingks) Portion of the pharynx that lies posterior to the mouth; it is continuous above with the nasopharynx and below with the laryngopharynx.

osmosis (os-mō′sis) [G. *osmos,* thrusting or an impulsion] Diffusion of solvent (water) through a selectively permeable membrane from a region of higher water concentration to one of lower water concentration.

osmotic pressure (os-mot′ik) Force required to prevent the movement of water across a selectively permeable membrane.

ossification (os′i-fi-kā′shŭn) [L. *os,* bone + *facio,* to make] Bone formation.

osteoblast (os′tē-ō-blast) [G. *osteo,* bone] Cell that makes bone.

osteoclast (os′tē-ō-klast) [bone eating] Cell that digests and removes bone.

osteocyte (os′tē-ō-sīt) [G. *osteon,* bone + *kytos,* cell] Mature bone cell surrounded by bone matrix.

osteon (os′tē-on) Single central canal, with its contents, and the associated lamellae and osteocytes surrounding it. Also called a haversian system.

otolith (ō′tō-lith) [G. *ous*, ear + *lithos*, stone] Small protein and calcium carbonate weight in the maculae of the vestibule.

ovary (ō′vă-rē) One of two female reproductive glands located in the pelvic cavity; produces the oocyte, estrogen, and progesterone.

ovulation (ov′ū-lā′shŭn) Release of an oocyte from the mature follicle.

oxidative metabolism (ok-si-dā′tiv mĕ-tab′ō-lizm) Metabolism in which oxygen is required to produce ATP.

oxygen deficit (ok′sē-jen) Amount of oxygen required to convert the lactic acid produced during anaerobic respiration to glucose and to replenish creatine phosphate stores.

oxytocin (ok′sĭ-tō′sin) [G., swift birth] Peptide hormone, secreted by the posterior pituitary gland, that increases uterine contraction and stimulates milk ejection from the mammary glands.

palate (pal′ăt) Roof of the oral cavity; consists of the anterior bony part, the hard palate, and the posterior soft palate that is composed mainly of skeletal muscle and connective tissue.

pancreas (pan′krē-as) Elongated gland extending from the duodenum to the spleen; consists of a head, a body, and a tail. There is an exocrine portion, which secretes digestive enzymes, which are carried by the pancreatic duct to the duodenum, and pancreatic islets, which secrete insulin and glucagon.

pancreatic duct (pan-krē′at′ik) Duct of the pancreas; it joins the common bile duct to empty into the duodenum.

pancreatic islet (i′let) Cellular mass in the tissue of the pancreas; composed of different cell types that constitute the endocrine portion of the pancreas and are the source of insulin and glucagon.

papilla (pă-pil′ă), pl. papillae (pă-pil′ē) [L., nipple] Small, nipplelike process; projection of the dermis, containing blood vessels and nerves, into the epidermis; projection on the surface of the tongue.

papillary muscle (pap′i-l-ār-ē) Raised area of cardiac muscle in the ventricle to which the chordae tendineae attach.

paracrine (par′ă-krin) [G. *para*, alongside + *krino*, to separate] Kind of hormone function in which the effects of the hormone are restricted to the local environment.

parafollicular cell (par-ă-fo-lik′ū-lăr) Cell type scattered in a network of loose connective tissue between the thyroid follicles of the thyroid gland; secretes calcitonin.

paranasal sinus (par-ă-nā′săl) Air-filled cavity within certain skull bones that connects to the nasal cavity; the four sets of paranasal sinuses are the frontal, maxillary, sphenoidal, and ethmoidal.

parasympathetic (par-ă-sim-pa-thet′ik) [G. *para*, alongside + *sympathetic*] Subdivision of the autonomic nervous system with preganglionic neurons in the brainstem and sacral part of the spinal cord; involved in involuntary functions, such as digestion, defecation, and urination.

parathyroid gland (par-ă-thī′royd) One of four glandular masses embedded in the posterior surface of the thyroid gland; secretes parathyroid hormone.

parathyroid hormone (PTH) (hōr′mōn) Hormone produced by the parathyroid gland; increases bone breakdown and blood calcium levels.

parietal (pă-rī′ĕ-tăl) [L. *paries*, wall] Relating to the wall of any cavity; parietal serous membranes are in contact with the walls of cavities. The parietal bones form part of the skull.

parietal peritoneum (pĕ′rĭ-tō-nē′ŭm) [L., wall] Portion of the serous membranes of the abdominal cavity lining the inner surface of the body wall.

parotid gland (pă-rot′id) Largest of the salivary glands; one of a pair of salivary glands located anterior and inferior to each ear.

partial pressure Pressure exerted by a single gas in a mixture of gases.

partial-thickness burn Burn that damages only the epidermis (first-degree burn) or the epidermis and part of the dermis (also called a second-degree burn).

parturition (par-toor-ish′ŭn) [L. *parturio*, to be in labor] Childbirth; the delivery of a baby at the end of pregnancy.

patella (pa-tel′ă) [L. *patina*, shallow disk] Kneecap.

pectoral (pek′tŏ-răl) [L. *pectoralis*, breastbone] Relating to the chest.

pedicle (ped′ĭ-kl) [L. *pedicellus*, foot] Portion of a vertebra that extends from the body to the transverse process.

pelvic cavity (pel′vik) Space completely surrounded by the pelvic bones.

pepsin (pep′sin) [G. *pepsis*, digestion] Principal digestive enzyme produced by the stomach; digests proteins into smaller peptide chains.

peptidase (pep′ti-dās) Enzyme capable of breaking peptide chains into smaller chains and amino acids.

peptide bond (pep′tīd) Covalent chemical bond between adjacent amino acids in a polypeptide chain.

pericardial cavity (per-i-kar′dē-ăl) [G. *peri-*, around + *kardia*, the heart] Space between the visceral and parietal pericardia, filled with pericardial fluid; a cavity that surrounds the heart.

pericardial fluid Serous fluid within the pericardial cavity.

pericardial membrane Serous membranes associated with the heart.

pericardium (per-i-kar′dē-ŭm) [G. *pericardion*, the membrane around the heart] Membrane consisting of the epicardium and parietal pericardium (of the serous layers) and the outer fibrous pericardium; also called the pericardial sac.

perilymph (per′i-limf) [*peri* + G. *lympha*, clear fluid] Fluid between the bony labyrinth and the membranous labyrinth of the inner ear.

perimetrium (per-i-mē′trē-ŭm) Outer layer of the uterus; also called the serous layer.

perimysium (per′-i-mis′ē-ŭm, per′-i-miz′ē-ŭm) [*peri-* + G. *mys*, muscle] Fibrous sheath enveloping each of the skeletal muscle fascicles.

perineum (per′-i-nē′ŭm) Area inferior to the pelvic diaphragm between the thighs; extends from the coccyx to the pubis.

periodontal (per′ē-ō-don′tăl) [*peri-* + G. *odous*, tooth] Referring to structures surrounding the tooth, primarily in the alveolus.

periosteum (per-ē-os′tē-ŭm) [*peri-* + G. *osteon*, bone] Thick, double-layered connective tissue sheath covering the entire surface of a bone, except the articular surface, which is covered with cartilage.

peripheral nervous system (PNS) Part of the nervous system not surrounded by the skull or vertebral column; consists of nerves and ganglia.

peristaltic waves (per-i-stal′tik) [*peri-* + G. *stalsis*, constriction] Waves of relaxation followed by waves of contraction moving along a tube; propel food along the digestive tract.

peritoneal cavity (per′i-tō-nē′ăl) [to stretch over] Space between the visceral and parietal peritoneum filled with peritoneal fluid; cavity that surrounds many abdominopelvic organs.

peritoneal membrane Serous membrane associated with the peritoneal cavity.

peritubular capillary (per′ĭ-too′bū-lăr) Capillary network in the cortex of the kidney; associated with the distal and proximal convoluted tubules.

peroxisome (per-ok′si-sōm) Membrane-bound body similar to a lysosome in appearance but often smaller and irregular in shape; contains enzymes that either decompose or synthesize hydrogen peroxide.

Peyer patch Named for Swiss anatomist Johann Peyer (1653–1712). Collection of lymphatic nodules in the distal half of the small intestine and in the appendix.

pH scale Measure of the hydrogen ion concentration of a solution; the scale extends from 0 to 14.0—a pH of 7.0 is neutral, a pH of less than 7 acidic, and a pH of greater than 7 basic.

phagocytosis (fag´ō-sī-tō´sis) [G. *phagein,* to eat + L. *kytos,* cell + *osis,* condition] Ingestion and digestion by cells of substances, such as other cells, bacteria, cell debris, and foreign particles.

pharynx (far´ingks) [G. *pharynx,* throat] Joint openings of the digestive tract and the windpipe. The part of the digestive and respiratory tracts superior to the larynx and esophagus and inferior and posterior to the oral and nasal cavities.

phenotype (fē´nō-tīp) [G. *phaino,* to display + *typos,* model] Characteristic observed in the individual resulting from expression of the genotype.

pheromones (fer´ō-mōnz) [G. *pherē,* to carry + *hormāo,* to excite] Chemical signals secreted by an individual into the environment and perceived by a second individual of the same, or similar, species, producing a change in sexual or social behavior of that individual.

phlebitis (fle-bī´tis) Inflammation of a vein.

phospholipid (fos-fō-lip´id) Lipid with phosphorus resulting in a molecule with a polar and a nonpolar end; main component of cell membranes.

physiology (fiz-ē-ol´ō-jē) [G. *physis,* nature + *logos,* study] Scientific discipline that deals with the processes or functions of living things.

pia mater (pī´ă mā´ter, pē´ă ma´ter) [L., affectionate mother] Innermost meningeal layer; tightly attached to the brain and spinal cord.

pineal gland (pin´ē-ăl) [L. *pineus,* pinecone-shaped] Small endocrine gland attached to the dorsal surface of the diencephalon; may influence the onset of puberty and may play a role in some long-term cycles.

pinocytosis (pī´nō-sī-tō´sis) [G. *pineo,* to drink + *kytos,* cell; *osis,* condition] Cell drinking; uptake of liquid by a cell.

pituitary dwarf (dwōrf) Individual of short stature, of relatively normal proportion, as a result of insufficient growth hormone secreted from the anterior pituitary gland.

pituitary gland (pi-too´i-tār-rē) [L. *pituita,* phlegm or a thick mucous secretion] Endocrine gland attached to the hypothalamus by the infundibulum; secretes hormones that influence the function of several other glands and tissues.

placenta (plă-sen´tă) Structure derived from embryonic and maternal tissues by which the embryo and fetus are attached to the uterus.

plasma (plaz´mă) Fluid portion of blood; blood minus the formed elements.

plasma membrane Cell membrane; outermost component of the cell, surrounding and binding the rest of the cell contents.

plasmin (plaz´min) Enzyme that breaks down the fibrin in blood clots; derived from plasminogen.

platelet (plāt´let) Minute fragment of cells derived from megakaryocytes; plays an important role in preventing blood loss.

platelet plug Accumulation of platelets that stick to connective tissue and to one another and prevent blood loss from damaged blood vessels.

pleural (ploor´ăl) [G., a rib or cavity] **cavity** Space between the visceral and parietal pleura, filled with pleural fluid; a cavity that surrounds each lung.

pleural membrane Serous membrane associated with the lungs.

plexus (plek´sŭs) [L., a braid] Intertwining of nerves or blood vessels.

PMS See *premenstrual syndrome.*

pneumothorax (noo-mō-thōr´aks) Presence of air in the pleural cavity.

podocyte (pod´ō-sīt) [Fr. *pous, podos,* foot + G. *kytos,* a hollow (cell)] Epithelial cell of Bowman's capsule attached to the outer surface of the glomerular capillary basement membrane; forms part of the filtration membrane.

polar body Oocyte receiving little cytoplasm; results from the first and the second meiotic division.

polar covalent bond Chemical bond in which electrons are shared unequally between two atoms.

polarize To create a difference in potential (charge) between two points, as between the inside and outside of a cell membrane.

polycythemia (pol´ē-sī-thē´mē-ă) [G. *polys,* many + *kytos,* cell] Increase in red blood cell numbers above the normal value.

polysaccharide (pol-ē-sak´ă-rīd) [many sugars] Many monosaccharides chemically bound together, such as glycogen and starch.

pons (ponz) [L., bridge] Part of the brainstem between the medulla oblongata and midbrain; contains nerve tracts between the cerebrum and cerebellum, as well as ascending and descending tracts.

portal system (pōr´tăl) System of vessels in which blood, after passing through one capillary bed, is conveyed through a second capillary network.

positive feedback Mechanism by which any deviation from an ideal normal value or set point is made greater.

posterior (pos-tēr´ē-ŏr) [L. *posterus,* following] That which follows; in humans, toward the back.

posterior horn Posterior extension of spinal cord gray matter; contains neuron cell bodies that receive input from primary sensory neurons and relay that input to the brain; also called the dorsal horn.

posterior pituitary gland Posterior portion of the pituitary gland, which consists of processes of nerve cells that have their cell bodies located in the hypothalamus; secretes oxytocin and antidiuretic hormone.

postganglionic (pōst´gang-glē-on´ik) Autonomic neurons whose cell bodies are located outside the central nervous system and that receive synaptic stimulation from preganglionic autonomic neurons.

preganglionic (prē´gang-glē-on´ik) Autonomic neurons whose cell bodies are located in the central nervous system and that synapse with postganglionic neurons.

preload (prē´lōd) Degree to which the ventricular wall is stretched at the end of diastole; increases as the venous return increases.

premenstrual syndrome (PMS) (prē-men´stroo-al sin´drōm) In some women of reproductive age, the regular monthly experience of physiological and emotional distress, usually during the few days preceding menses, typically involving fatigue, edema, irritability, tension, anxiety, and depression.

prenatal period (prē-nā´tăl) [L. *prae,* before + *natalis,* relating to birth] Period before birth.

prepuce (prē´poos) In the male, a free fold of skin that almost completely covers the glans penis; the foreskin; in the female, a fold of mucous membrane that covers the clitoris.

primary response Immune response that occurs as a result of the first exposure to an antigen; results in the production of antibodies and memory cells.

prime mover Muscle that plays the principal role in accomplishing a movement.

primitive streak Shallow groove in the ectodermal surface of the embryonic disk; cells migrating through the streak become mesoderm.

process (pros´es, prō´ses) Projection on a bone.

product (prod´ŭkt) Substance produced in a chemical reaction.

progesterone (prō-jes´ter-ōn) Hormone secreted primarily by the corpus luteum and the placenta; aids in growth and development of female reproductive organs and secondary sexual characteristics; causes growth and maturation of the endometrium of the uterus during the menstrual cycle.

prolactin (prō-lak´tin) [L. *pro,* precursor + *lact,* milk] Hormone of the anterior pituitary gland that stimulates the secretion of milk.

pronation (prō-nā´shŭn) [L. *pronare,* to bend forward] Rotation, as of the forearm, starting in the anatomical position, so that the anterior surface faces posteriorly.

prophase (prō´fāzs) [G. *prophasis,* to foreshadow] First stage of mitosis or meiosis, consisting of contraction and increase in thickness of the chromosomes.

proprioceptive neurons (prō´prē-ō-sep´tiv) [L. *proprius,* one's own + *capio,* to take] Nerves that innervate the joints and tendons and provide information about the position of the body and its various parts.

prostaglandin (pros-tă-glan′din) Class of physiologically active substances present in many tissues; effects include vasodilation, stimulation and contraction of uterine smooth muscle, and promotion of inflammation and pain.

prostate gland (pros′tāt) [G. *prostates,* one standing before] Gland that surrounds the beginning of the urethra in the male. The secretion of the gland is a milky fluid that is discharged into the urethra as part of the semen.

protein (prō′tēn) [G. *proteios,* primary] Large molecule consisting of long sequences of amino acids (polypeptides) linked by peptide bonds.

proteoglycan (prō′tē-ō-glī′kan) [G. *proteo,* protein + *glycan,* polysaccharide] Macromolecule consisting of numerous polysaccharides attached to a common protein core; attracts and retains large amounts of water.

proteolytic (prō′tē-ō-lit′ik) Enzyme capable of digesting proteins or polypeptides.

proton (prō′ton) [G. *protos,* first] Positively charged particle in the nuclei of atoms.

provitamin (prō-vīt′ă-min) Substance that can be converted into a vitamin.

proximal (prok′si-măl) [L. *proximus,* nearest] Closer to the point of attachment to the body than another structure.

proximal convoluted tubule Convoluted portion of the nephron that extends from Bowman's capsule to the descending limb of the loop of Henle.

pterygoid (ter′ĭ-goyd) [G. *pteryx,* wing] Wing-shaped structure; two of the muscles of mastication, attached to wing-shaped, bony projections.

PTH See *parathyroid hormone.*

puberty (pū′ber-tē) [L. *pubertas,* grown up] Series of events that transform a child into a sexually mature adult; involves an increase in the secretion of all reproductive hormones.

pudendal cleft (pū-den′dal) [L. *pudeo,* to feel ashamed] Cleft between the labia majora.

pudendum (pū-den′dŭm), pl. pudenda (pū-den′da) External genitals, especially the female genitals. See *vulva.*

pulmonary capacity (pŭl′mō-nār-ē) Sum of two or more pulmonary volumes.

pulmonary circulation Blood flow through the system of blood vessels that carry blood from the right ventricle of the heart to the lungs and back from the lungs to the left atrium.

pulmonary semilunar valve Semilunar valve at the base of the pulmonary trunk where it exits from the right ventricle.

pulmonary trunk Large, elastic artery that carries blood from the right ventricle of the heart to the right and left pulmonary arteries.

pulmonary volume Lung volume, measured by spirometry; deviations from a normal value can be used to diagnose certain lung diseases; the pulmonary volumes are the tidal volume, inspiratory reserve volume, expiratory reserve volume, and residual volume.

pulp (pŭlp) [L. *pulpa,* flesh] Soft tissue inside a tooth, consisting of connective tissue, blood vessels, nerves, and lymphatic vessels.

pulse (pŭls) Pressure wave that travels rapidly along the arteries when blood is ejected from the left ventricle into the aorta.

pulse pressure Difference between systolic and diastolic pressures.

pupil (pū′pĭl) [L. *pupa,* a doll, so called because you can see a little reflection, or doll, in the pupil of another person's eye] Opening in the iris of the eye through which light enters.

Purkinje fiber (pŭr-kĭn′jē) Named for Bohemian anatomist/physiologist Johannes Purkinje (1787–1869). A specialized cardiac muscle fiber that conducts action potentials through cardiac muscle; forms part of the conduction system of the heart.

pus (pŭs) Product of inflammation, consisting of a liquid containing white blood cells, dead cells, and cell fragments.

pyloric sphincter (pī-lōr′ik) [G., gatekeeper] Thickened ring of smooth muscle at the distal end of the stomach.

pyrogen (pī′rō-jen) Chemical released by microorganisms, neutrophils, monocytes, and other cells that stimulates fever production by acting on the hypothalamus.

pyruvic acid (pī-roo′vik) Three-carbon end product of glycolysis; two pyruvic acid molecules are produced from each glucose molecule.

quadrant (kwăh′drant) [L. *quadrans,* a quarter] One-quarter of a circle; the abdomen is divided into right upper, right lower, left upper, and left lower quadrants by a horizontal and a vertical line intersecting at the umbilicus.

RBC See *red blood cell.*

reactant (rē-ak′tant) Substance taking part in a chemical reaction.

receptor (rē-sĕp′tōr) [L., receiver] Protein molecule on the cell surface or within the cytoplasm that binds to a specific factor, such as a drug, a hormone, an antigen, or a neurotransmitter; one of the sensory nerve endings in the skin, deep tissues, viscera, and special sense organs.

recessive (rē-ses′iv) In genetics, a gene that may not be expressed phenotypically because of the expression of a contrasting dominant gene.

rectum (rek′tŭm) [L. *rectus,* straight] Last, straight part of the large intestine; between the colon and the anal canal.

rectus (rek′tŭs) Straight.

red blood cell (RBC) Biconcave disk that contains hemoglobin, which transports oxygen and carbon dioxide; red blood cells do not have a nucleus.

reflex (rē′fleks) Automatic response to a stimulus; does not require conscious thought.

reflex arc Consists of a sensory receptor, an afferent (sensory) neuron, an association neuron, an efferent (motor) neuron, and an effector organ.

regeneration (rē′jen-er-ā′shŭn) Tissue repair in which the damaged cells are replaced by cells of the same type as those damaged.

releasing hormone Hormone that is released from neurons in the hypothalamus and flows through the hypothalamic-pituitary portal system to the anterior pituitary gland; regulates the secretion of hormones from the cells of the anterior pituitary gland.

renal capsule (rē′năl) Connective tissue capsule that surrounds each kidney.

renal corpuscle (kōr′pŭs-l) Structure composed of a Bowman's capsule and its glomerulus.

renal pyramid (pir′ă-mid) Cone-shaped structure that extends from the renal sinus, where the apex is located, into the cortex of the kidney, where the base is located.

renal sinus (sī′nŭs) Cavity central to the medulla of the kidney that is filled with adipose tissue and contains the renal pelvis.

renin (rē′nin, ren′in) Enzyme secreted by the kidney that converts the plasma protein angiotensinogen to angiotensin I.

replacement Tissue repair in which the damaged cells are replaced by cells of a type different from those damaged.

respiration (res-pi-rā′shŭn) [L. *respiratio,* to breathe] Process in which oxygen is used to oxidize organic fuel molecules, providing a source of energy as well as carbon dioxide and water; includes ventilation, gas exchange, transport of oxygen and carbon dioxide in the blood, gas exchange between the blood and the tissues, and cell metabolism.

respiratory membrane (res′pi-ră-tōr-ē, rē-spīr′ă-tōr-ē) Membrane in the lungs across which gas exchange occurs with blood; consists of a thin layer of fluid, the alveolar epithelium, a basement membrane of the alveolar epithelium, interstitial space, the basement membrane of the capillary endothelium, and the capillary endothelium.

respiratory system Nose, nasal cavity, pharynx, larynx, trachea, bronchi, and lungs.

resting membrane potential Charge difference across the membrane of a resting cell (i.e., a cell that has not been stimulated to produce an action potential).

rete testis (rē′tē) Network of canals at the termination of the straight portion of the seminiferous tubules.

reticular formation (rē-tik′ū-lăr) [L. *rete*, net] Loose network of neuron cell bodies scattered throughout the brainstem; involved in the regulation of cycles such as the sleep-wake cycle.

retina (ret′i-nă) [L. *rete*, a net] Inner, light-sensitive tunic of the eye; nervous tunic.

retinaculum (ret-i-nak′ū-lŭm) [L., band, bracelet, halter, to hold back] Dense regular connective tissue sheath holding down the tendons at the wrist, ankle, or other sites.

retinal (rět′i-năl) Relating to the retina; retinaldehyde most commonly referring to the all-*trans* form (all-*trans*-retinal).

retroperitoneal (re′trō-per′i-tō-nē′ăl) Located behind the parietal peritoneum; includes the kidneys, the adrenal glands, the pancreas, portions of the intestines, and the urinary bladder.

reversible reaction (rē-ver′si-bl) Chemical reaction in which the reaction can proceed from reactants to products or from products to reactants; the amount of reactants relative to products is constant at equilibrium.

rhodopsin (rō-dop′sin) [G. *rhodon*, rose or red color + *opsin*, protein portion of rhodopsin] Purplish red protein in the external segment of the rods of the retina. Action of light converts it to opsin and all-*trans*-retinal.

ribonucleic acid (rī′bō-noo-klē′ik) (RNA) Type of nucleic acid containing the sugar ribose; involved in protein synthesis.

ribosomal RNA (rRNA) (ri′bō-sōm-ăl) RNA that is associated with certain proteins to form ribosomes.

ribosome (ri′bō-sōm) [ribose, a specific sugar] Small, spherical, cytoplasmic organelle where protein synthesis occurs.

right lymphatic duct Lymphatic duct that empties into the right subclavian vein; drains the right side of the head and neck, the right upper thorax, and the right upper limb.

RNA See *ribonucleic acid*.

rod Photoreceptor cell in the retina of the eye with a rod-shaped photoreceptive process; very light-sensitive cell that is important in dim light.

rotator cuff (rō-tā′tōr, rō-tā′tōr) Four deep muscles that attach the humerus to the scapula.

rRNA See *ribosomal RNA*.

ruga (roo′gă) Ridge or fold in the mucous membrane of the stomach.

SA See *sinoatrial*.

sagittal plane (saj′i-tăl) [L. *sagitta*, the flight of an arrow] Plane running vertically through the body and dividing it into right and left parts.

saliva (să-lī′vă) Fluid containing enzymes and mucus; produced by the salivary glands and released into the oral cavity.

salivary gland (sal′i-vār-rē) Gland opening into the mouth and producing saliva.

salt Molecule consisting of a positively charged ion other than hydrogen and a negatively charged ion other than hydroxide.

sarcolemma (sar′kō-lem′ă) [G. *sarx*, flesh, muscle; *lemma*, husk] Cell membrane of a muscle fiber.

sarcomere (sar′kō-mēr) [sarco- + G. *meros*, part] Part of a myofibril formed of actin and myosin myofilaments, extending from Z disk to Z disk; the structural and functional unit of a muscle.

sarcoplasm (sar′kō-plazm) [sarco- + *plasma*, a thing formed] Cytoplasm of a muscle fiber.

sarcoplasmic reticulum (sar′kō-plaz′mik re-tik′ū-lŭm) Endoplasmic reticulum of a muscle fiber.

scapula (skap′ū-lă) Shoulder blade.

Schwann cell Named for German histologist/physiologist Theodor Schwann (1810–1882). Neuroglial cell forming myelin sheaths around axons in the peripheral nervous system.

sciatic (sī-at′ik) [Fr. *ischion*, the hip joint] Ischiadic or sciatic nerve.

sclera (sklēr′ă) [L. *skleros*, hard] Dense, white, opaque posterior four-fifths of the fibrous tunic of the eye; white of the eye.

scoliosis (skō-lē-ō′sis) [G., a crookedness] Abnormal lateral curvature of the spine.

scrotum (skrō′tum) Musculocutaneous sac containing the testes.

sebaceous gland (sē-bā′shŭs) [L. *sebum*, tallow] Gland of the skin that produces sebum; usually associated with a hair follicle.

sebum (sē′bŭm) [L., tallow] Oily, white, fatty substance produced by the sebaceous glands; lubricates hair and the surface of the skin.

secondary response See *memory response*.

secretin (se-krē′tin) Hormone released from the epithelium of the duodenum; inhibits gastric secretion.

sella turcica (sel′ă tŭr′sĭ-kă) [L., saddle, Turkish] Saddle-shaped depression in the inner surface of the skull where the pituitary gland is located.

semen (sē′men) [L., seed] Penile ejaculate; thick, yellowish-white, viscous fluid containing sperm cells and secretions of the testes, seminal vesicles, prostate gland, and bulbourethral glands.

semicircular canal (sem′ē-sir′kū-lăr) One of three canals in each temporal bone; involved in the detection of motion.

semilunar valve (sem-ē-loo′năr) One of two valves in the heart composed of three crescent-shaped cusps that prevent blood flow into the ventricles following ejection; located at the beginning of the aorta and pulmonary trunk.

seminal vesicle (sem′i-năl ves′i-kl) One of two glandular structures that empty into the ejaculatory ducts; its secretion is one of the components of semen.

seminiferous tubule (sem′i-nif′er-ŭs) Tubule in the testis in which sperm cells develop.

sensory neuron Neuron that extends from sensory receptors in the periphery to the central nervous system.

septum (sep′tŭm) [L. *saeptum*, a partition or wall] Thin wall dividing two cavities or masses of softer tissue.

serosa (se-rō′să) Smooth, outermost covering of an organ where it faces a cavity and is not surrounded by connective tissue.

serous membrane (sēr′ŭs) Thin sheet consisting of epithelium and connective tissue that lines cavities not opening to the outside of the body; does not contain glands but does secrete serous fluid.

serum (sēr′ŭm) Fluid portion of blood after the removal of fibrin and formed elements.

sex chromosome Chromosome other than an autosome; responsible for sex determination.

sinoatrial (SA) node (sī′nō-ā′trē-ăl) Mass of specialized cardiac muscle fibers in the right atrium near the opening of the superior vena cava that acts as the "pacemaker" of the cardiac conduction system.

sliding filament model Mechanism by which actin and myosin myofilaments slide over one another during muscle contraction.

solute (sol′ūt, sō′loot) [L. *solutus*, dissolved] Dissolved substance in a solution.

solution (sō-loo′shŭn) Homogeneous mixture formed when a solute dissolves in a solvent (liquid).

solvent (sol′vent) [L. *solvens*, dissolve] Liquid that holds another substance in solution.

somatic motor (sō-mat′ik) [G. *soma*, body or bodily] Type of motor (efferent) neuron of the peripheral nervous system that innervates skeletal muscle.

somesthetic (sō′mes-thet′ik) [G. *soma*, body + *aisthesis*, sensation] Consciously perceived.

somesthetic cortex Part of the cerebral cortex involved with the conscious perception and localization of general body sensations.

spermatid (sper′mă-tid) Cell in the late stage of the development of the sperm cell (male gamete). It is haploid and is derived from the secondary spermatocyte.

spermatocyte (sper′mă-tō′sīt) Cell arising from a spermatogonium and destined to give rise to spermatozoa.

spermatogenesis (sper′mă-tō-jen′ĕ-sis) Formation and development of sperm cells.

spermatogonium (sper′mă-tō-gō′nē-ŭm), pl. spermatogonia (sper′mă-tō-gō′nē-ă) Most peripheral germ cell in the seminiferous tubules scattered between the sustentacular cells; divide by mitosis and some form primary spermatocytes.

spermatozoon (sper′mă-tō-zō′on), pl. spermatazoa (sper′mă-tō-zō′ă) [G. *sperma*, seed + *zoon*, animal] Male gamete, or sex cell, composed of a head, midpiece, and tail; contains the genetic information transmitted by the male; sperm cell.

sperm cell [G. *sperma*, seed] Male reproductive cell; see also *spermatozoon*.

sphenoid (sfē′noyd) [G. *sphenoeides*, wedge + *eidos*, resemblance] Sphenoid bone or relating to the sphenoid bone.

sphygmomanometer (sfig′mō-mă-nom′ĕ-ter) [G. *sphygmos*, pulse; *manos + metron*, measure] Instrument for measuring blood pressure consisting of an arm sleeve and an inflating bulb with a device attached for measuring pressure in the arm sleeve.

spina bifida (spī′nă bif′ĭ-dă, bī′fĭ-dă) Defect in the spinal column, consisting in absence of the vertebral arches, through which the spinal membranes, with or without spinal cord tissue, may protrude.

spinal cord Portion of the central nervous system extending from the foramen magnum at the base of the skull to the second lumbar vertebra; consists of a central gray portion and a peripheral white portion.

spinal nerve Peripheral nerve exiting from the spinal cord.

spirometer (spī-rom′ĕ-ter) [L. *spiro*, to breathe + G. *metron*, measure] Meter used for measuring the volume of respiratory gases; usually consists of a counterbalanced, cylindrical bell sealed by dipping into a circular trough of water.

spirometry (spī-rom′ĕ-trē) Process of making pulmonary measurements with a spirometer.

spleen (splēn) Large lymphatic organ in the left upper part of the abdominal cavity, between the stomach and the diaphragm; composed of white and red pulp; responds to foreign substances in the blood, destroys worn-out red blood cells, and is a reservoir for blood.

spongy bone [L., grating or lattice] Bone with a latticelike appearance.

squamous (skwā′mŭs) [L. *squama*, a scale] Scalelike, flat.

stapes (stā′pēz) [L., stirrup] The third of the three middle ear bones; attached to the oval window; the stirrup.

Starling's law of the heart Named for English physiologist Ernest Starling (1866–1927). Force of contraction of cardiac muscle is a function of the length of its muscle fibers at the end of diastole; the greater the degree of filling of the heart (the greater the venous return), the greater the force of contraction of the cardiac muscle.

stem cell Single population of cells that differentiate to give rise to the formed elements of blood.

stenosed valve (sten′ōzd) Valve that has its opening narrowed or partially closed.

sternum (ster′nŭm) [L. *sternon*, chest] Breastbone.

steroid (stēr′oyd, ster′oyd) Large family of lipids, including some hormones, vitamins, and cholesterol.

stethoscope (steth′ō-skōp) [G. *stetho-*, chest + *skopeo*, to view] Instrument originally devised for aid in hearing the respiratory and cardiac sounds in the chest and now used in hearing other sounds in the body as well.

strabismus (stra-biz′mŭs) [G. *strabismos*, a squinting] Lack of parallelism of the visual axes of the eyes.

stratum (strat′ŭm), pl. **strata** (strat′tă) [L., bed cover, layer] Layer of tissue.

stratum basale (bā-săl′ē) Deepest layer of the epidermis; consists of columnar cells that undergo mitotic divisions.

stratum corneum (kōr′nē-ŭm) Most superficial layer of the epidermis; consists of dead squamous cornified cells that have undergone keratinization.

stria, pl., **striae** (strī′e) [L., channel, furrow] Bands of thin, wrinkled skin, becoming red and white, that occur commonly on the abdomen, buttocks, and thighs at puberty and/or during and following pregnancy and result from overextension of the skin.

stroke volume Volume of blood ejected from either the right or the left ventricle during each heartbeat.

styloid (stī′loyd) [G. *stylos*, a stake or pen] Slender, pencil-shaped process.

subarachnoid space (sŭb-ă-rak′noyd) Fluid-filled space below the arachnoid layer covering the brain and spinal cord; contains cerebrospinal fluid.

subcutaneous (sŭb-koo-tā′nē-ŭs) [L. *sub*, under + *cutis*, skin] Under the skin; same tissue as the hypodermis.

sublingual gland (sŭb-ling′gwăl) One of a pair of salivary glands located below the tongue.

submandibular gland (sŭb-man-dib′ū-lăr) One of a pair of salivary glands located below the mandible.

submucosa (sŭb-moo-kō′să) Layer of connective tissue deep to the mucous membrane.

sulcus (sool′kŭs), pl. **sulci** (sŭl′sī) [L., ditch] Groove on the surface of the brain between gyri.

superficial (soo-per-fish′ăl) [L. *superficialis*, surface] Toward or on the surface.

superior (soo-pēr′ē-ōr) [L., higher] Up, or higher, with reference to the anatomical position.

superior vena cava (vē′nă kā′vă) Blood vessel that receives blood from the head, neck, and upper limbs and empties into the right atrium of the heart.

supination (soo′pi-nā′shŭn) [L. *supino*, to place something on its back] Rotation of the forearm so that the anterior surface is anterior; that is, the forearm is in the anatomical position.

surfactant (ser-fak′tănt) Mixture of lipoprotein molecules produced by the secretory cells of the alveolar epithelium of the lung; reduces water surface tension.

sustentacular cell (sŭs-ten-tak′ū-lăr) Cell in the wall of the seminiferous tubules to which spermatogonia and spermatids are attached; also called a Sertoli cell.

suture (soo′choor) [L. *surtura*, a seam] Fibrous joint between flat bones of the skull.

sweat gland (swet) Usually, a secretory organ that produces a watery secretion, called sweat, that is released onto the surface of the skin; some sweat glands, however, produce an organic secretion.

sympathetic (sim-pă-thet′ik) [G. *sympatheo*, to feel with + *pathos*, suffering] Subdivision of the autonomic nervous system with preganglionic nerve cell bodies located in the thoracic and lumbar regions of the spinal cord; generally involved in preparing the body for immediate physical activity.

synapse (sin′aps) [G. *syn*, together + *haptein*, to clasp] Junction between a nerve cell and another nerve cell, muscle cell, or gland cell; in a chemical synapse, chemicals are released from the nerve cell as a result of an action potential in the nerve cell, the chemicals cross the cleft between the cells, and they cause a response in the postsynaptic cell.

synapsis (si-nap′sis) Pairing of homologous chromosomes during prophase of the first meiotic division.

synergist (sin′er-jist) Muscle that works with another muscle to cause a movement.

synovial cavity (si-nō′vē-ăl) [G. *syn*, coming together + *ovia*, resembling egg albumin] Cavity surrounding articulating bones of a freely movable or synovial joint; contains synovial fluid.

synovial fluid Somewhat viscous substance serving as a lubricant in movable joints, tendon sheaths, and bursae.

synovial joint Freely movable joint.

synovial membrane Membrane that lines the inside of a joint cavity; produces synovial fluid.

synthesis reaction (sin′thĕ-sis) Combination of atoms, ions, or molecules to form a new, larger molecule.

systemic circulation (sis-tem′ik) Blood flow through the system of blood vessels that carry blood from the left ventricle of the heart to the tissues of the body and back from the body to the right atrium.

systemic inflammation Inflammation that occurs in many areas of the body; in addition to the symptoms of local inflammation, can include increased neutrophil numbers in the blood, fever, and shock.

systole (sis′tō-lē) [G., a contracting] Contraction of the heart chambers, during which blood leaves the chambers; usually refers to ventricular contraction.

systolic pressure (sis-tol′ik) Maximum arterial blood pressure reached during ventricular systole.

target tissue Tissue on which a hormone acts.

tarsal bone (tar′săl) [G. *tarsos*, sole of foot] Bone of the instep of the foot.

taste bud Sensory structure mostly on the tongue; functions as a taste receptor.

tectorial membrane (tek-tōr′ē-ăl) [L., a covering] Membrane attached to the spiral lamina and extending over the hair cells; hairs of the hair cells have their tips embedded in the membrane.

telophase (tel′ō-fāz) [G. *telos*, end + *phasis*, an appearance] Final stage of mitosis or meiosis that begins when migration of chromosomes to the poles of the cells has been completed.

temporal (tem′pŏ-răl) [L. *tempus*, time] Indicating the temple; the temple of the head is so named because it is there that the hair first begins turning white, indicating the passage of time.

tendinous intersection (ten′di-nŭs) One of the bands of connective tissue crossing the rectus abdominis muscle, subdividing it and attaching it to adjacent connective tissue.

tendon (ten′dŏn) Tough connective tissue band connecting a muscle to bone.

teniae coli (tē′nē-ē kō′lī) [G. *tainia*, band, tapeworm + *coli*, colon] Segmented, longitudinal smooth muscle layer of the colon.

testis (tes′tis), pl. **testes** (test′tēz) One of two male reproductive glands in the scrotum; produces testosterone and sperm cells.

testosterone (tes′tos′tĕ-rōn) Steroid hormone secreted primarily by the testes; aids in spermatogenesis, controls maintenance and development of male reproductive organs and secondary sexual characteristics, and influences sexual behavior.

tetanus (tet′ă-nŭs) [L. *tetanus*, convulsive tension] Sustained muscular contraction caused by a series of nerve stimuli repeated so rapidly that the individual contractions are fused, producing a sustained tetanic contraction; also a disease marked by painful tonic muscular contractions, caused by the neurotoxin of *Clostridium tetani* action on the central nervous system.

tetany (tet′ă-nē) Condition in muscle contraction in which there is no relaxation between muscle twitches.

tetraiodothyronine (tet′ră-ī-ō-dō-thī′rō-nēn) (T₄) One of the thyroid hormones; contains four iodine atoms; also called thyroxine.

thalamus (thal′ă-mŭs) [G., a bedroom] Large mass of gray matter making up the bulk of the diencephalon; involved in the relay of sensory input to the cerebrum.

thoracic cavity (thō-ras′ik) Space bounded by the neck, the thoracic wall, and the diaphragm.

thoracic duct Largest lymphatic vessel in the body; drains the left side of the head and neck, the left upper thorax, the left upper limb, and the inferior half of the body into the left subclavian vein.

thorax (thō′raks) [G., breastplate] Chest; the upper part of the trunk between the neck and the abdomen.

thrombocyte (throm′bō-sīt) [*thrombos-*, clot + G. *kytos*, cell] Cell fragment involved in platelet plug and clot formation; also called a platelet.

thrombosis (throm′bō′sis) [G. *thrombos*, clot] Formation or presence of a clot (thrombus) inside a blood vessel.

thrombus (throm′bŭs) [G. *thrombos*, clot] Clot within the cardiovascular system.

thymosin (thī′mō-sin) Hormone secreted from the thymus that helps activate the immune system.

thymus (thī′mŭs) [G. *thymos*, sweetbread] Bilobed lymphatic organ located in the inferior neck and superior mediastinum; involved with the maturation of T cells.

thyroid cartilage (thī′royd) [G. *thyreoeides*, shield] Largest laryngeal cartilage; forms the laryngeal prominence, or Adam's apple.

thyroid follicle One of many small spheres with walls consisting of cuboidal epithelial cells in the thyroid gland; filled with proteins to which thyroid hormones are attached until they are secreted.

thyroid gland Endocrine gland located inferior to the larynx and consisting of two lobes connected by a narrow band; secretes the thyroid hormones.

thyroid hormone Any hormone secreted by the thyroid gland, especially those, such as thyroxine, that contain iodine and regulate metabolism and the maturation of tissues.

thyroid-stimulating hormone (TSH) Hormone released from the hypothalamus that stimulates thyroid hormone secretion from the thyroid gland.

thyroxine (thī-rok′sēn, thī-rok′sin) See *tetraiodothyronine*.

tissue (tish′ū) [L. *texo*, to weave] Collection of cells with similar structure and function and the substances between the cells.

tissue repair Substitution of viable cells for damaged or dead cells by regeneration or replacement.

tonsil (ton′sil) Collection of lymphoid tissue; usually refers to large collections of lymphoid tissue beneath mucous membranes of the oral cavity and pharynx; lingual, pharyngeal, and palatine tonsils.

trabecula (tră-bek′ū-lă) [L. *trabs*, beam] Beam or plate of spongy bone or other tissue.

trachea (trā′kē-ă) [G. *tracheia arteria*, rough artery] Air tube extending from the larynx into the thorax, where it divides to form bronchi; has 16–20 C-shaped pieces of cartilage in its walls.

tracheostomy (trā′kē-os′tō-mē) Incision into the trachea.

tract (trakt) Nerve tract; a bundle of neuron cell processes (axons) in the central nervous system, usually having a common function.

transfer RNA (tRNA) RNA that attaches to individual amino acids and transports them to the ribosomes, where they are connected to form a protein polypeptide chain.

transverse plane (trans-vers′) Plane separating the body into superior and inferior parts.

transverse section Cut made at right angles to the long axis of an organ.

trapezius (tra-pē′zē-ŭs) Back muscle, shaped like a trapezium (a four-sided geometric figure in which no two sides are parallel), that rotates the scapula.

triacylglycerol (trī-as′il-glis′er-ol) See *triglyceride*.

triceps brachii (trī′seps brā′kē-ī) Three-headed muscle in the posterior arm that extends the forearm.

tricuspid valve (trī-kŭs′pid) Valve closing the opening between the right atrium and right ventricle of the heart.

triglyceride (trī-glis′er-īd) Common type of lipid, or fat, with three fatty acids bound to a glycerol molecule; also called a triacylglycerol.

trigone (trī′gōn) [L. *trigonium*, triangle] Triangular, smooth area at the base of the urinary bladder between the openings of the two ureters and that of the urethra.

triiodothyronine (trī-ī′ō-dō-thī′rō-nēn) (T₃) One of the thyroid hormones; contains three iodine atoms.

tRNA See *transfer RNA*.

trochanter (trō′kanter) [G., a runner] One of the large tubercles of the proximal femur.

trophoblast (trō′fō-blast) [G. *trophe*, nourishment + *blastos*, germ] Outer part of the blastocyst; enters the uterus and becomes the embryonic portion of the placenta.

trypsin (trip′sin) Enzyme released from the pancreas that digests proteins.

TSH See *thyroid-stimulating hormone*.

tubercle (too′ber-kl) Lump or knob on a bone.

tuberosity (too′ber-os′i-tē) Lump on a bone, usually larger than a tubercle.

tubular reabsorption Movement of materials, by means of diffusion or active transport, from the filtrate within a nephron into the blood.

tubular secretion Movement of materials, by means of active transport, from the blood into the filtrate of a nephron.

tumor (too′mŏr) Swelling, one of the cardinal signs of inflammation, or a new growth of tissue in which the multiplication of cells is uncontrolled and progressive; see also *neoplasm*.

tunic (too′nik) [L., coat] Layer or coat; one of the three enveloping layers of the wall of the eye; the three tunics are the fibrous, vascular, and nervous tunics; one of the three layers of blood vessels: tunica intima, tunica media, and tunica adventitia.

tunica adventitia (too′ni-kă ad-ven-tish′ă) Outermost fibrous coat of a vessel or an organ that is derived from the surrounding connective tissue.

tunica intima (in′ti-mă) Innermost layer of a blood or lymphatic vessel; consists of endothelium and a small amount of connective tissue.

tunica media Middle, usually muscular, coat of an artery or another tubular structure.

tympanic membrane (tim-pan′ik) [drumlike] Cellular membrane that covers the inner opening of the external auditory canal and separates the middle and external ears; vibrates in response to sound waves; the eardrum.

ulcer (ŭl′ser) [L. ulcus, a sore] Lesion on the surface of the skin or a mucous membrane, such as in the stomach or intestine, caused by a superficial loss of tissue, usually with inflammation.

umbilical cord (ŭm-bil′i-kăl) [L., navel] Cord connecting the fetus to the placenta; contains two umbilical arteries, which originate from the embryo's internal iliac arteries, that carry blood from the embryo to the placenta, and one umbilical vein, that carries blood back to the fetus.

umbilical vein Vein in the umbilical cord of the fetus by which the fetus receives nourishment from the placenta; becomes the round ligament of the liver in the adult.

upper motor neuron Motor neuron located in the cerebral cortex and synapsing with a lower motor neuron in the brainstem or spinal cord.

ureter (ū-rē′ter, ū′re-ter) [G. oureter, urinary canal] Tube conducting urine from the kidney to the urinary bladder.

urethra (ū-rē′thră) Duct leading from the urinary bladder, discharging the urine externally.

uterus (ū′ter-ŭs) Hollow muscular organ in which the fertilized oocyte develops into a fetus.

utricle (ū′trĭ-kl) Larger of the two membranous sacs in the vestibule of the labyrinth. The semicircular canals arise from it.

uvula (ū′vū-lă) [L. uva, grape] Small, grapelike appendage at the posterior margin of the soft palate.

vaccine (vak′sēn, vak-sēn′) Preparation of killed microorganisms, altered microorganisms, or derivatives of microorganisms intended to produce immunity; usually administered by injection, but sometimes ingestion is preferred.

vagina (vă-jī′nă) [L., sheath] Genital canal in the female, extending from the uterus to the vulva.

variable region Part of an antibody that combines with an antigen; responsible for the specificity of the antibody.

varicose (văr′ĭ-kōs) vein Vein that is so dilated that the cusps of the valves are no longer capable of preventing backflow of blood; usually the veins in the lower legs or the hemorrhoidal veins.

vasoconstriction (vă′sō-kon-strik′shŭn) Decreased diameter of blood vessels.

vasodilation (vă′sō-dī-lā′shŭn) Increased diameter of blood vessels.

vasomotor center (vă-sō-mō′ter) Area of the lower pons and upper medulla oblongata that continually transmits a low frequency of action potentials through sympathetic neurons to smooth muscle in blood vessels; can cause vasoconstriction and vasodilation.

vasomotor tone Partial constriction of blood vessels in the periphery, which results from relatively constant sympathetic stimulation.

vasopressin (vă-sō-pres′in) [L. vaso, blood vessel + pressum, to press down] Peptide hormone, related to oxytocin, secreted from the posterior pituitary gland. In large doses, it causes contraction of blood vessel smooth muscle; see also antidiuretic hormone.

vein (vān) Blood vessel that carries blood toward the heart.

venous return (vē′nŭs) Volume of blood returning to the heart.

ventilation (ven-ti-lā′shŭn) Movement of air in and out of the lungs.

ventral (ven′trăl) [L. venter, belly] In humans, synonymous with anterior.

ventral root Motor (efferent) root of a spinal nerve.

ventricle (ven′tri-kl) [L. venter, belly] Cavity; in the brain, one of four cavities filled with cerebrospinal fluid; one of two chambers of the heart that pump blood into arteries; there are a left and a right ventricle.

vernix caseosa (ver′niks kā′sē-ō′să) Epithelial cells and sebaceous matter that cover the skin of the fetus.

vesicle (ves′i-kl) [L. vesicula, blister or bladder] Small, membrane-bound sac containing material to be transported across the cell membrane.

vestibular fold (ves-tib′ū-lăr) [L., entrance hall] False vocal fold.

vestibule (ves′ti-bool) Small cavity or space at the entrance of a canal; see also vulva.

villus (vil′ŭs), pl. villi (vil′ī) [L., shaggy hair] Projection of the mucous membrane in the small intestine that increases surface area.

visceral (vis′er-ăl) [L. viscus, the soft parts, internal organs] Relating to the internal organs.

visceral peritoneum (per′i-tō-nē′ŭm) [L., organ] Part of the serous membrane in the abdominal cavity covering the surface of some abdominal organs.

vitamin (vīt′ă-min) [L. vita, life + amine, from ammonia] One of a group of organic substances, present in minute amounts in natural foods, that are essential to normal metabolism; insufficient amounts in the diet may cause deficiency diseases.

vitamin D Fat-soluble vitamin produced from a precursor molecule in skin exposed to ultraviolet light; increases calcium and phosphate uptake in the intestine.

vitreous humor (vit′rē-ŭs) Transparent, jellylike substance that fills the posterior compartment of the eye; helps maintain pressure within the eye and holds the lens and retina in place.

vocal fold (vō′kăl) One of the ligaments that extends from the posterior surface of the thyroid cartilage to the paired cartilages of the larynx; the superior pair are the false vocal folds, and the inferior pair are the true vocal folds.

vulva (vŭl′vă) [L., a wrapper or covering, seed covering, womb] External genitalia of the female; the mons pubis, labia majora and minora, the clitoris, the vestibule and its glands, the opening of the urethra, and the opening of the vagina.

white blood cell (WBC) Round, nucleated blood cell involved in immunity; includes neutrophils, basophils, eosinophils, lymphocytes, and monocytes; also called a leukocyte.

X-linked Trait caused by a gene on the X chromosome.

yolk sac (yōk, yōlk) Highly vascular endodermal layer surrounding the yolk of an embryo.

zona pellucida (zō′nă pe-loo′sid-ă) [L. zone, girdle + pellucidus, passage of light] Extracellular coat surrounding the oocyte; appears translucent.

zygomatic (zī′gō-mat′ik) [G. zygon, yoke] Referring to the zygomatic, or cheek, bone; the zygomatic arch is a bony arch created by the junction of the zygomatic and temporal bones.

zygomaticus muscle (zī′gō-mat′i-kŭs) Muscle originating on the zygomatic bone and inserting onto the corner of the mouth; involved in smiling.

zygote (zī′gōt) [G. zygotos, yoked] Single-celled, diploid product of fertilization, resulting from the union of a sperm cell and an oocyte.

Credits

Photographs

RF; **12A (left):** ©Hank Morgan/ Science Source; **12A (right):** ©SPL/Science Source.

Chapter 13
Opener: ©Silvia Otte/Taxi/Getty Images; **13.2:** ©Ed Reschke/ Photolibrary/Getty Images.

Chapter 14
Opener: ©McGraw-Hill Education/Jill Braaten; **14.6b:** ©Trent Stephens; **p. 406:** ©Chris Ryan/age fotostock RF; **14A:** ©BSIP/Science Source.

Chapter 15
Opener: ©BSIP/UIG via Getty Images; **15.2b:** ©R. T. Hutchings;

15.4b: ©CNRI/Science Source; **p. 436:** ©Thinkstock/ Jupiterimages RF; **15A:** ©Chris Cole/Riser/Getty Images; **15B (left):** ©Carolina Biological Supply Company/Phototake; **15B (right):** ©Image Source/ Getty Images RF.

Chapter 16
Opener: ©SW Productions/ Photodisc/Getty Images RF; **16.10c:** ©Victor Eroschenko; **16.14d:** ©SPL/Getty Images RF; **16.21b:** ©CNRI/SPL/Science Source; **p. 470:** ©Comstock Images RF; **16A:** ©PictureNet/ Corbis RF.

Chapter 17
Opener, 17.1: USDA; **17.2a:** ©David A. Tietz/Editorial Image, LLC RF; **17.2b:** USDA; **17.11:** ©M.M. Sweet/Flickr/Getty Images RF.

Chapter 18
Opener: ©Adam Crowley/Getty Images RF; **18.3b:** ©McGraw-Hill Education/Rebecca Gray, photographer/Don Kincaid, dissections; **Page 520 iPad:** ©DreamPictures/Blend Images LLC RF.

Chapter 19
Opener: ©Thierry Berrod, Mona Lisa Production/Science

Source; **19Aa-b:** ©McGraw-Hill Education/Jill Braaten; **19Ac:** ©Aaron Haupt/Science Source; **19Ad:** ©McGraw-Hill Education/ Jill Braaten; **19Ae:** ©Martin Shields/Alamy; **p. 552:** ©Brand X Pictures/PunchStock RF.

Chapter 20
Opener: ©Getty Images RF; **20.2 (all):** ©Dr. Yorgus Nikas/ Jason Burns/Phototake; **20.8:** ©John Giannicchi Science Source; **20.13a-b:** ©Petit Format/ Nestle Science Source; **20.13c:** ©Tissuepix Science Source; **20.18:** ©CNRI/SPL/Science Source.

Index

Selected Abbreviations

α alpha

ACE   angiotensin-converting enzyme

acetyl-CoA   acetyl coenzyme A

ACh   acetylcholine

ADH   antidiuretic hormone

ADP   adenosine diphosphate

ANH   atrial natriuretic hormone

ANS   autonomic nervous system

apo E   apolipoprotein E

ATP   adenosine triphosphate

AV   atrioventricular beta

$\overline{B}COP$ blood colloid osmotic pressure

BMI   body mass index

BMR   basal metabolic rate

BP   blood pressure

BPG   2,3-bisphosphoglycerate

bpm   beats per minute

BUN   blood urea nitrogen

$C_6H_{12}O_6$ glucose

$Ca_{10}(PO_4)_6(OH)_2$ hydroxyapatite

Ca^{2+}   calcium ion

cal   calorie

cAMP   cyclic adenosine monophosphate

CBC   complete blood count

cGMP   cyclic guanosine monophosphate

CH_3COOH   acetic acid

Cl^-   chloride ion

CNS   central nervous system

CO   cardiac output

CO   carbon monoxide

CO_2   carbon dioxide

—COOH   carboxyl group

COX-1   cyclooxygenase-1

CP   capsule pressure

CRH   corticotrophin-releasing hormone

CSF   cerebrospinal fluid

DAG   diacylglycerol

DNA   deoxyribonucleic acid

ECG or EKG   electrocardiogram

EEG   electroencephalogram

EGF   epidermal growth factor

ENS   enteric nervous system

EPSP   excitatory postsynaptic potential

FAD   flavin adenine dinucleotide

$FADH_2$ reduced flavin adenine diphosphate

Fe^{2+}   iron ion

FEV_1   forced expiratory volume in one second

FGF   fibroblast growth factor

FSH   follicle-stimulating hormone gamma

$\overline{g}$ gram

GABA   gamma-aminobutyric acid

GCP   glomerular capillary pressure

GDP   guanosine diphosphate

GFR   glomerular filtration rate

GH   growth hormone

GHIH   growth hormone-inhibiting hormone

GHRH   growth hormone-releasing hormone

GnRH   gonadotropin-releasing hormone

GTP   guanosine triphosphate

H^+   hydrogen ion

H_2CO_3   carbonic acid

H_2O   water

H_2O_2   hydrogen peroxide

$H_2PO_4^-$ dihydrogen phosphate ion

HCG   human chorionic gonadotropin

HCl   hydrochloric acid

HCO_3^-   bicarbonate ion

HDL   high-density lipoprotein

Hg   mercury

HIV   human immunodeficiency virus

HLA   human leukocyte antigen

HPO_4^{2-}   monohydrogen phosphate ion

HR   heart rate

I^-   iodide ion

ICSH   interstitial cell-stimulating hormone

IFP   interstitial fluid pressure

Ig   immunoglobulin

IP_3   inositol triphosphate

IPSP   inhibitory postsynaptic potential

IU   international units

K^+   potassium ion

kcal   kilocalorie

kg   kilogram

L   liter

LDL   low-density lipoprotein

LH   luteinizing hormone

LHRH   luteinizing hormone-releasing hormone

MAC   membrane attack complex

MALT   mucosa-associated lymphoid tissue

MAO   monoamine oxidase

MAP   mean arterial pressure

mEq   milliequivalent

Mg^{2+}   magnesium ion

MHC   major histocompatibility complex

mOsm   milliosmole

mRNA   messenger ribonucleic acid

mV   millivolt

Na^+   sodium ion

NaCl   sodium chloride

NAD^+   nicotinamide adenine dinucleotide

NADH   reduced nicotinamide adenine dinucleotide

$NaHCO_3$ sodium bicarbonate

NaOH   sodium hydroxide

NFP   net filtration pressure

—NH_2   amine group

NH_3   ammonia

NH_4^+   ammonium ion

NK cells   natural killer cells

NO   nitric oxide

O_2   oxygen

OH^-   hydroxide ion

PAH   *para*-aminohippuric acid

P_{ALV}   alveolar pressure

P_B   barometric air pressure

PGE   prostaglandin E

PGF   prostaglandin F

P_i   inorganic phosphate

PIF   prolactin-inhibiting factor

PIH   prolactin-inhibiting hormone

PIP_2   phosphoinositol

PMNs   polymorphonuclear neutrophils

PNS   peripheral nervous system

PO_4^{3-}   phosphate ion

P_{PL}   pleural pressure

PR   peripheral resistance

PRF   prolactin-releasing factor

PRH   prolactin-releasing hormone

PTH   parathyroid hormone

RANKL   receptor for activation of nuclear factor kappa B ligand

RAS   reticular activating system

RBC   red blood count

RDA   recommended daily allowances

RDI   reference daily intake

RhoGAM Rh_o (D) immune globulin

RMP   resting membrane potential

RNA   ribonucleic acid

SA   sinoatrial

SV   stroke volume

T_3   triiodothyronine

T_4   tetraiodothyronine

TF   tissue factor

TGF-_   transforming growth factor beta

TMJ   temporomandibular joint

TNF   tumor necrosis factor

TRH   thyroid-releasing hormone

TSH   thyroid-stimulating hormone

V_A   alveolar ventilation

VLDL   very low-density lipoprotein

vWF   von Willebrand factor

WBC   white blood count

Prefixes, Suffixes, and Combining Forms

The ability to break down medical terms into separate components or to recognize a complete word depends on mastery of the combining forms (roots or stems) and the prefixes and suffixes that alter or modify their meanings. Common prefixes, suffixes, and combining forms are listed below in boldface type, followed by the meaning of each form and an example illustrating its use.

a-, an- without, lack of: *a*phasia (lack of speech), *an*aerobic (without oxygen)

ab- away from: *ab*ductor (leading away from)

-able capable: vi*able* (capable of living)

acou- hearing: *acou*stics (science of sound)

acr- extremity: *acr*omegaly (large extremities)

ad- to, toward, near to: *ad*renal (near the kidney)

adeno- gland: *adeno*ma (glandular tumor)

-al expressing relationship: neur*al* (referring to nerves)

-algia pain: gastr*algia* (stomach pain)

angio- vessel: *angio*graphy (radiography of blood vessels)

ante- before, forward: *ante*cubital (before elbow)

anti- against, reversed: *anti*peristalsis (reversed peristalsis)

arthr- joint: *arthr*itis (inflammation of a joint)

-ary associated with: urin*ary* (associated with urine)

-asis condition, state of: homeost*asis* (state of staying the same)

auto- self: *auto*lysis (self breakdown)

bi- twice, double: *bi*cuspid (two cusps)

bio- live: *bio*logy (study of living)

-blast bud, germ: fibro*blast* (fiber-producing cell)

brady- slow: *brady*cardia (slow heart rate)

-c expressing relationship: cardia*c* (referring to heart)

carcin- cancer: *carcin*ogenic (causing cancer)

cardio- heart: *cardio*pathy (heart disease)

cata- down, according to: *cata*bolism (breaking down)

cephal- head: *cephal*ic (toward the head)

-cele hollow: blasto*cele* (hollow cavity inside a blastocyst)

cerebro- brain: *cerebro*spinal (referring to brain and spinal cord)

chol- bile: a*chol*ic (without bile)

cholecyst- gallbladder: *cholecyst*okinin (hormone causing the gallbladder to contract)

chondr- cartilage: *chondr*ocyte (cartilage cell)

-cide kill: bacteri*cide* (agent that kills bacteria)

circum- around, about: *circum*duction (circular movement)

-clast smash, break: osteo*clast* (cell that breaks down bone)

co-, com-, con- with, together: *co*enzyme (molecule that functions with an enzyme), *com*misure (coming together), *con*vergence (to incline together)

contra- against, opposite: *contra*lateral (opposite side)

crypto- hidden: *crypto*rchidism (undescended or hidden testes)

cysto- bladder, sac: *cysto*cele (hernia of a bladder)

-cyte-, cyto- cell: erythro*cyte* (red blood cell), *cyto*skeleton (supportive fibers inside a cell)

de- away from: *de*hydrate (remove water)

derm- skin: *derm*atology (study of the skin)

di- two: *di*ploid (two sets of chromosomes)

dia- through, apart, across: *dia*pedesis (ooze through)

dis- reversal, apart from: *dis*sect (cut apart)

-duct- leading, drawing: ab*duct* (lead away from)

-dynia pain: masto*dynia* (breast pain)

dys- difficult, bad: *dys*mentia (bad mind)

e- out, away from: *e*viscerate (take out viscera)

ec- out from: *ec*topic (out of place)

ecto- on outer side: *ecto*derm (outer skin)

-ectomy cut out: append*ectomy* (cut out the appendix)

-edem- swell: myo*edem*a (swelling of a muscle)

em-, en- in: *em*pyema (pus in), *en*cephalon (in the brain)

-emia blood: an*emia* (deficiency of blood)

endo- within: *endo*metrium (within the uterus)

entero- intestine: *entero*itis (inflammation of the intestine)

epi- upon, on: *epi*dermis (on the skin)

erythro- red: *erythro*cyte (red blood cell)

eu- well, good: *eu*phoria (well-being)

ex- out, away from: *ex*halation (breathe out)

exo- outside, on outer side: *exo*genous (originating outside)

extra- outside: *extra*cellular (outside the cell)

-ferent carry: af*ferent* (carrying to the central nervous system)

-form expressing resemblance: fusi*form* (resembling a fusion)

gastro- stomach: *gastro*dynia (stomach ache)

-genesis produce, origin: patho*genesis* (origin of disease)

gloss- tongue: hypo*gloss*al (under the tongue)

glyco- sugar, sweet: *glyco*lysis (breakdown of sugar)

-gram a drawing: myo*gram* (drawing of a muscle contraction)

-graph instrument that records: myo*graph* (instrument for measuring muscle contraction)

hem- blood: *hem*opoiesis (formation of blood)

hemi- half: *hemi*plegia (paralysis of half of the body)

hepato- liver: *hepato*itis (inflammation of the liver)

hetero- different, other: *hetero*zygous (different genes for a trait)

hist- tissue: *hist*ology (study of tissues)

homeo-, homo- same: *homeo*stasis (state of staying the same), *homo*logous (alike in structure or origin)

hydro- wet, water: *hydro*cephalus (fluid within the head)

hyper- over, above, excessive: *hyper*trophy (overgrowth)

hypo- under, below, deficient: *hypo*tension (low blood pressure)

-ia, -id expressing condition: neuralg*ia* (pain in nerve), flacc*id* (state of being weak)

-iatr- treat, cure: ped*iatr*ics (treatment of children)

-im not: *im*permeable (not permeable)

in- in, into: *in*jection (forcing fluid into)

infra- below, beneath: *infra*orbital (below the eye)

inter- between: *inter*costal (between the ribs)

intra- within: *intra*ocular (within the eye)

-ism condition, state of: dimorph*ism* (condition of two forms)